HISTORY OF THE WORLD

Other works by the author

French Revolution Documents I (edited)

Europe 1880–1945

The Mythology of the Secret Societies

The Paris Commune from the Right

The Age of Revolution and Improvement

The Triumph of the West

HISTORY
OF THE
WORLD

J.M. ROBERTS

New York
OXFORD UNIVERSITY PRESS
1993

This edition first published in Great Britain by Helicon Publishing Limited
42 Hythe Bridge Street
Oxford OX1 2EP, England

Previous editions published by Hutchinson & Co. Ltd. (1976, 1987)

Published in the United States of America by Oxford University Press, Inc.,
200 Madison Avenue,
New York, New York 10016, U.S.A.

Library of Congress Cataloging-in-Publication Data
Roberts, J. M. (John Morris). 1928-
History of the world / J.M. Roberts.
p. cm. Includes index.
ISBN 0-19-521043-3
1. World history. I. Title.
D20.R65 1993 909—dc20 93-14431

2 4 6 8 9 7 5 3 1

Printed in the United States of America
on acid-free paper

CONTENTS

List of Maps ix

Preface xi

BOOK I BEFORE HISTORY – BEGINNINGS

Introduction 3
1 The Foundations 4
2 *Homo Sapiens* 15
3 The Possibility of Civilization 24

BOOK II THE FIRST CIVILIZATIONS

Introduction 31
1 Early Civilized Life 32
2 Ancient Mesopotamia 38
3 Ancient Egypt 52
4 Intruders and Invaders: The Dark Ages of the Ancient Near East 71
A Complicating World 71
Early Civilized Life in the Aegean 76
The Near East in the Ages of Confusion 85
5 The Beginnings of Civilization in Eastern Asia 95
Ancient India 95
Ancient China 105
6 The Other Worlds of the Ancient Past 118
7 The End of the Old World 128

BOOK III THE CLASSICAL MEDITERRANEAN

Introduction 133
1 The Roots of One World 134
2 The Greeks 137
3 Greek Civilization 149
4 The Hellenistic World 168
5 Rome 180
6 The Roman Achievement 194
7 Jewry and the Coming of Christianity 206
8 The Waning of the Classical West 219
9 The Elements of a Future 237

BOOK IV THE AGE OF DIVERGING TRADITIONS

Introduction 251
1 Islam and the Re-making of the Near East 252
2 The Arab Empires 264
3 Byzantium and Its Sphere 275
4 The Disputed Legacies of the Near East 295
5 The Making of Europe 312
6 India 337
7 Imperial China 352
8 Japan 369
9 Worlds Apart 379
10 Europe: the First Revolution 390
The Church 391
Principalities and Powers 402
Working and Living 408
11 New Limits, New Horizons 417
Europe Looks Outward 417
The European Mind 427

BOOK V THE MAKING OF THE EUROPEAN AGE

Introduction 435
1 A New Kind of Society: Early Modern Europe 436
2 Authority and Its Challengers 452
3 The New World of Great Powers 476
4 Europe's Assault on the World 500
5 World History's New Shape 521
6 Ideas Old and New 535

BOOK VI THE GREAT ACCELERATION

Introduction 555
1 Long-Term Change 556
2 Political Change in an Age of Revolution 574
3 Political Change: A New Europe 594
4 Political Change: the Anglo-Saxon World 613
5 The European World Hegemony 630
6 European Imperialism and Imperial Rule 648
7 Asia's Response to a Europeanizing World 662

BOOK VII THE END OF THE EUROPEANS' WORLD

Introduction 689
1 Strains in the System 690
2 The Era of the First World War 706
3 A New Asia in the Making 731
4 The Ottoman Heritage and the Western Islamic Lands 745
5 The Second World War 756
6 The Shaping of a New World 777

BOOK VIII THE LATEST AGE

	Introduction	795
1	Perspectives	797
	Population Changes	797
	Plenty	801
	The Management of Nature	808
	Ideas, Attitudes, Authority	821
2	The Politics of the New World	829
	Cold War Beginnnings	829
	Asian Revolution	834
	Inheritors of Empire: the Middle East and Africa	847
	Latin America	860
3	Crumbling Certainties	867
	Superpower Difficulties	867
	Two Europes	879
	New Challenges to the Cold War World Order	885
4	The End of an Era	895

Epilogue: In the Light of History 914

Illustration Sources 923

Index 928

LIST OF MAPS

Page
11 Sites of Some Celebrated Discoveries of Hominid Fossils
26 Early Sites of Farming
36 The Fertile Crescent
56 Ancient Egypt
98 The Indus Valley
108 China – Physical
119 Climatic Changes in the Sahara
126 European Megalithic Monuments
130 Civilizations of the Near East
138 The Greek World of the Aegean
146 The Persian Empire of the Achaemenids
153 The Peloponnesian War
170 Alexander's March to the East
174 The Hellenistic World Soon After 200 BC
178 The Mediterranean c. 600 BC
181 Southern Italy 509–272 BC
186 The Punic Wars – Major Events
191 Roman Expansion
200 Major Roads, Cities and Garrisons of the Empire in the Age of the Antonines
205 Judaism in the Ancient World
213 Paul's Missions
223 The Sassanid Empire c. 400
230 The Making of the Eastern Roman Empire
233 Völkerwanderung
244 Justinian's Empire 527–565
253 Central Asia
257 China under the Han Dynasty
259 Seventh-century Arabia
261 The Early Spread of Islam
267 Islamic Iberia c. 1050
273 Islam Beyond the Arab World until 1800
277 The Byzantine Empire c. 1265 and c. 1354
286 The Growth of Venice as a Mediterranean Power
289 Kiev Rus
293 South-eastern Europe about 1400
299 The Mongol Empires
309 Ottoman Expansion
315 Charlemagne's Europe
322 Christendom before the Islamic Conquest
329 The Medieval Empire
334 Christendom in the Eleventh Century
338 Moslem India
348 Moghul India

398 European Universities Founded before 1500
407 German Eastward Expansion
421 The Crusader Wars
444 European Trading Stations and Possessions in Africa and Asia c. 1750
459 Reformation and Counter-Reformation Europe
483 Europe (Treaty of Westphalia 1648)
486 The Beginning of the Ottoman Retreat in Europe
489 Russian Expansion 1500–1800
506 The Growth of British Power in India 1783–1804
508 Exploration of the Americas
515 British Atlantic Trade in the 1770s
518 Economic Resources of the British American Colonies in the Eighteenth Century
523 Christian Missionary Activity in Africa and Asia in the Nineteenth Century
528 Africa in the Early Modern Era
560 The Slavery Problem in the United States
579 The Emergence and Consolidation of the USA
591 Napoleonic Europe
601 Europe in 1815
608 Russian Expansion to 1905
610 Europe in 1914
617 The Winning of the Far West
619 The American Civil War 1861–65
629 The British Empire (and protected territories) 1815–1914
639 South America after Independence
652 British India 1858–1947
654 Africa in 1880
657 Partitioned Africa: Areas of European Domination in 1914
664 Manchu China
676 Japanese Expansion 1895–1942
683 Major Religions of Asia in the Early Twentieth Century
684 Imperial Expansion in South-East Asia 1850–1914
691 Migration from Europe in the Nineteenth Century
698 Ottoman Decline and the Emergence of Modern Turkey 1683–1923
720 The Great War 1914–18
725 Russia in 1918
736 China 1918–49
766 Europe During the War of 1939–45
792 Proposed UN Partition of Palestine 1947/Israel 1948–67/Israel 1967–75
802 Worldwide Life Expectancy/Population Density/Gross Domestic Product/Daily Calorie Intake
806 World Energy Resources
830 Post-War Germany and Central Europe
844 The Post-War Recovery of Eastern Asia
850 The Post-Ottoman Near and Middle East
855 Decolonization in Africa and Asia
861 Post-War Latin America
882 Post-War Europe – Economic and Military Blocs
896 The Soviet Union and its Successors

PREFACE

'Turmoil everywhere, so it's back to the drawing board'. So ran the title given by the editor of *The Times* in 1989 to a light-hearted article setting out the case for revising this *History* and preparing a new edition of it. Eighteen months or so later, the last day of 1991 hardly seems to be one on which to labour the point that historical perspectives can quickly change, though that they do is the best reason for this new edition. Yet I have found that revision has still left much of the book unchanged. Perhaps this is unsurprising; though there is new material to include (and 31st December 1991 seems to me a good symbolic closing date) neither my general approach nor my judgements about the proportions of my story have altered, nor has the standpoint from which I wrote it. Most of my original preface might stand, therefore, with only minor alteration, in much the same words as in earlier revisions. It may, though, be worth repeating that world history is inseparably part of the way we see things. Even if the limits of our own historical horizon are not very far away, we strive to make sense of events by getting them 'in perspective', and in fact make judgements about world history all the time. Our minds are not going to be kept empty of them, and those who do not have decent history in their heads will have bad. Most men and women have some notions, however inadequate, about the world came to be what it is, and such ideas are less dangerous if they are made explicit. Perhaps professional historians are now readier to accept that than even a few years ago.

Readers of the first edition of this book will know that one way in which I have tried to increase our awareness of such judgements and make them a little easier to criticize is by recalling the sheer weight of the inherited past. Historical inertia is easily under-rated. This is not just a matter of what we can see. Ruins and beefeaters are picturesque, but for the most part less important than much mental and institutional history lost to sight in the welter of day-to-day events. It is already beginning to require an effort to remember that what was called the 'Cold War' dominated much of the 1950s and 1960s, recent as that is. The historical forces moulding the outlook of Americans, Russians and Chinese for centuries before the words capitalism and communism were invented are easier still to overlook. Distant history still clutters up our lives, and thinking. Even our choice of the dates around which we construct our accounts of the past are shaped by history: calendars are cultural artifacts, after all.

From the start, I tried in this book to balance the attention given to the effects of historical inertia by another great fact, mankind's unique power to produce change. Many people find this easier to recognize than the way past history inhibits human freedom. Evidence of the acceleration of change, its growth in scale and its wider and wider geographical spread, has continued to accumulate in recent years and much of it shows a continuing increase in conscious power to master the world of nature. Lately, though, this mastery has been clouded. The enthusiasm once felt for technical and intellectual achievement has even given way to disfavour. The Great Depression, Auschwitz and Hiroshima were followed by pollution, fear of overpopulation and the threat of war with ever more frightful weapons – to name only a handful of twentieth-century evils. Many people now seem to distrust those Promethean visions of man which were in the past so easily distorted into an optimism which assumed that inevitable success lay ahead.

Personally, I doubt whether any balanced judgement about the significance of much that has recently happened is ever struck solely on the basis of a knowledge of historical facts. Optimism and pessimism still seem to me to be for the most part a matter of temperament. And even if my impression is wrong, it does not seem to me that many safe predictions can follow from such facts as history provides. How many of us could rightly have judged the future of the USSR when the first edition of this book appeared? We can only make judgements, not necessary inductions. They do not force us to conclude either that we are now facing problems specially recalcitrant, or, on the other hand, that we are not. Just as when I wrote my first preface to this book, the odds still seem to me to be that the world organized as we know it certainly cannot last much longer, but that ordered and civilized life will go on in most places where it already exists. We have no reason to suppose that the outcome will be any more intolerable than, say, the results of changes forced on traditional Asia and Africa within the last century by the coming of Western technology (and many people may reasonably argue that this would be intolerable enough). Nevertheless, my main ideas have not changed. The words with which I ended this book seventeen years ago seem as apposite as ever, and I had no difficulty in deciding that they should still find a place in this new edition.

For those readers who have never looked into world history before, it may be worth adding that I have sought in these pages to tell a unified story and not to bring together a new collection of accounts of traditionally important themes. Beyond this, this book was shaped also by a wish to avoid detail and to set out instead major historical processes, their comparative scale and relations with one another. I have not tried to write continuous histories of all major countries or fields of human activity. The place for a comprehensive account of facts about the past is an encyclopaedia; I have assumed that my readers can get at one (and at dictionaries and atlases). Consequently, some topics of great scholarly interest and some of which we are much aware because of the glamour of what they have left behind are passed over briefly or even ignored in what follows. Though we still gape in amazement at the ruins of Yucatan and Zimbabwe and wonder over the statues of Easter Island, and intrinsically desirable though knowledge of the societies which produced these things may be, they are peripheral to world history; the early centuries of black Africa or the story of pre-Columbian America are for the same reason only lightly sketched in these pages. What Europeans later brought back from and did to those continents is a different matter.; it has shaped and continues to shape our lives, even if only in small degree. But nothing in the history of black Africa or the Americas between very remote times and the coming of the Europeans to those continents affected the great world-forming cultural traditions in which the legacies of, say, the Buddha, the Hebrew prophets, Plato, and Confucius were for centuries (as they still are today) living and shaping influences on millions of people.

Even in writing a selective account, it is easy to feel swamped by the huge quantities of printed evidence and massed scholarly monographs which now fill historical libraries. Fortunately, there is not the slightest chance of mastering all that might be read that is relevant to world history and I have an easy conscience. But bibliography can skew judgement; there is always a temptation to write about something which has given rise to much academic debate. What matters, though, is that a topic is important, not that we may be lucky enough to have information about it. Napoleon, however crucial for France and even Europe, seemed to me safer to pass over briefly than, say, the Chinese Revolution. In the most recent period of history it is more than ever important to distinguish the wood from the trees and not to mention something simply because it turns up every day in the newspaper or on television.

That medium, nonetheless, has influenced me in one respect since my first edition. An opportunity to make a series of films for the BBC under the title of *The Triumph of the West*, and the writing of a book about it both forced me to think again about the röle of Europe and its civilization in world history. The more I studied it, the less I felt misgivings about the recognition I had given to that civilization in this account. The more I thought about it, the more the centrality of Europe's röle in the making of the modern world stood out. Of course, I have striven not to be trapped by a 'Eurocentric' viewpoint and, if I have confirmed, I have not increased the weight I gave to that theme when writing my first drafts twenty or so years ago. The impulses of my own historical inheritance were bound to influence my choice of themes, organization and chronological arrangement, of course, and I still cannot believe (to quote Lord Actons's ideal of historical objectivity) that 'nothing shall reveal the country, the religion, or the party' to which I belong, nor that I could provide (as he hoped) an account of Waterloo to satisfy French, English, German and Dutch alike (not that I had time and space here to spend on such a theme). I hope nevertheless that an effort to remain aware of my assumptions and their limitations may have made it possible to provide what he termed a history 'which is distinct from the combined history of all countries', and which nevertheless displays the variety of the great cultural traditions which give it much of its structure.

The reader should blame no one but the author for what seems inadequate or erroneous nor, indeed, for anything else in this book. Yet many other people helped to bring it into being. In its latest form, it has been especially shaped by the tireless scrutiny of my editors, Adam Sisman and Anne-Lucie Norton, to whom I owe much, and in another important sense, it has been made possible only by the efforts of a team of ladies who have typed and re-typed my drafts – Mrs Moira Wise, Miss Clare Bass, Mrs Joan Barton and Mrs Lesley Walsh. To all of them I am most grateful. I wish also to thank Mr Eric Smith especially for his close eye and unflagging attention to the maps. There were many others to whose informed influence and suggestions I owed much over many years; I explained in my first preface that they were too numerous to be acknowledged individually, though I was happy to record then the debt I owed to them and now gladly do so again. To that I must add further thanks owed to many correspondents who have written to me in the last decade and a half offering specific criticism, suggestion and encouragement, too numerous though they, too, are for me to write down their names here. If I do not seem always to have taken much notice of what they have said, I can assure them they are mistaken in that impression. Those whose help was more easily identifiable, because I deliberately consulted them, and to whom I continue to feel special gratitude were listed in the old preface; that record stands, and there is no need to repeat it here. More names must now, nevertheless, be added to theirs and I wish to record my warm thanks to Sir Bryan Cartledge, Dr C.A Grocock and Mr R. Inskeep, to Professors Vassos Karagheorgis, H.W. Arndt and Elisabeth Vrba and to Mr M.M. Roberts, all of whom gave me valuable observations and suggestions. I must acknowledge, too, how stimulating I found the books of Professor E.L. Jones in preparing this new edition. None of these, of course, bears any responsibility for the final outcome; that rests entirely with me. As for my debt to my wife, I should not wish to alter what I have already written about that in any respect save one, by correcting a misprint which slid into the last preface to this book; in an indispensable sense, this book is hers.

J.M.R.
Funchal, 31 December 1991

Where they are known, dates of birth and death (and, in the case of rulers, regnal dates) for all persons mentioned by name in the text are given in the index.

HISTORY OF THE WORLD

The skull of homo erectus pekinensis *– so called because of the discoveries near* Peking *of his remains.*

BOOK I

BEFORE HISTORY – BEGINNINGS

When does History begin? It is tempting to reply 'In the beginning', but like many obvious answers, this soon turns out to be unhelpful. As a great Swiss historian once pointed out in another connexion, history is the one subject where you cannot begin at the beginning. We can trace the chain of human descent back to the appearance of vertebrates, or even to the photosynthetic cells and other basic structures which lie at the start of life itself. We can go back further still, to almost unimaginable upheavals which formed this planet and even to the origins of the universe. Yet this is not 'history'.

Commonsense helps here: history is the story of mankind, of what it has done, suffered or enjoyed. We all know that dogs and cats do not have histories, while human beings do. Even when historians write about a natural process beyond human control, such as the ups and downs of climate, or the spread of a disease, they do so only because it helps us to understand why men and women have lived (and died) in some ways rather than others.

This suggests that all we have to do is to identify the moment at which the first human beings step out from the shadows of the remote past. It is not quite as simple as that, though. First, we have to know what we are looking for, but most attempts to define humanity on the basis of observable characteristics prove in the end arbitrary and cramping, as long arguments about 'ape-men' and 'missing links' have shown. Physiological tests help us to classify data but do not identify what is or is not human. That is a matter of a definition about which disagreement is possible. Some people have suggested that human uniqueness lies in language, yet other primates possess vocal equipment similar to our own; when noises are made with it which are signals, at what point do they become speech? Another famous definition is that man is a tool-maker, but observation has cast doubt on our uniqueness in this respect, too, long after Dr Johnson scoffed at Boswell for quoting it to him.

What is surely and identifiably unique about the human species is not its possession of certain faculties or physical characteristics, but what it has done with them – its achievement, or history, in fact. Humanity's unique achievement is its remarkably intense level of activity and creativity, its cumulative capacity to create change. All animals have ways of living, some complex enough to be called cultures. Human culture alone is progressive; it has been increasingly built by conscious choice and selection within it as well as by accident and natural pressure, by the accumulation of a capital of experience and knowledge which man has exploited. Human history began when the inheritance of genetics and behaviour which had until then provided the only way of dominating the environment was first broken through by conscious choice. Of course, human beings have always only been able to make their history within limits. These limits are now very wide indeed, but they were once so narrow that it is impossible to identify the first step which took human evolution away from the determination of nature. We have for a long time only a blurred, story, obscure both because the evidence is fragmentary and because we cannot be sure exactly what we are looking for.

1
The Foundations

The roots of history lie in the prehuman past and it is hard to grasp just how long ago that was. If we think of a century on our calendar as a minute on some great clock recording the passage of time, then white Europeans began to settle in the Americas only about five minutes ago. Slightly less than fifteen minutes before that, Christianity appeared. Rather more than an hour ago a people settled in southern Mesopotamia who were soon to evolve the oldest civilization known to us. This is already well beyond the furthest margin of written record; according to our clock people began writing down the past much less than an hour ago, too. Some six or seven hours further back on our scale and much more remote, we can discern the first recognizable human beings of a modern physiological type already established in western Europe. Behind them, anything from a fortnight to three weeks earlier, appear the first traces of creatures with some manlike characteristics whose contribution to the evolution which followed is still in debate.

How much further back into a growing darkness we need go in order to understand the origins of Man is debatable, but it is worth considering for a moment even larger tracts of time simply because so much happened in them which, even if we cannot say anything very precise about it, shaped what followed. This is because humanity was to carry forward into historical times certain possibilities and limitations, and they were settled long ago, in a part even more remote than the much shorter period of time – three or four million years or so – in which creatures with at least some claim to human qualities are known to have existed. Though it is not our direct concern, we need to try to understand what was in the baggage of advantages and disadvantages with which Man alone among the primates emerged after these huge tracts of time as a change-maker. Virtually all the physical and much of the mental formation we still take for granted was by then determined, fixed in the sense that some possibilities were excluded and others were not. The crucial process is the evolution of manlike creatures as a distinct branch among the primates, for it is at this fork in the line, as it were, that we begin to look out for the station at which we get off for History. It is here that we can hope to find the first signs of that positive, conscious, impact upon environment which marks the first stage of human achievement.

The bedrock of the story is the earth itself. Changes recorded in fossils of flora and fauna, in geographical forms and geological strata, narrate a drama of epic scale lasting hundreds of millions of years. During them the shape of the world changed out of recognition many times. Great rifts opened and closed in its surface, coasts rose and fell; at times huge areas were covered with a long-since vanished vegetation. Many species of plants and animals emerged and proliferated. Most died out. Yet these 'dramatic' events happened with almost unimaginable slowness. Some lasted millions of years; even the most rapid took centuries. The creatures who lived while they were going on could no more have perceived them than could a twentieth-century butterfly, in its three weeks or so of life, sense the rhythm of the seasons. Yet slowly the earth was taking shape as a collection of habitats permitting different strains to survive. Meanwhile, biological evolution inched forwards with almost inconceivable slowness.

Climate was the first great pace-maker of change. About forty million years ago – an early enough point at which to begin to grapple with our story – a long warm climatic phase began to draw to a close. It had favoured the great reptiles and during it Antarctica had separated from Australia. There were no ice-fields then in any part of the globe. As the world grew colder and the new climatic conditions restricted their habitat, the great reptiles disappeared (though some have argued that other factors than environmental change were decisive). But the new conditions suited other animal strains which were already about, among them some mammals whose tiny ancestors had appeared two hundred million years or so earlier. They now inherited the earth, or a considerable part of it. With many breaks in sequence and accidents of selection on the way, these strains were themselves to evolve into the mammals which occupy our own world – ourselves included.

Crudely summarized, the main lines of this evolution were probably determined for millions of years by astronomical cycles. As the earth's position changed in relation to the sun, so did climate. A huge pattern emerges, of recurrent swings of temperature. The extremes which resulted, of climatic cooling on the one hand and aridity on the other, choked off some possible lines of development. Conversely, in other times, and in certain places, the onset of appropriately benign conditions allowed certain species to flourish and encouraged their spread into new habitats. The only major sub-division of this immensely long process which concerns us comes very recently (in prehistoric terms), slightly less than four million years ago. There then began a period of climatic changes which we believe to have been more rapid and violent than any observed in earlier times. 'Rapid', we must again remind ourselves, is a comparative term; these changes took tens of thousands of years. Such a pace of change, though, looks very different from the millions of years of much steadier conditions which lay in the past.

Scholars have long talked about 'Ice Ages', each lasting between fifty and a hundred thousand years, which covered big areas of the northern hemisphere (including much of Europe, and America as far south as modern New York) with great ice sheets, sometimes a mile or more thick. They have now distinguished some seventeen to nineteen (there is argument about the exact number) such 'glaciations' since the onset of the first, over three million years ago. We live in a warm period following the most recent of them, which came to an end some ten thousand years ago. Evidence about these glaciations and their effects is now available from all oceans and continents and they provide the backbone for prehistoric chronology. To the external scale which the Ice Ages provide we can relate such clues as we have to the evolution of humanity.

The Ice Ages make it easy to see how climate determined life and its evolution in prehistoric times, but to emphasize their dramatic direct effects is misleading. No doubt the slow onset of the ice was decisive and often disastrous for what lay in its path. Many of us still live in landscapes shaped by its scouring and gougings thousands of centuries ago. The huge inundations which followed the retreat of the ice as it melted must also have been locally catastrophic, destroying the habitats of creatures which had adapted to the challenge of arctic conditions. Yet they also created new opportunities. After each glaciation new species spread into the areas uncovered by the thaw. Beyond regions directly affected, though, the effects of the glaciations may have been even more important for the global story of evolution. Changes in environment followed cooling and warming thousands of miles from the ice itself; and the outcome had its own determining force. Both aridification and the spread of grassland, for instance, charged the possibilities of spreading themselves open to existing species. Some of those species form part of the human evolutionary story, and all the most important stages in that evolution of so far observed have been located in Africa, far from the ice-fields.

Climate can still be very important today, as contemplation of the disasters brought by drought can show. But such effects, even when they affect millions of people, are not so fundamental as the slow transformation of the basic geography of the world and its supplies of food which climate wrought in prehistoric times. Until very recently climate determined where and how men lived. It made technique very important (and still does): the possession in early times of a skill such as fishing or fire-making could make new environments available to branches of the human family fortunate enough to possess such skills, or able to discover and learn them. Different food-gathering possibilities in different habitats meant different chances of a varied diet and, eventually, of progressing from gathering to hunting, and then to growing. Long before the Ice Ages, though, and even before the appearance of the creatures from which humanity was to evolve, climate was setting the stage for humanity and thus shaping, by selection, the eventual genetic inheritance of humanity itself.

One more backward glance is useful before plunging into the still shallow (though gradually deepening) pool of evidence. Fifty-five million or so years ago, primitive mammals were of two main sorts. One, rodent-like, remained on the ground; the other took to the trees. In this way the competition of the two families for resources was lessened and strains of each survived to people the world with the creatures we know today. The second group were the prosimians. We are among their descendants, for they were the ancestors of the first primates.

It is best not to be too impressed by talk about 'ancestors' in any but the most general sense. Between the prosimians and ourselves lie millions of generations and many evolutionary blind alleys. It is important none the less that our remotest identifiable ancestors lived in trees because what survived in the next phase of evolution were genetic strains best suited to the special uncertainties and accidental challenges of the forest. That environment put a premium on the capacity to learn. Those survived whose genetic inheritance could respond and adapt to the surprising, sudden danger of deep shade, confused visual patterns, treacherous handholds. Strains prone to accident in such conditions were wiped out. Among those which prospered (genetically speaking) were some species with long digits which were to develop into fingers and, eventually, the oppositional thumb, and other forerunners of the apes already embarked upon an evolution towards three-dimensional vision and the diminution of the importance of the sense of smell.

The prosimians were little creatures. Tree-shrews still exist which give us some idea of what they were like; they were far from being monkeys, let alone men. Yet for millions of years they carried the traits which made humanity possible. During this time geography counted for much in their evolution, by imposing limits on contact between different strains, sometimes effectively isolating them, and thus increasing differentiation. Changes would not happen quickly but it is likely that fragmentations of the environment caused by geographical disturbance led to the isolation of zones in which, little by little, the recognizable ancestors of many modern mammals appeared. Among them are the first monkeys and apes. They do not seem to go back more than thirty-five million years or so.

They represent a great evolutionary stride. Both families had much greater manipulative dexterity than any predecessor. Within them, species distinct in size or acrobatic quality began to evolve. Physiological and psychological evolution blur in such matters. Like the development of better and stereoscopic vision, the growth of manipulative power seems to imply a growth of consciousness. Perhaps some of these creatures could already distinguish different colours. The brains of the first primates were already much more complex than those of any of their predecessors; they were bigger, too. Somewhere the brain of one or more of these strains became complex enough and its physical powers sufficiently developed for the animal to cross the line at which the world as a mass of undifferentiated sensations becomes at least in part a world of objects. Whenever this happened it was a decisive step towards mastering the world by using it, instead of reacting automatically to it.

Some twenty-five or thirty million years ago, as desiccation began to reduce the area of the forests, competition for diminishing forest resources became fiercer. Environmental challenge and opportunity appeared where the trees and the grasslands met. Some primates not powerful enough to hold on to their forest homes were able, because of some genetic quality, to penetrate the savannahs in search of food and could meet the challenge and exploit the opportunities. Probably they had a posture and movement marginally more like that of men than, say, that of the gorillas or chimpanzees. An upright stance and the capacity to move easily on two feet make it possible to carry burdens, among them food. The dangerous open savannah could then be explored and its resources withdrawn from it to a safer home base. Most animals consume their food where they find it; man does not. Freedom to use the forelimbs for something other than locomotion or fighting also suggests other possibilities. We do not know what was the first 'tool', but other primates than man have been seen to pick up objects which come to hand and wave them as a deterrent, use them as weapons, or investigate and expose possible sources of food with their aid.

The next step in the argument is enormous, for it takes us to the first glimpse of a member of the biological family to which both man and the great apes belong. The evidence is fragmentary, but suggests that some fifteen or sixteen million years ago a very successful species was widespread throughout Africa, Europe and Asia. Probably he was a tree-dweller and certainly he was not very large – he may have weighed about 40 pounds. Unfortunately, the evidence is such as to leave him isolated in time. We have no direct knowledge of his immediate forbears or descendants, but some kind of fork in the road of primate evolution had occurred. While one branch was to lead to the great apes and chimpanzees, the other led to human beings. This line has been named 'hominid'. But the first hominid fossil (found in Kenya) is dated only some four and a half to five million years ago, so that for about ten million years the record is obscure. During that time big geological and geographical changes must have favoured and disfavoured many new evolutionary patterns.

The first hominid fossil may not belong to it, but one evolutionary line which emerged from this upheaval was a small African hominid, subsequently named *Australopithecus*. The earliest fossil identified with this genus is more than four million years old and was found in Ethiopia. Evidence of other species of 'australopithecines' (as they are usually termed) found as far apart as Kenya and the Transvaal can be dated to various periods over the next two million years and has had a great impact upon archaeological thinking. In the last quarter-century, something like three million years has been added to the period in which the search for human origins goes on, thanks to the australopithecine discoveries. Great uncertainty and much debate still surrounds them, but if the human species has a common ancestor it seems most likely that it belonged to a species of this genus. It is with *Australopithecus*, though, and with what, for want of a better word, we must call its 'contemporaries' of other species, that the difficulties of distinguishing between apes, manlike apes and other creatures with some human characteristics first appear in their full complexity. The questions raised are still becoming in some ways more difficult to deal with. No simple picture has yet emerged and discoveries are still being made.

We have most evidence about *Australopithecus*. But there came to live contemporaneously with some australopithecine species other, more manlike, creatures, to whom the genus name *Homo* has been given. *Homo* was no doubt related to *Australopithecus*, but is first clearly identifiable as distinct about two million years ago on certain African sites; remains attributed to one of his species, though, have been dated by radio-activity to some million and a half years before that. Recently, to make confusion worse, the remains of an even bigger hominid have turned up near Lake Rudolf in northern Kenya. About five feet tall, with a brain about twice the size of a modern chimpanzee's, he has the undignified name of '1470 man', that being the number attached to his relics in the catalogue of the Kenya museum where they are to be found.

Where specialists disagree and may be expected to go on arguing about such fragmentary evidence as we have (all that is left of two million or so years of hominid life could be put on a big table), laymen had better not dogmatize. It is clear enough, though, that we can be fairly certain about the extent to which some characteristics later observable in humans already existed more than two million years ago. We know, for instance, that the australopithecines, though smaller than modern humans, had legbones and feet which were manlike rather than apelike. We know they walked upright and could run and carry loads for long distances as apes could not. Their hands showed a flattening at the finger-tips characteristic of those of men. These are stages far advanced on the road of human physique even if the actual descent of our species is from some other branch of the hominid tree.

It is to early members of the genus *Homo* (sometimes distinguished as *Homo habilis*) nonetheless, that we owe our first relics of tools. Tool-using is not confined to men, but the making of tools has long been thought of as a human characteristic. It is a notable step in winning a livelihood from the environment. Tools found in Ethiopia are the oldest which we have (about two and a half million years old) and they consist of stones crudely fashioned by striking flakes off pebbles to give them an edge. The pebbles seem often to have been carried purposefully and perhaps selectively to the site where they were prepared. Conscious creation of implements had begun. Simple pebble choppers of the same type from later times turn up all over the Old World of prehistory; about one million years ago, for example, they were in use in the Jordan valley. In Africa, therefore, begins the flow of what was to prove the biggest single body of evidence about prehistoric man and his precursors and the one which has provided most information about their distribution and cultures. A site at the Olduvai Gorge in Tanzania has provided the traces of the first identified building, a windbreak of stones which has been dated 1.9 million years ago, as well as evidence that its inhabitants were meat-eaters, in the form of bones smashed to enable the marrow and brains to be got at and eaten raw.

Olduvai prompts a tempting speculation. The bringing of stones and meat to the site combines with other evidence to suggest that the children of early hominids could not easily cling to their mother for long foraging expeditions as do the offspring of other primates. It may be that this is the first trace of the human institution of the home base.

Examples from our best line of evidence for a million years or so of men and manlike creatures – their technology. The scraper is from Libya, the hand-axe from Hampshire: thousands of years separate their manufacture. They are both very different and very much the same, speaking for the same prolonged phase of human development, the hundreds of millenia during which different traditions of shaping and working stone are our best guide to the variety of what was going forward.

Among primates, only humans have them: places where females and children normally stay while the males search for food to bring back to them. Such a base also implies the shady outlines of sexual differentiation in economic roles. It might even register the achievement already of some degree of forethought and planning, in that food was not devoured to gratify the immediate appetite on the spot where it was taken (as is the case with most primates), but reserved for family consumption elsewhere. Whether hunting, as opposed to scavenging, took place is another question, but the meat of large animals was consumed at a very early date at Olduvai.

Yet such exciting evidence only provides tiny and isolated islands of hard fact. It cannot be presumed that the East African sites were necessarily typical of those which sheltered and made possible the emergence of humanity; we know about them only because conditions there allowed the survival and subsequent discovery of early hominid remains. Nor, though the evidence may incline that way, can we be sure that *any* of these hominids is a direct ancestor of humanity; they may all only be precursors. What can be said is that these creatures show remarkable evolutionary efficiency in the creative manner we associate with human beings, and suggest the uselessness of categories such as ape-men (or men-apes) – and that few scholars would now be prepared to say categorically that we are not directly descended from *Homo habilis*, the species first identified with tool-using.

It is also easy to believe that the invention of the home base made biological survival easier. It would have made possible brief periods of rest and recovery from the hazards posed by sickness and accident, thus sidestepping, however slightly, the process of evolution by physical selection. Together with their other advantages, it may help to explain how examples of the genus *Homo* were able to leave traces of themselves throughout most of the world outside the Americas and Australasia in the next million or so years. But we do not certainly know whether this was through the spread of one stock, or because similar creatures evolved in different places. It is generally held, though, that tool-making was carried to Asia and India (and perhaps to Europe) by migrants originally from East Africa. The establishment and survival in so many different places of these hominids must show a superior capacity to grapple with changing conditions, but in the end we do not know what was the behavioural secret which suddenly (speaking once more in terms of prehistoric time) released that capacity and enabled them to spread over the landmass of Africa and Asia. No other mammal settled so widely and successfully before our own branch of the human family, which was eventually to occupy the whole planet, a unique biological achievement.

The next clear stage in human evolution is nothing less than a revolution in physique. After a divergence between hominids and more ape-like creatures which may have occurred more than four million years ago, it took rather less than two million years for one successful family of hominids to increase its brain size to about twice that of *Australopithecus*. One of the most important stages of this process and some of the most crucial in the evolution of Man were already reached in a species called *Homo erectus* which was widespread and successful a quarter of a million years ago. It had been in existence for at least half a million years and perhaps even longer (the oldest specimen so far identified may be about a million and half years old). This species lasted much longer, that is to say, than has (so far) *Homo sapiens*, the branch of hominids to which we belong. Many signs once more point to an African origin and thence to a spread through Europe and Asia (where *Homo erectus* was first found). Apart from fossils, a special tool helps to plot the distribution of the new species by defining areas into which *Homo erectus* did not spread as well as those into which he did. This is the so-called 'hand-axe' of stone whose main use seems to have been for skinning and cutting up large animals (its use as an axe seems unlikely, but the name is established). There can be no doubt of the success of *Homo erectus* as a genetic product.

When we finish with *Homo erectus* there is no precise dividing line (there never is in human prehistory, a fact it is only too easy to overlook or forget), but we are already dealing with a creature who has added to the upright stance of his predecessors a brain of the same order of magnitude as that of modern man. Though we still know little of the way in which the brain is organized, there is, allowing for body size, a rough correlation between size and intelligence. It is reasonable, therefore, to attribute great importance to the selection of strains with bigger brains and to reckon this a huge advance in the story of the slow accumulation of human characteristics.

Bigger brains meant bigger skulls and other changes, too. An increase in ante-natal size requires changes in the female pelvis to permit the birth of offspring with larger heads, and another consequence was a longer period of growth after birth; physiological evolution in the female was not sufficient to provide ante-natal accommodation to any point approaching physical maturity. Human children need maternal care long after birth. Prolonged infancy and immaturity in their turn imply prolonged dependency: it is a long time before such infants gather their own food. It may be with the offspring of *Homo erectus* that there began that long extension of the period of immaturity whose latest manifestation is the maintenance of young people by society during long periods of higher education.

Biological change also meant that care and nurture came gradually to count for more than large litters in ensuring the survival of the species. This in turn implied further and sharper differentiation in the roles of the sexes. Females were being pinned down much more by maternity at a time when food-gathering techniques seem to have become more elaborate and to demand arduous and prolonged cooperative action by males – perhaps because bigger creatures needed more and better food. Psychologically, too, the change may be significant. A new emphasis on the individual is one concomitant of prolonged infancy. Perhaps it was intensified by a social situation in which the importance of learning and memory was becoming more and more important and skills more complex. About this point the mechanics of what is going forward begin to slip from our grasp (if, indeed, they were ever in it). We are somewhere near the area in which the genetic programming of the hominids is infringed by learning. This is the beginning of the great change from the natural physical endowment to tradition and culture - and eventually to conscious control – as evolutionary selectors, though we may never be able to say where precisely this change occurs.

Another important physiological change is the loss of oestrus by the female hominid. We do not know when this happened, but after it had been completed her sexual rhythm was importantly differentiated from that of other animals. Man is the only animal in which the mechanism of the oestrus (the restriction of the female's sexual attractiveness and receptivity to the limited periods in which she is on heat) has entirely disappeared. It is easy to see the evolutionary connexion between this and the prolongation of infancy: if female hominids had undergone the violent disruption of their ordinary routine which the oestrus imposes, their offspring would have been periodically exposed to a neglect which would have made their survival impossible. The selection of a genetic strain which dispensed with oestrus, therefore, was essential to the survival of the species; such a strain must have been available, though the process in which it emerged may have taken a million or a million and a half years because it cannot have been effected consciously.

Such a change has radical implications. The increasing attractiveness and receptivity of females to males make individual choice much more significant in mating. The selection of a partner is less shaped by the rhythm of nature; we are at the start of a very long and obscure road which leads to the idea of sexual love. Together with prolonged infant dependency, the new possibilities of individual selection point ahead also to the stable and enduring family unit of father, mother and offspring, an institution unique to mankind. Some have even speculated that incest taboos (which are in practice well-nigh universal, however much the precise identification of the prohibited relationships may vary) originate in the recognition of the dangers presented by socially immature but sexually adult young males for long periods in close association with females who are always potentially sexually receptive.

In such matters it is best to be cautious. After all, modern communities do not seem to be very good at population control though they have much better information available. The evidence takes us only a very little way. Moreover, it is drawn from a very long span of time: examples of *Homo erectus* have been identified as active from at least a million and a half years ago, and to have continued to appear for another million. This would have given time for considerable physical, psychological and technological evolution. The earliest forms of *Homo erectus* may not have been much like the last, some of whom have been classified by some scientists as archaic forms of the next evolutionary stage of the hominid line. Yet all reflexions support the general hypothesis that the changes in hominids observable while *Homo erectus* occupies the centre of our stage were especially important in defining the arcs within which humanity was to evolve. He had unprecedented capacity to manipulate his environment, feeble though his handhold on it may seem to us. Besides the hand-axes which make possible the observation of his cultural traditions, late forms of *Homo erectus* left behind the earliest surviving traces of constructed dwellings (huts, sometimes fifty feet long, built of branches, with stone-slab or skin floors), the earliest worked wood, the first wooden spear and the earliest container, a wooden bowl. Creation on such a scale hints strongly at a new level of mentality, at a conception of the object formed before manufacture is begun, and perhaps an idea of process. Some have argued far more. In the repetition of simple forms, triangles, ellipses and ovals, in huge numbers of examples of stone tools, there has been discerned intense care to produce regular shapes which does not seem proportionate to any small gain in efficiency which may have been achieved. Can there be discerned in this the first tiny budding of the aesthetic sense?

The greatest of prehistoric technical and cultural advances was made when someone learnt how to manage fire. Until recently, the earliest available evidence of its use came from China, and probably from between three and five hundred thousand years ago. But very recent discoveries in the Transvaal have provided evidence convincing to many scholars that hominids there were using fire well before that. It remains fairly certain that *Homo erectus* never learnt how to make fire and, that even his successors did not for a long time possess this skill. That he knew how to use it, on the other hand, is indisputable. The importance of this knowledge is attested by the folklore of many later peoples; in almost all of them a heroic figure or magical beast first seizes fire. A violation of the supernatural order is implied: in the Greek legend Prometheus steals the fire of the gods. This is suggestive, not solid, but perhaps the first fire was taken from outbreaks of natural gas or volcanic activity. Culturally, economically, socially and technologically, fire was a revolutionary instrument, though we must again remember that a prehistoric 'revolution'

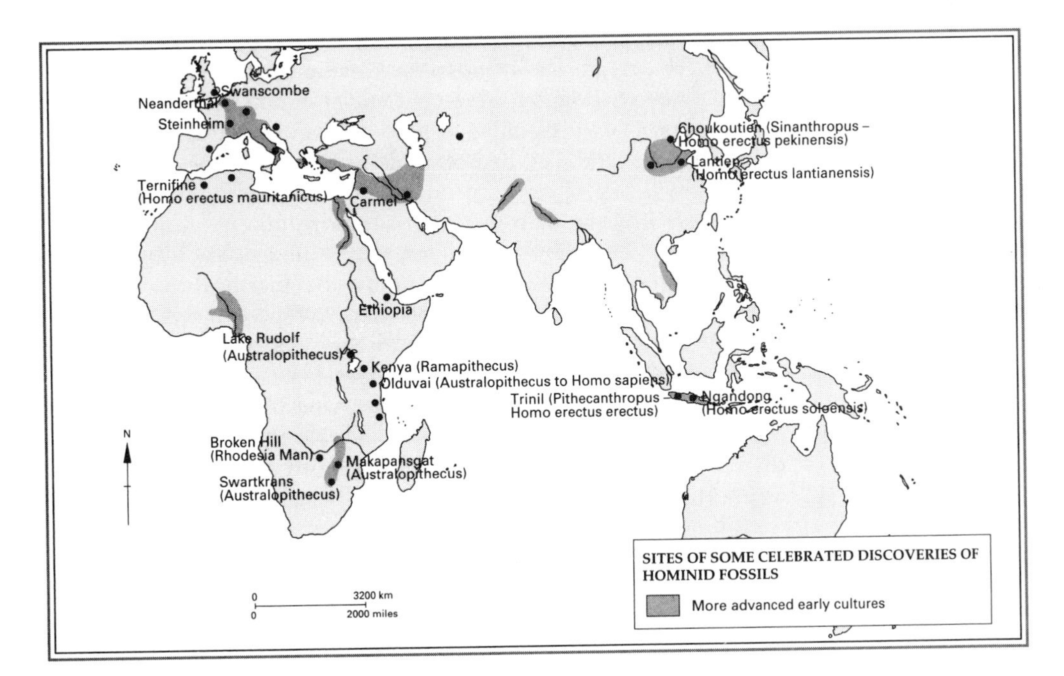

SITES OF SOME CELEBRATED DISCOVERIES OF HOMINID FOSSILS

took millennia. It brought the possibility of warmth and light and therefore of a double extension of Man's environment, into the cold and into the dark. In physical terms one obvious expression of this was the occupation of caves. Animals could now be driven out and kept out by fire (and perhaps the seed lies here of the use of fire to drive big game in hunting). Technology could move forward: spears could be hardened in fires and cooking became possible, indigestible substances such as seeds becoming sources of food and distasteful or bitter plants edible. This must have stimulated attention to the variety and availability of plant life; the science of botany was stirring without anyone knowing it.

Fire must have influenced mentality more directly, too. It was another factor strengthening the tendency to conscious inhibition and restraint, and therefore their evolutionary importance. The focus of the cooking fire as the source of light and warmth had also the deep psychological power which it still retains. Around the hearths after dark gathered a community almost certainly already aware of itself as a small and meaningful unit against a chaotic and unfriendly background. Language – of whose origins we as yet know nothing – would have been sharpened by a new kind of group intercourse. The group itself would be elaborated, too, in its structure. At some point, fire-bearers and fire specialists appeared, beings of awesome and mysterious importance, for on them depended, it might be, life and death. They carried and guarded the great liberating tool, and the need to guard it must sometimes have made them masters. Yet the deepest tendency of this new power always ran toward the liberation of early man. Fire began to break up the iron rigidity of night and day and even the discipline of the seasons. It thus carried further the breakdown of the great objective natural rhythms which bound our fireless ancestors. Behaviour could be less routine and automatic. There is even a discernible possibility of leisure.

Big-game hunting was the other great achievement of *Homo erectus*. Its origins must lie far back in the scavenging which turned vegetarian hominids into omnivores. Meat-eating provided concentrated protein. It released meat-eaters from the incessant nibbling of so many vegetarian creatures, and so permitted economies of effort. It is one of the first signs that the capacity for conscious restraint is at work when food is being carried home to be shared tomorrow rather than consumed on the spot today. At the beginning of the archaeological record, an elephant and perhaps a few giraffes and buffaloes were among the beasts whose scavenged meat was consumed at Olduvai, but for a long time the bones of smaller animals vastly preponderate in the rubbish. By about three hundred thousand years ago the picture is wholly altered.

This may be where we can find a clue to the way by which *Australopithecus* and his relatives were replaced by the bigger, more efficient *Homo erectus*. A new food supply permits larger consumption but also imposes new environments: game has to be followed if meat-eating becomes general. As the hominids become more or less parasitic upon other species there follows further exploration of territory and new settlements, too, as sites particularly favoured by the mammoth or woolly rhinoceros are identified. Knowledge of such facts has to be learnt and passed on; technique has to be transmitted and guarded, for the skills required to trap, kill and dismember the huge beasts of antiquity were enormous in relation to anything which preceded them. What is more, they were cooperative skills: only large numbers could carry out so complex an operation as the driving – perhaps by fire – of game to a killing-ground favourable because of bogs in which a weighty creature would flounder, or because of a precipice, well-placed vantage points, or secure platforms for the hunters. Few weapons were available to supplement natural traps and, once dead, the victims presented further problems. With only wood, stone and flint, they had to be cut up and removed to the home base. Once carried home, the new supplies of meat mark another step towards the provision of leisure as the consumer is released for a time from the drudgery of ceaselessly rummaging in his environment for small, but continuously available, quantities of nourishment.

It is very difficult not to feel that this is an epoch of crucial significance. Considered against a background of millions of years of evolution, the pace of change, though still unbelievably slow in terms of later societies, is quickening. This is not man as we know him, but these are beginning to be man-like creatures: the greatest of predators is beginning to stir in his cradle. Something like a true society, too, is dimly discernible, not merely in the complicated cooperative hunting enterprises, but in what this implies in passing on knowledge from generation to generation. Culture and tradition are slowly taking over from genetic mutation and natural selection as the primary sources of change among the hominids. It is the groups with the best 'memories' of effective techniques which will carry forward evolution. The importance of experience was very great, for knowledge of methods which were likely to succeed rested upon it, not (as increasingly in modern society) on experiment and analysis. This fact alone would have given new importance to older men and women. They knew how things were done and what methods worked and they did so at a time when the home base and big-game hunting made their maintenance by the group easier. They would not have been very old, of course. It is unlikely that many lived more than forty years.

Selection also favoured those groups whose members not only had good memories, but the increasing power to reflect upon it given by speech. We know very little about the prehistory of language. Modern types of language can only have appeared long after *Homo erectus* disappeared. Yet some sort of communication must have been used in big-game hunting, and all primates make meaningful signals. How early hominids communicated may never be known, but one plausible suggestion is that they began by breaking up calls akin to those of other animals into particular sounds capable of rearrangement. This would give the possibility of different messages and may be the remote taproot of grammar. What is certain is that a great acceleration of evolution would follow the appearance of groups able to pool experience, to practise and refine skills, to elaborate ideas through language. Once more, we cannot separate one process from others: better vision, an increased physical capacity to deal with the world as a set of discrete objects and the multiplication of artifacts by using tools were all going on simultaneously over the hundreds of thousands of years in which language was evolving. Together they contributed to a growing extension of mental capacity until one day conceptualization became possible and abstract thought appeared.

It remains true, though, that if nothing very general can be confidently said about the behaviour of hominids before man, still less can anything very precise. We move in a fog, dimly apprehending for a moment creatures now more, now less, man-like and familiar. Their minds, we can be sure, are almost inconceivably unlike our own as instruments for the registration of the outside world. Yet when we look at the range of the attributes of *Homo erectus* it is his human, not pre-human, characteristics which are most striking. Physically, he has a brain of an order of magnitude comparable to our own. He makes tools (and does so within more than one technical tradition), builds shelters, takes over natural refuges by exploiting fire, and sallies out of them to hunt and gather his food. He does this in groups with a discipline which can sustain complicated operations; he therefore has some ability to exchange ideas by speech. The basic biological units of his hunting groups probably prefigure the nuclear family of man, being founded on the institutions of the home base and a sexual differentiation of activity. There may even be some complexity of social organization in so far as fire-bearers and gatherers or old creatures whose memories made them the data banks of their 'societies' could be supported by the labour of others. There has to be some social organization to permit the sharing of cooperatively obtained food, too. There is nothing to be usefully added to an account such as this by pretending to say where exactly can be found a prehistorical point or dividing line at which such things had come to be, but subsequent human history is unimaginable without them. When a sub-species of *Homo erectus*, perhaps possessing slightly larger and more complex brains than others, evolved into *Homo sapiens* it did so with an enormous achievement and heritage already secure in its grasp. Whether we choose to call it human or not hardly matters.

The first representations of which was to become an enduring artistic genre, the female nude, survive from what was done by sculptors twenty thousand years ago in what we now call Gravettian culture. Scattered across lands from Siberia to western France, archaeologists have found grotesque, ugly, but powerful figurines and reliefs, the sexual characteristics highly exaggerated, all notion of individual identity lost in their brute force. This example, called the Laussel Venus, comes from the Dordogne. Does it represent a goddess? Was it some sort of talisman whose help was sought to avoid the extinction of mankind by representing elemental woman, the continuator of the species? We cannot know. We can only guess that so widely and obsessively reproduced an image must have had a meaning more powerful than any of the surviving images of men from that age.

2
Homo Sapiens

The appearance of *Homo sapiens* is momentous: here, at last, is recognizable humanity, however raw in form. Yet this evolutionary step is another abstraction. It is the beginning of the main drama and the end of the prologue, but we cannot usefully ask precisely when this happens. It is a process, not a point, and it is not a process occurring everywhere at the same rate. All we have to date it are a few physical relics of early men of types recognizably modern or closely related to the modern. Some of them may well overlap by thousands of years the continuing life of earlier hominids. Some may represent false starts and dead ends, for human evolution must have continued to be highly selective. Though much faster than in earlier times, this evolution is still very slow: we are dealing with something that took place over perhaps two hundred thousand years in which we do not know when our first true 'ancestor' appeared (though the place was almost certainly Africa). It is not ever easy to pose the right questions; the physiological and technical and mental lines at which we leave *Homo erectus* behind are matters of definition.

The few early human fossils have provoked much argument. Two famous European skulls seem to belong to the period between two Ice Ages about two hundred thousand years ago, an age climatically so different from ours that elephants browsed in a semi-tropical Thames valley and the ancestors of lions prowled about in what would one day be Yorkshire. The 'Swanscombe' skull, named after the place where it was found, shows its possessor to have had a big brain (about 1300 cc) but in other ways not much to resemble modern man: if 'Swanscombe man' was *Homo sapiens*, then he represents a very early version. The other skull, that of 'Steinheim man', differs in shape from that of *Homo sapiens* but again held a big brain. Perhaps they are best regarded as the forerunners of early prototypes of *Homo sapiens*, though creatures still living (as their tools show) much like *Homo erectus*.

The next Ice Age then brings down the curtain. When it lifts, a hundred and thirty thousand or so years ago, in the next warm period, human remains again appear. There has been much argument about what they show but it is indisputable that there has been a great step forward. At this point we are entering a period where there is a fairly dense though broken record. Its unravelling can begin in Europe. Creatures we must now call humans lived there just over a hundred thousand years ago. There are caves in the Dordogne area which were occupied on and off for some fifty thousand years after that. The cultures of these peoples therefore survived a period of huge climatic change; the first traces of them belong to a warm interglacial period and the last run out in the middle of the last Ice Age. This is an impressive continuity to set against what must have been great variation in the animal population and vegetation near these sites; to survive so long, such cultures must have been very resourceful and adaptive.

For all their essential similarity to ourselves, though, the peoples who created these cultures are still physiologically distinguishable from modern Man. The first discovery of their remains was at Neanderthal in Germany (because of this, humans of this type are usually called Neanderthal men) and it was of a skull so curiously shaped that it was for a long time thought to be that of a modern idiot. Scientific analysis still leaves much about it unexplained. But it is now thought that *Homo sapiens neanderthalis* (as the Neanderthal is scientifically classified) has its ultimate origin in an early expansion out of

Africa of advanced forms of *Homo erectus*, possibly seven hundred thousand years ago. Across many intervening genetic stages, there emerged a population of pre-Neanderthals, from which, in turn, the extreme form evolved whose striking remains were found in Europe (and, so far, nowhere else). This special development has been interpreted by some as a Neanderthal sub-species, perhaps cut off by some accident of glaciation. Evidence of other Neanderthalers has turned up elsewhere, in Morocco, in the northern Sahara, at Mount Carmel in Palestine and elsewhere in the Near East and Iran. They have also been traced in Central Asia and China, where the earliest specimens may go back something like two hundred millenia. Evidently, it was for a long time a highly successful species.

Eighty thousand years ago, the artifacts of Neanderthal man had spread all over Eurasia. They show differences of technique and form. But technology from over a hundred thousand years ago and associated with other forms of 'anatomically modern humans', as scholars term other creatures evolved from advanced forms of *Homo erectus*, has been identified in parts of Africa. Moreover, it was more widely spread than that of Neanderthal man. The primeval cultural unity had thus already fragmented, and distinct cultural traditions were beginning to emerge. From the start, there is a kind of provincialism within humanity.

Neanderthal man, like other species which specialists refer to as anatomically modern, walked erect and had a big brain. Though in other ways more primitive than the sub-species to which we belong, *Homo sapiens sapiens* (as the guess about the first skull suggests), he represents none the less a great evolutionary stride and shows a new mental sophistication we can still hardly grasp, let alone measure. One striking example is his use of technology to overcome environment: we know from the evidence of skin-scrapers he used to dress skins and pelts that Neanderthal man wore clothes (though none have survived; the oldest clothed body yet discovered, in Russia, has been dated to about thirty-five thousand years ago). Even this important advance in the manipulation of environment, though, is nothing like so startling as the appearance in Neanderthal culture of formal burial. The act of burial itself is momentous for archaeology; graves are of enormous importance because of the artifacts of ancient society they preserve. Yet the Neanderthal graves provide more than this: they also contain the first evidence of ritual or ceremony.

This may mean a great deal and it is very difficult to control speculation. Perhaps some early totemism explains the ring of horns within which a Neanderthal child was buried near Samarkand. Conjecture is stimulated, too, by a fleeting vision of the primitive community in northern Iraq which went out one day to gather the masses of wild flowers and grasses which eventually lay under and surrounded the dead companion it wished so to honour. Some have suggested that careful burial may reflect a new concern for the individual which was one result of the greater interdependence of the group in the renewed Ice Ages. This could have intensified the sense of loss when a member died and might also point to something more. A skeleton of a Neanderthal man who had lost his right arm years before his death has been found. He must have been very dependent on others, and was sustained by his group in spite of his handicap.

More hazardous still is the suggestion that ritualized burial implies some view of an after-life. If this were true, it would testify to a huge power of abstraction in the hominids and the origins of one of the greatest and most enduring myths, that life is an illusion, that reality lies invisible elsewhere, that things are not what they seem. Without going so far, it is at least possible to agree that a momentous change is under way. Like the hints of rituals involving animals which Neanderthal caves also offer here and there, careful burial may mark a new attempt to dominate the environment. The human brain must already have been capable of discerning questions it wanted to answer and perhaps of providing answers in the shape of rituals. Slightly, tentatively, clumsily – however we describe it and still in the shallows though it may be – the human mind is afloat; the greatest voyage of exploration has begun.

Neanderthal man also provides our first evidence of another great institution, warfare. It may have been practised in connexion with cannibalism, which was directed apparently to the eating of the brains of victims. Analogy with later societies suggests that here again we have the start of some conceptualizing about a soul or spirit; such acts are sometimes directed to acquiring the magical or spiritual power of the vanquished. Whatever the magnitude of the evolutionary step which the Neanderthals represent, however, they failed in the end as a sub-species. After long and widespread success they were not in the end to be the inheritors of the earth. Effectively, Neanderthal survivors were genetically 'vanquished' by the strain of *Homo sapiens* which was in the end to be dominant, and about the reasons for this we know nothing. Nor can we know to what extent, if at all, it was mitigated by some genetic transmission through the mingling of stocks.

The successor to Neanderthal man had a smaller face, lighter skull and straighter limbs. This was *Homo sapiens sapiens*, the outstandingly successful sub-species which was to spread world-wide in about twenty thousand years. To it we all belong. Scholars suggest that its origin is to be sought in another expansion out of Africa, about a hundred thousand years ago, which led to the establishment of anatomically modern humans over much of the Levant, the Near East and south-eastern Europe in the next fifty thousand years. Why expansion should have taken this direction rather than another is not known, but it has been suggested that warm climates away from glacial environments had allowed hunters to develop their techniques to a high level, while tropical conditions put different sorts of obstacle in the way of these further south. However this may be, when a shift to drier climates becomes observable, the coastal corridor between the mountains and the sea in what is today Israel and the Lebanon seems especially significant in the appearance of modern Man.

Palaeo-anthropologists are cautious. They do not like to assert that fossil remains more than thirty thousand or so years old are those of *Homo sapiens sapiens*. Nevertheless, it is clear that from about fifty thousand years ago to the end of the last Ice Age in about 9000 BC we are at last considering plentiful evidence of men of modern type. This period is normally referred to as the 'Upper' Palaeolithic, a name derived from the Greek for 'old stones'. It corresponds, roughly, to the more familiar term 'Stone Age', but, like other contributions to the chaotic terminology of prehistory, there are difficulties in using such words without careful qualification.

To separate 'Upper' and 'Lower' Palaeolithic is easy; the division represents the physical fact that the topmost layers of geological strata are the most recent and that, therefore, fossils and artifacts found among them are later than those found at lower levels. The Lower Palaeolithic is therefore the designation of an age more ancient than the Upper. Almost all the artifacts which survive from the Palaeolithic are made from stone; none is made from metal, whose appearance makes it possible to follow a terminology used by the Roman poet Lucretius by labelling what comes after the Stone Age as the Bronze and Iron Ages.

These are, of course, cultural and technological labels; their great merit is that they direct attention to the activities of man. At one time tools and weapons are made of stone, then of bronze, then of iron. None the less, these terms have disadvantages, too. The obvious one is that within the huge tracts of time in which stone artifacts provide the largest significant body of evidence, we are dealing for the most part with hominids. They had, in varying degree, some, but not all, human characteristics; many stone tools are not made by men. Increasingly, too, the fact that this terminology originated in European archaeology created difficulties as more and more evidence accumulated about the rest of the world which did not really fit in. A final disadvantage is that it blurs important distinctions within periods even in Europe. The result has been its further refinement. Within the Stone Age scholars have distinguished (in sequence) the Lower, Middle and Upper Palaeolithic and then the Mesolithic and the Neolithic (the last of which blurs the division attributed by the older schemes to the coming of metallurgy). The period down to the end of the last Ice Age in Europe is also sometimes called the Old Stone Age, another complication, because here we have yet another principle of classification, simply that provided by chronology. *Homo sapiens sapiens* appears in Europe roughly at the beginning of the Upper Palaeolithic, somewhere about forty thousand years ago although anatomically modern humans only shadowily distinct from him had appeared much earlier in Africa. It is in Europe, though, that the largest quantity of skeletal remains has been found, and it is on this evidence that the distinction of the species has been based.

For this period in Europe much has been done to classify and group in sequence cultures identified by their implements. The climate was not constant; though usually cold, there were important fluctuations, probably including the sharp onset of the coldest conditions for a million years somewhere about twenty thousand years ago. Such climatic variations still exercised great determinative force on the evolution of society. It was perhaps thirty thousand years ago that they made it possible for human beings for the first time to enter the Americas, crossing from Asia somewhere in the region of what is now the Bering Strait by a link provided by ice or, perhaps, by land left exposed because the ice-caps contained so much of what is now sea-water and sea-level was much lower. They moved southwards for thousands of years as they followed the game which had drawn them to the last uninhabited continent. The Americas were from the first peopled by immigrants. But the ice sheets also retreated and as they did huge transformtions occurred to coasts, routes and food supplies. This was all as it had been for ages, but there was this time a crucial difference. Man was there. A new order of intelligence was available to use new and growing resources in order to cope with environmental change. The change to history, when conscious human action to control environment will increasingly be effective, is under way.

This may seem a big claim in the light of the resources early men possessed, judging by their tool kits and weaponry. Yet they already represent a huge range of capacities if we compare them with their predecessors'. The basic tools of *Homo sapiens* were stone, but they were made to serve many more precise purposes than earlier tools and were made in a different way, by striking flakes from a carefully prepared core. Their variety and elaboration are another sign of the growing acceleration of human evolution. New materials came into use in the Upper Palaeolithic, too, as bone and antler were added to the wood and flint of earlier workshops and armouries. These provided new possibilities of manufacture; the bone needle was a great step in the elaboration of clothing, pressure flaking enabled some skilled workmen to carry the refinement of their flint blades to a point at which it seems non-utilitarian, so delicately thinned have they become. The first man-made material, a mixture of clay with powdered bone, also makes its appearance. Weapons especially are improved. The tendency which can be seen towards the end of the Upper Palaeolithic for small flint implements to appear more frequently and for them to be more regularly geometrical suggests the making of more complex weapon points. In the same era come the invention and spread of the spear-thrower, the bow and arrow, and the barbed harpoon, used first on mammals and later to catch fish. The last shows an extension of hunting – and therefore of resources – to water. Long before this, perhaps six hundred thousand years ago, hominids had gathered molluscs for food in China and doubtless elsewhere. With harpoons and, perhaps, more perishable implements such as nets and lines, new and richer aquatic sources of food (some created by the temperature changes of the last Ice Ages) could now be exploited, and this led to achievements in hunting, possibly connected with the growth of forests in post-glacial phases and with a new dependence on and knowledge of the movements of reindeer and wild cattle.

It is tempting to see support of this in the most remarkable and mysterious evidence of all which has survived the men of the Upper Palaeolithic: their art. It is the first of whose existence we can be sure. Earlier men or even manlike creatures may have scratched patterns in the mud, daubed their bodies, moved rhythmically in the dance or spread flowers in patterns, but of such things we know nothing, because of them, if they ever happened, nothing has survived. Some creature took the trouble to accumulate little hoards of red ochre some forty or sixty thousand years ago, but the purpose of doing so is unknown. It has been suggested that two indentations on a Neanderthal gravestone are the earliest surviving art, but the first plentiful and assured evidence comes in Europe thirty-five thousand years ago. It then swells dramatically until we find ourselves in the presence of a conscious art whose greatest technical and aesthetic achievements appear, without warning or forerunner, already almost mature. They continue so for thousands of years until this art vanishes. Just as it has no ancestor, it leaves no descendant, though it seems to have employed many of the basic processes of the visual arts still in use today.

Perhaps one of the first essays in European realism, from a cave near Lascaux in the Dordogne; the hunter falls back before the bison which charges him, his throwing-stick useless beside him on the ground, the spear having found its mark in the belly of the enraged beast whose entrails drag on the ground.

Its isolation both in space and time must be ground for suspicion that there is more to be discovered. Caves in Africa abound with prehistoric paintings and carvings dated as far back as twenty-seven thousand years ago and continuing to be added to well into the reign of England's Queen Victoria; in Australia there was cave painting at least twenty thousand years ago. Palaeolithic art is not, therefore, confined to Europe, but what has been discovered there has, so far, been studied much more intermittently. We do not yet know enough about the dating of cave paintings in other parts of the world, nor about the uniqueness of the conditions which led to the preservation in Europe of objects which may have had parallels elsewhere. Nor do we know what may have disappeared; there is a vast field of possibilities of what may have been produced in gesture, sound or perishable materials which cannot be explored. None the less, the art of western Europe in the Upper Palaeolithic, all qualifications made, has a colossal and solid impressiveness which is unique.

Most of it has been found in a relatively small area of south-western France and northern Spain and consists of three main bodies of material: small figures of stone, bone or, occasionally, clay (usually female), decorated objects (often tools and weapons) and the painted walls and roofs of caves. In these caves (and in the decoration of objects) there is an overwhelming preponderance of animal themes. The meaning of these designs, above all in the elaborate sequences of the cave paintings, has intrigued scholars. Obviously, many of the beasts so carefully observed were central to a hunting economy. At least in the French caves, too, it now seems highly probable that a conscious order exists in the sequences in which they are shown. But to go further in the argument is still very hard. Clearly, art in Upper Palaeolithic times has to carry much of a burden later carried by writing, but what its messages mean is still obscure. It seems likely that the paintings were connected with religious or magical practice: African rock painting has been convincingly shown to be linked to magic and shamanism and the selection of such remote and difficult corners of caves as those in which the European paintings have been traced is by itself strongly suggestive that some special rite was carried out when they were painted or gazed upon. (Artificial light, of course, was needed in these dark corners.) The origins of religion have been hinted at in Neanderthal burials and appear even more strongly in those of the Upper Palaeolithic peoples which are often elaborate; here, in their art, is something where inferences are even harder to resist. Perhaps it provides the first surviving relics of organized religion.

The birth, maturity and death of the earliest artistic achievement of mankind in Europe occupies a very long period, perhaps of thirty thousand years. Somewhere about thirty-five thousand years ago appear decorated and coloured objects, often of bone and ivory. Then, fifteen millenia or so later, we reach the first figurative art and, soon afterwards, the peak of the prehistoric aesthetic achievement, the great painted and incised cave 'sanctuaries' (as they have been called), with their processions of animals and mysterious repeated symbolic shapes. This high phase lasted about five thousand years, a startlingly long time for the maintenance of so consistent a style and content. So long a period – almost as long as the whole history of civilization on this planet – illustrates the slowness with which tradition changed in ancient times and its imperviousness to outside influence. Perhaps it is an index, too, of the geographical isolation of prehistoric cultures. The last phase of this art which has been discerned takes the story down to about 9000 BC; in it, the stag more and more replaces other animals as subject matter (no doubt thus reflecting the disappearance of the reindeer and the mammoth as the ice retreated) before a final burst of richly decorated tools and weapons brings Europe's first great artistic achievement to an end. The age which followed produced nothing approaching it in scale or quality; its best surviving relics are a few decorated pebbles. Six thousand years were to pass before the next great art.

For all the splendour of this art, we know little about its collapse. The light is never more than dim in the Upper Palaeolithic and the darkness closes in rapidly – which is to say, of course, over thousands of years. Nevertheless, the impression left by the violence of the contrast between what was before and what came after produces a sense of shock. So relatively sudden an extinction is a mystery. We have no precise dates or even precise sequences: nothing ended in one year or another. There was only a gradual closing down of artistic activity over a long time which seems in the end to have been absolute. Some scholars have blamed climate. Perhaps, they argue, the whole phenomenon of cave art was linked to efforts to influence the movements or abundance of the great game herds on which the hunting peoples relied. As the last Ice Age ebbed and each year the reindeer retreated a little, men sought new and magical techniques to manipulate them, but gradually as the ice sheets withdrew more and more, an environment to which they had successfully adapted disappeared. As it did, so did the hope of influencing nature. *Homo sapiens* was not powerless; far from it, he could adapt, and did, to a new challenge. But for a time one cultural impoverishment at least, the abandonment of his first art, was a consequence of adaptation.

It is easy to see much that is fanciful in such speculation, but difficult to restrain excitement over such an astonishing achievement. People have spoken of the great cave sequences as 'cathedrals' of the Palaeolithic world and such metaphors are justified if the level of achievement and the scale of the work undertaken is measured against what evidence we have of the earlier triumphs of man. With the first great art, the hominids are now left far behind and we have unequivocal evidence of the power of the human mind.

Much else that is known of the Upper Palaeolithic confirms the sense that the crucial genetic changes are behind and that evolution is now a mental and social phenomenon. The distribution of major racial divisions in the world which last down to early modern times is already broadly fixed by the end of the Upper Palaeolithic. Geographical and climatic divisions had produced specializations within *Homo sapiens* in skin pigment, hair characteristics, the shape of the skull and the bone structure of the face. In the earliest Chinese relics of *Homo sapiens* the Mongoloid characteristics are discernible. All the main racial groups are established by 10000 BC, broadly speaking in the areas they dominated until the great resettlement of the Caucasian stocks which was one aspect of the rise of European civilization to world domination after AD 1500. The world was filling up during the Old Stone Age. Men at last penetrated the virgin continents. Somewhere between thirty and fifteen thousand years ago Mongoloid peoples spread over the Americas; the earliest Australian Palaeolithic site has been dated about thirty-five thousand years ago and *Homo sapiens* probably reached that continent by water transport from south-east Asia.

Yet the Upper Palaeolithic world was still a very empty place. Calculations suggest that twenty thousand humans lived in France in Neanderthal times, possibly fifty thousand twenty millenia ago. There were then perhaps ten million humans in the whole world. 'A human desert swarming with game' is one scholar's description of it. They lived by hunting and gathering, and a lot of land was needed to support a family.

However questionable such figures may be, if they are agreed to be of this order of magnitude it is not hard to see that they still mean very slow cultural change. Greatly accelerated though Man's progress in the Old Stone Age may be and much more versatile though he is becoming, he is still taking thousands of years to transmit his learning across the barriers of geography and social division. A man might, after all, live all his life without meeting anyone from another group or tribe, let alone another culture. The divisions which already existed between different groups of *Homo sapiens* open a historical era whose whole tendency was towards the cultural distinction, if not isolation, of one group from another, and this was to increase human variety until reversed by technical and political forces in very recent times.

About the groups in which Upper Palaeolithic man lived there is still much unknown. What is clear is that they were both larger in size than in former times and also more settled. The earliest remains of buildings come from the hunters of the Upper Palaeolithic who inhabited what are now Czechoslovakia and southern Russia. In about 10000 BC in parts of France some clusters of shelters seem to have contained anything from four to six hundred people, but judging by the archaeological record, this was unusual. Something like the tribe probably existed, therefore, though about its organization and hierarchies it is virtually impossible to speak. All that is clear is that there was a continuing sexual specialization in the Old Stone Age as hunting grew more elaborate and its skills more demanding, while settlements provided new possibilities of vegetable gathering by women.

Cloudy though its picture is, none the less, the earth at the end of the Old Stone Age is in important respects one we can recognize. There were still to be geological changes (the English Channel was only to make its latest appearance in about 7000 BC, for example) but we have lived in a period of comparative topographical stability which has preserved the major shapes of the world of about 9000 BC. That world was already firmly Man's world. The descendants of the primates who came out of the trees had, by the acquisition of their tool-making skills, by using natural materials to make shelters and by domesticating fire, by hunting and exploiting other animals, long achieved an important measure of independence of natural rhythms. This had brought them to a high enough level of social organization to undertake important cooperative works. Their needs had provoked economic differentiation between the sexes. Grappling with these and other material problems had led to the transmission of ideas by speech, to the invention of ritual practices and ideas which lie at the roots of religion, and, eventually, to a great

Boys discovered this cave at Lascaux in 1940, after it had been lost for thousands of years. Scholars still argue about the meaning of the great cave paintings of the Upper Palaeolithic; whether they are to be interpreted as elements in a magical ritual or as, in some dim sense, the beginnings of something conceivable as 'pure' art, are still debated. What remains unambiguous is the power of the impression such a cave can still produce.

art. It has even been argued that Upper Palaeolithic man had a lunar calendar. Man as he leaves prehistory is already a conceptualizing creature, equipped with intellect, with the power to objectify and abstract. It is very difficult not to believe that it is this new strength which explains Man's capacity to make the last and greatest stride in prehistory, the invention of agriculture.

3
The Possibility of Civilization

Homo sapiens has existed for at least ten and perhaps twenty times as long as the civilization he has created. The waning of the last Ice Age allowed the long march to civilization to be completed and is the immediate prelude to History. Within five or six thousand years a succession of momentous changes took place; the most important of them was an increase in food supply. Nothing so sharply accelerated human development or had such widespread results until the changes called industrial revolution which have been going on for the last two and a half centuries.

One scholar summed up these changes which mark the end of prehistory with a similar phrase, the 'Neolithic revolution'. Here is another little tangle of potentially misleading terminology, though the last we need consider in prehistory. Archaeologists follow the Palaeolithic era by the Mesolithic and that by the Neolithic (some add a fourth, the Chalcolithic, by which they mean a phase of society in which artifacts of stone and copper are in simultaneous use). The distinction between the first two is really of moment only to the specialist, but all these terms describe cultural facts; they identify sequences of artifacts which show growing resources and capacities. Only the term 'Neolithic' need concern us. It means, at its strictest, a culture in which ground or polished stone tools replace chipped ones (though other criteria are sometimes added to this). This may not seem so startling a change as to justify the excitement over the Neolithic which has been shown by some prehistorians, far less talk of a 'Neolithic revolution'. In fact, though the phrase is still sometimes used it is unsatisfactory because it has had to cover too many different ideas. None the less, it was an attempt to pin down an important and complex change which took place with many local variations. It is still worth while, therefore, to try to understand what made the Neolithic so important.

We can start by noting that even in the narrowest technological sense, the Neolithic phase of human development does not begin, flower or end everywhere at the same time. In one place it may last thousands of years longer than in another and its beginnings are separated from what went before not by a clear line but by a mysterious zone of cultural change. Then, within it, not all societies possess the same range of skills and resources; some discover how to make pottery, as well as polished stone tools, others go on to domesticate animals and begin to gather or raise cereal crops. Slow evolution is the rule and not all societies had reached the same level by the time literate civilization appears. Nevertheless, Neolithic culture is the matrix from which civilization appears and provides the preconditions on which it rests, and they are by no means limited to the production of the highly finished stone tools which gave the phase its name.

We must also qualify the word 'revolution' when discussing this change. Though we leave behind the slow evolutions of the Pleistocene and move into an accelerating era of prehistory, there are still no clear-cut divisions. They are pretty rare in later history; even when they try to do so, few societies ever break with their past. What we can observe is a slow but radical transformation of human behaviour and organization over more and more of the world, not a sudden new departure. It is made up of several crucial changes which make the last period of prehistory identifiable as a unity, whatever we call it.

At the end of the Upper Palaeolithic, Man existed physically much as we know him. He was, of course, still to change somewhat in height and weight, most obviously in those areas of the world where he gained in stature and life expectancy as nutrition improved. In the Old Stone Age it was still unlikely that a man or a woman would reach an age of forty and if they did then they were likely to live pretty miserable lives, in our eyes prematurely aged, tormented by arthritis, rheumatism and the casual accidents of broken bones or rotting teeth. This would only slowly change for the better. The shape of the human face would go on evolving, too, as diet altered. (It seems to be only after AD 1066 that the edge-to-edge bite gave way among Anglo-Saxons to the overbite which was the ultimate consequence of a shift to more starch and carbohydrate, a development of some importance for the later appearance of the English.)

The physical types of men differed in different continents, but we cannot presume that capacities did. In all parts of the world *Homo sapiens sapiens* was showing great versatility in adapting his heritage to the climatic and geographical upheavals of the ebbing phase of the last Ice Age. In the beginnings of settlements of some size and permanence, in the elaboration of technology and in the growth of language and the dawn of characterization in art lay some of the rudimentary elements of the compound which was eventually to crystallize as civilization. But much more than these were needed. Above all, there had to be the possibility of some sort of economic surplus to daily requirements.

This was hardly conceivable except in occasional, specially favourable areas of the hunting and gathering economy which sustained all human life and was the only one known to human beings until about ten thousand years ago. What made it possible was the invention of agriculture.

The importance of this was so great that it does seem to justify a strong metaphor and the 'farming revolution' or 'food-gathering revolution' are terms whose meaning is readily clear. They single out the fact which explains why the Neolithic era could provide the circumstances in which civilizations could appear. Even a knowledge of metallurgy, which was spreading in some societies during their Neolithic phases, is not so fundamental. Farming truly revolutionized the conditions of human existence and it is the main thing to bear in mind when considering the meaning of Neolithic, a meaning once concisely summarized 'a period between the end of the hunting way of life and the beginning of a full metal-using economy, when the practice of farming arose and spread through most of Europe, Asia and North Africa like a slow-moving wave'.

The essentials of agriculture are the growing of crops and the practice of animal husbandry. How these came about and at what places and times is more mysterious. Some environments must have helped more than others; while some peoples pursued game across plains uncovered by the retreating ice, others were intensifying the skills needed to exploit the new, prolific river valleys and coastal inlets rich in edible plants and fish. The same must be true of cultivation and herding. On the whole, the Old World of Africa and Eurasia was better off in domesticable animals than what would later he called the Americas. Not surprisingly, then, agriculture began in more than one place and in different forms. It seems likely that the earliest instance based on the cultivation of primitive forms of millet and rice, occurred in south-east Asia, somewhere about 10000 BC.

Because of what was to happen later and because of the accidents of historical survival and the direction taken by scholarly effort, much more is known about early agriculture in the Near East than its possible precursors in the Far East. There are good reasons for continuing to regard it as a crucial zone, even if it is not given pride of place in terms of chronology. Both the predisposing conditions and the evidence point to the region later called the 'Fertile Crescent' as especially significant, the whole arc of territory running northward from Egypt through Palestine and the Levant, through Anatolia to the hills between Iran and the south Caspian to enclose the river valleys of Mesopotamia. Much of it now looks very different from the same area's lush landscape when the climate was at its best, five thousand or so years ago. Wild barley and a wheatlike cereal then grew in southern Turkey and emmer, a wild wheat, in the Jordan valley. Egypt enjoyed enough rain for the hunting of big game well into historical times, and elephants were still to be found in Syrian forests in 1000 BC. The whole region today is fertile by comparison with the desert which is its boundary, but in prehistoric times it was even more favoured. The cereal grasses which are the ancestors of later crops have been traced back furthest in these lands. There is evidence of the harvesting, though not necessarily of the cultivating, of wild grasses in Asia Minor in about 9500 BC. There, too, the afforestation which followed the end of the last Ice Age seems to have presented a manageable challenge; population pressure might well have stimulated attempts to extend living-space by clearing and planting when hunting-gathering areas became overcrowded. From this region the new foods and the techniques for planting and harvesting them seem to have spread into Europe in about 6000 BC. Within the region, of course, contacts were relatively easier than outside it; a date as early as 8000 BC has been given to discoveries of bladed tools found in south-west Iran but made from obsidian which came from Anatolia. But diffusion was not the only process at work. Agriculture later appeared in the Americas, seemingly without any import of techniques from outside.

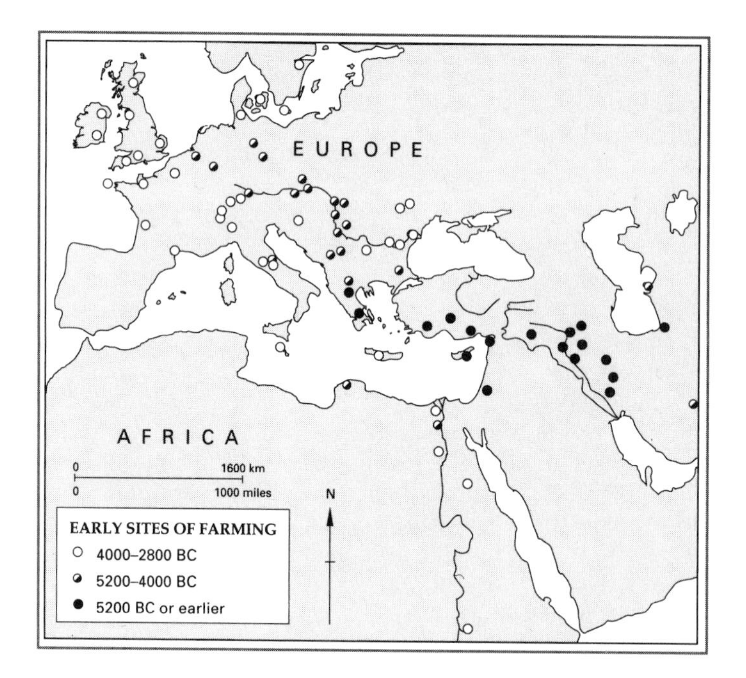

The jump from gathering wild cereals to planting and harvesting them seems marginally greater than that from driving game for hunting to herding, but the domestication of animals was almost as momentous. The first traces of the keeping of sheep come from northern Iraq, in about 9000 BC. Over such hilly, grassy areas the wild forbears of the Jersey cow and the Sussex White pig roamed untroubled for thousands of years except by occasional contact with their hunters. Pigs, it is true, could be found all over the Old World, but sheep and goats were especially plentiful in Asia Minor and a region running across much of Asia itself. From their systematic exploitation would follow the control of their breeding and other economic and technological innovations. The use of skins and wool opened new possibilities; the taking of milk launched dairying. Riding and the use of animals for traction would come later. So would domestic poultry.

The story of mankind is now far past the point at which the impact of such changes can be easily grasped. Suddenly, with the coming of agriculture, the whole material fabric on which subsequent human history was to be based flashes into view, though not into existence. It began the greatest of man's transformations of the environment. In a hunting-gathering society thousands of acres are needed to support a family, whereas in primitive agricultural society about twenty-five acres is enough. In terms of population growth alone, a huge acceleration became possible.

An assured or virtually assured food surplus also meant settlements of a new solidity. Bigger populations could live on smaller areas and true villages could appear. Specialists not engaged in food production could be tolerated and fed more easily while they practised their own skills. Before 9000 BC there was a village (and perhaps a shrine) at Jericho. A thousand years later it had grown to some eight to ten acres of mud-brick houses with substantial walls.

It is a long time before we can discern much of the social organization and behaviour of such communities. It seems possible that at this time, as much as at any other, local divisions of mankind were decisive. Physically, men were more uniform than ever, but culturally they were diversifying as they grappled with different problems and appropriated different resources. The adaptability of different branches of *Homo sapiens* in the conditions left behind after the retreat of the last Ice Age is very striking and produced variations in experience unlike those following earlier glaciations. They lived for the most part in isolated, settled traditions, in which the importance of routine was overwhelming. This would give new stability to the divisions of culture and race which had appeared so slowly throughout Palaeolithic times. It would take much less time in the historical future which lay ahead for these local peculiarities to crumble under the impact of population growth, speedier communication and the coming of trade – a mere ten thousand years, at most. Within the new farming communities it seems likely that distinctions of role multiplied and new collective disciplines had to be accepted. For some people there must have been more leisure (though for others actually engaged in the production of food, leisure may well have diminished). It certainly seems likely that social distinctions became more marked. This may be connected with new possibilities as surpluses became available for barter which led eventually to trade.

The same surpluses may also have encouraged Man's oldest sport after hunting, warfare. New prizes must have made raids and conquest more tempting. Perhaps, too, a conflict, which was to have centuries of vitality before it, finds its origins here - that between nomads and settlers. Political power may have its origin in the need to organize protection for crops and stock from human predators. We may even speculate that the dim roots of the notion of aristocracy are to be sought in the successes (which must have been frequent) of hunter-gatherers, representatives of an older social order, in exploiting the vulnerability of the settlers, tied to their areas of cultivation, by enslaving them. Hunting was long to be the sport of kings and mastery of the animal world was an attribute of the first heroes of whose exploits we have records in sculpture and legend. None the less, though the just prehistoric world must have been lawless and brutal, it is worth remembering that there was an offsetting factor: the world was still not very full. The replacement of hunter-gatherers by farmers did not have to be a violent process. The ample space and thin populations of Europe on the eve of the introduction of farming may explain the lack of archaeological evidence of struggle. It was only slowly that growing populations and pressure on the new farming resources increased the likelihood of competition.

In the long run metallurgy changed things as much as did farming, but it was to be a very much longer run. Immediately, it made a less rapid and fundamental difference. This is probably because the deposits of ore first discovered were few and scattered: for a long time there was just not much metal around. The first of whose use we find evidence is copper (which rather weakens the attractiveness of the old term 'Bronze Age' for the beginning of metal-using culture). At some time between 6000 and 7000 BC it was being hammered into shape without heating at Çatal Hüyük, in Anatolia, though the earliest known metal artifacts date from about 4000 BC and are beaten copper pins found in Egypt. Once the technique of blending copper with tin (which was in use in Mesopotamia soon after 3000 BC) to produce bronze was discovered, a metal was available which was both relatively easy to cast and retained a much better cutting edge. On this much could be built; from it much derived, among other results the quite new importance of ore-bearing areas. In its turn, this was to give a new twist to trade, to markets and to routes. Further complications, of course, followed the coming of iron, which appeared after some cultures had indisputably evolved into civilizations – another reflexion of the way in which the historical and prehistoric eras run so untidily into one another. Its obvious military value springs to the eye, but it had just as much importance when turned into agricultural tools. This is looking a long way ahead, but it made possible a huge extension of living-space and food-producing soil: however successfully he burned woodland and scrub, Neolithic man could only scratch at heavy soils with an antler or wooden pick. Turning them over and digging deep began to be possible only when the invention of ploughing (in the Near East in about 3000 BC) brought animal muscle-power to the assistance of human, and when iron tools became common.

It is already clear how quickly – the term is legitimate against the background of earlier prehistory even if it takes thousands of years in some places – interpenetration and interplay begin to influence the pace and direction of change. Long before these processes have exhausted their effects in some areas, too, the first civilizations are in being. Prehistorians used to argue whether innovations were diffused from a single source or appeared spontaneously and independently in different places, but so complex a background has made this seem a waste of time and energy. Both views, if put forward in an unqualified way, seem untenable. To say that in one place, and in one place only, all the conditions for the appearance of new phenomena existed and that these were then simply diffused elsewhere was as implausible as saying that in widely differing circumstances of geography, climate and cultural inheritance exactly the same inventions could be thrown up, as it were, time and time again. What we can observe is a concentration of factors in the Near East which made it at one crucial moment immeasurably the most concentrated, active and important centre of new developments. It does not mean that similar individual developments may not have occurred elsewhere: pottery, it seems, was first produced in Japan in about 10000 BC, and agriculture evolved in America perhaps as early as 5000 BC in complete isolation from the Old World.

This means that the prologue to human history comes to an end in a ragged, untidy way; once again, there is no neat dividing line. At the end of prehistory and on the eve of the first civilizations we can discern a world of human societies more differentiated than ever before and more successful than ever in mastering different environments and surviving. Some will continue into history. It is only within the last century or so that the Ainus of northern Japan have disappeared, taking with them a life that is said to have been very similar to one they lived fifteen thousand years ago. Englishmen and Frenchmen who went to North America in the sixteenth century AD found hunter-gatherers there who must have lived much as their own ancestors had done ten thousand years before. Plato and Aristotle were to live and die before prehistory in America gave way to the appearance of the great Mayan civilization of Yucatán, and prehistory lasted for Eskimos and Australian aborigines until the nineteenth century.

No crude divisions of chronology, therefore, will help in unravelling so interwoven a pattern. But its most important feature is clear enough: by 6000 or 5000 BC, there existed in at least one area of the Old World all the essential constituents of civilized life. Their deepest roots lay hundreds of thousands of years further back, in ages dominated by the slow rhythm of genetic evolution. Through the Upper Palaeolithic eras the pace of change had quickened by a huge factor as culture slowly became more important, but this was as nothing to what was to follow. Civilization was to bring conscious attempts on a quite new scale to control and organize men and their environment. It builds on a basis of cumulative mental and technological resources and the feedback from its own transformations further accelerates the process of change. Ahead lies faster development in every field, in the technical control of environment, in the elaboration of mental patterns, in the changing of social organization, in the accumulation of wealth, in the growth of population.

It is important to get our perspective in this matter right. From some modern points of view the centuries of the European Middle Ages look like a long slumber. No medievalist would agree, of course, but a twentieth-century man who is impressed by the rapidity of the change which encompasses him and the relative immobility of medieval society ought to reflect that the art which develops from the Romanesque of Charlemagne's Aachen to the Flamboyant of fifteenth-century France was revolutionized in five or six centuries; in a period about ten times as long, the first known art, that of Upper Palaeolithic Europe, shows, by comparison, insignificant stylistic change. Further back, the pace is even slower as the long persistence of early tool types shows. Still more fundamental changes are even less easy to comprehend. So far as we know, the last twelve thousand years register nothing new in human physiology comparable to the colossal transformations of the early Pleistocene which are registered for us in a handful of relics of a few of nature's experiments, yet those took hundreds of thousands of years.

In part, the contrast in the rate of change is the one with which we began, that between Nature and Man as makers of change. Man increasingly chooses for himself and even in prehistory the story of change is therefore increasingly one of conscious adaptation. So the story will continue into historical times, more intensively still. This is why the most important part of the story of Man is the story of consciousness; when, long ago, it broke the genetic slow march, it made everything else possible. Nature and nurture are there from the moment that Man is first identifiable; perhaps they can never be quite disentangled, but man-made culture and tradition are increasingly the determinants of change.

Two reflexions ought, none the less, to be made to balance this indisputable fact. The first is that Man has almost certainly not shown any improvement in innate capacity since the Upper Palaeolithic. His physique has not changed fundamentally in forty thousand years or so and it would be surprising if his mental capacity had done so. So short a time could hardly suffice for genetic changes comparable to those of earlier eras. The rapidity with which humanity has achieved so much since prehistoric times can be accounted for quite simply: there are many more of us upon whose talents humanity can draw and, more important still, human achievements are essentially cumulative. They rest upon a heritage itself accumulating at, as it were, compound interest. Primitive societies had far less inherited advantage in the bank. This makes the magnitude of their greatest steps forward all the more amazing.

If this is speculative, the second reflexion need not be: Man's genetic inheritance not only enables him to make conscious change, to undertake an unprecedented kind of evolution, but also controls and limits him. The irrationalities of this century show the narrow limits of Man's capacity for conscious control of his destiny. To this extent, he is still determined, still unfree, still a part of a nature which produced his unique qualities in the first place only by evolutionary selection. It is not easy to separate this part of his inheritance, either, from the emotional shaping he has received from the processes through which he evolved. That shaping still lies deep at the heart of all our aesthetic and affective life. Man must live with an inbuilt dualism. To deal with it has been the aim of most of the great philosophies and religions and the mythologies by which we still live, but they are themselves moulded by it. As we move from prehistory to history it is important not to forget that its determining effect still proves much more resistant to control than those blind prehistoric forces of geography and climate which were so quickly overcome. Nevertheless, Man at the edge of history is already the creature we know, Man the change-maker.

Cretan religion has attracted much speculation; what goddess can have been represented by this little faience statuette from Knossos? More can perhaps be inferred from it about Cretan dress than about Cretan mentality: other archaeological finds suggest that she wears much the same costume as that of court ladies in Minoan times.

BOOK II
THE FIRST CIVILIZATIONS

Ten thousand years ago, the physical shape of the world was much what it is today. The outlines of the continents were broadly those we know and the major natural barriers and channels of communication have remained constant ever since. By comparison with the upheavals of the hundreds of millennia preceding the end of the last Ice Age, climate, too, was from this time stable; from this point the historian need only regard its short-term fluctuations. Ahead there lay the age (in which we still live) in which most change was going to be man-made.

Civilization has been one of the great accelerators of such change. It began at least seven times says one historian, meaning by that that he can distinguish at least seven occasions on which particular mixes of human skills and natural facts came together to make possible a new order of life based on the exploitation of nature. Though all these beginnings fall within a span of three thousand years or so – barely a moment by comparison with the vast scale of prehistory – they were neither simultaneous, nor equally successful. They turned out very differently, some of them racing ahead to lasting achievements while others declined or disappeared, even if after spectacular flowerings. Yet all of them signified an increase in the rate and scale of change dramatic by comparison with anything achieved in earlier times.

Some of these early civilizations are still real foundations of our own world. Some of them, on the other hand, now exercise little or no influence, except perhaps upon our imaginations and emotions when we contemplate the relics which are all that is now left of them. None the less, together they determined much of the cultural map of the world down to this day because of the power of the traditions which sprang from them even when their achievements in ideas, social organization or technology had long been forgotten. The establishment of the first civilizations took place between about 3500 BC *and 500* BC *and it provides the first of the major chronological divisions of world history.*

1
Early Civilized Life

For as long as we know there has been at Jericho a never-failing spring feeding what is still a sizable oasis. No doubt it explains why people have lived there on and off for about ten thousand years. Farmers clustered about it in late prehistoric times; its population may then have numbered two or three thousand. Before 6000 BC it had great water tanks which suggest provision for big needs, possibly for irrigation, and there was a massive stone tower which was part of elaborate defences long kept in repair. Clearly its inhabitants thought they had something worth defending; they had property. Jericho was a considerable place.

For all that, it was not the beginnings of a civilization; too much was still lacking and it is worth considering for a moment at the outset of the era of civilization just what it is we are looking for. It is a little like the problem of pinning down in time the first human beings. There is a shaded area in which we know the change occurs, but we can still disagree about the point at which a line has been crossed. All over the Near East around 5000 BC farming villages provided the agricultural surpluses on which civilization could eventually be raised. Some of them have left behind evidence of complex religious practice and elaborate painted pottery, one of the most widespread forms of art in the Neolithic era. Somewhere about 6000 BC brick building was going on in Turkey at Çatal Hüyük, a site only slightly younger than Jericho. But by civilization we usually mean something more than ritual, art or the presence of a certain technology, and certainly something more than the mere agglomeration of human beings in the same place.

It is a little like speaking of 'an educated man': everyone can recognize one when they see him, but not all educated men are recognized as such by all observers, nor is a formal qualification (a university degree, for example) either a necessary or infallible indicator. Dictionary definitions are of no help in pinning down 'civilization', either. That of the *Oxford English Dictionary* is indisputable but so cautious as to be useless: 'a developed or advanced state of human society'. What we have still to make up our minds about is *how far* developed or advanced and along what lines.

Some have said that a civilized society is different from an uncivilized society because it has a certain attribute – writing, cities, monumental building have all been suggested. But agreement is difficult and it seems safer not to rely on any such single test. If, instead, we look at examples of what everyone has agreed to call civilizations and not at the marginal and doubtful cases, then it is obvious that what they have in common is complexity. They have all reached a level of elaboration which allows much more variety of human action and experience than even a well-off primitive community. Civilization is the name we give to the interaction of human beings in a very creative way, when, as it were, a critical mass of cultural potential and a certain surplus of resources have been built up. In civilization this releases human capacities for development at quite a new level and in large measure the development which follows is self-sustaining.

This is somewhat abstract and it is time to turn to examples. Somewhere about 3500 BC is the starting-point of the story of civilizations and it will be helpful to set out a rough overall chronology right at the start. We begin with the first recognizable civilization in Mesopotamia. The next example is in Egypt, where civilization is observable at a slightly later date, perhaps about 3100 BC. Another marker in the Near East is Minoan civilization in Crete, appearing about 2000 BC, and from that time we can disregard

The invention of writing was momentous. This limestone tablet is one of the earliest documents we have of the stages by which it happened. It was found in southern Mesopotamia and has been dated to about 3500 BC. The little pictures are of hands, feet and some kind of sledge (perhaps used for threshing grain); they mark a big stride towards using signs to stand for things, a process already foreshadowed in symbols on pottery. Later, the Egyptians clung to such 'pictographic' forms of writing, elaborating them into the 'hieroglyphics' which recorded their life and business for thousands of years. By then, the Mesopotamian tradition had taken another direction, away from picture-writing towards patterns of signs made of identical marks in different arrangements which could be easily formed by a reed-stalk stamp on a soft clay tablet. This process took about fifteen hundred years after these tablets were made.

questions of priorities in this part of the world: it is already a complex of civilizations in interplay with one another. Meanwhile, by that time, and perhaps about 2500 BC, another civilization has appeared in India and it is at least in a measure literate. China's first civilization starts later, towards the middle of the second millennium BC. Later still come the meso-Americans. Once we are past about 1500 BC, though, only this last example is sufficiently isolated for interaction not to be a big part of explaining what happens. From that time, there are no civilizations to be explained which appear without the stimulus, shock or inheritance provided by others which have appeared earlier. For the moment, then, our preliminary sketch is complete enough at this point.

About these first civilizations (whose appearance and shaping is the subject-matter of the next few chapters) it is very difficult to generalize. Of course they all show a low level of technological achievement, even if it is astonishingly high by comparison with that of their uncivilized predecessors. To this extent their shape and development were still determined much more than those of our own civilization by their setting. Yet they had begun to nibble at the restraints of geography. The topography of the world was already much as it is today; the continents were settled in the forms they now have and the barriers and channels to communication they supplied were to be constants, but there was a growing technological ability to exploit and transcend them. The currents of wind and water which directed early maritime travel have not changed, and even in the second millennium BC men were learning to use them and to escape from their determining force.

This suggests, correctly, that at a very early date the possibilities of human interchange were very considerable and this makes it very unwise to dogmatize about civilization appearing in any standard way in different places. Arguments have been put forward about favourable environments, river valleys for example: obviously, their rich and easily cultivated soils could support fairly dense populations of farmers in villages which would then grow to form the first cities. This was decisive in Mesopotamia, Egypt, the Indus valley and China. But cities and civilizations have also arisen away from river valleys, in meso-America, Minoan Crete and, later, in Greece. With the last two, there is the strong likelihood of important influence from the outside, but Egypt and the Indus valley, too, were in touch with Mesopotamia at a very early date in their evolution. Evidence of such contact led at one time to the view put forward a few years ago that we should look for one central source of civilization from which all others came. This is not now very popular. There is not only the awkward case of civilization in the isolated Americas to deal with, but great difficulty in getting the time-table of the supposed diffusion right as more and more knowledge of early chronology is acquired by the techniques of radio-carbon dating.

The most satisfactory answer appears to be that civilization was likely always to result from the coming together of a number of factors predisposing a particular area to throw up something dense enough to be recognized later as civilization, but that different environments, different influences from outside and different cultural inheritances from the past mean that men did not move in all parts of the world at the same pace or even towards the same goals. The idea of a standard pattern of social 'evolution' was discredited even before the idea of 'diffusion' from a common civilizing source. Clearly, a favourable geographical setting was essential; in the first civilizations everything rested on the existence of an agricultural surplus. But another factor was just as important – the capacity of the peoples on the spot to take advantage of an environment or rise to a challenge, and here external contacts may be as important as tradition. China seems at first sight almost insulated from the outside, but even there possibilities of contact existed. The way in which different societies generate the critical mass of elements necessary to civilization therefore remains very hard to pin down.

It is easier to say something generally true about the marks of early civilization than about the way it happened. Again, no absolute and universal statements are plausible. Civilizations have existed without writing, useful as it is for storing and using experience. More mechanical skills have been very unevenly distributed, too: the meso-Americans carried out major building operations with neither draught animals nor the wheel, and the Chinese could cast iron nearly fifteen hundred years before Europeans. Nor have all civilizations followed the same patterns of growth; there are wide disparities between their staying-power, let alone their successes.

None the less, early civilizations, like later ones, seem to have a common positive characteristic in that they change the human scale of things. They bring together the cooperative efforts of more men and women than in earlier societies and usually do this by physically bringing them together in larger agglomerations, too. Our word 'civilization' suggests, in its Latin roots, a connexion with urbanization. Admittedly, it would be a bold man who was willing to draw a precise line at the moment when the balance tipped from a dense pattern of agricultural villages clustered around a religious centre or a market to reveal the first true city. Yet it is perfectly reasonable to say that more than any other institution the city has provided the critical mass which produces civilization and that it has fostered innovation better than any other environment so far. Inside the city the surpluses of wealth produced by agriculture made possible other things characteristic of civilized life. They provided for the upkeep of a priestly class which elaborated a complex religious structure, leading to the construction of great buildings with more than merely economic functions, and eventually to the writing down of literature. Much bigger resources than in earlier times were thus allocated to something other than immediate consumption and this meant a storing of enterprise and experience in new forms. The accumulated culture gradually became a more and more effective instrument for changing the world.

This little picture of a scene at a Sumerian dairy was executed in limestone and shell set into bitumen. It probably decorated the walls of a temple dedicated to a fertility goddess often symbolized by a cow.

One change is quickly apparent: in different parts of the world men grew more rapidly more unlike one another. The most obvious fact about early civilizations is that they are startlingly different in style, but because it is so obvious we usually overlook it. The coming of civilization opens an era of ever more rapid differentiation – of dress, architecture, technology, behaviour, social forms and thought. The roots of this obviously lie in prehistory, when there already existed men with different life-styles, different patterns of existence, different mentalities, as well as different physical characteristics. With the emergence of the first civilizations this becomes much more obvious, but is no longer merely the product of the natural endowment as environment, but of the creative power of civilization itself. Only with the rise to dominance of Western technology in the twentieth century has this variety begun to diminish. From the first civilizations to our own day there have always been alternative models of society available even if they knew little of one another.

Much of this variety is very hard to recover. All that we can do in some instances is to be aware that it is there. At the beginning there is still little evidence about the life of the mind except institutions so far as we can recover them, symbols in art and ideas embodied in literature. In them lie presuppositions which are the great coordinates around which a view of the world is built – even when the people holding that view do not know they are there (History is often the discovery of what people did not know about themselves). Many of them are irrecoverable, and even when we can begin to grasp the shapes which defined the world of men living in the old civilizations, a constant effort of imagination must be made to avoid the danger of falling into anachronism which surrounds us on every side. Even literacy does not reveal very much of the minds of creatures so like and yet so unlike ourselves.

It is in the Near East that the stimulating effects of different cultures upon one another first become obvious and no doubt it is much of the story of the appearance of the earliest civilizations there. A turmoil of racial comings and goings for three or four thousand years both enriched and disrupted this area, where our history must begin. The Fertile Crescent was to be for most of historic times a great crucible of cultures, a zone not only of settlement but of transit, through which poured an ebb and flow of people and ideas. In the end this produced a fertile interchange of institutions, language and belief from which stems much of human thought and custom even today.

Why this began to happen cannot exactly be explained, but the overwhelming presumption must be that the root cause was over-population in the lands from which the intruders came. Over-population is at first sight a curious notion to apply to a world whose total population in about 4000 BC has been estimated only at between eighty and ninety millions. In the next four thousand years it grew by about fifty per cent to about one hundred and thirty millions; this implies an annual increase almost imperceptible by comparison with those we take for granted. It shows both the relative slowness with which our species added to its power to exploit the natural world and how much and how soon the new possibilities of civilization had already reinforced Man's propensity to multiply and prosper by comparison with prehistoric times.

Such growth was still slight by later standards because it was always based on a very fragile margin of resources and it is this fragility which justifies talk of over-population. Drought or desiccation could dramatically and suddenly destroy an area's capacity to feed itself and it was to be thousands of years before food could easily be brought from elsewhere. The immediate results must often have been famine, but in the longer run there were others more important. The disturbances which resulted were the prime movers of early history; climatic change was still at work as a determinant, though now in much more local and specific ways. Droughts, catastrophic storms, even a few decades of marginally lower or higher temperatures, could force peoples to get on the move and so help to bring on civilization by throwing together peoples of different tradition. In collision and cooperation they learnt from one another and so increased the total potential of their societies.

The peoples who are the actors of early history in the Near East all belonged to the light-skinned human family (sometimes confusingly termed Caucasian) which is one of the three major ethnic classifications of the species *Homo sapiens* (the others being Negroid and Mongoloid). Linguistic differences make it possible to distinguish them further. All the peoples in the Fertile Crescent of early civilized times can be assigned

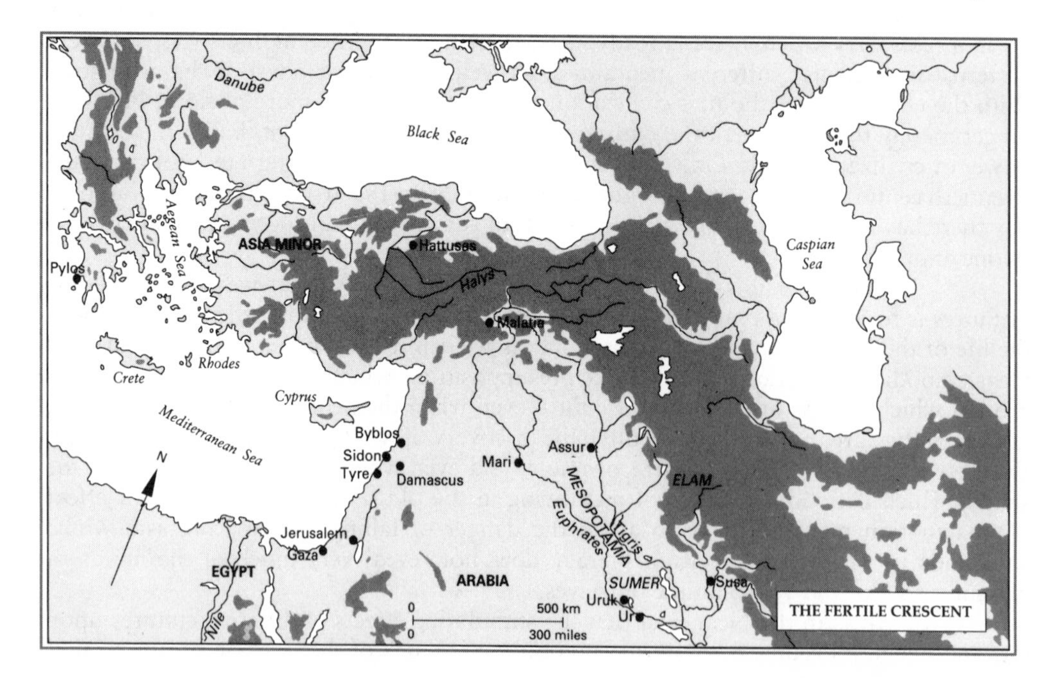

THE FERTILE CRESCENT

either to the Hamitic stocks who evolved in Africa north and north-east of the Sahara, to the Semites of the Arabian peninsula, to the Indo-Europeans who, from southern Russia, had spread also by 4000 BC into Europe and Iran, or to the true 'Caucasians' of Georgia. These are the *dramatis personae* of early Near Eastern history. Their historic centres all lay round the zone in which agriculture and civilization appear at an early date. The wealth of as well-settled an area must have attracted peripheral peoples.

By about 4000 BC most of the Fertile Crescent was occupied by Caucasians. Probably Semitic peoples had already begun to penetrate it by then, too; their pressure grew until by the middle of the third millennium BC (long after the appearance of civilization) they would be well established in central Mesopotamia, across the middle sections of the Tigris and Euphrates. The interplay and rivalry of the Semitic peoples with the Caucasians, who were able to hang on to the higher lands which enclosed Mesopotamia from the north-east, is one continuing theme scholars have discerned in the early history of the area. By 2000 BC the peoples whose languages form part of what is called the Indo-European group have also entered on the scene, and from two directions. One of these peoples, the Hittites, pushed into Anatolia from Europe, while their advance was matched from the east by that of the Iranians. Between 2000 BC and 1500 BC branches of these sub-units dispute and mingle with the Semitic and Caucasian peoples in the Crescent itself, while the contacts of the Hamites and Semites lie behind much of the political history of old Egypt. This scenario is, of course, highly impressionistic. Its value is only that it helps to indicate the basic dynamism and rhythms of the history of the ancient Near East. Much of its detail is still highly uncertain (as will appear) and little can be said about what maintained this fluidity. None the less, whatever its cause, this wandering of peoples was the background against which the first civilization appeared and prospered.

2
Ancient Mesopotamia

The best case for the first appearance of something which is recognizably civilization has been made for the southern part of Mesopotamia, the seven-hundred-mile-long land formed by the two river valleys of the Tigris and Euphrates. This end of the Fertile Crescent was thickly studded with farming villages in Neolithic times. Some of the oldest settlements of all seem to have been in the extreme south where deposits from centuries of drainage from up-country and annual floodings had built up a soil of great richness. It must always have been much easier to grow crops there than elsewhere provided that the water supply could be made continuously and safely available; this was possible, for though rain was slight and irregular, the river bed was often above the level of the surrounding plain. A calculation has been made that in about 2500 BC the yield of grain in southern Mesopotamia compared favourably with that of the best Canadian wheat-fields today. Here, at an early date, was the possibility of growing more than was needed for daily consumption, the surplus indispensable to the appearance of town life. Furthermore, fish could be taken from the nearby sea.

Such a setting was a challenge, as well as an opportunity. The Tigris and Euphrates could suddenly and violently change their beds: the marshy, low-lying land of the delta had to be raised above flood level by banking and ditching and canals had to be built to carry water away. Thousands of years later, techniques could still be seen in use in Mesopotamia which were probably those first employed to form the platforms of reed and mud on which were built the first homesteads of the area. These patches of cultivation would be grouped where the soil was richest. The drains and irrigation channels they needed could be managed properly only if they were managed collectively. No doubt the social organization of reclamation was another result. However it happened, the seemingly unprecedented achievement of making land from watery marsh must have been the forcing house of a new complexity in the way men lived together.

As the population rose, more land was taken to grow food. Sooner or later men of different villages would have come face to face with others intent on reclaiming marsh which had previously separated them from one another. Different irrigation needs may even have brought them into contact before this. There was a choice: to fight or to cooperate. Each meant further collective organization and a new agglomeration of power. Somewhere along this path it made sense for men to band together in bigger units than hitherto for self-protection or management of the environment. One physical result is the town, mud-walled at first to keep out floods and enemies, raised above the waters on a platform. It was logical for the local deity's shrine to be the place chosen: he stood behind the community's authority. It would be exercised by his chief priest, who became the ruler of a little theocracy competing with others.

Something like this explains the difference between southern Mesopotamia in the third and fourth millennia BC and the other zones of Neolithic culture with which by then it had already been long in contact. There is plenty of evidence in the existence of pottery and characteristic shrines of links between Mesopotamia and the Neolithic cultures of Anatolia, Assyria and Iran. They all had much in common. But only in one relatively small area did a pattern of village life common to much of the Near East begin to grow faster and harden into something else. From that background emerges the first true

Between four and five thousand years ago, little statues like these were carved by Mesopotamians for shrines where they were to pray and worship on behalf of those they represented. Perhaps they were statues of the sculptors and their families, perhaps those of wealthy men who commissioned them. The second is more likely: stone was scarce and must have been expensive in Sumer (the tallest of these figures is only about thirty inches high). These examples were discovered in what had been a pit under the floor of a temple at Tel Asmar, on a tributary of the Tigris. Besides throwing light on the way the Mesopotamians worshipped, they tell us something of the daily appearance of men and women of early civilization.

urbanism, that of Sumer, and the first observable civilization.

Sumer is an ancient name for southern Mesopotamia, which then extended about a hundred miles less to the south than it does now. The people who lived there may have been Caucasians, unlike their Semitic neighbours to the south-west and like their northern neighbours the Elamites who lived on the other side of the Tigris. Scholars are still divided about when the Sumerians - that is, those who spoke the language later called Sumerian - arrived in the area: they may have been there since about 4000 BC. But since we know the population of civilized Sumer to be a mixture of races, perhaps including the earlier inhabitants of the region, with a culture which mixed foreign and local elements, it does not much matter.

Sumerian civilization had deep roots. The people long shared a way of life not very different from that of their neighbours. They lived in villages and had a few important cult centres which were continuously occupied. One of these, at a place called Eridu, probably originated in about 5000 BC. It grew steadily well into historic times and by the middle of the fourth millennium there was a temple there which some have thought to have provided the original model for Mesopotamian monumental architecture, though nothing is now left of it but the platform on which it rested. Such cult centres began by serving those who lived near them. They were not true cities, but places of devotion and pilgrimage. They may have had no considerable resident populations, but they were usually the centres around which cities later crystallized and this helps to explain the close relationship religion and government always had in ancient Mesopotamia. Well before 3000 BC some such sites had very big temples indeed; at Uruk (which is called Erech in the Bible) there was an especially splendid one, with elaborate decoration and impressive pillars of mud brick, eight feet in diameter.

Pottery is among the most important evidence linking pre-civilized Mesopotamia with historic times. It provides one of the first clues that something culturally important is going forward which is qualitatively different from the evolutions of the Neolithic. The so-called Uruk pots (the name is derived from the site where they were found) are often duller, less exciting than earlier ones. They are, in fact, mass-produced, made in standard form on a wheel. The implication of this is strong that when they came to be produced there already existed a population of specialized craftsmen; it must have been maintained by an agriculture sufficiently rich to produce a surplus exchanged for their creations. It is with this change that the story of Sumerian civilization can conveniently be begun.

It lasts about thirteen hundred years (roughly from 3300 to 2000 BC), which is about as much time as separates us from the age of Charlemagne. At the beginning comes the invention of writing, possibly the only invention of comparable importance to the invention of agriculture before the age of steam. Writing had been preceded by the invention of cylinder seals, on which little pictures were incised to be rolled on to clay; pottery may have degenerated, but these seals were one of the great Mesopotamian artistic achievements. The earliest writings are in the form of pictograms or simplified pictures (a step towards non-representative communication), on clay tablets usually baked after they had been inscribed with a reed stalk. The earliest are in Sumerian and it can be seen that they are memoranda, lists of goods, receipts; their emphasis is economic and they cannot be read as continuous prose. The writing on these early notebooks and ledgers evolved slowly towards cuneiform, a way of arranging impressions stamped on clay by the wedge-like section of a chopped-off reed. With this the break with the pictogram form is complete. Signs and groups of signs come at this stage to stand for phonetic and possibly syllabic elements and are all made up of combinations of the same basic wedge shape. It was more flexible as a form of communication by signs than anything used hitherto and Sumer reached it soon after 3000 BC.

A fair amount is known about the Sumerian language. A few of its words have survived to this day; one of them is the original form of the word 'alcohol', which is suggestive. But its greatest interest is its appearance in written forms. Literacy must have been both unsettling and stabilizing. On the one hand it offered huge new possibilities of communicating; on the other it stabilized practice because the consultation of records became possible. It made much easier the complex operations of irrigating lands, harvesting and storing crops, which were fundamental to a growing society. Writing made for more efficient exploitation of resources. It also immensely strengthened government and emphasized its links with the priestly castes who at first monopolized literacy. Interestingly, one of the earliest uses of seals appears to be connected with this, since they were used somehow to certify the size of crops at their receipt in the temple. Perhaps they record at first the operations of an economy of centralized redistribution, where men brought their due produce to the temple and received there the food or materials they themselves needed.

Besides such records, the invention of writing opens more of the past to the historian in another way. He can at last begin to deal in hard currency when talking about mentality because writing preserves literature. The oldest story in the world is the Epic of Gilgamesh. Its most complete version, it is true, goes back only to the seventh century BC, but the tale itself appears in Sumerian times and is known to have been written down soon after 2000 BC. Gilgamesh was a real person, ruling at Uruk. He became also the first individual and hero in world literature, appearing in other poems, too. He is the first person whose name must appear in this book. To a modern reader the most striking part of the Epic is the coming of a great flood which obliterates mankind except for a favoured family who survive by building an ark; from them springs a new race to people the world after the flood has subsided. This was not part of the Epic's oldest versions, but a separate poem telling a story which turns up in many Near-Eastern forms, though its incorporation is easily understandable. Lower Mesopotamia must always have had much trouble with flooding which would undoubtedly put a heavy strain on the fragile system of irrigation on which its prosperity depended. Floods were the type, perhaps, of general disaster, and must have helped to foster the pessimistic fatalism which some scholars have seen as the key to Sumerian religion.

This sombre mood dominates the Epic. Gilgamesh does great things in his restless search to assert himself against the iron laws of the gods which ensure human failure, but they triumph in the end. Gilgamesh, too, must die.

> *The heroes, the wise men, like the new moon have their waxing and waning. Men will say, 'Who has ever ruled with might and with power like him?' As in the dark month, the month of shadows, so without him there is no light. O Gilgamesh, this was the meaning of your dream. You were given the kingship, such was your destiny; everlasting life was not your destiny.*

Apart from this mood and its revelation of the religious temperament of a civilization, there is much information about the gods of ancient Mesopotamia in the Epic. But it is hard to get at history through it, let alone relate it to the historical Gilgamesh. In particular, attempts to identify a single, cataclysmic flood by archaeological means have not been convincing, though plentiful evidence of recurrent flooding is available. From the water eventually emerges the land: perhaps, then, what we are being given is an account of the creation of the world, of genesis. In the Hebrew Bible earth emerges from the waters at God's will and this account was the one which was to satisfy most educated Europeans for a thousand years. It is fascinating to speculate that we may owe so much of our own intellectual ancestry to a mythical reconstruction by the Sumerians of their own pre-history when farming land had been created out of the morass of the Mesopotamian delta. But it is only speculation; caution suggests we remain satisfied merely to note the undeniable close parallels between the Epic and one of the best of the Bible stories, that of Noah's Ark.

This story hints at the possible importance of the diffusion of Sumerian ideas in the Near East long after the focus of its history had moved away to upper Mesopotamia. Versions and parts of the Epic – to stick to that text alone for a moment – have turned up in the archives and relics of many peoples who dominated parts of this region in the second millennium BC. Though later to be lost to sight until rediscovery in modern times, Gilgamesh was for two thousand years or so a name to which literature in many languages could knowingly refer, somewhat in the way, say, that European authors until recently could take it for granted that an allusion to classical Greece would be understood by their readers. The Sumerian language lived on for centuries in temples and scribal schools, much as Latin lived on for the learned in the muddle of vernacular cultures in Europe after the collapse of the western classical world of Rome. The comparison is suggestive, because literary and linguistic tradition embodies ideas and images which impose, permit and limit different ways of seeing the world; they have, that is to say, their own historic weight.

Probably the most important ideas kept alive by the Sumerian language were religious. Cities like Ur and Uruk were the seedbed of ideas which, after transmutation into other religions in the Near East during the first and second millennia BC, were four thousand years later to be influential world wide, albeit in almost unrecognizably different forms. There is, for example, in the Gilgamesh Epic an ideal creature of nature, the man Enkidu; his Fall from his innocence is sexual, a seduction by a harlot, and thereafter, though the outcome for him is civilization, he loses his happy association with the natural world. Literature makes it possible to observe such hints at the mythologies of other and later societies. In literature, men begin to make explicit the meanings earlier hidden in obscure relics of sacrificial offerings, clay figures and the ground plans of shrines and temples. In earliest Sumer these already reveal an organization of human discourse with the supernatural much more complex and elaborate than anything elsewhere at so early a date. Temples had been the focus of the early cities and they grew bigger and more splendid (in part, because of a tradition of building new ones on mounds enclosing their predecessors). Sacrifices were offered in them to ensure good crops. Later their cults elaborated, temples of still greater magnificence were built as far north as Assur, three hundred miles away up the Tigris, and we hear of one built with cedars brought from the Lebanon and copper from Anatolia.

No other ancient society at that time gave religion quite so prominent a place or diverted so much of its collective resources to its support. It has been suggested that this was because no other ancient society left men feeling so utterly dependent on the will of the gods. Lower Mesopotamia in ancient times was a flat, monotonous landscape of mudflats, marsh, water. There were no mountains for the gods to dwell in like men, only the empty heavens above, the remorseless summer sun, the overturning winds against which there was no protection, the irresistible power of flood-water, the blighting attacks of drought. The gods dwelt in these elemental forces, or in the 'high places' which alone dominated the plains, the brick-built towers and Ziggurats remembered in the biblical Tower of Babel. The Sumerians, not surprisingly, saw themselves as a people created to labour for the gods.

By about 2250 BC a pantheon of gods more or less personifying the elements and natural forces had emerged in Sumer. It was to be the backbone of Mesopotamian religion. This is the beginning of theology. Originally, each city had its particular god. Possibly helped by political changes in the relations of the cities, they were in the end organized into a kind of hierarchy which both reflected and affected men's views of human society. The gods of Mesopotamia in the developed scheme are depicted in human form. To each of them was given a special activity or role; there was a god of the air, another of the water, another of the plough. Ishtar (as she was later known under her Semitic name) was the goddess of love and procreation, but also of war. At the top of the hierarchy were three great male gods, whose roles are not easy to disentangle, Anu, Enlil and Enki. Anu was father of the gods. Enlil was at first the most prominent; he was 'Lord Air', without whom nothing could be done. Enki, god of wisdom and of the sweet waters that literally meant life to Sumer, was a teacher and life-giver, who maintained the order Enlil had shaped.

These gods demanded propitiation and submission in elaborate ritual. In return for this and for living a good life they would grant prosperity and length of days, but not more. In the midst of the uncertainties of Mesopotamian life, some feeling that a possible access to protection existed was essential. Men depended on the gods for reassurance in a capricious universe. The gods - though no Mesopotamian could have put it in these terms – were the conceptualization of an elementary attempt to control environment, to resist the sudden disasters of flood and dust-storm, to assure the continuation of the cycle of the seasons by the repetition of the great spring festival when the gods were again married and the drama of Creation was re-enacted. After that, the world's existence was assured for another year.

One of the great demands which men later came to make of religion was that it should help them to deal with the inevitable horror of death. The Sumerians and those who inherited their religious ideas can hardly have derived much comfort from their beliefs, in so far as we can apprehend them; they seem to have seen the world of life after death as a gloomy, sad place. It was 'The house where they sit in darkness, where dust is their food and clay their meat, they are clothed like birds with wings for garments, over bolt and door lie dust and silence.' In it lies the origin of the later notions of Sheol, of Hell. Yet at least one ritual involved virtual suicide, for a Sumerian king and queen of the middle of the third millennium were followed to their tombs by their attendants who were then buried with them, perhaps after taking some soporific drink. This could suggest that the dead were going somewhere where a great retinue and gorgeous jewellery would be as important as on earth.

An impression left by a Mesopotamian cylinder seal from the early centuries of the third millenium. The seal was rolled over soft clay to form a continuous pattern like a frieze. This one, called the 'Gilgamesh' seal, represents a recurrent pattern of animals, men, or mysterious bearded, semi-human bulls in combat. Others show a hero holding helpless or at bay two animals, one on each side, and this was to be a pattern cropping up again and again in the art of many different traditions and peoples.

There were important political aspects to Sumerian religion. All land belonged ultimately to the gods; the king, probably a king-priest as much as a warrior-leader in origin, was but their vicar. No human tribunal, of course, existed to call him to account. The vicariate also meant the emergence of a priestly class, specialists whose importance justified economic privilege which could permit the cultivation of special skills and knowledge. In this respect, too, Sumer was the origin of a tradition, that of the seers, soothsayers, wise men of the East. They also had charge of the first organized system of education, based on memorizing and copying in the cuneiform script.

Among the by-products of Sumerian religion were the first true likenesses of human beings in art. In particular at one religious centre, Mari, there seems to have been something of a fondness for portraying human figures engaged in ritual acts. Sometimes they are grouped in processions; thus is established one of the great themes of pictorial art. Two others are also prominent: war and the animal world. Some have detected in the early portraiture of the Sumerians a deeper significance. They have seen in them the psychological qualities which made the astonishing achievements of their civilization possible, a drive for pre-eminence and success. This, again, is speculative. What we can also see for the first time in Sumerian art is much of a daily life in earlier times hidden from us. Given the widespread contacts of Sumer and its basic similarity of structure to other, neighbouring peoples, it is not too much to infer that we can begin to see something of life much as it was lived over a large area of the ancient Near East.

Seals, statuary and painting reveal a people often clad in a kind of furry – goat-skin or sheep-skin? – skirt, the women sometimes throwing a fold of it over one shoulder. The men are often, but not always, clean-shaven. Soldiers wear the same costume and are only distinguishable because they carry weapons and sometimes wear a pointed leather cap. Luxury seems to have consisted in leisure and possessions other than dress, except for jewellery, of which quantities have survived. Its purpose often seems to be the indication of status and it symptomizes a society of growing complexity. There survives, too, a picture of a drinking-party; a group of men sit in armchairs with cups in their hands while a musician entertains them. At such moments Sumer seems less remote.

Sumerian marriage had much about it which would have been familiar to later societies. The crux of the matter was the consent of the bride's family. Once arranged to their satisfaction, a new monogamous family unit was established by the marriage which was recorded in a sealed contract. Its head was the patriarchal husband, who presided over both his relatives and his slaves. It is a pattern which was until very recently observable in most parts of the world. Yet there are interesting nuances. Legal and literary evidence suggest that even in early times Sumerian women were less down-trodden than their sisters in many later Near-Eastern societies. Semitic and non-Semitic traditions may diverge in this. Sumerian stories of their gods suggest a society very conscious of the dangerous and even awe-inspiring power of female sexuality; the Sumerians were the first people to write about passion.

It is not easy to relate such things to institutions, but Sumerian law (whose influence can be traced in post-Sumerian times well past 2000 BC) gave women important rights. A woman was not a mere chattel; even the slave mother of a free man's children had rights which could be protected at law. Divorce arrangements provided for women as well as men to seek separations and for the equitable treatment of divorced wives. Though a wife's adultery was punishable by death, while a husband's was not, this difference is to be understood in the light of concern over inheritance and property. It was not until long after Sumerian times that Mesopotamian law begins to emphasize the importance of virginity and to impose the veil on respectable women. Both were signs of a hardening and more cramping role for them.

The Sumerians also demonstrated great technical inventiveness. Other peoples would owe much to them. It was they who laid the foundations of mathematics, establishing the technique of expressing number by position as well as by sign (as we, for example, can reckon the figure 1 as one, one-tenth, ten or several other values, according to its relation to the decimal point), and they arrived at a method of dividing the circle into six equal segments. They knew about the decimal system, too, but did not exploit it.

By the end of their history as an independent civilization they had learnt to live in big groups; one city alone is said to have had thirty-six thousand males. This made big demands on building skill, and even more were made by the large monumental structures. Lacking stone, southern Mesopotamians had first built in reeds plastered with mud, then with bricks made from the mud and dried in the sun. Their brick technology was advanced enough by the end of the Sumerian period to make possible very large buildings with columns and terraces; the greatest of its monuments, the Ziggurat of Ur, had an upper stage over a hundred feet high and a base two hundred feet by a hundred and fifty. The earliest surviving potter's wheel was found at Ur; this was the first way in which man made use of rotary motion and on it rested the large scale production of pottery which made it a man's trade and not, like earlier pottery, a woman's. Soon, by 3000 BC, the wheel was being used for transport. Another invention of the Sumerians was glass, and specialized craftsmen were casting in bronze early in the third millennium BC.

This innovation raises further questions: where did the raw material come from? There is no metal in southern Mesopotamia. Moreover, even in earlier times, during the Neolithic, the region must have obtained from elsewhere the flint and obsidian it needed for the first agricultural implements. Clearly a widespread network of contacts abroad is in the background, above all with the Levant and Syria, huge distances away, but also with Iran and Bahrein, down the Persian Gulf. Before 2000 BC Mesopotamia was obtaining goods – though possibly indirectly - from the Indus valley. Together with the evidence of documentation (which reveals contacts with India before 2000 BC), it makes an impression of a dimly emerging international trading system already creating important patterns of interdependence. When, in the middle of the third millennium, supplies of tin from the Near East dried up, Mesopotamian bronze weapons had to give way to unalloyed copper ones.

The whole of this was sustained on an agriculture which was from an early date complicated as well as rich. Barley, wheat, millet and sesame were grains grown in quantity; the first may have been the main crop, and no doubt explains the frequent evidence of the presence of alcohol in ancient Mesopotamia. In the easy soil of the flood beds iron tools were not needed to achieve intensive cultivation; the great contribution of technology here was in the practice of irrigation and the growth of government. Such skills accumulated slowly; the evidence of Sumerian civilization was left to us by fifteen hundred years of history.

So far this huge stretch of time has been discussed almost as if nothing happened during it, as if it were an unchanging whole. Of course it was not. Whatever reservations are made about the slowness of change in the ancient world and though it may now seem to us very static, these were fifteen centuries of great change for the Mesopotamians – history, in the true sense. Scholars have recovered much of the story, but this is not the place to set it out in detail, especially as much of it is still debated, much of it remains obscure and even its dating is for much of the time only approximate. All that is needed here is to relate the first age of Mesopotamian civilization to its successors and to what was going on elsewhere at the same time.

Three broad phases can be marked out in the history of Sumer. The first, lasting from about 3360 BC to 2400 BC, has been called its archaic period. Its narrative content is a matter of wars between city-states, their waxings and wanings. Fortified cities and the application of the wheel to military technology in clumsy four-wheeled chariots are some of the evidence of this. Towards the middle of this nine-hundred-year phase, local dynasties begin to establish themselves with some success. Originally, Sumerian society seems to have had some representative, even democratic basis, but a growth of scale led to the emergence of kings distinct from the first priestly rulers; probably they began as warlords appointed by cities to command their forces who did not give up their power when the emergency which called them forth had passed. From them stemmed dynasties which fought one another. The sudden appearance of a great individual then opens a new phase.

He was Sargon I, a king of the Semitic city of Akkad who conquered the Sumerian cities between 2400 and 2350 BC and inaugurated an Akkadian supremacy. There exists a sculpted head which is probably of him; if it is, it is one of the first royal portraits. He was the first of a long line of empire-builders; he has been thought to have sent his troops as far as Egypt and Ethiopia. His rule was not based on the relative superiority of one city state to another; he set up a unified empire integrating the cities into a whole. His people were among those which for thousands of years pressed in on

At Mari and one or two other places, shaving the head while growing a luxuriant beard seems to have been fashionable. This portrait statue comes from the first half of the third millenium BC, *when the Sumerian tradition of sculpture was becoming more naturalistic and individual - this is a likeness of a steward, Ibihil, one of a few models whose name we know.*

the civilizations of the river valleys from outside. They took over from its culture what they wanted but imposed themselves by force and left behind a new style of Sumerian art marked by the theme of royal victory.

Akkadian empire was not the end of Sumer but its second main phase. Though itself an interlude, it was important as an expression of a new level of organization. By Sargon's time a true state has appeared. The division between secular and religious authority which had appeared in old Sumer was fundamental. Though the supernatural still interpenetrated daily life at every level, lay and priestly authority had diverged. The evidence is physically apparent in the appearance of palaces beside the temples in the Sumerian cities; the authority of the gods lay behind their occupants, too.

Obscure though the turning of the notables of early cities into kings remains, the evolution of professional soldiery probably played a part in it. Disciplined infantry, moving in a phalanx with overlapping shields and levelled spears, appear on monuments from Ur. In Akkad there is something of a climax to early militarism. Sargon, it was boasted, had 5400 soldiers eating before him in his palace. This, no doubt, was the end of a process which built power on power; conquest provided the resources to maintain such a force. But the beginnings may again have lain originally in the special challenges and needs of Mesopotamia. As population rose, one chief duty of the ruler must have been to mobilize labour for big works of irrigation and flood control. The power to do this could also provide soldiers and as weapons became more complex and expensive, professionalism would be more likely. One source of Akkadian success was that they used a new weapon, the composite bow made of strips of wood and horn.

The Akkadian hegemony was relatively short. After two hundred years, under Sargon's great-grandson, it was overthrown, apparently by mountain peoples called Gutians, and the last phase of Sumer, called 'neo-Sumerian' by scholars, began. For another two hundred years or so, until 2000 BC, hegemony again passed to the native Sumerians. This time its centre was Ur and, though it is hard to see what it meant in practice, the first king of the Third Dynasty of Ur who exercised this ascendancy called himself King of Sumer and Akkad. Sumerian art in this phase showed a new tendency to exalt the power of the prince; the tradition of popular portraiture of the archaic period almost vanished. The temples were built again, bigger and better, and the kings seem to have sought to embody their grandeur in the ziggurats. Administrative documents show that the Akkadian legacy was strong, too; neo-Sumerian culture show many Semitic traits and perhaps the aspiration to wider kingship reflects this inheritance. The provinces which paid tribute to the last successful kings of Ur stretched from Susa, on the frontiers of Elam on the lower Tigris, to Byblos on the coast of Lebanon.

This was the sunset of the first people to achieve civilization. Of course they did not disappear, but their individuality was about to be merged in the general history of Mesopotamia and the Near East. Their great creative era was behind them and has focused our attention on a relatively small area; the horizons of history are about to expand. Enemies abounded on the frontiers. In about 2000 BC, the Elamites came and Ur fell to them. Why, we do not know. There had been intermittent hostility between the peoples for a thousand years and some have seen in this the outcome of a struggle to control the routes of Iran which could guarantee access to the highlands where lay minerals the Mesopotamians needed. At all events, it was the end of Ur. With it disappeared the distinctive Sumerian tradition, now merged in the swirling currents of a world of more than one civilization. It would now be only visible from time to time in patterns made by others. For fifteen centuries or so Sumer had built up the subsoil of civilization in Mesopotamia, just as its precivilized forerunners had built up the physical subsoil on which it itself rested. It left behind writing, monumental buildings, an idea of justice and legalism and the roots of a great religious tradition. It is a considerable record and the seed of much else. The Mesopotamian tradition had a long life ahead of it and every side of it was touched by the Sumerian legacy.

While the Sumerians had been building up their civilization, their influence had contributed to changes elsewhere. All over the Fertile Crescent new kingdoms and peoples had been appearing. They were stimulated or taught by what they saw in the south and by the empire of Ur, as well as by their own needs. The diffusion of civilized ways was already rapid. This makes it very hard to delineate and categorize the main processes of these centuries in a clear-cut way. Worse still, the Near East was for long periods a great confusion of peoples, moving about for reasons we often do not understand. The Akkadians themselves had been one of them, pushing up originally from the great Semitic reservoir of Arabia to finish in Mesopotamia. The Gutians, who took part in the Akkadians' overthrow, were Caucasians. The most successful of all of these peoples were the Amorites, a Semitic stock which had spread far and wide and joined the Elamites to overthrow the armies of Ur and destroy its supremacy. They had established themselves in Assyria, or upper Mesopotamia, in Damascus, and in Babylon in a series of kingdoms which stretched as far as the coast of Palestine. Southern Mesopotamia, old Sumer, they continued to dispute with the Elamites. In Anatolia their neighbours were the Hittites, an Indo-European people which crossed from the Balkans in the third millennium. At the edges of this huge confusion stood another old civilization, Egypt, and the vigorous Indo-European peoples who had filled up Iran. The picture is a chaos; the area is a maelstrom of races pushing into it from all sides. Patterns grow hard to distinguish.

One convenient landmark is provided by the appearance of a new empire in Mesopotamia, one which has left behind a famous name: Babylon. Another famous name is inseparably linked to it, that of one of its kings, Hammurabi. He would have a secure place in history if we knew nothing of him except his reputation as a law-giver; his code is the oldest statement of the legal principle of an eye for an eye. He was also the first ruler to unify the whole of Mesopotamia, and though the empire was short-lived the city of Babylon was to be from his time the symbolic centre of the Semitic peoples of the south. It began with the triumph of one Amorite tribe over its rivals in the confused period following the collapse of Ur. Hammurabi may have become ruler in 1792 BC; his successors held things together until sometime after 1600 BC, when the Hittites destroyed Babylon and Mesopotamia was once more divided between rival peoples who flowed into it from all sides.

At its height the first Babylonian empire ran from Sumeria and the Persian Gulf north to Assyria, the upper part of Mesopotamia. Hammurabi ruled the cities of Nineveh and Nimrud on the Tigris, Mari high on the Euphrates, and controlled that river up to the point at which it is nearest to Aleppo. Seven hundred or so miles long and about a hundred miles wide, this was a great state, the greatest, indeed, to appear in the region up to this time, for the empire of Ur had been a looser, tributary affair. It had an elaborate administrative structure, and Hammurabi's code of laws is justly famous, though it owes something of its pre-eminence to chance. As probably happened to earlier collections of judgements and rules which have only survived in fragments, Hammurabi's was cut in stone and set up in the courtyard of temples for the public to consult. But at greater length and in a more ordered way than earlier collections it assembled some 282 articles, dealing comprehensively with a wide range of questions: wages, divorce, fees for medical attention and many other matters. This was not legislation, but a declaration of existing law, and to speak of a 'code' may be misleading unless this is remembered. Hammurabi assembled rules already current; he did not create those laws *de novo*. This body of 'common law' long provided one of the major continuities of Mesopotamian history.

The family, land and commerce seem to be the main concerns of this compilation of rules. It gives a picture of a society already far beyond regulation by the ties of kindred, local community and the government of village headmen. By Hammurabi's time the judicial process had emerged from the temple and non-priestly courts were the rule. In them sat the local town notables and from them appeals lay to Babylon and the king himself. Hammurabi's stele (the stone pillar on which his code was carved) clearly stated that its aim was to assure justice by publishing the law:

Let the oppressed man who has a cause
Come into the presence of my statue
And read carefully my inscribed stele.

Sadly, perhaps, its penalties seem to have harshened, by comparison with older Sumerian practice, but in other respects, such as the laws affecting women, Sumerian tradition survived in Babylon.

The code's provisions in respect of property included laws about slaves. Babylon, like every other ancient civilization and many of modern times, rested on slavery. Very possibly the origin of slavery is conquest; certainly slavery was the fate which probably awaited the loser of any of the wars of early history and his women and children, too. But by the time of the first Babylonian empire, regular slave-markets existed and there was a steadiness of price which indicates a fairly regular trade. Slaves from certain districts were especially prized for their reliable qualities. Though the master's hold on the slave was virtually absolute, some Babylonian slaves enjoyed remarkable independence, engaging in business and even owning slaves on their own account. They had legal rights, if narrow ones.

It is hard to assess what slavery meant in practice in a world lacking the assumption which we take for granted that chattel slavery cannot be justified. Generalities dissolve in the light of evidence about the diversity of things slaves might do; if most lived hard lives, then so, probably, did most men. Yet it is hard to feel anything but pity for the lives of captives being led away to slavery before conquering kings on scores of memorials from the 'golden standard' of Ur in the middle of the third millennium to the stone reliefs of Assyrian conquests fifteen hundred years later. The ancient world rested civilization on a great exploitation of man by man; if it was not felt to be very cruel, this is only to say that no other possible way of running things was conceivable.

Babylonian civilization in due time became a legend of magnificence. The survival of one of the great images of city life – the worldly, wicked city of pleasure and consumption – in the name 'Babylon' was a legacy which speaks of the scale and richness of its civilization, though it owes most to a later period. Yet enough remains, too, to see the reality behind this myth, even for the first Babylonian empire. The great palace of Mari is an outstanding example; walls in places forty feet thick surrounded courtyards, three hundred or so rooms forming a complex drained by bitumen-lined pipes running thirty feet deep. It covered an area measuring 150 by over 200 yards and is the finest evidence of the authority the monarch had come to enjoy. In this palace, too, were found great quantities of clay tablets whose writing reveals the business and detail which government embraces by this period.

Many more tablets survive from the first Babylonian empire than from its predecessors or immediate successors. They provide the detail which enables us to know this civilization better, it has been pointed out, than we know some European countries of a thousand years ago. They contribute evidence of the life of the mind in Babylon, too. It was then that the Epic of Gilgamesh took the shape in which we know it. The Babylonians gave cuneiform script a syllabic form, thus enormously increasing its flexibility and usefulness. Their astrology pushed forward the observation of nature and left another myth behind, that of the wisdom of the Chaldeans, a name sometimes misleadingly given to the Babylonians. Hoping to understand their destinies by scanning the stars, the Babylonians built up a science, astronomy, and established an important series of observations which was another major legacy of their culture. It took centuries to accumulate after its beginnings in Ur but by 1000 BC the prediction of lunar eclipses was possible and within another two or three centuries the path of the sun and some of the planets had been plotted with remarkable accuracy against the positions of the apparently fixed stars. This was a scientific tradition reflected in Babylonian mathematics, which has passed on to us the sexagesimal system of Sumer in our circle of 360 degrees and the hour of sixty minutes. The Babylonians also worked out mathematical tables and an algebraic geometry of great practical utility.

Astronomy began in the temple, in the contemplation of celestial movements announcing the advent of festivals of fertility and sowing, and Babylonian religion held close to the Sumerian tradition. Like the old cities, Babylon had a civic god, Marduk; gradually he elbowed his way to the front among his Mesopotamian rivals. This took a long time. Hammurabi said (significantly) that Anu and Enlil, the Sumerian gods, had conferred the headship of the Mesopotamian pantheon upon Marduk, much as they had bidden him to rule over all men for their good. Subsequent vicissitudes (sometimes accompanied by the abduction of his statue by invaders) obscured Marduk's status, but after the twelfth century BC it was usually unquestioned. Meanwhile, Sumerian tradition remained alive well into the first millennium BC in the use of Sumerian in the Babylonian liturgies, in the names of the gods and the attributions they enjoyed. Babylonian cosmogony began, like that of Sumer, with the creation of the world from watery waste (the name of one god meant 'silt') and the eventual fabrication of Man as the slave of the gods. In one version, gods turned men out like bricks, from clay moulds. It was a world picture suited to absolute monarchy, where kings exercised power like that of gods over the men who toiled to build their palaces and sustained a hierarchy of officials and great men which mirrored that of the heavens.

Hammurabi's achievement did not long survive him. Events in northern Mesopotamia indicated the appearance of a new power even before he formed his

The external facing and one of the stairways of the great Ziggurat at Ur excavated by Sir Leonard Woolley, the greatest of Sumerian archaeologists. Built by Urammu, the founder of the last and most successful dynasty of rulers in Ur, it was a colossal achievement, to be venerated and maintained by the biblical Nebuchadnezzar, fifteen hundred years after it was built.

empire. Hammurabi had overthrown an Amorite kingdom which had established itself in Assyria at the end of the hegemony of Ur. This was a temporary success. There followed nearly a thousand years during which Assyria was to be a battleground and prize, eventually overshadowing a Babylon from which it was separated; the centre of gravity of Mesopotamian history had decisively moved northwards from old Sumer. The Hittites who were establishing themselves in Anatolia in the last quarter of the third millennium BC, were pushing slowly forwards in the next few centuries; during this time they took up the cuneiform script, which they adapted to their own Indo-European language. By 1700 BC they ruled the lands between Syria and the Black Sea. Then, one of their kings turned southwards against a Babylonia already weakened and shrunken to the old land of Akkad. His successor carried the advance to completion; Babylon was taken and plundered and Hammurabi's dynasty and achievement finally came to an end. But then the Hittites withdrew and other peoples ruled and disputed Mesopotamia for a mysterious four centuries of which we know little except that during them the separation of Assyria and Babylonia which was to be so important in the next millennium was made final.

In 1162 BC the statue of Marduk was again taken away from Babylon by Elamite conquerors. By that time, a very confused era has opened and the focus of world history has shifted away from Mesopotamia. The story of the Assyrian empire still lies ahead, but its background is a new wave of migrations in the thirteenth and twelfth centuries BC which involve other civilizations far more directly and deeply than the successors of the Sumerians. Those successors, their conquerors and displacers, none the less built on the foundations laid in Sumer. Technically, intellectually, legally, theologically, the Near East, which by 1000 BC was sucked into the vortex of world politics – the term is by then not too strong – still bore the stamp of the makers of the first civilization. Their heritage would pass in strangely transmuted forms to others in turn.

3
Ancient Egypt

Mesopotamia was not the only great river valley to cradle a civilization, but the only early example to rival it in the antiquity and staying-power of what was created was that of Egypt. For thousands of years after it had died, the physical remains of the first civilization in the Nile valley fascinated men's minds and stirred their imaginations; even the Greeks were bemused by the legend of the occult wisdom of a land where gods were half men, half beasts, and people still waste their time trying to discern a supernatural significance in the arrangement of the Pyramids. Ancient Egypt has always been our greatest visible inheritance from antiquity.

The richness of its remains is one reason why we know more about Egyptian than about much of Mesopotamian history. In another way, too, there is an important difference between these civilizations: because Sumerian civilization appeared first, Egypt could benefit from its experience and example. Exactly what this meant has been much debated. Mesopotamian contributions have been seen in the motifs of early Egyptian art, in the presence of cylinder seals at the outset of Egyptian records, in similar techniques of monumental building in brick and in the debt of hieroglyph, the pictorial writing of Egypt, to early Sumerian script. That there were important and fruitful connexions between early Egypt and Sumer seems incontestable, but how and when the first encounter of the Nile peoples with Sumer came about will probably never be known. It seems at least likely that when it came, Sumerian influence was transmitted by way of the peoples of the delta and lower Nile. In any case, these influences operated in a setting which always radically differentiated Egyptian experience from that of any other centre of civilization. This was provided by the Nile itself, the heart of Egypt's prehistory, as of its history.

Egypt was defined by the Nile and the deserts which flanked it; it was the country the river watered, one drawn-out straggling oasis. In prehistoric times it must also have been one great marsh, six hundred miles long, and, except in the delta, never more than a few miles wide. From the start the annual floods of the river were the basic mechanism of the economy and set the rhythm of life on its banks. Farming gradually took root in the beds of mud deposited higher and higher year by year, but the first communities must have been precarious and their environment semi-aquatic; much of their life has been irrecoverably swept away to the delta silt-beds. What remain of the earliest times are things made and used by the peoples who lived on the edge of the flood areas or on occasional rocky projections within it or at the valley sides. Before 4000 BC they began to feel the impact of an important climatic change. Sand drifted in from the deserts and desiccation set in. Armed with elementary agricultural techniques, these people could move down to work the rich soils of the flood-plain.

From the start, therefore, the river was the bringer of life to Egypt. It was a benevolent deity whose never-failing bounty was to be thankfully received, rather than the dangerous, menacing source of sudden, ruinous inundations, amid which the men of Sumer struggled to make land out of a watery waste. It was a setting in which agriculture (though it established itself later than in the Levant or Anatolia) gave a quick and rich return and perhaps made possible a population 'explosion' which released its human and natural resources. Although, as signs of contact in the fourth millennium BC show, Sumerian experience may have been available as a fertilizing element, it cannot be said that it was decisive; there always existed a potential for civilization in the Nile valley and it may have needed no external stimulus to discharge it. It is at least obvious, when Egyptian civilization finally emerged, that it is unique, unlike anything we can find elsewhere.

The deepest roots of this civilization have to be pieced together from archaeology and later tradition. They reveal Hamitic peoples in Upper Egypt (the south, that is, up the Nile) in Neolithic times. From about 5000 BC such peoples were hunting, fishing, gathering crops and finally embarking on purposeful cultivation in the valley. They lived in villages grouped round market centres and seem to have belonged to clans which had animals as symbols or totems; these they copied on their pottery. This was the basis of the eventual political organization of Egypt which began with the emergence of clan chiefs controlling the regions inhabited by their followers.

At an early stage these peoples already had several important technological accomplishments to their credit, though they do not seem such advanced farmers as those of other parts of the ancient Near East. They knew how to make papyrus boats, how to work hard materials such as basalt, and how to hammer copper into small articles for daily use. They were, that is to say, pretty accomplished well before the dawn of written record, with specialist craftsmen and, to judge by their jewels, well-marked distinctions of class or status. Then, somewhere about the middle of the fourth millennium, there is an intensification of foreign influences, apparent first in the north, the delta. Signs of trade and contact with other regions multiply, notably with Mesopotamia, whose influence is shown in the art of this era. Meanwhile, hunting and occasional farming give way to a more intense cultivation. In art, the bas-relief appears which is to be so important later in the Egyptian tradition; copper goods become more plentiful. Everything seems suddenly to be emerging at once, almost without antecedents, and to this epoch belongs the basic political structure of the future kingdom.

It was twofold; at some time in the fourth millennium there solidified two kingdoms, one northern, one southern, one of Lower and one of Upper Egypt. This is interestingly different from Sumer; there were no city-states. Egypt seems to move straight from pre-civilization to the government of large areas. Egypt's early 'towns' were the market-places of agriculturalists; the agricultural communities and clans coalesced into groups which were the foundation of later provinces. Egypt was to be a united political whole seven hundred years before Mesopotamia, and even later she would have only a restricted experience of city life.

Of the kings of the two Egypts we know little until about 3200 BC, but we may guess that they were the eventual winners in centuries of struggles to consolidate power over larger and larger groups of people. It is about the same time that the written record begins and because writing is already there at the beginning of the Egyptian story, a much more historical account of the development of its civilization can be put together than in the case of Sumer. In Egypt writing was used from its first appearance not merely as an administrative and economic convenience but to record events on monuments and relics intended to survive.

In about 3200 BC, the records tell us, a great king of Upper Egypt, Menes, conquered the north. Egypt was thus unified in a huge state six hundred miles long, running up the river as far as Abu Simbel. It was to be even bigger and to extend even further up the great river which was its heart, and it was also to undergo disruption from time to time, but this is effectively the beginning of a civilization which was to survive into the age of classical Greece and Rome. For nearly three thousand years - one and a half times the life of Christianity – Egypt was a historical entity, for much of it a source of wonder and focus of admiration. In so long a period much happened and we by no means know all of it. Yet it is the stability and conservative power of Egyptian civilization which is the most striking thing about it, not its vicissitudes.

Roughly speaking, that civilization's greatest days were over by about 1000 BC. Before that date, Egyptian history can most easily be visualized in five big divisions. Three of these are called respectively the Old, Middle and New Kingdoms; they are separated by two others called the First and Second Intermediate periods.

Very roughly, the three 'kingdoms' are periods of success or at least of consolidated government; the two intermediate stages are interludes of weakness and disruption from external and internal causes. The whole scheme can be envisaged as a kind of layer cake, with three tiers of different flavours separated by two of somewhat formless jam.

This is by no means the only way of understanding Egyptian history, nor for all purposes the best. Many scholars use an alternative way of setting out ancient Egyptian chronology in terms of thirty-one dynasties of kings, a system which has the great advantage of being related to objective criteria; it avoids perfectly proper but awkward disagreements about whether (for example) the first dynasties should be put in the 'Old Kingdom' or distinguished as a separate 'archaic' period, or about the line to be drawn at the beginning or end of the intermediate era. None the less, the five-part scheme is sufficient for our purposes, if we also distinguish an archaic prelude. A dating in one recent presentation, together with the dynastic synchronization, is as follows:

Dynasties	
I–II	Protodynastic *c.* 3200-2665 BC
III–VIII	Old Kingdom 2664-2155 BC
IX–XI	First Intermediate 2154-2052 BC
XII	Middle Kingdom 2052-1786 BC
XIII–XVII	Second Intermediate 1785-1554 BC
XVIII–XX	New Kingdom 1554-1075 BC

R. A. Parker's table in *The Legacy of Egypt*, 2nd edn, ed. J.R. Harris (Oxford, 1971), pp. 24-5.

This takes us down to the time at which, as in Mesopotamian history, there is something of a break as Egypt is caught up in a great series of upheavals originating outside its own boundaries to which the overworked word 'crisis' can reasonably be applied. True, it is not until several more centuries have passed that the old Egyptian tradition really comes to an end. Some modern Egyptians insist on a continuing sense of identity among Egyptians since the days of the Pharaohs. None the less, somewhere about the beginning of the first millennium is one of the most convenient places at which to break the story, if only because the greatest achievements of the Egyptians were by then behind them.

These were above all the work of and centred in the monarchical state. The state form itself was the expression of Egyptian civilization. It was focused first at Memphis whose building was begun during the lifetime of Menes and which was the capital of the Old Kingdom. Later, under the New Kingdom, the capital was normally at Thebes, though there were also periods of uncertainty about where it was. Memphis and Thebes were great religious centres and palace complexes; they did not really progress beyond this to true urbanism. The absence of cities earlier was politically important, too. Egypt's kings had not emerged like Sumer's as the 'big men' in a city-state community which originally deputed them to act for it. Nor were they simply men who like others were subject to gods who ruled all men, great or small. The tension of palace with temple was missing in Egypt and when Egyptian kingship emerges it is unrivalled. The Pharaohs were to be gods, not servants of gods.

It was only under the New Kingdom that the title 'pharaoh' came to be applied personally to the king. Before that it indicated the king's residence and his court. None the less, at a much earlier stage Egyptian monarchs already had the authority which was so to impress the ancient world. It is expressed in the size with which they are depicted on the earliest monuments. This they inherited ultimately from prehistoric kings who had a special sanctity because of their power to assure prosperity through successful agriculture. Such powers are attributed to some African rainmaker-kings even today; in ancient Egypt they focused upon the Nile. The Pharaohs were believed to control its annual rise and fall: life itself, no less, to the riparian communities. The first rituals of Egyptian kingship which are known to us are concerned with fertility, irrigation and land reclamation. The earliest representations of Menes show him excavating a canal.

Under the Old Kingdom the idea appears that the king is the absolute lord of the land. Soon he is venerated as a descendant of the gods, the original lords of the land. He becomes a god, Horus, son of Osiris, and takes on the mighty and terrible attributes of the divine maker of order; the bodies of his enemies are depicted hanging in rows like dead gamebirds, or kneeling in supplication lest (like less fortunate enemies) their brains be ritually dashed out. Justice is 'what Pharaoh loves', evil 'what Pharaoh hates'; he is divinely omniscient and so needs no code of law to guide him. Later, under the New Kingdom, the Pharaohs were to be depicted with the heroic stature of the great warriors of other contemporary cultures; they are shown in their chariots, mighty men of war, trampling down their enemies and confidently slaughtering beasts of prey. Perhaps a measure of secularization can be inferred in this change, but it does not remove Egyptian kingship from the region of the sacred and awesome. 'He is a god by whose dealings one lives, the father and mother of all men, alone by himself, without an equal', wrote one of the chief civil servants of the Pharaoh as late as about 1500 BC. Until the Middle Kingdom, only he had an after-life to look forward to. Egypt, more than any other Bronze Age state, always stressed the incarnation of the god in the king, even when that idea was increasingly exposed by the realities of life in the New Kingdom and the coming of iron. Then, the disasters which befell Egypt at the hands of foreigners would make it impossible to continue to believe that Pharaoh was god of all the world.

The landscape of Egyptian agriculture has been vividly preserved for us in painted scenes on the walls of tombs. This vintage scene is from one at Thebes, for centuries Egypt's cultural and religious centre of gravity.

But long before this the Egyptian state had acquired another institutional embodiment and armature, an elaborate and impressive hierarchy of bureaucrats. At its apex were viziers, provincial governors and senior officials who came mainly from the nobility; a few of the greatest of these were buried with a pomp that rivalled that of the Pharaohs. Less eminent families provided the thousands of scribes needed to staff and service an elaborate government directed by the chief civil servants. The ethos of this bureaucracy can be sensed through the literary texts which list the virtues needed to succeed as a scribe: application to study, self-control, prudence, respect for superiors, scrupulous regard for the sanctity of weights, measures, landed property and legal forms. The scribes were trained in a special school at Thebes, where not only the traditional history and literature and command of various scripts were taught, but, it seems, surveying, architecture and accountancy also.

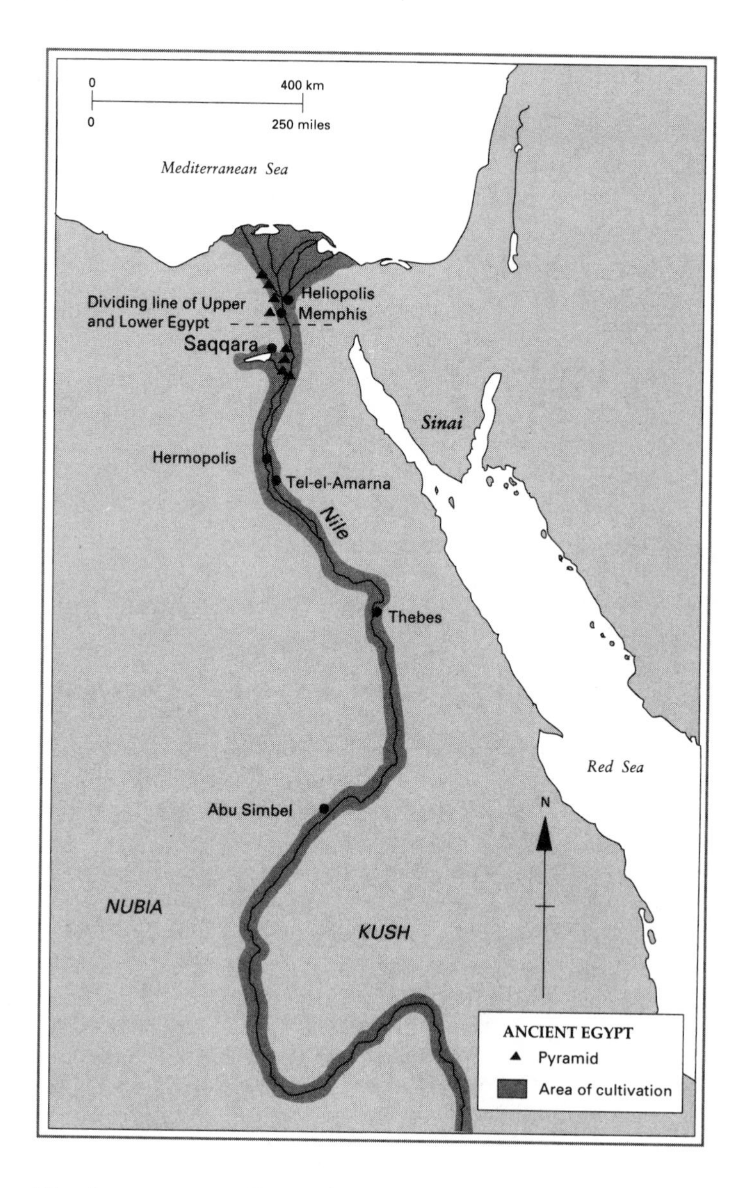

The bureaucracy directed a country most of whose inhabitants were peasants. They cannot have lived wholly comfortable lives, for they provided both the labour for the great public works of the monarchy and the surplus upon which a noble class, the bureaucracy and a great religious establishment could subsist. Yet the land was rich and was increasingly mastered by irrigation techniques established in a pre-dynastic period; these were probably one of the earliest manifestations of the unsurpassed capacity to mobilize collective effort which was to be one of the hallmarks of Egyptian government. Vegetables, barley, emmer were the main crops of the fields laid out along the irrigation channels; the diet they made possible was supplemented by poultry, fish and game (all of which figure plentifully in Egyptian art). Cattle were in use for traction and ploughing at least as early as the Old Kingdom. With little change this agriculture remained the basis of life in Egypt until modern times; it was sufficient to make her the granary of the Romans.

On the surplus of this agriculture there also rested Egypt's own spectacular form of conspicuous consumption, a range of great public works in stone unsurpassed in antiquity. Houses and farm buildings in ancient Egypt were built in the mud brick already used before dynastic times: they were not meant to outface eternity. The palaces, tombs and memorials of the Pharaohs were a different matter; they were built of the stone abundantly available in some parts of the Nile valley. Though they were carefully dressed with first copper and then bronze tools and often elaborately incised and painted, the technology of utilizing this material was far from complicated. Egyptians invented the stone column, but their great building achievement was not so much architectural and technical as social and administrative. What they did was based on an unprecedented and almost unsurpassed concentration of human labour. Under the direction of a scribe, thousands of slaves and sometimes regiments of soldiers were deployed to cut and manhandle into position the huge masses of Egyptian building. With only such elementary assistance as was available from levers and sleds - no winches, pulleys, blocks or tackle existed – and by the building of colossal ramps of earth, a succession of still-startling buildings was produced.

The step pyramid at Saqqara was conceived as a staircase to heaven and built as a monument to a Third Dynasty King in about 2650 BC by Imhotep. He was an administrator and engineer of genius who for the first time fully exploited techniques for the handling and shaping of stone which were just coming to maturity. The enclosure around the two-hundred-foot-high pyramid contains many examples of early Egyptian building style. They have survived because they were made in stone, to outface death and serve the king in his afterlife, instead of the brick, wood and mud of earlier buildings.

They began under the Third Dynasty. The most famous are the pyramids, the tombs of kings, at Saqqara, near Memphis. One of these, the 'Step Pyramid', was the masterpiece of the first architect whose name is recorded, Imhotep, chancellor to the king. His work was so impressive that he was later to be defied - as the god of medicine – as well as being revered as astronomer, priest and sage. The beginning of building in stone was attributed to him and it is easy to believe that the building of something so unprecedented as the two-hundred-foot-high pyramid was seen as evidence of godlike power. It and its companions rose without peer over a civilization which until then lived only in dwellings of mud. A century or so later, blocks of stone of fifteen tons apiece were used for the pyramid of Cheops, and it was at this time (during the Fourth Dynasty) that the greatest pyramids were completed at Giza. Cheops' Pyramid was twenty years in the building; the legend that 100 000 men were employed upon it is now thought an exaggeration but many thousands must have been and the huge quantities of stone (between five and six million tons) were brought from as much as 500 miles away. This colossal construction is perfectly orientated and its sides, 750 feet long, vary by less than eight inches – only about 0.09 per cent. It is not surprising that the Pyramids later figured among the Seven Wonders of the World, nor that they alone from those Wonders survive. They were the greatest evidence of the power and self-confidence of the pharaonic state. Nor, of course, were they the only great monuments of Egypt. Each of them was only the dominant feature of a great complex of buildings which made up together the residence of the king after death. At other sites there were great temples, palaces, the tombs of the Valley of the Kings.

These huge public works were in both the real and figurative sense the biggest things the Egyptians left to posterity. They make it less surprising that the Egyptians were later also to be reputed to have been great scientists: people could not believe that these huge monuments did not rest on the most refined mathematical and scientific skill. Yet this is invalid as an inference as well as in fact untrue. Though Egyptian surveying was highly skilled, it was not until our own day that a more than elementary mathematical skill became necessary to engineering; it was certainly not needed for the erection of the pyramids. What was requisite was outstanding competence in mensuration and the manipulation of certain formulae for calculating volumes and weights, and this was as far as Egyptian mathematics went, whatever later admirers believed. Modern mathematicians do not think much of the Egyptians' theoretical achievement and they certainly did not match the Babylonians in this art. They worked with a decimal numeration which at first sight looks modern, but it may be that their only significant contribution to later mathematics was the invention of unit fractions.

No doubt a primitive mathematics is a part of the explanation of the sterility of the Egyptians' astronomical endeavours - another field in which posterity, paradoxically, was to credit them with great things. Their observations were accurate enough to permit the forecasting of the rise of the Nile and the ritual alignment of buildings, it is true, but their theoretical astronomy was valueless. Here again they were left far behind by the Babylonians. The inscriptions in which Egyptian astronomical science was recorded were to command centuries of respect from astrologers, but their scientific value was low and their predictive quality relatively short-term. The one solid work which rested on the Egyptians' astronomy was the calendar. They were the first people to establish the solar year of 365¼ days and they divided it into twelve months, each of three 'weeks' of ten days, with five extra days at the end of the year. This arrangement, it may be remarked, was revived in 1793 when the French revolutionaries sought to replace the Christian calendar by one more rational.

The calendar, though it owed much to the observation of stars, must have reflected also in its remoter origins observation of the great pulse at the heart of Egyptian life, the flooding of the Nile. This gave the Egyptian farmer a year of three seasons, each of approximately four months, one of planting, one of flood, one of harvest. But the Nile's endless cycle also influenced Egypt at deeper levels.

The structure and solidity of the religious life of ancient Egypt greatly struck other peoples. Herodotus believed that the Greeks had acquired the names of their gods from Egypt; he was wrong, but it is interesting that he should have thought so. Later, the cults of Egyptian gods were seen as a threat by the Roman emperors; they were forbidden, but the Romans had eventually to tolerate them, such was their appeal. Mumbo-jumbo and charlatanry with an Egyptian flavour could still take in cultivated Europeans in the eighteenth century; a more amusing and innocent expression of the fascination of the myth of ancient Egypt can still be seen in the rituals of the Shriners, the secret fraternities of respectable American businessmen who parade about the streets of small towns improbably attired in fezzes and baggy trousers on great occasions. There was, indeed, a continuing vigour in Egyptian religion which, like other sides of Egyptian civilization, long outlived the political forms that had sustained and sheltered it.

Yet it remains something with which it is peculiarly difficult to come to grips. Words like 'vigour' can be misleading; religion in ancient Egypt was much more a matter of an all-pervasive framework, as much taken for granted as the circulatory system of the human body, than of an independent structure such as what later came to be understood as a church. It was not consciously seen as a growing, lively force: it was, rather, one aspect of reality, a description of an unchanging cosmos. But this, too, may be a misleading way of putting it. An important book about the world outlook of early Mesopotamians and Egyptians has the suggestive title *Before Philosophy*; we have

Egyptian goldsmiths at work, pictured on a Theban tomb of a Pharaoh's first minister during the Eighteenth Dynasty.

to remember that concepts and distinctions which we take for granted in assessing (and even talking about) the mentalities of other ages did not exist for the men whose minds we seek to penetrate. The boundary between religion and magic, for example, hardly mattered for the ancient Egyptian, though he might be well aware that each had its proper efficacy. It has been said that magic was always present as a kind of cancer in Egyptian religion; the image is too evaluative, but expresses the intimacy of the link. Another distinction lacking to ancient Egypt was the one most of us make automatically between the name and the thing. For the ancient Egyptian, the name was the thing; the real object we separate from its designation was identical with it. So might be other images. The Egyptians lived in symbolism as fishes do in water, taking it for granted, and we have to break through the assumptions of a profoundly unsymbolic age to understand them.

A whole world view is therefore involved in appreciating the meaning and role of religion in ancient Egypt. At the outset there is overwhelming evidence of its importance; for almost the whole duration of their civilization, the ancient Egyptians show a remarkably uniform tendency to seek through religion a way of penetrating the variety of the flow of ordinary experience so as to reach a changeless world most easily understood through the life the dead lived there. Perhaps the pulse of the Nile is to be detected here, too; each year it swept away and made new, but its cycle was ever recurring, changeless, the embodiment of a cosmic rhythm. The supreme change threatening men was death, the greatest expression of the decay and flux which was their common experience. Egyptian religion seems from the start obsessed with it: its most familiar embodiments, after all, are the mummy and the grave-goods from funeral chambers preserved in our museums. Under the Middle Kingdom it came to be believed that all men, not just the king, could expect life in another world. Accordingly, through ritual and symbol, through preparation of the case he would have to put to his judges in the afterworld, a man might prepare for the afterlife with a reasonable confidence that he would achieve the changeless well-being it offered in principle. The Egyptian view of the afterlife was, therefore, unlike the gloomy version of the Mesopotamians; men could be happy in it.

The struggle to assure this outcome for so many men across so many centuries gives Egyptian religion a heroic quality. It is the explanation, too, of the obsessively elaborate care shown in preparing tombs and conducting the deceased to his eternal resting-place. Its most celebrated expression is the building of the Pyramids and the practice of mummification. It took seventy days to carry out the funerary rites and mummification of a king under the Middle Kingdom.

The Egyptians believed, it appears, that after death a man could expect judgement before Osiris; if the verdict was favourable, he would live in Osiris' kingdom, if not, he was abandoned to a monstrous destroyer, part crocodile, part hippopotamus. This did not mean, though, that in life human beings need do no more than placate Osiris, for the Egyptian pantheon was huge. About two thousand gods existed and there were several important cults. Many of them originated in the prehistoric animal deities. Horus, the falcon god, was also god of the dynasty and probably arrived with the mysterious invaders of the fourth millennium BC. These animals underwent a slow but incomplete humanization; artists stick their animal heads on to human bodies. These totemlike creatures were rearranged in fresh patterns as the Pharaohs sought through the consolidation of their cults to achieve political ends. In this way the cult of Horus was consolidated with that of Amon-Re, the sun-god, of whom the Pharaoh came to be regarded as the incarnation. This was the official cult of the great age of pyramid-building and by no means the end of the story. Horus later underwent another transformation, to appear as the offspring of Osiris, the central figure of a national cult, and his consort Isis. This goddess of creation and love was probably the most ancient of all – her origins, like those of other Egyptian deities, go back to the pre-dynastic era, and she is one development of the ubiquitous mother-goddess of whom evidence survives from all over the Neolithic Near East. She was long to endure, her image, the infant Horus in her arms, surviving into the Christian iconography of the Virgin Mary.

Egyptian religion is an immensely complicated theme. Different places had different cults and there were even occasional variations of a doctrinal and speculative kind. The most famous of these was the attempt of a fourteenth-century pharaoh to establish the cult of Aton, another manifestation of the sun, in which has been discerned the first monotheistic religion. Yet there is a recurring sense of a striving after synthesis, even if it is often the expression of dynastic or political interest. Much of the history of Egyptian religion must be, if we could only decipher it, the story of ebbings and flowings about the major cults: politics, in fact, rather than religion.

Not only the pharaohs were interested. The institutions which maintained these beliefs were in the hands of a hereditary priestly class, initiated into the rituals to whose inner sancta the ordinary worshipper almost never penetrated. The cult statues at the shrine of the temple were rarely seen except by the priests. As time passed, they acquired important vested interests in the popularity and well-being of their cults.

The gods loom large in the subject-matter of ancient Egyptian art, but it contains much more besides. It was based on a fundamental naturalism of representation which, however restrained by conventions of expression and gesture, gives two millennia of classical Egyptian art at first a beautiful simplicity and later, in a more decadent period, an endearing charm and approachability. It permitted a realistic portrayal of scenes of everyday life. The rural themes of farming, fishing and hunting are displayed in them; craftsmen are shown at work on their products and scribes at their duties. Yet neither content nor technique is in the end the most striking characteristic of Egyptian art, but its enduring style. For some two thousand years, artists were able to work satisfyingly within the same classical tradition. Its origins may owe something to Sumer and it showed itself later able to borrow other foreign influences yet the strength and solidity of the central and native tradition never wavers. It must have been one of the most impressive visual features of Egypt to a visitor in ancient times; what he saw was all of a piece. If we exempt the work of the Upper Palaeolithic, of which we know so very little, it is the longest and strongest continuous tradition in the whole history of art.

It did not prove to be transplantable. Perhaps the Greeks took the column from ancient Egypt, where it had its origins in the mud-plastered bundle of reeds of which a reminiscence survives in fluting. What is clear apart from this is that although the monuments of Egypt continuously fascinated artists and architects of other lands, the result, even when they exploited them successfully for their own purposes, was always superficial and exotic. Egyptian style never took root anywhere else; it pops up from time to time down the ages as decoration and embellishment – sphinxes and serpents on furniture, an obelisk here, a cinema there. Only one great integral contribution was made by Egyptian art to the future, the establishment for the purposes of the huge incised and painted figures on the walls of tombs and temples of the classical canons of proportion of the human body which were to pass through the Greeks to western art. Artists were still to be fascinated by these as late as Leonardo, although by then the contribution was theoretical, not stylistic.

Another great artistic achievement not confined to Egypt, though exceptionally important there, was calligraphic. It seems that Egyptians deliberately took the Sumerian invention of representing sounds rather than things, but rejected cuneiform. They invented, instead, hieroglyphic writing. Instead of the device of arranging the same basic shape in different ways which had been evolved in Mesopotamia, they deliberately chose lifelike little pictures or near-pictures. It was much more decorative than cuneiform, but also much harder to master. The first hieroglyphs appear before 3000 BC; the last example of which we know was written in AD 394. Nearly 4000 years is an impressively long life for a calligraphy. But the uninitiated could still not read it for another fourteen and a half centuries after its disappearance, until a French scholar deciphered the inscription on the 'Rosetta stone' brought back to France after its discovery by scientists accompanying a French army in Egypt. None of the classical writers of antiquity who wrote about Egypt ever learnt to read hieroglyph, it seems, though enormous interest was shown in it. Yet it now seems likely that hieroglyph had importance in world as well as in Egyptian history because it was a model for Semitic scripts of the second millennium BC and thus came to be a remote ancestor of the modern Latin alphabet, which has spread round the world in our own times.

In the ancient world the ability to read hieroglyph was the key to the position of the priestly caste and, accordingly, a closely guarded professional secret. From pre-dynastic times it was used for historical record and as early as the First Dynasty the invention of papyrus – strips of reed-pith, laid criss-cross and pounded together into a homogeneous sheet – provided a convenient medium for its multiplication. Here was a real contribution to the progress of mankind. This invention had much greater importance for the world than hieroglyph; cheaper than skin (from which parchment was made) and more convenient (though more perishable) than clay tablets or slates of stone, it was the most general basis of correspondence and record in the Near East until well into the Christian era, when the invention of paper reached the Mediterranean world from the Far East (and even paper took its name from papyrus). Soon after the appearance of papyrus, writers began to paste sheets of it together into a long roll: thus the Egyptians invented the book, as well as the material on which it could first be written and a script which is an ancestor of our own. It may be our greatest debt to the Egyptians, for a huge proportion of what we know of antiquity comes to us directly or indirectly via papyrus.

Undoubtedly, the rumoured prowess of her religious and magical practitioners and the spectacular embodiment of a political achievement in art and architecture largely explains Egypt's continuing prestige. Yet if her civilization is looked at comparatively, it seems neither very fertile nor very responsive. Technology is by no means an infallible test – nor one easy to interpret – but it suggests a people slow to adopt new skills, reluctant to innovate once the creative jump to civilization had been made. Stone architecture is the only major innovation for a long time after the coming of literacy. Though papyrus and the wheel were known under the First Dynasty, Egypt had been in contact with Mesopotamia for getting on for two thousand years before she adopted the well-sweep, by then long in use to irrigate land in the other river valley.

Perhaps the weight of routine was insuperable, given the background of the unchanging reassurance provided by the Nile. Though Egyptian art records workmen organized in teams for the subdivision of manufacturing processes down to a point which faintly suggests the modern factory, many important devices came to Egypt only much later than elsewhere. There is no definite evidence of the presence of the potter's wheel before the Old Kingdom; for all the skill of the goldsmith and coppersmith, bronze-making does not appear until well into the second millennium BC and the lathe has to wait for the Hellenistic age. The bow-drill was almost the only tool for the multiplication and transmission of energy available to the mass of Egyptian craftsmen.

Only in medicine is there indisputable originality and achievement and it can be traced back at least as far as the Old Kingdom. By 1000 BC Egyptian pre-eminence in this art was internationally and justifiably recognized. While Egyptian medicine was never wholly separable from magic (magical prescriptions and amulets survive in great numbers), it had an appreciable content of rationality and pure empirical observation. It extended as far as a knowledge of contraceptive techniques. Its indirect contribution to subsequent history was great, too, whatever its efficiency in its own day; much of our knowledge of drugs and of plants furnishing *materia medica* was first established by the Egyptians and passed from them, eventually, through the Greeks to the scientists of medieval Europe. It is a considerable thing to have initiated the use of a remedy as long effective as has been castor oil. Here Egypt left Mesopotamia far behind.

What can be concluded about the health of the ancient Egyptians is another matter. They do not seem to have been so worried about alcoholic over-indulgence as the Mesopotamians, but it is not easy to infer anything from that. Some scholars have said there was an exceptionally high rate of infant mortality and hard evidence of a negative kind exists for some diseases of adults; whatever the explanation, the many mummified bodies surviving reveal no instance of cancer, rickets, or syphilis. On the other hand, the delibitating disease called schistosomiasis, carried by blood flukes and so prevalent in Egypt today, seems to have been prevalent there already in the second millennium.

Of course, none of this throws much light on ancient Egyptian medical practice. Such evidence as we have of prescriptions and recommended cures suggests that these were a mixed bag, no better and no worse than most of those deployed in other great centres of civilization at any time before the present. Considerable preservative skill was attributed to the practitioners of mummification, though unjustifiably. Curiously, the products of their art were later themselves regarded as of therapeutic value; powdered mummy was for centuries a sovereign cure for many ills in Europe and possessed other properties as well, as is suggested to readers of *Othello*.

The court of Ramses II, who added this to the existing temple of Amon at Luxor.

Most Egyptians were peasants, a consequence of Egypt long remaining non-urbanized as Mesopotamia did not. The picture of Egyptian life presented by its literature and art reveals a population living in the countryside, using little towns and temples as service centres rather than dwelling places. Egypt was for most of antiquity a country of a few great cult and administrative centres such as Thebes or Memphis and the rest nothing more than villages and markets. Life for the poor was hard, but not unremittingly so. The major burden must have been conscript labour services. When these were not exacted by Pharaoh, then the peasant would have considerable leisure at those times when he waited for the flooding Nile to do its work for him. The agricultural base was rich enough, too, to sustain a complex and variegated society with a wide range of craftsmen. About their activities we know more than of those of their Mesopotamian equivalents, thanks to stone-carvings and paintings. The great division of this society was between the educated, who could enter the state service, and the rest. Slavery was important, but, it appears, less fundamental an institution than elsewhere in the ancient Near East.

Tradition in later times remarked upon the seductiveness and accessibility of Egyptian women. With other evidence it helps to give an impression of a society in which women were more independent and enjoyed higher status than elsewhere. Some weight must be given to an art which depicts court ladies clad in the fine and revealing linens which the Egyptians came to weave, exquisitely coiffured and jewelled, wearing the carefully applied cosmetics to whose provision Egyptian commerce gave much attention. We should not lean too strongly on this, but our impression of the way in which women of the Egyptian ruling class were treated is important, and it is one of dignity and independence. The Pharaohs and their consorts – and other noble couples – are sometimes depicted, too, with an intimacy of mood found nowhere else in the art of the ancient Near East before the first millennium BC and suggestive of a real emotional equality; it can hardly be accidental that this is so.

The beautiful and charming women who appear in many of the paintings and sculptures may reflect also a certain political importance for their sex which was lacking elsewhere. The throne theoretically and often in practice descended through the female line. An heiress brought to her husband the right of succession; hence there was much anxiety about the marriage of princesses. Many royal marriages were of brother and sister, without apparently unsatisfactory genetic effects; some Pharaohs married their daughters, but perhaps to prevent anyone else marrying them rather than to ensure the continuity of the divine blood. Such a standing must have made royal ladies influential personages in their own right. Some exercised important power and one even occupied the throne, being willing to appear ritually bearded and in a man's clothes, and taking the title of Pharaoh. True, it was an innovation which seems not to have been wholly approved.

There is also much femininity about the Egyptian pantheon, notably in the cult of Isis, which is suggestive. Literature and art stress a respect for the wife and mother which goes beyond the confines of the circle of the notabilities. Both love stories and scenes of family life reveal what was at least thought to be an ideal standard for society as a whole and it emphasizes a tender eroticism, relaxation and informality, and something of an emotional equality of men and women. Some women were literate and there is even an Egyptian word for a female scribe, but there were, of course, not many occupations open to women except those of priestess or prostitute. If they were well-off, however, they could own property and their legal rights seem in most respects to have been akin to those of women in the Sumerian tradition. It is not easy to generalize over so long a period as that of Egyptian civilization but such evidence as we have from ancient Egypt leaves an impression of a society with a potential for personal expression by women not found among many later peoples until modern times.

So impressive is the solidity and material richness of Egyptian civilization in retrospect, so apparently unchanging, that it is even more difficult than in the case of Mesopotamia to keep in perspective what were its relations with the world outside or the ebb and flow of authority within the Nile valley. There are huge tracts of time to account for – the Old Kingdom alone, on the shortest reckoning, has a history two and a half times as long as that of the United States – and much happened under the Old Kingdom. The difficulty is to be sure exactly what it was that was going on and what was its importance. For nearly a thousand years after Menes, Egypt's history can be considered in virtual isolation. It was to be looked back upon as a time of stability when Pharaohs were impregnable. Yet under the Old Kingdom there has been detected a decentralization of authority; provincial officers show increasing importance and independence. The

Pharaoh, too, still had to wear two crowns and was twice buried, once in Upper and once in Lower Egypt; this division was still real. Relations with neighbours were not remarkable, though a series of expeditions was mounted against the peoples of Palestine towards the end of the Old Kingdom. The First Intermediate period which followed saw the position reversed and Egypt was invaded, rather than the invader. No doubt weakness and division helped Asian invaders to establish themselves in the valley of the lower Nile; there is a strange comment that 'the high born are full of lamentation but the poor are jubilant ... squalor is throughout the land ... strangers have come into Egypt'. Rival dynasties appeared near modern Cairo; the grasp of Memphis flagged.

The next great period of Egyptian history was the Middle Kingdom, effectively inaugurated by the powerful Amenemhet I who reunified the kingdom from his capital at Thebes. For about a quarter-millennium after 2000 BC, Egypt enjoyed a period of recovery whose repute may owe much to the impression (which comes to us through the records) of the horrors of the Intermediate period. Under the Middle Kingdom there was a new emphasis on order and social cohesion. The divine status of the Pharaoh subtly changes: not only is he God, but it is emphasized that he is descended from gods and will be followed by gods. The eternal order will continue unshaken after bad times have made men doubt. It is certain, too, that there was expansion and material growth. Great reclamation work was achieved in the marshes of the Nile. Nubia, to the south, between the first and third cataracts, was conquered and its goldmines fully exploited. Egyptian settlements were founded even farther south, too, in what was later to be a mysterious kingdom called Kush. Trade leaves more elaborate traces than ever before and the copper mines of the Sinai were now exploited again. Theological change also followed – there was something of a consolidation of cults under the god Amon-Re which reflected political consolidation. Yet the Middle Kingdom ended in political upheaval and dynastic competition.

The Second Intermediate period of roughly two hundred years was marked by another and far more dangerous incursion of foreigners. These were the Hyksos, possibly a Semitic people, who used the military advantage of the iron-fitted chariot to establish themselves in the Nile delta as overlords to whom the Theban dynasties paid tribute. Not much is known about them. Seemingly, they took over Egyptian conventions and methods, and even maintained the existing bureaucrats at first, but this did not lead to assimilation. Under the Eighteenth Dynasty the Egyptians evicted the Hyksos in a war of peoples; this was the start of the New Kingdom, whose first great success was to follow up victory in the years after 1570 BC by pursuing the Hyksos into their strongholds in south Canaan. In the end, the Egyptians occupied much of Syria and Palestine.

The New Kingdom in its prime was internationally so successful and has left such rich physical memorials that it is difficult not to think that the Hyksos domination must have had a cathartic or fertilizing effect. There was under the Eighteenth Dynasty almost a renaissance of the arts, a transformation of military techniques by the adoption of Asiatic devices such as the chariot, and, above all, a huge consolidation of royal authority. During it a female, Hatshepsut, for the first time occupied the throne in a reign notable for the expansion of Egyptian commerce, or so her mortuary temple seems to show. The next century or so brought further imperial and military glory, her consort and successor, Thotmes III, carrying the limits of Egyptian empire to the Euphrates. Monuments recording the arrival of tribute and slaves or marriages with Asiatic princesses testify to an Egyptian pre-eminence matched at home by a new richness of decoration in the temples and the appearance of a sculpture in the round which produced busts and statues generally regarded as the peak of Egyptian artistic achievement. Foreign influences also touched Egyptian art at this time; they came from Crete.

Towards the end of the New Kingdom, the evidence of multiplied foreign contacts begins to show something else: the context of Egyptian power had already changed importantly. The crucial area was the Levant coast which even Thotmes III had taken seventeen years to subdue. He had to leave unconquered a huge empire ruled by the Mitanni which dominated eastern Syria and northern Mesopotamia. His successors

Akhnaton (or Amenhotep IV), one of the most celebrated kings of ancient Egypt, but famous also for the disorder he introduced to Egyptian life through the promotion of the cult of Aton, the Sun God, bitterly opposed by the priests of Amon.

changed tack. A Mitanni princess married a pharaoh and to protect Egyptian interests in this area the New Kingdom came to rely on the friendship of her people. Egypt was being forced out of the isolation which had long protected her. But the Mitanni were under growing pressure from the Hittites, to the north, one of the most important of the peoples whose ambitions and movements break up the world of the Near East more and more in the second half of the second millennium BC.

We know a lot about the preoccupations of the New Kingdom at an early stage in this process because they are recorded in one of the earliest collections of diplomatic correspondence, for the reigns of Amenhotep III and IV (*c.* 1400-1362 BC). Under the first of these kings, Egypt reached its peak of prestige and prosperity. It was the greatest era of Thebes. Amenhotep was fittingly buried there in a tomb which was the largest ever prepared for a king, though nothing of it remains but the fragments of the huge statues the Greeks later called the colossi of Memnon (a legendary hero, whom they supposed to be Ethiopian).

Musicians and dancers entertain guests at a feast in an Egyptian house under the New Kingdom.

Amenhotep IV succeeded his father in 1379 BC. He attempted a religious revolution, the substitution of a monotheistic cult of the sun-god Aton for the ancient religion. To mark his seriousness, he changed his name to Akhnaton and founded a new city at Amarna, 300 miles north of Thebes, where a temple with a roofless sanctuary open to the sun's rays was the centre of the new religion. Although there can be no doubt of Akhnaton's seriousness of purpose and personal piety, his attempt must have been doomed from the start, given the religious conservatism of Egypt, then there may have been political motives for his persistence. Perhaps he was trying to recover power usurped by the priests of Amon-Re. Whatever the explanation, the opposition Akhnaton provoked by this religious revolution helped to cripple him on other fronts. Meanwhile, Hittite pressure was producing clear signs of strain in the Egyptian dependencies; Akhnaton could not save the Mitanni who lost all their lands west of the Euphrates to the Hittites in 1372 and dissolved in civil war which foreshadowed their kingdom's disappearance thirty years or so later. The Egyptian sphere was crumbling. There were other motives, perhaps, than religious outrage for the later exclusion of Akhnaton's name from the official list of kings.

His successor bore a name which is possibly the most famous to descend to us from ancient Egypt and a significant one: Amenhotep IV had changed his to Akhnaton because he wished to erase the reminiscence of the cult of the old god Amon; his successor and son-in-law changed his name from Tutankhaton to register the restoration of the old cult of Amon and the overthrow of the attempted religious reform. It may have been gratitude for this that led to the magnificent burial in the Valley of the Kings which was given to Tutankhamon after only a short and otherwise unremarkable reign.

When he died, the New Kingdom had two centuries of life ahead, but their atmosphere is one of only occasionally interrupted and steadily accelerating decline. Symptomatically, Tutankhamon's widow arranged to marry a Hittite prince (though he was murdered before the ceremony could take place). Later kings made efforts to recover lost ground and sometimes succeeded; the waves of conquest rolled back and forth over Palestine and at one time a pharaoh took a Hittite princess as a bride as his predecessors had taken princesses from other peoples. But there were yet more new enemies appearing; even a Hittite alliance was no longer a safeguard. The Aegean was in uproar, the islands 'poured out their people all together' and 'no land stood before them', say the Egyptian records. These sea peoples were eventually beaten off, but the struggle was hard.

At some time during these turbulent years a small Semitic people, called by the Egyptians 'Hebrews', left the delta and (according to their later tradition) followed their leader Moses out of Egypt into the deserts of Sinai. From about 1150 BC the signs of internal disorganization, too, are plentiful. One king, Ramses III, died as a result of a conspiracy in the harem; he was the last to achieve some measure of success in offsetting the swelling tide of disaster. We hear of strikes and economic troubles under his successors; there is the ominous symptom of sacrilege in a generation of looting of the royal tombs at Thebes. The pharaoh is losing his power to priests and officials and the last of the Twentieth Dynasty, Ramses XI, was in effect a prisoner in his own palace. The age of Egypt's imperial power was over. So in fact was that of the Hittites, and of other empires of the end of the second millennium. Not only Egypt's unquestioned power, but the world which was the setting of her glories, was passing away.

Undoubtedly, it is in changes affecting the whole ancient world that much of the explanation of the decline of Egypt must be sought, yet it is impossible to resist the feeling that the last centuries of the New Kingdom expose weaknesses present in Egyptian civilization from the beginning. These are not easy to discern at first sight; the spectacular heritage of Egypt's monuments and a history counted not in centuries but in millennia stagger the critical sense and stifle scepticism. Yet the creative quality of Egyptian civilization seems, in the end, strangely to miscarry. Colossal resources of labour are massed under the direction of men who, by the standards of any age, must have been outstanding civil servants, and the end is the creation of the greatest tombstones the world has ever seen. Craftsmanship of exquisite quality is employed, and its masterpieces are grave-goods. A highly literate elite utilizing a complex and subtle language and a material of unsurpassed convenience uses them copiously, but has no philosophical or religious idea comparable to those of Greek or Jew to give to the world. It is difficult not to sense an ultimate sterility, a nothingness, at the heart of this glittering *tour de force*.

In the other scale must be placed the sheer staying-power of ancient Egyptian civilization; after all, it worked for a very long time, a spectacular fact. Though it underwent at least two phases of considerable eclipse, it recovered from them, seemingly unchanged. Survival on such a scale is a great material and historical success; what remains obscure is why it should have stopped there. Egypt's military and economic power in the end made little permanent difference to the world. Her civilization was never successfully spread abroad. Perhaps this is because its survival owed much to its setting. If it was a positive success to create so rapidly institutions which with little fundamental change could last so long, this could probably have been done by any ancient civilization enjoying such a degree of immunity from intrusion. China was to show impressive continuity, too.

A huge building, the mortuary temple of Amenhotep III, once stood on the site in the Theban plain now marked only by these two towering statues – each over sixty feet high and now known as the Colossi of Memnon.

It is important also to remember once more how slow and imperceptible all social and cultural change was in early times. Because we are used to change, we must find it difficult to sense the huge inertia possessed by any successful social system (one, that is, which enables men to grapple effectively with their physical and mental environment) in almost any age before the most recent. In the ancient world the sources of innovation were far fewer and far more occasional than now. The pace of history is rapid in ancient Egypt if we think of prehistoric times; it seems glacially slow if we reflect how little daily life must have changed between Menes and Thotmes III, a period of more than fifteen hundred years and therefore comparable to that which separates us from the end of Roman Britain. Marked change could only come from sudden and overwhelming natural disaster (and the Nile was a reliable safeguard), or invasion or conquest (and Egypt long stood at the edge of the battleground of peoples in the Near East, affected only occasionally by their comings and goings). Only very slowly could technology or economic forces exert such pressures for change as we take for granted. As for intellectual stimuli, these could hardly be strong in a society where the whole apparatus of a cultural tradition was directed to the inculcation of routine.

In the end, speculation about the nature of Egyptian history tends always to revert to the great natural image of the Nile. It was always present to the Egyptian eyes, so prominent, perhaps, that it could not be seen for the colossal and unique influence it was, for no context broader than its valley needed consideration. While in the background the incomprehensible (but in the end world-making) wars of the Fertile Crescent rage across the centuries, the history of Old Egypt goes on for thousands of years, virtually a function of the remorseless, beneficent flooding and subsidence of the Nile. On its banks a grateful and passive people gathers the richness it bestows. From it could be set aside what they thought necessary for the real business of living: the proper preparation for death.

4
Intruders and Invaders: The Dark Ages of the Ancient Near East

Mesopotamia and Egypt are the foundation-stones of written history. For a long time the first two great centres of civilization dominate chronology and may conveniently be dealt with more or less in isolation. But obviously their story is not the whole story of the ancient Near East, let alone that of the ancient world. Soon after 2000 BC the movements of other peoples were already breaking it up into new patterns. A thousand years later, other centres of civilization were in existence elsewhere and we are well into the historical era.

Unfortunately for the historian, there is no simple and obvious unity to this story even in the Fertile Crescent, which for a long time continued to show more creativity and dynamism than any other part of the world. There is only a muddle of changes whose beginnings lie far back in the second millennium and which go on until the first of a new succession of empires emerges in the ninth century BC. The sweeping political upheavals which stud this confusion are hard even to map in outline, let alone to explain; fortunately, their details do not need to be unravelled here. History was speeding up and civilization was providing men with new opportunities. Rather than submerge ourselves in the flood of events, it is worth while to try to grasp some of the change-making forces at work.

A COMPLICATING WORLD

The most obvious of these continue to be the great migrations of peoples. Their fundamental pattern does not change much for a thousand years or so after 2000 BC nor does the ethnic cast of the drama. The basic dynamic was provided by the pressure of Indo-European peoples on the Fertile Crescent from both east and west. Their variety and numbers grow greater; their names need not be remembered here but some of them bring us to the remote origins of Greece. Meanwhile, Semitic peoples dispute with the Indo-Europeans the Mesopotamia valley; with Egypt and the mysterious 'Peoples of the Sea' they fight over Sinai, Palestine and the Levant. Another branch of the Indo-European race is established in Iran – and from it will eventually come the greatest of all the empires of the ancient past, that of sixth-century Persia. Still another branch pushes out into India. These movements must explain much of what lies behind a shifting pattern of empires and kingdoms stretching across the centuries. By the standards of modern times some of them were quite long-lived; from about 1600 BC a people called Kassites from Caucasia ruled in Babylon for four and a half centuries, which is a duration comparable to that of the entire history of British overseas empire. Yet, by the standards of Egypt such polities are the creatures of a moment, born today and swept away tomorrow.

It would indeed be surprising if they had not proved fragile in the end, for many other new forces were also at work which multiplied the revolutionary effects of the wanderings of peoples. One of them which has left deep traces is improvement in military technique. Fortification and, presumably, siege-craft had already reached a fairly high level in Mesopotamia by 2000 BC. Among the Indo-European peoples who nibbled at the civilization these skills protected were some with recent nomadic origins; perhaps for that reason they were able to revolutionize warfare in the field, though they long remained unskilled in siege-craft. Their introduction of the two-wheeled war chariot and the cavalryman transformed operations in open country. The soldiers of Sumer are depicted trundling about in clumsy four-wheeled carts, drawn by asses; probably these were simply a means of moving generals about or getting a leader into the mêlée, so that spear and axe could be brought to bear. The true chariot is a two-wheeled fighting vehicle drawn by horses, the usual crew being two, one man driving, the other using it as a platform for missile weapons, especially the composite bow formed of strips of horn. The Kassites were probably the first people to exploit the horse in this way and their rulers seem to have been of Indo-European stock. Access to the high pastures to the north and east of the Fertile crescent opened to them a reserve of horses in the lands of the nomads. In the river valleys horses were at first rare, the prized possessions of kings or great leaders, and the barbarians therefore enjoyed a great military and psychological superiority. Eventually, though, chariots were used in the armies of all the great kingdoms of the Near East; they were too valuable a weapon to be ignored. When the Egyptians expelled the Hyksos, they did so by, among other things, using this weapon against those who had conquered them with it.

Warfare was changed by riding horses, too. A cavalryman proper not only moves about in the saddle but fights from horseback; it took a long time for this art to be developed, for managing a horse and a bow or a spear at the same time is a complex matter. Horse-riding came from the Iranian highlands, where it may have been practised as early as 2000 BC. It spread through the Near East and Aegean well before the end of the next millennium. Later, after 1000 BC, there appeared the armoured horseman, charging home and dominating foot-soldiers by sheer weight and impetus. This was the beginning of a long era in which heavy cavalry were a key weapon, though their full value could only be exploited centuries later, when the invention of the stirrup gave the rider real control of his horse.

During the second millennium BC chariots came to have parts made of iron; soon they had hooped wheels. The military advantages of this metal are obvious and it is not surprising to find its uses spreading rapidly through the Near East and far beyond, in spite of attempts by those who had iron to restrict it. At first, these were the Hittites. After their decline iron-working spread rapidly, not only because it was a more effective metal for making arms, but because iron ore, though scarce, was more plentiful than copper or tin. It was a great stimulus to economic as well as military change. In agriculture, iron-using peoples could till heavy soils which had remained impervious to wood or flint. But there was no rapid general transfer to the new metal; iron supplemented bronze, as bronze and copper had supplemented stone and flint in the human tool-kit, and did so in some places more rapidly than others. Already in the eleventh century BC iron was used for weaponry in Cyprus (some have argued that steel was produced there, too) and from that island iron spread to the Aegean soon after 1000 BC. That date can serve as a rough division between the Bronze and Iron Ages, but is no more than a helpful prop to memory. Though iron implements became more plentiful after it, parts of what we may call the 'civilized world' long went on living in a Bronze Age culture. Together with the 'Neolithic' elsewhere, the Bronze Age lives on well into the first millenium BC, fading away only slowly like the smile on the face of the Cheshire cat. For a long time, after all, there was very little iron to go round.

Metallurgical demand helps to explain another change, a new and increasingly complex inter-regional and long-distance trade. It is one of those complicating inter-reactions which seem to be giving the ancient world a certain unity just before its disruption at the end of the second millenium BC. Tin, for example, so important a commodity, had to be brought from Mesopotamia and Afghanistan, as well as Anatolia, to what we should now call 'manufacturing' centres. The copper of Cyprus was another widely-traded commodity and the search for more of it gave Europe, at the margins of ancient history though she was, a new importance. Mine-shafts in what is now Yugoslavia were sunk sixty and seventy feet below ground to get at copper even before 4000 BC. Perhaps it is not surprising that some European peoples later came to display high levels of metallurgical skill, notably in the beating of large sheets of bronze and in the shaping of iron (a much more difficult material to work than bronze until temperatures high enough to cast it were available.)

Long-range commerce turns on transport. At first, the carriage of goods was a matter of asses and donkeys; the domestication of camels in the middle of the second millenium BC made possible the caravan trade of Asia and the Arabian peninsula which was later to seem to be of ageless antiquity, and opened an environment hitherto almost impenetrable, the waterless desert. Except among nomadic peoples, wheeled transport probably had only local importance, given the poor quality of early roads. Early carts were drawn by oxen or asses; they may have been in service in Mesopotamia about 3000 BC, in Syria around 2250 BC, in Anatolia two or three hundred years later and in mainland Greece about 1500 BC.

On the walls of a temple Ramses III celebrates his victory over the 'sea-peoples' at the beginning of the twelfth century BC. *This fragment shows Egyptian vessels in action against the invaders, who may have included peoples later termed Philistines and Achaeans.*

For goods in quantity, water transport was already likely to be cheaper and simpler than transport by land; this was to be a constant of economic life until the coming of the steam railway. Long before caravans began to bring up to Mesopotamia and Egypt the gums and resins of the south Arabian coasts, ships were carrying them up the Red Sea and merchants were moving back and forth in trading vessels across the Aegean. Understandably, it was in maritime technology that some of the most important advances in transport were made.

We know that Neolithic peoples could make long journeys by sea in dug-out canoes and there is even some evidence of navigation from the seventh millennium. The Egyptians of the Third Dynasty had put a sail on a sea-going ship; the central mast and square sail were the beginning of seamanship relying on anything but human energy. Improvements of rigging came slowly over the next two millennia. It has been thought that these made some approach to the fore-and-aft rigging which was necessary if ships were to sail closer to the wind, but for the most part the ships of antiquity were square-rigged. Because of this, the direction of prevailing winds was decisive in setting patterns of sea-borne communication. The only other source of energy was human: the invention of the oar is an early one and it provided the motive power for long sea crossings as well as for close handling. It seems likely, though, that oars were used more frequently in warships, and sail in what it is at a very early date possible to call merchantmen. By the thirteenth century BC, ships capable of carrying more than 200 copper ingots were sailing about the eastern Mediterranean, and within a few centuries more, some of these ships were being fitted with watertight decks.

Even in recent times goods have been exchanged or bartered and no doubt this was what trade meant for most of antiquity. Yet a great step was taken when money was invented. This seems to have happened in Mesopotamia, where values of account were being given in measures of grain or silver before 2000 BC. Copper ingots seem to have been treated as monetary units throughout the Mediterranean in the late Bronze Age. The first officially sealed means of exchange which survives comes from Cappadocia in the form of ingots of silver of the late third millennium BC: this was a true metal currency. Yet though money is an important invention and one which was to spread, we have to wait until the seventh century BC for the first coins. Refined monetary devices (and Mesopotamia had a credit system and bills of exchange in early times) may help to promote trade, but they are not indispensable. Peoples in the ancient world could get along without them. The Phoenicians, a trading people of legendary skill and acumen, did not have a currency until the sixth century BC; Egypt, a centrally controlled economy and of legendary wealth, did not adopt a coinage until two centuries after that, and Celtic Europe, for all its trade in metal goods, did not coin money until two centuries later still.

Meanwhile, men exchanged goods without money, though it is hard to be sure quite what this means. Although there was an important rise in the volume of goods moved about the world, by 1000 BC or so, not all of this was what would now be termed ‘trade’. Economic organization in ancient times is for a long time very obscure. Any specialized function – pottery-making, for example – implies a machinery which on the one hand distributes its products and, on the other, ensures subsistence to the specialist by redistributing to him and his fellows the food they need to survive, and perhaps other goods. But this does not require ‘trade’, even in the form of barter. Many peoples in historic times have been observed operating such distribution through their chiefs: these men presided over a common store, ‘owning’, in a sense, everything the community possessed, and doling out such shares from it as were required to keep society working smoothly. This may be what lay behind the centralization of goods and supplies in Sumerian temples; it would also explain the importance of the recording and sealing of consignments deposited there and hence the early association of writing with accounting.

As for economic exchange between communities, confident generalization about its earliest stages is even more hazardous. Once into the era of historical record, we can see many activities going on which involve the transfer of commodities, not all of them aimed at monetary gain. Payment of tribute, symbolic or diplomatic gifts between rulers, votive offerings, were some of the forms it took. We should not rush to be over-definite; right down to the nineteenth century AD the Chinese empire conceived its foreign trade in terms of tribute from the outside world and the pharaohs had a way of translating trade with the Aegean into similar notions, to judge by tomb paintings. In the ancient world, such transactions might include the transfer of standard objects such as tripods or vessels of a certain weight or rings of uniform size which therefore present at an early date some of the characteristics of currency. Sometimes such things were useful; sometimes they were merely tokens. All that is wholly certain is that the movement of commodities increased and that much of this increase in the end took the form of the profitable exchanges we now think of as commerce.

New towns must have helped. They sprang up all over the old Near East no doubt in part because of population growth. They register the successful exploitation of agricultural possibilities but also a growing parasitism. The literary tradition of the alienation of countrymen from the city is already there in the Old Testament. Yet city life also offered a new intensity of cultural creativity, a new acceleration of civilization.

One sign is the spreading of literacy. In about 2000 BC, literacy was still largely confined to the river-valley civilizations and the areas they influenced. Cuneiform had spread throughout Mesopotamia and two or three languages were written in it; in Egypt the monumental inscriptions were hieroglyphic and day-to-day writing was done on papyrus in a simplified form called hieratic. A thousand years or so later, the picture had changed. Literate peoples were then to be found all over the Near East, and in Crete and Greece, too. Cuneiform had been adapted to yet more languages with great success; even the Egyptian government adopted it for its diplomacy. Other scripts were being invented, too. One, in Crete, takes us to the edge of modernity, for it reveals a people in about 1500 BC whose language was Greek. With the adoption of a Semitic alphabet, the Phoenician, the medium of the first western literature was in existence by about 800 BC, and so, perhaps, was its first surviving expression, in what were later called the works of Homer.

Such themes make nonsense of chronology; they register changes lost to sight if history is pinned too closely to specific countries. Yet individual countries and their peoples, though subject to general forces and in more and more frequent contact, also become increasingly distinct. Literacy pins down tradition; in its turn, tradition expresses communal self-consciousness. Presumably tribes and peoples have always felt their identity; such awareness is much strengthened when states take on more continuing and institutionalized forms. The dissolution of empires into more viable units is a familiar story from Sumer to modern times, but some areas emerge time and time again as enduring nuclei of tradition. Even in the second millennium BC, states are getting more solid and show greater staying-power. They were still far from achieving that extensive and continuing control of their peoples whose possibilities have only fully been revealed in modern times. Yet even in the most ancient records there seems to be an unchecked trend towards a greater regularity in government and greater institutionalizing of power. Kings surround themselves with bureaucracies and tax-collectors find the resources for larger and larger enterprises. Law becomes a widely accepted idea; wherever it penetrates, there is a limitation, even if at first only implicit, of the power of the individual and an increase of that of the law-giver. Above all, the state expresses itself in military power; the problem of feeding, equipping and administering standing professional armies is solved by 1000 BC.

When such things happen, the story of governmental and social institutions begins to escape from the general categories of early civilization. In spite of a new cosmopolitanism made possible by easier intercourse and cross-fertilizing, societies take very diverse paths. In the life of the mind, the most conspicuous expression of diversity is religion. While some have discerned in the pre-classical era a tendency towards simpler, monotheistic systems, the most obvious fact is a huge and varied pantheon of local and specialized deities, mostly coexisting tolerantly, with only an occasional indication that one god is jealous of his distinction.

There is a new scope for differentiation in other expressions of culture, too. Before civilization began, art had already established itself as an autonomous activity not necessarily linked to religion or magic (often so linked though it continued to be). The first literature has already been mentioned and of other sides of the mind we also begin to see something. There is the possibility of play; gaming-boards appear in Mesopotamia, Egypt, Crete. Perhaps men were already gambling. Kings and noblemen hunted with passion, and in their palaces were entertained by musicians and dancers. Among sports, boxing seems to go back into Bronze Age Crete, an island where a unique and probably ritualistic sport of bull-leaping was also practised.

In such matters it is more obvious than anywhere else that we need not pay much heed to chronology, far less to particular dates, even when we can be sure of them. The notion of an individual civilization is less and less helpful over the area with which we have so far been concerned, too. There is too much interplay for it to bear the weight it can do in Egypt and Sumer. Somewhere between about 1500 and 800 BC big changes took place which ought not to be allowed to slip through the mesh of a net woven to catch the history of the first two great civilizations. In the confused, turbulent Near East and eastern Mediterranean of the centuries around 1000 BC a new world different from that of Sumer and the Old Kingdom was in the making.

EARLY CIVILIZED LIFE IN THE AEGEAN

A new interplay of cultures brought many changes to peoples on the fringe of the Near East but civilization in the Aegean islands was rooted in the Neolithic as it was elsewhere. The first metal object found in Greece – a copper bead – has been dated to about 4700 BC, and European as well as Asian stimuli may have been at work. Crete is the largest of the Greek islands. Several centuries before 2000 BC towns with a regular layout were being built there by an advanced people who had been there through Neolithic times. They may have had contacts with Anatolia which spurred them to exceptional achievements, but the evidence is indecisive. They could well have arrived at civilization for themselves. At any rate, for about a thousand years they built the houses and tombs by which their culture is distinguished and these did not change much in style. By about 2500 BC there were important towns and villages on the coasts, built of stone and brick; their inhabitants practised metal-working and made attractive seals and jewels. At this stage, that is to say, the Cretans shared much of the culture of mainland Greece and Asia Minor. They exchanged goods with other Aegean communities. There then came a change. About five hundred years later they began to build the series of great palaces which are the monuments of what we call Minoan civilization; the greatest of them, Knossos, was first built about 1900 BC. Nothing quite as impressive appears anywhere else among the islands

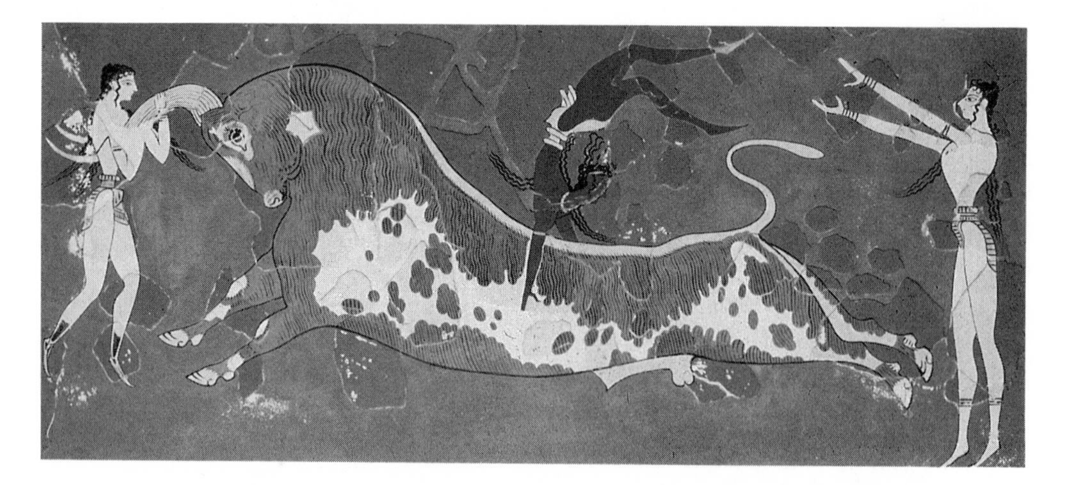

This damaged but reconstructed fresco from Knossos appears to show different stages of the acrobatic bull-leaping practised there. Its ritual content remains unknown, but the bull played a central part in Minoan mythology, perhaps as a representation of the elemental force of earthquakes not uncommon in the Aegean. More than a thousand years later, Greeks at Ephesus still celebrated a feast at which Poseidon, the 'earth-shaker' and god of the sea, was addressed as a bull and had black bulls sacrificed to him.

and it exercized a cultural hegemony over more or less the whole of the Aegean.

Minoan is a curious name; it is taken from the name of a King Minos who, although celebrated in legend, may never have existed. Much later, the Greeks believed – or said – that he was a great king in Crete who lived at Knossos, parleyed with the gods, and married Pasiphae, the daughter of the sun. Her monstrous offspring, the Minotaur, devoured sacrificial youths and maids sent as tribute from Greece at the heart of a labyrinth eventually penetrated successfully by the hero Theseus, who slew him. This is a rich and suggestive theme and has excited scholars who believe it can throw light on Cretan civilization, but there is no proof that King Minos ever existed. It may be that, as legend suggests, there was more than one of that name, or that his name was a titular identification of several Cretan rulers. He is one of those fascinating figures who, like King Arthur, remain just beyond the borders of history and inside those of mythology.

Minoan, then, simply means the civilization of people who lived in Bronze Age Crete; it has no other connotation. This civilization lasted some six hundred years, but only the outlines of a history can be put together. They reveal a people living in towns linked in some dependence on a monarchy at Knossos. For three or four centuries they prosper, exchanging goods with Egypt and the Greek mainland, and subsisting on a native agriculture. It may have been this which explains Minoan civilization's leap forward. Crete seems then, as today, to have been better for the production of olives and vines, two of the great staples of later mediterranean agriculture, than either the other islands or mainland Greece. It seems likely, too, that she raised large numbers of sheep and exported wool. Whatever its precise forms, Crete experienced an important agricultural advance in late Neolithic times, which led not only to better cereal-growing but, above all, to the cultivation of the olive and vine. They could be grown where grains could not and their discovery changed the possibilities of Mediterranean life. Immediately they permitted a larger population. On this much else could then be built because new human resources were available, but it also made new demands, for organization and government, for the regulation of a more complex agriculture and the handling of its produce.

Whether or not this explains the appearance of Minoan civilization, its peak came about 1600 BC. A century or so later, the Minoan palaces were destroyed. The mystery of this end is tantalizing. At about the same time the major towns of the Aegean islands were destroyed by fire, too. There had been earthquakes in the past; perhaps this was another of them. Recent scholarship identifies a great eruption in the island of Thera at a suitable time; it could have been accompanied by tidal waves and earthquakes in Crete, seventy miles away, and followed by the descent of clouds of ash which blighted Cretan fields. Some people have preferred to think of a rising against the rulers who lived in the palaces. Some have discerned signs of a new invasion, or postulated some great raid from the sea which carried off booty and prisoners, destroying a political power for ever by the damage it inflicted, by leaving no new settlers behind. None of these can be conclusively established. It is only possible to guess about what happened and the view which does least violence to the lack of evidence is that there was a natural cataclysm originating in Thera which broke the back of Minoan civilization.

Whatever the cause, this was not the end of early civilization in Crete, for Knossos was occupied for another century or so by people from the mainland. Nevertheless, though there were still some fairly prosperous times to come, the ascendancy of the indigenous civilization of Crete was, in effect, over. Far a time, it seems, Knossos still prospered. Then, early in the fourteenth century BC it, too, was destroyed by fire. This had happened before, but this time it was not rebuilt. So ends the story of early Cretan civilization.

Fortunately, its salient characteristics are easier to understand than the detail of its history. The most obvious is its close relationship with the sea. More than a thousand years later, Greek tradition said that Minoan Crete was a great naval power exercising political hegemony in the Aegean through her fleet. This idea has been much blown upon by modern scholars anxious to reduce what they believe to be an anachronistic conception to more plausible proportions and it certainly seems misleading to see behind this tradition the sort of political power later exercised through their navies by such states as fifth-century Athens or nineteenth-century Great Britain. The Minoans may have had a lot of ships, but they were unlikely to be specialized at this early date and there is no hope in the Bronze Age of drawing a line between trade, piracy and counter-piracy in their employment. Probably there was no permanent Cretan 'navy' in a public sense at all. Nevertheless, the Minoans felt sufficiently sure of the protection the sea gave them – and this must have implied some confidence in their ability to dominate the approaches to the natural harbours, most of which are on the north coast – to live in towns without fortifications, built near to the shore on only slightly elevated ground. We do not have to look for a Cretan Nelson among their defenders; that would be silly. But we can envisage a Cretan Hawkins or Drake, combining trade, freebooting and protection of the home base.

The Minoans thus exploited the sea as other peoples exploited their natural environments. The result was an interchange of products and ideas which shows once more how civilization can accelerate where there is the possibility of cross-fertilization. Minoans had close connexions with Syria before 1550 BC and traded as far west as Sicily, perhaps further. Someone took their goods up the Adriatic coasts. Even more important was their penetration of Greece. The Minoans may well have been the most important single conduit through which the goods and ideas of the first civilizations reached Bronze Age Europe. Certain Cretan products begin to turn up in Egypt in the second millennium BC and this was a major outlet; the art of the New Kingdom shows Cretan influence. There was even, some scholars think, an Egyptian resident for some time at Knossos, presumably to watch over well-established interests, and it has been argued that Minoans fought with the Egyptians against the Hyksos. Cretan vases and metal goods have been found at several places in Asia Minor: these are the things which survive, but it has been argued that a wide range of other products – timber, grapes, oil, wood, metal vases and even opium – were supplied by the Minoans to the mainland. In return, they took metal from Asia Minor, alabaster from Egypt, ostrich eggs from Libya. It was a complex trading world.

Together with a prosperous agriculture it made possible a civilization of considerable solidity, long able to recover from natural disaster, as the repeated rebuilding of the palace at Knossos seems to show. The palaces are the finest relics of Minoan civilization, but the towns were well built too, and had elaborate piped drains and sewers. This was technical achievement of a high order; early in the sequence of palaces at Knossos the bathing and lavatory provision is on a scale unsurpassed before Roman times. Other cultural achievement was less practical, though artistic rather than intellectual; Minoans seem to have taken their mathematics from Egypt and left it at that. Their religion went under with them, apparently leaving nothing to the future, but the Minoans had an important contribution to make to the style of another civilization on the Greek mainland. Art embodied Minoan civilization at its highest and remains its most spectacular legacy. Its genius was pictorial and reached a climax in palace frescoes of startling liveliness and movement. Here is a really original style, influential across the seas, in Egypt and in Greece. Through other palatial arts, too, notably the working of gems and precious metals, it was to shape fashion elsewhere.

Minoan art provides a little evidence about the Cretans' style of life because it is often representational. They seem to have dressed scantily, the women often being depicted bare-breasted; the men are beardless. There is an abundance of flowers and plants to suggest a people deeply and readily appreciative of nature's gifts; they do not give the impression that the Minoans found the world an unfriendly place. Their relative wealth – given the standards of ancient times – is attested by the rows of huge and beautiful oil-jars found in their palaces. Their concern for comfort and what cannot but be termed elegance comes clearly through the dolphins and lilies which decorate the apartments of a Minoan queen.

Archaeology has also provided evidence of a singularly unterrifying religious world, though this does not, perhaps, take us very far since we have no texts. Though we have representations of the Cretans' gods and goddesses, it is not easy to be sure who they are. Nor can we penetrate their rituals, beyond registering the frequency of sacrificial altars, double-headed axes, and the apparent centering of Minoan cults in a female figure. She is perhaps a Neolithic fertility figure such as was to appear again and again as the embodiment of female sexuality: the later Astarte and Aphrodite. In Crete she appears elegantly skirted, bare-breasted, standing between lions and holding snakes. Whether there was a male god, too, is less clear. But the appearance of bulls' horns in many places and of frescoes of these noble beasts is suggestive if it is linked to later Greek legend (Minos' mother, Europa, had been seduced by Zeus in the shape of a bull; his wife Pasiphae enjoyed a monstrous coition with a bull from which was born the half-bull, half-man Minotaur), and to the obscure but obviously important rites of bull-leaping. It is striking that whatever it was, Cretan religion does not seem gloomy; pictures of sports and dancing or delicate frescoes and pottery do not suggest an unhappy people.

The Greeks always looked back on prehistoric Crete as a mysterious land of legend. This vase of about 500 BC shows the hero Theseus, about to ward off a boulder hurled at him by the Minotaur, the monster at the heart of the labyrinth of King Minos.

The political arrangements of this society are obscure. The palace was not only a royal residence, but in some sense an economic centre – a great store – which may perhaps best be understood as the apex of an advanced form of exchange based on redistribution by the ruler. The palace was also a temple, but not a fortress. In its maturity it was the centre of a highly organized structure whose inspiration may have been Asian; knowledge of the literate empires of Egypt and Mesopotamia was available to a trading people. One source of our knowledge of what Minoan government was trying to do is a huge collection of thousands of tablets which are its administrative records. They indicate rigid hierarchy and systematized administration, but not how this worked in practice. However effective government was, the only thing the records certainly show is what it aspired to, a supervision far closer and more elaborate than anything conceivable by the later Greek world. If there are any analogies, they are again with the Asian empires and Egypt.

At present, the tablets tell us only of the last phase of Minoan civilization because many of them cannot be read. The weight of scholarly opinion now inclines to the view put forward a few years ago that the script of a great mass of them found at Knossos is used to write Greek and that they date from about 1450 to 1375 BC. This confirms the archaeological evidence of the arrival of successful invaders from the mainland at about this time and of their supersession of the native rulers. The tablets are their documents, and the script in which they are written has been termed 'Linear B'. The earlier written records are found at first in hieroglyph, with some symbols borrowed from Egypt, and then in another script (not yet deciphered) termed 'Linear A' and used from perhaps as early as 1700 BC. Almost certainly it was wholly non-Greek. It seems likely that the incoming Greeks took over pre-existing Minoan administrative practice and put down records, such as were already kept, in their own tongue. The earlier tablets probably contain, therefore, information very like the later, but it is about Crete before the coming of the Greek-speaking invaders who presided over the last phase and mysterious end of Minoan civilization.

Successful invasion from the mainland was itself a sign that the conditions which had made this civilization possible were crumbling away in the troubled times of the closing Bronze Age. Crete for a long time had no rival to threaten her coasts. Perhaps the Egyptians had been too busy; in the north there had long been no possible threat. Gradually, the second of these conditions had ceased to hold. Stirring on the mainland were the same Indo-European peoples who have already cropped up in so many places in this story. Some of them penetrated Crete again after the final collapse of Knossos; they were apparently successful colonists who exploited the lowlands and drove away the Minoans and their shattered culture to lonely little towns of refuge where they disappear from the stage of world history.

Ironically, only two or three centuries before this, Cretan culture had exercised something like hegemony in Greece, and Crete was always to hang about mysteriously at the back of the Greek mind, a lost and golden land. A direct transfusion of Minoan culture to the mainland had taken place through the first Achaean peoples (the name usually given to these early Greek-speakers) who came down into Attica and the Peloponnese and established towns and cities there in the seventeenth and eighteenth centuries BC. They entered a land long in contact with Asia, whose inhabitants had already contributed to the future one enduring symbol of Greek life, the fortification of the high place of the town, or *acropolis*. The new arrivals were culturally hardly superior to those they conquered, though they brought with them the horse and war-chariot. They were barbarians by comparison with the Cretans, with no art of their own. More aware of the role of violence and war in society than were the islanders (no doubt because they did not enjoy the protection of the sea and had a sense of continuing pressure from the homelands from which they had come), they fortified their cities heavily and built castles. Their civilization had a military style. Sometimes they picked sites which were to be the later centres of Greek city-states; Athens and Pylos were among them. They were not very large, the biggest containing at most not more than a few thousand people. One of the most important was at Mycenae, which gave its name to the civilization that finally spread over Bronze Age Greece in the middle of the second millennium.

It left some splendid relics, for it was very rich in gold; strongly influenced by Minoan art, it was also a true synthesis of Greek and indigenous cultures on the mainland. Its institutional basis seems to have been rooted in the patriarchal ideas found among many of the Indo-European peoples, but there is more to it than this. The bureaucratic aspiration revealed by the Knossos tablets and by others from Pylos in the western Peloponnese of about 1200 BC suggests currents of change flowing back from conquered Crete towards the mainland. Each considerable city had a king. The one at Mycenae, presiding over a society of warrior landowners whose tenants and slaves were the aboriginal peoples, may have been at an early date the head of some sort of federation of kings. There is suggestive evidence in Hittite diplomatic records which points to some political unity in Mycenaean Greece. Below the kings, the Pylos tablets show a close supervision and control of community life and also important distinctions between officials and, more fundamentally, between slave and free. What cannot be known is just what such differences meant in practice. Nor can we see much of the economic life which lay at the root of Mycenaean culture, beyond its centralization in the royal household, as in Crete.

For this almost spherical flask of about 1500 BC, ten inches or so in diameter, a Cretan artist found a decorative motif in the fronds of seaweed and octopus so plentiful on the shores of the Aegean Islands. It is an outstanding example of the relating of a naturalistic design to the form of the artwork, but Cretan art was for a long time rich in observations, particularly of marine subjects.

Whatever its material basis, the culture represented most spectacularly at Mycenae had by 1400 BC spread all over mainland Greece and to many of the islands. It was a whole, though well-established differences of Greek dialect persisted and distinguished one people from another down to classical times. Mycenae replaced the Minoan trading supremacy in the Mediterranean with its own. It had trading posts in the Levant and was treated as a power by Hittite kings. Sometimes Mycenaean pottery exports replaced Minoan, and there are even examples of Minoan settlements being followed by Mycenaean.

The Mycenaean empire, if the term is permissible, was at its height in the fifteenth and fourteenth centuries BC. For a while, the weakness of Egypt and the crumbling of the Hittite power favoured it; for a time a small people enriched by trade had disproportionate importance while great powers waned. Mycenaean colonies were established on the shores of Asia Minor; trade with other Asian towns, notably Troy, prospered. But there are signs of a flagging from about 1300 BC. War seems to have been one answer; Achaeans took important parts in attacks on Egypt at the end of the century and it now seems that a great raid by them which was immortalized as the Siege of Troy took place about 1200 BC. The troubled background to these events was a series of dynastic upheavals in the Mycenaean cities themselves.

What can be called the Dark Ages of the Aegean were about to close in and they are as obscure as what was happening in the Near East at about the same time. When Troy fell, new barbarian invasions of mainland Greece had already begun. At the very end of the thirteenth century the great Mycenaean centres were destroyed, perhaps by earthquakes, and the first Greece broke up into disconnected centres. As an entity Mycenaean civilization collapsed, but not all the Mycenaean centres were abandoned, though their life continued at a lower level of achievement. The kingly treasures disappear, the palaces are not rebuilt. In some places the established resident peoples hung on successfully for centuries; elsewhere they were ruled as serfs or driven out by new conquerors, Indo-Europeans from the north who had been on the move from about a century before the fall of Troy. It does not seem likely that these new peoples always settled the lands they ravaged, but they swept away the existing political structures and the future would be built on their kinships, not on the Mycenaean institutions. There is a picture of confusion as the Aegean Dark Age deepens; only just before 1000 BC are there a few signs that a new pattern – the ground-plan of classical Greece - is emerging.

Legendary accounts of this attribute much to one particular group among the newcomers, the Dorians. Vigorous and bold, they were to be remembered as the descendants of Heracles. Though it is very dangerous to argue back from the presence of later Greek dialects to identifiable and compact groups of early invaders, tradition makes them the speakers of a tongue, Doric, which lived on into the classical age as a dialect setting them apart. In this case, tradition has been thought by scholars to be justified. In Sparta and Argos, Dorian communities which would be future city-states established themselves. But other peoples also helped to crystallize a new civilization in this obscure period. The most successful were those later identified as speakers of 'Ionic' Greek, the Ionians of the Dark Age. Setting out from Attica (where Athens had either survived or assimilated the invaders who followed Mycenae), they took root in the Cyclades and Ionia, the present Turkish coast of the Aegean. Here, as migrants and pirates, they seized or founded towns, if not on islands, almost always on or near the coast, which were the future city-states of a seafaring race. Often the sites they chose had already been occupied by the Mycenaeans. Sometimes - at Smyrna, for example – they displaced earlier Greek settlers.

This is a confusing picture at best and for much of it there is only fragmentary evidence. Yet from this turmoil there would slowly re-emerge the unity of civilization enjoyed by the Bronze Age Aegean. At first, though, there were centuries of disruption and particularism, a new period of provincialism in a once cosmopolitan world. Trade flagged and ties with Asia languished. What replaced them was the physical transference of people, sometimes taking centuries to establish new settled patterns, but in the end setting out the ground-plan of a future Greek world.

Immediately, there was a colossal setback in civilized life which should remind us how fragile it could be in ancient times. Its most obvious sign was a depopulation between 1100 and 1000 BC so widespread and violent that some scholars have sought explanations in a sudden cataclysm – plague, perhaps, or a climatic change such as might have suddenly and terribly reduced the small cultivable area of the Balkan and Aegean hillsides. Whatever the cause, the effects are to be seen also in a waning of elegance and skill; the carving of hard gems, the painting of frescoes and the making of the fine pottery all come to a stop. Such cultural continuity as the age permitted must have been largely mental, a matter of songs, myths and religious ideas.

Of this troubled time a very little is dimly and remotely reflected in the bardic epics later set down in writing in the *Iliad* and the *Odyssey*. They include material transmitted for generations by recitation, whose origins lie in tradition near-contemporary with the events they purport to describe, though later attributed to one poet, Homer. Exactly what is reflected, though, is much harder to agree about; the consensus has recently been that it is hardly anything for Mycenaean times, and little more for what immediately followed them. The central episode of the *Iliad*, the attack on Troy, is not what matters here, though the account probably reflects a real preponderance of Achaean initiative in the settlement of Asia Minor. What survives is a little social and conceptual information carried incidentally by the poems. Though Homer gives an impression of some special pre-eminence attributed to the Mycenaean king, this is information about the

This silver bowl came originally from Cyprus, the site of important Phoenician settlements, whose art, as the sphinxes, Egyptian figures and Assyrian warriors on this example show, draws on a wide diversity of motifs.

post-Mycenaean Aegean of the eighth century, when recovery from the Dark Ages begins. It reveals a society whose assumptions are those of barbarian warlords rather than those of kings commanding regular armies or supervising bureaucracies like those of Asia. Homer's kings are the greatest of great nobles, the heads of large households, their acknowledged authority tempered by the real power of truculent near-equals and measured by their ability to impose themselves; their lives are troubled and exacting. The atmosphere is individualistic and anarchic: they are more like a band of Viking leaders than the rulers commemorated by the Mycenaean tablets. Whatever reminiscences of detail may survive from earlier times (and these have sometimes been confirmed in their accuracy by excavation) and however many reflexions of later society they eventually contained, the poems only fitfully illuminate a primitive society, still in confusion, settling down perhaps, but neither so advanced as Mycenae had been, nor even dimly foreshadowing what Greece was to become.

The new civilization which was at last to emerge from the centuries of confusion owed much to the resumption of intercourse with the East. It was very important that the Hellenes (the name by which the invaders of Greece came to be distinguished from their predecessors) had spread out into the islands and on to the Asian mainland; they provided many points of contact between two cultural worlds. But they were not the only links between Asia and Europe. Seeds of civilization were always carried about by the go-betweens of world history, the great trading peoples.

One of them, another seafaring race, had a long and troubled history, though not so long as its legends said; the Phoenicians claimed that they had arrived in Tyre in about 2700 BC. This may be treated on a level with stories about the descent of the Dorian kings from Heracles. None the less, they were already settled on the coast of the modern Lebanon in the second millennium BC when the Egyptians got their supplies of cedar-wood from them. The Phoenicians were a Semitic people. Like the Arabs of the Red Sea, they became seafarers because geography urged them to look outwards rather than inland. They lived in the narrow coastal strip which was the historic channel of communication between Africa and Asia. Behind them was a shallow hinterland, poor in agricultural resources, cut up by hills running down from the mountains to the sea so that the coastal settlements found it difficult to unite. There were parallels with the experience of later Greek states tempted to the sea in similar circumstances and in each case the result was not only trade but colonization.

Weak at home – they came under the sway of Hebrew, Egyptian and Hittite in turn – it cannot be entirely coincidental that the Phoenicians emerge from the historical shadows only after the great days of Egypt, Mycenae and the Hittite empire. They, too, prospered in others' decline. It was after 1000 BC, when the great era of Minoan trade was long past, that the Phoenician cities of Byblos, Tyre and Sidon had their brief golden age. Their importance then is attested by the biblical account of their part in the building of Solomon's Temple; 'thou knowest', says Solomon, 'that there is not among us any that can skill to hew timber like unto the Sidonians', and he paid up appropriately (1 Kings v, 6). This is perhaps evidence of a uniquely large and spectacular public works contract in ancient times, and there is copious later material to show the continuing importance of Phoenician enterprise. Ancient writers often stressed their reputation as traders and colonizers. They even exchanged goods with the savages of Cornwall, and must have been navigators of some skill to get so far. Phoenician dyes were long famous and much sought after down to classical times. No doubt commercial need stimulated their inventiveness; it was at Byblos (from which the Greeks were to take their name for a book) that the alphabet later adopted by the Greeks was invented. This was a great step, making a more widespread literacy possible, but no remarkable Phoenician literature survives, while Phoenician art tends to reflect their role of the middleman, borrowing and copying from Asian and Egyptian models, perhaps as the customer demanded.

Trade was the Phoenician occupation and did not at first require settlement overseas. Yet they came to base themselves more and more on colonies or trading stations, sometimes where Mycenaeans had traded before them. There were in the end some twenty-five of these up and down the Mediterranean, the earliest set up at Kition (the modern Larnaca) in Cyprus at the end of the ninth century BC. Some colonies may have followed earlier Phoenician commercial activity on the spot. They may also reflect the time of troubles which overtook the Phoenician cities after a brief phase of independence at the beginning of the first millennium. In the seventh century Sidon was razed to the ground and the daughters of the king of Tyre were carried off to the harem of the Assyrian Ashurbanipal. Phoenicia was then reduced to its colonies elsewhere in the Mediterranean and little else. Yet their establishment may also have reflected anxiety at a wave of Greek colonization in the west which threatened the supply of metal, especially of British tin and Spanish silver. This could explain the Phoenician foundation of Carthage a century earlier; it was to become the seat of a power more formidable by far than Tyre and Sidon had ever been and went on to establish its own chain of colonies. Further west, beyond the Straits of Gibraltar, Cadiz was already known to Phoenicians who called there while looking for an Atlantic trade further north.

The Phoenicians were among the most important traffickers in civilization but so, willy-nilly, had been others, the Mycenaeans by their diffusion of a culture and the Hellenes by their stirring up of the ethnic world of the Aegean. The Minoans had been something more; true originators, they not only took from the great established centres of culture, but remade what they took before diffusing it again. These peoples help to shape a more rapidly changing world. One important side-effect, of which little has yet been said, was the stimulation of continental Europe. The search for minerals would take explorers and prospectors further and further into that barbarian unknown. Already in the second millennium there are the first signs of a complicated future; beads found at Mycenae were manufactured in Britain from Baltic amber. Trade was always slowly at work, eating away isolation, changing peoples' relations with one another, imposing new shapes on the world. But it is hard to relate this story to the stirring of the ethnic pot in the Aegean, let alone to the troubled history of the Asian mainland from the second millennium BC.

THE NEAR EAST IN THE AGES OF CONFUSION

'Confusion' is a matter of perspective. For about eight hundred years from, say, the end of Knossos, the history of the Near East is indeed very confused if our standpoint is that of world history. What was essentially going on were disputes about control of the slowly-growing wealth of the best-defined agricultural region of the ancient world (the empires which came and went could not find resources in the desert and steppe area on the borders of the Near East which could justify their conquest) and in that story it is hard to find any continuing thread. Invaders came and went rapidly, some of them leaving new communities behind them, some setting up new states to replace those they overthrew. This could hardly have been grasped by those to whom these events would only have come home occasionally, and suddenly, when (for instance) their homes were burned, their wives and daughters raped, their sons carried off to slavery – or, less dramatically, when they discovered that a new governor was going to levy higher taxes. Such events would be upsetting enough – if a stronger word is not required. On the other hand, millions of people must also in those times have lived out their lives unaware of any change more dramatic than the arrival one day in their village of the first iron sword or sickle; hundreds of communities lived within a pattern of ideas and institutions unchanged for many generations. This is an important reservation. It must not be forgotten when we stress the dynamism and violence of the Near East's history during the transition from the Bronze to Iron Ages, an era already considered from the standpoint of the peoples of the Aegean.

On the mainland, wandering peoples moved about in a zone where there were well-established centres of government and population, powerful and long-lasting political structures, numerous hierarchies of specialists in administration, religion and learning. These partly explain why the coming of new peoples obliterates less of what had already been achieved than in the Aegean. Another conservative force was the long contact many of the barbarians had already had with civilization in this region. It left them wanting not to destroy it but to enjoy its fruits themselves. These two forces helped in the long run to diffuse civilization further and to produce the increasing cosmopolitanism of a large and confused, but civilized and interconnected, Near East.

The story begins very early, somewhere back towards the beginning of the second millennium BC with the arrival in Asia Minor of the Hittites. Perhaps they belonged to the same group of peoples as the Minoans, at any rate they were established in Anatolia at about the same time that Minoan civilization was rising to its greatest triumphs. They were far from being primitive barbarians. They had a legal system of their own and absorbed much of what Babylon could teach. They long enjoyed a virtual monopoly of iron in Asia; this not only had great agricultural importance but, together with their mastery of fortification and the chariot, gave the Hittites a military superiority which was the scourge of Egypt and Mesopotamia. The raid which cut down Babylon in about 1590 BC was something like the high-water mark of the first Hittite 'empire'. A period of eclipse and obscurity followed. Then, in the first half of the fourteenth century, came a renaissance of power. This second and even more splendid era saw a Hittite hegemony which stretched at one brief moment from the shores of the Mediterranean to the Persian Gulf. It dominated all of the Fertile Crescent except Egypt and successfully challenged even that great military power while almost ceaselessly at war with the Myceneans. But like other empires it crumbled after a century or so, the end coming in about 1200 BC.

The culmination and collapse of this great organizing effort at the beginning of 'dark ages' for Greece and the Aegean has two interesting features. The first is that the Hittites by this time no longer enjoyed a monopoly of iron; by about 1000 BC it is to be found in use all over the Near East and its diffusion must surely be part of the story of the swing of power against the Hittites. The other interesting feature is a coincidence with the rhythm of migrations, for it seems that the great diffusers of iron technology were the Indo-European peoples who from about 1200 BC were throwing everything into turmoil. The disappearance of Troy, which never recovered from the Achaean destruction, has been thought of great strategic importance in this respect; the city seems to have played until this time a leading rôle in an alliance of powers of Asia Minor who had held the line against the barbarians from the north. After its overthrow, no other focus for resistance appeared. There is a closeness of timing which some have thought too pronounced to be merely coincidental between the collapse of the last Hittite power and the attacks of 'sea peoples' recorded in the Egyptian records. The particular conquerors of the Hittites were a people from Thrace called the Phrygians.

The 'sea peoples' were one more symptom of the great folk movements of the era. Armed with iron, from the beginning of the twelfth century BC they were raiding the mainland of the East Mediterranean basin, ravaging Syrian and Levantine cities. Some of them may have been 'refugees' from the Mycenaean cities who moved first to the Dodeconese and then to Cyprus. One group among them, the Philistines, settled in Canaan in about 1175 BC and are commemorated still by a modern name derived from their own: Palestine. But Egyptians were the major victim of the sea peoples. Like the Vikings of the northern seas two thousand years later, sea-borne invaders and raiders plunged down on the delta again and again, undeterred by occasional defeat, at one time even wresting it from Pharaoh's control. Egypt was under great strain. In the early eleventh century, she broke apart and was disputed between two kingdoms. Nor were the sea peoples Egypt's only enemies. At one point, a Libyan fleet appears to have raided the delta, though it was drawn off. In the south, the Nubian frontier did not yet present a problem, but round about 1000 BC an independent kingdom emerged in the Sudan which would later be troublesome. The tidal surge of barbarian peoples was wearing away the old structures of the Near East just as it had worn away Mycenaean Greece.

This is far enough into the welter of events to make it clear that we have entered an age both too complex and too obscure for straightforward narration. Mercifully, there soon appear two threads through the turmoil. One is an old theme renewed, that of the continuing Mesopotamian tradition about to enter its last phase. The other is quite new. It begins with an event we cannot date and know only through tradition recorded centuries later, but which probably occurred during the testing time imposed on Egypt by the sea peoples. Whenever and however it happened, a turning-point had been reached in world history when there went out of Egypt people the Egyptians called Hebrews and the world later called Jews.

For many people during many centuries, mankind's history before the coming of Christianity was the history of the Jews and what they recounted of the history of others. Both were written down in the books called the Old Testament, the sacred writings of the Jewish people, subsequently diffused world-wide in many languages by the Christian missionary impulse and the invention of printing. They were to be the first people to arrive at an abstract notion of God and to forbid his representation by images. No people has produced a greater historical impact from such comparatively insignificant origins and resources, origins so insignificant indeed that it is still difficult to be sure of very much about them in spite of huge efforts.

Making bricks in Egypt, as shown on a wall painting from a Theban tomb. Some of the workers have features unlike those of Egyptians and we know that many such labourers were foreigners and prisoners of war. They were called by the Egyptians apiru *– a word scholars have connected with the later appearance of the name Hebrew.*

These origins lie among the Semitic, nomadic peoples of Arabia, whose pre-historic and historic tendency was to press into the richer lands of the Fertile Crescent nearest to their original homes. The first stage of their story of which history must take notice is the age of the patriarchs, whose traditions are embodied in the biblical accounts of Abraham, Isaac and Jacob. There do not seem to be good grounds for denying that men who were the origins of these gigantic and legendary figures actually existed. If they did, they lived in about 1800 BC and their story is a part of the confusion following the end of Ur. The Bible states that Abraham came from Ur to Canaan; this is quite plausible and would not conflict with what we know of the dispersal of Amorite and other tribes in the next four hundred years. Those among them who were to be remembered as the descendants of Abraham became known in the end as 'Hebrews', a word meaning 'wanderer' which does not appear before Egyptian writings and inscriptions of the fourteenth or thirteenth centuries BC, long after their first settlement in Canaan. Though this word is not wholly satisfactory, it is probably the best name to give the tribes with which we are concerned at this time. It is a better term to identify this group than 'Jews', and for all the traditional associations gathered round that word by centuries of popular usage it is best to reserve it (as scholars usually do) for a much later era than that of the patriarchs.

It is in Canaan that Abraham's people are first distinguishable in the Bible, They are depicted as pastoralists, organized tribally, quarrelling with neighbours and kinsmen over wells and grazing, still liable to be pushed about the Near East by the pressures of drought and hunger. One group among them went down into Egypt, we are told, perhaps in the early seventeenth century BC; it was to appear in the Bible as the family of Jacob. As the story unfolds in the Old Testament, we learn of Joseph, the great son of Jacob, rising high in Pharaoh's service. At this point we might hope for help from Egyptian records. It has been suggested that this happened during the Hyksos ascendancy, since only a period of large-scale disturbance could explain the improbable pre-eminence of a foreigner in the Egyptian bureaucracy. It may be so, but there is no evidence to confirm or disprove it. There is only tradition, as there is only tradition for all Hebrew history until about 1200 BC. This tradition is embodied in the Old Testament; its texts only took this present form in the seventh century BC, perhaps eight hundred years after the story of Joseph, though older elements can be and have been distinguished in them. As evidence, it stands in something like the relation to Jewish origins in which Homer stands to those of Greece.

None of this would matter very much, and certainly would not interest anyone except professional scholars, were it not for events which occurred from one to three thousand years later. Then, the destinies of the whole world were swayed by the Christian and Islamic civilizations whose roots lay in the religious tradition of a tiny, not very easily identifiable Semitic people for centuries hardly distinguishable from many similar wanderers by the rulers of the great empires of Mesopotamia and Egypt. This was because the Hebrews somehow arrived at a unique religious vision.

Throughout the world of the ancient Near East it is possible to see at work forces which were likely to make monotheistic religious views more appealing. The power of local deities was likely to be questioned after contemplation of the great upheavals and disasters which time and time again swept across the region after the first Babylonian empire. The religious innovations of Akhnaton and the growing assertiveness of the cult of Marduk have both been seen as responses to such a challenge. Yet only the Hebrews and those who came to share their beliefs were able to push the process home, transcending polytheism and localism to arrive at a coherent and uncompromising monotheism.

The timing of this process is very difficult to establish but its essential steps were not complete before the eighth century BC. In the earliest times at which Hebrew religion could be distinguished it was probably polytheistic, but also monolatrous – that is to say, that like other Semitic peoples the tribes who were the forerunners of the Jews believed that there were many gods, but worshipped only one, their own. The first stage of refinement was the idea that the people of Israel (as the descendants of Jacob came to be called) owed exclusive allegiance to Yahweh, the tribal deity, a jealous God, who had made a covenant with his people to bring them again to the promised land, the Canaan to which Yahweh had already brought Abraham out of Ur, and which remains a focus of racial passion right down to the present. The covenant was a master idea. Israel was assured that *if* it *did* something, then something desirable *would* follow. This is very unlike the religious atmosphere of Mesopotamia or Egypt.

The exclusive demands of Yahweh opened the way to monotheism, for when the time came for this the Israelites felt no respect for other gods which might be an obstacle for such evolution. Nor was this all. At an early date Yahweh's nature was already different from that of other tribal gods. That no graven image was to be made of him was the most distinctive feature of his cult. At times, he appears as other gods, in an immanent dwelling place, such as a temple made with hands, or even in manifestations of nature, but, as the Israelite religion developed, he could be seen as transcendent deity:

'the LORD is in his holy temple, the LORD's throne is in heaven'

Psalm xi, 4

says a hymn. He had created everything, but existed independently of his creation, a universal being.

'Whither shall I go from thy spirit? or whither shall I flee from thy presence?' asked the Psalmist.

Psalm cxxxix, 7

The creative power of Yahweh was something else differentiating the Jewish from the Mesopotamian tradition. Both saw Man's origins in a watery chaos; 'the earth was without form, and void; and darkness was upon the face of the deep', says the book of Genesis. For the Mesopotamian, no pure creation was involved; somehow, matter of some sort had always been there and the gods only arranged it. It was different for the Hebrew; Yahweh had already created the chaos itself. He was for Israel what was later described in the Christian creed, 'maker of all things, by whom all things are made'. Moreover, he made Man in his own image, as a companion, not as a slave; Man was the culmination and supreme revelation of His creative power, a creature able to know good from evil, as did Yahweh Himself. Finally, Man moved in a moral world set by Yahweh's own nature. Only He was just; man-made laws might or might or might not reflect His will, but He was the only author of right and justice.

The implications of such ideas were to take centuries to clarify and millennia to show their full weight. At first, they were well wrapped up in the assumptions of a tribal society looking for a god's favour in war. Much in them reflected the special experience of a desert-dwelling people. Later Jewish tradition placed great emphasis on its origins in the exodus from Egypt, a story dominated by the heroic and mysterious figure of Moses. Clearly, when the Hebrews came to Canaan, they were already consciously a people, grouped round the cult of Yahweh. The biblical account of the wanderings in Sinai probably reports the crucial time when this national consciousness was forged. But the biblical tradition is again all that there is to depend upon and it was only recorded much later. It is certainly credible that the Hebrews should at last have fled from harsh oppression in a foreign land – an oppression which could, for example, reflect burdens imposed by huge building operations. Moses is an Egyptian name and it is likely that there existed a historical original of the great leader who dominates the biblical story by managing the exodus and holding the Hebrews together in the wilderness. In the traditional account, he founded the Law by bringing down the Ten Commandments from his encounter with Yahweh. This was the occasion of the renewal of the covenant by Yahweh and his people at Mount Sinai, and it may be seen as a formal return to its traditions by a nomadic people whose cult had been eroded by long sojourn in the Nile delta. Unfortunately, the exact role of the great religious reformer and national leader remains impossible to define and the Commandments themselves cannot be convincingly dated until much later than the time when he lived.

Yet though the biblical account cannot be accepted as it stands, it should be treated with respect as our only evidence for much of Jewish history. It contains much that can be related to what is known or inferred from other sources. Archaeology comes to the historians' help only with the arrival of the Hebrews in Canaan. The story of conquest told in the book of Joshua fits evidence of destruction in the Canaanite cities in the thirteenth century BC. What we know of Canaanite culture and religion also fits the Bible's account of Hebrew struggles against local cult practice and a pervasive polytheism. Palestine was disputed between two religious traditions and two peoples throughout the twelfth century and this, of course, again illustrates the collapse of Egyptian power, since this crucial area could not have been left to be the prey of minor Semitic peoples had the monarchy's power still been effective. It now seems likely that the Hebrews attracted to their support other nomadic tribes, the touchstone of alliance being adherence to Yahweh. After settlement, although the tribes quarrelled with one another, they continued to worship Yahweh and this was for some time the only uniting force among them, for tribal divisions formed Israel's only political institution.

The Hebrews took as well as destroyed. They were clearly in many ways less advanced culturally than the Canaanites and they took over their script. They borrowed their building practice, too, though without always achieving the same level of town life as their predecessors. Jerusalem was for a long time a little place of filth and confusion, not within striking-distance of the level reached by the town life of the Minoans long before. Yet in Israel lay the seeds of much of the future history of the human race.

Settlement in Palestine had been essentially a military operation and military necessity provoked the next stage in the consolidation of a nation. It seems to have been the challenge from the Philistines (who were obviously more formidable opponents than the Canaanites) which stimulated the emergence of the Hebrew kingship at some time about 1000 BC. With it appears another institution, that of the special distinction of the prophets, for it was the prophet Samuel who anointed (and thus, in effect, designated) both Saul, the first king, and his successor, David. When Saul reigned, the Bible tells us, Israel had no iron weapons, for the Philistines took care not to endanger their supremacy by permitting them. None the less, the Jews learnt the management of iron from their enemies; the Hebrew words for 'knife' and 'helmet' both have Philistine roots. Ploughshares did not exist, but, if they had, they could have been beaten into swords.

Saul won victories, but died at last by his own hand and his work was completed by David. Of all Old Testament individuals, David is outstandingly credible both for his strengths and weaknesses. Although there is no archaeological evidence that he existed, he lives still as one of the great figures of world literature and was a model for kings for two thousand years. The literary account, confused though it is, is irresistibly convincing. It tells of a noble-hearted but flawed and all-too-human hero who ended the Philistine peril and reunited the kingdom which had split at Saul's death. Jerusalem became Israel's capital and David then imposed himself upon the neighbouring peoples. Among them were the Phoenicians who had helped him against the Philistines, and this was the end of Tyre as an important independent state.

Yet it was David's son and successor, Solomon, who was the first king of Israel to achieve major international standing. He gave his army a chariot arm, launched expeditions to the south against the Edomites, allied with Phoenicia and built a navy. Conquest and prosperity followed.

> *'And Solomon reigned over all kingdoms from the river [Euphrates] unto the land of the Philistines, and unto the border of Egypt. . . and Judah and Israel dwelt safely, every man under his vine and under his fig tree, from Dan even to Beersheba, all the days of Solomon.'*
>
> I Kings iv, 21, 25.

Again, this sounds like the exploitation of possibilities available to the weak when the great are in decline; the success of Israel under Solomon is further evidence of the eclipse of the older empires and it was matched by the successes of other now-forgotten peoples of Syria and the Levant who constituted the political world depicted in the obscure struggles recorded in the Old Testament. Most of them were descendants of the old Amorite expansion. Solomon was a king of great energy and drive and the economic and technical advances of the period were also notable. He was an entrepreneur ruler of the first rank. The legendary 'King Solomon's Mines' have been said to reflect the activity of the first copper refinery of which there is evidence in the Near East, though this is disputed. Certainly the building of the Temple (after Phoenician models) was only one of many public works, though perhaps the most important. David had given Israel a capital, thus increasing the tendency to political centralization. He had planned a temple and when Solomon built it the worship of Yahweh was given a more splendid form than ever before and an enduring focus.

A tribal religion had successfully resisted the early dangers of contamination by the fertility rites and polytheism of the agriculturalists among whom the Hebrews had settled in Canaan. But there was always a threat of backsliding which would compromise the covenant. With success came other dangers, too. A kingdom meant a court, foreign contacts and – in Solomon's day - foreign wives who cherished the cults of their own gods. Denunciation of the evils of departing from the law by going a-whoring after the fertility gods of the Philistines had been the first role of the prophets; a new luxury gave them a social theme as well.

The prophets brought to its height the Israelite idea of God. They were not soothsayers such as the Near East already knew (though this is probably the tradition which formed the first two great prophets, Samuel and Elijah), but preachers, poets, political and moral critics. Their status depended essentially on the conviction they could generate in themselves and others, that God spoke through them. Few preachers have had such success. In the end Israel would be remembered not for the great deeds of her kings but for the ethical standards announced by her prophets. They shaped the connexions of religion with morality which were to dominate not only Judaism but Christianity and Islam.

A Jewish king makes obeisance to the Assyrian ruler Shalmaneser III in 841 BC. *Jehu, the king shown here, may be the first Jew of whom a picture survives. Above the figures is an inscription in cuneiform.*

The prophets evolved the cult of Yahweh into the worship of a universal God, just and merciful, stern to punish sin but ready to welcome the sinner who repented. This was the climax of religious culture in the Near East, a point after which religion could be separated from locality and tribe. The prophets also bitterly attacked social injustice. Amos, Isaiah and Jeremiah went behind the privileged priestly caste to do so, denouncing religious officialdom directly to the people. They announced that all men were equal in the sight of God, that kings might not simply do what they would; they proclaimed a moral code which was a given fact, independent of human authority. Thus the preaching of adherence to a moral law which Israel believed was god-given became also a basis for a criticism of existing political power. Since the law was not made by man it did not ostensibly emerge from that power; the prophets could always appeal to it as well as to their divine inspiration against king or priest. It is not too much to say that, if the heart of political liberalism is the belief that power must be used within a moral framework independent of it, then its tap-root is the teaching of the prophets.

Most of the prophets after Samuel spoke against a troubled background, which they called in evidence as signs of backsliding and corruption. Israel had prospered in the eclipse of paramount powers, when kingdoms came and went with great rapidity. After Solomon's death in 935 BC Hebrew history had ups as well as downs, but broadly took a turn for the worse. There had already been revolts; soon the kingdom split. Israel became a northern kingdom, built on ten tribes gathered together around a capital at Samaria; in the south the tribes of Benjamin and Judah still held Jerusalem, capital of the kingdom of Judah. The Assyrians obliterated Israel in 722 BC and the ten tribes disappeared from history in mass deportations. Judah lasted longer. It was more compact and somewhat less in the path of great states; it survived until 587 BC, when Jerusalem's walls and Temple were razed by a Babylonian army. The Judaeans, too, then suffered deportations, many of them being carried away to Babylon, to the great experience of the Exile, a period so important and formative that after it we may properly speak of 'the Jews', the inheritors and transmitters of a tradition still alive and easily traced. Once more great empires had established their grip in Mesopotamia and gave its civilization its last flowering. The circumstances which had favoured the appearance of a Jewish state had disappeared. Fortunately for the Jews, the religion of Judah now ensured that this did not mean that their national identity was doomed too.

Since the days of Hammurabi, the peoples of the Mesopotamian valley had been squeezed in a vice of migratory peoples. For a long time its opposing jaws had been the Hittites and the Mitanni, but from time to time others had ruled in Assur and Babylon. When, in due course, the Hittites also crumbled, ancient Mesopotamia was the seat of no great military power until the ninth century BC, though such a sentence conceals much. One Assyrian king briefly conquered Syria and Babylon early in the eleventh century; he was soon swept away by a cluster of pushful Semitic tribes whom scholars call Aramaeans, followers of the old tradition of expansion into the fertile lands from the desert. Together with a new line of Kassite kings in Babylon they were the awkward and touchy neighbours of the reduced kings of Assyria for two hundred years or so – for about as long as the United States has existed. Though one of these Semitic peoples was called the Chaldees and therefore subsequently gave its name somewhat misleadingly to Babylonia, there is not much to be remarked in this story except further evidence of the fragility of the political constructions of the ancient world.

Shape only begins to reappear in the turmoil of events in the ninth century BC when Mesopotamia recovered. Then, the Old Testament tells us, Assyrian armies were once more on the move against the Syrian and Jewish kingdoms. After some successful resistance the Assyrians came back again and again, and they conquered. This was the beginning of a new, important and unpleasant phase of Near Eastern history. A new Assyrian empire was in the making. In the eighth century it was moving to its apogee, and Nineveh, the capital high up the Tigris, which had replaced the ancient centre of Assur, became the focus of Mesopotamian history as Babylon had once been. Assyrian empire was a unity in a way that other great empires were not; it did not rely on the vassalization of kings and the creation of tributaries. Instead, it swept native rulers away and installed Assyrian governors. Often, too, it swept away peoples. One of its characteristic techniques was mass deportation; the Ten Tribes of Israel are the best-remembered victims.

Assyrian expansion was carried forward by repeated and crushing victory. Its greatest successes followed 729 BC, when Babylon was seized. Soon after, Assyrian armies destroyed Israel, Egypt was invaded, its kings were confined to Upper Egypt and the delta was annexed. By then Cyprus had submitted, Cilicia and Syria had been conquered. Finally, in 646 BC, Assyria made its last important conquest, part of the land of Elam, whose kings dragged the Assyrian conqueror's chariot through the streets of Nineveh. The consequences were of great importance for the whole Near East. A standardized system of government and law spanned the whole area. Conscript soldiers and deported populations were moved about within it, sapping its provincialism. Aramaic spread widely as a common language. A new cosmopolitanism was possible after the Assyrian age.

This great formative power is commemorated in monuments of undeniable impressiveness. Sargon II (721-705 BC) built a great palace at Khorsabad, near Nineveh, which covered half a square mile of land and was embellished with more than a mile of sculpted reliefs. The profits of conquest financed a rich and splendid court. Ashurbanipal (668-626 BC) also left his monuments (including obelisks carried off to Nineveh from Thebes), but he was a man with a taste for learning and antiquities and his finest relic is what survives of the great collection of tablets he made for his library. In it he accumulated copies of all that he could discover of the records and literature of ancient Mesopotamia. It is to these copies that we owe much of our knowledge of Mesopotamian literature, among them the Epic of Gilgamesh in its fullest edition, a translation made from Sumerian. The ideas that moved this civilization are thus fairly accessible from literature as well as from other sources. The frequent representation of Assyrian kings as hunters may be a part of the image of the warrior-king, but may also form part of a conscious identification of the king with legendary conquerors of nature who had been the heroes of a remote Sumerian past.

The stone reliefs which commemorate the great deeds of Assyrian kings also repeat, monotonously, another tale, that of sacking, enslavement, impalement, torture, and the Final Solution of mass deportation. Assyrian empire had a brutal foundation of conquest and intimidation. It was made possible by the creation of the best army seen until that time. Fed by conscription of all males and armed with iron weapons, it also had siege artillery able to breach walls until this time impregnable, and even some mailed cavalry. It was a coordinated force of all arms. Perhaps, too, it had a special religious fervour. The god Assur is shown hovering over the armies as they go to battle and to him kings reported their victories over unbelievers.

Whatever the fundamental explanation of Assyrian success, it quickly waned. Possibly, empire put too great a strain on Assyrian numbers. The year after Ashurbanipal died, the empire began to crumble, the first sign being a revolt in Babylon. The rebels were supported by the Chaldeans and also by a great new neighbour, the kingdom of the Medes, now the leading Iranian people. Their entrance as a major power on the stage of history marks an important change; for a long time the Medes had been distracted by having to deal with yet another wave of barbarian invaders from the north, the Scythians, who poured down into Iran from the Caucasus (and at the same time down the Black Sea coast towards Europe). They were light cavalrymen, fighting with the bow from horseback, and it took time to come to terms with them in the seventh century. This was, in fact, the first major eruption into western Asia of a new force in world history, nomadic peoples straight from Central Asia. Like all other great invasions, too, the Scythian advance pushed other peoples before them (the kingdom of Phrygia was overrun by one of these). Meanwhile, the last of the political units of the Near East based on the original Caucasian inhabitants was gobbled up by Scyths, Medes or Assyrians. All this took a century and more, but amounted to a great clearing of the stage. The instability and fragmentation of the periphery of the Fertile Crescent had long favoured Assyria; it ceased to do so when Scyths and Medes joined forces. This pushed Assyria over the edge and gave the Babylonians independence again; Assyria passes from history with the sack of Nineveh by the Medes in 612 BC.

This thunderbolt was not quite the end of the Mesopotamian tradition. Assyria's collapse left the Fertile Crescent open to new masters. The north was seized by the Medes, who pushed across Anatolia until halted at the borders of Lydia and at last drove the Scyths back into Russia. An Egyptian pharaoh made a grab at the south and the Levant, but was defeated by a Babylonian king, Nebuchadnezzar, who gave Mesopotamian civilization an Indian summer of grandeur and a last Babylonian empire which more than any other captured the imagination of posterity. It ran from Suez, the Red Sea and Syria across the border of Mesopotamia and the old kingdom of Elam (by then ruled by a minor Iranian dynasty called the Achaemenids). If for nothing else, Nebuchadnezzar would be remembered as a great conqueror. He destroyed Jerusalem in 587 BC after a Jewish revolt and carried off the tribes of Judah into captivity, using them as he used other captives, to carry out the embellishment of his capital, whose 'hanging gardens' or terraces were to be remembered as one of the Seven Wonders of the World. He was the greatest king of his time, perhaps of any time until his own.

The glory of the empire came to a focus in the cult of Marduk, which was now at its zenith. At a great New Year festival held each year all the Mesopotamian gods – the idols and statues of provincial shrines – came down the rivers and canals to take counsel with Marduk at his temple and acknowledge his supremacy. Borne down a processional way three-quarters of a mile long (which was, we are told, probably the most magnificent street of antiquity) or landed from the Euphrates nearer to the temple, they were taken into the presence of a statue of the god which, Herodotus reported two centuries later, was made of two and a quarter tons of gold. No doubt he exaggerated, but it was indisputably magnificent. The destinies of the whole world, whose centre was this temple, were then debated by the gods and determined for another year. Thus theology reflected political reality. The re-enacting of the drama of creation was the endorsement of Marduk's eternal authority, and this was an endorsement of the absolute monarchy of Babylon. The king had responsibility for assuring the order of the world and therefore the authority to do so.

It was the last flowering of the Mesopotamian tradition and was soon to end. More and more provinces were lost under Nebuchadnezzar's successors. Then came an invasion in 539 BC by new conquerors from the east, the Persians, led by the Achaemenids. The passage from worldly pomp and splendour to destruction had been swift. The book of Daniel telescopes it in a magnificent closing scene, Belshazzar's feast. 'In that night,' we read, 'was Belshazzar the king of the Chaldeans slain. And Darius the Median took the kingdom.' Unfortunately, this account was only written three hundred years later and it was not quite like that. Belshazzar was neither Nebuchadnezzar's son, nor his successor, as the book of Daniel says, and the king who took Babylon was called Cyrus. None the less, the emphasis of the Jewish tradition has a dramatic and psychological truth. In so far as the story of antiquity has a turning-point, this is it. An independent Mesopotamian tradition going back to Sumer is over. We are at the edge of a new world. A Jewish poet summed it up exultantly in the book of Isaiah, where Cyrus appears as a deliverer to the Jews:

> *'Sit thou silent, and get thee into darkness, O daughter of the Chaldeans:*
> *for thou shalt no more be called, The lady of kingdoms.'*

Daniel v, 30-31, Isaiah xlvii, 5, xlv, 1-2

5
The Beginnings of Civilization in Eastern Asia

From the beginnings to the most recent times the centre of gravity of world history has usually swung about between the Atlantic and Iran. Yet (also until the most recent times) what went on there had little direct impact elsewhere. Much of the life of other parts of the world long remained virtually impervious to the influence of its civilizations and two areas were especially resistant: India and China. By 1000 BC civilizations had appeared in these countries which were, in spite of peripheral contacts, quite independent of the Near East. They were the foundations of major and enduring cultural traditions long to outlive those of Mesopotamia and Egypt and they would each enjoy a huge sphere of influence.

ANCIENT INDIA

Even now, ancient India is still visible and accessible to us in a very direct sense. At the beginning of this century, some Indian communities still lived as all our primeval ancestors must once have lived, by hunting and gathering. The bullock-cart and the potter's wheel of many villages today are, as far as can be seen, identical with those used four thousand years ago. A caste-system whose main lines were set by about 1000 BC still regulates the lives of millions, and even of some Indian Christians and Moslems. Gods and goddesses whose cults can be traced to the Stone Age are still worshipped at village shrines.

In some ways, then, ancient India is with us still as is no other ancient civilization. Yet though such examples of the conservatism of Indian life are commonplace, the country that contains them contains many other things too. The hunter-gatherers of the early twentieth century were the contemporaries of other Indians used to travelling in railway trains. The diversity of Indian life is enormous, but wholly comprehensible given the size and variety of its setting. The sub-continent is, after all, about the size of Europe and is divided into regions clearly distinguished by climate, terrain and crops. There are two great river valleys, the Indus and Ganges systems, in the north; between them lie desert and arid plains, and to the south the highlands of the Deccan, largely forested. When written history begins, India's racial complexity, too, is already very great: scholars identify six main ethnic groups. Many others were to arrive later and make themselves at home in the Indian sub-continent and society, too. All this makes it hard to find a focus.

Yet Indian history has a unity in the fact of its enormous power to absorb and transform forces playing on it from the outside. This provides a thread to guide us through the patchy and uncertain illumination of its early stages which is provided by archaeology and texts long transmitted only by word of mouth. Its basis is to be found in another fact, India's large measure of insulation from the outside world by geography. In spite of her size and variety, until the oceans began to be opened up in the sixteenth and seventeenth centuries India had only to grapple with occasional, though often irresistible, incursions by alien peoples. To the north and north-west she was protected by some of the highest mountains in the world; to the east lay belts of jungle. The other two sides of the sub-continent's great triangle opened out into the huge expanses of the Indian ocean. This natural definition not only channelled and restricted communication with the outside world; it also gave India a distinctive climate. Much of India does not lie in the tropics, but none the less that climate is tropical. The mountains keep away the icy winds of Central Asia; the long coasts open themselves to the rain-laden clouds which roll in from the oceans and cannot go beyond the northern ranges. The climatic clock is the annual monsoon, bringing the rain during the hottest months of the year. It is still the central prop of an agricultural economy.

Few pieces of sculpture have attracted so much speculation as this soapstone figure of a priest or deity from Mohenjo-Daro. The ornamentation of the cloak perhaps reflects a theme to be found in Mesopotamia and Persia, the arrangement of the same garment anticipates the later style in which robes were worn by Buddhist monks, the modelling of the head and features suggests faint reminiscences of Sumer. Nothing about the relationship of sculpture from Mohenjo-Daro – of which this is one of only a few fragmentary examples – to the sculpture of other times and places can safely by deduced from it.

Protected in some measure from external forces though she has always been before modern times, India's north-western frontier is more open than her others to the outside world. Baluchistan and the frontier passes were the most important zones of encounter between India and other peoples right down to the seventeenth century AD; in civilized times even India's contacts with China were first made by this roundabout route (though it is not quite as roundabout as Mercator's familiar projection makes it appear). At times, this north-western region has fallen directly under foreign sway, which is suggestive when we consider the first Indian civilizations; we do not know much about the way in which they arose but we know that Sumer and Egypt antedated them. Mesopotamian records of Sargon I of Akkad report contacts with a 'Meluhha' which scholars have believed to be the Indus valley, the alluvial plains forming the first natural region encountered by the traveller once he has entered India. It was there, in rich, heavily forested countryside, that the first Indian civilizations appeared at the time when, farther west, the great movements of Indo-European peoples were beginning to act as the levers of history. There may have been more than one stimulus at work.

The evidence also shows that agriculture came later to India than to the Near East. It, too, can first be traced in the sub-continent in its north-west corner. There is archaeological evidence of domesticated animals in Baluchistan as far back as 3700 BC. By 3000 BC there are signs of settled life on the alluvial plains and parallels with other river-valley cultures begin to appear. Wheel-thrown pottery and copper implements begin to be found. All the signs are of a gradual build-up in intensity of agricultural settlements until true civilization appears as it did in Egypt and Sumer. But there is the possibility of direct Mesopotamian influence in the background and, finally, there is at least a reasonable inference that already India's future was being shaped by the coming of new peoples from the north. At a very early date the complex racial composition of India's population suggests this, though it would be rash to be assertive about it.

When at last indisputable evidence of civilized life is available, the change is startling. One scholar speaks of a cultural 'explosion'. There may have been one crucial technological step, the invention of burnt brick (as opposed to the sun-baked mud brick of Mesopotamia) which made flood control possible in a flat plain lacking natural stone. Whatever the process, the outcome was a remarkable civilization which stretched over a half-million square miles of the Indus valley, an area greater than either the Sumerian or Egyptian.

This civilization has been given the name 'Harappan', because one of its great sites is the city of Harappa on a tributary of the Indus. There is another such site at Mohenjo-Daro; others are being discovered. Together they reveal human beings highly organized and capable of carefully regulated collective works on a scale equalling those of Egypt and Mesopotamia. There were large granaries in the cities, and weights and measures seem to have been standardised over a large area. It is a clear that a well-developed culture was established by 2250 BC and lasted for something like four hundred years with very little change.

The two cities which are its greatest monuments may have contained more than thirty thousand people each. This says much for the agriculture which sustained them; the region was then far from being the arid zone it later became. Mohenjo-Daro and Harappa were between two and two and a half miles in circumference and the uniformity and complexity of their building speaks for a very high degree of administrative and organizational skill. They each had a citadel and a residential area; streets of houses were laid out on rectangular grid plans and made of bricks of standardized sizes. Both the elaborate and effective drainage systems and the internal layout of the houses show a strong concern for bathing and cleanliness; in some streets of Harappa nearly every house has a bathroom. Perhaps it is not fanciful to see in this some of the first manifestations of what has become an enduring feature of Indian religion, the bathing and ritual ablutions which are still important to Hindus.

Before 2000 BC these cities were trading far afield and living an economic life of some complexity. A great dockyard, connected by a mile-long canal to the sea at Lothal, four hundred miles south of Mohenjo-Daro, suggests the importance of an external trade which reached, through the Persian Gulf, as far north as Mesopotamia. In the Harappan cities themselves evidence survives of specialized craftsmen drawing their materials from a wide area and subsequently sending out again across its length and breadth the products of their skills. This civilization had cotton cloth (the first of which we have evidence) which was plentiful enough to wrap bales of goods for export whose cordage was sealed with seals found at Lothal. These seals are part of our evidence for Harappan literacy; a few inscriptions on fragments of pottery are all that supplements them and provides the first traces of Indian writing. The seals, of which about 2500 survive, provide some of our best clues to Harappan ideas. The pictographs on the seals run from right to left. Animals often appear on them and may represent six seasons into which the year was divided. Many 'words' on the seals remain unreadable, but it now seems at least likely that they are part of a language akin to the Dravidian tongues still used in southern India.

Ideas and techniques from the Indus spread throughout Sind and the Punjab, and down the west coast of Gujarat. The process took centuries and the picture revealed by archaeology is too confused for a consistent pattern to emerge. Where its influence did not reach – the Ganges valley, the other great silt-rich area where large populations could live, and the south-east – different cultural processes were at work, but they have left nothing so spectacular behind them. Some of India's culture must derive from other sources; there are traces elsewhere of Chinese influence. But it is hard to be positive. Rice, for example, began to be grown in India in the Ganges valley; we simply do not know where it came from, but one possibility is China or South-East Asia, on whose coasts it was grown from about 3000 BC. Two thousand years later, this crucial item in Indian diet was used over most of the north.

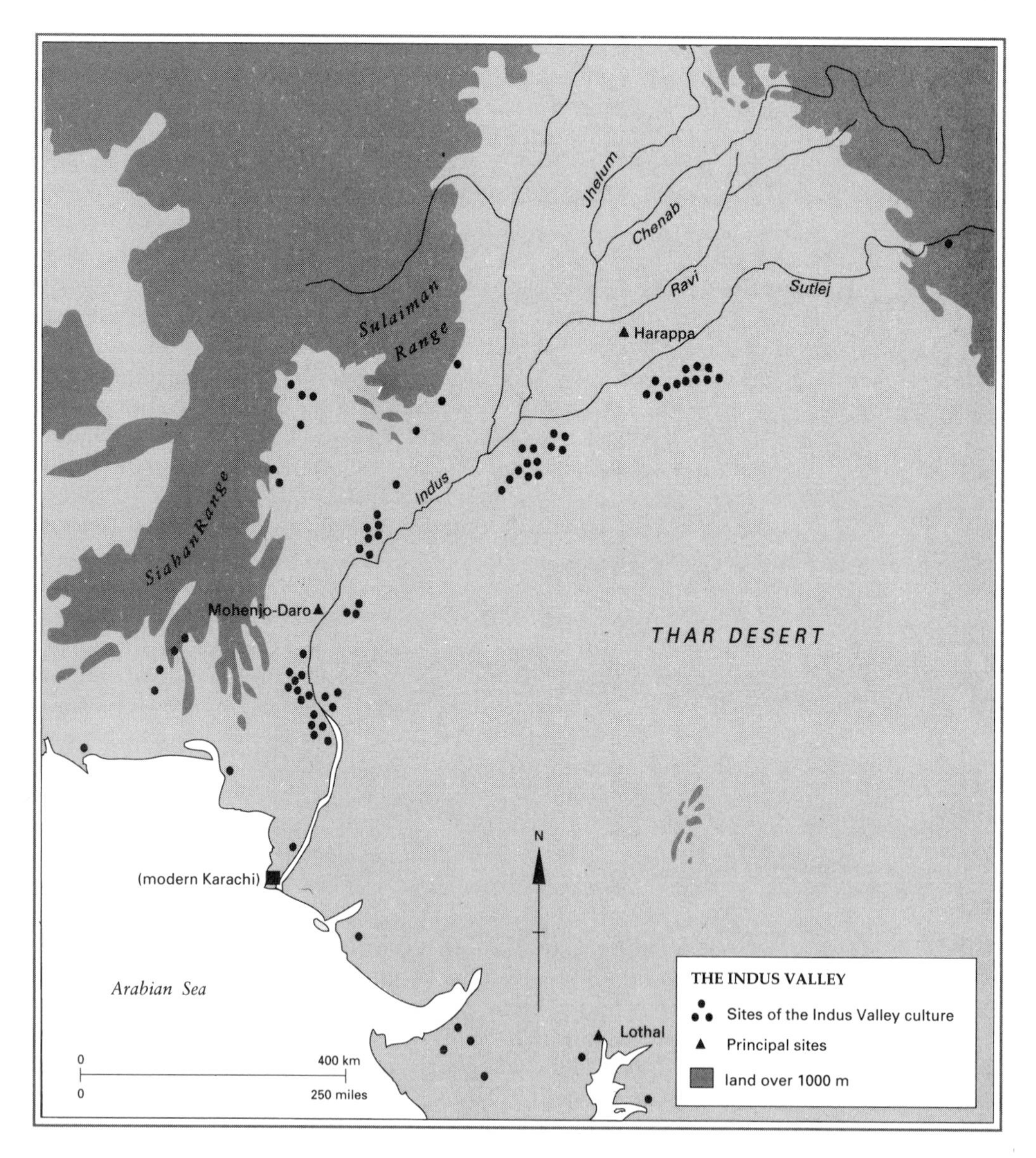

Nor do we know why the first Indian civilizations began to decline, though their passing can be roughly dated. The devastating floods of the Indus or uncontrollable alterations of its course may have wrecked the delicate balance of the agriculture on its banks. The forests may have been destroyed by tree-felling to provide fuel for the brick-kilns on which Harappan building depended. But perhaps there were also other agencies at work. Skeletons, possibly those of men killed where they fell, were found in the streets of Mohenjo-Daro. Harappan civilization seems to end in the Indus valley about 1750 BC and this coincides strikingly with the irruption into Indian history of one of its great creative forces, the invading Aryans, though scholars do not favour the idea that invaders destroyed the Indian valley cities. Perhaps the newcomers entered a land already devastated by over-exploitation and nature disorders.

Strictly speaking, 'Aryan' is a linguistic term, like 'Indo-European'. None the less, it has customarily and conveniently been used to identify one group of those Indo-European peoples whose movements make up so much of the dynamic of ancient history in other parts of the Old World after 2000 BC. At about the time when other Indo-Europeans were flowing into Iran, somewhere about 1750 BC, a great influx of Aryans began to enter India from the Hindu Kush. This was the beginning of centuries during which waves of these migrants washed deeper and deeper into the Indus valley and the Punjab and eventually reached the upper Ganges. They did not obliterate the native peoples, though the Indus valley civilization crumbled. No doubt much violence marked their coming, for the Aryans were warriors and nomads, armed with bronze weapons, bringing horses and chariots, but they settled and there are plenty of signs that the native populations lived on with them, keeping their own beliefs and practices alive. There is much archaeological evidence of the fusion of Harappan with later ways. However qualified, this was an early example of the assimilation of cultures which was always to characterize Indian society and was eventually to underly classical Hinduism's remarkable digestive power.

It seems clear that the Aryans brought to India no culture so advanced as that of the Harappans. It is a little like the story of the coming of Indo-Europeans in the Aegean. Writing, for example, disappears and does not emerge again until the middle of the first millennium BC; cities, too, have to be re-invented and when they are again to be found they lack the elaboration and order of their Indus valley predecessors. Instead, the Aryans appear to have slowly given up their pastoral habits and settled into agricultural life, spreading east and south from their original settlement areas in a sprawl of villages. This took centuries. Not until the coming of iron was it complete and the Ganges valley colonized; iron implements made cultivation easier. Meanwhile, together with this physical opening up of the northern plains, Aryan culture had made two decisive contributions to Indian history, in its religious and in its social institutions.

The Aryans laid the foundations of the religion which has been the heart of Indian civilization. Theirs centred on sacrificial concepts; through sacrifice the process of creation which the gods achieved at the beginning of time was to be endlessly repeated. Agni, the god of fire, was very important, because it was through his sacrificial flames that men could reach the gods. Great importance and standing was given to the *brahmans*, the priests who presided over these ceremonies. There was a pantheon of gods of whom two of the most important were Varuna, god of the heavens, controller of natural order and the embodiment of justice, and Indra, the warrior god who, year after year, slew a dragon and thus released again the heavenly waters which came with the breaking of the monsoon. We learn about them from the *Rig-Veda*, a collection of more than a thousand hymns performed during sacrifice, collected for the first time in about 1000 BC but certainly accumulated over centuries. It is one of our most important sources for the history not only of Indian religion but also of Aryan society.

The *Rig-Veda* seems to reflect an Aryan culture as it has been shaped by settlement in India and not Aryan culture as it had existed at earlier times. It is, like Homer, the eventual written form of a body of oral tradition, but quite different in being much less difficult to use as a historical source, since its status is much more certain. Its sanctity made its memorization in exact form essential, and though the *Rig-Veda* was not to be written down until after AD 1300, it was then still almost certainly largely uncorrupted from its original form. Together with later Vedic hymns and prose works, it is our best source for Aryan India, whose archaeology is cramped for a long time because building materials less durable than the brick of the Indus valley cities were used in its towns and temples.

There is a suggestion again of the world of Homer in the world revealed by the *Rig-Veda*, which is one of Bronze Age barbarians. Some archaeologists now believe they can identify in the hymns references to the destruction of the Harappan cities. Iron is not mentioned and appears only to have come to India after 1000 BC (there is argument about how late and from what source). The setting of the hymns is a land which stretches from the western banks of the Indus to the Ganges, inhabited by Aryan peoples and dark-skinned native inhabitants. These formed societies whose fundamental units were families and tribes, but the legacies of these were less enduring than the pattern of Aryan social organization which eventually emerges and which we call caste.

About the early history of this vast and complicated subject and its implications it is impossible to speak with assurance. Once the rules of caste were written down, they appeared as a hard and solid structure, incapable of variation. Yet this did not happen until caste had been in existence for hundreds of years, during which it was still flexible and evolving. Its root appears to be a recognition of the fundamental class-divisions of a settled agricultural society, a warrior-aristocracy (*kshatriyas*), priestly *brahmans* and the ordinary peasant-farmers (*vaishyas*). These are the earliest divisions of Aryan society which can be observed and seem not to have been exclusive; movement between them was possible. The only unleapable barrier in early times seems to have been that between non-Aryans and Aryans; one of the words used to denote the aboriginal inhabitants of India by Aryans was *dasa*, which came eventually to denote 'slave'. To the occupational categories was soon added a fourth category for non-Aryans. Clearly it rested on a wish to preserve racial integrity. These were the *shudras*, or 'unclean', who might not study or hear the Vedic hymns.

This structure has been elaborated almost ever since. Further divisions and subdivisions appeared as society became more complex and movements within the original threefold structure took place. In this the *brahmans*, the highest class, played a crucial role. Landowners and merchants came to be distinguished from farmers; the first were called *vaishyas*, and *shudras* became cultivators. Marriage and eating taboos were codified. This process gradually led to the appearance of the caste system as we know it. A vast number of castes and sub-castes slowly inserted themselves into the system. Their obligations and demands eventually became a primary regulator of Indian society, perhaps the only significant one in many Indians' lives. By modern times there were thousands of *jatis* – local castes with members restricted to marrying within them, eating only food cooked by fellow-members, and obeying their regulations. Usually, too, a caste limited those who belonged to it to the practice of one craft or profession. For this reason (as well as because of the traditional ties of tribe, family and locality and the distribution of wealth) the structure of power in Indian society right down to the present day has had much more to it than formal political institutions and central authority.

In early times Aryan tribal society threw up kings, who emerged, no doubt, because of military skill. Gradually, some of them acquired something like divine sanction, though this must always have depended on a nice balance of relations with the *brahman* caste. But this was not the only political pattern. Not all Aryans accepted this evolution. By about 600 BC, when some of the detail of early Indian political history at last begins to be dimly discernible through a mass of legend and myth, two sorts of political communities can be discerned, one non-monarchical, tending to survive in the hilly north, and one monarchical, established in the Ganges valley. This reflected centuries of steady pressure by the Aryans towards the east and south during which peaceful settlement and intermarriage seem to have played as big a part as conquest. Gradually, during this era, the centre of gravity of Aryan India had shifted from the Punjab to the Ganges valley as Aryan culture was adopted by the peoples already there.

As we emerge from the twilight zone of the Vedic kingdoms, it is clear that they established something like a cultural unity in northern India. The Ganges valley was by the seventh century BC the great centre of Indian population. It may be that the cultivation of rice made this possible. A second age of Indian cities began there, the first of them market-places and centres of manufacture, to judge by the way they brought together specialized craftsmen. The great plains, together with the development of armies on a larger and better-equipped scale (we hear of the use of elephants), favoured the consolidation of larger political units. At the end of the seventh century BC, northern India was organized in sixteen kingdoms, though how this happened and how they were related to one another is still hard to disentangle from their mythology. None the less, the existence of coinage and the beginnings of writing make it likely that they had governments of growing solidity and regularity.

The processes in which they emerged are touched on in some of the earliest literary sources for Indian history, the *Brahmanas*, texts composed during the period when Aryan culture came to dominate the Ganges valley (*c*. 800-600 BC). But more about them and the great names involved can be found in later documents, above all in two great Indian epics, the *Ramayana* and the *Mahabharata*. The present texts are the result of constant revision from about 400 BC to AD 400, when they were written down as we know them for the first time, so their interpretation is not easy. In consequence, it

The main streets of Mohenjo-Daro, thirty to forty feet wide, ran north and south, probably because of the direction of the prevailing winds. The city was better ordered and planned than earlier Mesopotamian cities or later Indian ones. Even the carefully laid-out drains running along these streets were provided with inspection holes. It was a rich well-ordered place, apparently investing its wealth in its comfort and material well-being, rather than in temples and palaces.

remains hard to get at the political and administrative reality behind, say, the kingdom of Magadha, based on southern Bihar, which emerged eventually as the preponderant power and was to be the core of the first historical empires of India. On the other hand (and possibly more importantly), the evidence is clear that the Ganges valley was already what it was to remain, the seat of empire, its cultural domination assured as the centre of Indian civilization, the future Hindustan.

The later Vedic texts and the general richness of the Aryan literary record make it all too easy to forget the existence of half the sub-continent. Written evidence tends to confine Indian history down to this point (and even after) to the history of the north. The state of archaeological and historical scholarship also reflects and further explains the concentration of attention on northern India. There is just much more known about it in ancient times than about the south. But there are also better and less accidental justifications for such an emphasis. The archaeological evidence shows, for example, a clear and continuing cultural lag in this early period between the area of the Indus system and the rest of India (to which, it may be remarked, the river was to give its name). Enlightenment (if it may be so expressed) came from the north. In the south, near modern Mysore, settlements roughly contemporaneous with Harappa show no trace of metal, though there is evidence of domesticated cattle and goats. Bronze and copper only begin to appear at some time after the Aryan arrival in the north. Once outside the Indus system, too, there are no contemporary metal sculptures, no seals and fewer terracotta figures. In Kashmir and eastern Bengal there are strong evidences of Stone Age cultures with affinities with those of south China, but it is at least clear that, whatever the local characteristics of the Indian cultures with which they were in contact and within the limits imposed by geography, first Harappan and then Aryan civilization were dominant. They gradually asserted themselves towards Bengal and the Ganges valley, down the west coast towards Gujarat, and in the central highlands of the sub-continent. This is the pattern of the Dark Age, and when we reach that of history, there is not much additional light. The survival of Dravidian languages in the south shows the region's persistent isolation.

Topography explains much of it. The Deccan has always been cut off from the north by jungle-clad mountains, the Vindhya. Internally, too, the south is broken and hilly, and this did not favour the building of large states as did the open plains of the north. Instead, south India remained fragmented, some of its peoples persisting, thanks to their inaccessibility, in the hunting and gathering cultures of a tribal age. Others, by a different accident of geography, turned to the seas – another contrast with the predominantly agrarian empires of the north.

Millions of people must have been affected by the changes so far described. Estimates of ancient populations are notoriously unreliable. India's has been put at about 25 millions in 400 BC, which would be roughly a quarter of the whole population of the world at that time. The importance of India's early history nevertheless lies in the way it laid down patterns still shaping the lives of even larger numbers today, rather than in its impact on big populations in antiquity. This is above all true of religion. Classical Hinduism crystallized in the first millennium BC. A great world religion, Buddhism, was also launched then in India; it was eventually to dominate wide areas of Asia. What men do is shaped by what they believe they can do; it is the making of a culture that is the pulse of Indian history, not the making of a nation or an economy, and to this culture religion was central.

The deepest roots of the Indian religious and philosophical synthesis go very deep indeed. One of the great popular cult figures of the Hindu pantheon today is Shiva, in whose worship many early fertility cults have been brought together. A seal from Mohenjo-Daro already shows a figure who looks like an early Shiva, and stones like the *lingam* found in modern temples, the phallic cult-object which is his emblem, have been found in the Harappan cities. There is some presumptive evidence therefore for the view that worship of Shiva may be the oldest surviving religious cult in the world. Though he has assimilated many important Aryan characteristics, he is pre-Aryan and survives in all his multi-faceted power, still an object of veneration in the twentieth century. Nor is Shiva the only possible survival from the remote past of Indus civilization. Other Harappan seals seem to suggest a religious world centred about a mother-goddess and a bull. The bull survives to this day, the Nandi of countless village shrines all over Hindu India (and newly vigorous in his latest incarnation, as the electoral symbol of the Congress Party).

Vishnu, another focus of modern popular Hindu devotion, is much more an Aryan. Vishnu joined hundreds of local gods and goddesses still worshipped today to form the Hindu pantheon. Yet his cult is far from being either the only or the best evidence of the Aryan contribution to Hinduism. Whatever survived from the Harappan (or even pre-Harappan) past, the major philosophical and speculative traditions of Hinduism stem from Vedic religion. These are the Aryan legacy. To this day, Sanskrit is the language of religious learning; it transcends ethnic divisions, being used in the Dravidian-speaking south as much as in the north by the brahman. It was a great cultural adhesive and so was the religion it carried. The Vedic hymns provided the nucleus for a system of religious thought more abstract and philosophical than primitive animism. Out of Aryan notions of hell and paradise, the House of Clay and the World of the Fathers, there gradually evolved the belief that action in life determined human destiny. An immense, all-embracing structure of thought slowly emerged, a world view in which all things are linked in a huge web of being. Souls might pass through different forms in this immense whole; they might move up or down the scale of being, between castes, for example, or even between the human and animal worlds. The idea of transmigration from life to life, its forms determined by proper behaviour, was linked to the idea of purgation and renewal, to the trust in liberation from the transitory, accidental and apparent, and to belief in the eventual indentity of soul and absolute being in *Brahma*, the creative principle. The duty of the believer was the observation of *Dharma* – a virtually untranslatable concept, but one which embodies something of the western ideas of a natural law of justice and something of the idea that men owed respect and obedience to the duties of their station.

These developments took a long time. The steps by which the original Vedic tradition began its transformation into classical Hinduism are obscure and complicated. At the centre of the early evolution had been the *brahmans* who long controlled religious thought because of their key role in the sacrificial rites of Vedic religion. The brahmanical class appears to have used its religious authority to emphasize its seclusion and privilege. To kill a brahman soon became the gravest of crimes; even kings could not contend with their powers. Yet they seem to have come to terms with the gods of an older world in early times; it has been suggested that it may have been the infiltration of the brahmanical class by priests of the non-Aryan cults which ensured the survival and later popularity of the cult of Shiva.

The sacred *Upanishads*, texts dating from about 700 BC, mark the next important evolution towards a more philosophical religion. They are a mixed bag of about two hundred and fifty devotional utterances, hymns, aphorisms and reflexions of holy men pointing to the inner meaning of the traditional religious truths. They give much less emphasis to personal gods and goddesses than earlier texts and also include some of the earliest ascetic teachings which were to be so visible and striking a feature of Indian religion, even if only practised by a small minority. The *Upanishads* met the need felt by some men to look outside the traditional structure for religious satisfaction. Doubt appears to have been felt about the sacrificial principle. New patterns of thought had begun to appear at the beginning of the historical period and uncertainty about traditional beliefs is already expressed in the later hymns of the *Rig-Veda*. It is convenient to mention such developments here because they cannot be understood apart from the Aryan and pre-Aryan past. Classical Hinduism was to embody a synthesis of ideas like those in the *Upanishads* (pointing to a monistic conception of the universe) with the more polytheistic popular tradition represented by the *brahmans*.

Abstract speculation and asceticism were often favoured by the existence of monasticism, a stepping-aside from material concerns to practise devotion and contemplation. The practice appeared in Vedic times. Some monks threw themselves into ascetic experiment, others pressed speculation very far and we have records of intellectual systems which rested on outright determinism and materialism. One very successful cult which did not require belief in gods and expressed a reaction against the formalism of the brahmanical religion was Jainism, a creation of a sixth-century teacher who, among other things, preached a respect for animal life which made agriculture or animal husbandry impossible. Jains therefore tended to become merchants, with the result that in modern times the Jain community is one of the wealthiest in India. But much the most important of the innovating systems was the teaching of the Buddha, the 'enlightened one' or 'aware one' as his name may be translated.

It has been thought significant that the Buddha, like some other religious innovators, was born in one of the states to the northern edge of the Ganges plain where the orthodox, monarchical pattern emerging elsewhere did not establish itself. This was early in the sixth century BC. Siddhartha Gautama was not a *brahman*, but a prince of the warrior class. After a comfortable and gentlemanly upbringing he found his life unsatisfying and left home. His first recourse was asceticism. Seven years of this proved to him that he was on the wrong road. He began instead to preach and teach. His reflexions led him to propound an austere and ethical doctrine, whose aim was liberation from suffering by achieving higher states of consciousness. This was not without parallels in the teaching of the *Upanishads*.

An important part in this was to be played by *yoga*, which was to become one of what were termed the 'Six Systems' of Hindu philosophy. The word has many meanings but in this context is roughly translatable as 'method' or 'technique'. It sought to achieve truth through meditation after a complete and perfect control of the body had been attained. Such control was supposed to reveal the illusion of personality which, like all else in the created world, is mere flux, the passage of events, not identity. This system, too, had already been sketched in the *Upanishads* and was to become one of the aspects of Indian religion which struck visitors from Europe most forcibly. The Buddha taught his disciples so to discipline and shed the demands of the flesh that no obstacle should prevent the soul from attaining the blessed state of *Nirvana* or self-annihilation, freedom from the endless cycle of rebirth and transmigration, a doctrine urging men not to do something, but to be something – in order not to be anything. The way to achieve this was to follow an Eightfold path of moral and spiritual improvement. All this amounts to a great ethical and humanitarian revolution.

The Buddha apparently had great practical and organizing ability which, together with his unquestionable personal quality, quickly made him a popular and successful teacher. He sidestepped, rather than opposed, the brahmanical religion and this must have smoothed his path. The appearance of communities of Buddhist monks gave his work an institutional setting which would outlive him. He also offered a role to those not satisfied by traditional practice, in particular to women and to low-caste followers, for caste was irrelevant in his eyes. Finally, Buddhism was non-ritualistic, simple and atheistic. It soon underwent elaboration, and, some would say, speculative contamination, and like all great religions it assimilated much preexisting belief and practice, but by doing so it retained great popularity.

Yet Buddhism did not supplant brahmanical religion and for two centuries or so was confined to a relatively small part of the Ganges valley. In the end, too – though not until well into the Christian era – Hinduism was to be the victor and Buddhism would dwindle to a minority belief in India. But it was to become the most widespread religion in Asia and a potent force in world history. It is the first world religion to spread beyond the society in which it was born, for the older tradition of Israel had to wait for the Christian era before it could assume a world role. In its native India, Buddhism was to be important until the coming of Islam. The teaching of the Buddha marks, therefore, a recognizable epoch in Indian history; it justifies a break in its exposition. By his day, an Indian civilization still living today and still capable of enormous assimilative feats stood complete in its essentials. This was a huge fact; it would separate India from the rest of the world.

Much of the achievement of early civilization in India remains intangible.

There is a famous figure of a beautiful dancing-girl from Mohenjo-Daro, but ancient India before the Buddha's time did not produce great art on the scale of Mesopotamia, Egypt, or Minoan Crete, far less their great monuments. Marginal in her technology, she came late – though how much later than other civilizations cannot be exactly said – to literacy, too. Yet the uncertainties of much of India's early history cannot obscure the fact that her social system and her religions have lasted longer than any other great creations of the mind. Even to guess at what influence they exercised through the attitudes they encouraged, diffused through centuries in pure or impure forms, is rash. Only a negative dogmatism is safe; so comprehending a set of world views, institutions so careless of the individual, philosophy so assertive of the relentless cycles of being, so lacking in any easy ascription of responsibility for good and evil, cannot but have made a history very different from that of men reared in the great Semitic traditions. And these attitudes were formed and settled for the most part a thousand years before Christ.

ANCIENT CHINA

The most striking fact of China's history is that it has gone on for so long. For about two and a half thousand years there has been a Chinese nation using a Chinese language. Its government as a single unit has long been taken to be normal, in spite of intervals of division and confusion. China has had a continuing experience of civilization rivalled in duration only by that of ancient Egypt. This experience is the key to Chinese historical identity; it is as much cultural as political. The example of India shows how much more important culture can be than government, and China's makes the same point in a different way; in China, culture made unified government easier. Somehow at a very early date she crystallized certain institutions and attitudes which were to endure because they suited her circumstances. Some of them seem even to transcend the revolution of the twentieth century.

We must begin with the land itself, and at first sight it does not suggest much that makes for unity. The physical theatre of Chinese history is vast. China is bigger than the United States and now contains three to four times as many people. The Great Wall which guarded the northern frontier was made up of between 2500 and 3000 miles of fortifications and has never been completely surveyed. From Peking to Hong Kong, more or less due south, is 1200 miles as the crow files. This huge expanse contains many climates and many regions. One great distinction stands out among them, that between northern and southern China. In summer the north is scorching and arid while the south is humid and used to floods; the north looks bare and dustblown in the winter, while the south is always green. This is not all that this distinction implies. One of the major themes of early Chinese history is of the spread of civilization, sometimes by migration, sometimes by diffusion, from north to south, of the tendency of conquest and political unification to take the same broad direction, and of the continual stimulation and irrigation of northern civilization by currents from the outside, from Mongolia and Central Asia.

China's major internal divisions are set by mountains and rivers. There are three great river valleys which drain the interior and run across the country roughly from west to east. They are, from north to south, the Hwang-Ho, or Yellow River, the Yangtze and the Hsi. It is surprising that a country so vast and thus divided should form a unity at all. Yet China is isolated, too. One scholar thinks the country a world by itself since long before the Pleistocene. Much of China is mountainous and except in the extreme south and northeast her frontiers still sprawl across and along great ranges and plateaux. The headwaters of the Yangtze, like those of the Mekong, lie in the high Kunlun, north of Tibet. These highland frontiers are great insulators. The arc they form is broken only where the Yellow River flows south into China from inner Mongolia and it is on the banks of this river that the story of civilization in China begins.

Skirting the Ordos desert, itself separated by another mountain range from the desolate wastes of the Gobi, the Yellow River opens a sort of funnel into north China.

Through it have flowed people and soil; the *loess* beds of the river valley, easily worked and fertile, laid down by wind from the north, are the basis of the first Chinese agriculture. Once this region was richly forested and well watered, but it became colder and more desiccated in one of those climatic transformations which are behind so much primeval social change. To Chinese pre-history overall, of course, there is a bigger setting than one

The Great Wall of China. Still without a complete archaeological survey, it is first distinguishable as a continuity over one thousand miles long in an immense undertaking by the first Ch'in emperor to link existing walls of individual states.

river valley. 'Peking man' turns up as a fire-user about six hundred thousand years ago, and there are Neanderthal traces in all three of the great river basins. The trail from these forerunners to the dimly discernible cultures which are their successors in early Neolithic times leads us to a China already divided into two cultural zones, with a meeting place and mixing area on the Yellow River. It is impossible to separate the tangle of cultural interconnexions already detectable by that time. But there was no even progress towards a uniform or united culture; even in early historical times, we are told, 'the whole of China ... was teeming with Neolithic survivals'. Against this varied background emerged settled agriculture; nomads and settlers were to coexist in China until our own day. Rhinoceros and elephant were still hunted in the north not long before 1000 BC.

As in other parts of the world, the coming of agriculture meant a revolution. It has been argued that peoples who lived in the semi-tropical coastal areas of South-East Asia and south China were clearing forests to make fields as far back as 10,000 BC. Certainly they exploited vegetation to provide themselves with fibres and food. But this is still a topic about which much more needs to be known. A much better record exists in northern China where ground just above the flood level of the Yellow River begins to yield evidence of agriculture from about 5000 BC. Somewhat like that of early Egypt, it seems to have been exhaustive or semi-exhaustive. The land was cleared, used for a few years, and then left to revert to nature while the cultivators turned attention elsewhere. From this area agriculture can be seen later to spread both north to Manchuria and to the south. It has been called the 'nuclear area of North China'. Within it there soon appeared complex cultures which combined with agriculture the use of jade and wood for carving, the domestication of silk-worms, the making of ceremonial vessels in forms which were to become traditional and perhaps even the use of chopsticks. In other words, this was in Neolithic times already the home of much that is characteristic of the later Chinese tradition in the historic area.

Ancient writers recognized the importance of this revolutionary social change and legends identified a specific inventor of agriculture, yet very little can be inferred confidently or clearly about social organization at this stage. Perhaps because of this there has been a persistent tendency among Chinese to idealize it. Long after private property had become widespread it was assumed that 'under heaven every spot is the sovereign's ground' and this may reflect early ideas that all land belonged to the community as a whole. The Chinese Marxists have maintained this tradition, discerning in the archaeological evidence a golden age of primitive Communism preceding a descent into slave and feudal society. Argument is unlikely to convince those interested in the question one way or the other. Ground seems to be firmer in attributing to these times the appearance of a clan structure and totems, with prohibitions on marriage within the clan. Kinship in this form is almost the first institution which can be seen to have survived to be important in historical times. The evidence of the pottery, too, suggests some new complexity in social roles. Already things were being made which cannot have been intended for the rough and tumble of everyday use; a stratified society seems to be emerging before we reach the historical era.

One material sign of a future China already obvious at this stage is the widespread use of millet, a grain well adapted to the sometimes arid farming of the north. It was to be the basic staple of Chinese diet until about a thousand years ago and sustained a society which in due course arrived at literacy, at a great art of bronze-casting based on a difficult and advanced technology, at the means of making exquisite pottery far finer than anything made anywhere else in the world, and, above all, at an ordered political and social system which identifies the first major age of Chinese history. But it must be remembered once more that the agriculture which made this possible was for a long time confined to north China and that many parts of this huge country only took up farming when historical times had already begun.

The narrative of early times is very hard to recover, but can be outlined with some confidence. It has been agreed that the story of civilization in China begins under rulers from a people called the Shang, the first name with independent evidence to support it in the traditional list of dynasties which was for a long time the basis of Chinese chronology. From the late eighth century BC we have better dates, but we still have no chronology for early Chinese history as well founded as, say, that of Egypt. It is more certain that somewhere about 1700 BC (and a century each way is an acceptable margin of approximation) a tribe called the Shang, which enjoyed the military advantage of the chariot, imposed itself on its neighbours over a sizable stretch of the Yellow River valley. Eventually, the Shang domain was a matter of about 40,000 square miles in northern Honan; this made it somewhat smaller than modern England, though its cultural

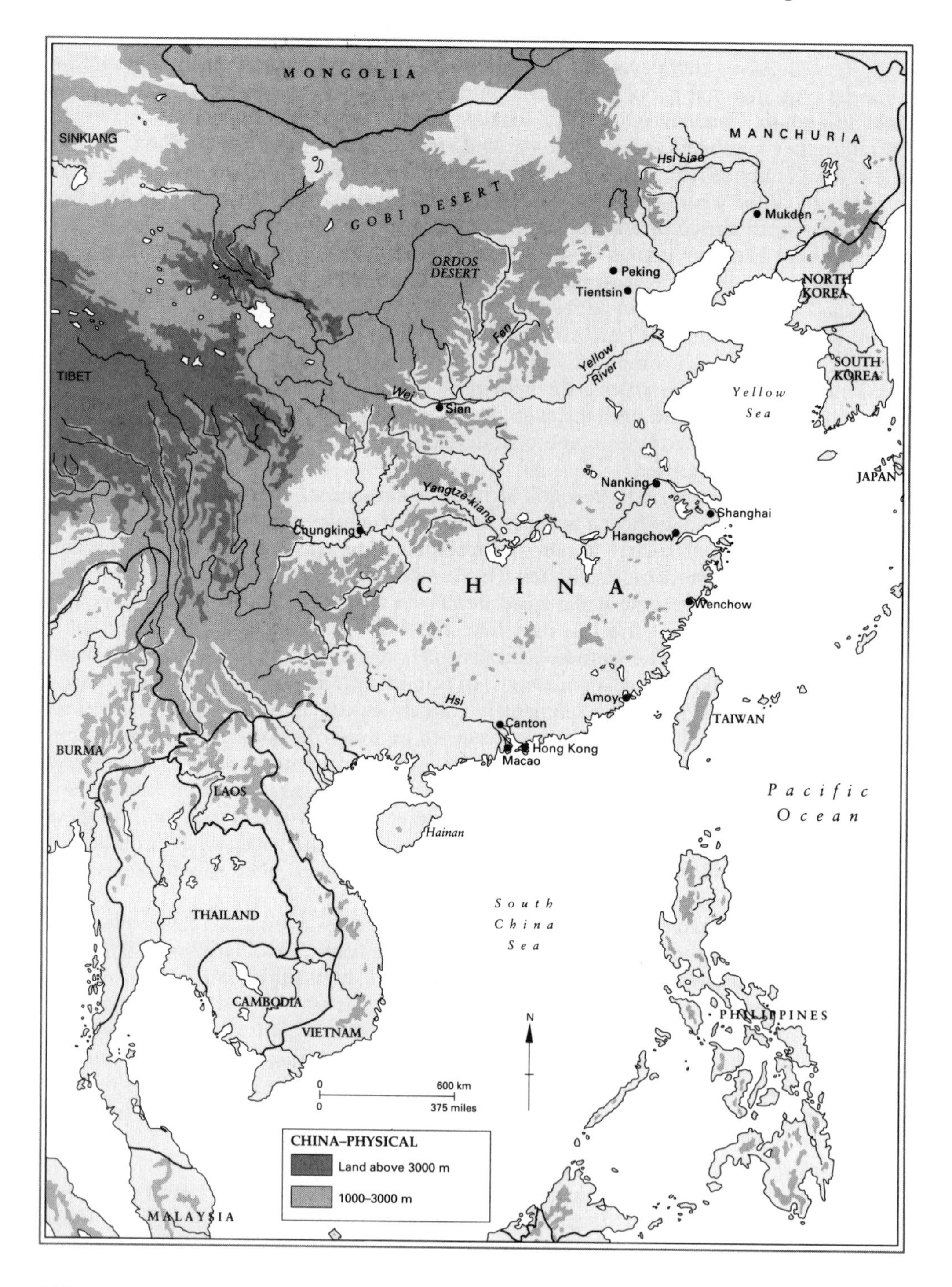

influences reached far beyond its periphery, as evidence from as far away as south China, Chinese Turkestan and the north-eastern coast shows.

Shang kings lived and died in some state; slaves and human sacrificial victims were buried with them in deep and lavish tombs. Their courts had archivists and scribes, for this was the first truly literate culture east of Mesopotamia. This is one reason for distinguishing between Shang civilization and Shang dynastic paramountcy; this people showed a cultural influence which certainly extended far beyond any area they could have dominated politically. The political arrangements of the Shang domains themselves seem to have depended on the uniting of landholding with obligations to a king; the warrior landlords who were the key figures were the leading members of aristocratic lineages with semi-mythical origins. Yet Shang government was advanced enough to use scribes and had a standardized currency. What it could do when at full stretch is shown in its ability to mobilize large amounts of labour for the building of fortifications and cities.

Shang China succumbed in the end to another tribe from the west of the valley, the Chou. A probable date is 1027 BC. Under the Chou, many of the already elaborate governmental and social structures inherited from the Shang were preserved and further refined. Burial rites, bronze-working techniques and decorative art also survived in hardly altered forms. The great work of the Chou period was the consolidation and diffusion of this heritage. In it can be discerned the hardening of the institutions of a future Imperial China which would last two thousand years.

The Chou thought of themselves as surrounded by barbarian peoples waiting for the benevolent effects of Chou tranquillization (an idea, it may be remarked, which still underlay the persistent refusal of Chinese officials two thousand years later to regard diplomatic missions from Europe as anything but respectful bearers of tribute). Chou supremacy in fact rested on war, but from it flowed great cultural consequences. As under the Shang, there was no truly unitary state and Chou government represented a change of degree rather than kind. It was usually a matter of a group of notables and vassals, some more dependent on the dynasty than others, offering in good times at least a formal acknowledgement of its supremacy and all increasingly sharing in a common culture. Political China (if it is reasonable to use such a term) rested upon big estates which had sufficient cohesion to have powers of long survival and in this process their original lords turned into rulers who could be called kings, served by elementary bureaucracies.

This system collapsed from about 700 BC, when a barbarian incursion drove the Chou from their ancestral centre to a new home farther east, in Honan. The dynasty did not end until 256 BC, but the next distinguishable epoch dates from 403 to 221 BC and is significantly known as the Period of the Warring States. In it, historical selection by conflict grew fierce. Big fish ate little fish until one only was left and all the lands of the Chinese were for the first time ruled by one great empire, the Ch'in, from which the country was to get its name. This is matter for discussion elsewhere; here it is only apropos in that it registers an epoch in Chinese history.

Reading about these events in the traditional Chinese historical accounts can produce a slight feeling of beating the air, and historians who are not experts in Chinese studies may perhaps be forgiven if they cannot trace over this period of some fifteen hundred years or so any helpful narrative thread in the dimly discernible struggles of kings and over-mighty subjects. They should be; after all, scholars have not yet provided one. Nevertheless, two basic processes were going on for most of this time which were very important for the future and which give the period some unity, through their detail is elusive. The first of these was a continuing diffusion of culture outwards from the Yellow River basin.

To begin with, Chinese civilization was a matter of tiny islands in a sea of barbarism. Yet by 500 BC it was the common possession of scores, perhaps hundreds, of 'states' scattered across the north, and it had also been carried into the Yangtze valley. This had long been a swampy, heavily forested region very different from the north and inhabited by far more primitive peoples. Chou influence – in part thanks to military expansion – irradiated this area, and helped to produce the first major culture and state of the Yangtze valley, the Ch'u civilization. Although owing much to the Chou, it had many distinctive linguistic, calligraphic, artistic and religious traits of its own. By the end of the Period of Warring States we have reached the point at which the stage of Chinese history is about to be much enlarged.

One of the first images of a human face – from pre-Shang Honan. It may represent a shaman or magician wearing some kind of professional costume (the ruff suggests something special), but this is speculation. It is at least as probable that the form was determined by the shape of the clay pot for which this was the lid. Other such pots to which human faces form lids have been found.

The second of these fundamental and continuing processes under both Shang and Chou was the establishment of landmarks in institutions which were to survive until modern times. Among them was a fundamental division of Chinese society into a landowning nobility and the common people. Most of these were peasants, making up the vast majority of the population and paying for all that China produced in the way of civilization and state power. What little we know of their countless lives can be quickly said; even less can be discovered than about the anonymous masses of toilers at the base of every other ancient civilization. There is one good physical reason for this: the life of the Chinese peasant was an alternation between his mud hovel in the winter and an encampment where he lived during the summer months to guard and tend his growing crops. Neither has left much trace. For the rest, he appears sunk in the anonymity of his community (he does not belong to a clan), tied to the soil, occasionally taken from it to carry out other duties and to serve his lord in war or hunting. His depressed state is expressed by the classification of modern Chinese communist historiography which lumps Shang and Chou together as 'Slavery Society' preceding the 'Feudal Society' which comes next.

Though Chinese society was to grow much more complex by the end of the Warring States period, this distinction of common people from the nobly born remained. There were important practical consequences: the nobility, for example, were not subject to punishments – such as mutilation – inflicted on the commoner; it was a survival of this in later times that the gentry were exempt from the beatings which might be visited on the commoner (though, of course, they might suffer appropriate and even dire punishment for more serious crimes). The nobility long enjoyed a virtual monopoly of wealth, too, which outlasted its earlier monopoly of metal weapons. None the less, these were not the crucial distinctions of status, which lay elsewhere, in the nobleman's special religious standing through a monopoly of certain ritual practices. Only noblemen could share in the cults which were the heart of the Chinese notion of kinship. Only the nobleman belonged to a family – which meant that he had ancestors. Reverence for ancestors and propitiation of their spirits had existed before the Shang, though it does not seem that in early times many ancestors were thought likely to survive into the spirit world. Possibly the only ones lucky enough to do so would be the spirits of particularly important persons; the most likely, of course, were the rulers themselves, whose ultimate origin, it was claimed, was itself godly.

The family emerged as a legal refinement and subdivision of the clan, and the Chou period was the most important one in its clarification. There were about a hundred clans, within each of which marriage was forbidden. Each was supposed to be founded by a hero or a god. The patriarchal heads of the clan's families and houses exercised special authority over its members and were all qualified to carry out its rituals and thus influence spirits to act as intermediaries with the powers which controlled the universe on the clan's behalf. These practices came to identify persons entitled to possess land or hold office. The clan offered a sort of democracy of opportunity at this level: any of its members could be appointed to the highest place in it, for they were all qualified by the essential virtue of a descent whose origins were godlike. In this sense, a king was only *primus inter pares*, a patrician outstanding among all patricians.

The family absorbed enormous quantities of religious feeling and psychic energy; its rituals were exacting and time-consuming. The common people, not sharing in this, found a religious outlet in maintaining the worship of nature gods. These always got some attention from the élite, too, the worship of mountains and rivers and the propitiating of their spirits being an important imperial duty from early times, but they were to influence the central developments of Chinese thought less than similar notions in other religions.

Religion had considerable repercussions on political forms. The heart of the ruling house's claim to obedience was its religious superiority. Through the maintenance of ritual, it had access to the goodwill of unseen powers, whose intentions might be known from the oracles. When these had been interpreted, the ordering of the agricultural life of the community was possible, for they regulated such matters as the time of sowing or harvesting. Much turned, therefore, on the religious standing of the king; it was of the first importance to the state. This was reflected in the fact that the Chou displacement of the Shang was religious as well as military. The idea was introduced that there existed a god superior to the ancestral god of the dynasty and that from him there was derived a mandate to rule. Now, it was claimed, he had decreed that the mandate should pass to other hands. This was the introduction of another idea fundamental to the Chinese conception of government and it was to be closely linked to the notion of a cyclic history, marked by the repeated rise and fall of dynasties. Inevitably, it provoked speculation about what might be the signs by which the recipient of the new mandate should be recognized. Filial piety was one, and to this extent, a conservative principle was implicit. But the Chou writers also introduced an idea rendered not very comfortably into English by the word 'virtue'. Clearly, its content remained fluid; disagreement and discussion were therefore possible.

In its earliest forms the Chinese 'state' – and one must think over long periods of more than one coexisting – seems little more than an abstraction from the idea of the ruler's estate and the necessity to maintain the rituals and sacrifices. The records do not leave an impression of a very busy monarchy. Apart from the extraordinary decisions of peace or war, the king seems to have had little to do except fulfil his religious duties, hunt, and initiate building projects in the palace complexes which appear as early as Shang times, though there are indications of Chou kings also undertaking (with the labour of prisoners) extensive agricultural colonization. For a long time the early Chinese rulers did without any very considerable bureaucracy. Gradually a hierarchy of ministers emerged who regulated court life, but the king was a landowner who for the most part needed only bailiffs, overseers and a few scribes. No doubt much of his life was spent on the move about his lands. The only other aspect of his activity which needed expert support was the supernatural. Out of this much was to grow, not least the intimate connexion between rule in China and the determination of time and the calendar, both very important in agricultural societies. These were based on astronomy, and though this came to have a respectable basis in observation and calculation, its origins were magical and religious.

In the earliest days all the great decisions of state, and many lesser ones, were taken by consulting oracles. This was done by engraving turtle shells or the shoulder-blades of certain animals with written characters and then applying to them a heated bronze pin so as to produce cracks on the reverse side. The direction and length of these cracks in relation to the characters would then be considered and the oracle read accordingly. This was an enormously important practice from the point of view of historians, for such oracles were kept, presumably as records. They also provide us with evidence for the foundation of Chinese language, for the characters on the oracle bones (and some early bronzes) are basically those of classical Chinese. The Shang had about 5000 such characters, though not all can be read. Nevertheless, the principles of this writing show a unique consistency; while other civilizations gave up pictographic characterization in favour of phonetic systems, the Chinese language grew and evolved, but remained essentially within the pictographic framework. Already under the Shang, moreover, the structure of the language was that of modern Chinese – monosyllabic and depending on word order, not on the inflection of words, to convey meaning. The Shang, in fact, already used a form of Chinese.

Writing was to remain high on the scale of Chinese arts and has always retained some trace of the religious respect given to the first characters. Only a few years ago, examples of Mao Tse-tung's calligraphy were widely reproduced during his ascendancy and were used to enhance his prestige. This reflects the centuries during which writing remained the jealously guarded privilege of the élite. The readers of the oracles, the so-called *shih*, were the primitive form of the later scholar-gentry class; they were indispensable experts, the possessors of hieratic and arcane skills. Their monopoly was to pass to the much larger class of the scholar-gentry in later times. The language thus remained the form of communication of a relatively small élite which not only found its privileges rooted in its possession but also had an interest in preserving it against corruption or variation. It was of enormous importance as a unifying and stabilizing force because written Chinese became a language of government and culture transcending divisions of dialect, religion and region. Its use by the élite tied the country together.

Several great determinants of future Chinese history had thus been settled in outline by the end of the Chou period. That end came after increasing signs of social changes which were affecting the operation of the major institutions. This is not surprising; China long remained basically agricultural, and change was often initiated by the pressure of population upon resources. This accounts for the impact of the introduction of iron, probably in use by about 500 BC. As elsewhere a sharp rise in agricultural production (and therefore in population) followed. The first tools which have been found come from the fifth century BC; iron weapons came later. At an early date, too, tools were made by casting, for iron moulds for sickle blades have been found dating from the fourth or fifth centuries. Chinese technique in handling the new metal was thus advanced in very early times. Whether by development from bronze casting or by experiments with pottery furnaces which could produce high temperatures, China somehow arrived at the casting of iron at about the same time as knowledge of how to forge it. Exact precedence is unimportant; what is noteworthy is that sufficiently high temperatures for casting were not available elsewhere for another nineteen centuries or so.

Another important change under the later Chou was a great growth of cities. They tended to be sited on plains near rivers, but the first of them had probably taken their shape and location from the use of landowners' temples as centres of administration for their estates. This drew to them other temples, those of the popular nature gods, as communities collected about them. Then, under the Shang, a new scale of government begins to make itself felt; we find stamped-earth ramparts, specialized aristocratic and court quarters and the remains of large buildings. At Anyang, a Shang capital in about 1300 BC, there were metal foundries and potters' kilns as well as palaces and a royal graveyard. By late Chou times, the capital Wang Ch'eng is surrounded by a rectangle of earth walls each nearly three kilometres long.

There were scores of cities by 500 BC and their prevalence implies an increasingly varied society. Many of them had three well-defined areas: a small enclosure where the aristocracy lived, a larger one inhabited by specialized craftsmen and merchants, and the fields outside the walls which fed the city. A merchant class was another important development. It may not have been much regarded by the landowners but well before 1000 BC a cowry shell currency was used which shows a new complexity of economic life and the presence of specialists in trade. Their quarters and those of the craftsmen were distinguished from those of the nobility by walls and ramparts round the latter, but they, too, fell within the walls of the city – a sign of a growing need for defence. In the commercial streets of cities of the Warring States Period could be found shops selling jewellery, curios, food and clothing, as well as taverns, gambling houses and brothels.

The heart of Chinese society, none the less, still beat to the slow rhythms of the countryside. The privileged class which presided over the land system showed unmistakable signs of a growing independence of its kings as the Chou period came to an end. Landowners originally had the responsibility of providing soldiers to the king and development in the art of war helped to increase their independence. The nobleman had always had a monopoly of arms; this was already significant when, in Shang times, Chinese weaponry was limited for the most part to the bow and the bronze halberd. As time passed only noblemen could afford the more expensive weapons, armour and horses which increasingly came into use. The warrior using a chariot as a platform for archery before descending to fight the last stage of the battle on foot with bronze weapons evolved in the last centuries of the pre-Christian era into a member of a team of two or three armoured warriors, moving with a company of sixty or seventy attendants and supporters, accompanied by a battle-wagon carrying the heavy armour and new weapons like the cross-bow and long iron sword which were needed at the scene of action. The nobleman remained the key figure under this system as in earlier times.

As historical records become clearer, it can be seen that economic supremacy was rooted in customary tenure which was very potent and far reaching. Ownership of estates – theoretically all granted by the king – extended not only to land but to carts, livestock, implements and, above all, people. Labourers could be sold, exchanged, or left by will. This was another basis of a growing independence for the nobility, but it also gave fresh importance to distinctions within the landowning class. In principle, estates were held by them in concentric circles about the king's own demesne, according to their closeness to the royal line and, therefore, according to the degree of closeness of their relations with the spirit world. By about 600 BC, it seems clear that this had effectively reduced the king to dependence on the greatest princes. There appear a succession of protectors of the royal house; kings could only resist the encroachments of these oriental Bolingbrokes and Warwicks in so far as the success of any one of them inevitably provoked the jealousy of others, and because of the kingly religious prestige which still counted for much with the lesser nobility. The whole late Chou period was marked by grave disorder and growing scepticism, though, about the criteria by which the right to rule was recognizable. The price of survival for the princes who disputed China was the elaboration of more effective governments and armed forces, and often they welcomed innovators prepared to set aside tradition.

In the profound and prolonged social and political crisis of the last, decaying centuries of the Chou and the Period of Warring States (433–221 BC), there was a burst of speculation about the foundations of government and ethics. The era was to remain famous as the time of the 'Hundred Schools', when wandering scholars moved about from patron to patron, expounding their teachings. One sign of this new development was the appearance of a school of writers known as the 'Legalists'. They urged that law-making power should replace ritual observances as the principle of organization of the state; there should be one law for all, ordained and vigorously applied by one ruler. The aim of this was the creation of a wealthy and powerful state. This seemed to many of their opponents to be little more than a cynical doctrine of power, but the Legalists were to have important successes in the next few centuries because kings, at least, liked their ideas. The debate went on for a long time. In this debate the main opponents of the Legalists were the followers of the teacher who is the most famous of all Chinese thinkers, Confucius. It is convenient to call him by that name, though it is only a latinized version of his Chinese name, K'ung-fu-tzu, and was given to him by Europeans in the seventeenth century, more than two thousand years after his birth in the middle of the sixth century BC. He was to be more profoundly respected in China than any other philosopher. What he said – or was said to have said – shaped his countrymen's thinking for two thousand years and was to be paid the compliment of bitter attack by the first post-Confucian Chinese state, the Marxist republic of today.

Confucius came from a *shih* family. He was a member of the lesser nobility who had spent some time as a minister of state and an overseer of granaries. When he could not find a ruler to put into practice his recommendations for just government he turned to meditation and teaching; his aim was to present a purified and more abstract version of the doctrine he believed to lie at the heart of the traditional practices and thus to revive personal integrity and disinterested service in the governing class. He was a reforming conservative, seeking to teach his pupils the essential truths of a system materialized and obscured by routine. Somewhere in the past, he thought, lay a mythical age when each man knew his place and did his duty; to return to that was Confucius' ethical goal. He advocated the principle of order – the attribution to everything of its correct place in the great whole of experience. The practical expression of this was the strong Confucian predisposition to support the institutions likely to ensure order – the family, hierarchy, seniority – and due reverence for the many nicely graded obligations between men.

This was teaching likely to produce men who would respect the traditional culture, emphasize the value of good form and regular behaviour, and seek to realize their moral obligations in the scrupulous discharge of duties. It was immediately successful in that many of Confucius' pupils won fame and worldly success (though his teaching deplored the conscious pursuit of such goals, urging, rather, a gentlemanly self-effacement). But it was also successful in a much more fundamental sense, since generations of Chinese civil servants were later to be drilled in the precepts of behaviour and government which he laid down. 'Documents, conduct, loyalty and faithfulness', four precepts attributed to him as his guidance on government, helped to form reliable, disinterested and humane civil servants for hundreds of years, even if not always with uniform success.

Confucian texts were later to be treated with something like religious awe. His name gave great prestige to anything with which it was associated. He was said to have compiled some of the texts later known as the Thirteen Classics, a collection which only took its final form in the thirteenth century AD. Rather like the Old Testament, they were a somewhat miscellaneous collection of old poems, chronicles, early state documents, moral sayings and an early cosmogony called the Book of Changes, but they were used for centuries in a unified and creative way to mould generations of Chinese civil servants and

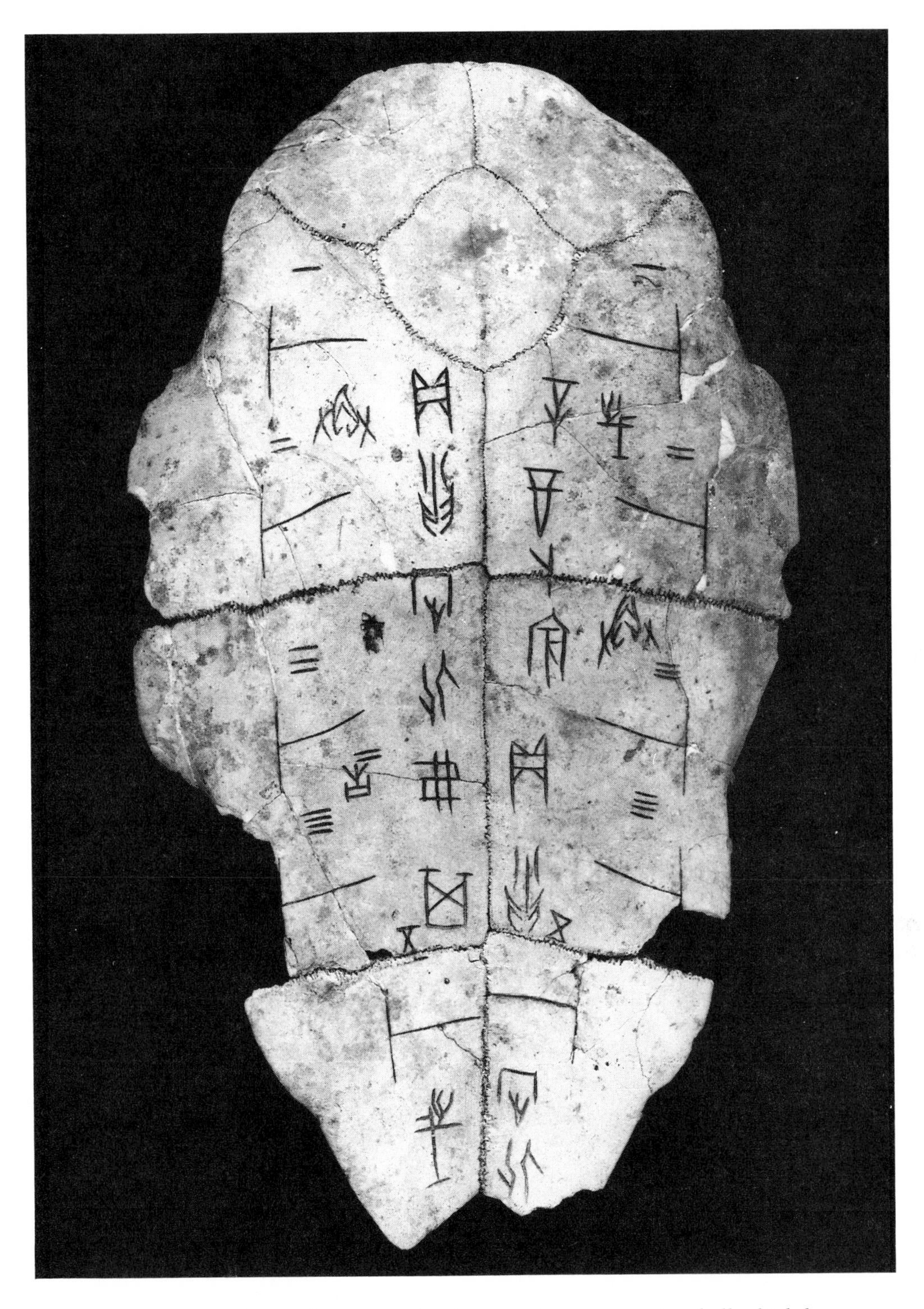

An especially fine and complete example of an oracle on a tortoise shell which has survived almost intact from Shang times. After cracks had been observed in answer to the questions previously engraved on the shell or bone, these answers were then themselves also inscribed; thus a permanent record was available for consultation. Archaeologists have now recovered more than a hundred thousand fragments of the Shang oracles. They have provided examples of some five thousand of the first Chinese characters (about a third of them have been interpreted).

rulers in the precepts which were believed to be those approved by Confucius (the parallel with the use of the Bible, at least in Protestant countries, is striking here, too). The stamp of authority was set upon this collection by the tradition that Confucius had selected it and that it must therefore contain doctrine which digested his teaching. Almost incidentally it also reinforced still more the use of the Chinese in which these texts were written as the common language of Chinese intellectuals; the collection was another tie pulling a huge and varied country together in a common culture.

It is striking that Confucius had so little to say about the supernatural. In the ordinary sense of the word he was not a 'religious' teacher (which probably explains why other teachers had greater success with the masses). He was essentially concerned with practical duties, an emphasis he shared with several other Chinese teachers of the fourth and fifth centuries BC. Possibly because the stamp was then so firmly taken, Chinese thought seems less troubled by agonized uncertainties over the reality of the actual or the possibility of personal salvation than other, more tormented, traditions. The lessons of the past, the wisdom of former times and the maintenance of good order came to have more importance in it than pondering theological enigmas or seeking reassurance in the arms of the dark gods.

Yet for all his great influence Confucius was not the only maker of Chinese intellectual tradition. In part, the tone of Chinese intellectual life is perhaps not attributable to any individual's teaching, but shares something with other oriental philosophies in its emphasis upon the meditative and reflective mode rather than the methodical and interrogatory which is more familiar to Europeans. The mapping of knowledge by systematic questioning of the mind about the nature and extent of its own powers was not to be a characteristic activity of Chinese philosophers. This does not mean they inclined to other-worldliness and fantasy, for Confucianism was emphatically practical. Unlike the ethical sages of Judaism, Christianity and Islam, those of China tended always to turn to the here and now, to pragmatic and secular questions, rather than to theology and metaphysics.

This can also be said of systems rivalling Confucianism which were evolved to satisfy Chinese needs. One was the teaching of Mo-Tzu, a fifth-century thinker, who preached an active creed of universal altruism; men were to love strangers like their own kinsmen. Some of his followers stressed this side of his teaching, others a religious fervour which encouraged the worship of spirits and had greater popular appeal. Lao-Tse, another great teacher (though one whose vast fame conceals the fact that we know virtually nothing about him), was supposed to be the author of the text which is the key document of the philosophical system later called Taoism. This was much more obviously competitive with Confucianism, for it advocated the positive neglect of much that Confucianism upheld; respect for the established order, decorum and scrupulous observance of tradition and ceremonial, for example. Taoism urged submission to a conception already available in Chinese thought and familiar to Confucius, that of the *Tao* or 'way', the cosmic principle which runs through and sustains the harmoniously ordered universe. The practical results of this were likely to be political quietism and non-attachment; one ideal held up to its practitioners was that a village should know that other villages existed because it would hear cockerels crowing in the mornings, but should have no further interest in them, no commerce with them and no political order binding them together. Such an idealization of simplicity and poverty was the very opposite of the empire and prosperity Confucianism upheld.

Still another and later sage, the fourth-century Mencius, taught men to seek the welfare of mankind. The following of a moral code based on this principle would assure that Man's fundamentally beneficent nature would be able to operate; this was more a development of Confucian teaching than a departure from it. But all schools of Chinese philosophy had to take account of Confucian teaching, so great was its prestige and influence. Eventually, with Buddhism (which had not reached China by the end of the Warring States Period) and Taoism, Confucianism was habitually referred to as one of the 'three teachings' which were the basis of Chinese culture.

The total effect of such views is imponderable, but enormous. It is hard to say how many people were directly affected by such doctrines and, in the case of Confucianism its great period of influence lay still in the remote future at the time of Confucius' death. Yet Confucianism's importance for the directing élites of China was to be immense. It set standards and ideals to China's leaders and rulers whose eradication was to prove impossible until our own day. Moreover, some of its precepts - filial piety, for example – filtered down to popular culture through stories and the traditional motifs of art. It thus further solidified a civilization many of whose most striking features were well entrenched by the third century BC. Certainly its teachings accentuated the preoccupation with the past among China's rulers which was to give a characteristic bias to Chinese historiography, and it may also have had a damaging effect on scientific enquiry. Evidence suggests that after the fifth century BC a tradition of astronomical observation which had permitted the prediction of lunar eclipses fell into decline. Some scholars have seen the influence of Confucianism as part of the explanation of this.

China's great schools of ethics are one striking example of the way in which almost all the categories of her civilization differ from those of our own tradition and, indeed, from those of any other civilization of which we have knowledge. Its uniqueness is not only a sign of its comparative isolation, but also of its vigour. Both are displayed in its art, which is what now remains of ancient China that is most immediately appealing and accessible. Of the architecture of the Shang and Chou, not much survives; their building was often in wood, and the tombs do not reveal very much. Excavation of cities, on the other hand, reveals a capacity for massive construction; the wall of one Chou capital was made of pounded earth thirty feet high and forty thick.

Smaller objects survive much more plentifully and they reveal a civilization which even in Shang times is capable of exquisite work, above all in its ceramics, unsurpassed in the ancient world. A tradition going back to Neolithic times lay behind them. Pride of place must be given none the less to the great series of bronzes which begin in early Shang times and continue thereafter uninterruptedly. The skill of casting sacrificial containers, pots, wine-jars, weapons, tripods was already at its peak as early as 1600 BC. And it is argued by some scholars that the lost-wax method, which made new triumphs possible, was also known in the Shang era. Bronze casting appears so suddenly and at such a high level of achievement that people long sought to explain it by transmission of the technique from outside. But there is no evidence for this and the most likely origin of Chinese metallurgy is from locally evolved techniques in several centres in the late Neolithic.

None of the bronzes reached the outside world in early times, or at least there has been no discovery of them elsewhere which can be dated before the middle of the first millennium BC. Nor are there many discoveries outside China at earlier dates of the other things to which Chinese artists turned their attention, to the carving of stone or the appallingly hard jade, for example, into beautiful and intricate designs. Apart from what she absorbed from her barbaric nomadic neighbours China not only had little to learn from the outside until well into the historical era, it seems, but had no reason to think that the outside world – if she knew of it – wanted to learn much from her.

6
The Other Worlds of the Ancient Past

So far in this account huge areas of the world have hardly been mentioned. Though Africa has priority in the story of the evolution and spread of humanity and though the entry of men to the Americas and Australasia calls for remark, once those remote events have been touched upon, the beginnings of history focus attention elsewhere. The homes of the creative cultures which have dominated the story of civilization were the Near East and Aegean, India and China. In all these areas some meaningful break in rhythm can be seen somewhere in the first millennium BC; there are no neat divisions, but there is a certain rough synchrony which makes it reasonable to divide their histories in this era. But for the great areas of which nothing has so far been said, such a chronology would be wholly unrevealing.

This is, in the main, because none of them had achieved levels of civilization comparable to those already reached in the Mediterranean and Asia by 1000 BC. Remarkable things had been done by then in western Europe and the Americas, but when they are given due weight there still remains a qualitative gap between the complexity and resources of the societies which produced them and those of the ancient civilizations which were to found durable traditions. The interest in the ancient history of these areas lies rather in the way they illustrate that varied roads might lead towards civilization and that different responses might be demanded by different environmental challenges than in what they left as their heritage. In one or two instances they may allow us to reopen arguments about what constitutes 'civilization', but for the period of which we have so far spoken the story of Africa, of the Pacific peoples, of the Americas and western Europe is not history but still prehistory. There is little or no correspondence between its rhythms and what was going on in the Near East or Asia, even when there were (as in the case of Africa and Europe though not of the Americas) contacts with them.

Africa is a good place to start, because that is where the human story first began. Historians of Africa, sensitive to any slighting or imagined slighting of their subject, like to dwell upon Africa's importance in pre-history. As things earlier in this book have shown, they are quite right to do so; most of the evidence for the life of the earliest hominids is African. With the Upper Palaeolithic and the Neolithic the focus moves elsewhere. Much continues to happen in Africa but the period of its greatest creative influence on the rest of the world is over.

Why this happened we cannot say, but there is a strong possibility that the primary force may have been a change of climate. Even recently, say in about 3000 BC, the Sahara supported animals such as elephants and hippopotami which have long since disappeared there; more remarkably, it was the home of pastoral peoples herding cattle, sheep and goats. In those days, what is now desert and arid canyon was fertile savannah intersected and drained by rivers running down to the Niger and by another system seven hundred and fifty miles long, running into Lake Tchad. The peoples who lived in the hills where these rivers rose have left a record of their life in rock painting and engraving very different from the earlier cave art of Europe which depicted little but animal life and only an occasional human. This record also suggests that the Sahara was then a meeting place of Negroid and what some have called 'Europoid' peoples, those who were, perhaps, the ancestors of later Berbers and Tuaregs. One of these peoples seems to have made its way down from Tripoli with horses and chariots and perhaps to have conquered the

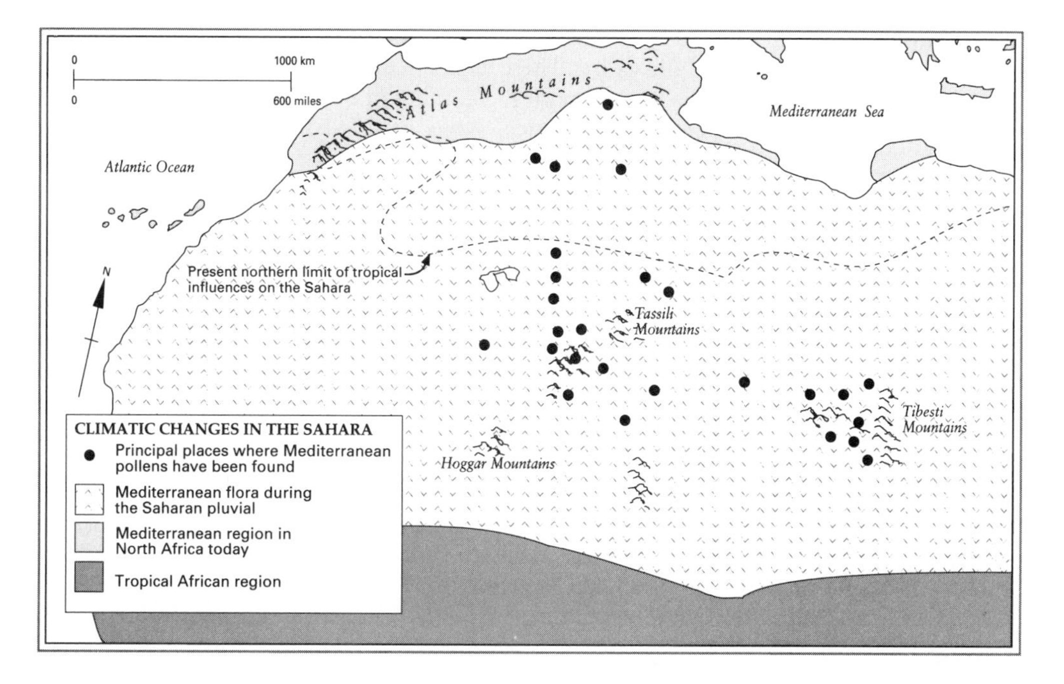

pastoralists. Whether they did so or not, their presence and that of the Negroid peoples of the Sahara show that Africa's vegetation was once very different from that of later times: horses need grazing. Yet when we reach historical times the Sahara is already desiccated, the sites of a once prosperous people are abandoned, the animals have gone.

Perhaps, therefore, it is climate which drives us back upon Egypt as the beginning of African history. Yet Egypt exercised little creative influence beyond the limits of the Nile valley. Though there were contacts with other cultures, it is not easy to penetrate them. Presumably the Libyans of Egyptian records were the sort of people who are shown with their chariots in the Sahara cave-paintings, but we do not certainly know. When the Greek historian Herodotus came to write about Africa in the fifth century BC, he found little to say about what went on outside Egypt. His Africa was a land defined by the Nile, which he took to run south roughly parallel to the Red Sea and then to swing west along the borders of Libya. South of the Nile there lay for him in the east the Ethiopians, in the west a land of deserts, without inhabitants. He could obtain no information about it, though a travellers' tale spoke of a dwarfish people who were sorcerers. Given his sources, this was topographically by no means an unintelligent construction, but Herodotus had grasped only a third or a quarter of the ethnic truth. The Ethiopians, like the old inhabitants of upper Egypt, were members of the Hamitic peoples who make up one of three racial groups in Africa at the end of the Stone Age later distinguished by anthropologists. The other two were the ancestors of the modern Bushmen, inhabiting, roughly, the open areas running from the Sahara south to the Cape, and the Negroid group, eventually dominant in the central forests and West Africa. (Opinion is divided about the origin and distinctiveness of a fourth group, the Pygmies.) To judge by the stone tools, cultures associated with Hamitic or proto-Hamitic peoples seem to have been the most advanced in Africa before the coming of farming. This was, except in Egypt, a slow evolution and in Africa the hunting and gathering cultures of prehistory have coexisted with agriculture right down to modern times.

The same growth which occurred elsewhere when food began to be produced in quantity soon changed African population patterns, first by permitting the dense settlements of the Nile valley which were the preliminary to Egyptian civilization, then by building up the Negroid population south of the Sahara, along the grasslands separating desert and equatorial forest in the second and first millennia BC. This seems to reflect a spread of agriculture southwards from the north. It also reflects the discovery of nutritious crops better suited to tropical conditions and other soils than the wheat and barley which flourished in the Nile valley. These were the millets and rice of the savannahs. The forest areas could not be exploited until the coming of other plants suitable to them from South-East Asia and eventually America. None of this happened before the birth of Christ. Thus was established one of the major characteristics of African history, a divergence of cultural trends within the continent.

By that time, iron had come to Africa and it had already produced the first exploitation of African ores. This occurred in the first independent African state other than Egypt of which we have information, the kingdom of Kush, high up the Nile, in the region of Khartoum. This had originally been the extreme frontier zone of Egyptian activity. After Nubia had been absorbed, the Sudanese principality which existed to its south was garrisoned by the Egyptians, but in about 1000 BC it emerged as an independent kingdom, showing itself deeply marked by Egyptian civilization. Probably its inhabitants were Hamitic people and its capital was at Napata, just below the Fourth Cataract. By 730 BC Kush was strong enough to conquer Egypt and five of its kings ruled as the pharaohs known to history as the Twenty-Fifth or 'Ethiopian' Dynasty. None the less, they could not arrest the Egyptian decline. When the Assyrians fell on Egypt, the Kushite dynasty ended. Though Egyptian civilization continued in the kingdom of Kush, a pharaoh of the next dynasty invaded it in the early sixth century BC. After this, the Kushites, too, began to push their frontiers further to the south and in so doing their

The 'Hallstatt' culture was named after a place in upper Austria where a great cemetery of an iron-using Celtic people was found. In the seventh and sixth centuries BC, these people were in touch with Mediterranean civilization, probably by way of what was later Marseilles and the Rhône valley. They may have traded salt from their mines with Mediterranean peoples. This bronze cauldron not only shows the metallurgical skill of Iron Age Europe, but employs decorations and forms derived from the Aegean. In this example they are interestingly combined with the very un-Mediterranean device of riveting to the bottom of the cauldron a cow and her calf to serve as a handle.

kingdom underwent two important changes. It became more Negroid, its language and literature reflecting a weakening of Egyptian trends, and it extended its territory over new territories which contained both iron ore and the fuel needed to smelt it. The technique of smelting had been learnt from the Assyrians. The new Kushite capital at Meroe became the metallurgical centre of Africa. Iron weapons gave the Kushites the advantages over their neighbours which northern peoples had enjoyed in the past over Egypt, and iron tools extended the area which could be cultivated. On this basis was to rest some three hundred years of prosperity and civilization in the Sudan, though later than the age we are now considering.

It is clear that the history of man in the Americas is much shorter than that in Africa or, indeed, in any other part of the world except Australasia. Something like thirty thousand years ago, Mongoloid peoples crossed into North America from Asia. Over the next few thousand years they filtered slowly southwards. Cave-dwellers have been traced in the Peruvian Andes as many as eighteen thousand years ago. The Americas contain very varied climates and environments; it is scarcely surprising, therefore, that archaeological evidence shows there were almost equally varied patterns of life, based on different opportunities for hunting, food-gathering and fishing. What they learnt from one another is probably undiscoverable. What is indisputable is that some of these cultures arrived at the invention of agriculture independently of the Old World.

Disagreement is still possible about when precisely this happened because, paradoxically, a great deal is known about the early cultivation of plants at a time when the scale on which this took place cannot reasonably be called agriculture. It is, nevertheless, a change which comes later than in the Fertile Crescent. Maize began to be cultivated in Mexico in about 5000 BC, but had been improved by 2000 BC in Meso-america into something like the plant we know today. This is the sort of change which made possible the establishment of large settled communities. Farther south, potatoes and manioc (another starchy root vegetable) also begin to appear at about this time and a little later there are signs that maize has spread southwards from Mexico. Everywhere, though, change is gradual; to talk of an 'agricultural revolution' is even less appropriate in the Americas than in the Near East.

Farming, villages, weaving and pottery all appear in Central America in the second millennium BC and towards the end of it come the first stirrings of the culture which produced the first recognized American civilization, that of the Olmecs of the eastern Mexican coast. It was focused, it seems, on important ceremonial sites with large earth pyramids. At these sites have been found colossal monumental sculpture and fine carvings of figures in jade. The style of this work is highly individual. It concentrates on human and jaguar-like images, sometimes fusing them. For several centuries after 800 BC it seems to have prevailed right across Central America as far south as what is now El Salvador. But it retains its mystery, appearing without antecedents or warning in a swampy, forested region which makes it hard to explain in economic terms. We do not know why civilization which elsewhere required the relative plenty of the great river valleys should in America spring from such unpromising soil.

Olmec civilization transmitted something to the future, for the gods of the later Aztecs were to be descendants of those of the Olmecs. It may also be that the early hieroglyphic systems of Central America originate in Olmec times, though the first survivals of the characters of these systems follow only a century or so after the disappearance of Olmec culture in about 400 BC. Again, we do not know why or how this happened. Much further south, in Peru, a culture called Chavin (after a great ceremonial site) also appeared and survived a little later than Olmec civilization to the north. It, too, had a high level of skill in working stone and spread vigorously only to dry up mysteriously.

What should be thought of these early lunges in the direction of civilization is very hard to see. Whatever their significance for the future, they are millennia behind the appearance of civilization elsewhere, whatever the cause of that may be. When the Spanish landed in the New World nearly two thousand years after the disappearance of Olmec culture they would still find most of its inhabitants working with stone tools. They would also find complicated societies (and the relics of others) which had achieved prodigies of building and organization far out-running, for example, anything Africa could offer after the decline of ancient Egypt. It is clear only that there are no unbreakable sequences in these matters.

The only other area where a startlingly high level of achievement in stone-working was reached was western Europe. This has led enthusiasts to claim it as another seat of early 'civilization', almost as if its inhabitants were some sort of depressed class needing historical rehabilitation. Europe has already been touched upon as a supplier of metals to the ancient Near East. Yet, though much that we now find interesting was happening there in prehistoric times, it does not provide a very impressive or striking story. In the history of the world, prehistoric Europe has little except illustrative importance. To the great civilizations which rose and fell in the river valleys of the Near East, Europe was largely an irrelevance. It sometimes received the impress of the outside world but contributed only marginally and fitfully to the process of historic change. A parallel might be Africa at a later date, interesting for its own sake, but not for its positive contribution to world history. It was to be a very long time before men would even be able to conceive that there existed a geographical, let alone a cultural, unity corresponding to the later idea of Europe. To the ancient world, the northern lands where the barbarians came from before they appeared in Thrace were irrelevant (and most of them probably came from further east anyway). The north-western hinterland was only important because it occasionally disgorged commodities wanted in Asia and the Aegean.

The still mysterious Chavin culture of Peru evolved a high degree of metal-working skill. This plaque is one of the earliest gold objects to survive from the Americas. The design is made up of disconnected conventional signs originally forming the face of a god. Peruvians may have been metal workers two thousand years before Mexicans emerged from the Stone Age, though their civilization reached its peak in the last centuries BC.

There is not much to say, therefore, about prehistoric Europe, but in order to get a correct perspective, one more point should be made. Two Europes must be separated. One is that of the Mediterranean coasts and their peoples. Its rough boundary is the line which delimits the cultivation of the olive. South of this line, literate, urban civilization comes fairly quickly once we are into the Iron Age, and apparently comes by direct contact with more advanced areas. By 800 BC the coasts of the western Mediterranean were already beginning to experience fairly continuous contact with the East. The Europe north and west of this line is a different matter. In this area literacy was never achieved in

A remarkable series of prehistoric tombs approached through passages lined with slabs of stone, sometimes decorated like this one, can be found in Spain, western France and Ireland. They date from the second millennium BC and similarities of design have led some to think that they must be expressions of a common culture or technological source. This example is from Brittany.

antiquity, but was imposed much later by conquerors. It long resisted cultural influences from the south and east - or at least did not offer a favourable reception to them – and it is for two thousand years important not for its own sake but because of its relationship to other areas. Its role was not entirely passive: the movements of its peoples, its natural resources and skills all at times impinged marginally on events elsewhere. But in 1000 BC – to take an arbitrary date – or even at the beginning of the Christian era, Europe has little of its own to offer the world except its minerals, and nothing which represents cultural achievement on the scale reached by the Near East, India or China. Europe's age was still to come; hers would be the last great civilization to appear.

This was not because the continent's natural endowment was unfavourable. It contains a disproportionately large area of the world's land naturally suitable for cultivation. It would be surprising if this had not favoured an early development of agriculture and this the archaeological evidence shows. The relative ease of simple agriculture in Europe may have had a negative effect on social evolution; in the great river valleys men *had* to work collectively to control irrigation and exploit the soil if they were to survive, while in much of Europe an individual family could scratch a living on its own. There is no need to fall into extravagant speculation about the origins of western individualism in order to recognize that there is something very distinctive and potentially very important.

A superb electrum torc from Snettisham in Norfolk. Perhaps it resembles one which an ancient historian tells us was worn by the legendary Boadicea when she rode into battle. It was probably made within the half-century or so before Julius Caesar landed in England and is a fine example of the skill of the Celtic jewellers at its peak, combining as it does the different techniques of moulding and plaiting wire.

Recently, much earlier dates have been given to the evolution of prehistoric European farming – and metallurgy – than was once thought probable. Before 4000 BC there were farming settlements in France and the British Isles. Perhaps a thousand years before this copper was being worked in the Balkans – some two thousand years before the appearance of recognizable civilization in Sumer. This early dating is not just of interest to scholars. It weakens the case once generally accepted, that Europe acquired its major cultural advances in prehistoric times by diffusion from the Aegean and Near East. The dates are now hard to fit. It seems certain that agriculture and copper-working were arrived at independently in Europe; this emphasizes once again the relative isolation of the area in ancient times, though important transfers to it from the outside could take place. Though agriculture appeared there spontaneously, it seems that the most important cereals were brought in thousands of years later from the Near East.

Most of the north-western and western parts of Europe were occupied in about 3000 BC by peoples sometimes termed western Mediterranean, who were gradually squeezed during the third millennium by Indo-Europeans from the east. By about 1800 BC the resulting cultures seem to have fragmented sufficiently distinctly for us to identify among them the ancestors of the Celts, the most important of prehistoric European

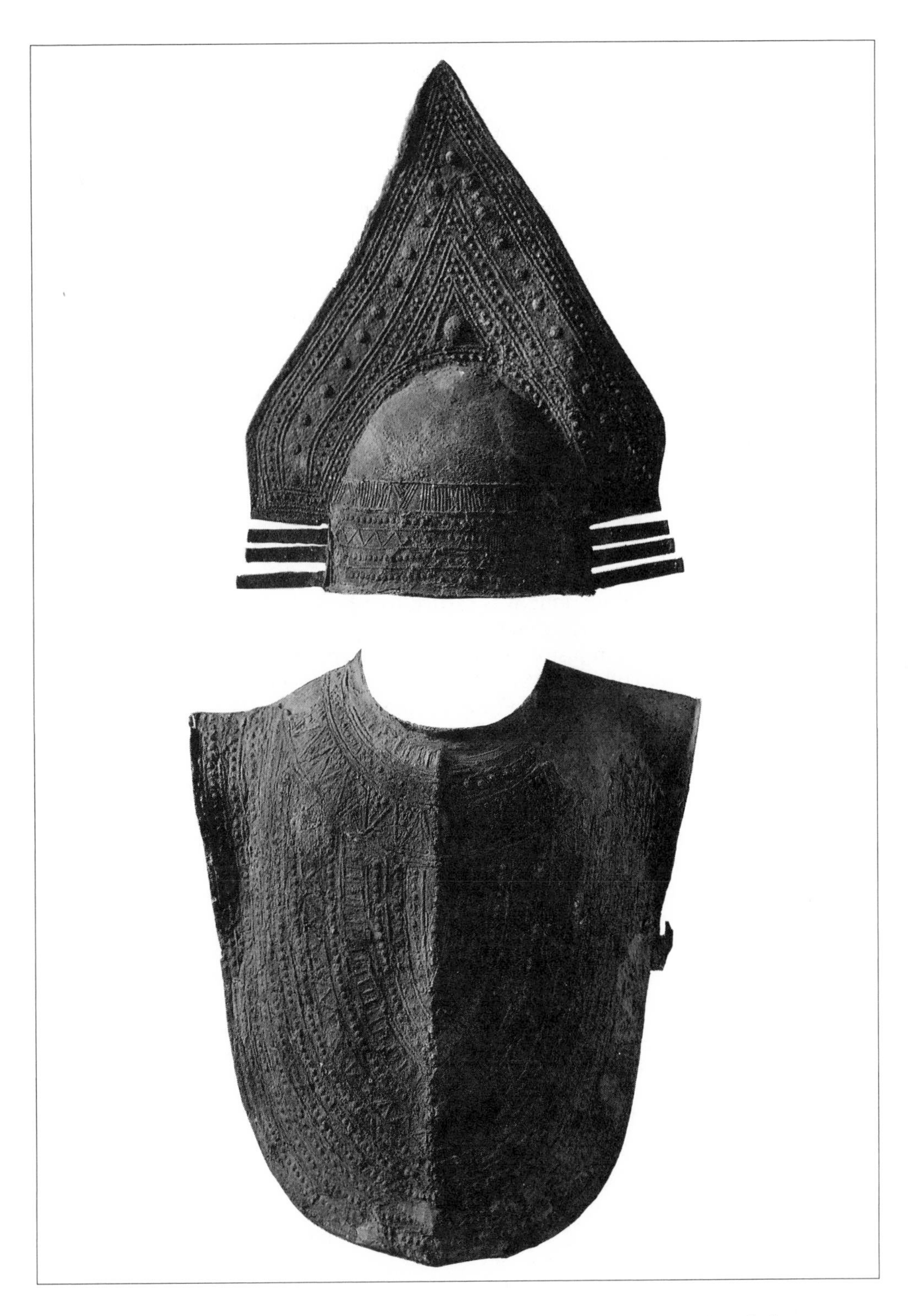

The Villanovans may have come originally from the across the Alps, already bringing with them the elements of their later metallurgical prowess. Once established in Italy their culture rapidly shows the influence of models and styles from the Aegean and the Eastern Mediterranean. Among the objects which have survived there are splendid crested helmets (sometimes used as lids for the funeral urns in which they placed the ashes of their cremated dead) and highly ornamented body-armour.

peoples, a society of warriors rather than traders or prospectors. They had wheeled transport. One enterprising group got to the British Isles and have some claim to being the first north-European sea-travellers. There is much disagreement about how far Celtic influence is to be traced, but it will not much disfigure the truth if we think of Europe divided in about 1800 BC into three groups of peoples. The ancestors of the Celts then occupied most of modern France, Germany, the Low Countries and upper Austria. To their eastward were the future Slavs, to the north (in Scandinavia) the future Teutonic tribes. Outside Europe, in northern Scandinavia and northern Russia, are the Finns, a non-Indo-European race.

Except in the Balkans and Thrace, the movements of these peoples affected the older centres of civilization only in so far as they affected access to the resources of the areas into which they moved. This was above all a matter of minerals and skills. As the demands of the Near East civilizations grew, so did Europe's importance. The first centre of metallurgy to develop there had been in the Balkans. Developments in southern Spain, Greece and the Aegean and central Italy followed by 2000 BC. In the later Bronze Age, metal- working was advanced to high levels even in places where no local ores were available. We have here one of the earliest examples of the emergence of crucial economic areas based on the possession of special resources. Copper and tin shaped the penetration of Europe and also its coastal and river navigation because these commodities were needed and were only available in the Near East in small quantities. Europe was the major primary producer of the ancient metallurgical world, as well as a major manufacturer. Metal-working was carried to a high level and produced beautiful objects long before that of the Aegean, but it is possibly an argument against exaggerated awe about material factors in history that this skill, even when combined with a bigger supply of metals after the collapse of Mycenaean demand, did not release European culture for the achievement of a full and complex civilization.

Ancient Europe had, of course, one other art form which remains impressive. It is preserved in the thousands of megalithic monuments to be found stretching in a broad arc from Malta, Sardinia and Corsica, round through Spain and Brittany to the British Isles and Scandinavia. They are not peculiar to Europe but are more plentiful there, and appear to have been erected earlier – from about 3500 BC – there than in other continents. 'Megalith' is a word derived from the Greek for 'large stone' and many of the stones used are very large indeed. Some of these monuments are tombs,

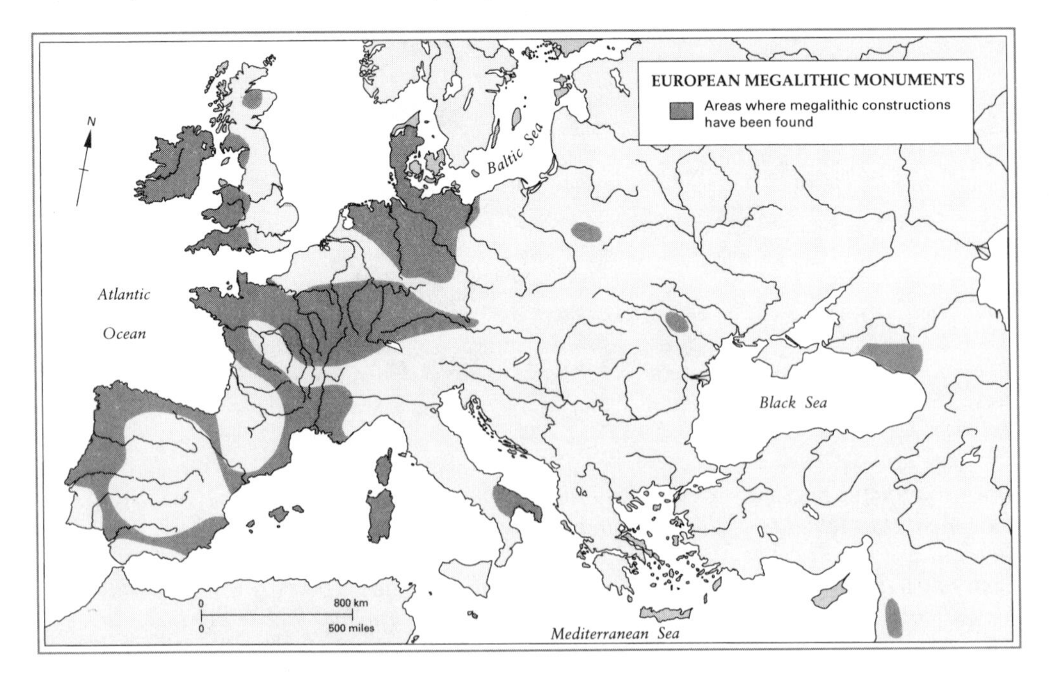

roofed and lined with slabs of stone, some are stones standing singly, or in groups. Some of them are laid out in patterns which run for miles across country; others enclose small areas like groves of trees. The most complete and striking megalithic site is Stonehenge, in southern England, now dated to about 2000 BC. What such places originally looked like is hard to guess or imagine. Their modern austerity and weathered grandeur may well be misleading; great places of human resort are not like that when in use and it is more likely that the huge stones were daubed in ochres and blood, hung with skins and fetishes. They may well often have looked more like totem-poles than the solemn, brooding shapes we see today. Except for the tombs, it is not easy to say what these works were for, though it has been argued that some were giant clocks or huge solar observatories, aligned to the rising and setting of sun, moon, and stars at the major turning-points of the astronomical year. Careful observation underlay such work, even if it fell far short in detail and precision of what was done by astronomers in Babylon and Egypt.

These relics represent huge concentrations of labour and argue for well-developed social organization. Stonehenge contains several blocks weighing about fifty tons apiece and they had to be brought some eighteen miles to the site before being erected. There are some eighty pieces of stone there weighing about five tons which came 150 miles or so from the mountains of Wales. The peoples who put up Stonehenge without the help of wheeled vehicles, like those who built the carefully lined tombs of Ireland, the lines of standing stones of Brittany or the dolmens of Denmark, were capable of work on a scale approaching that of ancient Egypt, therefore, though without its fineness or any means of recording their purposes and intentions except these great constructions themselves. Such skill, coupled with the fact of the monuments' distribution in a long chain within short distances of the sea, has suggested that their explanation might lie in what was learnt from wandering stonemasons from the East, perhaps from Crete, Mycenae, or the Cyclades, where the technique of dressing and handling such masses was understood. But recent advances in dating have once again removed a plausible hypothesis; Stonehenge was probably complete before Mycenaean times, megalithic tombs in Spain and Brittany antedate the pyramids, and Malta's mysterious temples with their huge carved blocks of building stone were there before 3000 BC. Nor do the monuments have to form part of any one process of distribution, even in the north-west. They may all have been achieved more or less in isolation by four or five cultures made up of relatively small and simple agricultural societies in touch with one another, and the motives and occasions of their building may have been very different. Like its agriculture and metallurgy, prehistoric Europe's engineering and architecture arose independently of the outside world.

For all their considerable achievements, the Europeans of ancient times seem strangely passive and unresisting when they finally appear in regular contact with advanced civilization. Their hesitations and uncertainties may have resembled those of other primitive peoples meeting advanced societies at later dates - eighteenth-century Africans for example. But, in any case, regular contact only began shortly before the Christian era. Before then, the European peoples seem to have exhausted their energies in grappling with an environment which, though easily worked to satisfy modest needs, required the coming of iron to make it fully exploitable. Though far more advanced than their contemporaries in America, or in Africa south of the Nile valley, they never reached the stage of urbanization. Their greatest cultural achievements were decorative and mechanical. At best, in their metallurgy, the ancient Europeans serviced other civilizations' needs. Beyond that, they would only provide the stocks which would receive the impress of civilization later.

Only one group of western barbarians had a more positive contribution to make to the future. South of the olive-line an Iron Age people of central Italy had already during the eighth century BC established trading contacts with Greeks further south in Italy and with Phoenicia. We call them Villanovans, after one of the sites where they lived. In the next two hundred years they adopted Greek characters for writing their language. By then they were organized in city-states, producing art of high quality. These were the Etruscans. One of their city-states would one day be known as Rome.

7
The End of the Old World

Of what was going on in India and China and its importance for the future the rulers of the Mediterranean and Near Eastern peoples knew hardly anything. Some of them, listening to traders, may have had a dim perception of a barbarian northern and north-western Europe. Of what happened beyond the Sahara and of the existence of the Americas they knew nothing. Yet their world was to expand rapidly in the first millennium BC and, equally and perhaps even more obviously, it was to become more integrated as its internal communications grew more complex and efficient. A world of a few highly distinctive and almost independent civilizations was giving way to one where larger and larger areas shared in the same achievements of civilization – literacy, government, technology, organized religion, city life – and, under their influence, changed more and more rapidly as the interplay of different traditions increased. It is important not to think of this in terms too abstract or grandiose. It is not only registered by art and speculative thought, but also by much that is more down-to-earth. Small things show it as well as great. On the legs of the huge statues at Abu Simbel, seven hundred miles up the Nile, sixth-century Greek mercenaries in the Egyptian army cut inscriptions which recorded their pride in coming that far, just as two thousand five hundred years later English county regiments would leave their badges and names cut into the rocks of the Khyber Pass.

There is no clear chronological line to be drawn in this increasingly complicated world. If one exists it has already been several times crossed before we reach the eve of the classical age of the West. The military and economic drive of the Mesopotamians and their successors, the movements of the Indo-Europeans, the coming of iron and the spread of literacy thoroughly mixed up the once-clear patterns of the Near East well before the appearance of a Mediterranean civilization which is the matrix of our own. Nevertheless, there is a sense in which it becomes manifest that an important boundary was crossed somewhere early in the first millennium BC. The greatest upheavals of the *Völkerwanderung* in the ancient Near East were then over. The patterns set there in the late Bronze Age would still be modified locally by colonization and conquest, but not for another thousand years by big comings and goings of peoples. The political structures left behind from antiquity would be levers of the next era of world history in a zone which stretched from Gibraltar to the Indus. Civilization within this area would more and more be a matter of interplay, borrowing and cosmopolitanism. The framework for this was provided by the great political change of the middle of the first millennium BC, the rise of a new power, Persia, and the final collapse of the Egyptian and Babylonian-Assyrian traditions.

The story of Egypt is the easiest to summarize, for it records little except decline. She has been called a 'Bronze Age anachronism in a world that steadily moved away from her' and her fate seems to be explained by an inability to change or adapt. She survived the first attacks of the iron-using peoples and had beaten off the Peoples of the Sea at the beginning of the age of turmoil. But this was the last big achievement of the New Kingdom: thereafter the symptoms are unmistakably those of a machine running down. At home kings and priests disputed power while Egypt's suzerainty beyond her borders declined to a shadow. A period of rival dynasties was briefly followed by a reunification which again took an Egyptian army to Palestine, but by the end of the eighth century a dynasty of Kushite invaders had established itself; in 671 BC it was ejected from Lower Egypt by the Assyrians. Ashurbanipal sacked Thebes. As Assyrian power ebbed, there was again an illusory period of Egyptian 'independence'. By this time, evidence of a new world towards which Egypt had to make more than political concessions can be seen in the establishment of a school for Greek interpreters and of a Greek trading enclave with special privileges at Naucratis in the delta. Then again, in the sixth century, Egypt went down to defeat first at the hands of the forces of Nebuchadnezzar (588 BC) and sixty years later, before the Persians (525 BC), to become a province of an empire which was to set boundaries for a new synthesis and would for centuries dispute world supremacy with new powers appearing in the Mediterranean. It was not quite the end of Egyptian independence, but from the fourth century BC to the twentieth AD, she was to be ruled by foreigners or immigrant dynasties and passes from view as an independent nation. The last bursts of Egyptian recovery show little innate vitality. They express, rather, temporary relaxations of the pressures upon her which always, in the end, were followed by their resumption. The Persian threat was the last of these and was fatal.

Once again, the starting-point is a migration. On the high plateau which is the heart of modern Iran there were settlements in 5000 BC, but the word 'Iran' (which does not appear until about AD 600) in its oldest form means 'land of the Aryans' and it is somewhere around 1000 BC, with an irruption of Aryan tribes from the north, that the history of the Persian empire begins. In Iran, as in India, the impact of the Aryans was to prove ineffaceable and founded a long-enduring tradition. Among their tribes two, especially vigorous and powerful, have been remembered by their biblical names as the Medes and Persians. The Medes moved west and north-west to Media; their great age came at the beginning of the sixth century, after they had overthrown Assyria, their neighbour. The Persians went south towards the Gulf, establishing themselves in Khuzistan (on the edge of the Tigris valley and in the old kingdom of Elam) and Fars, the Persia of the ancients.

Oral tradition preserves a story of legendary kings more important for the light it throws on later Persian attitudes to kingship than as history. It was none the less from the Persian dynasty of the Achaemenids that there descended the first king of a united Persia – anachronistic though this term is. He was Cyrus, the conqueror of Babylon. In 549 BC he humbled the last independent king of the Medes and thenceforth the boundaries of conquest rolled outwards, swallowing Babylon and advancing through Asia Minor to the sea, dropping down into Syria and Palestine. Only in the east (where he was eventually killed fighting the Scythians) did Cyrus find it difficult to stabilize his frontiers, though he crossed the Hindu Kush and set up some sort of supremacy over the region of Gandhara, north of the Jhelum.

This was the largest empire the world had seen until that time. Its style was different from its predecessors; the savagery of the Assyrians seems muted. At least brutality was not celebrated in official art and Cyrus was careful to respect the institutions and ways of his new subjects. The result was a diverse empire, but a powerful one, commanding loyalties of a kind lacking to its predecessors. There are some notable religious symptoms; the protection of Marduk was solicited for Cyrus's assumption of the Babylonian kingship and at Jerusalem he launched the rebuilding of the Temple. A Jewish prophet saw in his victories God's hand, named him the Lord's anointed and gloated over the fate of the old enemy, Babylon.

> *'Let now the astrologers, the stargazers, the monthly prognosticators, stand up, and save thee from these things that shall come upon thee.'*
>
> Isaiah xlv, 1, xlvii, 1-13.

Cyrus's success owed much to the material resources of his kingdom. It was rich in minerals, above all in iron, and in the high pastures of the valleys lay a great reserve of horses and cavalrymen. Yet it is impossible to resist the conclusion that sheer personal ability also counted for much; Cyrus lives as a world-historical figure, recognized as such by other would-be conquerors who were to strive in the next few centuries to emulate him. He based his government upon provincial governors who were the forbears of the later Persian satraps, and required from his subject provinces little beyond tribute – usually in gold, which replenished the treasuries of Persia – and obedience.

Thus began the empire which, though with setbacks a plenty, provided for nearly two centuries a framework for the Near East, sheltering a great cultural tradition which grew to nourish itself both from Asia and Europe. Large areas knew longer periods of peace under it than for centuries and it was in many ways a beautiful and gentle civilization. Greeks were told already by Herodotus that the Persians loved flowers and there are many things we could do without more easily than the tulip, which we owe to them. Cyrus's son added Egypt to the empire; yet he died before he could deal with a pretender to the throne whose attempts encouraged Medes and Babylonians to seek to recover their independence. The restorer of Cyrus's heritage was a young man who claimed Achaemenid descent, Darius.

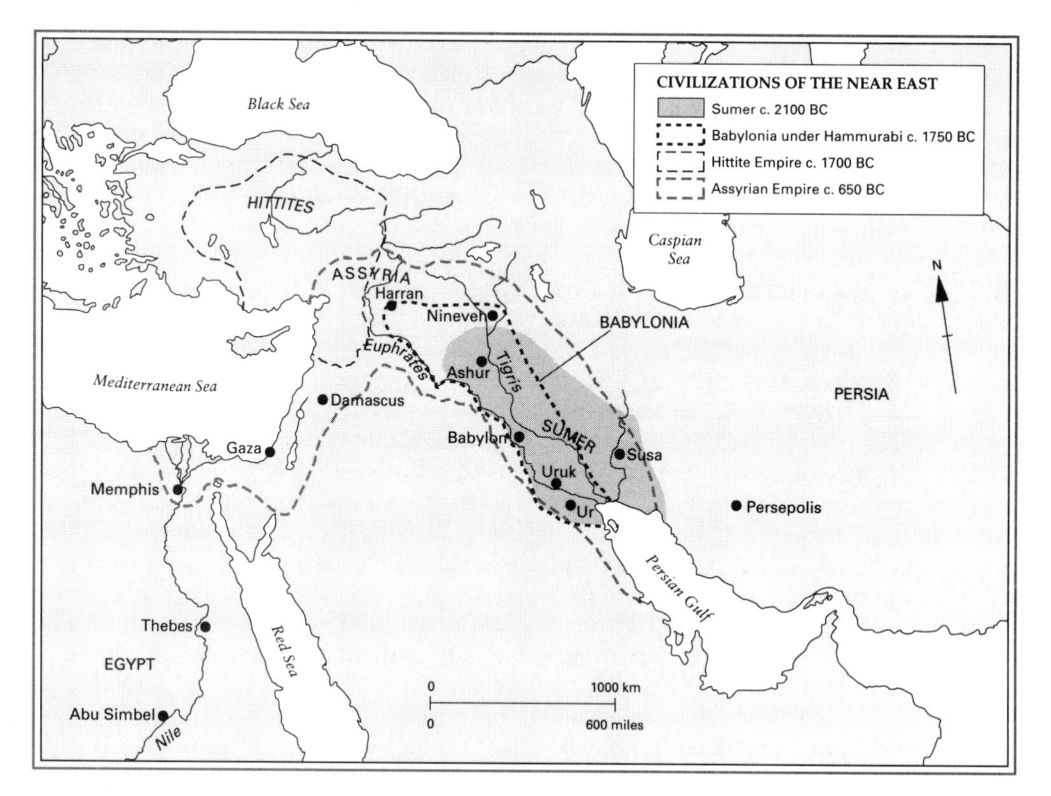

Darius (who reigned 522-486) did not achieve all he wished. His work, none the less, rivalled that of Cyrus. His own inscription on the monument recording his victories over rebels may be thought justified by what he did: 'I am Darius the Great King, King of Kings, King in Persia', a recitation of an ancient title whose braggadocio he adopted. In the east the boundaries of the empire were carried further into the Indus valley. In the west they advanced to Macedonia, though they were checked there, and in the north Darius failed, as Cyrus before him, to make much headway against the Scythians. Inside the empire a remarkable work of consolidation was undertaken. Decentralization was institutionalized with the division of the empire into twenty provinces, each under a satrap who was a royal prince or great nobleman. Royal inspectors surveyed their work and their control of the machine was made easier by the institution of a royal secretariat to conduct correspondence with the provinces, and Aramaic, the old *lingua franca* of Assyrian empire, became the administrative language. It was well adapted to the conduct of affairs because it was not written in cuneiform but in the Phoenician alphabet. The bureaucracy rested on better communications than any yet seen, for much of the provincial tribute was invested in road-building. At their best these roads could convey messages at two hundred miles a day.

A monument to this imperial achievement was to have been a great new capital at Persepolis, where Darius himself was buried in a rock tomb cut into the cliff face. Intended as a colossal glorification of the king, it remains impressive even when it seems pompous. Persepolis was in the end a collective creation; later kings added their palaces to it and embodied in it the diversity and cosmopolitanism of the empire. Assyrian colossi, man-headed bulls and lions guarded its gates as they had done those of Nineveh. Up its staircases marched stone warriors bearing tribute; they are a little less mechanical than the regimented Assyrians of earlier sculpture, but only a little. The decorative columns recall Egypt, but it is an Egyptian device transmitted through Ionian stone-cutters and sculptors. Greek details are to be found also in the reliefs and decoration and a similar mixture of reminiscences is to be found in the royal tombs not far away. They recall the Valley of the Kings in their conception while their cruciform entrances speak of something else. Cyrus's own tomb, at Pasargadae, had also been marked by Greek design. A new world is coming to birth.

These monuments fittingly express the continuing diversity and tolerance of Persian culture. It was one always open to influence from abroad and would continue to be. Persia took up not only the language of those she conquered, but also sometimes their ideas. She also contributed. Vedic and Persian religion mingled in Gandhara, where stood the Indian city the Greeks called Taxila, but both, of course, were Aryan. The core of Persian religion was sacrifice and centred on fire. By the age of Darius the most refined of its cults had evolved into what has been called Zoroastrianism, a dualist religion accounting for the problem of evil in terms of the struggle of a good with an evil god. Of its prophet, Zoroaster, we know little, but it seems that he taught his disciples to uphold the cause of the god of light with ritual and moral behaviour; ahead lay a messianic deliverance, the resurrection of the dead and life everlasting after judgement. This creed spread rapidly through western Asia with Persian rule, even though it was probably never more than the cult of a minority. It would influence Judaism and the oriental cults which were to be part of the setting of Christianity; the angels of Christian tradition and the notion of the hellfire which awaited the wicked both came from Zoroaster.

It is too early to speak of the interplay of Asia and Europe, but there are few more striking examples of the interplay of reciprocal influences which marks the end of the ancient world. We can mark an epoch. Right across the Old World, Persia suddenly pulled peoples into a common experience. Indians, Medes, Babylonians, Lydians, Greeks, Jews, Phoenicians, Egyptians were for the first time all governed by one empire whose eclecticism showed how far civilization had already come. The era of civilization embodied in distinct historical entities was over in the Near East. Too much had been shared, too much diffused for the direct successors of the first civilizations to be any longer the building blocks of world history. Indian mercenaries fought in the Persian armies; Greeks in those of Egypt. City-dwelling and literacy were widespread through the Near East. Men lived in cities around much of the Mediterranean, too. Agricultural and metallurgical techniques stretched even beyond that area and were to be spread further as the Achaemenids transmitted the irrigation skill of Babylon to Central Asia and brought rice from India to be planted in the Near East. When Asian Greeks came to adopt a currency it would be based on the sexagesimal numeration of Babylon. The base of a future world civilization was in the making.

Early in the sixth century BC *an anonymous sculptor carved this herdsman, one of the earliest Attic statues to be found on the Athenian Acropolis. An inscription tells us that the subject may have been Rhombos, who to judge by his dress was probably a wealthy man. Yet he has been caught in a pose which is simple and even consciously so, one evocative of the pastoral and rural background of classical civilization. Similar figures, carrying lambs (or even rams) on their shoulders, have been found as far away as Crete and were dated a century or so earlier than this one.*

BOOK III
THE CLASSICAL MEDITERRANEAN

Measured in years, more than half the story of civilization is already over by about 500 BC. We are still nearer to that date than were the men who lived then to their first civilized predecessors. In the three thousand or so years between them, humanity had come a long way; however imperceptibly slow the changes of daily life in them had been, there is an enormous qualitative gap between Sumer and Achaemenid Persia. By the sixth century, a great period of foundation and acceleration was already over. From the western Mediterranean to the coasts of China a variety of cultural traditions had established themselves. Distinct civilizations had taken root in them, some firmly and deeply enough to survive into our own era. Some of them lasted, moreover, with little but superficial and temporary change for hundreds or even thousands of years. Virtually isolated, they contributed little to mankind's shared life outside their own areas. For the most part, even the greatest centres of civilization were indifferent to what lay outside their spheres for at least two thousand years after the fall of Babylon except when troubled by an occasional invasion. Only one of the civilizations already discernible by the sixth century BC *in fact showed much potential for expanding beyond its cradle, that of the eastern Mediterranean. It was the youngest of them but was to be very successful, lasting for over a thousand years without a break in its tradition. Even this is less remarkable than what it left behind, though, for it was the seedbed of almost all that played a dynamic part in shaping the world we still inhabit.*

1
The Roots of One World

The appearance of a new civilization in the eastern Mediterranean owed much to older Near Eastern and Aegean traditions. From the start we confront an amalgam of Greek speech, a Semitic alphabet, ideas whose roots lie in Egypt and Mesopotamia, reminiscences of Mycenae. Even when this civilization matured it still showed the diversity of its origins. It was never to be a simple, monolithic whole and in the end was very complex indeed. For all that integrated it and gave it unity, it was always hard to delimit, a cluster of similar cultures around the Mediterranean and Aegean, their frontier zones blurring far outwards into Asia, Africa, barbarian Europe and southern Russia. Even when its boundaries with them were clear, other traditions always played upon Mediterranean civilization and received much from it.

This civilization also varied in time. It showed greater powers of evolution than any of its predecessors. Even when they had undergone important political changes their institutions remained fundamentally intact, while Mediterranean civilization displays a huge variety of transient political forms and experiments. In religion and ideology, whereas other traditions tended to develop without violent changes or breaks, so that civilization and religion were virtually coterminous, the one living and dying with the other, Mediterranean civilization begins in a native paganism and ends by succumbing to an exotic import, Christianity; it was a revolutionized Judaism which produced the first world religion. This was a huge change and it transformed this civilization's possibilities of influencing the future.

Of all the forces making for its crystallization, the most fundamental was the setting itself, the Mediterranean basin. It was both a collecting area and a source; currents flowed easily into it from the lands of the old civilizations and from this central reservoir they also flowed back to where they came from and northwards into the barbarian lands. Though it is large and contains a variety of peoples, this basin has well-defined general characteristics. Most of its coasts are narrow plains behind which quickly rise fairly steep and enclosing mountain ranges, broken by few important river valleys. Those who lived on the coasts tended to look along them and outward across the sea, rather than behind them to their hinterland. This, combined with a climate they all shared, made the spreading of ideas and techniques within the Mediterranean natural for enterprising peoples.

The Romans, with reason, named the Mediterranean *Mare Magnum*, the Great Sea. It was the outstanding geographical fact of their world, the centre of classical maps. Its surface was a great uniting force for those who knew how to use it, and by 500 BC maritime technology was advanced enough to make this possible except in winter. Prevailing winds and currents determined the exact routes of ships whose only power was provided by sails or oars, but any part of the Mediterranean was accessible by water from any other. The upshot was a littoral civilization, with a few languages spoken widely within it. It had specialized trading centres, for exchanges of materials were easy by sea, but the economy rested firmly on the growing of wheat and barley, olives and vines, mainly for local consumption. The metals increasingly needed by this economy could be brought in from outside. The deserts to the south were held at bay further from the coast and for perhaps thousands of years North Africa was richer than it now is, more heavily wooded, better watered, and more fertile. The same sort of civilization therefore tended to appear all round the Mediterranean. Such a difference between Africa and Europe as we take for granted did not exist until after 500 AD.

The outward-looking peoples of this littoral civilization created a new world. The great valley civilizations had not colonized, they had conquered. Their peoples looked inward to the satisfaction of limited aims under local despots. Many later societies, even within the classical world, were to do the same, but there is a discernible change of tempo and potential from the start, and eventually Greeks and Romans grew corn in Russia, worked tin from Cornwall, built roads into the Balkans and enjoyed spices from India and silk from China.

Individual artists begin to emerge from the anonymous flow of early art in the sixth century. One master of potting and the painting of pots was Exekias of Athens whose signature appears on this cup decorated with the ship of Dionysus. The god of wine lies back in his ship holding a drinking-vessel, perhaps contemplating the vine which decorates its mast and breaks out into bunches of grapes over his head, or perhaps amused by the leaping dolphins into which (legend said) he had transformed a gang of pirates he had encountered on his voyage.

About this world we know a great deal, partly because is left behind a huge archaeological and monumental legacy. Much more important, though, is the new richness of written material. With this, we enter the era of full literacy. Among other things, we confront the first true works of history; important as were to be the great folk records of the Jews, the narrations of a cosmic drama built about the pilgrimage of one people through time, they are not critical history. In any case, they, too, reach us through the classical Mediterranean world. Without Christianity, their influence would have been limited to Israel; through it, the myths they presented and the possibilities of meaning they offered were to be injected into a world with four hundred years of what we can recognize as critical writing of history already behind it. Yet the work of ancient historians, important as it is, is only a tiny part of the record. Soon after 500 BC, we are in the presence of the first complete great literature, ranging from drama to epic, lyric hymn, history and epigram, though what is left of it is only a small part – seven out of more than a hundred plays by its greatest tragedian, for example. Nevertheless, it enables us to enter the mind of a civilization as we can enter that of none earlier.

Even for Greece, of course, the source of this literature, and *a fortiori* for other and more remote parts of the classical world, the written record is not enough on its own. The archaeology is indispensable, but it is all the more informative because literary sources are so much fuller than anything from the early past. The record they offer us is for the most part in Greek or Latin, the two languages which provided the intellectual currency of Mediterranean civilization. The persistence in English, the most widely used of languages today, of so many words drawn from them is by itself almost enough evidence to show this civilization's importance to its successors (all seven nouns in the last sentence but one, for example, are based on Latin words). It was through writings in these languages that later men approached this civilization and in them they detected the qualities which made them speak of what they found simply as '*the* classical world'.

This is a perfectly proper usage, provided we remember that the men who coined it were heirs to the traditions they saw in it and stood, perhaps trapped, within its assumptions. Other traditions and civilizations, too, have had their 'classical' phases. What it means is that men see in some part of the past an age setting standards for later times. Many later Europeans were to be hypnotized by the power and glamour of classical Mediterranean civilization. Some men who lived in it, too, thought that they, their culture and times were exceptional, though not always for reasons we should now find convincing. Yet it *was* exceptional; vigorous and restless, it provided standards and ideals, as well as technology and institutions, on which huge futures were to be built. In essence, the unity later discerned by those who admired the Mediterranean heritage was a mental one.

Inevitably, there was to be much anachronistic falsification in some of the later efforts to study and utilize the classical ideal, and much romanticization of a lost age, too. Yet even when this is discounted, and when the classical past has undergone the sceptical scrutiny of scholars, there remains a big indissoluble residue of mental achievement which somehow places it on our side of a mental boundary, while the great empires of Asia lie beyond it. With whatever difficulty and possibility of misconstruction, the mind of the classical age is recognizable and comprehensible in a way perhaps nothing earlier can be. 'This', it has been well said, 'is a world whose air we can breathe.'

The role of the Greeks was pre-eminent in making this world and with them its story must begin. They contributed more than any other single people to its dynamism and to its mythical and inspirational legacy. The Greek search for excellence defined for later men what excellence was and their achievement remains difficult to exaggerate. It is the core of the process which made classical Mediterranean civilization.

2
The Greeks

In the second half of the eighth century BC, the clouds which have hidden the Aegean since the end of the Bronze Age begin to part a little. Processes and sometimes events become somewhat more discernible. There is even a date or two, one of which is important in the history of a civilization's self-consciousness: in 776 BC, according to later Greek historians, the first Olympian games were held. After a few centuries the Greeks would count from this year as we count from the birth of Christ.

The people who gathered for that and later festivals of the same sort recognized by doing so that they shared a culture. Its basis was a common language; Dorians, Ionians, Aeolians all spoke Greek. What is more they had done so for a long time; the language was now to acquire the definition which comes from being written down, an enormously important development, making possible, for example, the recording of the traditional oral poetry which was said to be the work of Homer. Our first surviving inscription in Greek characters is on a jug of about 750 BC. It shows how much the renewal of Aegean civilization owed to Asia. The inscription is written in an adaptation of Phoenician script; Greeks were illiterate until their traders brought home this alphabet. It seems to have been used first in the Peloponnese, Crete and Rhodes; possibly these were the first areas to benefit from the renewal of intercourse with Asia after the Dark Ages. The process is mysterious and can probably never be recovered, but somehow the catalyst which precipitated Greek civilization was contact with the East.

Who were the Greek-speakers who attended the first Olympiad? Though it is the name by which they and their descendants are still known, they were not called Greeks; that name was only given them centuries later by the Romans. The word they would have used was the one we render in English as 'Hellenes'. First used to distinguish invaders of the Greek peninsula from the earlier inhabitants, it became the name of all the Greek-speaking peoples of the Aegean. This was the new conception and the new name emerging from the Dark Ages and there is more than a verbal significance to it. It expressed a consciousness of a new entity, one still emerging and one whose exact meaning would always remain uncertain. Some of the Greek-speakers had in the eighth century already long been settled and their roots were lost in the turmoil of the Bronze Age invasions. Some were much more recent arrivals. None came as Greeks; they became Greeks by being there, all round the Aegean. Language identified them and wove new ties between them. Together with a shared heritage of religion and myth, it was the most important constituent of being Greek, always and supremely a matter of common culture.

Yet such ties were never politically effective. They were unlikely to make for unity because of the size and shape of the theatre of Greek history, which was not what we now call Greece, but was, rather, the whole Aegean. The wide spread of Minoan and Mycenaean influences in earlier civilized times had foreshadowed this, for between the scores of its islands and the shores which closed about them it was easy to voyage during much of the year. The explanation of the appearance of Greek civilization at all may well be largely a matter of this geography. The past certainly counted for something, too, but Minoan Crete and Mycenae probably left less to Greece than Anglo-Saxon England left to a later Great Britain. The setting was a much more important factor than history. It offered a specially dense distribution of economically viable communities using the same language and easily accessible not only to one another but to older centres of civilization in the Near

East. Like the old river valleys – but for different reasons – the Aegean was a propitious place; civilization could appear there.

Much of the Aegean was settled by Greeks as a consequence of limitations and opportunities which they found on the mainland. Only in very small patches did its land and climate combine to offer the chance of agricultural plenty. For the most part, cultivation was confined to narrow strips of alluvial plain which had to be dry-farmed framed by rocky or wooded hills; minerals were rare, there was no tin, copper or iron. A few valleys ran direct to the sea and communication between them was usually difficult. All this inclined the inhabitants of Attica and the Peloponnese to look outward to the sea, on the surface of which movement was much easier than on land. None of them, after all, lived more than forty miles from it.

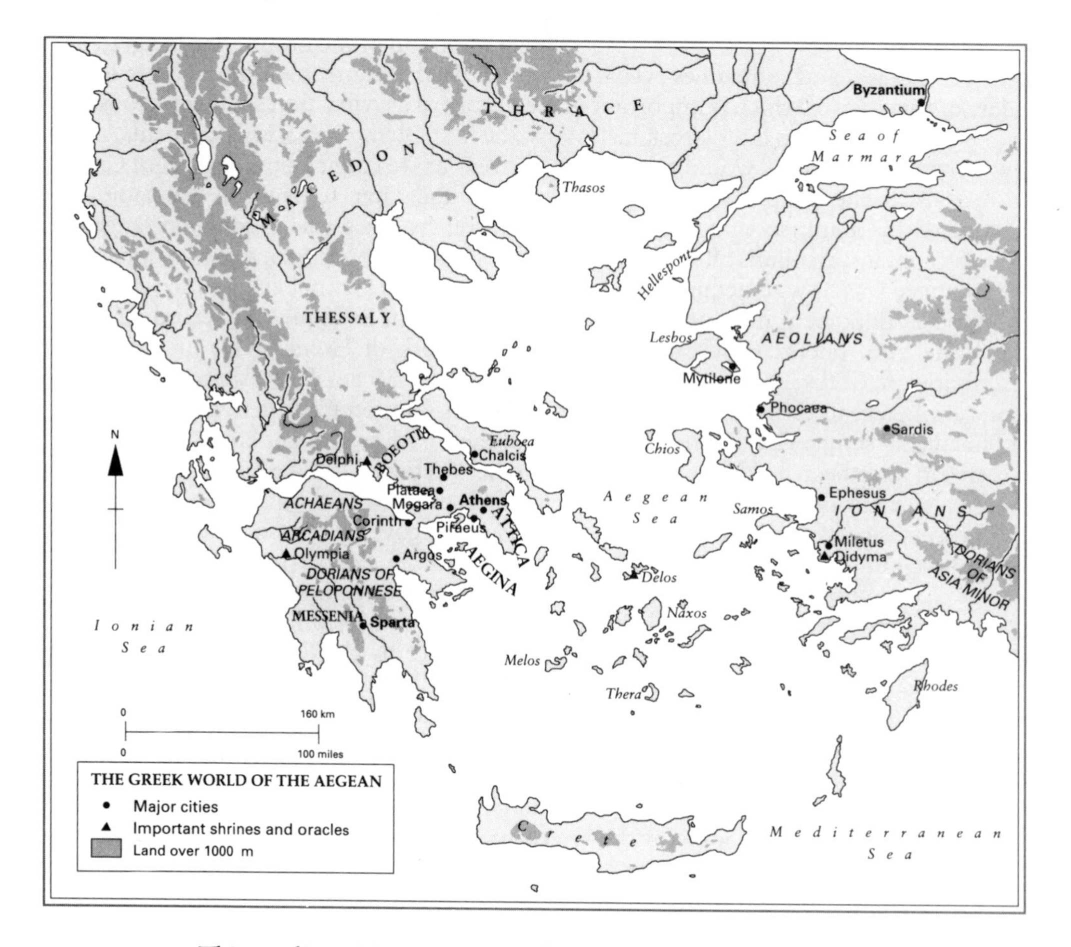

This predisposition was intensified as early as the tenth century by a growth of population which brought greater pressure on available land. Ultimately this led to a great age of colonization; by the end of it, in the sixth century, the Greek world stretched far beyond the Aegean, from the Black Sea in the east to the Balearics, France and Sicily in the west and Libya in the south. But this was the result of centuries during which forces other than population pressure had also been at work. While Thrace was colonized by agriculturalists looking for land, other Greeks settled in the Levant or south Italy in order to trade, whether for the wealth it would bring or for the access it offered to the metals they needed and could not find in Greece. Some Black Sea Greek cities seem to be where they are because of trade, some because of their farming potential. Nor were traders and farmers the only agents diffusing Greek ways and teaching Greece about the outside world. The historical records of other countries show us a flow of Greek mercenaries from the sixth century (when they fought for the Egyptians against the Assyrians) onwards. All these facts were to have important social and political repercussions on the homeland, but before considering them there is much to be said about what kind of civilization it was which was being diffused in this way and was absorbing, by way of return, what others had to give.

Quarrel violently among themselves though they did, and cherish the traditional and emotional distinctions of Boeotian, or Dorian, or Ionian though they might, the Greeks were always very conscious that they were different from other peoples. This could be practically important; Greek prisoners of war, for example, were in theory not to be enslaved, unlike 'barbarians'. This word expressed self-conscious Hellenism in its essence but is more inclusive and less dismissive than it is in modern speech; the barbarians were the rest of the world, those who did not speak an intelligible Greek (dialect though it might be) but who made a sort of 'bar-bar' noise which no Greek could understand. The great religious festivals of the Greek year, when people from many cities came together, were occasions to which only the Greek-speaker was admitted.

Religion was the other foundation of Greek identity. The Greek pantheon is enormously complex, the amalgam of a mass of myths created by many communities over a wide area at different times, often incoherent or even self-contradictory until ordered by later, rationalizing minds. Some were imports, like the Asian myth of golden, silver, bronze and iron ages. Local superstition and belief in such legends was the bedrock of the Greek religious experience. Yet it was a religious experience very different from that of other peoples in its ultimately humanizing tendency. Greek gods and goddesses, for all their supernatural standing and power, are remarkably human. They express the man-centred quality of later Greek civilization. Much as it owed to Egypt and the East, Greek mythology and art usually presents its gods as better, or worse, men and women, a world away from the monsters of Assyria and Babylonia, or from Shiva the many-armed. Whoever is responsible, this was a religious revolution; its converse was the implication that men could be godlike. This is already apparent in Homer; perhaps he did as much as anyone to order the Greek supernatural in this way and he does not give much space to popular cults. He presents the gods taking sides in the Trojan war in postures all too human. They compete with one another; while Poseidon harries the hero of the *Odyssey*, Athena takes his part. A later Greek critic grumbled that Homer 'attributed to the gods everything that is disgraceful and blameworthy among men: theft, adultery and deceit'. It was a world which operated much like the actual world.

The *Iliad* and *Odyssey* have already been touched upon because of the light they throw on prehistory; they were also shapers of the future. They are at first sight curious objects for a people's reverence. The *Iliad* gives an account of a short episode from a legendary long-past war; the *Odyssey* is more like a novel, narrating the wandering of one of the greatest of all literary characters, Odysseus, on his way home from the same struggle. That, on the face of it, is all. But they came to be held to be something like sacred books. If, as seems reasonable, the survival rate of early copies is thought to give a true reflexion of relative popularity, they were copied more frequently than any other text of Greek literature. Much time and ink have been spent on argument about how they were composed. It now seems most likely that they took their present shape in Ionia slightly before 700 BC The Greeks referred to their author without qualification as 'the poet' (a sufficient sign of his standing in their eyes) but some have found arguments for thinking the two poems are the work of different men. For our purpose, it is unimportant whether he was one author or not; the essential point is that someone took material presented by four centuries of bardic transmission and wove it into a form which acquired stability and in this sense these works are the culmination of the era of Greek heroic poetry. Though they were probably written down in the seventh century, no standard version of these poems was accepted until the sixth; by then they were already regarded as the authoritative account of early Greek history, a source of morals and models, and the staple of literary education. Thus they became not only the first documents of Greek self-consciousness, but the embodiment of the fundamental values of classical civilization. Later they were to be even more than this: together with the Bible, they are the source of western literature.

Human though Homer's gods might be, the Greek world had also a deep respect for the occult and mysterious. It was recognized in such embodiments as omens and oracles. The shrines of the oracles of Apollo at Delphi or at Didyma in Asia Minor were places of pilgrimage and the sources of respected if enigmatic advice. There were ritual cults which practised 'mysteries' which re-enacted the great natural processes of germination and growth at the passage of the seasons. Popular religion does not loom large in the literary sources, but it was never wholly separated from 'respectable' religion. It is important to remember this irrational subsoil, given that the achievements of the Greek élite during the later classical era are so impressive and rest so importantly on rationality and logic; the irrational was always there and in the earlier, formative period with which this chapter is concerned, it loomed large.

The literary record and accepted tradition also reveal something, if nothing very precise, of the social and (if the word is appropriate) political institutions of early Greece. Homer shows us a society of kings and aristocrats, but by his day this was already anachronistic. The title of king sometimes lived on, and in one place, Sparta, where there were always two kings at once, it had a shadowy reality which sometimes was effective, but by historical times power had passed from monarchs to aristocracies in almost all the Greek cities. The council of the Areopagus at Athens is an example of the sort of restricted body which usurped the kingly power in many places. Such ruling élites rested fundamentally on land; their members were the outright owners of the estates which provided not only their livelihood but the surplus for the expensive arms and horses which made leaders in war. Homer depicts such aristocrats behaving with a remarkable degree of independence of his kings; this probably reflects the reality of his own day. They were the only people who counted; other social distinctions have little importance in these poems. Thersites is properly chastised for infringing the crucial line between Gentlemen and the Rest.

A military aristocracy's preoccupation with courage may also explain a continuing self-assertiveness and independence in Greek public life; Achilles, as Homer presents him, was as prickly and touchy a fellow as any medieval baron. To this day a man's standing in his peers' eyes is what many Greeks care about more than anything else and their politics have often reflected this. It was to prove true during the classical age when time and time again individualism wrecked the chances of cooperative action. The Greeks were never to produce an enduring empire, for it could only have rested on some measure of subordination of the lesser to the greater good, or some willingness to accept the discipline of routine service. This may have been no bad thing, but meant that for all their Hellenic self-consciousness the Greeks could not unite even their homeland into one state.

Below the aristocrats of the early cities were the Other Ranks of a still not very complex society. Freemen worked their own land or sometimes for others. Wealth did not change hands rapidly or easily until money made it available in a form more easily transferred than land. Homer measured value in oxen and seems to have envisaged gold and silver as elements in a ritual of gift-giving, rather than as means of exchange. This was the background of the later idea that trade and menial tasks were degrading; an aristocratic view lingered on. It helps to explain why in Athens (and perhaps elsewhere) commerce was long in the hands of metics, foreign residents who enjoyed no civic privilege, but who provided the services Greek citizens would not provide for themselves.

Slavery, of course, was taken for granted, though much uncertainty surrounds the institution. It was clearly capable of many different interpretations. In archaic times, if that is what Homer reflects, most slaves were women, the prizes of victory, but the slaughter of male prisoners later gave way to enslavement. Large-scale plantation slavery such as that of Rome or the European colonies of modern times was unusual. Many Greeks of the fifth century who were freemen owned one or two slaves and one estimate is that about one in four of the population was a slave when Athens was most prosperous. They could be freed; one fourth-century slave became a considerable banker. They were also often well treated and sometimes loved. One has become famous: Aesop.

But they were not free and the Greeks thought that absolute dependence on another's will was intolerable for a free man though they hardly ever developed this notion into positive criticism of slavery. It would be anachronistic to be surprised at this. The whole world outside Greece, too, was organized on the assumption that slavery would go on. It was the prevailing social institution almost everywhere well into Christian times and it still survives. It is hardly cause for comment, therefore, that Greeks took it for granted. There was no task that slavery did not sustain for them, from agricultural labour to teaching (our word 'pedagogue' originally meant a slave who accompanied a well-born boy to school). A famous Greek philosopher later tried to justify this state of affairs by arguing that there were some human beings who were truly intended to be slaves by nature, since they had been given only such faculties as fitted them to serve the purposes of more enlightened men. To modern ears this does not seem a very impressive argument, but in the context of the way Greeks thought about nature and man there was more to it than simple rationalization of prejudice.

To the Greeks the expedition to Troy and the great deeds of the heroes who fought there were history, and their own history, too. The combat between Hector and Achilles was one of the great episodes of the story, and it is depicted on this early fifth-century vase. The other figures shown are divinities – Apollo, the patron of Hector, the Trojan hero, and Athene, protector of Achilles.

Slaves may and foreign residents must have been among the many channels by which the Greeks continued to be influenced by the Near East long after civilization had re-emerged in the Aegean. Homer had already mentioned the *demiourgoi*, foreign craftsmen who must have brought with them to the cities of the Hellenes not only technical skill but the motifs and styles of other lands. In later times we hear of Greek craftsmen settled in Babylon and there were many examples of Greek soldiers serving as mercenaries to foreign kings. When the Persians took Egypt in 525 BC, Greeks fought on each side. Some of these men must have returned to the Aegean, bringing with them new ideas and impressions. Meanwhile, there was all the time a continuing commercial and diplomatic intercourse between the Greek cities in Asia and their neighbours.

The multiplicity of day-to-day exchanges resulting from the enterprise of the Greeks makes it very hard to distinguish native and foreign contributions to the culture of archaic Greece. One tempting area is art; here, just as Mycenae had reflected Asian models, so the animal motifs which decorate Greek bronze work, or the postures of goddesses such as Aphrodite, recall the art of the Near East. Later, the monumental architecture and statuary of Greece was to imitate Egypt's, and Egyptian antiquities shaped the styles of the things made by Greek craftsmen at Naucratis. Although the final product, the mature art of classical Greece, was unique, its roots lie far back in the renewal of ties with Asia in the eighth century. What is not possible to delineate quickly is the slow subsequent irradiation of a process of cultural interplay which was by the sixth century working both ways, for Greece was by then both pupil and teacher. Lydia, for example, the kingdom of the legendary Croesus, richest man in the world, was Hellenized by its tributary Greek cities; it took its art from them and, probably more important, the alphabet, indirectly acquired via Phrygia. Thus Asia received again what Asia had given.

Well before 500 BC, this civilization is so complex that it is easy to lose touch with the exact state of affairs at any one time. By the standards of its contemporaries, early Greece was a rapidly changing society, and some of its changes are easier to see than others. One important development towards the end of the seventh century seems to have been a second and more important wave of colonization, often from the eastern Greek cities. Their colonies were a response to agrarian difficulties and population pressure at home. There followed an upsurge of commerce: new economic relationships appearing as trade with the non-Greek world became easier. Part of the evidence is an increased circulation of silver. The Lydians had been the first to strike true coins – tokens of standard weight and imprint - and in the sixth century money began to be widely used in both foreign and internal trade; only Sparta resisted its introduction. Specialization became a possible answer to land shortage at home. Athens assured the grain imports she needed by specializing in the output of great quantities of pottery and oil; Chios exported oil and wine. Some Greek cities became notably more dependent on foreign corn, in particular, from Egypt on the Greeks colonies of the Black Sea.

Commercial expansion meant not only that land was no longer the only important source of wealth, but also that more men could buy the land which was so important in status. This began a revolution both military and political. The old Greek ideal of warfare had been single combat, a form of fighting natural to a society whose warriors were aristocrats, riding or driving to the field of battle to meet their equals while less well-armed inferiors brawled about them. The new groups of wealthy men could afford the armour and arms which provided a better military instrument, the regiment of 'hoplites', the heavy-armed infantry who were to be for two centuries the backbone of Greek armies and give them superiority. They would prevail by disciplined cohesion, rather than by individual derring-do.

The hoplite wore helmet and body-armour and carried a shield. His main weapon was the spear, which he did not throw, but with which he thrust and stabbed in the mêlée which followed a charge by an ordered formation of spearmen whose weight gave it its effect. Such tactics could work only on relatively level ground, but it was such ground that was usually being contested in Greek wars, for the agriculture on which a Greek city depended could be devastated by seizure of the little plains of the valley floor where most of its crops were grown. On such terrain, the hoplites would charge as a mass, with the aim of sweeping away defenders by their impact. They depended completely on their power to act as a disciplined unit. This both maximized the effect of the charge and enabled them to prevail in the hand-to-hand fighting which followed, because each hoplite had to rely for protection on his right-hand side by the shield of his neighbour. To keep an ordered line was therefore crucial. The Spartans were in particular admired for their expertise in performing the preliminary evolutions which preceded such an encounter and for retaining cohesion as a group once the scrimmage had begun.

The ability to act collectively was the heart of the new warfare. Though bigger numbers now took part in battles numbers were no longer all that counted, as three centuries of Greek success against Asian armies were to prove. Discipline and tactical skill began to matter more and they implied some sort of regular training, as well as a social widening of the warrior group. More men thus came to share in the power which comes from a near-monopoly of the means of exercising force.

This was not the only crucial innovation of these years. It was then, too, that the Greeks invented politics; the notion of running collective concerns by discussion of possible choices in a public setting is theirs. The magnitude of what they did lives on in the language we still use, for 'politics' and 'political' are terms derived from the Greek word for city, *polis*. This was the framework of Greek life. It was much more than a mere agglomeration of people living in the same place for economic reasons. That it was more is shown by another Greek turn of speech: they did not speak of Athens doing this, or Thebes doing that, but of the Athenians and the Thebans. Bitterly divided though it might often be, the *polis* – or, as for convenience it can be called, the city-state – was a community, a body of men conscious of shared interests and common goals.

Such collective agreement was the essence of the city-state; those who did not like the institutions of the one they lived in could look for alternatives elsewhere. This helped to produce a high degree of cohesiveness, but also a narrowness; the Greeks never long transcended the passion for local autonomy (another Greek word) and the city-state characteristically looked outwards defensively and distrustfully. Gradually, it acquired its protecting gods, its festivals and its liturgical drama which connected living men with the past and educated them in its traditions and laws. Thus it came to be an organism living in time, spanning generations. But at its root lay the hoplite ideal of disciplined, cooperative action in which men stood shoulder to shoulder with their neighbours, relying on them to support them in the common cause. In early days the citizen body – those, that is to say, who constituted the politically effective community – was confined to the hoplites, those who could afford to take their place in the ranks on which the defence of the city-state depended. It is not surprising that in later times Greek reformers who were worried about the results of political extremism would often turn hopefully to the hoplite class when looking for a stable, settled foundation for the *polis*.

At the roots of city-states lay also other facts: geography, economics, kinship. Many of them grew up on very ancient sites, settled in Mycenaean times; others were newer, but almost always the territory of a city-state was one of the narrow valleys which could provide just enough for its maintenance. A few were luckier: Sparta sat in a broad valley. A few were specially handicapped: the soil of Attica was poor and Athens would have to feed its citizens on imported grain in consequence. Dialect intensified the sense of independence latent in the mountains separating a city from its neighbours. In it was preserved a sense of common tribal origin which lived on in the great public cults.

By the beginning of historical times these forces had already generated intense feelings of community and individuality which made it virtually impossible for Greeks to transcend the city-state: a few shadowy leagues and confederations did not count for much. Within the city the involvement of citizens in its life was close; we might find it excessive. Yet because of its scale the city-state could do without elaborate bureaucracies; the citizen body, always much smaller than the whole population, could always assemble at one meeting place. There was no likelihood that a city-state could or would aspire to a minute bureaucratic regulation of affairs; anything like this would probably have been beyond the capacity of its institutions. If we judge by the evidence of Athens, the state of which we know most because it recorded so much in stone, the distinction between administration, judgement and law-making was not as we know it; as in the Europe of the Middle Ages, an executive act might be clothed as a decision of a court interpreting established law. Law-courts were, formally speaking, only sections of the assembly of the citizens.

The size and qualification of the membership of this body determined the constitutional character of the state. Upon it depended, more or less, the authorities of day-to-day government, whether magistrates or courts. There was nothing like the modern permanent civil service. True, it is still risky to generalize about such matters. There were over a hundred and fifty city-states and about many of them we know nothing; of most of the rest we know only a little. Obviously there were important differences between

Not much pottery survives from Sparta – for her citizens were debarred from manual labour and looked down on craftsmen. What has survived is for the most part from the sixth century; some was exported and includes pieces of high quality. This example has a certain commercial interest, for it represents the king of Cyrene in North Africa supervising the weighing and storing of bales of a medicinal plant which was one of his kingdom's most important exports. Workmen call out the weights, and his pets play round his chair as he watches.

the ways in which they ran their affairs; in the fourth century BC, Aristotle made a great collection of their constitutions and there would not have been much point in a political scientist doing this unless they were significantly different from one another. But the detail of what went on is hard to discern, even in the few cases where we have good information.

As for the history of political forms, the origins are usually buried in legends as informative as the story of Hengist and Horsa to the historian of England. Even Homer is unhelpful about the city-state; he hardly mentions it because his subject is warrior bands. Yet when the historical age dawns the city-state is there, ruled by aristocracies. The forces which determined the broad lines of its later evolution have already been touched upon. New wealth meant new men, and the new men battered away at the existing élites to get admission to citizenship. The aristocracies which had supplanted the kings themselves became objects of rivalry and attack. The new men sought to replace them by governments less respectful of traditional interests; the result was an age of rulers the Greeks called tyrants. They were often moneyed, but their justification was their popularity; they were strong men who set aside the aristocracies. The later sinister connotations of the word 'tyrant' did not then exist; many tyrants must have seemed benevolent despots. They brought peace after social struggles probably intensified by a new crisis arising from pressure on land. Peace favoured economic growth, as did the usually good relations the tyrants enjoyed with one another. The seventh century was their golden age. Yet the institution did not long survive. Few tyrannies lasted two generations. In the sixth century the current turned almost everywhere towards collective government; oligarchies, constitutional governments, even incipient democracies began to emerge.

Athens was an outstanding example. For a long time it seems that Attica, though poor, had sufficient land for Athens to escape the social pressures which in other states led to the colonization movement. In other ways, too, her economy early reflected a special vigour; even in the eighth century her pottery suggests that Athens was something of a commercial and artistic leader. In the sixth, though, she too was racked by conflict between rich and poor. A soon legendary lawgiver, Solon, forbade the enslavement of debtors by wealthy creditors (which had the effect of leading men to turn to greater dependence on chattel slaves, since debt bondage could no longer guarantee a labour force). Solon also encouraged farmers to specialise. Oil and wine (and their containers) became staple Athenian exports and grain was kept at home. Simultaneously, a series of reforms (also attributed to Solon) gave to the newly enriched equality with the old landed class and provided for a new popular council to prepare business for the *ecclesia*, the general assembly of all citizens.

Such changes did not at once quiet Athens' divisions. An age of tyrants only closed with the expulsion of the last in 510 BC. Then there at last began to operate the institutions whose paradoxical outcome was to be the most democratic government in Greece, though one over a state which held more slaves than any other. All political decisions were taken in principle by majority vote of the *ecclesia* (which also elected the important magistrates and military commanders). Ingenious arrangements provided for the organization of the citizens in units which would prevent the emergence of sectional factions representing city-dwellers as against farmers or merchants. It was the beginning of a great age, one of prosperity, when Athens would consciously foster festivals and cults looking beyond the city and offered something to all Greeks. This was something of a bid for leadership.

Much has long been made of the contrast between Athens and her great rival, Sparta. Unlike Athens, Sparta met the pressures upon her not by modifying her institutions but by resisting change. She embodied the most conservative approach to the problem, solving it for a long time by rigid social discipline at home and by conquest among her neighbours which allowed her to meet the demand for land at others' expense. A very early consequence was a fossilizing of the social structure. So tradition-bound was she that it was alleged that her legendary law-giver, Lycurgus, had even forbidden her to write down her laws; they were driven home in the minds of the Spartiates by a rigorous training all undergone in youth, boys and girls alike.

Sparta had no tyrants. Her effective government appears to have been shared between a council of old men and five magistrates called 'ephors', while the two hereditary kings had special military powers. These oligarchs were in the last resort answerable to the assembly of the Spartiates (of whom, according to Herodotus, there were early in the fifth century about five thousand). Sparta was, therefore, a large aristocracy whose origin, ancient writers agreed, was the hoplite class. Society remained agricultural; no commercial class was allowed to appear and when the rest of Greece took up the use of money, around 600 BC, Sparta stood out and permitted only an iron currency for internal use. Spartiates were not supposed to own silver or gold until the fourth century. Sparta even stood aside from the colonizing movement; she launched only one enterprise of this sort.

This produced a sort of militarized egalitarianism often admired by later puritans, and an atmosphere strongly suggestive, for good and ill, of the aspirations of the more old-fashioned English public school. Though the passing of time and the position of kings slightly softened their practice, Spartiates knew no great distinctions of wealth or comfort. Until well into classical times they avoided dressing differently and ate at communal messes. Their conditions of life were, in a word, 'spartan', reflecting the idealization of military virtues and strict discipline. The details are often strikingly unpleasant as well as curious. Marriage, for example, was a ceremony for which the bride's hair was cropped and she was dressed as a boy. It was followed by a simulated rape, after which the couple did not live together, the man continuing to live with his companions in a male dormitory and eating in messes with them. It is interesting that Sparta exported nursemaids to other Greek states (later parallels will again occur to the reader). She had no artistic or cultural achievement to speak of and her internal politics remain mysterious.

Possibly Spartan politics were simplified or muted by Sparta's gravest problem, the division between the citizen commune and the rest. The bulk of the inhabitants of the Spartan state were not citizens. Some were freemen, but most were helots, serf-like workers bound to the land who shared with the free peasants the task of producing the food consumed at the Spartiates' communal meals. Originally the helot population may have been the native population enslaved by the Dorian invasions, but they were like later serfs in being tied to land rather than being the chattels of individual owners. Certainly their number was later swollen by conquest, above all by the annexation in the eighth century of the plain of Messenia, which disappeared from Greek history as an independent state for more than three hundred years. As a result, a cloud hung over the Spartan achievement, the fear of a helot revolt, and it was remarked by other Greeks. It hobbled the Spartans in their relations with other states. Increasingly they feared to have their army abroad lest its absence should tempt revolt at home. Sparta was always on sentry-go and the feared enemy was at home.

Sparta and Athens were to quarrel fatally in the fifth century and this has led them to be seen as always the poles of the political world of ancient Greece. They were not, of course, the only models available, and herein lies one of the secrets of the Greek achievement. It would draw upon a richness of political experience and data far greater than anything seen in the world until this time. This experience would provide the first systematic reflexions upon the great problems of law, duty, and obligation which have exercized men's minds ever since, largely in terms set by the classical Greeks. In pre-classical times, speculation on such themes is almost nonexistent. The weight of custom and the limitations of local experience sufficiently explain this.

The city-state was the shared inheritance and experience of the Greeks, but they knew of other types of political organization through contacts made in the course

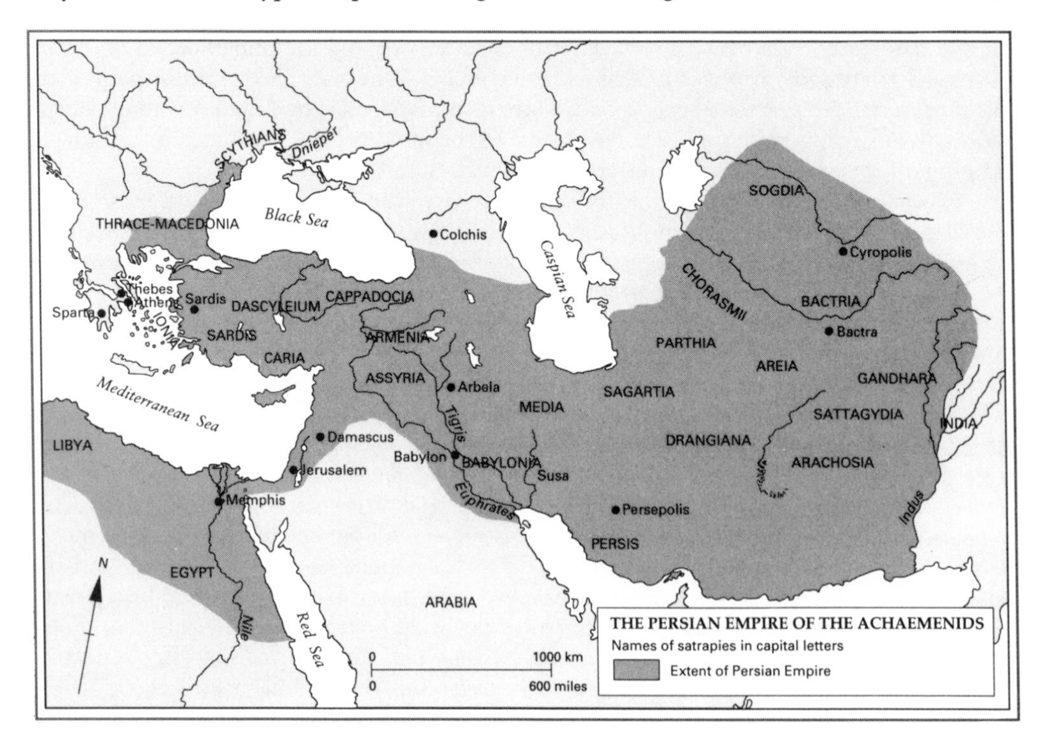

of trade and because of the exposed nature of many of their own settlements. The Greek world had frontier regions where conflict was likely. In the west they once seemed to be pushing ahead in an almost limitless expansion, but two centuries of striking advance came to an end round about 550 BC, when Carthaginian and Etruscan power prescribed a limit. The first settlements – once again, at sites sometimes used centuries earlier by Minoans and Mycenaeans – show that trade mattered as much as agriculture in their foundation. Their main strength lay in Sicily and in southern Italy, an area significantly to be called *Magna Graecia* in later classical times. The richest of these colonies was Syracuse, founded by Corinthians in 733 BC and eventually the dominating Greek state in the west. It had the best harbour in Sicily. Beyond this colonial area, settlements were made in Corsica and southern France (at Massilia, the later Marseilles) while some Greeks went to live among the Etruscans and Latins of central Italy. Greek products have turned up even as far afield as Sweden and Greek style has been seen in sixth-century fortification in Bavaria. More impalpable influence is hard to pin down, but a Roman historian believed that Greek example first civilized the barbarians of what was later to be France and set them not only to tilling their fields, but to cultivating the vine. If so, posterity owes Greek commerce a debt indeed.

This vigorous expansion seems to have provoked Phoenician envy and imitation. It led the Phoenicians to found Carthage and the Carthaginians to seize footholds in western Sicily. Eventually they were able to close down Greek trade in Spain. Yet they could not turn the Greek settlers out of Sicily any more than the Etruscans could drive them from Italy. The decisive battle in which the Syracusans routed a Carthaginian force was in 480 BC.

This was a date of even greater significance for Greek relations with Asia, where the Greek cities of Asia Minor had often been at loggerheads with their neighbours. They had suffered much from the Lydians until they came to terms with the Lydian king, Croesus of legendary wealth, and paid him tribute. Before this, Greece already influenced Lydian fashions; some of Croesus' predecessors had sent offerings to the shrine at Delphi. Now the Hellenization of Lydia went even more quickly ahead. None the less, a much more formidable opponent loomed up even further east: Persia.

The Greek struggle with Persia is the climax of the early history of Greece and the inauguration of the classical age. Because the Greeks made so much of their long conflict with the Persians it is easy to lose sight of the many ties that linked their cultures. The Persian fleets – and to a lesser extent, Persian armies – launched against the Peloponnese had thousands of Greeks, mainly from Ionia, serving in them. Cyrus had employed Greek stone-cutters and sculptors and Darius had a Greek physician. Probably the war did as much to create as to feed the antagonism, however deep the emotional revulsion proclaimed by the Greeks for a country which treated its kings like gods.

The origins of the war lay in the great expansion of Persia under the Achaemenids. In about 540 BC, the Persians overthrew Lydia (and that was the end of Croesus, who was supposed to have provoked the assault by an incautious interpretation of an utterance of the Delphic oracle, which said that if he went to war with Persia, he would destroy a great empire but not which one). This brought Greeks and Persians face to face; elsewhere, the tide of Persian conquest rolled on. When the Persians took Egypt they damaged Greek traders' interests there. Next, the Persians crossed to Europe and occupied the cities of the coast as far west as Macedon; across the Danube and they failed, and soon retired from Scythia. At this point there was something of a pause. Then, in the first decade of the fifth century, the Asian Greek cities revolted against Persian suzerainty, encouraged, perhaps, by Darius' failure against the Scythians. The mainland cities, or some of them, decided to help. Athens and Eretria sent a fleet to Ionia. In the subsequent operations the Greeks burnt Sardis, the former capital of Lydia and the seat of the western satrapy of the Persian empire. But the revolt failed in the end and left the mainland cities facing an enraged opponent.

Things did not usually happen very quickly in the ancient world, and large-scale expeditions still take a long time to prepare, but almost as soon as the Ionian revolt was crushed the Persians sent a fleet against the Greeks; it was wrecked off Mount Athos. A second attempt, in 490 BC, sacked Eretria but then came to grief at the hands of the Athenians in a battle whose name has become legendary: Marathon.

Though this was an Athenian victory, the leader in the next phase of the struggle with Persia was Sparta, the strongest of the city-states on land. Out of the Peloponnesian League, an alliance whose origins had been domestic in that its aim had been to assure Sparta's future by protecting her from the need to send her army abroad, there devolved upon Sparta something like national leadership. When the Persians came again, ten years later, almost all the Greek states accepted this – even Athens, whose strengthening of her fleet had made her the preponderant power of the League at sea.

The Greeks said, and no doubt believed, that the Persians came again (in 480 BC, through Thrace) in millions; if, as now seems more likely, there were in fact well under a hundred thousand of them, this was still an overwhelming enough disproportion for the defenders of Greece. The Persian army moved slowly along the coast and down towards the Peloponnese, accompanied by a huge fleet which hung on its flanks. Yet the Greeks had important advantages in their better-armed and trained heavy infantry, a terrain which nullified the Persian cavalry superiority, and morale.

This time the crucial battle was at sea. It followed another legendary episode, the overwhelming of Leonidas the Spartan king and his three hundred at the pass of Thermopylae, after which Attica had to be abandoned to the Persians. The Greeks retired to the isthmus of Corinth, their fleet massed in the bay of Salamis near Athens. Time was on their side. It was autumn; a winter which would catch the Persians unprepared would soon be coming and Greek winters are severe. The Persian king threw his numerical advantage away by deciding to engage the Greek fleet in the narrow waters of Salamis. His fleet was shattered and he began a long retreat to the Hellespont. The next year the army he had left behind was defeated at Plataea and the Greeks won another great sea fight, at Mycale on the other side of the Aegean, on the same day. This was the end of the Persian War.

It was a great moment in Greek history, perhaps the greatest, and Sparta and Athens had covered themselves with glory. The liberation of Asiatic Greece followed. It opened an age of huge self-confidence for the Greeks. Their outward drive was to continue until its culmination in a Macedonian empire a century and a half later. The sense of Greek identity was at its height, and men looking back at these heroic days were to wonder later if some great chance to unite Greece as a nation had not then been missed for ever. Perhaps, too, it was something more, for in the repulse of Asia by Greece lay the beginnings of a distinction between Europe and Asia whose reality would not appear for centuries but which would eventually lead men to look back anachronistically at Marathon and Salamis as the first time that Europe was saved.

3
Greek Civilization

Victory over the Persians opened the greatest age of Greek history. Some have spoken of a 'Greek miracle', so high do the achievements of classical civilization appear. Yet those achievements had as their background a political history so embittered and poisoned that it ended in the extinction of the institution which sheltered Greek civilization, the city-state. Complicated though it is in detail, the story can easily be summarized.

For thirty years after Plataea and Mycale the war with Persia dragged on, but as a background to a more important theme, a sharpening rivalry between Athens and Sparta. Survival assured, the Spartans had gone home with relief, anxious about their helots. This left Athens undisputed leader of those states which wanted to press ahead with the liberation of other cities from the Persians. A confederation called the Delian League was formed which was to support a common fleet to fight the Persians and command of it was given to an Athenian. As time passed, the members contributed not ships but money. Some did not wish to pay up as the Persian danger declined. Athenian intervention to make sure that they did not default increased and grew harsher. Naxos, for example, which tried to leave the alliance, was besieged back into it. The League was turning gradually into an Athenian empire and the signs were the removal of its headquarters from Delos to Athens, the use of the tribute money for Athenian purposes, the imposition of resident Athenian magistrates and the transfer of important legal cases to Athenian courts. When peace was made with Persia in 449 BC, the League continued, though its excuse had gone. At its peak, over one hundred and fifty states were paying tribute to Athens.

Sparta had welcomed the first stages of this process, happy to see others take up commitments outside her own borders. Like other states, Sparta only gradually became aware of a changing situation. When they did, this had much to do with the fact that Athenian hegemony increasingly affected the internal politics of the Greek states. They were often divided about the League, the richer, tax-paying citizens resenting the tribute, while the poorer did not; they did not have to find the money to pay it. When Athenian interventions occurred they were sometimes followed by internal revolution, the result of which was often imitation of Athenian institutions. Athens was herself living through struggles which steadily drove her in the direction of democracy. By 460 BC, the issue at home was really settled, so that irritation over her diplomatic behaviour soon came to have an ideological flavour. Other things, too, may have added to an irritation with Athens. She was a great trading state and another big trading city, Corinth, felt herself threatened. The Boeotians were directly the subjects of Athenian aggression, too. The materials thus accumulated for a coalition against Athens, and Sparta eventually took the lead in it by joining in war against Athens begun in 460. Fifteen years of not very determined fighting followed and then a doubtful peace. It was only after almost another fifteen years, in 431 BC, that there began the great internal struggle which was to break the back of classical Greece, the Peloponnesian War.

It lasted, with interruptions, twenty-seven years, until 404 BC. Essentially it was a struggle of land against sea. On one side was the Spartan league, with Boeotia, Macedon (an unreliable ally) and Corinth as Sparta's most important supporters; they held the Peloponnese and a belt of land separating Athens from the rest of Greece. Athens' allies were scattered round the Aegean shore, in the Ionian cities and the islands, the area

The Acropolis was the site of the most sacred place in Athens, the home of the divine guardians of the city. Under Pericles' leadership there began to be built upon it (from money subscribed to resist Persia by the allies of Athens) the astonishing range of buildings which made it the greatest shrine of Greek classical architecture. The most celebrated of its buildings is the Parthenon, first a temple to Athene, later an Orthodox church, a Catholic cathedral, an Islamic mosque and finally a gunpowder magazine until it blew up in 1687. Then began the era of ruin and spoliation which lasted until very recent times.

she had dominated since the days of the Delian League. Strategy was dictated by the means available. Sparta's army, clearly, was best used to occupy Athenian territory and then exact submission. The Athenians could not match their enemies on land. But they had the better navy. This was in large measure the creation of a great Athenian statesman and patriot, the demagogue Pericles. On the fleet he based a strategy of abandoning the Athenian countryside to annual invasion by the Spartans – it was in any case never capable of feeding the population – and withdrawing the inhabitants to the city and its port, the Piraeus, to which it was linked by two walls some five miles long, two hundred yards apart. There the Athenians could sit out the war, untroubled by bombardment or assault, which were beyond the capacities of Greek armies. Their fleet, still controlling the sea, would assure they were fed in war as in peace, by imported corn, so that blockade would not be effective.

Things did not work as well as this, because of plague within the city and the absence of leadership after Pericles' death in 429 BC, but the basic sterility of the first ten years of the war rests on this strategical deadlock. It brought peace for a time in 421 BC, but not a lasting one. Athenian frustrations found an outlet in the end in a scheme to carry the war further afield.

In Sicily lay the rich city of Syracuse, the most important colony of Corinth, herself the greatest of Athens' commercial rivals. To seize Syracuse would deeply wound an enemy, finish off a major grain-supplier to the Peloponnese, and provide immense booty. With this wealth Athens could hope to build and man a yet bigger fleet and thus achieve a final and unquestioned supremacy in the Greek world – perhaps the mastery of the Phoenician city of Carthage and a western Mediterranean hegemony, too. The result was the disastrous Sicilian Expedition of 415–413 BC. It was decisive, but as a death-blow to the ambitions of Athens. Half her army and all her fleet were lost; a period of political upheaval and disunion began at home. Finally, the defeat crystallized once more the alliance of Athens' enemies.

The Spartans now sought and obtained Persian help in return for a secret undertaking that the Greek cities of mainland Asia should again become vassals of Persia (as they had been before the Persian War). This enabled them to raise the fleet which could help the Athenian subject cities who wanted to shake off her imperial control. Military and naval defeat undermined morale in Athens. In 411 BC an unsuccessful revolution briefly replaced the democratic régime with an oligarchy. Then there were more disasters, the capture of the Athenian fleet, and, finally, blockade. This time starvation was effective. In 404 BC Athens made peace and her fortifications were slighted.

Formally the story ends here, for what followed was implicit in the material and psychological damage the leading states of Greece had done to one another in these bitter years. There followed a brief Spartan hegemony during which she attempted to prevent the Persians cashing the IOU on the Greek Asian cities, but this had to be conceded after a war which brought a revival of Athenian naval power and the rebuilding of the Long Walls. In the end, Sparta and Persia had a common interest in preventing a renaissance of Athenian power and made peace in 387 BC. The settlement included a joint guarantee of all the other Greek cities except those of Asia. Ironically, the Spartans soon became as hated as the Athenians had been. Thebes took the leadership of their enemies. At Leuctra, in 371 BC, to the astonishment of the rest of Greece, the Spartan army was defeated. It marked a psychological and military epoch in something of the same way as the battle of Jena in Prussian history over two thousand years later. The practical consequences made this clear, too; a new confederation was set up in the Peloponnese as a counterweight to Sparta on her very doorstep and the foundation of a revived Messenia in 369 BC was another blow. The new confederation was a fresh sign that the day of the city-state was passing. The next half-century would see it all but disappear, but 369 BC is far enough to take the story for the moment.

Such events would be tragic in the history of any country. The passage from the glorious days of the struggle against Persia to the Persians' almost effortless recouping of their losses, thanks to Greek divisions, is a rounded drama which must always grip the imagination. Another reason why such intense interest has been given to it is that it was the subject-matter of an immortal book, Thucydides' *History of the Peloponnesian War*, the first work of contemporary as well as of scientific history. But the fundamental explanation why these few years should fascinate us when greater struggles do not is because we feel that at the heart of the jumble of battles, intrigues, disasters and glory still lies an intriguing and insoluble puzzle: was there a squandering of real opportunities after Mycale, or was this long anticlimax simply a dissipation of an illusion, circumstances having for a moment seemed to promise more than in fact was possible?

The war years have another startling aspect, too. During them there came to fruition the greatest achievement in civilization the world had ever seen. Political and military events then shaped that achievement in certain directions and in the end limited it and determined what should continue to the future. This is why the century or so of this small country's history whose central decades are those of the war is worth as much attention as the millennial empires of antiquity.

At the outset we should recall how small a plinth supported Greek civilization. There were many Greek states, certainly, and they were scattered over a large expanse of the Aegean, but even if Macedonia and Crete were included, the land-surface of Greece would fit comfortably into England without Wales or Scotland - and of it only about one-fifth could be cultivated. Of the states, most were tiny, containing not more than 20,000 souls at most; the biggest might have had 300,000, thus just surpassing (but not by much) the size of, say, Bournemouth. Within them only a small élite took part in civic life and the enjoyment of what we now think of as Greek civilization.

The other thing to be clear about at the outset is that civilization's essence. The Greeks were far from underrating comfort and the pleasures of the senses. The physical heritage they left behind set the canons of beauty in many of the arts for two thousand years. Yet in the end the Greeks are remembered as poets and philosophers; it is an achievement of the mind that constitutes their claim on our attention. This has been recognized implicitly in the idea of classical Greece, a creation of later ages rather than of the Greeks themselves. Certainly some Greeks of the fifth and fourth centuries BC saw themselves as the bearers of a culture which was superior to any other available, but the force of the classical ideal lies in its being a view from a later age, one which looked back to Greece and found there standards by which to assess itself. Later generations saw these standards above all in the fifth century, in the years following victory over the Persians, but there is a certain distortion in this. There is also an Athenian bias in such a view, for the fifth century was the apogee of Athenian cultural success. Nevertheless, to distinguish classical Greece from what went before – usually named 'archaic' or 'pre-classical' – makes sense. The fifth century has an objective unity because it saw a special heightening and intensification of Greek civilization, even if that civilization was ineradicably tied to the past, ran on into the future and spilled out over all the Greek world.

That civilization was rooted still in relatively simple economic patterns; essentially, they were those of the preceding age. No great revolution had altered it since the introduction of money and for three centuries or so there were only gradual or specific changes in the direction or materials of Greek trade. Some markets opened, some closed, but that was all; the technical arrangements grew slightly more elaborate as the years went by. And trade between countries and cities was the most advanced economic sector. Below this level, the Greek economy was still nothing like as complicated as would be now taken for granted. Barter, for example, persisted for everyday purposes well into the era of coinage. It also speaks for relatively simple markets, with only limited demands made on them by the consumer. The scale of manufacture, too, was small. It has been suggested that at the height of the craze for the best Athenian pottery not more than 150 craftsmen were at work making and painting it. We are not dealing with a world of factories; most craftsmen and traders probably worked as individuals with a few employees and slaves. Even great building projects, such as the embellishment of Athens, reveal subcontracting to small groups of workers. The only exception may have been in mining, where the silver mines of Laurium in Attica seem to have been worked by thousands of slaves, though the arrangements under which this was done – the mines belonged to the state and were in some way sublet – remain obscure. The heart of the economy almost everywhere was subsistence agriculture. In spite of the specialized demand and production of an Athens or a Miletus (which had something of a name as a producer of woollens) the typical community depended on the production by small farmers of the grain, olives, vines and timber needed by the home market.

Such men were the typical Greeks. Some were rich, most of them were probably poor by modern standards, but even now the Mediterranean climate makes a relatively low income more tolerable than it would be elsewhere. Commerce on any scale, and other kinds of entrepreneurial activity, were likely to be mainly in the hands of metics. They might have considerable social standing and were often rich men, but (for example) in Athens they could not acquire land without special permission, though they were liable for military service (which gives us a little information about their numbers, for at the beginning of the Peloponnesian War there were some 3000 who could afford the arms and armour needed to serve as hoplite infantry). The other male inhabitants of the city-state who were not citizens were either freemen or slaves.

Women, too, were excluded from citizenship, though it is hazardous to generalize any further about their legal rights. In Athens, for example, they could neither inherit nor own property, though both were possible in Sparta, nor could they undertake

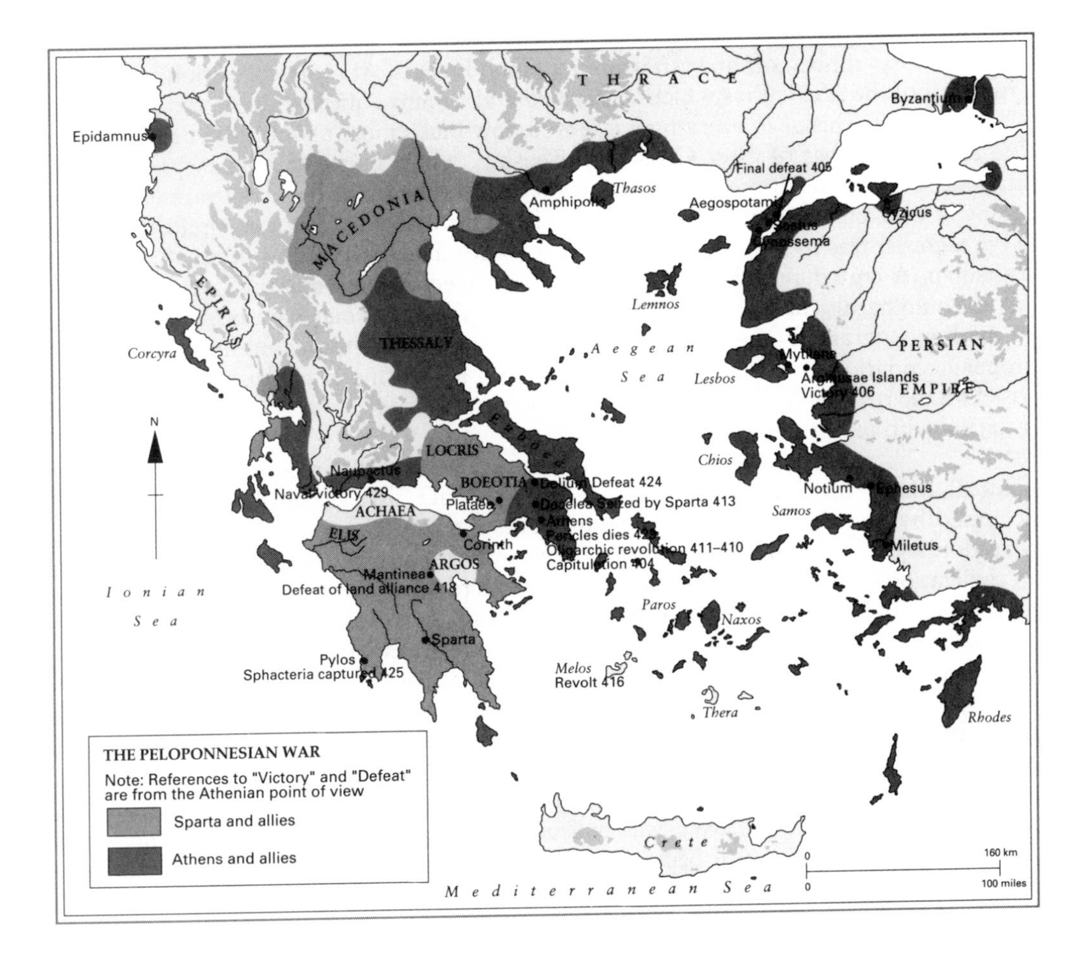

a business transaction if more than the value of a bushel of grain was involved. Divorce at the suit of the wife was, it is true, available to Athenian women, but it seems to have been rare and was probably practically harder to obtain than it was for men, who seem to have been able to get rid of wives fairly easily. Literary evidence suggests that wives other than those of rich men lived, for the most part, the lives of drudges. The social assumptions that governed all women's behaviour were very restrictive; even women of the upper classes stayed at home in seclusion for most of the time. If they ventured out, they had to be accompanied; to be seen at a banquet put their respectability in question. Entertainers and courtesans were the only women who could normally expect a public life; they could enjoy a certain celebrity, but a respectable woman could not. Significantly, in classical Greece girls were thought unworthy of education. Such attitudes suggest the primitive atmosphere of the society out of which they grew, one very different from, say, Minoan Crete among its predecessors, or later Rome.

So far as sexuality is revealed by literature, Greek marriage and parenthood could produce deep feeling and as high a mutual regard between individual men and women as in our own societies. One element in it which is nowadays hard to weigh up exactly was a tolerated and even romanticized male homosexuality. Convention regulated this. In many Greek cities, it was acceptable for young upper-class males to have love-affairs with older men (interestingly, there is much less evidence in Greek literature of homosexual love between men of the same age). This was not thought to disqualify them for subsequent heterosexual marriage. Something must be allowed for fashion in this, and all societies can provide examples of homosexual relationships which suit many men at one stage of their lives; those of the ancient Greeks have attracted undue attention, perhaps because of the absence of inhibitions and controls which made the expression of homosexual affection improper in other societies and because the general prestige of their civilization has rubbed off on even its minor embodiments. At root, it may only have been a function of the restrictions which segregated and circumscribed the lives of free women.

In this as in everything else we know much more about the behaviour of an élite than about that of most Greeks. Citizenship, which must often have spanned very different social levels in practice, is a category too big to permit generalizations. Even in democratic Athens the kind of man who rose in public life and of whom, therefore, we read in the records, was usually a landowner; he was not likely to be a businessman, far less a craftsman. A craftsman might be important as a member of his group in the assembly, but he could hardly make his way to leadership. Businessmen may have been handicapped by the long-engrained conviction of upper-class Greeks that trade and industry were no proper occupations for a gentleman, who should ideally live a life of cultivated leisure based on the revenues of his own lands. This was a view which was to pass into European tradition with important effect.

Social history therefore blurs into politics. The Greek preoccupation with political life – the life of the *polis* – and the fact that classical Greece is neatly delimited by two distinct political epochs (that of the Persian Wars and that of a new, Macedonian, empire) makes it easy to appreciate the importance of Greek political history to civilization. Yet to reconstruct it in any complete sense is impossible, Many, perhaps most, English parishes have records richer than those we can recover for most of the city-states of Greece. What can be discovered from the evidence is much of the history of Athens, quite a lot of that of a few other states, almost nothing of many, and a fairly full narrative of their relations with one another. Together, these facts provide us with a pretty clear picture of the political context of classical Greek civilization, but uncertainty about many of its details.

Athens dangerously dominates this picture. There are big risks in arguing too readily from Athens to what was typical. What we know most about we often tend to think most important and because some of the greatest of fifth-century Greeks were Athenians and Athens is one pole of the great story of the Peloponnesian War, scholars have given its history enormous attention. Yet we also know that Athens was – to take only two points – both big and a commercial centre; it must, therefore, have been very untypical in important ways.

The temptation to over-value Athens' cultural importance is less dangerous. Such a primacy was, after all, recognized at the time. Though many of the greatest Greeks were not Athenians, and many Greeks rejected the Athenians' claims to superiority, the Athenians felt themselves the leaders of Greece. Only a few of the most scrupulous among them hesitated to use the tribute of the Delian League for embellishing its leading city. Thus were built the buildings whose ruins still crown the Acropolis, the Parthenon and Propylaea, but, of course, the money spent on them was available just because so many Greek states recognized Athens' paramountcy. This reality is what the tribute lists record. When on the eve of the Peloponnesian War Pericles told his countrymen that their state was a model for the rest of Greece there was an element of propaganda in what he said, but there was also conviction.

Solid grounds for the importance traditionally given to Athens ought, indeed, to be suggested *a priori* by the basic facts of geography. Her position recalls the tradition that she played an ill-defined but seemingly important role in the Ionian plantation of the Aegean and Asia Minor. Easy access to this region, together with poor agricultural resources, made her a trading and maritime power early in the sixth century. Thanks to this she was the richest of the Greek cities; at the end of it the discovery of the silver deposits of Laurium gave her the windfall with which to build the fleet of Salamis. From the fleet came her undisputed pre-eminence in the Aegean and thence, eventually, the tribute which refreshed her treasury in the fifth century. The peak of her power and wealth was reached just before the Peloponnesian War, in the years when creative activity and patriotic inspiration reached their height. Pride in the extension of empire was then linked to a cultural achievement which was truly enjoyed by the people.

Commerce, the navy, ideological confidence and democracy are themes as inseparably and traditionally interwoven in the history of fifth-century Athens as of late nineteenth-century England, though in very different ways. It was widely recognized at the time that a fleet of ships whose movement depended ultimately upon about two hundred paid oarsmen apiece was both the instrument of imperial power and the preserve of the democracy. Hoplites were less important in a naval state than elsewhere, and no expensive armour was needed to be an oarsman, who would be paid by the tribute of the League or the proceeds of successful warfare – as it was hoped, for example, the Sicilian Expedition would prove. Imperialism was genuinely popular among Athenians who would expect to share its profits, even if only indirectly and collectively, and not to have to bear its burdens. This was an aspect of Athenian democracy which was given much attention by its critics.

Thales, one of the founders of Greek science: a Roman copy of a Greek original, but since Thales lived before the age of classical portraiture his appearance remains conjectural.

Attacks on Athenian democracy began in early times and have continued ever since. They have embodied as much historical misrepresentation as have over-zealous and idealizing defences of the same institutions. The misgivings of frightened conservatives who had never seen anything like it before are understandable, for democracy emerged at Athens unexpectedly and at first almost unobserved. Its roots lay in sixth-century constitutional changes which replaced the organizing principle of kinship with that of locality; in theory and law, at least, local attachment came to be more important than the family you belonged to. This was a development which appears to have been general in Greece and it put democracy on the localised institutional basis which it has usually had ever since. Other changes followed from this. By the middle of the fifth century all adult males were entitled to take part in the assembly and through it, therefore, in the election of major administrative officers. The powers of the Areopagus were steadily reduced; after 462 BC it was only a law-court with jurisdiction over certain offences. The other courts were at the same time rendered more susceptible to democratic influence by the institution of payment for jury-service. As they also conducted much administrative business, this meant a fair amount of popular participation in the daily running of the city. Just after the Peloponnesian War, when times were hard, pay was also offered for attendance at the assembly itself. Finally, there was the Athenian belief in selecting by lot; its use for the choice of magistrates told against hereditary prestige and power.

At the root of this constitution lay distrust of expertise and entrenched authority and confidence in collective common sense. From this derived, no doubt, the relative lack of interest Athenians showed in rigorous jurisprudence – argument in an Athenian court was occupied much more with questions of motive, standing and substance, than with questions of law-and the importance they gave to the skills of oratory. The effective political leaders of Athens were those who could sway their fellow-citizens by their words. Whether we call them demagogues or orators does not matter; they were the first politicians seeking power by persuasion.

Towards the end of the fifth century, though even then by no means usually, some such men came from families outside the traditional ruling class. The continuing importance of old political families was nevertheless an important qualification of the democratic system. Themistocles at the beginning of the century and Pericles when the war began were members of old families, their birth making it proper for them even in the eyes of conservatives to take the lead in affairs; the old ruling classes found it easier to accept democracy because of this practical qualification of it. There is a rough parallel in the grudging acceptance of Whig reform by nineteenth-century English aristocrats; government in Athens as in Victorian England remained for a long time in the hands of men whose forefathers might have expected to rule the state in more aristocratic days. Another tempering qualification was provided by the demands of politics on time and money. Though jurors and members of the assembly might be paid, the fee for attendance was small; it seems to have been prompted, too, by the need to make sure of a quorum, which does not suggest that the assembly found it easy to get the mass of the citizens to attend. Many of them must have lived too far away and it has been calculated that not more than about one in eight of them were present at the usual statutory meetings, of which some forty were held each year. These facts tend to be lost to sight both in the denunciation and the idealization of Athenian democracy and they go some way to explaining its apparent mildness. Taxation was light and there was little discriminatory legislation against the rich such as we would now associate with democratic rule and such as Aristotle said would be the inevitable result of the rule of the poor.

Even in its emergent period Athenian democracy was identified with adventure and enterprise in foreign policy. Popular demand lay behind support for the Greek cities of Asia in their revolt against Persia. Later, for understandable reasons, it gave foreign policy an anti-Spartan bias. The struggle against the Areopagus was led by Themistocles, the builder of the Athenian fleet of Salamis, who had sensed a potential danger from Sparta from the moment the Persian War was over. Thus the responsibility for the Peloponnesian War and for its exacerbation of the factions and divisions of all the other cities of Greece came to be laid at the door of democracy. It not only brought disaster upon Athens itself, its critics pointed out, but exported to or at least awoke in all the Greek cities the bitterness of faction and social conflict. Oligarchy was twice restored in Athens – not that it helped matters – and by the end of the century faith in Athenian decmocracy was grievously weakened. Thucydides could take his history only down to 411 BC but it closes in misgiving and disillusion over his native city – which had exiled him – and Plato was to imprint for ever upon the Athenian democrats the stigma of the execution of Socrates in 399 BC.

If Athenian democracy's exclusion of women, metics and slaves is also placed in the scale, the balance against it seems heavy; to modern eyes, it looks both narrow and disastrously unsuccessful. Yet it should not disqualify Athens for the place she later won in the regard of posterity. Anachronistic and invalid comparisons are too easy; Athens is not to be compared with ideals still imperfectly realized after two thousand years, but with her contemporaries. For all the survival of the influence of the leading families and the practical impossibility that even a majority of its members would turn up to any particular meeting of the assembly, more Athenians were engaged in self-government than was the case in any other state. Athenian democracy more than any other institution brought about the liberation of men from the political ties of kin which is one of the great Greek achievements. Many who could not have contemplated office elsewhere could experience in Athens the political education of taking responsible decisions which is the heart of political culture. Men of modest means could help to run the institutions which nurtured and protected Athens' great civilized achievement. They listened to arguments of an elevation and thoughtfulness which makes it impossible to dismiss them as mere rhetoric; they must surely have weighed them seriously *sometimes*. Just as the physical divisions between the old Greek communities fostered a variety of experience which led in the end to a break with the world of god-given rulers and a grasp of the idea that political arrangements could be consciously chosen, so the stimulus of participation in affairs worked on unprecedentedly large numbers of men in classical Athens, not only in the assembly, but in the daily meetings of the peoples council which prepared its business. Even without the eligibility of all citizens to office Athenian democracy would still have been the greatest instrument of political education contrived down to that time.

It is against that background that the errors, vanities and misjudgements of Athenian politics must be seen. We do not cease to treasure the great achievements of British political culture because of the shallowness and corruptness of much of twentieth-century democracy. Athens may be judged, like any political system, by its working at its best; under the leadership of Pericles it was outstanding. It left behind the myth of the individual's responsibility for his own political fate. We need myths in politics and have yet to find a better.

The Athenians, in any case, would have been uninterested in many modern criticisms of their democracy. Its later defenders and attackers have both often fallen into another sort of anachronism, that of misinterpreting the goals Greeks thought worth achieving. Greek democracy, for example, was far from being dominated, as is ours, by the mythology of cooperativeness, and cheerfully paid a larger price in destructiveness than would be welcomed today. There was a blatant competitiveness in Greek life apparent from the Homeric poems onwards. Greeks admired men who won and thought men should strive to win. The consequent release of human power was colossal, but also dangerous. The ideal expressed in the much-used word which we inadequately translate as 'virtue' illustrates this. When Greeks used it, they meant that people were able, strong, quick-witted, just as much as just, principled, or virtuous in a modern sense. Homer's hero, Odysseus, frequently behaved like a rogue, but he is brave and clever and he succeeds; he is therefore admirable. To show such quality was good; it did not matter that the social cost might sometimes be high. The Greek was concerned with 'face'; his culture taught him to avoid shame rather than guilt and the fear of shame was never far from the fear of public evidence of guilt. Some of the explanation of the bitterness of faction in Greek politics lies here; it was a price willingly paid.

When all is said, Athenian democracy must be respected above all for what it cradled, a series of cultural triumphs which are peaks even in the history of Greek civilization. These were public facts. The art of Athens was applauded and sustained by many men; the tragedies were tested not by the takings of a box-office but by judges interpreting a public taste vigorously expressed. The sculptor Phidias worked to beautify the city and not for an individual patron. And as democracy degenerated, so it seems, there was a waning of artistic nerve. This was a loss to the whole of Greece.

The achievement which made Greece teacher of Europe (and through her of the world) is too rich and varied to generalize about even in long and close study; it is impossible to summarize in a page or so. But there is a salient theme which emerges in it: a growing confidence in rational, conscious enquiry. If civilization is advance towards the control of mentality and environment by reason, then the Greeks did more for it than any of their predecessors. They invented the philosophical question as part and parcel of one

Dionysus, depicted in about 500 BC, holds in his hands the vine-staff and the empty cup which symbolize his role as god of wine. To the Greeks he was also much more – the focus of a major religious cult. Probably Dionysus was in origin a nature god but his cult evolved into an interesting and psychologically fruitful complement to that of Apollo, the embodiment above all of the qualities of harmony, rationality, equity. The cult of Dionysus gave expressions to another side of human nature and an important one, for it stressed abandon, frenzy, even violence. It was celebrated in the winter months when Apollo was supposed to be absent from his shrine at Delphi; then troops of young women, Maenads as they were called, would follow the priests of Dionysus to remote places in the mountains for the secret rites of the cult, dancing in ecstasy, impervious to cold and exposure, destroying animals which fell in their way and, perhaps, human intruders too.

of the great intuitions of all time, that a coherent and logical explanation of things could be found, that the world did not ultimately rest upon the meaningless and arbitrary fiat of gods or demons. Put like that, of course, it is not an attitude which could be or was grasped by all, or even most, Greeks. It was an attitude which had to make its way in a world permeated with irrationality and superstition. Nevertheless, it was a revolutionary and beneficial idea. It looked forward to the possibility of a society where such an attitude would be generalized; even Plato, who thought it impossible that most men could share it, gave to the rulers of his ideal state the task of rational reflexion as the justification both of their privileges and of the discipline laid upon them. The Greek challenge to the weight of irrationality in social and intellectual activity tempered its force as it had never been tempered before. For all the subsequent exaggeration and myth-making about it, the liberating effect of this emphasis was felt again and again for thousands of years. It was the greatest single Greek achievement.

This was so big a revolution in modes of thought in the Aegean that it now obscures its own scale. So remarkable are the works of the Greek intellectuals and so large have they loomed that it requires effort to penetrate through them to the values of the world from which they emerged. It is made a little easier because no such revolution is ever complete. A look at the other side of the coin reveals that most Greeks continued to live in cocoons of traditional irrationality and superstition; even those who were in a position to understand something of the speculations which were opening new mental worlds rarely accepted the implications. A continuing respect was shown to the old public orthodoxies. It was impiety in late fifth-century Athens, for example, to deny belief in the gods. One philosopher believed that the sun was a red-hot disc; it did not protect him that he had been the friend of Pericles when he said so, and he had to flee. It was at Athens, too, that public opinion was convulsed, on the eve of the Sicilian Expedition, by the mysterious and ominous mutilation of certain public statues, the 'Hermae', or busts of Hermes. The disasters which followed were attributed by some to this sacrilege. Socrates, the Athenian philosopher who became, thanks to his pupil Plato, the archetypal figure of the man of intellect, and left as a maxim the view that 'the unexamined life is not worth living', offended the pieties of his state and was condemned to die for it by his fellow-citizens; he was also condemned for questioning received astronomy. It does not seem that similar trials took place elsewhere, but they imply a background of popular superstition which must have been more typical of the Greek community than the presence of a Socrates.

In spite of such important historical residues, Greek thought, more than that of any earlier civilization, always reflected changes of emphasis and fashion. They arose from its own dynamism and did not always lead to a greater ability to grapple with nature and society rather than surrender to them, but sometimes to dead ends and blind alleys, to exotic and extravagant fantasies. Greek thought is not monolithic; we should think not of a bloc with a unity pervading all its parts, but of a historical continuum extending across three or four centuries, in which different elements are prominent at different times and which is hard to assess.

One reason for this is that Greek categories of thought – the way, so to speak, in which they laid out the intellectual map before beginning to think about its individual components in detail at all – are not our own, though often deceptively like them. Some of those we use did not exist for the Greeks and their knowledge led them to draw different boundaries between fields of enquiry from those which we take for granted. Sometimes this is obvious and presents no difficulties; when a philosopher, for example, locates the management of the household and its estate (economics) as a part of a study of what we should call politics, we are not likely to misunderstand him. In more abstract topics it can cause trouble.

One example is to be found in Greek science. For us, science seems to be an appropriate way of approaching the understanding of the physical universe, and its techniques are those of empirical experiment and observation. Greek thinkers found the nature of the physical universe just as approachable through abstract thought, through metaphysics, logic and mathematics. It has been said that Greek rationality actually came in the end to stand in the way of scientific progress, because enquiry followed logic and abstract deduction, rather than the observation of nature. Among the great Greek philosophers, only Aristotle gave prominence to collecting and classifying data, and he did this for the most part only in his social and biological studies. This is one reason for not separating the history of Greek science and philosophy too violently. They are a whole, the product of scores of cities and developing across four centuries or so in time.

Their beginnings constitute a revolution in human thought and it has already taken place when there appear the earliest Greek thinkers of whom we have information. They lived in the Ionian city of Miletus in the seventh and sixth centuries BC. Important intellectual activity went on there and in other Ionian cities right down to the remarkable age of Athenian speculation which begins with Socrates. No doubt the stimulus of an Asian background was important here as in so many other ways in getting things started; it may also have been significant that Miletus was a rich place; early thinkers seem to have been rich men who could afford the time to think. None the less, the early emphasis on Ionia gives way before long to a spectrum of intellectual activity going on all over the Greek world. The western settlements of *Magna Graecia* and Sicily were crucial in many sixth- and fifth-century developments and primacy in the later Hellenistic age was to go to Alexandria. The whole Greek world was involved in the success of the Greek mind and even the great age of Athenian questioning should not be given exaggerated standing within it.

In the sixth century BC. Thales and Anaximander launched at Miletus the conscious speculation about the nature of the universe which shows that the crucial boundary between myth and science has been crossed. Egyptians had set about the practical manipulation of nature and had learned much inductively in the process, while Babylonians had made important measurements. The Miletan school made good use of this information, and possibly took more fundamental cosmological notions from the old civilizations, too; Thales is said to have held that the earth had its origin in water. Yet the Ionian philosophers soon went beyond their inheritance. They set out a general view of the nature of the universe which replaced myth with impersonal explanation. This is more impressive than the fact that the specific answers they put forward were in the end to prove unfruitful. The Greek analysis of the nature of matter is an example. Although an atomic theory was adumbrated which was over two thousand years before its time, this was by the fourth century rejected in favour of a view, based on that of the early Ionian thinkers, that all matter was composed of four 'elements' - air, water, earth, fire – which combined in different proportions in different substances. This theory subsequently dominated western science down to the Renaissance. It was of enormous historical importance because of the boundaries it set and the possibilities it opened. It was also, of course, erroneous.

This should be firmly kept in place as a secondary consideration at this point. What mattered about the Ionians and the school they founded was what has rightly been called their 'astonishing' novelty. They pushed aside gods and demons from the understanding of nature. Time was to overwhelm some of what they had done, it is true. In Athens in the late fifth century more than a temporary alarm in the face of defeat and danger has been seen in the condemnation as blasphemous of views far less daring than those of Ionian thinkers two centuries before. One of them had said 'If the ox could paint a picture, his god would look like an ox'; a few centuries later, classical Mediterranean civilization has lost much of such perceptiveness. Its early appearance is the most striking sign of the vigour of Greek civilization.

Not only popular superstition swamped such ideas. Other philosophical tendencies also played a part. One coexisted with the Ionian tradition for a long time and was to have much longer life and influence. Its crux was the view that reality was immaterial, that, as Plato later put it in one of its most persuasive expressions, in life we experience only the images of pure Form and Ideas, the heavenly embodiments of true reality, which can only be apprehended by thought, a matter not only of systematic speculation, but of intuition, too. For all its immateriality, this kind of thinking also had its roots in Greek science, though not in the speculations of the Ionians about matter but in the activities of mathematicians.

Some of their greatest advances were not to be made until long after Plato's death, when they would round off what is the biggest single triumph of Greek thought, its establishment of most of the arithmetic and geometry which served western civilization down to the seventeenth century. Every schoolboy used to know the name of Pythagoras, who lived at Crotone in southern Italy in the middle of the sixth century and may be said to have founded the deductive proof. Fortunately or unfortunately, he did more than this. He discovered the mathematical basis of harmonics by studying a vibrating string and he became especially interested in the relationship of numbers and geometry. His approach to them was semi-mystical; Pythagoras, like many mathematicians, was a religiously minded man who is said to have celebrated the satisfactory conclusion of his famous proof by sacrificing an ox. His school – there was a secret Pythagorean 'Brotherhood' – later came to hold that the ultimate nature of the universe was mathematical and numerical. 'They fancied that the principles of mathematics were the principles of all things,' reported Aristotle, somewhat disapprovingly, yet his own teacher, Plato, had been greatly influenced by this belief, and by the scepticism of Parmenides, an early fifth-century

Pythagorean, about the world known to the senses. Numbers seemed more attractive than the physical world; they possessed both the defined perfection and the abstraction of the Idea which embodied reality.

Beside the Parthenon on the Acropolis at Athens stands a small temple called the Erechtheum, or house of Erechtheus, one of the legendary kings of Athens. It contained a small wooden image of the goddess Athene supposed to have been placed there by the gods, and was a shrine as sacred as the Parthenon itself – from some points of view it was even more sacred. The south porch contains probably the finest examples of the female figures called Caryatids and used by the Greeks (as they had been earlier used in the Near East) as a variant on the usual columns for the support of architectural loads. How they came to be named is not exactly known, but it may have been as the result of a joke comparing them to the (allegedly) robust ladies of Caryae, a district of Sparta.

Pythagorean influence on Greek thought is an immense subject; fortunately, it need not be summarized. What matters here is its ultimate repercussions in a view of the universe which, because it was constructed on mathematical and deductive principles, rather than from observation, fixed astronomy on the wrong lines for nearly two thousand years. From it came the vision of a universe built up of successively enclosing spheres on which moved sun, moon and planets in a fixed and circular pattern about the earth. The Greeks noticed that this did not seem to be the way the heavens moved in practice. But – to summarize crudely – appearances were saved by introducing more and more refinements into the basic scheme while refusing to scrutinize the principles from which it was deduced. The final elaborations were not achieved until work in the second century AD by a famous Alexandrian, Ptolemy. These efforts were remarkably successful, and only a few dissentients demurred (which shows that other intellectual results were possible in Greek science). For all the inadequacies of Ptolemy's system, predictions of planetary movement could be made which would still serve as adequate guides for oceanic navigation in the age of Columbus, even if they rested on misconceptions which sterilized cosmological thinking until his day.

Both the theory of the four elements and the development of Greek astronomy illustrate the deductive bias of Greek thought and its characteristic weakness, its urge to set out a plausible theory to account for the widest possible range of experience without submitting it to the test of experiment. It affected most fields of thought which we now think to be covered by science and philosophy. Its fruits were on the one hand argument of unprecedented rigour and acuteness and on the other an ultimate scepticism about sense-data. Only the Greek doctors, led by the fifth-century Hippocrates, made much of empiricism.

In the case of Plato – and, for good or ill, philosophical discussion has been shaped more by him and his pupil Aristotle than by any other two men – this bias may have been reinforced by his low opinion of what he observed. By birth an aristocratic Athenian, he turned away from the world of practical affairs in which he had hoped to take part, disillusioned with the politics of the Athenian democracy, and, in particular, with its treatment of Socrates, whom it had condemned to death. From Socrates Plato had learnt not only his Pythagoreanism but an idealist approach to ethical questions, and a technique of philosophical enquiry. The Good, he thought, was discoverable by enquiry and intuition; it was reality. It was the greatest of a series of 'ideas' - Truth, Beauty, Justice were others – which were not ideas in the sense that at any moment they had shape in anyone's mind (as one might say 'I have an idea about that'), but were real entities, enjoying a real existence in a world fixed and eternal, of which such ideas were the elements. This world of changeless reality, thought Plato, was hidden from us by the senses, which deceived us and misled us. But it was accessible to the soul, which could understand it by the use of reason.

Such ideas had a significance going far beyond technical philosophy. In such ideas (like those of Pythagoras) can be found, for example, traces of a familiar later idea, that of puritanism, that man is irreconcilably divided between the soul, of divine origin, and the body which imprisons it. Not reconciliation, but the victory of one or another, must be the outcome. It was an idea which would pass into Christianity with enormous effect. Immediately, too, Plato had an intensely practical concern since he believed that knowledge of the Ideal world of universals and reality could be helped or hindered by the arrangements under which men lived. He set out his views in a series of dialogues between Socrates and people who came to argue with him. They were the first textbooks of philosophical thinking and the one we call *The Republic* was the first book in which anyone had ever set out a scheme for a society directed and planned to achieve an ethical goal. It describes an authoritarian state (reminiscent of Sparta) in which marriages would be regulated to produce the best genetic results, families and private property would not exist, culture and the arts would be censored and education carefully supervised. The few who ruled this state would be those of sufficient intellectual and moral stature to fit them for the studies which would enable them to realize the just society in practice by apprehending the Ideal world. Like Socrates, Plato held that wisdom was the understanding of reality and he assumed that to see truth ought to make it impossible not to act in accordance with it. Unlike his teacher, he held that for most people education and the laws should impose exactly that unexamined life which Socrates had thought not worth living.

The Republic and its arguments were to provoke centuries of discussion and imitation, but this was true of almost all Plato's work. As a twentieth-century English philosopher put it, practically all subsequent philosophy in the West was a series of footnotes to Plato. In spite of Plato's distaste for what he saw about him and the prejudice it engendered in him, he anticipated almost all the great questions of philosophy, whether they concerned morals, aesthetics, the basis of knowledge, or the nature of mathematics, and he set out his ideas in great works of literature, which have always been read with pleasure.

The Academy which Plato founded has some claim to be the first university. From it emerged his pupil Aristotle, a thinker more comprehensive and balanced, less sceptical of the possibilities of the actual, and less adventurous than he. Aristotle never altogether rejected his master's teaching but he departed from it in fundamental ways. He was a great classifier and collector of data (with a special interest in biology) and did not reject sense experience as did Plato. Indeed, he sought both firm knowledge and happiness in the world of experience, rejecting the notion of universal ideas and arguing inductively from facts to general laws. Aristotle was so rich a thinker and interested in so many sides of experience that his historical influence is as hard to delimit as that of Plato. What he wrote provided a framework for the discussion of biology, physics, mathematics, logic, literary criticism, aesthetics, psychology, ethics and politics for two thousand years. He provided ways of thinking about these subjects and approaches to them which were elastic and capacious enough eventually to contain Christian philosophy. He also founded a science of deductive logic which was not displaced until the end of the nineteenth century. It is a vast achievement, different in kind but not less important than that of Plato.

Aristotle's political thinking was in one sense in agreement with Plato's: the city-state was the best conceivable social form, but required reform and purification to work properly, he thought. But beyond this point he diverged greatly from his master. Aristotle saw the proper working of the *polis* as being that which would give each of its parts the role appropriate to it and that was essentially for him a matter of understanding what led in most existing states to happiness. In formulating an answer, he made use of a Greek idea to which his teaching was to give long life, that of the Mean, the idea that excellence lay in a balance between extremes. The empirical facts seemed to confirm this and Aristotle assembled greater quantities of such evidence in a systematic form than any predecessor, it seems; but in stressing the importance of facts about society, he had been anticipated by another Greek invention, that of history.

This was another major achievement. In most countries, chronicles or annals which purport simply to record successions of events precede history. In Greece, this was not so. Historical writing in Greek emerged from poetry. Amazingly, it at once reached its highest level in its first embodiments, two books by masters which were never equalled by their successors. The first of them, Herodotus, has reasonably been termed 'the father of history'. The word – *historie* – existed before him; it meant enquiry. Herodotus gave it an added meaning, that of enquiry about events in time, and in putting down the results wrote the first prose work of art in a European language which survives. His stimulus was a wish to understand a near-contemporary fact, the great struggle with Persia. He accumulated information about the Persian Wars and their antecedents by reading a huge mass of the available literature and by interrogating people on his travels and assiduously recording what he was told and read. For the first time, these things became the subject of more than a chronicle. The result is his *Histories*, a remarkable account of the Persian empire, with, built into it, much information about early Greek history and a sort of world survey, followed by an account of the Persian Wars down to Mycale. He spent much of his life travelling, having been born (it was traditionally said) in the Dorian town of Halicarnassus in south-west Asia Minor in 484 BC. At one point he came to Athens where he remained for a few years living as a metic, and while there he may have been rewarded for public recitations of his work. He went later to a new colony in south Italy; there he completed his work and died, a little after 430 BC. He therefore knew something by experience of the whole spread of the Greek world and travelled in Egypt and elsewhere as well. Thus wide experience lay behind his great book, an account scrupulously based on witnesses, even if Herodotus sometimes treated them somewhat credulously.

It is usually conceded that one of the superiorities of Thucydides, Herodotus' greater successor, was his more rigorous approach to reports of fact and his attempts to control them in a critical way. The result is a more impressive intellectual achievement, though its austerity throws into even stronger relief the charm of Herodotus' work. Thucydides' subject was even more contemporary, the Peloponnesian War. The choice reflected deep personal involvement and a new conception. Thucydides was a member of a leading Athenian family (he served as a general until disgraced for an alleged failure in command) and he wanted to discover the causes which had brought his city and Greece into their dreadful plight. He shared with Herodotus a practical motive, for he thought (as most Greek historians were to do after him) that what he found out would have practical value, but he sought not merely to describe, but to explain. The result is one of the most striking pieces of historical analysis ever written and the first ever to seek to penetrate through different levels of explanation. In the process he provided a model of disinterested judgement to future historians, for his Athenian loyalties rarely obtrude. The book was not completed – it takes the story only to 411 BC – but the overall judgement is concise and striking: 'the growth of Athens' power and Sparta's fear was, in my view, the cause which compelled them to go to war'.

The invention of history is itself evidence of the new intellectual range of the literature created by the Greeks. It is the first complete one known to man. The Jewish is almost as comprehensive, but contains neither drama nor critical history, let alone the lighter *genres*. But Greek literature shares with the Bible a primacy shaping the whole of subsequent western writing. Besides its positive content, it imposed the major forms of literature and the first themes of a criticism by which to judge them.

From the beginning, as Homer shows, it was closely linked to religious belief and moral teaching. Hesiod, a poet who probably lived in the late eighth century and is usually considered to be the first Greek poet of the post-epic age, consciously addressed himself to the problem of justice and the nature of the gods, thus confirming the tradition that literature was for more than enjoyment and setting out one of the great themes of Greek literature for the next four centuries. For the Greeks, poets were always likely to be seen as teachers, their work suffused with mystical overtones, inspiration. Yet there were to be many poets, many styles of poetry in Greek. The first which can be distinguished is writing in a personal vein which was to the taste of aristocratic society. But as private patronage became concentrated during the era of the tyrants, so it passed slowly into the collective and civic area. The tyrants deliberately fostered the public festivals which were to be vehicles of the greatest specimens of Greek literary art, the tragedies. The drama's origins lie everywhere in religion and its elements must have been present in every civilization. The ritual of worship is the first theatre. Yet there, too, the Greek achievement was to press this towards conscious reflexion on what was going forward; more was to be expected of the audience than passive resignation or orgiastic possession. The didactic impulse emerges in it.

The first form of the Greek drama was the dithyramb, the choral song recited at the festivals of Dionysus, together with dance and mime. In 535 BC, we are told, this was the subject of a crucial innovation, when Thespis added to it an individual actor whose speech was some kind of antiphone to the chorus. Further innovation and more actors followed and within a hundred years we have reached the full, mature theatre of Aeschylus, Sophocles and Euripides. Of their work thirty-three plays survive (including one complete trilogy), but we know that more than three hundred different tragedies were performed in the fifth century. In this drama the religious undertone is still there, though not so much in the words as in the occasions at which they would have been performed. The great tragedies were sometimes performed in trilogies at civic festivals attended by citizens who were already familiar with the basic stories (often mythological) they had come to see. This, too, suggests the educational effect. Probably most Greeks never saw a play by Aeschylus; certainly an infinitesimally small number by comparison with the number of modern Englishmen who have seen a play of Shakespeare. None the less, those who were not too busy on their farms, or too far away, provided a large audience.

More men than in any other ancient society were able to scrutinize and reflect upon the content of their own moral and social world. What they expected was a revealing emphasis in familiar rites, a new selection from their meaning, and this is what

The inspiration of Greek sculpture came first from Asia, then from Egypt. The basic posture of the statues known as kouroi – youths – reflects this. The four-square stance is meant to be seen from the front; one leg is in advance of the other, as if in a walking position. The rigidity of the posture, the clenched fists and set smile, all characterized this type of statue from its first appearance in the Greek islands (this example is from Melos) until it gave way to a more relaxed, naturalistic manner at the beginning of the fifth century. Such statues stood over graves or as dedications to gods; they were not portraits, but figures only one degree removed in their nature from the anonymous cult figures of shrines.

the great dramatists mostly gave them, even if some plays went beyond this and some even, at favourable moments, satirized the social pieties. It was not, of course, a naturalistic picture that was presented, but the operation of the laws of a heroic, traditional world and their agonizing impact on individuals caught in their working. In the second half of the fifth century Euripides had even begun to use the conventional tragic form as a vehicle for questioning conventional assumptions; thus he inaugurated a technique to be exploited in the western theatre by authors as late and as different as Gogol and Ibsen. The framework provided by plot, though, was familiar, and at its heart lay a recognition of the weight of inexorable law and *nemesis*. The acceptance of this setting may be thought, in the last resort, to be testimony to the irrational rather than the rational side of the Greek mind. Yet it was a long way from the state of mind in which the congregation of an eastern temple fearfully or hopefully witnessed the round of unchanging ritual and sacrifice.

In the fifth century the scope of the theatre was also broadening in other ways. This was when Attic comedy developed as a form in its own right, and found in Aristophanes its first great manipulator of men and events for others' amusement. His material was often political, almost always highly topical, and frequently scurrilous. His survival and success is the most striking evidence we possess of the tolerance and freedom of Athenian society. A hundred years later, we have almost reached the modern world in a fashion for plays about the intrigues of slaves and troubled love-affairs. It has not the impact of Sophocles, but it can still amuse and remains a near-miracle, for nothing like it had been there two hundred years before. The rapidity with which Greek literature grew after the age of epic poetry and its enduring power is evidence of Greek powers of innovation and mental development which is easy to appreciate even when we cannot explain it.

Literature at the end of the classical age still had a long and important life ahead when the city-states disappeared. It had a growing audience, for Greek was to become both *lingua franca* and an official language over all the Near East and much of the Mediterranean. It was not to reach again the heights of Athenian tragedy, but it was still to show us masterpieces. The sense of decline in the visual arts is more apparent. Here, above all in monumental architecture and the nude, Greece had again set standards for the future. From the first borrowings from Asia a wholly original architecture was evolved, the classical style whose elements are still consciously evoked even by the austerities of twentieth-century builders. Within a few hundred years it spread over much of the world from Sicily to India; in this art, too, the Greeks were cultural exporters.

They were in one respect favoured by geology, for Greece contained much high-quality stone. Its durability is attested by the magnificence of the relics we look at today. Yet there is an illusion in this. The purity and austerity with which fifth-century Athens speaks to us in the Parthenon conceals its image in Greek eyes. We have lost the garish statues of gods and goddesses, the paint and ochre and the clutter of monuments, shrines and *stelae* that must have encumbered the Acropolis and obscured the simplicity of its temples. The reality of many great Greek centres may have been more like, say, modern Lourdes; a jumble of untidy little shrines cluttered by traders, booths, and the rubbish of superstition is the impression gained when approaching the Temple of Apollo at Delphi (though we must make allowances for the contribution made by the archaeologist to this impression).

None the less, this qualification made, the erosion of time has allowed a beauty of form which is almost unequalled to emerge from the superficial experience. There is no possibility here of discounting the interplay of judgement of the object with standards of judgement which derive ultimately from the object itself. It remains simply true that to have originated an art that has spoken so deeply and powerfully to men's minds across such ages is itself not easily interpreted except as evidence of an unsurpassed artistic greatness and an astonishing skill in giving it expression.

This quality is also present in Greek sculpture. Here, too, the presence of good stone was an advantage, and the original influence of oriental, often Egyptian, models important. Like pottery, the eastern models once absorbed, sculpture evolved towards greater naturalism. The supreme subject of the Greek sculptors was the human form, portrayed no longer as a memorial or cult object, but for its own sake. Again it is not always possible to be sure of the finished statue the Greeks saw; these figures were often gilded, painted or decorated with ivory and precious stones. Some bronzes have undergone looting or melting down, so that the preponderance of stone may itself be misleading. Their evidence, though, records a clear evolution. We begin with statues of gods and of young men and women whose identity is often unknown, simply and symmetrically presented in poses not too far removed from those of the orient. In the classical figures of the fifth century, naturalism begins to tell in an uneven distribution of weight and the abandonment of the simple frontal stance and to evolve towards the mature, human style of Praxiteles and the fourth century in which the body - and for the first time the female nude – is treated.

A great culture is more than a mere museum and no civilization can be reduced to a catalogue. For all its élite quality, the achievement and importance of Greece comprehended all sides of life; the politics of the city-state, a tragedy of Sophocles and a statue by Phidias are all part of it. Later ages grasped this intuitively, happily ignorant of the conscientious discrimination which historical scholarship made possible between periods and places. This was a fruitful error, because in the end what Greece was to be thought to be was as important to the future as what she was. The meaning of the Greek experience was to be represented and reinterpreted, and ancient Greece was to be rediscovered and reconsidered and, in different ways, reborn and re-used, for more than two thousand years. For all the ways in which reality had fallen short of later idealization and for all the strength of ties with past, Greek civilisation was quite simply the most important extension of Man's grasp of his own destiny down to that time. Within four centuries, Greece invented philosophy, politics, most of arithmetic and geometry, and the categories of western art. It would be enough, even if her errors, too, had not been so fruitful. Europe has drawn interest on the capital Greece laid down ever since, and through Europe the rest of the world has traded on the same account.

4
The Hellenistic World

The history of Greece rapidly becomes less interesting after the fifth century. It is also less important. What remains important is the history of Greek civilization and the shape of this, paradoxically, was determined by a kingdom in northern Greece which some said was not Greek at all: Macedon. In the second half of the fourth century it created an empire bigger than any yet seen, the legatee of both Persia and the city-states. It organized the world we call Hellenistic because of the preponderance and uniting force within it of a culture. Greek in inspiration and language. Yet Macedon was a barbarous place, perhaps centuries behind Athens in the quality of its life and culture.

The story begins with the decline of Persian power. Persian recovery in alliance with Sparta had masked important internal weaknesses. One of them is commemorated by a famous book, the *Anabasis* of Xenophon, the story of the long march of an army of Greek mercenaries back up the Tigris and across the mountains to the Black Sea after an unsuccessful attempt on the Persian throne by a brother of the king. This was only a minor and subsidiary episode in the important story of Persian decline, an offshoot of one particular crisis of internal division. Throughout the fourth century that empire's troubles continued, province after province (among them Egypt which won its independence as early as 404 and held it for sixty years) slipping out of control. A major revolt by the western satraps took a long time to master and though in the end imperial rule was restored the cost had been great. When at last reimposed, Persian rule was often weak.

One ruler tempted by the possibilities of this decline was Philip II of Macedon, a not very highly regarded northern state whose power rested on a warrior aristocracy; it was a rough, tough society, its rulers still somewhat like the warlords of Homeric times, their power resting more on personal ascendancy than institutions. Whether this was a state which was a part of the world of the Hellenes was disputed; some Greeks thought Macedonians barbarians, though their kings claimed descent from Greek houses (one going back to Heracles) and their claim was generally recognized. Philip himself sought status; he wanted Macedon to be thought of as Greek. When he became regent of Macedon in 359 BC he began a steady acquisition of territory at the expense of other Greek states. His ultimate argument was an army which became by the end of his reign the best-trained and organized in Greece. The Macedonian military tradition had emphasized heavy, armoured cavalry, and this continued to be a major arm. Philip added to this tradition the benefit of lessons about infantry he had drawn while a hostage at Thebes in his youth. From hoplite tactics he evolved a new weapon, the sixteen-deep phalanx of pikemen. The men in its ranks carried pikes twice as long as a hoplite spear and they operated in a more open formation, pike shafts from the second and third ranks running between men in the front to present a much denser array of weapons for the charge. Another advantage of the Macedonians was a grasp of siege-warfare techniques not shown by other Greek armies; they had catapults which made it possible to force a besieged town's defenders to take cover while battering-rams, mobile towers and mounds of earth were brought into play. Such things had previously been seen only in the armies of Assyria and their Asian successors. Finally, Philip ruled a fairly wealthy state, its riches much increased once he had acquired the goldmines of Mount Pangaeum, though he spent so much that he left huge debts.

For more than two thousand years the story of Alexander the great conqueror fascinated men who dreamed they might carry out exploits as astonishing. No doubt it added to its glamour that Alexander's physical image had to be evoked from representations of him of doubtful authenticity. This one is at least second-hand – a copy in mosaic by a Roman artist of a Hellenistic painting long since perished. The original may have been near-contemporary (it may even have been commissioned by Cassander, the king of Macedon, murderer of Alexander's mother, wife and heir); by the time it was embodied in this version it is more an idealization of a soldier, pictured in the vigour of his onslaught on the Persians, than a portrait. The huge mosaic at Pompeii (sixteen feet by eight) from which this detail is taken was no doubt a copy of one painting in a series which represented the whole cycle of Alexander's almost immediately legendary life.

He used his power first to ensure the effective unification of Macedon itself. Within a few years the infant king for whom he was regent was deposed and Philip was elected king. Then he began to look to the south and north-east. In these areas expansion sooner or later meant encroachment upon the interests and position of Athens. Her allies in Rhodes, Cos, Chios and Byzantium placed themselves under Macedonian patronage. Another, Phocis, went down in a war in which Athens had egged her on but failed to give effective support. Although Demosthenes, the last great agitator of Athenian democracy, made himself a place in history, which is recalled by the word 'philippic', by warning his countrymen of the dangers they faced, he could not save them. When a war between others and Macedon (355–346 BC) at last ended, Philip had won not only Thessaly, but had established himself in central Greece and controlled the pass of Thermopylae.

His situation favoured designs on Thrace and this implied a return of Greek interest towards Persia. One Athenian writer advocated a Hellenic crusade to exploit Persia's weakness (in opposition to Demosthenes who continued to denounce the Macedonian 'barbarian'), and once more plans were made to liberate the Asian cities, a notion attractive enough to bear fruit in a reluctant League of Corinth formed by the major Greek states other than Sparta in 337 BC. Philip was its president and general and it was somewhat reminiscent of the Delian League; the apparent independence of its members was a sham, for they were Macedonian satellites. Though the culmination of

Philip's work and reign (he was assassinated the following year), it had only come into being after Macedon had defeated the Athenians and Thebans in 338 BC. The terms of peace imposed by Philip were not harsh, but the League had to agree to go to war with Persia under Macedonian leadership. There was one more kick of Greek independence after Philip's death, but his son and successor Alexander crushed the Greek rebels as he did others in other parts of his kingdom. Thebes was then razed to the ground and its population enslaved (335 BC).

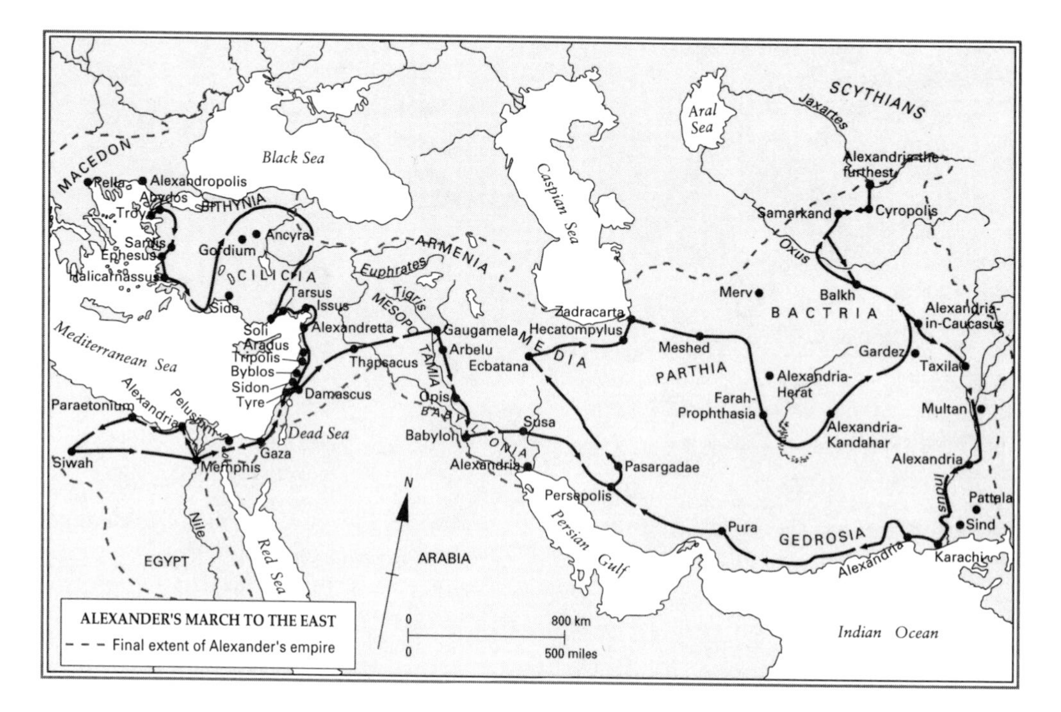

This was the real end of four centuries of Greek history. During them civilization had been created and sheltered by the city-state, one of the most successful political forms the world has ever known. The immediate future for Greece was Macedonian governors and garrisons. Not for the first time nor the last, the future seemed to belong to the bigger battalions, the bigger organizations. Mainland Greece was from this time a political backwater. Like his father, Alexander sought to conciliate the Greeks by giving them a large measure of internal self-government in return for adherence to his foreign policy. This was always to leave some Greeks, notably the Athenian democrats, unreconciled. When Alexander died, Athens once more tried to organize an anti-Macedonian coalition. The results were disastrous. A part of the price of defeat was the replacement of democracy by oligarchy at Athens (322 BC); Demosthenes fled to an island off the coast, seeking sanctuary in the temple of Poseidon there, but poisoned himself when the Macedonians came for him. A Macedonian governor henceforth ruled the Peloponnese.

Alexander's reign had thus begun with difficulties, but once they were surmounted, he could turn his attention to Persia. In 334 BC he crossed to Asia at the head of an army of which a quarter was drawn from Greece. There was more than idealism in this; aggressive war might also be prudent, for the fine army left by Philip had to be paid if it was not to present a threat to a new king, and conquest would provide the money. He was twenty-two years old and before him lay a short career of conquest so brilliant that it would leave his name a myth down the ages and provide a setting for the widest expansion of Greek culture. He drew the city-states into a still wider world.

The story is simple to summarize. Legend says that after crossing to Asia Minor he cut the Gordian Knot. He then defeated the Persians at the battle of Issus. This was followed by a campaign which swept south through Syria, destroying Tyre on the way, and eventually to Egypt, where Alexander founded the city still bearing his name. In every battle he was his own best soldier and he was wounded several times in the mêlée. He pushed into the desert, interrogated the oracle at Siwah and then went back into Asia to inflict a second and decisive defeat on Darius III in 331 BC. Persepolis was sacked and burnt and Alexander proclaimed successor to the Persian throne; Darius was murdered by one of his satraps the next year. On went Alexander, pursuing the Iranians of the north-east into Afghanistan (where Kandahar, like many cities elsewhere, commemorates his name) and penetrating a hundred miles or so beyond the Indus into the Punjab. Then he turned back because his army would go no further. It was tired and having defeated an army with 200 elephants may have been disinclined to face a further 5,000 reported to be waiting for it in the Ganges valley. Alexander returned to Babylon. There he died in 323 BC, thirty-two years old and just ten years after he had left Macedon.

Both his conquests and their organization in empire bear the stamp of individual genius; the word is not too strong, for achievement on this scale is more than the fruit of good fortune, favourable historical circumstance or blind determinism. Alexander was a creative mind, but self-absorbed, obsessed with his pursuit of glory, and something of a visionary. With great intelligence he combined almost reckless courage; he believed his mother's ancestor to be Homer's Achilles and strove to emulate the hero. He was ambitious as much to prove himself in men's eyes - or perhaps those of his forceful and repellent mother – as to win new lands. The idea of the Hellenic crusade against Persia undoubtedly had reality for him, but he was also, for all his admiration of the Greek culture of which he had learnt from his tutor Aristotle, too egocentric to be a missionary, and his cosmopolitanism was grounded in an appreciation of realities. His empire had to be run by Persians as well as Macedonians. Alexander himself married first a Bactrian and then a Persian princess, and accepted – unfittingly, thought some of his companions – the homage which the East rendered to rulers it thought to be godlike. He was also at times rash and impulsive; it was his soldiers who finally made him turn back at the Indus, and the ruler of Macedon had no business to plunge into battle with no attention for what would happen to the monarchy if he should die without a successor. Worse still, he killed a friend in a drunken brawl and he may have arranged his father's murder.

Alexander lived too short a time either to ensure the unity of his empire in the future or to prove to posterity that even he could not have held it together for long. What he did in this time is indubitably impressive. The foundation of twenty-five 'cities' is by itself a considerable matter, even if some of them were only spruced-up strongpoints; they were keys to the Asian land routes. The integration of east and west in their government was still more difficult, but Alexander took it a long way in ten years. Of course, he had little choice; there were not enough Greeks and Macedonians to conquer and govern the huge empire. From the first he ruled through Persian officials in the conquered areas and after coming back from India he began the reorganization of the army in mixed regiments of Macedonians and Persians. His adoption of Persian dress and his attempt to exact prostration – an obligatory kow-tow like that which so many Europeans in recent times found degrading when it was asked for by Chinese rulers – from his compatriots as well as from Persians, also antagonized his followers, for they revealed his taste for oriental manners. There were plots and mutinies; they were not successful, and his relatively mild reprisals do not suggest that the situation was ever very dangerous for Alexander. The crisis was followed by his most spectacular gesture of cultural integration when, himself taking Darius' daughter as a wife (in addition to his Bactrian princess, Roxana), he then officiated at the mass wedding of 9,000 of his soldiers to eastern women. This was the famous 'marriage of East and West', an act of state rather than of idealism, for the new empire had to be cemented together if it was to survive.

What the empire really meant in cultural interplay is more difficult to assess. There was certainly a wider physical dispersal of Greeks. But the results of this were only to appear after Alexander's death, when the formal framework of empire collapsed and yet the cultural fact of a Hellenistic world emerged from it. We do not in fact know very

A statue of Aphrodite found on the isle of Melos, and possibly the best-known work of sculpture surviving from the Hellenistic age.

much about life in Alexander's empire and it must be unlikely, given its brief duration, the limitations of ancient government and a lack of will to embark upon fundamental change, that most of its inhabitants found things very different in 323 BC from what they had been ten years before.

Alexander's impact was made in the east. He did not reign long enough to affect the interplay of the western Greeks with Carthage which was the main preoccupation of the later fourth century in the west. In Greece itself things stayed quiet until his death. It was in Asia that he ruled lands no Greeks had ruled before. In Persia he had proclaimed himself heir to the Great King and rulers in the northern satrapies of Bithynia, Cappadocia and Armenia did him homage.

Weak as the cement of the Alexandrine empire must have been, it was submitted to intolerable strain when he died without a competent heir. His generals fell to fighting for what they could get and keep, and the empire was dissolving even before the birth of his posthumous son by Roxana. She had already murdered his second wife, so when she and her son died in the troubles any hope of direct descent vanished. In forty-odd years of fighting it was settled that there would be no reconstitution of Alexander's empire. There emerged instead in the end a group of big states, each of them a hereditary monarchy. They were founded by successful soldiers, the *diadochi*, or 'Successors'.

Ptolemy Soter, one of Alexander's best generals, had at once seized power in Egypt at his master's death and to it he subsequently conveyed the valuable prize of Alexander's body. Ptolemy's descendants were to rule the province for nearly three hundred years until the death of Cleopatra in 30 BC. Ptolemaic Egypt was the longest-lived and richest of the successor states. Of the Asian empire, the Indian territories and some of Afghanistan passed out of Greek hands altogether, being ceded to an Indian ruler in return for military help. The rest of it was by 300 BC a huge kingdom of one and a half million square miles and perhaps thirty million subjects, stretching from Afghanistan to Syria, the site of its capital, Antioch. This vast domain was ruled by the descendants of Seleucus, another Macedonian general. Attacks by migrating Celts from northern Europe (who had already invaded Macedonia itself) led to its partial disruption early in the third century BC and part of it thenceforth formed the kingdom of Pergamon, ruled by a dynasty called the Attalids, who pushed the Celts further into Asia Minor. The Seleucids kept the rest, though they were to lose Bactria in 225 BC, where descendants of Alexander's soldiers set up a remarkable Greek kingdom. Macedon, under another dynasty, the Antigonids, strove to retain a control of the Greek states contested in the Aegean by the Ptolemaic fleet and in Asia Minor by the Seleucids. Once again, about 265 BC, Athens made a bid for independence but failed.

These events are complicated, but not very important for our purpose. What mattered more was that for about sixty years after 280 BC the Hellenistic kingdoms lived in a rough balance of power, preoccupied with events in the eastern Mediterranean and Asia and, except for the Greeks and Macedonians, paying little attention to events further west. This provided a peaceful setting for the greatest extension of Greek culture and this is why these states are important. It is their contribution to the diffusion and growth of a civilization that constitute their claim on our attention, not the obscure politics and unrewarding struggles of the *diadochi*.

Greek was now the official language of the whole Near East; even more important, it was the language of the cities, the foci of the new world. Under the Seleucids the union of Hellenistic and oriental civilization to which Alexander may have aspired began to be a reality. They urgently sought Greek immigrants and founded new cities wherever they could as a means of providing some solid framework for their empire and of Hellenizing the local population. The cities were the substance of Seleucid power, for beyond them stretched a heterogeneous hinterland of tribes, Persian satrapies, vassal princes. Seleucid administration was still based fundamentally upon the satrapies; the theory of absolutism was inherited by the Seleucid kings from the Achaemenids just as was their system of taxation. Yet it is not certain what this meant in practice and the east seems to have been less closely governed than Mesopotamia and Asia Minor, where Hellenistic influence was strongest and the capital lay. The size of the Hellenistic cities here far surpassed those of the older Greek emigrations; Alexandria, Antioch and the new capital city, Seleucia, near Babylon, quickly achieved populations of between one and two hundred thousand.

This reflected economic growth as well as conscious policy. The wars of Alexander and his successors released an enormous booty, much of it in bullion, accumulated by the Persian empire. It stimulated economic life all over the Near East, but also brought the evils of inflation and instability. Nevertheless, the overall trend was towards greater wealth. There were no great innovations, either in manufacture or in the tapping of new natural resources. The Mediterranean economy remained much what it had always been except in scale, but Hellenistic civilization was richer than its predecessors and population growth was one sign of this.

Its wealth sustained governments of some magnificence, raising large revenues and spending them in spectacular and sometimes commendable ways. The ruins of the Hellenistic cities show expenditure on the appurtenances of Greek urban life; theatres and gymnasia abound, games and festivals were held in all of them. This probably did not much affect the native populations of the countryside who paid the taxes and some of them resented what would now be called 'westernization'. None the less, it was a solid achievement. Through the cities the east was Hellenized in a way which lasted until the coming of Islam. Soon they produced their own Greek literature.

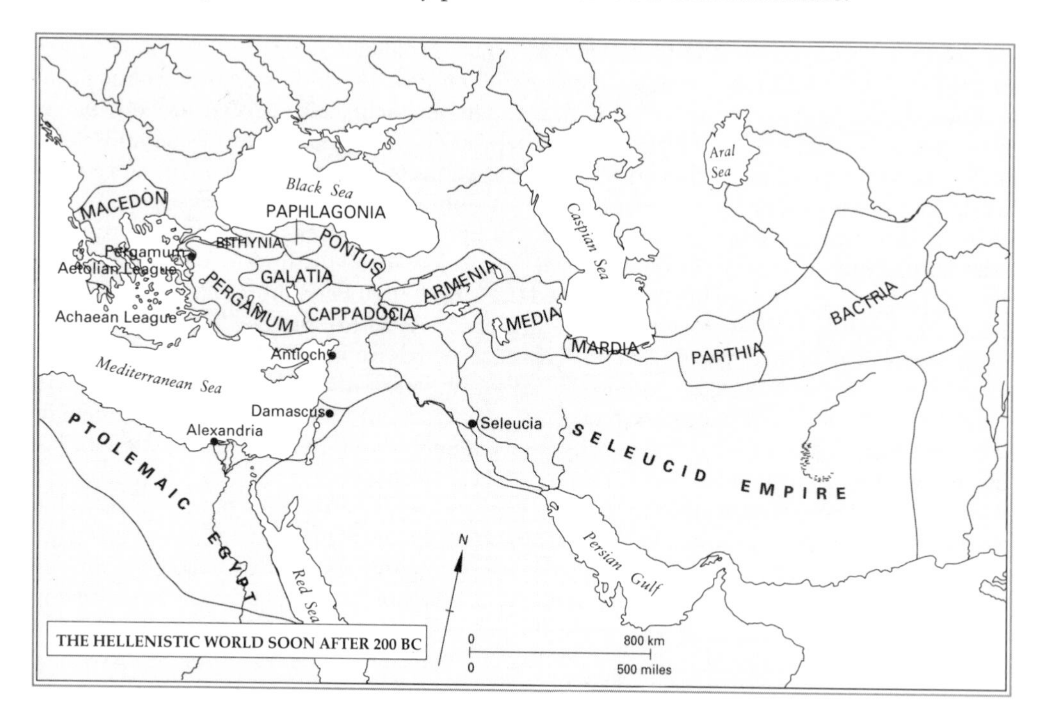

THE HELLENISTIC WORLD SOON AFTER 200 BC

Yet though this was a civilization of Greek cities, it was in spirit unlike that of the past, as some Greeks noted sourly. The Macedonians had never known the life of the city-state and their creations in Asia lacked its vigour; the Seleucids founded scores of cities but maintained the old autocratic and centralized administration of the satrapies above that level. Bureaucracy was highly developed and self-government languished. Ironically, besides having to bear the burden of disaster in the past, the cities of Greece itself, where a flickering tradition of independence lived on, were the one part of the Hellenistic world which actually underwent economic and demographic decline.

Though the political nerve had gone, city culture still served as a great transmission system for Greek ideas. Large endowments provided at Alexandria and

Pergamon the two greatest libraries of the ancient world. Ptolemy I also founded the Museum, a kind of institute of advanced study. In Pergamon a king endowed schoolmasterships and it was there that people perfected the use of parchment (*pergamene*) when the Ptolemies cut off supplies of papyrus. In Athens the Academy and the Lyceum survived and from such sources the tradition of Greek intellectual activity was everywhere refreshed. Much of this activity was academic in the narrow sense that it was in essence commentary on past achievement, but much of it was also of high quality and now seems lacking in weight only because of the gigantic achievements of the fifth and fourth centuries. It was a tradition solid enough to endure right through the Christian era, though much of its content has been irretrievably lost. Eventually, the world of Islam would receive the teaching of Plato and Aristotle through what had been passed on by Hellenistic scholars.

Hellenistic civilization preserved the Greek tradition in nothing more successfully than in science, and here Alexandria, the greatest of all Hellenistic cities, was pre-eminent. Euclid was the greatest systematizer of geometry, defining it until the nineteenth century, and Archimedes, who is famous for his practical achievements in the construction of war-machines in Sicily, was probably Euclid's pupil. Another Alexandrian, Eratosthenes, was the first man to measure the size of the earth, and yet another, Hero, is said to have invented a steam engine and certainly used steam to transmit energy. It is inconceivable that the state of contemporary metallurgy could ever have made the widespread application of this discovery practicable, which probably explains why we hear no more of it. The point is of general relevance; the intellectual achievements of the ancient world (and of European medieval civilization later) often pushed up to the limits of existing technical skills but could not be expected to go beyond them; further progress had to wait for better instrumentation. Another Hellenistic Greek, Aristarchus of Samos, got so far as to say that the earth moved round the sun, though his views were set aside by contemporaries and posterity because they could not be squared with Aristotelian physics which stated the contrary; the truth or falsity of both views remained untested experimentally. In hydrostatics, it is true, Archimedes made great strides (and invented the windlass, too) but the central achievement of the Greek tradition was always mathematical, not practical, and in Hellenistic times it reached its apogee with the theory of conic sections and ellipses and the founding of trigonometry.

These were important additions to humanity's toolkit. Yet they were less distinct from what went before than was Hellenistic moral and political philosophy. It is tempting to find the reason for this in the political change from the city-state to larger units. It was still in Athens that the philosophy of the age found its greatest centre and Aristotle had hoped to reinvigorate the city-state; in the right hands, he thought, it could still provide the framework for the good life. The unhappy last age of the city-state after the Peloponnesian War and the size and impersonality of the new monarchies must have soon sapped such confidence. In them, the old patriotic impulse of the city-states had dried up. Efforts were made to find other ways of harnessing public loyalty and emotion. Perhaps because of the need to impress non-Greeks, perhaps because they felt the positive attraction of the world beyond Greek culture, the new monarchs buttressed themselves more and more with oriental cults attached to the person of the ruler whose origins went back into the Mesopotamian and Egyptian past. Extravagant titles were employed but perhaps much of this was flattery: 'Soter', as Ptolemy I was called, meant 'Saviour'. The Seleucids allowed themselves to be worshipped, but the Ptolemies outdid them; they took over the divine status and prestige of the Pharaohs (and practice, too, to the extent of marrying their sisters). Meanwhile, the real basis of the Hellenistic states was bureaucracy unchecked by traditions of civic independence – since the Seleucids had founded or refounded most of the Greek cities in Asia, what they had given they could take back – and armies of Greek and Macedonian mercenaries which relieved them of dependence on native troops. Powerful and awe-inspiring though they might be, there was little in such structures to capture their very mixed subjects' loyalties and emotions.

Twenty feet high, this huge statue found near Luxor evokes in its style and stance the images of the Egyptian kings. Yet its face is curiously un-Egyptian. Made in Ptolemaic times, it is now thought to represent Alexander the Great.

Probably the erosion of those emotions had gone too far even before Alexander. The triumph of Greek culture was deceptive. Language went on being used, but with a different meaning. Greek religion, for example, a great force for unity among Hellenes, rested not on ecclesiastical institutions but on respect for the Homeric gods and goddesses and the behaviour they exemplified. Beyond this, there were the city cults and official mysteries. This had already begun to change, possibly as early as the fifth century, when, under the impact of the prolonged war, the Olympian gods began to lose the respect paid to them. There was more than one cause of this. The rationalism of much Greek fourth-century philosophy is as much a part of the story as the rise of new fears. With the Hellenistic age another influence is felt, that of a pervasive irrationality, of the pressure of fortune and fate. Men sought reassurance in new creeds and faiths. The popularity of astrology was one symptom. All this only came to its climax as late as the first century BC, 'the period', says one scholar, 'when the tide of rationalism, which for the past hundred years had flowed ever more sluggishly, has finally expended its force and begins to retreat'. This is perhaps further ahead than we need look at this point in the story, but one thing about this reversal is striking at an early date. Swamped as the Hellenistic world was with mysteries and crazes of all kinds, from the revival of Pythagorean mysticism to the raising of altars to dead philosophers, traditional Greek religion was not a beneficiary. Its decay had already gone too far. The decline of Delphi, remarked from the third century, was not arrested.

This collapse of a traditional religious framework of values was the background to philosophical change. The study of philosophy was still vigorous in Greece itself and even there its Hellenistic development suggests that men were falling back upon personal concerns, contracting out of societies they could not influence, seeking shelter from the buffets of fate and the strain of daily life. It seems somewhat familiar. One example was Epicurus, who sought the good in an essentially private experience of pleasure. Contrary to later misinterpretations, he meant by this something far from self-indulgence. For Epicurus, pleasure was psychological contentment and the absence of pain - a view of pleasure somewhat austere to modern eyes. But symptomatically its importance is considerable because it reveals a shift in men's preoccupations towards the private and personal. Another form of this philosophic reaction advocated the ideals of renunciation and non-attachment. The school known as the Cynics expressed contempt for convention and sought release from dependence on the material world. One of them, Zeno, a Cypriot, who lived at Athens, began to teach a doctrine of his own in a public place, the *stoa Poikile*. The place gave its name to those he taught, the Stoics. They were to be among the most influential of philosophers because their teaching was readily applicable to daily life. Essentially the Stoics taught that life should be lived to fit the rational order they discerned running through the universe. Man could not control what happened to him, they said, but he could accept what was sent by fate, the decree of the divine will in which they believed. Virtuous acts, accordingly, should not be performed for their likely consequences, which might well be unfortunate or thwarted, but for their own sake, because of their intrinsic value.

In stoicism, which was to have great success in the Hellenistic world, lay doctrine which gave the individual a new ground for ethical confidence at a time when neither *polis* nor traditional Greek religion retained their authority. Stoicism also had the potential for a long life, because it applied to all men, who, it taught, were all alike: this was the seed of an ethical universalism which gradually transcended the old distinction between Greek and barbarian, as it would any other distinction between reasonable men. It spoke to a common humanity and actually produced a condemnation of slavery, an amazing step in a world built by forced labour. It was to be a fecund source for thinkers for two thousand years. Soon its ethic of disciplined common sense was to have great success at Rome.

Philosophy thus showed the symptoms of the eclecticism and cosmopolitanism which strike the eye in almost every other aspect of Hellenistic culture. Perhaps their most obvious expression was the adaptation of Greek sculpture to the monumental statuary of the East which produced such monsters as the hundred-foot-high Colossus of Rhodes; yet in the end eclecticism and cosmopolitanism appeared everywhere, in the aspirations of the Stoics just as in the exotic oriental cults which displaced the Greek gods. It was the scientist Eratosthenes who said that he saw all good men as fellow-countrymen and the remark expresses the new spirit which was Hellenism at its best.

The political framework of this world was bound in the end to change, because sources of change grew up beyond its circumference. One early omen was the appearance of a new threat in the east, the kingdom of Parthia. By the middle of the third century BC the weakness imposed by the Seleucid kingdom's concentration of population and wealth in its western half was leading to over-preoccupation with relations with the other Hellenistic states. The north-east was threatened – as always - by nomads from the steppes, but government was distracted from this danger by the need to supply money and resources for quarrels with Ptolemaic Egypt. The temptation to a remote satrap to strike out on his own as a warlord was often irresistible. Scholars contest the details, but one of the satrapies in which this happened was Parthia, an important area to the south-east of the Caspian. It was to become more important still as the centuries passed for it lay across the caravan route to Central Asia by which the western classical world and China came to be remotely in touch, the Silk Road.

Who were the Parthians? They were originally the Parni, one of those Indo-European nomadic peoples who emerged from Central Asia to create and re-create a political unity in the highlands of Iran and Mesopotamia. They became a byword for a military skill then peculiar to them: the discharging of arrows by mounted horsemen. They did not build nearly five hundred years of political continuity only on this, though. They also inherited an administrative structure which the Seleucids were left by Alexander, who had taken it from the Persians. Indeed, in most things the Parthians seemed inheritors, not originators; their great dynasty used Greek for its official documents, and they seem to have had no law of their own but to have readily accepted existing practice, whether Babylonian, Persian or Hellenistic.

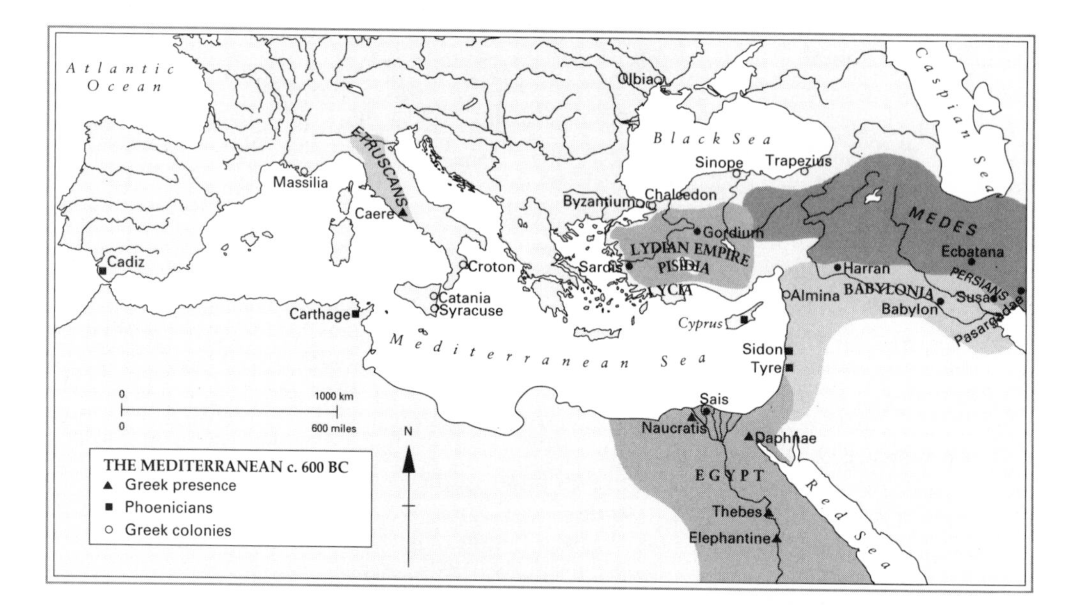

Much about their early history remains obscure. There was a kingdom, whose centre remains undiscovered, in Parthia in the third century BC, but the Seleucids do not seem to have reacted strongly to it. In the second century, when the Seleucid monarchy was much more disastrously engaged in the west, two brothers, the younger of whom was Mithridates I, established a Parthian empire which at his death stretched from Bactria (another fragment of the Seleucid inheritance which had been finally separated from it

MAIDEN CASTLE *Still the most haunting and impressive relic of the British Iron Age peoples, Maiden Castle, in Dorset, enclosed a site of forty-five acres within a huge double belt of earth ramparts and ditches. Its most remote origins lay in a neolithic settlement of the fourth millenium* BC, *but the great defensive system began to be built only in the third century* BC. *The Roman forces under Claudius took it soon after their arrival in the island, and a few years later the site was abandoned by its inhabitants who probably migrated to the newly founded town of Durnovaria (the modern Dorchester) a few miles away.*

SPHINX *Guarding the approach to the pyramids of Kephren, the Sphinx is the best-known and largest statue of the Old Kingdom. The body is sixty metres long. Even after the ravages of time – and French cannonballs – it produces an impression of outstanding grandeur. The sphinx, a mythical beast, has the body of a lion and the head of a man. This is unlike what came to be the conventional representation of the gods of Egypt, whose animal heads were placed on human bodies.*

at about the same time as Parthia) in the east to Babylonia in the west. Consciously reminiscent of those who had gone before, Mithridates described himself on his coins as the 'great king'. There were setbacks after his death but his namesake Mithridates II recovered lost ground and went even further. The Seleucids were now confined to Syria. In Mesopotamia the frontier of his empire was the Euphrates and the Chinese opened diplomatic relations with him. The coins of the second Mithridates bore the proud Achaemenid title, 'King of Kings', and the inference is reasonable that the Arsacid dynasty to which Mithridates belonged was now being consciously related to the great Persian line. Yet the Parthian state seems a much looser thing than the Persian. It is more reminiscent of a feudal grouping of nobles about a warlord than a bureaucratized state.

On the Euphrates, Parthia was eventually to meet a new power from the West. Less remote from it than Parthia and therefore with less excuse, even the Hellenistic kingdoms had been almost oblivious of the rise of Rome, this new star of the political firmament, and went their way almost without regard for what was happening in the West. The western Greeks, of course, knew more about it, but they long remained preoccupied with the first great threat they had faced, Carthage, a mysterious state which almost may be said to have derived its being from hostility to the Greeks. Founded by Phoenicians somewhere about 800 BC, perhaps even then to offset Greek commercial competition on the metal routes, Carthage had grown to surpass Tyre and Sidon in wealth and power. But she long remained a city-state, using alliance and protection rather than conquests and garrisons, her citizens preferring trade and agriculture to fighting. Unfortunately, the native documentation of Carthage was to perish when, finally, the city was razed to the ground and we know little of its own history.

Yet it was clearly a formidable commercial competitor for the western Greeks. By 480 BC these had been confined commercially to little more than the Rhône valley, Italy and, above all, Sicily. This island, and one of its cities, Syracuse, was the key to the Greek west. Syracuse for the first time protected Sicily from the Carthaginians when she fought and beat them in the year of Salamis. For most of the fifth century Carthage troubled the western Greeks no more and the Syracusans were able to turn to supporting the Greek cities of Italy against the Etruscans. Then Syracuse was the target of the ill-fated Sicilian Expedition from Athens (415–413 BC) because she was the greatest of the western Greek states. The Carthaginians came back after this, but Syracuse survived defeat to enjoy soon afterwards her greatest period of power, exercized not only in the island, but in southern Italy and the Adriatic. During most of it she was at war with Carthage. There was plenty of vigour in Syracuse; at one moment she all but captured Carthage, and another expedition added Corcyra (Corfu) to her Adriatic possessions. But soon after 300 BC it was clear that Carthaginian power was growing while Syracuse had also to face a Roman threat in mainland Italy. The Sicilians fell out with a man who might have saved them, Pyrrhus of Epirus, and by mid-century the Romans were masters of the mainland.

There were now three major actors in the arena of the West, yet the Hellenistic east seemed strangely uninterested in what was going forward (though Pyrrhus was aware of it). This was perhaps short-sighted, but at this time the Romans did not see themselves as world conquerors. They were as much moved by fear as by greed in entering on the Punic Wars, from which they would emerge victors. Then they would turn east. Some Hellenistic Greeks were beginning to be aware by the end of the century of what might be coming. A 'cloud in the west' was one description of the struggle between Carthage and Rome viewed from the Hellenized east. Whatever its outcome, this struggle was bound to have great repercussions for the whole Mediterranean. None the less, the east was to prove in the event that it had its own strengths and powers of resistance. As one Roman later put it, Greece would take her captors captive, Hellenizing yet more barbarians.

5
Rome

All round the western Mediterranean shores and across wide tracts of western Europe, the Balkans and Asia Minor, relics can still be seen of a great achievement, the empire of Rome. In some places – Rome itself, above all – they are very plentiful. The explanation why they are there is a thousand years of history. If we no longer look back on the Roman achievement as our ancestors after did, feeling dwarfed by it, we can still be puzzled and even amazed that men could do so much. Of course, the closer the scrutiny historians give to those mighty remains and the more scrupulous their sifting of the documents which explain Roman ideals and Roman practice, the more we realize that Romans were not, after all, superhuman. The grandeur that was Rome sometimes looks more like tinsel and the virtues its publicists proclaimed sound as much like cant as do many political slogans of today. Yet when all is said and done, there remains an astonishing and solid core of creativity. In the end, Rome remade the setting of Greek civilization. Thus Romans settled the shape of the first civilization embracing all the West. This was a self-conscious achievement. Romans who looked back on it when it was later crumbling about them still felt themselves to be Romans like those who had built it up. They were, though only in the sense that they believed it. Yet that was the most important sense. For all its material impressiveness and occasional grossness, the core of the explanation of the Roman achievement was an idea, the idea of Rome itself, the values it embodied and imposed, the notion of what was one day to be called *Romanitas*.

It was believed to have deep roots. Romans said their city was founded by one Romulus in 753 BC. We need not take this seriously, but the legend of the foster-mother wolf which suckled both Romulus and his twin, Remus, is worth a moment's pause; it is a good symbol of early Rome's debt to a past that was dominated by the people called Etruscans, among whose cults has been traced a special reverence for the wolf.

In spite of a rich archaeological record with many inscriptions and much scholarly effort to make sense of it the Etruscans remain a mysterious people. All that has been delineated with some certainty is the general nature of Etruscan culture, not its history or chronology. Different scholars have argued that Etruscan civilization came into existence at a wide range of different times, stretching from the tenth to the seventh century BC. Nor have they been able to agree about where the Etruscans came from; one hypothesis points to immigrants from Asia just after the end of the Hittite empire, but several other possibilities have their supporters. All that is obvious is that they were not the first Italians. Whenever they came to the peninsula and wherever from, Italy was then already a confusion of peoples.

There were probably still at that time some aboriginal natives among them whose ancestors had been joined by Indo-European invaders in the second millennium BC. In the next thousand years some of these Italians developed advanced cultures. Iron-working was going on in about 1000 BC. The Etruscans probably adopted the skill from the peoples there before them, possibly from a culture which has been called Villanovan (after an archaeological site near modern Bologna). They brought metallurgy to a high level and vigorously exploited the iron deposits of Elba, off the coast of Etruria. With iron weapons, they appear to have established an Etruscan hegemony which at its greatest extent covered the whole central peninsula, from the valley of the Po down to

Campania. Its organization remains obscure, but Etruria was probably a loose league of cities governed by kings. The Etruscans were literate, using an alphabet derived from Greek which may have been acquired from the cities of *Magna Graecia* (though hardly anything of their writing can be understood), and they were relatively rich.

In the sixth century BC the Etruscans were installed in an important bridgehead on the south bank of the river Tiber. This was the site of Rome, one of a number of small cities of the Latins, an old-established people of the Campania. Through this city something of the Etruscan legacy was to survive to flow into and eventually be lost in the European tradition. Near the end of the sixth century BC Rome broke away from Etruscan dominion during a revolt of the Latin cities against their masters. Until then, the city had been ruled by kings, the last of whom, tradition later said, was expelled in 509 BC. Whatever the exact date, this was certainly about the time at which Etruscan power, over-strained by struggle with the western Greeks, was successfully challenged by the Latin peoples, who thereafter went their own ways. Nevertheless, Rome was to retain much from her Etruscan past. Through it she had first had access to the Greek civilization with which she continued to live in contact both by land and sea. Rome was a focus of important land and water routes, high enough up the Tiber for her to bridge it, but not so high that she could not be reached by sea-going vessels. Fertilization by Greek influence was perhaps her most important inheritance, but Rome also carried forward many Etruscan institutions. One was the way she organized her people in 'centuries' for military purposes; more superficial but striking instances were her gladiatorial games, civic triumphs and reading of auguries – a consultation of the entrails of sacrifices in order to discern the shape of the future.

The republic was to last for more than four hundred and fifty years and even after that its institutions survived in name. Romans always harped on continuity and their loyal adherence (or reprehensible non-adherence) to the good old ways of the early republic. There was some reality in such claims, much as there is, for example, in the claims made for the continuity of parliamentary government in Great Britain or for the wisdom of the founding fathers of the United States in agreeing a constitution which still operates successfully. Yet, of course, great changes took place as the centuries passed. They eroded the institutional and ideological continuities and historians still argue about how to interpret them. Yet for all these changes Rome's institutions made possible a Roman Mediterranean and a Roman empire stretching far beyond it which was to be the cradle of Europe and Christianity. Thus Rome, like Greece (which reached many later men only through Rome), shaped much of the modern world. It is not just in a physical sense that men still live among her ruins.

Broadly speaking the changes of republican times were symptoms and results of two main processes. One was of decay; gradually the republic's institutions ceased to work. They could no longer contain political and social realities and, in the end, this destroyed them, even when their names survived. The other was the extension of Roman rule first beyond the city and then beyond Italy. For about two centuries both processes went on rather slowly.

Internal politics were rooted in arrangements originally meant to make impossible the return of monarchy. Constitutional theory was concisely expressed in the motto carried by the monuments and standards of Rome until well into imperial times: *SPQR*, the abbreviation of the Latin words for 'the Roman Senate and People'. Theoretically, ultimate sovereignty always rested with the people, which acted through a complicated set of assemblies attended by all citizens in person (of course, not all inhabitants of Rome were citizens). This was similar to what went on in many Greek city-states. The general conduct of business was the concern of the Senate; it made laws and regulated the work of elected magistrates. It was in the form of tensions between the poles of Senate and people that the most important political issues of Roman history were usually expressed.

Somewhat surprisingly, the internal struggles of the early republic seem to have been comparatively bloodless. Their sequence is complicated and sometimes mysterious, but their general result was that they gave the citizen body as a whole a greater say in the affairs of the republic. The Senate, which concentrated political leadership, had come by 300 BC or so to represent a ruling class which was an amalgamation of the old patricians of prerepublican days with the wealthier members of the *plebs*, as the rest of the citizens were termed. The Senate's members constituted an oligarchy, self-renewing though some were usually excluded in each census (which took place once every five years). Its core was a group of noble families whose origins might be plebeian, but among whose ancestors were men who had held the office of consul, the highest of the magistracies.

Two consuls had replaced the last kings at the end of the sixth century BC. Appointed for a year, they ruled through the Senate and were its most important officers. They were bound to be men of experience and weight, for they had to have passed through at least two subordinate levels of elected office, as *quaestors* and *praetors*, before they were eligible. The quaestors (of whom there were twenty elected each year) also automatically became members of the Senate. These arrangements gave the Roman ruling élite great cohesiveness and competence; for progress to the highest office was a matter of selection from a field of candidates who had been well tested and trained in office. That this constitution worked well for a long time is indisputable. Rome was never short of able men. What it masked was the natural tendency of oligarchy to decay into faction, for whatever victories were won by the plebs, the working of the system ensured that it was the rich who ruled and the rich who disputed the right to office among themselves. Even in the electoral college which was supposed to represent the whole people, the *comitia centuriata*, organization gave an undue proportion of influence to the wealthy.

These figures of a man and his wife reclining on a couch have something of the feel of archaic Greek art about them. They come, in fact, from Cerveteri in Italy: they form the lid of an Etruscan sarcophagus made in about 500 BC.

Plebs, in any case, is a misleadingly simple term. The word stood for different social realities at different times. Conquest and enfranchisement slowly extended the boundaries of citizenship. Even in early times they ran well beyond the city and its environs as other cities were incorporated in the republic. At that time, the typical citizen was a countryman. The basis of Roman society was always agricultural and rural. It is significant that the Latin word for money, *pecunia*, is derived from the word for a flock of sheep or herd of cattle and that the Roman measure of land was the *iugerum*, the extent that could be ploughed in a day by two oxen. Land and the society it supported were related in changing ways during the republic, but always its base was the rural population. The later preponderance in men's minds of the image of imperial Rome, the great parasitic city, obscures this.

The free citizens who made up the bulk of the population of the early republic were therefore peasants, some much poorer than others. They were legally grouped in complicated arrangements whose roots were sunk in the Etruscan past. Such distinctions were economically insignificant, though they had constitutional importance for electoral purposes, and tell us less about the social realities of republican Rome than distinctions made by the Roman census between those able to equip themselves with the arms and armour needed to serve as soldiers, those whose only contribution to the state was to breed children (the *proletarii*) and those who were simply counted as heads, because they neither owned property nor had families. Below them all, of course, were the slaves.

There was a persistent tendency, accelerating rapidly in the third and second centuries BC, for many of the plebs who in earlier days had preserved some independence through possession of their own land to sink into poverty. Meanwhile, the new aristocracy increased its relative share of land as conquest brought it new wealth. This was a long-drawn-out process, and while it went on, new subdivisions of social interest and political weight appeared. Furthermore, to add another complicating factor, there grew up the practice of granting citizenship to Rome's allies. The republic in fact saw a gradual enlargement of the citizen class but a real diminution of its power to affect events.

This was not only because wealth came to count for so much in Roman politics. It was also because everything had to be done at Rome, though there were no representative arrangements which could effectively reflect the wishes of even those Roman citizens who lived in the swollen city, let alone those scattered all over Italy. What tended to happen instead was that threats to refuse military service or to withdraw altogether from Rome and found a city elsewhere enabled the plebs to restrict somewhat the powers of Senate and magistrates. After 366 BC, too, one of the two consuls had to be a plebeian and in 287 BC the decisions of the plebeian assembly were given overriding force of law. But the main restriction on the traditional rulers lay in the ten elected Tribunes of the People, officers chosen by popular vote, who could initiate legislation or veto it (one veto was enough) and were available night and day to citizens who felt themselves unjustly treated by a magistrate. The tribunes had most weight when there was great social feeling or personal division in the Senate, for then they were courted by the politicians. In the earlier republic and often thereafter, the tribunes, who were members of the ruling class and might be nobles, worked for the most part easily enough with the consuls and the rest of the Senate. The administrative talent and experience of this body and the enhancement of its prestige because of its leadership in war and emergency could hardly be undermined until there were social changes grave enough to threaten the downfall of the republic itself.

The constitutional arrangements of the early republic were thus very complicated, but effective. They prevented violent revolution and permitted gradual change. Yet they would be no more important to us than those of Thebes or Syracuse, had they not made possible and presided over the first phase of victorious expansion of Roman power. The story of the republic's institutions is important for even later periods, too, because of what the republic itself became. Almost the whole of the fifth century was taken up in mastering Rome's neighbours and her territory was doubled in the process. The other cities of the Latin League were next subordinated; when some of them revolted in the middle of the fourth century they were forced back into it on harsher terms. It was a little like a land version of the Athenian empire a hundred years before; Roman policy was to leave her 'allies' to govern themselves, but they had to subscribe to Roman foreign policy and supply contingents to the Roman army. In addition, Roman policy favoured established dominant groups in the other Italian communities, and Roman aristocratic families multiplied their personal ties with them. The citizens of those communities were also admitted to rights of citizenship if they migrated to Rome. Etruscan hegemony in central Italy, the richest and most developed part of the peninsula, was thus replaced by Roman.

Roman military power grew as did the number of subjected states. The republic's own army was based on conscription. Every male citizen who owned property was obliged to serve if called and the obligation was heavy, sixteen years for an infantryman and ten for cavalry. The army was organized in legions of 5,000 which fought at first in solid phalanxes with long pike-like spears. It not only subdued Rome's neighbours, but also beat off a series of fourth-century incursions by Gauls from the north, though on one occasion they sacked Rome itself (390 BC). The last struggles of this formative period came at the end of the fourth century when the Romans conquered the Samnite peoples of the Abruzzi. Effectively, the republic could now tap allied manpower from the whole of central Italy.

Rome was now at last face to face with the western Greek cities. Syracuse was by far the most important of them. Early in the third century the Greeks asked the assistance of a great military leader of mainland Greece, Pyrrhus, King of Epirus, who campaigned against both the Romans and the Carthaginians (280–275 BC), but achieved only the costly and crippling victories to whose type he gave his name. He could not destroy the Roman threat to the western Greeks. Within a few years they were caught up willynilly in a struggle between Rome and Carthage in which the whole western Mediterranean was at stake – the Punic Wars.

They form a duel of more than a century. Their name comes from the Roman rendering of the word Phoenician and, unfortunately, we have only the Roman version of what happened. There were three bursts of fighting, but the first two settled the question of preponderance. In the first (264–241 BC) the Roman began naval warfare on a large scale for the first time. With their new fleet they took Sicily and established themselves in Sardinia and Corsica. Syracuse abandoned an earlier alliance with Carthage and western Sicily and Sardinia became the first Roman provinces, a momentous step, in 227 BC.

Even recently, the backbone of the population of Italy was its peasantry. These farmers were depicted on a bronze vessel used for the ashes of the cremated dead in a cemetery of the fifth century BC *near Bologna.*

This was only round one. As the end of the third century approached, the final outcome was still not discernible and there is still argument about which side, in this touchy situation, was responsible for the outbreak of the second Punic War (218–201 BC), the greatest of the three. It was fought in a greatly extended theatre, for when it began the Carthaginians were established in Spain. Some of the Greek cities there had been promised Roman protection. When one of them was attacked and sacked by a Carthaginian general, Hannibal, the war began. It is famous for Hannibal's great march to Italy and passage of the Alps with an army including elephants and for its culmination in the crushing Carthaginian victories of Lake Trasimene and Cannae (217 and 216 BC), where a Roman army twice the size of Hannibal's was destroyed. At this point Rome's grasp on Italy was badly shaken; some of her allies and subordinates began to look at Carthaginian power with a new respect. Virtually all the south changed sides, though central Italy remained loyal. With no resources save her own exertions and the great advantage that Hannibal lacked the numbers needed to besiege Rome, Rome hung on and saved herself. Hannibal campaigned in an increasingly denuded countryside far from his base.

The Romans mercilessly destroyed Capua, a rebellious ally, without Hannibal coming to help her and then boldly embarked upon a strategy of striking at Carthage in her own possessions, especially in Spain. In 209 BC 'New Carthage' (Cartagena) was taken by the Romans. When an attempt by Hannibal's younger brother to reinforce him was beaten off in 207 BC the Romans transferred their offensives to Africa itself. There, at last, Hannibal had to follow them to meet his defeat at Zama in 202 BC, the end of the war.

This battle settled more than a war; it decided the fate of the whole western Mediterranean. Once the Po valley was absorbed early in the second century, Italy was, whatever the forms, henceforth a single state ruled from Rome. The peace imposed on Carthage was humiliating and crippling. Roman vengeance pursued Hannibal himself and drove him to exile at the Seleucid court. Because Syracuse had once more allied with Carthage during the war, her presumption was punished by the loss of her independence; she was the last Greek state in the island. All Sicily was now Roman, as was southern Spain, where another province was set up.

Nor was this all. These events opened the way to the east. At the end of the second Punic War it is tempting to imagine Rome at a parting of the ways. On the one hand lay the alternative of moderation and the maintenance of security in the west, on the other of expansion and imperialism in the east. Yet this over-simplifies reality. Eastern and western issues were already too entangled to sustain so simple an antithesis. As early as 228 BC the Romans had been admitted to the Greek Isthmian games; it was a recognition, even if only formal, that for some Greeks they were already a civilized power and part of the Hellenistic world. Through Macedon, that world had already been involved directly in the wars of Italy, for Macedon had allied with Carthage; Rome had therefore taken the side of Greek cities opposed to Macedon and thus began to dabble in Greek politics. When a direct appeal for help against Macedon and the Seleucids came from Athens, Rhodes and a king of Pergamon in 200 BC, the Romans were already psychologically ready to commit themselves to eastern enterprise. It is unlikely, though, that any of them saw that this could be the beginning of a series of adventures from which would emerge a Hellenistic world dominated by the republic.

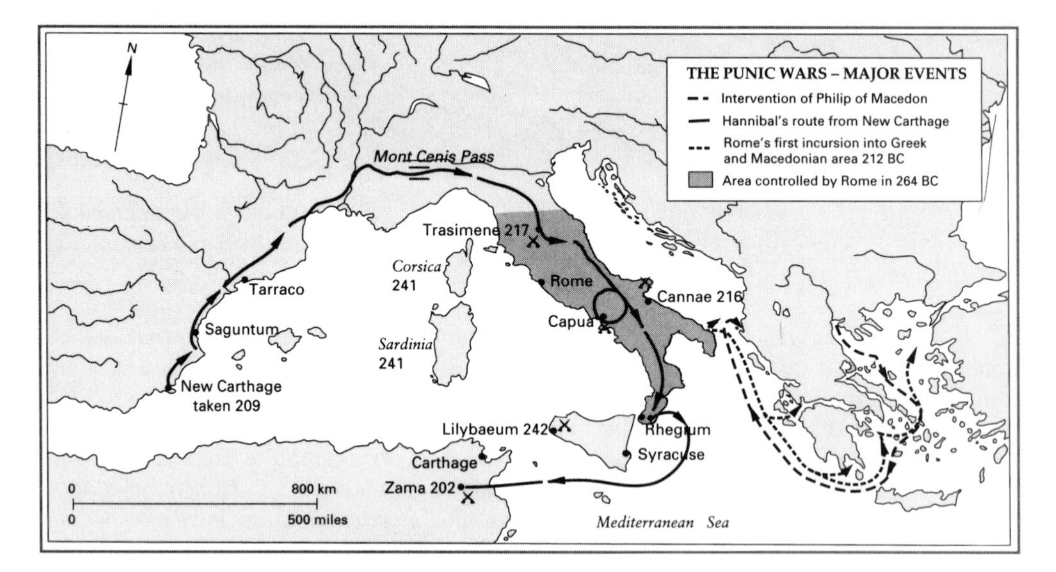

Another change in Roman attitudes was not yet complete, but was beginning to be effective. When the struggle with Carthage began, most upper-class Romans probably saw it as essentially defensive. Some went on fearing even the crippled enemy left after Zama. The call of Cato in the middle of the next century - 'Carthage must be destroyed' – was to be famous as an expression of an implacable hostility arising from fear. None the less, the provinces won by war had begun to awake men's minds to other possibilities and soon supplied other motives for its continuation. Slaves and gold from Sardinia, Spain and Sicily were soon opening the eyes of Romans to what the rewards of empire might be. These countries were not treated like mainland Italy, as allies, but as resource pools to be administered and taxed. A tradition grew up under the republic, too, of generals distributing some of the spoils of victory to their troops.

The twists and turns are complicated, but the main stages of Roman expansion in the east in the second century BC are obvious enough. The conquest and reduction of Macedon to a province was accomplished in a series of wars ending in 148 BC; the phalanxes were not what they had been, nor Macedonian generalship. On the way the cities of Greece had also been reduced to vassalage and forced to send hostages to Rome. An intervention by a Syrian king led to the first passage of Roman forces to Asia Minor; next came the disappearance of the kingdom of Pergamon, Roman hegemony in the Aegean and the establishment of the new province of Asia in 133 BC. Elsewhere, the conquest of the remainder of Spain except the north-west, the organization of a tributary confederacy in Illyria, and the provincial organization of southern France in 121 BC, meant that the coasts from Gibraltar to Thessaly were all under Roman rule. Finally, the chance long sought by the enemies of Carthage came in 149 BC with the start of the third and last Punic War. Three years later the city *was* destroyed, ploughs were run over its site and a new Roman province, Africa, existed in its stead.

Thus was the empire made by the republic. Like all empires, but perhaps more obviously than any earlier one, its appearance owed as much to chance as to design. Fear, idealism and eventually cupidity were the mingled impulses which sent the legions further and further afield. Military power was the ultimate basis of Roman empire, and it was kept up by expansion. Numbers were decisive in overcoming Carthaginian experience and tenacity and the Roman army was large. It could draw upon an expanding pool of first-class manpower available from allies and satellites, and republican rule brought order and regular government to new subjects. The basic units of the empire were its provinces, each ruled by a governor with proconsular powers whose posting was formally for one year. Beside him stood a taxing officer.

Empire inevitably had political consequences at home. In the first place it made it even more difficult to ensure popular participation – that is, the participation of poor citizens – in government. Prolonged warfare reinforced the day-to-day power and the moral authority of the Senate, and it must be said that its record was a remarkable one. Yet the expansion of territory carried even further shortcomings already apparent in the extension of Roman rule over Italy. Serious and novel problems arose. One was posed by the new opportunities war and empire gave to generals and provincial governors. The fortunes to be made, and made quickly, were immense; not until the days of the Spanish *conquistadores* or the British East India Company were such prizes so easily available to those in the right place at the right time. Much of this was legal; some was simply looting and theft. Significantly, in 149 BC a special court was created to deal with illegal extortion by officials. Whatever its nature, access to this wealth could only be obtained through participation in politics, for it was from the Senate that governors were chosen for the new provinces and it was the Senate which appointed the tax-gatherers who accompanied them from among the wealthy but non-noble class of *equites*, or 'knights'.

Another constitutional weakness arose because the principle of annual election of magistrates had more and more frequently to be set aside in practice. War and rebellion in the provinces provided emergencies which consuls elected for their political skill might well find beyond them. Inevitably, proconsular power fell into the hands of those who could deal with emergencies effectively, usually proven generals. It is a mistake to think of the republic's commanders as professional soldiers in the modern sense; they were members of the ruling class who might expect in a successful career to be civil servants, judges, barristers, politicians and even priests. One key to the administrative proficiency of Rome was its acceptance of the principle of non-specialization in its rulers. None the less, a general who stayed years with his army became a different sort of political animal from the proconsuls of the early republic who commanded an army for one

There survive many likenesses of Julius Caesar, several made soon after his death. Almost alone among the portraits of him, this one is believed to date from his lifetime.

The leader among Caesar's assassins (and reputedly a descendant of the Brutus who founded the Republic) was Brutus; like other politicians and rulers of the ancient world, he used coins as vehicles for propaganda. This one commemorates the day of Caesar's death as a day of liberation – a cap of liberty stands between two daggers - and may be the beginning of a tradition of idolizing republican tyrannicide which runs down to modern times.

campaign and then returned to Rome and politics. Paradoxically, it was a weakness that the provincial governorships were themselves annual. In that lay a temptation to make hay while the sun shone. If this was one way by which irresponsibility crept into the administrative structure, there was a corresponding tendency for successful generals long in the field to draw to themselves the loyalty soldiers owed to the republic. Finally, there was even a kind of socialized corruption, for all Roman citizens benefited from an empire which made possible their exemption from any direct taxation; the provinces were to pay for the homeland. Awareness of such evils lies behind much moralizing condemnation and talk of decline which arose in the first century BC, when their impact became fatal.

Another change brought by empire was a further spread of Hellenization. Here there are difficulties of definition. In some measure, Roman culture was already Hellenized before conquest went beyond Italy. The republic's conscious espousal of the cause of the Greek cities' independence of Macedon was a symptom. On the other hand, whatever Rome already possessed, there was much that could be hers only after more direct contact with the Hellenized world. In the last resort, Rome looked to many Greeks like another barbarian power, almost as bad as Carthage. There is symbolism in the legend of the death of Archimedes, struck down while pondering geometrical problems in the sand, by the sword of a Roman soldier who did not know who he was.

With empire the contact became direct and the flow of Hellenistic influence manifold and frequent. Later ages were to wonder at the Roman passion for baths; the habit was one they had learnt from the Hellenized East. The first Roman literature was translated Greek drama and the first Latin comedies were imitations of Greek models. Art began to flow to Rome through pilfering and looting, but Greek style – above all its architecture – was already familiar from the western cities. There was a movement of people, too. One of the thousand hostages sent to Rome from the Greek cities in the middle of the second century BC was Polybius, who provided Rome with its first scientific history in the tradition of Thucydides. His history of the years 220–146 BC was a conscious exploration of a phenomenon which he felt to mark a new epoch, Rome's success in overthrowing Carthage and conquering the Hellenistic world. He first among historians recognized a complement to the earlier civilizing work of Alexander in the new unity given to the Mediterranean by Rome. He also admired the disinterested air Romans appeared to bring to imperial government – a reminder to be set against the Romans' own denunciation of their wickednesses under the late republic.

Rome's greatest triumph rested on the bringing of peace and it was a second great Hellenistic age in which men could travel from one end to another of the Mediterranean without hindrance. The essential qualities of the structure which sustained it were already there under the republic, above all in the cosmopolitanism encouraged by Roman administration, which sought not to impose a uniform pattern of life but only to collect taxes, keep the peace and regulate the quarrels of men by a common law. The great achievements of Roman jurisprudence lay still far ahead, but the early republic in about 450 BC launched Roman law on its history of definition by the consolidation of the Twelve Tables which little Roman boys lucky enough to go to school had still to get by heart hundreds of years later. On them was eventually built a framework within which many cultures might survive to contribute to a common civilization.

It is convenient to finish the story of the spread of the rule of the republic to its limits before considering how such success in the end proved fatal. Transalpine Gaul (southern France) was a province in 121 BC but (like north Italy) it remained troubled from time to time by the incursions of Celtic tribes. The Po valley was given provincial status as Cisalpine Gaul in 89 BC and nearly forty years later (51 BC) the rest of Gaul – roughly northern France and Belgium – was conquered and with that the Celtic danger effectively came to an end. Meanwhile there had been further conquests in the east. The last king of Pergamon had bequeathed his kingdom to Rome in 133 BC. There followed the acquisition of Cilicia in the early first century BC, and then a series of wars with Mithridates, King of Pontus, a state on the Black Sea. The outcome was the reorganization of the Near East, Rome being left with possession of a coast running from Egypt to the Black Sea, all of which was divided between client kingdoms or provinces (one of them named 'Asia'). Finally, Cyprus was annexed in 58 BC.

Ironically, the counterpoint of this continuing and apparently irresistible success abroad was growing strife at home. The crux of the matter was the restriction of access to office to members of the ruling class. Electoral institutions and political conventions had come to work differently because of two grave long-term problems. The first was the gradual impoverishment of the Italian peasant who had been the typical figure of the early republic. It had several causes, but the root of the matter was the terrible cost of the second Punic War. Not only had conscripted soldiers been absent for long years of almost continuous campaigns, but the physical damage to southern Italy was enormous. Meanwhile, those who were lucky enough to amass wealth in imperial enterprise laid it out in the only good investment available, land. The effect in the long run was to concentrate property in large estates usually worked by slaves made cheaper by the wars; there was no place on them for the smallholder, who now had to make his way to the city and fend for himself as best he could, a Roman citizen in name, but a proletarian in the making. Yet as a citizen he still had a vote. To those with wealth and political ambition he became someone to buy or to intimidate. Since the road to lucrative office lay through popular elections, the politics of the republic could hardly fail increasingly to reflect the power of money. This, too, had repercussions far and wide in Italy. Once votes had a price, the citizen proletariat of Rome was unlikely to welcome their continual devaluation by extending civic rights to other Italians, even though Rome's allies had to put up with conscription.

The second problem was a change in the army. The legions had more than four hundred years' history under the republic and their evolution can hardly be condensed in a simple formula, but if one is to be sought, it is perhaps best to say that the army became increasingly professional. After the Punic Wars it was impossible any longer to rely solely on soldiers fighting in such time as they could spare from farming. The burden of conscription had always been heavy and became unpopular. When campaigns carried men further and further afield for year after year, and as garrisons had sometimes to remain for decades in conquered provinces, even the Roman pool of manpower showed signs of drying up. In 107 BC a formal change registered what was happening: the property qualification for service was abolished. This was the work of a consul called Marius, who thus solved the problem of recruitment, for after this there were usually enough poor volunteers for conscription to be unnecessary. Military service still continued to be restricted to citizens, but there were many of these; in the end, though, service itself was to confer citizenship. Another innovation of Marius was to give the legions their 'eagles', the standards so important to their *esprit de corps*, something between an idol and a modern regimental badge. Such changes gradually turned the army into a new kind of political force, available to a man like Marius who was an able general and much called upon for service in the provinces. He actually exacted a personal oath of allegiance from one army under his own command.

The widening gap of rich and poor in central Italy as peasant farming gave way to large estates bought (and stocked with slaves) with the spoils of empire, and the new possibilities open to political soldiers, proved fatal to the republic in the end. As the end of the second century BC, the Gracchi brothers, Tribunes of the People, sought to do something about the social problem in the only way open to an agrarian economy, by land reform, as well as reducing senatorial power and giving the *equites* a bigger role in government. They tried, in effect, to spread the wealth of empire, but their attempts only ended in their deaths. This itself marked the raising of the stakes in politics; in the last century of the republic factional bitterness reached its peak because politicians knew their lives might be forfeit. It also saw the beginning of what has been called the Roman revolution, for the conventions of Roman politics were set aside when Tiberius Gracchus (the elder brother), then consul, persuaded the plebs to unseat the tribune who had vetoed his land-bill and thus announced that he would not accept the traditional circumvention of the popular will by the prerogative of a tribune to use his veto.

The final plunge of the republic into confusion was precipitated in 112 BC by a new war when a north African king massacred a great number of Roman businessmen. Not long afterwards a wave of barbarian invaders in the north threatened Roman rule in Gaul. The emergency brought forward the consul Marius, who dealt successfully with the enemies of the republic, but at the cost of further constitutional innovation, for he was elected to the consulship for five years in succession. He was, in fact, the first of a series of warlords who were to dominate the last century of the republic, for other wars rapidly followed. Demand grew for the extension of Roman citizenship to the other Latin and Italian states. In the end these allies (*socii*) revolted in what is somewhat misleadingly called the 'Social War' in 90 BC. They were only pacified with concessions which made nonsense of the notion that the Roman popular assemblies were the ultimate sovereign; citizenship was extended to most of Italy. Then came new Asian wars – from which emerged another general with political ambitions, Sulla. There was civil war, Marius died after once more being consul, and Sulla returned to Rome in 82 BC to launch a dictatorship (voted by the Senate) with a ruthless 'proscription' of his opponents (a posting of their names which signified that anyone who could do so was entitled to kill them), an assault on the popular powers of the constitution and an attempted restoration of those of the Senate.

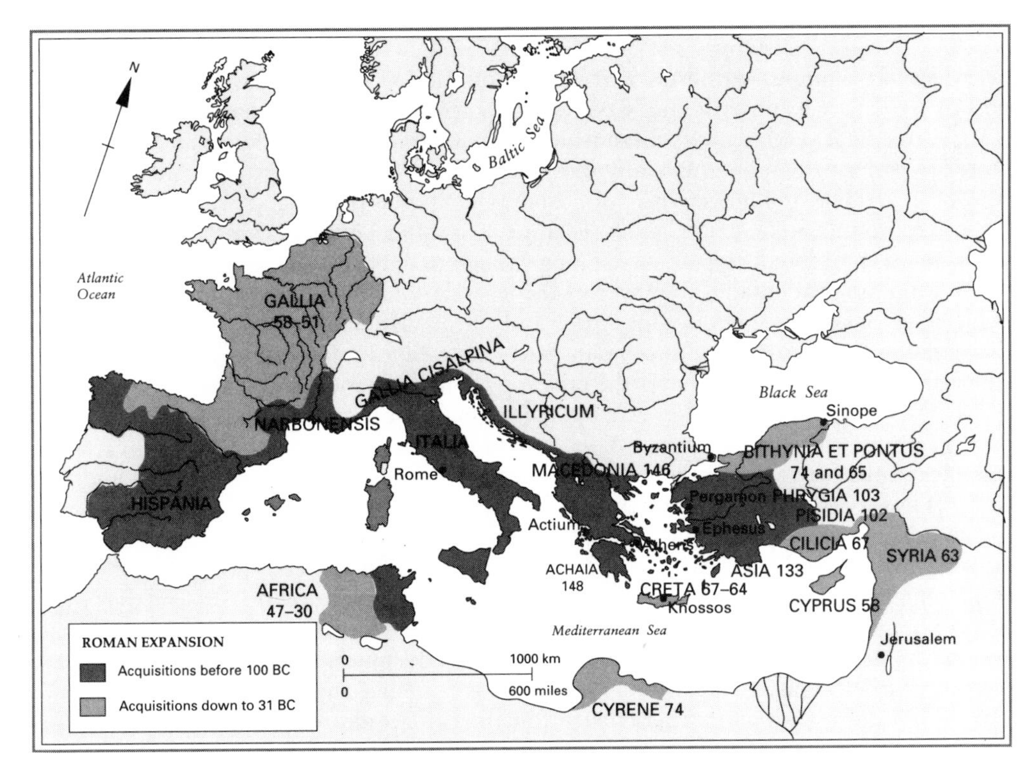

One former supporter and protégé of Sulla was a young man whose name has passed into English as Pompey. Sulla had advanced his career by giving him posts normally held only by consuls and in 70 BC he was elected to that office, too. He left for the east three years later to eliminate piracy from the Mediterranean and went on to conquer huge Asian territories in the wars against Pontus. Pompey's youth, success and outstanding ability began to make him feared as a potential dictator. But the interplay of Roman politics was complicated. As the years passed, disorder increased in the capital and corruption in ruling circles. Fears of dictatorship were intensified, but the fears were those of one oligarchic faction among several and it was less and less clear where the danger lay. Moreover one danger went long disregarded before people awoke to it.

In 59 BC another aristocrat, the nephew of Marius' wife, had been elected consul. This was the young Julius Caesar. For a time he had cooperated with Pompey. The consulship led him to the command of the army of Gaul and a succession of brilliant campaigns in the next seven years ending in its complete conquest. Though he watched politics closely, these years kept Caesar away from Rome where gangsterism, corruption and murder disfigured public life and discredited the Senate. After them he was enormously rich and had a loyal, superbly experienced and confident army looking to him for the leadership which would give them pay, promotion and victory in the future. He was also a cool, patient and ruthless man. There is a story of him joking and playing at dice with some pirates who captured him. One of his jokes was that he would crucify them when he was freed. The pirates laughed, but crucify them he did.

Some senators suddenly became alarmed when this formidable man wished to remain in Gaul in command of his army and the province, although its conquest was complete, retaining command until the consular election. His opponents strove to get him recalled to face charges about illegalities during his consulship. Caesar then took the step which, though neither he nor anyone else knew it, was the beginning of the end of the republic. He led his army across Rubicon, the boundary of his province, beginning a march which brought him in the end to Rome. This was in January 49 BC. It was an act of treason, though he claimed to be defending the republic against its enemies.

In its extremity the Senate called Pompey to defend the republic. Without forces in Italy, Pompey withdrew across the Adriatic to raise an army. The consuls and most of the Senate went with him. Civil war was now inevitable. Caesar marched quickly to Spain to defeat seven legions there which were loyal to Pompey; they were then mildly treated in order to win over as many of the soldiers as possible. Ruthless and even cruel though he could be, mildness to his political opponents was politic and prudent; he did not propose to imitate Sulla, said Caesar. Then he went after Pompey, chasing him to Egypt, where he was murdered. Caesar stayed long enough to dabble in an Egyptian civil war and became, almost incidentally, the lover of the legendary Cleopatra. Then he went back to Rome, to embark almost at once for Africa and defeat a Roman army there which opposed him. Finally, he returned again to Spain and destroyed a force raised by Pompey's sons. This was in 45 BC, four years after the crossing of the Rubicon.

Brilliance like this was not just a matter of winning battles. Brief though Caesar's recent visits to Rome had been, he had organized his political support carefully and packed the Senate with his men. The victories brought him great honours and real power. He was voted dictator for life and became in effect a monarch in all but name. His power he used without much regard for the susceptibilities of politicians and without showing an imaginativeness which suggests his rule would have been successful in the long term, though he imposed order in the Roman streets, and undertook steps to end the power of the money-lenders in politics. To one reform in particular the future of Europe was to owe much – the introduction of the Julian calendar. Like much else we think of as Roman, it came from Hellenistic Alexandria, where an astronomer suggested to Caesar that the year of 365 days, with an extra day each fourth year, would make it possible to emerge from the complexities of the traditional Roman calendar. The new calendar began on 1 January 45 BC.

Fifteen months later Caesar was dead, struck down in the Senate on 15 March 44 BC at the height of his success. His assassins' motives were complex. The timing was undoubtedly affected by the knowledge that he planned a great eastern campaign against the Parthians. Were he to join his army, it might be to return again in triumph, more unassailable than ever. There had been talk of a kingship; a Hellenistic despotism was envisaged by some. The complicated motives of his enemies were given respectability by the distaste some felt for the flagrant affront to republican tradition in the *de facto* despotism of one man. Minor acts of disrespect for the constitution antagonized others and in the end his assassins were a mixed bag of disappointed soldiers, interested oligarchs and offended conservatives.

His murderers had no answer to the problems which Caesar had not had the time and their predecessors had so conspicuously failed to solve. Nor could they protect themselves for long. The republic was pronounced restored, but Caesar's acts were confirmed. There was a revulsion of feeling against the conspirators who soon had to flee the city. Within two years they were dead and Julius Caesar was proclaimed a god. The republic was moribund, too. Damaged fatally long before the crossing of the Rubicon, the heart had gone out of its constitution whatever attempts were made to restore it. Yet its myths, its ideology and forms lived on in a Romanized Italy. Romans could not bring themselves to turn their backs on the institutional heritage and admit that they had done with it. When eventually they did, they had already ceased in all but name to resemble the Romans of the republic.

6
The Roman Achievement

If the Greek contribution to civilization was essentially mental and spiritual, that of Rome was structural and practical; its essence was the empire itself. Though no man is an empire, not even the great Alexander, its nature and government were to an astonishing degree the creation of one man of outstanding ability, Julius Caesar's great-nephew and adopted heir, Octavian. Later he was celebrated as Caesar Augustus. An age has been named after him; his name gave an adjective to posterity. Sometimes one has the feeling that he invented almost everything that characterized imperial Rome, from the new Praetorian Guard, which was the first military force stationed permanently in the capital, to the taxation of bachelors. One reason for this impression (though only one) is that he was a master of public relations;

In about AD 10, this gem was carved from onyx as a contribution to the imperial mythology then still in the making. Augustus, accompanied by the goddess Roma, is shown sitting in state, presumably contemplating the glory of his position while another figure places on his head a crown representing the civilized world. At the side, Tiberius steps from his chariot after his victory over the warlike Pannonians, while soldiers erect a trophy below. This remarkable work has been attributed to Augustus' official engraver and so no doubt expresses themes the emperor approved.

significantly, more representations of him than of any other Roman emperor have come down to us.

Though a Caesar, Octavian came of a junior branch. From Julius (whom he succeeded as the age of eighteen) he inherited aristocratic connexions, great wealth and military support. For a time he cooperated with one of Caesar's henchmen, Mark Antony, in a ferocious series of proscriptions to destroy the party which had murdered the great dictator. Mark Antony's departure to win victories in the east, failure to do so and injudicious marriage to Cleopatra, Julius Caesar's sometime mistress, gave Octavian further opportunities. He fought in the name of the republic against a threat that Antony might make a proconsular return bringing oriental monarchy in his baggage-train. The victory of Actium (31 BC) was followed by the legendary suicides of Antony and Cleopatra; the kingdom of the Ptolemies came to an end and Egypt too was annexed a province of Rome.

This was the end of civil war. Octavian returned to become consul. He had every card in his hand and judiciously refrained from playing them, leaving it to his opponents to recognize his strength. In 27 BC he carried out what he called a republican restoration with the support of a Senate whose republican membership, purged and weakened by civil war and proscription, he reconciled to his real primacy by his careful preservation of forms. He re-established the reality of his great-uncle's power behind a facade of republican piety. He was *imperator* only by virtue of his command of the troops of the frontier provinces – but that was where the bulk of the legions were. As old soldiers of his and his great-uncle's armies returned to retirement, they were duly settled on smallholdings and were appropriately grateful. His consulship was prolonged from year to year and in 27 BC he was given the honorific title of Augustus, the name by which he is remembered. At Rome, though, he was formally and usually called by his family name, or was identified as *princeps*, first citizen. As the years passed Augustus' power still grew. The Senate accorded him a right of interference in those provinces which it formally ruled (that is, those where there was no need to keep a garrison army). He was voted the tribunician power. His special status was enhanced and formalized by a new recognition of his state or *dignitas*, as the Romans called it; he sat between the two consuls after his resignation from that office in 23 BC and his business was given precedence in the agenda of the Senate. Finally, in 12 BC he became *Pontifex maximus*, the head of the official cult, as his great-uncle had been. The forms of the republic with their popular elections and senatorial elections were maintained, but Augustus said who should be elected.

The political reality masked by this supremacy was the rise to domination within the ruling class of men who owed their position to the Caesars. But the new élites were not to be allowed to behave like the old. The Augustan benevolent despotism regularized the provincial administration and army by putting them into obedient and salaried hands. The conscious resuscitation of republican tradition and festivals had a part to play in this, too. Augustan government was heavily tinged with concern for moral revival; the virtues of ancient Rome seemed to some to live again. Ovid, a poet of pleasure and love, was packed off to exile in the Black Sea when a sexual scandal at the edge of the imperial family provided an excuse. When to this official austerity is added the peace which marked most of the reign and the great visible monuments of the Roman architects and engineers, the reputation of the Augustan age is hardly surprising. After his death in AD 14 Augustus was deified as Julius Caesar had been.

Augustus intended to be succeeded by a member of his own family. Although he respected republican forms (and they were to endure with remarkable tenacity) Rome was now really a monarchy and this was demonstrated by the succession of five members of the same family. Augustus' only child was a daughter; his immediate successor was his adopted stepson, Tiberius, one of his daughter's three husbands. The last of his descendants to reign was Nero, who died in AD 68.

The rulers of the classical world did not usually live easy lives. Some Roman emperors had great mirrors installed at the corners of the corridors of their palaces so that would-be assassins could not lurk around them. Tiberius himself may have not

After almost nineteen centuries, the water-supply of the Spanish city of Segovia is still carried to it by this Roman aqueduct, borne by 128 arches which raise the water-channel to a hundred feet above ground at its highest point. Such monuments understandably made barbarians gape as at the works of gods. The Romans were the greatest engineers yet to set about extending the human exploitation of the material world. Not for centuries was anything of such comparable scale and utility again to be built in Europe.

have died a natural death, and none of his four successors did. The fact is significant of the weaknesses inherent in Augustus' legacy. There was still scope for pinpricks from a Senate which formally continued to appoint the first magistrate and always room for intrigue and cabal about the court and imperial household. Yet the Senate could never hope to recover authority, for the ultimate basis of power was always military. If there was confusion and indecision at the centre, then the soldiers would decide. This was what happened in the first great burst of civil war to shake the empire, in the year of the Four Emperors, AD 69, from which there emerged Vespasian, the grandson of a centurion and far from an aristocrat. The first magistracy had passed out of the hands of the great Roman families.

When Vespasian's younger son was murdered in AD 96 this upstart house came to an end. Its successor was an elderly senator, Nerva. He solved the problem of succession by breaking with attempts to ensure natural dynastic continuity. Instead, he institutionalized the practice of adoption to which Augustus had been driven. The result was a succession of four emperors, Trajan, Hadrian, Antoninus Pius and Marcus Aurelius, who gave the empire a century of good government; it has been named (after the third of them) the age of the Antonines. All of them came of families with provincial roots; they were evidence of the degree to which the empire was a cosmopolitan reality, the framework of the post-Hellenistic world of the West, and not merely the property of the Italian-born. Adoption made it easier to find candidates upon whom army, provinces and Senate could agree, but this golden age came to an end with a reversion to the hereditary principle, the succession of Commodus, son of Marcus Aurelius. He was murdered in AD 192, and a new AD 69 followed when, in the following year, there were again four emperors, each acclaimed by his own army. The Illyrian army prevailed in the end, imposing an African general. Other and later emperors were to be the nominees of soldiers too; bad times lay ahead.

By this time, the emperors ruled a far larger area than had Augustus. In the north Julius Caesar had carried out reconnaissances into Britain and Germany, but had left Gaul with the Channel and the Rhine as its frontiers. Augustus pressed into Germany, and also up to the Danube from the south. The Danube eventually became the frontier of the empire, but incursions beyond the Rhine were less successful and the frontier was not stabilized on the Elbe as Augustus had hoped. Instead, a grave shock had been given to Roman confidence in AD 9 when the teutonic tribes led by Arminius (in whom later Germans were to see a national hero) destroyed three legions. The ground was never recovered, nor the legions, for their numbers were thought so ill-omened that they never again appear in the army lists. Eight remained stationed along the Rhine, the most strongly held part of the frontier because of the dangers which lay beyond it.

Elsewhere, Roman rule still advanced. In AD 43 Claudius began the conquest of Britain, which was carried to its furthest enduring limit when Hadrian's wall was built across the north as an effective boundary forty or so years later. In AD 42 Mauretania had become a province. In the east Trajan conquered Dacia, the later Romania, in AD 105, but this was more than a century and a half after a quarrel which was to be long-lasting had opened in Asia.

Rome had first faced Parthia on the Euphrates when Sulla's army campaigned there in 92 BC. Nothing of importance followed until thirty years later when Roman armies began to advance against Armenia. Two spheres of influence overlapped there and Pompey at one moment arbitrated between the Armenian and Parthian kings in a boundary dispute. Then, in 54 BC, the Roman politician Crassus launched an invasion of Parthia across the Euphrates. Within a few weeks he was dead and a Roman army of forty thousand destroyed. It was one of the worst military disasters of Roman history. Evidently there was a new great power in Asia. The Parthian army had more than good mounted archers to it by this time. It also had heavy cavalry of unrivalled quality, the cataphracts, mail-clad horsemen with their mounts mailed, too, charging home with a heavy lance. The fame of their great horses even awoke the envy of the distant Chinese.

After this, the eastern frontier on the Euphrates was to remain undisturbed for a century, but the Parthians did not endear themselves to Rome. They dabbled in the politics of the civil war, harassing Syria and encouraging unrest among the Palestinian Jews. Mark Antony had to retreat in disgrace and distress to Armenia after losing thirty-five thousand men in a disastrous campaign against them. But Parthia suffered from internal divisions, too, and in 20 BC Augustus was able to obtain the return of the Roman standards taken from Crassus and thankfully set aside any need to attack Parthia for reasons of honour. Yet the likelihood of conflict persisted, both because of the sensitivity with which each power regarded Armenia and because of the instability of Parthia's dynastic politics. One emperor, Trajan, conquered the Parthian capital of Ctesiphon and fought his way down to the Persian Gulf, but his successor Hadrian wisely conciliated the Parthians by handing back much of his conquest.

It was the Roman boast that their new subjects all benefited from the extension to them of the *Pax Romana*, the imperial peace which removed the threats of barbarian incursion or international strife. The claim has to be qualified by recognition of the violence with which many subject peoples resisted Roman rule, and the bloodshed this cost, but there is something in it. Within the frontiers there was order and peace as never before. In some places this permanently changed the patterns of settlement as new cities were founded in the east or descendants of Caesar's soldiers were settled in new military colonies in Gaul. Sometimes there were even more far-reaching results. The adoption of the Rhine frontier permanently affected the history of Europe by its division of the Germanic peoples. Meanwhile, everywhere, as things settled down, a gradual romanization of the local notables occurred. They were encouraged to share a common civilization whose spread was made easier by the new swiftness of communication along the roads whose main purpose was the movement of the legions. Napoleon could not move couriers faster from Paris to Rome than could the emperors of the first century AD.

The empire was a huge area and required the solution of problems of government which had not been faced by Greeks or solved by Persians. A complex bureaucracy appeared, with remarkable scope. To cite one small example, the records of all officers of centurion rank and above (company commanders upwards, as it were) were centralized at Rome. The corps of provincial civil servants was the administrative armature, sustained by a practical reliance for many places upon the army, which did much more than merely fight. Bureaucracy was controlled by the adoption of fairly limited aims. These were above all fiscal; if the taxes came in, then Roman rule did not want to interfere in other ways with the operation of local custom. Rome was tolerant. It would provide the setting within which the example of its civilization would wean barbarians from their native ways. The reform of the administrators had begun under Augustus. The Senate still appointed to many posts on an annual basis, but the emperor's *legati* who acted for him in the frontier provinces held office at his pleasure. All the evidence is that whatever the means were by which it was achieved, the administration underwent a notable improvement under the empire by comparison with the corruption of the last century of the republic. It was much more centralized and integrated than the satrapy system of Persia.

The cooperation of the subject peoples was tempted with a bait. First the republic and then the empire had been extended by granting citizenship to wider and wider numbers of Rome's subjects. It was an important privilege; among other things, as the Acts of the Apostles remind us, it carried with it rights of appeal from local courts to the emperor at Rome. On the granting of citizenship could be based the winning of the loyalties of local notables; more and more non-Romans make their appearance in the Senate and at Rome as the centuries pass. Finally, in AD 212 citizenship was granted to all free subjects of the empire.

This was an outstanding instance of Roman digestive power. The empire and the civilization it carried were unashamedly cosmopolitan. The administrative framework contained an astonishing variety of contrasts and diversities. They were held together not by an impartial despotism exercised by a Roman élite or a professional bureaucracy, but by a constitutional system which took local élites and romanized them. From the first century AD the senators themselves included only a dwindling number of men of Italian descent. Roman tolerance in this was diffused among other peoples. The empire was never a racial unity whose hierarchies were closed to non-Italians. Only one of its peoples, the Jews, felt strongly about the retention of their distinction within it and that distinction rested on religion.

Already Hellenistic civilization had achieved a remarkable mixing of East and West; now Rome continued the process over an even wider area. The element in the new cosmopolitanism which was most obvious was, indeed, the Greek, for the Romans themselves made much of their inheritance from the Greeks, though it was the Greeks of the Hellenistic era with whom they were most at home. All educated Romans were bilingual and this illustrates the tradition upon which they drew. Latin was the official language and always remained the language of the army; it was spoken widely in the West and to judge by the military records, literacy in it was high. Greek was the *lingua franca* in the eastern provinces, understood by all officials and merchants, and used in the courts if the litigants wished. Educated Romans grew up to read the Greek classics and drew from them their standards; the creation of a literature which could stand on an equal footing with the older was the laudable ambition of most Roman writers. In the first century AD they got nearest to this and the coincidence of a cultural and an imperial achievement is striking in Virgil, the conscious renewer of the epic tradition who was also the poet of imperial mission.

It may be that in this lies one clue to explain the peculiar feel of Roman culture. Perhaps it is the obviousness and pervasiveness of the Greek background which does much to deprive it of the air of novelty. Its weight was accentuated by the static, conservative concern of Roman thinkers. Between them, their attention was absorbed almost exclusively by the two foci provided by the Greek inheritance and the moral and political traditions of the republic. Both lived on curiously and somewhat artificially in a material setting which more and more ceased to fit them. Formal education changed little in practice and content from century to century, for example. Livy, the great Roman historian, sought again to quicken republican virtues in his history, but not to criticize and reinterpret them. Even when Roman civilization was irreversibly urban the (almost extinct) virtues of the independent peasant continued to be celebrated and rich Romans longed (they said) to get away from it all to the simple life of the countryside. Roman sculpture only provided again what Greeks had already done better. The philosophies of Rome were Greek, too. Epicureanism and Stoicism held the centre of the stage; neo-Platonism was innovatory, but came from the East, as did the mystery religions which were eventually to provide Roman men and women with something their culture could not give them.

Only in two practical fields were the Romans to be great creators, law and engineering. The achievements of the lawyers were relatively late; it was in the second and early third centuries AD that the jurisconsults began the accumulation of commentary which would be so valuable a legacy to the future when codification passed their work to medieval Europe. In engineering – and Romans did not distinguish it from architecture - the quality of their achievement is more immediately impressive. It was a source of pride to the Romans and one of the few things in which they were sure they outstripped the Greeks. It was based on cheap labour: at Rome it was slaves and in the provinces often the unemployed legions on garrison duty in peaceful times who carried out the great works of hydraulic engineering, bridging and road-building. But more was involved than material factors. The Romans virtually founded town-planning as an art and administrative skill west of the Indus, and their inventions of concrete and the vaulted dome revolutionized the shapes of buildings. For the first time the interiors of buildings became more than a series of surfaces for decoration. Volumes and lighting became part of the subject-matter of architecture; the later Christian basilicas were to be the first great expressions of a new concern with the spaces inside buildings.

Roman technical accomplishment was stamped on an area stretching from the Black Sea in the east to Hadrian's Wall in the north and the Atlas mountains in the south. The capital, of course, contained some of its most spectacular relics. There, the wealth of empire expressed itself in a richness of finish and decoration nowhere else so concentrated. When the marble facings were intact, and paint and stucco moulding relieved the sheer mass of stone, Rome must have had some of the appeal to the imagination earlier possessed by Babylon. There was an ostentation about it which

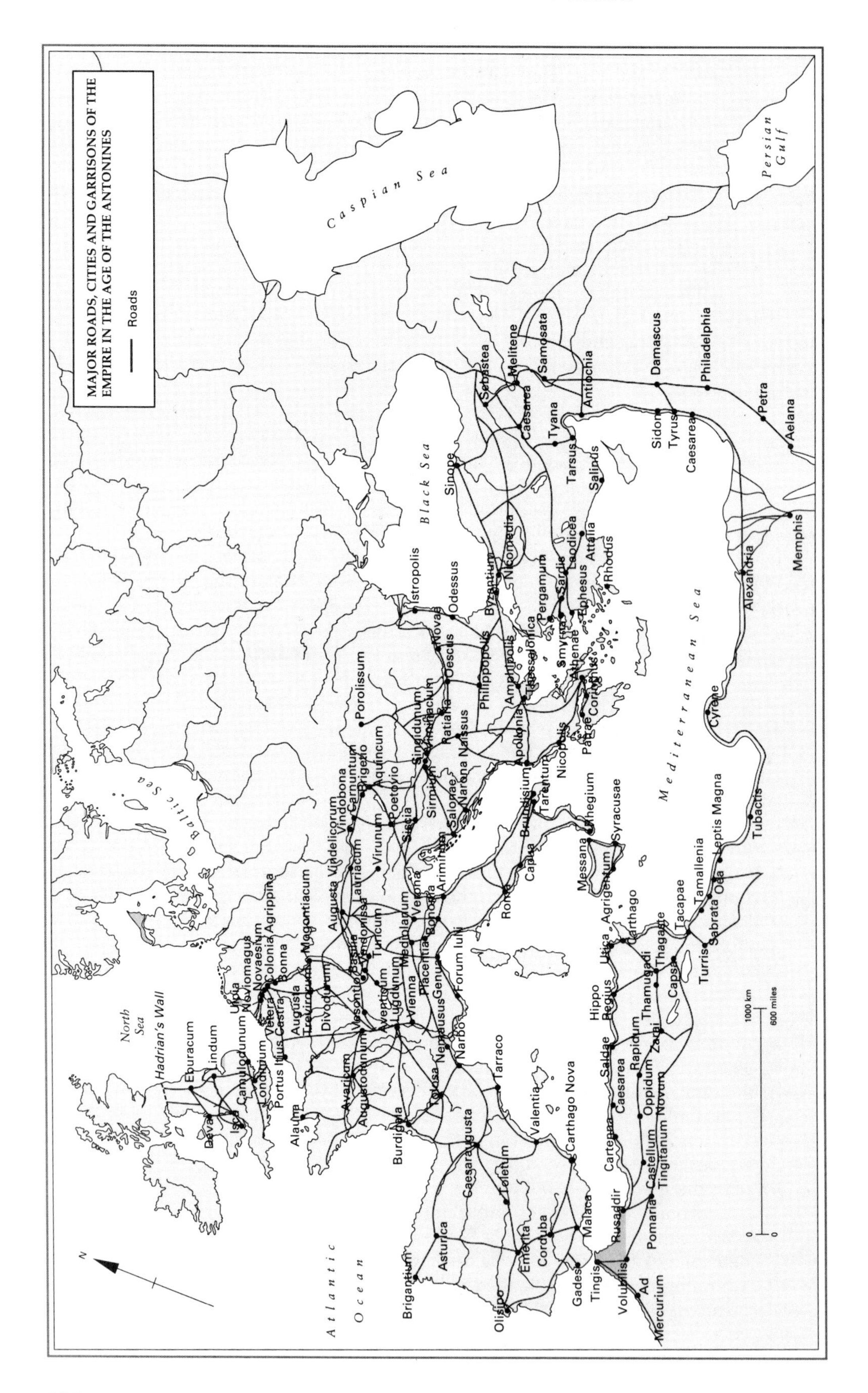
MAJOR ROADS, CITIES AND GARRISONS OF THE EMPIRE IN THE AGE OF THE ANTONINES
Roads
Caspian Sea
Persian Gulf
Black Sea
Baltic Sea
North Sea
Atlantic Ocean
Mediterranean Sea
Hadrian's Wall
Eburacum
Lindum
Deva
Isca
Camulodunum
Londinium
Portus Itius
Ulpia
Noviomagus
Novaesium
Vetera
Colonia Agrippina
Bonna
Castra
Augusta Treverorum
Mogontiacum
Divodurum
Augusta Vindelicorum
Vindobona
Carnuntum
Brigetio
Aquincum
Porolissum
Lauriacum
Virunum
Poetovio
Siscia
Singidunum
Sirmium
Viminacium
Ratiaria
Oescus
Novae
Istropolis
Odessus
Naissus
Salonae
Narona
Vesontio
Basilia
Vindonissa
Turicum
Mediolanum
Verona
Bononia
Ariminum
Placentia
Genua
Vienna
Lugdunum
Aventicum
Augustodunum
Avaricum
Alauna
Burdigala
Tolosa
Nemausus
Narbo
Forum Iulii
Tarraco
Caesaraugusta
Valentia
Carthago Nova
Toletum
Asturica
Brigantium
Emerita
Corduba
Malaca
Gades
Olisipo
Rome
Capua
Brundisium
Tarentum
Messana
Rhegium
Syracusae
Agrigentum
Nicopolis
Apollonia
Amphipolis
Thessalonica
Philippopolis
Byzantium
Nicomedia
Pergamum
Sardis
Smyrna
Ephesus
Athenae
Corinthus
Patrae
Laodicea
Attalia
Rhodus
Sinope
Sebastea
Melitene
Samosata
Caesarea
Tyana
Tarsus
Antiochia
Salamis
Damascus
Philadelphia
Sidon
Tyrus
Caesarea
Petra
Aelana
Memphis
Alexandria
Cyrene
Tubactis
Leptis Magna
Oea
Sabrata
Tamallenia
Tacapae
Turris
Capsa
Thagaste
Carthago
Utica
Hippo Regius
Thamugadi
Zarai
Saldae
Rapidum
Caesarea
Oppidum Novum
Cartenna
Castellum Tingitanum
Rusaddir
Pomaria
Tingis
Volubilis
Ad Mercurium
1000 km
600 miles
0
N

spoke of a certain vulgarity, too, and in this again it is not hard to sense a difference of quality between Rome and Greece; Roman civilization has a grossness and materiality inescapable in even its greatest monuments.

In part this was the simple expression of the social realities on which the empire rested; Rome, like all the ancient world, was built on a sharp division of rich and poor, and in the capital itself this division was an abyss not concealed but consciously expressed. The contrasts of wealth were flagrant in the difference between the sumptuousness of the houses of the new rich, drawing to themselves the profits of empire and calling on the services of perhaps scores of slaves on the spot and hundreds on the estates which maintained them, and the swarming tenements in which the Roman proletariat lived. Romans found no difficulty in accepting such divisions as part of the natural order; for that matter, few civilizations have ever much worried about them before our own, though few displayed them so flagrantly as imperial Rome. Unfortunately, though easy to recognize, the realities of wealth in Rome still remain curiously opaque to the historian. The finances of only one senator, the younger Pliny, are known to us in any detail.

The Roman pattern was reflected in all the great cities of the empire. It was central to the civilization which Rome sustained everywhere. The provincial cities stood like islands of Graeco-Roman culture in the aboriginal countrysides of the subject-peoples. Due allowance made for climate, they reflected a pattern of life of remarkable uniformity, displaying Roman priorities. Each had a forum, temples, a theatre, baths, whether added to old cities, or built as part of the basic plan of those which were refounded. Regular grid-patterns were adopted as ground plans. The government of the cities was in the hands of local bigwigs, the *curiales* or city-fathers who at least until Trajan's time enjoyed a very large measure of independence in the conduct of municipal affairs, though later a tighter supervision was to be imposed on them. Some of these cities, such as Alexandria or Antioch, or Carthage (which the Romans refounded), grew to a very large size. The greatest of all cities was Rome itself, eventually containing more than a million people.

In this civilization the omnipresence of the amphitheatre is a standing reminder of the brutality and coarseness of which it was capable. It is important not to get this out of perspective, just as it is important not to infer too much about 'decadence' from the much-quoted works of would-be moral reformers. One disadvantage under which the repute of Roman civilization has laboured is that it is one of the few before modern times in which we have very much insight into the popular mind through its entertainments, for the gladiatorial games and the wild-beast shows were emphatically mass entertainment in a way in which the Greek theatre was not. Popular relaxation is in any era hardly likely to be found edifying by the sensitive, and the Romans institutionalized its least attractive aspects by building great centres for their shows, and by permitting the mass-entertainment industry to be used as a political device; the provision of spectacular games was one of the ways in which a rich man could bring to bear his wealth to secure political advancement. Nevertheless, when all allowances are made for the fact that we cannot know how, say, the ancient masses of Egypt or Assyria amused themselves, we are left with the uniqueness of the gladiatorial spectacle; it was an exploitation of cruelty as entertainment on a bigger scale than ever before and one unrivalled until the twentieth-century cinema. It was made possible by the urbanization of Roman culture, which could deliver larger mass audiences than ever. The ultimate roots of the 'games' were Etruscan, but their development sprang from a new scale of urbanism and the exigencies of Roman politics.

Another aspect of the brutality at the heart of Roman society was, of course, far from unique: the omnipresence of slavery. As in Greek society, it was so varied in its expression that it cannot be summarized in a generalization. Many slaves earned wages, some bought their freedom and the Roman slave had rights at law. The growth of large plantation estates, it is true, provided examples of a new intensification of it in the first century or so, but it would be hard to say Roman slavery was worse than that of other ancient societies. A few who questioned the institution were very untypical: moralists reconciled themselves to slave-owning as easily as later Christians.

The loser: a bronze head of a Gallic chief of the first century AD.

Much of what we know about popular mentality before modern times is known through religion. Roman religion was a very obvious part of Roman life, but that may be misleading if we think in modern terms. It had nothing to do with individual salvation and not much with individual behaviour; it was above all a public matter. It was a part of the *res publica*, a series of rituals whose maintenance was good for the state, whose neglect would bring retribution. There was no priestly caste set apart from other men (if we exclude one or two antiquarian survivals in the temples of a few special cults) and priestly duties were the task of the magistrates who found priesthood a useful social and political lever. Nor was there creed or dogma. What was required of Romans was only that the ordained services and rituals should be carried out in the accustomed way; for the proletarian this meant little except that he should not work on a holiday. The civic authorities were everywhere responsible for the rites, as they were responsible for the maintenance of the temples. The proper observances had a powerfully practical purpose: Livy reports a consul saying the gods 'look kindly on the scrupulous observance of religious rites which has brought our country to its peak'. Men genuinely felt that the peace of Augustus was the *pax deorum*, a divine reward for a proper respect for the gods which Augustus had reasserted. Somewhat more cynically, Cicero had remarked that the gods were needed to prevent chaos in society. This, if different, was also an expression of the Roman's practical approach to religion. It was not insincere or disbelieving; the recourse to diviners for the interpretation of omens and the acceptance of the decisions of the augurs about important acts of policy would alone establish that. But it was unmysterious and down-to-earth in its understanding of the official cults.

The content of these was a mixture of Greek mythology and festivals and rites derived from primitive Roman practice and therefore heavily marked by agricultural preoccupations. One which lived to deck itself out in the symbols of another religion was the December Saturnalia, which is with us still as Christmas. But the religion practised by Romans stretched far beyond official rites. The most striking feature of the Roman approach to religion was its eclecticism and cosmopolitanism. There was room in the empire for all manner of belief, provided it did not contravene public order or inhibit adherence to the official observances. For the most part, the peasants everywhere pursued the timeless superstitions of their local nature cults, townsmen took up new crazes from time to time, and the educated professed some acceptance of the classical pantheon of Greek gods and led the people in the official observances. Each clan and household, finally, sacrificed to its own god with appropriate special rituals at the great moments of human life: childbirth, marriage, sickness and death. Each household had its shrine, each street-corner its idol.

Under Augustus there was a deliberate attempt to reinvigorate old belief, which had been somewhat eroded by closer acquaintance with the Hellenistic East and about which a few sceptics had shown cynicism even in the second century BC. After Augustus, emperors always held the office of chief priest (*pontifex maximus*) and political and religious primacy were thus combined in the same person. This began the increasing importance and definition of the imperial cult itself. It fitted well the Roman's innate conservatism, his respect for the ways and customs of his ancestors. The imperial cult linked respect for traditional patrons, the placating or invoking of familiar deities and the commemoration of great men and events, to the ideas of divine kingship which came from the East, from Asia. It was there that altars were first raised to Rome or the Senate, and there that they were soon reattributed to the emperor. The cult spread through the whole empire, though it was not until the third century AD that the practice was wholly respectable at Rome itself, so strong was republican sentiment. But even there the strains of empire had already favoured a revival of official piety which benefited the imperial cult.

This was not all that came from the East. By the second century, the distinction of a pure Roman religious tradition from others within the empire is virtually impossible. The Roman pantheon, like the Greek, was absorbed almost indistinguishably into a mass of beliefs and cults, their boundaries blurred and fluid, merging imperceptibly over a scale of experience running from sheer magic to the philosophical monotheism popularized by the stoic philosophies. The intellectual and religious world of the empire was omnivorous, credulous and deeply irrational. It is important here not to be over-impressed by the visible practicality of the Roman mind; practical men are often superstitious. Nor was the Greek heritage understood in an altogether rational way; its philosophers were seen by the first century BC as men inspired, holy men whose mystical teaching was the most eagerly studied part of their works, and even Greek civilization had always rested on a broad basis of popular superstition and local cult practice. Tribal gods swarmed throughout the Roman world.

All this boils down to a large measure of practical criticism of the ancient Roman ways. Obviously, they were no longer enough for an urban civilization, however numerically preponderant the peasants on which it rested. Many of the traditional festivals were pastoral or agricultural in origin, but occasionally even the god they invoked was forgotten. City-dwellers gradually came to need more than piety in a more and more puzzling world. Men grasped desperately at anything which could give meaning to the world and some degree of control over it. Old superstitions and new crazes benefited. The evidence can be seen in the appeal of the Egyptian gods, whose cults flooded through the empire as its security made travel and intercourse easier (they were even patronized by an emperor, the Lybian Septimius Severus). A civilized world of greater complexity and unity than any earlier was also one of greater and greater religiosity and a curiousness almost boundless. One of the last great teachers of pagan antiquity, Apollonius of Tyana, was said to have lived and studied with the Brahmans of India. Men were looking about for new saviours long before one was found in the first century AD.

Another symptom of eastern influence was the popularization of mysteries, cults which rested upon the communication of special virtues and powers to the initiated by secret rites. The sacrificial cult of Mithras, a minor Zoroastrian deity especially favoured by soldiers, was one of the most famous. Almost all the mysteries register impatience with the constraints of the material world, an ultimate pessimism about it and a preoccupation with (and perhaps a promise of survival after) death. In this lay their power to provide a psychological satisfaction no longer offered by the old gods and never really possessed by the official cult. They drew individuals to them; they had the appeal that was later to draw men to Christianity, which in its earliest days was often seen, significantly, as another mystery.

That Roman rule did not satisfy all Roman subjects all the time was even true in Italy itself when as late as 73 BC, in the disorderly last age of the Republic, a great slave revolt required three years of military campaigning and was punished with the crucifixion of 6,000 slaves along the roads from Rome to the south. In the provinces revolt was endemic, always likely to be provoked by a particular burst of harsh or bad government. Such was the famous rebellion of Boadicea in Britain, or the earlier Pannonian revolt under Augustus. Sometimes such troubles could look back to local traditions of independence, as was the case at Alexandria where they were frequent. In one particular instance, that of the Jews, they touched chords not unlike those of later nationalism. The spectacular Jewish record of disobedience and resistance goes back beyond Roman rule to 170 BC, when they bitterly resisted the 'westernizing' practices of the Hellenistic kingdoms which first adumbrated policies later to be taken up by Rome. The imperial cult made matters worse. Even Jews who did not mind Roman tax-gatherers and thought that Caesar should have rendered unto him what was Caesar's were bound to draw the line at the blasphemy of sacrifice at his altar. In AD 66 came a great revolt; there were others under Trajan and Hadrian. Jewish communities were powder-barrels. Their sensitivity makes somewhat more understandable the unwillingness of a Procurator of Judaea in about AD 30 to press hard for the strict observance of the legal rights of an accused man when Jewish leaders demanded his death.

Taxes kept the empire going. Although not heavy in normal times, when they paid for administration and police quite comfortably, they were a hated burden and one augmented, too, from time to time, by levies in kind, requisitioning and forced recruiting. For a long time, they drew on a prosperous and growing economy. This was not only a matter of such lucky imperial acquisitions as the gold-mines of Dacia. The growth in the circulation of trade and the stimulus provided by the new markets of the great frontier encampments also favoured the appearance of new industry and suppliers. The huge numbers of wine jars found by archaeologists are only an indicator of what must have been a vast commerce – of foodstuffs, textiles, spices, which have left fewer traces. Yet the economic base of empire was always agriculture. This was not rich by modern standards, for its techniques were primitive; no Roman farmer ever saw a windmill and watermills were still rare when the empire ended in the West. For all its idealization, rural life was a harsh and laborious thing. To it too, therefore, the *pax Romana* was essential: it meant that taxes could be found from the small surplus produced and that lands would not be ravaged.

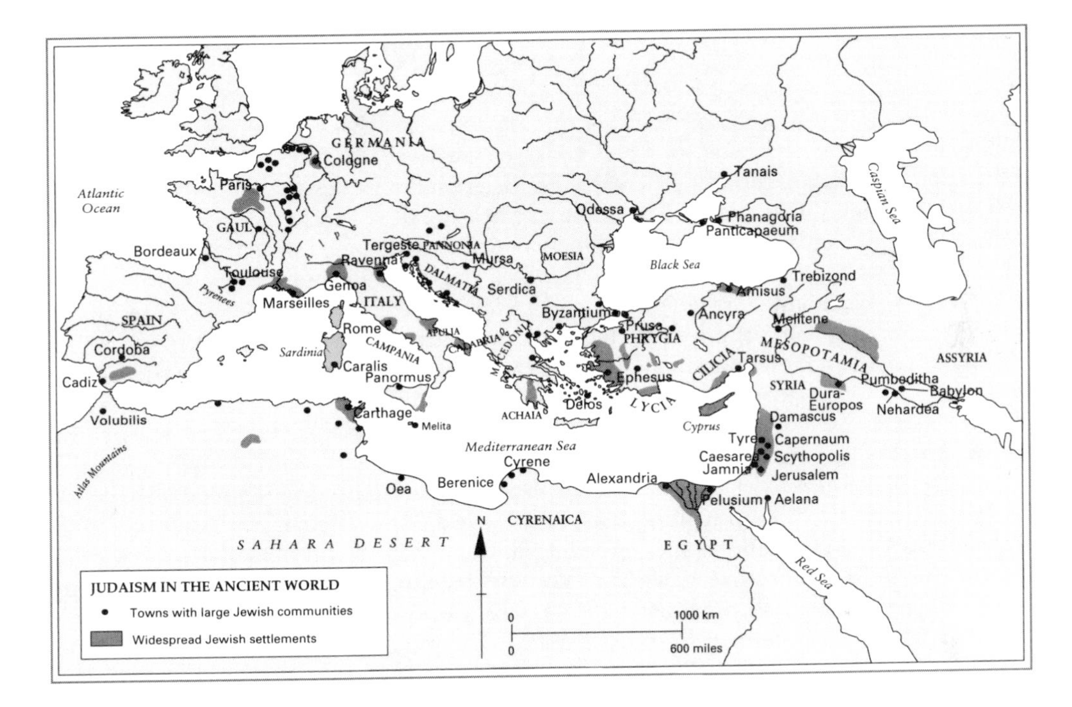

In the last resort almost everything seems to come back to the army, on which the Roman peace depended, yet it was an instrument which changed over six centuries as much as did the Roman state itself. Roman society and culture were always militaristic, yet the instruments of that militarism changed. From the time of Augustus the army was a regular long-service force, no longer relying even formally upon the obligation of all citizens to serve. The ordinary legionary served for twenty years, four in reserve, and he more and more came from the provinces as time went by. Surprising as it may seem, given the repute of Roman discipline, volunteers seem to have been plentiful enough for letters of recommendation and the use of patrons to be resorted to by would-be recruits. The twenty-eight legions which were the normal establishment after the defeat in Germany were distributed along the frontiers, about 160,000 men in all. They were the core of the army, which contained about as many men again in the cavalry, auxiliaries and other arms. The legions continued to be commanded by senators (except in Egypt) and the central issue of politics at the capital itself was still access to opportunities such as this. For, as had become clearer and clearer as the centuries passed, it was in the camps of the legions that the heart of the empire lay, though the Praetorian Guard at Rome sometimes contested their right to choose an emperor. Yet the soldiers made only part of the history of the empire. Quite as much impact was made on it, in the long run, by the handful of men who were the followers and disciples of the man the Procurator of Judaea had handed over to execution.

7
Jewry and the Coming of Christianity

Few readers of this book are likely to have heard of Abgar, far less of his east Syrian kingdom, Osrhoene; both were unknown to the writer until he was well embarked upon this book. Yet this little-known and obscure monarch is a landmark: he was long believed to be the first Christian king. In fact, the story of his conversion is a legend; it seems to have been under his descendant, Abgar VIII (or IX, so vague is our information), that Osrhoene became Christian at the end of the second century AD. The conversion may not even have included the king himself, but this did not trouble hagiographers. They placed Abgar at the head of a long and great tradition; in the end it was to incorporate virtually the whole history of monarchy in Europe. From there, in turn, it was to spread to influence rulers in other parts of the world.

All these monarchs would behave differently because they saw themselves as Christian, yet, important though it was, this is only a tiny part of the difference Christianity has made to history. Until the coming of industrial society, in fact, it is the only historical phenomenon we have to consider whose implications, creative power and impact are comparable with the great determinants of prehistory in shaping the world we live in. Christianity grew up within the classical world of the Roman empire, fusing itself in the end with its institutions and spreading through its social and mental structures to become our most important legacy from that civilization. Often disguised or muted, its influence runs through all the great creative processes of the last fifteen hundred years; almost incidentally, it defined Europe. We are what we are today because a handful of Jews saw their teacher and leader crucified and believed he rose again from the dead.

The Jewishness of Christianity is fundamental and was probably its salvation (to speak in purely human terms), for the odds against the historical survival, let alone world-wide success, of a small sect centred upon a holy man in the Roman eastern empire were enormous. Judaism was a matrix and protecting environment for a long time as well as the source of the most fundamental Christian ideas. In return, Jewish ideas and myths were to be generalized through Christianity to become world forces. At the heart of these was the Jewish view that history was a meaningful story, providentially ordained, a cosmic drama of the unfolding design of the one, omnipotent God for His chosen people. Through His covenant with that people could be found guidance for right action, and it lay in adherence to His law. The breaking of that law had always brought punishment; it had come to the whole people in the deserts of Sinai and by the waters of Babylon. This great drama was the inspiration of Jewish historical writing, in which the Jews of the Roman empire discerned the pattern which made their lives meaningful.

That mythological pattern was deeply rooted in Jewish historical experience, which, after the great days of Solomon, had been bitter, fostering an enduring distrust of the foreigner and an iron will to survive. Few things are in fact more remarkable in the life of this remarkable people than the simple fact of its continued existence. The Exile which began in 587 BC when Babylonian conquerors took many of the Jews away after the destruction of the Temple was the last crucial experience in the moulding of their national identity before modern times. It finally crystallized the Jewish vision of history. The exiles heard prophets like Ezekiel promise a renewed covenant; Judah had been punished for her sins by exile and the Temple's destruction, now God would turn His face again to her, she would return again to Jerusalem, delivered out of Babylon as Israel had been delivered out of Ur, out of Egypt. The Temple would be rebuilt. Perhaps only a minority of the Jews of the Exile heeded this, but it was a large one and it included Judah's religious and administrative élite, if we are to judge by the quality of those – again, probably a minority – who, when they could do so, returned to Jerusalem, a saving Remnant, according to prophecy.

Before then, the experience of the Exile had transformed Jewish life as well as confirming the Jewish vision. Scholars are divided as to whether the more important developments took place among the exiles or among the Jews who were left in Judah to lament what had happened. In one way or another, though, Jewish religious life was deeply stirred. The most important change was the implanting of the reading of the scriptures as the central act of Jewish religion. While the Old Testament was not to assume its final form for another three or four centuries, the first five books, or 'Pentateuch', traditionally ascribed to Moses, were substantially complete soon after the return from the Exile. Without the focus of cult practice at the Temple the Jews seem to have turned to weekly meetings to hear these sacred texts read and expounded. They contained the promise of a future and guidance to its achievement through maintenance of the Law, now given a new detail and coherence. This was one of the slow effects of the work of the interpreters and scribes who had to reconcile and explain the sacred books. In the end there was to grow out of these weekly meetings both the institution of the synagogue and a new liberation of religion from locality and ritual, however much and long Jews continued to pine for the restoration of the Temple. The Jewish religion could eventually be practised wherever Jews could come together to read the scriptures; they were to be the first of the peoples of a Book, and Christians and Moslems were to follow them. It made possible greater abstraction and universalizing of the vision of God.

There were narrowings, too. Although Jewish religion might be separated from the Temple cult, some prophets had seen the redemption and purification which must lie ahead as only to be approached through an even more rigid enforcement of what was now believed to be Mosaic law. Ezra brought back its precepts from Babylon and observances which had been in origin those of nomads were now imposed rigorously on an increasingly urbanized people. The self-segregation of Jews became much more important and obvious in towns; it was seen as a part of the purification which was needed that every Jew married to a gentile wife (and there must have been many) should divorce her.

This was after the Persian overthrow of Babylon. In 538 BC some of the Jews took the opportunity offered to them and came back to Jerusalem. The Temple was rebuilt and Judah became under Persian overlordship a sort of theocratic satrapy, effective administrative power being in the hands of the priestly aristocracy, which provided the political articulation of Jewish nationhood until Roman times.

With the ending of Persian rule, the age of Alexander's heirs brought new problems. After being ruled by the Ptolemies, the Jews eventually passed to the Seleucids. The social behaviour and thinking of the upper classes underwent the influence of Hellenization; this sharpened divisions by exaggerating contrasts of wealth and differences between townsmen and countrymen. It also separated the priestly families from the people, who remained firmly in the tradition of the Law and the Prophets, as expounded in the synagogues. It was against a king of Hellenistic Syria, Antiochus, and cultural 'westernization' approved by the priests, but resented as have been such processes in modern times by the masses, that the great Maccabaean revolt broke out (168–164 BC). Antiochus had tried to go too fast; not content with the steady erosion of Jewish insularity by Hellenistic civilization and the friction of example, he had interfered with Jewish rites and profaned the Temple. After the revolt had been repressed with difficulty (and guerrilla war went on long after), a more conciliatory policy was resumed by the Seleucid kings. It did not satisfy many Jews, who in 142 BC were able to take advantage of a favourable set of circumstances to win an independence which was to last for nearly eighty years. Then, in 63 BC, Pompey imposed Roman rule and there disappeared the last independent Jewish state in the Near East for nearly two thousand years.

Independence had not been a happy experience. A succession of kings drawn from the priestly families had thrown the country into disorder by innovation and high-handedness. They and the priests who acquiesced in their policies excited opposition. They were challenged in their authority by a new, more austere, school of interpreters, who clung to the Law, rather than the cult, as the heart of Judaism and gave it new and searchingly rigorous interpretation. These were the Pharisees, the representatives of a reforming strain which was time and time again to express itself in Jewry in protest against the danger of creeping Hellenization. They also accepted proselytism among non-Jews, teaching a belief in the resurrection of the dead and a divine Last Judgement; there was a mixture in their stance of national and universal aspiration and they drew out further the implications of Jewish monotheism.

Most of these changes took place in Judaea, the tiny rump of the once great kingdom of David; fewer Jews lived there in the time of Augustus than in the rest of the empire. From the seventh century onwards they had spread over the civilized world. The armies of Egypt, Alexander and the Seleucids all had Jewish regiments. Others had settled abroad in the course of trade. One of the greatest Jewish colonies was at Alexandria, where they had gathered from about 300 BC. The Alexandrian Jews were Greek-speakers; there the Old Testament was first translated into Greek and when Jesus was born there were probably more Jews there than in Jerusalem. At Rome there were another 50,000 or so. Such agglomerations increased the opportunities to proselytize and therefore the danger of friction between communities.

Jewry offered much to a world where traditional cults had waned. Circumcision and dietary restraints were obstacles, but were far outweighed for many a proselyte by the attractions of a code of behaviour of great minuteness, a form of religion not dependent on temples, shrines or priesthood for its exercise, and, above all, the assurance of salvation. A prophet whose teaching was ascribed by the Old Testament compilers to Isaiah, but who is almost certainly of the Exile, had already announced a message to bring light to the gentiles, and many of them had responded to that light long before the Christians, who were to promote it in a special sense. The proselytes could indentify themselves with the Chosen People in the great story which inspired Jewish historical writing, the only achievement in this field worthy of comparison with the Greek invention of scientific history, and one which gave meaning to the tragedies of the world. In their history the Jews discerned an unfolding pattern by which they were being refined in the fire for the Day of Judgement. A fundamental contribution of Jewry to Christianity would be its sense of the people apart, its eyes set on things not of this world; Christians were to go on to the idea of the leaven in the lump, working to redeem the world. Both myths were deeply rooted in Jewish historical experience and in the remarkable though simple fact of this people's survival at all.

The big communities of Jews and Jewish proselytes were important social facts to Roman governors, standing out not only because of their size but because of their tenacious separateness. Archaeological evidence of synagogues as special and separate buildings does not appear until well into the Christian era, but Jewish quarters in cities were distinct, clustering about their own synagogues and courts of law. While proselytizing was widespread and even some Romans were attracted by Jewish belief, there were also early signs of popular dislike of Jews at Rome itself. Rioting was frequent in Alexandria and easily spread to other towns of the Near East. This led to distrust on the part of authorities and (at least at Rome) to the dispersal of Jewish communities when things became difficult.

Judaea itself was regarded as a particularly ticklish and dangerous area and to this the religious ferment of the last century and a half BC had greatly contributed. In 37 BC the Senate appointed a Jew, Herod the Great, King of Judaea. He was an unpopular monarch. No doubt this was in part a matter of national distaste for a Roman nominee and a ruler anxious – with reason – to preserve the friendship of Rome, but it was exacerbated by the Hellenistic style of life at his court (though he was careful to display his loyalty to

The menorah or seven-branched candlestick symbolizes for Jews the seven days of creation. This carving of one (from a sarcophagus at Rome in about AD 300) is combined with classical figures and shows how assimilated members of the large Roman Jewish community could become. Conventional 'victories' support the menorah; the cupid-like figure represents winter, the other three seasons, once part of the design, having been lost.

the Jewish religion) and by the heavy taxes which he raised, some of them for grandiose building. Even if it were not for the legendary Massacre of the Innocents and his place in Christian demonology, Herod would not have had a good historical press. At his death, in 4 BC, his kingdom was divided between his three sons, an unsatisfactory arrangement which was superseded in AD 6, when Judaea became part of the Roman province of Syria governed from Caesarea. In AD 26 Pontius Pilate became Procurator, or governor, and was to hold the uncomfortable and exacting post for ten years.

It was a bad moment in the history of a turbulent province. Something of a climax to the excitements of nearly two centuries was being reached. The Jews were at loggerheads with their Samaritan neighbours and resented an influx of Greek-Syrians noticeable in the coastal towns. They detested Rome as the latest of a long line of conquerors and also because of its demands for taxes; tax-gatherers – the 'publicans' of the New Testament – were unpopular not just because of what they took but because they took it for the foreigner. But worse still, the Jews were also bitterly divided between themselves. The great religious festivals were often stained by bloodshed and rioting. Pharisees, for instance, were bitterly divided from Sadducees, the formalising representatives of the aristocratic priestly caste. Other sects rejected them both. One of the most interesting of them has become known to us only in recent years, through the discovery and reading of the Dead Sea Scrolls, in which it can be seen to have promised its adherents much that was also the promise of early Christianity. It looked forward to a last deliverance which would follow Judaea's apostasy and would be announced by the coming of a Messiah. Jews attracted by such teaching searched the writings of the Prophets for the prefigurings of these things. Others sought a more direct way. The Zealots looked to the nationalist resistance movement as the way ahead.

Into this electric atmosphere Jesus was born in about 6 BC, into a world in which thousands of his countrymen awaited the coming of a Messiah, a leader who would lead them to military or symbolic victory and inaugurate the last and greatest days of Jerusalem. The evidence for the facts of his life is contained in the records written down after his death in the Gospels, the assertions and traditions which the early Church

From the whole New Testament, no non-Jew is better remembered than Pontius Pilate, Roman governor of Judaea from AD *26 to 36. His historical existence has long been attested not only by the Gospels but by other literary sources; only in 1961, though, did this fragment of inscribed stone come to light and provide evidence of his existence in the form of his dedication of a building to the emperor.*

based on the testimony of those who had actually known Jesus. The Gospels are not by themselves satisfactory evidence but their inadequacies can be exaggerated. They were no doubt written to demonstrate the supernatural authority of Jesus and the confirmation provided by the events of his life for the prophecies which had long announced the coming of Messiah. This interested and hagiographical origin does not demand scepticism about all the facts asserted; many have inherent plausibility in that they are what might be expected of a Jewish religious leader of the period. They need not be rejected; much more inadequate evidence about far more intractable subjects has often to be employed. There is no reason to be more austere or rigorous in our canons of acceptability for early Christian records than for, say, the evidence in Homer which illuminates Mycenae. Nevertheless, it is very hard to find corroborative evidence of the facts stated in the Gospels in other records.

The picture of Jesus presented in them is of a man of modest though not destitute family, with a claim to royal lineage. Such a claim would no doubt have been

Thirteen hundred feet above the Dead Sea, the rock of Masada was chosen by Herod the Great as a site for a fortress containing palaces, arsenals, barracks and great water tanks. It was virtually a personal refuge for him and his family. This was in about 30 BC. *A century later, in* AD *73, this fortress was the scene of the siege endured by Jewish rebels whose revolt against the Romans had by then been crushed everywhere else. In the end, faced with defeat, the entire garrison and their families destroyed themselves, only two women and five children whom they had hidden surviving, reports Josephus. Excavated in the 1960s, the scene of this heroic struggle is now a shrine of Israeli patriotism.*

denied by his opponents if there had not been something in it. Galilee, where Jesus grew up, was something of a frontier area for Judaism, where it was most exposed to the contact with Syrian-Greeks which often irritated religious sensibilities. There preached in the neighbourhood a man called John, a prophet to whom crowds had flocked in the days before his arrest and execution. Scholars now believe John to have been connected with the Qumran community which left behind the Dead Sea Scrolls. One evangelist tells us that he was the cousin of Jesus; this is possibly true, but less important than the agreement of all the Gospels that John baptized Jesus as he baptized countless others who came to him fearing the approach of the Last Day. He is also said to have recognized in Jesus a teacher like himself and perhaps something more: 'Art thou He that cometh, or look we for another?' Jesus knew himself to be a holy man; his teaching and the evidence of his sanctity which was seen in miracles soon convinced the excited multitude to Jerusalem. His triumphal entry to the city was based on their spontaneous feeling. They followed him as they followed other great teachers in the hope of the Messiah that was to come. The end was a charge of blasphemy before the Jewish court and the relaxation of the letter of Roman law by a Roman governor in order to avoid further trouble in a violent city. Jesus was not a Roman citizen and for such men the extreme penalty was crucifixion after scourging. The inscription on the cross on which he was nailed said: 'Jesus of Nazareth, King of the Jews'; this made clear that a political act was envisaged, and that the significance of it should not go unmissed was ensured by posting the inscription in Latin, Greek and Hebrew. This was probably in AD 33, though AD 29 and AD 30 have also been put forward as dates. Shortly after his death, Jesus' disciples believed that he had risen from the dead, that they had seen him and his ascension into heaven, and that they had received a divine gift of power from him at Pentecost which should sustain them and their adherents until the Last Day. That would soon come, they also believed, and would bring back Jesus as the judge sitting at the right hand of God. All this the Gospels tell us.

If this was what the first Christians saw in Christ (as he came to be called, from the Greek word meaning 'the anointed one') there were also in his teaching other elements susceptible of far wider application. The reported devotional ideas of Jesus do not go beyond the Jewish observances; service in the Temple, together with private prayer, were all that he indicated. In this very real sense, he lived and died a Jew. His

'The Dead Sea Scrolls' are manuscripts discovered first in 1947 in caves and hiding-places around the Dead Sea. One of them, the largest (twenty-four feet if extended) contains almost the whole book of Isaiah and is thought to have been part of a monastic library belonging to a sect which settled near Qumran sometime after the Roman occupation of Palestine in 63 BC. It then stayed there until its adherents were dispersed or destroyed after the Jewish uprising of AD *66.*

moral teaching, though, focused upon repentance and deliverance from sin, and upon a deliverance available to all, and not just to Jews. Retribution had its part in Jesus' teaching (on this the Pharisees agreed with him), and, strikingly, most of the more terifying things said in the New Testament are attributed to him. Fulfilment of the Law was essential. Yet it was not enough; beyond observance lay the duties of repentance and restitution in the case of wrong done, even self-sacrifice. The law of love was the proper guide to action. Emphatically, Jesus rejected the role of the political leader. A political quietism was one of the meanings later discerned in a dictum which was to prove to be of terrible ambiguity: 'My kingdom is not of this world.'

Yet a Messiah who would be a political leader was expected by many. Others sought a leader against the Jewish religious establishment and therefore were potentially a danger to order even if they aimed only at religious purification and reform. Inevitably, Jesus, of the house of David, became a dangerous man in the eyes of the authorities. One of his disciples was Simon the Zealot, an alarming associate because he had been a member of an extremist sect. Many of Jesus' teachings encouraged feeling against the dominant Sadducees and Pharisees, and they in their turn strove to draw out any anti-Roman implication which could be discerned in what he said.

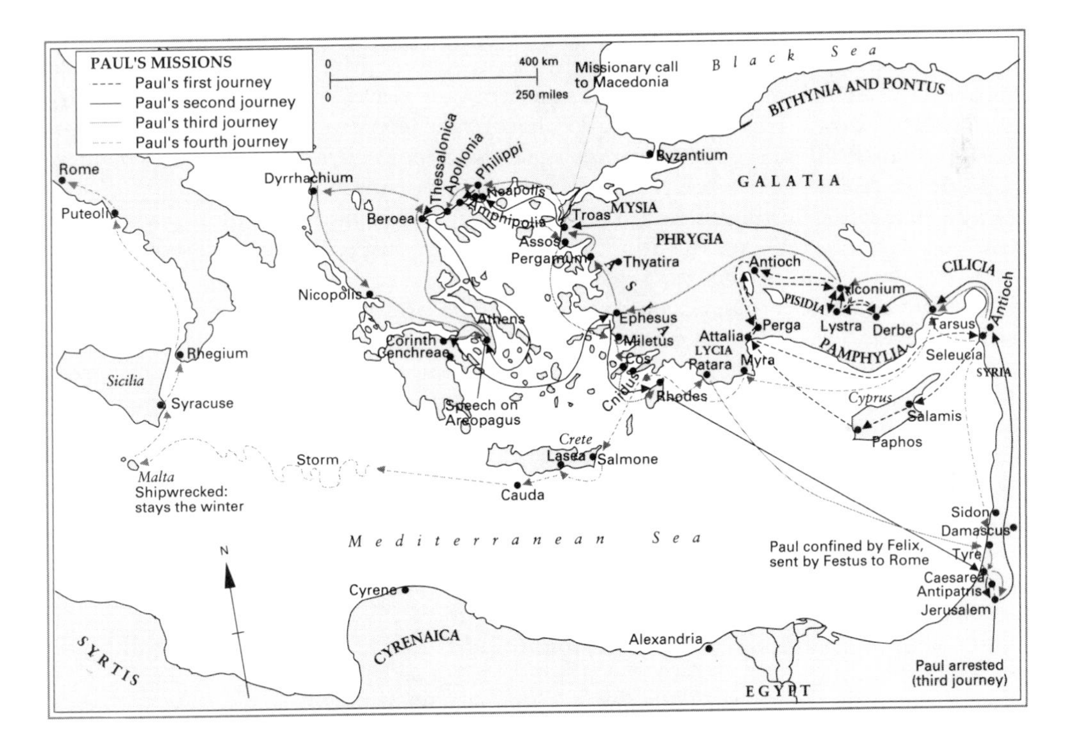

Such facts provide the background to Jesus' destruction and the disappointment of the people; they do not explain the survival of his teaching. He had appealed not only to the politically dissatisfied but to Jews who felt that the Law was no longer guide enough and to non-Jews who, though they might win second-class citizenship of Israel as proselytes, wanted something more to assure them of acceptance at the Last Day. Jesus had also attracted the poor and outcast; they were many in a society which offered enormous contrasts of wealth and no mercy to those who fell by the wayside. These were some of the appeals and ideas which were to yield in the end an astonishing harvest. Yet though they were effective in his own lifetime, they seemed to die with him. At his death his followers were only one tiny Jewish sect among many. But they believed that a unique thing had happened. They believed that Christ had risen from the dead, that they had seen him, and that he offered to them and those that were saved by his baptism the same overcoming of death and personal life after God's judgement. The generalization of this message and its presentation to the civilized world was achieved within a half-century of Jesus' death.

The conviction of the disciples led them to remain at Jerusalem, an important centre of pilgrimage for Jews from all over the Near East, and therefore a seminal centre for a new doctrine. Two of Jesus' disciples, Peter and Jesus' brother, James, were the leaders of the tiny group which awaited the imminent return of the Messiah, striving to prepare for it by penitence and the service of God in the Temple. They stood emphatically within the Jewish fold; only the rite of baptism, probably, distinguished them. Yet other Jews saw in them a danger; their contacts with Greek-speaking Jews from outside Judaea led to questioning of the authority of the priests. The first martyr, Stephen, one of this group, was lynched by a Jewish crowd. One of those who witnessed this was a Pharisee from Tarsus of the tribe of Benjamin, named Paul. It may have been that as a Hellenized Jew of the dispersion he was especially conscious of the need for orthodoxy. He was proud of his own. Yet he is the greatest influence in the making of Christianity after Jesus himself.

Somehow, Paul underwent a change of heart. From being a persecutor of the followers of Christ, he became one himself: it seems to have followed a sojourn of meditation and reflexion in the deserts of east Palestine. Then, in AD 47 (or perhaps earlier; dating Paul's life and travels is a very uncertain business), he began a series of missionary journeys which took him all over the eastern Mediterranean. In AD 49 an apostolic council at Jerusalem took the momentous decision to send him as a missionary to gentiles, who would not be required to undergo the circumcision which was the most important act of submission to the Jewish faith; it is not clear whether he, the council, or both in agreement were responsible. There were already little communities of Jews following the new teaching in Asia Minor, where it had been carried by pilgrims. Now these were given a great consolidation by Paul's efforts. His especial targets were Jewish proselytes, gentiles to whom he could preach in Greek and who were now offered full membership of Israel through the new covenant. The doctrine that Paul taught was new. He rejected the Law (as Jesus had never done), and strove to reconcile the essentially Jewish ideas at the heart of Jesus' teaching with the conceptual world of the Greek language. He continued to emphasize the imminence of the coming end of things, but offered all nations, through Christ, the chance of understanding the mysteries of creation and, above all, of the relationship of things seen and things invisible, of the spirit and the flesh, and of the overcoming of the second by the first. In the process, Jesus became more than a human deliverer who had overcome death, and was God Himself – and this was to shatter the mould of Jewish thought within which the faith had been born. There was no lasting place for such an idea within Jewry, and Christianity was now forced out of the Temple. The intellectual world of Greece was the first of many new resting-places it was to find as the centuries went by. A colossal theoretical structure was to be built on this change.

The Acts of the Apostles give plentiful evidence of the uproar which such teaching could cause and also of the intellectually tolerant attitude of the Roman administration when public order was not involved. But it often was. In AD 59, Paul had to be rescued from the Jews at Jerusalem by the Romans. When put on trial in the following year, he appealed to the emperor and to Rome he went, apparently with success. From that time he is lost to history; he may have perished in a persecution by Nero in AD 67.

The first age of Christian missions permeated the civilized world by sinking roots everywhere in the first place in the Jewish communities. The 'Churches' which emerged were administratively wholly independent of one another, though the community at Jerusalem was recognized to have an understandable primacy. There were to be found those who had seen the risen Christ and their successors. The only links of the Churches other than their faith were the institutional one of baptism, the sign of acceptance in the new Israel, and the ritual practice of the eucharist, the re-enactment of the rites performed by Jesus at his last supper with his disciples, the evening before his arrest. It has remained

The crushing of the Jewish Revolt of AD 66 was commemorated as a great triumph of Roman arms. On the Arch of Titus in Rome were placed reliefs, this one recording the bringing to Rome from the Temple of the Menorah, the table of the shewbread and the sacred trumpets, all of them spoils of war.

the central sacrament of the Christian Churches to this day.

The local leaders of the Churches exercised independent authority in practice, therefore, but this did not cover much. There was nothing except the conduct of the affairs of the local Christian community to be decided upon, after all. Meanwhile, Christians waited for the Second Coming. Such influence as Jerusalem had flagged after AD 70, when a Roman sack of the city dispersed many of the Christians there; after this time Christianity had less vigour within Judaea. By the beginning of the second century the communities outside Palestine were clearly more numerous and more important and had already evolved a hierarchy of officers to regulate their affairs. These identified the three later orders of the Church: bishops, presbyters and deacons. Their sacerdotal functions were at this stage minimal, and it was their administrative and governmental role which mattered.

The response of the Roman authorities to the rise of a new sect was largely predictable; its governing principle was that when no specific cause for interference existed, new cults were tolerated unless they awoke disrespect or disobedience to the empire. There was a danger at first that the Christians might be confounded with other Jews in a vigorous Roman reaction to Jewish nationalist movements which culminated in a number of bloody encounters, but their own political quietism and the announced hostility of other Jews saved them. Galilee itself had been in rebellion in AD 6 (perhaps a memory of it influenced Pilate's handling of the case of a Galilean among whose disciples was a Zealot) but a real distinction from Jewish nationalism came with the great Jewish rising of AD 66. This was the most important in the whole history of Jewry under the empire, when the extremists gained the upper hand in Judaea and took over Jerusalem. The Jewish historian Josephus has recorded the atrocious struggle which followed, the final storming of the Temple, the headquarters of resistance, and its burning after the Roman victory. Before this, the unhappy inhabitants had been reduced to cannibalism in their struggle to survive. Archaeology has recently revealed at Masada, a little way from the city, what may well have been the site of the last stand of the Jews before it, too, fell to the Romans in AD 73.

This was not the end of Jewish turbulence, but it was a turning-point. The extremists never again enjoyed such support and must have been discredited. The Law was now more than ever the focus of Jewishness, for the Jewish scholars and teachers (after this time, they are more and more designated as 'rabbis') had continued to unfold its meaning in centres other than Jerusalem while the revolt was in progress. Their good conduct may have saved these Jews of the dispersion. Later disturbances were never so important as had been the great revolt, though in AD 117 Jewish riots in Cyrenaica developed into full-scale fighting, and in AD 132 the last 'Messiah', Simon Bar Kochba, launched another revolt in Judaea. But the Jews emerged with their special status at law still intact. Jerusalem had been taken from them (Hadrian made it an Italian colony, which Jews might enter only once a year), but their religion was granted the privilege of having a special officer, a patriarch, with sovereignty over it, and they were allowed exemption from the obligations of Roman law which might conflict with their religious duties. This was the end of a volume of Jewish history. For the next eighteen hundred years Jewish history was to be the story of communities of the *diaspora* (dispersion), until a national state was again established in Palestine among the debris of another empire.

The nationalists of Judaea apart, Jews elsewhere in the empire were for a long time thereafter safe enough during the troubled years. Christians did less well, though their religion was not much distinguished from Judaism by the authorities; it was, after all, only a variant of Jewish monotheism with, presumably, the same claims to make. It was the Jews, not the Romans, who first persecuted it, as the Crucifixion itself, the martyrdom of Stephen and the adventures of Paul had shown. It was a Jewish King, Herod Agrippa, who, according to the author of the Acts of the Apostles, first persecuted the community at Jerusalem. It has even appeared plausible to some scholars that Nero, seeking a scapegoat for a great fire at Rome in AD 64, should have had the Christians pointed out to him by hostile Jews. Whatever the source of this persecution, in which, according to popular Christian tradition, St Peter and St Paul both perished, and which was accompanied by horrific and bloody scenes in the arena, it seems to have been for a long time the end of any official attention by Rome to the Christians. They did not take up arms against the Romans in the Jewish revolts, and this must have soothed official susceptibilities with regard to them.

When they emerge in the administrative records as worth notice by government it is in the early second century AD. This is because of the overt disrespect which Christians were by then showing in refusing to sacrifice to the emperor and the Roman deities. This was their distinction. Jews had a right to refuse; they had possessed a historic cult which the Romans respected – as they always respected such cults – when they took Judah under their rule. The Christians were now clearly seen as distinct from other Jews and were a recent creation. Yet the Roman attitude was that although Christianity was not legal it should not be the subject of general persecution. If, on the other hand, breaches of the law were alleged – and the refusal to sacrifice might be one – then the authorities should punish when the allegations were specific and shown in court to be well founded. This led to many martyrdoms, as Christians refused the well-intentioned attempts of Roman civil servants to persuade them to sacrifice or abjure their god, but there was no systematic attempt to eradicate the sect.

Indeed, the authorities' hostility was much less dangerous than that of the Christians' fellow-subjects. As the second century passed, there is more evidence of pogroms and popular attacks on Christians, who were not protected by the authorities since they followed an illegal religion. They may sometimes have been acceptable scapegoats for the administration or lightning-conductors diverting dangerous currents. It was easy for the popular mind of a superstitious age to attribute to Christians the offences to the gods which led to famine, flood, plague and other natural disasters. Other equally convincing explanations of these things were lacking in a world with no other technique of explaining natural disaster. Christians were alleged to practise black magic, incest, even cannibalism (an idea no doubt explicable in terms of misleading accounts of the eucharist). They met secretly at night. More specifically and acutely, though we cannot be sure of the scale of this, the Christians threatened by their control of their members the whole customary structure which regulated and defined the proper relations of parents and children, husbands and wives, masters and slaves. They proclaimed that in Christ there was neither bond nor free and that he had come to bring not peace but a sundering sword to families and friends. It is not hard, therefore, to understand the violent outbursts in the big provincial towns, such as that at Smyrna in AD 165, or Lyons in AD 177. They were the popular aspect of an intensification of opposition to Christianity which had an intellectual counterpart in the first attacks on the new cult by pagan writers.

Persecution was not the only danger facing the early Church. Possibly it was the least grave. A much more serious one was that it might develop into just another cult of the kind of which many examples could be seen in the Roman empire and, in the end, be engulfed like them in the magical morass of ancient religion. All over the Near East could be found examples of the 'mystery religions', whose core was the initiation of the believer into the occult knowledge of a devotion centred on a particular god (the Egyptian Isis was a popular one, the Persian Mithras another). Almost always the believer was offered the chance to identify himself with the divine being in a ceremony which involved a simulated death and resurrection and thus overcame mortality. Such cults offered, through their impressive rituals, the peace and liberation from the temporal which many craved. They were very popular.

Among the oriental cults which spread widely throughout the empire, one of the most successful was that of the ancient Egyptian goddess Isis. This Coptic fresco of the third century depicts her with her son Harpocrates, god of silence (hence his finger before his mouth). There is an obvious similarity to Christian iconography. Other parallels could be made; the myth of Osiris, consort of Isis, centred upon his death and resurrection.

That there was a real danger that Christianity might develop in this way is shown by the importance in the second century of the Gnostics. Their name derives from the Greek word *gnosis*, meaning 'knowledge': the knowledge the Christian Gnostics claimed was a secret, esoteric tradition, not revealed to all Christians but only to a few (one version said only to the Apostles and the sect to which it had subsequently descended). Some of their ideas came from Zoroastrian, Hindu and Buddhist sources which stressed the conflict of matter and spirit in a way which distorted the Judaeo-Christian tradition; some came from astrology and even magic. There was always a temptation in such a dualism, the attribution of evil and good to opposing principles and entities and the denial of the goodness of the material creation. The Gnostics were haters of this world and in some of their systems this led to the pessimism typical of the mystery cults; salvation was only possible by the acquisition of arcane knowledge, secrets of an initiated elect. A few Gnostics even saw Christ not as the saviour who confirmed and renewed a covenant but as one who delivered men from Yahweh's error. It was a dangerous creed in whatever form it came, for it cut at the roots of hopefulness which was the heart of the Christian revelation. It turned its back on the redemption of the here and now of which Christians could never wholly despair, since they accepted the Judaic tradition that God made the world and that it was good.

In the second century, with its communities scattered throughout the *diaspora* and their organizational foundations fairly firmly settled, Christianity thus seems to stand at a parting of ways, either of which could prove fatal to it. Had it turned its back on the implications of Paul's work and remained merely a Jewish heresy, it would at best have been reabsorbed eventually into the Judaic tradition; on the other hand, a flight from a Jewry which rejected it might have driven Christians into the Hellenistic world of the mystery cults or the despair of the Gnostics. Thanks to a handful of men, it escaped both.

The achievement of the Fathers of the Church who navigated these perils was, for all its moral and pietistic content, above all intellectual. They were stimulated by their danger. Irenaeus, who succeeded the martyred Bishop of Lyons in AD 177, provided the first great outline of Christian doctrine, a creed and definition of the scriptural canon. All of these set off Christianity from Judaism. But he wrote also against the background of the challenge of heretical beliefs. In AD 172 the first Council had met to reject Gnostic doctrines. Christian doctrine was squeezed into intellectual respectability by the need to resist the pressures of competitors. Heresy and orthodoxy were born twins. One of the pilots who steered an emerging Christian theology through this period was the prodigiously learned Clement of Alexandria, a Christian Platonist (perhaps born in Athens), through whom Christians were brought to an understanding of what the Hellenistic tradition might mean apart from the mysteries. In particular, he directed Christians to the thought of Plato. To his even greater pupil, Origen, he transmitted the thought that God's truth was a reasonable truth, a belief which could attract men educated in the stoic view of reality.

The intellectual drive of the early Fathers and the inherent social appeal of Christianity made it possible for it to utilize the huge possibilities of diffusion and expansion inherent in the structure of the classical and later Roman world. Its teachers could move freely and talk and write to one another in Greek. It had the great advantage of emerging in a religious age; the monstrous credulousness of the second century cloaks deep longings. They hint that the classical world is already running out of vigour; the Greek capital needed replenishment and one place to look for it was in new religions. Philosophy had become a religious quest and rationalism or scepticism appealed only to an infinitesimally small minority. Yet this promising setting was also a challenge to the Church; early Christianity has to be seen always in the context of thriving competitors. To be born in a religious age was a threat as well as an advantage. How successfully Christianity met the threat and seized its opportunity was to be seen in the crisis of the third century, when the classical world all but collapsed and survived only by colossal, and in the end mortal, concession.

8
The Waning of the Classical West

After 200 AD there are many signs that Romans were beginning to look back on the past in a new way. Men had always talked of golden ages in the past, indulging in a conventional, literary nostalgia. But the third century brought something new, a sense of conscious decline.

Historians have spoken of a 'crisis', but its most obvious expressions were in fact surmounted. The changes Romans carried out or accepted by the year 300 gave a new lease of life to much of classical Mediterranean civilization. They may even have been decisive in ensuring that it would in the end transmit so much of itself to the future. Yet the changes themselves took a toll, for some of them were essentially destructive of the spirit of that civilization. Restorers are often unconscious imitators. Somewhere about the beginning of the fourth century we can sense that the balance has tipped against the Mediterranean heritage. It is easier to feel it than to see what was the crucial moment. The signs are a sudden multiplication of ominous innovations – the administrative structure of the empire is rebuilt on new principles, its ideology is transformed, the religion of a once-obscure Jewish sect becomes established orthodoxy, and physically, large tracts of territory are given up to settlers from outside, alien immigrants. A century later still, and the consequence of these changes is apparent in political and cultural disintegration.

The ups and downs of imperial authority mattered a lot in this process. Classical civilization had come by the end of the second century AD to be coterminous with the empire. It was dominated by the conception of *romanitas*, the Roman way of doing things. Because of this, the weaknesses of the structure of government were fundamental to what was going wrong. The imperial office had long since ceased to be held, as Augustus had carefully pretended, by the agent of the Senate and people; the reality was a despotic monarch, his rule tempered only by such practical considerations as the placating of the Praetorian Guard on which he depended. A round of civil wars which followed the accession of the last, inadequate, Antonine emperor in 180 opened a terrible era. This wretched man, Commodus, was strangled by a wrestler at the bidding of his concubine and chamberlain in 192, but it solved nothing. From the struggles of four 'emperors' in the months following his death there finally emerged an African, Septimius Severus, married to a Syrian, who strove to base the empire again on heredity, attempting to link his own family with the Antonine succession and thus to deal with one fundamental constitutional weakness.

This was really to deny the fact of his own success. Severus, like his rivals, had been the candidate of a provincial army. Soldiers were the real emperor-makers throughout the third century and their power lay at the root of the empire's tendency to fragment. Yet the soldiers could not be dispensed with; indeed, because of the barbarian threat, now present on several frontiers simultaneously, the army had to be enlarged and pampered. Here was a dilemma to face emperors for the next century. Severus' son Caracalla, who prudently began his reign by bribing the soldiers heavily, was none the less murdered by them in the end.

In theory the Senate still appointed the emperor. In fact it had little effective power except in so far as it could commit its prestige to one of a number of contending

Stilicho, the last general of the western empire, was the son of a Vandal, and twice saved Italy from barbarian invasion with the aid of armies composed of Visigoths, Alans and even Huns. When in the end he turned against the eastern empire, the emperor Honorius, his godson and former pupil, permitted Stilicho's murder. Three months later, the Goths sacked Rome.

candidates. This was not much of an asset but still had some importance so long as the moral effect of maintaining the old forms was still significant. It was inevitable, though, that the arrangements should intensify the latent antagonism of Senate and emperor. Severus gave more power to officers drawn from the equestrian class; Caracalla inferred that a purge of the Senate would help and took this further step towards autocratic rule. More military emperors followed him; soon there was for the first time one who did not come from the senatorial ranks, though he was from the *equites*. Worse was to follow. In 235 Maximinus, a huge ex-ranker from the Rhine legions, contested the prize with an octogenarian from Africa who had the backing of the African army and, eventually, of the Senate. Many emperors were murdered by their troops; one died fighting his own commander-in-chief in battle (his conqueror subsequently being slain by the Goths after his betrayal to them by one of his other officers). It was a dreadful century; altogether, twenty-two emperors came and went during it and that number does not include mere pretenders (or such semi-emperors as Postumus, who for a while maintained himself in Gaul, thus pre-figuring a later division of the empire).

Though Severus' reforms had for a time improved matters, the fragility of his successors' position accelerated a decline in administration. Caracalla was the last emperor to try to broaden the basis of taxation by making all free inhabitants of the empire Roman citizens and thus liable to inheritance taxes, but no fundamental fiscal reform was attempted. Perhaps decline was inevitable, given the emergencies to be faced and the resources available. With irregularity and extemporization went growing rapacity and corruption as those with power or office used it to protect themselves. This reflected another problem, the economic weakness which the empire was showing in the third century.

Few generalizations are safe about what this meant to the consumer and supplier. For all its elaboration and organization around a network of cities, the economic life of the empire was overwhelmingly agrarian. Its bedrock was the rural estate, the *villa*, large or small, which was both the basic unit of production and also, in many places, of society. Such estates were the source of subsistence for all those lived on them (and that meant nearly all the rural population). Probably, therefore, most people in the countryside were less affected by the long-term swings of the economy than by the requisitioning and heavier taxation which resulted from the empire ceasing to expand; the armies had to be supported from a narrower base. Sometimes, too, the land was devastated by fighting. But peasants lived at subsistence level, had always been poor, and continued to be so, whether bond or free. As times got worse, some sought to bind themselves as serfs, which suggests an economy in which money was in retreat before payment in goods and services. It also probably reflects another impact of troubled times such as drove peasants to the towns or to banditry; men everywhere sought protection.

Requisitioning and higher taxation may in some places have helped to produce depopulation – though the fourth century provides more evidence of this than the third – and to this extent were self-defeating. In any case, they were likely to be inequitable, for many of the rich were exempt from taxation and the owners of the estates cannot have suffered much in inflationary times unless they were imprudent. The continuity of many of the great estate-owning families in antiquity does not suggest that the troubles of the third century bit deeply into their resources.

The administration and the army felt most of the effects of economic troubles, and particularly the major ill of the century, inflation. Its sources and extent are complex and still disputed. In part it derived from an official debasement of the coinage which was aggravated by the need to pay tribute in bullion to barbarians who from time to time were best placated by this means. But barbarian incursions themselves often helped to disrupt supply, and this again told against the cities, where prices rose. Because the soldiers' pay was fixed it fell in real value (this made them, of course, more susceptible to generals who offered lavish bribes). Although the overall impact is hard to assess, it has been suggested that money may have fallen during the century to about one-fiftieth of its value at the beginning.

The damage showed both in the towns and in imperial fiscal practice. From the third century onwards many towns shrank in size and prosperity; their early medieval successors were only pale reflexions of the important places they once had been. One cause was the increasing demands of the imperial tax-collectors. From the beginning of the fourth century the depreciation of coin led imperial officials to levy taxes in kind – they could often be used directly to supply local garrisons but were also the means of payment to civil servants – and this not only made the government more unpopular, but also the *curiales* or municipal office-holders who had the task of raising these impositions. By 300 they often had to be forced to take office, a sure sign that a once sought-after dignity had become a strenuous obligation. Some towns suffered from actual physical damage, too, especially those in the frontier regions. Significantly, as the third century wore on, towns well within the frontier began to rebuild (or build for the first time) walls for their protection. Rome began again to fortify itself soon after 270.

Meanwhile, the army steadily grew bigger. If the barbarians were to be kept out it had to be paid, fed and equipped. If the barbarians were not kept out there would be tribute to pay to them instead. And there was not only the barbarian to contend with. Only in Africa was the imperial frontier reasonably secure against Rome's neighbours (because there were no neighbours there who mattered). In Asia things were much grimmer. Ever since the days of Sulla a cold war with Parthia had flared up from time to time into full-scale campaigning. Two things prevented the Romans and Parthians from ever really setting down peacefully. One was the overlapping of their spheres of interest. This was most obvious in Armenia, a kingdom which was alternately a buffer and shuttlecock between them for a century and a half, but the Parthians also dabbled in the disturbed waters of Jewish unrest, another sensitive matter for Rome. The other factor making for disturbance was the temptation presented to Rome time and time again by Parthia's own internal dynastic troubles.

Such facts had led in the second century AD to intense fighting over Armenia, its details often obscure. Severus eventually penetrated Mesopotamia but had to withdraw; the Mesopotamian valleys were too far away. The Romans were trying to do too much and faced the classic problem of over-extended imperialism. But their opponents were tiring and at low ebb, too. Parthian written records are fragmentary, but the tale of exhaustion and growing incompetence emerges from a coinage declining into unintelligibility and blurred derivations from earlier Hellenized designs.

In the third century Parthia disappeared, but the threat to Rome from the East did not. A turning-point was reached in the history of the old area of Persian civilization. In about 225 a king called Ardashir (later known in the West as Artaxerxes) killed the last king of Parthia and was crowned in Ctesiphon. He was to re-create the Achaemenid empire of Persia under a new dynasty, the Sassanids; it would be Rome's greatest antagonist for more than four hundred years. There was much continuity here; the Sassanid empire was Zoroastrian, as Parthia had been, and evoked the Achaemenid tradition as Parthia had done.

Within a few years the Persians had invaded Syria and opened three centuries of struggle with the empire. In the third century there was not a decade without war. The Persians conquered Armenia and took one emperor (Valerian) prisoner. Then they were driven from Armenia and Mesopotamia in 297. This gave the Romans a frontier on the Tigris, but it was not one they could keep for ever. Neither could the Persians keep their conquests. The outcome was a long-drawn-out and ding-dong contest. A sort of equilibrium grew up in the fourth and fifth centuries and only in the sixth did it begin to break down. Meanwhile, commercial ties appeared. Though trade at the frontier was officially limited to three designated towns, important colonies of Persian merchants came to live in the great cities of the empire. Persia, moreover, lay across trade routes to India and China which were as vital to Roman exporters as to those who wanted oriental silk, cotton and spices. Yet these ties did not offset other forces. When not at war, the two empires tended to co-exist with cold hostility; their relations were complicated by communities and peoples settled on both sides of the frontier, and there was always the danger of the strategic balance being upset by a change in one of the buffer kingdoms – Armenia, for instance. The final round of open struggle was long put off, but came at last in the sixth century.

This is to jump too far ahead for the present; by then huge changes had taken place in the Roman empire which have still to be explained. The conscious dynamism of the Sassanid monarchy was only one of the pressures encouraging them. Another came from the barbarians along the Danube and Rhine frontiers. The origins of the folk-movements which propelled them forward in the third century and thereafter must be sought in a long development and are less important than the outcome. These peoples grew more insistent, acted in larger groupings and had, in the end, to be allowed to settle inside Roman territory. Here they were first engaged as soldiers to protect the empire against other barbarians and then, gradually, began to take a hand in running the empire themselves.

In 200 this still lay in the future; all that was clear then was that new pressures were building up. The most important barbarian peoples involved were the Franks and Alamanni on the Rhine and the Goths on the lower Danube. From about 230 the empire was struggling to hold them off but the cost of fighting on two fronts was heavy; his Persian entanglements soon led one emperor to make concessions to the Alamanni. When his immediate successors added their own quarrels to their Persian burdens the Goths took advantage of a promising situation and invaded Moesia, the province immediately south of the Danube, killing an emperor there *en passant* in 251. Five years later, the Franks crossed the Rhine. The Alamanni followed and got as far as Milan. Gothic armies invaded Greece and raided Asia and the Aegean from the sea. Within a few years the European dams seemed to give way everywhere at once

The scale of these incursions is not easy to establish. Perhaps the barbarians could never field an army of more than twenty or thirty thousand. But this was too much at any one place for the imperial army. Its backbone was provided by recruits from the Illyrian provinces; appropriately, it was a succession of emperors of Illyrian stock who turned the tide. Much of what they did was simple good soldiering and intelligent extemporization. They recognized priorities; the main dangers lay in Europe and had

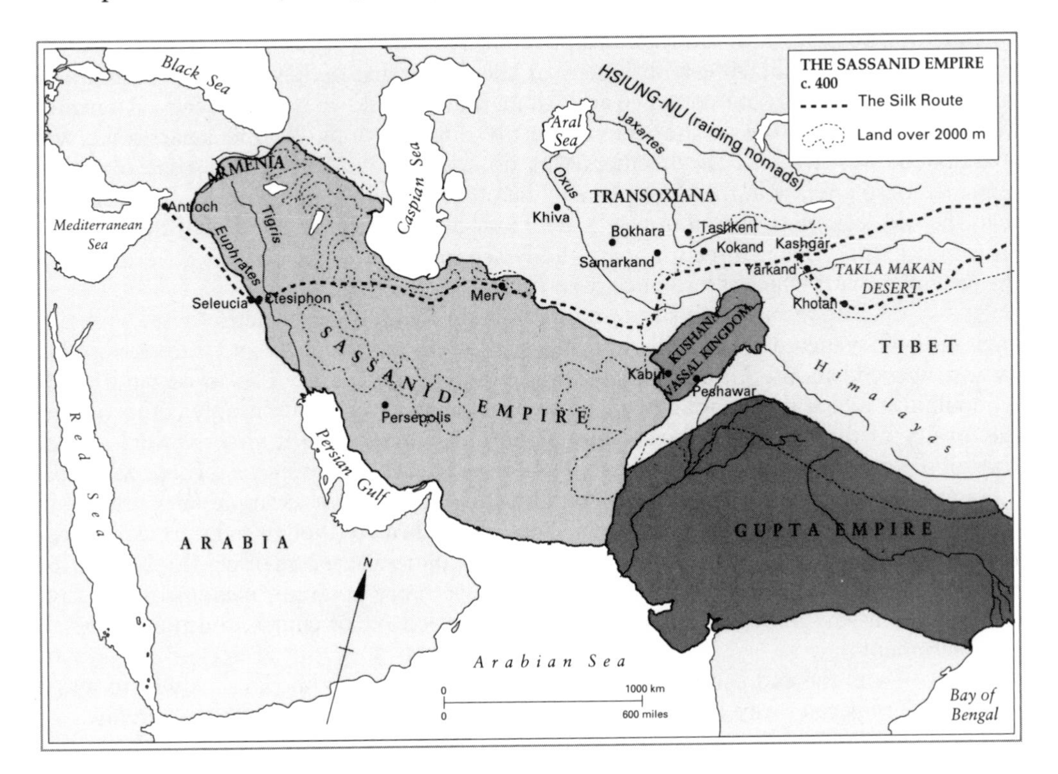

to be dealt with first. Alliance with Palmyra helped to buy time against Persia. Losses were cut; trans-Danubian Dacia was abandoned in 270. The army was reorganized to provide effective mobile reserves in each of the main danger areas. This was all the work of Aurelian, whom the Senate significantly called 'Restorer of the Roman empire'. But the cost was heavy. A more fundamental reconstruction was implicit if the work of the Illyrian emperors was to survive and this was the aim of Diocletian. A soldier of proven bravery, he sought to restore the Augustan tradition but revolutionized the empire instead.

Diocletian had an administrator's genius rather than a soldier's. Without being especially imaginative, he had an excellent grasp of organization and principles, a love of order and great skill in picking and trusting men to whom he could delegate. He was also energetic. Diocletian's capital was wherever the imperial retinue found itself; it moved about the empire, passing a year here, a couple of months there, and sometimes only a day or two in the same place. The heart of the reforms which emerged from this court was a division of the empire intended to deliver it both from the dangers of internal quarrels between pretenders in remote provinces and from the over-extension of its administrative and military resources. In 285 Diocletian appointed a co-emperor, Maximian, with responsibility for the empire west of a line running from the Danube to Dalmatia. The two *augusti* were subsequently each given a *caesar* as co-adjutor; these were to be both their assistants and successors, thus making possible an orderly transfer of power. In fact, the machine of succession only once operated as Diocletian intended, at his own abdication and that of his colleague, but the practical separation of administration in two imperial structures was not reversed. After this time all emperors had to accept a large measure of division even when there was nominally still only one of them.

There also now emerged explicitly a new conception of the imperial office. No longer was the title *princeps* employed; the emperors were the creation of the army, not the Senate, and were deferred to in terms recalling the semi-divine kingship of oriental courts. Practically, they acted through pyramidal bureaucracies. 'Dioceses' responsible directly to the emperors through their 'vicars' grouped provinces much smaller and about twice as numerous as the old ones had been. The senatorial monopoly of governmental power had long since gone; senatorial rank now meant in effect merely a social distinction (membership of the wealthy landowning class) or occupation of one of the important bureaucratic posts. Equestrian rank disappeared.

The military establishment of the Tetrarchy, as it was called, was much larger (and, therefore, more expensive) than that laid down originally by Augustus. The theoretical mobility of the legions, deeply dug into long occupied garrisons, was abandoned. The army of the frontiers was now broken up into units, some of which remained permanently in the same place while others provided new mobile forces smaller than the old legions. Conscription was reintroduced. Something like a half-million men were under arms. Their direction was wholly separated from the civilian government of the provinces with which it had once been fused.

The results of this system do not seem to have been exactly what Diocletian envisaged. They included a considerable measure of military recovery and stablization, but its cost was enormous. An army whose size doubled in a century had to be paid for by a population which had probably grown smaller. Heavy taxation not only compromised the loyalty of the empire's subjects and encouraged corruption; it also required a close control of social arrangements so that the tax base should not be eroded. There was great administrative pressure against social mobility; the peasant, for example, was obliged to stay where he was recorded at the census. Another celebrated (though so far as can be seen totally unsuccessful) example was the attempt to regulate wages and prices throughout the empire by a freeze. Such efforts, like those to raise more taxation, meant a bigger civil service, and as the number of administrators increased so, of course, did the overheads of government.

In the end Diocletian probably achieved most by opening the way to a new view of the imperial office itself. The religious aura which it acquired was a response to a real problem. Somehow, under the strain of continued usurpation and failure the empire had ceased to be unquestioningly accepted. This was not merely because of dislike of higher taxation or fear of its growing numbers of secret police. Its ideological basis had been eroded and it could not focus men's loyalties. A crisis of civilization was going on as well as a crisis of government. The spiritual matrix of the classical world was breaking up; neither state nor civilization was any longer to be taken for granted and they needed a new ethos before they could be.

An emphasis on the unique status of the emperor and his sacral role was one early response to this need. Consciously, Diocletian acted as a saviour, a Jupiter-like figure holding back chaos. Something in this spoke of affinities with those thinkers of the late classical world who saw life as a perpetual struggle of good and evil. Yet this was a vision not Greek or Roman at all, but oriental. The acceptance of a new vision of the emperor's relation to the gods, and therefore of a new conception of the official cult, did not bode well for the traditional practical tolerance of the Greek world. Decisions about worship might now decide the fate of the empire.

These possibilities shaped the history of the Christian Churches for both good and ill. In the end Christianity was to be the legatee of Rome. Many religious sects have risen from the position of persecuted minorities to become establishments in their own right. What sets the Christian Church apart is that this took place within the uniquely comprehensive structure of the late Roman empire, so that in both attached itself to and strengthened the lifeline of classical civilization, with enormous consequences not only for itself but for Europe and ultimately the world.

At the beginning of the third century missionaries had already carried the faith to the non-Jewish peoples of Asia Minor and North Africa. Particularly in North Africa, Christianity had its first mass successes in the towns; it long remained a predominantly urban phenomenon. But it was still a matter of minorities. Throughout the empire, the old gods and the local deities held the peasants' allegiance. By the year 300 Christians may have made up only about a tenth of the population of the empire. But there had already been striking signs of official favour and even concession. One emperor had been nominally a Christian and another had included Jesus Christ among the gods honoured privately in his household. Such contacts with the court illustrate an interplay of Jewish and classical culture which is an important part of the story of the process by which Christianity took root in the empire. Perhaps Paul of Tarsus, the Jew who could talk to Athenians in terms they understood, had launched this. Later, early in the second century, Justin Martyr, a Palestinian Greek, had striven to show that Christianity had a debt to Greek philosophy. This had a political point; cultural identification with the classical tradition helped to rebut the charge of disloyalty to the empire. If a Christian could stand in the ideological heritage of the Hellenistic world he could also be a good citizen, and Justin's rational Christianity (even though he was martyred for it in about 165) envisaged a revelation of the Divine Reason in which all the great philosophers and prophets had partaken, Plato among them, but which was only complete in Christ. Others were to follow similar lines, notably the learned Clement of Alexandria, who strove to integrate pagan scholarship with Christianity, and Origen (though his exact teaching is still debated because of the disappearance of many of his writings). A North African Christian, Tertullian, had contemptuously asked what the Academy had to do with the Church; he was answered by the Fathers who deliberately employed the conceptual armoury of Greek philosophy to provide a statement of the Faith which anchored Christianity to rationality as Paul had not done.

When coupled to its promise of salvation after death and the fact that the Christian life could be lived in a purposeful and optimistic way, such developments might lead us to suppose that Christians were by the third century confident about the future. In fact, favourable portents were much less striking than the persecutions so prominent in the history of the early Church. There were two great outbreaks. That of the middle of the century expressed the spiritual crisis of the establishment. It was not only economic strain and military defeat that were troubling the empire, but a dialectic inherent in Roman success itself: the cosmopolitanism which had been so much the mark of the empire was, inevitably, a solvent of the *romanitas* which was less and less a reality and more and more a slogan. The emperor Decius seems to have been convinced that the old recipe of a return to traditional Roman virtue and values could still work; it implied the revival of service to the gods whose benevolence would then be once more deployed in favour of the empire. The Christians, like others, must sacrifice to the Roman tradition, said Decius, and many did, to judge by the certificates issued to save them from persecution; some did not, and died. A few years later, Valerian renewed persecution on the same grounds, though his proconsuls addressed themselves rather to the directing personnel and the property of the Church – its buildings and books – than to the mass of believers. Thereafter, persecution ebbed, and the Church resumed its shadowy, tolerated existence just below the horizon of official attention.

Persecution had shown, nevertheless, that it would require great efforts and prolonged determination to eradicate the new sect; it may even have been already beyond the capacities of Roman government to carry out such an eradication. The exclusiveness and isolation of early Christianity had waned. Christians were increasingly prominent in local affairs in the Asian and African provinces. Bishops were often public figures with whom officials expected to do business; the development of distinct traditions within the Faith (those of the churches of Rome, Alexandria and Carthage being the most important) spoke for the degree to which it was rooted in local society and could express local needs.

Outside the empire, too, there had been signs that better times might lie ahead for Christianity. The local rulers of the client states under the shadow of Persia could not afford to neglect any source of local support. Respect for widely held religious views was at least prudent. In Syria, Cilicia and Cappadocia, Christians had been very successful in their missionary work and in some towns they formed a social élite. Simple superstition, too, helped to convince kings; the Christian god might prove powerful and it could hardly be damaging to insure against his ill-will. Thus Christianity's political and civic prospects improved.

Christians noted with some satisfaction that their persecutors did not prosper; the Goths slew Decius, and Valerian was said to have been skinned alive by the Persians (and stuffed). But Diocletian did not appear to draw any conclusions from this and in 303 launched the last great Roman persecution. It was not at first harsh. The main targets were Christian officials, clergy and the books and buildings of the Church. The books were to be handed over for burning, but for some time there was no death penalty for failing to sacrifice. (Many Christians none the less did sacrifice, the bishop at Rome among them.) Constantius, the Caesar of the West, did not enforce the persecution after 305 when Diocletian abdicated, though his eastern colleague (Diocletian's successor Galerius) felt strongly about it, ordering a general sacrifice on pain of death. This meant that persecution was worst in Egypt and Asia where it was kept up a few years longer. But before this it had been cut across by the complicated politics which led to the emergence of the emperor Constantine the Great.

This was the son of Constantius, who died in Britain in 306, a year after his accession as *Augustus*. Constantine was there at the time and although he had not been his father's Caesar he was hailed as emperor by the army at York. A troubled period of nearly two decades followed. Its intricate struggles demonstrated the failure of Diocletian's arrangements for the peaceful transmission of the empire and only ended in 324, when Constantine re-united the empire under one ruler.

By this time he had already addressed himself vigorously and effectively to its problems, though with more success as a soldier than as an administrator. Often with barbarian recruits, he built up a powerful field army distinct from the frontier guards; it was stationed in cities within the empire. This was a strategically sound decision which proved itself in the fighting power the empire showed in the East for the next two centuries. Constantine also disbanded the Praetorian Guard and created a new, German, bodyguard. He restored a stable gold currency and paved the way to the abolition of payments of taxes in kind and the restoration of a money economy. His fiscal reforms had more mixed results but attempted some readjustment of the weight of taxation so that more should be

THOLOS *Almost certainly, the idea of circular buildings in Greece derived ultimately from the round store huts of the early Aegeans. Mycenaens built round tombs and later* tholoi, *as circular constructions were called, were built in stone and marble for two or three centuries before the example, the Tholos at Delphi, was erected in about 390* BC. *Delphi was a great religious centre, where could be found the shrine of Apollo and the seat of the most renowned of Greek oracles, a place hallowed by association, superstition and myth, sited magnificently and dramatically high above the Gulf of Corinth.*

STELA *Some of the pagan traditions with which the early Church had to come to terms in the process of making a Christian Europe were very ancient indeed. Behind the Germanic invasions lay the Roman empire, and behind that, Celtic and even pre-Celtic Europe. At Lampan-Ploudalmézean, almost at the north-west tip of Brittany, an area littered with megaliths and menhirs, some missionary or priest assured the local people that the mysterious stela put there by pagan Celts over a thousand years earlier in the fifth or sixth century* BC *and worshipped by them could be a focus of Christian belief and worship. In due course, it acquired a medieval cross.*

borne by the rich. None of these things, though, so struck contemporaries as his attitude to Christianity.

Constantine gave the Church official houseroom. He thus played a more important part in shaping its future than any other Christian layman and was to be called the 'thirteenth Apostle'. Yet his personal relationship to Christianity was complicated. He grew up intellectually with the monotheistic predisposition of many late classical men and was in the end undoubtedly a convinced believer (it was not then unusual for Christians to do as he did and postpone baptism until their deathbed). But he believed from fear and hope, for the god he worshipped was a god of power. His first adherence was to the sun-god whose sign he bore and whose cult was already officially associated with that of the emperor. Then, in 312, on the eve of battle and as a result of what he believed to be a vision he ordered his soldiers to put on their shields a Christian monogram. This showed a willingness to show suitable respect to whatever gods there might be. He won the battle and thenceforth, though continuing publicly to acknowledge the cult of the sun, he began to show important favours to the Christians and their god.

A huge seated statue of Constantine, more than thirty feet tall, dominated his basilica in Rome. It represented a new kind of imperial iconography setting the emperor far above other men and foreshadowing the later exaltation of the emperor's image under the Christian empire of Byzantium.

One manifestation of this was an edict the following year which was issued by another of the contenders for the empire, after agreement with Constantine at Milan. It restored to Christians their property, and granted them the toleration that other religions enjoyed. The justification may reveal Constantine's own thinking as well as his wish to arrive at a satisfactory compromise formula with his colleague, for it explained its provisions by the hope 'that whatever divinity dwells in the heavenly seat may be appeased and be propitious towards us and to all who are placed under our authority'. Constantine went on to make considerable gifts of property to the churches, favouring, in particular, that of Rome. Besides providing important tax concessions to the clergy, he conferred, an unlimited right to receive bequests on the Church. Yet for years his coins continued to honour pagan gods, notably the 'Unconquered Sun'.

Constantine gradually came to see himself as having a quasi-sacerdotal role, and this was of the first importance in the further evolution of the imperial office. He saw himself as responsible to God for the well-being of the Church to which he more and more publicly and unequivocally adhered. After 320 the sun no longer appeared on his coins and soldiers had to attend church parades. But he was always cautious of the susceptibilities of his pagan subjects. Though he later despoiled temples of their gold while building splendid Christian churches and encouraging converts by preferment, he did not cease to tolerate the old cults.

In some of Constantine's work (like that of Diocletian) there was the development of things latent and implicit in the past, an extension of earlier precedents. This was true of his interventions in the internal affairs of the Church. As early as 272, the Christians of Antioch had appealed to the emperor to remove a bishop and Constantine himself in 316 tried to settle a controversy in North Africa by installing a bishop of Carthage against the will of a local sectarian group known as Donatists. Constantine came to believe that the emperor owed to God more than a grant of freedom to the Church or even an endowment. His conception of his role evolved towards that of the guarantor and, if need be, the imposer of the unity which God required as the price of His continuing favour. When he turned on the Donatists it was this view of his duty which gave them the unhappy distinction of being the first schismatics to be persecuted by a Christian government. Constantine was the creator of Caesaropapism, the belief that the secular ruler has divine authority to settle religious belief, and of the notion of established religion in Europe for the next thousand years.

Constantine's greatest act in the ordering of religion came just after he had formally declared himself a Christian in 324 (a declaration preceded by another victory over an imperial rival who had, interestingly, been persecuting Christians). This was the calling of the first ecumenical council, the Council of Nicaea. It met for the first time in 325, nearly 300 bishops being present, and Constantine presided over it. Its task was to settle the response of the Church to a new heresy, Arianism, whose founder, Arius, taught that the Son did not share the divinity of the Father. Though technical and theological, the nice issues to which this gave rise prompted enormous controversy. Grave scandal was alleged by Arius' opponents. Constantine sought to heal the division and the Council laid down a Creed which decided against the Arians, but went on in a second reunion to readmit Arius to communion after suitable declarations. That this did not satisfy all the bishops (and that there were few from the West at Nicaea) was less important than that Constantine had presided at this crucial juncture of proclaiming the emperor's enjoyment of special authority and responsibility. The Church was clothed in the imperial purple.

There were other great implications, too. Behind the hair-splitting of the theologians lay a great question both of practice and principle: in the new ideological unity given to the empire by the official establishment of Christianity, what was to be the place of diverging Christian traditions which were social and political, as well as liturgical and theological, realities? The churches of Syria and Egypt, for example, were strongly tinctured by their inheritance of thought and custom both from the Hellenistic culture and the popular religion of those regions. The importance of such considerations helps to explain why the practical outcome of Constantine's ecclesiastical policy was less than he had hoped. The Council did not produce an emollient formula to make easier a general reconciliation in a spirit of compromise. Constantine's own attitude to the Arians soon relaxed (in the end, it was to be an Arian bishop who baptized him as he lay dying) but the opponents of Arius, led by the formidable Athanasius, bishop of Alexandria, were relentless. The quarrel remained unsettled when Arius died, and Constantine's own death followed not long after. Yet Arianism was not to prosper in the East. Its last successes, instead, were won by Arian missionaries to the Germanic tribes of south-east Russia; borne by these barbarian nations, Arianism was to survive until the seventh century in the West-but this is to anticipate.

How much of the Church's rise was in the end inevitable it is hardly profitable to consider. Certainly – in spite of a North African Christian tradition which saw the state as an irrelevancy - something so positively important as Christianity could hardly have remained for ever unrecognized by the civil power. Yet someone had to begin. Constantine was the man who took the crucial steps which linked Church and empire for so long as the empire should last. His choices were historically decisive. The Church gained most, for it acquired the charisma of Rome. The empire seemed less changed. Yet Constantine's sons were brought up as Christians and even if the fragility of much in the new establishment was to appear soon after his death in 337, he had registered a decisive break with the tradition of classical Rome. Ultimately, unwittingly, he founded Christian Europe and, therefore, the modern world.

One of his decisions only slightly less enduring in its effects was his foundation, 'on the command of God', he said, of a city to rival Rome on the site of the old Greek colony of Byzantium at the entrance to the Black Sea. It was dedicated in 330 as Constantinople. Though his own court remained at Nicomedia and no emperor was to reside there permanently until another fifty years were past, Constantine was again shaping the future. For a thousand years Constantinople would be a Christian capital, unsullied by pagan rites. After that, for five hundred years more, it would be a pagan capital and the constant ambition of would-be successors to its traditions.

Once again, though, this is anticipating too much. We must return to the empire as Constantine left it, in Roman eyes still coterminous with civilization. Its frontiers ran for the most part along natural features which recognized, more or less, the demarcations of distinct geographical or historical regions. Hadrian's wall in Britannia was their northern limit; in continental Europe they followed the Rhine and Danube. The Black Sea coasts north of the mouths of the Danube had been lost to barbarians by 305 BC, but Asia Minor remained in the empire; it stretched as far east as the shifting boundary with Persia. Further south, the Levant coast and Palestine lay within a frontier which ran to the Red Sea. The lower Nile valley was still held by the empire and so was the North African coast; the African frontiers were the Atlas and the desert.

This unity was, for all Constantine's great work, in large measure an illusion. As the first experiments with co-emperors had shown, the world of Roman civilization had grown too big for a unified political structure, however desirable the preservation of the myth of unity might be. Growing cultural differentiation between a Greek-speaking East and a Latin-speaking West, the new importance of Asia Minor, Syria and Egypt (in all of which there were large Christian communities) after the establishment of Christianity and the continuing stimulus of direct contact with Asia in the East all drove the point home. After 364 the two parts of the old empire were only once more and then only briefly ruled by the same man. Their institutions diverged further and further. In the East the emperor was a theological as well as a juridical figure; the identity of Empire and Christendom and the emperor's standing as the expression of divine intention were unambiguous. The West, on the other hand, had by 400 already seen adumbrated the distinction of the roles of Church and State which was to father one of the most creative arguments of European politics. There was an economic contrast, too: the East was populous and could still raise great revenues, while the West was by 300 already unable to feed itself without Africa and the Mediterranean islands. It now seems obvious that two distinct civilizations were to emerge, but it was a long time before any of the participants could see that.

Instead, they saw something much more appalling: the western empire simply disappeared. By 500, when the boundaries of the eastern empire were still much what they had been under Constantine, and his successors were still holding their own against the Persians, the last western emperor had been deposed and his *insignia* sent to Constantinople by a barbarian king who claimed to rule as the eastern emperor's representative in the West.

This is striking: what, actually, had collapsed? What had declined or fallen? Fifth-century writers bewailed it so much that it is easy to have the impression, heightened by such dramatic episodes as sackings of Rome itself, that the whole of society fell apart. This was not so. It was the state apparatus which collapsed, some of its functions ceasing to be carried out, and some passing into other hands. This was quite enough to explain the alarm. Institutions with a thousand years of history behind them gave way within a half-century. It is hardly surprising that people have asked why ever since.

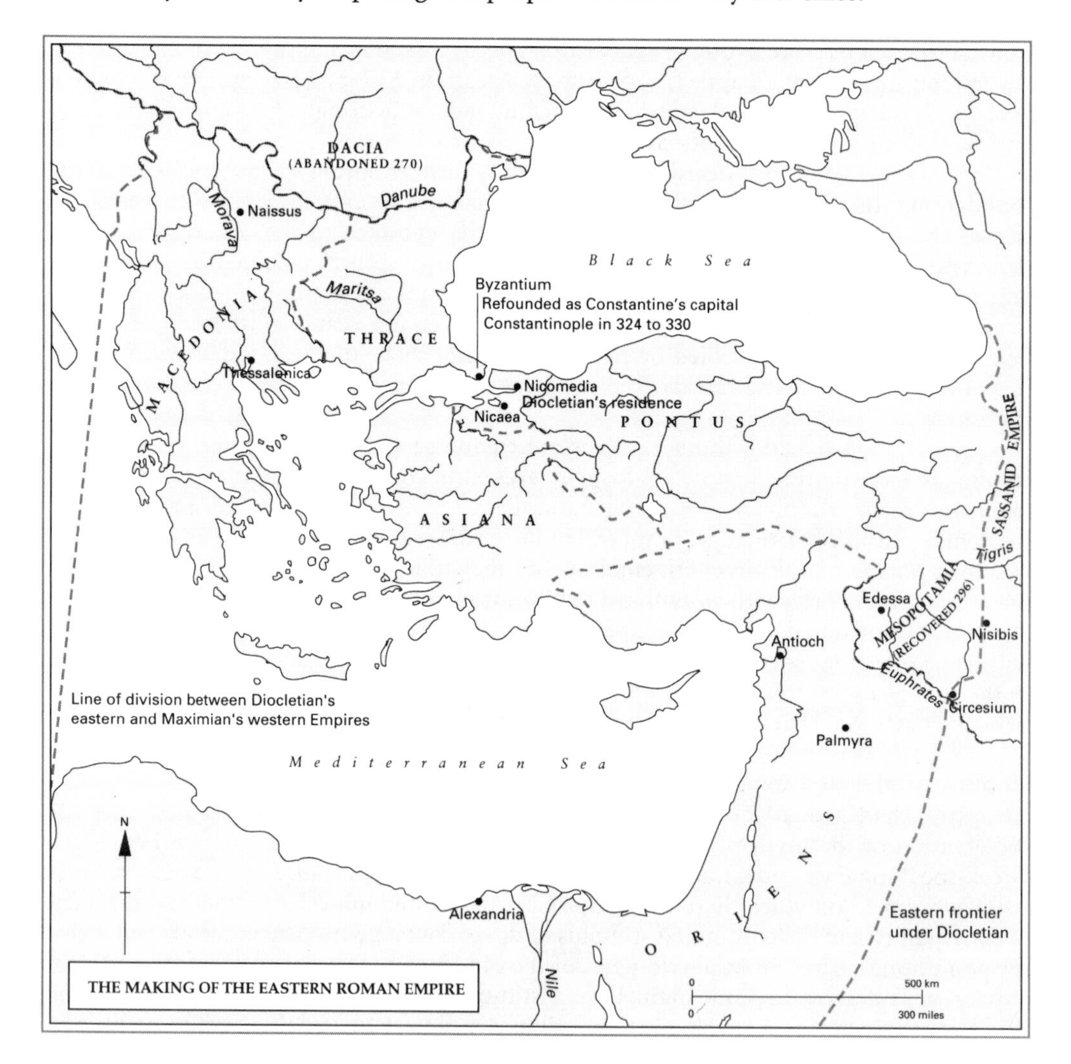

THE MAKING OF THE EASTERN ROMAN EMPIRE

One explanation is cumulative: the state apparatus in the West gradually seized up after the recovery of the fourth century. The whole concern became too big for the demographic, fiscal and economic base which carried it. The main purpose of raising revenue was to pay for the military machine, but it became more and more difficult to raise enough. There were no more conquests after Dacia to bring in new tribute. Soon the measures adopted to squeeze out more taxes drove rich and poor alike to devices for avoiding them. The effect was to make agricultural estates rely more and more upon meeting their own needs and becoming self-supporting, rather than producing for the market. Parallel with this went a crumbling of urban government as trade languished and the rich withdrew to the countryside.

The military result was an army recruited from inferior material, because better could not be paid for. Even the reform of dividing it into mobile and garrison forces had its defects, for the first lost their fighting spirit by being stationed at the imperial residence and becoming used to the pampering and privileges that went with city postings, while the second turned into settled colonists, unwilling to take risks which would jeopardize their homesteads. Another descent in the unending spiral of decline logically followed. A weaker army drove the empire to rely still more on the very barbarians the army was supposed to keep at bay. As they had to be recruited as mercenaries, soothing and conciliatory politics were needed to keep them sweet. This led the Romans to concede more to the barbarians just when the pressure of the Germanic folk-movements was reaching a new climax. Migration and the attractive prospect of paid service with the empire probably counted for much more in the barbarian contribution to imperial collapse than the simple desire for loot. The prospect of booty might animate a raiding-party but could hardly bring down an empire.

At the beginning of the fourth century Germanic peoples were stretched along the whole length of the frontier from the Rhine to the Black Sea, but it was in the south that the most formidable concentration was at that moment assembled. These were the Gothic peoples, Ostrogoth and Visigoth, who waited beyond the Danube. Some of them were already Christian, though in the Arian form. Together with Vandals, Burgundians and Lombards, they made up an east Germanic group. To the north were the west Germans: Franks, Alamanni, Saxons, Frisians and Thuringians. They would move into action in the second phase of the *Völkerwanderung* of the fourth and fifth centuries.

The crisis began in the last quarter of the fourth century. The pressure on more western barbarians of the Huns, a formidable nomadic people from central Asia, was mounting after 370. They overran the Ostrogothic territory, defeated the Alans and then turned on the Visigoths near the Dniester. Unable to hold them, the Visigoths fled for refuge to the empire. In 376 they were allowed to cross the Danube to settle within the frontier. This was a new departure. Earlier barbarian incursions had been driven out or absorbed. Roman ways had attracted barbarian rulers and their followers had joined Rome's army. The Visigoths, though, came as a people, perhaps 40,000 strong, keeping their own laws and religion and remaining a compact unit. The Emperor Valens intended to disarm them; it was not done and instead there was fighting. At the battle of Adrianople in 378 the emperor was killed and a Roman army defeated by the Visigoth cavalry. The Visigoths ravaged Thrace.

This was in more than one way a turning-point. Now whole tribes began to be enrolled as confederates – *foederati*, a word first used in 406 – and entered Roman territory to serve against other barbarians under their own chiefs. A temporary settlement with the Visigoths could not be maintained. The eastern empire was helpless to protect its European territories outside Constantinople, though when the Visigothic armies moved north towards Italy early in the fifth century, they were checked for a while by a Vandal general. By now the defence of Italy, the old heart of the empire, was entirely dependent on barbarian auxiliaries and soon even this was not enough; Constantinople might be held, but in 410 the Goths sacked Rome. After an abortive move to the south, with a view to pillaging Africa as they had pillaged Italy, the Visigoths again turned north, crossed the Alps into Gaul and eventually settled as the new kingdom of Toulouse in 419, a Gothic state within the empire, where a Gothic aristocracy shared its over-lordship with the old Gallo-Roman landlords.

These are confused events, difficult to follow, but there is still one other major movement of peoples which has to be noticed in order to explain the fifth-century re-making of the European racial and cultural map. In return for their settlement in Aquitania, the western emperor had succeeded in getting the Visigoths to promise that they would help him to clear Spain of other barbarians. Of these the most important were the Vandals. In 406 the Rhine frontier, denuded of soldiers sent to defend Italy against the Visigoths, had given way too and the Vandals and Alans had broken into Gaul. From there they made their way southward, sacking and looting as they went and crossing the Pyrenees to establish a Vandal state in Spain. Twenty years later they were tempted to Africa by a dissident Roman governor who wanted their help. Visigoth attacks encouraged them to leave Spain. By 439 they had taken Carthage. The Vandal kingdom of Africa now had a naval base. They were to stay there for nearly a century and in 455 they, too, crossed to sack Rome and leave their name to history as a synonym for mindless destructiveness. Terrible as this was, though, it was less important than the seizure of Africa, the mortal blow to the old western empire. It had now lost much of its economic base. Though great efforts could and would still be made in the West by eastern emperors, Roman rule there was on its last legs. The dependence on one barbarian against another was a fatal handicap. The cumulative impact of fresh pressure made recovery impossible. The protection of Italy had meant abandoning Gaul and Spain to the Vandals; their invasion of Africa had meant the loss of Rome's graingrowing provinces.

The collapse was completed in Europe in the third quarter of the century. It followed the greatest of the Hun assaults. These nomads had followed the Germanic tribes into the Balkans and central Europe after a preliminary diversion to ravage Anatolia and Syria. By 440 the Huns were led by Attila, under whom their power was at its height. From Hungary, where the great steppe corridor of Asia peters out, he drove west for the last time with a huge army of allies, but was defeated near Troyes in 451 by a 'Roman' army of Visigoths under a commander of barbarian origin. This was the end of the Hun threat; Attila died two years later, apparently scheming to marry the western emperor's sister and perhaps become emperor himself. A great revolt the following year by the Huns' subjects in Hungary finally broke them and they are thenceforth almost lost to sight. In Asia, their home, new confederations of nomads were forming to play a similar part in the future, but their story can wait.

The Huns had all but delivered the *coup de grâce* in the West; one emperor had sent the pope to intercede with Attila. The last western emperor was deposed by a Germanic warlord, Odoacer, in 476 and formal sovereignty passed to the eastern emperors. Though Italy, like the rest of the former western provinces, was henceforth a barbarian kingdom, independent in all but name, Italians regarded the emperor as their sovereign, resident in Constantinople though he might be.

The structure which had finally given way under these blows has in its last decades something of the Cheshire cat about it. It was fading away all the time; it is not particularly meaningful to pick one date rather than another as its end. It is unlikely that 476 seemed especially remarkable to contemporaries. The barbarian kingdoms were only a logical development of the reliance upon barbarian troops for the field army and their settlement as *foederati* within the frontiers. The barbarians themselves usually wanted no more, unless it was simple loot. Certainly they did not plan to replace imperial authority with their own. It is a Goth who is reported saying, 'I hope to go down to posterity as the restorer of Rome, since it is not possible that I should be its supplanter.' Other dangers were greater and more fundamental than barbarian swagger.

Socially and economically, the tale of the third century had been resumed in the fifth. Cities decayed and population fell. The civil service slid deeper into disorder as officials sought to protect themselves against inflation by taking payment for carrying out their duties. Though revenue declined as provinces were lost, the sale of offices somehow kept up the lavish expenditure of the court. But independence of action was gone. From being emperors whose power rested on their armies, the last emperors of the West declined through the stage of being the equals in negotiation of barbarian warlords whom they had to placate, to being their puppets, cooped up in the last imperial capital, Ravenna. Contemporaries had been right in this sense to see the sack of Rome in 410 as the end

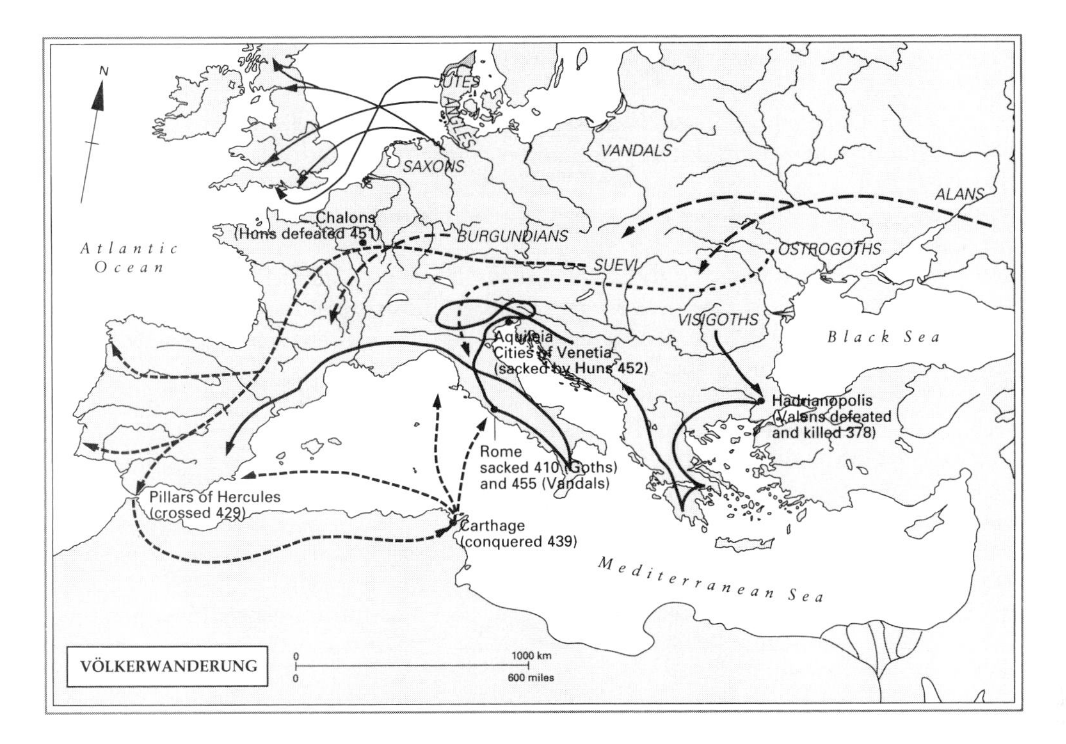

of an age, for then it was revealed that the empire could no longer preserve the very heart of *romanitas*. By then, there had been many other signs, too, of what was going on. The the last emperor of Constantine's house had tried during a brief reign (361–3) to restore the pagan cults; this had earned him historical fame (or, in Christian eyes, infamy) and, revealingly, the title 'the Apostate', but he was not successful. Believing that a restoration of the old sacrifices would ensure the return of prosperity, he had too little time to test the proposition. What is now perhaps more striking is the unquestioned assumption that religion and public life were inseparably intertwined, on which his policy was based and which commanded general agreement; it was an assumption whose origins were Roman, not Christian. Julian did not threaten Constantine's work and Theodosius, the last ruler of a united empire, at last forbade the public worship of the ancient gods in 380.

What this meant in practice is hard to say. In Egypt it seems to have been the final landmark in the process of overcoming the ancient civilization which had been going on for eight centuries or so. The victory of Greek ideas first won by the philosophers of Alexandria was now confirmed by the Christian clergy. The priests of the ancient cults were to be harried as pagans. Roman paganism found outspoken defenders still in the fifth century and only at the end of it were pagan teachers expelled from the universities at Athens and Constantinople. None the less a great turning-point had been reached; in principle the closed Christian society of the Middle Ages was now in existence.

Christian emperors soon set about developing it in a particular direction which became only too familiar by depriving Jews, the most easily identifiable of groups alien to the closed society, of their juridical equality with other citizens. Here was another turning-point. Judaism had long been the only monotheistic representative in the pluralistic religious world of Rome and now it was ousted by its derivative, Christianity. A prohibition on proselytizing was the first blow and others soon followed. In 425 the patriarchate under which Jews had enjoyed administrative autonomy was abolished. When pogroms occurred, Jews began to withdraw to Persian territory. Their growing alienation from the empire weakened it, for they could soon call upon Rome's enemies for help. Jewish Arab states which lay along trade-routes to Asia through the Red Sea, were able to inflict damage on Roman interests in support of their co-religionists, too. Ideological rigour came at a high price.

Theodosius' reign is also notable in Christian history because of his quarrel with St Ambrose, Bishop of Milan. In 390, after an insurrection at Thessalonica, Theodosius pitilessly massacred thousands of its inhabitants. To the amazement of contemporaries, the emperor was soon seen standing in penance for the deed in a Milan church. Ambrose had refused him communion. Superstition had won the first round of what was to prove a long battle for humanity and enlightenment. Other men of might were to be tamed by excommunication or its threat, but this was the first time the spiritual arm had been so exercised and it is significant that it happened in the western Church. Ambrose had alleged a higher duty for his office than that owed to the emperor. It is the inauguration of a great theme of western European history, the tension of spiritual and secular claims which was time and time again to pull it back into a progressive channel, the conflict of Church and State.

By then, a glorious century for Christianity was almost over. It had been a great age of evangelization, in which missionaries had penetrated as far afield as Ethiopia, a brilliant age of theology and, above all, the age of establishment. Yet the Christianity of the age has about it much which now seems repellent and bilious. Establishment gave Christians power they did not hesitate to use. 'We look on the same stars, the same heavens are above us all,' pleaded one pagan to St Ambrose, 'the same universe surrounds us. What matters it by what method each of us arrives at the truth?' But Symmachus asked in vain. East and West, the temper of the Christian Churches was intransigent and enthusiastic; if there was a distinction between the two, it lay between the Greeks' conviction of the almost limitless authority of a Christianized empire, blending spiritual and secular power, and the defensive, suspicious hostility to the whole secular world, state included, of a Latin tradition which taught Christians to see themselves as a saving remnant, tossed on the seas of sin and paganism in the Noah's Ark of the Church. Yet to be fair to the Fathers, or to understand their anxieties and fears, a modern observer has to recognize the compelling power of superstition and mystery in the whole late classical world. Christianity acknowledged and expressed it. The demons among whom Christians walked their earthly ways were real to them and to pagans alike, and a fifth-century pope consulted the augurs in order to find out what to do about the Goths.

This is part of the explanation of the bitterness with which heresy and schism were pursued. Arianism had not been finished off at Nicaea; it flourished among the Gothic peoples and Arian Christianity was dominant over much of Italy, Gaul and Spain. The Catholic Church was not persecuted in the Arian barbarian kingdoms, but it was neglected there and when everything depended on the patronage of rulers and the great, neglect could be dangerous. Another threat was the Donatist schism in Africa, which had taken on a social content and broke out in violent conflicts of town and country. In Africa, too, the old threat of Gnosticism lived again in Manichaeism which came to the West from Persia; another heresy, Pelagianism, showed the readiness of some Christians in Latinized Europe to welcome a version of Christianity which subordinated mystery and sacramentalism to the aim of living a good life.

Few men were better fitted by temperament or education to discern, analyse and combat such dangers than was St Augustine, the greatest of the Fathers. It was important that he came from Africa – that is to say, the Roman province of that name, which corresponded roughly to Tunisia and eastern Algeria – where he was born in 354. African Christianity had more than a century's life behind it by then but was still a minority affair. The African Church had a special temper of its own since the days of Tertullian, its great founding figure. Its roots did not lie in the Hellenized cities of the East, but in soil laid down by the religions of Carthage and Numidia which lingered on amid the Berber peasantry. The humanized deities of Olympus had never been at home in Africa. The local traditions were of remote gods dwelling in mountains and high places, worshipped in savage and ecstatic rituals (the Carthaginians are supposed to have practised child sacrifice).

The intransigent, violent temper of the African Christianity which grew up against this background was reflected to the full in Augustine's own personality. He responded to the same psychological stimuli and felt the need to confront the fact of an evil lurking in himself. One answer was available and popular. The stark dualism of Manichaeism had a very wide appeal in Africa; Augustine was a Manichee for nearly ten years. Characteristically, he then reacted against his errors with great violence.

Before adulthood and Manichaeism, Augustine's education had orientated him towards a public career in the western empire. That education was overwhelmingly Latin (Augustine probably spoke only that language and certainly found Greek difficult) and very selective. Its skills were those of rhetoric and it was in them that Augustine first won prizes, but as for ideas, it was barren. Augustine taught himself by reading; his first great step forward was the discovery of the works of Cicero, probably his first contact, though at secondhand, with the classical Athenian tradition.

Augustine's lay career ended in Milan (where he had gone to teach rhetoric) with his baptism as a Catholic by St Ambrose himself in 387. At that time Ambrose exercised an authority which rivalled that of the empire itself in one of its most important cities. Augustine's observation of this relation between religion and secular power confirmed him in views very different from those of Greek churchmen, who welcomed the conflation of lay and religious authority in the emperor which followed establishment. Augustine then returned to Africa, first to live as a monk at Hippo and then, reluctantly, to become its bishop. There he remained until his death in 430, building up Catholicism's position against the Donatists and almost by the way, thanks to vast correspondence and a huge literary output, becoming a dominant personality of the western Church.

In his lifetime Augustine was best known for his attacks on the Donatists and the Pelagians. The first was really a political question: which of two rival Churches was to dominate Roman Africa? The second raised wider issues. They must seem remote to our non-theologically minded age but on them turned much future European history. Essentially, the Pelagians preached a kind of stoicism; they were part of the classical world and tradition, dressed up in Christian theological language though it might be. The danger this presented – if it was a danger – was that the distinctiveness of Christianity would be lost and the Church simply become the vehicle of one strain in classical Mediterranean civilization, with the strengths and weaknesses which that implied. Augustine was uncompromisingly other-worldly and theological; for him the only possibility of redemption for mankind lay in the Grace which God conferred and no man could command by his works. In the history of the human spirit Augustine deserves a place for having laid out more comprehensively than any predecessor the lines of the great debate between predestination and free will, Grace and Works, belief and motive, which was to run for so long through European history. Almost incidentally, he established Latin Christianity firmly on the rock of the Church's unique power of access to the source of Grace through the sacraments.

This is now largely forgotten except by specialists. St Augustine (as he came to be) now enjoys instead some notoriety as one of the most forceful and insistent exponents of a distrust of the flesh which was long especially to mark Christian sexual attitudes and thereby the whole of western culture. He stands in strange company – with Plato, for example – as a Founding Father of puritanism. But his intellectual legacy was far richer than this suggests. In his writings can also be seen the foundations of much medieval political thought in so far as they are not Aristotelian or legalistic, and a view of history which would long dominate Christian society in the West and would affect it as importantly as the words of Christ himself.

The book now called *The City of God* contains the writing of Augustine which had most future impact. It is not so much a matter of specific ideas or doctrines – there is difficulty in locating his precise influence on medieval political thinkers, perhaps because there is much ambiguity about what he says – as of an attitude. He laid out in this book a way of looking at history and the government of men which became inseparable from Christian thinking for a thousand years and more. The subtitle of the book is *Against the Pagans*. This reveals his aim: to refute the reactionary and pagan charge that the troubles crowding in on the empire were to be blamed on Christianity. He was inspired to write by the Gothic sack of Rome in 410; his overriding aim was to demonstrate that the understanding of even such an appalling event was possible for a Christian and, indeed, could only be understand through the Christian religion, but his huge book swoops far and wide over the past, from the importance of chastity to the philosophy of Thales of Miletus, and expounds the civil wars of Marius and Sulla as carefully as the meaning of God's promises to David. It is impossible to summarize: 'It may be too much for some, too little for others,' said Augustine wryly in his last paragraph. It is a Christian interpretation of a whole civilization and what went to its making. Its most remarkable feature is its own central judgement: that the whole earthly tissue of things is dispensable, and culture and institutions – even the great empire itself – of no final value, if God so wills.

That God did so will was suggested by Augustine's central image of two cities. One was earthly, founded in men's lower nature, imperfect and made with sinful hands, however glorious its appearance and however important the part it might from time to time have to play in the divine scheme. Sometimes its sinful aspect predominates and it is clear that men must flee the earthly city – but Babylon, too, had had its part in the divine plan. The other city was the heavenly city of God, the community founded on the assurance of God's promise of salvation, a goal towards which mankind might make a fearful pilgrimage from the earthly city, led and inspired by the Church. In the Church was to be found both the symbol of the City of God and the means of reaching it. History had changed with the appearance of the Church: from that moment the struggle of good and evil was clear in the world and human salvation rested upon its defence. Such arguments would be heard long into modern times.

The two cities sometimes make other appearances in Augustine's argument, too. They are sometimes two groups of men, those who are condemned to punishment in the next world and those who are making the pilgrimage to glory. At this level the cities are divisions of the actual human race, here and now, as well as of all those since Adam who have already passed to judgement. But Augustine did not think that membership of the Church explicitly defines one group, the rest of humanity being the other. Perhaps the power of Augustine's vision was all the greater because of its ambiguities, dangling threads of argument and suggestion. The state was not *merely* earthly and wicked: it had its rôle in the divine scheme and government, in its nature, was divinely given. Much was later to be heard about that; the state would be asked to serve the Church by preserving it from its carnal enemies and by using its own power to enforce the purity of the Faith. Yet the Mandate of Heaven (as another civilization might put it) could be withdrawn and, when it was, even an event like the sack of Rome was only a landmark in the working of judgement on sin. In the end the City of God would prevail.

St Augustine escapes simple definition in his greatest book but perhaps he escapes it in every sense. Much remains to be said about him for which there is little room here. He was, for example, a careful and conscientious bishop, the loving pastor of his flock; he was also a persecutor with the dubious distinction of having persuaded the imperial government to use force against the Donatists. He wrote a fascinating spiritual study which, though profoundly misleading on the facts of his early life, virtually founded the literary genre of romantic and introspective autobiography. He could be an artist with words – Latin ones, not Greek (he had to ask St Jerome for help with Greek translation) – and a prize-winning scholar, but his artistry was born of passion rather than of craftsmanship and his Latin is often poor. Yet he was soaked in the classical Roman past. It was from the high ground of his mastery of this tradition that he looked out with the eyes of Christian faith to a cloudy, uncertain and, to other men, frightening future. He embodied two cultures more completely, perhaps, than any other man of those divided times and perhaps this is why, fifteen hundred years later, he still seems to dominate them.

9
The Elements of a Future

From the Germanic invasions grew in the end the first nations of modern Europe, but when the western empire disappeared the barbarian peoples did not occupy areas looking much like later states. They fall clearly into four major and distinctive groups. The northernmost, the Saxons, Angles and Jutes, were moving into the old Roman province of Britain from the fourth century onwards, well before the island was abandoned to its inhabitants when the last emperor to be proclaimed there by his soldiers crossed with his army to Gaul in 407. Britain was then contested between successive waves of invaders and the Romano-British inhabitants until there emerged from it at the beginning of the seventh century a group of seven Anglo-Saxon kingdoms fringed by a Celtic world consisting of Ireland, Wales and Scotland.

Though the first British still lived on in communities which seem to have survived sometimes to the tenth century, and perhaps longer, Romano-British civilization disappeared more completely than its equivalents anywhere else in the western empire. Even the language was to go; a Germanic tongue almost completely replaced it. We may have a fleeting glimpse of the last spasms of Romano-British resistance in the legend of King Arthur and his knights, which could be a reminiscence of the cavalry-fighting skills of the late imperial army, but that is all. Of administrative or spiritual continuity between this imperial province and the barbarian kingdoms there is virtually no trace. The imperial heritage of the future England was purely physical. It lay in the ruins of towns and villas, occasional Christian crosses, or the great constructions like Hadrian's wall which were to puzzle newcomers until they came at last to believe that they were the work of giants of superhuman power. Some of these relics, like the complex of baths built upon the thermal springs at Bath, disappeared from sight for hundreds of years until rediscovered by the antiquaries of the eighteenth and nineteenth centuries. The roads remained, sometimes serving for centuries as trade-routes even when their engineering had succumbed to time, weather and pillage. Finally, there were the natural immigrants who had come with the Romans and stayed: animals like the ferrets which often give an English country boy his first taste of the excitement of the hunt, or plants like the mustard which was to spice the roast beef that became a minor national mythology over a thousand years later. But of the things of the mind left by the Romans we have hardly a trace. Romano-British Christianity, whatever it may have been, disappeared and the keepers of the faith retired for a time to the misty fastnesses where there brooded the monks of the Celtic Church. It was another Rome which was to convert the English nation, not the empire. Before that, Germanic tradition would be the preponderant formative influence as nowhere else within the old imperial territory.

Across the Channel, things were very different. Much survived. After its devastation by the Vandals, Gaul continued to lie in the shadow of the Visigoths of Aquitaine. Their share in repelling the Huns gave them greater importance than ever. To the north-east of Gaul, nevertheless, lay German tribes which were to displace them from this superiority, the Franks. Unlike the Visigoths, they were not Arian and in part because of this the future was to belong to them. They were to have a bigger impact on the shaping of Europe than any other barbarian people.

The graves of the first Franks reveal a warrior society, divided into a hierarchy of ranks. More willing to settle than some other barbarians, they were

Eight miles off the west coast of Ireland lies the seven-hundred-foot-high rock of Skellig Michael. Some beehive cells for monks and a tiny church are the remains of an old Celtic monastery established there at an unknown but early date. From such precarious footholds at the edge of the civilized world came the Irish monks who were the first Christians to evangelize the former Roman province of Britain and the Celtic peoples of Scotland.

established in the fourth century in modern Belgium, between the Scheldt and the Meuse, where they became Roman *foederati*. Some of them moved on into Gaul. One group, settled at Tournai, threw up a ruling family subsequently called Merovingians; the third king (if this is the correct word) of this line was Clovis. His is the first great name in the history of the country known as Francia after the peoples which Clovis put together.

Clovis became ruler of the western Franks in 481. Though formally the subject of the emperor, he soon turned on the last Roman governors of Gaul and conquered lands far to the west and down to the Loire. Meanwhile the eastern Franks defeated the Alamanni and when Clovis had been elected their king, too, a united Frankish kingdom straddled the lower Rhine valley and northern France. This was the heartland of the Frankish state which was the heir to Roman supremacy in north Europe. Clovis married a princess from another Germanic people, the Burgundians, who had settled in the Rhône valley and the area running south-east to modern Geneva and Besançon. She was a Catholic, though her people were Arians, and at some time after their marriage (traditionally in 496) and after a battlefield conversion which is reminiscent of Constantine's, Clovis himself embraced Catholicism. This gave him the support of the Roman Church, the most important power still surviving from the empire in the barbarian lands, in what it now chose to regard as a religious war against the other Germanic peoples of Gaul. Catholicism was also the way to friendship with the Romano-Gaulish population. No doubt the conversion was political; it was also momentous. A new Rome was to rule in Gaul.

The Burgundians were Clovis' first victims, though they were not subjugated completely until after his death, when they were given Merovingian princes but kept an independent state structure. The Visigoths were tackled next; they were left only the south-eastern territories they held north of the Pyrenees (the later Languedoc and Roussillon and Provence). Clovis was now the successor of the Romans in all Gaul; the emperor recognized it by naming him a consul.

The Frankish capital was moved to Paris by Clovis and he was buried in the church he had built there, the first Frankish king not to be buried as a barbarian. But this was not the start of the continuous history of Paris as a capital. A Germanic kingdom was not what later times would think of as a state nor what a Roman would recognize. It was a heritage composed partly of lands, partly of kinship groups. Clovis' heritage was divided among his sons. The Frankish kingdom was not reunited until 558. A couple of years later it broke up again. Gradually, it settled down in three bits. One was Austrasia, with its capital at Metz and its centre of gravity east of the Rhine; Neustria was the western equivalent and had its capital at Soissons; under the same ruler, but distinct, was the kingdom of Burgundy. Their rulers tended to quarrel over the lands where these regions touched.

In this structure there begins to appear a Frankish nation no longer a collection of barbarian warbands, but peoples belonging to a recognizable state, speaking a Latin vernacular, and with an emerging class of landowning nobles. Significantly, from it there also comes a Christian interpretation of the barbarian role in history, the *History of the Franks*, by Gregory, Bishop of Tours, himself from the Romano-Gaulish aristocracy. Other barbarian peoples would produce similar works (the greatest, perhaps, is that written for England by the Venerable Bede) which sought to reconcile traditions in which paganism was still strong to Christianity and the civilized heritage. It must be said that Gregory presented a picture of the Franks after the death of his hero Clovis which was pessimistic; he thought the Frankish rulers had behaved so badly that their kingdom was doomed.

The Merovingians kept other barbarians out of Gaul, and took their lands north of the Alps from the Ostrogoths, where their greatest king was Theodoric. His right to rule in Italy, where he fought off other Germans, was recognized by the emperor in 497. He was utterly convinced of Rome's authority; he had an emperor as godfather and had been brought up at Constantinople until he was eighteen. 'Our royalty is an imitation of yours, a copy of the only Empire on earth', he once wrote to the emperor in Constantinople from his capital in Ravenna. On his coins appeared the legend 'Unvanquished Rome' (*Roma invicta*), and when he went to Rome, Theodoric held games in the old style in the circus. Yet technically he was the only Ostrogoth who was a Roman citizen, his authority accepted by the Senate; his countrymen were merely the mercenary soldiers of the empire. To civil offices he appointed Romans. One of them was his friend and adviser, the philosopher Boethius, who was to be possibly the most important single channel through which the legacy of the classical world passed to Medieval Europe.

Theodoric seems to have been a judicious ruler, maintaining good relations with other barbarian peoples (he married Clovis' sister) and enjoying some sort of primacy among them. But he did not share his own people's Arian faith, and religious division told against Ostrogothic power in the long run. Unlike the Franks, and in spite of their ruler's example they were not to ally with the Roman past and after Theodoric the Ostrogoths were expelled from Italy and history by the generals from the eastern empire. They left a ruined Italy, soon to be invaded by yet another barbarian people, the Lombards.

In the west Clovis had left the Visigoths virtually confined to Spain, from which they had driven the Vandals. Other Germanic peoples were already settled there. Its terrain presented quite special problems – as it has continued to do to all invaders and governments – and the Visigothic kingdom of Spain was not able to resist much more romanization than its founders had undergone in Gaul, where they had fused much less with existing society than had the Franks. The Visigoths – and there were not so very many of them, less than 100,000 at most – clustered about their leaders who spread out from Old Castile through the provinces; they then quarrelled so much that imperial rule was able to re-establish itself for more than a half-century in the south. Finally, the Visigothic kings turned to Catholicism and thus enlisted the authority of the Spanish bishops. In 587 begins the long tradition of Catholic monarchy in Spain.

What this adds up to is hard to say. Generalization is hazardous. Simple duration alone almost explains this; the Visigoths underwent three centuries of evolution between the creation of the kingdom of Toulouse and the end of their ascendancy in Spain. Much changed in so long a time. Though economic life and technology hardly altered, mental and institutional forms were undergoing radical, if slow, transformations in all the barbarian kingdoms. Soon it is not quite right to think of them still as merely such (except, perhaps, the Lombards). The Germanic tribesmen were a minority, often isolated in alien settings, dependent on routines long established by the particular environment for their living and forced into some sort of understanding with the conquered. The passage of their invasions must sometimes have seemed at close quarters like a flood tide, but when it had passed there were often only tiny, isolated pools of invaders left behind, here and there replacing the Roman masters, but often living alongside them and with them. Marriage between Roman and barbarian was not legal until the sixth century, but that was not much of a check. In Gaul the Franks took up its Latin, adding Frankish words to it. By the seventh century, western European society has already a very different atmosphere from that of the turbulent fifth.

None the less, a barbarian past left its imprint. In almost all the barbarian kingdoms society was long and irreversibly shaped by Germanic custom. This sanctioned a hierarchy reflected in the characteristic Germanic device for securing public order, the blood feud. Men – and women, and cattle, and property of all sorts – had in the most literal sense their price; wrongs done were settled by interesting a whole clan or family in the outcome if customary compensation were not forthcoming. Kings more and more wrote down and thus in a sense 'published' what such customs were. Literacy was so rare that there can have been no point in imagining devices such as the stele of Babylon or the white boards on which the decrees of Greek city-states were set out. Recording by a scribe on parchment for future consultation was all that could be envisaged. None the less, in this Germanic world lie the origins of a jurisprudence one day to be carried across oceans to new cultures of European stock. The first institution to open the way to this was the acceptance of kingly or collective power to declare what was to be recorded. All the Germanic kingdoms moved towards the writing down and codification of their law.

Where the early forms of public action are not religious or supernatural, they are usually judicial, and it is hardly surprising that, for example, the Visigothic court of Toulouse should have sought the skills of Roman legal experts. But this was only one form of a respect which almost every barbarian aristocracy showed for Roman tradition and forms. Theodoric saw himself as the representative of the emperor; his problem did not lie in identifying his own role, but in the need to avoid irritating his followers who could be provoked by any excess of romanization. Perhaps similar considerations weighed with Clovis before his conversion, which was an act of identification with Empire as well as with Church. At the level just below such heroic figures, both Frankish and Visigothic noblemen seem to have taken pleasure in showing themselves the heirs of Rome by writing to one another in Latin and patronizing light literature. There was a tie of interest with the Romans, too; Visigothic warriors sometimes found employment in putting down the revolts of peasants who menaced the Romano-Gaulish landowner as well as the invaders. Yet so long as Arianism stood in the way, there was a limit to the identification with *romanitas* possible for the barbarians. The Church, after all, was the supreme relic of empire west of Constantinople.

The eastern emperors had not seen these changes with indifference. But troubles in their own domains hamstrung them and in the fifth century their barbarian generals dominated them too. They watched with apprehension the last years of the puppet emperors of Ravenna but recognized Odoacer, the deposer of the last of them. They maintained a formal claim to rule over a single empire, east and west, without actually questioning Odoacer's independence in Italy until an effective replacement was available in Theodoric, to whom the title of patrician was given. Meanwhile, Persian wars

The Frankish warrior. Taken from a Saxon funeral stele of about 700, this is one of the earliest naturalistic renderings of a barbarian which survives.

and the new pressure of Slavs in the Balkans were more than enough to deal with. It was not until the accession of the Emperor Justinian in 527 that reality seemed likely to be restored to imperial government.

In retrospect Justinian seems something of a failure. Yet he behaved as people thought an emperor should; he did what most people still expected that a strong emperor would one day do. He boasted that Latin was his native tongue; for all the wide sweep of the empire's foreign relations, he could still think plausibly of reuniting and restoring the old empire, centred on Constantinople though it now had to be. We labour under the handicap of knowing what happened, but he reigned a long time and his contemporaries were more struck by his temporary successes. They expected them to herald a real restoration. After all, no one could really conceive a world without the empire. The barbarian kings of the West gladly deferred to Constantinople and accepted titles from it; they did not grasp at the purple themselves. Justinian sought autocratic power, and his contemporaries found the goal both comprehensible and realistic. There is a certain grandeur about his conception of his role; it is a pity that he should have been so unattractive a man.

Justinian was almost always at war. Often he was victorious. Even the costly Persian campaigns (and payments to the Persian king) were successful in the limited sense that they did not lose the empire much ground. Yet they were a grave strategic handicap; the liberation of his resources for a policy of recovery in the West which had been Justinian's aim in his first peace with the Persians always eluded him. Nevertheless, his greatest general, Belisarius, destroyed Vandal Africa and recovered that country for the empire (though it took ten years to reduce it to order). He went on to invade Italy and begin the struggle which ended in 554 with the final eviction of the Ostrogoths from Rome and the unification once more of all Italy under imperial rule, albeit an Italy devastated by the imperial armies as it had never been by the barbarians. These were great achievements, though badly followed up. More were to follow in southern Spain, where the imperial armies exploited rivalry between Visigoths and again set up imperial government in Córdoba. Throughout the western Mediterranean, too, the imperial fleets were supreme; for a century after Justinian's death, Byzantine ships moved about unmolested.

It did not last. By the end of the century most of Italy was gone again, this time to the Lombards, another Germanic people and the final extinguishers of imperial power in the peninsula. In eastern Europe, too, in spite of a vigorous diplomacy of bribery and missionary ideology, Justinian had never been successful in dealing with the barbarians. Perhaps enduring success there was impossible. The pressure from behind on these migrant peoples was too great and, besides, they could see great prizes ahead; 'the barbarians,' wrote one historian of the reign, 'having once tasted Roman wealth, never forgot the road that led to it'. By Justinian's death, in spite of his expensive fortress-building, the ancestors of the later Bulgars were settled in Thrace and a wedge of barbarian peoples separated west and east Rome.

Justinian's conquests, great as they were, could not be maintained by his successors in the face of the continuing threat from Persia, the rise of Slav pressure in the Balkans and, in the seventh century, a new rival, Islam. A terrible time lay ahead. Yet even then Justinian's legacy would be operative through the diplomatic tradition he founded, the building of a network of influences among the barbarian peoples beyond the frontier, playing off one against another, bribing one prince with tribute or a title, standing godparent to the baptized children of a second. If it had not been for the client princedoms of the Caucasus who were converted to Christianity in Justinian's day, or his alliance with the Crimean Goths (which was to last seven centuries), the survival of the eastern empire would have been almost impossible. In this sense, too, the reign sets out the ground-plan of a future Byzantine sphere.

Within the empire, Justinian left an indelible imprint. At his accession the monarchy was handicapped by the persistence of party rivalries which could draw upon popular support, but in 532 this led to a great insurrection which made it possible to strike at the factions and, though much of the city was burned, this was the end of domestic threats to Justinian's autocracy. It showed itself henceforth more and more consistently and nakedly.

Its material monuments were lavish; the greatest is the basilica of St Sophia itself (532–7), but all over the empire public buildings, churches, baths and new towns mark the reign and speak for the inherent wealth of the eastern empire. The richest and most civilized provinces were in Asia and Egypt; Alexandria, Antioch and Beirut were their great cities. A nonmaterial, institutional, monument of the reign was Justinian's codification of Roman law. In four collections a thousand years of Roman jurisprudence was put together in a form which gave it deep influence across the centuries and helped to shape the modern idea of the state. Justinian's efforts to win administrative and organizational reform were far less successful. It was not difficult to diagnose ills known to be dangerous as long before as the third century. But given the expense and responsibilities of Empire, permanent remedies were hard to find. The sale of offices, for example, was

After Justinian's re-conquest of Italy, much was done to beautify Ravenna, the last imperial capital there. This mosaic in the church of San Vitale date from this period and shows the empress approaching the altar with offerings, attended by suites. Theodora was a tiny woman, but in accordance with Byzantine convention she is shown taller than the attendants.

known to be an evil and Justinian abolished it, but then had to tolerate it when it crept back.

The main institutional response to the empire's problem was a progressive regimentation of its citizens. In part, this was in the tradition of regulating the economy which he had inherited. Just as peasants were tied to the soil, craftsmen were now attached to their hereditary corporations and guilds; even the bureaucracy tended to become hereditary. The resulting rigidity was unlikely to make imperial problems easier to solve.

It was unfortunate, too, that a quite exceptionally disastrous series of natural calamities fell on the east at the beginning of the sixth century: they go far to explain why it was hard for Justinian to leave the empire in better fettle than he found it. Earthquake, famine, plague devastated the cities and even the capital itself, where men saw phantoms in the streets. The ancient world was a credulous place, but tales of the emperor's capacity to take off his head and then put it on again, or to disappear from sight at will, already suggest that under these strains the mental world of the eastern empire was already slipping its moorings in classical civilization. Justinian was to make the separation easier by his religious outlook and policies, another paradoxical outcome, for it was far from what he intended. After it had survived for eight hundred years, he abolished the academy of Athens; he wanted to be a Christian emperor, not a ruler of unbelievers, and decreed the destruction of all pagan statues in the capital. Worse still, he accelerated the demotion of the Jews in civic status and the reduction of their freedom to exercise their religion. Things had already gone a long way by then. Pogroms had long been connived at and synagogues destroyed; now Justinian went on to alter the Jewish calendar and interfere with the Jewish order of worship. He even encouraged barbarian rulers to persecute Jews. Long before the cities of western Europe, Constantinople had a ghetto.

Justinian was all the more confident of the rightness of asserting imperial authority in ecclesiastical affairs because (like the later James I of England) he had a real taste for theological disputation. Sometimes the consequences were unfortunate; such an attitude did nothing to renew the loyalty to the empire of the Nestorians and Monophysites, heretics who had refused to accept the definitions of the precise relationship of God the Father to God the Son laid down in 451 at a council at Chalcedon. The theology of such deviants mattered less than the fact that their symbolic tenets were increasingly identified with important linguistic and cultural groups. The empire began to create its Ulsters. Harrying heretics intensified separatist feeling in parts of Egypt and Syria. In the former, the Coptic Church went its own way in opposition to Orthodoxy in the later fifth century and the Syrian Monophysites followed, setting up a 'Jacobite' church. Both were encouraged and sustained by the numerous and enthusiastic monks of those countries. Some of these sects and communities, too, had important connexions outside the empire, so that foreign policy was involved. The Nestorians found refuge in Persia and, though not heretics, the Jews were especially influential beyond the frontiers; Jews in Iraq supported Persian attacks on the empire and Jewish Arab states in the Red Sea interfered with the trade routes to India when hostile measures were taken against Jews in the empire.

Justinian's hopes of reuniting the western and eastern Churches were to be thwarted in spite of his zeal. A potential division between them had always existed because of the different cultural matrices in which each had been formed. The western Church had never accepted the union of religious and secular authority which was the heart of the political theory of the eastern empire; the empire would pass away as others had done (and the Bible told) and it would be the Church which would prevail against the gates of hell. Now such doctrinal divergences became more important, and separation had been made more likely by the breakdown in the West. A Roman pope visited Justinian and the emperor spoke of Rome as the 'source of priesthood', but in the end two Christian communions were first to go their own ways and then violently to quarrel. Justinian's own view, that the emperor was supreme, even on matters of doctrine, fell victim to clerical intransigence on both sides.

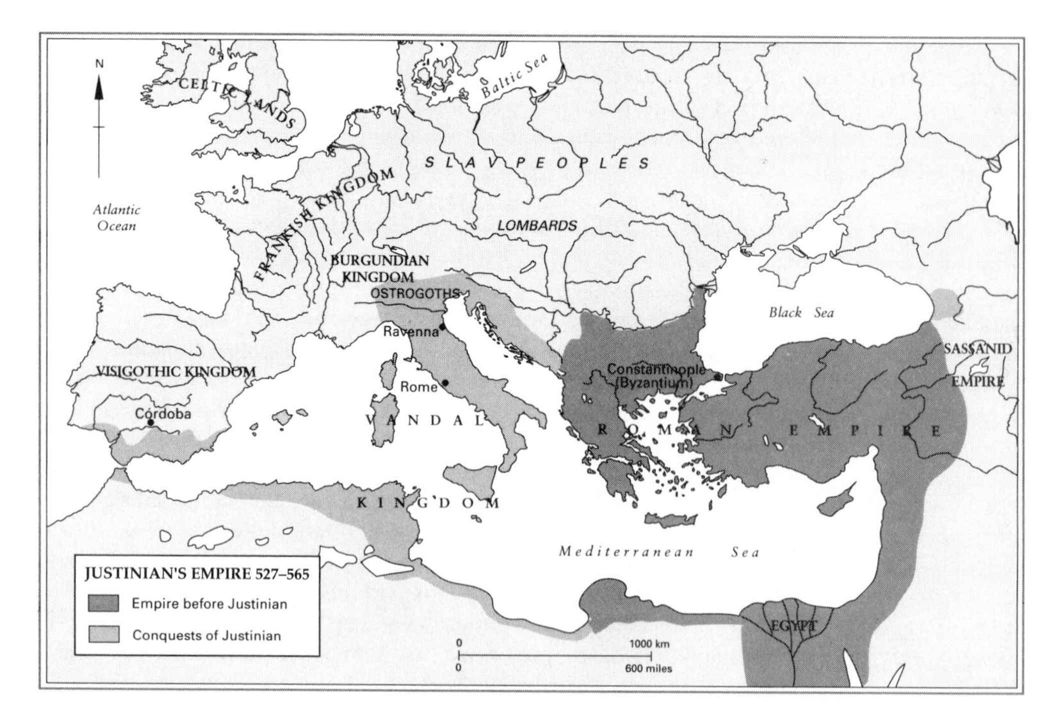

This seems to imply (as do so many others of his acts) that Justinian's real achievement was not that which he sought and temporarily achieved, the re-establishment of the imperial unity, but a quite different one, the easing of the path towards the development of a new, Byzantine civilization. After him, Byzantium was a reality, even if not yet recognized. It was evolving away from the classical world towards a style clearly related to it, but independent of it. This was made easier by contemporary developments in both eastern and western culture, by now overwhelmingly a matter of new tendencies in the Church.

As often in later history, the Church and its leaders had not at first recognized or welcomed an opportunity in disaster. They identified themselves with what was collapsing and understandably so. The collapse of Empire was for them the collapse of civilization; the Church in the West was, except for municipal authority in the impoverished towns, often the sole institutional survivor of *romanitas*. Her bishops were men with experience of administration, at least as likely as other local notables to be intellectually equipped to grapple with new problems. A semi-pagan population looked to them with superstitious awe and attributed to them near-magical power. In many places they were the last embodiment of authority left when imperial armies went away and imperial administration crumbled and they were lettered men among a new unlettered ruling class which craved the assurance of sharing the classical heritage. Socially, they were often drawn from the leading provincial families; that meant that they were sometimes great aristocrats and proprietors with material resources to support their spiritual role. Naturally, new tasks were thrust upon them.

This was not all. The end of the classical world also saw two new institutions emerge in the western Church which were to be lifelines in the dangerous rapids between a civilization which had collapsed and one yet to be born. The first was Christian monasticism, a phenomenon first appearing in the East. It was about 285 that a Copt, St Antony, retired to a hermit's life in the Egyptian desert. His example was followed by others who watched, prayed and strove with demons or mortified the flesh by fasting and more dubious disciplines. Some of them drew together in communities. In the next century this new form of spirituality established itself in a communal form in the Levant and Syria. From there, the idea spread to the West, to the Mediterranean coast of France. In a crumbling society such as fifth-century Gaul the monastic ideal of undistracted worship and service to God in prayer, within the discipline of an ascetic rule, was attractive to many men and women of intellect and character. Through it they could assure personal salvation. The communities attracted many from among the well-born who sought a refuge from a changing world. Unfriendly critics who hankered after the old Roman ideal of service to the state condemned them for shirking their proper responsibilities to society by withdrawing from it. Nor did churchmen always welcome what they saw as the desertion of some of the most zealous among their congregations. Yet many of the greatest churchmen of the age were monks and the institution prospered. Landowners founded communities or endowed existing ones with lands. There were some scandals and no doubt many compromises of principle in grappling with patrons and men of power.

One Italian monk, of whom we know little except his achievement and that he was believed to work miracles, found the state of monasticism shocking. This was St Benedict, one of the most influential men in the Church's history. In 529 he set up a monastery at Monte Cassino in southern Italy, giving it a new rule which he had compiled by sifting and selecting among others available. It is a seminal document of western Christianity and therefore of western civilization. It directed the attention of the monk to the community, whose abbot was to have complete authority. The community's purpose was not merely to provide a hotbed for the cultivation or the salvation of individual souls but that it should worship and live as a whole. The individual monk was to contribute to its task in the framework of an ordered routine of worship, prayer and labour. From the individualism of traditional monasticism a new human instrument was forged; it was to be one of the main weapons in the armoury of the Church.

Pope Leo The Great turns back Attila the Hun from Italy and the holy city of Rome: a legendary encounter depicted one thousand years later by Raphael. The heroic and anachronistic treatment of the subject (here depicted on a tapestry) was embodied in a fresco in the Vatican, where visitors would at once catch the analogy with a later pope who sought vigorously to defend Italy from those he called 'barbarians' – Raphael's patron, Julius II.

St Benedict did not set his sights too high and this was one secret of his success; the Rule was within the powers of ordinary men who loved God and his monks did not need to multilate either body or spirit. Its success in estimating their need was demonstrated by its rapid spread. Very quickly Benedictine monasteries appeared everywhere in the West. They became the key sources of missionaries and teaching for the conversion of pagan England and Germany. In the west, only the Celtic Church at its fringe clung to the older, eremitical model of the monkish life.

The Church's other new great support was the Papacy. The prestige of St Peter's see and the legendary guardianship of the Apostle's bones always gave Rome a special place among the bishoprics of Christendom. It was the only one in the West to claim descent from one of the Apostles. But in principle she had little else to offer; the western Church was a junior branch and it was in the Churches of Asia that the closest links with the Apostolic age could be asserted. Something more was required for the Papacy to begin its rise to the splendid pre-eminence which was taken for granted by the medieval world.

To begin with there was the city. Rome had been seen for centuries as the capital of the world, and for much of the world that had been true. Its bishops were the business colleagues of Senate and Emperor and the departure of the imperial court only left their eminence more obvious. The arrival in Italy of alien civil servants from the eastern empire whom the Italians disliked as much as they did the barbarians directed new attention to the Papacy as the focus of Italian loyalties. It was, too, a wealthy see, with an apparatus of government commensurate with its possessions. It generated administrative skill superior to anything to be found outside the imperial administration itself. This distinction stood out all the more clearly in times of trouble, when the barbarians lacked these skills. The see of Rome had the finest records of any; already in the fifth century papal apologists were exploiting them. The characteristically conservative papal stance, the argument that no new departures are being made but that old positions are being defended, is already present and was wholly sincere; popes did not see themselves as conquerors of new ideological and legal ground; but as men desperately trying to keep the small foothold the Church had already won.

This was the setting of the Papacy's emergence as a great historical force. The fifth-century Leo the Great was the first pope under whom the new power of the bishop of Rome was clearly visible. An emperor declared papal decisions to have the force of law and Leo vigorously asserted the doctrine that the popes spoke in the name of St Peter. He assumed the title *pontifex maximus* discarded by the emperors. It was believed that his intervention by visiting Attila had staved off the Hun attack on Italy; bishops in the West who had hitherto resisted claims for Rome's primacy became more willing to accept them in a world turned upside-down by barbarians. Still, though, Rome was a part of the state church of an empire whose religion Justinian saw as above all the emperor's concern.

The pope in whom the future medieval Papacy is most clearly revealed was also the first pope who had been a monk. In Gregory the Great, who reigned from 590 to 604, there thus came together the two great institutional innovations of the early Church. He was a statesman of great insight. A Roman aristocrat, loyal to the empire and respectful of the emperor, he was nevertheless the first pope who fully accepted the barbarian Europe in which he reigned; his pontificate at last reveals a complete break with the classical world. He saw as his duty the first great missionary campaign, one of whose targets was pagan England, to which he sent Augustine of Canterbury in 596. He struggled against the Arian heresy and was delighted by the conversion of the Visigoths to Catholicism. He was as much concerned with the Germanic kings as with the emperor in whose name he claimed to act, but was also the doughtiest opponent of the Lombards; for help against them he turned both to the emperor and, more significantly, the Franks. Yet the Lombards also made the pope, of necessity, a political power. Not only did they cut him off from the imperial representative at Ravenna but he had to negotiate with them when they stood before the walls of Rome. Like other bishops in the West who inherited civilian authority, he had to feed his city and govern it. Slowly Italians came to see the pope as successor to Rome as well as to St Peter.

In Gregory the classical-Roman heritage and the Christian are subsumed; he represented something new though he can hardly have seen it like that. Christianity had been a part of the classical heritage, yet it was now turning away from much of it and was distinct from it. Significantly, Gregory did not speak Greek; nor did he feel he needed to. There had already been signs of transformation in the Church's relations with the barbarians. With Gregory, one focus of this story has come at last to be Europe, not the Mediterranean basin. There were already sown in it the seeds of the future, though not of the near future; for most of the world's people the existence of Europe for the next thousand years or so is almost irrelevant. But a Europe is at last discernible, unimaginably different though it may be from what was to come.

It was also decisively different from the past. The ordered, literate, unhurried life of the Roman provinces had given way to a fragmented society with, encamped in it, a warrior aristocracy and their tribesmen, sometimes integrated with the earlier inhabitants,

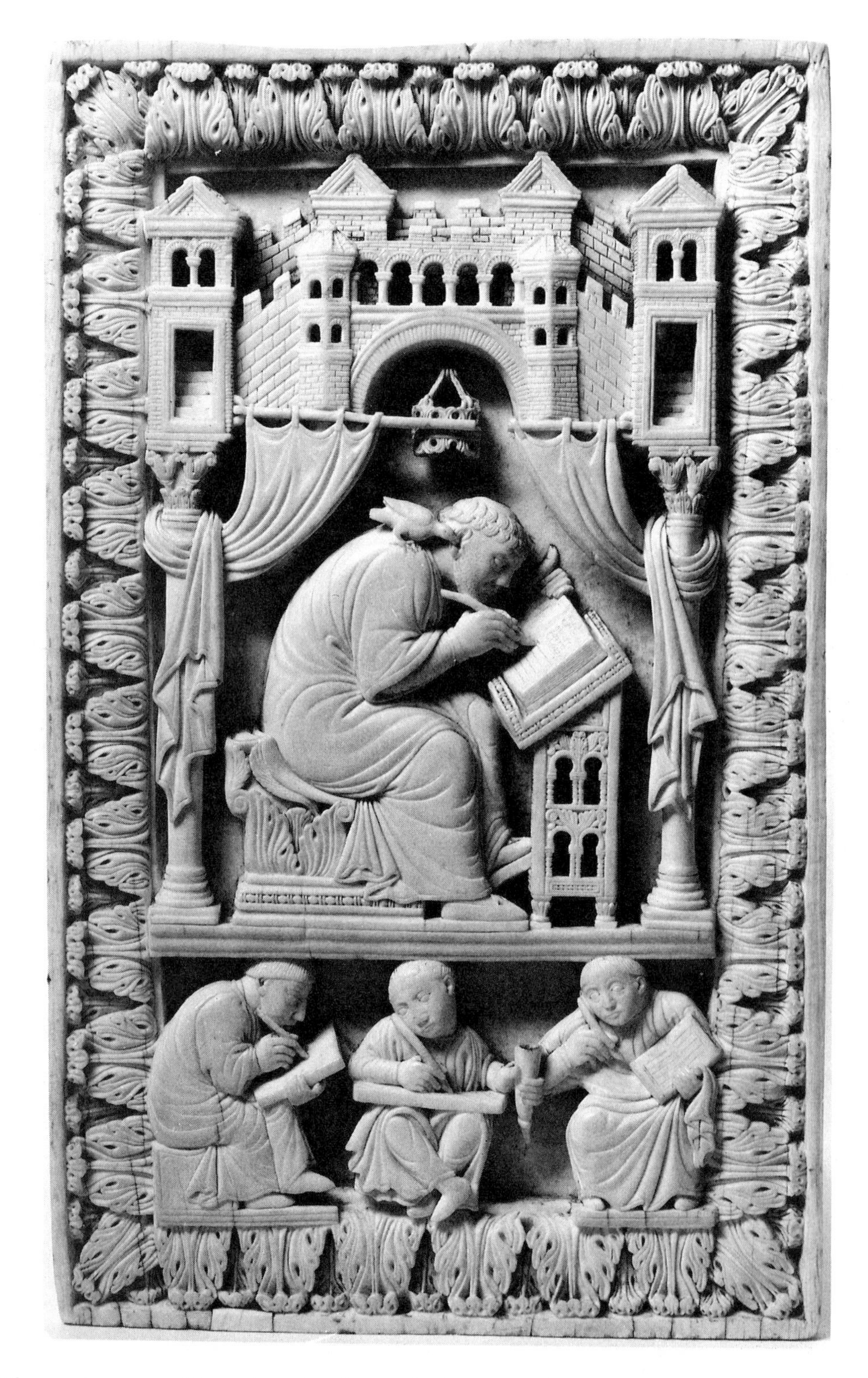

The tenth-century Rhenish artist who made this ivory chose for his subject St Gregory, writing under the influence of the Holy Spirit, here shown as a dove on his shoulder. Below him are the scribes of a monastic copying-school, one holding the inkhorn for the others. Gregory left a large number of writings which were to influence the future developments of the church after his death as much as he had done through the great decisions of his pontificate.

sometimes not. Their chiefs were called kings and were certainly no longer merely chiefs, any more than their followers, after nearly two centuries of involvement with what Rome had left behind, were mere barbarians. It was in 550 that a barbarian king – a Goth – for the first time represented himself on his coins decked in the imperial insignia. Through the impression wrought on their imaginations by the relics of a higher culture, through the efficacy of the idea of Rome itself and through the conscious and unconscious work of the Church, above all, these peoples were on their way to civilization and their art remains to prove it.

Of formal culture, they brought nothing with them to compare with antiquity. There was no barbarian contribution to the civilized intellect. Yet the cultural traffic was not all in one direction at less formal levels. The extent to which Christianity, or at least the Church, was still an elastic form must not be underestimated. Everywhere Christianity had to flow in the channels available and these were defined by layers of paganism, Germanic upon Roman upon Celtic. The conversion of a king like Clovis did not mean that his people made at once even a formal adherence to Christianity; some were still pagan after generations had passed, as their graves showed. But this conservatism presented opportunities as well as obstacles. The Church could utilize the belief in folk magic, or the presence of a holy site which could associate a saint with respect for age-old deities of countryside and forest. Miracles, knowledge of which was assiduously propagated in the saints' lives read aloud to pilgrims to their shrines, were the persuasive arguments of the age. Men were used to the magical interventions of the old Celtic deities or manifestations of Woden's power. For most men then, as for most of human history, the role of religion was not the provision of moral guidance or spiritual insight, but the propitiation of the unseen. Only over blood-sacrifice did Christianity draw the line between itself and the pagan past unambiguously; much other pagan practice and reminiscence it simply christened.

The process by which this came about has often been seen as one of decline and there are certainly reasonable arguments to be made to that effect. In material terms, barbarian Europe was an economically poorer place than the empire of the Antonines; all over Europe tourists gape still at the monuments of Rome's builders as our barbarian predecessors must have done. Yet out of this confusion something quite new and immeasurably more creative than Rome would emerge in due course. It was perhaps impossible for contemporaries to view what was happening in anything but apocalyptic terms. But some may have seen just a little beyond this, as the concerns of Gregory suggest.

Hagia Sophia, the church of the Holy Wisdom, better known as St Sophia, was for centuries the greatest of Christendom's temples. It was Justinian's architectural masterpiece. The dome is an extraordinary feat of engineering (though it collapsed on one occasion) and its three aisles a huge theatre for the religious ritual of the court and the hierarchy of the Church. It was not a place of worship for the commonalty; as one scholar has put it, 'the audience was God'. Its original appearance, bedecked with hangings of silk and gold and blazing with the colour of its mosaics and marbles is hard now to recapture, for what the modern visitor sees is a building first 'purged' by iconoclasts, then redecorated and embellished again over the centuries until the Ottoman conquest which turned it into a mosque, adding minarets to the outside and Koranic inscriptions, pulpit, chandeliers and railings to the interior. Nowadays a museum, it became also in its last phase a model for Moslem architects and the inspiration of one of their finest works, the great Blue mosque which stands about three-quarters of a mile away.

BOOK IV
THE AGE OF DIVERGING TRADITIONS

The 'Romans' of Justinian's day knew they were very different from other men and were proud of it. They belonged to a particular civilization; some of them, at least, thought it was the best conceivable. They were not unique in this. The same was true of men in other parts of the globe. Long before the birth of Christ, civilization was at work in every continent except Australasia, deepening and quickening the divisions opened in human behaviour in prehistoric times. Mankind's cultural variety even in the earliest historical times already escapes all but the finest net and when the classical Mediterranean world had at last cracked irreparably – 500 AD *will do as a rough marker – the world was full of contrasting cultures.*

Most of the globe's surface was then still without civilization, but what was civilized fell into relatively few zones in each of which powerful, distinctive, often self-conscious and largely independent traditions were at work. Their differences were to go on deepening for another thousand years or so, until by about 1500 mankind was probably more diversified than ever before or since. There was still no single dominant cultural tradition.

One result was that Chinese, Indian, western European and Islamic civilizations all lived independently long enough to leave ineradicable traces in the ground-plan of our world. They coexisted and part of the explanation, paradoxically, is that in one respect all these civilizations were much alike. Broadly speaking, they were all based on subsistence agriculture and all had to find their main sources of energy in wind, running water, and animal or human muscles. None of them could bring to bear overwhelming power to change the others. Everywhere, too, the weight of tradition was enormous; the unquestioned, if different, routines under which all mankind then lived would seem intolerable today. Of course, variety in cultural development shaped technology. It was to be a long time before Europeans could again undertake engineering on the Roman scale, yet the Chinese had already long before that discovered how to print with movable characters and knew about gunpowder. Nevertheless, the impact of such advantages or disadvantages was only marginal, largely because intercourse between traditions was difficult except in a few favoured areas. Yet the insulation of one civilization from another was never absolute; there was always some physical and mental interaction going on. The barriers between them resembled permeable membranes rather than impenetrable walls, though for the most part men in these times lived contentedly in traditional patterns, ignoring others following other ways a few hundred – or even a few score – miles from them.

This great era of cultural diversity spans a very long time. In some traditions we must go back to the third century BC *to resume the story, and the breaches in the defences which separated them from others only became irreparable well after 1500. Before then, most civilizations moved largely to rhythms of their own, only occasionally showing the effect of major disturbance from outside. One such disturbance which affected men from Spain to Indonesia, and from the river Niger to China, originated in the Near East, the zone of the oldest civilized traditions and the logical place to begin to examine this diverse world.*

1
Islam and the Re-making of the Near East

With relatively brief interruptions, great empires based in Iran hammered away at the West for a thousand years before 500. Wars can sometimes bring civilizations closer, and in the Near East two cultural traditions had so influenced one another that their histories, though distinct, are inseparable. Through Alexander and his successors, the Achaemenids had passed to Rome the ideas and style of a divine kingship whose roots lay in ancient Mesopotamia; from Rome they went on to flower in the Byzantine Christian empire which fought the Sassanids. Persia and Rome fascinated and, in the end, helped to destroy one another; their antagonism was a fatal commitment to both of them when their attention and resources were urgently needed elsewhere. In the end both succumbed.

The first Sassanid, Ardashir, or Artaxerxes, had a strong sense of continuing Persian tradition. He deliberately evoked memories of the Parthians and the Great King, and his successors followed him in cultivating them by sculpture and inscription. Ardashir claimed all the lands once ruled by Darius and went on himself to conquer the oases of Merv and Khiva, and invade the Punjab; the conquest of Armenia took another hundred and fifty years to confirm but most of it was in the end brought under Persian hegemony. This was the last reconstitution of the ancient Iranian empire which in the sixth century even dominated the Yemen.

Geographical and climatic variety always threatened this huge sprawl of territory with disintegration, but for a long time the Sassanids solved the problems of governing it. There was a bureaucratic tradition running back to Assyria to build on and a royal claim to divine authority. The tension between these centralizing forces and the interests of great families was what the political history of the Sassanid state was about. The resultant pattern was of alternating periods of kings encumbered or unsuccessful in upholding their claims. There were two good tests of this. One was their ability to appoint their own men to the major offices of state and resist the claims on them of the nobility. The other was their retention of control over claims on them of the nobility. The other was their retention of control over the succession. Some kings were deposed and though the kingship itself formally passed by nomination by the ruler, this gave way at times to a semi-electoral system in which the leading officers of state, soldiers and priests made a choice from the royal family.

The dignitaries who contested the royal power and often ruled in the satrapies came from a small number of great families which claimed descent from the Parthian Arsacids, the paramount chiefs of that people. They enjoyed large fiefs for their maintenance but their dangerous weight was balanced by two other forces. One was the mercenary army, largely officered by members of the lesser nobility, who were thus given some foothold against the greater. Its corps d'élite, the heavy-armed household cavalry, was directly dependent on the king. The other force was the priesthood.

Sassanid Persia was a religious as well as a political unity. Zoroastrianism had been formally restored by Ardashir, who gave important privileges to its priests, the *magi*. They led in due course to political power as well. Priests confirmed the divine nature of the kingship, had important judicial duties, and came, too, to supervise the collection of the land-tax which was the basis of Persian finance. The doctrines they taught seem to have varied considerably from the strict monotheism attributed to Zoroaster but focused on a creator, Ahura Mazda, whose viceroy on earth was the king. The Sassanids' promotion of the state religion was closely connected with the assertion of their own authority.

The ideological basis of the Persian state became even more important when the Roman empire became Christian. Religious differences began to matter much more; religious disaffection came to be seen as political. The wars with Rome made Christianity treasonable. Though Christians in Persia had at first been tolerated, their persecution became logical and continued well into the fifth century. Nor was it only Christians who were tormented. In 276 a Persian religious teacher called Mani was executed – by the particularly agonizing method of being flayed alive. He was to become known in the West under the Latin form of his name, Manichaeus, and the teaching attributed to him had a great future as a Christian heresy. Manichaeism brought together Judaeo-Christian beliefs and Persian mysticism and saw the whole cosmos as a great drama in which the forces of Light and Darkness struggled for domination. Those who apprehended this truth sought to participate in the struggle by practising austerities which would open to them the way to perfection and to harmony with the cosmic drama of salvation. Manichaeism sharply differentiated good and evil, nature and God; its fierce dualism appealed to some Christians who saw in it a doctrine coherent with what Paul had taught. St Augustine was a Manichee in his youth and Manichaean traces have been detected much later in the heresies of medieval Europe. Perhaps an uncompromising dualism has always a strong appeal to a certain cast of mind. However that may be, the distinction of being persecuted both by a Zoroastrian and a Christian monarchy preceded the spread of Manichaean ideas far and wide. Their adherents found refuge in central Asia and China, where Manichaeism appears to have flourished as late as the thirteenth century.

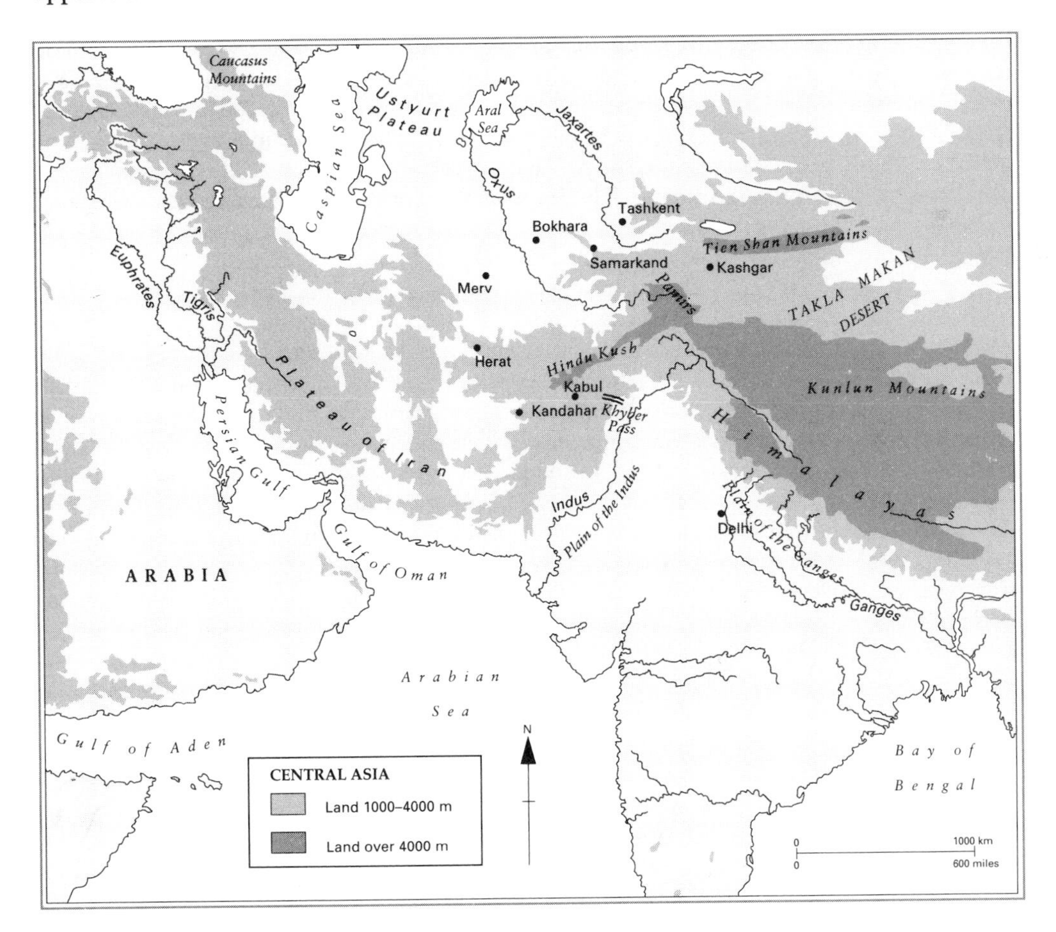

As for orthodox Christians in Persia, although a fifth-century peace stipulated that they should enjoy toleration, the danger that they might turn disloyal in the continual wars with Rome made this a dead letter. Only at the end of the century did a Persian king issue an edict of toleration and this was merely to conciliate the Armenians. It did not end the problem; Christians were soon irritated by the vigorous proselytizing of Zoroastrian enthusiasts. Further assurances by Persian kings that Christianity was to be tolerated do not suggest that they were very successful or vigorous in seeing that it was. Perhaps it was impossible against the political background: the exception which proves the rule is provided by the Nestorians, who *were* tolerated by the Sassanids, but this was just because they were persecuted by the Romans. They were, therefore, thought likely to be politically reliable.

Though religion and the fact that Sassanid power and civilization reached their peak under Chosroes I in the sixth century both help to give the rivalry of the empires something of the dimensions of a contest between civilizations, the renewed wars of that century are not very interesting. They offer for the most part a dull, ding-dong story, though they were the last round but one of the struggle of East and West begun by the Greeks and Persians a thousand years earlier. The climax to this struggle came at the beginning of the seventh century in the last world war of antiquity. Its devastations may well have been the fatal blow to the Hellenistic urban civilization of the Near East.

Chosroes II, the last great Sassanid, then ruled Persia. His opportunity seemed to have come when a weakened Byzantium – Italy was already gone and the Slavs and Avars were pouring into the Balkans – lost a good emperor, murdered by mutineers. Chosroes owed a debt of gratitude to the dead Maurice, for he had been restored to the Persian throne with his aid. He seized on the crime as an excuse and said he would avenge it. His armies poured into the Levant, ravaging the cities of Syria. In 615 they sacked Jerusalem, bearing away the relic of the True Cross which was its most famous treasure. The Jews, it may be remarked, often welcomed the Persians and seized the chance to carry out pogroms of Christians no doubt all the more delectable because the boot had for so long been on the other foot. The next year Persian armies went on to invade Egypt; a year later still, their advance-guards paused only a mile from Constantinople. They even put to sea, raided Cyprus and seized Rhodes from the empire. The empire of Darius seemed to be restored almost at the moment when, at the other end of the Mediterranean, the Roman empire was losing its last possessions in Spain.

This was the blackest moment for Rome in her long struggle with Persia, but a saviour was at hand. In 610 the imperial viceroy of Carthage, Heraclius, had revolted against Maurice's successor and ended that tyrant's bloody reign by killing him. In his turn he received the imperial crown from the Patriarch. The disasters in Asia could not at once be stemmed but Heraclius was to prove one of the greatest of the soldier emperors. Only sea-power saved Constantinople in 626, when the Persian army could not be transported to support an attack on the city by their Avar allies. Next year Heraclius broke into Assyria and Mesopotamia, the old disputed heartland of Near Eastern strategy. The Persian army mutinied, Chosroes was murdered and his successor made peace. The great days of Sassanid power were over. The relic of the True Cross – or what was said to be such – was restored to Jerusalem. The long duel of Persia and Rome was at an end and the focus of world history was to shift at last to another conflict.

The Sassanids went under in the end because they had too many enemies. The year 610 had brought a bad omen: for the first time an Arab force defeated a Persian army. But for centuries Persian kings had been much more preoccupied with enemies on their northern frontiers than with those of the south. They had to contend with the nomads of central Asia who have already made their mark on this narrative, but whose history it is hard to see either as a whole or in detail. None the less, one salient fact is clear – for nearly fifteen centuries these peoples provided an impetus in world history which was felt spasmodically and confusedly and had results ranging from the Germanic invasions of the West to the revitalizing of Chinese government in east Asia.

The best starting-point is geography. The place from which they came, 'central Asia', is not very well named. The term is imprecise. 'Land-locked Asia' might be better, for it is its remoteness from oceanic contact which distinguishes the crucial area. In the first place, this remoteness produced a distinctive and arid climate; secondly, it ensured until modern times an almost complete seclusion from external political pressure, though Buddhism, Christianity and Islam all showed that it was open to cultural influence from the outside.

One way to envisage the area is in a combination of human and topographic terms. It is that part of Asia which is suitable for nomads and it runs like a huge corridor from east to west for four thousand miles or so. Its northern wall is the Siberian forest mass; the southern is provided by deserts, great mountain ranges, and the plateaux of Tibet and Iran. For the most part it is grassy steppe, though the boundary with the desert fluctuates and it extends into it to important oases which have always been a distinctive part of its economy. They sheltered settled populations whose way of life both aroused the antagonism and envy of the nomads and also complemented it. The oases were most frequent and richest in the region of the two great rivers known to the Greeks as the Oxus and the Jaxartes. Cities rose there which were famous for their wealth and skills – Bokhara, Samarkand, Merv – and the trade-routes which bound distant China to the West passed through them.

No one knows the ultimate origins of the peoples of central Asia. They seem distinctive at the moment they enter history, but more for their culture than for their genetic stock. By the first millennium BC they were specialists in the difficult art of living on the move, following pasture with their flocks and herds and mastering the special skills this demanded. It is almost completely true that until modern times they remained illiterate and they lived in a mental world of demons and magic except when converted to the higher religions. They were skilled horsemen and especially adept in the use of the composite bow, the weapon of the mounted archer, which took extra power from its construction not from a single piece of wood but from strips of wood and horn. They could carry out elaborate weaving, carving and decoration, but, of course, did not build, for they lived in their tents.

The first to be named among these peoples were the Scythians, though it is not easy to say very precisely who they were. Scythians have been identified by archaeologists in many parts of Asia and Russia, and as far into Europe as Hungary. They seem to have had a long history of involvement in the affairs of the Near East. Some of them are reported harrying the Assyrian borders in the eighth century BC. Later they attracted the attention of Herodotus, who had much to say about a people who fascinated the Greeks. Possibly they were never really one people, but a group of related tribes. Some of them seem to have settled in south Russia long enough to build up regular relations with the Greeks as farmers, exchanging grain for the beautiful gold objects made by the Greeks of the Black Sea coasts which have been found in Scythian graves. But they also most impressed the Greeks as warriors, fighting in the way which was to be characteristic of the Asian nomads, using bow and arrow from horseback, falling back when faced with a superior force. They harassed the Achaemenids and their successors for centuries and shortly before 100 BC overran Parthia.

The Scythians can serve as an example of the way in which such peoples are set in motion, for they were responding to very distant impulses. They moved because other peoples were moving them. The balance of life in central Asia was always a nice one; even a small displacement of power or resources could deprive a people of its living-space and force it to long treks in search of a new livelihood. Nomads could not travel fast with flocks and herds, but seen from a background of long immunity their irruptions into settled land could seem dramatically sudden. It is through its large-scale periodic upheavals rather than the more or less continuous frontier raiding and pillaging that central Asia has made its impact on world history.

In the third century BC another nomadic people was at the height of its power in Mongolia, the Hsiung-Nu, in whom some recognize the first appearance on

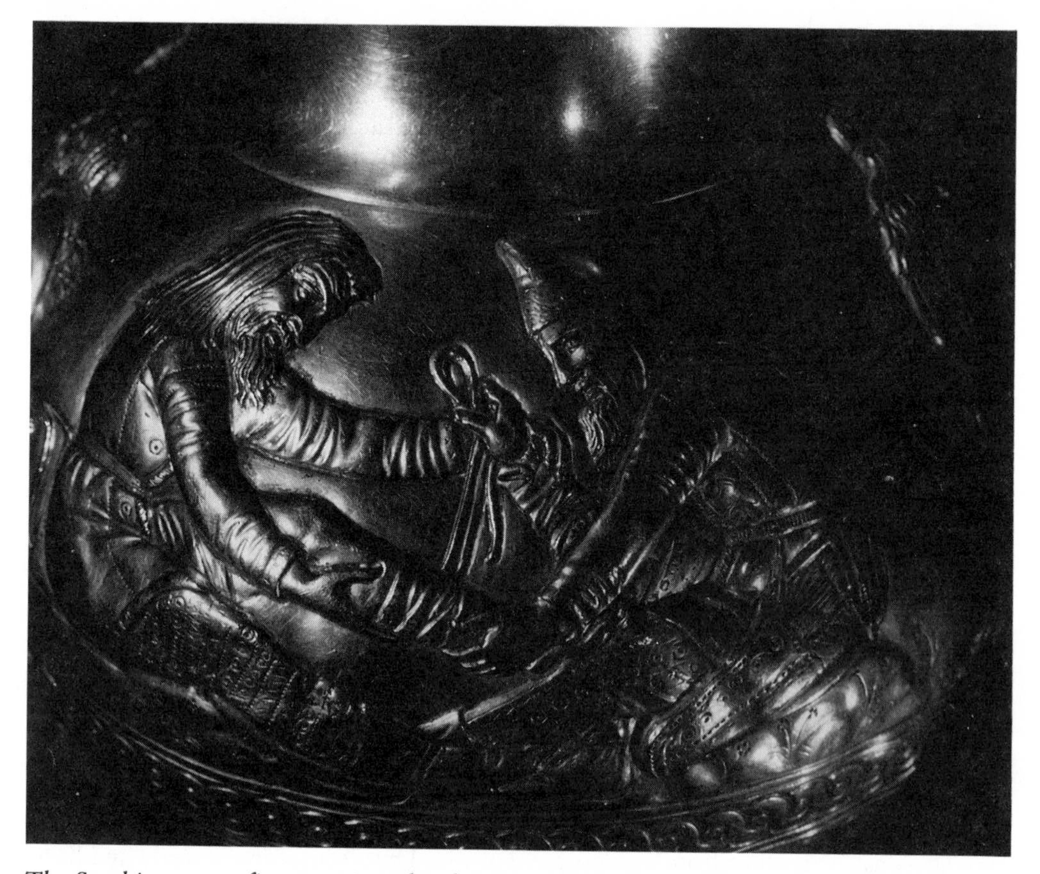

The Scythians were first mentioned in historical narrative by Herodotus and in the twentieth century were acclaimed by Russian enthusiasts as the original stock on which their national tradition was founded. Much, though, remains to be discovered about them. Archaeology has already made it clear that contact with the Greek communities of the Black Sea zone stimulated them to a high level of artistry in precious metals. Much of our knowledge of this gifted people comes from objects laid in their tombs, such as the vessels on which these reliefs appear.

the historical stage of those more familiar as Huns. For centuries they were a byword; all sources agree at least that they were most unpleasant opponents, ferocious, cruel and, unfortunately, skilled warriors. It was against them that the Chinese emperors built the Great Wall, a fourteen-hundred-mile-long insurance policy. Later Chinese governments none the less found it inadequate protection and suffered at the Huns' hands until they embarked on a forward policy, penetrating Asia so as to outflank the Hsiung-Nu. This led to a Chinese occupation of the Tarim basin up to the foothills of the Pamirs and the building on its north side of a remarkable series of frontier works. It was an early example of the generation of imperialism by suction; great powers can be drawn into areas of no concern to them except as sources of trouble. Whether or not this Chinese advance was the primary cause, the Hsiung-Nu now turned on their fellow nomads and began to push west. This drove before them another people, the Yueh-chih, who in turn pushed out of their way more Scyths. At the end of the line stood the post-Seleucid Greek state of Bactria; it disappeared towards 140 BC and the Scythians then went on to invade Parthia.

They also pushed into south Russia, and into India, but that part of the story may be set aside for a moment. The history of the central Asian peoples quickly takes the non-specialist out of his depth; experts are in much disagreement, but it is clear that there was no comparable major upheaval such as that of the third century BC

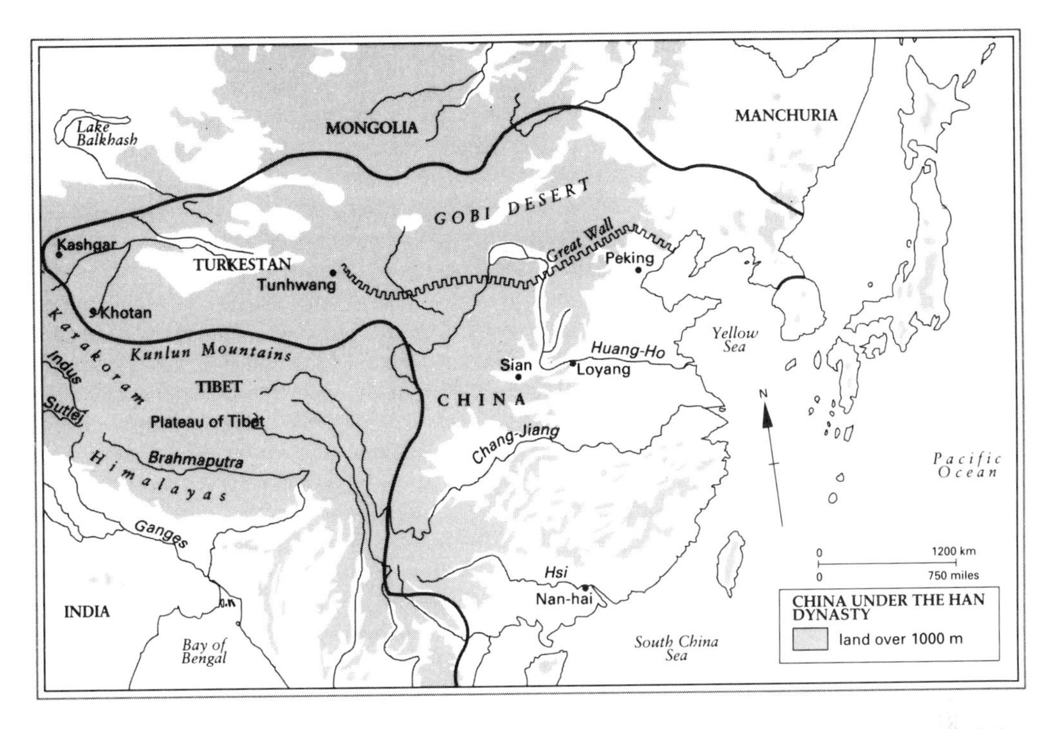

for another four hundred years or so. Then about AD 350 came the re-emergence of the Hsiung-Nu in history when Huns began to invade the Sassanid empire (where they were known as Chionites). In the north, Huns had been moving westwards from Lake Baikal for centuries, driven before more successful rivals as others had been driven before them. Some were to appear west of the Volga in the next century; we have already met them near Troyes in 451. Those who turned south were a new handicap to Persia in its struggle with Rome.

Only one more major people from Asia remains to be introduced, the Turks. Again, the first impact on the outside world was indirect. The eventual successors of the Hsiung-Nu in Mongolia had been a tribe called the Juan-Juan. In the sixth century its survivors were as far west as Hungary, where they were known as Avars; they are noteworthy for introducing a revolution in cavalry warfare to Europe by introducing there the stirrup, which had given them an important advantage. But they were only in Europe because in about 550 they had been displaced in Mongolia by the Turks, a clan of iron-workers who had been their slaves. Among them were tribes – Khazars, Pechenegs; Cumans – which played important parts in the later history of the Near East and Russia. The Khazars were Byzantium's allies against Persia, when the Avars were allies of the Sassanids. What is called the first Turkish empire seems to have been a loose dynastic connexion of such tribes running from the Tamir river to the Oxus. A Turkish khan sent emissaries to Byzantium in 568, roughly nine centuries before other Turks were to enter Constantinople in triumph. In the seventh century the Turks accepted the nominal suzerainty of the Chinese emperors, but by then a new element had entered Near Eastern history, for in 637 Arab armies overran Mesopotamia.

This follow-up to the blows of Heraclius announced the end of an era in Persian history. In 620 Sassanid rule stretched from Cyrenaica to Afghanistan and beyond; thirty years later it no longer existed. The Sassanid empire was gone, its last king murdered by his subjects in 651. More than a dynasty passed away, for the Zoroastrian state went down before a new religion as well as before the Arab armies and it was one in whose name the Arabs would go on to yet greater triumphs.

Islam has shown greater expansive and adaptive power than any other religion except Christianity. It has appealed to peoples as different and as distant from one another as Nigerians and Indonesians; even in its heartland, the lands of Arabic civilization between the Nile and the Hindu Kush, it encompasses huge differences of culture and climate. Yet none of the other great shaping factors of world history was based on fewer initial resources, except perhaps the Jewish religion. Perhaps significantly, the Jews' own nomadic origins lay in the same sort of tribal society, barbaric, raw and backward, which supplied the first armies of Islam. The comparison inevitably suggests itself for another reason, for Judaism, Christianity and Islam are the great monotheistic religions. None of them, in their earliest stages, could have been predicted to be world-historical forces, except perhaps by their most obsessed and fanatical adherents.

The history of Islam begins with Muhammad, but not with his birth, for its date is one of many things which are not known about him. His earliest Arabic biographer did not write until a century or so after he died and even his account survives only indirectly. What is known is that round about 570 Muhammad was born in the Hejaz of poor parents, and was soon an orphan. He emerges as an individual in young manhood preaching the message that there is one God, that He is just and will judge all men, who may assure their salvation by following His will in their religious observance and their personal and social behaviour. This God had been preached before, for he was the God of Abraham and the Jewish prophets, of whom the last had been Jesus of Nazareth.

Muhammad belonged to a minor clan of an important Bedouin tribe, the Quraysh. It was one of many in the huge Arabian peninsula, an area six hundred miles wide and over a thousand long. Those who lived there were subjected to very testing physical conditions; scorched in its hot season, most of Arabia was desert or rocky mountain. In much of it even survival was an achievement. But round its fringes there were little ports, the homes of Arabs who had been seafarers even in the second millennium BC. Their enterprise linked the Indus valley to Mesopotamia and brought the spices and gums of east Africa up the Red Sea to Egypt. The origins of these peoples and those who lived inland is disputed, but both language and the traditional genealogies which go back to Old Testament patriarchs suggest ties with other early Semitic pastoralists who were also ancestors of the Jews, however disagreeable such a conclusion may be to some Arabs today.

Arabia had not always been so uninviting. Just before and during the first centuries of the Christian era it contained a group of prosperous kingdoms. They survived until, possibly, the fifth century AD; both Islamic tradition and modern scholarship link their disappearance with the collapse of the irrigation arrangements of south Arabia. This produced migration from south to north, which created the Arabia of Muhammad's day. None of the great empires had penetrated more than briefly into the peninsula, and Arabia had undergone little sophisticating fertilization from higher civilizations. It declined swiftly into a tribal society based on nomadic pastoralism. To regulate its affairs, patriarchy and kinship was enough so long as the Bedouin remained in the desert.

At the end of the sixth century new changes can be detected. At some oases, population was growing. There was no outlet for it and this was straining traditional social practice. Mecca, where the young Muhammad lived, was such a place. It was important both as an oasis and as a pilgrim centre, for people came to it from all over Arabia to venerate a black meteoric stone, the Kaaba, which had for centuries been important in Arab religion. But Mecca was also an important junction of caravan routes between the Yemen and Mediterranean ports. Along them came foreigners and strangers. The Arabs were polytheists, believing in nature gods, demons and spirits, but as intercourse with the outside world increased, Jewish and Christian communities appeared in the area; there were Christian Arabs before there were Moslems.

At Mecca some of the Quraysh began to go in for commerce (another of the few early biographical facts we know about Muhammad is that in his twenties he was married to a wealthy Qurayshi widow who had money in the caravan business). But such developments brought further social strains as the unquestioned loyalties of tribal structure were compromised by commercial values. The social relationships of a pastoral society assumed noble blood and age to be the accepted concomitants of wealth and this was no longer always the case. Here were some of the formative psychological pressures working on the tormented young Muhammad. He began to ponder the ways of God to man. In the end he articulated a system which helpfully resolved many of the conflicts arising in his disturbed society.

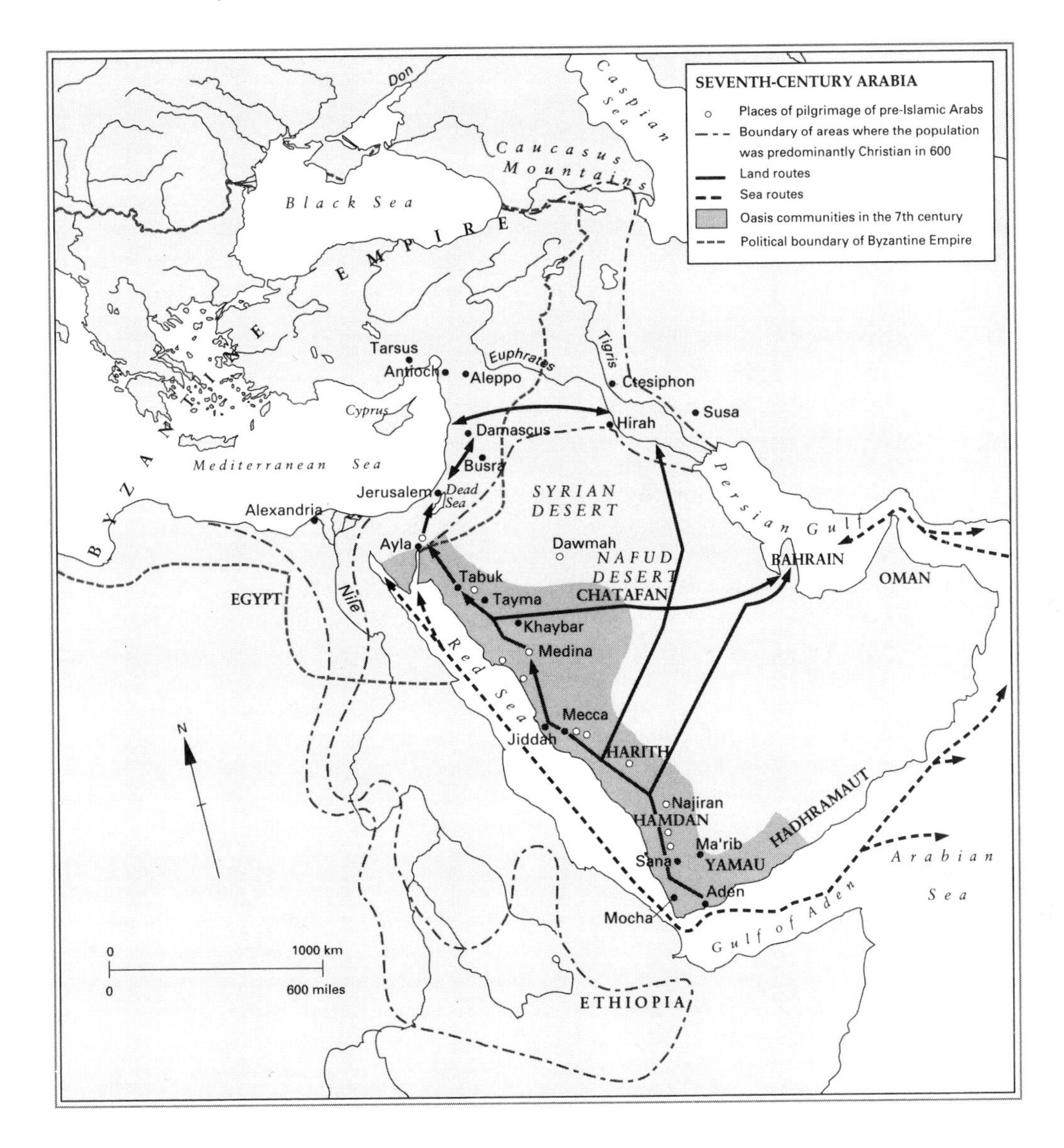

The roots of his achievement lay in the observation of the contrast between the Jews and Christians who worshipped the God familiar also to his own people as Allah, and the Arabs; Christians and Jews had a scripture for reassurance and guidance, and Muhammad's people had none. One day while he contemplated in a cave outside Mecca a voice came to him revealing his task:

Recite, in the name of the Lord, who created, Created man from a clot of blood.

For twenty-two years Muhammad was to recite and the result is one of the great formative books of mankind, the Koran. Its narrowest significance is still enormous and, like that of Luther's Bible or the Authorized Version, it is linguistic; the Koran crystallized a language. It was the crucial document of Arabic culture not only because of its content but because it was to propagate the Arabic tongue in a written form. But it is much more; it is a visionary's book, passionate in its conviction of divine inspiration;

vividly conveying Muhammad's spiritual genius and vigour. Though not collected in his lifetime, it was taken down by his entourage as delivered by him in a series of revelations; Muhammad saw himself as a passive instrument, a mouthpiece of God. The word *Islam* means submission or surrender. Muhammad believed he was to convey God's message to the Arabs as other messengers had earlier brought His word to other peoples. But Muhammad was sure that his position was special; though there had been prophets before him, their revelations heard (but falsified) by Jew and Christian, he was the final Prophet. Through him, Moslems were to believe, God spoke his last message to mankind.

The message demanded exclusive service for Allah. Tradition says that Muhammad on one occasion entered the Kaaba's shrine and struck with his staff all the images of the other deities which his followers were to wash out, sparing only that of the Virgin and Child (he retained the stone itself). His teaching began with the uncompromising preaching of monotheism in a polytheistic religious centre. He went on to define a series of observances necessary to salvation and a social and personal code which often conflicted with current ideas, for example in its attention to the status of the individual believer, whether man, woman or child. It can readily be understood that such teaching was not always welcome. It seemed yet another disruptive and revolutionary influence – as it was – setting its converts against those of their tribe who worshipped the old gods and would certainly go to hell for it. It might damage the pilgrim business, too (though in the upshot it improved it, for Muhammad insisted strictly on the value of pilgrimage to so holy a place). Finally, as a social tie it placed blood second to belief; it was the brotherhood of believers which was the source of community, not the kinship group.

It is not surprising that the leaders of his tribe turned on Muhammad. Some of his followers emigrated to Ethiopia, a monotheistic country already penetrated by Christianity. Economic boycott was employed against the recalcitrants who stayed. Muhammad heard that the atmosphere might be more receptive at another oasis about two hundred and fifty miles further north, Yathrib. Preceded by some two hundred followers, he left Mecca and went there in 622. This *Hegira*, or emigration, was to be the beginning of the Moslem calendar and Yathrib was to change its name, becoming the 'city of the prophet', Medina.

It, too, was an area unsettled by economic and social change. Unlike Mecca, though, Medina was not dominated by one powerful tribe, but was a focus of competition for two; moreover, there were other Arabs there who adhered to Judaism. Such divisions favoured Muhammad's leadership. Converted families gave hospitality to the immigrants. The two groups were to form the future élite of Islam, the 'Companions of the Prophet'. Muhammad's writings for them show a new direction in his concerns, that of organizing a community. From the spiritual emphasis of his Mecca revelations he turned to practical, detailed statements about food, drink, marriage, war. The characteristic flavour of Islam, a religion which was also a civilization and a community, was now being formed.

Medina was the base for subduing first Mecca and then the remaining tribes of Arabia. A unifying principle was available in Muhammad's idea of the *umma*, the brotherhood of believers. It integrated Arabs (and, at first, Jews) in a society which maintained much of the traditional tribal framework, stressing the patriarchal structure in so far as it did not conflict with the new brotherhood of Islam, even retaining the traditional primacy of Mecca as a place of pilgrimage. Beyond this it is not clear how far Muhammad wished to go. He had made approaches to Jewish tribesmen at Medina, but they had refused to accept his claims; they were therefore driven out, and a Moslem community alone remained, but this need not have implied any enduring conflict with either Judaism or its continuator, Christianity. Doctrinal ties existed in their monotheism and their scriptures even if Christians were believed to fall into polytheism with the idea of the Trinity. Nevertheless, Muhammad enjoined the conversion of the infidel and for those who wished there was a justification here for proselytizing.

Muhammad died in 632. At that moment the community he had created was in grave danger of division and disintegration. Yet on it two Arab empires were to be built, dominating successive historical periods from two different centres of gravity.

In each the key institution was the caliphate, the inheritance of Muhammad's authority as the head of a community, both its teacher and ruler. From the start, there was no tension of religious and secular authority in Islam, no 'Church and State' dualism such as was to shape Christian policies for a thousand years and more. Muhammad, it has been well said, was his own Constantine, prophet and sovereign in one. His successors would not prophesy as he had done, but they were long to enjoy his legacy of unity in government and religion.

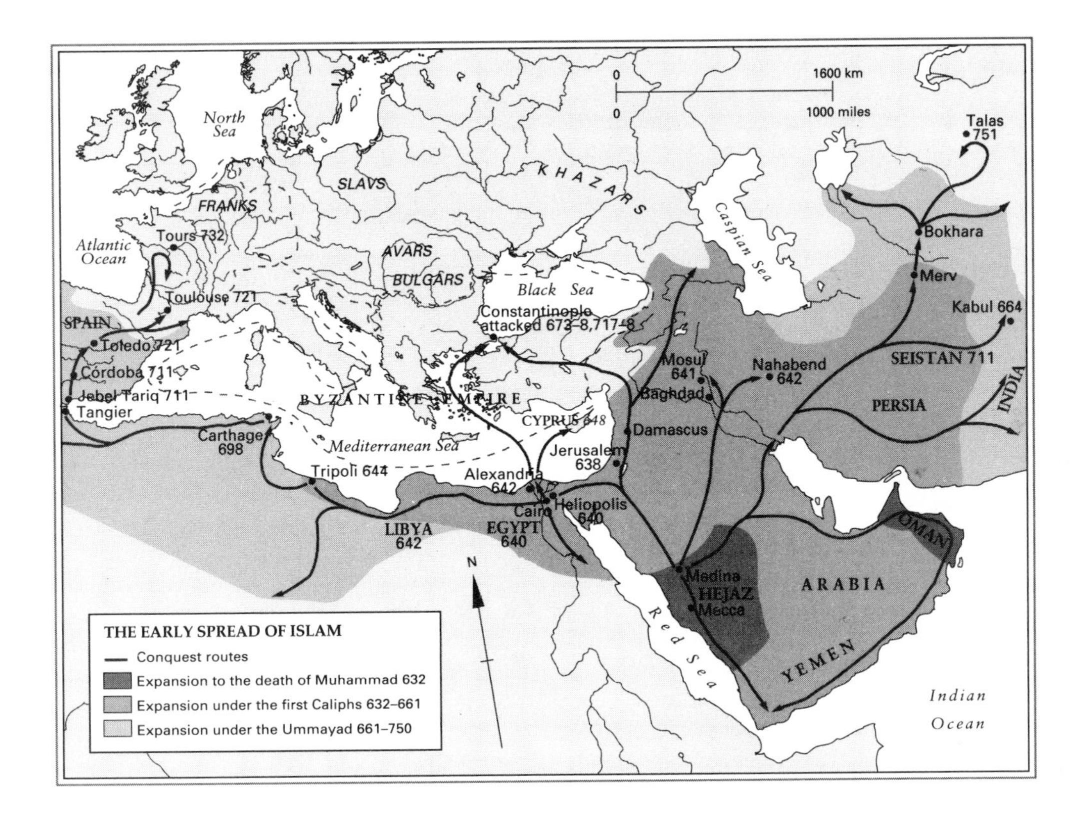

The first 'patriarchal' caliphs were all Quraysh, most of them related to the Prophet by blood or marriage. Soon, they were criticized for their wealth and status and were alleged to act as tyrants and exploiters. The last of them was deposed and killed in 661 after a series of wars in which conservatives contested what they saw as the deterioration of the caliphate from a religious to a secular office. The year 661 saw the beginning of the Umayyad caliphate, the first of the two major chronological divisions of Arab empire, focused on Syria, with its capital as Damascus. It did not bring struggle within the Arab world to an end for in 750 the Abbasid caliphate displaced it. The new caliphate lasted longer. Soon moved to a new location, Baghdad, it would survive nearly two centuries (until 946) as a real power and even longer as a puppet regime. Between them the two dynasties gave the Arab peoples three centuries of ascendancy in the Near East.

The first and most obvious expression of this was a great series of conquests in the first century of Islam which remade the world map from Gibraltar to the Indus. They had in fact begun immediately after the Prophet's death with the assertion of the first caliph's authority. Abu-Bakr set about conquering the unreconciled tribes of southern and eastern Arabia for Islam. But this led to fighting which spread to Syria and Iraq. Something analogous to the processes by which barbarian disturbances in central Asia rolled outward in their effects was at work in the overpopulated Arabian peninsula; this time there was a creed to give it direction as well as a simple love of plunder.

Once beyond the peninsula, the first victim of Islam was Sassanid Persia.

The challenge came just as she was under strain at the hands of the Heraclian emperors who were likewise to suffer from this new scourge. In 633 Arab armies invaded Syria and Iraq. Three years later the Byzantine forces were driven from Syria and in 638 Jerusalem fell to Islam. Mesopotamia was wrested from the Sassanids in the next couple of years, and at about the same time Egypt was taken from the empire. An Arab fleet was now created and the absorption of North Africa began. Cyprus was raided in the 630s and 640s; later in the century it was divided between the Arabs and the empire. At the end of the century the Arabs took Carthage, too. Meanwhile, after the Sassanids' disappearance the Arabs had conquered Khurasan in 655, Kabul in 664; at the beginning of the eighth century they crossed the Hindu Kush to invade Sind, which they occupied between 708 and 711. In the latter year an Arab army with Barber allies crossed the Straits of Gibraltar (its Berber commander, Tariq, is commemorated in that name, which means *Jebel Tariq*, or mount of Tariq) and advanced into Europe, shattering at last the Visigothic kingdom. Finally, in 732, a hundred years after the death of the Prophet, the Moslem army, deep in France, puzzled by over-extended communications and the approach of winter, turned back near Poitiers. The Franks who faced them and killed their commander claimed a victory; at any rate, it was the high water-mark of Arab conquest, though in the next few years Arab expeditions raided into France as far as the upper Rhône. Whatever brought it to an end (and possibly it was just because the Arabs were not much interested in European conquest, once away from the warm lands of the Mediterranean littoral), the Islamic onslaught in the West remains an astonishing achievement, even if Gibbon's vision of an Oxford teaching the Koran was never remotely near realization.

The Arab armies were at last stopped in the East, too, although at a cost of two sieges of Constantinople and the confining of the empire to the Balkans and Anatolia. From eastern Asia there is a report that an Arab force reached China in the early years of the eighth century; even if questionable, such a story is evidence of the conquerors' prestige. What is certain is that the frontier of Islam settled down along the Caucasus mountains and the Oxus after a great Arab defeat at the hands of the Khazars in Azerbaijan, and a victory in 751 over a Chinese army commanded by a Korean general on the river Talas, in the high Pamirs. On all fronts, in western Europe, central Asia, Anatolia and in the Caucasus, the tide of Arab conquest at last came to an end in the middle of the eighth century.

That tide had not flowed without interruption. There had been something of a lull in Arab aggressiveness during the intestine quarrelling just before the establishment of the Umayyad caliphate and there had been bitter fighting of Moslem against Moslem in the last two decades of the seventh century. But for a long time circumstances favoured the Arabs. Their first great enemies, Byzantium and Persia, had both had heavy commitments on other fronts and had been for centuries one another's fiercest antagonists. After Persia went under, Byzantium still had to contend with enemies in the west and to the north, fending them off with one hand while grappling with the Arabs with the other. Nowhere did the Arabs face an opponent comparable to the Byzantine empire nearer than China. Because of this, they pressed their conquests to the limit of geographical possibility or attractiveness, and sometimes their defeat showed they had overstretched themselves. Even when they met formidable opponents, though, the Arabs still had great military advantages. Their armies were recruited from hungry fighters to whom the Arabian desert had left small alternative; the spur of overpopulation was behind them. Their assurance in the Prophet's teaching that death on the battlefield against the infidel would be followed by certain removal to paradise was a huge moral advantage. They fought their way, too, into lands whose peoples were often already disaffected with their rulers; in Egypt, for example, Byzantine religious orthodoxy had created dissident and alienated minorities. Yet when all such influences have been totted up, the Arab success remains amazing. The fundamental explanation must lie in the movement of large numbers of men by a religious ideal. The Arabs thought they were doing God's will and creating a new brotherhood in the process; they generated an excitement in themselves like that of later revolutionaries. And conquest was only the beginning of the story of the impact of Islam on the world. In its range and complexity it can only be compared to that of Judaism or Christianity. At one time it looked as if Islam might be irresistible everywhere. That was not to be, but one of the great traditions of civilization was to be built on its conquests and conversions.

2
The Arab Empires

In 661 the Arab governor of Syria, Mu-Awiyah, set himself up as caliph after a successful rebellion and the murder (though not at his hands) of the caliph Ali, cousin and son-in-law of the Prophet. This ended a period of anarchy and division, and so, thought many Moslems, excused what he did. It was also the foundation of the Umayyad caliphate.

This usurpation gave political ascendancy among the Arab peoples to the aristocrats of the Quraysh, the very people who had opposed Muhammad at Mecca. Mu-Awiyah set up his capital at Damascus and later named his son crown prince, an innovation which introduced the dynastic principle. This was also the beginning of a schism within Islam, for a dissident group, the Shi'ites, henceforth claimed that the right of interpreting the Koran was confined to Muhammad's descendants. The murdered caliph, they said, had been divinely designated as *imam* to transmit his office to his descendants and was immune from sin and error. The Umayyad caliphs, correspondingly, had their own party of supporters, called Sunnites, who believed that doctrinal authority changed hands with the caliphate. Together with the creation of a regular army and a system of supporting it by taxation of the unbelievers, a decisive movement was thus made away from an Arab world solely of tribes. The site of the Umayyad capital, too, was important in changing the style of Islamic culture, as were the personal tastes of the first caliph. Syria was a Mediterranean state, but Damascus was roughly on the border between the cultivated land of the Fertile Crescent and the barren expanses of the desert; its life was fed by two worlds. To the desert-dwelling Arabs, the former must have been the more striking. Syria had a long Hellenistic past and both the caliph's wife and his doctor were Christians. While the barbarians of the West looked to Rome, Arabs were to be shaped by the heritage of Greece.

The first Umayyad speedily reconquered the East from dissidents who resisted the new régime and the Shi'ite movement was driven underground. There followed a glorious century whose peak came under the sixth and seventh caliphs between 685 and 705. Unfortunately we know little about the detailed and institutional history of Umayyad times. Archaeology sometimes throws light on general trends and reveals something of the Arabs' impact on their neighbours. Foreign records and Arab chroniclers log important events. Nevertheless, early Arab history produces virtually no archive material apart from an occasional document quoted by an Arab author. Nor did Islamic religion have a bureaucratic centre of ecclesiastical government. Islam had nothing remotely approaching in scope the records of the papacy, for example, though the analogy between the popes and the caliphs might reasonably arouse similar expectations. Instead of administrative records throwing light on continuities there are only occasional collections preserved almost by chance, such as a mass of papyri from Egypt, special accumulations of documents by minority communities such as the Jews, and coins and inscriptions. The huge body of Arabic literature in print or manuscript provides further details, but it is much more difficult to make general statements about the government of the caliphates with confidence than, say, similar statements about Byzantium.

It seems, none the less, that the early arrangements of the caliphates, inherited from the orthodox caliphs, were loose and simple – perhaps too loose, as the Umayyad defection showed. Their basis was conquest for tribute, not for assimilation, and the result was a series of compromises with existing structures. Administratively and politically, the early caliphs took over the ways of earlier rulers. Byzantine and Sassanid arrangements continued to operate; Greek was the language of government in Damascus, Persian in Ctesiphon, the old Sassanid capital, until the early eighth century. Institutionally, the Arabs left the societies they took over by and large undisturbed except by taxation. Of course, this does not mean that they went on just as before. In north-western Persia, for example, Arab conquest seems to have been followed by a decline in commerce and a drop in population, and it is hard not to associate this with the collapse of a complex drainage and irrigation system successfully maintained in Sassanid times. In other places, Arab conquest had less drastic effects. The conquered were not antagonized by having to accept Islam, but took their places in a hierarchy presided over by the Arab Moslems. Below them came the converted neo-Moslems of the tributary peoples, then the *dhimmi*, or 'protected persons' as the Jewish and Christian monotheists were called. Lowest down the scale came unconverted pagans or adherents of no revealed religion. In the early days the Arabs were segregated from the native population and lived as a military caste in special towns paid by the taxes raised locally, forbidden to enter commerce or own land.

This could not be kept up. Like the Bedouin customs brought from the desert, segregation was eroded by garrison life. Gradually the Arabs became landowners and cultivators, and so their camps changed into new, cosmopolitan cities such as Kufa or Basra, the great *entrepôt* of the trade with India. More and more Arabs mixed with the local inhabitants in a two-way relationship, as the indigenous élites underwent an administrative and linguistic arabization. The caliphs appointed more and more of the officials of the provinces and by the mid-eighth century Arabic was almost everywhere the language of administration. Together with the standard coinage bearing Arabic inscriptions it is the major evidence of Umayyad success in laying the foundations of a new, eclectic civilization. Such changes went fastest in Iraq, where they were favoured by prosperity as trade revived under the Arab peace.

The assertion of their authority by the Umayyad caliphs was one source of their troubles. Local bigwigs, especially in the eastern half of the empire, resented interference with their practical independence. Whereas many of the aristocracy of the former Byzantine territories emigrated to Constantinople, the élites of Persia could not; they had nowhere to go and had to remain, irritated by their subordination to the Arabs who left them much of their local authority. Nor did it help that the later Umayyad caliphs were men of poor quality, who did not command the respect won by the great men of the dynasty. Civilization softened them. When they sought to relieve the tedium of life in the towns they governed, they moved out into the desert, not to live again the life of the Bedouin, but to enjoy their new towns and palaces, some of them remote and luxurious, equipped as they were with hot baths and great hunting enclosures, and supplied from irrigated plantations and gardens.

There were opportunities here for the disaffected, among whom the *Shi'a*, the party of the Shi'ites, were especially notable. Besides their original political and religious appeal, they increasingly drew on social grievances among the non-Arab converts to Islam, particularly in Iraq. From the start, the Umayyad *régime* had distinguished sharply between those Moslems who were and those who were not by birth members of an Arab tribe. The numbers of the latter class grew rapidly; the Arabs had not sought to convert (and sometimes even tried to deter from conversion in early times) but the attractiveness of the conquering creed was powerfully reinforced by the fact that adherence to it might bring tax relief. Around the Arab garrisons, Islam had spread rapidly among the non-Arab populations which grew up to service their needs. It was also very successful among the local élites who maintained the day-to-day administration. Many of these neo-Moslems, the *mawali*, as they were called, eventually became soldiers, too. Yet they increasingly felt alienated and excluded from the aristocratic society of the pure Arabs. The puritanism and orthodoxy of the Shi'ites, equally alienated from the same society for political and religious reasons, made a great appeal to them.

Islamic mechanics built upon Hellenistic texts which transmitted the skills of such men as Archimedes and Hero of Alexandria. This elaborate machine, depicted in a fourteenth-century miniature (probably from Syria), is a clock. It indicated the passage of time by an elaborate series of movements by the little figures making it up and finally by the depositing of a little ball inside the elephant to mark each half-hour.

Increasing trouble in the east heralded the breakdown of Umayyad authority. In 749 a new caliph, Abu-al-Abbas, was hailed publicly in the mosque at Kufa in Iraq. This was the beginning of the end for the Umayyads. The pretender, a descendant of an uncle of the Prophet, announced his intention of restoring the caliphate to orthodox ways; he appealed to a wide spectrum of opposition including the Shi'ites. His full name was promising: it meant 'Shedder of Blood'. In 750 he defeated and executed the last Umayyad caliph. A dinner-party was held for the males of the defeated house; the guests were murdered before the first course, which was then served to their hosts. With this clearing of the decks began nearly two centuries during which the Abbasid caliphate ruled the Arab world, the first of them the most glorious.

The support the Abbasids enjoyed in the eastern Arab dominions was reflected by the shift of the capital to Iraq, to Baghdad, until then a Christian village on the Tigris. The change had many implications. Hellenistic influences were weakened; Byzantium's prestige seemed less unquestionable. A new weight was given to Persian influence which was both politically and culturally to be very important. There was a change in the ruling caste, too, and one sufficiently important to lead some historians to call it a social revolution. They were from this time Arabs only in the sense of being Arabic-speaking; they were no longer Arabian. Within the matrix provided by a single religion and a single language the élites which governed the Abbasid empire came from many peoples right across the Middle East. They were almost always Moslem but they were often converts or children of convert families. The cosmopolitanism of Baghdad reflected the new cultural atmosphere. A huge city, rivalling Constantinople, with perhaps a half-million inhabitants, it was a complete antithesis of the ways of life brought from the desert by the first Arab conquerors. A great empire had come again to the whole Middle East. It did not break with the past ideologically, though, for after dallying with other possibilities the Abbasid caliphs confirmed the Sunnite orthodoxy of their predecessors. This was soon reflected in the disappointment and irritation of the Shi'ites who had helped to bring them to power.

The Abbasids were a violent lot and did not take risks with their success. They quickly and ruthlessly quenched opposition and bridled former allies who might turn sour. Loyalty to the dynasty, rather than the brotherhood of Islam, was increasingly the basis of the empire and this reflected the old Persian tradition. Much was made of religion as a buttress to the dynasty, though, and the Abbasids persecuted nonconformists. The

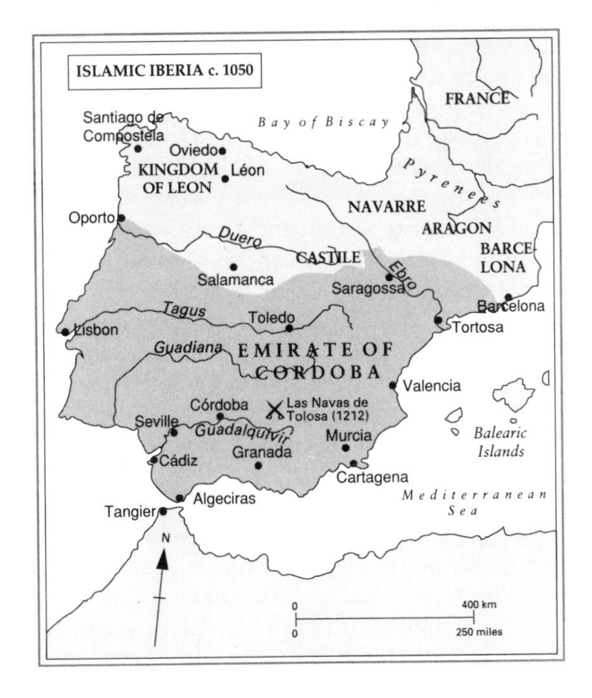

machinery of government became more elaborate. Here one of the major developments was that of the office of vizier (monopolized by one family until the legendary caliph Haroun-al-Raschid wiped them out). The whole structure became somewhat more bureaucratized, the land taxes raising a big revenue to maintain a magnificent monarchy. Nevertheless, provincial distinctions remained very real. Governorships tended to become hereditary, and, because of this, central authority was eventually forced on to the defensive. The governors exercised a greater and greater power in appointments and the handling of taxation. It is not easy to say what was the caliphate's real power, for it regulated a loose collection of provinces whose actual dependence was related very much to the circumstances of the moment. But of Abbasid wealth and prosperity at its height there can be no doubt. They rested not only on its great reserves of manpower and the large areas where agriculture was untroubled during the Arab peace, but also upon the favourable conditions it created for trade. A wider range of commodities circulated over a larger area then ever before. This revived commerce in the cities along the caravan routes which passed through the Arab lands from east to west. The riches of Haroun-al-Raschid's Baghdad reflected the prosperity they brought.

Islamic civilization in the Arab lands reached its peak under the Abbasids. Paradoxically, one reason was the movement of its centre of gravity away from Arabia and the Levant. Islam provided a political organization which, by holding together a huge area, cradled a culture which was essentially synthetic, mingling, before it was done, Hellenistic, Christian, Jewish, Zoroastrian and Hindu ideas. Arabic culture under the Abbasids had closer access to the Persian tradition and closer contact with India which brought to it renewed vigour and new creative elements.

One aspect of Abbasid civilization was a great age of translation into Arabic, the new *lingua franca* of the Middle East. Christian and Jewish scholars made available to Arab readers the works of Plato and Aristotle, Euclid and Galen, thus importing the categories of Greek thought into Arab culture. The tolerance of Islam for its tributaries made this possible in principle from the moment when Syria and Egypt were conquered, but it was under the early Abbasids that the most important translations were made. So much it is possible to chart fairly confidently. To say what this meant, of course, is more difficult, for though the texts of Plato might be available, it was the Plato of late Hellenistic culture, transmitted through interpretations by Christian monks and Sassanid academics.

The culture these sources influenced was predominantly literary; Arabic Islam produced beautiful buildings, lovely carpets, exquisite ceramics, but its great medium was the word, spoken and written. Even the great Arab scientific works are often huge prose compendia. The accumulated bulk of this literature is immense and much of it simply remains unread by western scholars. Large numbers of its manuscripts have never been examined at all. The prospect is promising; the absence of archive material for early Islam is balanced by a huge corpus of literature of all varieties and forms except the drama. How deeply it penetrated Islamic society remains obscure, though it is clear that educated people expected to be able to write verses and could enjoy critically the performances of singers and bards. Schools were widespread; the Islamic world was probably highly literate by comparison, for example, with medieval Europe. Higher learning, more closely religious in so far as it was institutionalized in the mosques or special schools of religious teachers, is more difficult to assess. How much, therefore, the potentially divisive and stimulating effect of ideas drawn from other cultures was felt below the level of the leading Islamic thinkers and scientists is hard to say, but potentially many seeds of a questioning and self-critical culture were there from the eighth century onwards. They seem not to have ripened.

Judged by its greatest men, Arabic culture was at its height in the East in the ninth and tenth centuries and in Spain in the eleventh and twelfth. Although Arab history and geography are both very impressive, its greatest triumphs were scientific and mathematical; we still employ the 'arabic' numerals which made possible written calculations with far greater simplicity than did Roman numeration and which were set out by an Arab arithmetician (although in origin they were Indian). This transmission function of Arabic culture was always important and characteristic but must not obscure its originality. The name of the greatest of Islamic astronomers, Al-Khwarizmi, indicates Persian Zoroastrian origins; it expresses the way in which Arabic culture was a confluence of sources. His astronomical tables, none the less, were an Arabic achievement, an expression of the synthesis made possible by Arab empire.

Alfonso the Wise, King of Castile, added Murcia to his kingdom and took Cadiz from the Arabs, but is remembered as a great king less for that than for his patronage of scholars and artists, and for the splendour and eclecticism of Spanish culture during his reign. Something of its spirit is expressed by this illumination from a book on chess by the king which shows an Arab nobleman and a Christian knight, his guest, peacefully at play. Christian culture in Spain was for centuries refreshed by currents from the Islamic world; the myth of the Reconquest was to obscure much of this from later Spaniards.

The translation of Arabic works into Latin in the later Middle Ages, and the huge repute enjoyed by Arab thinkers in Europe, testify to the quality of this culture. Of the works of Al-Kindi, one of the greatest of Arab philosophers, more survive in Latin than in Arabic, while Dante paid Ibn-Sina (Avicenna in Europe) and Averroes the compliment of placing them in limbo (together with Saladin, the Arab hero of the crusading epoch) when he allocated great men to their fate after death in his poem, and they were the only men of the Christian era whom he treated thus. The Persian practitioners who dominated Arabic medical studies wrote works which remained for centuries the standard textbooks of western training. European languages are still marked by Arabic words which indicate the special importance of Arabic study in certain areas: 'zero', 'cipher', 'almanac', 'algebra' and 'alchemy' are among them. The survival of a technical vocabulary of commerce, too – tariff, *douane*, magazine – is a reminder of the superiority of Arab commercial technique; the Arab merchants taught Christians how to keep accounts. Strikingly, this cultural traffic with Europe was almost entirely one way. Only one Latin text, it appears, was ever translated into Arabic during the Middle Ages, at a time when Arabic scholars were passionately interested in the cultural legacies of Greece, Persia and India. A single fragment of paper bearing a few German words with their Arabic equivalents is the only evidence from eight hundred years of Islamic Spain of any interest in western languages outside the peninsula. The Arabs regarded the civilization of the cold lands of the north as a meagre, unsophisticated affair, as no doubt it was. But Byzantium impressed them.

An Arabic tradition in visual art founded under the Umayyads also flourished under the Abbasids, but it was narrower in its scope than Islamic science. Islam came to forbid the making of likenesses of the human form or face; this was not scrupulously enforced, but it long inhibited the appearance of naturalistic painting or sculpture. Of course, it did not restrict architects. Their art developed very far within a style whose essentials had appeared at the end of the seventh century; it was at once in debt to the past and unique to Islam. The impression produced upon the Arabs by Christian building in Syria was the catalyst; from it they learnt, but they sought to surpass it, for believers should, they were sure, have places of worship better and more beautiful than the Christians' churches. Moreover, a distinctive architectural style could visibly serve as a separating force in the non-Moslem world which surrounded the first Arab conquerors of Egypt and Syria.

The Arabs borrowed Roman technique and Hellenistic ideas of internal space, but what resulted was distinctive. The oldest architectural monument of Islam is the Dome of the Rock built at Jerusalem in 691. Stylistically, it is a landmark in architectural history, the first Islamic building with a dome. It appears to have been built as a monument to victory over Jewish and Christian belief, but unlike the congregational mosques which were to be the great buildings of the next three centuries, the Dome of the Rock was a shrine glorifying and sheltering one of the most sacred places of Jew and Moslem alike; men believed that on the hill-top it covered Abraham had offered up his son Isaac in sacrifice and that from it Muhammad was taken up into heaven.

Soon afterwards came the Umayyad mosque at Damascus, the greatest of the classical mosques of a new tradition. As so often in this new Arab world, it embodied much of the past; a Christian basilica (which had itself replaced a temple of Jupiter) formerly stood on its site, and it was itself decorated with Byzantine mosaics. Its novelty was that it established a design derived from the pattern of worship initiated by the Prophet in his house at Medina; its essential was the *mihrab*, or alcove in the wall of the place of worship, which indicated the direction of Mecca.

Architecture and sculpture, like literature, continued to flourish and to draw upon elements culled from traditions all over the Near East and Asia. Potters strove to achieve the style and finish of the Chinese porcelain which came to them down the Silk Road. The performing arts were less cultivated and seem to have drawn little on other traditions, whether Mediterranean or Indian. There was no Arab theatre, though the storyteller, the poet, the singer and the dancer were esteemed. Arabic musical art is commemorated in European languages through the names of lute, guitar, and rebec; its

The great mosque of Córdoba remains one of the most remarkable buildings of the Arab world but is almost all that is left of the splendour of a once-great Arab city which had no rival west of Constantinople. It was built as a forest of short columns, probably because the architect had many short Roman pillars at his disposal, but none tall enough for a high ceiling. Later the Spanish built a church in the middle of it. In spite of this and though it is now difficult to imagine it in its glory as the spiritual centre of western Islam, the mosque still remains a building of surpassing beauty and mystery.

achievements, too, have been seen as among the greatest of Arabic culture, though they remain less accessible to western sensibility than those of the plastic and visual arts.

Many of the greatest names of this civilization were writing and teaching when its political framework was already in decay, even visibly collapsing. In part this was a matter of the gradual displacement of Arabs within the caliphate's élites, but the Abbasids in their turn lost control of their empire, first of the peripheral provinces and then of Iraq itself. As an international force they peaked early; in 782 an Arab army appeared for the last time before Constantinople. They were never to get so far again. Haroun-al-Raschid might be treated with respect by Charlemagne but the first signs of an eventually irresistible tendency to fragmentation were already there in his day.

In Spain, in 756, an Umayyad prince who had not accepted the fate of his house had proclaimed himself *emir*, or governor, of Córdoba. Others were to follow in Morocco and Tunisia. Meanwhile, El-Andalus acquired its own caliph only in the tenth century (until then its rulers remained emirs) but long before that was independent *de facto*. This did not mean that Umayyad Spain was untroubled. Islam had never conquered the whole peninsula and the Franks recovered the north-east by the tenth century. There were by then Christian kingdoms in northern Iberia and they were always willing to help stir the pot of dissidence within Arab Spain where a fairly tolerant policy towards Christians did not end the danger of revolt.

Yet El-Andalus prospered. The Umayyads developed their sea-power and contemplated imperial expansion not towards the north, at the expense of the Christians, but into Africa, at the expense of Moslem powers, even negotiating for alliance with Byzantium in the process. It was not until the eleventh and twelfth centuries, when the caliphate of Córdoba was in decline, that Spain's Islamic civilization reached its greatest beauty and maturity in a golden age of creativity which rivalled that of Abbasid Baghdad. This left behind great monuments as well as producing great learning and philosophy. The seven hundred mosques of tenth-century Córdoba numbered among them one which can still be thought the most beautiful building in the world. Arab Spain was of enormous importance to Europe, a door to the learning and science of the East, but one through which were also to pass more material goods as well: through it Christendom received knowledge of agricultural and irrigation techniques, oranges and lemons, sugar. As for Spain itself, the Arab stamp went very deep, as many students of the later, Christian, Spain have pointed out, and can still be observed in language, manners and art.

Another important breakaway within the Arab world came when the Fatimids from Tunisia set up their own caliph and moved their capital to Cairo in 973. The Fatimids were Shi'ites and maintained their government of Egypt until a new Arab invasion destroyed it in the twelfth century. Less conspicuous examples could be found elsewhere in the Abbasid dominions as local governors began to term themselves emir and sultan. The power base of the caliphs narrowed more and more rapidly; they were unable to reverse the trend. Civil wars among the sons of Haroun led to a loss of the support by the religious teachers and the devout. Bureaucratic corruption and embezzlement alienated the subject populations. Recourse to tax-farming as a way round these ills only created new examples of oppression. The army was increasingly recruited from foreign mercenaries and slaves; even by the death of Haroun's successor it was virtually Turkish. Thus, barbarians were incorporated within the structure of the caliphates as had been the western barbarians within the Roman empire. As time went by they took on a praetorian look and increasingly dominated the caliphs. And all the time popular opposition was exploited by the Shi'ites and other mystical sects. Meanwhile, the former economic prosperity waned. The wealth of Arab merchants was not to crystallize in a vigorous city life such as that of the late medieval West.

Abbasid rule effectively ended in 946 when a Persian general and his men deposed a caliph and installed a new one. Theoretically, the line of Abbasids continued, but in fact the change was revolutionary; the new Buwayhid dynasty lived henceforth in Persia. Arab Islam had fragmented; the unity of the Near East was once more at an end. No empire remained to resist the centuries of invasion which followed, although it was not until 1258 that the last Abbasid was slaughtered by the Mongols. Before that, Islamic unity had another revival in response to the Crusades, but the great days of Islamic empire were over.

The peculiar nature of Islam meant that religious authority could not long be separated from political supremacy; the caliphate was eventually to pass to the Ottoman Turks, therefore, when they became the makers of Near-Eastern history. They would carry the frontier of Islam still farther afield and once again deep into Europe. But their Arab predecessors' work was awe-inspiringly vast for all its ultimate collapse. They had destroyed both the old Roman Near East and Sassanid Persia, hemming Byzantium in to Anatolia. In the end, though, this would call western Europeans back into the Levant. The Arabs had also implanted Islam ineradicably from Morocco to Afghanistan. Its coming was in many ways revolutionary. It kept women, for example, in an inferior position, but gave them legal rights over property not available to women in many European countries until the nineteenth century. Even the slave had rights and inside the community of the believers there were no castes nor inherited status. This revolution was rooted in a religion which – like that of the Jews – was not distinct from other sides of life, but embraced them all; no words exist in Islam to express the distinctions of sacred and profane, spiritual and temporal, which our own tradition takes for granted. Religion *is* society for the Moslems, and the unity this has provided has outlasted centuries of political division. It was a unity both of law and of a certain attitude; Islam is not a religion of miracles (though it claims some), but of practice and intellectual belief.

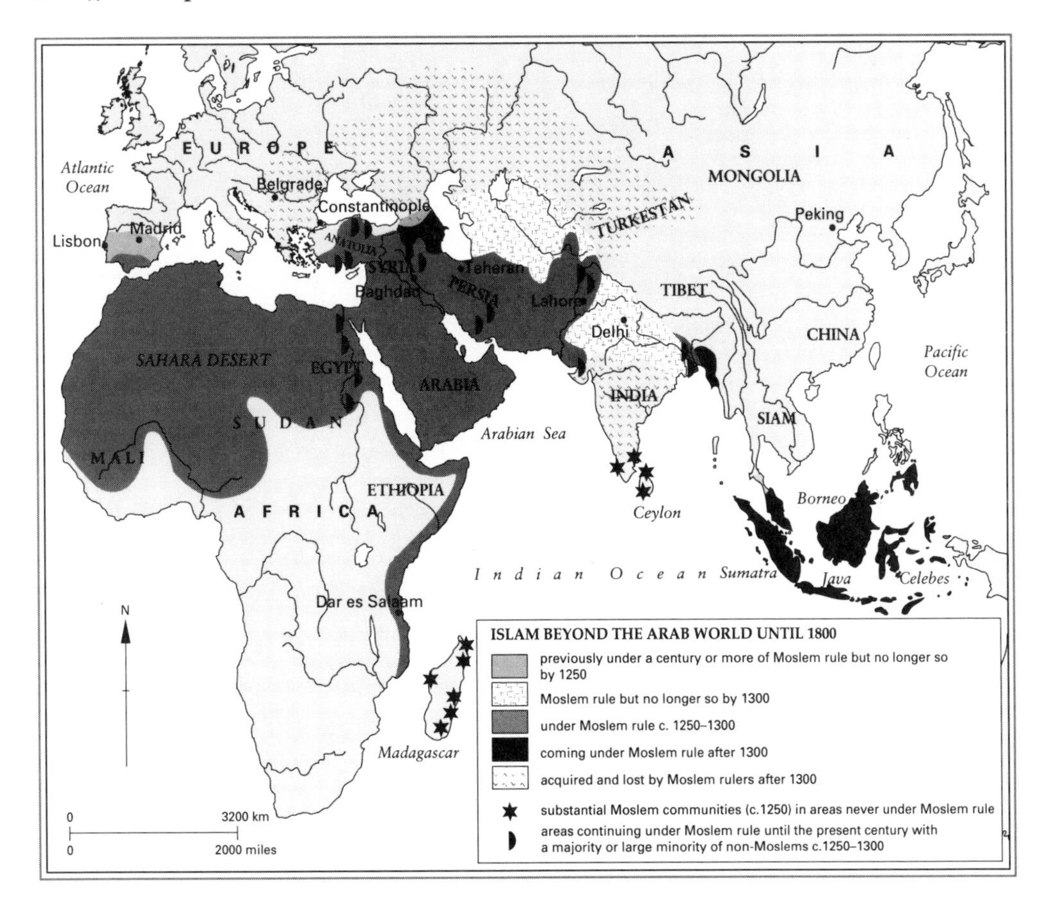

Besides having a great intellectual impact on Christendom, Islam also spread far beyond the world of Arab hegemony, to central Asia in the tenth century, India between the eighth and eleventh and in the eleventh beyond the Sudan and to the Niger. Between the twelfth and sixteenth centuries still more of Africa would become Moslem;

Islam remains today the fastest-growing faith of that continent. Thanks to the conversion of Mongols in the thirteenth century, Islam would also reach China. In the fifteenth and sixteenth centuries it spread across the Indian ocean to Malaya and Indonesia. Missionaries, migrants and merchants carried it with them, the Arabs above all, whether they moved in caravans into Africa or took their dhows from the Persian Gulf and Red Sea to the bay of Bengal. There would even be a last, final, extension of the faith in south-east Europe in the sixteenth and seventeenth centuries. It was a remarkable achievement for an idea at whose service there had been in the beginning no resources except those of a handful of Semitic tribes. But in spite of its majestic record no Arab state was ever again to provide unity for Islam after the tenth century. Even Arab unity was to remain only a dream, though one cherished still today.

3
Byzantium and its Sphere

In 1453, nine hundred years after Justinian, Constantinople fell to an infidel army. 'There has never been and there never will be a more dreadful happening,' wrote one Greek scribe. It was indeed a great event. No one in the West was prepared; the whole Christian world was shocked. More than a state, Rome itself was at an end. The direct descent from the classical Mediterranean civilization had been snapped at last; if few saw this in quite so deep a perspective as the literary enthusiasts who detected in it retribution for the Greek sack of Troy, it was still the end of two thousand years' tradition. And if the pagan world of Hellenistic culture and ancient Greece were set aside, a thousand years of Christian empire at Byzantium itself was impressive enough for its passing to seem an earthquake.

This is one of those subjects where it helps to know the end of the story before beginning it. Even in their decline Byzantine prestige and traditions had amazed strangers who felt through them the weight of an imperial past. To the end its emperors were *augusti* and its citizens called themselves 'Romans'. For centuries, St Sophia had been the greatest of Christian churches, the Orthodox religion it enshrined needing to make even fewer concessions to religious pluralism as previously troublesome provinces were swallowed by the Moslems. Though in retrospect it is easy to see the inevitability of decline and fall, this was not how the men who lived under it saw the eastern empire. They knew, consciously or unconsciously, that it had great powers of evolution. It was a great conservative *tour de force* which had survived many extremities and its archaic style was almost to the end able to cloak important changes.

None the less, a thousand years brought great upheavals in both east and west; history played upon Byzantium, modifying some elements in its heritage, stressing others, obliterating others, so that the empire was in the end very different from Justinian's while never becoming wholly distinct from it. There is no clear dividing line between antiquity and Byzantium. The centre of gravity of the empire had begun to shift eastward before Constantine and when his city became the seat of world empire it was the inheritor of the pretensions of Rome. The office of the emperors showed particularly sharply how evolution and conservatism could combine. Until 800 there was no formal challenge to the theory that the emperor was the secular ruler of all mankind. When a western ruler was hailed as an 'emperor' in Rome that year, the uniqueness of the imperial purple of Byzantium was challenged, whatever might be thought and said in the East about the exact status of the new regime. Yet Byzantium continued to cherish the fantasy of universal empire; there would be emperors right to the end and their office was one of awe-inspiring grandeur. Still theoretically chosen by Senate, army and people, they had none the less an absolute authority. While the realities of his accession might determine for any particular emperor the actual extent of his power – and sometimes the dynastic succession broke under the strains – he was *autocrat* as a western emperor never was. Respect for legal principle and the vested interests of bureaucracy might muffle the emperor's will in action, but it was always supreme in theory. The heads of the great departments of state were responsible to no one but him. This authority explains the intensity with which Byzantine politics focused at the imperial court, for it was there, and not through corporate and representative institutions such as evolved slowly in the West, that authority could be influenced.

Honorius III, a vigorous and intelligent pope, carried on the good work of his predecessor Innocent III in asserting the authority and role of the Roman see. He also approved the foundation of the Dominican order and confirmed by a bull the revised rule of the Franciscans, founded under Innocent III, a ceremony commemorated in this sculpture.

Autocracy had its harsh side. The *curiosi* or secret police informers who swarmed through the empire were not there for nothing. But the nature of the imperial office also laid obligations on the emperor. Crowned by the Patriarch of Constantinople, the emperor had the enormous authority, but also the responsibilities, of God's representative upon earth. The line between lay and ecclesiastical was always blurred in the East where there was nothing like the western opposition of Church and State as a continuing challenge to unchecked power. Yet in the Byzantine scheme of things there was a continuing pressure upon God's vice-regent to act appropriately, to show *philanthropia*, a love of mankind, in his acts. The purpose of the autocratic power was the preservation of mankind and of the conduits by which it drew the water of life – orthodoxy and the Church. Appropriately most of the early Christian emperors were canonized – just as pagan emperors had been deified. Other traditions than the Christian also affected the office, as this suggested. Byzantine emperors were to receive the ritual prostrations of oriental tradition and the images of them which look down from their mosaics show their heads surrounded by the nimbus in which the last pre-Christian emperors were depicted, for it was part of the cult of the sun god. (Some representations of Sassanid rulers have it, too.) It was, none the less, above all as a Christian ruler that the emperor justified his authority.

The imperial office itself thus embodied much of the Christian heritage of Byzantium. That heritage also marked the eastern empire off sharply from the West at many other levels. There were, in the first place, the ecclesiastical peculiarities of what came to be called the Orthodox Church. Islam, for example, was sometimes seen by the eastern clergy less as a pagan religion than a heresy. Other differences lay in the Orthodox view of the relationship of clergy to society; the coalescence of spiritual and lay was important at many levels below the throne. One symbol of it was the retention of a married clergy; the Orthodox priest, for all his presumed holiness, was never to be quite the man apart his western and Catholic colleague became. This suggests the great role of the Orthodox Church as a cementing force in society down to modern times. Above all, no sacerdotal authority as great as that of the papacy would emerge. The focus of authority was the emperor, whose office and responsibility towered above the equally ranked bisnops. Of course, so far as social regulation went, this did not mean that Orthodoxy was more tolerant than the Church of the medieval West. Bad times were always liable to be interpreted as evidence that the emperor had not been doing his

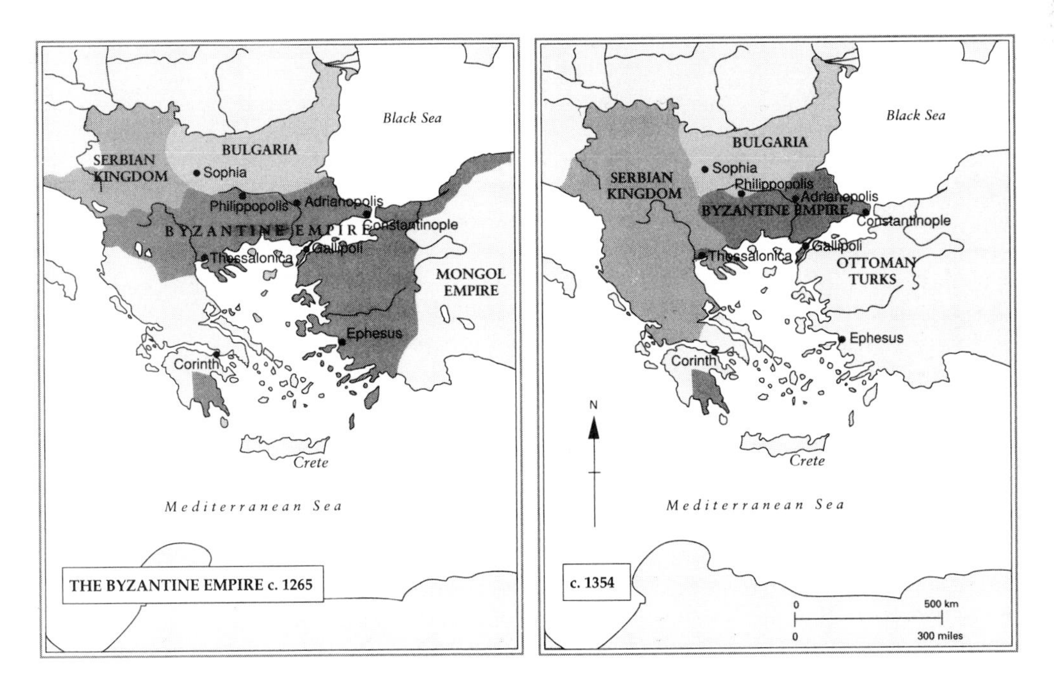

Christian duty – which included the harrying of such familiar scapegoats as Jews, heretics and homosexuals.

Distinction from the West was in part a product of political history, of the gradual loosening of contact after the division of the empires, in part a matter of an original distinction of style. The Catholic and Orthodox traditions were on divergent courses from early times, even if at first the divergence was only slight. At an early date Latin Christianity was somewhat estranged by the concessions the Greeks had to make to Syrian and Egyptian practice. Yet such concessions had also kept alive a certain polycentrism within Christendom. When Jerusalem, Antioch and Alexandria, the other three great patriarchates of the East, fell into Arab hands, the polarization of Rome and Constantinople was accentuated. Gradually, the Christian world was ceasing to be bilingual; a Latin west came to face a Greek east. It was at the beginning of the seventh century that Latin finally ceased to be the official language of the army and of justice, the two departments where it had longest resisted Greek. That the bureaucracy was Greek-speaking was to be very important. When the eastern Church failed among Moslems, it opened a new missionary field and won much ground among the pagans to the north. Eventually, south-eastern Europe and Russia were to owe their evangelizing to Constantinople. The outcome – among many other things – was that the Slav peoples would take from their teachers not only a written language based on Greek, but many of their most fundamental political ideas. And because the West was Catholic, its relations with the Slav world were sometimes hostile, so that the Slav peoples came to view the western half of Christendom with deep reservations. This lay far in the future and takes us further afield than we need to go for the present.

The distinctiveness of the eastern Christian tradition could be illustrated in many ways. Monasticism, for example, remained closer to its original forms in the East and the importance of the Holy Man has always been greater there than in the more hierarchically aware Roman Church. The Greeks, too, seem to have been more disputatious than Latins; the Hellenistic background of the early Church had always favoured speculation and the eastern Churches were open to oriental trends, always susceptible to the pressures of many traditional influences. Yet this did not prevent the imposition of dogmatic solutions to religious quarrels.

Some of these were about issues which now seem trivial or even meaningless. Inevitably, a secular age such as our own finds even the greatest of them difficult to fathom simply because we lack a sense of the mental world lying behind them. It requires an effort to recall that behind the exquisite definitions and logic-chopping of the Fathers lay a concern of appalling importance, nothing less than that mankind should be saved from damnation. A further obstacle to understanding arises for the diametrically opposed reason that theological differences in eastern Christianity often provided symbols and debating forms for questions about politics and society, about the relationship of national and cultural groups to authority, much as hair-splitting about the secular theology of Marxist-Leninism was to mask practical differences between twentieth-century communists. There is more to these questions than appears at first sight and much of it affected world history just as powerfully as the movements of armies or even peoples. The slow divergence of the two main Christian traditions is of enormous importance; it may not have originated in any sense in theological division, but theological disputes propelled divergent traditions yet further apart. They created circumstances which make it more and more difficult to envisage an alternative course of events.

One episode provides an outstanding example, the debate on Monophysitism, a doctrine which divided Christian theologians from about the middle of the fifth century. The significance of the theological issue is at first sight obscure to our post-religious age. It originated in an assertion that Christ's nature while on earth was single; it was wholly divine, instead of dual (that is, both divine and human), as had generally been taught in the early Church. The delicious subtleties of the long debates which this view provoked must, perhaps regrettably, be bypassed here. It is sufficient only to notice that there was an important non-theological setting of the uproar of Aphthartodocetists, Corrupticolists and Theopaschitists (to name a few of the contesting schools). One element in it was the slow crystallization of three Monophysite Churches separated from eastern Orthodoxy and Roman Catholicism. These were the Coptic Church of Egypt and Ethiopia, the Syrian Jacobite and the Armenian Churches; they became, in a sense, national Churches in their countries. It was in an endeavour to reconcile such groups and consolidate the unity of the empire in the face of first the Persian and then the Arab threat that the emperors were drawn into theological dispute; there was more to it, that is to say, than the special responsibility of the office first revealed by Constantine's presiding at the Council of Nicaea. The emperor Heraclius, for example, did his best in the early seventh century to produce a compromise formula to reconcile the disputants over Monophysitism. It took the form of a new theological definition soon called Monothelitism, and on it, for a time, agreement seemed likely, though it was in the end condemned as Monophysitism under a new name.

Meanwhile, the issue had pushed East and West still further apart in practice. Though, ironically, the final theological outcome was agreement in 681, Monophysitism had produced a forty-year schism between Latins and Greeks as early as the end of the fifth century. This was healed, but then came the further trouble under Heraclius. The empire had to leave Italy to its own devices when threatened by the Arab onslaught but both pope and emperor were now anxious to show a common front. This partly explains the pope's endorsement of Monothelitism (on which Heraclius had asked his view so as to quieten the theological misgivings of the Patriarch of Jerusalem). Pope Honorius, successor of Gregory the Great, supported Heraclius and so enraged the anti-Monophysites that almost half a century later he achieved the distinction (unusual among popes) of being condemned by an ecumenical council at which even the western representatives at the council joined in the decision. At a crucial moment of danger Honorius had done much damage. The sympathies of many eastern churchmen in the early seventh century had been alienated still further from Rome by his imprudent action.

The Byzantine inheritance was not only imperial and Christian. It also owed debts to Asia. These were not merely a matter of the direct contacts with alien civilizations symbolized by the arrival of Chinese merchandise along the Silk Road, but also of the complex cultural inheritance of the Hellenistic East. Naturally, Byzantium preserved the prejudice which confused the idea of 'barbarians' with that of peoples who did not speak Greek, and many of its intellectual leaders felt they stood in the tradition of Hellas. Yet the Hellas of which they spoke was one from which the world had long been cut off except through the channels of the Hellenistic East. When we look at that area it is hard to be sure how deep Greek roots went there and how much nourishment they owed to Asiatic sources. The Greek language, for example, seems in Asia Minor to have been used mainly by the few who were city-dwellers. Another sign comes from the imperial bureaucracy and leading families, which reveal more and more Asian names as the centuries go by. Asia was bound to count for more after the losses of territory the empire suffered in the fifth and sixth centuries, for these pinned it increasingly into only a strip of mainland Europe around the capital. Then the Arabs hemmed it in to Asia Minor, bounded in the north by the Caucasus and in the south by the Taurus. On the edges of this, too, ran a border always permeable to Moslem culture. The people who lived on it naturally lived in a sort of marcher world, but sometimes there are indications of deeper external influence than this upon Byzantium. The greatest of all the Byzantine ecclesiastical disputes, that over Iconoclasm, had its parallels almost contemporaneously within Islam.

The most characteristic features of a complicated inheritance were set in the seventh and eighth centuries: an autocratic tradition of government, the Roman myth, the guardianship of eastern Christianity and practical confinement to the East. There had by then begun to emerge from the late Roman empire the medieval state which was sketched under Justinian. Yet of these crucial centuries we know little. Some say that no adequate history of Byzantium in that era can be written, so poor are the sources and so skimpy the present state of archaeological knowledge. At the start of this disturbed period the empire's assets are clear enough. It had at its disposal a great accumulation of diplomatic and bureaucratic skills, a military tradition and enormous prestige. Once its commitments could be reduced in proportion, its potential tax resources were considerable and so were its reserves of manpower. Asia Minor was a recruiting ground which relieved the eastern empire of the need to rely upon Germanic barbarians as had been necessary in the West. It had a notable war-making technology; the 'Greek fire' which was its secret weapon was used powerfully against ships which might attack the capital. The situation of Constantinople, too, was a military asset. Its great walls, built in the fifth century, made it hard to attack by land without heavy weapons unlikely to be available to barbarians; at sea the fleet could prevent a landing.

What was less secure in the long run was the social basis of the empire. It was always to be difficult to maintain the smallholding peasantry and prevent powerful provincial landlords from encroaching on their properties. The law courts would not always protect the small man. He was, too, under economic pressure from the steady expansion of church estates. These forces could not easily be offset by the imperial practice of making grants to smallholders on condition that they supplied military service. But this was a problem whose dimensions were only to be revealed with the passage of centuries; the short-term prospects gave the emperors of the seventh and eighth centuries quite enough to think about.

They were over-extended. In 600 the empire still included the North African coast, Egypt, the Levant, Syria, Asia Minor, the far coast of the Black Sea beyond Trebizond, the Crimean coast and that from Byzantium up to the mouths of the Danube. In Europe there were Thessaly, Macedonia and the Adriatic coast, a belt of territory across central Italy, enclaves in the toe and heel of the peninsula, and finally the islands of Sicily, Corsica and Sardinia. Given the empire's potential enemies and the location of its resources, this was a strategist's nightmare. The story of the next two centuries was to be of the return again and again of waves of invaders. Persians, Avars, Arabs, Bulgars and Slavs were to harry the main body of the empire, while in the West the territories won back by the generals of Justinian were almost all soon taken away again by Arabs and Lombards. Eventually, the West, too, was to reveal itself as a predator; that the eastern empire for centuries absorbed much of the punishment which might otherwise have fallen on the West, would not save her. The result of this was that the eastern empire faced continual warfare. In Europe it meant fighting up to the very walls of Constantinople; in Asia it meant wearisome campaigning to dispute the marches of Asia Minor.

This challenge was offered to a state which, even at the beginning of the seventh century, already had only a very loose control over its domain and depended for much of its power on a penumbra of influence, diplomacy, Christianity and military prestige. Its relations to its neighbours might be seen in more than one way; what looks to a later eye like blackmail paid by every emperor from Justinian to Basil II to menacing barbarians was in the Roman tradition bounty to subject allies and *foederati*. Its diversity of peoples and religions was masked by official ideology. Its Hellenization was often superficial. The reality was expressed in the willingness with which many of the Christian communities of Syria welcomed the Arab, as, later, many of those in Anatolia were to welcome the Turk. Here, religious persecution came home to roost. Moreover, Byzantium numbered no great power among her allies. In the troubled seventh and eighth centuries the most important friendly power was the Khanate of Khazaria, a huge, but loose, state founded by nomads who by 600 dominated the other peoples of the Don and Volga valleys. This established them across the Caucasus, the strategic land bridge which they thus barred to Persians and Arabs for two centuries. At its widest the Khazar state ran round the Black Sea coast to the Dniester and northwards to include the Upper Volga and Don. Byzantium made great efforts to keep the goodwill of the Khazars and seems to have tried, but failed, to convert them to Christianity. What exactly happened is a mystery, but the Khazar leaders, while tolerating Christianity and several other cults, were apparently converted to Judaism in about 740, possibly as a result of Jewish immigration from Persia after the Arab conquest and probably as a conscious act of diplomacy. As Jews they were not likely to be sucked either into the spiritual and political orbit of the Christian empire, or into that of the Caliphs. Instead they enjoyed diplomatic relations and trade with both.

The first great hero of the Byzantine struggle for survival was Heraclius, who strove to balance the threats in Europe with alliances and concessions so that he could campaign vigorously against the Persians. Successful though he eventually was, the Persians had by then done appalling damage to the empire in the Levant and Asia Minor before their expulsion. They have been believed by some scholars to be the real destroyers of the Hellenistic world of great cities; the archaeology is mysterious still, but after Heraclius' victory there are signs that once great cities lay in ruins, that some were reduced to little more than the acropolis which was their core and that population fell sharply. It was, then, on a structure much of which was already badly shaken that the Arab onslaughts fell – and they were to continue for two centuries. Before Heraclius died in 641 virtually all his military achievements had been overturned. Some of the emperors of his line were men of ability, but they could do little more than fight doggedly against a tide flowing strongly against them. In 643 Alexandria fell to the Arabs and that was the end of Greek rule in Egypt. Within a few years they had lost North Africa and Cyprus. Armenia, that old battleground, went in the next decade and finally the high-water mark of Arab success came with the five years of attacks on Constantinople (673–8); it may have been Greek fire that saved the capital from the Arab fleet. Before this, in spite of a personal visit by the emperor to Italy, no progress had been made in recovering the Italian and Sicilian territories taken by Arabs and Lombards. And so the century went on, with yet another menace appearing in its last quarter as Slavs pressed down into Macedonia and Thrace and another race, the Bulgars, themselves one day to be Slavicized, crossed the Danube.

The century ended with a revolt in the army and the replacement of one emperor by another. All the symptoms suggested that the eastern empire would undergo the fate of the West, the imperial office becoming the prize of the soldiers. A succession of beastly or incompetent emperors at the beginning of the eighth century let the Bulgars come to the gates of Constantinople and finally brought about a second siege of the capital by the Arabs in 717. But this was a true turning-point, though it was not to be the last Arab appearance in the Bosphorus. In 717 there had already come to the throne one of the greatest Byzantine emperors, the Anatolian Leo III. He was a provincial official who had successfully resisted Arab attacks on his territory and who had come to the capital to defend it and force the emperor's abdication. His own elevation to the purple followed and was both popular and warmly welcomed by the clergy. This was the foundation of the Isaurian dynasty, so-called from their place of origin; it was an indication of the way in which the élites of the eastern Roman empire were gradually transformed into those of Byzantium, an oriental monarchy.

The eighth century brought the beginning of a period of recovery, though with setbacks. Leo himself cleared Anatolia of the Arabs and his son pushed back the frontiers to those of Syria, Mesopotamia and Armenia. From this time, the frontiers with the caliphate had rather more stability than hitherto, although each campaigning season brought border raids and skirmishes. From this achievement – in part attributable, of course, to the relative decline in Arab power – opened out a new period of progress and expansion which lasted until the early eleventh century. In the West little could be done. Ravenna was lost and only a few toeholds remained in Italy and Sicily. But in the East the empire expanded again from the base of Thrace and Asia Minor which was its heart. A chain of 'themes', or administrative districts, was established along the fringe of the Balkan peninsula; apart from them, the empire had no foothold there for two centuries. In the tenth century Cyprus, Crete and Antioch were all recovered. Byzantine forces at one time crossed the Euphrates and the struggle for north Syria and the Taurus continued. The position in Georgia and Armenia was improved.

In eastern Europe the Bulgar threat was finally contained after reaching its peak at the beginning of the tenth century, when the Bulgars had already been converted to Christianity. Basil II, who has gone down in history as *Bulgaroctonos*, the 'slayer of Bulgars', finally destroyed their power at a great battle in 1014 which he followed up by blinding 15 000 of his prisoners and sending them home to encourage their countrymen. The Bulgar ruler is said to have died of shock. Within a few years Bulgaria was a Byzantine province, though it was never to be successfully absorbed. Shortly afterwards the last conquests of Byzantium were made when Armenia passed under its rule.

The other side of the long Byzantine struggle with the Bulgarians: the emperor Basil II, 'slayer of Bulgars'. At his feet grovel Bulgarian princes: Michael, standard-bearer of the heavenly host, steadies his spear and Gabriel places a crown upon his head. Above them all, the figure of Christ is seen, extending a heavenly crown to his champion. The psalter from which the painting is taken was made in Constantinople in the early eleventh century.

The overall story of these centuries is therefore one of advance and recovery. It was also one of the great periods of Byzantine culture. Politically there was an improvement in domestic affairs in that, by and large, the dynastic principle was observed between 820 and 1025. The Isaurian dynasty had ended badly in an empress who was followed by another series of short reigns and irregular successions until Michael II, the founder of the Phrygian dynasty, succeeded a murdered emperor in 820. His house was replaced in 867 by the Macedonian dynasty, under whom Byzantium reached its summit of success. Where there were minorities the device of a co-emperor was adopted to preserve the dynastic principle.

One major source of division and difficulty for the empire in the earlier part of this period was, as so often before, religion. This plagued the empire and held back its recovery because it was so often tangled with political and local issues. The outstanding example was a controversy which embittered feelings for over a century, the campaign of the iconoclasts.

The depicting of the saints, the Blessed Virgin and God Himself had come to be one of the great devices of Orthodox Christianity for focusing devotion and teaching. In late antiquity such images, or icons, had a place in the West, too, but to this day they occupy a special place in Orthodox churches where they are displayed in shrines and on special screens to be venerated and contemplated by the believer. They are much more than mere decoration, for their arrangement conveys the teachings of the Church and (as one authority has said) provides 'a point of meeting between heaven and earth', where the faithful amid the icons can feel surrounded by the whole invisible Church, by the departed, the saints and angels, and Christ and His mother themselves. It is hardly surprising that something concentrating religious emotion so intensely should have led in paint or mosaic to some of the highest achievements of Byzantine (and, later, Slav) art.

Icons had become prominent in eastern churches by the sixth century. There followed two centuries of respect for them and in many places growing popular devotion to them, but then their use came to be questioned. Interestingly, this happened just after the caliphate had mounted a campaign against the use of images in Islam, but it cannot be inferred that the iconoclasts took their ideas from Moslems. The critics of the icons claimed that they were idols, perverting the worship due to God to the creations of men. They demanded their destruction or expunging and set to work with a will with whitewash, brush and hammer.

Leo III favoured such men. There is still much that is mysterious about the reason why imperial authority was thrown behind the iconoclasts, but he acted on the advice of bishops, and Arab invasions and volcanic eruptions were no doubt held to indicate God's disfavour. In 730, therefore, an edict forbade the use of images in public worship. A persecution of those who resisted followed; enforcement was always more marked at Constantinople than in the provinces. The movement reached its peak under Constantine V and was ratified by a council of bishops in 754. Persecution became fiercer, and there were martyrs, particularly among monks, who usually defended icons more vigorously than did the secular clergy. But iconoclasm was always dependent on imperial support; there were ebbings and flowings in the next century. Under Leo IV and Irene, his widow, persecution was relaxed and the 'iconophiles' (lovers of icons) recovered ground, though this was followed by renewed persecution. Only in 843, on the first Sunday of Lent, a day still celebrated as a feast of Orthodoxy in the eastern Church, were the icons finally restored.

What was the meaning of this strange episode? There was a practical justification, in that the conversion of Jews and Moslems was said to be made more difficult by Christian respect for images, but this does not take us very far. Once again, a religious dispute cannot be separated from factors external to religion, but the ultimate explanation probably lies in a sense of religious precaution, and given the passion often shown in theological controversy in the eastern empire, it is easy to understand how the debate became embittered. No question of art or artistic merit arose: Byzantium was not like that. What was at stake was the feeling of reformers that the Greeks were falling into idolatry in the extremity of their (relatively recent) devotion to icons and that the Arab disasters were the first rumblings of God's thunder; a pious king, as in the Israel of the Old Testament, could yet save the people from the consequences of sin by breaking the idols. This was easier in that the process suited the mentalities of a faith which felt itself at bay. It was notable that iconoclasm was particularly strong in the army. Another fact which is suggestive is that icons had often represented local saints and holy men; they were replaced

Iconoclasts at work. Whitewash is applied to an image of Christ in this ninth-century illustration from a manuscript.

by the uniting, simplifying symbols of eucharist and cross, and this says something about a new, monolithic quality in Byzantine religion and society from the eighth century onwards. Finally, iconoclasm was also in part an angry response to a tide which had long flowed in favour of the monks who gave such prominence to icons in their teaching. As well as a prudent step towards placating an angry God, therefore, iconoclasm represented a reaction of centralized authority, that of emperor and bishops, against local pieties, the independence of cities and monasteries, and the cults of holy men.

Iconoclasm offended many in the western Church but it showed more clearly than anything yet how far Orthodoxy now was from Latin Christianity. The western Church had been moving, too; as Latin culture was taken over by the Germanic peoples, it drifted away in spirit from the churches of the Greek east. The iconoclast synod of bishops had been an affront to the papacy, which had already condemned Leo's supporters. Rome viewed with alarm the emperor's pretensions to act in spiritual matters. Thus iconoclasm drove deeper the division between the two halves of Christendom. Cultural differentiation had now gone very far – not surprisingly when it could take two months by sea to go from Byzantium to Italy and by land a wedge of Slav peoples soon stood between two languages.

Contact between East and West could not be altogether extinguished at the official level. But here, too, history created new divisions, notably when the Pope crowned a Frankish king 'emperor' in 800. This was a challenge to the Byzantine claim to be the legatee of Rome. Distinctions within the western world did not much matter in Constantinople; the Byzantine officials identified a challenger in the Frankish realm and thereafter indiscriminately called all westerners 'Franks', the usage which was to spread as far as China. The two states failed to cooperate against the Arab and offended one another's susceptibilities. The Roman coronation may itself have been in part a response to the assumption of the title of emperor at Constantinople by a woman, Irene, an unattractive mother who had blinded her own son. But the Frankish title was only briefly recognized in Byzantium; later emperors in the West were regarded there only as kings. Italy divided the two Christian empires, too, for the remaining Byzantine lands there came to be threatened by Frank and Saxon as much as they had ever been by Lombards. In the tenth century the manipulation of the papacy by Saxon emperors made matters worse.

Of course the two Christian worlds could not altogether lose touch. One German emperor of the tenth century had a Byzantine bride and German art of the tenth century was much influenced by Byzantine themes and techniques. But it was just the difference of two cultural worlds that made such contacts fruitful, and as the centuries went by, the difference became more and more palpable. The old aristocratic families of Byzantium were replaced gradually by others drawn from Anatolian and Armenian stocks. Above all, there was the unique splendour and complication of the life of the imperial city itself, where religious and secular worlds seemed completely to interpenetrate one another. The calendar of the Christian year was inseparable from that of the court; together they set the rhythms of an immense theatrical spectacle in which the rituals of both Church and State displayed to the people the majesty of the empire. There was some secular art, but the art constantly before men's eyes was overwhelmingly religious. Even in the worst times it had a continuing vigour, expressing the greatness and omnipresence of God, whose vice-regent was the emperor. Ritualism sustained the rigid etiquette of the court about which there proliferated the characteristic evils of intrigue and conspiracy. The public appearance of even the Christian emperor could be like that of the deity in a mystery cult, preceded by the raising of several curtains from behind which he dramatically emerged. This was the apex of an astonishing civilization which showed half the world for perhaps half a millennium what true empire was. When a mission of pagan Russians came to Byzantium in the tenth century to examine its version of the Christian religion as they had examined others, they could only report that what they had seen in Hagia Sophia had amazed them. 'There God dwells among men,' they said.

What was happening at the base of the empire, on the other hand, is not easy to say. There are strong indications that population fell in the seventh and eighth centuries; this may be connected both with the disruptions of war and with plague. At the same time there was little new building in the provincial cities and the circulation of the coinage diminished. All these things suggest a flagging economy, as does more and more interference with it by the state. Imperial officials sought to ensure that its primary needs would be met by arranging for direct levies of produce, setting up special organs to feed the cities and by organizing artisans and tradesmen bureaucratically in guilds and corporations. Only one city of the empire retained its economic importance throughout, and that was the capital itself, where the spectacle of Byzantium was played out at its height. Trade never dried up altogether in the empire and right down to the twelfth century there was still an important transit commerce in luxury goods from Asia to the West; its position alone guaranteed Byzantium a great commercial role and stimulation for the artisan industries which provided other luxuries to the West. Finally, there is evidence across the whole period of the continuing growth in power and wealth of the great landowners. The peasants were more and more tied to their estates and the later years of the empire see something like the appearance of important local economic units based on the big landholdings.

This economy was able to support both the magnificence of Byzantine civilization at its height and the military effort of recovery under the ninth-century emperors. Two centuries later an unfavourable conjuncture once more overtaxed the empire's strength and opened a long era of decline. It began with a fresh burst of internal and personal troubles. Two empresses and a number of short-lived emperors of poor quality weakened control at the centre. The rivalries of two important groups within the Byzantine ruling class got out of hand; an aristocratic party at court whose roots lay in the provinces was entangled in struggles with the permanent officials, the higher bureaucracy. In part this reflected also a struggle of a military with an intellectual élite. Unfortunately, the result was that the army and navy were starved of the funds they needed by the civil servants and were rendered incapable of dealing with new problems.

At one end of the empire these were provided by the last barbarian migrants of the West, the Christian Normans, now moving into south Italy and Sicily. In Asia Minor they arose from Turkish pressure. Already in the eleventh century a Turkish sultanate of Rum was established inside imperial territory (hence its name, for 'Rum' signified 'Rome'), where Abbasid control had slipped into the hands of local chieftains. After a shattering defeat by the Turks at Manzikert in 1071 Asia Minor was virtually lost, and this was a terrible blow to Byzantine fiscal and manpower resources. The caliphates with which the emperors had learnt to live were giving way to fiercer enemies. Within the empire there was a succession of Bulgarian revolts in the eleventh and twelfth centuries and there

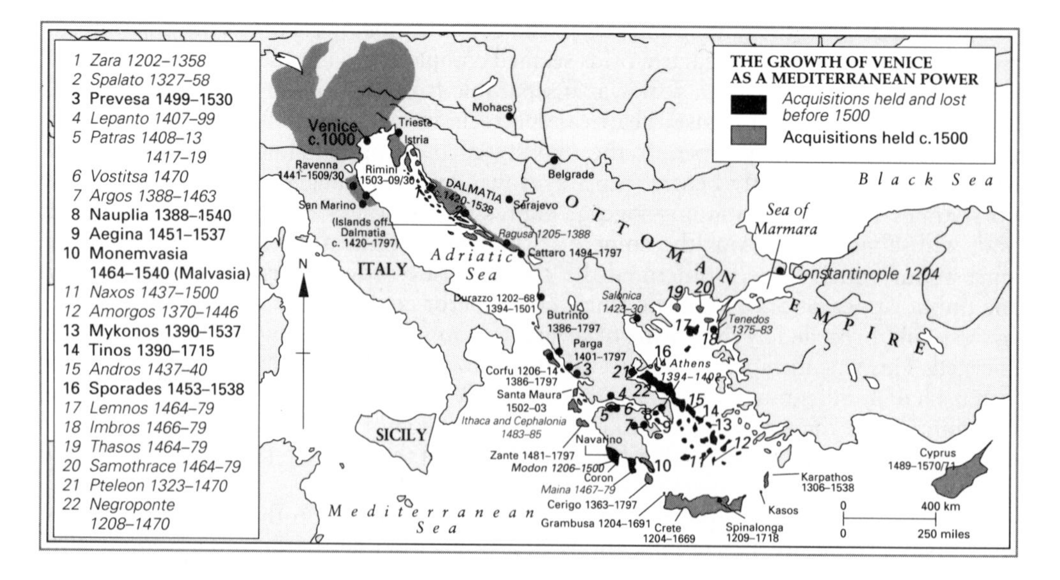

spread widely in that province the most powerful of the dissenting movements of medieval Orthodoxy, the Bogomil heresy, a popular movement drawing upon hatred of the Greek higher clergy and their Byzantinizing ways.

A new dynasty, the Comneni, once again rallied the empire and managed to hold the line for another century (1081–1185). They pushed back the Normans from Greece and they fought off a new nomadic threat from south Russia, the Pechenegs, but could not crack the Bulgars or win back Asia Minor and had to make important concessions to do what they did. Some concessions were to their own magnates; some were to allies who would in turn prove dangerous.

To one of these, the Republic of Venice, once a satellite of Byzantium, concessions were especially ominous, for her whole *raison d'être* had come to be aggrandizement in the eastern Mediterranean. She was the major beneficiary of Europe's trade with the East and at an early time had developed a specially favoured position. In return for help against the Normans in the eleventh century, the Venetians were given the right to trade freely throughout the empire; they were to be treated as subjects of the emperor, not as foreigners. Venetian naval power grew rapidly and, as the Byzantine fleet fell into decline, it was more and more dominant. In 1123 the Venetians destroyed the Egyptian fleet and thereafter were uncontrollable by their former suzerain. One war was fought with Byzantium, but Venice did better from supporting the empire against the Normans and from the pickings of the Crusades. Upon these successes followed commercial concessions and territorial gains and the former mattered most; Venice, it may be said, was built on the decline of the empire, which was an economic host of huge potential for the Adriatic parasite – in the middle of the twelfth century there were said to be 10 000 Venetians living at Constantinople, so important was their trade there. By 1204 the Cyclades, many of the other Aegean islands, and much of the Black Sea coasts belonged to them: hundreds of communities were to be added to those and Venetianized in the next three centuries. The first commercial and maritime empire since ancient Athens had been created.

The appearance of the Venetian challenge and the persistence of old ones would have been embarrassing enough for the Byzantine emperors had they not also faced new trouble at home. In the twelfth century revolt became more common. This was doubly dangerous when the West was entering upon enterprise in the East in the great and complex movement which is famous as the Crusades. The western view of the Crusades need not detain us here; from Byzantium these irruptions from the West looked more and more like new barbarian invasions. In the twelfth century they left behind four crusading states in the former Byzantine Levant as a reminder that there was now another rival in the field in the Near East. When the Moslem forces rallied under Saladin, and there was a resurgence of Bulgarian independence at the end of the twelfth century, the great days of Byzantium were finally over.

The fatal blow came in 1204, when Constantinople was at last taken and sacked, but by Christians, not the pagans who had threatened it so often. A Christian army which had gone east to fight the infidel in a fourth crusade was turned against the empire by the Venetians. It terrorized and pillaged the city (this was when the bronze horses of the Hippodrome were carried off to stand, as they still do, in front of St Mark's Cathedral in Venice), and enthroned a prostitute in the patriarch's seat in St Sophia. East and West could not have been more brutally distinguished; the act was to live in Orthodox memory as one of infamy. The 'Franks', as the Greeks called them, did not see Byzantium as a part of their civilization, nor, perhaps, as even a part of Christendom, for a schism had existed in effect for a century and a half. Though they were to abandon Constantinople and the emperors would be restored in 1261 the Franks would not again be cleared from the old Byzantine territories until a new conqueror came along, the Ottoman Turk. Meanwhile, the heart had gone out of Byzantium, though it had still two centuries in which to die. The immediate beneficiaries were the Venetians and Genoese to whose history the wealth and commerce of Byzantium was now annexed.

The legacy of Byzantium – or a great part of it – was on the other hand already secured to the future, though not, perhaps, in a form in which the eastern Roman would have felt much confidence or pride. It lay in the rooting of Orthodox Christianity among the Slav peoples. This was to have huge consequences, with many of which we still live. The Russian state and the other modern Slav nations would not have been incorporated into Europe and would not now be reckoned as part of it, if they had not been converted to Christianity in the first place.

Much of the story of how this happened is still obscure, and what is known about the Slavs before Christian times is even more debatable. Though the ground-plan of the Slav peoples of today was established at roughly the same time as that of western Europe, geography makes for confusion. Slav Europe covers a zone where nomadic invasions and the nearness of Asia still left things very fluid long after barbarian society had settled down in the west. Much of the central and south-eastern European landmass is mountainous. There, river valleys channelled the distribution of stocks. Most of modern Poland and European Russia, on the other hand, is a vast plain. Though for a long time covered in forests, it provided neither obvious natural lodgements nor insuperable barriers to settlements. In its huge spaces, rights were disputed for many centuries. By the end of the process, at the beginning of the thirteenth century, there had emerged in the East a number of Slav peoples who would have independent historical futures. The pattern thus set has persisted down to our own day.

There had also come into existence a characteristic Slav civilization, though not all Slavs belonged wholly to it and in the end the peoples of Poland and modern Czechoslovakia were to be more closely tied by culture to the West than to the East. The state structures of the Slav world would come and go, but two of them, those evolved by the Polish and Russian nations, proved particularly tenacious and capable of survival in organized form. They would have much to survive, for the Slav world was at times – notably in the thirteenth and twentieth centuries – under pressure as much from the West as from the East. Western aggressiveness is another reason why the Slavs retained a strong identity of their own.

The story of the Slavs has been traced back at least as far as 2000 BC when this ethnic group appears to have been established in the eastern Carpathians. For two thousand years they spread slowly both west and east, but especially to the east, into modern Russia. From the fifth to the seventh century AD Slavs from both the western and eastern groups began to move south into the Balkans. Perhaps their direction reflects the power of the Avars, the Asiatic people who, after the ebbing of the Hun invasions, lay like a great barrier across the Don, Dnieper and Dniester valleys, controlling south Russia as far as the Danube and courted by Byzantine diplomacy.

Throughout their whole history the Slavs have shown remarkable powers of survival. Harried in Russia by Scythians and Goths, in Poland by Avars and Huns, they none the less stuck to their lands and expanded them; they must have been tenacious agriculturists. Their early art shows a willingness to absorb the culture and techniques of others; they learnt from masters whom they outlasted. It was important, therefore, that in the seventh century there stood between them and the dynamic power of Islam a barrier of two peoples, the Khazars and the Bulgars. These strong peoples also helped to channel the gradual movement of Slavs into the Balkans and down to the Aegean. Later it was to run up the Adriatic coast and was to reach Moravia and central Europe, Croatia, Slovenia and Serbia. By the tenth century Slavs must have been numerically dominant throughout the Balkans.

In this process the first Slav state to emerge was Bulgaria, though the Bulgars were not Slavs, but stemmed from tribes left behind by the Huns. Some of them gradually became Slavicized by intermarriage and contact with Slavs; these were the western Bulgars, who were established in the seventh century on the Danube. They cooperated with the Slav peoples in a series of great raids on Byzantium; in 559 they had penetrated the defences of Constantinople and camped in the suburbs. Like their allies, they were pagans. Byzantium exploited differences between Bulgar tribes and a ruler from one of them was baptized in Constantinople, the Emperor Heraclius standing godfather. He used the Byzantine alliance to drive out the Avars from what was to be Bulgaria. Gradually, the Bulgars were diluted by Slav blood and influence. When a Bulgar state finally appears at the end of the century we can regard it as Slav. In 716 Byzantium recognized its independence; now an alien body existed on territory long taken for granted as part of the empire. Though there were alliances, this was a thorn in the side of Byzantium which helped to cripple her attempts at recovery in the West. At the beginning of the ninth century the Bulgars killed an emperor in battle (and made a cup for their king from his skull); no emperor had died on campaign against the barbarians since 378.

A turning-point – though not the end of conflict – was reached when the Bulgars were converted to Christianity. After a brief period during which, significantly, he dallied with Rome and the possibility of playing her off against Constantinople, another Bulgarian prince accepted baptism in 865. There was opposition among his people, but from this time Bulgaria was Christian. Whatever diplomatic gain Byzantine statesmen may have hoped for, it was far from the end of their Bulgarian problem. None the less, it is a landmark, a momentous step in a great process, the christianizing of the Slav peoples. It was also an indication of how this would happen: from the top downwards, by the conversion of their rulers.

What was at stake was a great prize, the nature of the future Slav civilization. Two great names dominate the beginning of its shaping, those of the brothers St Cyril

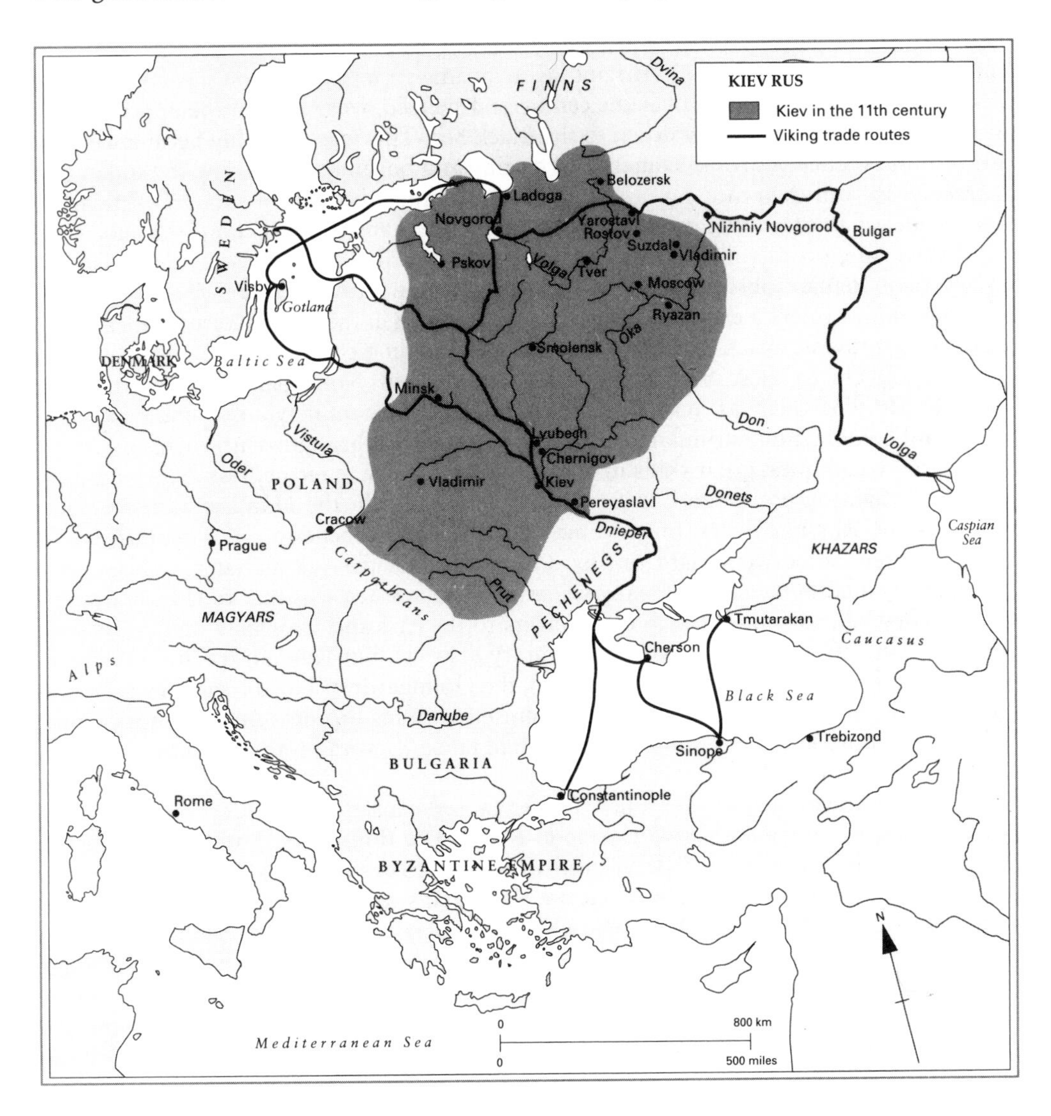

and St Methodius, priests still held in honour in the Orthodox communion. Cyril had earlier been on a mission to Khazaria and their work must be set in the overall context of the ideological diplomacy of Byzantium; Orthodox missionaries cannot neatly be distinguished from Byzantine diplomatic envoys, and these churchmen would have been hard put to it to recognize such a distinction. But they did much more than convert a dangerous neighbour. Cyril's name is commemorated still in the name of the Cyrillic alphabet which he devised. It was rapidly diffused through the Slav peoples, soon reaching Russia, and it made possible not only the radiation of Christianity but the crystallization of Slav culture. That culture was potentially open to other influences, for Byzantium was not its only neighbour, but eastern Orthodoxy was in the end the deepest single influence upon it.

From the Byzantine point of view a still more important conversion was to follow, though not for more than a century. In 860 an expedition with 200 ships raided Byzantium. The citizens were terrified. They listened tremblingly in St Sophia to the prayers of the patriarch: 'a people has crept down from the north ... the people is fierce and has no mercy, its voice is as the roaring sea ... a fierce and savage tribe ... destroying everything, sparing nothing'. It might have been the voice of a western monk invoking divine protection from the sinister longships of the Vikings, and understandably so, for Vikings in essence these raiders were. But they were known to the Byzantines as Rus (or Rhos) and the raid marks the tiny beginnings of Russia's military power.

As yet, there was hardly anything that could be called a state behind it. Russia was still in the making. Its origins lay in an amalgam to which the Slav contribution was basic. The east Slavs had over the centuries dispersed over much of the upper reaches of the river valleys which flow down to the Black Sea. This was probably because of their agricultural practice, a primitive matter of cutting and burning, exhausting the soil in two or three years and then moving on. By the eight century there were enough of them for there to be signs of relatively dense inhabitation, perhaps of something that could be called town life, on the hills near Kiev. They lived in tribes whose economic and social arrangements remain obscure, but this was the basis of future Russia. We do not know who their native rulers were, but they seem to have lived in the defended stockades which were the first towns, exacting tribute from the surrounding countryside.

On to these Slav tribes fell the impact of Norsemen who became their overlords or sold them as slaves in the south. These Scandinavians combined trade, piracy and colonization, stimulated by land-hunger. They brought with them important commercial techniques, great skills in navigation and the management of their longships, formidable fighting power and, it seems, no women. As in the Humber and the Seine, they used the Russian rivers, much longer and deeper, to penetrate the country which was their prey. Some went right on; by 846 we hear of the 'Varangians', as they were called, at Baghdad. One of their many sallies in the Black Sea was that to Constantinople in 860. They had to contend with the Khazars to the east and may have first established themselves in Kiev, one of the Khazar tributary districts, but Russian traditional history begins with their establishment in Novgorod, the Holmgardr of Nordic saga. Here, it was said, a prince called Rurik had established himself with his brothers in about 860. By the end of the century another Varangian prince had taken Kiev and transferred the capital of a new state to that town.

The appearance of a new power caused consternation but provoked action in Byzantium. Characteristically, its response to a new diplomatic problem was cast in ideological terms; there seems to have been an attempt to convert some Rus to Christianity and one ruler may have succumbed. But the Varangians retained their northern paganism – their gods were Thor and Woden – while their Slav subjects, with whom they were increasingly mingled, had their own gods, possibly of very ancient Indo-European origins; in any case, these deities tended to merge as time passed. Soon there were renewed hostilities with Byzantium. Oleg, a prince of the early tenth century, again attacked Constantinople while the fleet was away. He is said to have brought his fleet ashore and to have put it on wheels to outflank the blocked entrance to the Golden Horn. However he did it, he was successful in extracting a highly favourable treaty from Byzantium in 911. This gave the Russians unusually favourable trading privileges and made clear the enormous importance of trade in the life of the new principality. Half a century or so after the legendary Rurik, it was a reality, a sort of river-federation centred on Kiev and linking the Baltic to the Black Sea. It was pagan, but when civilization and Christianity came to it, it would be because of the easy access to Byzantium which water gave to the young principality, which was first designated as Rus in 945. Its unity was still very loose. An incoherent structure was made even less rigid by the Vikings' adoption of a Slav principle which divided an inheritance. Rus princes tended to move around as rulers among the centres of which Kiev and Novgorod were the main ones. Nevertheless, the family of Kiev became the most important.

During the first half of the tenth century the relation between Byzantium and Kiev Rus was slowly ripening. Below the level of politics and trade a more fundamental re-orientation was taking place as Kiev relaxed its links with Scandinavia and looked more and more to the south. Varangian pressure seems to have been diminishing, and this may have had something to do with the success of Norsemen in the West, where one of their rulers, Rollo, had been granted in 911 land later to be known as the duchy of Normandy. Yet it was a long time before there were closer ties between Kiev and Byzantium. One obstacle was the caution of Byzantine diplomacy, still quite as concerned in the early tenth century to fish in troubled waters by negotiating with the wild tribes of the Pechenegs as to placate the Rus whose territories they harried. The Pechenegs had already driven to the west the Magyar tribes which had previously formed a buffer between the Rus and the Khazars and more trouble could be expected there. Nor did Varangian raids come to an end, though there was something of a turning-point when the Rus fleet was driven off by Greek fire in 941. A treaty followed which significantly reduced the trading privileges granted thirty years earlier. But the reciprocity of interests was emerging more clearly as Khazaria declined and the Byzantines realized that Kiev might be a valuable ally against Bulgaria. Signs of contact multiplied; Varangians appeared in the royal guard at Constantinople and Rus merchants came there more frequently. Some are believed to have been baptized.

Christianity, though sometimes despising the merchant, has often followed the trader's wares. There was already a church in Kiev in 882, and it may have been there for foreign merchants. But nothing seems to have followed from this. There is little evidence of Russian Christianity until the middle of the next century. Then, in 945, the widow of a Kievan prince assumed the regency on behalf of his successor, her son. This was Olga. Her son was Sviatoslav, the first prince of Kiev to bear a Slav and not a Scandinavian name. Later, Olga made a state visit to Constantinople. She may have been secretly baptized a Christian before this, but she was publicly and officially converted on this visit in 957, the emperor himself attending the ceremonies in St Sophia. Because of its diplomatic overtones it is difficult to be sure exactly how to understand this event. Olga had, after all, also sent to the West for a bishop, to see what Rome had to offer. Furthermore, there was no immediate practical sequel. Sviatoslav, who reigned from 962 to 972, turned out to be a militant pagan, like other Viking military aristocrats of his time. He clung to the gods of the north and was doubtless confirmed in his belief by his success in raiding Khazar lands. He did less well against the Bulgars, though, and was finally killed by the Pechenegs.

This was a crucial moment. Russia existed but was still Viking, poised between eastern and western Christianity. Islam had been held back at the crucial period by Khazaria, but Russia might have turned to the Latin West. Already the Slavs of Poland had been converted to Rome and German bishoprics had been pushed forward to the east in the Baltic coastlands and Bohemia. The separation, even hostility, of the two great Christian Churches was already a fact, and Russia was a great prize waiting for one of them.

In 980 a series of dynastic struggles ended with the victorious emergence of the prince who made Russia Christian, Vladimir. It seems possible that he had been brought up as a Christian, but at first he showed the ostentatious paganism which became a Viking warlord. Then he began to enquire of other religions. Legend says that he had their different merits debated before him; Russians treasure the story that Islam was rejected by him because it forbade alcoholic drink. A commission was sent to visit the Christian Churches. The Bulgarians, they reported, smelt. The Germans had nothing to offer. But Constantinople had won their hearts. There, they said in words often to be quoted, 'we knew not whether we were in heaven or earth, for on earth there is no such vision nor beauty, and we do not know how to describe it; we know only that there God dwells among men'. The choice was accordingly made. Around about 986–8 Vladimir accepted Orthodox Christianity for himself and his people.

It was a turning-point in Russian history and culture, as Orthodox churchmen have recognized ever since. 'Then the darkness of idolatry began to leave us, and the dawn of orthodoxy arose,' said one, eulogizing Vladimir a half-century or so later. Yet for all the zeal Vladimir showed in imposing baptism on his subjects (by physical force if necessary), it was not only enthusiasm which influenced him. There were diplomatic dimensions to the choice, too. Vladimir had been giving military help to the emperor and now he was promised a Byzantine princess as a bride. This was an unprecedented acknowledgement of the standing of a prince of Kiev. The emperor's sister was available because Byzantium needed the Rus alliance against the Bulgars. When things did not go smoothly, Vladimir put on the pressure by occupying Byzantine possessions in the Crimea. The marriage then soon took place. Kiev was worth a nuptial mass to Byzantium, though Vladimir's choice was decisive of much more than diplomacy. Two hundred years later his countrymen acknowledged this: Vladimir was canonized. He had made the single decision which, more than any other, determined Russia's future.

Probably tenth-century Kiev Rus had in many ways a richer culture than most of western Europe could offer. Its towns were great trading centres, channelling goods into the Near East where Russian furs and beeswax were prized. This commercial emphasis reflects another difference: in western Europe he self-contained, subsistence economy of the manor had emerged as the institution bearing the strain of the collapse of the classical economic world. Without the western manor, Russia would also be without the western feudal nobleman. A territorial aristocracy would take longer to emerge in Russia than in Catholic Europe to remain Russian nobles were for a long time very much the companions and followers of a war-leader. Some of them opposed Christianity and paganism hung on in the north for decades. As in Bulgaria, the adoption of Christianity was a political act with internal as well as external dimensions and though the capital of a Christian principality, Kiev was not yet the centre of a Christian nation. The monarchy had to assert itself against a conservative alliance of aristocracy and paganism. Lower down the social scale, in the towns, the new faith gradually took root, at first thanks to Bulgarian priests, who brought with them the liturgy of the south Slav Church and the Cyrillic alphabet which created Russian as a literary language. Ecclesiastically, the influence of Byzantium was strong and the Metropolitan of Kiev was usually appointed by the Patriarch of Constantinople.

Kiev became famous for the magnificence of its churches; it was a great time of building in a style showing Greek influence. Unhappily, being of wood, few of them survive. But the repute of this artistic primacy reflects Kiev's wealth. Its apogee came under Jaroslav 'the Wise', when one western visitor thought she rivalled Constantinople. Russia was then culturally as open to the outside world as it was ever to be for centuries. In part this reflected Jaroslav's military and diplomatic standing. He exchanged diplomatic missions with Rome while Novgorod received the merchants of the German Hanse. Having himself married a Swedish princess, he found husbands for the womenfolk of his family in kings of Poland, France and Norway. A harried Anglo-Saxon royal family took refuge at his court. Links with western courts were never to be so close again. Culturally, too, the first fruits of the Byzantine implantation on Slav culture were being gathered. Educational foundation and legal creation reflected this. From this reign comes also one of the first great Russian works of literature, *The Primary Chronicle*, an interpretation of Russian history with a political purpose. Like much other early Christian history, it sought to provide a Christian and historical argument for what had already been done by Christian princes, in this case the unification of Russia under Kiev. It stressed the Slav heritage and offered an account of Russian history in Christian terms.

The weaknesses of Kiev Rus lay in the persistence of a rule of succession which almost guaranteed division and dispute at the death of the major prince. Though one other eleventh-century prince managed to assert his authority and hold foreign enemies at bay, the Kiev supremacy waned after Jaroslav. The northern princedoms showed greater autonomy; Moscow and Novgorod were, eventually, the two most important among them, though another 'grand' princedom to match Kiev's was established at Vladimir in the second half of the thirteenth century. In part this shift of the centre of gravity of Russia's history reflects a new threat to the south in the pressure of the Pechenegs, now reaching its peak.

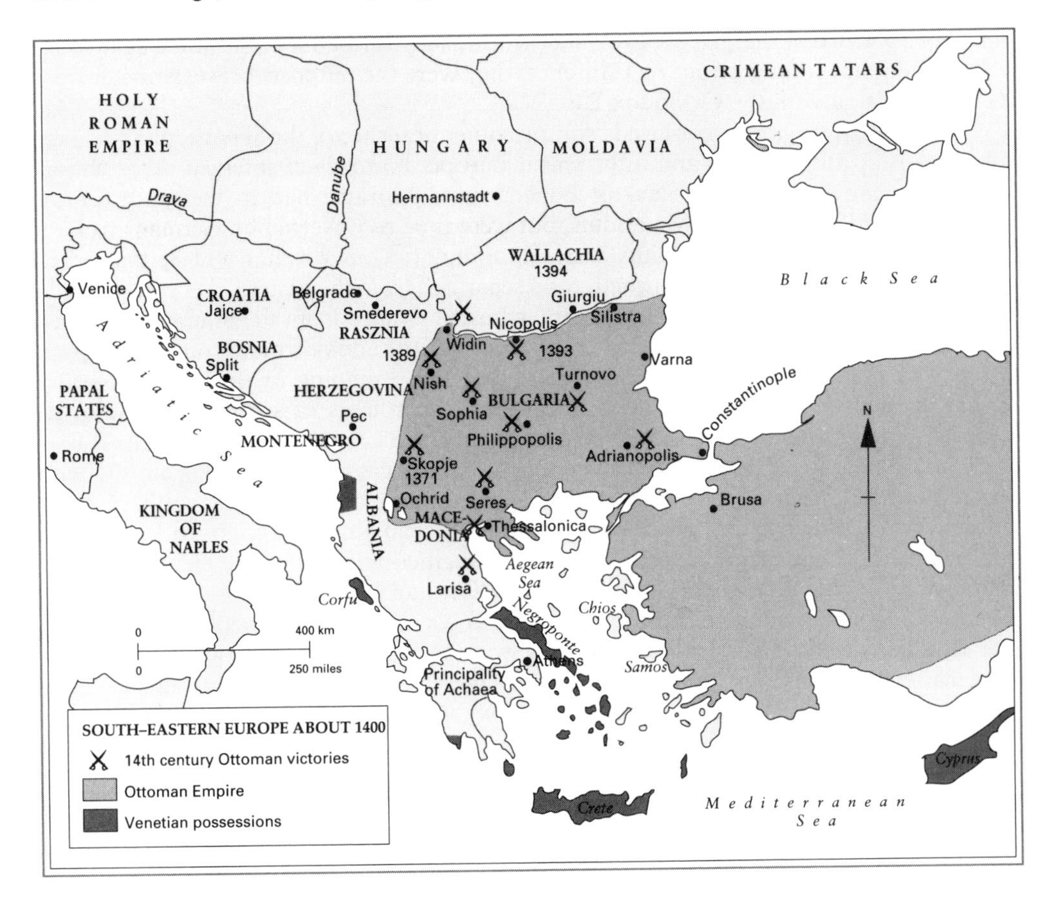

This was a momentous change. In these northern states, the beginnings of future trends in Russian government and society can be discerned. Slowly, grants from the princes were transforming the old followers and boon-companions of the warlord kings into a territorial nobility. Even settled peasants began to acquire rights of ownership and inheritance. Many of those who worked the land were slaves, but there was no such pyramid of obligations as constituted the territorial society of the medieval West. Yet these changes unrolled within a culture whose major direction had been settled by the Kiev period of Russian history.

Another enduring national entity which began to crystallize at about the same time as Russia was Poland. Its origins lay in a group of Slav tribes who appear at the outset, in the tenth century, struggling against pressure from the Germans in the west. It may well have been politics, therefore, that dictated the choice of Christianity as a religion by Poland's first historically recorded ruler, Mieszko I. The choice was not, as in Russia's case, the eastern Orthodox Church. Mieszko plumped for Rome. Poland, therefore, would be linked throughout her history to the west as would be Russia to the East. This conversion, in 966, opened a half-century of rapid consolidation for the new state. A vigorous successor began the creation of an administrative system and extended his lands to the Baltic in the north and through Silesia, Moravia and Cracow in the west. One German emperor recognized his sovereignty in 1000 and in 1025 he was crowned King of Poland as Boleslav I. Political setbacks and pagan reactions dissipated much of what he had done and there were grim times to come, but Poland was henceforth a historical reality. Moreover, three of the dominating themes of her history had also made their appearance: the struggle against German encroachment from the west, the identification with the interests of the Roman Church, and the factiousness and independence of the nobles towards the Crown. The first two of these do much to account for Poland's unhappy history, for they tugged her in different directions. As Slavs, Poles guarded the glacis of the Slav world; they formed a breakwater against the tides of Teutonic immigration. As Catholics, they were the outposts of western culture in its confrontation with the Orthodox East.

During these confused centuries other branches of the Slav peoples had been pushing on up the Adriatic and into central Europe. From them emerged other nations with important futures. The Slavs of Bohemia and Moravia had in the ninth century been converted by Cyril and Methodius, but were then reconverted by Germans to Latin Christianity. The conflict of faiths was important, too, in Croatia and Serbia, where another branch settled and established states separated from the eastern Slav stocks first by Avars, and then by Germans and Magyars, whose invasions from the ninth century were especially important in cutting off central European Orthodoxy from Byzantine support.

A Slav Europe therefore existed at the beginning of the twelfth century. It was divided, it is true, by religion and into distinct areas of settlement. One of the peoples settled in it, the Magyars, who had crossed the Carpathians from south Russia, were not Slav at all. The whole of the area was under growing pressure from the west, where politics, crusading zeal and land-hunger all made a drive to the east irresistibly attractive to Germans. The greatest Slav power, Kievan Russia, developed less than its full potential; it was handicapped by political fragmentation after the eleventh century and harried in the next by the Cumans. By 1200 it had lost its control of the Black Sea river route; Russia had retreated to the north and was becoming Muscovy. Bad times for the Slavs lay ahead. A hurricane of disasters was about to fall upon Slav Europe, and for that matter on Byzantium. It was in 1204 that the crusaders sacked Constantinople and the world power which had sustained Orthodoxy was eclipsed. Worse still was to come. Thirty-six years later the Christian city of Kiev fell to a terrible nomadic people. These were the Mongols.

4
The Disputed Legacies of the Near East

Byzantium was not the only temptation to the predators prowling about the Near East; indeed, she survived their attentions longer than her old enemy the Abbasid caliphate. The Arab empire slipped into decline and disintegration and from the tenth century we enter an age of confusion which makes any brief summary of what happened a despairing exercise. There was no take-off into sustained growth such as the flowering of commerce and the emergence of moneyed men outside the ruling and military hierarchies might have seemed to promise. Rapacious and arbitrary expectations by government may be the basic explanation. In the end, for all the comings and goings of rulers and raiders, nothing disturbed the foundations of Islamic society. The whole area from the Levant to the Hindu Kush was pervaded for the first time in history by a single culture and it was to endure. Within that zone, the Christian inheritance of Rome hung on as a major cultural force only until the eleventh century, bottled up beyond the Taurus in Asia Minor. After that, Christianity declined in the Near East to become only a matter of the communities tolerated by Islam.

The stability and deep-rootedness of Islamic social and cultural institutions were enormously important. They far transcended the weaknesses – which were mainly political and administrative – of the semi-autonomous states which emerged to exercise power under the formal supremacy of the caliphate in its decadent period. About them little need be said. Interesting to Arabists though they are, they need be noted here rather as convenient landmarks than for their own sake. The most important and strongest of them was ruled by the Fatimid dynasty which controlled Egypt, most of Syria and the Levant, and the Red Sea coast. This territory included the great shrines of Mecca and Medina and therefore the profitable and important pilgrim trade. On the borders of Anatolia and northern Syria another dynasty, the Hamdanid, stood between the Fatimids and the Byzantine empire, while the heartland of the caliphate, Iraq and western Iran, together with Azerbaijan, was ruled by the Buwayhid. Finally, the north-eastern provinces of Khurasan, Sijistan and Transoxiana had passed to the Samanids. Listing these four groupings of power far from exhausts the complications of the unsettled Arab world of the tenth century, but it provides all the background now needed to narrate the unrolling of the process by which two new empires appeared within Islam, one based on Anatolia and one on Persia.

The thread is provided by a central Asian people already introduced into this story, the Turks. Some of them had been granted a home by the Sassanids in their last years in return for help. In those days the Turkish 'empire', if that is the right word for their tribal confederation, ran right across Asia; it was their first great era. Like that of other nomadic peoples, this ascendancy soon proved to be transient. The Turks faced at the same time inter-tribal divisions and a resurgence of Chinese power and it was on a divided and disheartened people that there had fallen the great Arab onslaught. In 667 the Arabs invaded Transoxiana and in the next century they finally shattered the remains of the Turkish empire in western Asia. They were only stopped at last in the eighth century by the Khazars, another Turkish people. Before this the eastern Turkish confederation had broken up.

In spite of this collapse what had happened was very important. For the first time a nomadic polity of sorts had spanned Asia and it had lasted for more than a century. All four of the great contemporary civilizations, China, India, Byzantium, and Persia, had felt bound to undertake relations with the Turkish khans, whose subjects had learned much from these contacts. Among other things, they acquired the art of writing; the first surviving Turkish inscription dates from the early eighth century. Yet in spite of this, for long stretches of Turkish history we must rely upon other people's accounts and records, for no Turkish authority seems to go back beyond the fifteenth century and the archaeological record is sporadic.

A Mongol warrior, plaiting the tail of his mount (from a fifteenth-century Persian painting).

This, combined with the fragmentation of the Turkish tribes, makes for obscurity until the tenth century. Then came the collapse of the T'ang dynasty in China, a great event which offered important opportunities to the eastern and Sinicized Turks, just at the moment when signs of weakness were multiplying in the Islamic world. One was the emergency of the Abbasid successor states. Turkish slaves or 'Mamelukes' had long served in the caliphates' armies; now they were employed as mercenaries by the dynasties which tried to fill their vacuum of power. But the Turkish peoples themselves were again on the move by the tenth century. In the middle of it a new dynasty re-established Chinese power and unity; perhaps it was this which provided the decisive impetus for another of the long shunting operations by which central Asian peoples jostled one another forward to other lands. Whatever the cause, a people called the Oghuz Turks were in the van of those who pressed into the north-eastern lands of the old caliphate and set up their own new states there. One clan among them were the Seljuks. They were notable because they were already Moslem. In 960 they had been converted by the assiduous missionary efforts of the Samanids, when still in Transoxiana.

Many of the leaders of the new Turkish régimes were former slave soldiers of the Arab-Persians; one such group were the Ghaznavids, a dynasty who briefly built a huge dominion which stretched into India (this was also the first post-Abbasid régime to choose its generals as *sultans*, or heads of state). But they were in their turn pushed aside as new nomadic invaders arrived. The Oghuz came in sufficient numbers to produce a major change in the ethnic composition of Iran and also in its economy. In another way, too, their arrival means a deeper change than any preceding one and opened a new phase of Islamic history. Because of what the Samanids had done, some of the Oghuz Turks were already Moslem and respected what they found. There now began the translation into Turkish of the major works of Arabic and Persian scholarship which was to give the Turkish peoples access to Arab civilization as never before.

Early in the eleventh century the Seljuks crossed the Oxus, too. This was to lead to the creation of a second Turkish empire, which lasted until 1194, and, in Anatolia, to 1243. After evicting the Ghaznavids from eastern Iran, the Seljuks turned on the Buwayhids and seized Iraq, thus becoming the first central Asian invaders of historical times to penetrate further than the Iranian plateau. Perhaps because they were Sunnites they seem to have been readily welcomed by many of the former subjects of the Shi'ite Buwayhid. They went on, though, to much greater deeds than this. After occupying Syria and Palestine they invaded Asia Minor, where they inflicted on the Byzantines one of the worst defeats of their history at Manzikert in 1071. Significantly, the Seljuks called the sultanate they set up there the Sultanate of Rum, for they saw themselves henceforth as the inheritors of the old Roman territories. That Islam should have a foothold inside the old Roman empire touched off crusading zeal in the West; it also opened Asia Minor to the settlement of Turks.

In many ways, then, the Seljuks played an outstanding historic role. Not only did they begin the conversion of Asia Minor from Christianity to Islam, but they provoked the crusades and long bore the brunt of resisting them, too. This cost them heavily on other fronts. By the mid-twelfth century Seljuk power was already dwindling in the Iranian lands. Nevertheless, the Seljuk empire lasted long enough to make possible a final crystallization over the whole Islamic heartlands of a common culture and of institutions which this time included Turkish peoples.

This was less because Seljuk government innovated than because it recognized social (and in Islam that meant religious) realities. The essence of the Seljuk structure was tribute rather than administrative activity. It was something of a confederation of tribes and localities and was no more capable of standing up to long-term stress than its predecessors. The central apparatus of the empire was its armies and what was necessary to maintain them; locally, the notables of the *ulema*, the teachers and religious leaders of Islam, ruled. They provided a consolidation of authority and social custom which would survive the caliphates and become the cement of Islamic society all over the Middle East. They would run things until the coming of nationalism in the twentieth century. For all the divisions of schools within the *ulema*, it provided at local levels a common cultural and social system which ensured that the loyalty of the masses would be available to new régimes which replaced one another at the top and might have alien origins. It provided political spokesmen who could assure satisfaction at the local level and legitimize new régimes by their support.

This produced one of the most striking differences between Islamic and Christian society. Religious élites were the key factor in the *ulema*; they organized the locally, religiously based community, so that bureaucracy, in the western sense, was not needed. Within the political divisions of the Islamic world in the age of the caliphates' decadence these élites provided its social unity. The Seljuk pattern spread over the Arabic world, and was maintained under the successor empires. Another basic institution was the use of slaves, a few as administrators, but many in the armies. Though the Seljuks granted some great fiefs in return for military service, it was the slaves – often Turkish – who provided the real force on which the régime rested, its armies. Finally, it relied also on the maintenance, where possible, of the local grandee, Persian or Arab.

The declining years of the Seljuk régime exposed the weaknesses in this structure. It depended heavily for its direction upon the availability of able individuals supported by tribal loyalties. But the Turks were thin on the ground and could not keep their subjects' loyalties if they did not succeed. When the first wave of Moslem settlement in Anatolia was spent, that area was still only superficially Turkish, and Moslem towns stood in the middle of a countryside linguistically distinct; local language was not arabized as it was further south and the submergence of the Greek culture of the area was only very slowly achieved. Further east, the first Moslem lands to be lost were lost to pagans in the twelfth century; a nomad ruler (widely supposed in the West to be a Christian king, Prester John, on his way from central Asia to help the crusaders) took Transoxiana from the Seljuks.

The crusading movement was in part a response to the establishment of Seljuk power. The Turks, perhaps because of their late conversion to Islam, were less tolerant than the Arabs. They began to trouble Christian pilgrims going to the holy places. The other causes which promoted the crusades belong rather to European than to Islamic history and can be dealt with elsewhere, but by 1100 the Islamic world felt itself on the defensive even though the Frankish threat was not yet grave. Still, the reconquest of Spain had begun, and the Arabs had already lost Sicily. The first crusade (1096–9) was favoured by Moslem divisions which enabled the invaders to establish four Latin states in the Levant: the kingdom of Jerusalem, and its three fiefs, the county of Edessa, the principality of Antioch and the county of Tripoli. They were not to have much of a future, but in the early twelfth century their presence seemed ominous to Islam. The crusaders' success provoked Moslem reaction and a Seljuk general seized Mosul as a centre from which he built up a new state in northern Mesopotamia and Syria. He recaptured Edessa (1144); his son saw the possibilities of exploiting the Christians' alienation of the local Moslem population by bad treatment. It was a nephew of this prince, Saladin, who seized power in Egypt in 1171, declaring the Fatimid caliphate at an end.

Saladin was a Kurd. He came to be seen as the hero of the Moslem reconquest of the Levant and he remains a captivating figure even after strenuous efforts by unromantic and sceptical scholars to cut through the image of the *beau idéal* of Saracenic chivalry. The fascination he exercised over the minds of his Christian contemporaries was rooted in paradoxes which must have had real educational force. He was indisputably a pagan, yet he was good, a man of his word and just in his dealings; he was chivalrous, yet of a world that did not know the knightly ideal. (This puzzled some Frenchmen so much that they were forced to believe he had in fact been knighted by a Christian captive and that he baptized himself on his deathbed.) On a more mundane level, Saladin's first great triumph was the recapture of Jerusalem (1187), which provoked a new, and third, crusade (1189–92). This could achieve little against him, though it further intensified the irritation of Moslems who now began to show a quite new and unprecedented bitterness and ideological hostility towards Christianity. Persecution of Christians followed and with it began the slow but irreversible decline of the formerly large Christian populations of the Moslem lands.

Saladin founded a dynasty, the Abbuyid sultans, which ruled the Levant (outside the crusader enclaves), Egypt and the Red Sea coast. It lasted until it was replaced by rulers drawn from its own palace guards, the Turkish Mamelukes. These were to be the destroyers of the remaining crusader conquests in Palestine. The revival of the caliphate which followed at Cairo (it was given to a member of the Abbasid house) is of small significance in comparison with this. It registered, nevertheless, that so far as Islam still had a preponderant power and a cultural focus, both were now to be found in Egypt. Baghdad was never to recover.

The Mamelukes had another great achievement to their credit by this time. It was they who finally halted the tide of a conquest far more threatening than that of the Franks, when it had been rising for more than half a century. This was the onslaught of the Mongols. Their history makes nonsense of chronological and territorial divisions. In an astonishingly short time this nomadic people drew into their orbit China, India, the Near East and Europe and left ineffaceable marks behind them. Yet there is no physical focus for their history except the felt tents of their ruler's encampment; they blew up like a hurricane to terrify half a dozen civilizations, slaughtered and destroyed on a scale the twentieth century alone has emulated, and then disappeared almost as suddenly as they came. They demand to be considered alone as the last and most terrible of the nomadic conquerors.

Twelfth-century Mongolia is as far back as a search for their origins need go. A group of peoples speaking the languages of the family called Mongol who had long demanded the attention of Chinese governments then lived there. Generally, China played off one of them against another in the interests of its own security. They were barbarians, not much different in their cultural level from others who have already crossed these pages. Two tribes among them, the Tatars and that which became known as the Mongols,

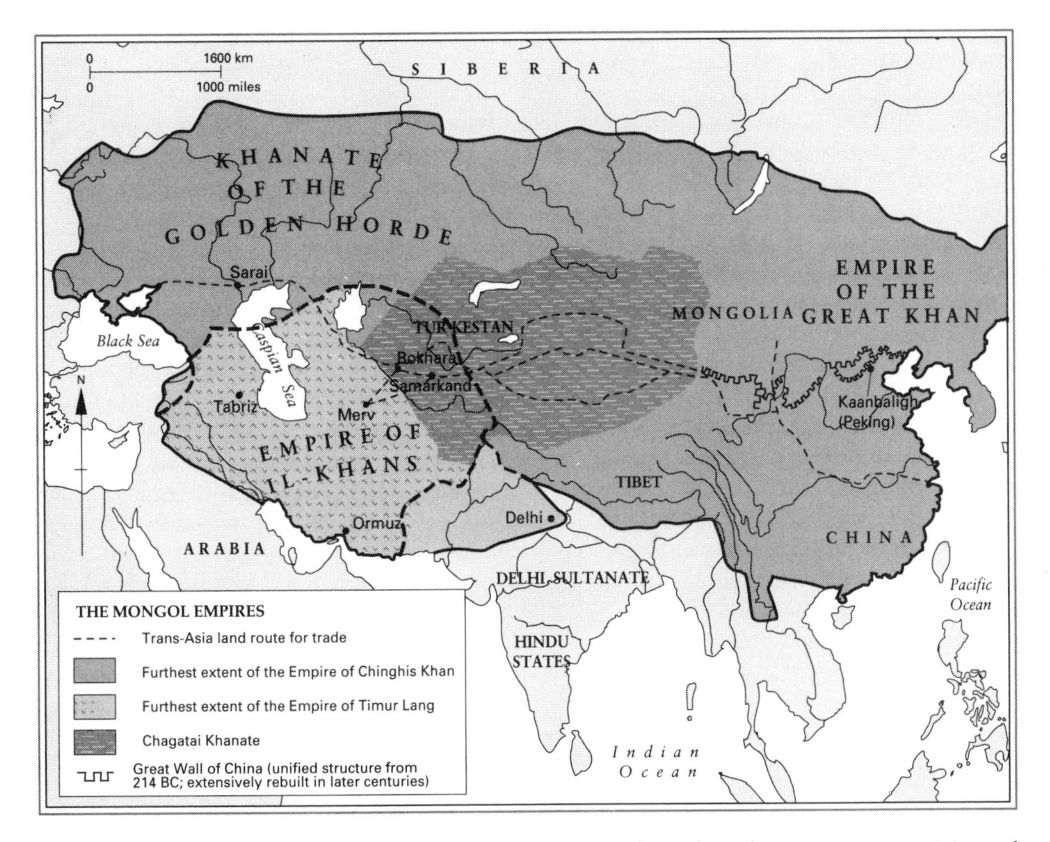

competed and on the whole the Tatars had the best of it. They drove one young Mongol to extremes of bitterness and self-assertion. The date of his birth is uncertain, but in the 1190s he became khan to his people. A few years later he was supreme among the Mongol tribes and was acknowledged as such by being given the title of Chinghis Khan. By an Arabic corruption of this name he was to become known in Europe as Genghis Khan. He extended his power over other peoples in central Asia and in 1215 defeated (though he did not overthrow) the Chin state in northern China and Manchuria. This was only the beginning. By the time of his death, in 1227, he had become the greatest conqueror the world has ever known.

He seems unlike all earlier nomad warlords. Chinghis genuinely believed he had a mission to conquer the world. Conquest, not booty or settlement, was his aim and what he conquered he often set about organizing in a systematic way. This led to a structure which deserves the name 'empire' more than do most of the nomadic polities.

He was superstitious, tolerant of religions other than his own paganism, and, said a Persian historian, 'used to hold in esteem beloved and respected sages and hermits of every tribe, considering this a procedure to please God'. Indeed, he seems to have held that he was himself the recipient of a divine mission. This religious eclecticism was of the first importance, as was the fact that he and his followers (except for some Turks who joined them) were not Moslem, as the Seljuks had been when they arrived in the Near East. Not only was this a matter of moment to Christians and Buddhists – there were both Nestorians and Buddhists among the Mongols – but it meant that the Mongols were not identified with the religion of the majority in the Near East.

In 1218 Chinghis Khan turned to the west and the era of Mongol invasions opened in Transoxiana and northern Iran. He never acted carelessly, capriciously, or without premeditation, but it may well be that the attack was provoked by the folly of a Moslem prince who killed his envoys. From there Chinghis went on to a devastating raid into Persia followed by a swing northward through the Caucasus into south Russia, and returned, having made a complete circuit of the Caspian.

All this was accomplished by 1223. Bokhara and Samarkand were sacked with massacres of the townspeople which were meant to terrify others who contemplated resistance. (Surrender was always the safest course with the Mongols and several minor peoples were to survive with nothing worse than the payment of tribute and the arrival of a Mongol governor.) Transoxiana never recovered its place in the life of Islamic Iran after this. Christian civilization was given a taste of Mongol prowess by the defeat of the Georgians in 1221 and of the southern Russian princes two years later. Even these alarming events were only the overture to what was to follow.

Chinghis died in the East in 1227, but his son and successor returned to the West after completing the conquest of northern China. In 1236 his armies poured into Russia. They took Kiev and settled on the lower Volga, from which they organized a tributary system for the Russian principalities they had not occupied. Meanwhile they raided Catholic Europe. The Teutonic knights, the Poles and the Hungarians all went down before them. Cracow was burnt and Moravia devastated. A Mongol patrol crossed into Austria, while the pursuers of the king of Hungary chased him through Croatia and finally reached Albania before they were recalled.

The Mongols left Europe because of dissensions of their leaders and the arrival of the news of the death of the khan. A new one was not chosen until 1246. A Franciscan friar attended the ceremony (he was there as an emissary of the pope); so did a Russian grand duke, a Seljuk sultan, the brother of the Abbuyid sultan of Egypt, an envoy from the Abbasid caliph, a representative of the king of Armenia, and two claimants to the Christian throne of Georgia. The election did not solve the problems posed by dissension among the Mongols and it was not until another Great Khan was chosen (after his predecessor's death had ended a short reign) that the stage was set for another Mongol attack.

This time it fell almost entirely upon Islam, and provoked unwarranted optimism among Christians who noted also the rise of Nestorian influence at the Mongol court. The area nominally still subject to the caliphate had been in a state of disorder since Chinghis Khan's campaign. The Seljuks of Rum had been defeated in 1243 and were not capable of asserting authority. In this vacuum, relatively small and local Mongol forces could be effective and the Mongol empire relied mainly upon vassals among numerous local rulers.

The campaign was entrusted to the younger brother of the Great Khan and began with the crossing of the Oxus on New Year's Day 1256. After destroying the notorious sect of the Assassins *en route*, he moved on Baghdad, summoning the caliph to surrender. The city was stormed and sacked and the last Abbasid caliph murdered – because there were superstitions about shedding his blood he is supposed to have been rolled up in a carpet and trampled to death by horses. It was a black moment in the history of Islam as, everywhere, Christians took heart and anticipated the overthrow of their Moslem overlords. When, the following year, the Mongol offensive was launched against Syria, Moslems were forced to bow to the cross in the streets of a surrendered Damascus and a mosque was turned into a Christian church. The Mamelukes of Egypt were next on the list for conquest when the Great Khan died. The Mongol commander in the West favoured the succession of his younger brother, Kubilai, far away in China. But he was distracted and withdrew many of his men to Azerbaijan to wait on events. It was on a weakened army that the Mamelukes fell at the Goliath Spring near Nazareth on 3 September 1260. The Mongol general was killed, the legend of Mongol invincibility was shattered and a turning-point in world history was reached. For the Mongols the age of conquest was over and that of consolidation had begun.

The unity of Chinghis Khan's empire was at an end. After civil war the legacy was divided among the princes of his house, under the nominal supremacy of his grandson Kubilai, Khan of China, who was to be the last of the Great Khans. The Russian khanate was divided into three: the khanate of the Golden Horde ran from the Danube to the Caucasus and to the east of it lay the 'Cheibanid' khanate in the north (it was named after its first khan) and that of the White Horde in the south. The khanate of Persia included much of Asia Minor, and stretched across Iraq and Iran to the Oxus. Beyond that lay the khanate of Turkestan. The quarrels of these states left the Mamelukes free to mop up the crusader enclaves and to take revenge upon the Christians who had compromised themselves by collaboration with the Mongols.

In retrospect it is still far from easy to understand why the Mongols were so successful for so long. In the west they had the advantage that there was no single great power such as Persia or the eastern Roman empire had been, to stand up to them, but in the east they defeated China, undeniably a great imperial state. It helped, too, that they faced divided enemies; Christian rulers toyed with the hope of using Mongol power against the Moslem and even against one another, while any combination of the civilizations of the west with China against the Mongols was inconceivable given Mongol control of communication between the two. Their tolerance of religious diversity, except during the period of implacable hatred of Islam, also favoured the Mongols; those who submitted peacefully had little to fear. Would-be resisters could contemplate the ruins of Bokhara or Kiev, or the pyramids of skulls where there had been Persian cities; much of the Mongol success must have been a result of the sheer terror which defeated many of their enemies before they ever came to battle. In the last resort, though, simple military skill explained their victories. The Mongol soldier was tough, well-trained and led by generals who exploited all the advantages which a fast-moving cavalry arm could give them. Their movement was in part the outcome of the care with which reconnaissance and intelligence work was carried out before a campaign. The discipline of their cavalry and their mastery of the techniques of siege warfare (which, none the less, the Mongols preferred to avoid) made them much more formidable than a horde of nomadic freebooters. As conquests continued, too, the Mongol army was recruited by specialists among its captives; by the middle of the thirteenth century there were many Turks in its ranks.

Though his army's needs were simple, the empire of Chinghis Khan and, in somewhat less degree, of his successors was an administrative reality over a vast area. One of the first innovations of Chinghis was the reduction of Mongol language to writing, using the Turkish script. This was done by a captive. Mongol rule always drew willingly upon the skills made available to it by its conquests. Chinese civil servants organized the conquered territories for revenue purposes; the Chinese device of paper money, when introduced by the Mongols into the Persian economy in the thirteenth century, brought about a disastrous collapse of trade, but the failure does not make the example of the use of alien techniques less striking.

In so great an empire, communications were the key to power. A network of post-houses along the main roads looked after rapidly moving messengers and agents. The roads helped trade, too, and for all their ruthlessness to the cities which resisted them, the Mongols usually encouraged rebuilding and the revival of commerce, from the taxation of which they sought revenue. Asia knew a sort of *Pax Mongolica*. Caravans were protected against nomadic bandits by the policing of the Mongols, poachers turned gamekeepers.

In 1526 Suleiman the Magnificent inflicted a shattering defeat on the Christian Hungarian army at Mohacs, south of Budapest. In this Turkish painting, Hungarian knights flee before the victorious Ottoman army.

The most successful nomads of all, they were not going to let other nomads spoil their game. Land trade was as easy between China and Europe during the Mongol era as at any time; Marco Polo is the most famous of Europe's travellers to the Far East in the thirteenth century and by the time he went there the Mongols had conquered China, but before he was born his father and uncle had begun travels in Asia which were to last years. They were both Venetian merchants and were sufficiently successful to set off again almost as soon as they got back, taking the young Marco with them. By sea, too, China's trade was linked with Europe, through the port of Ormuz on the Persian Gulf, but it was the land-routes to the Crimea and Trebizond which carried most of the silks and spices westward and provided the bulk of Byzantine trade in its last centuries. The land-routes depended on the Khans, and, significantly, the merchants were always strong supporters of the Mongol régime.

In its relations with the rest of the world, the Mongol empire came to show the influence of China in its fundamental presuppositions. The khans were the representatives on earth of the one sky god, Tengri; his supremacy had to be acknowledged, though this did not mean that the practice of other religions would not be tolerated. But it did mean that diplomacy in the western sense was inconceivable. Like the Chinese emperors whom they were to replace, the khans saw themselves as the upholders of a universal monarchy; those who came to it had to come as suppliants. Ambassadors were the bearers of tribute, not the representatives of powers of equal standing. When in 1246 emissaries from Rome conveyed papal protests against the Mongol treatment of Christian Europe and a recommendation that he should be baptized, the new Great Khan's reply was blunt: 'If you do not observe God's command, and if you ignore my command, I shall know you as my enemy. Likewise I shall make you understand.' As for baptism, the pope was told to come in person to serve the khan. It was not an isolated message, for another pope had the same reply from the Mongol governor of Persia a year later: 'If you wish to keep your land, you must come to us in person and thence go on to him who is master of the earth. If you do not, we know not what will happen: only God knows.'

The cultural influences playing upon the Mongol rulers and their circle were not only Chinese. There is much evidence of the importance of Nestorian Christianity at the Mongol court and it encouraged European hopes of a rapprochement with the khans. One of the most remarkable western visitors to the khan, the Franciscan William of Roebruck, was told just after New Year 1254, by an Armenian monk, that the Great Khan would be baptized a few days later, but nothing came of it. William went on, however, to win a debate before him, defending the Christian faith against Moslem and Buddhist representatives and coming off best. This was, in fact, just the moment at which Mongol strength was being gathered for the double assault on world power, against Sung China and the Moslems, which was finally checked in Syria by the Mamelukes in 1260.

Not that this was the end of attempts by Mongols to conquer the Levant. None was successful, though; the Mongols' quarrels among themselves had given the Mamelukes a clear field for too long. Logically, Christians regretted the death of Hulugu, the last khan to pose a real threat to the Near East for decades. After him a succession of Il-khans, or subordinate khans, ruled in Persia, preoccupied with their quarrels with the Golden and White Hordes. Gradually Persia recovered under them from the invasions it had suffered earlier in the century. As in the East, the Mongols ruled through locally recruited administrators and were tolerant of Christians and Buddhists, though not, at first, of Moslems. There was a clear sign of a change in the relative position of Mongol and European when the Il-khans began to suggest to the pope that they should join in an alliance against the Mamelukes.

When Kubilai Khan died in China in 1294 one of the few remaining links that held together the Mongol empire had gone. In the following year an Il-khan called Ghazan made a momentous break with the Mongol tradition; he became a Moslem. Since then the rulers of Persia have always been Moslem. But this did not do all that might have been hoped and the Il-khan died young, with many problems unsolved. To embrace Islam had been a bold stroke, but it was not enough. It had offended many Mongols and in the last resort the khans depended upon their captains. Nevertheless, the contest with the Mamelukes was not yet abandoned. Though in the end unsuccessful, Ghazan's armies took Aleppo in 1299; he was prayed for in the Ummayad mosque at Damascus the next year. He was the last khan to attempt to realize the plan of Mongol conquest of the Near East set out a half-century before, but was frustrated in the end when the Mamelukes defeated the last Mongol invasion of Syria in 1303. The Il-khan died the following year.

As in China, it soon appeared in Persia that Mongol rule had enjoyed only a brief Indian summer of consolidation before it began to crumble. Ghazan was the last Il-khan of stature. Outside their own lands, his successors could exercise little influence; the Mamelukes terrorized the old allies of the Mongols, the Christian Armenians, and Anatolia was disputed between different Turkish princes. There was little to hope for from Europe, where the illusion of the crusading dream had been dissipated. As Mongol states crumbled, one last flash of the old terror in the West came with a conqueror who rivalled even Chinghis.

In 1369 Timur Lang, or Timur the lame, became ruler of Samarkand. For thirty years the history of the Il-khans had been one of civil strife and succession disputes; Persia was conquered by Timur in 1379. Timur (who has passed into English literature, thanks to Marlowe, as Tamberlane) aspired to rival Chinghis. In the extent of his conquests and the ferocity of his behaviour he did; he may even have been as great a leader of men. None the less, he lacked the statesmanship of his predecessors. Of creative art he was barren. Though he ravaged India and sacked Delhi (he was as hard on his fellow-Moslems as on Christians), thrashed the Khans of the Golden Horde, defeated Mameluke and Turk alike and incorporated Mesopotamia as well as Persia in his own domains, he left little behind. His historic role was, except in two respects, almost insignificant. One negative achievement was the almost complete extinction of Asiatic Christianity in its Nestorian and Jacobite form. This was hardly in the Mongol tradition, but Timur was as much a Turk by blood as a Mongol and knew nothing of the nomadic life of central Asia from which Chinghis came, with its willingness to indulge Christian clergy. His sole positive achievement was unintentional and temporary: briefly, he prolonged the life of Byzantium. By a great defeat of an Anatolian Turkish people, the Ottomans, in 1402, he prevented them for a while from going in to the kill against the eastern empire.

This was the direction in which Near Eastern history had been moving ever since the Mongols had been unable to keep their grip on Seljuk Anatolia. The spectacular stretch of Mongol campaigning – from Albania to Java – makes it hard to sense this until Timur's death, but then it was obvious. Before that, the Mongols had already been overthrown in China. Timur's own legacy crumbled, Mesopotamia eventually becoming the emirate of the attractively named Black Sheep Turks, while his successors for a while still hung on to Persia and Transoxiana. By the middle of the fifteenth century, the Golden Horde was well advanced in its break-up. Though it could still terrorize Russia, the Mongol threat to Europe was long over.

By then, Byzantium was at its last gasp. For more than two centuries she fought a losing battle for survival, and not merely with powerful Islamic neighbours. It was the West which had first reduced Byzantium to a tiny patch of territory and had sacked her capital. After the mortal wound of 1204 she was only a small Balkan state. A Bulgarian king had seized the opportunity of that year to assure his country's independence and this was one of several ephemeral successor states which made their appearance. Furthermore, on the ruins of Byzantine rule there was established a new western European maritime empire, that of Venice, the cuckoo in the nest which had been in the first place bribed to enter it. This former client, though having to sustain a bitter commercial and political rivalry with another Italian city-state, Genoa, had by the middle of the fourteenth century taken for herself from the Byzantine heritage the whole Aegean complex of islands, with Rhodes, Crete, Corfu and Chios.

In 1261 the Byzantines had regained possession of their own capital from the Franks. They did so with the help of a Turkish power in Anatolia, the Osmanlis.

Two factors might still benefit the empire; the crucial phase of Mongol aggression was past (though this could hardly have been known and Mongol attacks continued to fall on peoples who cushioned her from them), and in Russia there existed a great Orthodox power which was a source of help and money. But there were also new threats and these outweighed the positive factors. Byzantine recovery in Europe in the later thirteenth century was soon challenged by a Serbian prince with aspirations to empire. He died before he could take Constantinople, but he left the empire with little but the hinterland of the capital and a fragment of Thrace. Against the Serbs, the empire once more called on Osmanli help. Already firmly established on the Asian shores of the Bosphorus, the Turks took a toehold in Europe at Gallipoli in 1333.

At Samarkand stands the Gur-i-Mir, the burial place of Timur Lang. Commissioned by him as the tomb of a favourite nephew, it became his own and a part of a complex of buildings which grew to be a family mausoleum. Craftsmen were brought from Isfahan to work upon this masterpiece of Timurid art, covered with coloured tiles and surmounted by a dome of unusual magnificence.

The best that the last eleven emperors, the Palaeologi, could manage in these circumstances was a rearguard action. They lost what was left of Asia Minor to the Osmanlis in 1326 and it was there that the fatal danger lay. In the eastern Black Sea they had an ally in the Greek empire of Trebizond, a great trading state which was just to outlive Byzantium itself, but in Europe they could hope for little. The ambitions of the Venetians and Genoese (who by now dominated even the trade of the capital city itself), and the King of Naples, gave Byzantium little respite. One emperor desperately accepted papal primacy and reunion with the Roman Church; this policy did little except antagonize his own clergy and his successor abandoned it. Religion still divided Christendom.

As the fourteenth century wore on, the Byzantines had a deepening sense of isolation. They felt abandoned to the infidel. An attempt to use western mercenaries from Catalonia only led to their attacking Constantinople and setting up yet another breakaway state, the Catalan duchy of Athens, in 1311. Occasional victories when an island or a province was retaken did not offset the general tendency of these events, nor the debilitating effect of occasional civil war within the empire. True to their traditions, the Greeks managed even in this extremity to invest some of these struggles with a theological dimension. On top of all this, the plague in 1347 wiped out a third of what was left of the empire's population.

In 1400, when the emperor travelled the courts of western Europe to drum up help (a little money was all he got) he ruled only Constantinople, Salonica and the Morea. Many in the West now spoke of him, significantly, as 'emperor of the Greeks', forgetting he was still titular emperor of the Romans. The Turks surrounded the capital on all sides, and had already carried out their first attack on it. There was a second in 1422. John VIII made a last attempt to overcome the strongest barrier to cooperation with the West. He went in 1439 to an ecumenical council sitting in Florence and there accepted papal primacy and union with Rome. Western Christendom rejoiced; the bells were rung in all the parish churches of England. But the Orthodox East scowled. The council's formula ran headlong against its tradition; too much stood in the way – papal authority, the equality of bishops, ritual and doctrine. The most influential Greek clergy had refused to attend the council; the large number who did all signed the formula of union except one (he, significantly, was later canonized) but many of them recanted when they went home. 'Better,' said one Byzantine dignitary, 'to see in the city the power of the Turkish turban than that of the Latin tiara.' Submission to the pope was for most Greeks a renegade act; they were denying the true Church, whose tradition Orthodoxy had conserved. In Constantinople itself priests known to accept the council were shunned; the emperors were loyal to the agreement but thirteen years passed before they dared to proclaim the union publicly at Constantinople. The only benefit from the submission was the pope's support for a last crusade (which ended in disaster in 1441). In the end the West and East could not make common cause. The infidel was, as yet, battering only at the West's outermost defences. France and Germany were absorbed in their own affairs; Venice and Genoa saw their interest might lie as much in conciliation of the Turk as in opposition to him. Even the Russians, harried by Tatars, could do little to help Byzantium, cut off as they were from direct contact with her. The imperial city, and little else, was left alone and divided within itself to face the Ottomans' final effort.

Who were the Osmanlis, or, as they became known in Europe, the Ottomans? They were one of the Turkish peoples who had emerged from the collapse of the sultanate of Rum. When the Seljuks arrived they found on the borderlands between the dissolved Abbasid caliphate and the Byzantine empire a number of Moslem marcher lords, petty princes called *ghazis*, sometimes Turkish by race, lawless, independent and the inevitable beneficiaries of the ebbing of paramount power. Their existence was precarious, and the Byzantine empire had absorbed some of them in its tenth-century recovery, but they were hard to eliminate. Many survived the Seljuk era and benefited from the Mongol destruction of the Seljuks at a time when Constantinople was in the hands of the Latins. One of these *ghazis* was Osman, a Turk who may have been an Oghuz. But his appeal lay in his leadership and enterprise, and men gathered to him. His quality is shown by the transformation of the world *ghazi*: it came to mean 'warrior of the faith'. Fanatical frontiersmen, his followers seem to have been distinguished by a certain spiritual *élan*. Some of them were influenced by a particular mystical tradition within Islam. They also developed highly characteristic institutions of their own. They had a military organization somewhat like that of merchant guilds or religious orders in medieval Europe and it has been suggested that the West learnt in these matters from the Ottomans. Their situation on a curious borderland of cultures, half-Christian, half-Islamic, must also have been provoking. Whatever its ultimate source, their staggering record of conquest rivals that of Arab and Mongol. They were in the end to reassemble under one ruler the territory of the old eastern Roman empire and more.

The first Ottoman to take the title of Sultan did so in the early fourteenth century. This was Orkhan, Osman's son. Under him began the settlement of conquered

Mehmet II, the Ottoman conqueror, painted by the Venetian artist Gentile Bellini in 1480.

lands which was eventually to be the basis of Ottoman military power. Like his foundation of the 'Janissaries', the 'New Army' of infantry which he needed to fight in Europe, the change marked an important stage in the evolution of Ottoman empire away from the institutions of a nomadic people of natural cavalrymen. Another sign that things were settling down was Orkhan's issue of the first Ottoman coinage. At his death he ruled the strongest of the post-Seljuk states of Asia Minor as well as some European lands. Orkhan was important enough to be three times called upon by the Byzantine emperor for help and he married one of the emperor's daughters.

His two successors steadily ate up the Balkans, conquering Serbia and Bulgaria. They defeated another 'crusade' against them in 1396 and went on to take Greece. In 1391 they began their first siege of Constantinople, which they maintained successfully for six years. Meanwhile, Anatolia was absorbed by war and diplomacy. There was only one bad setback, the defeat by Timur which brought on a succession crisis and almost dissolved the Ottoman empire. The advance was then resumed and the Venetian empire now began to suffer, too. But for Byzantine and Turk alike, the struggle was essentially a religious one and its heart was the possession of the thousand-year-old Christian capital, Constantinople.

It was under Mehmet II, named the Conqueror, that in 1453 Constantinople fell to the Turks and the western world shuddered. It was a great feat of arms, depleted though the resources of Byzantium were, and supremely Mehmet's achievement, for he had persisted against all obstacles. The age of gunpowder was now well under way and he had a Hungarian engineer build him a gigantic cannon whose operation was so cumbersome that it could only be moved by a hundred oxen and fired only seven times a day (the Hungarian's assistance had been turned down by the Christians though the fee he asked was a quarter of what Mehmet gave him). It was a failure. Mehmet did better with orthodox methods, driving his soldiers forward ruthlessly, cutting them down if they flinched from the assault. Finally, he carried seventy ships overland to get them behind the imperial squadron guarding the Horn.

The last attack began early in April 1453. After nearly two months, on the evening of 28 May, Roman Catholics and Orthodox alike gathered in St Sophia and the fiction of the religious reunion was given its last parade. The Emperor Constantine XI, eightieth in succession since his namesake, the great first Constantine, took communion and then went out to die worthily, fighting. Soon afterwards, it was all over. Mehmet entered the city, went straight to St Sophia and there set up a triumphant throne. The church which had been the heart of Orthodoxy was made a mosque.

This was only a step, great as it was; the banner of Ottoman success was to be raised yet higher. The invasion of Serbia in 1459 was almost at once followed by the conquest of Trebizond. Unpleasant though this may have been for the inhabitants, it would merit only a footnote to the roll of Turkish conquest were it not also the end of Hellenism. At this remote spot on the south-eastern coast of the Black Sea in 1461 the world of Greek cities made possible by the conquest of Alexander the Great gave its last gasp. It marked an epoch as decisively as the fall of Constantinople, which a humanist pope bewailed as 'the second death of Homer and Plato'. From Trebizond, Turkish conquest rolled on. In the same year the Turks occupied the Peloponnese. Two years later they took Bosnia and Herzegovina. Albania and the Ionian islands followed in the next twenty years. In 1480 they captured the Italian port of Otranto and held it for nearly a year. In 1517 Syria and Egypt were conquered. They took longer to pick up the remainder of the Venetian empire, but at the beginning of the sixteenth century Turkish cavalry were near Vicenza. In 1526 at Mohacs they wiped out the army of the Hungarian king in a defeat which is remembered still as the black day of Hungarian history. Three years later they besieged Vienna for the first time. In 1571 Cyprus fell to them and nearly a century later Crete. By this time they were deep into Europe. They again besieged Vienna in the seventeenth century; their second failure to take it was the high-water mark of Turkish conquest. But they were still conquering new territory in the Mediterranean as late as 1715. Meanwhile, they had taken Kurdistan from Persia, with whom they had hardly ceased to quarrel since the appearance of a new dynasty there in 1501, and had sent an army as far south as Aden.

The Ottoman empire was of unique importance to Europe. It is one of the big differences marking off the history of its eastern from that of its western half. It was crucial that the Church survived and was tolerated in the Ottoman empire. That preserved the heritage of Byzantium for its Slav subjects (and, indeed, ended any threat to the supremacy of the patriarch at Constantinople either from the Catholics or from national Orthodox churches in the Balkans). Outside the former empire, only one important focus of Orthodoxy remained; it was crucial that the Orthodox Church was now the heritage of Russia. The establishment of the Ottoman empire for a time sealed off Europe from the Near East and the Black Sea and, therefore, in large measure from the land routes to Asia. The Europeans had really only themselves to blame; they had never been (and were never to be) able to unite effectively against the Turks. Byzantium had been left to her fate. 'Who will make the English love the French? Who will unite Genoese and Aragonese?' asked a fifteenth-century pope despairingly; not long after, one of his successors was sounding out the possibilities of Turkish help against France. Yet the challenge had awoken another sort of response, for even before the fall of Constantinople Portuguese ships were picking their way southwards down the African coast to look for a new route to the spices of the East and, possibly, an African ally to take the Turk in the flank from the south. People had mused over finding a way round the Islamic barrier since the thirteenth century, but the means had long been inadequate. By one of history's ironies they were just about to become available as Ottoman power reached its menacing peak.

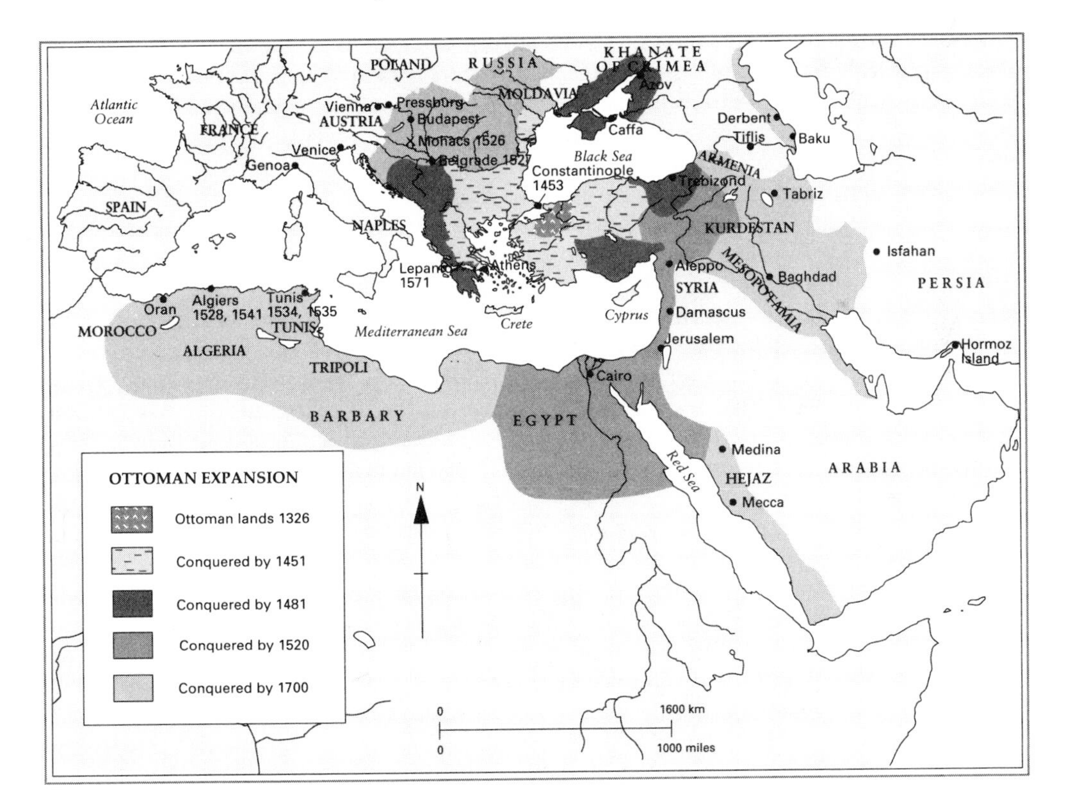

Behind the Ottoman frontiers a new multi-racial policy was organized. Mehmet was a man of wide, if volatile, sympathies and later Turks found it hard to understand his forbearance to the infidel. He was a man who could slaughter a boy, the godson of the emperor, because his sexual advances were refused, but he allowed a band of Cretans who would not surrender to sail away after the fall of Constantinople because he admired their courage. He seems to have wanted a multi-religious society. He brought back Greeks to Constantinople from Trebizond and appointed a new patriarch under whom the Greeks were eventually given a kind of self-government. The Turkish record towards Jew and Christian was better than that of Spanish Christians towards Jew and Moslem.

Thus the Ottomans reconstructed a great power in the eastern Mediterranean. While they rebuilt something like the Byzantine empire, another power was emerging in Persia which was also reminiscent of the past, this time of the empire of the Sassanids.

Between 1501 and 1736 the Safavid dynasty ruled Persia. Like their predecessors, the Safavids were not themselves Persian. Since the days of the Sassanids, conquerors had come and gone. The continuities of Persian history were meanwhile provided by culture and religion. Persia was defined by geography, by its language and by Islam, not by the maintenance of national dynasties. The Safavids were originally Turk, *ghazis* like the Osmanlis and succeeded, like them, in distancing possible rivals. The first ruler they gave to Persia was Ismail, a descendant of the fourteenth-century tribal ruler who had given his name to the line.

At first, Ismail was only the most successful leader of a group of warring Turkish tribes rather like those further west, exploiting similar opportunities. The Timurid inheritance had been in dissolution since the middle of the fifteenth century. In 1501 Ismail defeated the people known as the White Sheep Turks, entered Tabriz and proclaimed himself shah. Within twenty years he had carved out an enduring state and had also embarked upon a long rivalry with the Ottomans.

This rivalry had a religious dimension, for the Safavids were Shi'ites. When in the early sixteenth century the caliphate passed to the Ottomans they became the leaders of Sunnite Moslems who saw the caliphs as the proper interpreters and governors of the faith. The Shi'ites were therefore automatically anti-Ottoman. Ismail's establishment of the sect in Persia thus gave a new distinctiveness to Persia's civilization and this was to prove of great importance in preserving it.

His immediate successors had to fight off the Turks several times before a peace was made in 1555 which left Persia intact and opened Mecca and Medina to Persian pilgrims. There were domestic troubles, too, and fighting for the throne, but in 1587 there came to it one of the most able of Persian rulers, Shah Abbas the Great. Under his rule the Safavid dynasty was at its zenith. Politically and militarily he was very successful, defeating the Uzbeks and the Turks and taming the old tribal loyalties which had weakened his predecessors. He had important advantages: the Ottomans were distracted in the West, the potential of Russia was sterilized by internal troubles and Moghul India was past its peak. He was clever enough to see that Europe could be enrolled against the Turk. Yet a favourable conjuncture of international forces did not lead to schemes of world conquest. The Safavids did not follow the Sassanid example. They never took the offensive against Turkey except to recover earlier loss and they did not push north through the Caucasus to Russia, or beyond Transoxiana.

Persian culture enjoyed a spectacular flowering under Shah Abbas, who built a new capital at Isfahan. Its beauty and luxury astounded European visitors. Literature flourished. The only ominous note was religious. The shah insisted on abandoning the religious toleration which had until now characterized Safavid rule and imposed conversion to Shi'ite views. This did not at once mean the imposition of an intolerant system; that would only come later. But it did mean that Safavid Persia had taken a significant step towards decline and towards the devolution of power into the hands of religious officials.

After Shah Abbas' death in 1629 events rapidly took a turn for the worse. His unworthy successor did little about this, preferring to withdraw to the seclusion of the harem and its pleasures, while the traditional splendour of the Safavid inheritance cloaked its actual collapse. The Turks took Baghdad again in 1638. In 1664 came the first portents of a new threat: Cossack raids began to harry the Caucasus and the first Russian mission arrived in Isfahan. Western Europeans had already long been familiar with Persia. In 1507 the Portuguese had established themselves in the port of Ormuz where Ismail levied tribute on them. In 1561 an English merchant reached Persia overland from Russia and opened up Anglo-Persian trade. In the early seventeenth century his connexion was well established and by then Shah Abbas had Englishmen in his service. This was the result of his encouragement of relations with the West, where he hoped to find support against the Turk.

The growing English presence was not well received by the Portuguese. When the East India Company opened operations they attacked its agents, but unsuccessfully. A little later the English and Persians joined forces to eject the Portuguese from Ormuz. By this time other European countries were becoming interested, too. In the second half of the seventeenth century the French, Dutch and Spanish all tried to penetrate the Persian trade. The shahs did not rise to the opportunity of playing off one set of foreigners against another.

At the beginning of the eighteenth century Persia was suddenly exposed to a double onslaught. The Afghans revolted and established an independent Sunnite state; religious antagonism had done much to feed their sedition. From 1719 to 1722 the Afghans were at war with the last Safavid shah. He abdicated in that year and an Afghan, Mahmud, took the throne, thus ending Shi'ite rule in Persia. The story must none the less be taken a little further forward, for the Russians had been watching with interest the progress of Safavid decline. The Russian ruler had sent embassies to Isfahan in 1708 and 1718. Then, in 1723, on the pretext of intervention in the succession, the Russians seized Derbent and Baku and obtained from the defeated Shi'ites promises of much more. The Turks decided not to be left out and, having seized Tiflis, agreed in 1724 with the Russians upon a dismemberment of Persia. That once great state seemed to be ending in nightmare. In Isfahan a massacre of possible Safavid sympathizers was carried out by orders of a shah who had now gone mad. There was, before long, to be a last Persian recovery by the last great Asiatic conqueror, Nadir Kali. But though he might restore Persian empire, the days when the Iranian plateau was the seat of a power which could shape events far beyond its borders were over until the twentieth century, and then it would not be armies which gave Iran its leverage.

5
The Making of Europe

In comparison with Byzantium or the caliphate, Europe west of the Elbe was for centuries after the Roman collapse an almost insignificant backwater. Its boundaries were soon far narrower than had been those of western Christianity. Its inhabitants felt themselves a beleaguered remnant and in a sense so they were. Islam cut them off from Africa and the Near East and Arab raids tormented their southern coasts. From the eighth century the seemingly inexplicable violence of the Norse peoples we call Vikings fell like a flail time and time again on the northern coasts, river valleys and islands. In the ninth century the eastern front was harried by the pagan Magyars. Europe was formed in a hostile, heathen world.

The foundations of a new civilization had to be laid in barbarism and backwardness, which only a handful of men was available to tame and cultivate. No city in the west could approach in magnificence Constantinople, Córdoba, Baghdad or Ch'ang-an. Europe would long be a cultural importer. It took centuries before its architecture could compare with that of the classical past, of Byzantium or the Asian empires, and when it emerged it did so by borrowing the style of Byzantine Italy and the pointed arch of the Arabs. For just as long, no science, no school in the West could match those of Arab Spain or Asia. Nor could the western Christendom produce an effective political unity or theoretical justification of power such as the eastern empire and the caliphates; for centuries even the greatest European kings were hardly more than barbarian warlords to whom men clung for protection and in fear of something worse.

Had it come from Islam, that something might well have been better. At times, such an outcome must have seemed possible, for the Arabs established themselves not only in Spain but in Sicily, Corsica, Sardinia and the Balearics; men long feared they might go further. They had more to offer than the Scandinavian barbarians, yet the northerners left more of a mark in the end on the kingdoms established by earlier migrants. As for Slavic Christendom and Byzantium, both were culturally sundered from Catholic Europe and able to contribute little to Europe. Yet they were a cushion which just saved Europe from the full impact of eastern nomads and of Islam. A Moslem Russia would have meant a very different history for the West.

Roughly speaking, western Christendom before AD 1000 meant half the Iberian peninsula, all modern France and Germany west of the Elbe, Bohemia, Austria, the Italian mainland and England. At the fringes of this area lay barbaric, but Christian, Ireland and Scotland, and, just at the end of these centuries, the Scandinavian kingdoms. To this area the word 'Europe' began to be applied in the tenth century; a Spanish chronicle even spoke of the victors of 732 as 'European'. The area they occupied was all but landlocked; though the Atlantic was wide open, there was almost nowhere to go in that direction once Iceland was settled by the Norwegians, while the western Mediterranean, the highway to other civilizations and their trade, was an Arab lake. Only a thin channel of seaborne communication with an increasingly alien Byzantium brought Europe some relief from its introverted, narrow existence. Men grew used to privation rather than opportunity. They huddled together under the rule of a warrior class which they needed for their protection.

In fact, the worst was over in the tenth century. The Magyars were checked, the Arabs were beginning to be challenged at sea, and the northern barbarians were on the road to Christianity. Though, as the portentous date 1000 approached, men thought that the end of the world might be at hand, that year can serve, very approximately, as the marker of an epoch. Not only had the pressures upon Europe begun to relax, but the lineaments of a later, expanding Europe were already hardening. Her basic political and social structure was set and her Christian culture had already much of its peculiar flavour. The eleventh century was to begin an era of revolution and adventure, for which the centuries sometimes called the Dark Ages had provided raw materials. As a way to understand how this happened, a good starting-point is the map.

Before the eleventh century, three great changes had begun which were to shape the European map we know. One was a cultural and psychological shift away from the Mediterranean, the focus of classical civilization. Between the fifth and eighth centuries, the centre of European life, in so far as there was one, moved to the valley of the Rhine and its tributaries. By preying on the sea-lanes to Italy and by its distraction of Byzantium in the seventh and eighth centuries, Islam, too, helped to throw back the West upon this heartland of a future Europe. The second change was more positive, a gradual advance of Christianity and settlement in the east. Though far from complete by 1000, the advance guards of Christian civilization had by then long been pushed out well beyond the old Roman frontier. The third change was the slackening of barbarian pressure. The Magyars were checked in the tenth century; the Norsemen who were eventually to provide rulers in England, northern France, Sicily and some of the Aegean came from the last wave of Scandinavian expansion, which was in its final phase in the early eleventh century. Europe was no longer to be just a prey to others. True, even two hundred years later, when the Mongols menaced her, it must have been difficult to feel this. None the less, by 1000 she was ceasing to be wholly plastic.

Western Christendom can be considered in three big divisions. In the central area built round the Rhine valley the future France and the future Germany were to emerge. Then there was a west Mediterranean littoral civilization, embracing at first Catalonia, the Languedoc and Provence. With time and the recovery of Italy from the barbarian centuries, this extended itself further to the east and south. A third Europe was the somewhat varied periphery in the west, north-west and north where there were to be found the first Christian states of northern Spain, which emerged from the Visigothic period, England, with its independent Celtic and semi-barbarous neighbours, Ireland, Wales and Scotland, and lastly the Scandinavian states. We must not be too categorical about such a picture. There were areas one might allocate to one or the other of these three regions, such as Aquitaine, Gascony and sometimes Burgundy. Nevertheless, these distinctions are real enough to be useful. Historical experience, as well as climate and race, made these regions significantly different, yet of course most men living in these areas would not have known in which one they lived; they would certainly have been more interested in differences between them and their neighbours in the next village than of those between their region and its neighbour. Dimly aware that they were a part of Christendom, very few of them would have had even an approximate conception of what lay in the awful shadows beyond that comforting idea.

The origin of the heartland of the medieval West was the Frankish heritage. It had fewer towns than the south and they mattered little; a settlement like Paris was less troubled by the collapse of commerce than, say, Milan. Life centered on the soil, and aristocrats were successful warriors turned landowners. From this base, the Franks began the colonization of Germany, protected the Church and hardened and passed on a tradition of kingship whose origins lay somewhere in the magical powers of Merovingian rulers. But for centuries, state structures were fragile things, dependent on strong kings, for ruling was a personal activity.

Frankish ways and institutions did not help. After Clovis, though there was dynastic continuity, a succession of impoverished and therefore feeble kings led to more independence for landed aristocrats, who warred with one another; they had the wealth which could buy power. One family from Austrasia came to overshadow the Merovingian

A tenth-century history of the councils of Spain, the 'Codex Vigilanus' (or codex of Vigila) described the history of Spain, as the chroniclers of other former barbarian peoples described theirs, as a story of Christian monarchy. The Codex ends with a page showing five kings, three Visigothic, two kings of Leon, with the wife of the one who commissioned it. The bottom row depicts some of the very few individual artists who emerge from the anonymity of early medieval art, the scribe Vigila himself (in the centre) with an assistant and a pupil.

royal line. It produced Charles Martel, the soldier who turned the Arabs back at Tours in 732 and the supporter of St Boniface, the evangelizer of Germany. This is a considerable double mark to have left on European history (St Boniface said he could not have succeeded without Charles' support) and it confirmed the alliance of Martel's house with the Church. His second son, Pepin the Short, was chosen king by the Frankish nobles in 751. Three years later, the pope came to France and anointed him king as Samuel had anointed Saul and David.

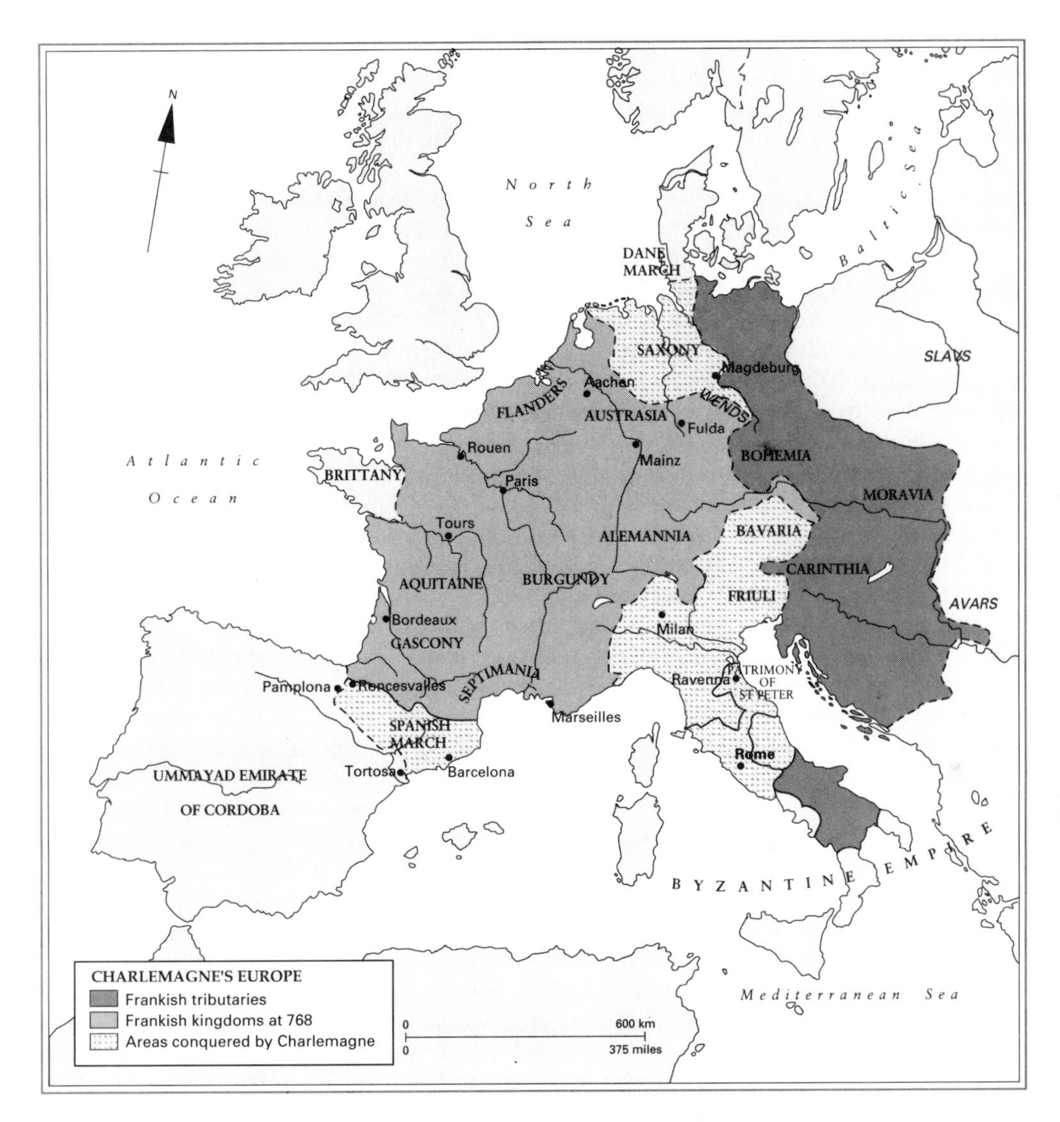

The papacy needed a powerful friend. The pretensions of the emperor in Constantinople were a fiction and in Roman eyes he had fallen into heresy, in any case, through taking up iconoclasm. To confer the title of Patrician on Pepin, as Pope Stephen did, was really a usurpation of imperial authority, but the Lombards were terrorizing Rome. The papacy drew the dividend on its investment almost at once. Pepin defeated the Lombards and in 756 established the Papal States of the future by granting Ravenna 'to St Peter'. This was the beginning of eleven hundred years of the Temporal Power, the secular authority enjoyed by the pope over his own dominions as a ruler like any other ruler. A Romano-Frankish axis was created. From it stemmed the reform of the Frankish Church, further colonization and missionary conversion in Germany (where wars were waged against the pagan Saxons), the throwing back of the Arabs across the Pyrenees and the conquest of Septimania and Aquitaine. These were big gains for the Church. It is hardly surprising to find Pope Hadrian I no longer dating official documents by the regnal year of the emperor at Byzantium, and minting coins in his own name. The Papacy had a new basis for independence. Nor did the new magic of anointing benefit only kings. Though it could replace or blur mysteriously with the old Merovingian thaumaturgy and raise kings above common men in more than their power, the pope gained the subtle implication of authority latent in the power to bestow the sacral oil.

Pepin, like all Frankish kings, divided his land at his death but the whole Frankish heritage was united again in 771 in his elder son. This was Charlemagne, crowned emperor in 800. The greatest of the Carolingians, as the line came to be called, he was soon a legend. This increases the difficulties, always great in medieval history, of penetrating a man's biography. Charlemagne's actions speak for certain continuing prepossessions. He was obviously still a traditional Frankish warrior-king; he conquered and his business was war. What was more novel was the seriousness with which he took the Christian sanctification of this role. He took his duties seriously, too, in patronizing learning and art; he wanted to magnify the grandeur and prestige of his court by filling it with evidence of Christian learning.

Territorially, Charlemagne was a great builder, overthrowing the Lombards in Italy and becoming their king; their lands, too, passed into the Frankish heritage. For thirty years he hammered away in campaigns on the Saxon March and achieved the conversion of the Saxon pagans by force. Fighting against the Avars, Wends and Slavs brought him Carinthia and Bohemia and, perhaps as important, the opening of a route down the Danube to Byzantium. To master the Danes, the Dane Mark (March) was set up across the Elbe. Charlemagne pushed into Spain early in the ninth century and instituted the Spanish March across the Pyrenees down to the Ebro and the Catalonian coast. But he did not put to sea; the Visigoths had been the last western European sea-power.

Thus he put together a realm bigger than anything in the West since Rome. Historians have been arguing almost ever since about what its reality was and about what Charlemagne's coronation by the pope on Christmas Day, 800, and his acclamation as emperor, actually meant. 'Most pious Augustus, crowned by God, the great and peace-giving Emperor' ran the chart at the service – but there already was an emperor whom everybody acknowledged to be such: he lived in Constantinople. Did a second ruler with the title mean that there were two emperors of a divided Christendom, as in later Roman times? Clearly, it was a claim to authority over many peoples; by this title, Charlemagne said he was more than just a ruler of Franks. Perhaps Italy mattered most in explaining it, for among the Italians a link with the imperial past might be a cementing factor as nowhere else. An element of papal gratitude – or expediency – was involved, too; Leo III had just been restored to his capital by Charlemagne's soldiers. Yet Charlemagne is reported to have said that he would not have entered St Peter's had he known what the pope intended to do. He may have disliked the pope's implied arrogation of authority. He may have foreseen the irritation the coronation would cause at Constantinople. He must have known that to his own people, the Franks, and to many of his northern subjects he was more comprehensible as a traditional Germanic warrior-king than as the successor of Roman emperors, yet before long his Seal bore the legend *Renovatio Romani imperii*, a conscious reconnexion with a great past.

In fact, Charlemagne's relations with Byzantium were troubled, though his title was a few years later recognized as valid in the West in return for a concession to Byzantium of sovereignty over Venice, Istria and Dalmatia. With another great state, the Abbasid caliphate, Charlemagne had somewhat formal but not unfriendly relations; Haroun-al-Raschid is said to have given him a cup bearing a portrait of Chosroes I, the king under whom Sassanid power and civilization was at its height (perhaps it is significant that it is from Frankish sources that we learn of these contacts; they do not seem to have struck the Arab chroniclers as important enough to mention). The Umayyads of Spain were different; they were marked down as the enemies of a Christian ruler because near enough to be a threat. To protect the faith from pagans was a part of Christian kingship but in government this kingship had other expressions. For all his support and protection, the Church was firmly subordinate to Charlemagne's authority. He presided over the Frankish synods, pronouncing upon dogmatic questions as authoritatively as had Justinian, and seems to have hoped for an integrated reform of the Frankish Church and the Roman, imposing upon them both the Rule of St Benedict. In such a scheme there is the essence of the later European idea that a Christian king is responsible not only for the protection of the Church but for the quality of the religious life within his dominions. Charlemagne also used the Church as an instrument of government, ruling through bishops.

Further evidence of religion's special importance to Charlemagne lies in the tone of the life of his court at Aachen. He strove to beautify its physical setting with architecture and decorative treasures. There was, of course, much to be done. The ebbing of economic life and of literacy meant that a Carolingian court was a primitive thing by comparison with Byzantium – and possibly even in comparison with those of some of the early barbarian kingdoms which were sometimes open to influence from a more cultivated world, as the appearance of Coptic themes in early barbarian art attests. When Charlemagne's men brought materials and ideas to beautify Aachen from Ravenna, Byzantine art, too, moved more freely into the north European tradition and classical models still influenced his artists. But it was its scholars and scribes who made Charlemagne's court most spectacular. It was an intellectual centre. From it radiated the impulse to copy texts in a new refined and reformed hand called Carolingian minuscule which was to be one of the great instruments of culture in the West. Charlemagne had hoped to use it to supply an authentic copy of the Rule of St Benedict to every monastery in his realm, but the major expression of a new manuscript potential was first evident in the copying of the Bible. This had a more than religious aim, for the scriptural story was to be interpreted as a justification of Carolingian rule. The Jewish history of the Old Testament was full of examples of pious and anointed warrior-kings. The Bible was the major text in the monastic libraries which now began to be assembled throughout the Frankish lands.

Copying and the diffusion of texts went on for a century after the original impulse had been given at Aachen and were the core of what modern scholars have called 'the Carolingian Renaissance'. It had none of the pagan connotations of that word as it was used of a later revival of learning which focused attention on the classical past, for it was emphatically Christian. Its whole purpose was the training of clergy to raise the level of the Frankish Church and carry the faith further to the east. The leading men in the beginnings of this transmission of sacred knowledge were not Franks. There were several Irishmen and Anglo-Saxons in the palace school at Aachen and among them the outstanding figure was Alcuin, a cleric from York, a great centre of English learning. His most famous pupil was Charlemagne himself, but he had several others and managed the palace library. Besides writing books of his own he set up a school at Tours, where he became abbot, and began to expound Boethius and Augustine to the men who would govern the Frankish church in the next generation.

Alcuin's pre-eminence is as striking a piece of evidence as any of the shift in the centre of cultural gravity in Europe, away from the classical world and to the north. But others than his countrymen were involved in teaching, copying and founding the new monasteries which spread outwards into east and west Francia; there were Franks, Visigoths, Lombards and Italians among them, too. One of these, a layman called Einhard, wrote a life of the emperor from which we learn such fascinating human details as the fact that he could be garrulous, that he was a keen hunter and that he passionately loved swimming and bathing in the thermal springs which explain his choice of Aachen as a residence. Charlemagne comes to life in Einhard's pages as an intellectual, too, speaking Latin as well, we are told, as Frankish, and understanding Greek. This is made more credible because we hear also of his attempts to write, keeping notebooks under his pillow so that he could do so in bed, 'but', Einhard says, 'although he tried very hard, he had begun too late in life'.

From this account and from his work a remarkably vivid picture can be

u autem dn̄e miserere mei & resuscita me
& retribuam eis
n hoc cognoui qm̄ uoluisti me
qm̄ non gaudebit inimicus meus sup me
e autem propter innocentiā suscepisti
& confirmasti me in conspectu tuo in
enedictus dn̄s ds̄ israhel
a sáeculo & in saeculum fiat fiat
42.
uem admodum desiderat
ceruus ad fontes aquarum
ita desiderat anima mea ad te ds̄
itiuit anima mea ad dm̄
fortē uiuum quando
ueniā & ap parebo ante faciē di

A psalter copied at a French Abbey in about 825, written in the new, easily-read hand of the Carolingian copyists.

formed of a dignified, majestic figure, striving to make the transition from warlord to ruler of a great Christian empire, and having remarkable success in his own lifetime in so doing. Clearly his physical presence was impressive (he probably towered over most of his entourage), and men saw in him the image of a kingly soul, gay, just and magnanimous, as well as that of the heroic paladin of whom poets and minstrels would be singing for centuries. His authority was a more majestic spectacle than anything seen to that time in barbarian lands. When his reign began, his court was still peripatetic, it normally ate its way from estate to estate throughout the year. When Charlemagne died, he left a palace and a treasury established at the place where he was to be buried. He had been able to reform weights and measures, and had given to Europe the division of the pound of silver into 240 pennies (*denarii*) which was to survive in the British Isles for eleven hundred years. But his power was also very personal. This may be inferred from the efforts he made to prevent his noblemen from replacing tribal rulers by setting down into hereditary positions of their own, and from the repeated issuing of 'capitularies' or instructions to his servants (a sign that his wishes were not carried out). In the last resort, even a Charlemagne could only rely on personal rule, and that meant a monarchy based on his own domain and its produce and on the big men close enough to him for supervision. These vassals were bound to him by especially solemn oaths, but even they began to give trouble as he grew older.

Charlemagne thought in traditional Frankish terms of his territorial legacy. He made plans to divide it and only the accident of sons dying before him ensured that the empire passed undivided to the youngest, Louis the Pious, in 814. With it came the imperial title (which Charlemagne gave to his son) and the alliance of monarchy and papacy; two years after his succession the pope crowned Louis at a second coronation. Partition was only delayed by this. Charlemagne's successors had neither his authority nor his experience, nor perhaps an interest in controlling fissiparous forces. Regional loyalties were forming around individuals and a series of partitions finally culminated in one between three of Charlemagne's grandsons, the Treaty of Verdun of 843, which had great consequences. It gave a core kingdom of Frankish lands centred on the western side of the Rhine valley and containing Charlemagne's capital, Aachen, to Lothair, the reigning emperor (thus it was called Lotharingia) and added to this the kingdom of Italy. North of the Alps, this united Provence, Burgundy, Lorraine and the lands between the Scheldt, Meuse, Saône and Rhône. To the east lay a second block of lands of Teutonic speech between the Rhine and the German Marches; it went to Louis the German. Finally, in the west, a tract of territory including Gascony, Septimania and Aquitaine, and roughly the equal of the rest of modern France, went to a half-brother of these two, Charles the Bald.

This settlement was not long untroubled, but it was decisive in a broad and important way; it effectively founded the political distinction of France and Germany, whose roots lay in west and east Francia. Between them it set up a third unit with much less linguistic, ethnic, geographical and economic unity. Lotharingia was there in part because three sons had to be provided for. Much future Franco-German history was going to be about the way in which it could be divided between neighbours bound to covet it and therefore likely to grow apart from one another in rivalry.

No royal house could guarantee a continuous flow of able kings, nor could they for ever buy loyalty from their supporters by giving away lands. Gradually, and like their predecessors, the Carolingians declined in power. The signs of break-up multiplied, an independent kingdom of Burgundy appeared and people began to dwell on the great days of Charlemagne, a significant symptom of decay and dissatisfaction. The histories of west and east Franks diverged more and more.

In west Francia the Carolingians lasted just over a century after Charles the Bald. By the end of his reign Brittany, Flanders, and Aquitaine were to all intents and purposes independent. The west Frankish monarchy thus started the tenth century in a weak position and it had the attacks of Vikings to deal with as well. In 911 Charles III, unable to expel the Norsemen, conceded lands in what was later Normandy to their leader, Rollo. Baptized the following year, Rollo set to work to build the duchy for which he did homage to the Carolingians; his Scandinavian countrymen continued to arrive and settle there until the end of the tenth century, yet somehow they soon became French in speech and law. After this, the unity of the west Franks fell even more rapidly apart. From confusion over the succession there emerged a son of a count of Paris who steadily built up his family's power around a domain in the Ile de France. This was to be the core of the later France. When the last Carolingian ruler of the west Franks died in 987, this man's son, Hugh Capet, was elected king. His family was to rule for nearly four hundred years. For the rest, the west Franks were divided into a dozen or so territorial units ruled by magnates of varying standing and independence.

Among the supporters of Hugh's election was the ruler of the east Franks. Across the Rhine, the repeated division of their heritage had quickly proved fatal to the

The emperor Otto III receives the homage of the four parts of the empire, Sclavonia, Germania, Gallia and Roma. Like the leading personage in a Byzantine painting, he is drawn larger than the accompanying figures.

Carolingians. When the last Carolingian king died in 911 there emerged a political fragmentation which was to characterize Germany down to the nineteenth century. The assertiveness of local magnates combined with stronger tribal loyalties than in the west to produce a half-dozen powerful dukedoms. The ruler of one of these, Conrad of Franconia, was chosen as king by the other dukes, somewhat surprisingly. They wanted a strong leader against the Magyars. The change of dynasty made it advisable to confer some special standing on the new ruler; the bishops therefore anointed Conrad at his coronation. He was the first ruler of the east Franks so to be treated. But Conrad was not successful against the Magyars; he lost and could not win back Lotharingia and he strove, with the support of the Church, to exalt his own house and office. Almost automatically, the dukes gathered their peoples about them to safeguard their own independence. The four whose distinction mattered most were the Saxons, the Bavarians, the Swabians and the Franconians (as the east Franks became known). Regional differences, blood and the natural pretensions of great nobles stamped on Germany in Conrad's reign the pattern of its history for a thousand years: a tug-of-war between central authority and local power not to be resolved in the long run in favour of the centre as elsewhere, though in the tenth century it looked otherwise for a while. Conrad faced ducal rebellion but nominated one of the rebels his successor and the dukes agreed. In 919, Henry 'the Fowler' (as he was called), Duke of Saxony, became king. He and his descendants, the 'Saxon emperors', or Ottonians, ruled the eastern Franks until 1024.

Henry the Fowler avoided the ecclesiastical coronation. He had great family properties and the tribal loyalties of the Saxons on his side and brought the magnates into line by proving himself a good soldier. He won back Lotharingia from the west Franks, created new Marches on the Elbe after victorious campaigns against the Wends, made Denmark a tributary kingdom and began its conversion, and, finally, he defeated the Magyars. His son, Otto I, thus had a goodly inheritance and made good use of it. In disciplining the dukes, he continued his father's work. In 955 he inflicted on the Magyars a defeat which ended for ever the danger they had presented. Austria, Charlemagne's east March, was recolonized. Though he faced some opposition, Otto made a loyal instrument out of the German church; it was an advantage of the Saxon emperors that in Germany, unlike west Francia, churchmen tended to look with favour to the monarchy for protection against predatory laymen. A new archiepiscopal province, Magdeburg, was organized to direct the bishoprics established among the Slavs. With Otto ends, it has been said, the period of mere anarchy in central Europe; under him, certainly, we have the first sense of something we might call Germany. But Otto's ambition did not stop there.

In 936 Otto had been crowned at Aachen, Charlemagne's old capital. Not only did he accept the ecclesiastical service and anointing which his father had avoided, but he afterwards held a coronation banquet at which the German dukes served him as his vassals. This was in the old Carolingian style. Fifteen years later he invaded Italy, married the widow of a claimant to the crown of Italy, and assumed it himself. Yet the pope refused him an imperial coronation. Ten years later, in 962, Otto was back in Italy again in response to an appeal by the pope for help, and this time the pope crowned him.

Thus was revived the Roman and the Carolingian ideal of empire. The German and Italian crowns were united again in what would one day be known as the Holy Roman Empire and would last nearly a thousand years. Yet it was not so wide an empire as Charlemagne's, nor did Otto dominate the Church as Charlemagne had done. For all his strength (and he deposed two popes and nominated two others) Otto was the Church's protector who thought he knew what was best for it, but he was not its governor. Nor was the structure of the empire very solid; it rested on the political manipulation of local magnates rather than on administration.

Nevertheless, the Ottonian empire was a remarkable achievement. Otto's son, the future Otto II, married a Byzantine princess. Both he and Otto III had reigns troubled by revolt, but successfully maintained the tradition established by Otto the Great of exercising power south of the Alps. Otto III made a cousin pope (the first German to sit in the chair of St Peter) and followed him by appointing the first French pope. Rome seemed to captivate him and he settled down there. Like both his immediate predecessors, he called himself *augustus* but in addition his seals revived the legend 'Renewal of the Roman empire' – which he equated with the Christian empire. Half Byzantine by birth, he saw himself as a new Constantine. A diptych of a gospel-book painted nearly at the end of the tenth century shows him in state, crowned and orb in hand, receiving the homage of four crowned women: they are Sclavonia (Slavic Europe), Germany, Gaul and Rome. His notion of a Europe organized as a hierarchy of kings serving under the emperor was eastern. In this there was megalomania as well as genuine religious conviction; the real basis of Otto's power was his German kingship, not the Italy which obsessed and detained him. Nevertheless, after his death in 1002, he was taken to Aachen, as he had ordered, to be buried beside Charlemagne.

He left no heir, but the direct Saxon line was not exhausted; Henry II, who was elected after a struggle, was a great-grandson of Henry the Fowler. But his coronation at Rome masked the reality; he was a German ruler, not emperor of the West, at heart. His seal's inscription read 'Renewal of the kingdom of the Franks' and his attention was focused on pacification and conversion in east Germany. Though he made three expeditions to Italy, Henry relied there not on government but on politics, the playing off of factions against one another. With him the Byzantine style of the Ottonian empire began to wane.

Thus the eleventh century opened with the idea of western empire still capable of beguiling monarchs, but with the Carolingian inheritance long since crumbled into fragments. They set out the lines of European history for ages to come. The idea of Germany barely existed but the country was a political reality, even if still inchoate. The curious federal structure which was to emerge from the German middle ages was to be the last refuge of the imperial idea in the West, the Holy Roman Empire. Meanwhile, in France, too, the main line of the future was settled, though it could not have been discerned at the time. West Francia had dissolved into a dozen or so major units over which the suzerainty of the Capetians was for a long time feeble. But they had on their side a centrally placed royal domain, including Paris and the important diocese of Orleans, and the friendship of the Church. These were advantages in the hands of able kings, and able kings would be forthcoming in the next three centuries.

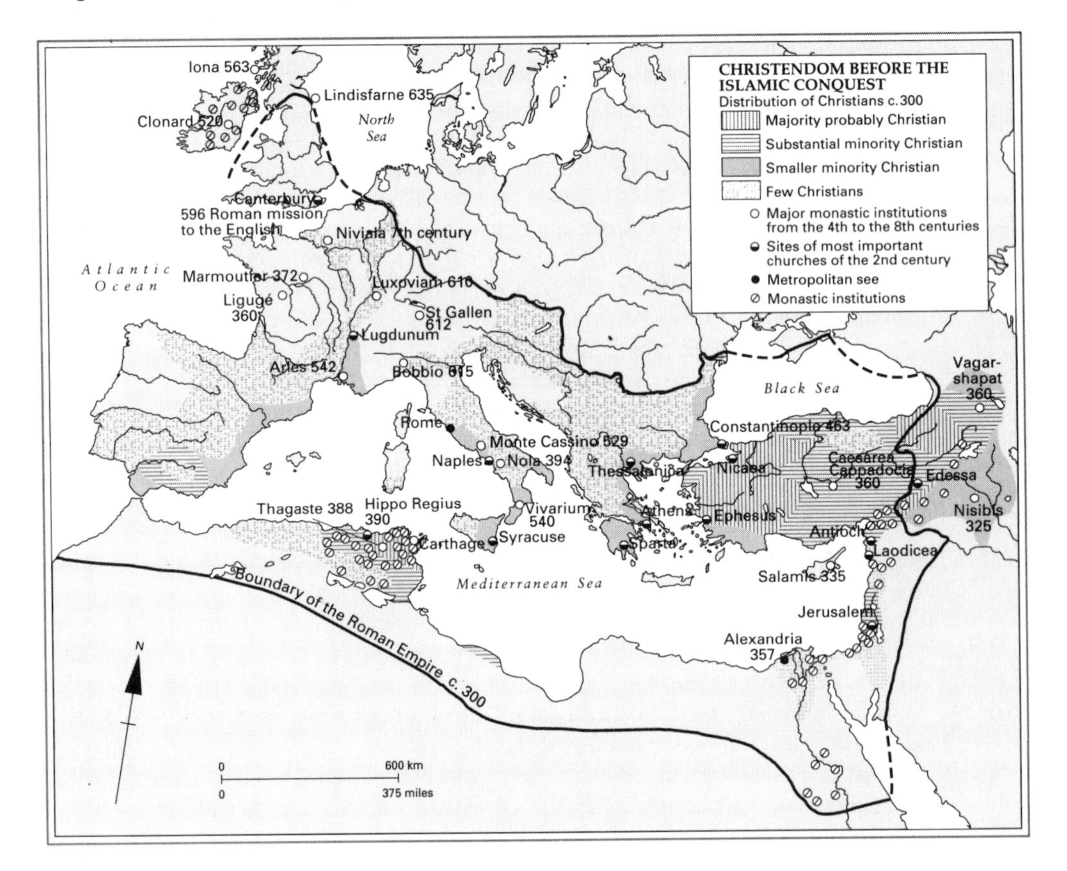

The other major component of the Carolingian heritage had been Italy. It had gradually become more and more distinct from the territories north of the Alps; since the seventh century it had been evolving away from the possibility of integration with northern Europe and back towards re-emergence as a part of Mediterranean Europe. By the middle of the eighth century, much of Italy had been subjugated by the Lombards. This barbarian people had settled down in the peninsula and had adopted an Italianate speech, but they remained an aggressive minority, whose social tensions demanded release in frequent wars of conquest, and they had shaped the Catholicism they had adopted to their own needs and institutions. In spite of the theoretical survival of the legal claims of the eastern emperors, the only possible balancing power to them in Italy until the eighth century was the pope. When the Lombard principalities began to consolidate under a vigorous monarchy, this was no longer enough; hence the evolution of papal diplomacy towards alliance with the Carolingians. Once the Lombard kingdom had been destroyed by Charlemagne, there was no rival in the peninsula to the Papal States, though after the waning of the Carolingians' power the popes had to face both the rising power of the Italian magnates and their own Roman aristocracy. The western Church was at its lowest ebb of cohesion and unity and the Ottonians' treatment of the papacy showed how little power it had. An anarchic Italian map was another result of this situation. The north was a scatter of feudal statelets. Only Venice was very successful; for two hundred years she had been pushing forward in the Adriatic and her ruler had just assumed the title of duke. She is perhaps better regarded as a Levantine and Adriatic rather than a Mediterranean power. City-states which were republics existed in the south, at Gaeta, Amalfi, Naples. Across the middle of the peninsula ran the Papal States. Over the whole fell the shadow of Islamic raids as far north as Pisa, while emirates appeared at Taranto and Bari in the ninth century. These were not to last, but the Arabs completed the conquest of Sicily in 902 and were to rule it for a century and a half with profound effects.

One of the most beautiful machines of its age, the Viking ship. This one, buried as the bier of a prince, may have originally belonged to a king, for it was exceptionally finely carved and decorated.

The Arabs shaped the destiny of the other west Mediterranean coasts of Europe, too. Not only were they established in Spain, but even in Provence they had more or less permanent bases (one of them being St Tropez). The inhabitants of the European coasts of the Mediterranean had, perforce, a complex relationship with the Arabs, who appeared to them both as free-booters and as traders; the mixture was not unlike that observable in the Viking descents except that the Arabs showed little tendency to settle. Southern France and Catalonia were areas in which Frankish had followed Gothic conquest, but many factors differentiated them from the Frankish north. The physical reminiscences of the Roman past were plentiful in these areas and so was a Mediterranean agriculture. Another distinctive characteristic was the appearance of a family of Romance languages in the south, of which Catalan and Provençal were the most enduring.

In 1000 AD, the peripheral Europe of the north barely included Scandinavia, if Christianity is the test of inclusion. Missionaries had been at work for a long time but the first Christian monarchs only appear there in the tenth century and not until the next were all Scandinavian kings Christian. Long before that, pagan Norsemen had changed the history of the British Isles and the northern fringe of Christendom.

For reasons which, as in the case of many other folk-movements, are by no means clear, but are possibly rooted in over-population, the Scandinavians began to move outwards from the eighth century onwards. Equipped with two fine technical instruments, a longboat which oars and sails could take across seas and up shallow rivers and a tubby cargo-carrier which could shelter large families, their goods and animals for six or seven days at sea, they thrust out across the water for four centuries, and left behind a civilization which in the end stretched from Greenland to Kiev. Not all sought the same things. The Norwegians who struck out to Iceland, the Faroes, Orkney and the far west wanted to colonize. The Swedes who penetrated Russia and survive in the records as Varangians were much busier in trade. The Danes did most of the plundering and piracy the Vikings are remembered for. But all these themes of the Scandinavian migrations wove in and out of one another. No branch of these peoples had a monopoly of any one of them.

The Viking colonization of remote islands was their most spectacular achievement. They wholly replaced the Picts in the Orkneys and the Shetlands and from them extended their rule to the Faroes (previously uninhabited except for a few Irish monks and their sheep) and the Isle of Man. Offshore, the Viking lodgement was more lasting and profound than on the mainland of Scotland and Ireland, where settlement began in the ninth century. Yet the Irish language records their importance by its adoption of Norse words in commerce, and the Irish map marks it by the situation of Dublin, founded by the Vikings and soon turned into an important trading-post. The most successful colony of all was Iceland. Irish hermits had anticipated Vikings there, too, and it was not until the end of the ninth century that they came in large numbers. By 930 there may have been 10000 Norse Icelanders, living by farming and fishing, in part for their own subsistence, in part to produce commodities such as salt fish which they might trade. In that year the Icelandic state was founded and the *Thing* (which romantic antiquarians later saw as the first European 'parliament') met for the first time. It was more like a council of the big men of the community than a modern representative body and it followed earlier Norwegian practice, but Iceland's continuous historical record is in this respect a remarkable one.

Colonies in Greenland followed in the tenth century; there were to be Norsemen there for five hundred years. Then they disappeared, probably because the settlers were wiped out by Eskimos pushed south by an advance of the ice. Of discovery and settlement further west we can say much less. The Sagas, the heroic poems of medieval Iceland, tell us of the exploration of 'Vinland', the land where Norsemen found the wild vine growing, and of the birth of a child there (whose mother subsequently returned to Iceland and went abroad again as far as Rome as a pilgrim before settling into a highly sanctified retirement in her native land). There are reasonably good grounds to believe that a settlement discovered in Newfoundland is Norse. But we cannot at present go much further than this in uncovering the traces of the predecessors of Columbus.

In western European tradition, the colonial and mercantile activities of the Vikings were from the start obscured by their horrific impact as marauders. Certainly, they had some very nasty habits, spread-eagling among them, but so did most barbarians. Some exaggeration must therefore be allowed for, especially because our main evidence comes from the pens of churchmen doubly appalled, both as Christians and as victims, by attacks on churches and monasteries; as pagans, of course, Vikings saw no special sanctity in the concentrations of precious metals and food so conveniently provided by such places, and found them especially attractive targets. Nor were the Vikings the first people to burn monasteries in Ireland.

None the less, however such considerations are weighed, it is indisputable that the Viking impact on northern and western Christendom was very great and very terrifying. They first attacked England in 793, the monastery of Lindisfarne being their victim; the attack shook the ecclesiastical world (yet the monastery lived on another eighty years). Ireland they raided two years later. In the first half of the ninth century the Danes began a harrying of Frisia which went on regularly year after year, the same towns being plundered again and again. The French coast was then attacked; in 842 Nantes was sacked with a great massacre. Within a few years a Frankish chronicler bewailed that 'the endless flood of Vikings never ceases to grow.' Towns as far inland as Paris, Limoges, Orleans, Tours and Angoulême were attacked. The Vikings had become professional pirates. Soon Spain suffered and the Arabs, too, were harassed; in 844 the Vikings stormed Seville. In 859 they even raided Nimes and plundered Pisa, though they suffered heavily at the hands of an Arab fleet on their way home.

At its worst, think some scholars, the Viking onslaught came near to destroying civilisation in West Francia; certainly the West Franks had to endure more than their cousins in the east and the Vikings helped to shape the differences between a future France and a future Germany. In the west their ravages threw new responsibilities on local magnates, while central and royal control crumbled away and men looked more and more towards their local lord for protection. When Hugh Capet came to the throne, it was very much as *primus inter pares* in a recognisably feudal society.

Not all the efforts of rulers to meet the Viking threat were failures. Charlemagne and Louis the Pious did not, admittedly, have to face attacks as heavy and persistent as their successors, but they managed to defend the vulnerable ports and

According to Germanic tradition, a king should reign with counsel from his great men. In Anglo-Saxon England the witanagemot *joins the king in judging a criminal.*

river-mouths with some effectiveness. The Vikings could be (and were) defeated if drawn into full-scale field engagements and, though there were dramatic exceptions, the main centres of the Christian West were on the whole successfully defended. What could not be prevented were repeated small-scale raids on the coasts. When the Vikings learnt to avoid pitched battles, the only way to deal with them was to buy them off and Charles the Bald began paying them tribute so that his subjects should be left in peace.

This was the beginning of what the English called Danegeld. Their island had soon become a major target, to which Vikings began to come to settle as well as to raid. A small group of kingdoms had emerged there from the Germanic invasions; by the seventh century many of Romano-British descent were living alongside the communities of the new settlers, while others had been driven back to the hills of Wales and Scotland. Christianity continued to be diffused by Irish missionaries from the Roman mission which had established Canterbury. It competed with the older Celtic Church until 664, a crucial date. In that year a Northumbrian king at a synod of churchmen held at Whitby pronounced in favour of adopting the date of Easter set by the Roman church. It was a symbolic choice, determining that the future England would adhere to the Roman traditions, not the Celtic.

From time to time, one or other of the English kingdoms was strong enough to have some sway over the others. Yet only one of them could successfully stand up to the wave of Danish attacks from 851 onwards which led to the occupation of two-thirds of the country. This was Wessex and it gave England its first national hero who is also an historical figure, Alfred the Great.

As a child of four, Alfred had been taken to Rome by his father and was given consular honours by the pope. The monarchy of Wessex was indissolubly linked with Christianity and Carolingian Europe. It defended the faith against paganism as much as England against an alien people. In 871 Alfred inflicted the first decisive defeat on a Danish army in England. Significantly, a few years later the Danish king agreed not only to withdraw from Wessex but to accept conversion as a Christian. This registered that the Danes were in England to stay (they had settled in the north) but also that they might be divided from one another. Soon Alfred was leader of all the surviving English kings; eventually, none was left but he. He recovered London and when he died in 899 the worst period of Danish raids was over and his descendants were to rule a united country. Even the settlers of the Danelaw, the area marked to this day by Scandinavian place-names and fashions of speech as that of Danish colonization defined by Alfred, accepted their rule. Nor was this all. Alfred had also founded a series of strongholds ('burghs') as a part of a new system of national defence by local levies. They not only gave his successors bases for the further reduction of the Danelaw but set much of the pattern of early medieval urbanization in England; on them were built towns whose sites are still inhabited today. Finally, with tiny resources, Alfred deliberately undertook the cultural and intellectual regeneration of his people. The scholars of his court, like those of Charlemagne, proceeded by way of copying and translation: the Anglo-Saxon nobleman and cleric were intended to learn of Bede and Boethius in their own tongue.

Alfred's innovations were a creative effort of government unique in Europe, and marked the beginning of a great age for England. The shire structure took shape and boundaries were established which lasted until 1974. The English Church was soon to experience a remarkable surge of monasticism, the Danes were held in a united kingdom through a half-century's turbulence. It was only when ability failed in Alfred's line that the Anglo-Saxon monarchy came to grief and a new Viking offensive took place. Colossal sums of Danegeld were paid until a Danish king (this time a Christian) overthrew the English king and then died, leaving a young son to rule his conquest. This was the celebrated Canute, under whom England was briefly part of a great Danish empire (1006–35). There was a last great Norwegian invasion of England in 1066, but it was shattered at the battle of Stamford Bridge. By that time, all the Scandinavian monarchies were Christian and Viking culture was being absorbed into Christian forms. It left many evidences of its individuality and strength in both Celtic and continental art.

Under Moslem rule, Spanish Christians produced what was later called Mozarabic or 'arabized' art. One of its most astonishing productions is a collection of illuminated manuscripts of a popular work of devotion and scholarship, a commentary on the Apocalypse by an eighth-century monk, Beatus of Liebana. About twenty copies survive and this illustration of the text 'Behold, he cometh with clouds; and every eye shall see him' is from an eleventh-century version.

Its institutions survive in Iceland and other islands. The Scandinavian legacy is strongly marked for centuries in English language and social patterns, in the emergence of the duchy of Normandy, and, above all, in the literature of the Sagas. Yet where they entered settled lands, the Norsemen gradually merged with the rest of the population. When the descendants of Rollo and his followers turned to the conquest of England in the eleventh century they were really Frenchmen and the war-song they sang at Hastings was about Charlemagne the Frankish paladin. They conquered an England where the men of the Danelaw were by then English. Similarly, the Vikings lost their distinctiveness as an ethnic group in Kiev Rus and Muscovy.

The only other western peoples of the early eleventh century who call for remark because of the future that lay before them were those of the Christian states of northern Spain. Geography, climate and Moslem division had all helped Christianity's survival in the peninsula and in part defined its extent. In the Asturias and Navarre Christian princes or chieftains still hung on early in the eighth century. Aided by the establishment of the Spanish March by Charlemagne and its subsequent growth under the new Counts of Barcelona, they nibbled away successfully at Islamic Spain while it was distracted by civil war and religious schism. A kingdom of Léon emerged in the Asturias to take its place beside a kingdom of Navarre. In the tenth century, however, it was the Christians who fell out with one another and the Arabs who again made headway against them. The blackest moment came at the very end of the century when a great Arab conqueror, Al-Mansur, took Barcelona, Léon, and in 998 the shrine of Santiago de Compostela itself at which St James the Apostle was supposed to be buried. The triumph was not long-lived, for here, too, what had been done to found Christian Europe proved ineradicable. Within a few decades Christian Spain had rallied as Islamic Spain fell into disunion. In the Iberian peninsula as elsewhere, the age of expansion which this inaugurated belongs to another historical era, but was based on long centuries of confrontation with another civilization. For Spain, above all, Christianity was the crucible of nationhood.

The Iberian example suggests just how much of the making of the map of Europe is the making of the map of the Church, but an emphasis only on successful missions and ties with powerful monarchs is misleading. There was much more to early Christian Europe and the Christian life than this. The western Church provides one of the great success stories of history, yet its leaders between the end of the ancient world and the eleventh or twelfth century long felt isolated and embattled in a pagan or semi-pagan world. Increasingly at odds with, and finally almost cut off from, eastern Orthodoxy, it is hardly surprising that western Christianity developed an aggressive intransigence almost as a defensive reflex. It was another sign of its insecurity. Nor was it threatened merely by enemies without. Inside western Christendom, too, the Church felt at bay and beleaguered. It strove in the middle of still semi-pagan populations to keep its teaching and practice intact while christening what it could of a culture with which it had to live, judging nicely the concession which could be made to local practice or tradition and distinguishing it from a fatal compromise of principle. All this it had to do with a body of clergy of whom many, perhaps most, were men of no learning, not much discipline and dubious spirituality. Perhaps it is not surprising that the leaders of the Church sometimes overlooked the enormous asset they enjoyed in being faced by no spiritual rival in western Europe after Islam was turned back by Charles Martel; they had to contend only with vestigial paganism and superstition, and these the Church knew how to use. Meanwhile, the great men of this world surrounded it, sometimes helpfully, sometimes hopefully, always a potential and often a real threat to the Church's independence of the society it had to strive to save.

Inevitably, much of the history which resulted is the history of the papacy. It is the central and best-documented institution of Christianity. Its documentation is part of the reason why so much attention has been given to it, a fact that should provoke reflexion about what can be known about religion in these centuries. Though papal power had alarming ups and downs, the division of the old empire meant that if there was anywhere in the West a defender of the interests of religion, it was Rome, for it had no ecclesiastical rival. After Gregory the Great it was obviously implausible to maintain the theory of one Christian Church in one empire, even if the imperial exarch did stay at Ravenna. The last emperor who came to Rome did so in 663 and the last pope to go to Constantinople went there in 710. Then came iconoclasm, which inflicted grievous ideological damage. When Ravenna fell to the renewed advance of the Lombards, Pope Stephen set out for Pepin's court, not that of Byzantium. There was no desire to break with the eastern empire, but

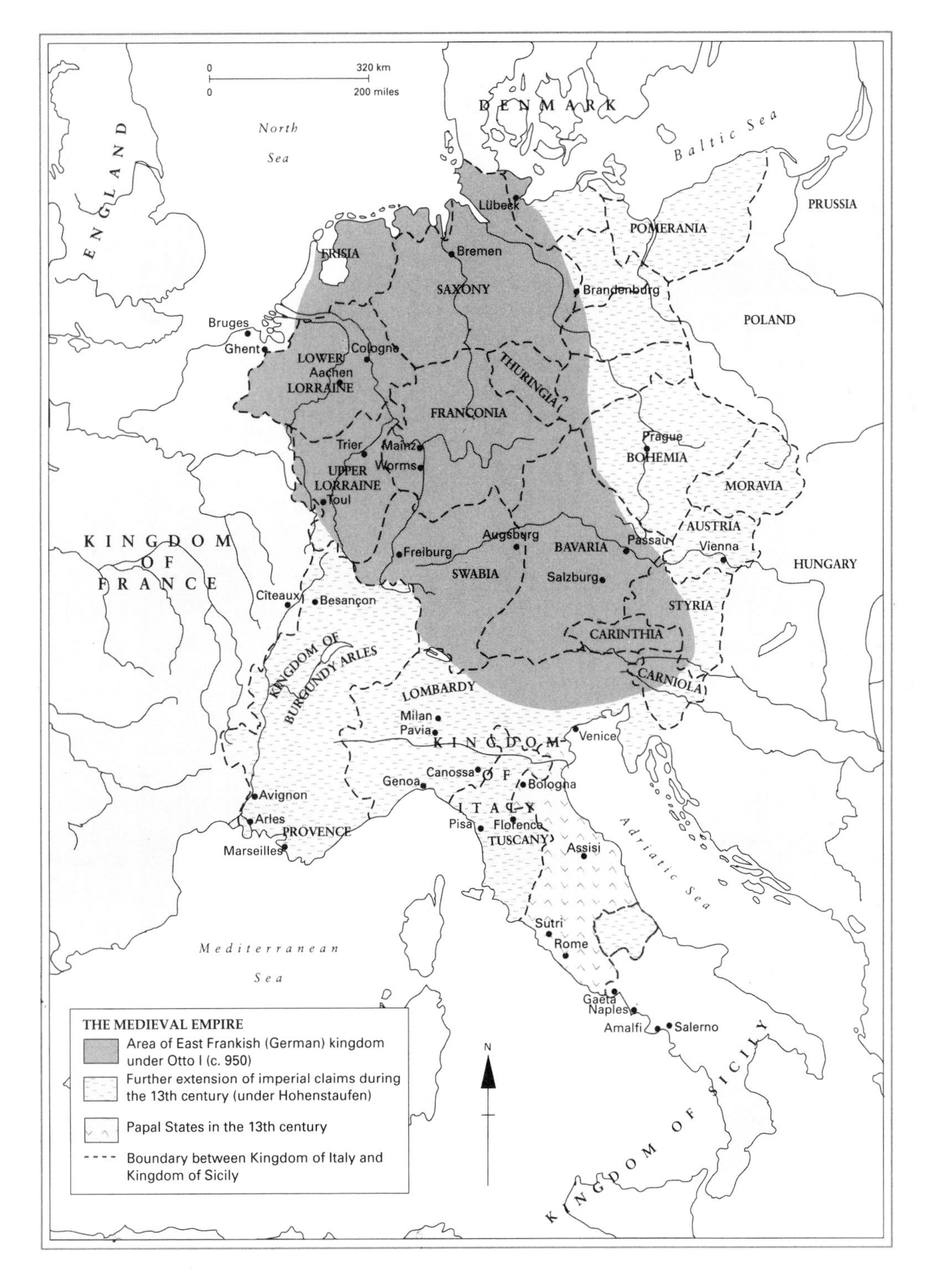

Frankish armies could offer protection no longer available from the east. Protection was needed, too, for the Arabs menaced Italy from the beginning of the eighth century, and, increasingly, the native Italian magnates became obstreperous in the ebbing of Lombard hegemony.

There were some very bad moments in the two and a half centuries after Pepin's coronation. Rome seemed to have very few cards in its hands and at times only to have exchanged one master for another. Its claim to primacy was a matter of the respect due to the guardianship of St Peter's bones and the fact that the see was indisputably the only apostolic one in the West: a matter of history rather than of practical power. For a long time the popes could hardly govern effectively even within the temporal domains, for they had neither adequate armed forces nor a civil administration. As great Italian property-owners, they were exposed to predators and blackmail. Charlemagne was only the first, and perhaps he was the most high-minded, of several emperors who made clear to the papacy their views of the respective standing of pope and emperor as guardians of the Church. The Ottonians were great makers and unmakers of popes. The successors of St Peter could not welcome confrontations, for they had too much to lose.

Yet there was another side to the balance sheet, even if it was slow to reveal its full implications. Pepin's grant of territory to the papacy would in time form the nucleus of a powerful Italian territorial state. In the pope's coronation of emperors there rested veiled claims, perhaps to the identification of rightful emperors; significantly, as time passed, popes withdrew from the imperial coronation ceremony (as from that of English and French kings) the use of the chrism, the specially sacred mixture of oil and balsam for the ordination of priests and the coronation of bishops, substituting simple oil. Thus was expressed a reality long concealed but easily comprehensible to an age used to symbols: the pope conferred the crown and the stamp of God's recognition on the emperor. Perhaps, therefore, he could do so conditionally. Leo's coronation of Charlemagne, like Stephen's of Pepin, may have been expedient, but it contained a potent seed. When, as often happened, personal weaknesses and succession disputes disrupted the Frankish kingdoms, Rome might gain ground.

More immediately and practically, the support of powerful kings was needed for the reform of local Churches and the support of missionary enterprise in the East. For all the jealousy of local clergy, the Frankish Church changed greatly; in the tenth century what the pope said mattered a great deal north of the Alps. From the *entente* of the eighth century there emerged gradually the idea that it was for the pope to say what the Church's policy should be and that the individual bishops of the local Churches should not pervert it. A great instrument of standardization was being forged. It was there in principle when Pepin used his power as a Frankish king to reform his countrymen's Church and did so on lines which brought it into step with Rome on questions of ritual and discipline, and further away from Celtic influences.

The balance of advantage and disadvantage long tipped to and fro, the boundaries of the effective powers of the popes ebbing and flowing. Significantly, it was after a further sub-division of the Carolingian heritage so that the crown of Italy was separated from Lotharingia that Nicholas I pressed most successfully the papal claims. A century before, a famous forgery, the 'Donation of Constantine', purported to show that Constantine had given to the Bishop of Rome the former dominion exercised by the empire in Italy; Nicholas addressed kings and emperors as if this theory ran everywhere in the West. He wrote to them, it was said, 'as though he were lord of the world', reminding them that he could appoint and depose. He used the doctrine of papal primacy against the emperor of the East, too, in support of the Patriarch of Constantinople. This was a peak of pretension which the papacy could not long sustain in practice, for it was soon clear that force at Rome would decide who should enjoy the imperial power the pope claimed to confer. Nicholas' successor, revealingly, was the first pope to be murdered. None the less, the ninth century laid down precedents, even if they could not yet be consistently followed.

Especially in the collapse of papal authority in the tenth century, when the throne became the prey of Italian factions whose struggles were occasionally cut across by the interventions of the Ottonians, the day-to-day work of safeguarding Christian interests could only be in the hands of the bishops of the local Churches. They had to respect the powers that were. Seeking the cooperation and help of the secular rulers, they often moved into positions in which they were all but indistinguishable from royal servants. They were under the thumbs of their secular rulers just as, often, the parish priest was under the thumb of the local lord – and had to share his ecclesiastical proceeds in consequence. This humiliating dependency was later to lead to some of the sharpest papal interventions in the local Churches.

The bishops did much good, too; in particular, they encouraged missionaries. This had a political side to it. In the eighth century the Rule of St Benedict was well-established in England. A great Anglo-Saxon missionary movement, whose outstanding figures were St Willibrord in Frisia and St Boniface in Germany, followed. Largely independent of the east Frankish bishops, the Anglo-Saxons asserted the supremacy of Rome; their converts tended therefore to look directly to the throne of St Peter for religious authority. Many made pilgrimages to Rome. This papal emphasis died away in the later phases of evangelizing the East, or, rather, became less conspicuous because of the direct work of the German emperors and their bishops. Missions were combined with conquest and new bishoprics were organized as governmental devices.

Another great creative movement, that of reform in the tenth century, owed something to the episcopate but nothing to the papacy. It was a monastic movement which enjoyed the support of some rulers. Its essence was the renewal of monastic ideals; a few noblemen founded new houses which were intended to recall a degenerate monasticism to its origins. Most of them were in the old central Carolingian lands, running down from Belgium to Switzerland, west into Burgundy and east into Franconia, the area from which the reform impulse radiated outwards. At the end of the tenth century it began to enlist the support of princes and emperors. Their patronage in the end led to fear of lay dabbling in the affairs of the Church but it made possible the recovery of the papacy from a narrowly Italian and dynastic nullity.

The most celebrated of these foundations was the Burgundian abbey of Cluny, founded in 910. For nearly two and a half centuries it was the heart of reform in the Church. Its monks followed a revision of the Benedictine rule and evolved something quite new, a religious order resting not simply on a uniform way of life, but on a centrally disciplined organization. The Benedictine monasteries had all been independent communities, but the new Cluniac houses were all subordinate to the abbot of Cluny itself; he was the general of an army of (eventually) thousands of monks who only entered their own monasteries after a period of training at the mother house. At the height of its power, in the middle of the twelfth century, more than three hundred monasteries throughout the West – and even some in Palestine – looked for direction to Cluny, whose abbey contained the greatest church in western Christendom after St Peter's at Rome.

This is to look too far ahead for the present. Even in its early days, though, Cluniac monasticism was disseminating new practices and ideas throughout the Church. This takes us beyond questions of ecclesiastical structure and law, though it is not easy to speak with certainty of all aspects of Christian life in the early Middle Ages. Religious history is especially liable to be falsified by records which sometimes make it very difficult to see spiritual dimensions beyond the bureaucracy. They make it clear, though, that the Church was unchallenged, unique, and that it pervaded the whole fabric of society. It had something like a monopoly of culture. The classical heritage had been terribly damaged and narrowed by the barbarian invasions and the intransigent other-worldliness of early Christianity: 'What has Athens to do with Jerusalem?' Tertullian had asked, but this intransigence had subsided. By the tenth century what had been preserved of the classical past had been preserved by churchmen, above all by the Benedictines and the copiers of the palace schools who transmitted not only the Bible but Latin compilations of Greek learning. Through their version of Pliny and Boethius a slender line connected early medieval Europe to Aristotle and Euclid.

In the second half of the tenth century a school of copying and illumination flowered at Winchester under the influence of currents from continental Europe. The newly founded Benedictine abbey – the New Minster – received a charter which is the oldest manuscript made for the house and contains this miniature showing King Edgar offering to Christ the charter he has granted to the monks.

Literacy was virtually coterminous with the clergy. The Romans had been able to post their laws on boards in public places, confident that enough literate people existed to read them; far into the Middle Ages, even kings were normally illiterate. The clergy controlled virtually all access to such writing as there was. In a world without universities, only a court or church school offered the chance of letters beyond what might be offered, exceptionally, by an individual cleric-tutor. The effect of this on all the arts and intellectual activity was profound; culture was not just related to religion but took its rise only in the setting of overriding religious assumptions. The slogan 'art for art's sake' could never have made less sense than in the early Middle Ages. History, philosophy, theology, illumination, all played their part in sustaining a sacramental culture, but, however narrowed it might be, the legacy they transmitted, in so far as it was not Jewish, was classical.

In danger of dizziness on such peaks of cultural generalisation, it is salutary to remember that we can know very little directly about what must be regarded both theologically and statistically much more important than this and, indeed, as the most important of all the activities of the Church. This is the day-to-day business of exhorting, teaching, marrying, baptizing, shriving and praying, the whole religious life of the secular clergy and laity which centred about the provision of the major sacraments. The Church was in these centuries deploying powers which often cannot have been distinguished clearly by the faithful from those of magic. It used them to drill a barbaric world into civilization. It was enormously successful and yet we have almost no direct information about the process except at its most dramatic moments, when a spectacular conversion or baptism reveals by the very fact of being recorded that we are in the presence of the untypical.

Of the social and economic reality of the Church we know much more. The clergy and their dependants were numerous and the Church controlled much of society's wealth. The Church was a great landowner. The revenues which supported its work came from its land and a monastery or chapter of canons might have very large estates. The roots of the Church were firmly sunk in the economy of the day and to begin with that implied something very primitive indeed.

Difficult though it is to measure exactly, there are many symptoms of economic regression in the West at the end of antiquity. Not everyone felt the setback equally. The most developed economic sectors went under most completely. Barter replaced money and a money economy emerged again only slowly. The Merovingians began to coin silver, but for a long time there was not much coin – particularly coin of small denominations – in circulation. Spices disappeared from ordinary diet; wine became a costly luxury; most people ate and drank bread and porridge, beer and water. Scribes turned to parchment, which could be obtained locally, rather than papyrus, now hard to get; this turned out to be an advantage, for minuscule was possible on parchment, and had not been on papyrus, which required large, uneconomical strokes, but none the less it reflects difficulties within the old Mediterranean economy. Though recession often confirmed the self-sufficiency of the individual estate, it ruined the towns. The universe of trade also disintegrated from time to time because of war. Contact was maintained with Byzantium and further Asia, but the western Mediterranean's commercial activity dwindled during the seventh and eighth centuries as the Arabs seized the North African coast. Later, thanks again to the Arabs, it was partly revived (one sign was a brisk trade in slaves, many of whom came from eastern Europe, from the Slav peoples who thus gave their name to a whole category of forced labour). In the north, too, there was a certain amount of exchange with the Scandinavians, who were great traders. But this did not matter to most Europeans, for whom life rested on agriculture.

Subsistence was for a long time to be almost all that they could hope for. That it was the main concern of the early medieval economy is one of the few safe generalizations about it. Animal manure or the breaking of new and more fertile ground were the only ways of improving a return on seed and labour which was by modern standards derisory. Only centuries of laborious husbandry could change this. The animals who lived with the stunted and scurvy-ridden human tenants of a poverty-stricken landscape were themselves undernourished and undersized, yet for fat, the luckier peasant depended upon the pig, or, in the south, on oil. Only with the introduction in the tenth century of plants yielding food of higher protein content did the energy return from the soil begin to improve. There were some technological innovations, notably the diffusion of mills and the adoption of a better plough, but when production rose it did so for the most part because new land was brought into cultivation. And there was much to exploit. Most of France and Germany and England was still covered with forest and waste.

The economic relapse at the end of antiquity left behind few areas where towns thrived. The main exception was Italy, where some commercial relations with the outside world always persisted. Elsewhere, towns did not begin much to expand again until after 1100; even then, it would be a long time before western Europe contained a city comparable with the great centres of the classical Islamic and Asian civilizations. Almost universally in the West the self-sufficient agricultural estate was for centuries the rule. It fed and maintained a population probably smaller than that of the ancient world in the same area, though even approximate figures are almost impossible to establish. At any rate, there is no evidence of more than a very slow growth of population until the eleventh century. The population of western Europe may then have stood at about forty million – fewer than live in the United Kingdom today.

In this world, possession of land or access to it was the supreme determinant of the social order. Somehow, slowly, but logically, the great men of western society, while continuing to be the warriors they had always been in barbarian societies, became landowners too. With the dignitaries of the Church and their kings, they were the ruling class. From the possession of land came not only revenue by rent and taxation, but jurisdiction and labour service, too. Landowners were the lords, and gradually their

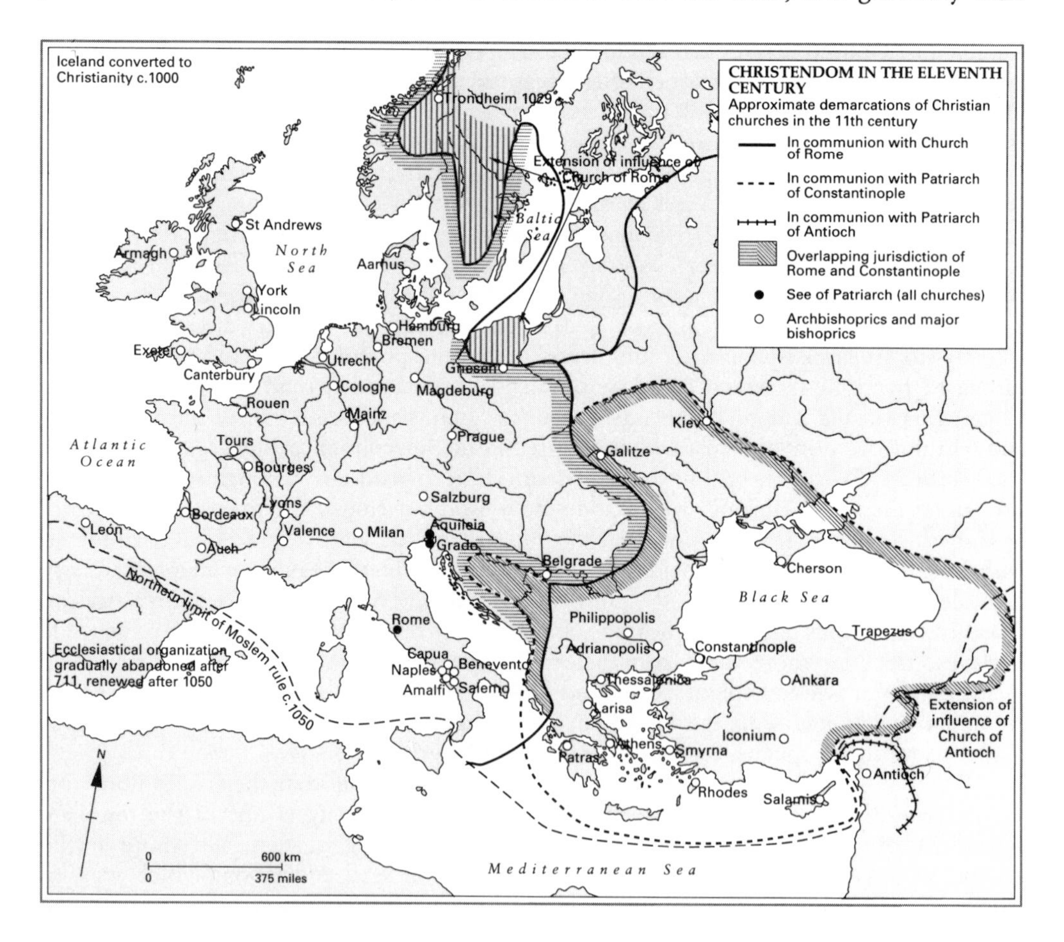

hereditary status was to loom larger and their practical prowess and skill as warriors was to be less emphasized (though in theory it long persisted) as the thing that made them noble.

The lands of some of these men were granted to them by a king or great prince. In return they were expected to repay the favour by turning out when required to do him military service. Moreover, administration had to be decentralized after imperial times; barbarian kings did not have the bureaucratic and literate resources to rule directly over great areas. Thus the grant of exploitable economic goods in return for specific obligations of service was very common, and this idea was what lay at the heart of what later historians, looking back at the European Middle Ages, called feudalism.

Many tributaries flowed into this stream. Both Roman and Germanic custom favoured the elaboration of such an idea. It helped, too, that in the later days of the empire, or in the troubled times of Merovingian Gaul, it had become common for men to 'commend' themselves to a great lord for protection; in return for his protection they offered him a special loyalty and service. This was a usage easily assimilated to the practices of Germanic society. Under the Carolingians, the practice began of 'vassals' of the king doing him homage; that is to say, they acknowledged with distinctive ceremonies, often public, their special responsibilities of service to him. He was their llord; they were his men. The old loyalties of the blood-brotherhood of the warrior-companions of the barbarian chief began to blend with notions of commendation in a new moral ideal of loyalty, faithfulness and reciprocal obligation. Vassals then bred vassals and one lord's man was another man's lord. A chain of obligation and personal service might stretch in theory from the king down through his great men and their retainers to the lowest of the free. And, of course, it might produce complicating and conflicting demands. A king could be another king's vassal in respect of some of his lands. Below the free were the slaves, more numerous perhaps in southern Europe than in the north and everywhere showing a tendency to evolve marginally upwards in status to that of the serf – the unfree man, born tied to the soil of his manor, but nevertheless, not quite without rights of any kind.

Some people later spoke as if the relationship of lord and man could explain the whole of medieval society. This was never so. Though much of the land of Europe was divided into fiefs – the *feuda* from which 'feudalism' takes its name – which were holdings bearing obligation to a lord, there were always important areas, especially in southern Europe, where the 'mix' of Germanic overlay and Roman background did not work out in the same way. Much of Italy, Spain and southern France was not 'feudal' in this sense. There were also always some freeholders even in more 'feudal' lands, an important class of men, more numerous in some countries than others, who owed no service for their lands but owned them outright.

For the most part, nevertheless, contractual obligations based on land set the tone of medieval civilization. Corporations, like men, might be lords or vassals; a tenant might do homage to the abbot of a monastery (or the abbess of a nunnery) for the manor he held of its estates, and a king might have a cathedral chapter or a community of monks as one of his vassals. There was much room for complexity and ambiguity in the feudal order. But the central fact of an exchange of obligations between superior and inferior ran through the whole structure and does more than anything else to make it intelligible to modern eyes. Lord and man were bound to one another reciprocally: 'Serfs, obey your temporal lords with fear and trembling; lords, treat your serfs according to justice and equity' was a French cleric's injunction which concisely summarized a principle in a specific case. On this rationalization rested a society of growing complexity which it long proved able to interpret and sustain.

It also justified the extraction from the peasant of the wherewithal to maintain the warrior and build his castle. From this grew the aristocracies of Europe. The military function of the system which supported them long remained paramount. Even when personal service in the field was not required, that of the vassal's fighting-men (and later of his money to pay fighting-men) would be. Of the military skills, that which was most esteemed (because it was the most effective) was that of fighting in armour on horse-back. At some point in the seventh or eighth century the stirrup was adopted; from that time the armoured horseman had it for the most part his own way on the battlefield until the coming of weapons which could master him. From this technical superiority emerged the knightly class of professional cavalrymen, maintained by the lord either directly or by a grant of a manor to feed them and their horses. They were the source of the warrior aristocracy of the Middle Ages and of European values for centuries to come. Yet for a long time, the boundaries of this class were ill-defined and movement into (and out of) it was common.

Political realities often militated against theory. In the intricate web of vassalage, a king might have less control over his own vassals than they over theirs. The great lord, whether lay magnate or local bishop, must always have loomed larger and more important in the life of the ordinary man than the remote and probably never-seen king or prince. In the tenth and eleventh centuries there are everywhere examples of kings obviously under pressure from great men. The country where this seemed to present the least trouble was Anglo-Saxon England, whose monarchical tradition was the strongest of any. But pressure was not always effective against even a weak king if he were shrewd. He had, after all, other vassals, and if wise he would not antagonize all of them at once. Furthermore, his office was unique. The anointing of the Church confirmed its sacred, charismatic authority. Kings were set apart in the eyes of most men by the special pomp and ceremony which surrounded them and which played as important a part in medieval government as does bureaucratic, paper in ours. If in addition a king had the advantage of large domains of his own, then he stood an excellent chance of having his way.

Not always in the technical and legal sense, but in common sense, kings and great magnates were the only men who enjoyed much freedom in early medieval society. Yet even they led lives cramped and confined by the absence of much that we take for granted. There was nothing much to do, after all, except pray, fight, hunt and run your estate; there were no professions for men to enter, except that of the Church, and small possibility of innovation in the style or content of daily life. Women's choices were even more restricted, and so they were for men as one went further down the social scale. Only with the gradual revival of trade and urban life as the economy expanded was this to change. Obviously, dividing lines are of almost no value in such matters, but it is not really until after 1100 that important economic expansion begins, and only then that we have the sense of moving out of a society still semi-barbarous, whose pretensions to civilization are sometimes negligible.

6
India

Though accompanied and advised by scholars and savants, Alexander the Great had only hazy ideas of what he would find in India; he seems to have thought the Indus part of the Nile and that beyond it lay more of Ethiopia. A fair amount had long been known by the Greeks about the Indian north-west, the seat of the Persian satrapy of Gandhara. But beyond that all was darkness. So far as political geography is concerned, the obscurity has remained; the relations between and, for that matter, the nature of the states of the Ganges valley at the time of Alexander's invasion are still hard to get at. A kingdom of Magadha, based on the lower river and exercising some sort of hegemony over the rest of the valley, had been the most important political unit in the subcontinent for two centuries or more, but not much is known about its institutions or history. Indian sources say nothing of Alexander's arrival in India and as the great conqueror never penetrated beyond the Punjab we can learn from Greek accounts of his day only of his disruption of the petty kingdoms of the north-west, not about the heartland of Indian power.

Under the Seleucids more reliable information became available in the West about what lay beyond the Punjab. This new knowledge roughly coincides with the rise of a new Indian power, the Maurya empire, and here the India of historical record really begins. One of our informants is a Greek ambassador, Megasthenes, sent to India by the Seleucid king in about 300 BC. Fragments of his account of what he saw were preserved long enough for later writers to quote him at length. As he travelled as far as Bengal and Orissa and was respected both as a diplomat and as a scholar, he met and interrogated many Indians. Some later writers found him a credulous and unreliable reporter; they dwelt upon his tales of men who subsisted on odours instead of food and drink, of others who were cyclopean or whose feet were so large that they used them to shelter from the sun, of pygmies and men without mouths. Such tales were, of course, nonsense. But they were not necessarily without foundation. They may well represent only the highly developed awareness shown by Aryan Indians of the physical differences which marked them off from neighbours or remote acquaintances from central Asia or the jungles of Burma. Some of these must have looked very strange indeed, and some of their behaviour was, no doubt, also very strange in Indian eyes. Others among these tales may dimly reflect the curious ascetic practices of Indian religion which have never ceased to impress outsiders and usually improve in the telling. Such tales need not discredit the teller, therefore they do not mean that other things he reports must be wholly untrue. They may even have a positive value if they suggest something of the way in which Megasthenes' Indian informants saw the outside world.

He describes the India of a great ruler, Chandragupta, founder of the Maurya line. Something is known about him from other sources. The ancients believed that he had been inspired to conquest by having as a youth seen Alexander the Great during his invasion of India. However this may be, Chandragupta usurped the Magadha throne in 321 BC and on the ruins of that kingdom built a state which encompassed not only the two great valleys of the Indus and Ganges, but most of Afghanistan (taken from the Seleucids) and Baluchistan. His capital was at Patna, where Chandragupta inhabited a magnificent palace. It was made of wood; archaeology still cannot help us much at this stage of Indian history. From Megasthenes' account it might be inferred that Chandragupta exercised a sort of monarchical presidency but Indian sources seem to reveal a bureaucratic state, or at least something that aspired to be one. What it was like in practice is hard to see. It had been built from political units formed in earlier times, many of which had been republican or popular in organization, and many of these were connected to the emperor through great men who were his officers; some of these, nominally subjects, must often have been very independent in practice.

About the empire's inhabitants, too, Megasthenes is informative. Besides providing a long list of different peoples, he distinguished two religious traditions (one

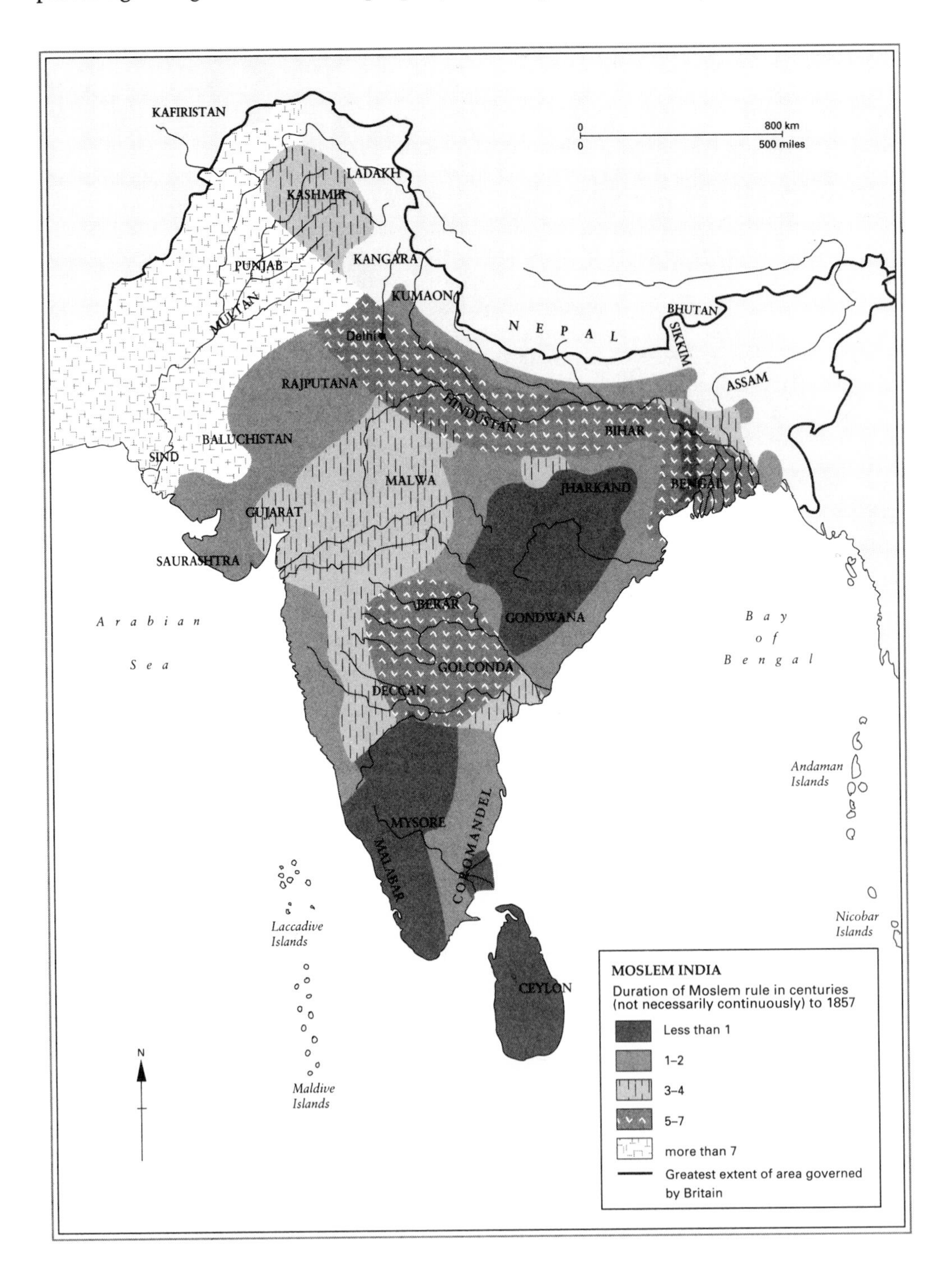

the Brahmanical and the other apparently Buddhist), mentioned the rice-eating habits of Indians and their abstention from wine except for ritual purposes, said much about the domestication of elephants, and remarked on the fact (surprising to Greek eyes) that in India there were no slaves. He was wrong, but excusably so. Though Indians were not bought and sold in absolute servitude, there were those bound to labour for their masters and legally incapable of removal. Megasthenes also reported that the king diverted himself by hunting, which was done from raised platforms, or from the backs of elephants – much as tigers still are shot today.

Chandragupta is said to have spent his last days in retirement with Jains, ritually starving himself to death in a retreat near Mysore. His son and successor turned the expansive course of empire already shown by his father to the south. Maurya power began to penetrate the dense rain-forests east of Patna, and to push down the eastern coast. Finally, under the third Maurya, the conquest of Orissa gave the empire control of the land and sea routes to the south and the subcontinent acquired a measure of political unity not matched in extent for over two thousand years. The conqueror who achieved this was Asoka, the ruler under whom a documented history of India at last begins to be possible.

From Asoka's era survive many inscriptions bearing decrees and injunctions to his subjects. The use of this means of propagating official messages and the individual style of the inscriptions both suggest Persian and Hellenistic influence, and India under the Mauryas was certainly more continually in touch with the civilizations to the West than ever before. At Kandahar, Asoka left inscriptions in both Greek and Aramaic.

Such evidence reveals a government capable of much more than that sketched by Megasthenes. A royal council ruled over a society based on caste. There was a royal army and a bureaucracy; as elsewhere the coming of literacy was an epoch in government as well as in culture. There seems also to have been a large secret police, or internal intelligence service. Besides raising taxes and maintaining communication and irrigation services, this machine, under Asoka, undertook the promotion of an official ideology. Asoka had himself been converted to Buddhism early in his reign. Unlike Constantine's conversion, his did not precede but followed a battle whose cost in suffering appalled Asoka. Be that as it may, the result of his conversion was the abandonment of the pattern of conquest which had marked Asoka's career until then. Perhaps this is why he felt no temptation to campaign outside the subcontinent – a limitation which, however, he shared with most Indian rulers, who never aspired to rule over barbarians and one which, of course, was only evident when he had completed the conquest of India.

The most remarkable consequence of Asoka's Buddhism has usually been thought to be expressed in the recommendations he made to his subjects in the rock-inscriptions and pillars dating from this part of his reign (roughly after 260 BC). They really amounted to a complete new social philosophy. Asoka's precepts have the overall name of *Dhamma*, a variant of a Sanskrit word meaning 'Universal Law', and their novelty has led to much anarchronistic admiration of Asoka's modernity by Indian politicians of this century. Asoka's ideas are, none the less, striking. He enjoined respect for the dignity of all men, and, above all, religious toleration and non-violence. His precepts were general rather than precise and they were not laws. But their central themes are unmistakable and they were intended to provide principles of action. While Asoka's own bent and thinking undoubtedly made such ideas agreeable to him, they suggest less a wish to advance the ideas of Buddhism (this is something Asoka did in other ways) than a wish to allay differences; they look very much like a device of government for a huge, heterogeneous and religiously divided empire. Asoka was seeking to establish some focus for a measure of political and social unity spanning all India, which would be based on men's interests as well as upon force and spying. 'All men', said one of his inscriptions, 'are my children.'

This may also explain his pride in what might be called his 'social services', which sometimes took forms appropriate to the climate 'on the roads I have had banyan trees planted', he proclaimed, 'which will give shade to beasts and men'. The value of this apparently simple device would have been readily apparent to those who toiled and travelled in the great Indian plains. Almost incidentally, improvements also smoothed the path of trade, but like the wells he dug and the rest-houses he set up at nine-mile intervals, the banyan trees were an expression of *Dhamma*. Yet *Dhamma* does not appear to have succeeded, for we hear of sectarian struggles and the resentment of priests.

Asoka did better in promoting simple Buddhist evangelization. His reign brought the first great expansion of Buddhism, which had prospered, but had remained hitherto confined to north-eastern India. Now Asoka sent missionaries to Burma who did well; in Ceylon others did better still, and from his day the island was predominantly Buddhist. Those sent, more optimistically, to Macedonia and Egypt were less successful though Buddhist teaching left its mark on some of the philosophies of the Hellenistic world and some Greeks were converted.

The vitality of Buddhism under Asoka may in part explain signs of reaction in the Brahmanical religion. It has been suggested that a new popularization of certain cults which dates from about this time may have been a conscious brahman response to challenge. Notably, the third and second centuries BC brought a new prominence to the cults of two of the most popular avatars of Vishnu. One is the proteiform Krishna, whose legend offers vast possibilities of psychological identification to the worshipper, and the other Rama, the embodunent of the benevolent king, good husband and son, a family god. It was in the second century BC, too, that the two great Indian epics, the *Mahabharata* and the *Ramayana*, began to take their final form. The first of these was extended by a long passage which is now the most famous work of Indian literature and its greatest poem, the *Bhagavad Gita*, or 'Song of the lord'. It was to become the central testament of Hinduism, weaving around the figure of Vishnu/Krishna the ethical doctrine of duty in the performance of the obligations laid upon one by membership of one's class (*dharma*) and the recommendation that works of devotion, however meritorious, might be less efficacious than love of Krishna as a means to release into eternal happiness.

These were important facts for the future of Hinduism, but were to develop fully only over a period which ran on far past the crumbling of the Mauryan empire, which began soon after Asoka's death. Such a disappearance is so dramatically impressive – and the Mauryan empire had been so remarkable a thing – that, though we are tempted to look for some special explanation, yet perhaps there is only a cumulative one. In all ancient empires except the Chinese, the demands made on government eventually outgrew the technical resources available to meet them: when this happened, they broke up. The Mauryas had done great things. They conscripted labour to exploit large areas of wasteland, thereby both feeding a growing population and increasing the tax base of the empire. They undertook great irrigation works which survived them for centuries. Trade prospered under Maurya rule, if we may judge from the way northern pottery spread throughout India in the third century BC. They kept up a huge army and a diplomacy which ranged as far afield as Epirus. The cost, however, was great. The government and army were parasitical upon an agricultural economy which could not be limitlessly expanded. There was a limit to what it could pay for. Nor, though bureaucracy seems at this distance to have been centralized in principle, was it likely to have been very effective, let alone flawless. Without a system of control and recruitment to render it independent of society, it fell at one end into the control of the favourites of the monarch on whom all else depended and at the other into the gift of local élites who knew how to seize and retain power.

One political weakness was rooted deep in pre-Maurya times. Indian society had already sunk its anchors in the family and the institutions of caste. Here, in social institutions rather than in a dynasty or an abstract notion of a continuing state (let alone a nation), was the focus of Indian loyalties. When an Indian empire began to crumble under economic, external or technical pressures, it had no unthinking popular support to fall back upon. This is a striking indication of the lack of success of Asoka's attempts to provide ideological integument for his empire. What is more, India's social institutions, and especially caste, in its elaborated forms, imposed economic costs. Where functions were inalterably allocated by birth, economic aptitude was held back. So was ambition. India had a social system which was bound to cramp the possibilities of economic growth.

The assassination of the last Maurya was followed by a Ganges dynasty of Brahmanical origin and thereafter the story of India for five hundred years is once more one of political disunity. References in Chinese sources become available from the end of the second century BC, but it cannot be said that they have made agreement between scholars about what was happening in India any easier, even the chronology is still largely conjectural. Only the general processes stand out.

The most important of these is a new succession of invasions of India from the historic north-western routes. First came Bactrians, descendants of the Greeks left behind by Alexander's empire on the upper Oxus, where by 239 BC they had formed an independent kingdom standing between India and Seleucid Persia. Our knowledge of this mysterious realm is largely drawn from its coins and has grave gaps in it, but it is known that a hundred years later the Bactrians were pushing into the Indus valley. They were the foremost in a current which was to flow for four centuries. A complex series of movements was in train whose origins lay deep in the nomadic societies of Asia. Among those who followed the Indo-Greeks of Bactria and established themselves at different times in the Punjab were Parthians and Scythians. One Scythian king, according to legend, received St Thomas the apostle at this court.

One important people came all the way from the borders of China and left behind them the memory of another big Indian empire, stretching from Benares beyond the mountains to the caravan routes of the steppes. These were the Kushanas. Historians still argue about how they are related to other nomadic peoples, but two things about them seem clear enough. The first is that they (or their rulers) were both enthusiastically Buddhist and also patronized some Hindu sects. The second was that their political interests were focused in central Asia, where their greatest king died fighting.

The Kushana period brought strong foreign influences once more into Indian culture, often from the West, as the Hellenistic flavour of its sculpture, particularly of the Buddha, shows. It marks an epoch in another way, for the depicting of the Buddha was something of an innovation in Kushan times. The Kushanas carried it very far and the Greek models gradually gave way to the forms of Buddha familiar today. This was one expression of a new complicating and developing of Buddhist religion. One thing which was happening was that Buddhism was being popularized and materialized; Buddha was turning into a god. But this was only one among many charges. Millenarianism, more emotional expressions of religion and more sophisticated philosophical systems were all interplaying with one another. To distinguish Hindu or Buddhist 'orthodoxy' in this is somewhat artificial.

In the end the Kushanas succumbed to a greater power. Bactria and the Kabul valley were taken by Artaxerxes early in the third century AD. Soon after, another Sassanid king took the Kushana capital of Peshawar – and such statements make it easy to feel impatient with the narrative they provide. Contemplating them, the reader may well feel with Voltaire: 'What is it to me if one king replaces another on the banks of the Oxus and Jakartes?' It is like the fratricidal struggles of Frankish kings, or of the Anglo-Saxon kingdoms of the Heptarchy, on a slightly larger scale. It is indeed difficult to see much significance in this ebb and flow beyond its registration of two great constants of Indian history, the importance of the north-western frontier as a cultural conduit and the digestive power of Hindu civilization. None of the invading peoples could in the end resist the assimilative power India always showed. New rulers were before long ruling Hindu kingdoms (whose roots went back possibly beyond Maurya times to political units of the fourth and fifth centuries BC), and adopting Indian ways.

Invaders never penetrated far to the south. After the Maurya break-up, the Deccan long remained separate and under its own Dravidian rulers. Its cultural distinction persists even today. Though Aryan influence was stronger there after the Maurya era and Hinduism and Buddhism were never to disappear, the south was not again truly integrated politically with the north until the coming of the British Raj.

India's most famous building and the supreme achievement of Islamic architects in the sub–continent, the Taj Mahal, built by Shah Jehan as a tomb for his favourite wife.

In this confusing period not all India's contacts with outsiders were violent. Trade with Roman merchants grew so visibly that Pliny blamed it (wrongly) for draining gold out of the empire. We have little hard information, it is true, except about the arrival of embassies from India to negotiate over trade but the remark suggests that one feature of India's trade with the West was already established; what Mediterranean markets sought were luxuries which only India could supply and there was little they could offer in return except bullion. This pattern held until the nineteenth century. There are also other interesting signs of intercontinental contacts arising from trade. The sea is a uniter of the cultures of trading communities; Tamil words for commodities turn up in Greek and Indians from the south had traded with Egypt since Hellenistic times. Later, Roman merchants lived in southern ports where Tamil kings kept Roman bodyguards. Finally, it seems likely that whatever the truth may be about the holy apostle Thomas, Christianity appeared in India first in the western trading ports, possibly as early as the first century AD.

Political unity did not appear again even in the north until hundreds of years had passed. A new Ganges valley state, the Gupta empire, was then the legatee of five centuries of confusion. Its centre was at Patna, where a dynasty of Gupta emperors established itself. The first of these, another Chandra Gupta, began to reign in 320, and within a hundred years north India was once more for a time united and relieved of external pressure and incursion. It was not so big an empire as Asoka's, but the Guptas preserved theirs longer. For some two centuries north India enjoyed under them a sort of Antonine age, later to be imagined with nostalgia, India's classical period.

The Gupta age brought the first great consolidation of an Indian art. From the earlier times little has survived before the perfection of stone-carving under the Mauryas. The columns which are its major monuments were the culmination of a native tradition of stonework. For a long time stone-carving and building still showed traces of styles evolved in an age of wood construction, but techniques were well advanced before the arrival of Greek influence, once thought to be the origin of Indian stone sculpture. What the Greeks brought were new artistic motifs and techniques from the West. If we are to judge by what survives, the major deployment of these influences was found in Buddhist sculpture until well into the Christian era. But before the Gupta era, a rich and indigenous tradition of Hindu sculpture too, had been established and from this time India's artistic life is mature and self-sustaining. In Gupta times there began to be built the great numbers of stone temples (as distinct from excavated and embellished caves) which are the great glories both of Indian art and architecture before the Moslem era.

Gupta civilization was also remarkable for its literary achievement. Again, the roots are deep. The standardization and systematization of Sanskrit grammar just before Maurya times opened the path to a literature which could be shared by the élite of the whole subcontinent. Sanskrit was a tie uniting north and south in spite of their cultural differences. The great epics were given their classical form in Sanskrit (though they were also available in translations into local languages) and in it wrote the greatest of Indian poets, Kalidasa. He was also a dramatist, and in the Gupta era there emerges from the shadowy past the Indian theatre whose traditions have been maintained and carried into the popular Indian film of the twentieth century.

Intellectually, too, the Gupta era was a great one. It was in the fifth century that Indian arithmeticians invented the decimal system. A layman can perhaps glimpse the importance of this more readily than he can that of the Indian philosophical resurgence of the same period. The resurgence was not confined to religious thought, but what can be gathered from it about general attitudes or the direction of culture seems highly debatable. In a literary text such as the *Kama Sutra* a western observer may be most struck by the prominence given in it to the acquisition of techniques whose use, however stimulating to the individual, can at most have absorbed only a small fraction of the interest and time of a tiny élite. A negative point is perhaps safest: neither the emphasis on *dharma* of the Brahmanical tradition, nor the ascetic severities of some Indian teachers, nor the frank acceptance of sensual pleasure suggested by many texts beside the *Kama Sutra* have anything in common with the striving, militant puritanism so strong in both the Christian and Islamic traditions. Indian civilization moved to very different rhythms from those further west; here, perhaps, lay its deepest strength and the explanation of its powers of resistance to alien cultures.

In the Gupta era Indian civilization came to its mature, classical form. Chronology derived from politics is a hindrance here; important developments flow across the boundaries of any arbitrary period. Nevertheless, in Gupta culture we can sense the presence of the fully evolved Hindu society. Its outstanding expression was a caste system which by then had come to overlay and complicate the original four-class division of Vedic society. Within castes which locked them into well-defined groups for marriage and, usually, to their occupations, most Indians lived a life close to the land. The cities were for the most part great markets or great centres of pilgrimage. Most Indians were, as they are now, peasants, whose lives were lived within the assumptions of a religious culture already set in its fundamental form in pre-Maurya times. Some of its later developments have been mentioned already; others run on past the Gupta period and will have to be discussed elsewhere. Of their vigour and power there can be no doubt; with centuries of further elaboration ahead, they were already expressed in Gupta times in a huge development of carving and sculpture which manifest the power of popular religion and take their place alongside the stupas and Buddhas of pre-Gupta times as an enduring feature of the Indian landscape. Paradoxically, India, largely because of its religious art, is a country where we have perhaps more evidence about the mind of the men of the past than we have about their material life. We may know little about the precise way in which Gupta taxation actually weighed on the peasant (though we can guess), but in the contemplation of the endless dance of the gods and demons, the forming and dissolving patterns of animals and symbols, we can touch a world still alive and to be found in the village shrines and juggernauts of our own day. In India as nowhere else, there is some chance of access to the life of the uncounted millions whose history should be recounted in such books as this, but which usually escapes us.

In the climax of Hindu civilization between Gupta times and the coming of Islam, the fertility of Indian religion, the soil of Indian culture, was hardly troubled by political events. One symptom was the appearance by 600 or thereabouts of an important new cult which quickly took a place it was never to lose in the Hindu worship, that of the mother-goddess Devi. Some have seen in her an expression of a new sexual emphasis which marked both Hinduism and Buddhism. Her cult was part of a general effervescence of religious life, lasting a couple of centuries or more, for a new popular emotionalism is associated with the cults of Shiva and Vishnu at about the same time. Dates are not very helpful here; we have to think of continuing change during the whole of the centuries corresponding to those of the early Christian era whose result was the final evolution of the old Brahmanical religion into Hinduism.

From it there emerged a spectrum of practice and belief offering something for all needs. It ran from the philosophic system of *Vedanta*, an abstract creed stressing the unreality of the factual and material and the desirability of winning disengagement from it in true knowledge of reality – *brahma* – to the crudities of the village shrines at which local deities were worshipped which had been easily assimilated to the cults of Shiva or Vishnu by the belief that these two leading deities might appear in more than one incarnation. Religious effervescence thus found expression antithetically in the simultaneous growth of image worship and the rise of new austerity. Animal sacrifice had never stopped. It was one of the things now endorsed by a new strictness of conservative religious practice. So was a new rigidity of attitudes towards women and their intensified subordination. The religious expression of this was an upsurge of child marriage and the practice called *suttee*, or self-immolation of widows on their husbands' funeral pyres.

Yet the richness of Indian culture is such that this coarsening of religion was accompanied also by the development to their highest pitch of the philosophical tradition of the Vedanta, the culmination of Vedic tradition, and the new development of *Mahayana* Buddhism, which asserted the divinity of the Buddha. The roots of the latter went back to early deviations from the Buddha's teaching on contemplation, purity and non-attachment. These deviations had favoured a more ritualistic and popular religious approach and also stressed a new interpretation of the Buddha's role. Instead of merely being understood as a teacher and an example, Buddha was now seen as the greatest of *bodhisattvas*, saviours who, entitled to the bliss of self-annihilation themselves, nevertheless rejected it to remain in the world and teach men the way to salvation.

To become a *bodhisattva* gradually became the aim of many Buddhists. In part, the efforts of a Buddhist council summoned by the Kushan ruler Kanishka had been directed towards re-integrating two tendencies in Buddhism which were increasingly divergent. This had not been successful. *Mahayana* Buddhism (the word means 'great vehicle') focused upon a Buddha who was effectively a divine saviour who might be worshipped and followed in faith, one manifestation of a great, single heavenly Buddha who begins to look somewhat like the undifferentiated soul behind all things found in Hinduism. The disciplines of austerity and contemplation Gautama had taught were now increasingly confined to a minority of orthodox Buddhists, the followers of *Mahayana* winning conversions among the masses. One sign of this was the proliferation in the first and second centuries AD of statues and representations of the Buddha, a practice hitherto restrained by the Buddha's prohibition of idol-worship. *Mahayana* Buddhism eventually replaced earlier forms in India, and spread also along the central Asian trade routes through Central Asia to China and Japan. The more orthodox tradition did better in south-eastern Asia and Indonesia.

Hinduism and Buddhism were thus both marked by changes which broadened their appeal. The Hindu religion prospered better, though there is a regional factor at work here; since Kushan times, the centre of Indian Buddhism had been the north-west, the region most exposed to the devastations of the Hun raiders. Hinduism prospered most in the south. Both the north-west and the south, of course, were zones where cultural currents intermingled most easily with those from the classical Mediterranean world, in the one across land and in the other by sea.

These changes provoke a sense of culmination and climax. They matured only shortly before Islam arrived in the subcontinent, but early enough for a philosophical outlook to have solidified which has marked India ever since and has shown astonishing invulnerability to competing views. At its heart was a vision of endless cycles of creation and reabsorption into the divine, a picture of the cosmos which predicated a cyclic and

Madras was long to be associated in Europe with its cottons. This painted hanging of the seventeenth century provides and early example of the Indian's vision of the European.

not a linear history. What difference this made to the way Indians have actually behaved – right down to the present day – is a huge subject, and almost impossible to grasp. It might be expected to lead to passivity and scepticism about the value of practical action, yet this is very debatable. Few Christians live lives logically wholly coherent with their beliefs and there is no reason to expect Hindus to be more consistent. The practical activity of sacrifice and propitiation in Indian temples survives still. Yet the direction of a whole culture may none the less be determined by the emphasis of its distinctive modes of thought and it is difficult not to feel that much of India's history has been determined by a world outlook which stressed the limits rather than the potential of human action.

For the background to Islam in India we have to return to 500 or so. From about that time, northern India was once again divided in obedience both to the centrifugal tendencies which afflicted early empires and to the appearance of a mysterious invasion of 'Hunas'. Were they perhaps Huns? Certainly they behaved like them, devastating much of the north-west, sweeping away many of the established ruling families. Across the mountains, in Afghanistan, they mortally wounded Buddhism, which had been strongly established there. In the subcontinent itself, this anarchic period did less fundamental damage. Though the northern plains had broken up again into warring kingdoms, Indian cities do not seem to have been much disturbed and peasant life recovers quickly from all but the worst blows. Indian warfare appears rapidly to have acquired important and effective conventional limits on its potential for destructiveness. The state of affairs over much of the north at this time seems in some ways rather like that of some European countries during the more anarchic periods of the Middle Ages, when feudal relationships more or less kept the peace between potentially competitive grandees but could not completely contain outbreaks of violence which were essentially about different forms of tribute.

Meanwhile, Islam had come to India. It did so first through Arab traders on the western coasts. Then, in 712 or thereabouts, Arab armies conquered Sind. They got no further, gradually settled down and ceased to trouble the Indian peoples. A period of calm followed which lasted until a Ghaznavid ruler broke deep into India early in the eleventh century with raids which were destructive, but again did not produce radical change. Indian religious life for another two centuries moved still to its own rhythms, the most striking changes being the decline of Buddhism and the rise of Tantrism, a semi-magical and superstitious growth of practices promising access to holiness by charms and ritual. Cults centred on popular festivals at temples also prospered, no doubt in the absence of a strong political focus in post Gupta times. Then came a new invasion of central Asians.

These invaders were Moslems and were drawn from the complex of Turkish peoples. Theirs was a different sort of Islamic onslaught from earlier ones, for they came to stay, not just to raid. They first established themselves in the Punjab in the eleventh century and then launched a second wave of invasions at the end of the twelfth century which led within a few decades to the establishment of Turkish sultans at Delhi who ruled the whole of the Ganges valley. Their empire was not monolithic. Hindu kingdoms survived within it on a tributary basis, as Christian kingdoms survived to be tributaries of Mongols in the West. The Moslem rulers, perhaps careful of their material interests, did not always support their co-religionists of the *Ulema* who sought to proselytize, and were willing to persecute (as the destruction of Hindu temples shows).

The heartland of the first Moslem empire in India was the Ganges Valley. The invaders rapidly overran Bengal and later established themselves on the west coast of India and the tableland of the Deccan. Further south they did not penetrate and Hindu society survived there largely unchanged. In any case, their rule was not to last long even in the north. In 1398 Timur Lang's army sacked Delhi after a devastating approach march which was made all the speedier, said one chronicler, because of the Mongols' desire to escape from the stench of decay arising from the piles of corpses they left in their wake. In the troubled waters after this disaster, generals and local potentates struck out for themselves and Islamic India fragmented again. None the less, Islam was by now established in the subcontinent, the greatest challenge yet seen to India's assimilative powers, for its active, prophetic, revelatory style was wholly antithetical both to Hinduism and to Buddhism (though Islam, too, was to be subtly changed by them).

New Sultans emerged at Delhi but long showed no power to restore the former Islamic empire. Only in the sixteenth century was it revived by a prince from outside, Babur of Kabul. On his father's side he descended from Timur and on his mother's from Chinghis, formidable advantages and a source of inspiration to a young man schooled in adversity. He quickly discovered he had to fight for his inheritance and there can have been few monarchs who, like Babur, conquered a city of the importance of Samarkand at the age of fourteen (albeit to lose it again almost at once). Even when legend and anecdote are separated, he remains, in spite of cruelty and duplicity, one of the most attractive figures among great rulers, munificent, hardy, courageous, intelligent and sensitive. He left a remarkable autobiography, written from notes made throughout his life, which was to be treasured by his descendants as a source of inspiration and guidance. It displays a ruler who did not think of himself as Mongol in culture, but Turkish in the tradition of those peoples long settled in the former eastern provinces of the Abbasid caliphate. His taste and culture were formed by the inheritance of the Timurid princes of Persia; his love of gardening and poetry came from that country and fitted easily into the setting of an Islamic India whose court cultures were already much influenced by Persian models. Babur was a bibliophile, another Timurid trait; it is reported that when he took Lahore he went at once to his defeated adversary's library to choose texts from it to send as gifts to his sons. He himself wrote, among other things, a forty-page account of his conquests in Hindustan, noting its customs and caste structure and, even more minutely, its wildlife and flowers.

This young prince was called in to India by Afghan chiefs, but had his own claims to make to the inheritance of the Timurid line in Hindustan. This was to prove the beginning of Moghul India; Moghul was the Persian word for Mongol, though it was not a word Babur applied to himself. Originally, those whose discontent and intrigue called him forward had only aroused in him the ambition of conquering the Punjab, but he was soon drawn further. In 1526 he took Delhi after the Sultan had fallen in battle. Soon Babur was subduing those who had invited him to come to India while at the same time conquering the infidel Hindu princes who had seized an opportunity to renew their own independence. The result was an empire which in 1530, the year of his death, stretched from Kabul to the borders of Bihar. Babur's body, significantly, was taken as he directed to Kabul, where it was buried, in his favourite garden with no roof over his tomb, in the place he had always thought of as home.

The reign of Babur's son, troubled by his own instability and inadequacy and by the presence of half-brothers anxious to exploit the Timurid tradition which, like the Frankish, prescribed the division of a royal inheritance, showed that the security and consolidation of Babur's realm could not be taken for granted. For five years of his reign, he was driven from Delhi, though he returned there to die in 1555. His heir, Akbar born during his father's distressed wanderings (but enjoying the advantages of a very auspicious horoscope and the absence of rival brothers), thus came to the throne as a boy. He inherited at first only a small part of his grandfather's domains, but was to build from them an empire recalling that of Asoka, winning the awed respect of Europeans, who called him 'the Great Moghul'.

Akbar had many kingly qualities. He was brave to the point of folly – his most obvious weakness was that he was headstrong – enjoying as a boy riding his own fighting elephants and preferring hunting and hawking to lessons (one consequence was that, uniquely in Babur's line he was almost illiterate). He once killed a tiger with his sword in single combat and was proud of his marksmanship with a gun (Babur had introduced firearms to the Moghul army). Yet he was also, like his predecessors, an admirer of learning and all things beautiful. He collected books and in his reign Moghul architecture and painting came to their peak, a department of court painters being maintained at his expense. Above all, he was statesmanlike in his handling of the problems posed by religious difference among his subjects.

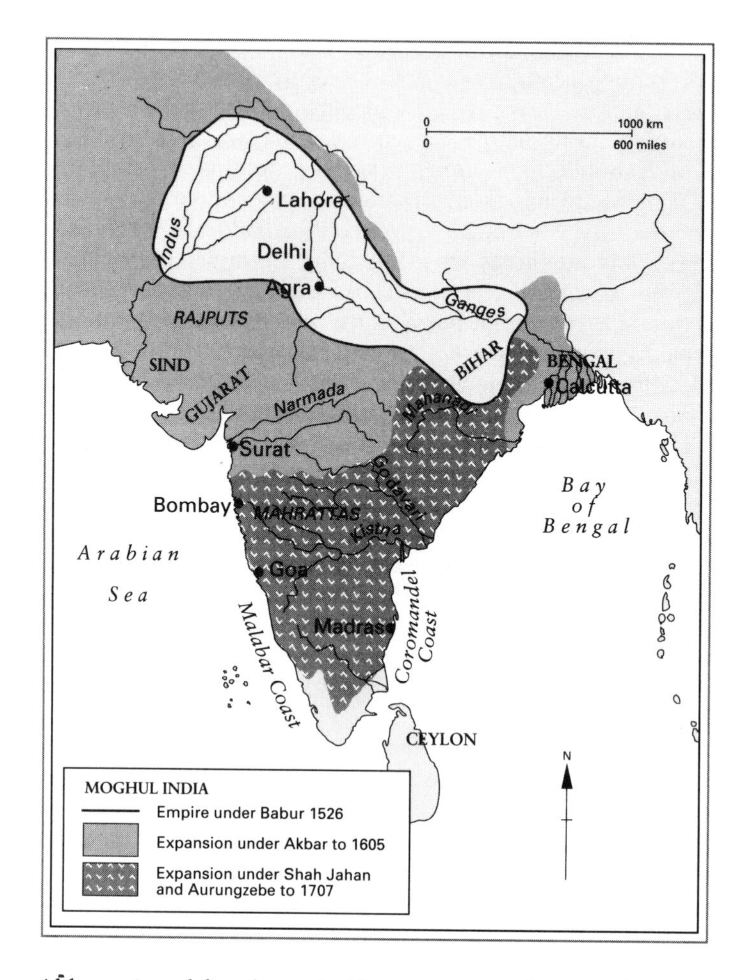

Akbar reigned for almost half a century, until 1605, thus just overlapping at each end the reign of his contemporary, Queen Elizabeth I of England. Among his first acts on reaching maturity was to marry a Rajput princess who was, of course, a Hindu. Marriage always played an important part in Akbar's diplomacy and strategy, and this lady (the mother of the next emperor) was the daughter of the greatest of the Rajput kings and therefore an important catch. None the less, something more than policy may be seen in the marriage. Akbar had already permitted the Hindu ladies of his harem to practise the rites of their own religion within it, an unprecedented act for a Moslem ruler. Before long, he abolished the poll-tax on non-Moslems; he was going to be the emperor of all religions, not a Moslem fanatic. Akbar even went on to listen to Christian teachers; he invited the Portuguese who had appeared on the west coast to send missionaries learned in their faith to his court and three Jesuits duly arrived there in 1580. They disputed vigorously with Moslem divines before the emperor and received many marks of his favour, though they were disappointed in their long-indulged hope of his conversion. He seems, in fact, to have been a man of genuine religious feeling and eclectic mind; he went so far as to try to institute a new religion of his own, a sort of mishmash of Zoroastrianism, Islam and Hinduism. It had little success except among prudent courtiers and offended some.

However this is interpreted, it is evident that the appeasement of non-Moslems would ease the problems of government in India. Babur's advice in his memoirs to conciliate defeated enemies pointed in this direction too, for Akbar launched himself on a career of conquest and added many new Hindu territories to his empire. He rebuilt the unity of northern India from Gujarat to Bengal and began the conquest of the Deccan. The empire was governed by a system of administration much of which lasted well into the era of the British Raj, though Akbar was less an innovator in government than the confirmer and establisher of institutions he inherited. Officials ruled in the emperor's name and at his pleasure, they had the primary function of providing soldiers as needed and raising the land tax, now reassessed on an empire-wide and more flexible system devised by a Hindu finance minister which seems to have had an almost unmatched success in that it actually led to increases in production which raised the standard of living in Hindustan. Among other reforms which were notable in intention if not in effect was the discouragement of *suttee.*

Above all, Akbar stabilized the regime. He was disappointed in his sons and quarrelled with them, yet the dynasty was solidly based when he died. There were revolts nevertheless. Some of them seem to have been encouraged by Moslem anger at Akbar's apparent falling-away from the faith. Even in the 'Turkish' era the sharpness of the religious distinction between Moslem and non-Moslem had somewhat softened as invaders settled down in their new country and took up Indian ways. One earlier sign of assimilation was the appearance of a new language, Urdu, the tongue of the camp. It was the *lingua franca* of rulers and ruled, with a Hindi structure and a Persian and Turkish vocabulary. Soon there were signs that the omnivorous power of Hinduism would perhaps even incorporate Islam; a new devotionalism in the fourteenth and fifteenth centuries had spread through popular hymns an abstract, almost monotheistic, cult, of a God whose name might be Rama or Allah, but who offered love, justice and mercy to all men. Correspondingly, some Moslems even before Akbar's reign had shown interest in and respect for Hindu ideas. There was some absorption of Hindu ritual practice. Soon it was noticeable that converts to Islam tended to revere the tombs of holy men: these became places of resort and pilgrimage which satisfied the scheme of a subordinate focus of devotion in a monotheistic religion and thus carried out the functions of the minor and local deities who had always found a place in Hinduism.

Another important development before the end of Akbar's reign was the consolidation of India's first direct relations with Atlantic Europe. Links with Mediterranean Europe may already have been made slightly easier by the coming of Islam; from the Levant to Delhi a common religion provided continuous, if distant, contact. European travellers had turned up from time to time in India and its rulers had been able to attract the occasional technical expert to their service, though they were few after the Ottoman conquests. But what was now about to happen was to go much further and would change India for ever. The Europeans who now arrived would be followed by others in increasing numbers and they would not go away.

The process had begun when a Portuguese admiral reached Malabar at the end of the fifteenth century. Within a few years his countrymen had installed themselves as traders – and behaved sometimes as pirates at Bombay and on the coast of Gujarat. Attempts to dislodge them failed in the troubled years following Babur's death and in the second half of the century the Portuguese moved round to found new posts in the Bay of Bengal. They made the running for Europeans in India for a long time. They were liable, none the less, to attract the hostility of good Moslems because they brought with them pictures and images of Christ, His mother and the saints, which smacked of idolatry. Protestants were to prove less irritating to religious feeling when they arrived. The British age in India was still a long way off, but with rare historical neatness the first British East India Company was founded on 31 December 1600, the last day of the sixteenth century. Three years later the Company's first emissary arrived at Akbar's court at Agra and by then Elizabeth I, who had given the merchants their charter of incorporation, was dead. Thus at the end of the reigns of two great rulers, came the first contact between two countries whose historical destinies were to be entwined so long and with such enormous effect for them both and for the world. At that moment no hint of such a future could have been sensed. The English then regarded trade in India as less interesting than that with other parts of Asia. The contrast between the two realms, too, is fascinating: Akbar's empire was one of the most powerful in the world, his court one of the most sumptuous and he and his successors ruled over a civilization more glorious and spectacular than anything India had known since the Guptas, while Queen Elizabeth's kingdom, barely a great power, even in European terms, was crippled by debt and contained fewer people than modern Calcutta. Akbar's successor was contemptuous of the presents sent to him by James I a few years later. Yet the future of India lay with the subjects of the queen.

The Moghul emperors continued in Babur's line in direct descent, though not without interruption, until the middle of the nineteenth century. After Akbar, so great was the dynasty's prestige that it became fashionable in India to claim Mongol descent. Only the three rulers who followed Akbar matter here, for it was under Jahangir and Shah Jahan that the empire grew to its greatest extent in the first half of the seventeenth century and under Aurungzebe that it began to decay in the second. The reign of Jahangir was not so glorious as his father's, but the empire survived his cruelty and alcoholism, a considerable test of its administrative structure. The religious toleration established by Akbar also survived intact. For all his faults, too, Jahangir was a notable promoter of the arts, above all of painting. During his reign there becomes visible for the first time the impact of European culture in Asia, through artistic motifs drawn from imported pictures and prints. One of these motifs was the halo or nimbus given to Christian saints and, in Byzantium, to emperors. After Jahangir all Moghul emperors were painted with it.

Shah Jahan began the piecemeal acquisition of the Deccan sultanates though he had little success in campaigns in the north-west and failed to drive the Persians from Kandahar. In domestic administration there was a weakening of the principle of religious toleration, though not sufficiently to place Hindus at a disadvantage in government service; administration remained multi-religious. Although the emperor decreed that all newly built Hindu temples should be pulled down, he patronized Hindu poets and musicians. At Agra, Shah Jahan maintained a lavish and exquisite court life. It was there, too, that he built the most celebrated and the best-known of all Islamic buildings, the Taj Mahal, a tomb for his favourite wife; it is the only possible rival to the mosque of Cordoba for the title of the most beautiful building in the world. She had died soon after Shah Jahan's accession and for over twenty years his builders were at work. It is the culmination of the work with arch and dome which is one of the most conspicuous Islamic legacies to Indian art and the greatest monument of Islam in India. The waning of Indian representational sculpture after the Islamic invasions had its compensations. Shah Jahan's court brought also to its culmination a great tradition of miniature painting.

Below the level of the court, the picture of Moghul India is far less attractive. Local officials had to raise more and more money to support not only the household expenses and campaigns of Shah Jahan but also the social and military élites who were essentially parasitic on the producing economy. Without regard for local need or natural disaster, a rapacious tax-gathering machine may at times have been taking from the peasant producer as much as half his income. Virtually none of this was productively invested. The flight of peasants from the land and rise of rural banditry is a telling symptom of the suffering and resistance these exactions provoked. Yet even Shah Jahan's demands probably did the empire less damage than the religious enthusiasm of his third son, Aurungzebe, who set aside three brothers and imprisoned his father to become emperor in 1658. He combined, disastrously, absolute power, distrust of his subordinates and a narrow religiosity. To have succeeded in reducing the expenses of his court is not much of an offsetting item in the account. New conquests were balanced by revolts against Moghul rule which were said to owe much to Aurungzebe's attempt to prohibit the Hindu religion and destroy its temples and to his restoration of the poll-tax on non-Moslems. The Hindu's advancement in the service of the state was less and less likely; conversion became necessary for success. A century of religious toleration was cancelled and one result was the alienation of many subjects' loyalties.

Among other results, this helped to make it impossible finally to conquer the Deccan, which has been termed the ulcer which ruined the Moghul empire. As under Asoka, North and South India could not be united. The Mahrattas, the hillmen who were the core of Hindu opposition, constituted themselves under an independent ruler in 1674. They allied with the remains of the Moslem armies of the Deccan sultans to resist the Moghul armies in a long struggle which threw up a heroic figure who has become something of a paladin in the eyes of modern Hindu nationalists. This was Shivagi, who built from fragments a Mahratta political identity which soon enabled him to exploit the tax-payer as ruthlessly as the Moghuls had done. Aurungzebe was continuously campaigning against the Mahrattas down to his death in 1707. There followed a grave crisis for the régime, for his three sons disputed the succession. The empire almost at once began to break up and a much more formidable legatee than the Hindu or local prince was waiting in the wings – the European.

Perhaps the negative responsibility for the eventual success of the Europeans in India is Akbar's, for he did not scotch the serpent in the egg. Shah Jahan, on the other hand, destroyed the Portuguese station on the Hooghly, though Christians were later tolerated at Agra. Strikingly, Moghul policy never seems to have envisaged the building of a navy, a weapon used formidably against the Mediterranean Europeans by the Ottomans. One consequence was already felt under Aurungzebe, when coastal shipping and even the pilgrim trade to Mecca were in danger from the Europeans. On land, the Europeans had been allowed to establish their toeholds and bridgeheads. After beating a Portuguese squadron, the English won their first west-coast trading concession early in the seventeenth century. Then, in 1639, on the Bay of Bengal and with the permission of the local ruler, they founded at Madras the first settlement of British India, Fort St George. The headstones over their graves in its little cemetery still commemorate the first English who lived and died in India, as would thousands more for over three centuries.

The English later fell foul of Aurungzebe, but got further stations at Bombay and Calcutta before the end of the century. Their ships had maintained the paramountcy in trade won from the Portuguese, but a new European rival was also in sight by 1700. A French East India Company had been founded in 1664 and soon established its own settlements.

A century of conflict lay ahead, but not only between the newcomers. Europeans were already having to make nice political choices because of the uncertainties aroused when Moghul power was no longer as strong as it once had been. Relations had to be opened with his opponents as well as with the emperor, as the English in Bombay discovered, looking on helplessly while a Mahratta squadron occupied one island in Bombay harbour and a Moghul admiral the one next to it. In 1677 an official sent back a significant warning to his employers in London: 'the times now require you to manage your general commerce with your sword in your hands.' By 1700 the English were well aware that much was at stake.

With that date we are into the era in which India is increasingly caught up in events not of her own making, the era of world history, in fact. Little things show it as well as great; in the sixteenth century the Portuguese had brought with them chilli, potatoes and tobacco from America. Indian diet and agriculture were already changing. Soon maize, pawpaws and pineapple were to follow. The story of Indian civilizations and rulers can be broken once this new connexion with the larger world is achieved. Yet it was not the coming of the European which ended the great period of Moghul empire; that was merely coincidental, though it was important that newcomers were there to reap the advantages. No Indian empire had ever been able to maintain itself for long. The diversity of the subcontinent and the failure of its rulers to find ways to tap indigenous popular loyalty are probably the main explanation. India remained a continent of exploiting ruling élites and producing peasants upon whom they battened. The 'states', if the term can be used, were only machinery for transferring resources from producers to parasites. The means by which they did this destroyed the incentive to save – to invest productively.

India was, by the end of the seventeenth century, ready for another set of conquerors. They were awaiting their cue, already on stage, but as yet playing hardly more than bit parts. Yet in the long run the European tide, too, would recede. Unlike early conquerors, though Europeans were to stay a long time, they were not to be overcome by India's assimilative power as their predecessors had been. They would go away defeated, but would not be swallowed. And when they went they would leave a deeper imprint than any of their predecessors because they would leave true state structures behind.

7
Imperial China

One explanation of the striking continuity and independence of Chinese civilization is obvious: China was remote, inaccessible to alien influence, far from sources of disturbance in other great civilizations. Empires came and went in both countries, but Islamic rule made more difference to India than any dynasty's rise or fall made to China. It was also endowed with an even greater capacity to assimilate alien influence, probably because the tradition of civilization rested on different foundations in each country. In India the great stabilizers were provided by religion and a caste system inseparable from it. In China it rested on the culture of an administrative élite which survived dynasties and empires and kept China on the same course.

One thing we owe to this élite is the maintenance of written records from very early times. Thanks to them, Chinese historical accounts provide an incomparable documentation, crammed with often reliable facts, though the selection of them was dominated by the assumptions of a minority, whose preoccupations they reflect. The Confucian scholars who kept up the historical records had a utilitarian and didactic aim: they wanted to provide a body of examples and data which would make easier the maintenance of traditional ways and values. Their histories emphasize continuity and the smooth flow of events. Given the needs of administration in so huge a country this is perfectly understandable; uniformity and regularity were clearly to be desired. Yet such a record leaves much out. It remains very difficult even in historical times – and much more difficult than in the classical Mediterranean world – to recover the concerns and life of the vast majority. Moreover, official history may well give a false impression both of the unchanging nature of Chinese administration and of the permeation of society by Confucian values. For a long time, the assumptions behind the Chinese administrative machine can only have been those of a minority, even if they came in the end to be shared by many Chinese and accepted, unthinkingly and even unknowingly, by most.

The official culture was extraordinarily self-sufficient. Such outside influences as played upon it did so with little effect and this remains impressive. The fundamental explanation, again, is geographic isolation. China was much further removed from the classical West than the Maurya and Gupta empires. She had little intercourse with it even indirectly, although until the beginning of the seventh century Persia, Byzantium and the Mediterranean depended upon Chinese silk and valued her porcelain. Always, too, China had complicated and close relations with the people of central Asia; yet, once unified, she had for many centuries on her borders no great states with whom relations had to be carried on. This isolation was, if anything, to increase as the centre of gravity of Western civilization moved west and north and as the Mediterranean was more and more cut off from East Asia first by the inheritors of the Hellenistic legacy (the last and most important of which was Sassanid Persia), and then by Islam.

China's history between the end of the period of Warring States and the beginning of the T'ang in 618 has a backbone of sorts in the waxing and waning of dynasties. Dates can be attached to these, but there is an element of the artificial, or at least a danger of being over-emphatic, in using them. It could take decades for a dynasty to make its power a reality over the whole empire and even longer to lose it. With this reservation, the dynastic reckoning can still be useful. It gives us major divisions of Chinese history down to this century which are called after the dynasties which reached their peaks during them. The first three which concern us are the Ch'in, the Han, and the Later Han.

The Ch'in ended the disunity of the period of Warring States. They came from a western state still looked upon by some as barbarous as late as the fourth century BC. Nevertheless, the Ch'in prospered, perhaps in part because of a radical reorganization carried out by a legalist-minded minister in about 356 BC; perhaps also because of their soldiers' use of a new long iron sword. After swallowing Szechwan, the Ch'in claimed the status of a kingdom in 325 BC. The climax of Ch'in success was the defeat of their last opponent in 221 BC and the unification of China for the first time in one empire under the dynasty which gives the country its name.

Principal Chinese Dynasties

SHANG ?1523–?1027
CHOU ?1027–?256
CH'IN 221–206
(having annihilated CHOU in 256 and other rival states afterwards)
FORMER HAN 206 BC–AD9
HSIN AD 9–23
LATER HAN 25–220
WEI 220–265 SHU 221–263 WU 222–280
WESTERN CHIN 265–316
SIXTEEN KINGDOMS 304–439 EASTERN CHIN 317–420
LIU SUNG 420–479
NORTHERN WEI 386–581 SOUTHERN CH'I 479–502
WESTERN WEI 535–557 EASTERN WEI 534–550 LIANG 502–557
NORTHERN CHOU 557–581 NORTHERN CH'I 550–577 CH'EN 557–589
SUI 581–618
T'ANG 618–907
FIVE DYNASTIES 906–960 TEN KINGDOMS 907–979
NORTHERN HAN 951–979
(reckoned as one of the Ten Kingdoms)
SUNG 960–1126
(the extreme north of China being rules by the LIAO 947–1125)
CHIN 1126–1234 **SOUTHERN SUNG** 1127–1279
YUAN 1279–1368
(having succeeded the CHIN in North China in 1234)
MING 1366–1644
CH'ING 1644–1912

This was a great achievement. China from this time may be considered the seat of a single, self-conscious civilization. There had been earlier signs that such an outcome was likely. Given the potential of their own Neolithic cultures, the stimuli of cultural diffusion and some migration from the north, the first shoots of civilization had appeared in several parts of China before 500 BC. By the end of the Warring States Period some of them showed marked similarities which offset the differences between them. The political unity achieved by Ch'in conquest over a century was in a sense the logical corollary of a cultural unification already well under way. Some have even claimed that a sense of Chinese nationality can be discerned before 221 BC; if so, it must have made conquest itself somewhat easier.

Fundamental administrative innovations by the Ch'in were to survive that dynasty's displacement after less than twenty years by the Han, who ruled for two hundred years (206 BC – AD 9), to be followed after a brief interlude by the almost equally creative Later Han dynasty (AD 25–220). Though they had their ups and downs, the Han emperors showed unprecedented strength. Their sway extended over almost the whole of modern China, including southern Manchuria and the south-eastern province of Yueh. The Later Han went on to create an empire as big as that of their Roman contemporaries. They faced an old threat from Mongolia and a great opportunity towards the south. They handled both with skill aided by the tactical superiority given their armies by the new crossbow. This weapon was probably invented soon after 200 BC and was both more powerful and more accurate than the bows of the barbarians, who did not for a long time have the ability to cast the bronze locks required. It was the last major achievement of Chinese military technology before the coming of gunpowder.

In Mongolia at the beginning of Han times lived the Hsiung-Nu, whom we have already met as the forerunners of the Huns. The Han emperors drove them north of the Gobi desert and then seized control of the caravan routes of central Asia, sending armies far west into Kashgaria in the first century BC. They even won tribute from the Kusharas, whose own power straddled the Pamirs. To the south, they occupied the coasts as far as the Gulf of Tonkin; Annam accepted their suzerainty and Indo-China has been regarded by Chinese statesmen as part of their proper sphere ever since. To the north-east they penetrated Korea. All this was the work of the later, or 'eastern', Han whose capital was at Loyang. From there they continued to press forward in Turkestan and raised tribute from the oases of central Asia. One general in AD 97 may have got as far as the Caspian.

Tentative diplomatic encounters with Rome in Han times suggest that expansion gave China much more contact with the rest of the world. Before the nineteenth century this was in the main by land, and besides the silk trade which linked her regularly with the Near East (caravans were leaving for the West with silk from about 100 BC), China also developed more elaborate exchanges with her nomadic neighbours. Sometimes this was within the fictional framework of tribute acknowledged in turn by gifts, sometimes within official monopolies which were the foundation of great merchant families. Nomadic contacts may explain one of the most astonishing works of Chinese art, the great series of bronze horses found in tombs at Wu-Wei. These were only one among many fine works of Han bronze-workers; they evidently broke more readily with tradition than the Han potters, who showed more antiquarian respect for past forms. At a different level, though, Han pottery provides some of the very few exploitations in art of the subject-matter of the daily life of most Chinese in the form of collections of tiny figures of peasant families and their livestock.

This was a brilliant culture, centred on a court with huge, rich palaces built in the main of timber – unhappily, for the result is that they have disappeared, like the bulk of the Han collections of paintings on silk. Much of this cultural capital was dissipated or destroyed during the fourth and fifth centuries, when the barbarians returned to the frontiers. Failing at last to provide China's defence from her own manpower, the Han emperors fell back on a policy tried elsewhere, that of bringing within the Wall some of the tribes who pressed on it from outside and then deploying them in its defence. This raised problems of relations between the newcomers and the native Chinese. The Han emperors could not prolong for ever their empire, and after four hundred years China once more dissolved into a congeries of kingdoms.

Some of these had barbarian dynasties, but in this crisis there is observable for the first time China's striking powers of cultural digestion. Gradually the barbarians were swallowed by Chinese society, losing their own identity and becoming only another kind of Chinese. The prestige which Chinese civilization enjoyed among the peoples of Central Asia was already very great. There was a disposition among the uncivilized to see China as the centre of the world, a cultural pinnacle, somewhat in the way in which the Germanic peoples of the West had seen Rome. One Tatar ruler actually imposed Chinese customs and dress on his people by decree in 500. The central Asian threat was not over; far from it, there appeared in Mongolia in the fifth century the first Mongol empire. Nonetheless, when the T'ang, a northern dynasty, came to receive the mandate of heaven in 618 China's essential unity was in no greater danger than it had been at any time in the preceding two or three centuries.

Political disunity and barbarian invasion had not damaged the foundations of Chinese civilization, which entered its classical phase under the T'ang. Among those foundations, the deepest continued to lie in kinship. Throughout historical times the clan retained its importance because it was the mobilized power of many linked families, enjoying common institutions of a religious and sometimes of an economic kind. The diffusion and ramification of family influence were all the easier because China did not have primogeniture; the paternal inheritance was usually divided at death. Over a social ocean in which families were the fish that mattered presided one Leviathan, the state. To it and to the family the Confucians looked for authority; those institutions were unchallenged by others, for in China there were no entities such as Church or communes which confused questions of right and government so fruitfully in Europe.

This celebrated bronze horse buried at Kansu in north-west China in the second century AD may have been meant to be a magical creature; it seems to be galloping through the air, momentarily startling the swallow brushed by its foot. Similar models were often made of real horses and were buried with their owners.

The state's essential characteristics were all in place by T'ang times. They were to last until this century and the attitudes they built up linger on still. In their making, the consolidating work of the Han had been especially important, but the office of the emperor, holder of the mandate of heaven, could be taken for granted even in Ch'in times.

The comings and goings of dynasties did not compromise the standing of the office since they could always be ascribed to the withdrawal of the heavenly mandate. The emperor's liturgical importance was, if anything, enhanced by the inauguration under the Han of a sacrifice only he could make. Yet his position also changed in a positive sense. Gradually, a ruler who was essentially a great feudal magnate, his power an extension of that of the family or the manor, was replaced by one who presided over a centralized and bureaucratic state.

This had begun a long way back. Already in Chou times a big effort was made to build canals for transport. Great competence in organization and large human resources were required for this and only a potent state could have deployed them. A few centuries later the first Ch'in emperor was able to link together the existing sections of the Great Wall in 1400 miles of continuous barrier against the barbarians (legendarily, his achievement cost a million lives and that story, too, is revealing of the way the empire was seen). His dynasty went on to standardize weights and measures and impose a degree of disarmament on its subjects while itself putting in the field perhaps a million soldiers. The Han were able to impose a monopoly of coining and standardized the currency. Under them, too, entry to the civil service by competitive examination began; though it was to fade out again, not to be resumed until T'ang times, it was very important. Territorial expansion had required more administrators. The resulting bureaucracy was to survive many periods of disunion (a proof of its vigour) and remained to the end one of the most striking and characteristic institutions of imperial China. It was probably the key to China's successful emergence from the era when collapsing dynasties were followed by competing petty and local states which broke up the unity already achieved. It linked China together by an ideology as well as by administration. The civil servants were trained and examined in the Confucian classics; under the Han, legalism finally lost its grip after a lively ideological struggle. Literacy and political culture were thus wedded in China as nowhere else.

The scholars had been deeply offended by the Ch'in. Though a few of them had been favoured and gave the dynasty advice, there had been a nasty moment in 213 BC when the emperor turned on scholars who had criticized the despotic and militaristic character of his régime. Books were burned and only 'useful' works on divination, medicine or agriculture were spared; more than four hundred scholars perished. What was really at stake is not clear; some historians have seen this attack as an offensive aimed at 'feudal' tendencies opposed to Ch'in centralization. If so, it was far from the end of the confusion of cultural and political struggle with which China has gone on mystifying foreign observers even in this century. Whatever the sources of this policy, the Han took a different tack and sought to conciliate the intellectuals.

This led first to the formalization of Confucian doctrine into what quickly became an orthodoxy. The canonical texts were established soon after 200 BC. True, Han Confucianism was a syncretic matter; it had absorbed much of legalism. But the important fact was that Confucianism had been the absorbing force. Its ethical precepts remained dominant in the philosophy which formed China's future rulers. In AD 58 sacrifices to Confucius were ordered in all government schools. Eventually, under the T'ang, administrative posts were confirmed to those trained in this orthodoxy. For over a thousand years it provided China's governors with a set of moral principles and a literary culture doggedly acquired by rote-learning. The examinations they underwent were designed to show which candidates had the best grasp of the moral tradition discernible in the classical texts as well as to test mechanical abilities and the capacity to excel under pressure. It made them one of the most effective and ideologically homogeneous bureaucracies the world has ever seen and also offered great rewards to those who successfully made the values of Confucian orthodoxy their own.

The official class was in principle distinguished from the rest of society only by educational qualification (the possession of a degree, as it were). Most civil servants came from the land-owning gentry, but they were set apart from them. Their office once achieved by success in the test of examination, they enjoyed a status only lower than that of the imperial family, and great material and social privileges besides. Officials' duties were general rather than specific, but they had two crucial annual tasks, the compilation of the census returns and the land registers on which Chinese taxation rested. Their other main work was judicial and supervisory, for local affairs were very much left to local gentlemen acting under the oversight of about two thousand or so district magistrates from the official class. Each of these lived in an official compound, the *yamen*, with his clerks, runners and household staff about him.

The gentry undertook a wide range of quasi-governmental and public service activities which were both an obligation of the privileged class and also an insurance of much of its income. Local justice, education, public works were all part of this. The gentry also often organized military forces to meet local emergencies and even collected the taxes, from which it might recoup its own expenses. Over the whole of these arrangements and the official class itself, there watched a state apparatus of control, checking and reporting on a bureaucracy bigger by far than that of the Roman empire and at its greatest extent ruling a much large area.

This structure had huge conservative power. Crisis only threatened legal authority, rarely the social order. The permeation of governmental practice by the agreed ideals of Confucian society was rendered almost complete by the examination system. Moreover, though it was very hard for anyone not assured of some wealth to support himself during the long studies necessary for the examination – writing in the traditional literary forms itself took years to master – the principle of competition ensured that a continuing search for talent was not quite confined to the wealthier and established gentry families; China was a meritocracy in which learning always provided some social mobility. From time to time there were corruption and examples of the buying of places, but such signs of decline usually appear towards the end of a dynastic period. For the most part, the imperial officials showed remarkable independence of their background. They were not supposed to act on such assumptions of obligation to family and connexion as characterized the public servants drawn from the eighteenth-century English gentry. The civil servants were the emperor's men; they were not allowed to own land in the province where they served, serve in their own provinces, or have relatives in the same branch of government. They were not the representatives of a class, but a selection from it, an independently recruited élite, renewed and promoted by competition. They made the state a reality.

Imperial China was thus not an aristocratic polity; political power did not pass by descent within a group of noble families, though noble birth was socially important. Only in the small closed circle of the court was hereditary access to office possible, and there it was a matter of prestige, titles and standing, rather than of power. To the imperial counsellors who had risen through the official hierarchy to its highest levels and had become more than officials, the only rivals of importance were the court eunuchs. These creatures were often trusted with great authority by the emperors because, by definition, they could not found families. They were thus the only political force escaping the restraints of the official world.

Clearly, in the Chinese state there was little sense of the European distinction between government and society. Official, scholar and gentleman were usually the same man, combining many roles which in Europe were increasingly to be divided between governmental specialists and the informal authorities of society. He combined them, too, within the framework of an ideology which was much more obviously central to society than any to be found elsewhere than perhaps in Islam. The preservation of Confucian values was not a light matter, nor satisfiable by lip-service. The bureaucracy maintained those values by exercising a moral supremacy somewhat like that long exercised by the clergy in the West – and in China there was no Church to rival the state. The ideas which inspired it were profoundly conservative; the predominant administrative task was seen to be the maintenance of the established order; the aim of Chinese government was to oversee, conserve and consolidate, and occasionally to innovate in practical matters by carrying out large public works. Its overriding goals were regularity and the maintenance of common standards in a huge and diverse empire, where many district magistrates were divided from the people in their charge even by language. In achieving its conservative aims, the bureaucracy was spectacularly successful and its ethos survived intact across all the crises of the dynasties.

Below the Confucian orthodoxy of the officials and gentry, it is true, other creeds were important. Even some who were high in the social scale turned to Taoism or Buddhism. The latter was to be very successful after the Han collapse, when disunity gave it an opportunity to penetrate China. In its Mahayana variety it posed more of a threat

An eighteenth century album contains this painting of an examination, a traditional procedure long unchanged in China. Magistrates are here shown at their re-examination in the Confucian classics. The emperor himself (a T'ang monarch of the eighth century) assesses the results, on which promotion or dismissal was based.

to China than any other ideological force before Christianity, for, unlike Confucianism, it posited the rejection of worldly values. It was never to be eradicated altogether, in spite of persecution under the T'ang; attacks on it were, in any case, probably mounted for financial rather than ideological reasons. Unlike the persecuting Roman empire, the Chinese state was more interested in property than in the correction of individual religious eccentricity. Under the fiercest of the persecuting emperors (who is said to have been a Taoist) over four thousand monasteries were dissolved, and over a quarter of a million monks and nuns dispersed from them. Nevertheless, in spite of such material damage to Buddhism, Confucianism had to come to terms with it. No other foreign religion influenced China's rulers so strongly until Marxism in the twentieth century; even some emperors were Buddhists.

Well before this, Taoism had developed into a mystical cult (borrowing something from Buddhism in the process) appealing both to those who sought personal immortality and to those who felt the appeal of a quietistic movement as an outlet from the growing complexity of Chinese life. As such it would have enduring significance. Its recognition of the subjectivity of human thought gives it an appearance of humility which some people in different cultures with more self-confident intellectual changes find attractive today. Such religious and philosophical notions, important as they were, touched the life of the peasant directly only a little more than Confucianism, except in debased forms. A prey to the insecurities of war and famine, his outlet lay in magic or superstition. What little can be discerned of his life suggests that it was often intolerable, sometimes terrible. A significant symptom is the appearance under the Han of peasant rebellion, a phenomenon which became a major theme of Chinese history, punctuating it almost as rhythmically as the passing of dynasties. Oppressed by officials acting either on behalf of an imperial government seeking taxes for its campaigns abroad or in their own interest as grain speculators, the peasants turned to secret societies, another recurrent theme. Their revolts often took religious forms. A millenarian, Manichaean strain has run through Chinese revolution, bursting out in many guises, but always positing a world dualistically divided into good and evil, the righteous and the demons. Sometimes this threatened the social fabric, but the peasants were rarely successful for long.

Chinese society therefore changed very slowly. In spite of some important cultural and administrative innovations, the lives of most Chinese were for centuries little altered in style or appearance. The comings and goings of the dynasties were accounted for by the notion of the mandate of heaven and although great intellectual achievements were possible, China's civilization already seemed self-contained, self-sufficient, stable to the point of immobility. No innovation compromised the fundamentals of a society more closely woven into a particular governmental structure than anything in the West. This structure proved quite competent to contain such changes as did take place and to regulate them so as not to disturb the traditional forms.

One visibly important change was a continuing growth of commerce and towns which made it easier to replace labour service by taxation. Such new resources could be tapped by government both to rule larger areas effectively and to provide a series of great material monuments. They had already permitted the Ch'in to complete the Great Wall, which later dynasties were further to extend, sometimes rebuilding portions of it. It still astonishes the observer and far outranks the walls of Hadrian and Antoninus. Just before the inauguration of the T'ang, too, at the other end of this historical epoch, a great system of canals was completed which linked the Yangtze valley with the Yellow River valley to the north, and Hangchow to the south. Millions of labourers were employed on this and on other great irrigation schemes. Such works are comparable in scale with the Pyramids and surpass the great cathedrals of medieval Europe. They imposed equally heavy social costs, too, and there were revolts against conscription for building and guard duties.

It was a state with great potential and a civilization with impressive achievements already to its credit which entered its mature phase in 618. For the next thousand years, as for the previous eight hundred, its formal development can be linked to the comings and goings of the dynasties which provide a chronological structure (T'ang, 618–907; Sung, 960–1126; Mongol ascendancy, 1234–1368; Ming, 1368–1644; Manchu or Ch'ing, 1644–1912). Many historical themes overrun these divisions. One is the history of population. There was an important shift of the demographic centre of gravity towards the south during the T'ang period; henceforth most Chinese were to live in the Yangtze valley rather than the old Yellow River plain. The devastation of the southern forests and exploitation of new lands to grow rice fed them, but new crops became available, too. Together they made possible an overall growth of population which accelerated under the Mongols and the Ming. Estimates have been made that a population of perhaps eighty million in the fourteenth century more than doubled in the next two hundred years, so that in 1600 there were about 160 million subjects of the empire. This was a huge number, given populations elsewhere, but there was still great increase to come.

The weight of this fact is great. Apart from the enormous importance it gives to China in world population history, it puts in perspective the great manifestations of Chinese culture and imperial power, which rested on the huge mass of desperately poor peasants utterly unconcerned with such things. For the most part their lives were confined to their villages; only a few could hope to escape from this, or can have envisaged doing so. Most could have dreamed only of obtaining the precarious, but best, security available to them: the possession of a little land. Yet this became more and more difficult as numbers grew and, gradually, all available land was occupied. It was farmed more and more intensively in smaller and smaller plots. The one way out of the trap of famine was rebellion. At a certain level of intensity and success this might win support from the gentry and officials, whether from prudence or sympathy. When that happened, the end of a dynasty was probably approaching, for Confucian principles taught that, although rebellion was wrong if a true king reigned, a government which provoked rebellion and could not control it ought to be replaced for it was *ipso facto* illegitimate. At the *very* end of this road lay the success of a twentieth-century Chinese revolution based on the peasants.

For many centuries, population pressure, a major fact of modern China's history, made itself felt to the authorities only in indirect and obscured ways, when, for instance, famine or hunger drove men to rebellion. A much more obvious threat came from the outside. Essentially the problem was rather like that of Rome, an overlong frontier beyond which lay barbarians. T'ang influence over them was weakened when central Asia succumbed to Islam. Like their Roman predecessors, too, the later T'ang emperors found that reliance on soldiers could be dangerous. There were hundreds of military rebellions by local warlords under the T'ang and any rebellion's success, even if short-lived, had a multiplier effect, tending to disrupt administration and damage the irrigation arrangements on which food (and therefore internal peace) depended. A regime thought of as a possible ally by Byzantium, which had sent armies to fight the Arabs and received ambassadors from Haroun-al-Raschid, was a great world power. In the end, though, unable to police their frontier effectively, the T'ang went under in the tenth century, and China collapsed again into political chaos. The Sung who emerged from it had to face an even graver external threat, the Mongols, and were in due course swallowed after the barbarian dynasty which had evicted them from north China had itself been engulfed by the warriors of Chinghis Khan.

During the whole of this time, the continuity and recuperative power of the bureaucracy and the fundamental institutions of society kept China going. After each dynastic change, the inheritors of power, even if from outside, turned to the relatively smaller number of officials (an estimate for the eighteenth century gives less than 30,000 civil and military officers actually in post). They thus drew into the service of each new government the unchanging values of the Confucian system, which were strengthened, if narrowed, by disaster. Only a small number of especially crucial matters were ever expected to be the reserved province of the imperial government. Confucian teaching supported this distinction of spheres of action and made it easy for the dynasty to change without compromising the fundamental values and structure of society. A new dynasty would have to turn to the officials for its administration and to the gentry for most of its officials who, in their turn, could get some things done only on the gentry's terms.

Recurrent disunity did not prevent China's rulers, sages and craftsmen from bringing Chinese civilization to its peak in the thousand years after the T'ang inauguration. Some have placed the classical age as early as the seventh and eighth centuries, under the T'ang themselves, while others discern it under the Sung. Such judgements usually rest on the art-form considered, but even Sung artistic achievement was in any case a culmination of development begun under the T'ang, between whom and the Han much more of a break in style is apparent. It was in fact the last important break in the continuity of Chinese art until this century.

T'ang culture reflects the stimulus of contacts with the outside world, but especially with central Asia, unprecedentedly close under this dynasty. The capital was then at Ch'ang-an, in Shensi, a western province. Its name means 'long-lasting peace' and to this city at the end of the Silk Road came Persians, Arabs and central Asians who made it one of the most cosmopolitan cities in the world. It contained Nestorian churches, Zoroastrian temples, Moslem mosques, and was probably the most splendid and luxurious capital of its day, as the objects which remain to us show. Many of them reflect Chinese recognition of styles other than their own – the imitation of Iranian silverware, for example – while the flavour of a trading entrepôt is preserved in the pottery figures of horsemen and loaded camels which reveal the life of central Asia swirling in the streets of Ch'ang-an. These figures were often finished with the new polychromatic glazes achieved by T'ang potters; their style was imitated as for away as Japan and Mesopotamia. The presence of the court was as important in stimulating such craftsmanship as the visits of merchants from abroad, and from tomb-paintings something of the life of the court aristocracy can be seen. The men relax in hunting, attended by central Asian retainers; the women, vacuous in expression, are luxuriously dressed and, if servants, elaborately equipped with fans, cosmetic boxes, back-scratchers and other paraphernalia of the boudoir. Great ladies, too, favour central Asian fashions borrowed from their domestic staff.

The history of women, though, is the history of one of those other Chinas always obscured by the bias of the documentation towards the official culture. We hear little of them, even in literature, except in sad little poems and love stories. Yet presumably they must have made up about half of the population, or perhaps slightly less, for in hard times girl babies were exposed by poor families to die. That fact, perhaps, characterizes women's place in China until very recent times even better than the more familiar and superficially striking practice of foot-binding, which produced grotesque deformations and could leave a high-born lady almost incapable of walking. Another China still all but excluded from the historical evidence by the nature of the established tradition was that of the peasants. They become shadowly visible only as numbers in the census returns and as eruptions of revolt; after the Han pottery figures, there is little in Chinese art to reveal them, and certainly nothing to match the uninterrupted recording of the life of the common man in the fields which runs from medieval European illumination, through the vernacular literature to the Romantics, and into the peasant subjects of the early Impressionists.

Official culture also excluded the tenth or so of the Chinese population who lived in the cities, some of which grew as time passed to become the biggest in the world. Ch'ang-an, when the T'ang capital, is said to have had two million inhabitants. No eighteenth-century European city was as big as contemporary Canton or Peking, which were even larger. Such huge cities housed societies of growing complexity. Their development fostered a new commercial world; the first Chinese paper money was issued in 650. Prosperity created new demands, among other things for a literature which did not confine itself to the classical models and in colloquial style far less demanding than the elaborate classical Chinese. City life thus gradually secreted a literate alternative to the official culture, and because it was literate, it is the first part of unofficial China to which we have some access. Such popular demand could be satisfied because of two enormously important inventions, that of paper in the second century BC, and of printing before AD 700. This derived from the taking of rubbed impressions from stone under the Han. Printing from wood blocks was taking place under the T'ang and movable type appeared in the eleventh century AD. Soon after this large numbers of books were published in China, long before they appeared anywhere else. In the cities, too, flourished popular poetry and music which abandoned the classical tradition.

The culture of Ch'ang-an never recovered from its disruption by rebellion in 756, only two years after the foundation of an Imperial Academy of Letters (about nine hundred years before any similar academy in Europe). After this the dynasty was in decline. The Sung ascendancy produced more great pottery; the earlier, northern phase of Sung history was marked by work still in the coloured, patterned tradition, while

On the site of an ancient shrine, the wooden Temple of Heaven of Peking as it now appears is really the outcome of an eighteenth-century repair programme. Its basic design, though, goes back to pre-Han times, an example of the conservatism of Chinese architecture exemplified in the ground plan of the 'forbidden city' of Peking itself which followed the gridiron street-plan prescribed by the ancient books of architecture. Properly speaking, the 'forbidden city' was the Palace compound only; the Temple of Heaven lay outside this central area.

southern Sung craftsmen came to favour monochromatic, simple products. Significantly, they attached themselves to another tradition: that of the forms evolved by the great bronze-casters of earlier China. For all the beauty of its ceramics, though, Sung is more notable for some of the highest achievements of Chinese painting, their subject-matter being, above all, landscape. As a phase of Chinese development, though, the Sung era is more remarkable still for a dramatic improvement in the economy.

In part this can be attributed to technological innovation – gunpowder, movable type and the sternpost all can be traced to the Sung era – but it was also linked to the exploitation of technology already long available. Technological innovation may indeed have been as much a symptom as a cause of a surge in economic activity between the tenth and thirteenth centuries which appears to have brought most Chinese a real rise in incomes in spite of continuing population growth. For once in the pre-modern world economic growth seems for a long period to have outstripped demographic trends. One change making this possible was certainly the discovery and adoption of a rice variety which permitted two crops a year to be taken from well-irrigated land and one from hilly ground only watered in the spring. The evidence of rising production in a different sector of the economy has been dramatically distilled into one scholar's calculation that within a few years of the battle of Hastings, China was producing nearly as much iron as the whole of Europe six centuries later. Textile production, too, underwent dramatic development (notably through the adoption of water-driven spinning machinery) and it is possible to speak of Sung 'industrialization' as a recognisable phenomenon.

It is not easy (the evidence is still disputed) to say why this remarkable burst of growth took place. Undoubtedly there was a real input to the economy by public – that is, governmental – investment in public works, above all, communications. Prolonged periods of freedom from foreign invasion and domestic disorder also must have helped, though the second benefit may be explained as much by economic growth as the other way round. The main explanation, though, seems likely to be an expansion in markets and the rise of a money economy which owed something to factors already mentioned, but which rested fundamentally on the great expansion in agricultural productivity. So long as this kept ahead of population increase, all was well. Capital became available to utilize more labour, and to tap technology by investment in machines. Real incomes rose.

It is even harder to say why, after temporary and local regression at the end of the Sung era, and the resumption of economic growth, this intensive growth, which made possible rising consumption by greater numbers, came to an end. Nonetheless, it did, and was not resumed. Instead, average real incomes in China stabilized for something like five centuries, as production merely kept pace with population growth. (After that time, incomes began to fall, and continued to do so to a point at which the early twentieth-century Chinese peasant could be described as a man standing neck-deep in water, whom even ripples could drown.) But the economic relapse after Sung times is not the only factor to be taken into account in explaining why China did not go on to produce a dynamic, progressive society. In spite of printing, the mass of Chinese remained illiterate down to the present century. China's great cities, for all their growth and commercial vitality, produced neither the freedom and immunities which sheltered men and ideas in Europe, nor the cultural and intellectual life which in the end revolutionized European civilization, nor effective questioning of the established order. Even in technology, where China achieved so much so soon, there is a similar strange gap between intellectual fertility and revolutionary change. The Chinese could invent (they had a far more efficient wheelbarrow than other civilizations), but once Chou times were over, it was the use of new land and the introduction of new crops rather than technical change which raised production. Other examples of a low rate of innovation are even more striking. Chinese sailors already had the magnetic compass in Sung times, but though naval expeditions were sent to Indonesia, the Persian Gulf, Aden and East Africa in the fifteenth century, their aim was to impress those places with the power of the Ming, not to accumulate information and experience for further voyages of exploration and discovery. Masterpieces had been cast in bronze in the second millennium BC and the Chinese knew how to cast iron fifteen hundred years before Europeans, yet much of the engineering potential of this metallurgical tradition was unexplored even when iron production rose so strikingly. What he called 'a sort of black stone' was burnt in China when Marco Polo was there; it was coal, but there was to be no Chinese steam engine.

This list could be much lengthened. Perhaps the explanation lies in the very success of Chinese civilization in pursuit of a different goal, the assurance of continuity and the prevention of fundamental change. Neither officialdom nor the social system favoured the innovator. Moreover, pride in the Confucian tradition and the confidence generated by great wealth and remoteness made it difficult to learn from the outside. This was not because the Chinese were intolerant. Jews, Nestorian Christians, Zoroastrian Persians, and Arab Moslems long practised their own religion freely, and the last even made some converts, creating an enduring Islamic minority. Contacts with the West multiplied, too, later under Mongol rule. But what has been called a 'neo-Confucian' movement was by then already manifesting tendencies of defensive hostility, and formal tolerance had never led to much receptivity in Chinese culture.

Invasion by the Mongols showed China's continuing seductive power over its conquerors. By the end of the thirteenth century, all China had been overrun by them – and this may have cost the country something like thirty million lives, or well over a quarter of its whole population in 1200 – but the centre of gravity of the Mongol empire had moved from the steppes to Peking, Kubilai's capital. This grandson of Chinghis was the last of the Great Khans and after his time Mongol China can be considered Chinese, not Mongol; Kubilai adopted a dynastic life in 1271 and the remainder of the Mongol era is recorded as that of the Yunan dynasty. China changed Mongols more than Mongols changed China, and the result was the magnificence reported by the amazed Marco Polo. Kubilai made a break with the old conservatism of the steppes, the distrust of civilization and its works, and his followers slowly succumbed to Chinese culture in spite of their initial distrust of the scholar officials. They were, after all, a tiny minority of rulers in an ocean of Chinese subjects; they needed collaborators to survive. Kubilai spent nearly all his life in China, though his knowledge of Chinese was poor.

But the relationship of Mongol and Chinese was long ambiguous. Like the British in nineteenth-century India who set up social conventions to prevent their assimilation by their subjects, so the Mongols sought by positive prohibition to keep themselves apart. Chinese were forbidden to learn the Mongol language or marry Mongols. They were not allowed to carry arms. Foreigners, rather than Chinese, were employed in administration where possible, a device paralleled in the western khanates of the Mongol empire: Marco Polo was for three years an official of the Great Khan; a Nestorian presided over the imperial bureau of astronomy; Moslems from Transoxiana administered Yunan. For some years, too, the traditional examination system was suspended. Some of the persistent Chinese hostility to the Mongols may be explained by such facts, especially in the south. When Mongol rule in China collapsed, seventy years after Kubilai's death, there appeared an, if possible, even more exaggerated respect for tradition and a renewed distrust of foreigners among the Chinese ruling class.

The short-run achievement of the Mongols was, nonetheless, very impressive. It was most obvious in the re-establishment of China's unity and the realization of its potential as a great military and diplomatic power. The conquest of the Sung south was not easy, but once it was achieved (in 1279) Kubilai's resources were more than doubled (they included an important fleet) and he began to rebuild the Chinese sphere of influence in Asia. Only in Japan was he totally unsuccessful. In the south, Vietnam was invaded (Hanoi was three times captured) and after Kubilai's death Burma was occupied for a time. These conquests were not, it is true, to prove long-lasting and they resulted in tribute rather than prolonged occupation. In Java, too, success was qualified; a landing was made there and the capital of the island taken in 1292, but it proved impossible to hold. There was also further development of the maritime trade with India, Arabia and the Persian Gulf which had been begun under the Sung.

Since it failed to survive, the Mongol régime cannot be considered wholly successful, but this does not take us far. Much that was positive was done in just over a century. Foreign trade flourished as never before. Marco Polo reports that the poor of Peking were fed by the largesse of the Great Khan, and it was a big city. A modern eye finds something attractive, too, about the Mongols' treatment of religion. Only Moslems were hindered in the preaching of their doctrine; Taoism and Buddhism were positively encouraged, for example by relieving Buddhist monasteries of taxes (this, of course, meant heavier impositions on others, as any state support for religion must; the peasants paid for religious enlightenment).

In the fourteenth century, natural disasters combined with Mongol exactions to produce a fresh wave of rural rebellions, the telling symptom of a dynasty in decline. They may have been made worse by Mongol concessions to the Chinese gentry. Giving landlords new rights over their peasants can hardly have won the régime popular support. Secret societies began to appear again and one of them, the 'Red Turbans', attracted support from gentry and officials. One of its leaders, Chu Yan-chang, a monk, seized Nanking in 1356. Twelve years later he drove the Mongols from Peking and the Ming era began. Yet like many other Chinese revolutionary leaders Chu Yan-chang gradually became an upholder of the traditional order. The dynasty he founded, though it presided over a great cultural flowering and managed to maintain the political unity of China which was to last from Mongol times to the twentieth century, confirmed China's conservatism and isolation. In the early fifteenth century the maritime expeditions by great fleets came to an end. An imperial decree forbade Chinese ships to sail beyond coastal waters or individuals to travel abroad. Soon, Chinese shipyards lost the capacity to build the great ocean-going junks; they did not even retain their specifications. The great voyages of the eunuch Cheng Ho, a Chinese Vasco Da Gama, were almost forgotten. At the same time, the merchants who had prospered under the Mongols were harassed.

In the end the Ming dynasty ran to seed. A succession of emperors virtually confined to their palaces while favourites and imperial princes disputed around them the enjoyment of the imperial estates registered the decline. Except in Korea, where the Japanese were beaten off at the end of the sixteenth century, the Ming could not maintain the peripheral zones of Chinese empire. Indo-China fell away from the Chinese sphere, Tibet went more or less out of Chinese control and in 1544 the Mongols burnt the suburbs of Peking.

Under the Ming, too, came the first Europeans to seek more than a voyage of trade or discovery. In 1557 Portuguese established themselves at Macao. They had little to offer which China wanted, except silver; but missionaries followed and the official tolerance of Confucian tradition gave them opportunities they successfully exploited. They became very influential at the Ming court and in the early seventeenth century Chinese officials began to feel alarmed. The Portuguese were ordered back to Macao. By then, besides the mechanical toys and clocks which the missionaries added to the imperial collections, their scientific and cosmographical learning had begun to interest Chinese intellectuals. The correction of the Chinese calendar, which one Jesuit carried out, was of great importance, for the authenticity of the emperor's sacrifices depended on accurate dating. From the Jesuits the Chinese learnt also to cast heavy cannon, another useful art.

Early in the seventeenth century, the Ming needed any military advantages they could procure. They were threatened from the north by a people living in Manchuria, a province to which they later gave its name, but who were not known as Manchu until after their conquest of China. The way was opened to them in the 1640s by peasant revolt and an attempted usurpation of the Chinese throne. An imperial general asked the Manchu to help him and they came through the Wall, but only to place their own dynasty, the Ch'ing, on the throne in 1644 (and incidentally wipe out the general's own clan). Like other barbarians and semi-barbarians, the Manchu had long been fascinated by the civilization they threatened and were already somewhat sinicized before their arrival. They were familiar with the Chinese administrative system, which they had imitated at their own capital of Mukden, and found it possible to cooperate with the Confucian gentry as they extended their grip on China. The attachment of Manchu inspectors stimulated the bureaucracy who needed to change little in their ways except to conform to the Manchu practice of wearing pigtails (thus was introduced what later struck Europeans as one of the oddest features of Chinese life).

The cost of Manchu conquest was high. Some twenty-five million people perished. Yet recovery was rapid. China's new power was already spectacularly apparent under the Emperor K'ang-hsi, who reigned from 1662 to 1722. This roughly corresponded to the reign of Louis XIV of France, whose own exercises in magnificence and aggrandizement took different forms but showed curious parallels on the other side of the world. K'ang-hsi was capable of a personal violence which the Sun King would never have permitted himself (he once attacked two of his sons with a dagger) but for all the difference in the historical backgrounds which formed them, there is a similarity in their

A depiction of scholars from a scroll of Sung times.

style of rule. Jesuit observers speak of K'ang-hsi's 'nobility of soul' and the description seems to have been prompted by more than the desire to flatter, and justified by more than his patronage. He was hard-working, scrutinizing with a close eye the details of business (and its manner, for he would painstakingly correct defective calligraphy in the memorials placed before him), and, like Louis, he refreshed himself by indulging his passion for hunting.

Characteristically, though K'ang-hsi was unusual among the Chinese emperors in admiring European skill (he patronized the Jesuits for their scientific knowledge), the merits of his reign were set firmly within accepted tradition; he identified himself with the enduring China. He rebuilt Peking, destroyed during the Manchu invasion, carefully restoring the work of the Ming architects and sculptors. It was as if Versailles had been put up in the Gothic style or London rebuilt in Perpendicular after the Great Fire. K'ang-hsi's principles were Confucian and he had classical works translated into Manchu. He sought to respect ancient tradition and assured his Chinese subjects their usual rights; they continued to rise to high office in the civil service in spite of its opening to Manchus, and K'ang-hsi appointed Chinese generals and viceroys. In the style of his personal life the emperor was, if not austere, at least moderate. He enjoyed the bracing life of the army and on campaigns lived simply; in Peking the pleasures of the palace were deliberately reduced and the emperor relaxed from the burdens of state with a harem of a mere three hundred girls.

K'ang-hsi extended imperial control to Formosa, occupied Tibet, mastered the Mongols and made them quiescent vassals. This was something of a turning-point, as final as anything can be in history; from this time the nomadic peoples of Central Asia at last begin gradually to recede before the settler. Further north, in the Amur valley, another new historical chapter opened when, in 1685, a Chinese army attacked a Russian post at Albazin. Negotiations led to the withdrawal of the Russians and the razing of their fort. The treaty of Nershinsk which settled matters contained among its clauses one which prescribed that boundary posts should be set up with inscriptions not only in Russian, Manchu, Chinese and Mongolian, but also in Latin. The suggestion had been made by a French Jesuit who was a member of the Chinese delegation and, like the establishment of a frontier line at all, was a symptom of new Chinese relationships to the outside world, relationships developing faster, perhaps, than any Chinese knew. The treaty was far from being the final settlement of accounts between China and the only European power with which she shared a land frontier but it quietened things for a time. Elsewhere, Manchu conquest continued to unroll; later in the eighteenth century Tibet was again invaded and vassal status reimposed on Korea, Indo-China and Burma. These were major feats of arms.

At home, peace and prosperity marked the last years of Manchu success. It was a silver age of the high classical civilization which some scholars believe to have reached its peak under the later Ming. If it did, it could still produce much beauty and scholarship under the Manchu. Great efforts of compilation and criticism, initiated and inspired by K'ang-hsi himself, opened a hundred years of transcription and publication which not only spawned such monsters as a five thousand-volume encyclopedia, but also collections of classical editions now given canonical form. In K'ang-hsi's reign, too, the imperial kilns began a century of technical advance in enamelling which produced exquisite glazes.

Yet however admirable, and however the emphasis is distributed between its various expressions in different arts, Manchu China's civilization was still, like that of its predecessors, the civilization of an élite. Although there was at the same time a popular culture of great vigour, the Chinese civilization which Europeans were struck by was as much the property of the Chinese ruling class as it always had been, a fusion of artistic, scholarly and official activity. Its connexion with government still gave it a distinctive tone and colour. It remained profoundly conservative, not only in social and political matters but even in its aesthetic. The art it esteemed was based on a distrust of innovation and originality; it strove to imitate and emulate the best, but the best was always past. The traditional masterpieces pointed the way. Nor was art seen as the autonomous expression of aesthetic activity. Moral criteria were brought to the judgement of artistic work and these criteria were, of course, the embodiments of Confucian values. Restraint, discipline, refinement, and respect for the great masters were the qualities admired by the scholar-civil servant who was also artist and patron.

Whatever appearances might suggest at first sight, therefore, Chinese art was no more directed towards escape from conventional life and values than that of any other culture before the European nineteenth century. This was also paradoxically apparent in its traditional exaltation of the amateur and the disapprobation it showed towards professionals. The man most esteemed was the official or landowner who was able to execute with sureness and apparent lack of effort works of painting, calligraphy or literature. Brilliant amateurs were greatly admired and in their activities, Chinese art escapes from its anonymity; we often know such artists' names. Its beautiful ceramics and textiles, on the other hand, are the products of tradesmen whose names are lost, often working under the direction of civil servants. Artisans were not esteemed for originality; the craftsman was encouraged to develop his skill not to the point of innovation but towards technical perfection. Central direction of large bodies of craftsmen within the precincts of the imperial palace only imposed upon these arts all the more firmly the stamp of traditional style. Even a brilliant explosion of new technical masteries at the imperial kilns during the reign of K'ang-hsi still expressed itself within the traditional canons of restraint and simplicity.

The final paradox is the most obvious and by the eighteenth century it seems starkly apparent. For all her early technological advances China never arrived at a mastery of nature which could enable her to resist Western intervention. Gunpowder is the most famous example; the Chinese had it before anyone else, but could not make guns as good as those of Europe, nor even employ effectively those made for them by European craftsmen. Chinese sailors had long had the use of the mariner's compass and a cartographical heritage which produced the first grid map, but they were only briefly exploring navigators. They neither pushed across the Pacific like the more primitive Melanesians, nor did they map it, as did later the Europeans. For six hundred years or so before Europe had them, the Chinese made mechanical clocks fitted with the escapement which is the key to successful time-keeping by machines, yet the Jesuits brought with them an horological technology far superior to the Chinese when they arrived in the sixteenth century. The list of unexploited intellectual triumphs could be much lengthened, by important Chinese innovations in hydraulics, for example, but there is no need to do so. The main point is clear. Somehow, a lack of interest in the utilisation of invention was rooted in a Confucian social system which, unlike that of Europe, did not regard as respectable association between the gentleman and the technician.

Pride in a great cultural tradition long continued to make it very hard to recognize its inadequacies. This made learning from foreigners – all barbarians, in Chinese eyes – very difficult. To make things worse, Chinese morality prescribed contempt for the soldier and for military skills. In a period when external threats would multiply, China was, therefore, dangerously cramped in her possibilities of response. Even under K'ang-hsi there were signs of new challenges ahead. In his old age he had to restore Manchu power in Tibet, when Mongol tribes had usurped it. The Russians were by 1700 installed in Kamchatka, were expanding their trade on the caravan routes and were soon to press on into the Trans-Caspian region. Even peace and prosperity had a price, for they brought faster population growth. Here, unsolved because unrecognized and perhaps insoluble, was another problem to upset the stability of the order authorized by the mandate of heaven. By 1800 there were over three hundred, perhaps even four hundred, million Chinese, and already signs were appearing of what such an increase might portend.

8
Japan

There was a time when Englishmen liked to think of Japan as the Great Britain of the Pacific. The parallel was developed at many levels; some were less plausible than others, but there was an indisputable hard nugget of reality in the facts of geography. Both are island kingdoms whose peoples' destinies have been shaped deeply by the sea. Both, too, live close to neighbouring land masses whose influence on them could not but be profound. The Straits of Tsushima which separate Korea from Japan are about five times as wide as the Straits of Dover, it is true, and Japan was able to maintain an isolation from the Asian *terra firma* far more complete than any England could hope for from Europe. Nevertheless, the parallel can be pressed a good way and its validity is shown by the excitement which the Japanese have always shown about the establishment of a strong power in Korea; it rivals that of the British over the danger that the Low Countries might fall into unfriendly hands.

Even before Japan emerges in her own historical records, in the eighth century AD, there was Japanese-held territory in the Korean peninsula. In those days, Japan was a country divided up among a number of clans, presided over by an emperor with an ill-defined supremacy and an ancestry traced back to the Sun Goddess. The Japanese did not occupy the whole of the territory of modern Japan, but lived in the main on the southern and central islands. Here were the mildest climate and the best agricultural prospects. In prehistoric times, the introduction of rice-growing and the fishing potential of Japanese waters had already made it possible for this mountainous country to feed a disproportionately large population, but pressure on land was to be a recurrent theme of Japanese history.

In 645 a political crisis in the dominant clan brought about its downfall and a new one arose, the Fujiwara. It was to preside over a great age of Japanese civilization and to dominate the emperors. There was more than political significance in the change. It also marked a conscious effort to redirect Japanese life along paths of renewal and reform. The direction could only be sought from the guidance offered by the highest example of civilization and power of which the Japanese were aware, and possibly the finest in the world at that time, that of imperial China, which was also an example of expanding, menacing power.

Its continuing and often changing relationship with China is another theme of Japanese history. Both peoples were of Mongoloid stock, though some Caucasoids whose presence it is difficult to account for also form a part of the Japanese ethnic heritage (these, the Ainus, were, at the beginning of the historical era, mainly to be found in the north-east). In prehistoric times Japan appears to have followed in the wake of the civilization of the mainland; bronze artifacts, for example, appear in the islands only in the first century or so BC. Such innovations in the last millennium BC may owe something to immigrants displaced by the Chinese as they moved southwards on the mainland. But the first references to Japan in the Chinese records (in the third century AD) still depict a country not much affected by mainland events and Chinese influence was not very marked until the centuries following the Han collapse. Then, a vigorous Japanese intervention in Korea seems to have opened the way to closer contact. It was subsequently fostered by the movement of Buddhist students. Confucianism, Buddhism, and iron technology all came to Japan from China. There were attempts to bring about administrative changes on Chinese lines. Above all, Chinese writing had been brought to Japan and its characters were used to provide a written form of the native language. Yet cultural attraction and dependence had not meant political submission.

The Japanese central administration was already well-developed in scope and scale at the beginning of the period of centralization and major efforts of reform were made in the seventh and eighth centuries. Yet, in the end, Japan evolved not in the direction of a centralized monarchy but of what might be termed, in a western analogy, feudal

From very ancient times, female images in clay were among the objects of Japanese devotion. With the coming of Buddhism the practice of carving them in wood began, as well as the giving of a new composure and tranquility to the figures. This is a ninth-century carving of Nakatasuhime, one of the supposed progenitors of the Japanese people.

anarchy. For almost nine hundred years it is hard to find a political thread to Japanese history. Its social continuity is much more obvious. From the beginnings of the historical era, even down to the present day, the keys to the continuity and toughness of Japanese society have been the family and the traditional religion. The clan was an enlarged family, and the nation the most enlarged family of all. In patriarchal style, the emperor presided over the national family as did a clan leader over his clan or, even, the small farmer over his family. The focus of family and clan life was participation in the traditional rites, the religion known as Shinto, whose essence was the worship at the proper times of certain local or personal deities. When Buddhism came to Japan it was easily conjoined with this traditional way.

The institutional coherence of old Japan was less marked than its social unity. The emperor was its focus. From the beginning of the eighth century, though, the emperor's power was more and more eclipsed and so, in spite of the efforts of an occasional vigorous individual, it remained until the nineteenth century. This eclipse arose in part from the activities of the would-be reformers of the seventh century, for one of them was the founder of the great Fujiwara clan. In the next hundred years or so, his family tied itself closely to the imperial household by marriage. As children were frequently brought up in the household of their mother's family, the clan could exercise a crucial influence upon future emperors while they were children. In the ninth century the chief of the Fujiwara was made regent for the emperor – who was an adult – and for most of what is called the 'Heian' period (794–1185: the name comes from that of the capital city, the modern Kyoto), that clan effectively controlled central government through marriage alliances and court office, its leaders acting in the emperor's name. The power of the Fujiwara did something to disguise the decline of the royal authority, but, in fact, the imperial clan was tending to become simply one among several which existed in the shade of the Fujiwara, each of them governing its own estates more or less independently.

The displacement of the emperor became much more obvious after the passing of the power of the Fujiwara. The 'Kamakura' period (1185–1333) was so called because power passed to a clan whose estates were in the area of that name and the bypassing of the imperial court, which remained at Heian, became much more obvious. It was early in the Kamakura period that there appeared the first of a series of military dictators who bore the title of *shōgun*. These ruled in the emperor's name but in fact with a large independence. The emperor lived on the revenues of his own estates, and as long as he acquiesced in the shogun's intentions he would have military power behind him; when he did not, he would be overruled.

This eclipse of the imperial power was so different from what had occurred in China, the model of the seventh-century reformers, that the explanation is not easy to see. It was complex. There was a steady progression through the centuries from the exercise of a usurped central authority in the emperor's name to the virtual disappearance of any central authority at all. No doubt there was a fundamental bias in the traditional clan loyalties of Japanese society and the topography of Japan which would have told against any central power; remote valleys provided lodgements for great magnates. But other countries have met these problems successfully: the Hanoverian governments of eighteenth-century Great Britain tamed the Scottish highlands with punitive expeditions and military roads. A more specific explanation can be seen in the way in which the land reforms of the seventh century, which were the key to political change, were in practice whittled away by the clans with influence at court. Some of these exacted privileges and exemptions, as did some land-holding religious institutions. The commonest example of the abuses which resulted from this was the granting of tax-free manors to noblemen who were imperial court officials by way of payment for carrying out their duties. The Fujiwara themselves were unwilling to check this practice. At a lower level, smaller proprietors would then seek to commend themselves and their land to a powerful clan in order to get assured tenure in return for rent and an obligation to provide service. The double result of such developments was to create a solid base for the power of local magnates while starving the central administrative structure of support from taxation. Taxes (in the form of a share of the crops) went not to the imperial administration but to the person to whom a manor had been granted.

Minamoto Yoritomo, the unscrupulous, brave and resourceful warrior who established in Kamakura a power which was the base of the first bakufu, *and who was the first of the shōguns. This is a contemporary portrait; we have no such likenesses of the lords of early medieval Europe, some of whom must have been not dissimilar in their ambitions for their families – however inferior in their successes.*

Such a civil service as existed, unlike the Chinese, was firmly reserved to the aristocracy. Not being recruited by competition, it could not provide a foothold for a group whose interests might be opposed to the hereditary noble families. In the provinces, posts just below the highest level tended to go to the local notables, only the most senior appointments being reserved to civil servants proper.

No one planned that this should happen. Nor did anyone plan a gradual transition to military rule, whose origins lay in the need to make some of the families of the frontier districts responsible for defence against the still unsubdued Ainu peoples. Slowly the prestige of the military clans drew to their leaders the loyalties of men seeking security in troubled times. And, indeed, there was a need for such security. Provincial dissidence began to express itself in outbreaks in the tenth century. In the eleventh there was clearly discernible an emerging class of manorial officers on the great estates. They enjoyed the real management and use of the lands of their formal masters and felt loyalties to the warrior clans in an elementary tie of service and loyalty. In this situation the Minamoto clan rose to a dominance which recreated central government in the early Kamakura period.

In one way these struggles were a luxury. The Japanese could indulge them because they lived in an island-state where no foreign intruder was ever more than occasionally threatening. Amongst other things, this meant that there was no need for a national army which might have mastered the clans. Although she came near to it in 1945, Japan has never been successfully invaded, a fact which has done much to shape the national psychology. The consolidation of the national territory was for the most part achieved in the ninth century when the peoples of the north were mastered and, after this, Japan rarely faced any serious external threat to her national integrity, though her relations with other states underwent many changes.

In the seventh century the Japanese had been ousted from Korea and this was the last time for many centuries that they were physically installed there. It was the beginning of a phase of cultural subservience to China which was matched by an inability to resist her on the mainland. Japanese embassies were sent to China in the interests of trade, good relations and cultural contact, the last one in the first half of the ninth century. Then, in 894, another envoy was appointed. His refusal to serve marks something of an epoch, for he gave it as his reason that China was too much disturbed and distracted by internal problems and that she had, in any case, nothing to teach the Japanese. Official relations were not resumed until the Kamakura period.

There were exploratory gestures in the thirteenth century. They did not prevent the expansion of irregular and private trade with the mainland in forms some of which looked much like freebooting and piracy. It may have been this which did much to provoke the two attempted Mongol invasions of 1274 and 1281. Both retired baffled, the second after grievous losses by storm – the *Kamikaze*, or 'divine wind', which came to be seen in much the same light as the English saw the storms which shattered the Armada – and this was of the greatest moment in strengthening the belief which Japanese came to hold in their own invincibility and national greatness. Officially, the Mongols' motive had been the Japanese refusal to recognize their claim to inherit the Chinese pretensions to empire and to receive tribute from them. In fact, this conflict once more killed off the recently revived relations with China; they were not taken up again until the coming of Ming rule. By then the reputation of the Japanese as pirates was well established. They ranged far and wide through the Asian seas just as Drake and his companions ranged the Spanish Main. They had the support of many of the feudal lords of the south and it was almost impossible for the *shōguns* to control them even when they wished (as they often did) to do so for the sake of good relations with the Chinese.

The collapse of the Kamakura shogunate in 1333 brought a brief and ineffective attempt to restore real power to the emperor, which ended when confronted with the realities of the military power of the clans. In the ensuing period neither *shōgun* nor emperor often enjoyed assured power. Until the end of the sixteenth century civil warfare was almost continuous. Yet these troubles did not check the consolidation of a Japanese cultural achievement which remains across the centuries a brilliant and moving spectacle and still shapes Japanese life and attitudes even in an era of industrialism. It is an achievement notable for its power to borrow and adopt from other cultures without sacrificing its own integrity or nature.

Even at the beginning of the historical era, when the prestige of T'ang art makes the derivative nature of what is done in Japan very obvious, there was no merely passive acceptance of a foreign style. Already in the first of the great periods of high

Japanese culture, in the eighth century, this is apparent in Japanese painting and a poetry already written in Japanese, though men for centuries still wrote works of art or learning in Chinese (it had something of the status long held by Latin in Europe). At this time, and still more during the climax of the Fujiwara ascendancy, Japanese art other than religious architecture was essentially a court art, shaped by the court setting and the work and enjoyment of a relatively narrow circle. It was hermetically sealed from the world of ordinary Japan by its materials, subject-matter and standards. The great majority of Japanese would never even see the products of what can now be discerned as the first great peak of Japanese culture. The peasant wove hemp and cotton; his womenfolk would no more be likely to touch the fine silks whose careful gradations of colour established the taste displayed by a great court lady's twelve concentric sleeves than he would be to explore the psychological complexities of the Lady Murasaki's subtle novel, the *Tale of Genji*, a study as compelling as Proust and almost as long. Such art had the characteristics to be expected of the art of an élite insulated from society by living in the compound of the imperial palace. It was beautiful, refined, subtle, and sometimes brittle, insubstantial and frivolous. But it already found a place for an emphasis which was to become traditional in Japan, that on simplicity, discipline, good taste and love of nature.

The culture of the Heian court attracted criticism from provincial clan leaders who saw in it an effete and corrupting influence, sapping both the independence of the court nobles and their loyalty to their own clans. From the Kamakura period, a new subject-matter – the warrior – appears in both literature and painting. Yet, as the centuries passed, a hostile attitude to traditional arts changed into one of respect and during the troubled centuries the warring magnates showed by their own support for them that the central canons of Japanese culture were holding fast. It was protected more and more by an insularity and even a cultural arrogance confirmed by the defeat of the Mongol invasions. A new, military, element, too, was added to this culture during the centuries of war, in part originating in criticism of the apparently effete court circles but then blending with their traditions. It was fed by the feudal ideal of loyalty and self-sacrificing service, by the warrior ideals of discipline and austerity, and by an aesthetic arising out of them. One of its characteristic expressions was an offshoot of Buddhism, Zen. Gradually there emerged a fusion of the style of the high nobility with the austere virtues of the *samurai* warrior which was to run through Japanese life down to the present day. Buddhism also left a visible mark on the Japanese landscape in its temples and the great statues of the Buddha himself. Overall, the anarchy was the most creative of all periods of Japanese culture, for in it there appeared the greatest landscape painting, the culmination of the skill of landscape gardening and the arts of flower arrangement, and the *Nō* drama.

In particular areas, the lawlessness of these centuries often inflicted grave social and economic damage. As was long to be the case, most Japanese were peasants: they might suffer terribly from an oppressive lord, from banditry, or the passage of an army of retainers from a rival fief. Yet such damage was nationally insignificant, it seems. In the sixteenth century a great burst of castle-building testifies to the availability of substantial resources, there was a prolonged expansion of the circulation of copper coinage, and Japanese exports – particularly the exquisite examples of the work of the swordsmiths – began to appear in the markets of China and south-eastern Asia. By 1600 Japan's population stood at about eighteen million. Both its slow growth (it had somewhat more than trebled in five centuries) and its substantial urban component rested on a steady improvement in agriculture which had been able to carry the costs of civil strife and lawlessness as well. It was a healthy economic position.

Sooner or later the Europeans were bound to come to find out more about the mysterious islands which produced such beautiful things. The first were the Portuguese who stepped ashore from Chinese ships, probably in 1543. Others followed in the next few years and in their own ships. It was a promising situation. Japan was virtually without a central government to undertake the regulation of intercourse with foreigners and many of the southern magnates were themselves highly interested in competing for foreign trade. Nagasaki, then a little village, was opened to the newcomers by one of them in 1570. This nobleman was a zealous Christian and had already built a church there; in 1549 the first Christian missionary had arrived, St Francis Xavier. Nearly forty years later Portuguese missionaries were forbidden, so much had the situation changed, though the ban was not at once enforced.

Among other things brought by the Portuguese to Japan were new food crops originally from the Americas – sweet potatoes, maize, sugar cane. They also brought muskets. The Japanese soon learnt to make them. This new weapon played an important part in assuring that the baronial wars of 'feudal' Japan came to an end, as did those of medieval Europe, with the emergence of a preponderant power, a brilliant, humbly-born

On her balcony, in the moonlight, the Lady Murasaki composes another chapter of the Tale of Genji, *the greatest work of Japanese literature and an unsurpassed exploration of the human heart and the life of love. The painting is of the seventeenth century.*

soldier-dictator, Hideyoshi. His successor was one of his henchmen, a member of the Tokugawa family. In 1603 he revived and assumed the old title of *shōgun* and so inaugurated a period of Japanese history known as the 'great peace', which lasted until a revolutionary change in 1868 but was itself an immensely creative period, in which Japan changed significantly.

During the Tokugawa shogunate, for two and a half centuries, the emperor passed even further into the wings of Japanese politics and was firmly kept there. Court gave way to camp; the shogunate rested on a military overlordship. The *shōguns* themselves changed from being outstandingly important feudal lords to being in the first place hereditary princes and in the second the heads of a stratified social system over which they exercised viceregal powers in the name of the emperor and on his behalf. This régime was called the *bakufu* – the government of the camp. The *quid pro quo* provided by Ieyasu, the first Tokugawa *shōgun*, was order and the assurance of financial support for the emperor.

The key to the structure was the power of the Tokugawa house itself. Ieyasu's origins had been pretty humble, but by the middle of the seventeenth century the clan appears to have controlled about one-quarter of Japan's rice-growing land. The feudal lords became in effect vassals of the Tokugawa, linked to the clan by a variety of ties. The term 'centralized feudalism' has been coined to label this system. Not all the lords, or *daimyo*, were connected to the *shōgun* in the same way. Some were directly dependent, being vassals with a hereditary family attachment to the Tokugawa family. Others were related to it by marriage, patronage or business. Others, less reliable, formed an outer category of those families which had only at length submitted. But all were carefully watched. The lords lived alternately at the *shōgun's* court or on their estates; when they were on their estates, their families lived as potential hostages of the *shōgun* at Edo, the modern Tokyo, his capital.

Below the lords was a society strictly and legally separated into hereditary classes and the maintenance of this structure was the primary goal of the régime. The noble *samurai* were the lords and their retainers, the warrior rulers who dominated society and gave it its tone as did the gentry bureaucrats of China. They followed a spartan, military ideal symbolized by the two swords they carried, and were allowed to use on commoners guilty of disrespect. *Bushido*, their creed, stressed above all the loyalty owed by a man to his lord. The original links of the retainers with the land were virtually gone by the seventeenth century and they lived in the castle towns of their lords. The other classes were the peasants, the artisans, and the merchants, the lowest in the social hierarchy because of their non-productive character; the self-assertive ethos of the merchant which emerged in Europe was unthinkable in Japan, in spite of the vigour of Japanese trade. As the aim of the whole system was stability, attention to the duties of one's station and confinement to them was determinedly enforced. Hideyoshi himself had supervised a great sword hunt whose aim was to take away these weapons from those who were not supposed to have them – the lower classes. Whatever the equity of this, it must have told in favour of order. Japan wanted stability and her society accordingly came to emphasize the things that could ensure it: knowing one's place, discipline, regularity, scrupulous workmanship, stoical endurance. At its best it remains one of humanity's most impressive social achievements.

One weakness of this system it shared with the Chinese; it presumed effective insulation from external stimuli to change. It was for a long time threatened by the danger of a relapse into internal anarchy; there were plenty of discontented *daimyo* and restless swordsmen about in seventeenth-century Japan. By then, one obvious external danger came from Europeans. They had already brought to Japan imports which would have profound effect. Among them the most obvious were firearms, whose powerfully disruptive impact went beyond that which they achieved on their targets, and Christianity. This faith had at first been tolerated and even welcomed as something tempting traders from outside. In the early seventeenth century the percentage of Japanese Christians in the population was higher than it has ever been since. Soon, it has been estimated, there were over half a million of them. Nevertheless, this happy state of affairs did not last. Christianity has always had great subversive potential. Once this was grasped by Japan's rulers, a savage persecution began. It not only cost the lives of thousands of Japanese martyrs, who often suffered cruel deaths, but brought trade with Europe almost to an end. The English left and the Spanish were excluded in the 1620s. After the Portuguese had undergone a similar expulsion they rashly sent an embassy in 1640 to argue the toss;

almost all of its members were killed. Japanese had already been forbidden to go abroad, or to return if they were already there, and the building of large ships was banned. Only the Dutch, who promised not to proselytize and were willing to trample on the cross, kept up Japan's henceforth tiny contact with Europe. They were allowed a trading station on a tiny island in Nagasaki harbour.

Swordsmanship was highly esteemed in Japan, as were the great fencing-masters, while the beautiful weapons forged by the Japanese smiths remained treasured and revered possessions.

After this, there was no real danger of foreigners exploiting internal discontent. But there were other difficulties. In the settled conditions of the 'great peace', military skill declined. The *samurai* retainers sat about in the castle towns of their lords, their leisure broken by little except the ceremonial parade in outdated armour which accompanied a lord's progress to Edo. When the Europeans came back in the nineteenth century with up-to-date weapons, Japan's military forces would be technically unable to match them.

This could, perhaps, hardly have been foreseen. Nor could another result of the general peace in which internal trade prospered. The Japanese economy became more dependent on money. Old relationships were weakened by this and new social stresses appeared. Payment in cash forced lords to sell most of the tax rice which was their subsistence to pay for their visits to the capital. At the same time, the market became a national one. Merchants did well: some of them soon had money to lend their rulers. Gradually the warriors became dependent on the bankers. Besides feeling a shortage of cash, those rulers found themselves sometimes embarrassed by their inability to deal with economic change and its social repercussions. If retainers were to be paid in coin, they might more easily transfer loyalty to another paymaster. Towns were growing, too, and by 1700 Osaka and Kyoto both had more than 300,000 inhabitants, while Edo may have had 800,000. Other changes were bound to follow such growth. Price fluctuations in the rice market of the towns sharpened hostility towards the wealthy dealers.

Here we face the great paradox of Tokugawa Japan. While its rulers slowly came to show less and less ability to contain new challenges to traditional ways, those challenges stemmed from a fundamental fact – economic growth – which in historical perspective now appears the dominant theme of the era. Under the Tokugawa, Japan was developing fast. Between 1600 and 1850 agricultural production approximately doubled, while the population rose by less than half. Since the régime was not one which was able to skim off the new wealth for itself, it remained in society as savings for investment by those who saw opportunities, or went into a rising standard of living for many Japanese.

Dispute continues about the explanation of what seems to have been a successful stride to self-sustaining economic growth of a kind which was elsewhere to appear only in Europe. Some are obvious and have been touched upon: the passive advantages conferred by the seas around Japan which kept out invaders such as the steppe-borne nomads who time and again harried the wealth-producers of mainland Asia. The shogunate's own great peace ended feudal warfare and was another bonus. Then there were positive improvements to agriculture which resulted from more intensive cultivation, investment in irrigation, the exploitation of the new crops brought (originally from the Americas) by the Portuguese. But at this point the enquiry is already touching on reciprocal effects: the improvement of agriculture was possible because it became profitable to the producer, and it was profitable because social and governmental conditions were of a certain kind. Enforced residence of noblemen and their families at Edo not only put rice on the market (became the nobles had to find cash), but created a new huge urban market at the capital which sucked in both labour (because it supplied employment) and goods which it became more and more profitable to produce. Regional specialization (in textile manufacture, for example) was favoured by disparities in the capacity to grow food: most of Japanese industrial and handicraft production was, as in early industrial Europe, to be found in rural areas. Government helped, too; in the early years of the shogunate there was organized development of irrigation, standardizing of weights and currency. But for all its aspirations to regulate society, the government of the *bakufu* in the end probably favoured economic growth because it lacked power. Instead of an absolute monarchy, it came to resemble a balance-of-power system of the great lords, and able to maintain itself only so long as there was no foreign invader to disturb it. As a result it could not obstruct the path to economic growth and divert resources from producers who could usefully employ them. Indeed, the economically quasiparasitical *samurai* actually underwent a reduction in their share of the national income at a time when producers' shares were rising. It has been suggested that by 1800 the *per capita* income and life expectancy of the Japanese was much the same as that of their British contemporaries.

Much of this has been obscured by more superficial but strikingly apparent features of the Tokugawa era. Some of these, of course, were important, but at a different level. The new prosperity of the towns created a clientele for printed books and the coloured wood-block prints which were later to excite European artists' admiration. It also provided the audiences for the new *kabuki* theatre. Yet brilliant though it often was, and successful, at the deepest economic level (if undesignedly) as it was, it is not clear that the Tokugawa system could have survived much longer even without the coming of a new threat from the West in the nineteenth century. Towards the end of the period there were signs of uneasiness. Japanese intellectuals began to sense that somehow their isolation had preserved them from Europe but also had cut them off from Asia. They were right. Japan had already made for herself a unique historical destiny and it would mean that she faced the West in a way very different from the subjects of Manchu or Moghul.

9
Worlds Apart

Africa and the Americas moved towards civilization to rhythms very different from those operating elsewhere. Of course, this was not quite so true of Africa as of the Americas, which were long cut off by the oceans from all but fleeting contacts with the rest of the world. The Africans, by contrast, lived in a continent much of which was gradually Islamicized, and for a long time had at least peripheral encounters with first Arab and then European traders. These were of growing importance as time went by, though they did not suck Africa completely into the mainstream of world history until the late nineteenth century. This isolation, combined with an almost complete dependence for much of the story on archaeological evidence, makes much African and American history an obscure business.

African history before the coming of European trade and exploration is largely a matter of an internal dynamic we can barely discern, but we may presume folk-movements to have played a large part in it. There are many legends of migration and they always speak of movement from the north to the south and west. In each case, scholars have to evaluate the legend in its context, and with help from reference in Egyptian records, travellers' tales and archaeological discovery, but the general tendency is striking. It seems to register a general trend, the enrichment and elaboration of African culture in the north first and its appearance in the south only much later.

The kingdom of Kush, whose connexions with Egypt have been noted, is a convenient beginning. By the fifth century BC the Kushites had lost control of Egypt and retreated once more to Meroe, their capital in the south, but they had centuries of flourishing culture still ahead of them. From Egypt, probably, they had brought with them a hieroglyph (claims are now being made to have penetrated it). Certainly they diffused their knowledge to the south and west in the Sudan, where notable metallurgical skills were later to flourish among the Nubians and Sudanese. In the last few centuries BC iron-working appears south of the Sahara, in central Nigeria. Its importance was recognized by its remaining the closely-guarded secret of kings, but so valuable a skill slowly travelled southwards. By about the twelfth century AD it had penetrated the south-east and the pygmies and the bushmen of the south were the only Africans then still living in the Stone Age.

Probably the greatest difference made by the spread of iron-working was to agriculture. It made possible a new penetration of the forests and better tilling of the soil (which may be connected with the arrival of new food-crops from Asia in the early Christian era), and so led to new folk-movements and population growth. Hunting and gathering areas were broken up by the coming of herdsmen and farmers who can be discerned already by about AD 500 in much of east and south-east Africa, in modern Zimbabwe and the Transvaal. Yet those Africans did not acquire the plough. Possibly the reason lies in the lack in most of the continent south of Egypt of an animal resistant enough to African diseases to draw one. One area where there were ploughs was Ethiopia, and there animals could be bred successfully, as the early use of the horse indicates. Horses were also bred for riding in the southern Sahara.

This suggests once again the important limiting factor of the African environment. Most of the continent's history is the story of response to influences from the outside – iron-working and new crops from the Near East, Asia, Indonesia and the Americas; steam engines and medicine from nineteenth-century Europe. These made it possible gradually to grapple with African nature. Without them, Africa south of the Sahara seems almost inert under the huge pressures exercised upon it by geography, climate and disease. It remained (with some exceptions) for the most part tied to a shifting agriculture, not achieving an intensive one; this was a positive response to difficult conditions but could not sustain more than a slow population growth. Nor did southern Africa arrive at the wheel; so it lagged behind in transport, milling and pottery.

Probably the first civilization to influence Black Africa was that of Egypt, which spread southward through the Sudanese kingdom of Meroe. This lion-headed god on a native tablet inscribed in Meriotic is one relic of that.

The story was different north of the Equator. Much Kushite history waits, in the most literal sense, to be uncovered, for few of the major cities have yet been excavated. It is known that in about AD 300 Kush was overthrown by Ethiopians. They were not then the unique people they were to become, with kings claiming descent from Solomon and for centuries the only Christian people in Africa outside Egypt. They were converted to Christianity by Copts only later in the fourth century; at that time they were still in touch with the classical Mediterranean world. But the Islamic invasions of Egypt placed between them and it a barrier which was not breached for centuries, during which the Ethiopians battled for survival against pagan and Moslem, virtually isolated from Rome or Byzantium. An Amharic-speaking people, they were the only literate non-Islamic African nation.

The only other place in Africa where Christianity established itself was in the Roman north. Here it had been a vigorous, if minority, cult. The violence of its dissensions and the pursuit of the Donatists as heretics probably explain its weakness when the Arab invasions brought it face to face with Islam. Except in Egypt, Christianity was extinguished in the Africa of the Arab states. Islam, on the other hand, was and has remained enormously successful in Africa. Borne by Arab invasion it spread in the eleventh century right across to the Niger and western Africa. Arab sources therefore provide our main information about the non-literate African societies which stretched across the Sudan and Sahara after the passing of Kush. They were often trading communities and may reasonably be thought of as city-states; the most famous was Timbuctoo, impoverished by the time Europeans finally got there, but in the fifteenth century important enough to be the site of what has been described as an Islamic university. Politics and economics are still as closely intertwined in Africa as in any part of the world, and it is not surprising that the early kingdoms of black Africa should have appeared and prospered at the end of important trade routes where there was wealth to tap. Merchants liked stability.

Another African state, the earliest recorded by the Arabs, had a name later taken by a modern nation: Ghana. Its origins are obscure, but may well have lain in the assertion of its supremacy by a people in the late pre-Christian era who had the advantage of iron weapons and horses. However this may be, the Ghana recorded by Arab chroniclers and geographers is already an important kingdom when it appears in the records in the eighth century AD. At its greatest extent, Ghana spanned an area about five hundred miles across the region framed to the south by the upper reaches of the Niger and Senegal and protected to the north by the Sahara. The Arabs spoke of it as 'the land of gold'; the gold came from the upper Senegal and the Ashanti, and was passed by Arab traders up to the Mediterranean by trans-Saharan routes or through Egypt. The most important other commodities traded across the Sahara were salt and slaves. Ghana collapsed during the twelfth and thirteenth centuries.

Its eclipse was followed by the pre-eminence of Mali, a kingdom whose ruler's wealth caused a sensation when in 1307 he made a pilgrimage to Mecca and another source of a name for a twentieth-century African state. Mali was even bigger than Ghana, taking in the whole Senegal basin and running about a thousand miles inland from the coast at the beginning of the fourteenth century. The Mali ruler is said to have had ten thousand horses in his stables. This empire broke up in the sixteenth century after defeat by the Moroccans. Other states were to follow. But although in some cases the Arab records speak of African courts attended by men of letters there is no native documentation which enables us to reach these peoples. Clearly they remained pagan while their rulers belonged to the Islamic world. It may be that the dissolution of Ghana owed something to dissent caused by conversions to Islam. Arab reports make it obvious that the Islamic cult was associated with the ruler in the Sudanese and Saharan states but had also still to accommodate traditional practice from the pagan past – rather as early Christianity in Europe accepted a similar legacy. Nor did social custom always adapt itself to Islam: Arabic writers expressed shocked disapproval at the public nakedness of Mali girls.

Africa further south of the Sahara is even harder to get at. At the roots of the history which determined its structure on the eve of its absorption into world events was a folk-migration of the negroid peoples who speak languages of the group called Bantu. This is a term somewhat like 'Indo-European', referring to identifiable linguistic characteristics, not genetic qualities. The detailed course of this movement is, of course, still highly obscure but its beginnings lie in eastern Nigeria, where there were early Bantu-speakers. From there they took their language and agriculture south, first into the Congo basin. There followed a rapid spread, round about the beginning of the Christian era, over most of southern Africa. This set the ethnic pattern of modern Africa.

Some peoples, speaking the language the Arabs called 'Swahili'(from the Arabic word meaning 'of the coast'), established towns on the east African coasts which were linked to mysterious kingdoms in the interior. This was before the eighth century AD, when the Arabs began to settle in these towns and turn them into ports. The Arabs called the region the land of the Zanz (from which was later to come the name of Zanzibar) and said that its peoples prized iron above gold. It is probable that these polities had some kind of trading relations with Asia even before Arab times; who the inter-mediaries were it is not possible to say, but they may have been Indonesians such as those who colonized Madagascar. The Africans had gold and iron to offer for luxuries and they also began the implantation of new crops from Asia, cloves and bananas among them.

Even a vague picture of the working of these states is hard to arrive at. Monarchy was by no means the rule in them and a sense of the importance of ties of kin seems to have been the only widespread characteristic of black African polities. Organization must have reflected the needs of particular environments and the possibilities presented by particular resources. Yet kingship was widely diffused. Again, the earliest signs are northern, in Nigeria and Benin. By the fifteenth century there are kingdoms in the region of the great eastern lakes and we hear of the kingdom of the Kongo, on the lower Congo River. There are not many signs of organization on this scale and African states were for a long time not to produce bureaucratized administration or standing armies. The powers of kings must have been limited, not only by custom and respect for tradition, but by the lack of resources to bind men's allegiance beyond the ties imposed by kinship and respect. No doubt this accounts for the transitory and fleeting nature of many of these 'states'. Ethiopia was an untypical African country.

Yet some remarkable traces remain of these dim and shadowy kingdoms. A high level of culture in the east African interior in about the twelfth century is demonstrated by the remains of mine-workings, roads, rock paintings, canals and wells; these were the product of a technology which archaeologists have called 'Azanian'. It was the achievement of an advanced Iron Age culture. Agriculture had been practised in the region since about the beginning of the Christian era. On the basis it provided, it was possible to exploit the gold which was for a long time easily accessible in what is now Zimbabwe. Only simple techniques were needed at first; large quantities could be obtained by little more than scratching the surface. This drew traders to the area – Arabs first, and later Portuguese – but also other Africans as migrants. The search for gold had in the end to be taken underground as the most easily available supplies ran out.

None the less there was a rich enough supply to support a 'state' lasting four centuries. It produced the only significant building in stone in southern Africa. There are relics of it in hundreds of places in modern Zimbabwe, but the most famous is at the place itself called by that name (which means only 'stone houses'). From about 1400 this was a royal capital, the burial place of kings and a sacred site for worship. So it remained until it was sacked in about 1830 by another African people. The Portuguese of the sixteenth century had already reported a great fortress built of dry-stone masonry but only in the nineteenth century have we records by Europeans of what we know to be this site. They were amazed to find massive walls and towers in carefully shaped stone, laid in courses without mortar but with great accuracy. There was disinclination to believe that Africans could have produced anything so impressive; some suggested the Phoenicians should have the credit and a few romantics toyed with the idea that Zimbabwe had been put there by the masons of the Queen of Sheba. Today, remembering the world of other Iron Age peoples in Europe and the civilizations of America, such hypotheses do not seem necessary. The Zimbabwe ruins may reasonably be attributed to the Africans of the fifteenth century.

Advanced as East Africa was, its peoples failed to arrive at literacy for themselves; like the early Europeans, they were to acquire it from other civilizations. Perhaps the absence of a need for careful records of land, or of crops which could be stored, is a part of the explanation. Whatever the reason, the absence of literacy was a handicap in acquiring and diffusing information and in consolidating government. It was also a cultural impoverishment: Africa would not have a native tradition of learned men from whom would come scientific and philosophic skill. On the other hand, the artistic capacity of black Africa was far from negligible, as the achievement of Zimbabwe, or the bronzes of Benin which captivated later Europeans, show.

Islam had been at work in Africa for nearly eight hundred years (and before that there had been the influence of Egypt on its neighbours) by the time the Europeans arrived in America, to discover civilizations which had achieved much more than those of Africa and appeared to have done so without stimuli from the outside. This has seemed so improbable to some people that much time has been spent investigating and discussing the possibility that the elements of civilization were implanted in the Americas by trans-Pacific voyagers a very long time ago. Most scholars find the evidence inconclusive. If there was such a contact in remote times, it had long since ceased. There is no unequivocal trace of connexion between the Americas and any other continent between the time when the first Americans crossed the Bering Straits and the landings of Vikings. There is then none thereafter until the Spanish arrived at the end of the fifteenth century. To an even greater degree than Africa, and for a longer time, we must assume the Americas to have been cut off from the rest of the world.

Eight thousand feet high in the Andes, the Inca city of Machu Picchu was never reached by the Spaniards. It was only discovered in 1911. No written records survive to explain its history, but it seems to have been a religious centre, built in the fifteenth century AD.

Their isolation accounts for the fact that even in the nineteenth century pre-agricultural peoples still survived in North America. On the eastern plains of the modern United States there were 'Indians' (as Europeans later came to call them) practising agriculture before the arrival of Europeans, but further west other communities were then still hunting and gathering. They would go on doing so, though with important changes of techniques as first the horse and metal, brought by Europeans, and then firearms were added to their technical equipment. Further west still, there were peoples on the west coast who fished or collected their subsistence on the seashore, again in ways fixed since time immemorial. Far to the north, a *tour de force* of specialization has enabled the Eskimos to live with great efficiency in an all but intolerable environment; this pattern survives in its essentials even today. Yet although the Indian cultures of North America are respectable achievements in their overcoming of environmental challenge, they are not civilization. For the American achievement in indigenous civilization it is necessary to go south of the Rio Grande. Here were to be found a series of major civilizations linked by common dependence on the cultivation of maize and by possessing pantheons of nature gods, but strikingly different in other ways.

In Meso-america the Olmec foundation proved very important. The calendars, hieroglyphics and the practice of building large ceremonial sites which mark so much of the region in later times may all be ultimately derived from it; the gods of Meso-america were already known in Olmec times, too. Between the beginning and the fourth centuries of the Christian era the successors of the Olmecs built the first great American city, Teotihuacan, in what it now Mexico. It was for two or three centuries a major trading centre and probably of outstanding religious importance, for it contained a huge complex of pyramids and great public buildings. Mysteriously, it was destroyed in about the seventh century, possibly by one of a series of waves of invaders moving southwards into the valley of central Mexico. These movements began an age of migration and warfare which was to last until the coming of the Spaniards, and produced several brilliant regional societies.

The most remarkable of them was the Mayan society of Yucatán, Guatemala and Honduras. Even its setting was extraordinary. Virtually all the great Mayan sites lie in tropical rain-forest, a setting which when cleared is briefly immensely fertile, but demands big efforts under heavy attacks from animals, insects, climate and disease. Yet in such a setting the Maya not only maintained themselves for many centuries with a primitive agricultural technique (they lacked ploughs and metal tools and after burning and clearing land could only use it for a couple of seasons before moving on to another site), but also raised stone buildings comparable to those of ancient Egypt.

Many Mayan sites may remain undiscovered in the jungle, but enough have now been found to reconstruct an outline of Mayan history and society. Mayan culture begins to appear from about AD 100, blossoming into a great period between 600 and 900, when its finest buildings, sculpture and pottery were produced. The Maya did not build cities, but great ceremonial complexes, combinations of temples, pyramids, tombs and ritual courts, usually uninhabited except by priests and their attendants. This in part explains their impressiveness; like the ziggurats and pyramids the great Mayan towers were meant to impress the beholder with a sense of the remoteness of the gods and the importance of those who climbed the long flights of steps to commune with them. Whatever their social organization, the Maya at this stage appear to have been governed theocratically. Religious practice appears to have consisted of the performance of regular acts of intercession and worship in a cycle calculated from a calendar derived from astronomical observation. Many scholars have found this the only Mayan achievement worthy of comparison with the buildings and it was indeed a great feat of mathematics. Through the calendar, enough of Mayan thinking can be grasped to make it evident that this people's religious leaders had an idea of time much vaster than that of any other civilization of which we have knowledge; they calculated an antiquity of hundreds of thousands of years. They may even have arrived at the idea that time has no beginning.

Stone hieroglyph and three surviving books tell us something about this calendar and also help to provide a chronology; the Maya of the classical era used to put up dated monuments every twenty years to record the passage of time. The last of them is dated to 928. By then, Mayan civilization had reached its peak. For all the skill of its builders and craftsmen in jade and obsidian, it had considerable limitations. The makers of the great temples never achieved the arch, nor could they employ carts in their operations for the Maya never discovered the wheel, while the religious world in whose shadows they lived was peopled by two-headed dragons, jaguars and grinning skulls.

Mayan civilization was, that is to say, a very specialized achievement, requiring huge investments of labour in economically unproductive building. The Egyptians had done the same, but also much more. Remarkable though this was – for it arose out of a formidably difficult context – it may be the fundamental reason for a decline which begins to be apparent after the tenth century: perhaps Mayan civilization was overloaded at an early date. Soon after its beginning a people from the valley of Mexico, probably Toltec, seized Chichen Itza, the greatest Mayan site, and from this time the jungle centres of the south began to be abandoned. The invaders brought metal with them and also the Mexican practice of sacrificing prisoners of war. Their gods begin to appear in sculpture at the Mayan sites. Seemingly, there was also a shift of power from priests to secular rulers among the Mayans and a cultural recession marked by cruder pottery and sculpture; hieroglyph declined, too. Chichen Itza was abandoned in the thirteenth century after being sacked and the Mayan capital removed to a site defended by five miles of wall; the earlier sites had been usually undefended. This capital, too, was sacked, apparently by a rising among the Mayan peasantry in about 1460, and with this Mayan civilization may be considered at an end. In the sixteenth century Yucatán passed into the hands of the Spanish, though only in 1699 did the last Mayan stronghold fall to them.

The Spaniards were only in the most formal sense the destroyers of Mayan civilization. It had already collapsed from within by the time they arrived. Explanation is not easy, given our information, and it is tempting to fall back on metaphor: Mayan civilization was the answer to a huge challenge and could meet it for a time, but only at the cost of a narrow specialization and burdens colossal in relation to its resources. The result, as in the case of the other American civilizations, was a virtually complete extinction; for not quite the same reasons as elsewhere, but equally decisively, Mayan traditions came to an end, leaving behind no technology, no living style, no literature, no institutions except at the most elementary level. What it left, instead, were wondrous ruins which would bemuse and fascinate those who had later to try to explain them.

While Mayan society was in its final decay, one of the last peoples to arrive in the valley of Mexico won an hegemony there which amazed the Spanish more than anything they later found in Yucatán. These were the Aztecs, who had entered the valley in about AD 1350, overthrowing the Toltecs who then exercised supremacy there. They settled in two villages on marshy land at the edge of Lake Texcoco; one of these was called Tenochtitlan and it was to be the capital of an Aztec empire which expanded in less than two centuries to cover the whole of central Mexico. Aztec expeditions went far south into what was later the republic of Panama but showed no diligence in settlement. The Aztecs were warriors and preferred an empire of tribute: their army gave them the obedience of some thirty or so minor tribes or states which they left more or less alone, provided the agreed tribute was forthcoming. The gods of these peoples were given the compliment of inclusion in the Aztec pantheon.

The centre of Aztec civilization was Tenochtitlan, the capital they had built up from the village. It stood in Lake Texcoco on a group of islands connected to the lake shores by causeways, one of which was five miles long and took eight horsemen abreast. The Spanish left excited descriptions of this city: its magnificence, said one, exceeded that of Rome or Constantinople. It probably contained about a hundred thousand inhabitants at the beginning of the sixteenth century and to its maintenance went what was received from the subject peoples. By comparison with European cities, it was an astonishing place, filled with temples and dominated by huge artificial pyramids, yet its magnificence seems to have been derivative, for the Aztecs exploited the skills of their subjects. Not a single important invention or innovation of Mexican culture can confidently be assigned to the post-Toltec period. The Aztecs controlled, developed and exploited the civilization that they found.

When the Spanish arrived in the early sixteenth century, the Aztec empire

A Mayan relief once forming part of the decoration of the lintel of a house. It shows a form of ritual self-torture or, perhaps, penance. Before a priest-like figure kneels a suppliant who draws through his or her tongue a cord studded with thorns.

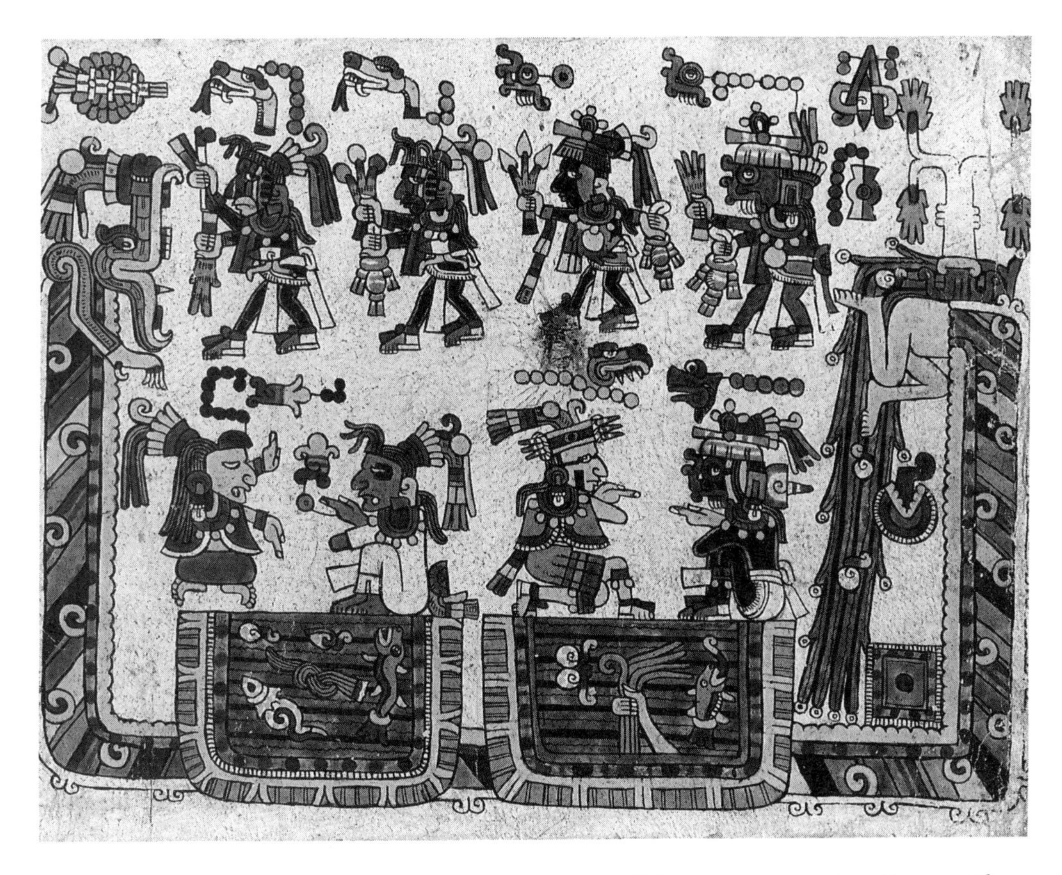

Among the Indian peoples who fell under the sway of the Aztecs were the Mixtecs. This example of their manuscript was sent to the emperor Charles V by Cortés.

was still expanding. Not all of its subject peoples were completely subdued, but Aztec rule ran from coast to coast. At its head was a semi-divine but elected ruler, chosen from a royal family. He directed a highly ordered and centralized society, making heavy demands on its members for compulsory labour and military service, but also providing them with an annual subsistence. It was a civilization pictographically literate, highly skilled in agriculture and the handling of gold, but knowing nothing of the plough, iron-working or the wheel. Its central rituals – which greatly shocked the Spaniards – included human sacrifice; no less than twenty thousand victims were killed at the dedication of the great pyramid of Tenochtitlan. Such holocausts re-enacted a cosmic drama which was the heart of Aztec mythology; it taught that the gods had been obliged to sacrifice themselves to give the sun the blood it needed as food.

This religion struck Europeans by its revolting details – the tearing out of victims' hearts, the flayings and ceremonial decapitations – but its bizarre and horrific accompaniments were less significant than its profound political and social implications. The importance of sacrifice meant that a continual flow of victims was needed. As these were usually supplied by prisoners of war – and because death in battle was also a route to the paradise of the sun for the warrior – a state of peace in the Aztec empire would have been disastrous from a religious point of view. Hence, the Aztecs did not really mind that their dependencies were only loosely controlled and that revolts were frequent. Subject tribes were allowed to keep their own rulers and governments so that punitive raids could be made upon them at the slightest excuse. This ensured that the empire could not win the loyalty of the subject peoples; they were bound to welcome the Aztec collapse when it came. Religion was also to affect in other ways the capacity to respond to the threat from Europeans, notably in the Aztecs' desire to take prisoners for sacrifice rather than to kill their enemies in battle, and in their belief that one day their great god, Quetzalcoatl, white-skinned and bearded, would return from the east, where he had gone after instructing his people in the arts.

Altogether, for all its aesthetic impressiveness and formidable social efficiency, the feel of Aztec civilization is harsh, brutal and unattractive. Few civilizations of which we know much have pressed so far their demands on their members. It seems to have lived always in a state of tension, a pessimistic civilization, its members uneasily aware that collapse was more than a possibility.

To the south of Mexico and Yucatán lay several other cultures distinct enough in their degree of civilization but none of them was so remarkable as the most distant, the Andean civilization of Peru. The Mexican peoples still lived for the most part in the Stone Age; the Andeans had got much further than this. They had also created a true state. If the Maya excelled among the American cultures in the elaborate calculations of their calendar the Andeans were far ahead of their neighbours in the complexity of their government. The imagination of the Spaniards was captured by Peru even more than by Mexico, and the reason was not simply its immense and obvious wealth in precious metals, but its apparently just, efficient and highly complex social system. Some Europeans soon found accounts of it attractive, for it required an almost total subordination of the individual to the collective.

This was the society ruled by the Incas. In the twelfth century a people from Cuzeo began to extend its control over earlier centres of civilization in Peru. Like the Aztec, they began as neighbours of those longer civilized than themselves; they were barbarians who soon took over the skills and fruits of higher cultures. At the end of the fifteenth century the Incas ruled a realm extending from Ecuador to central Chile, their conquest of the coastal areas being the most recent. This was an astonishing feat of government, for it had to contend with the natural obstacles provided by the Andes. The Inca state was held together by about ten thousand miles of roads passable in all weathers by chains of runners who bore messages either orally or recorded in *quipu*, a code of knots in coloured cords. With this device elaborate records were kept. Though preliterate, the Andean empire was formidably totalitarian in the organization of its subjects' lives. The Incas became the ruling caste of the empire, its head becoming *Sapa Inca* – the 'only Inca'. His rule was a despotism based on the control of labour. The population was organized in units of which the smallest was that of ten heads of families. From these units, labour service and produce were exacted. Careful and tight control kept population where it was needed; removal or marriage outside the local community was not allowed. All produce was state property; in this way agriculturists fed herdsmen and craftsmen and received textiles in exchange (the llama was the all-purpose beast of Andean culture, providing wool as well as transport, milk and meat). There was no commerce. Mining for precious metals and copper resulted in an exquisite adornment of Cuzco which amazed the Spaniards when they came to it. Tensions inside this system were not dealt with merely by force, but by the resettlement of loyal populations in a disaffected area and a strict control of the educational system in order to inculcate the notables of conquered peoples with the proper attitudes.

Like the Aztecs, the Incas organized and exploited the achievements of culture which they found already to hand, though less brutally. Their aim was integration rather than obliteration and they tolerated the cults of conquered peoples. Their own god was the sun. The absence of literacy makes it hard to penetrate the mind of this civilization, but it is noticeable that, though in a different way, the Peruvians seem to have shared the Aztecs' preoccupation with death. Accidents of climate, as in Egypt, favoured its expression in rites of mummification; the dry air of the high Andes was as good a preservative as the sand of the desert. Beyond this it is not easy to say what divisions among the conquered peoples persisted and were expressed in the survival of tribal cults. When a challenge appeared from Europe it became apparent that Inca rule had not eliminated discontent among its subjects, for all its remarkable success.

All the American civilizations were in important and obvious ways very different from those of Asia or Europe. A complete literacy escaped them, though the

Incas had good enough record-keeping processes to run complex governmental structures. Their technologies, though they had certain skills at a high level, were not so developed as those already long known elsewhere. Though these civilizations provided satisfactory settings and institutions for cultures of intense (but limited) power, the contribution of the indigenous Americans to the world's future was not to be made through them, therefore. It had in fact already been made before they appeared, through the obscure, unrecorded discoveries of primitive cultivators who had first discovered how to exploit the ancestors of tomatoes, maize, potatoes and squash. In so doing they had unwittingly made a huge addition to the resources of mankind. The glittering civilizations built on that in the Americas, though, were fated in the end to be no more than beautiful curiosities in the margin of world history, ultimately without progeny.

10
Europe: the First Revolution

Few terms have such misleading connotations as 'the Middle Ages'. A wholly Eurocentric usage, meaning nothing in the history of other traditions, the phrase embodies the negative idea that no interest attaches to certain centuries except their position in time. They were first singled out and labelled by men in the fifteenth and sixteenth centuries who wanted to recapture a classical antiquity long cut off from them. In that remote past, they thought, men had done and made great things; a sense of rebirth and quickening of civilization upon them, they could believe that in their own day great things were being done once more. But in between two periods of creativity they saw only a void – *Medio Evo*, *Media Aetas*, the Middle Ages – defined just by falling in between other ages, and in itself dull, uninteresting, barbaric.

It was not long before people could see a little more than this to a thousand or so years of European history. One way in which they gained perspective was by looking for the origins of what they knew; seventeenth-century Englishmen talked about a 'Norman Yoke' supposedly laid on their ancestors and eighteenth-century Frenchmen idealized their aristocracy by attributing its origins to Frankish Conquest. Such reflexions, nonetheless, were very selective; in so far as the Middle Ages were thought of as a whole it was still, even two hundred years ago, usually with contempt. Then, quite suddenly, came a great change. Men started to idealize those lost centuries as vigorously as their forebears had ignored them. Europeans began to fill out their picture of the past with historical novels about chivalry and their countryside with mock baronial castles inhabited by cotton-spinners and stockbrokers. More important, a huge effort of scholarship was then brought to bear on the records of these times. This was an improvement, but still left impediments to understanding, some of which are still with us. Men came to idealize the unity of medieval Christian civilization and the seeming stability of its life, but in so doing blurred the huge variety within it.

Consequently, it is still very hard to be sure we understand the European Middle Ages, though one crude distinction in this great tract of time nevertheless seems obvious enough. The centuries between the end of antiquity and the year 1000 or so now look very much like an age of foundation. Certain great markers then laid out the patterns of the future, though change was slow and its staying power still uncertain. Then, in the eleventh century, a change of pace can be sensed. New developments quicken and become discernible. It becomes clear, as time passes, that they are opening the way to something quite different. An age of adventure and revolution is beginning in Europe, and it will go on until European history merges with the first age of global history.

This makes it hard to say when the Middle Ages 'end'. In many parts of Europe, they were still going strong at the end of the eighteenth century, the moment at which Europe's first independent offshoot had just come into existence across the Atlantic. Even in the new United States, too, there many people who then, like millions of Europeans, were still gripped by a supernatural view of life, and traditional religious views about it, much as medieval men and women had been five hundred years earlier. Many Europeans then still lived lives which in their material dimensions were still those of their medieval forerunners, too. Yet at that moment, the Middle Ages were in some countries long over in any important sense. Old institutions had gone or were crumbling, taking unquestioned traditions of authority with them. In many places, something we can recognize as the life of the modern world was already going on. This first became possible, then likely, and finally unavoidable in what can now be seen as Europe's second formative phase, and the first of her revolutionary eras.

THE CHURCH

The Church is a good place to begin. By 'the Church' as an earthly institution, Christians mean the whole body of the faithful, lay and cleric alike. In this sense the Church came to be the same thing as European society during the Middle Ages. By 1500 only a few Jews, visitors and slaves stood apart from the huge body of people who (at least formally) shared Christian beliefs. Europe was Christian. Explicit paganism had disappeared from the map between the Atlantic coasts of Spain and the eastern boundaries of Poland. This was a great qualitative as well as a quantitative change. The religious beliefs of Christians were the deepest spring of a whole civilization which had matured for hundreds of years and was not yet threatened seriously by division or at all by alternative mythologies. Christianity had come to define Europe's purpose and to give its life a transcendent goal. It was also the reason why Europeans first became conscious of themselves as members of a particular society.

Nowadays, non-Christians are likely to think of something else as 'the Church'. People use the word to describe ecclesiastical institutions, the formal structures and organizations which maintain the life of worship and discipline of the believer. In this sense, too, the Church had come a long way by 1500. Whatever qualifications and ambiguities hung about them, its successes were huge; its failures might be great, too, but within the Church there were plenty of men who confidently insisted on the Church's power (and duty) to put them right. The Roman Church which had been a backwater of ecclesiastical life in late antiquity was, long before the fall of Constantinople, the possessor and focus of unprecedented power and influence. It had not only acquired new independence and importance but also had given a new temper to the Christian life since the eleventh century. Christianity then had become both more disciplined and more aggressive. It also become more rigid: many doctrinal and liturgical practices dominant until this century are less than a thousand years old – they were set up, that is to say, when more than half the Christian era was already over.

The most important changes took roughly from 1000 to 1250, and they constituted a revolution. Their beginnings lay in the Cluniac movement. Four of the first eight abbots of Cluny were later canonized: seven of them were outstanding men. They advised popes, acted as their legates, served emperors as ambassadors. They were men of culture, often of noble birth, sprung from the greatest families of Burgundy and the West Franks (a fact which helped to widen Cluny's influence) and they threw their weight behind the moral and spiritual reform of the Church. Leo IX, the pope with whom papal reform really begins, eagerly promoted Cluniac ideas. He spent barely six months of his five years' pontificate at Rome, moving about, instead, from synod to synod in France and Germany, correcting local practice, checking interference with the Church by lay magnates, punishing clerical impropriety, imposing a new pattern of ecclesiastical discipline. Greater standardization of practice within the Church was one of the first results. It begins to look more homogeneous.

Another outcome was the founding of a second great monastic order, the Cistercians (so named after the place of their first house, at Cîteaux), by monks dissatisfied with Cluny and anxious to return to the original strictness of the Benedictine rule, in particular by resuming the practical and manual labour Cluny had abandoned. A Cistercian monk, St Bernard, was to be the greatest leader and preacher of both Christian reform and crusade in the twelfth century, and his Order had widespread influence both on monastic discipline and upon ecclesiastical architecture. It too pushed the Church towards greater uniformity and regularity.

The success of reform was also shown in the fervour and moral exaltation of the crusading movement, often a genuinely popular manifestation of religion. But new ways also aroused opposition, some of it among churchmen themselves. Bishops did not always like papal interference in their affairs and parochial clergy did not always see a need to change inherited practices which their flock accepted (clerical marriage, for example). The most spectacular opposition to ecclesiastical reform came in the great quarrel which has gone down in history as the Investiture Contest. The attention given to it has been perhaps slightly disproportionate and, some would say, misleading. The central episodes lasted only a half-century or so and the issue was by no means clear-cut. The very distinction of Church and State implicit in some aspects of the quarrel was in anything like the modern sense still unthinkable to medieval man. The specific administrative and legal

St Etienne at Nevers: one of the most magnificent and unspoiled Romanesque churches of France, built between 1063 and 1097 to serve as a chapel for a Cluniac priory.

practices at issue were by and large quite soon the subject of agreement and many clergy felt more loyalty to their lay rulers than to the Roman Pope. Much of what was at stake, too, was very material. What was in dispute was the sharing of power and wealth within the ruling classes who supplied the personnel of both royal and ecclesiastical government in Germany and Italy, the lands of the Holy Roman Empire. Yet other countries were touched by similar quarrels – the French in the late eleventh century, the English in the early twelfth – because there was a transcendent theoretical principle at stake which did not go away: what was the proper relationship of lay and clerical authority?

The most public battle of the Investiture struggle was fought just after the election of Pope Gregory VII in 1073. Hildebrand (Gregory's name before his election: hence the adjective 'Hildebrandine' sometimes used of his policies and times) was a far from attractive person, but a pope of great personal and moral courage. He had been one of Leo IX's advisers and fought all his life for the independence and dominance of the Papacy within western Christendom. He was an Italian, but not a Roman, and this, perhaps, explains why before he was himself pope he played a prominent part in the transfer of papal election to the college of cardinals, and the exclusion from it of the Roman lay nobility. When reform became a matter of politics and law rather than morals and manners (as it did during his twelve years' pontificate) Hildebrand was likely to provoke rather than avoid conflict. He was a lover of decisive action without too nice a regard for possible consequences.

Perhaps strife was already inevitable. At the core of reform lay the ideal of an independent Church. It could only perform its task, thought Leo and his followers, if free from lay interference. The Church should stand apart from the state and the clergy should live lives different from laymen's lives: they should be a distinct society within Christendom. From this ideal came the attacks on simony (the buying of preferment), the campaign against the marriage of priests, and a fierce struggle over the exercise of hitherto uncontested lay interference in appointment and promotion. This last gave its name to the long quarrel over lay 'investiture': who rightfully appointed to a vacant bishopric, the temporal ruler or the Church? The right was symbolized in the act of giving his ring and staff to the new bishop.

Further potential for trouble lay in more mundane issues. Perhaps the emperors were bound to find themselves in conflict with the Papacy sooner or later, once it ceased to be in need of them against other enemies, for they inherited big, if shadowy claims of authority from the past which they could hardly abandon without a struggle. In Germany the Carolingian tradition had subordinated the Church to a royal protection which easily blurred into domination. Furthermore, within Italy the empire had allies, clients and interests to defend. Since the tenth century, both the emperors' practical control of the Papacy and their formal authority had declined. The new way of electing popes left the emperor with a theoretical veto and no more. The working relationship, too, had deteriorated in that some popes had already begun to dabble in troubled waters by seeking support among the emperor's vassals.

The temperament of Gregory VII was no emollient in this delicate situation. Once elected, he took his throne without imperial assent, simply informing the emperor of the fact. Two years later he issued a decree on lay investiture. Curiously, what it actually said has not survived, but its general content is known: Gregory forbade any layman to invest a cleric with a bishopric or other ecclesiastical office and excommunicated some of the emperor's clerical councillors on the grounds that they had been guilty of simony in purchasing their preferment. To cap matters, Gregory summoned Henry IV to Rome to appear before him and defend himself against charges of misconduct.

Henry responded at first through the Church itself; he got a German synod to declare Gregory deposed. This earned him excommunication, which would have mattered less had he not faced powerful enemies in Germany who now had the pope's support. The result was that Henry had to give way. To avoid trial before the German bishops presided over by Gregory (who was already on his way to Germany), Henry came in humiliation to Canossa, where he waited in the snow barefoot until Gregory

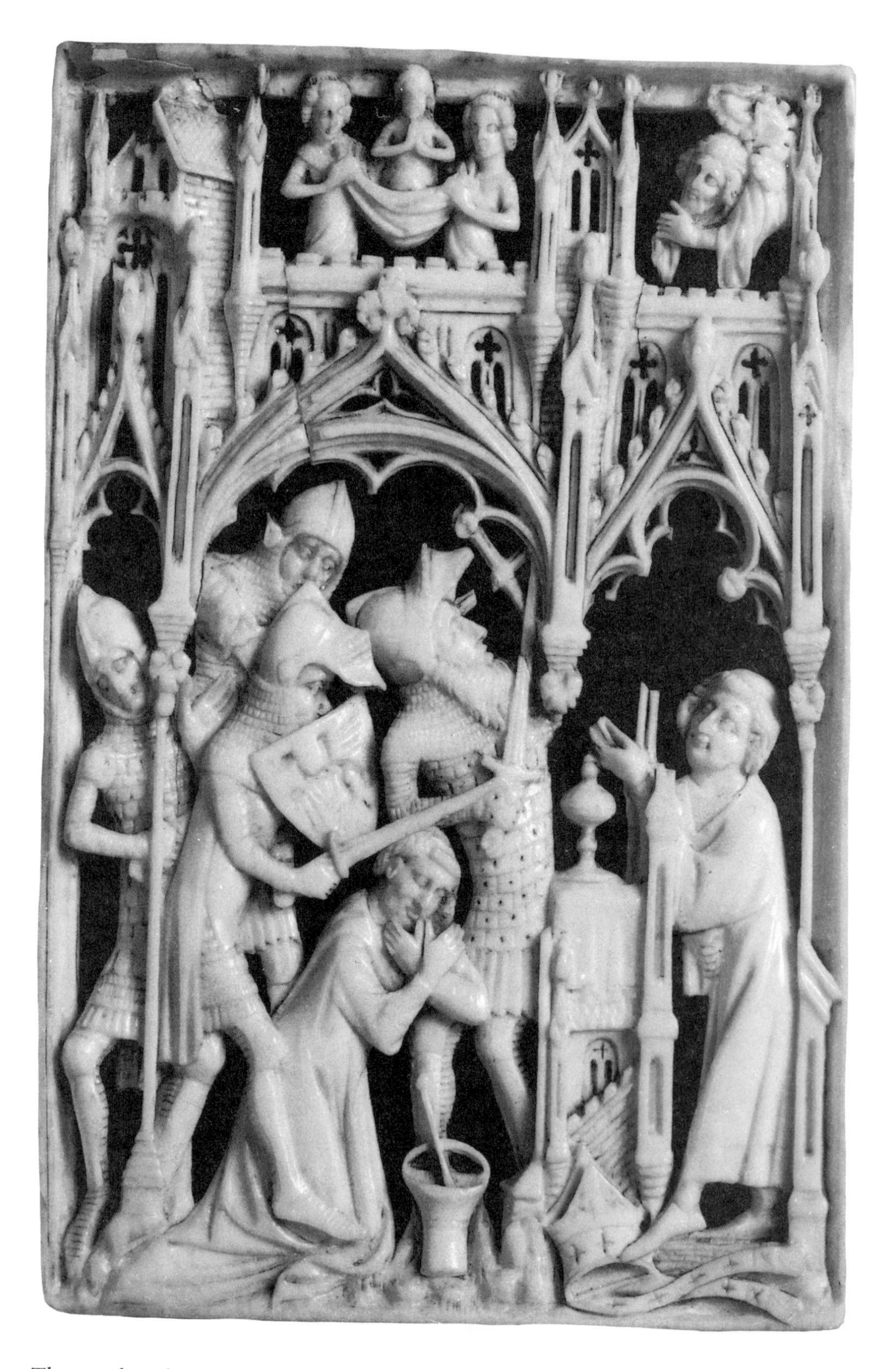

The murder of Becket, an incident in the medieval struggle of kings and churchmen, and a martyrdom which made Canterbury a place of pilgrimage for western Christendom.

would receive his penance in one of the most dramatic of all confrontations of lay and spiritual authority. But Gregory had not really won. Not much of a stir was caused by Canossa at the time. The pope's position was too extreme; he went beyond canon law to assert a revolutionary doctrine, that kings were but officers who could be removed when the pope judged them unfit or unworthy. This was almost unthinkably subversive to men whose moral horizons were dominated by the idea of the sacredness of oaths of fealty; it foreshadowed later claims to papal monarchy but was bound to be unacceptable to any king.

Investiture ran on as an issue for the next fifty years. Gregory lost the sympathy he had won through Henry's bullying and it was not until 1122 that another emperor agreed to a concordat which was seen as a papal victory, though one diplomatically disguised. Yet Gregory had been a true pioneer; he had differentiated clerics and laymen as never before and had made unprecedented claims for the distinction and superiority of papal power. More would be heard of them in the next two centuries. Though his immediate successors acted less dramatically than he, they steadily pressed papal claims to papal advantage. Urban II used the first crusade to become the diplomatic leader of Europe's lay monarchs; they looked to Rome, not the empire. Urban also built up the Church's administrative machine; under him emerged the curia, a Roman bureaucracy which corresponded to the household administrations of the English and French kings. Through it the papal grip on the Church itself was strengthened. In 1123, a historic date, the first ecumenical council was held in the West and its decrees were promulgated in the pope's own name. And all the time, papal jurisprudence and jurisdiction ground away; more and more legal disputes found their way from the local church courts to papal judges, whether resident at Rome, or sitting locally.

Prestige, dogma, political skill, administrative pressure, judicial practice and the control of more and more benefices all buttressed the new ascendancy of the papacy in the Church. By 1100 the groundwork was done for the emergence of a true papal monarchy. As the investiture contest receded, secular princes were on the whole well-disposed to Rome and it appeared that no essential ground had been lost by the Papacy. There was indeed a spectacular quarrel in England over the question of clerical privilege and immunity from the law of the land which would be an issue of the future; immediately, it led to the murder (and then the canonization) of Becket, the archbishop of Canterbury. But on the whole, the large legal immunities of clergy were not much challenged.

Under Innocent III papal pretensions to monarchical authority reached a new theoretical height. True, Innocent did not go quite so far as Gregory. He did not claim an absolute plenitude of temporal power everywhere in western Christendom, but he said that the papacy had by its authority transferred the empire from the Greeks to the Franks. Within the Church his power was limited by little but the inadequacies of the bureaucratic machine through which he had to operate. Yet papal power was still often deployed in support of the reforming ideas – which shows that much remained to be done. Clerical celibacy became more common and more widespread. Among new practices which were pressed on the Church in the thirteenth century was that of frequent individual confession, a powerful instrument of control in a religiously-minded and anxiety-ridden society. Among doctrinal innovations, the theory of transubstantiation, that by a mystical process the body and blood of Christ were actually present in the bread and wine used in the communion service, was imposed from the thirteenth century onwards.

The final christening of Europe in the central Middle Ages was a great spectacle. Monastic reform and papal autocracy were wedded to intellectual effort and the deployment of new wealth in architecture to make this the next peak of Christian history after the age of the Fathers. It was an achievement whose most fundamental work lay, perhaps, in intellectual and spiritual developments, but it became most visible in stone. What we think of as 'Gothic' architecture was the creation of this period. It produced the European landscape which, until the coming of the railway, was dominated or punctuated by a church tower or spire rising above a little town. Until the twelfth century the major buildings of the Church were usually monastic; then began the building of the astonishing series of cathedrals, especially in northern France and England, which remains one of the great glories of European art and, together with castles, constitutes the major architecture of the Middle Ages. There was great popular enthusiasm, it seems, for these huge investments, though it is difficult to penetrate to the mental attitudes behind them. Analogies might be sought in the feeling of twentieth-century enthusiasts for space exploration, but this omits the supernatural dimension of these great buildings. They were both offerings to God and an essential part of the instrumentation of evangelism and education on earth. About their huge naves and aisles moved the processions of relics and the crowds of pilgrims who had come to see them. Their windows were filled with the images of the biblical story which was the core of European culture; their façades were covered with the didactic representations of the fate awaiting just and unjust. Christianity achieved in them a new publicity and collectiveness. Nor is it possible to assess the full impact of these great churches on the imagination of medieval Europeans without reminding ourselves how much greater was the contrast their splendour presented to the reality of everyday life than any imaginable today.

A fifteenth-century picture of thirteenth-century heretics about to die by burning before the French king.

The power and penetration of organized Christianity were further reinforced by new religious orders. Two were outstanding, the mendicant Franciscans and Dominicans, who in England came to be called respectively the Grey and Black Friars, from the colours of their habits. The Franciscans were true revolutionaries: their founder, St Francis of Assisi, left his family to lead a life of poverty among the sick, the needy and the leprous. The followers who soon gathered about him eagerly took up a life directed towards the imitation of Christ's poverty and humility. There was at first no formal organization and Francis was never a priest, but Innocent III, shrewdly seizing the opportunity of patronizing this potentially divisive movement instead of letting it escape from control, bade them elect a Superior. Through him the new fraternity owed and maintained rigorous obedience to the Holy See. They could provide a counterweight to local episcopal authority because they could preach without the licence of the bishop of the diocese. The older monastic orders recognized a danger and opposed the Franciscans, but the friars prospered, despite internal quarrels about their organization. In the end they acquired a considerable administrative structure, but they always remained peculiarly the evangelists of the poor and the mission field.

The Dominicans were founded to further a narrower end. Their founder was a Castilian priest who went to preach in the Languedoc to heretics, the Albigensians. From his companions grew a new preaching order; when Dominic died in 1221 his seventeen followers had become over five hundred friars. Like the Franciscans, they were mendicants vowed to poverty, and like them, too, they threw themselves into missionary work. But their impact was primarily intellectual and they became a great force in a new institution of great importance, just taking shape, the first western universities. Dominicans came also to provide many of the personnel of the Inquisition, an organization to combat heresy which appeared in the early thirteenth century. From the fourth century onwards, churchmen had urged the persecution of heretics. Yet the first papal condemnation of them did not come until 1184. Only under Innocent III did persecution come to be the duty of Catholic kings. The Albigensians were certainly not Catholic, but there is some doubt whether they should really be regarded even as Christian heretics. Their beliefs reflect Manichaean doctrines. They were dualists; some of whom rejected all material creation as evil. Like those of many later heretics, heterodox religious views were taken to imply aberration or at least nonconformity in social and moral practices. Innocent III seems to have decided to persecute the Albigensians after the murder of a papal legate in the Languedoc and in 1209 a crusade was launched against them. It attracted many laymen (especially from northern France) because of the chance it offered for a quick grab at the lands and homes of the Albigensians, but it also marked a great innovation: the joining of state and Church in western Christendom to crush by force dissent which might place either in danger. It was for a long time an effective device, though never completely so.

In judging the theory and practice of medieval intolerance it must be remembered that the danger in which society was felt to stand from heresy was appalling: its members might face everlasting torment. Yet persecution did not prevent the appearance of new heresies again and again in the next three centuries, because they expressed real needs. Heresy was, in one sense, an exposure of a hollow core in the success which the Church had so spectacularly achieved. Heretics were living evidence of dissatisfaction with the outcome of a long and often heroic battle. Other critics would also make themselves heard in due course and different ways. Papal monarchical theory provoked counter-doctrine; thinkers would argue that the Church had a defined sphere of activity which did not extend to meddling in secular affairs. As men became more conscious of national communities and respectful of their claims, this would seem more and more appealing. The rise of mystical religion, too, was another phenomenon always tending to slip outside the ecclesiastical structure. In movements like the Brethren of the Common Life, following the teachings of the mystic Thomas à Kempis, laymen created religious practices and devotional forms which often escaped from clerical control.

Such movements expressed the great paradox of the medieval Church. It had risen to a pinnacle of power and wealth. It deployed lands, tithes, and papal taxation in the service of a magnificent hierarchy whose worldly greatness reflected the glory of God and whose lavish cathedrals, great monastic churches, splendid liturgies, learned foundations and libraries embodied the devotion and sacrifices of the faithful. Yet the point of this huge concentration of power and grandeur was to preach a faith at whose heart lay the glorification of poverty and humility and the superiority of things not of this world.

The worldliness of the Church drew increasing criticism. It was not just that a few ecclesiastical magnates lolled back upon the cushion of privilege and endowment to gratify their appetites and neglect their flocks. There was also a more subtle corruption inherent in power. The identification of the defence of the faith with the triumph of an institution had given the Church an increasingly bureaucratic and legalistic face. The point had arisen as early as the days of St Bernard; even then, there were too many ecclesiastical lawyers, it was said. By the mid-thirteenth century legalism was blatant. The papacy itself was soon criticized. At the death of Innocent III the Church of comfort and of the sacraments was already obscured behind the granite face of centralization. The claims of religion were confused with the assertiveness of an ecclesiastical monarchy demanding freedom from constraint of any sort. It was already difficult to keep the government of the Church in the hands of men of spiritual stature; Martha was pushing Mary aside, because administrative and legal gifts were needed to run a machine which more and more generated its own purposes. A higher authority than that of the pope, some argued, lay in an ecumenical council.

In 1294 a hermit of renowned piety was elected pope. The hopes this roused were quickly dashed. Celestine V was forced to resign within a few weeks, seemingly unable to impose his reforming wishes on the *curia*. His successor was Boniface VIII. He has been called the last medieval pope because he embodied all the pretensions of the papacy at its most political and its most arrogant. He was by training a lawyer and by temperament far from a man of spirituality. He quarrelled violently with the kings of England and France and in the Jubilee of 1300 had two swords carried before him to symbolize his possession of temporal as well as spiritual power. Two years later he asserted that a belief in the sovereignty of the pope over every human being was necessary to salvation.

Under him the long battle with kings came to a head. Nearly a hundred years before, England had been laid under interdict by the pope; this terrifying sentence forbade the administration of any of the sacraments while the king remained unrepentant and unreconciled. Men and women could not have their children baptized or obtain

EUROPEAN UNIVERSITIES FOUNDED BEFORE 1500
Borders are shown at the end of 1991

absolution for their own sins, and those were fearful deprivations in a believing age. King John had been forced to yield. A century later, things had changed. Bishops and their clergy were often estranged from Rome, which had undermined their authority, too. They could sympathize with a stirring national sense of opposition to the papacy whose pretensions reached their peak under Boniface. When the kings of France and England rejected his authority they found churchmen to support them. They also had resentful Italian noblemen to fight for them. In 1303 some of them (in French pay) pursued the old pope to his native city and there seized him with, it was said, appalling physical indignity. His fellow-townsmen released Boniface and he was not (like Celestine, whom he had put in prison) to die in confinement, but die he did, no doubt of shock, a few weeks later.

This was only the beginning of a bad time for the Papacy and, some would claim, for the Church. For more than four centuries it was to face recurrent and mounting waves of hostility which, though often heroically met, ended by calling Christianity itself in question. Even by the end of Boniface's reign, the legal claims he had made were almost beside the point; no one stirred to avenge him. Now spiritual failure increasingly drew fire; henceforth the Papacy was to be condemned more for standing in the way of reform than for claiming too much of kings. For a long time, though, criticism had important limits. The notion of autonomous, self-justified criticism was unthinkable in the Middle Ages: it was for failures in their traditional religious task that churchmen were criticized.

In 1309, a French pope brought the papal *curia* to Avignon, a town belonging to the king of Naples but overshadowed by the power of the French kings whose lands overlooked it. There was to be a preponderance of French cardinals, too, during the papal residence at Avignon (which lasted until 1377). The English and Germans soon believed the popes had become the tool of the French kings and took steps against the independence of the Church in their own territories. The imperial electors declared that their vote required no approval or confirmation by the pope and that the imperial power came from God alone.

At Avignon the popes lived in a huge palace, whose erection was a symbol of their decision to stay away from Rome, and whose luxury was a symbol of growing worldliness. The papal court was of unexampled magnificence, attended by a splendid train of servitors and administrators paid for by ecclesiastical taxation and misappropriation. Unfortunately the fourteenth century was a time of economic disaster; a much reduced population was being asked to pay more for a more costly (and, some said, extravagant) Papacy. Centralization continued to breed corruption – the abuse of the papal rights to appoint to vacant benefices was an obvious instance – and accusations of simony and pluralism had more and more plausibility. The personal conduct of the higher clergy was more and more obviously at variance with apostolic ideals. A crisis arose among the Franciscans themselves, some of the brothers, the 'spirituals', insisting that they take seriously their founder's rule of poverty, while their more relaxed colleagues refused to give up the wealth which had come to their order. Theological issues became entangled with this dispute. Soon there were Franciscans preaching that Avignon was Babylon, the scarlet whore of the Apocalypse, and that the Papacy's overthrow was at hand, while a pope, asserting that Christ Himself had respected property, condemned the ideal of apostolic poverty and unleashed the Inquisition against the 'spirituals'. They were burned for their preachings, but not before they had won audiences.

Thus the exile in Avignon fed a popular anti-clericalism and anti-papalism different from that of kings exasperated against priests who would not accept their jurisdiction. Many of the clergy themselves felt that rich abbeys and worldly bishops were a sign of a Church that had become secularized. This was the irony that tainted the legacy of Gregory VII. Criticism eventually rose to the point at which the Papacy returned to Rome in 1377, only to face the greatest scandal in the history of the Church, a 'Great Schism'. Secular monarchs set on having quasi-national churches in their own realms, and the college of twenty or so cardinals, manipulating the Papacy so as to maintain their own revenues and position, together brought about the election of two popes, the second by the French cardinals alone. For thirty years popes at Rome and Avignon simultaneously claimed the headship of the Church. Eight years afterwards there was a third contender as well. As the schism wore on, the criticism directed against the Papacy became more and more virulent. 'Antichrist' was a favourite term of abuse for the claimant to the patrimony of St Peter. It was complicated by the involvement of secular rivalries, too. For the Avignon pope, broadly, there stood as allies France, Scotland, Aragon and Milan; the Roman was supported by England, the German emperors, Naples and Flanders.

Yet the schism at one moment seemed to promise renovation and reformation. The instrument to which reformers turned was an ecumenical or general council of the Church; to return to the days of the apostles and the Fathers for a means to put the papal house in order sounded good sense to many Catholics. Unfortunately, it did not turn out well. Four councils were held. The first, held at Pisa in 1409, struck out boldly, proclaiming the deposition of both popes and choosing another. This meant there were now three pretenders to the chair of St Peter; moreover, when the new one died after a few months, another was elected whose choice was said to be tainted by simony (this was the first John XXIII, now no longer recognized as a pope and the victim of one of Gibbon's most searing judgements). The next council (Constance, 1414–18) removed John (though he had summoned it), got one of his competitors to abdicate and then deposed the third pretender. At last there could be a fresh start; the schism was healed. In 1417 a new pope was elected, Martin V. This was a success, but some people had hoped for more; they had sought reform and the council had been diverted from that. Instead it had devoted its time to heresy, and support for reform dwindled once the unity of the Papacy was restored. After another council (Siena, 1423–4) had been dissolved by Martin V for urging reform ('that the Supreme Pontiff should be called to account was perilous', he declared), the last met at Basle (1431–49), but was ineffective long before its

Eve, mother of all men, came to play an ambiguous part in Christian thinking and one more complicated than she had ever sustained in Jewish legend. She was remembered not only as our originator but as the agent of the Fall of Man and so she is portrayed, apple in hand and seductive, in this carving made for the great twelfth-century Burgundian cathedral at Autun.

dissolution. The conciliar movement had not achieved the desired reform and papal power was restored. The principle that there existed an alternative conciliar source of authority inside the Church was always henceforth regarded with suspicion at Rome. Within a few years it was declared heresy to appeal from the pope to a general council.

The Church had not risen to the level of the crisis now upon it. The Papacy had maintained its superiority, but its victory was only partial; secular rulers had reaped the benefits of anti-papal feeling in new freedoms for national Churches. As for the moral authority of Rome, that had clearly not been restored and one result would be a more damaging movement for reform three-quarters of a century later. The Papacy now began to look more and more Italian, and so it was to remain. There were some dismal popes to come in the next two centuries, but that did less damage to the Church than the evolution of their See towards becoming just one more Italian state.

Heresy, always smouldering, had burst out in a blaze of reforming zeal during the conciliar period. Two outstanding men, Wyclif and Hus, focused the discontents to which schism had given rise. They were first and foremost reformers, though Wyclif was a teacher and thinker rather than a man of action. Hus, a Bohemian, became the leader of a movement which involved national as well as ecclesiastical issues; he exercised huge influence as a preacher in Prague. He was condemned by the council of Constance for heretical views on predestination and property and was burned in 1415. The great impulse given by Wyclif and Hus flagged as their criticisms were muffled, but they had tapped the nerve of national anti-papalism which was to prove so destructive of the unity of the western Church. Catholics and Hussites were still disputing Bohemia in bitter civil wars twenty years after Hus' death. Meanwhile, the Papacy itself made concessions in its diplomacy with the lay monarchies of the fifteenth century.

Religious zeal in the fifteenth century more and more appeared to bypass the central apparatus of the Church. Fervour manifested itself in a continuing flow of mystical writing and in new fashions in popular religion. A new obsession with the agony of the passion was reflected in art; new devotions to saints, a craze for flagellation, outbreaks of dancing frenzy all show a heightened excitability. An outstanding example of the appeal and power of a popular preacher can be seen in Savonarola, a Dominican whose immense success made him for a time moral dictator of Florence in the 1490s. But religious favour often escaped the formal and ecclesiastical structures. In the fourteenth and fifteenth centuries much of the emphasis of popular religion was individual and devotional. Another impression of the inadequacy of both vision and machinery within the hierarchies is to be found, too, in a neglect of missionary work outside Europe.

All in all, the fifteenth century leaves a sense of withdrawal, an ebbing after a big effort which had lasted nearly two centuries. Yet to leave the medieval Church with that impression uppermost in our minds would be to risk a grave misunderstanding of a society made more different from our own by religion than by any other factor. Europe was still Christendom and was so even more consciously after 1453. Within its boundaries, almost the whole of life was defined by religion. The Church was for most men and women the only recorder and authenticator of the great moments of their existence – their marriages, their children's births and baptisms, their deaths. Many of them wholly gave themselves up to it; a much greater proportion of the population became monks and nuns than is the case today, but though they might think of withdrawal to the cloister from a hostile everyday, what they left behind was no secular world such as ours, wholly distinct from and indifferent to the Church. Learning, charity, administration, justice and huge stretches of economic life all fell within the ambit and regulation of religion. Even when men attacked churchmen, they did so in the name of the standards the Church had itself taught them and with appeals to the knowledge of God's purpose it gave them. Religious myth was not only the deepest spring of a civilization, it was still the life of all men. It defined human purpose and did so in terms of a transcendent good. Outside the Church, the community of all believers, lay only paganism. The devil – conceived in a most material form – lay in wait for those who strayed from the path of grace. If there were some bishops and even popes among the errant, so much the worse for them. Human frailty could not compromise the religious view of life. God's justice would be shown and He would divide sheep from goats in the Day of Wrath when all things would end.

PRINCIPALITIES AND POWERS

Most people today are used to the idea of the state. It is generally agreed that the world's surface is divided between impersonal organizations working through officials marked out in special ways, and that such organizations provide the final public authority for any given area. Often, states are thought in some way to represent people or nations. But whether they do or not, states are the building blocks from which most of us would construct a political account of the modern world.

None of this would have been intelligible to a European in 1000; five hundred years later much of it might well have been, depending on who the European was. The process by which the modern state emerged, though far from complete by 1500, is one of the markers which delimit the modern era of history. The realities had come first, before principles and ideas. From the thirteenth century onwards many rulers, usually kings, were able for a variety of reasons to increase their power over those they ruled. This was often because they could keep up large armies and arm them with the most effective weapons. Iron cannons were invented in the early fourteenth century; bronze followed, and in the next century big cast-iron guns became available. With their appearance, great men could no longer brave the challenges of their rulers from behind the walls of their castles. Steel crossbows, too, gave a big advantage to those who could afford them. Many rulers were by 1500 well on the way to exercising a monopoly of the use of armed force within their realms. They were arguing more, too, about the frontiers they shared, and this expressed more than just better techniques of surveying. It marked a change in emphasis within government, from a claim to control persons who had a particular relationship to the ruler to one to control people who lived in a certain area. Territorial was replacing personal dependence.

Over such territorial agglomerations, royal power was increasingly exercised directly through officials who, like weaponry, had to be paid for. A kingship which worked through vassals known to the king, who did much of his work for him in return for his favours and who supported him in the field when his needs went beyond what his own estates could supply, gave way to one in which royal government was carried out by employees, paid for by taxes (more and more in cash, not kind), the raising of which was one of their most important tasks. The parchment of charters and rolls began by the sixteenth century to give way to the first trickles and rivulets of what was to become the flood of modern bureaucratic paper.

Such a sketch hopelessly blurs this immensely important and complicated change. It was linked to every side of life, to religion and the sanctions and authority it embodied, to the economy, the resources it offered and the social possibilities it opened or closed, to ideas and the pressure they exerted on still plastic institutions. But the upshot is not in doubt. Somehow, Europe was beginning by 1500 to organize itself differently from the Europe of Carolingians and Ottonians. Though personal and local ties were to remain for centuries overwhelmingly the most important ones for most Europeans, society was institutionalized in a different way from that of the days when even tribal loyalties still counted. The relationship of lord and vassal which, with the vague claims of pope and emperor in the background, so long seemed to exhaust political thought, gave way to an idea of princely power over all the inhabitants of a domain which, in extreme assertions (such as that of Henry VIII of England that a prince knew no external superior save God) was really quite new.

Necessarily, such a change neither took place everywhere in the same way nor at the same pace. By 1800 France and England would have been for centuries unified in a way that Germany and Italy were still not. But wherever it happened, the centre of the process was usually the steady aggrandizement of royal families. Kings enjoyed great advantages. If they ran their affairs carefully they had a more solid power base for their power despite their often, usually large, domains than had noblemen in their smaller estates. The kingly office had a mysterious aura about it, reflected in the solemn circumstances of coronations and anointings. Royal courts and laws seemed to promise a more independent, less expensive justice than could be got from the local feudal lords. Kings could therefore appeal not only to the resources of the feudal structure at whose head – or somewhere near it – they stood, but also to other forces outside. One of these which was slowly revealed as of growing importance was the sense of nationhood.

Longbow and crossbow gave way only slowly to what Shakespeare called 'villainous saltpetre', but the coming of the first guns heralded the end of a long era of warfare during which men could rely only on the extra power given to missile weapons by the tension of cord, wood or steel, or the acceleration of sling or throwing-stick. Early guns, like the mortar shown in this fifteenth-century manuscript, were crude, unreliable and inaccurate, but they rapidly improved and by 1500 were essential elements in any royal armoury.

This is another idea which modern man takes for granted, but we must be careful not to antedate it. No medieval state was national in our sense. Nevertheless, by 1500 the subjects of the kings of England and France could often think of themselves as different from aliens who were not their fellow-subjects, even if they might also regard people who lived in the next village as virtually foreigners. Even two hundred years earlier this sort of distinction was being made between those born within and those born outside the realm and the sense of community of the native-born was steadily enhanced. One symptom was the appearance of belief in national patron saints; though churches had been dedicated to him under the Anglo-Saxon kings, only in the fourteenth century did St George's red cross on a white background became a kind of uniform for English soldiers when he was recognized as official protector of England (his exploit in killing the dragon had only been attributed to him in the twelfth century and may be the result of mixing him up with a legendary Greek hero, Perseus). Another was the writing of national histories (already foreshadowed by the Dark Age histories of the Germanic peoples) and the discovery of national heroes. In the twelfth century a Welshman more or less invented the mythological figure of Arthur, while an Irish chronicler of the same period built up an unhistorical myth of the High King Brian Boru and his defence of Christian Ireland against the Vikings. Above all, there was more vernacular literature. First Spanish and Italian, then French and English began to break through the barrier set about literary creativity by Latin. The ancestors of these tongues are recognizable in twelfth-century romances such as the Song of Roland which transformed a defeat of Charlemagne by Pyrenean mountaineers into the glorious stand of his rearguard against the Arabs, or the Poem of the Cid, the epic of a Spanish national hero. With the fourteenth century came Dante, Langland and Chaucer, each of them writing in a language which we can read with little difficulty.

We must not exaggerate the immediate impact. For centuries yet, family, local community, religion or trade were still to be the focus of most men's loyalties. Such national institutions as they could have seen growing among them would have done little to break into this conservatism; in few places was it more than a matter of the king's justices and the king's tax-gathers – and even in England, in some ways the most national of late medieval states, many people might never have seen either. The rural parishes and little towns of the Middle Ages, on the other hand, were real communities, and in ordinary times provided enough to think about in the way of social responsibilities. We really need another word than 'nationalism' to suggest the occasional and fleeting glimpses of a community of the realm which might once in a while touch a medieval man, or even the irritation which might suddenly burst out in a riot against the presence of foreigners, whether workmen or merchants. (Medieval anti-semitism, of course, had different roots.) Yet such hints of national feeling occasionally reveal the slow consolidation of support for new states in western Europe.

The first of them to cover anything like the areas of their modern successors were England and France. A few thousand Normans had come over from France after the invasion of 1066 to Anglo-Saxon England to form a new ruling class. Their leader, William the Conqueror, gave them lands, but retained more for himself (the royal estates were larger than those of his Anglo-Saxon predecessors) and asserted an ultimate lordship over the rest: he was to be lord of the land and all men held what they held either directly or indirectly of him. He inherited the prestige and machinery of the old English monarchy, too, and this was important, for it raised him decisively above his fellow-Norman warriors. The greatest of them became William's earls and barons, the lesser ones among them knights, ruling England at first from the wooden and earth castles which they spread over the length of the land.

They had conquered one of the most civilized societies in Europe, which went on under the Anglo-Norman kings to show unusual vigour. A few years after the Conquest, English government carried out one of the most remarkable administrative acts of the Middle Ages, the compilation of Domesday Book, a huge survey of England for royal purposes. The evidence was taken from juries in every shire and hundred and its minuteness deeply impressed the Anglo-Saxon chronicler who bitterly noted ('it is shameful to record, but did not seem shameful for him to do') that not an ox, cow or pig escaped the notice of William's men. In the next century there was rapid, even spectacular, development in the judicial strength of the Crown. Though minorities and weak kings from time to time led to royal concessions to the magnates, the essential integrity of the monarchy was not compromised. The constitutional history of England is for five hundred years the story of the authority of the Crown. This owed much to the fact that England was separated from possible enemies, except to the north, by water; it was hard for foreigners to interfere in her domestic politics and the Normans were to remain her last successful invaders.

For a long time, though, the Anglo-Norman kings were more than kings of an island state. They were heirs of a complex inheritance of possessions and feudal dependencies which at its furthest stretched far into south-western France. Like their followers, they still spoke Norman French. The loss of most of their 'Angevin' inheritance (the name came from Anjou) at the beginning of the twelfth century was decisive for France as well as for England. A sense of nationhood was nurtured in each of them by their quarrels with one another.

The Capetians had hung on grimly to the French crown. From the tenth century to the fourteenth their kings succeeded one another in unbroken hereditary succession. They added to the domain lands which were the basis of royal power. The Capetians' lands were rich, too. They fell in the heartland of modern France, the cereal – growing area round Paris called the *Île de France*, which was for a long time the only part of the country bearing the old name of Francia, thus commemorating the fact that it was a fragment of the old kingdom of the Franks. The domains of the first Capetians were thus distinguished from the other west Carolingian territories, such as Burgundy; by 1300 their vigorous successors had expanded 'Francia' to include Bourges, Tours, Gisors and Amiens. By then the French kings had also acquired Normandy and other feudal dependencies from the kings of England.

This is a reminder that in the fourteenth century (and later) there were still great fiefs and feudal principalities in what is now France which make it improper to think of the Capetian kingdom as a monolithic unity. Yet it was a unity of sorts, though much rested on the personal tie. During the fourteenth century that unity was greatly enhanced by a long struggle with England, remembered by the misleading name of the Hundred Years' War. In fact, English and French were only sporadically at war between 1337 and 1453. Sustained warfare was difficult to keep up; it was too expensive. Formally, though, what was at stake was the maintenance by the kings of England of territorial and feudal claims on the French side of the Channel; in 1350 Edward III had quartered his arms with those of France. There were therefore always likely to be specious grounds to start fighting again, and the opportunities it offered to English noblemen for booty and ransom money made war seem a plausible investment to many of them.

For England, these struggles supplied new elements to the infant mythology of nationhood (largely because of the great victories won at Crécy and Agincourt) and generated a long-lived distrust of the French. The Hundred Years' War was important to the French monarchy because it did something to check feudal fragmentation and broke down somewhat the barriers between Picard and Gascon, Norman and French. In the long run, too, French national mythology benefited; its greatest acquisition was the story and example of Joan of Arc whose astonishing career accompanied the turning of the balance of the long struggle against the English, though few Frenchmen at the time knew she existed. The two long-term results of the war which mattered most were that Crécy soon led to the English conquest of Calais and that England was the loser in the long run. Calais was to be held by the English for two hundred years and opened Flanders, where a cluster of manufacturing towns was ready to absorb English wool and later cloth exports, to English trade. England's ultimate defeat meant that her territorial connexion with France was virtually at an end by 1500 (though in the eighteenth century George III was still entitled 'King of France'). Once more England became almost an island. After 1453 French kings could push forward with the consolidation of their state undisturbed by the obscure claims of England's kings from which the wars had sprung. They could settle down to establish their sovereignty over their rebellious magnates at their leisure. In each country, war in the long run strengthened the monarchy.

Progress towards a future national consolidation was also to be seen in Spain. She achieved a measure of unity by the end of the fifteenth century which was mythologically underpinned by the Reconquest. The long struggle against Islam gave Spanish nationhood a quite special flavour from the start because of its intimate connexion with Christian faith and fervour; the Reconquest was a crusade uniting men of different origins. Toledo had been a Christian capital again in the mid-twelfth century. A hundred years later, Seville belonged to the kingdom of Castile and the crown of Aragon ruled the great Arab city of Valencia. In 1340, when the last great Arab offensive was defeated, success brought the threat of anarchy as the turbulent nobles of Castile strove to assert themselves. The monarchy took the burghers of the towns into alliance. The establishment of stronger personal rule followed the union of the crowns of Aragon and Castile by the marriage in 1479 of *Los Reyes Católicos*, 'the Catholic Monarchs', Ferdinand of Aragon and Isabella of Castile. This made easier both the final expulsion of the Moors and the eventual creation of one nation, though the two kingdoms long remained formally and legally separate. Only Portugal in the peninsula remained outside the framework of a new

Spain; she clung to an independence often threatened by her powerful neighbour.

Little sign of the ground-plans of future nations was to be found in Italy and Germany. Potentially, the claims of the Holy Roman Emperors were an important and broad base for political power. Yet after 1300 they had lost virtually all the special respect due to their title. The last German to march to Rome and force his coronation as emperor did so in 1328, and it proved an abortive effort. A long thirteenth-century dispute between rival emperors was one reason for this. Another was the inability of the emperors to consolidate monarchical authority in their diverse dominions.

In Germany, the domains of successive imperial families were usually scattered and disunited. The imperial election was in the hands of great magnates. Once elected, emperors had no special capital city to provide a centre for a nascent German nation. Political circumstances led them more and more to devolve such power as they possessed. Important cities began to exercise imperial powers within their territories. In 1356, a document traditionally accepted as a landmark in German constitutional history (though only a registration of established fact), the Golden Bull, named seven electoral princes who acquired the exercise of almost all the imperial rights in their own lands. Their jurisdiction, for example, was henceforth absolute; no appeals lay from their courts to the emperor. What persisted in this situation of attenuated imperial power was a reminiscence

The Hundred Years War

The name conventionally applied to a period of intermittent Anglo-French struggle in pursuit of English claims to the French crown. After performing homage for his lands in Aquitaine to the King of France, the English King, Edward III, quarrelled with his overlord which led to open hostilities and in

1339	Edward III proclaimed himself King of France, in right of his mother. There follow
1340	English victories at Sluys (naval, 1340) and Crécy (1346), and the capture of Calais (1347).
1355–6	Raids by the Black Prince across France from south-west and French defeat at Poitiers.
1360	Treaty of Brétigny ends first phase of war. Edward given an enlarged, sovereign duchy of Aquitaine.
1369	The French re-open the conflict, the English fleet defeated at La Rochelle (1372) and loss of Aquitaine. Steady decline of English position follows.
1399	Deposition of Richard II (married 1396 to daughter of Charles VI of France) renews French hostility.
1405–6	French landing in Wales and attack on English lands in Guienne.
1407	Outbreak of civil war in France, exploited by English.
1415	Henry V re-asserts claim to French throne. Alliance with Burgundy and defeat of French at Agincourt, followed by re-conquest of Normandy (1417–19).
1420	Treaty of Troyes confirms conquest of Normandy, marriage of Henry V to daughter of the King of France and his recognition as regent of France.
1422	Death of both Henry V and Charles VI of France. Infant Henry VI succeeds to the English throne; continuation of war successfully by English until
1429	Intervention of Jeanne d'Arc saves Orléans; Charles VII crowned at Reims.
1430	Henry VI crowned King of France.
1436	Loss of Paris after collapse of Anglo-Burgundian alliance.
1444	Treaty of Tours: England concedes duchy of Maine.
1449	The treaty of Tours is broken by the English, resulting in the collapse of English resistance under concerted French pressure.
1453	English defeat at Castillon ends English effort to reconquer Gascony; English left with only Calais and the Channel Islands and the struggle peters out in their abortive expeditions of 1474 and 1492.
1558	Loss of Calais to France (but the title of King of France is retained by English kings down to George III – and the French coat of arms is displayed in the *Times* newspaper's device until 1932).

of the mythology which would still prove a temptation to vigorous princes.

An Austrian family, the house of Habsburg, eventually succeeded to the imperial throne. Theirs was to be a great name; Habsburgs were to provide emperors almost without break from the accession of Maximilian I, who became emperor in 1493 and opened his house's greatest era, to the end of the empire in 1806. And even then they were to survive another century as the rulers of a great state. They began with an important advantage: as German princes went they were rich. But their most important resources only became available after a marriage which in the end brought them the inheritance of the duchy of Burgundy, the most affluent of all fifteenth-century European states and one including much of the Netherlands. Other inheritances and marriages would add Hungary and Bohemia to their possessions. For the first time since the thirteenth century, it seemed possible that an effective political unity might be imposed on Germany and central Europe; Habsburg family interest in uniting the scattered dynastic territories now had a possible instrument in the imperial dignity.

By that time the empire had virtually ceased to matter in Italy. The struggle to preserve it there had long been tangled with Italian politics: the contestants in feuds which tormented Italian cities called themselves Guelph and Ghibelline long after those names ceased to mean, as they once did, allegiance respectively to pope or emperor. After the fourteenth century there was no imperial domain in Italy and emperors hardly went there except to be crowned with the Lombard crown. Imperial authority was delegated to 'vicars' who made of their vicariates units almost as independent as the electorates of Germany. Titles were given to these rulers and their vicariates, some of which lasted until the nineteenth century; the duchy of Milan was one of the first. But other Italian states had different origins. Besides the Norman south, there were the republics, of which Venice, Genoa and Florence were the greatest.

The city republics represented the outcome of two great trends sometimes interwoven in early Italian history, the 'communal' movement and the rise of commercial wealth. In the tenth and eleventh centuries, in much of north Italy, general assemblies of the citizens had emerged as effective governments in many towns. They described themselves sometimes as *parliamenta* or, as we might say, town meetings, and represented municipal oligarchies who profited from a revival of trade beginning to be felt from 1100 onwards. In the twelfth century the Lombard cities took the field against the emperor and beat him. Thereafter they ran their own internal affairs.

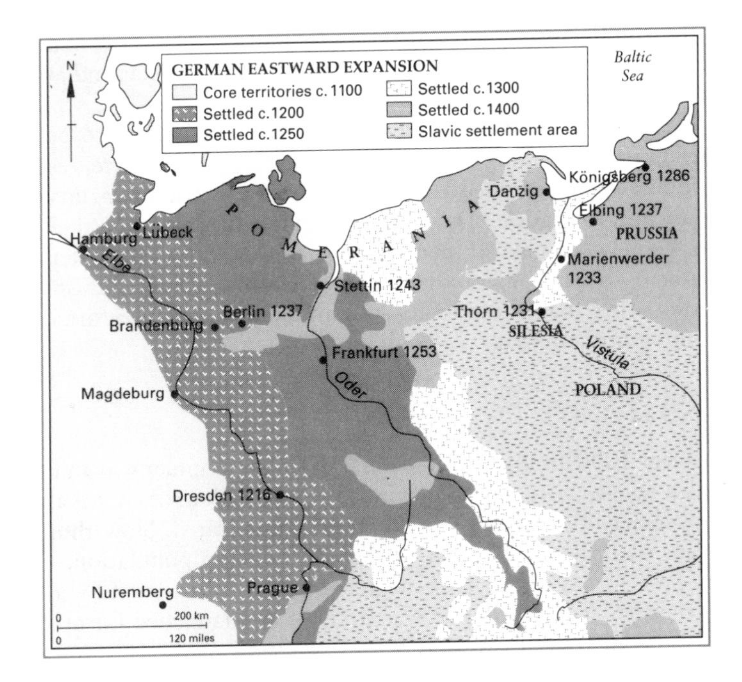

The greatest beneficiary of the revival of trade after 1100 was Venice and she contributed much to it. Formally a dependency of Byzantium, she was long favoured by the detachment from the troubles of the European mainland given to her by her position on a handful of islands in a shallow lagoon. Men already fled to her from the Lombards. Besides offering security, geography also imposed a destiny; Venice, as its citizens loved later to remember, was wedded to the sea, and a great festival of the Republic long commemorated it by the symbolic act of throwing a ring into the waters of the Adriatic. Venetian citizens were forbidden to acquire estates on the mainland and instead turned their energies to commercial empire overseas. Venice became the first west European city to live by trade. She was also the most successful of the pillagers of the eastern empire and fought and won a long struggle with Genoa for commercial supremacy in the East. Yet there was enough to go round: Genoa, Pisa and the Catalan ports all prospered with the revival of Mediterranean trade with the East.

Much of the political ground-plan of modern Europe was, therefore, in being by 1500. Portugal, Spain, France and England were recognizable in their modern form, though in Italy and Germany, where vernacular language defined nationhood, there was no correspondence between the nation and the state. That institution, too, was still far from enjoying the firmness it later acquired. The kings of France were not kings of Normandy but dukes. Different titles symbolized different legal and practical powers in different provinces. There were many such complicated survivals; constitutional relics everywhere cluttered up the idea of monarchical sovereignty, and they could provide excuses for rebellion. One explanation of the success of Henry VII, the first of the Tudors, was that by judicious marriages he drew much of the remaining poison from the bitter struggle of great families which had bedevilled the English Crown in the fifteenth-century Wars of the Roses, but there were still to be feudal rebellions to come.

One limitation on monarchical power had appeared which has a distinctly modern look. In the fourteenth and fifteenth centuries can be found the first examples of the representative, parliamentary bodies which are so characteristic of the modern state. The most famous of them all, the English parliament, was the most mature by 1500. Their origins are complex and have been much debated. One root is Germanic tradition, which imposed on a ruler the obligation of taking counsel from his great men and acting on it. The Church, too, was an early exponent of the representative idea, using it, among other things, to obtain taxation for the Papacy. It was a device which united towns with monarchs, too: in the twelfth century representatives from Italian cities were summoned to the diet of the empire. By the end of the thirteenth century most countries had seen examples of representatives with full powers being summoned to attend assemblies which princes had called to find new ways of raising taxation.

This was the nub of the matter. New resources had to be tapped by the new (and more expensive) state. Once summoned, princes found representative bodies had other advantages. They enabled voices other than those of the magnates to be heard. They provided local information. They had a propaganda value. On their side, the early parliaments (as we may loosely call them) of Europe were discovering that the device had advantages for them, too. In some of them the thought arose that taxation needed consent and that someone other than the nobility had an interest and therefore ought to have a voice in the running of the realm.

WORKING AND LIVING

From about the year 1000 another fundamental change was under way in Europe: it began to get richer. As a result, more men slowly acquired a freedom of choice almost unknown in earlier times; society became more varied and complicated. Slow though it was, this was a revolution; wealth at last began to grow faster than population. This was by no means obvious everywhere to the same degree and was punctuated by a bad setback in the fourteenth century. Yet the change was decisive and launched Europe on a career of economic growth lasting to our own day.

One crude but by no means misleading index is the growth of population. Only approximate estimates can be made but they are based on better evidence than is available for any earlier period. The errors they contain are unlikely much to distort the overall trend. They suggest that a Europe of about forty million people in 1000 rose to sixty million or so in the next two centuries. Growth then seems to have further accelerated to reach a peak of about seventy-three million around 1300, after which there is indisputable evidence of decline. The total population is said to have gone down to about fifty million by 1360 and only to have begun to rise in the fifteenth century. Then it began to go up again, and overall growth has been uninterrupted ever since.

Of course, the rate of increase varied even from village to village. The Mediterranean and Balkan lands did not succeed in doubling their population in five centuries and by 1450 had relapsed to levels only a little above those of 1000. The same appears to be true of Russia, Poland and Hungary. Yet France, England, Germany and Scandinavia probably trebled their populations before 1300 and after bad setbacks in the next hundred years still had twice the population of the year 1000. Contrasts within countries could be made, too, sometimes between areas very close to one another, but the general effect is indisputable: population grew overall as never before, but unevenly, the north and west gaining more than the Mediterranean, Balkan and eastern Europe.

The explanation lies in food supply, and therefore in agriculture. It was for a long time the only possible major source of new wealth. More food was obtained by bringing more land under cultivation and by increasing its productivity. Thus began the rise in food production which has gone on ever since. Europe had great natural advantages (which she has retained) in her moderate temperatures and good rainfall and these, combined with a physical relief whose predominant characteristic is a broad northern plain, have always given her a large area of potentially productive agricultural land. Huge areas of it still wild and forested in 1000 were brought into cultivation in the next few centuries.

Land was not short in medieval Europe and a growing population provided the labour to clear and till it. Though slowly, the landscape changed. The huge forests were gradually cut into as villages pushed out their fields. In some places, new colonies were deliberately established by landlords and rulers. The building of a monastery in a remote spot – as many were built – was often the beginning of a new nucleus of cultivation or stock-raising in an almost empty desert of scrub and trees. Some new land was reclaimed from sea or marsh. In the east, much was won in the colonization of the first German *Drang nach Osten*. Settlement there was promoted as consciously as it was later to be promoted in Elizabethan England in the first age of North American colonization.

The breaking in of fresh land slowed down by about 1300. There were even signs of over-population. The first big increase in the cultivated and grazed area was over, and an indispensable increase in productivity had occurred. Some say that in places it roughly doubled output. In part it was the slow result of more cultivation, the effect of regular fallows and cropping, of the slow enrichment of the soil, but new crops had been introduced, too. Although grain-growing was still the main business of the cultivator in northern Europe, the appearance of beans and peas of various sorts in larger quantities from the tenth century onwards meant that more nitrogen was being returned to the soil. Cause and effect are difficult to disentangle in economic history; other suggestive signs of change go along with these. In the thirteenth century the first manuals of agricultural practice appear and the first agricultural bookkeeping, a monastic innovation. More specialized cultivation brought a tendency to employ wage labourers instead of serfs carrying out obligatory work. By 1300 it is likely that most household servants in England were recruited and paid as free labour and probably a third or so of the peasants as well. The bonds of servitude were relaxing and a money economy was spreading slowly into the countryside.

Some peasants benefited, but increased wealth usually went to the landlord who took most of the profits. Most still lived poor and cramped lives, eating coarse bread and various grain-based porridges, seasoned with vegetables and only occasionally fish

December was the month for killing pigs, whose salted or smoked meat was the best that could be had during the long winter months, and whose blood and offal could be turned into sausages and stored. In the margin of this fifteenth-century picture can be seen a little of a lighter side of medieval life – little boys playing some kind of tug-of-war on their sledges.

or meat. Calculations suggest the peasant consumed about two thousand calories daily (almost exactly the figure calculated for the average daily intake of a Sudanese in 1988), and this had to sustain him for very laborious work. If he grew wheat he did not eat its flour, but sold it to the better-off, keeping barley or rye for his own food. He had little elbow-room to better himself. Even when his lord's legal grip through bond labour became less firm, the lord still had practical monopolies of mills and carts, which the peasants needed to work the land. 'Customs', or taxes for protection, were levied without regard to distinctions between freeholders and tenants and could hardly be resisted.

More cash crops for growing markets gradually changed the self-sufficient manor into a unit producing for sale. Its markets were to be found in towns which grew steadily between 1100 and 1300; urban population increased faster than rural. This is a complicated phenomenon. The new town life was in part a revival going hand in hand with the revival of trade, in part a reflexion of growing population. It is a chicken-and-egg business to decide which comes first. A few new towns grew up around a castle or a monastery. Sometimes this led to the establishment of a market. Many new towns, especially in Germany, were deliberately settled as colonies. On the whole long-established towns grew bigger – Paris may have had about eighty thousand inhabitants in 1340 and Venice, Florence and Genoa were probably comparable – but few were so big. Fourteenth-century Germany had only fifteen towns of more than ten thousand inhabitants, and London, with about thirty-five thousand was then by far the biggest English city. Of the great medieval towns, only those in the south had been important Roman centres (though many in the north, of course, had, like London, Roman nuclei). New cities tended to be linked distinctively to economic possibilities. They were markets, or lay on great trade routes such as the Meuse and Rhine, or were grouped in an area of specialized production such as Flanders, where already in the late twelfth century Ypres, Arras and Ghent were famous as textile towns, or Tuscany, also a cloth-producing region. Wine was one of the first agricultural commodities to loom large in international trade and this underpinned the early growth of Bordeaux. Ports often became the metropolitan centres of maritime regions, as did Genoa and Bruges.

The commercial revival was most conspicuous in Italy, where trade with the outside world was resumed, above all, by Venice. In that great commercial centre banking for the first time separated itself from the changing of money. By the middle of the twelfth century, whatever the current state of politics, Europeans enjoyed continuing trade not only with Byzantium but with the Arab Mediterranean. Beyond those limits, an even wider world was involved. In the early fourteenth century trans-Saharan gold from Mali relieved a bullion shortage in Europe. By then, Italian merchants had long been at work in central Asia and China. They sold slaves from Germany and central Europe to the Arabs of Africa and the Levant. They bought Flemish and English cloth and took it to Constantinople and the Black Sea. In the thirteenth century the first voyage was made from Italy to Bruges; before this the Rhine, Rhône and overland routes had been used. Roads were built across Alpine passes. Trade fed on trade and the northern European fairs drew other merchants from the north-east. The German towns of the Hanse, the league which controlled the Baltic, provided a new outlet for the textiles of the West and the spices of the East.

In such ways, European economic geography was revolutionized. In Flanders and the Low Countries economic revival soon began to generate a population big enough to stimulate agricultural innovation. Everywhere, towns which could escape from the cramping monopolies of the earliest manufacturing centres enjoyed the most rapid new prosperity. One visible result was a great wave of building. It was not only a matter of the houses and guildhalls of newly prosperous cities; it left a glorious legacy in Europe's churches, not only the great cathedrals, but the scores of magnificent parish churches of little English towns.

Building was a major expression of medieval technology. The architecture of a cathedral posed engineering problems as complex as those of a Roman aqueduct; in solving them, the engineer was slowly to emerge from the medieval craftsmen. Medieval technology was not in a modern sense science-based, but it achieved much by the accumulation of experience and reflexion on it. Possibly its most important achievement was the harnessing of other forms of energy to do the work of muscles and, therefore, to deploy muscle-power more effectively and productively. Winches, pulleys and inclined planes thus eased the shifting of heavy loads, but change was most obvious in agriculture, where metal tools had been becoming more common since the tenth century. The iron plough had made available the heavier soils of valley lands; since it required oxen to pull it the evolution of a more efficient yoke followed and with it more efficient traction. The whipple-tree and the shoulder collar for the horse also made possible bigger loads. There were not many such innovations, but they were sufficient to effect a considerable increase in the cultivators' control of the land. They also imposed new demands. Using horses meant that more grain had to be grown to feed them, and this led to new crop rotations.

Another innovation was the spread of milling; both windmills and watermills, first known in Asia, were widely spread in Europe even by 1000. In the centuries to come they were put to more and more uses. Wind often replaced muscle-power in milling foodstuffs, as it had already done in the evolution of better ships; water was used when possible to provide power for other industrial operations. It drove hammers both for cloth-fulling and for forging (here the invention of the crank was of the greatest importance), an essential element in a great expansion of Europe's metallurgical industry in the fifteenth century, and one closely connected with rising demand for an earlier technological innovation of the previous century, artillery. Water-driven hammers were also used in paper-making. The invention of printing soon gave this industry an importance which may even have surpassed that of the new metal-working of Germany and Flanders. Print and paper had their own revolutionary potential, too, because books made the diffusion of techniques faster and easier in the growing pool of craftsmen and artifiers able to use such knowledge. Some innovations were simply taken over from other cultures; the spinning-wheel came to medieval Europe from India (though the application of a treadle to it to provide drive with the foot seems to have been a European invention of the sixteenth century).

Whatever qualifications are needed, it is clear (if only from what was to follow) that by 1500 a technology was available which was already embodied in a large capital investment. It was making the accumulation of further capital for manufacturing enterprises easier than ever before. The availability of this capital must have been greater, moreover, as new devices eased business. Medieval Italians invented much of modern accountancy as well as new credit instruments for the financing of international trade. The bill of exchange appears in the thirteenth century and with it and the first true bankers we are at the edge of modern capitalism. Limited liability appears at Florence in 1408. Yet though such a change from the past was by implication colossal, it is easy to get it out of proportion if we do not recall its scale. For all the magnificence of its palaces, the goods shipped by medieval Venice in a year could all have fitted comfortably into one large modern ship.

The change was also precarious. For centuries, economic life was fragile, never far from the edge of collapse. Medieval agriculture, in spite of such progress as had been made, was appallingly inefficient. It abused the land and exhausted it. Little was consciously put back into it except manure. As new land became harder to find, family holdings got smaller; probably most European households farmed less than eight acres in 1300. Only in a few places (the Po valley was one) was there a big investment in collective irrigation or improvement. Above all, agriculture was vulnerable to weather; two successive bad harvests in the early fourteenth century reduced the population of Ypres by a tenth. Local famine could rarely be offset by imports. Roads had broken down since Roman times, carts were crude and for the most part goods had to be carried by packhorse or mule. Water transport was cheaper and swifter, but could rarely meet the need. Commerce could have its political difficulties, too; the Ottoman onslaught brought a gradual recession in eastern trade in the fifteenth century. Demand was small enough for a very little change to determine the fate of cities: cloth production at Florence and Ypres fell by two-thirds in the fourteenth century.

It is very difficult to generalize but about one thing there is no doubt: a great and cumulative setback occurred during that time. There was a sudden rise in mortality, not occurring everywhere at the same time, but notable in many places after a series of bad harvests round about 1320. This started a slow decline of population which suddenly became a disaster with the onset of attacks of epidemic disease. These are often called by the name of one of them, the 'Black Death' of 1348–50 and the worst single attack. It was of bubonic plague, but no doubt it masked many other killing diseases which swept Europe with it and in its wake. Europeans died of typhus, influenza and smallpox, too; all contributed to a great demographic disaster. In some areas a half or a third of the population may have died; over Europe as a whole the total loss has been calculated as a quarter. A papal enquiry put the figure at more than forty million. Toulouse was a city of thirty thousand in 1335 and a century later only eight thousand lived there; fourteen hundred died in three days at Avignon.

There was no universal pattern, but all Europe shuddered under these blows. In extreme cases a kind of collective madness broke out. Pogroms of Jews were a common expression of a search for scapegoats or those guilty of spreading the plague; the burning of witches and heretics was another. The European psyche bore a scar for the rest of the Middle Ages, which were haunted by the imagery of death and damnation in painting, carving and literature. The fragility of settled order illustrated the precariousness of the balance of food and population. When disease killed enough people, agricultural production would collapse; then the inhabitants of the towns would die of famine if they were not already dying of plague. Probably a plateau of productivity had already been reached by about 1300. Both available techniques and easily accessible new land for cultivation had reached a limit and some have seen signs of population pressure treading close upon resources even by that date. From this flowed the huge setback of the fourteenth century and then the slow recovery in the fifteenth.

It is scarcely surprising that an age of such colossal dislocations and disasters should have been marked by violent social conflicts. Everywhere in Europe the fourteenth and fifteenth centuries brought peasant risings. The French *jacquerie* of 1358 which led to over thirty thousand deaths and the English Peasants' Revolt of 1381 which for a time captured London were especially notable. The roots of rebellion lay in the ways in which landlords had increased their demands under the spur of necessity and in the new demands of royal tax collectors. Combined with famine, plague and war they made an always miserable existence intolerable. 'We are made men in the likeness of Christ, but you treat us like savage beasts', was the complaint of English peasants who rebelled in 1381. Significantly, they appealed to the Christian standards of their civilization; the demands of medieval peasants were often well formulated and effective but it is anachronistic to see in them a nascent socialism.

Demographic disaster on such a scale paradoxically made things better for some poor men. One obvious and immediate result was a severe shortage of labour; the pool of permanently under-employed had been brutally dried up. A rise in real wages followed. Once the immediate impact of the fourteenth-century disasters had been absorbed the standard of living of the poor may have slightly risen, for the price of cereals tended to fall. The tendency for the economy, even in the countryside, to move on to a money basis was speeded up by the labour shortage. By the sixteenth century, serf labour and servile status had both receded a long way in western Europe, particularly in England. This weakened the manorial structure and the feudal relationships clustered about it. Landlords were also suddenly confronted with a drop in their rent incomes. In the previous two centuries the habits of consumption of the better-off had become more expensive. Now property-owners suddenly ceased to grow more prosperous. Some landlords could adapt. They could, for example, switch from cultivation which required much labour to sheep-running which required little. In Spain there were still even possibilities of taking in more land and living directly off it. Moorish estates were the reward of the soldier of the Reconquest. Elsewhere, many landlords simply let their poorer land go out of cultivation.

The results are very hard to pin down, but they were bound to stimulate further and faster social change. Medieval society changed dramatically, and sometimes in oddly assorted ways, between the tenth century and the sixteenth. Even at the end of that age, though, it seems still almost unimaginably remote. Its obsession with status and hierarchy is one index of this. Medieval European man was defined by his legal status. Instead of being an individual social atom, so to speak, he was the point at which a number of coordinates met. Some of them were set by birth and the most obvious expression of this was the idea of nobility. The noble society which was to remain a reality in some places until the twentieth century was already present in its essentials in the thirteenth. Gradually, warriors had turned into landowners. Descent then became important because there were inheritances to argue about. One indicator was the rise of the sciences of heraldry and genealogy, which have since had a profitable life right down to our own day. New titles appeared as distinctions within the nobility ripened. The first English duke was created in 1337, an expression of the tendency to find ways of singling out the greater magnates from among their peers. Symbolic questions of precedence became of intense interest; rank was at stake. From this rose the dread of disparagement, the loss of status which might follow for a woman from an unequal marriage or for a man from contamination by a lowly occupation. For centuries it was to be assumed that only arms, the Church or the management of his own estates were fit occupations for a nobleman in northern Europe. Trade, above all, was closed to him except through agents. Even when, centuries later, this barrier gave way, hostility to retail trade was the last thing to be abandoned by those who cared about these things. When a sixteenth-century French king called his Portuguese cousin 'the grocer king' he was being rude as well as witty and no doubt his courtiers appreciated the sneer.

The values of the nobility were, at bottom, military. Through their gradual refinement there emerged slowly the notions of honour, loyalty, disinterested self-sacrifice which were to be held up as models for centuries to well-born boys and girls. The ideal of chivalry articulated these values and softened the harshness of a military code. It was blessed by the Church, which provided religious ceremonies to accompany the bestowal of knighthood and the knights' acceptance of Christian duties. The heroic figure who came supremely to embody the notion was the mythological English King Arthur, whose cult spread to many lands. It was to live on in the ideal of the gentleman and gentlemanly conduct, however qualified in practice.

Of course, it never worked as it should have done. But few great creative myths do; neither did the feudal theory of dependence, nor does democracy. The pressures of war and, more fundamentally, economics, were always at work to fragment and confuse social obligations. The increasing unreality of the feudal concept of lord and vassal was one factor favouring the growth of kingly power. The coming of a money economy made further inroads, service had increasingly to be paid for in cash, and rents became more important than the services that had gone with them. Some sources of feudal income remained fixed in terms made worthless by changes in real prices. Lawyers evolved devices which enabled new aims to be realized within a 'feudal' structure more and more unreal and worm-eaten.

Medieval nobility had been for a long time very open to new entrants, but usually this became less and less true as time passed. In some places attempts were actually made to close for ever a ruling caste. Yet European society was all the time generating new kinds of wealth and even of power which could not find a place in the old hierarchies and challenged them. The most obvious example was the emergence of rich merchants. They often bought land; it was not only the supreme economic investment in a world where there were few, but it might open the way to a change of status for which landownership was either a legal or social necessity. In Italy merchants sometimes themselves became the nobility of trading and manufacturing cities. Everywhere, though, they posed a symbolic challenge to a world which had, to begin with, no theoretical place for them. Soon, they evolved their own social forms, guilds, mysteries, corporations, which gave new definitions to their social rôle.

The rise of the merchant class was almost a function of the growth of towns; that is to say that merchants were inseparably linked with the most dynamic element in medieval European civilization. Unwittingly, at least at first, towns and cities held within their walls much of the future history of Europe. Though their independence varied greatly in law and practice, there were parallels in other countries to the Italian communal movement. Towns in the German east were especially independent, which helps to explain the appearance there of the powerful Hanseatic League of more than a hundred and fifty free towns. The Flemish towns also tended to enjoy a fair degree of freedom: French and English towns usually had less. Yet lords everywhere sought the support of towns against kings, while kings sought the support of townsmen and their wealth against overmighty subjects. They gave towns charters and privileges. The walls which surrounded the medieval city were the symbol as well as a guarantee of its immunity. The landlords' writ did not run in them and sometimes their anti-feudal implication was even more explicit: villeins, for example, could acquire their freedom in some towns if they lived in them for a year and a day. 'The air of the town makes for free' said a German proverb. The communes and within them the guilds were associations of free men for a long time isolated in a world unfree. The burgher – the *Bourgeois*, the dweller in bourg or borough – was a man who stood up for himself in a universe of dependence.

Much of the history behind this remains obscure because it is for the most part the history of obscure men. The wealthy merchants who became the typically dominant figures of the new town life and fought for their corporate privileges are visible enough, but their humbler predecessors are usually not. In earlier times a merchant can have been little but the pedlar of exotica and luxuries which the medieval European estate could not provide for itself. Ordinary commercial exchange for a long time hardly needed a middleman: craftsmen sold their own goods and cultivators their own crops. Yet somehow in the towns there emerged men who dealt between them and the countryside, and their successors were to be men using capital to order in advance the whole business of production for the market.

In the blossoming of urban life lies buried also much which made European history different from that of other continents. Neither in the ancient world (except, perhaps, classical Greece) nor in Asia or America, did city life develop the political and social power it came to show in Europe. One reason was the absence of destructively parasitic empires of conquest to eat away at the will to betterment; Europe's enduring political fragmentation made rulers careful of the geese which laid the golden egg they needed to compete with their rivals. A great sack of a city was a noteworthy event in the European Middle Ages; it was the inescapable and recurrent accompaniment of warfare in much of Asia. This, of course, could not be the whole story. It also must have mattered that, for all its obsession with status, Europe had no caste system such as that of India, no stultifying ideological homogeneity so intense as China's. Even when rich, the city-dwellers of other cultures seem to have aquiesced in their own inferiority. The merchant, the craftsman, the lawyer and the doctor had rôles in Europe, though, which at an early date made them more than simple appendages of landed society. Their society was not closed to change and self-advancement; it offered routes to self-improvement other than the warrior's or the court favourite's. Townsmen were equal and free, even if some were more equal than others.

It need not surprise us that practical, legal and personal freedom was much greater for men than for women (though there were still those of both sexes who were legally unfree at the bottom of society). Whether they were of noble or common blood, women suffered, by comparison with their menfolk, from important legal and social disabilities, as they did in every civilization which had ever existed. Their rights of inheritance were often restricted; they could inherit a fief, for example, but could not enjoy personal lordship, and had to appoint men to carry out the obligations that went with it. In all classes below the highest there was much drudgery to be done by women; until this century European peasant women worked on the land as women do in Africa and Asia today.

There were theoretical elements in the subjection of women and a large contribution was made to them by the Church. In part this was a matter of its traditionally hostile stance towards sexuality. Its teaching had never been able to find any justification for sex except the link with the reproduction of the species. Woman being seen as the origin of Man's Fall and a standing temptation to concupiscence, the Church threw its weight behind the domination of society by men. Yet this is not all there is to be said. Other societies have done more to seclude and oppress women than Christendom, and the Church at least offered women the only respectable alternative to domesticity available until modern times; the history of the female religious is studded with outstanding women of learning, spirituality and administrative gifts. The position of at least a minority of well-born women, too, was marginally bettered by the idealization of women in the chivalric codes of behaviour of the thirteenth and fourteenth centuries. There lay in this a notion of romantic love and an entitlement to service, a stage towards a higher civilization.

Yet such ideas can have affected very few. Among themselves, medieval European women were more equal before death than would be rich and poor in Asia today, but, then, so were men. Women lived less long than men, it seems, and frequent confinements and a high mortality rate no doubt explain this. Medieval obstetrics remained, as did other branches of medicine, rooted in Aristotle and Galen; there was nothing better available. But men died young, too. Aquinas lived only to forty-seven and philosophy is not nowadays thought to be physically exacting. This was about the age to which a man of twenty in a medieval town might normally expect to survive: he was lucky to have got as far already and to have escaped the ferocious toll of infant mortality which imposed an average life of about thirty-three years and a death rate about twice that of modern industrial countries. Judged by the standards of antiquity, so far as they can be grasped, this was of course by no means bad.

This reminds us of one last novelty in the huge variety of the Middle Ages; they left behind the means for us to measure just a little more of the dimensions of human life. From these centuries come the first collections of facts upon which reasoned estimates can be made. When in 1087 William the Conqueror's officers rode out into England to interrogate its inhabitants and to record its structure and wealth in Domesday Book, they were unwittingly pointing the way to a new age. Other collections of data, usually for tax purposes, followed in the next few centuries. Some have survived, together with the first accounts which reduce farming and business to quantities. Thanks to them historians can talk with a little more confidence about late medieval society than about that of any earlier time.

11
New Limits, New Horizons

In the Near East Europeans were until very recently called 'Franks', a word first used in Byzantium to mean western Christians. It caught on elsewhere and was still being used in various distortions and mispronunciations from the Persian Gulf to China a thousand years later. This is more than just a historical curiosity; it is a helpful reminder that non-Europeans were struck from the start by the unity, not the diversity, of the western peoples and long thought of them as one.

EUROPE LOOKS OUTWARDS

The roots of this idea can be seen even in the remote beginnings of Europe's long and victorious assault on the world, when a relaxation of pressure on her eastern land frontier and northern coasts at last began to be felt. By AD 1000 or so, the barbarians were checked; then they began to be Christianized. Within a short space of time Poland, Hungary, Denmark and Norway came to be ruled by Christian kings. One last great threat, the Mongol onslaught, still lay ahead, it is true, but that was at that time unimaginable. By the eleventh century, too, the rolling back of Islam had already begun. The Islamic threat to southern Europeans diminished because of the decline into which the Abbasid caliphate had fallen in the eighth and ninth centuries.

The struggle with Islam was to continue vigorously until the fifteenth century. It was given unity and fervour by religion, the deepest source of European self-consciousness. Christianity bound men together in a great moral and spiritual enterprise. But that was only one side of the coin. It also provided a licence for the predatory appetites of the military class which dominated lay society. They could spoil the pagans with clear consciences. The Normans were in the vanguard, taking south Italy and Sicily from the Arabs, a task effectively complete by 1100. (Almost incidentally they swallowed the last Byzantine possessions in the West as well.) The other great struggle in Europe against Islam was the epic of Spanish history, the Reconquest, whose climax came in 1492, when Granada, the last Moslem capital of Spain, fell to the armies of the Catholic Monarchs.

The Spaniards had always seen the Reconquest as a religious cause, and as such it had drawn warriors from all over Europe since its opening in the eleventh century. It had benefited from the same religious revival and quickening of vigour in the West which expressed itself in a succession of great enterprises in Palestine and Syria. The Crusades, as they came to be called, went on for more than two centuries, and though they were to be unsuccessful in their aim of delivering the Holy Land from Islamic rule they would leave profound marks not only on the Levant, but also on European society and psychology. The first four were the most important. The earliest and most successful was launched in 1096. Within three years the crusaders recaptured Jerusalem, where they celebrated the triumph of the Gospel of Peace by an appalling massacre of their prisoners, women and children included. The second crusade (1147–9), in contrast, *began* with a successful massacre (of Jews in the Rhineland), but thereafter, though the presence of an emperor and a king of France gave it greater importance than its predecessor, it was a disaster. It failed to recover Edessa, the city whose loss had largely provoked it, and did much to discredit St Bernard, its most fervent advocate (though it had a by-product of some importance when

The Reconquest: the Moors go down before James I of Aragon in 1235. When this fifteenth-century altar-piece was painted, the Reconquest had become the central and unifying theme of the Spaniard's view of their history.

an English fleet took Lisbon from the Arabs and it passed into the hands of the King of Portugal). Then in 1187 Saladin recaptured Jerusalem for Islam. The third crusade which followed (1189–92) was socially the most spectacular. A German emperor (drowned in the course of it) and the kings of England and France all took part. They quarrelled and the crusaders failed to recover Jerusalem. No great monarch answered Innocent III's appeal to go on the next crusade, though many land-hungry magnates did. The Venetians financed the expedition, which left in 1202. It was at once diverted by interference in the dynastic troubles of Byzantium, which suited the Venetians, who helped to recapture Constantinople for a deposed emperor. There followed the terrible sack of the city in 1204

TIMGAD *At Timgad in Algeria there stands the ruins of the city built in 100* AD *by the emperor Trajan as a colony of retired soldiers – a striking example of the conscious urbanism of Roman planning.*

EASTER ISLAND *A small island about two thousand miles east of Peru was visited by a Dutch expedition in 1722 which gave it the name Easter Island. The Dutch found the inhabitants seemingly worshipping huge statues of their gods, whose likenesses were first brought home for Europeans to ponder over fifty years later, when a painter on Cook's second voyage recorded them. Cook arrived, in fact, at the beginning of sixty to seventy years of turmoil and strife on the island which brought with it the over-turning of all the statues. Now, some of them have been re-erected and something is understood of the techniques with which these huge monoliths – some over twenty feet high – were sited and stood erect. The story of the peoples who put these up is still obscure and conjectural. One theory is that such skill in working huge pieces of stone must have come from Peru, the nearest culture with advanced stone-working techniques, and possibly by way of the Marqueras islands to the north. But when? What seems certain is that there were Easter Islanders already on the site as the Roman empires crumbled in the late fourth century* AD.

and that was the end of the fourth crusade, whose monument was the establishment of a 'Latin Empire' at Constantinople which survived there only for a half-century.

Several more crusades set out in the thirteenth century, but though they helped to put off a little longer the dangers which faced Byzantium, crusading to the holy land was dead as an independent force. Its religious impulse could still move men, but the first four crusades had too often shown the unpleasant face of greed. They were the first examples of European overseas imperialism, both in their characteristic mixture of noble and ignoble aims and in their abortive settler colonialism. Whereas in Spain, and on the pagan marches of Germany, Europeans were pushing forward a frontier of settlement, they tried in Syria and Palestine to transplant western institutions to a remote and exotic setting as well as to seize lands and goods no longer easily available in the West. They did this with clear consciences because their opponents were infidels who had by conquest installed themselves in Christianity's most sacred shrines. 'Christians are right, pagans are wrong', said the Song of Roland and that probably sums up well enough the average crusader's response to any qualms about what he was doing.

The brief successes of the first crusade had owed much to a passing phase of weakness and anarchy in the Islamic world, and the feeble transplants of the Frankish states and the Latin empire of Constantinople soon collapsed. But there were important and permanent results. As we noted, the crusades had embittered still further the division of western from eastern Christendom: the first warriors to sack Constantinople had been crusaders. Secondly, the crusaders had cherished and intensified the sense of unbridgeable ideological separation between Islam and Christianity. The Crusades both expressed and helped to forge the special temper of western Christianity, giving it a militant tone and an aggressiveness which would make its missionary work more potent in the future when it would have technological superiority on its side as well, but also more ruthless. In it lay the roots of a mentality which, when secularized, would power the world-conquering culture of the modern era. The Reconquest was scarcely to be complete before the Spanish would look to the Americas for the battlefield of a new crusade.

Krak des Chevaliers, the great Syrian crusader fortress, an affirmation of staggering confidence by its builders, few as they were, and far from home in an alien land.

Yet Europe was not impervious to Islamic influence. In these struggles she imported and invented new habits and institutions. Wherever they encountered Islam, whether in the crusading lands, Sicily or Spain, western Europeans found things to admire. Sometimes they took up luxuries not to be found at home, silk clothes, the use of perfumes and new dishes. One habit acquired by some crusaders was that of taking more frequent baths. This may have been unfortunate, for it added the taint of religious infidelity to a habit already discouraged in Europe by the association of bath-houses with sexual licence. Cleanliness had not yet achieved its later quasi-automatic association with godliness.

One institution crystallizing the militant Christianity of the high Middle Ages was the military order of knighthood. It brought together soldiers who professed vows as members of a religious order and of an accepted discipline to fight for the faith. Some of these orders became very rich, owning endowments in many countries. The Knights of St John of Jerusalem (who are still in existence) were to be for centuries in the forefront of the battle against Islam. The Knights Templar rose to such great power and prosperity that they were destroyed by a French king who feared them, and the Spanish military orders of Calatrava and Santiago were in the forefront of the Reconquest.

The Crusades

Conventionally, the 'Crusades' are a series of expeditions directed from western Chistendom to the Holy Land whose aim was to recover the Holy Places from their Islamic rulers. Those who took part were assured by papal authority of certain spiritual rewards – indulgences (remission of time spent in purgatory after death) and the status of martyr in the event of death on the expedition. The first four were the most important and made up what is usually thought of as the Crusading era.

1095	Urban II proclaims the *First Crusade* at the Council of Clermont. It culminated in
1099	the capture of Jerusalem and foundation of the Latin Kingdom.
1144	The Seljuk Turks capture the (Christian) city of Edessa, whose fall inspires St. Bernard's preaching of a new Crusade (1146).
1147–49	The *Second Crusade*, a failure (its only significant outcome was the capture of Lisbon by an English fleet and its transfer to the King of Portugal).
1187	Saladin reconquers Jerusalem for Islam.
1189	Launching of *Third Crusade* which fails to recover Jerusalem, though
1192	Saladin allows pilgrims access to the Holy Sepulchre.
1202	*Fourth Crusade*, the last of the major crusades, which culminates in the capture and sack of Constantinople by the crusaders (1204) and establishment of a 'Latin Empire' there.
1212	The so-called 'Children's Crusade'.
1216	The Fifth Crusade captures Damietta in Egypt, soon again lost.
1228–9	The emperor Frederick II (excommunicate) undertakes a 'crusade' and recaptures Jerusalem, crowning himself king.
1239–40	'Crusades' by Theobold of Champagne and Richard of Cornwall.
1244	Jerusalem retaken for Islam.
1248–54	Louis IX of France leads crusade to Egypt where he is taken prisoner, ransomed, and goes on pilgrimage to Jerusalem.
1270	Louis IX's second crusade, against Tunis, where he died.
1281	Acre, the last Christian foothold in the Levant, falls to Islam.

There were many other expeditions to which the title of 'crusade' was given, sometimes formally. Some were directed against non-Christians (Moorish Spain and the Slav peoples), some against heretics (e.g. the Albigenses), some against monarchs who had offended the Papacy. There were also further futile expeditions to the near East. In 1464 Pius II failed to obtain support for what proved to be a last attempt to mount a further Crusade to that region.

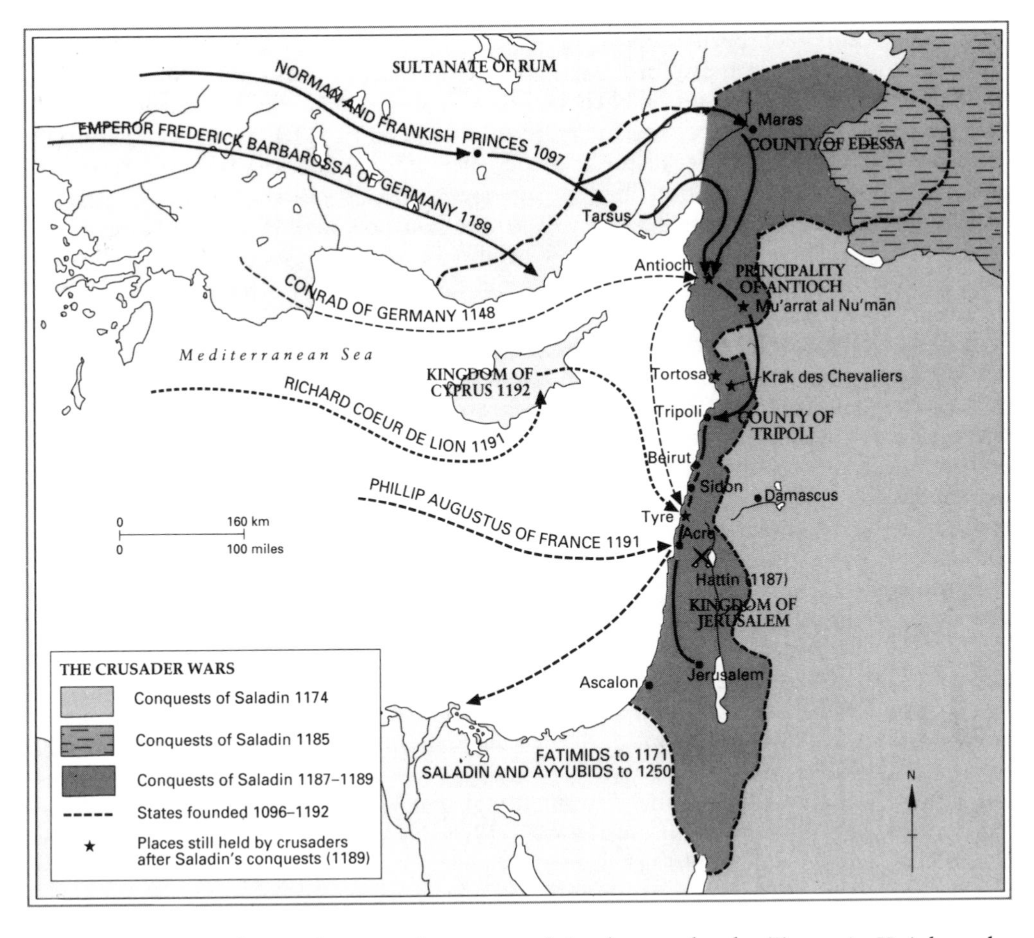

Another military order operated in the north, the Teutonic Knights, the warrior monks who were the spearhead of Germanic penetration of the Baltic and Slav lands. There, too, missionary zeal combined with greed and the stimulus of poverty to change both the map and the culture of a whole region. The colonizing impulse which failed in the Near East had lasting success further north. German expansion eastwards was a huge folk-movement, a centuries-long tide of men and women clearing forest, planting homesteads and villages, founding towns, building fortresses to protect them and monasteries and churches to serve them. When the crusades were over, and the narrow escape from the Mongols had reminded Europe that it could still be in danger, this movement went steadily on. Out on the Prussian and Polish marches, the soldiers, among whom the Teutonic Knights were outstanding, provided its shield and cutting-edge at the expense of the native peoples. This was the beginning of a cultural conflict between Slav and Teuton which persisted down to the twentieth century. The last time that the West threw itself into the struggle for Slav lands was in 1941: many Germans saw 'Barbarossa' (as Hitler's attack on Russia was named in memory of a medieval emperor) as another stage in a centuries-old civilizing mission in the East. In the thirteenth century a Russian prince, Alexander Nevsky, grand duke of Novgorod, had to beat off the Teutonic Knights (as Russians were carefully reminded by a great film in 1937) at a moment when he also faced the Tatars on another front.

While the great expansion of the German East between 1100 and 1400 made a new economic, cultural and racial map, it also raised yet another barrier to the union of the two Christian traditions. Papal supremacy in the West made the Catholicism of the late medieval period more uncompromising and more unacceptable than ever to Orthodoxy. From the twelfth century onwards Russia was more and more separated from western Europe by her own traditions and special historical experience. The Mongol capture of Kiev in 1240 had been a blow to eastern Christianity as grave as the sack of Constantinople in 1204. It also broke the princes of Muscovy. With Byzantium in decline and the Germans and Swedes on their backs, they were to pay tribute to the Mongols and their Tatar successors of the Golden Horde for centuries. This long domination by a nomadic people was another historical experience sundering Russia from the West.

Tatar domination had its greatest impact on the southern Russian principalities, the area where the Mongol armies had operated. A new balance within Russia appeared; Novgorod and Moscow acquired new importance after the eclipse of Kiev, though both paid tribute to the Tatars in the form of silver, recruits and labour. Their emissaries, like other Russian princes, had to go to the Tatar capital at Sarai on the Volga, and make their separate arrangements with their conquerors. It was a period of the greatest dislocation and confusion in the succession patterns of the Russian states. Both Tatar policy and the struggle to survive favoured those which were most despotic. The future political tradition of Russia was thus now shaped by the Tatar experience as it had been by the inheritance of imperial ideas from Byzantium. Gradually Moscow emerged as the focus of a new centralizing trend. The process can be discerned as early as the reign of Alexander Nevsky's son, who was prince of Muscovy. His successors had the support of the Tatars, who found them efficient tax-gatherers. The Church offered no resistance and the metropolitan archbishopric was transferred from Vladimir to Moscow in the fourteenth century.

Meanwhile, a new challenge to Orthodoxy had arisen in the West. A Roman Catholic but half-Slav state had emerged which was to hold Kiev for three centuries. This was the medieval duchy of Lithuania, formed in 1386 in a union by marriage which incorporated the Polish kingdom and covering much of modern Poland, Prussia, the Ukraine and Moldavia. Fortunately for the Russians, the Lithuanians fought the Germans, too; it was they who shattered the Teutonic Knights at Tannenberg in 1410. Harassed by the Germans and the Lithuanians to the west, Muscovy somehow survived by exploiting divisions within the Golden Horde.

The fall of Constantinople brought a great change to Russia; eastern Orthodoxy had now to find its centre there, and not in Byzantium. Russian churchmen soon came to feel that a complex purpose lay in such awful events. Byzantium, they believed, had betrayed its heritage by seeking religious compromise at the Council of Florence. 'Constantinople has fallen', wrote the Metropolitan of Moscow, 'because it has deserted the true Orthodox faith. ... There exists only one true Church on earth, the Church of Russia.' A few decades later, at the beginning of the sixteenth century, a monk could write to the ruler of Muscovy in a quite new tone: 'Two Romes have fallen, but the third stands and a fourth will not be. Thou art the only Christian sovereign in the world, the lord of all faithful Christians.'

The end of Byzantium came when other historical changes made Russia's emergence from confusion and Tatar domination possible and likely. The Golden Horde was rent by dissension in the fifteenth century. At the same time, the Lithuanian state began to crumble. These were opportunities, a ruler who was capable of exploiting them came to the throne of Muscovy in 1462. Ivan the Great (Ivan III) gave Russia something like the definition and reality won by England and France from the twelfth century onwards. Some have seen him as first national ruler of Russia. Territorial consolidation was the foundation of his work. When Muscovy swallowed the republics of Pskov and Novgorod, his authority stretched at least in theory as far as the Urals. The oligarchies which had ruled them were deported, to be replaced by men who held lands from Ivan in return for service. The German merchants of the Hanse who had dominated the trade of these republics were expelled, too. The Tatars made another onslaught on Moscow in 1481 but were beaten off, and two invasions of Lithuania gave Ivan much of White Russia and Little Russia in 1503. His successor took Smolensk in 1514.

Ivan the Great was the first Russian ruler to take the title of 'Tsar'. It was a conscious evocation of an imperial past, a claim to the heritage of the Caesars, the word in which it originated. In 1472 Ivan married a niece of the last Greek emperor. He was called 'autocrat by the grace of God' and during his reign the double-headed eagle was adopted, which was to remain part of the insignia of Russian rulers until 1917. This gave a further Byzantine colouring to Russian monarchy and Russian history, which became still more unlike that of western Europe. By 1500 western Europeans already recognized a distinctive kind of monarchy in Russia; Basil, Ivan's successor, was acknowledged to have a despotic power over his subjects greater than that of any other Christian rulers over theirs.

Much of Europe's future seems already discernible by 1500. A great process of definition and realization had been going on for centuries. Europe's land limits were now filled up; in the East further advance was blocked by the consolidation of Christian Russia, in the Balkans by the Ottoman empire of Islam. The first, crusading, wave of overseas expansion was virtually spent by about 1250. With the onset of the Ottoman threat in the fifteenth century, Europe was again forced on the defensive in the eastern Mediterranean and Balkans. Those unhappy states with exposed territories in the East, such as Venice, had to look after them as best they could. Meanwhile, others were taking a new look at their oceanic horizons. A new phase of western Europe's relations with the rest of the world was about to open.

In 1400 it had still seemed sensible to see Jerusalem as the centre of the world. Though the Vikings had crossed the Atlantic, men could still think of a world which, though spherical, was made up of three continents, Europe, Asia and Africa, around the shores of one land-locked sea, the Mediterranean. A huge revolution lay just ahead which for ever swept away such views and the route to it lay across the oceans, because elsewhere advance was blocked. Europe's first direct contacts with the East had been on land rather than on water. The caravan routes of central Asia were their main channel and brought goods west to be shipped from Black Sea or Levant ports. Elsewhere, ships rarely ventured far south of Morocco until the fifteenth century. Then, a mounting wave of maritime enterprise becomes noticeable. With it, the age of true world history was beginning.

One explanation of it was the acquisition of new tools and skills. Different ships and new techniques of long-range navigation were needed for oceanic sailing and they became available from the fourteenth century onwards, thus making possible the great effort of exploration which has led to the fifteenth century being called 'the Age of Reconnaissance'. In ship design there were two crucial changes. One was specific, the adoption of the stern-post rudder; though we do not know exactly when this happened, some ships had it by 1300. The other was a more gradual and complex process of improving rigging. This went with a growth in the size of ships. A more complex maritime trade no doubt spurred such developments. By 1500, the tubby medieval 'cog' of northern Europe, square-rigged with a single sail and mast, had developed into a ship carrying up to three masts, with mixed sails. The main-mast still carried square-rigging, but more than one sail; the mizzen-mast had a big lateen sail borrowed from the Mediterranean tradition; a fore-mast might carry more square-rigged sails, but also newly invented fore-and-aft jib sails attached to a bowsprit. Together with the lateen sail aft, these head-sails made vessels much more manoeuvrable; they could be sailed much closer to the wind.

Once these innovations were absorbed, the design of ships which resulted was to remain essentially unchanged (though refined) until the coming of steam propulsion. Though he would have found them small and cramped, Columbus's ships would have been perfectly comprehensible machines to a nineteenth-century clipper captain. Since they carried guns, though tiny ones by comparison with what was to come, they would also have been comprehensible to Nelson.

By 1500 some crucial navigational developments had also taken place. The Vikings had first shown how to sail an oceanic course. They had better ships and navigational skill than anything previously available in the West. Using the Pole Star and the sun, whose height above the horizon in northern latitudes at midday had been computed in tables by a tenth-century Irish astronomer, they had crossed the Atlantic by running along a line of latitude. Then, with the thirteenth century, there is evidence of two great innovations. At that time the compass came to be commonly used in the

Mediterranean (it already existed in China, but it is not certain that it was transmitted from Asia to the West), and in 1270 there appears the first reference to a chart, one used in a ship engaged on a crusading venture. The next two centuries gave birth to modern geography and exploration. Spurred by the thought of commercial prizes, by missionary zeal and diplomatic possibilities, some princes began to subsidize research. In the fifteenth century they came to employ their own cartographers and hydrographers. Foremost among these princes was the brother of the King of Portugal, Henry, 'the Navigator' as English-speaking scholars were later to call him (unsuitably, for he never navigated anything).

The Portuguese had a long Atlantic coast. They were land-locked by Spain, and virtually barred from the Mediterranean trade by the experience and armed force with which the Italians guarded it. Almost inevitably, it seems, they were bound to push out into the Atlantic and they had already started to familiarize themselves with northern waters when Prince Henry began to equip and launch a series of maritime expeditions. His initiative was decisive. From a mixture of motives, he turned his countrymen southward. Gold and pepper, it was known, were to be found in the Sahara; perhaps the Portuguese could discover where. Perhaps, too, there was a possibility of finding an ally here to take the Turk in the flank, the legendary Prester John. Certainly there were converts, glory and land to be won for the Cross. Henry, for all that he did so much to launch Europe on the great expansion which transformed the globe and created one world, was a medieval man to the soles of his boots. He cautiously sought papal authority and approval for his expeditions. He had gone crusading in North Africa, taking with him a fragment of the True Cross. This was the beginning of the age of discovery, and its heart was systematic, government-subsidized research, but it was rooted in the world of chivalry and crusade

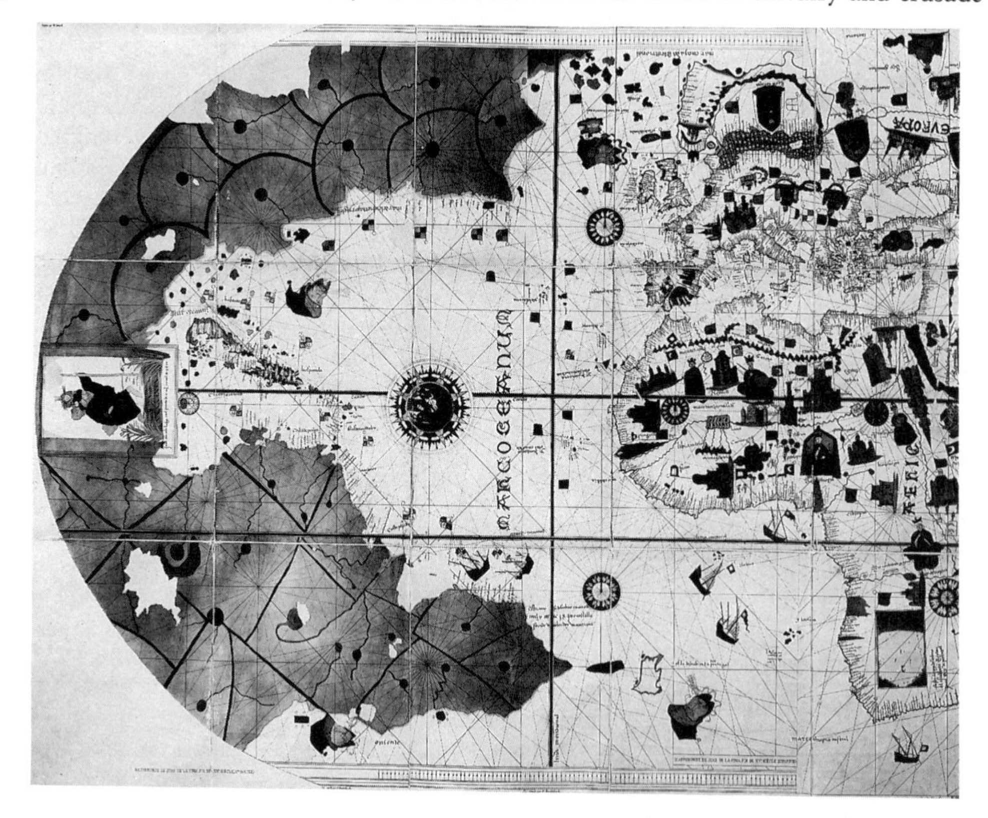

The first map to show Columbus' discoveries in the New World depicts Cuba, correctly, as an island. Together with the rest of the crew of 1495, the map's draughtsman, Juan de la Cosa, renounced the oath made to Columbus that it was part of the Asian mainland. St Christopher, who appears at the head of the map, was Columbus' patron saint.

which had shaped Henry's thinking. He is an outstanding example of a man who wrought much more than he knew.

The Portuguese pushed steadily south. They began by hugging the African coast, but some of the bolder among them reached the Madeiras and began to settle there already in the 1420s. In 1434 one of their captains passed Cape Bojador, an important psychological obstacle whose overcoming was Henry's first great triumph; ten years later they rounded Cape Verde and established themselves in the Azores. By then they had perfected the *caravel*, a ship which used new rigging to tackle head winds and contrary currents on the home voyage by going right out into the Atlantic and sailing a long semi-circular course home. In 1445 they reached Senegal. Their first fort was built soon after. Henry died in 1460, but by then his countrymen were ready to continue further south. In 1473 they crossed the Equator and in 1487 they were at the Cape of Good Hope. Ahead lay the Indian Ocean; Arabs had long traded across it and pilots were available. Beyond it lay even richer sources of spices. In 1498 Vasco da Gama dropped anchor at last in Indian waters.

By that time, another sailor, the Genoese Columbus, had crossed the Atlantic to look for Asia, confident in the light of Ptolemaic geography that he would soon come to it. He failed. Instead he discovered the Americas for the Catholic Monarchs of Spain. In the name of the 'West Indies' the modern map commemorates his continuing belief that he had accomplished the discovery of islands off Asia by his astonishing venture, so different from the cautious, though brave, progress of the Portuguese towards the East round Africa. Unlike them, but unwittingly, he had in fact discovered an entire continent, though even on the much better-equipped second voyage which he made in 1493 he explored only its islands. The Portuguese had reached a known continent by a new route. Soon (though to his dying day Columbus refused to admit it, even after two more voyages and experience of the mainland) it began to be realized that what he had discovered might not be Asia after all. In 1494 the historic name 'New World' was first applied to what had been found in the western hemisphere. (Not until 1726, though, was it to be realized that Asia and America were not joined together in the region of the Bering Straits.)

The two enterprising Atlantic nations tried to come to understandings about their respective interests in a world of widening horizons. The first European treaty about trade outside European waters was made by Portugal and Spain in 1479; now they went on to delimit spheres of influence. The pope made a temporary award, based on a division of the world between them along a line a hundred leagues west of the Azores, but this was overtaken by the treaty of Tordesillas in 1494 which gave to Portugal all the lands east of a line of longitude running 370 leagues west of Cape Verde and to Spain those west of it. In 1500 a Portuguese squadron on the way to the Indian Ocean ran out into the Atlantic to avoid adverse winds and to its surprise struck land which lay east of the treaty line and was not Africa. It was Brazil. Henceforth Portugal had an Atlantic as well as an Asian destiny. Though the main Portuguese effort still lay to the east, an Italian in Portuguese service, Amerigo Vespucci, soon afterwards ran far enough to the south to show that not merely islands but a whole new continent lay between Europe and Asia by a western route. Before long it was named after him, America, the name of the southern continent later being extended to the northern, too.

In 1522, thirty years after Columbus' landfall in the Bahamas, a ship in the Spanish service completed the first voyage round the world. Its commander had been Magellan, a Portuguese, who got as far as the Philippines, where he was killed, having discovered and sailed through the straits named after him. With this voyage and its demonstration that all the great oceans were interconnected, the prologue to the European age can be considered over. Just about a century of discovery and exploration had changed the shape of the world and the course of history. From this time the nations with access to the Atlantic would have opportunities denied to the land-locked powers of central Europe and the Mediterranean. In the first place this meant Spain and Portugal, but they would be joined and surpassed by France, Holland and, above all, England, a collection of harbours incomparably placed at the centre of the newly enlarged hemisphere, all of them easily accessible from their shallow hinterland, and within easy striking distance of all the great European sea routes of the next two hundred years.

The enterprise behind these changes had only been possible because of a growing substratum of maritime skill and geographical knowledge. The new and characteristic figure of this movement was the professional explorer and navigator. Many of the earliest among them were, like Columbus himself, Italian. New knowledge, too, underlay not only the conception of these voyages and their successful technical performance, but also allowed Europeans to see their relationship with the world in a new way. To sum the matter up, Jerusalem ceased to be centre of the world; the maps men began to draw, for all their crudity, are maps which show the basic structure of the real globe.

In 1400 a Florentine had brought back from Constantinople a copy of Ptolemy's Geography. The view of the world it contained had been virtually forgotten for a thousand years. In the second century AD Ptolemy's world already included the Canaries, Iceland and Ceylon, all of which found a place on his maps, along with the misapprehension that the Indian ocean was totally enclosed by land. Translation of his text, misleading as it was, and the multiplication of copies first in manuscript and then in print (there were six editions between 1477, when it was first printed, and 1500) was a great stimulus to better map-making. The Atlas – a collection of engraved and printed

A near-contemporary celebration of Magellan's discovery of the straits that bear his name on the voyage (sponsored by Charles V) on which he met his death. Flanked by the volcanoes of Tierra del Fuego, urged on by the gods of sea and sky, the great navigator is shown sailing through the straits, a new sort of hero, the explorer for exploration's sake.

maps bound in a book – was invented in the sixteenth century; more men than ever could now buy or consult a picture of their world. With better projections, navigation was simpler, too. Here the great figure was a Dutchman, Gerhard Kremer, who is remembered as Mercator. He was the first man to print on a map the word 'America' and he invented the projection which is still today the most familiar – a map of the world devised as if it were an unrolled cylinder, with Europe at its centre. This solved the problem of providing a flat surface on which to read direction and courses without distortion, even if it posed problems in the calculation of distances. The Greeks of the fourth century BC had known the world was a globe and the making of terrestrial and celestial globes was another important branch of the geographical revolution (Mercator made his first globe in 1541).

The most striking thing about this progression is its cumulative and systematic nature. European expansion in the next phase of world history would be conscious and directed as it had never been before. Europeans had long wanted land and gold; the greed which lay at the heart of enterprise was not new. Nor was the religious zeal which sometimes inspired them and sometimes cloaked their springs of action even from the actors themselves. What was new was a growing confidence derived from knowledge and success. Europeans stood in 1500 at the beginning of an age in which their energy and confidence would grow seemingly without limit. The world did not come to them, they went out to it and took it.

The scale of such a break with the past was not to be seen at once. In the Mediterranean and Balkans, Europeans still felt threatened and defensive. Navigation and seamanship still had far to go – not until the eighteenth century, for example, would there be available a time-keeper accurate enough for exact sailing. But the way was opening to new relationships between Europe and the rest of the world, and between European countries themselves. Discovery would be followed by conquest. A world revolution was beginning. An equilibrium which had lasted a thousand years was dissolving. As the next two centuries unrolled, thousands of ships would put out year after year, day after day, from Lisbon, Seville, London, Bristol, Nantes, Antwerp and many other European ports, in search of trade and profit in other continents. They would sail to Calicut, Canton, Nagasaki. In time, they would be joined by ships from places where Europeans had established themselves overseas – from Boston and Philadelphia, Batavia and Macao. And during all that time, not one Arab dhow was to find its way to Europe. It was 1848 before a Chinese junk was brought to the Thames. Only in 1867 would a Japanese vessel cross the Pacific to San Francisco, long after the great sea-lanes had been established by Europeans.

THE EUROPEAN MIND

In 1500 Europe is clearly recognizable as the centre of a new civilization; before long that civilization was to spread to other lands, too. Its heart was still religion. The institutional implications of this have already been touched upon; the Church was a great force of social regulation and government, whatever vicissitudes its central institution had suffered. But it was also the custodian of culture and the teacher of all men, the vehicle and vessel of civilization itself.

Since the thirteenth century the burden of recording, teaching and study so long borne by the monks had been shared by friars and, more important still, by a new institution, in which friars sometimes played a big part, the university. Bologna, Paris and Oxford were the first universities; by 1400 there were fifty-three more. They were new devices both for concentrating and directing intellectual activity and for education. One result was the revivifying of the training of the clergy. Already in the middle of the fourteenth century half the English bishops were graduates. But this was not the only reason why universities had been set up. The Emperor Frederick II founded the University of Naples to supply administrators for his south Italian kingdom; and when in 1264 Walter de Merton, an English bishop and royal servant, founded the first college at Oxford, among his purposes was that of providing future servants for the crown.

The universities' importance for the future of Europe, though, was greater than this, though it could not have been forseen and proved in respect incalculable. Their existence assured that when laymen came to be educated in substantial numbers, they too would be formed by an institution under the control of the Church and suffused with religion. Furthermore, universities would be a great uniting, cosmopolitan force. Their lectures were given in Latin, the language of the Church and the lingua franca of educated men until this century. Its former pre-eminence is still commemorated in the vestigial Latin of university ceremonies and the names of degrees.

Law, medicine, theology and philosophy all benefited from the new institution. Philosophy had all but disappeared into theology in the early medieval period. Only one important figure stands out, John Scotus Erigena, an Irish thinker and scholar of the ninth century. Then, as direct translation from Greek to Latin began in the twelfth century, European scholars could read for themselves works of classical philosophy. The texts became available from Islamic sources. As the works of Aristotle and Hippocrates were turned into Latin they were at first regarded with suspicion. This persisted until well into the thirteenth century, but gradually a search for reconciliation between the classical and Christian accounts of the world got under way and it became clear, above all because of the work of two Dominicans, Albertus Magnus and his pupil Thomas Aquinas, that reconciliation and synthesis were indeed possible. So it came about that the classical heritage was recaptured and rechristened in western Europe. Instead of providing a contrasting and critical approach to the theocentric culture of Christendom it was incorporated with it. The classical world began to be seen as the forerunner of the Christian. For centuries man would turn for authority in matters intellectual to religion or to the classics. Of the latter it was Aristotle who enjoyed unique prestige. If it could not make him a saint, the Church at least treated him as a kind of prophet.

The immediate evidence was the remarkable systematic and rationalist achievement of medieval scholasticism, the name given to the intellectual effort to penetrate the meaning of Christian teaching. Its strength lay in its embracing sweep, displayed nowhere more brilliantly than in the *Summa Theologica* of Aquinas which has been judged, contrastingly, both its crowning achievement and a brittle synthesis. It strove to account for all phenomena. Its weakness lay in its unwillingness to address itself to observation and experiment. Christianity gave the medieval mind a powerful training in logical thinking, but only a few men, isolated and untypical, could dimly see the possibility of breaking through authority to a truly experimental method.

Nevertheless, within the Christian cultural achievement the first signs of liberation from the enclosed world of the early Middle Ages can be seen. Paradoxically, Christendom owed them to Islam, though for a long time there was deep suspicion and fear in the attitudes of ordinary men towards Arab civilization. There was also ignorance (before 1100, one medievalist has pointed out, there is no evidence that anyone in northern Europe had ever heard the name of Muhammad). Not until 1143 was a Latin translation of the Koran available. Easy and tolerant relationships between the Faithful and the Infidel (both sides thought in the same terms) were possible only in a few places. In Sicily and Spain, above all, the two cultures could meet. There the great work of translation of the twelfth and thirteenth centuries took place. The Emperor Frederick II was regarded with the deepest suspicion because although he persecuted heretics he was known to welcome Jews and Saracens to his court at Palermo. Toledo, the old Visigothic capital, was another especially important centre. In such places scribes copied and recopied the Latin texts of the bestsellers of the next six centuries. Euclid's works began a career of being copied, recopied and then printed which may well have meant that in the end they surpassed the success of any book except the Bible – at least until the twentieth century – and became the foundation of mathematics teaching in western Europe until the nineteenth century. In such ways the Hellenistic world began again to irrigate the thought of the West.

Roughly speaking, the Islamic transmission of antiquity began with astrology, astronomy and mathematics, subjects closely linked to one another. Ptolemy's astronomy reached the West by this route and was found a satisfactory basis for cosmology and navigation until the sixteenth century. Islamic cartography was in fact more advanced than European for most of the Middle Ages, and Arab sailors used the magnet for navigation well before their European counterparts (yet it was the latter who were to carry through the great oceanic discoveries). The astrolabe had been a Greek invention, but its use was spread in the West by Arab writings. When Chaucer wrote his treatise on its use, he took as his model an earlier Arab one. The arrival from Arab sources of a new numeration and the decimal point (both of Indian origin) was perhaps most important of all; the latter's usefulness in simplifying calculation can be easily tested by trying to write sums in Roman numerals.

Of the sciences of observation other than astronomy, the most important to come to the West from Islam was medicine. Besides providing access to the medical works of Aristotle, Galen, and Hippocrates (direct translation from the Greek was not begun until after 1100), Arabic sources and teachers also brought into European practice a huge body of therapeutic, anatomical and pharmacological knowledge built up by Arab physicians. The prestige of Arab learning and science made easier the acceptance of more subtly dangerous and subversive ideas; Arab philosophy and theology, too, began to be studied in the West. In the end, even European art seems to have been affected by Islam, for the invention of perspective, which was to transform painting, is said to have come from thirteenth-century Arab Spain. Europe offered little in exchange except the technology of gunnery.

A medieval lecture-room, carved in a relief on the tomb of a professor of Law at Bologna, the first great centre of legal scholarship in western Europe.

To no other civilization did Europe owe so much in the Middle Ages as to Islam. For all their dramatic and exotic interest, the travels of a Marco Polo or the missionary wanderings of friars in central Asia did little to change the West. The quantity of goods exchanged with other parts of the world was still tiny, even in 1500. Technically, Europe owed for certain to the Far East only the art of making silk (which had reached her from the eastern empire) and paper which, though made in China in the second century AD, took until the thirteenth to reach Europe and then did so again by way of

Arab Spain. Nor did ideas reach Europe from nearer Asia, unless like Indian mathematics they had undergone refinement in the Arabic crucible. Given the permeability of Islamic culture, it seems less likely that this was because in some sense Islam insulated Europe from the Orient by imposing a barrier between them, than because China and India could not make their impact felt across such huge distances. They had hardly done so, after all, in pre-Christian antiquity, when communications had been no more difficult.

The reintegration of classical and Christian, though manifested in work like that of Aquinas, was an answer, ten centuries late, to Tertullian's jibing question about what Athens had to do with Jerusalem. In one of the supreme works of art of the Middle Ages – some would judge *the* supreme – the *Divine Comedy* of Dante, the importance of the re-attachment of the world of Christendom to its predecessor is already to be seen. Dante describes his journey through Hell, Purgatory and Paradise, the universe of Christian truth. Yet his guide is not a Christian, but a pagan, the classical poet Virgil. This role is much more than decorative; Virgil is an authoritative guide to truth, for before Christ, he foretold Him. The classical poet has become a prophet to stand beside those of the Old Testament. Though the notion of a link with antiquity had never quite disappeared (as attempts by enthusiastic chroniclers to link the Franks or the Britons to the descendants of the Trojans had shown) there is in Dante's attitude something marking an epoch. It is an acceptance of the classical world by Christendom, and this, for all the scholastic clutter of its surroundings, was decisive in making possible a change which has usually been seen as more radical, the great revival of humanistic letters of the fifteenth and sixteenth centuries.

One central figure of that moment of cultural history was Erasmus of Rotterdam, sometime a monk and later, as the foremost exponent of classical studies of his day, the correspondent of most of the great humanists. Yet he still saw his classics as the entrance to the supreme study of scripture and his most important book was an edition of the Greek New Testament. The effects of printing a good text of the Bible were, indeed, to be revolutionary, but Erasmus had no intention of overthrowing religious order, for all the vigour and wit with which he had mocked and teased puffed-up churchmen, and for all the provocation to independent thought which his books and letters provided. His roots lay in the piety of a fifteenth-century mystical movement in the Low Countries called the *devotio moderna*, not in pagan antiquity.

Some of the men who began to cultivate the study of classical authors, and to invoke explicitly pagan classical ideals, invented the notion of the Middle Ages to emphasize their sense of novelty. They in their turn were spoken of as men of a 're-birth' of a lost tradition, a 'Renaissance' of classical antiquity. Yet they were formed in the culture which the great changes in Christian civilization from the twelfth century onwards had made possible. To speak of Renaissance may be helpful if we keep in mind the limitations of the context in which we use the word, but it falsifies history if we take it to imply a transformation of culture marking a radical break with medieval Christian civilization. The Renaissance is and was a useful myth, one of those ideas which help men to master their own bearings and therefore to act more effectively. Whatever the Renaissance may be, there is no clear line in European history which separates it from the Middle Ages – however we like to define them.

What can be noticed almost everywhere, though, is a change of emphasis. It shows especially in the relation of the age to the past. Men of the thirteenth century, like those of the sixteenth, portrayed the great men of antiquity in the garb of their own day. Alexander the Great at one time looks like a medieval king; later, Shakespeare's Caesar wears not a toga but doublet and hose. There is, that is to say, no real historical sense in either of these pictures of the past, no awareness of the immense differences between past and present men and things. Instead, history was seen at best as a school of examples. The difference between the two attitudes is that in the medieval view antiquity could also be scrutinized for the signs of a divine plan, evidence of whose existence once more triumphantly vindicated the teachings of the Church. This was St Augustine's legacy and what Dante accepted. But by 1500 something else was also being discerned in the past, equally unhistorical, but, men felt, more helpful to their age and predicament. Some saw a classical inspiration, possibly even pagan, distinct from the Christian, and a new attention to classical writings was one result.

The idea of Renaissance is especially linked to innovation in art. Medieval Europe had seen much of this; it seems more vigorous and creative than any of the other great centres of civilized tradition from the twelfth century onwards. In music, drama and poetry new forms and styles were created which move us still. By the fifteenth century, though, it is already clear that they can in no sense be confined to the service of God. Art is becoming autonomous. The eventual consummation of this change was the major aesthetic expression of the Renaissance, transcending by far its stylistic innovations, revolutionary though these were. It is the clearest sign that the Christian synthesis and the ecclesiastical monopoly of culture are breaking up. The slow divergence of classical and Christian mythology was one expression of it; others were the appearance of the Romance and Provencal love poetry (which owed much to Arabic influence), the deployment of the Gothic style in secular building such as the great guildhalls of the new cities, or the rise of a vernacular literature for educated laymen of which perhaps the supreme example is Chaucer's *Canterbury Tales.*

Such changes are not easily dated, because acceptance did not always follow rapidly on innovation. In literature, there was a particularly severe physical restriction on what could be done because of a long-enduring shortage of texts. It was not until well into the sixteenth century that the first edition of Chaucer's complete works was printed and published. By then a revolution in thinking was undoubtedly under way, of which all the tendencies so far touched on form parts, but which was something much more than the sum of them and it owes almost everything to the coming of the printed book. Even a vernacular text such as the *Canterbury Tales* could not reach a wide public until printing made large numbers of copies easily available. When this happened, the impact of books was vastly magnified. This was true of all classes of book, poetry, history, philosophy, technology and, above all, the Bible itself. The effect was the most profound change in the diffusion of knowledge and ideas since the invention of writing; it was the greatest cultural revolution of these centuries.

The new technique owed nothing to stimulus from China, where it was already practised in a different form, except very indirectly, through the availability of paper. From the fourteenth century, rags were used in Europe to make paper of good quality and this was one of the elements which contributed to the printing revolution. Others were the principle of printing itself (the impressing of images on textiles had been practised in twelfth-century Italy), the use of cast metal for typefaces instead of wood (already used to provide blocks for playing-cards, calendars and religious images), the availability of oil-based ink, and, above all, the use of movable metal type. It was the last invention which was crucial. Although the details are obscure, and experiments with wood letters were going on at the beginning of the fifteenth century in Haarlem, there seems to be no good reason not to credit it to the man whose name has traditionally been associated with it, Johannes Gutenberg, the diamond polisher of Mainz. In about 1450 he and his colleagues brought the elements of modern printing together and in 1455 there appeared what is agreed to be the first true book printed in Europe, the Gutenberg Bible.

Gutenberg's own business career was by then a failure; something prophetic of a new age of commerce appears in the fact that he was probably under-capitalized. The accumulation of equipment and type was an expensive business and a colleague from whom he borrowed money took him to court for his debts. Judgement went against Gutenberg, who lost his press, so that the Bible, when it appeared, was not his property. (Happily, the story does not end there; Gutenberg was in the end ennobled by the Archbishop of Mainz, in recognition of what he had done.) But he had launched a revolution. By 1500, it has been calculated, some thirty-five thousand separate editions of books – *incunabula*, as they were called – had been published. This probably means between fifteen and twenty million copies; there may well have been already at that date fewer copies of books in manuscript in the whole world. In the following century there were between a hundred and fifty and two hundred thousand separate editions and perhaps ten times as many copies printed. Such a quantitative change merges into one which is qualitative; the culture which resulted from the coming of printing with movable type was as different from any earlier one as it is from one which takes radio and television for granted. The modern age was the age of print.

It is interesting but natural that the first printed European book should have been the Bible, the sacred text at the heart of medieval civilization. Through the printing press, knowledge of it was to be diffused as never before and with incalculable results. In 1450 it would have been very unusual for a parish priest to own a Bible, or even to have easy access to one. A century later, it was becoming likely that he had one, and in 1650 it would have been remarkable if he had not. The first German Bible was printed in 1466; Italian and French translations followed before the end of the century, but Englishmen had to wait for a New Testament printed in their language until 1526. Into the diffusion of sacred texts – of which the Bible was only the most important – pious laymen and churchmen alike poured resources for fifty to sixty years; presses were even set up in monastic houses. Meanwhile, grammars, histories, and, above all, the classical authors now edited by the humanists also appeared in increasing numbers. Another innovation from Italy was the introduction of simpler, clearer typefaces modelled upon the manuscript of Florentine scholars who were themselves copying Carolingian minuscule.

The impact could not be contained. The domination of the European conciousness by printed media would be the outcome. With some prescience the pope suggested to bishops in 1501 that the control of printing might be the key to preserving the purity of the faith. But more was involved than any specific threat to doctrine, important as that might be. The nature of the book itself began to change. Once a rare work of art, whose mysterious knowledge was accessible only to a few, it became a tool and artifact for the many. Print was to provide new channels of communication for governments and a new medium for artists (the diffusion of pictorial and architectural style in the sixteenth century was much more rapid and widespread than ever before because of the growing availability of the engraved print) and would give a new impetus to the diffusion of technology. A huge demand for literacy and therefore education would be stimulated by it. No single change marks so clearly the ending of one era and the beginning of another.

It is very hard to say exactly how such changes affected Europe's role in the coming era of world history. By 1500, there was certainly much to give confidence to the few Europeans who were likely to think at all about these things. The roots of their civilization lay in a religion which taught them they were a people voyaging in time, their eyes on a future made a little more comprehensible and perhaps a little less frightening by contemplation of past perils navigated and awareness of a common goal. As a result Europe was to be the first civilization aware of time not as endless (though perhaps cyclical) pressure, but as continuing change in a certain direction, as progress. The chosen people of the Bible, after all, were *going* somewhere; they were not simply people to whom inexplicable things happened which had to be passively endured. From the simple acceptance of change was before long to spring the will to live in change which was the peculiarity of modern man. Secularized and far away from their origins, such ideas could be very important; the advance of science soon provided an example. In another sense, too, the Christian heritage was decisive for, after the fall of Byzantium, Europeans believed that they alone possessed it (or in effect alone, for there was little sense among ordinary folk of what Slav, Nestorian or Coptic Christianity might be). It was an encouraging idea for men who stood at the threshold of centuries of unfolding power, discovery and conquest. Even with the Ottomans to face, Europe in 1500 was no longer just the beleaguered fortress of the Dark Ages, but a stronghold from which men were beginning to sally forth in counter-attack. Jerusalem had been abandoned to the infidel, Byzantium had fallen. Where should be the new centre of the world?

The men of the Dark Ages who had somehow persevered in adversity and had built a Christian world from the debris of the past and the gifts of the barbarians had thus wrought infinitely more than they could have known. Yet such implications required time for their development; in 1500 there was still little to show that the future belonged to the Europeans. Such contacts as they had with other peoples by no means demonstrated the clear superiority of their own way. Portuguese in West Africa might manipulate black men to their own ends and relieve them of their gold dust and slaves, but in Persia or India they stood in the presence of great empires whose spectacle often dazzled them. The men of 1500 were thus and in many other ways not modern men. We cannot without an effort understand them, even when they speak Latin, for their Latin had overtones and associations we are bound to miss; it was not only the language of educated men, but the language of religion.

In the half-light of a dawning modernity the weight of that religion remains the best clue to the reality of Europe's first civilization. Religion was one of the most impressive reinforcements of the stability of a culture which has been considered in this book almost entirely from an important but fundamentally anachronistic perspective, that of change. Except in the shortest term, change was not something most Europeans would have been aware of in the fifteenth century. For all men, the deepest determinant of their lives was still the slow but ever-repeated passage of the seasons, a rhythm which set the pattern of work and leisure, poverty and prosperity, of the routines of home, workshop and study. English judges and university teachers still work to a year originally divided by the need to get in the harvest. On this rhythm were imposed those of religion itself. When the harvest was in the Church blessed it and the calendar of the Christian year provided the more detailed timetable to which men lived. Some of it was very old, even pre-Christian; it had been going on for centuries and could hardly be imagined otherwise. It even regulated many people's days, for every three hours the religious were called to worship and prayer in thousands of monasteries and convents by the bell of their house. When it could be heard outside the walls, laymen set the pattern of their day by it, too. Before there were striking clocks, only the bell of the parish church, cathedral or monastery supplemented the sun or the burning of a candle as a record of passing time, and it did so by announcing the hour of another act of worship.

It is only in a very special, long perspective that we can rightly speak of centuries during which this went on and on as ones of 'revolutionary' changes. Truly revolutionary as some changes were, even the most obvious of them, the growth of a town, an onset of plague, the displacement of one noble family by another, the building of a cathedral or the collapse of a castle, all took place in a remarkably unchanged setting. The shapes of the fields tilled by English peasants in 1500 were often still those visited by the men who wrote them down in Domesday Book, over four hundred years before, and when men went to visit the nuns of Lacock in order to wind up their house in the 1530s, they found, to their amazement, these aristocratic ladies still speaking among themselves the Norman-French commonly used in noble families three centuries earlier.

Such immense inertia must never be forgotten; it was made all the more impressive and powerful by the fleeting lives of most men and women of the Middle Ages. Only very deep in the humus of this society did there lie a future. Perhaps the key to that future's relationship with the past can be located in the fundamental Christian dualism of this life and the world to come, the earthly and the heavenly. This was to prove an irritant of great value, secularized in the end as a new critical instrument, the contrast of what is and what might be, of ideal and actual. In it, Christianity secreted an essence to be utilized against itself, for in the end it would make possible the independent critical stance, a complete break with the world Aquinas and Erasmus both knew. The idea of autonomous criticism would only be born very gradually, though; it can be traced in many individual adumbrations between 1300 and 1700, but they only go to show that, once again, sharp dividing lines between medieval and modern are matters of expository convenience, not of historical reality.

Vasco da Gama, explorer, whose achievements were rewarded by wealth, honours, and the vice-royalty of Portuguese India, where he died.

LION-STRANGLING *The techniques of silk-weaving originated in China, making their way in the early Christian era to Persia, the Levant and Byzantium. From there, they were to spread into Europe and, as they did so, they bore with them themes from ancient art. This example of Persian-style work in silk and gold thread from twelfth-century Islamic Spain depicts a royal hero dominating beasts (in this case lions), a motif traceable as far back as ancient Mesopotamian seals. It is sometimes called the 'Gilgamesh' motif – an incongruous item to end in the tomb of a medieval bishop of Vich, in north-eastern Catalonia, where it was found.*

JAPANESE SCROLL *On this scroll, Samurai warriors storm through one of the gates of Kyoto in the 'Heiji Insurrection' of 1156. Essentially, this was a struggle of two military houses for control of the Japanese imperial capital. The rivalry of Taira and Minamoto culminated in the month's fighting of the insurrection and the rout of the Minamoto. That was not, however, the end of the story – decades later the descendants of those who fought in 1156 were at one another's throats in the Gempei War (1180-1185) – a struggle commemorated in an epic literature which played a major part in the formation of the Samurai tradition, much as did the Arthurian Legend in the shaping of medieval Europe's chivalric ideal.*

THE MEDIEVAL WORLD ORDER *In 1355, Andrea Bonaiuti painted this fresco for the church of Santa Maria Novella in Florence. Against the symbolic background of the (then uncompleted) cathedral of Florence, he set side-by-side the figures of Pope and Emperor. In due order, they are flanked by great officers and magnates of church and state – ecclesiastics to the Pope's right, laymen to the Emperor's left. Christ's flock nestles at the feet of Christ's two Vicars on earth, guarded by dogs who protect them from the wolf of heresy. A pun is intended: these are* Domini canes – *'the Lord's dogs' – whose coats echo the black and white habits of the Dominicans in whose honour the fresco was commissioned. At the sides of the fresco cluster bishops, nuns, doctors on the one side, and, on the other, peasants, nobles and townsfolk (among them, it is believed, representations of Dante, Boccacio, the painter Cimabue and other notable Florentines).*

BOOK V

THE MAKING OF THE EUROPEAN AGE

After 1500 or so, there are many signs that a new age of world history is beginning. Some of them have already appeared in these pages; the discoveries in the Americas and the first shoots of European enterprise in Asia are among them. At the outset they provide hints about the dual nature of a new age – that it is increasingly an age of truly world history and that it is one whose story is dominated by the astonishing success of one civilization among many, that of Europe. These are two aspects of the same process; there is a more and more continuous and organic interconnexion between events in all countries, but it is largely to be explained by the efforts of Europeans. They eventually became masters of the globe and they used their mastery – sometimes without knowing it – to make the world one. As a result world history has for the last two or three centuries a growing identity and unity of theme.

In a famous passage, the English historian Macaulay once spoke of red men scalping one another on the shores of the Great Lakes so that a European king could rob his neighbour of a province he coveted. This was one striking side of the story we must now embark upon – the gradual entanglement of struggles the world over with one another in greater and greater wars – but politics, empire-building and military expansion were only a tiny part of what was going on. The economic integration of the globe was another part of the process; more important still was the spreading of common assumptions and ideas. The result was to be, in a cant phrase of today, 'One World' – of sorts. The age of independent or nearly independent civilizations has come to a close.

Men and nations are still so different from one another that this may seem at first sight a wildly misleading exaggeration. National, cultural and racial differences have not ceased to produce and inspire appalling conflicts; the history of the centuries since 1500 can be (and often is) written mainly as a series of wars and violent struggles and men in different countries obviously do not feel much more like one another than did their predecessors centuries ago. Yet they are much more alike than their ancestors of, say, the tenth century and show it in hundreds of ways ranging from the superficialities of dress to the forms in which they get their living and organize their societies. The origins, extent and limits of this change make up most of the story which follows. It is the outcome of something still going on in many places which we sometimes call modernization. For centuries it has been grinding away at differences between cultures and it is the deepest and most fundamental expression of the growing integration of world history. Another way of describing the process is to say that the world is Europeanized, for modernization is above all a matter of ideas and techniques which are European in origin. Whether 'modernization' is the same as 'Europeanization', though, can be left for discussion elsewhere; perhaps it is only a matter of verbal preferences. What is obvious is that, chronologically, it is with the modernization of Europe that the unification of world history begins. A great change in Europe was the starting-point of modern history.

1
A New Kind of Society: Early Modern Europe

'Modern history' is a familiar term, but it does not always mean the same thing. There was a time when modern history was what had happened since the 'ancient' history whose subject-matter was the story of the Jews, Greeks and Romans; this is a sense which, for example, is still used to define a course of study at Oxford which includes the Middle Ages. Then it came to be distinguished from 'medieval' history, too. Now a further refinement is often made, for historians have begun to make distinctions within it and sometimes speak of an 'early modern' period. By this, though, they are really drawing our attention to a process, for they apply it to the era in which the modern Atlantic world emerged from the tradition-dominated, agrarian, superstitious and confined western Christendom of the Middle Ages, and this took place at different times in different countries. In England it happened very rapidly; in Spain it was far from complete by 1800, while much of eastern Europe was still hardly affected by it even a century later. But the reality of the process is obvious, for all the irregularity with which it expressed itself. So is its importance, for it laid the groundwork for a European world hegemony.

A useful starting-point for thinking about what was involved is to begin with the simple and obvious truth that for most of human history most people's lives have been deeply and cruelly shaped by the fact that they have had little or no choice about the way in which they could provide themselves and their families with shelter and enough to eat. The possibility that things might be otherwise has only recently become a conceivable one to even a minority of the world's population and it became a reality for any substantial number of people only with changes in the economy of early modern Europe, for the most part, west of the Elbe.

We can follow some of these changes as we can follow no earlier ones because for the first time there is reasonably plentiful and continuous quantified data. In one important respect, historical evidence gets much more informative in the last four or five centuries: it becomes much more statistical. Measurement therefore becomes easier. The source of new statistical material was often government. For many reasons, governments wanted to know more and more about the resources or potential resources at their disposal. But private records, especially of business, also give us much more numerical data after 1500. The multiplication of copies as paper and printing became more common meant that the chance of their survival was enormously increased. Commercial techniques appeared which required publication of data in collated forms; the movements of ships, or reports of prices, for example. Moreover, as historians have refined their techniques, they have attacked even poor or fragmentary sources with much greater success than was possible even a few years ago.

All this has provided much knowledge of the size and shape of change in early modern Europe, though we must be careful not to exaggerate either the degree of precision such material permits or what can be learnt from it. For a long time the collection of good statistics was very difficult. Even quite elementary questions, about, for example, who lived in a certain place, were very difficult to answer accurately until recent times. One of the great aims of reforming monarchs in the eighteenth century was merely to carry out accurate listings of land within their states, cadastral surveys, as they were called, or even to find out how many subjects they had. It was only in 1801 that the first census was held in Great Britain – nearly eight centuries after Domesday Book. France did not have her first official census until 1876 nor the Russian empire her only one until 1897. Such delays are not really surprising. A census or a survey requires a complex and reliable administrative machine. It may arouse strong opposition (when governments seek new information, new taxes often follow). Such difficulties are enormously increased where the population is as illiterate as it was in much of Europe for the greater part of modern history.

New statistical material can also raise as many historical problems as it solves. It can reveal a bewildering variety of contemporary phenomena which often makes generalization harder; it has become much harder to say anything at all about the French peasantry of the eighteenth century since research revealed the diversity hidden by that simple term and that perhaps there was no such thing as *a* French peasantry, but only several different ones. Finally, too, statistics can illuminate facts while throwing no light at all on causes. Nevertheless after 1500 we are more and more in an age of measurement and the overall effect of this is to make it easier to make defensible statements about what was happening than in earlier times at other places.

Demographic history is the most obvious example. At the end of the fifteenth century European population was poised on the edge of growth which has gone on ever since. After 1500 we may crudely distinguish two phases. Until about the middle of the eighteenth century the increase of population was (except for notable local and temporary interruptions) relatively slow and steady; this roughly corresponds to 'early modern' history and was one of the things characterizing it. In the second phase the increase much accelerated and great changes followed. Only the first phase concerns us here, because it regulated the way in which modern Europe took shape. The general facts and trends within it are clear enough. Though they rely heavily on estimates, the figures are much better based than in earlier times, in part because there was almost continuous interest in population problems from the early seventeenth century onwards. This contributed to the foundation of the science of statistics (then called 'Political Arithmetic') at the end of the seventeenth century, mainly by Englishmen. They did some remarkable work, though their efforts only provided a tiny island of relatively rigorous method in a sea of guesses and inferences. Nevertheless the broad picture is clear. In 1500 Europe had about eighty million inhabitants, two centuries later she had less than one hundred and fifty million and in 1800 slightly less than two hundred million. Before 1750 Europe had grown fairly steadily at a rate which maintained her share of the world's population at about one-fifth until 1700 or so, but by 1800 she had nearly a quarter of the world's inhabitants.

Obviously, therefore, for a long time there were no such startling disparities as appeared later between the rate of growth in Europe and that elsewhere. It seems reasonable to conclude that this meant that in other ways, too, European and non-European populations were less different than they were to come to be after 1800. The usual age of death among Europeans, for example, still remained low. Before 1800 they were on the average always much younger than nowadays, because people died earlier. At birth a French peasant of the eighteenth century had a life expectancy of about twenty-two years and only a roughly one in four chance of surviving infancy. His chances were therefore much the same as those of an Indian peasant in 1950 or an Italian under imperial Rome. Comparatively few people would have survived their forties, and, since they were less well fed than we are, they would have looked old to us at that age, and probably rather small in stature and unhealthy-looking. As in the Middle Ages, women tended still to die before men. This meant that many men made a second or even a third marriage, not, as today, because of divorce, but because they were soon widowers. The average European couple had a fairly short married life. West of a line running roughly from the Baltic to the Adriatic, they had shorter marriages than east of it, moreover, because those who lived there tended to make their first marriage later in their twenties, and this was long to be a habit making for different population patterns east and west. Generally, though, if Europeans were well-off they could afford a fairly large family; the poor had smaller ones. There is strong inferential evidence both that some form of family limitation was already taking place in some places in the seventeenth century and that other methods of achieving it than abortion and infanticide were available and widely known. No doubt both cultural and economic facts are needed to explain this mysterious topic. It remains one of those areas where a largely illiterate society is almost impossible to penetrate historically. We can say very little with confidence about the material facts of early birth control and still less about its implications – if there were any – for the ways in which early modern Europeans thought about themselves and their control over their own lives.

Overall such demographic facts reflect the continuing economic predominance of agriculture. For a long time it produced only slightly more food than was needed and could only feed a slowly growing population. In 1500 Europe was still largely a rural continent of villages in which people lived at a pretty low level of subsistence. It would have seemed very empty to modern eyes. England's population, heavy in relation to area by comparison with the rest of the continent, was in 1800 only about a fifth of today's; in eastern Europe there were huge empty spaces for which population was eagerly sought by rulers who encouraged immigration in all sorts of ways. Yet the towns and cities managed to grow in number and size, one or two of them spectacularly faster than the population as a whole. Amsterdam reached a total of about 200,000 inhabitants in the eighteenth century. Paris probably doubled in size between 1500 and 1700, and rose to slightly less than a half-million. London shot ahead of Paris by going up from about 120,000 to nearly 700,000 in the same two centuries; in a much smaller English population this, of course, meant a much bigger shift to urban life. A significant new word came into use in English: suburbs. But it is not easy to generalize about medium-sized and smaller towns. Most were quite small, still under 20,000 in 1700, but the nine European cities of more than 100,000 in 1500 had become at least a dozen two hundred years later. Yet Europe's predominance in urbanization was not so marked in these centuries as it was to become and there were still many great cities in other continents. Mexico, for example, outdid all European cities of the sixteenth century with its population of 300,000.

Neither urbanization nor population growth was evenly spread. France remained the largest west European nation in these years; she had about 21 million inhabitants in 1700, when England and Wales had only about 6 million. But it is not easy to make comparisons because estimates are much less reliable for some areas than others and because boundary changes often make it hard to be sure what we are talking about under the same name at different times. Some certainly underwent checks and possibly setbacks in their population growth in a wave of seventeenth-century disasters. Spain, Italy and Germany all had bad outbreaks of epidemic disease in the 1630s and there were other celebrated local attacks such as the Great Plague of London of 1665. Famine was another sporadic and local check; we hear even of cannibalism in the middle seventeenth century in Germany. Poor feeding and the lower resistance it led to quickly produced disaster when coupled to the disruption of the economy which could follow a bad harvest. When accentuated by warfare, of which there was always a great deal in central Europe, the result could be cataclysmic. Famine and the diseases which followed armies about in their baggage-trains could quickly depopulate a small area. Yet this in part reflected the degree to which economic life was still localized; the converse was that a particular town might get off unscathed even in a campaigning zone if it escaped siege or sack, while only a few miles away another was devastated. The situation was always precarious until population growth began to be overtaken by productivity.

In this, as in so many things, different countries have different histories. A renewed expansion of agriculture seems to have got under way in the middle of the fifteenth century. One sign was the resumption of land which had reverted to waste in the depopulation of the fourteenth century. Yet this had made little headway in any but a few places before 1550 or so. It remained confined to them for a long time, though by then there had already been important improvements in techniques which raised the productivity of land, mainly by the application of labour, that is by intensive cultivation. Where their impact was not felt the medieval past long lingered in the countryside. Even the coming of money was slow in breaking into the near self-sufficiency of some communities. In eastern Europe serfdom actually extended its range when it was dying out elsewhere. Yet by 1800, taking Europe as a whole and a few leading countries in particular, agriculture was one of the two economic sectors where progress was most marked (commerce was the other). Overall, it had proved capable of sustaining a continuing rise of population at first very slowly, but at a quickening rate.

Agricultural progress increasingly took two main forms: orientation towards the market, and technical innovation. They were interconnected. A large population in the neighborhood meant a market and therefore an incentive. Even in the fifteenth century the inhabitants of the Low Countries were already leaders in the techniques of intensive cultivation. It was in Flanders, too, that better drainage opened the way to better pasture and to a larger animal population. Another area with relatively large town populations was the Po valley; in north Italy new crops were introduced into Europe from Asia. Rice, for example, an important addition to the European larder, appeared in the Arno and Po valleys in the fifteenth century. On the other hand not all crops enjoyed instant success. It took about two centuries for the potato, which came to Europe from America, to become established as a normal item of consumption in England, Germany and France, in spite of its obvious nutritional value and much promotional folklore stressing its qualities as an aphrodisiac and value in the treatment of warts.

From the Low Countries agricultural improvements spread in the sixteenth century to eastern England where they were slowly elaborated further. In the seventeenth century London became a corn-exporting port and in the next other Europeans came to England to learn how to farm. The eighteenth century also brought better husbandry and animal breeding. Such improvements led to yields on crops and a quality of livestock now taken for granted but until then unimaginable. The appearance of the countryside and its occupants was transformed. Agriculture provided the first demonstration of what might be done by even rudimentary science – by experiment, observation, record, and experiment again – to increase Man's control of his environment more rapidly than could the selection imposed by custom. Improvement favoured the reorganization of land in bigger farms, the reduction of the number of small holders except on land which specially favoured them, the employment of wage-labour, and high capital investment in buildings, drainage and machinery. The speed of change must not be exaggerated. One index of change in England was the pace of 'enclosure', the consolidation for private use of the open fields and common lands of the traditional village. It was only at the end of the eighteenth century and the beginning of the nineteenth that the Acts of Parliament authorizing this became frequent and numerous. The complete integration of agriculture with the market economy and the treatment of land simply as a commodity like any other would have to wait for the nineteenth century, even in England, the leader of world agriculture until the opening of the overseas cornlands. Yet by the eighteenth century the way ahead was beginning to appear.

This overall agricultural change in the end eliminated the recurrent dearths which so long retained their power to destroy demographic advance. Perhaps the last moment when European population seems to have pressed on resources so as to threaten another great calamity like that of the fourteenth century came at the end of the sixteenth century. In the next bad spell, in the middle decades of the following century, England and the Netherlands escaped the worst. Thereafter, famine and dearth become in Europe local and national events, still capable, it is true, of causing large-scale demographic damage, but gradually succumbing to the increasing availability of imported grain. Bad harvests, it has been said, made France 'one great hospital' in 1708–9, but that was in wartime. Later in the century some Mediterranean countries depended for their flour on corn from the Baltic lands. True, it would be a long time before import would be a sure resource; often it could not operate quickly enough, especially where land transport was required. Some parts of France and Germany were to suffer dearth even in the nineteenth century, and in the eighteenth century the French population grew faster than production so that

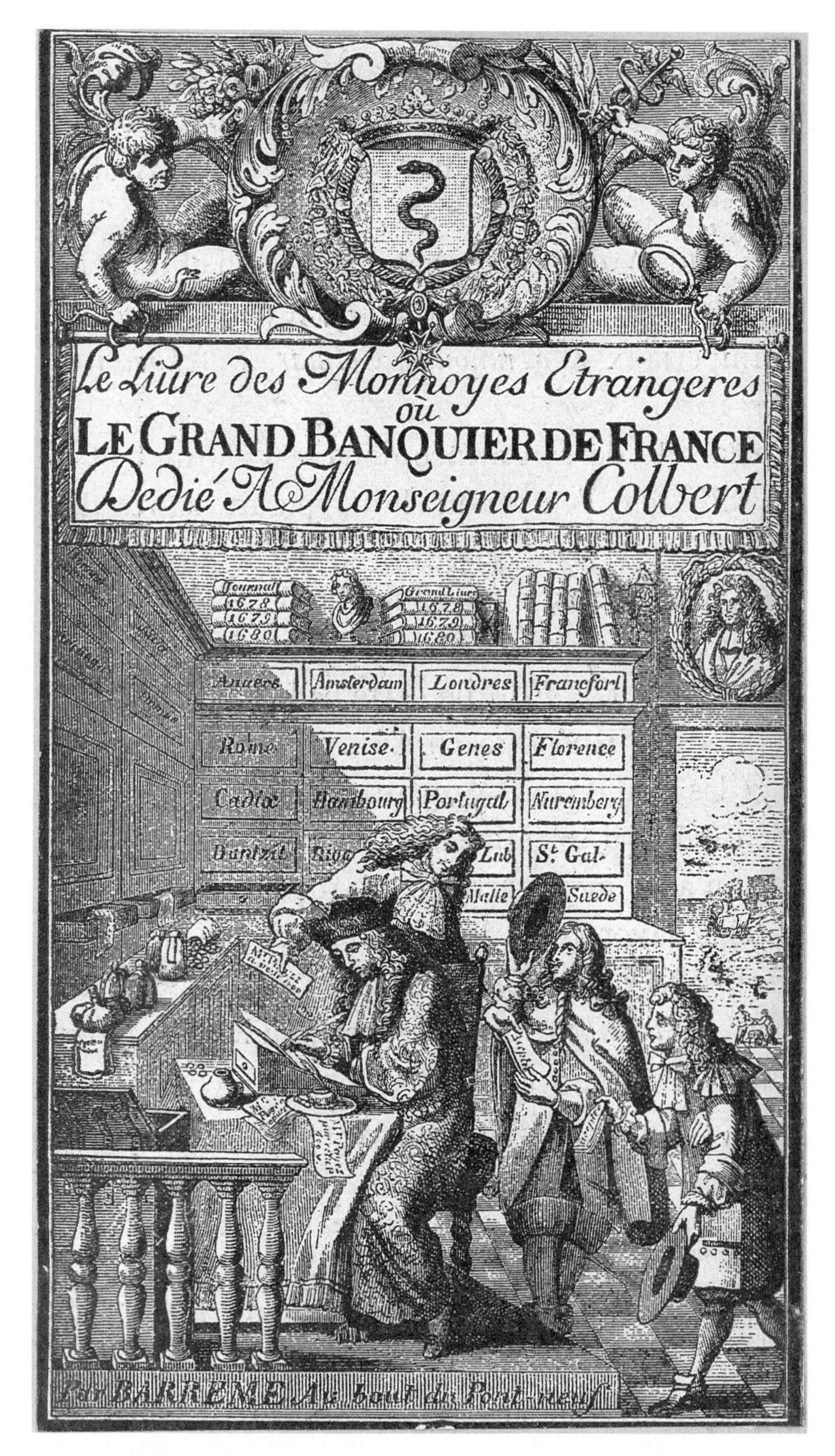

The demand for business manuals of all sorts provided a profitable field of publication to printers. They began to appear in large numbers from the seventeenth century; this example, from France, and dedicated to an outstanding finance minister, deals with the complicated science of the exchange of specie, an important part of the banker's skills in an age which still depended heavily for its business on the physical transfer of coin.

the standard of living of many Frenchmen then actually fell back. For the English rural labourer, though, some of that century was later looked back to as a golden age of plentiful wheaten bread and even meat on the table.

In the late sixteenth century one response to the obscurely felt pressure of an expanding population upon slowly growing resources had been the promoting of emigration. By 1800, Europeans had made a large contribution to the peopling of new lands overseas. Though far from so great a one as was to come, this has to be taken into account in their demographic history. In 1751 an American reckoned that North America contained a million persons of British origin; modern calculations are that about 250,000 British emigrants went to the New World in the seventeenth century, one and a half millions in the next. There were also Germans (about 200,000) there, and Frenchmen in Canada. By 1800 it seems reasonable to suppose that something like two million Europeans had gone to America north of the Rio Grande. South of it there were about 100,000 Spaniards and Portuguese.

Fear that there was not enough to eat at home helped to initiate these great migrations and reflected the continuing pre-eminence of agriculture in all thinking about economic life. There were important changes in three centuries in the structure and scale of all the main sectors of the European economy, but it was still true in 1800 (as it had been true in 1500) that the agricultural sector predominated even in France and England, the two largest western countries where commerce and manufacture had much progressed. Moreover, nowhere was anything but a tiny part of the population engaged in industry entirely unconnected with agriculture. Brewers, weavers, and dyers all depended on it, while many who grew crops or cultivated land also span, wove or dealt in commodities for the market. Apart from agriculture, it is only in the commercial sector that we can observe sweeping change. Here there is from the second half of the fifteenth century a visible quickening of tempo. Europe was regaining something like the commercial vigour first displayed in the thirteenth century. It showed in scale, technique and direction. Again there is a connexion with the growth of towns. They both needed and provided a living for specialists. The great fairs and markets of the Middle Ages still continued. So did medieval laws on usury and the restrictive practices of guilds. Yet a whole new commercial world came into existence before 1800.

It was already discernible in the sixteenth century when there began the long expansion of world commerce which was to last, virtually uninterrupted except briefly by war, until 1930, and then to be resumed again after the next world war. It started by carrying further the shift of economic gravity from southern to north-western Europe, from the Mediterranean to the Atlantic, which has already been remarked. One contribution to this was made by political troubles and wars such as ruined Italy in the early sixteenth century; others are comprised in tiny, short-lived but crucial pressures like the Portuguese harassment of Jews which led to so many of them going, with their commercial skills, to the Low Countries at about the same time. The great commercial success story of the sixteenth century was Antwerp's, though it collapsed after a few decades in political and economic disaster. In the seventeenth century Amsterdam and London surpassed it. In each case an important trade based on a well-populated hinterland provided profits for diversification into manufacturing industry, services and banking. The old banking supremacy of the medieval Italian cities passed first to Flanders and the German bankers of the sixteenth century and then, finally, to Holland and London. The Bank of Amsterdam and the Bank of England were already international economic forces in the seventeenth century. About them clustered other banks and merchant houses undertaking operations of credit and finance. Interest rates came down and the bill of exchange, a medieval invention, underwent an enormous extension of use and became the primary financial instrument of international trade.

This was the beginning of the increasing use of paper, instead of bullion. In the eighteenth century came the first European paper currencies and the invention of the cheque. Joint stock companies generated another form of negotiable security, their own shares. Quotation of these in London coffee-houses in the seventeenth century was

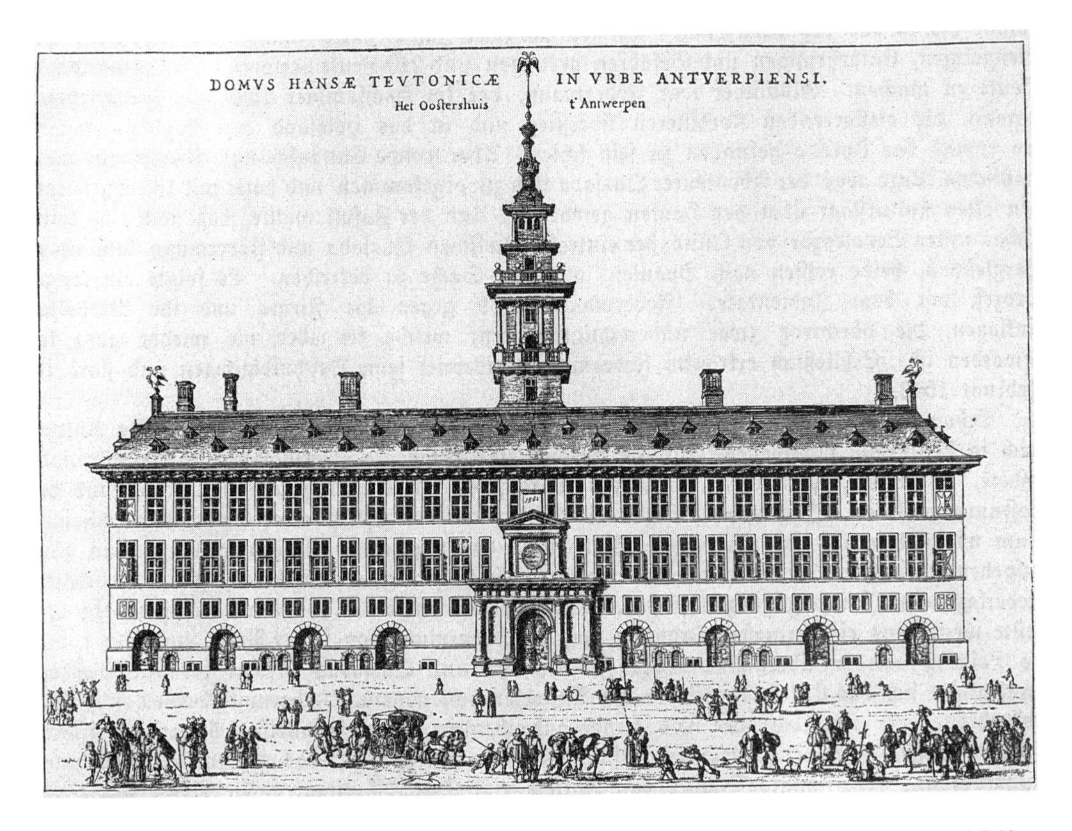

The hall of the Hanse merchants, the master of the rich Baltic trade, at Antwerp in 1563.

overtaken by the foundation of the London Stock Exchange. By 1800 similar institutions existed in many other countries. New schemes for the mobilization of capital and its deployment proliferated in London, Paris and Amsterdam. Lotteries and tontines at one time enjoyed a vogue; so did some spectacularly disastrous investment projects, of which the most notorious was the great English South Sea 'Bubble'. But all the time the world was growing more commercial, more used to the idea of employing money to make money, and was supplying itself with the apparatus of modern capitalism.

One effect quickly appeared in the much greater attention paid to commercial questions in diplomatic negotiation from the later seventeenth century and in the fact that countries were prepared to fight over them. The English and Dutch went to war over trade in 1652. This opened a long era during which they, the French and Spanish, fought again and again over quarrels in which questions of trade were important and often paramount.

Governments not only looked after their merchants by going to war to uphold their interests, but also intervened in other ways in the working of the commercial economy. One advantage they could offer was a grant of monopoly privileges to a company under a charter; this made the raising of capital easier by offering some security for a return. In the end people came to think that chartered companies might not be the best way of securing economic advantage and they fell into disfavour (enjoying a last brief revival at the end of the nineteenth century). None the less, such activities closely involved government and therefore the concerns of businessmen shaped both policy and law.

Occasionally the interplay of commercial development and society seems to throw light on changes with very deep implications indeed. One example came when a seventeenth-century English financier for the first time offered life insurance to the public. There had already begun the practice of selling annuities on a man's life. What was new was the application of actuarial science and the newly available statistics of 'political arithmetic' to this business. A reasonable calculation instead of a bet was now possible on a matter hitherto of awe-inspiring uncertainty and irrationality: death. With increasing refinement men would go on to offer (at a price) protection against a wide range of disaster. This would, incidentally, also provide another and very important device for the mobilization of wealth in large amounts for further investment. But the timing of the discovery of life insurance, at the start of what has sometimes been called the 'Age of Reason', suggests also that the dimensions of economic change are sometimes very far-reaching indeed. It was one tiny source and expression of a coming secularizing of the universe.

The most impressive structural development in European commerce was the sudden new importance to it of overseas trade from the second half of the seventeenth century onwards. This was part of the shift of economic activity from Mediterranean to northern Europe already observable before 1500. By the late seventeenth century, though the closed trade of Spain and Portugal with their transatlantic colonies was important, overseas commerce was dominated by the Dutch and their followers and increasingly successful rivals, the English. Dutch commerce grew out of the supply of salted herrings to European markets and the possession of a particularly suitable bulk-carrying vessel, the 'flute' or 'fly-boat'. With this the Dutch dominated the important Baltic trade from whose mastery they advanced to become the carriers of Europe. In the later seventeenth century they tended to be displaced gradually by the English, though they maintained a wide spread of colonies and trading stations, especially in the Far East, where they had succeeded the Portuguese in the domination of maritime commerce. The Atlantic was the basis of English supremacy at sea. Fish was important here, too. The English caught the immensely nutritious and therefore immensely valuable cod on the Newfoundland banks, dried it and salted it ashore, and then sold it in Mediterranean countries, where fish was in great demand because of the practice of fasting on Fridays. *Bacalao*, as it was called, can still be found on the tables of Portugal southern Spain, once the tourist coast is left behind. Gradually, both Dutch and English broadened and diversified their carrying trade and became dealers themselves, too. Nor was France out of the race; her overseas trade doubled in the first half of the seventeenth century.

Rising populations and some assurance of adequate transport (water was always cheaper than land carriage) slowly built up an international trade in cereals. Shipbuilding itself promoted the movement of such commodities as pitch, flax or timber, staples first of Baltic trade and later important in the economy of North America. More than European consumption was involved; all this took place in a setting of growing colonial empires. By the eighteenth century we are already in the presence of an oceanic economy and an international trading community which does business – and fights and intrigues for it – around the globe.

In this economy an important and growing part was played by slaves. Most of them were black Africans, the first of whom to be brought to Europe were sold at Lisbon in 1444. In Europe itself, slavery had by then all but withered away (though Europeans were still being enslaved and sold into slavery by Arabs and Turks). Now it was to undergo a vast extension in other continents. Within two or three years over a thousand more blacks had been sold by the Portuguese, who soon set up a permanent slaving station in West Africa. Such figures show the rapid discovery of the profitability of the new traffic but gave little hint of the scale of what was to come. What was already clear was the brutality of the business (the Portuguese quickly noted that the seizure of children usually ensured the docile captivity of the parents) and the complicity of Africans in it; as the search for slaves went further inland, it became simpler to rely on local potentates who would round up captives and barter them wholesale.

For a long time, Europe and the Portuguese and Spanish settlements in the Atlantic islands took almost all the slaves West Africa supplied. Then came a change. From the mid-sixteenth century African slaves were shipped across the Atlantic to Brazil, the Caribbean islands and the North American mainland. The trade thus entered upon a long period of dramatic growth whose demographic, economic and political consequences are still with us. African slaves are by no means the only ones important in modern history, nor were Europeans the only slavers. None the less, black slavery based on the buying of Africans from other Africans by Portuguese, Englishmen, Dutchmen and Frenchmen, and their sale to other Europeans in the Americas, is a phenomenon whose repercussions have been much more profound than the enslavement of Europeans by Ottomans or Africans by Arabs. Much of the labour which made American colonies possible and viable was supplied by black slaves, though for climatic reasons the slave population was not uniformly spread among them. Always the great majority of slaves worked in agriculture or domestic service: black craftsmen or, later, factory workers were unusual.

The slave trade was commercially very important, too. Huge profits were occasionally made – a fact which partly explains the crammed and pestilential holds of the slave-ships in which were confined the human cargoes. They rarely had a death-rate per voyage of less than 10 per cent and sometimes suffered much more appalling mortality. The supposed value of the trade made it a great and contested prize, though the normal return on capital has been much exaggerated. For two centuries it provoked diplomatic wrangling and even war as nation after nation sought to break into it or monopolize it. This testified to the trade's importance in the eyes of statesmen, whether it was economically justified or not.

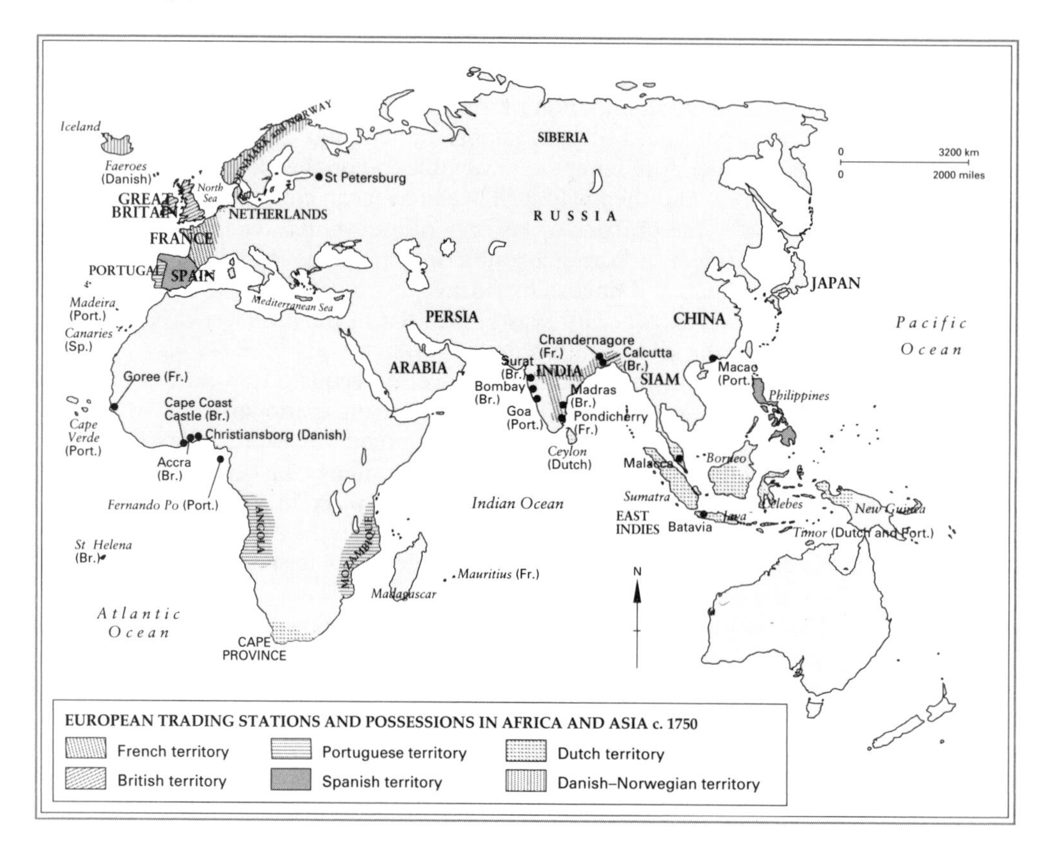

It was once widely held that the slave trade's profits provided the capital for European industrialization, but this no longer seems plausible. Industrialization was a slow process. Before 1800, though examples of industrial concentration could be found in several European countries, the growth of both manufacturing and extractive industry was still in the main a matter of the multiplication of small-scale artisan production and its technical elaboration, rather than of radically new methods and institutions. Europe had by 1500 an enormous pool of wealth to draw on in her large numbers of skilled craftsmen, already used to investigating new process and exploring new techniques. Two centuries of gunnery had brought mining and metallurgy to a high pitch. Scientific instruments and mechanical clocks testified to a wide diffusion of skill in the making of precision goods. Such advantages as these shaped the early pattern of the industrial age and soon began to reverse a traditional relationship with Asia. For centuries oriental craftsmen had astounded Europeans by their skill and the quality of their work. Asian textiles and ceramics had a superiority which lives in our everyday language: china, muslin, calico, shantung are still familiar words. Then, in the fourteenth and fifteenth centuries, supremacy in some forms of craftsmanship had passed to Europe, notably in mechanical and engineering skills. Asian potentates began to seek Europeans who could teach them how to make effective firearms; they even collected mechanical toys which were the commonplaces of European fairs. Such a reversal of roles was based on Europe's accumulation of skills in traditional occupations and their extension into new fields. This happened usually in towns; craftsmen often travelled from one to another, following demand. So much it is easy to see. What is harder is to see what it was in the European mind which pressed the European craftsman forward and also stimulated the interest of his social betters so that a craze for mechanical engineering is as important an aspect of the age of the Renaissance as is the work of its architects and goldsmiths. After all, this did not happen elsewhere.

Early industrial areas grew by accretion, often around the centres of established European industries (such as textiles or brewing) closely related to agriculture. This long continued to be true. These old trades had created concentrations of supporting industry. Antwerp had been the great port of entry to Europe for English cloth; as a result, finishing and dyeing establishments appeared there to work up further the commodities flowing through the port. In the English countryside, wool merchants shaped the early pattern of industrial growth by putting out to peasant spinners and weavers the raw materials they needed. The presence of minerals was another locating factor; mining and metallurgy were the most important industrial activities independent of agriculture. But industries could stagnate or even, sometimes, collapse. This seems to have happened to Italy. Its medieval industrial pre-eminence disappeared in the sixteenth century while that of the Flemish Low Countries and western and southern Germany – the old Carolingian heartland – lasted another century or so until it began to be clear that England, the Dutch Netherlands and Sweden were the new industrial leaders. In the eighteenth century Russia's extractive industries would add her to this list. By that time, too, other factors were beginning to enter the equation of economic development as organized science was being brought to bear on industrial techniques and state policy shaped industry both consciously and unconsciously.

The long-term picture of overall expansion and growth obviously requires much qualification. Dramatic fluctuations could easily occur even in the nineteenth century, when a bad harvest could lead to runs on banks and a contraction of demand for manufactured goods big enough to be called a slump. This reflected the growing development and integration of the economy. It could cause new forms of distress. Not long after 1500, for example, it began to be noticed that prices were rising with unprecedented speed. Locally this trend was sometimes very sharp indeed, doubling costs in a year. Though nothing like this rate was anywhere maintained for long the general effect seems to have been a roughly fourfold rise in European prices in a century. Given modern rates of inflation, this does not seem very shocking, but it was quite novel and bound to have great and grave repercussions. Among those who owned some property, some benefited and some suffered. Some landowners reacted by putting up rents and increasing as much as possible the yields from their feudal dues. Some had to sell out. In this sense, inflation made for social mobility, as it often does. Among the poor, the effects were usually harsh, for the price of agricultural produce shot up and money wages did not keep pace. Real wages therefore fell. This was sometimes made worse by local factors, too. In England, for example, high wool prices tempted landlords to enclose common land and thus remove it from common use in order to put sheep on it. The wretched peasant grazier starved and, thus, as one famous contemporary comment put it, 'sheep ate men'. Everywhere in the central third of the century there were popular revolts and a running disorder which reveal both the incomprehensibility and the severity of what was going on.

Everywhere it was the extremes of society which felt the pinch of inflation most sharply; to the poor it brought starvation while kings were pinched because they had to spend more than anyone else.

Much ink has been spent by historians on explaining this century-long price rise. They no longer feel satisfied with the explanation first put forward by contemporary observers, that the essential cause was a new supply of bullion which followed the opening of the New World mines by the Spanish; inflation was well under way before American bullion began to arrive in any significant quantity, even if gold later aggravated things. Probably the fundamental pressure always came from a population whose numbers were increasing when big advances in productivity still lay in the future. The rise in prices continued until the beginning of the seventeenth century. Then it began occasionally even to show signs of falling until a slower increase was resumed around 1700.

The twentieth century needs no reminders that social change can quickly follow economic change. We have little belief in the immutability of social forms and institutions. Three hundred years ago, many men and women believed them to be virtually God-given and the result was that although social changes took place in the aftermath of inflation (and, it must be said, for many other reasons) they were muffled and masked by the persistence of old forms. Superficially and nominally much of European society remained unchanged between 1500 and 1800 or thereabouts. Yet the economic realities underlying it changed a great deal. Appearances were deceptive.

Rural life had already begun to show this in some countries before 1500. As agriculture became more and more a matter of business (though by no means only because of that), traditional rural society had to change. Forms were usually preserved, and the results were more and more incongruous. Although feudal lordship still existed in France in the 1780s it was by then less a social reality than an economic device. The 'seigneur' might never see his tenants, might not be of noble blood, and might draw nothing from his lordship except sums of money which represented his claims on his tenants' labour, time and produce. Further east, the feudal relationship remained more of a reality. This in part reflected an alliance of rulers and nobles to take advantage of the new market for grain and timber in the growing population of western and southern Europe. They tied peasants to the land and exacted heavier and heavier labour services. In Russia serfdom became the very basis of society.

In England, on the other hand, even the commercialized 'feudalism' which existed in France had gone long before 1800, and noble status conferred no legal privilege beyond the rights to be summoned to a parliament (their other legal distinction was that like most of the other subjects of King George III, they could not vote in the election of a Member of Parliament). The English nobility was a tiny set; until the end of the eighteenth century the House of Lords had fewer than two hundred hereditary members, whose status could only be transmitted to their one direct heir. Consequently, there did not exist in England the large class of noble men and women, all enjoying extensive legal privileges separating them from the rest of the population, such as there was almost universally elsewhere in Europe. In France there were perhaps a quarter of a million nobles on the eve of the Revolution. All had important legal and formal rights; the corresponding legal order in England could comfortably have been assembled in the hall of an Oxford college and would have had rights correspondingly less impressive.

On the other hand, the wealth and social influence of English landowners was immense. Below the peerage stretched the ill-defined class of English gentlemen, linked at the top to the peers' families and disappearing at the other end into the ranks of prosperous farmers and merchants who were eminently respectable but not 'gentlefolk'. Its permeability was of enormous value in promoting cohesion and mobility. Gentlemanly status could be approached by enrichment, by professional distinction, or by personal merit. It was essentially a matter of a shared code of behaviour, still reflecting the aristocratic concept of honour, but one civilized by the purging away of its exclusiveness, its gothicisms and its legal supports. In the seventeenth and eighteenth centuries the idea of the gentleman became one of the formative influences of English history.

The mid-eighteenth-century house of commons, still sitting in the small St Stephen's chapel which was destroyed by fire in 1834, as it appeared to the statesmen of the age in which Great Britain became a world power.

It is true that ruling hierarchies differed in all countries. Contrasts could be drawn right across Europe. There would be nothing tidy about the result. None the less, a broad tendency towards social change which strained old forms is observable in many countries by 1700. In the most advanced countries it brought new ideas about what constituted status and how it should be recognized. Though not complete, there was a shift from personal ties to market relationships as a way of defining people's rights and expectations, and a shift from a corporate vision of society to an individualist one. This was most notable in the United Provinces, the republic which emerged in the Dutch Netherlands during this era. It was in effect ruled by merchants, particularly those of Amsterdam, the centre of Holland, its richest province. Here the landed nobility had never counted for as much as the mercantile and urban oligarchs.

Nowhere else in Europe had social change gone as far by 1789 as in Great Britain and the United Provinces. Elsewhere questioning of traditional status had barely begun. Figaro, the valet-hero of a notably successful eighteenth-century French comedy, jibed that his aristocratic master had done nothing to deserve his privileges beyond giving himself the trouble to be born. This was recognized at the time as a dangerous and subversive idea, but it need not have caused much alarm. Europe was still soaked in the assumptions of aristocracy (and was to be for a long time even after 1800). Degrees of exclusiveness varied, but the distinction between noble and non-noble remained crucial. Though alarmed aristocrats accused them of doing so, kings would nowhere ally with commoners against them even in the last resort. Kings were aristocrats, too; it was their trade, one of them said. Only the coming of a great revolution in France changed things much and then hardly at all outside that country before the end of the century. As the nineteenth century began, it looked as if most Europeans still respected noble blood. All that had changed was that not quite so many people still automatically thought it was a distinction which ought to be reflected in laws.

Just as men began to feel that to describe society in terms of orders with legally distinct rights and obligations no longer expressed its reality, so also they ceased to feel so sure that religion upheld a particular social hierarchy. It was still for a long time possible to believe that

The rich man in his castle,
the poor man at his gate,
God made them, high and lowly,
and ordered their estate

as an Ulsterwoman put in the nineteenth century, but this was not quite the same thing as saying that a fixed unchanging order was the expression of God's will. Even by 1800 a few people were beginning to think God rather liked the rich man to show the wisdom of God's way by having made his own way in the world rather than by inheriting his father's place. 'Government is a contrivance of human wisdom for the satisfaction of human wants,' said an eighteenth-century Irishman, and he was a conservative, too. A broad utilitarianism was coming to be the way more and more people assessed institutions in advanced countries, social institutions among them.

The old formal hierarchies were under most pressure where strain was imposed upon them by increasing economic mobility, by the growth of towns, by the rise of a market economy, by the appearance of new commercial opportunities and by the spread of literacy and awareness. Broadly speaking, three situations can be distinguished. In the East, in Russia, and almost to the same extent in Poland or east Prussia and Hungary, agrarian society was still so little disturbed by new developments that the traditional social pattern was not only intact but all but unchallenged at the end of the eighteenth century. In these landlocked countries, safe from the threats to the existing order implicit in the commercial development of maritime Europe, the traditional ruling classes usually not only retained their position but had often showed that they could actually enlarge their privileges. In a second group of countries, there was enough of a clash between the economic and social worlds which were coming into being and the existing order to provoke demands for change. When political circumstances permitted its resolution, these would demand satisfaction, though they could be contained for a time. France was the outstanding example, but in some of the German states, Belgium and parts of Italy there were signs of the same sort of strain. The third group of countries were those relatively open societies, such as England, the Netherlands and, across the sea, British North America, where the formal distinctions of society already meant less by comparison with wealth (or even talent), legal rights were widely diffused, economic opportunity was felt to be widespread, and wage-dependency was very marked. Even in the sixteenth century, English society seems much more fluid than that of continental countries and, indeed, when the North Americans came to give themselves a new constitution in the eighteenth century they forbade the conferring of hereditary titles. In these countries individualism had a scope almost untrammelled by law, whatever the real restraints of custom and opportunity.

It is only too easy in a general account such as this, though, to be over-precise, over-definite. Even the suggested rough tripartite division blurs too much. There were startling contrasts within societies we must misinterpret if we think of them as homogeneous. In the advanced countries there was still much that we should find strange, even antediluvian. The towns of England, France and Germany were for the most part little Barchesters, wrapped in a comfortable provincialism lorded over by narrow merchant oligarchies, successful guildsmen or cathedral chapters. Yet Chartres, contentedly rooted in its medieval countryside and medieval ways, its eighteenth-century population still the same size as five hundred years earlier, was part of the same country as Nantes or Bordeaux, thriving, bustling ports which were only two of several making up the dynamic sector of the French economy. Even the nineteenth century would find its immediate forebears unprogressive; far be it from us, therefore, to predicate the existence of a mature and clearly defined individualist and capitalist society wholly conscious of itself as such in any European country. What marked the countries we might call 'advanced' was a tendency to move further and faster in that direction than the great majority of the rest of the world.

Sometimes this won them admiration by would-be reformers. One great questioner of the *status quo*, Voltaire, was greatly struck by the fact that even in the early eighteenth century a great merchant could be as esteemed and respected in England as was a nobleman. He may have slightly exaggerated and he certainly blurred some important nuances, yet it is remarkable – and a part of the story of the rise of Great Britain to world power – that the political class which governed eighteenth-century England was a landed class and fiercely reflected landed values, yet constantly took care to defend the commercial interests of the country and accepted the leadership and guidance in this of the collective wisdom of the City of London. Though people went on talking of a political division between the 'moneyed' and the 'landed' interest, and though politics long remained a matter of disputed places and conflicting traditions within the landed class, interests which in other countries would have conflicted with these nevertheless prospered and were not alienated. The explanations must be complex. Some, like the commercialization of British agriculture, go far back into the history of the previous century; some, such as the growth of facilities for private investment in the government and commercial world, were much more recent.

The coincidence of the advanced social evolution of the Netherlands and Great Britain with their economic, and especially their commercial, success is striking. This was once largely attributed to their religion: as a result of a great upheaval within Christendom both had ceased to be dominated by the Catholic church. Anti-clericals in the eighteenth century and sociologists in the twentieth sought to explore and exploit this coincidence; Protestantism, it was said, provided an ethic for capitalism. This no longer seems plausible. There were too many Catholic capitalists, for one thing, and they were often successful. France and Spain were still important trading countries in the eighteenth century and the first seems to have enjoyed for much of it something like the same rates of growth as great Britain, though she was later to fall behind. They were both countries with Atlantic access, and so among those which had tended to show economic growth ever since the sixteenth century. Yet this is not an explanation which goes very far, either. Scotland – northern, Protestant and Atlantic – long remained backward, poor and feudal. There was more to the differences separating Mediterranean and eastern Europe from the north and west than simple geographical position and more than one factor to the explanation of differing rates of modernization. The progress of English and Dutch agriculture, for example, probably owes more to the relative scarcity of land in each country than to anything else.

The European East remained backward. The social and economic structure remained fundamentally unchanged until the nineteenth century. Deep-rooted explanations have been offered – that, for example, a shorter growing season and less rich soils then were to be found further west gave it from the start a poorer return on seed, and therefore handicapped it economically in the crucial early stages of agricultural growth. It had man-made handicaps, too. Settlement there had long been open to disturbance by Central Asian nomads, while on its southern flank lay the Balkans and the frontier with Turkey, for many centuries a zone of warfare, raiding and banditry. In some areas (Hungary for example) the effects of Turkish rule had been so bad as to depopulate the country. When it was reacquired for Europe, care was taken to tie the peasantry to the land. In the Russia which emerged from Muscovy in this period, too, the serf population grew larger as a proportion of the whole. Harsher law guaranteed their masters control of the peasants. In other eastern countries (Prussia was one), the powers of landlords over tenants were strengthened. This was more than just a kingly indulgence of aristocracies which might, if not placated, turn against royal authority. It was also a device for economic development. Not for the first time, nor the last, economic progress went with social injustice; serfdom was a way of making available one of the resources needed if land was to be made productive, just as forced labour was in many other countries at many other times.

One result which is still in some degree visible was a Europe divided roughly along the Elbe. To the west lay countries evolving slowly by 1800 towards more open social forms. To the east lay authoritarian governments presiding over agrarian societies where a minority of landholders enjoyed great powers over a largely tied peasantry. In this area towns did not often prosper as they had done for centuries in the West. They tended to be overtaxed islands in a rural sea, unable to attract from the countryside the labour they needed because of the extent of serfdom. Over great tracts of Poland and Russia

From the Middle Ages to the present century, pogroms of Jews have been endemic in some parts of Europe. A print of a notorious attack on the Frankfurt ghetto in 1614.

even a money economy barely existed. Much of later European history was implicit in this difference between east and west.

It was discernible in informal institutions, too, in the way, for example, in which women were treated, though here another division could be drawn, that between Mediterranean Europe and the north, which was in due course extended to run between Latin and North America. Formally and legally, little changed anywhere in these centuries; the legal status of women remained what it had been and this was only to be questioned right at the end of this period. Nevertheless, the real independence of women and, in particular, of upper-class women, does seem to have been enlarged in the more advanced countries. Even in the fifteenth century it had been remarked by foreigners that Englishwomen enjoyed unusual freedom. This lead does not seem to have diminished, but in the eighteenth century there are signs that in France, at least, a well-born woman could enjoy considerable real independence.

This was in part because the eighteenth century brought the appearance of a new sort of upper-class life, one which had room for other social gatherings than those of a royal court, and one increasingly independent of religious and family ritual. At the end of the seventeenth century we hear of men in London meeting in the coffee-houses from which the first clubs were to spring. Soon there appears the *salon*, the social gathering of friends and acquaintances in a lady's drawing-room which was especially the creation of the French; some eighteenth-century *salons* were important intellectual centres and show that it had become proper and even fashionable for a woman to show an interest in things of the mind other than religion. When Mme de Pompadour, the mistress of Louis XV, was painted, she chose to have included in the picture a book – Montesquieu's sociological treatise, *De l'esprit des lois*. But even when women did not aspire to blue stockings, the *salon* and the appearance of a society independent of the court presented them with a real, if limited, escape from the confinement of the family, which, together with religious and professional gatherings, had until then been virtually the only structures within which even men might seek social variety and diversion.

By the end of the eighteenth century we have arrived at the age of the female artist and novelist and of acceptance of the fact that spinsterhood need not mean retirement to a cloister. Where such changes came from is not easy to see. In the early years of the century the *Spectator* already thought it worth while to address itself to women readers as well as men, which suggests that we should look a fair way back. Perhaps it helped that the eighteenth century produced such conspicuous examples of women of great political influence – an English queen and four empresses (one Austrian and three Russian) all ruled in their own right, often with success. But it is not possible to say so with confidence for the prehistory of female emancipation still largely awaits study.

Finally, none of this touched the life of the overwhelming majority in even the most advanced societies of early modern Europe. There had not yet come into being the mass industrial jobs which would provide the first great force to prise apart the unquestioned certainties of traditional life for most men and women alike. Though they may have weighed most heavily in the primitive villages of Poland or in a southern Spain where Moorish influences had intensified the subordination and seclusion of women, those certainties were everywhere still heavy in 1800.

2
Authority and Its Challengers

In 1800 many Europeans still held ideas about social and political organization which would have been comprehensible and appropriate four hundred years earlier. The 'Middle Ages' no more came to a sudden end in this respect than in many others. Ideas about society and government which may reasonably be described as 'medieval' survived as effective forces over a wide area and during the centuries more and more social facts had been fitted into them. Broadly speaking, what has been called the 'corporate' organization of society, the grouping of men in bodies with legal privileges which protected their members and defined their status was still the rule in the eighteenth century in continental Europe. Over much of its central and eastern zones, as we have noted, serfdom had grown more rigid and more widespread. Many continuities in political institutions were obvious. The Holy Roman Empire still existed in 1800 as it had done in 1500; so did the temporal power of the pope. A descendant of the Capetians was still king of France (though he no longer came from the same branch of the family as in 1500 and, indeed, was in exile). Even in England and as late as 1820, a king's champion rode in full armour into Westminster Hall at the coronation banquet of King George IV, to uphold that monarch's title against all comers. In most countries it was still taken for granted that the state was a confessional entity, that religion and society were intertwined and that the authority of the Church was established by law. Although such ideas had been much challenged and in some countries had undergone grievous reverses, in this as in many matters the weight of history was still enormous in 1800 and only ten years earlier it had been even heavier.

When all this is acknowledged, it was nevertheless the general tendency of the three centuries between 1500 and 1800 to dissolve or at least weaken old social and political bonds characteristic of medieval government. Power and authority had instead tended to flow towards the central concentration provided by the state, and away from 'feudal' arrangements of personal dependence. (The very invention of the 'feudal' idea as a technical term of law was in fact the work of the seventeenth century and it suggests the age's need to define something whose reality was ebbing way.) The idea of Christendom, too, though still important in emotional, even subconscious ways, effectively lost any political reality in this period. Papal authority had begun to suffer at the hands of national sentiment in the age of the Schism and that of the Holy Roman emperors had been of small account since the fourteenth century. Nor did any new unifying principle emerged to integrate Europe. The test case was the Ottoman threat. Christian princes exposed to the Moslem onslaught might appeal to their fellow-Christians for help, popes might still use the rhetoric of crusade, but the reality, as the Turks well knew, was that Christian states would follow their own interest and ally with the infidel, if necessary. This was the era of *Realpolitik*, of the conscious subordination of principle and honour to intelligent calculation of the interests of the state. It is curious that in an age in which Europeans more and more agreed that greater distinctions of culture separated them (to their credit, they were sure) from other civilizations they paid little attention to institutions (and did nothing to create new ones) which acknowledged their essential unity. Only the occasional philosopher advocated the building of something which transcended the state. Perhaps, though, it is just in a new awareness of cultural superiority that the explanation lies. Europe was entering an age of triumphant expansion and did not need shared institutions to tell her so. Indeed, the authority of states, and therefore the power of their governments, waxed in these centuries.

Here it is important not to be misled by forms. For all the arguments about who should exercise it and a mass of political writing which suggested all sorts of limits on it, the general trend was towards acceptance of the idea of legislative sovereignty – that is, Europeans came to feel that, provided the authority of the state were in the right hands, there should be no restriction upon its power to make laws.

The last medieval emperor rides out in defence of Christendom much as he must have thought the great paladins who were his predecessors had done; Charles V, painted by Titian in one of his greatest works.

This was an enormous break with the thinking of the past. To a medieval man the idea that there might not be rights and rules above interference by any man, legal immunities and chartered freedoms inaccessible to change to subsequent lawmakers, fundamental laws which would always be respected, or laws of God which could never be contravened by those of men, would have been social and juridical, as well as theological, blasphemy. English lawyers of the seventeenth century floundered about in disagreement over what the fundamental laws of the land might be, but all thought some must exist.

A century later the leading legal minds of France were doing just the same. In the end, nevertheless, there emerged in both countries (as, to a greater or lesser degree, in most others) the acceptance of the idea that a sovereign, legally unrestrained lawmaking power was the characteristic mark of the state. Yet this took a long time. For most of the history of early modern Europe the emergence of the modern sovereign state was obscured by the fact that the most widely prevalent form of government was monarchy. Struggles about the powers of rulers make up much of European history in these centuries and sometimes it is hard to see exactly what is at stake. The claims of kings, after all, could be challenged on two quite distinct grounds: there was resistance based on the principle that it would be wrong for any government to have powers such as some monarchs claimed (and this might be termed the medieval or conservative defence of freedom) and there was resistance based on the principle that such powers could properly exist, but were being gathered into the wrong hands (and this can be called the modern or liberal defence of freedom). In practice, the two claims are often inextricably confused, but the confusion is itself a significant indicator of changing ideas.

Once away from legal principle, the strengthening of the state showed itself in the growing ability of kings and princes to get their way. One indicator was the nearly universal decline in the sixteenth and seventeenth centuries of the representative institutions which had appeared in many countries in the later Middle Ages. By 1789, most of western (if not central and Eastern) Europe was ruled by monarchs little hindered by representative bodies; the main exception was in Great Britain. Kings began in the sixteenth century to enjoy powers which would have seemed remarkable to medieval barons and burghers. The phenomenon is sometimes described as the rise of absolute monarchy. If we do not exaggerate a monarch's chances of actually carrying out his declared wishes (for many practical checks might exist to his power which were just as restricting as medieval immunities or a representative assembly), the term is acceptable. Everywhere, or almost everywhere, the relative strength of rulers *vis-à-vis* their rivals increased greatly from the sixteenth century onwards. New financial resources gave them standing armies and artillery to use against great nobles who could not afford them. Sometimes the monarchy was able to ally itself with the slow growth of a sense of nationhood in imposing order on the over-mighty. In many countries the late fifteenth century had brought a new readiness to accept royal government if it would guarantee order and peace. There were special reasons in almost every case, but nearly everywhere monarchs raised themselves further above the level of the greatest nobles and buttressed their new pretentions to respect and authority with cannon and taxation. The obligatory sharing of power with great subjects whose status entitled then *de facto* and sometimes *de jure* to office, ceased to weigh so heavily upon kings. England's Privy Council under the Tudors was a meritocracy as well as a gathering of magnates.

In the sixteenth and early seventeenth centuries this brought about the appearance of what some have called the 'Renaissance State'. This is a rather grandiose term for swollen bureaucracies, staffed by royal employees and directed by aspirations to centralization, but clear enough if we remember the implied antithesis: the medieval kingdom, whose governmental functions were often in large measure delegated to feudal and personal dependents or to corporations (of which the Church was the greatest). Of course, neither model of political organization existed historically in a pure form. There had always been royal officials, 'new men' of obscure origin, and governments today still delegate tasks to non-governmental bodies. There was no sudden transition to the modern 'state': it took centuries and often used old forms. In England, the Tudors seized on the existing institution of royal Justices of the Peace to weld the local squires into the structure of royal government. This was yet another stage in a long process of undermining seigneurial authority which elsewhere still had centuries of life before it. Even in England, too, as Tudor rebellions showed, noblemen had long to be treated with care if they were not to be fatally antagonized. Rebellion was not an exceptional but a continuing fact of life for the sixteenth-century statesman. Royal troops might prevail in the end, but no monarch wanted to be reduced to reliance on force. As a famous motto had it, artillery was the *last* argument of kings. The history of the French nobility's turbulence right down to the middle of the seventeenth century, of the effects of antagonizing local interest in England during the same period, of Habsburg attempts to unify their territories at the expense of local magnates, all show this. The United Kingdom had its last feudal rebellion in 1745; other countries then still had theirs to come.

Taxation, too, because of the danger of rebellion and the inadequacy of administrative machinery to collect it, could not be pressed very far, yet officials and armies had to be paid for. One way was to allow officials to charge fees or levy perquisities on those who needed their services. For obvious reasons, this was not a complete answer. The raising of greater sums by the ruler was therefore necessary. Something might still be done by exploiting royal domains. But all monarchs, sooner or later, were driven back to seek new taxation and it was a problem few could solve. There were technical problems here which could not be dealt with until the nineteenth century or even later, but for three centuries great fertility of imagination was to be shown in inventing new taxes. Broadly speaking, only consumption (through indirect taxes such as customs and excise or taxes on sale, or through requiring licenses and authorisations to trade which had to be paid for) or real property could be tapped by the tax-gatherer. Usually, this bore disproportionately upon the poorest, who spent a larger part of their small disposable income on necessities than the wealthy. Nor was it easy to stop a landowner from passing his tax burdens along to the man at the bottom of the property pyramid. Taxation, too, was particularly hindered by the surviving medieval idea of legal immunity. In 1500 it was generally accepted that there were areas, persons and spheres of action which were specially protected from invasion by the power of the rule. They might be defended by an irrevocable royal grant in past ages such as were the privileges of many cities, by contractual agreement such as the English *Magna Carta* was said to be, by immemorial custom, or by divine law. The supreme example was the Church. Its properties were not normally subject to lay taxation, it had jurisdiction in its courts of matters inaccessible to royal justice, and it controlled important social and economic institutions – marriage, for example. But a province, or a profession, or a family might also enjoy immunities, usually from royal jurisdiction or taxation. Nor was royal standing uniform. Even the French king was only a duke in Brittany and that made a difference to what he was entitled to do there. Such facts were the realities which the 'Renaissance State' had to live with. It could do no other than accept their survival, even if the future lay with the royal bureaucrats and their files.

In the early sixteenth century, a great crisis which shook western Christianity and destroyed forever the old medieval unity of the faith much accelerated the consolidation of royal power. The Protestant Reformation began as just one more dispute over religious authority, the calling in question of the papal claims whose formal and theoretical structure had successfully survived so many challenges. Thus, in origin at least, it was a thoroughly medieval phenomenon. But important as that was, that was not to be the whole story of the Reformation and far from exhausts its political significance. Given that it also detonated a cultural revolution, there is no reason to question its traditional standing as the start of modern history.

There was nothing new about demands for reform. The sense that Papacy and curia did not necessarily serve the interests of all Christians was well grounded by 1500. Some critics had already gone on from this to doctrinal dissent. The deep, uneasy devotional swell of the fifteenth century had expressed a search for new answers to spiritual questions but also a willingness to look for them outside the limits laid down by ecclesiastical authority. Heresy had never been blotted out, it had only been contained. Popular anti-clericalism was an old and widespread phenomenon. There had also appeared in the fifteenth century another current in religious life, perhaps more profoundly subversive than heresy, because, unlike heresy, it contained forces which might in the end cut at the roots of the traditional religious outlook itself. This was the learned, humanistic, rational, sceptical intellectual movement which, for want of a better word, we may call Erasmian after the man who embodied its ideals most clearly in the eyes of contemporaries, and who was the first Dutchman to play a leading role in European history. He was profoundly loyal to his faith; he knew himself to be a Christian and that meant, unquestionably, that he remained within the Church. But of that Church he had an ideal which embodied a vision of a possible reformation. He sought a simpler devotion and a purer pastorate. Though he did not challenge the authority of the Church or papacy, in a subtler way he challenged authority in principle, for his scholarly work had implications which were deeply subversive and so was the tone of the correspondence which he conducted with colleagues the length and breadth of Europe. They learnt from him to disentangle their logic and therefore the teaching of the faith from the scholastic mummifications of Aristotelian philosophy. In his Greek New Testament he made available a firm basis for argument on doctrine at a time when Greek was again becoming widespread. Erasmus, too, was the exposer of the spuriousness of texts on which bizarre dogmatic structures had been raised.

Yet Erasmus and others who shared his viewpoint did not attack religious authority outright or turn ecclesiastical into universal issues. They were good Catholics. Humanism, like heresy, discontent with clerical behaviour and the cupidity of princes, was something in the air at the beginning of the sixteenth century, waiting – as many things had long waited – for the man and the occasion which would make them into a religious revolution. No other term is adequate to describe what followed the unwitting act of a German monk. His name was Martin Luther and in 1517 he launched a movement which was to end by fragmenting a Christian unity intact in the West since the disappearance of the Arians.

Unlike Erasmus, the international man, Luther lived all his life except for brief absences in a small German town, Wittenberg, almost at the back of beyond on the Elbe. He was an Augustinian monk, deeply read in theology, somewhat tormented in spirit, who had already come to the conclusion that he must preach the Scriptures in a new light, to present God as a forgiving God, not a punitive one. This need not have made him a revolutionary; the orthodoxy of his views was never in question until he quarrelled with the papacy. He had been to Rome, and he had not liked what he saw there, for the papal city seemed a worldly place and its ecclesiastical rulers no better than they should be. This did not dispose him to feel warmly towards an itinerant Dominican, who came to Saxony only as a pedlar of indulgences. These documents were papal certificates that the possessor, in consideration of payment (which went towards the building of the new and magnificent St Peter's then rising in Rome), would be assured that some of the penalties incurred by him for sin would be remitted in the next world. Accounts of the preaching of this man were brought to Luther by peasants who had heard him and bought their indulgences. Research has made it clear that what had been said to them was not only misleading but outrageous; the crudity of the transaction promoted by the preacher displays one of the most unattractive faces of medieval Catholicism. It infuriated Luther, almost obsessed as he was by the overwhelming seriousness of the transformation necessary in a man's life before he could be sure of redemption. He formulated his protests against this and certain other papal practices in a set of ninety-five theses setting out his positive views. In the tradition of the scholarly disputation he posted them in Latin on the door of the castle church in Wittenberg on 21 October 1517. He had also sent the theses to the Archbishop of Mainz, primate of Germany, who passed them to Rome with a request that Luther be forbidden by his order to preach on this theme. By this time the theses had been put into German and the new information technology had transformed the situation; they were printed and circulated everywhere in Germany. So Luther got the debate he sought. Only the protection of Frederick of Saxony, the ruler of Luther's state, who refused to surrender him, kept him out of danger of his life. The delay in scotching the chicken of heresy in the egg was fatal. Luther's order abandoned him, but his university did not. Soon the papacy found itself confronted by a German national movement of grievance against Rome sustained and inflamed by Luther's own sudden discovery that he was a literary genius of astonishing fluency and productivity, the first to exploit the huge possibilities of the printed pamphlet.

A portrait of Luther by Nicholas Cranach.

Within two years, Luther was being called a Hussite. The Reformation had by then become entangled in German politics. Even in the Middle Ages would-be reformers had looked to secular rulers for help. This did not necessarily mean going outside the fold of the faith; the great Spanish churchman Ximenes had sought to bring to bear the authority of the Catholic Monarchs on the problems facing the Spanish Church. Kings tended not to like heretics; their duty was to uphold the true faith. Nevertheless, an appeal to lay authority could open the way to changes which went further perhaps, than, their authors had intended, and this, it seems, was the case with Luther. His arguments had rapidly carried him beyond the desirability and grounds of reform in practice to the questioning of papal authority and, then, of doctrine. The core of his early protests had not been theological. Nevertheless, he now rejected transubstantiation (replacing it with a view of the eucharist which is even more difficult to grasp) and preached that men were justified – that is, set aside for salvation – not by observance of the sacraments only ('works', as this was called), but by faith. This was, clearly, an intensely individualist position. It struck at the root of traditional teaching which saw no salvation possible outside the Church. (Yet, it may be noted, Erasmus, when asked for his view, would not condemn Luther; it was known, moreover, that he thought Luther to have said many valuable things.) In 1520 Luther was excommunicated. Before a wondering audience he burnt the bull of excommunication in the same fire as the books of canon law. He continued to preach and write. Summoned to explain himself before the imperial Diet, he refused to retract his views. Germany seemed on the verge of civil war. After leaving the Diet under a safe-conduct, he disappeared, kidnapped for his own safety by a sympathetic prince. In 1521 Charles V, the emperor, placed him under the Imperial Ban; Luther was now an outlaw.

Luther's doctrines, which he extended to condemnations of confession and absolution and clerical celibacy, by now appealed to many Germans. His followers spread them by preaching and by distributing his German translation of the New Testament. Lutheranism was also a political fact; the German princes, who entangled it in their own complicated relations with the emperor and his vague authority over them, ensured this. Wars ensued and the word 'Protestant' came into use. By 1555, Germany was irreparably divided into Catholic and Protestant states. This was recognized in agreement at the Diet of Augsburg that the prevailing religion of each state should be that of its ruler, the first European institutionalizing of religious pluralism. It was a curious concession for an emperor who saw himself as the defender of universal Catholicism. Yet it was necessary if he was to keep the loyalty of Germany's princes. In Catholic and Protestant Germany alike, religion now looked as never before to political authority to uphold it in a world of competing creeds.

By this time, other varieties of Protestantism had appeared. Some drew on social unrest. Luther soon had to distinguish his own teaching from the views of peasants who involved his name to justify rebellion against their masters. One radical group were the Anabaptists, persecuted by Catholic and Protestant rulers alike. At Munster in 1534 their leaders' introduction of communism of property and polygamy confirmed their opponents' fears and brought a ferocious suppression upon them. Of other forms of Protestantism, only one demands notice in so general an account as this. Calvinism was to be Switzerland's most important contribution to the Reformation, but it was the creation of a Frenchman, John Calvin. He was a theologian who formulated his essential doctrines while still a young man: the absolute depravity of man after the Fall of Adam and the impossibility of salvation except for those few, the Elect, predestined by God to salvation. If Luther, the Augustinian monk, spoke with the voice of Paul, Calvin evoked the tones of Augustine. It is not easy to understand the attractiveness of this gloomy creed. But to its efficacy, the history not only of Geneva, but of France, England, Scotland, the Dutch Netherlands, and British North America all witnessed. The crucial step was conviction of membership of the Elect. As the signs of this were outward adherence to the commandments of God and participation in the sacraments, it was less difficult to achieve such conviction than might be imagined.

Under Calvin, Geneva was not a place for the easy-going. He had drawn up the constitution of a theocratic state which provided the framework for a remarkable exercise in self-government. Blasphemy and witchcraft were punished by death, but this would not have struck contemporaries as surprising. Adultery, too, was a crime in most European countries and one punished by ecclesiastical courts. But Calvin's Geneva took this offence much more seriously and imposed the death penalty for that, too. Adulterous women were drowned, men beheaded (an apparent reversal of the normal penal practice of a male-dominated European society where women, considered weaker vessels morally and intellectually, were usually indulged with milder punishments than men). The most severe punishments, though, were reserved for those guilty of heresy.

From Geneva, where its pastors were trained, the new sect took root in France, where it won converts among the nobility and had more than 2,000 congregations by 1561. In the Netherlands, England and Scotland and, in the end, Germany, it challenged Lutheranism. It spread also to Poland, Bohemia and Hungary. Thus in its first century it showed a remarkable vigour, surpassing that of Lutheranism which, except in Scandinavia, was never strongly entrenched beyond the German lands which first adopted it.

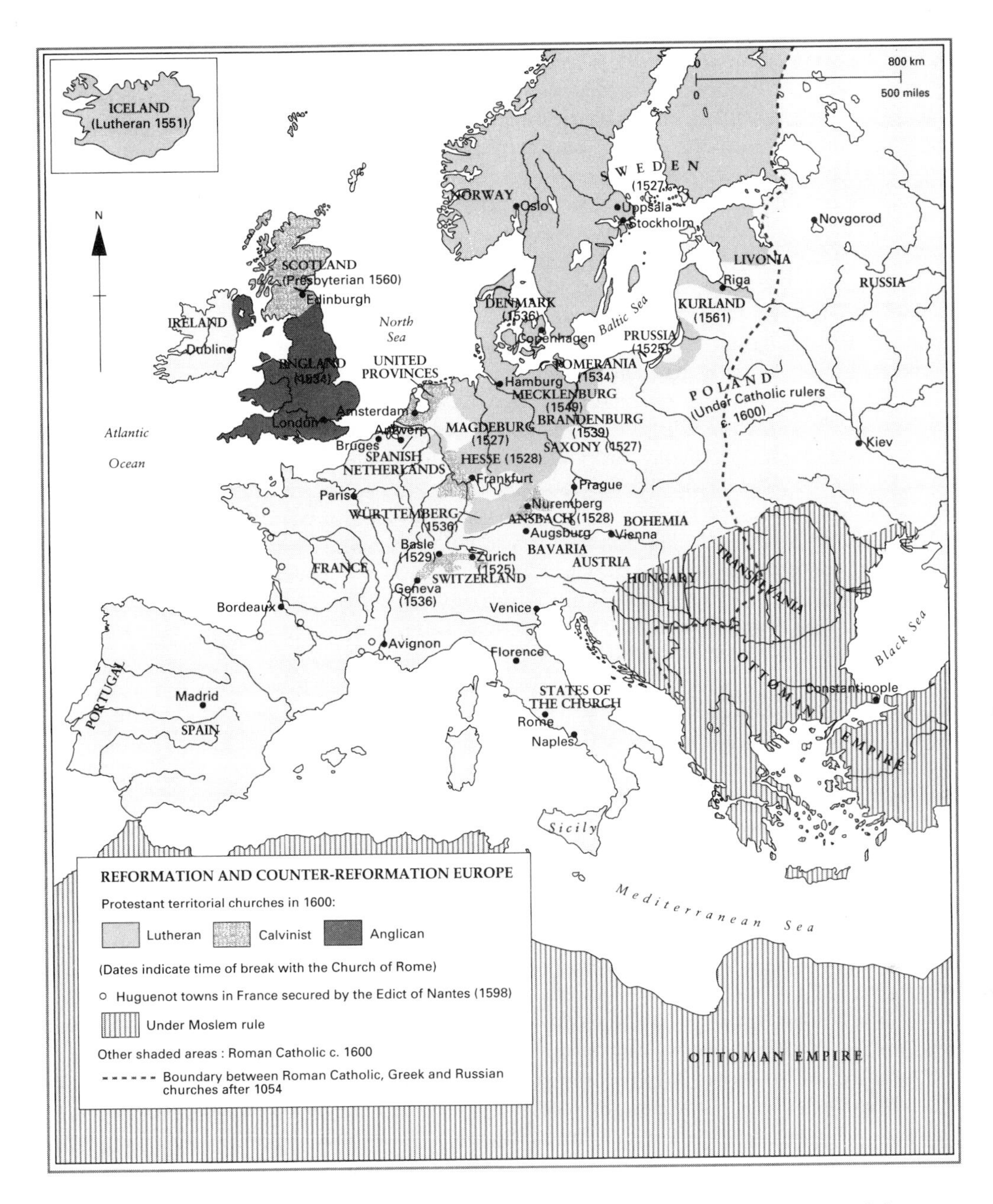

The Protestant Reformation still defies summary and simplification. Complex and deep-rooted in its origins, it also owed much to circumstance and was very rich and far-reaching in its effects. In Europe and the Americas it created new ecclesiastical cultures founded on the study of the Bible and preaching, to which it gave an importance sometimes surpassing that of the sacraments. It was to shape the lives of millions by accustoming them to a new and an intense scrutiny of private conduct and conscience (thus, ironically, achieving something long sought by Roman Catholics) and re-created the non-celibate clergy. Negatively, it slighted or at least called in question all existing ecclesiastical institutions and created new political forces in the form of churches which princes could now manipulate for their own ends – often against popes whom they saw simply as princes like themselves.

Curiously, neither Lutheranism nor Calvinism provoked the first rejection of papal authority by a nation-state. In England a unique religious change arose almost by accident. A new dynasty originating in Wales, the Tudors, had established itself at the end of the fifteenth century and the second king of this line, Henry VIII, became entangled with the papacy over his wish to dissolve the first of his six marriages in order to remarry and get an heir, an understandable preoccupation. This led to a quarrel and one of the most remarkable assertions of lay authority in the whole sixteenth century; it was also one fraught with significance for England's future. With the support of his parliament, which obediently passed the required legislation, Henry VIII proclaimed himself Head of the Church in England. Doctrinally, he conceived no break with the past; he was, after all, entitled Defender of the Faith by the pope because of a refutation of Luther from the royal pen (his descendant still bears his title). But the assertion of the royal Supremacy opened the way to an English Church separate from Rome. A vested interest in it was soon provided by a dissolution of monasteries and some other ecclesiastical foundations and the sale of property to their buyers among the aristocracy and gentry. Churchmen sympathetic to new doctrines sought to move the Church in England significantly towards continental Protestant ideas in the next reign. Popular reactions were mixed. Some saw this as the satisfaction of old national traditions of dissent from Rome; some resented innovations. From a confused debate and murky politics emerged one literary masterpiece, the Book of Common Prayer, and some martyrs both Catholic and Protestant. The latter were the first to suffer, for there was a reversion to papal authority (and the burning of Protestant heretics) under the fourth Tudor, the unfairly named and unhappy Bloody Mary, perhaps England's most tragic queen. By this time, moreover, the question of religion was thoroughly entangled with national interest and foreign policy, for the states of Europe drew apart more and more on religious grounds.

This was not all that was notable about the English Reformation which, like the German, was a landmark in the evolution of a national consciousness. It had been carried out by Act of Parliament and a constitutional question was implicit in the religious settlement: were there any limits to legislative authority? With the accession of Mary's half-sister, Elizabeth I, the pendulum swung back, though for a long time it was unclear how far. Yet Elizabeth insisted, and had parliament legislate, that she retain the essentials of her father's position; the English Church, or Church of England, as it may henceforth be called, claimed to be Catholic in doctrine but rested on the Royal Supremacy. More important still, because that Supremacy was recognized by Act of Parliament, England was soon to be at war with a Catholic king of Spain who was well known for his determination to root out heresy in the lands he subjugated. So another national cause was identified with that of Protestantism.

Reformation helped the English parliament to survive when other medieval representative bodies were going under before the new power of kings, though this was far from the whole story. A kingdom united since Anglo-Saxon times without provincial assemblies or 'estates' which might rival it made it much easier for parliament to focus national politics than any similar body elsewhere. A royal mistake helped, too; Henry VIII had squandered a great opportunity to achieve a sound basis for absolute monarchy when he rapidly liquidated the mass of property – about a fifth of the land of the whole kingdom – which he held briefly as a result of the dissolutions. Nevertheless, all such imponderables duly weighed, the fact that Henry chose to seek endorsement of his will from the national representative body in creating a national church still seems one of the most crucial decisions in parliament's history.

Catholic martyrs died under Elizabeth because they were judged traitors, not because they were heretics – but England was far less divided by religion than Germany and France. Sixteenth-century France was tormented and torn by Catholic and Calvinist interests. Each was in essence a group of noble clans, who fought for power in the Wars of Religion, of which nine have been distinguished between 1562 and 1598. At times their struggles brought the French monarchy very low; the nobility of France came near to winning the battle against the centralizing state. Yet, in the end, their divisions benefited a crown which could use one faction against another. The wretched population of France had to bear the brunt of disorder and devastation until there came to the throne in 1589 (after the murder of his predecessor) a member of a junior branch of the royal

Elizabeth I, greatest of the Tudors; she brought to her office a style and glamour which surpassed even those of her father, Henry VIII; in sagacity and caution she rivalled her grandfather, Henry VII. After her death, she became a folk memory against which were tested the acts of the Stuarts – who were usually found wanting.

family, Henry, king of the little state of Navarre, who became Henry IV of France and inaugurated the Bourbon line whose descendants still claim the French throne. He had been a Protestant, but accepted Catholicism as the condition of his succession, recognizing that Catholicism was the religion most Frenchmen would cling to. The Protestants were assured special guarantees which left them a state within a state, the possessors of fortified towns where the king's writ did not run; this very old-fashioned sort of solution assured protection for their religion by creating new immunities. Henry and his successors could then turn to the business of re-establishing the authority of a throne badly shaken by assassination and intrigue. But the French nobility were still far from tamed.

Long before this, religious struggles had been further inflamed by the wave of internal reassessment and innovation within the Roman Church which is called the Counter-Reformation. Its formal expression was a general council, the Council of Trent, summoned in 1543 and meeting in three sessions over the next eighteen years. Bishops from Italy and Spain dominated it. This mattered, for the Reformers challenged Catholicism little in Italy and not at all in Spain. The result was somewhat to accentuate the intransigence of the Council's decisions. They not only became the touchstone of orthodoxy in doctrine and discipline until the nineteenth century, providing a standard to which Catholic rulers could rally, but also initiated institutional change. Bishops were given more authority and parishes took on new importance. More remarkably still (though almost unnoticed), it answered by implication an old question about the headship of Catholic Europe; from this time, it indisputably lay with the Pope. Like the Reformation, the Counter-Reformation went beyond forms and legal principles. It expressed and gave direction to a new devotional intensity and rejuvenated the fervour of laity and clergy alike. It made attendance at mass each week obligatory and regulated baptism and marriage more strictly; it also ended the selling of indulgences by 'pardoners' – the very practice which had detonated the Lutheran explosion.

Papal authority was not the only source of Catholic reform nor was Counter-Reformation just a response to the Protestant challenge. The spirituality and spontaneous fervour already apparent among the faithful in the fifteenth century lay behind it, too. One of the most potent expressions of its spirituality, as well as an institution which was to prove enduring was the invention of a Spaniard, the soldier Ignatius Loyola. By a curious irony he had been a student at the same Paris college as Calvin in the early 1530s, but it is not recorded that they ever met. In 1534 he and a few companions took vows; their aim was missionary work and as they trained for it Loyola devised a rule for a new religious order. In 1540 it was recognized by the pope and named the Society of Jesus. The Jesuits, as they soon came to be called, were to have an importance in the history of the Church akin to that of the early Benedictines or the Franciscans of the thirteenth century. Their warrior-founder liked to think of them as the militia of the Church, utterly disciplined and completely subordinate to papal authority through their general, who lived in Rome. They transformed Catholic education. They were in the forefront of a renewed missionary effort which carried their members to every part of the world. In Europe their intellectual eminence and political skill raised them to high places in the courts of kings.

Yet though it brought new instruments to the support of papal authority, the Counter-Reformation (like the Reformation) could also strengthen the authority of lay rulers over their subjects. The new dependence of religion upon political authority – that is to say, upon organized force – further extended the grip of the political apparatus. This was most obvious in the Spanish kingdoms. Here two forces ran together to create an unimpeachably Catholic monarchy long before the Council of Trent. The Reconquest so recently completed had been a crusade; the title of the Catholic Monarchs itself proclaimed the identification of a political process with an ideological struggle. Secondly, the Spanish monarchy had the problem of absorbing suddenly great numbers of non-Christian subjects, both Moslem and Jew. They were feared as a potential threat to security in a multi-racial society. The instrument deployed against them was a new one: an Inquisition not, like its medieval forerunner, under clerical control, but under that of the Crown.

The vigour and self-confidence of Counter-Reformation Europe was expressed vividly in its art. Rubens, one of its master-painters, enthusiastically poured out his gifts in a flood of canvases of religious subjects (as in this picture of a miracle attributed to St Ignatius Loyola) as well as in classical and pagan subjects. Catholicism remained a religion of visual imagery while Protestantism turned to the word.

Established by papal bull in 1478, the Spanish Inquisition began to operate in Castile in 1480. The pope soon had misgivings; in Catalonia lay and ecclesiastical authority alike resisted, but to no avail. By 1516, when Charles I, the first ruler to hold both the thrones of Aragon and Castile, became king, the Inquisition was the only institution in the Spanish domains which, from a royal council, exercised authority in all of them, in the Americas, Sicily and Sardinia, as much as in Castile and Aragon. The most striking effects had already been the expulsion from them of the Jews and a severe regulation of the Moriscoes or converted Moors.

This gave Spain a religious unity unbreakable by a handful of Lutherans with whom the Inquisition found it easy to deal. The cost to Spain was in the end to be heavy. Yet already under Charles, a fervent Catholic, Spain was, in religion as in her secular life, aspiring to a new kind of centralized, absolutist monarchy, the Renaissance State *par excellence*, in fact. The residues of formal constitutionalism within the peninsula hardly affected this. Spain was a model for Counter-Reformation states elsewhere and one to be imposed upon much of Europe by force or example in the century after 1558, when Charles died after a retirement spent largely in his devotions in a remote monastery in Estremadura.

Of all European monarchs who identified himself with the cause of the Counter-Reformation and saw himself as the extirpator of heresy, none was more determined – and bigoted – than his son and successor, Philip II of Spain, widower of Mary Tudor. To him had come half Charles' empire: Spain, the Indies, Sicily and the Spanish Netherlands. (In 1581 he acquired Portugal, too, and it remained Spanish until 1640.) The results of his policies of religious purification in Spain have been variously interpreted. What is not open to dispute is the effect in the Spanish Netherlands, where they provoked the emergence of the first state in the world to break away from the old domination of monarchy and landed nobility.

What some call the 'Revolt of the Netherlands' and the Dutch the 'Eighty Years' War' has been, like many other events at the roots of nations, a great source of myth-making, some of it conscious. Even this, though, may have been less misleading than the assumption that because in the end a very modern sort of society emerged, it was a very 'modern' sort of revolt, dominated by a passionate struggle for religious toleration and national independence. Nothing was less true. The troubles of the Netherlands arose in a very medieval setting, the Old Burgundian inheritance of the lands of the richest state in northern Europe, the duchy which had passed to the Habsburgs by marriage. The Spanish Netherlands, seventeen provinces of very different sorts, formed part of it. The southern provinces, where many of the inhabitants spoke French, included the most urbanized part of Europe and the great Flemish commercial centre of Antwerp. They had long been troublesome and the Flemish towns had at one moment in the late fifteenth century seemed to be trying to turn themselves into independent city-states. The northern provinces were more agricultural and maritime. Their inhabitants showed a peculiarly tenacious feeling for their land, perhaps because they had actually been recovering it from the sea and making polders since the twelfth century.

North and South were to be the later Netherlands and Belgium, but this was inconceivable in 1554. Nor could a religious division between the two be then envisaged. Though the Catholic majority of the south grew somewhat as many Protestants emigrated northwards, the two persuasions were mixed upon both sides of a future boundary. Early sixteenth-century Europe was much more tolerant of religious divisions than it would be after the Counter-Reformation got to work.

Philip's determination to enforce the decrees of the Council of Trent explains something of what followed but the origins of trouble went back a long way. As the Spaniards strove to modernize the relations of central government and local communities, they did so with more up-to-date methods and perhaps less tact than the Burgundians had shown. Spanish royal envoys came into conflict first with the nobility of the southern provinces. As prickly and touchy as other nobilities of the age in defence of their symbolic 'liberties' – that is, privileges and immunities – they felt threatened by a monarch more

A Jesuit missionary at the Chinese court in the seventeenth century. He is shown with the astronomical instruments he took with him to China and which so impressed the sages of a culture that attached special importance to the accurate management of the calendar.

remote than the great Charles who, they felt, had understood them (he spoke their language), even if he was Charles' son. The Spanish commander, the Duke of Alva, they argued, was further violating local privilege by interfering with local jurisdictions in the pursuit of heretics. Catholic though they were, they had a stake in the prosperity of the Flemish cities where Protestantism had taken root and feared the introductions to them of the Spanish Inquisition. They were, too, as uneasy as other noblemen of the times about the pressures of inflation.

Resistance to Spanish government began in thoroughly medieval forms, in the Estates of Brabant, and for a few years the brutality of the Spanish army and the leadership of one of their number, William of Orange, united the nobles against their lawful ruler. Like his contemporary, Elizabeth Tudor, William (nicknamed the 'Silent' because of his reputed refusal to allow unguarded anger to escape him when he learnt of his ruler's determination to bring his heretic subjects to heel) was good at suggesting sympathy for popular causes. But there was always a potential rift between noblemen and Calvinist townsmen who had more at stake. Better political tactics by Spanish governors and the victories of the Spanish armies were in the end enough to force it open. The nobles fell back into line and thus, without knowing it, the Spanish armies defined modern Belgium. The struggle continued only in the northern provinces (though still under the political direction of William the Silent until his murder in 1584).

The Dutch (as we may now call them) had much at stake and were not encumbered as their southern co-religionists had been with the ambiguous dissatisfactions of the nobility. But they were divided among themselves; the provinces could rarely come easily to agreement. On the other hand, they could use the cry of religious freedom and a broad toleration to disguise their divisions. They benefited, too, from a great migration northward of Flemish capital and talent. Their enemies had difficulties; the Spanish army was formidable but could not easily deal with an enemy which retired behind its town walls and surrounded them with water by opening the dykes and flooding the countryside. The Dutch, almost by accident, transferred their main effort to the sea where they could do a great deal of damage to the Spanish on more equal terms. Spanish communications with the Netherlands were more difficult once the northern sea route was harried by the rebels. It was expensive to maintain a big army in Belgium by the long road up from Italy and even more expensive when other enemies had to be beaten off. That was soon the case. The Counter-Reformation had infected international politics with a new ideological element. Together with their interest in maintaining a balance of power on the continent and preventing the complete success of the Spanish, this led the English first to a diplomatic and then to a military and naval struggle against Spain which brought the Dutch allies.

The war created, almost fortuitously and incidentally, a remarkable new society, a loose federation of seven little republics with a weak central government, called the United Provinces. Soon, its citizens discovered a forgotten national past (much as decolonized Africans have done in this century) and celebrated the virtues of Germanic tribesmen dimly discernible in Roman accounts of rebellion; relics of their enthusiasm remain in the paintings commissioned by Amsterdam magnates depicting attacks upon Roman camps (this was in the era we remember for the work of Rembrandt). The distinctiveness of a new nation thus consciously created is now more interesting than such historical propaganda. Once survival was assured, the United Provinces enjoyed religious tolerance, great civic freedom and provincial independence; the Dutch did not allow Calvinism in the upper hand in government.

Later generations came to think they saw a similar linkage of religious and civic freedom in Elizabethan England; this was anachronistic, although comprehensible given the way English institutions were to evolve over the next century or so. Paradoxically, one part of this was a great strengthening of the legislative authority of the state, one which carried the limitation of privilege so far that at the end of the seventeenth century it was regarded with amazement by other Europeans. For a long time this cannot have seemed a likely outcome. Elizabeth had been an incomparable producer of the royal spectacle. When the myths of beauty and youth faded she had acquired the majesty of those who outlive their early counsellors. In 1603 she had been queen for forty-five years, the centre of a national cult fed by her own Tudor instinct for welding the dynasty's interest to patriotism, by poets of genius, by mundane devices such as the frequent travel (which kept down expenses, since she stayed with her nobility) which made her visible to her people, and by her astonishing skill with her parliaments. Nor did she persecute for religion's sake; she did not, as she put it, want to make 'windows into men's souls'. It is hardly surprising that the accession day of Good Queen Bess became a festival of patriotic opposition to government under her successors. Unhappily, she had no child to whom to bequeath the glamour she brought to monarchy, and she left an encumbered estate. Like all other rulers of her day, she never had a big enough income. The inheritance of debt did not help the first king of the Scotch house of Stuart, who succeeded her, James I. The shortcomings of the males of that dynasty are still difficult to write about with moderation; the Stuarts gave England four bad kings in a row. Still, James was neither as foolish as his son nor as unprincipled as his grandsons. It was probably his lack of tact and alien ways rather than more serious defects that did most to embitter politics in his reign.

In defence of the Stuarts, it can be agreed that this was not the only troubled monarchy. In the seventeenth century there was a roughly contemporaneous crisis of authority in several countries, and one curiously parallel to an economic crisis which was Europe-wide. The two may have been connected, but it is not easy to be sure what the nature of the connexion was. It is also interesting that these civil struggles coincided with the last phase of a period of religious wars which had been opened by the Counter-Reformation. We may at least assume that a contemporaneous breakdown of normal political life in a number of countries, notably England, France and Spain, owed something to the needs of governments forced to take part in them.

In England the crisis came to a head in civil war, regicide and the establishment of the only republic in English history. Historians still argue about where lay the heart of the quarrel and the point of no return in what became armed conflict between Charles I and his parliament. One crucial moment came when he found himself at war with one set of his subjects (for he was King of Scotland, as well as England), and had to call parliament to help him in 1640. Without new taxation, England could not be defended. But by then some of its members were convinced that there was a royal scheme to overturn from within the Church by law established and reintroduce the power of Rome. Parliament harried the king's servants (sending the two most conspicuous to the block). Charles decided in 1642 that force was the only way out and so the Civil War began. In it he was defeated. Parliament was uneasy, as were all men, for if you stepped outside the ancient constitution of King, Lords and Commons, where would things end? But Charles threw away his advantage by seeking a foreign invasion in his support (the Scots were to fight for him, this time). Those who dominated parliament had had enough and Charles was tried and executed. His son went into exile.

There followed in England an interregnum during which the dominant figure until his death in 1658 was one of the most remarkable of all Englishmen, Oliver Cromwell. He was a country gentleman who had risen in the parliamentary side's councils by his genius as a soldier. This gave him great power, for provided his army stood by him he could dispense with the politicians, but also imposed limitations on him, for he could not risk losing the army's support. The result was an English republic astonishingly fertile in new constitutional schemes, as Oliver cast about to find a way of governing through parliament without delivering England to an intolerant Protestantism. This was the Commonwealth.

The intolerance of some parliamentarians was one expression of a many-sided strain in English (and American) Protestantism which has been named Puritanism. It was an ill-defined but growing force in English life since Elizabeth's reign. Its spokesmen had originally sought only a particularly close and austere interpretation of religious doctrine and ceremony. There were Puritans inside the Anglican Church as well as among the critics who were impatient over its retention of much from the Catholic past, but it was to this second tendency that the name was more and more applied.

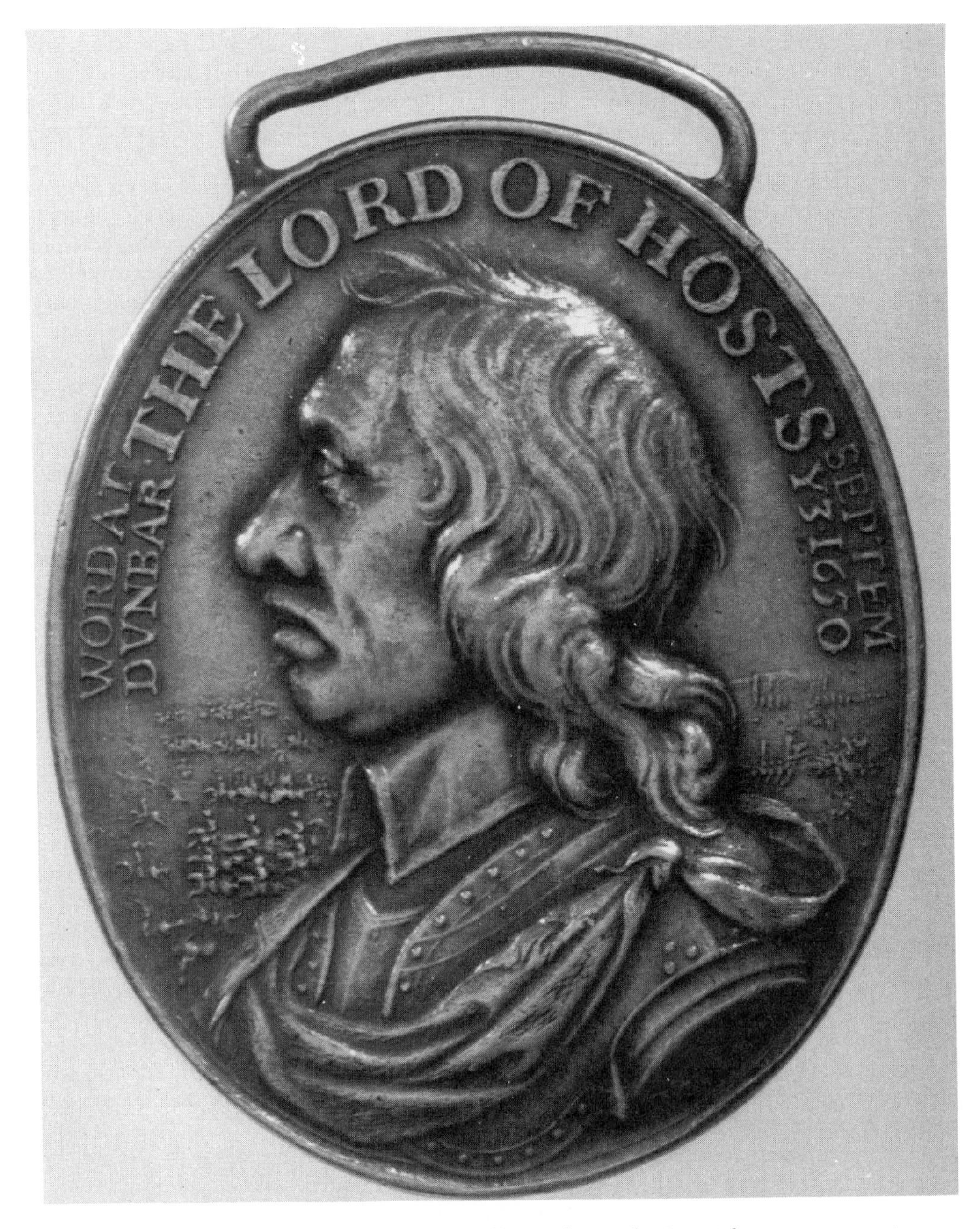

On 3 September 1650 at Dunbar, Cromwell overthrew the Scottish army supporting Charles II. Another victory at Worcester a year later to the day confirmed the triumph, marking the real end of the Civil War; soon, Scotland was united to England and the Commonwealth was safe – until its supporters fell out among themselves.

By the seventeenth century the epithet 'puritan' also betokened, besides rigid doctrine and disapproval of ritual, the reform of manners in a strongly Calvinistic sense. By the time of the republic, many who had been on the parliament's side in the Civil War appeared to wish to use its victory to impose Puritanism, both doctrinal and moral, by law not only on conservative and royalist Anglicans, but on dissenting religious minorities – Congregationalists, Baptists, Unitarians – which had found their voice under the Commonwealth. There was nothing politically or religiously democratic about Puritanism. Those who were of the Elect might freely choose their own elders and act as a self-governing community, but from outside the self-designated circle of the saved they looked (and were) an oligarchy claiming to know God's will for others, and therefore all the more unacceptable. It was a few, untypical minorities, not the dominant Protestant establishment, which threw up the democratic and levelling ideas which contributed so much to the great debate of the republican years.

The publication of more than twenty thousand books and pamphlets on political and religious issues would by itself have made the Civil War and Commonwealth years a great epoch in English political education. Unfortunately, once Oliver had died, the institutional bankruptcy of the republic was clear. Englishmen could not agree in sufficient numbers to uphold any new constitution. But most of them, it turned out, would accept the old device of monarchy. So the story of the Commonwealth ended with the restoration of the Stuarts in 1660. England in fact had her king back on unspoken conditions: in the last resort, Charles II came back because parliament said so, and believed he would defend the Church of England. Counter-Reformation Catholicism frightened Englishmen as much as did revolutionary Puritanism. Although the struggle of king and parliament was not over, there would be no absolute monarchy in England; henceforth the Crown was on the defensive.

Historians have argued lengthily about what the so-called 'English Revolution' expressed. Clearly, religion played a big part in it. Extreme Protestantism was given a chance to have an influence on the national life it was never again to have; this earned it the deep dislike of the Church of England and made political England anti-clerical for centuries. It was not without cause that the best English historian of the struggle has spoken of the 'Puritan Revolution'. But religion no more exhausts the meaning of these years than does the constitutional quarrel. Others have sought a class struggle in the Civil War. Of the interested motives of many of those engaged there can be no doubt, but it does not fit any clear general pattern. Still others have seen a struggle between a swollen 'Court', a governmental nexus of bureaucrats, courtiers and politicians, all linked to the system by financial dependence upon it, and 'Country', the local notables who paid for this. But localities often divided: it was one of the tragedies of the Civil War that even families could be split by it. It remains easier to be clear about the results of the English Revolution than about its origins or meaning.

Most continental countries were appalled by the trial and execution of Charles I, but they had their own bloody troubles. A period of conscious assertion of royal power in France by Cardinal Richelieu not only reduced the privileges of the Huguenots (as French Calvinists had come to be called) but had installed royal officials in the provinces as the direct representatives of royal power; these were the *intendants*. Administrative reform was an aggravation of the almost continuous suffering of the French people in the 1630s and 1640s. In the still overwhelmingly agricultural economy of France, Richelieu's measures were bound to hurt the poor most. Taxes on the peasant doubled and sometimes trebled in a few years. An eruption of popular rebellion, mercilessly repressed, was the result. Some parts of France, moreover, were devastated by the campaigns of the last phase of the great struggle for Germany and central Europe called the Thirty Years' War, the phase in which it became a Bourbon-Habsburg conflict. Lorraine, Burgundy and much of eastern France were reduced to ruins, the population of some areas declining by a quarter or a third. The claim that the French monarchy sought to impose new and (some said) unconstitutional taxation finally detonated political crisis under Richelieu's successors. The role of defender of the traditional constitution was taken up by special interests, notably the *parlement* of Paris, the corporation of lawyers who sat in and could plead before the first law court of the kingdom. In 1648 they led an insurrection in Paris (soon named the *Fronde*). A compromise settlement was followed after an uneasy interval by a second, much more dangerous *Fronde*, led this time by great noblemen. Though the *parlement* of Paris did not long maintain a united front with them, these men could draw on the anti-centralist feelings of the provincial nobility, as regional rebellions showed. Yet the Crown survived (and so did the *intendants*). In 1660 the absolute monarchy of France was still essentially intact.

In Spain, too, taxation provoked troubles. An attempt by a minister to overcome the provincialism inherent in the formally federal structure of the Spanish state led to revolt in Portugal (which had been absorbed into Spain with promises of respect for her liberties from Philip II), among the Basques, and in Catalonia. The last was to take twelve years to suppress. There was also a revolt in 1647 in the Spanish kingdom of Naples.

In all these instances of civic turbulence, demands for money provoked resistance. In the financial sense, then, the Renaissance State was far from successful. The appearance of standing armies in most states in the seventeenth century did not mark only a military revolution. War was a great devourer of taxes. Yet the burdens of taxation laid on Frenchmen seem far greater than those laid on Englishmen: why, then, did the French monarchy appear to suffer less from the 'crisis'? England, on the other hand, had civil war and the overthrow (for a time) of her monarchy without the devastation which went with foreign invasion. Nor were her occasional riots over high prices to be compared with the appalling bloodshed of the peasant risings of seventeenth-century France. In England, too, there was a specific challenge to authority from religious dissent. In Spain this was non-existent and in France it had been contained long before. The Huguenots, indeed, were a vested interest; but they saw their protector in the monarchy and therefore rallied to it in the *Frondes*. Regionalism was important in Spain, to a smaller extent in France where it provided a foothold for conservative interests threatened by governmental innovation, but seems to have played very little part in England.

1660, when the young Louis XIV assumed full powers in France and Charles II returned to England, was, in fact, something of a turning-point. France was not to prove ungovernable again until 1789 and was to show in the next half-century astonishing military and diplomatic power. In England there was never again to be, in spite of further constitutional troubles and the deposition of another king, a civil war. The last English rebellion, by an inadequate pretender and a few thousand deluded yokels in 1685, in no sense menaced the state. After 1660 there was an English standing army, after all. This makes it all the more striking, in retrospect, that men remained so unwilling to admit the reality of sovereignty. Englishmen solemnly legislated a series of defences of individual liberty in the Bill of Rights, yet even in 1689 it was hard to argue that what one king in Parliament had done, another could not undo. In France everyone agreed the king's power was absolute, yet lawyers went on saying that there were things he could not legally do.

One thinker at least, the greatest of English political philosophers, Thomas Hobbes, showed in his books, notably in the *Leviathan* of 1651, that he recognized the way society was moving. Hobbes argued that the disadvantages and uncertainties of not agreeing that someone should have the last word in deciding what was law clearly outweighed the danger that such power might be tyrannically employed. The troubles of his times deeply impressed him with the need to know certainly where authority was to be found. Even when they were not continuous, disorders were always liable to break out: as Hobbes put it (roughly), you do not have to live all the time under a torrential downpour to say that the weather is rainy. The recognition that legislative power – sovereignty – rested, limitless, in the state and not elsewhere, and that it could not be restricted by appeals to immunities, customs, divine law or anything else without the danger of falling into anarchy, was Hobbes' contribution to political theory, though he got small thanks for it and had to wait until the nineteenth century for due recognition. People often acted as though they accepted his views, but he was almost universally condemned.

Constitutional England was in fact one of the first states to operate on Hobbes' principles. By the early eighteenth century, Englishmen (Scotchmen were less sure, even when they came under the parliament at Westminster after the Act of Union of 1707) accepted in principle and sometimes showed in practice that there could be no limits except practical ones to the potential scope of law. This conclusion was to be explicitly challenged even as late as Victorian times, but was implicit when in 1688 England at last rejected the direct descent of the Stuart male line, pushed James II off the throne and put his daughter and her consort on it on conditions. With the creation

Evocations of power, secular and spiritual, surround the frontispiece of Leviathan, *the 'mortal God' whom Thomas Hobbes saw as the only guarantor of the conditions necessary to civilized life.*

of a contractual monarchy England at last broke with her *ancien régime* and began to function as a constitutional state. Effectively, centralized power was shared; its major component lay with a House of Commons which represented the dominant social interest, the landowning classes. The king still kept important powers of his own but his advisers, it soon became clear, must possess the confidence of the House of Commons. The legislative sovereign, the Crown in Parliament, could do anything by statute. No such immunity as still protected privilege in continental countries existed. The English answer to the danger posed by such a concentration of authority was to secure, by revolution if necessary, that the authority should only act in accordance with the wishes of the most important elements in society.

The year 1688 gave England a Dutch king, Queen Mary's husband, William III, to whom the major importance of the 'Glorious Revolution' of that year was that England could be mobilized against France, now threatening the independence of the United Provinces. There were too many complicated interests at work in them for the Anglo-French wars which followed to be interpreted in merely constitutional or ideological terms. Moreover, the presence of the Holy Roman Empire, Spain and various German princes in the shifting anti-French coalitions of the next quarter-century would certainly make nonsense of any neat contrast of political principle between the two sides. Nevertheless, it rightly struck some contemporaries that there was an ideological element buried somewhere in the struggle. England and Holland were more open societies than the France of Louis XIV. They allowed and protected the exercise of different religions. They did not censor the press but left it to be regulated by the laws which protected persons and the state against defamation. They were governed by oligarchies representing the effective possessors of social and economic power. France was at the opposite pole.

Under Louis XIV, absolute government reached its climax in France. It is not easy to pin his ambitions down in familiar categories; for him personal, dynastic and national greatness were hardly distinguishable. Perhaps that is why he became a model for all European princes. Politics was reduced effectively to administration; the royal councils, together with the royal agents in the provinces, the *intendants* and military commanders, took due account of such social facts as the existence of the nobility and local immunities, but the reign played havoc with the real independence of the political forces so powerful hitherto in France. This was the era of the establishment of royal power throughout the country and some later saw it as a revolutionary one; in the second half of the century the frame which Richelieu had knocked together was at last filled up by administrative reality. Louis XIV tamed aristocrats by offering them the most glamorous court in Europe; his own sense of social hierarchy made him happy to caress them with honours and pensions, but he never forgot the *Frondes* and controlled the nobility as had Richelieu. Louis' relatives were excluded from his council, which contained non-noble ministers on whom he could safely rely. The *parlements* were restricted to their judicial role; the French Church's independence of Roman authority was asserted, but only to bring it the more securely under the wing of the Most Christian King (as one of Louis' titles had it). As for the Huguenots, Louis was determined, whatever the cost, not to be a ruler of heretics; those who were not exiled were submitted to a harsh persecution to bring them to conversion.

The coincidence with a great age of French cultural achievement seems to make it hard for Frenchmen to recognize the harsh face of the reign of Louis XIV. He ruled a hierarchical, corporate, theocratic society which, even if up-to-date in methods, looked to the past for its goals. Louis even hoped to become Holy Roman Emperor. He refused to allow Descartes, the defender of religion, to be given religious burial in France because of the dangers of his ideas. Yet for a long time his kind of government seems to have been what most Frenchmen wanted. The process of effective government could be brutal, as Huguenots who were coerced into conversion by having soldiers billeted on them, or peasants reluctant to pay taxes who were visited by a troop of cavalry for a month or so, both knew. Yet life may have been better than life a few decades previously, in spite of some exceptionally hard years. The reign was the end of an era of disorder, not the start of one. France was largely free from invasion and there was a drop in the return expected from investment in land which lasted well into the eighteenth century. These were solid realities to underpin the glittering facade of an age later called the *Grand Siècle*.

Louis' European position was won in large measure by success in war (and by the end of the reign, he had undergone bad setbacks), but it was not only his armies and diplomacy which mattered. He carried French prestige to a peak at which it was long to remain because of the model of monarchy he presented; he was the perfect absolute monarch. The physical setting of the Ludovican achievement was the huge new palace of Versailles. Few buildings or the lives lived in them can have been so aped and imitated. In the eighteenth century Europe was to be studded with miniature reproductions of the French court, painfully created at the expense of their subjects by would-be *grands monarques* in the decades of stability and continuity which almost everywhere followed the upheavals of the great wars of Louis' reign.

There were between 1715 and 1740 no important international tensions to provoke internal change in states, nor were there great ideological divisions such as those of the seventeenth century, nor rapid economic and social development with their consequential strain. Not surprisingly, therefore, governments changed little and everywhere society seemed to settle down after a turbulent century or so. Apart from Great Britain, the United Provinces, the cantons of Switzerland and the fossil republics of Italy, absolute monarchy was the dominant state form. It remained so for most of the eighteenth century, sometimes in a style which came to be called 'enlightened despotism' – a slippery term, which neither has nor ever had a clear meaning any more than terms like 'Right' or 'Left' have today. What it indicates is that from about 1750 the wish to carry out practical reforms led some rulers to innovations which seemed to be influenced by the advanced thought of the day. Such innovations, when effective, were imposed none the less by the machinery of absolute monarchical power. If sometimes humanitarian, the policies of 'enlightened despots' were not necessarily politically liberal. They were, on the other hand, usually modern in that they undermined traditional social and religious authority, cut across accepted notions of social hierarchy or legal rights, and helped to concentrate lawmaking power in the state and assert its unchallenged authority over its subjects, who were treated increasingly as an aggregate of individuals rather than as members of a hierarchy of corporations.

Not surprisingly, it is almost impossible to find an example which in practice perfectly fulfils this general description, just as it is impossible to find a definition of a 'democratic' state today, or a 'fascist' state in the 1930s, which fits all examples. Among Mediterranean and southern countries, for example, Naples, Spain, Portugal and some other Italian states (and even at times the papal states) had ministers who sought economic reform. Some of these were stimulated by novelty; others – Portugal and Spain – turned to enlightened despotism as a way to recover lost status as great powers. Some encroached on the powers of the Church. Almost all of them served rulers who were part of the Bourbon family connexion. The involvement of one of the smallest of them, Parma, in a quarrel with the papacy led to a general attack in all of these countries on the right arm of the Counter-Reformation papacy, the Society of Jesus. In 1773 the pope was driven by them to dissolve the Society, a great symbolic defeat, as important for its demonstration of the strength of advanced anti-clerical principles even in Catholic Europe as for its practical effects.

Among these states only Spain had any pretension to great power status and she was in decline. Of the eastern enlightened despotisms, on the other hand, three out of four certainly had. The odd man out was Poland, the sprawling ramshackle kingdom where reform on 'enlightened' lines came to grief on constitutional rocks; the enlightenment was there all right, but not the despotism to make it effective. More successfully, Prussia, the Habsburg empire and Russia, all managed to sustain a facade of enlightenment while strengthening the state. Once more, the clue to change can be found in war, which cost far more than building even the most lavish replica of Versailles. In Russia modernization of the state went back to the earliest years of the century, when

Peter the Great sought to guarantee her future as a great power through technical and institutional change. In the second half of the century, the Empress Catherine II reaped many of the benefits of this. She also gave the regime a thin veneer of up-to-the-minute ideas by advertizing widely her patronage of letters and humanitarianism. This was all very superficial; the traditional ordering of society was unchanged. Russia was a conservative despotism whose politics were largely a matter of the struggles of noble factions and families. Nor did enlightenment much change things in Prussia, where there was a well-established tradition of efficient, centralized, economical administration embodying much of what reformers sought elsewhere. Prussia already enjoyed religious toleration but the Hohenzollern monarchy ruled a strongly traditional society virtually unchanged in the eighteenth century. The Prussian king was obliged to recognize – and willingly did so – that his power rested on the acquiescence of his nobles and he carefully preserved their legal and social privilege. Frederick II remained convinced that only noblemen should be given commissioned rank in his army and at the end of his reign there were more serfs in Prussian territory than there had been at the beginning.

Competition with Prussia was a decisive stimulus to reform in the Habsburg dominions. There were great obstacles in the way. The dynasty's territories were very diverse, in nationality, language, institutions; the emperor was King of Hungary, Duke of Milan, Archduke of Austria, to name only a few of his many titles. Centralization and greater administrative uniformity were essential if this variegated empire was to exercise its due weight in European affairs. Another problem was that, like the Bourbon states, but unlike Russia or Prussia, the Habsburg empire was overwhelmingly Roman Catholic. Everywhere the power of the Church was deeply entrenched; the Habsburg lands included most of those outside Spain where the Counter-Reformation had been most successful. The Church also owned huge properties; it was everywhere protected by tradition, canon law and papal policy, and it had a monopoly of education. Finally, the Habsburgs provided almost without interruption during these centuries the successive occupants of the throne of the Holy Roman Empire. In consequence they had special responsibilities in Germany.

This background was always likely to give modernization in the Habsburg dominions an 'enlightened' colour. Everywhere practical reform seemed to conflict with entrenched social power or the Church. The Empress Maria Theresia was herself by no means sympathetic to reform which had such implications, but her advisers were able to present a persuasive case for it when, after the 1740s, it became clear that the Habsburg monarchy would have to struggle for supremacy with Prussia. Once the road to fiscal and consequently administrative reform had been entered upon, it was in the end bound to lead to conflict between Church and State. This came to a climax in the reign of Maria Theresia's son and successor, Joseph II, a man who did not share the pieties of his mother and who was alleged to have advanced views. His reforms became especially associated with measures of secularization. Monasteries lost their property, religious appointments were interfered with, the right of sanctuary was removed and education was taken out of the hands of the clergy. So far as it went, this awoke angry opposition, but mattered less than the fact that by 1790 Joseph had antagonized to the point of open defiance the nobles of Brabant, Hungary and Bohemia. The powerful local institutions – estates and diets – through which those lands could oppose his policies paralysed government in many of Joseph's realms at the end of his reign.

Differences in the circumstances in which they were applied, in the preconceptions which governed them, in the success they achieved and in the degree to which they did or did not embody 'enlightened' ideas, all show how misleading is any idea that there was, anywhere, a 'typical' enlightened despotism to serve as a model. The government of France, clearly touched by reforming policies and aspirations, only confirms this. Obstacles to change had, paradoxically, grown stronger after the death of Louis XIV. Under his successor (whose reign began with a minority under a regent), the real influence of the privileged had grown and increasingly there grew up in the *Parlements* a tendency to criticize laws which infringed special interest and historic privilege. There was a new and growing resistance to the idea that there rested in the Crown any right of unrestricted legislative sovereignty. As the century wore on, France's international role imposed heavier and heavier burdens on her finances and the issue of reform tended to crystalize in the issue of finding new tax revenue – an exercise that was bound to bring on conflicts. Onto this rock ran most of the proposals for reform within the French monarchy.

Paradoxically, France was in 1789 the country most associated with the articulation and diffusion of critical and advanced ideas, yet also one of those where it seemed most difficult to put them into practice. But this issue posed problems Europe-wide in the traditional monarchies of the end of the eighteenth century. Wherever reform and modernization had been tried, the hazards of vested historical interest and traditional social structure threw obstacles in the way. In the last resort, it was unlikely that monarchical absolutism could have solved this problem anywhere. It could not question historical authority too closely for this was what it rested upon itself. Unrestricted legislative sovereignty seemed still in the eighteenth century to call too much in question. If historic rights were infringed, could not property be? This was a fair point, though Europe's most successful ruling class, the English, seemed not to fear that such a revolutionary idea was likely to be used against them.

With this important qualification, though, enlightened despotism, too, embodies the theme already set out – that at the heart of the complex story of political evolution in many countries over a period of three centuries, continuity lies in the growth of the power of the state. The occasional successes of those who tried to put the clock back almost always proved temporary. True, even the most determined reformers and the ablest statesmen had to work with a machinery of state which to any modern bureaucrat would seem woefully inadequate. Though the eighteenth-century state might mobilize resources much greater than had done its predecessors it had to do so with no revolutionary innovations of technique. Communications in 1800 depended just as they had done in 1500 on wind and muscle; the 'telegraph' which came into use in the 1790s was only a semaphore system, worked by pulling ropes. Armies could move only slightly faster than three centuries earlier, and if their weapons were improved, they were not improved out of recognition. No police force such as exists today existed in any country; income tax lay still in the future. The change in the power of the state came about because of changes in ideas and because of the development to greater efficiency of well-known institutions, rather than because of technology. In no major state before 1789 could it even be assumed that all its subjects would understand the language of government, while none, except perhaps Great Britain and the United Provinces, succeeded in so identifying itself with its subjects as to leave its government more concerned to protect them against foreigners than itself against them.

3
The New World of Great Powers

Among the institutions which took their basic shape in the fifteenth and sixteenth centuries and are still with us today are those of formal diplomacy. Rulers had long sent messages to one another and negotiated, but there were always many ways of doing this and of understanding what was going on. The Chinese, for example, used the fiction that their emperor was ruler of the world and that all embassies to him were therefore of the nature of petitions or tributes by subjects. Medieval kings had sent one another heralds, about whom a special ceremonial had grown up and whom special rules protected, or occasional missions of ambassadors. After 1500, it slowly became the practice to use in peacetime the standard device we still employ, of a permanently resident ambassador through whom all ordinary business is at least initially transacted and who has the task of keeping his own rulers informed about the country to which he is accredited.

Venetian commerce outlived the Byzantine empire which had cradled it in the Near East. A picture by Bellini of a Venetian embassy arriving in Cairo in the early sixteenth century.

The Venetian ambassadors were the first notable examples. It is not surprising that a republic so dependent on trade and the maintenance of regular relationships should have provided the first examples of the professional diplomat. More changes followed. Gradually, the hazards of the life of earlier emissaries were forgotten as diplomats were given a special status protected by privileges and immunities. The nature of treaties and other diplomatic forms also became more precise and regularized.

Procedure became more standardized. All these changes came about slowly, when they were believed to be useful. For the most part, it is true, the professional diplomat in the modern sense had not yet appeared by 1800, ambassadors were then still usually noblemen who could afford to sustain a representative role, not paid civil servants. None the less, the professionalization of diplomacy was beginning. It is another sign that after 1500 a new world of relationships between sovereign powers was replacing that of feudal ties between persons and the vague supremacies of Pope and Emperor.

The most striking characteristic of this new system is the assumption that the world is divided into sovereign states. This idea took time to emerge; sixteenth-century Europe was certainly not seen by contemporaries as a set of independent independent areas, each governed by a ruler of its own, belonging to it alone. Still less were its components thought to have in any but a few cases any sort of unity which might be called 'national'. That this was so was not only because of the survival of such museums of past practice as the Holy Roman Empire. It was also because the dominating principle of early modern Europe's diplomacy was dynasticism.

In the sixteenth and seventeenth centuries, the political units of Europe were less states than landed estates. They were accumulations of property put together over long or short periods by aggressiveness, marriage and inheritance – by the same processes and forces, that is to say, by which any private family's estate might be built up. The results were to be seen on maps whose boundaries continually changed as this or that portion of an inheritance passed from one ruler to another. The inhabitants had no more say in the matter than might the peasants living on a farm which changed hands. Dynasticism accounts for the monotonous preoccupation of negotiations and treaty-making with the possible consequences of marriages and the careful establishment and scrutiny of lines of succession.

Besides their dynastic interests, rulers also argued and fought about religion, and, increasingly, trade or wealth. Some of them acquired overseas possessions; this, too, became a complicating factor. Occasionally, the old principles of feudal superiority might still be invoked. There were also always map-making forces at work which fell outside the operation of these principles, such as settlement of new land or a rising national sentiment. Nevertheless, broadly speaking, most rulers in the sixteenth and seventeenth centuries saw themselves as the custodians of inherited rights and interests which they had to pass on. In this they behaved as was expected; they mirrored the attitudes of other men and other families in their societies. It was not only the Middle Ages which were fascinated by lineage, and the sixteenth and seventeenth centuries were the great age of genealogy.

In 1500 the dynastic map of Europe was about to begin a major transformation. For the next two centuries, two great families were to dispute much of Europe as they were already at that date disputing Italy. These were the house of Habsburg and the ruling house of France, first Valois, then after the accession of Henry IV in 1589, Bourbon. The one would come to be predominantly Austrian and the other's centre would always be France. But both would export rulers and consorts of rulers to many other countries. The heart of their quarrel when the sixteenth century began was the Burgundian inheritance. Each of them was then far from playing a wider European role. Indeed, there was not a great deal to distinguish them at that date in power – though much in antiquity – from other dynasties, the Welsh Tudors, for example, whose first ruler, Henry VII, had ascended the throne of England in 1485.

Only in England, France and Spain could there be discerned any real national cohesion and sentiment to sustain political unity. England, a relatively unimportant power, was a well-developed example. Insular, secluded from invasion and rid, after 1492, of continental appendages other than the seaport of Calais (finally lost only in 1558), her government was unusually centralized. The Tudors, anxious to assert the unity of the kingdom after the long period of disorder we call the 'Wars of the Roses', consciously associated national interest with that of the dynasty. Shakespeare quite naturally uses the language of patriotism (and, it may be remarked, says little about religious differences). France, too, had already come far along the road to national cohesion. The house of Valois-Bourbon had greater problems than the Tudors, though, in the continued survival of immunities and privileged enclaves within its territories over which its monarchs did not exercise full sovereignty as kings of France. Some of their subjects did not even speak French. Nevertheless, France was well on the way to becoming a national state.

So was Spain, though its two crowns were not united until the grandson of the Catholic Monarchs, Charles of Habsburg, became co-ruler with his insane mother in 1516 as Charles I. He had still carefully to distinguish the rights of Castile from those of Aragon, but Spanish nationality was made more self-conscious during his reign because, although at first popular, Charles obscured the national identity of Spain in a larger Habsburg empire and, indeed, sacrificed Spanish interest to dynastic aims and triumphs. The great diplomatic event of the first half of the century was his election in 1519 as Charles V, Holy Roman Emperor. He succeeded his grandfather Maximilian, who had sought his election, and careful marriages in the past had by then already made him the ruler of the furthest-flung territorial empire the world had ever seen, to which the imperial title supplied a fitting crown. From his mother he inherited the Spanish kingdoms, the newly discovered Americas and Sicily. From his father, Maximilian's son, came the Netherlands which had been part of the duchy of Burgundy, and from his grandfather the Habsburg lands of Austria and the Tyrol, with Franche-Comté, Alsace and a bundle of claims in Italy. This was the greatest dynastic accumulation of the age, and the crowns of Bohemia and Hungary were held by Charles' brother, Ferdinand, who was to succeed him as Emperor. Habsburg pre-eminence was the central fact of European politics for most of the sixteenth century. Its real and unreal pretensions are well shown in the list of Charles' titles when he ascended the imperial throne: 'King of the Romans; Emperor-elect; semper Augustus; King of Spain, Sicily, Jerusalem, the Balearic Islands, the Canary Islands, the Indies and the mainland on the far side of the Atlantic; Archduke of Austria; Duke of Burgundy, Brabant, Styria, Carinthia, Carniola, Luxemburg, Limburg, Athens and Patras; Count of Habsburg, Flanders and Tyrol; Count Palatine of Burgundy, Hainault, Pfirt, Roussillon; Landgrave of Alsace; Count of Swabia; Lord of Asia and Africa.'

Whatever this conglomeration was, it was not national. It fell, for practical purposes, into two main blocks: the Spanish inheritance, rich through the possession of the Netherlands and irrigated by a growing flow of bullion from the Americas, and the old Habsburg lands, demanding an active role in Germany to maintain the family's pre-eminence there. Charles, though, saw from his imperial throne much more than this. Revealingly, he liked to call himself 'God's standard-bearer' and campaigned like a Christian paladin of old against the Turk in Africa and up and down the Mediterranean. In his own eyes he was still the medieval emperor, much more than one ruler among many; he was leader of Christendom and responsible only to God for his charge. Certainly he had a far better claim to be called 'Defender of the Faith' than his Tudor rival Henry VIII, another aspirant to the Imperial throne. Germany, Spain and Habsburg dynastic interest were all to be sacrificed in some degree to Charles' vision of his role. Yet what he sought was impossible. The dream of making a reality of universal empire was beyond the powers of any man, given the strains imposed by the Reformation and the inadequate apparatus of sixteenth-century communication and administration. Charles, moreover, strove to rule personally, travelling ceaselessly in pursuit of this futile aim and thereby, perhaps, he ensured also that no part of his empire (unless it was the Netherlands) felt identified with his house. His aspiration reveals the way in which the medieval world still lived on, but also his anachronism.

The Holy Roman Empire was, of course, distinct from the Habsburg family possessions. It, too, embodied the medieval past, but at its most worm-eaten and unreal. Germany, where most of it lay, was a chaos supposedly united under the emperor and his tenants-in-chief, the imperial Diet. Since the Golden Bull the seven electors were virtually sovereign in their territories. There were also a hundred princes and more than fifty imperial cities, all independent. Another three hundred or so minor statelets and imperial vassals completed the patchwork which was what was left of the early medieval empire.

Claims under pressure: Pope (Clement VII) and Emperor (Charles V) are painted as co-equal rulers of a Christendom at a moment when the medieval unity expressed in that idea was about to disintegrate.

As the sixteenth century began, an attempt to reform this confusion and give Germany some measure of national unity failed; this suited the lesser princes and the cities. All that emerged were some new administrative institutions. Charles' election as emperor in 1519 was by no means a foregone conclusion; rightly, people feared that German interests in the huge Habsburg dominions might be over-ridden or neglected. Heavy bribery of the electors was needed before he prevailed over the King of France (the only other serious candidate, for nobody believed that Henry VIII, although a runner, would be able to pay enough). Habsburg dynastic interest was thereafter the only unifying principle at work in the Holy Roman Empire until its abolition in 1806.

Italy, one of the most striking geographical unities in Europe, was also still fragmented into independent states, most of them ruled by princely despots, and some of them dependencies of external powers. The pope was a temporal monarch in the states of the Church. A king of Naples of the house of Aragon ruled that country. Sicily belonged to his Spanish relatives. Venice, Genoa and Lucca were republics. Milan was a large duchy of the Po valley ruled by the Sforza family. Florence was theoretically a republic but from 1509 really a monarchy in the hands of the Medici, a former banking house. In north Italy the dukes of Savoy ruled Piedmont, on the other side of the Alps from their own ancestral lands. The divisions of the peninsula made it an attractive prey and a tangle of family relationships gave French and Spanish rulers excuses to dabble in affairs there. For the first half of the sixteenth century the main theme of European diplomatic history is provided by the rivalry of Habsburg and Bourbon, above all in Italy.

A series of Habsburg-Valois wars in Italy began in 1494 with a French invasion reminiscent of medieval adventuring and raiding, and lasted until 1559. They constitute a distinct period in the evolution of the European states system. Charles V's accession and the defeat of Francis in the imperial election brought out the lines of dynastic competition more clearly. There were altogether six so-called 'Italian' wars and they were more important than they might at first appear. To Charles the Emperor, they were a fatal distraction from the Reformation problem, and to Charles the King of Spain they were the start of a fatal draining of their country's power. To the French, they brought impoverishment and invasion, and to their kings, in the end, frustration, for Spain was left dominant in Italy. To the inhabitants of that country, the wars brought a variety of disasters. For the first time since the age of the barbarian invasions, Rome was sacked (in 1527, by a mutinous imperial army) and Spanish hegemony finally ended the great days of the city republics and brought about the emigration of skilled craftsmen. At one time, the coasts of Italy were raided by French and Turkish ships in concert; the hollowness of the unity of Christendom was revealed by a formal alliance of a French king with the Sultan.

Perhaps these were good years only for the Ottomans. Venice, usually left to face the Turks alone, watched her empire in the eastern Mediterranean begin to crumble away. Both Charles V and his son were defeated in African enterprises and the defeat of the Turks at Lepanto in 1571 was only a momentary setback for them; three years later they took back Tunis from the Spanish. The struggle with the Ottomans and the support of the Habsburg cause in Italy had by then overburdened even Spain's flow of silver from America, a recourse badly needed by her at home. In his last years, Charles V was crippled by debt.

He abdicated in 1556, just after the first settlement at Augsburg of the religious disputes of Germany, to be succeeded as emperor by his brother, who took the Austrian inheritance, and as ruler of Spain by his son, Philip II, a Spaniard born and bred. Charles had been born in the Netherlands and the ceremony which ended the great Emperor's reign took place there, in the Hall of the Golden Fleece; he was moved to tears as he left the assembly, leaning on the shoulder of a young nobleman, William of Orange. This division of the Habsburg inheritance marks the watershed of European affairs in the 1550s.

What followed was the blackest period of Europe's history for centuries. With a brief lull as it opened, European rulers and their people indulged in the seventeenth century in an orgy of hatred, bigotry, massacre, torture and brutality which has no parallel until the twentieth. The dominating facts of this period were the military pre-eminence of Spain, the ideological conflict opened by the Counter-Reformation, the paralysis of Germany and, for a long time, France, by internal religious quarrels, the emergence of new centres of power in England, the Dutch Netherlands and Sweden, and the first adumbrations of the overseas conflicts of the next two centuries. Only with the end of this period did it appear that the power of Spain had dwindled and that France had inherited her continental ascendancy.

The best starting-point is the Dutch Revolt. Like the Spanish Civil War of 1936–9 (but for much longer) it mixed up the relations of many outsiders in a confusion of ideological, political, strategic and economic issues. France could not be easy while Spanish armies might invade her from Spain, Italy and Flanders. England's involvement arose in other ways. Though she was Protestant, she was only just Protestant, and Philip tried to avoid an outright break with Elizabeth I. He was for a long time unwilling to sacrifice the chance of reasserting the English interests he had won by marriage to Mary Tudor, and thought he might retain them by marrying a second English queen. Moreover he was long distracted by campaigns against the Ottomans. But national and religious feeling were inflamed in England by Spanish responses to English piracy at the expense of the Spanish empire; Anglo-Spanish relations decayed rapidly in the 1570s and 1580s. Elizabeth overtly and covertly helped the Dutch, whom she did not want to see go under, but did so without enthusiasm; being a monarch, she did not like rebels. In the end, armed with papal approval for the deposition of Elizabeth, the heretic queen, a great

Spanish invasion effort was mounted in 1588. 'God blew and they were scattered' said the inscription on an English commemorative medal; bad weather completed the work of Spanish planning and English seamanship and gunnery (though not a ship on either side was sunk by gunfire) to bring the Armada to disaster. War with Spain went on long after its shattered remnants had limped back to Spanish harbours but a great danger was over. Also, almost incidentally, an English naval tradition of enormous importance had been born.

'God blew, and they were scattered'; runs the inscription of an English Armada medal. Many English Protestants must have felt a comfortable reassurance that they were on God's side.

James I strove sensibly to avoid a renewal of the conflict once peace had been made and succeeded, for all the anti-Spanish prejudices of his subjects. England was not sucked into the continental conflict when the revolt of the Netherlands, re-ignited after a Twelve Years' Truce, was merged into a much greater struggle, the Thirty Years' War. Its heart was a Habsburg attempt to rebuild the imperial authority in Germany by linking it with the triumph of the Counter-Reformation. This called in question the Peace of Augsburg and the survival of a religiously pluralistic Germany. Once again, cross-currents confused the pattern of ideological conflict. As Habsburg and Valois had disputed Italy in the sixteenth century, Habsburg and Bourbon disputed Germany in the next. Dynastic interest brought Catholic France into the field against the Catholic Habsburgs. Under the leadership of a cardinal, the 'eldest daughter of the Church', as France was claimed to be, allied with Dutch Calvinists and Swedish Lutherans to assure the rights of German princes. Meanwhile the unhappy inhabitants of much of central Europe had often to endure the whims and rapacities of quasi-independent warlords. Cardinal Richelieu has a better claim than any other man to be the creator of a foreign policy of stirring up trouble beyond the Rhine which was to serve France well for over a century. If anyone still doubted it, with him the age of *Realpolitik* and *raison d'état*, of simple, unprincipled assertion of the interest of the sovereign state, had clearly arrived.

The Peace of Westphalia which ended the Thirty Years' War in 1648 was in several ways a registration of change. Yet it showed traces still of the fading past. This makes it a good vantage-point. It was the end of the era of religious wars in Europe; for the last time European statesmen had as one of their main concerns in a general settlement the religious future of their peoples. It also marked the end of Spanish military supremacy and of the dream of reconstituting the empire of a Charles V. It closed, too, an era of Habsburg history. In Germany a new force had appeared in the Electorate of Brandenburg, with which later Habsburgs would contend, but the frustration of Habsburg aims in Germany had been the work of outsiders, Sweden and France. Here was the real sign of the future: a period of French ascendancy was beginning in Europe west of the Elbe.

A century and a half after Columbus, though, when Spain, Portugal, England, France and Holland all already had important overseas empires, these were apparently of no interest to the authors of the peace. England was not even represented at either of the centres of negotiation; she had hardly been concerned in what was at issue once the first phase of the war was over. Preoccupied by internal quarrels and troubled by her Scotch neighbours, her foreign policy was directed towards ends more extra-European than European – though it was these ends which soon led her to fight the Dutch (1552–4). Although Cromwell quickly restored peace, telling the Dutch there was room in the world for both of them to trade, English and Dutch diplomacy was already showing more clearly than that of other nations the influence of commercial and colonial interest.

French ascendancy was founded on solid natural advantages. France was the most populous state of western Europe and on this simple fact rested French military power until the nineteenth century; it would always require the assembling of great international forces to contain it. France, however miserably poor its inhabitants may seem to modern eyes, had great economic resources, and was able to sustain a huge efflorescence of power and prestige under Louis XIV. His reign began formally in 1643, but actually in 1661 when, at the age of twenty-two, he announced his intention of managing his own affairs. This assumption of supreme power was a great fact in international as well as French history; Louis was the most consummate exponent of the trade of kingship who has ever lived. Only for convenience may his foreign policy be distinguished from other aspects of his reign. The building of Versailles, for example, was not only the gratification of a personal taste, but an exercise in building a prestige essential to his diplomacy. Similarly, though they may be separated, his foreign and domestic policy were closely entwined with one another and with ideology. Louis wanted to improve the strategical shape of France's north-western frontier, but also (though he might buy millions of tulip bulbs a year from them for Versailles) he despised the Dutch as merchants, disapproved of them as republicans, and detested them as Protestants. Nor was that all. Louis was a legalistic man – kings had to be – and he felt easier when there existed legal claims good enough to give respectability to what he was doing. This was the complicated background to a foreign policy of expansion. Though in the end it cost his country dearly, it carried France to a pre-eminence from which she was to freewheel through half the eighteenth century, and created a legend to which Frenchmen still look back with nostalgia.

Louis' wish for an improved frontier meant conflict with Spain, still in possession of the Spanish Netherlands and the Franche-Comté. The defeat of Spain opened the way to war with the Dutch. The Dutch held their own, but the war ended in 1678 with a peace usually reckoned the peak of Louis' achievement in foreign affairs. He now turned to Germany. Besides territorial conquest, he sought the imperial crown and to obtain it was willing to ally with the Turk. A turning-point came in 1688, when William of Orange, the Stadtholder of Holland, took his wife Mary Stuart to England to replace her father on the English throne. From this time Louis had a new and persistent enemy across the Channel, instead of the complaisant Stuart kings. Dutch William could deploy the resources of the leading Protestant country and for the first time since the days of Cromwell, England fielded an army on the continent in support of a league of European states (even the pope joined it secretly) against Louis. King William's war (or the war of the League of Augsburg), brought together Spain and Austria, as well as the Protestant states of Europe, to contain the overweening ambition of the French king. The peace which ended it was the first in which he had to make concessions.

In 1700 Charles II of Spain died childless. It was an event Europe had long awaited, for he had been a sickly, feeble-minded fellow. Enormous diplomatic preparations had been made for his demise because of the great danger and opportunity

EUROPE
(TREATY OF WESTPHALIA 1648)
Brandenburg-Prussia
Austrian Habsburg
Spanish Habsburg
Swedish possessions
Venetian possessions
Ottoman Empire
Boundary of the Holy Roman Empire
0 800 km
0 500 miles
N
Atlantic Ocean
North Sea
Baltic Sea
Black Sea
Mediterranean Sea
IRELAND
SCOTLAND
Edinburgh
ENGLAND
London
NORWAY
SWEDEN
Stockholm
DENMARK
Copenhagen
ESTONIA
LIVONIA
Königsberg
Duchy of Prussia
Warsaw
POLAND
RUSSIA
Kiev
UKRAINE
CRIMEA
MOLDAVIA
WALLACHIA
TRANSYLVANIA
HUNGARY
Budapest
Belgrade
MONTENEGRO
Constantinople
OTTOMAN EMPIRE
Ionian Islands
Crete
UNITED NETHERLANDS
Amsterdam
Antwerp
SPANISH NETHERLANDS
Hamburg
Bremen
Munster
WEST-PHALIA
PALATINATE
Heidelberg
POMERANIA
BRANDENBURG
SAXONY
BOHEMIA
MORAVIA
Vienna
AUSTRIA
STYRIA
CARINTHIA
CARNIOLA
BAVARIA
TYROL
SWITZERLAND
Geneva
SAVOY
PIEDMONT
MILAN
Genoa
VENETIAN REPUBLIC
PAPAL STATES
Rome
Naples
KINGDOM OF THE TWO SICILIES
Palermo
Corsica
Sardinia
FRANCE
Paris
SPAIN
Madrid
PORTUGAL

which it must present. A huge dynastic inheritance was at stake. A tangle of claims arising from marriage alliances in the past meant that the Habsburg emperor and Louis XIV (who had passed his rights in the matter on to his grandson) would have to dispute the matter. But everyone was interested. The English wanted to know what would happen to the trade of Spanish America, the Dutch the fate of the Spanish Netherlands. The prospect of an undivided inheritance going either to Bourbon or Habsburg alarmed everybody. The ghost of Charles V's empire walked again. Partition treaties had therefore been made. But Charles II's will left the whole Spanish inheritance to Louis' grandson. Louis accepted it, setting aside the agreements into which he had entered. He also offended the English by recognizing the exiled Stuart Pretender as James III of England. A Grand Alliance of Emperor, United Provinces and England was soon formed, and there began the War of the Spanish Succession, twelve years' fighting, which eventually drove Louis to terms. By treaties signed in 1713 and 1714 and called the Peace of Utrecht, the crowns of Spain and France were declared for ever incapable of being united. The first Bourbon king of Spain took his place on the Spanish throne, though, taking with Spain the Indies but not the Netherlands, which went to the emperor as compensation and to provide a tripwire defence for the Dutch against further French aggression. Austria also profited in Italy. France made concessions overseas to Great Britain (as it was after the union of England with Scotland in 1707). The Stuart Pretender was expelled from France and Louis recognized the Protestant succession in England.

These important facts assured the virtual stabilization of western continental Europe until the French Revolution seventy-five years later. Not everyone liked it (the emperor refused to admit the end of his claim to the throne of Spain) but to a remarkable degree the major definitions of western Europe north of the Alps have remained what they were in 1714. Belgium, of course, did not exist, but the Austrian Netherlands occupied much of the same area as that country, and the United Provinces corresponds to the modern Netherlands. France would keep Franche-Comté and, except between 1871 and 1918, the Alsace and Lorraine which Louis XIV had won for her. Spain and Portugal would after 1714 remain separate within their present boundaries; they still had large colonial empires but were never again to be able to deploy their potential strength so as to rise out of the second rank of powers. Great Britain was the new great power in the west; since 1707, England no longer had to bother about the old northern threat, though once more attached by a personal connexion to the continent because after 1714 her rulers were also Electors of Hanover. South of the Alps, the dust took longer to settle. A still disunited Italy underwent another thirty-odd years of uncertainty, minor representatives of European royal houses shuffling around it from one state to another in attempts to tie up the loose ends and seize the left-overs of the age of dynastic rivalry. After 1748 there was only one important native dynasty left in the peninsula, that of Savoy, which ruled Piedmont on the south side of the Alps and the island of Sardinia. The papal states, it is true, could since the fifteenth century be regarded as an Italian monarchy, though only occasionally a dynastic one, and the decaying republics of Venice, Genoa and Lucca also upheld the tattered standard of Italian independence. Foreign rulers were installed in the other states.

Western political geography was thus set for a long time. Immediately, this owed much to the need felt by all statesmen to avoid for as long as possible another conflict such as that which had just closed. For the first time a treaty of 1713 declared the aim of the signatories to be the security of peace through a balance of power. So practical an aim was an important innovation in political thinking. There were good grounds for such realism; wars were more expensive than ever and even Great Britain and France, the only countries in the eighteenth century capable of sustaining war against other great powers without foreign subsidy, had been strained. But the end of the War of the Spanish Succession also brought effective settlements of real problems. A new age was opening. Outside Italy, most of the political map of the twentieth century was already visible in western Europe. Dynasticism was beginning to be relegated to the second rank as a principle of foreign policy. The age of national politics had begun, for some kings, at least, who could no longer separate the interests of their house from those of their nation.

East of the Rhine (and still more east of the Elbe) none of this was true. Great changes had already occurred there and many more were to come before 1800. But their origins have to be traced back a long way, as far as the beginning of the sixteenth century. At that time Europe's eastern frontiers were guarded by Habsburg Austria and the vast Polish-Lithuanian kingdom ruled by the Jagiellons which had been formed by marriage in the fourteenth century. They shared with the maritime empire of Venice the burden of resistance to Ottoman power, the supreme fact of east European politics at that moment. The phrase 'Eastern Question' had not then been invented; if it had been, men would have meant by it the problem of defending Europe against Islam. The Turks won victories and made conquests as late as the eighteenth century, though by then their last great effort was spent. For more than two centuries after the capture of Constantinople, nevertheless, they set the terms of eastern European diplomacy and strategy.

The capture of Constantinople was followed by more than a century of naval warfare and Turkish expansion to the west from which the main sufferer was Venice. While she long remained rich by comparison with other Italian states, Venice suffered a relative decline, first in military and then in commercial power. The first, which led to the second, was the result of a long, losing battle against the Turks. In 1479, they took the Ionian islands and imposed an annual charge for trade in the Black Sea. Though Venice acquired Cyprus two years later, and turned it into a major base, it was in its turn lost in 1571. By 1600, though still (thanks to her manufacturers) a rich state, Venice had lost her commercial leaderships; first Antwerp and then Amsterdam had replaced her. She was no longer a mercantile power at the level of the United Provinces or even England. Turkish success was interrupted in the early seventeenth century but then resumed; in 1669 the Venetians had to recognize that they had lost Crete. Meanwhile, Hungary was invaded in 1664. This was the last Turkish conquest of a European kingdom, though the Ukrainians soon acknowledged Turkish suzerainty and the Poles had to give up Podolia. In 1683 the Turks opened a siege of Vienna and Europe seemed in its greatest danger for over two centuries. In fact it was not. This was to be the last time Vienna was besieged, for the great days of Ottoman power were over. The effort which began with the conquest of Hungary had been the last heave.

Difficulties had long troubled the Ottomans. Their army was no longer abreast of the latest military technology: it lacked the field artillery which had become the decisive weapon of the seventeenth-century battlefield. At sea, the Turks clung to the old galley tactics of ramming and boarding and were less and less successful against the Atlantic nations' technique of using the ship as a floating artillery battery (unfortunately for themselves, the Venetians were conservative too). Turkish power was in any case badly overstretched. It was pinned down in Asia (where the conquest of Iraq from Persia in 1639 brought almost the whole Arab-Islamic world under Ottoman rule) as well as in Europe and Africa, and the strain was too much for a structure allowed to relax by inadequate or incompetent rulers. A great vizier had pulled things together in the middle of the century to make the last offensives possible. But there were weaknesses which he could not correct, for they were inherent in the nature of the empire itself.

More a military occupation than a political unity, Ottoman empire was dangerously dependent on subjects whose loyalty it could not win. The Ottomans usually respected the customs and institutions of non-Moslem communities, which were ruled under the *millet* system, through their own authorities. The Greek Orthodox, Armenians and Jews were the most important and each had their own arrangements, the Greek Christians having to pay a special poll-tax, for example, and being ruled, ultimately, by their own patriarch in Constantinople. At lower levels, such arrangements as seemed best were made with leaders of local communities for the support of the military machine which was the heart of the Ottoman structure. In the end this bred over-mighty subjects as pashas feathered their own nests amid incoherence and inefficiency. It gave the subjects of the sultan no sense of identification with his rule but, rather, alienated them from it while the Ottoman lands in Europe got poorer and poorer.

The year 1683, therefore, although a good symbolic date as the last time that Europe stood upon the defensive in her old bastion against Islam before going over to the attack, was a less dangerous moment than it looked. Afterwards the tide of Turkish power was to ebb almost without interruption until in 1918 it was once more confined to the immediate hinterland of Constantinople and the old Ottoman heartland, Anatolia. The relief of Vienna by the King of Poland, John Sobieski, was followed by the liberation of Hungary. The dethronement of an unsuccessful sultan in 1687 and his incarceration in a cage proved no cure for Turkish weakness. In 1699 Hungary became part of the Habsburg dominions again, devastated though it was. In the following century Transylvania, the Bukovina, and most of the Black Sea coasts would follow it out of Ottoman control.

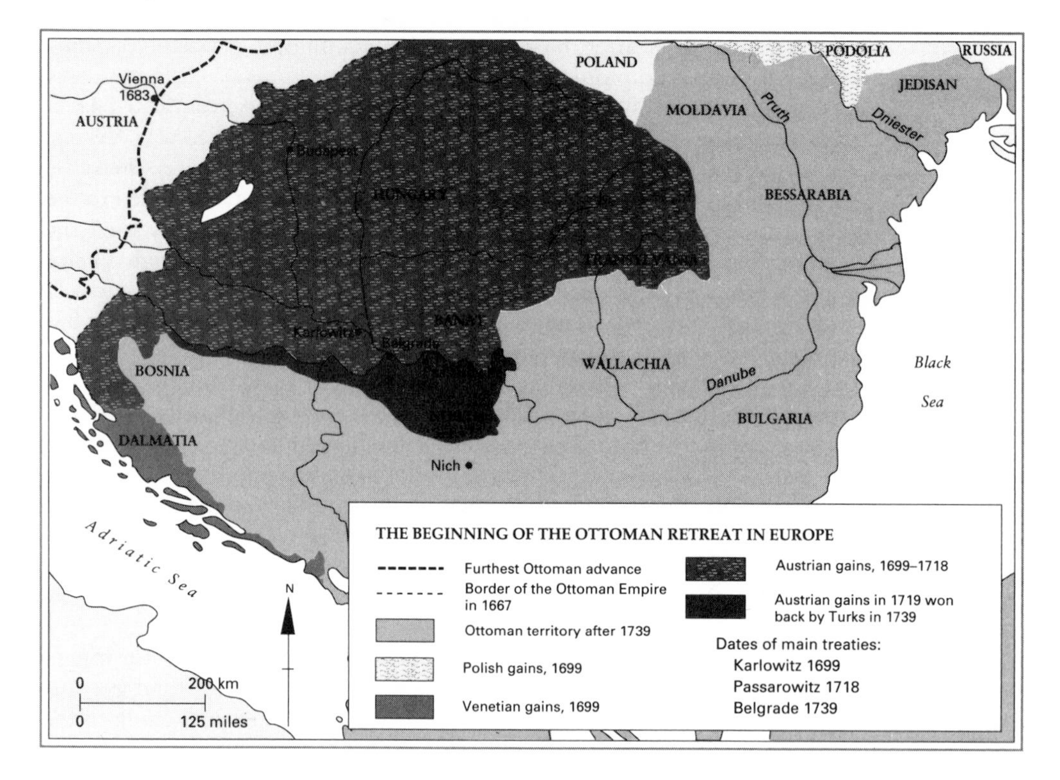

By 1800, the Russians had asserted a special protection over the Christian subjects of the Ottomans and had already tried promoting rebellion among them. In the eighteenth century, too, Ottoman rule ebbed in Africa and Asia; by the end of it, though forms might be preserved, the Ottoman caliphate was somewhat like that of the Abbasids in their declining days. Morocco, Algeria, Tunis, Egypt, Syria, Mesopotamia and Arabia were all in varying degrees independent or semi-independent.

It was not the traditional guardians of eastern Europe, the once-great Polish-Lithuanian commonwealth and the Habsburgs, who were the legatees of the Ottoman heritage, nor they who inflicted the most punishing blows as the Ottoman empire crumbled. The Poles were in fact nearing the end of their own history as an independent nation. The personal union of Lithuania and Poland had been turned into a real union of the two countries too late. In 1572, when the last king of the Jagiellon line died without an heir, the throne had become not only theoretically but actually elective. His successor was French and for the next century Polish magnates and foreign kings disputed each election, while their country was under grave and continuing pressure from Turks, Russians and Swedes. Poland prospered against these enemies only when they were embarrassed elsewhere. The Swedes descended on her northern territories during the Thirty Years' War and the last of the Polish coast was given up to them in 1660. Internal divisions had worsened, too; the Counter-Reformation brought religious persecution to the Polish Protestants and there were risings of Cossacks in the Ukraine and continuing serf revolts.

The election as king of the heroic John Sobieski was the last which was not the outcome of machinations by foreign rulers. He had won important victories and managed to preside over Poland's curious and highly decentralized constitution. The elected kings had very little legal power to balance that of the landowners. They had no standing army and could rely only on their own personal troops when factions among the gentry or magnates fell back on the practice of armed rebellion ('Confederation') to obtain their wishes. In the Diet, the central parliamentary body of the kingdom, a rule of unanimity stood in the way of any reform. Yet reform was badly needed, if a geographically ill-defined, religiously divided Poland, ruled by a narrowly selfish rural gentry, was to survive. Poland was a medieval community in a modernizing world.

John Sobieski could do nothing to change this. Poland's social structure was strongly resistant to reform. The nobility or gentry were effectively the clients of a few great families of extraordinary wealth. One clan, the Radziwills, owned estates half the size of Ireland and held a court which outshone that of Warsaw; the Potocki estates covered 6,500 square miles (roughly half the area of the Dutch Republic). The smaller landowners could not stand up to such grandees. Their estates made up less than a tenth of Poland in 1700. The million or so gentry who were legally the Polish 'nation' were for the most part poor, and therefore dominated by great magnates reluctant to surrender their power to arrange a confederation or manipulate a Diet. At the bottom of the pile were the peasants, some of the most miserable in Europe, engaged in unending battle with their landlords over feudal dues, over whom landlords still had in 1700 rights of life and death. The towns were powerless. Their total population was only half the size of the gentry and they had been devastated by the seventeenth-century wars. Yet Prussia and Russia also rested on backward agrarian and feudal infrastructures and survived. Poland was the only one of the three eastern states to go under completely. The principle of election blocked the emergence of Polish Tudors or Bourbons who could identify their own dynastic instincts of self-aggrandizement with those of the nation. Poland entered the eighteenth century under a foreign king, the Elector of Saxony, who was chosen to succeed John Sobieski in 1697, soon deposed by the Swedes, and then put back again on his throne by the Russians.

Russia was the coming new great power in the east. Her national identity had been barely discernible in 1500. Two hundred years later her potential was still only beginning to dawn on most western statesmen, though the Poles and Swedes were already alive to it. It now requires an effort to realize how rapid and astonishing was the appearance as a major force of what was to become one of the two most powerful states in the world. At the beginning of the European age, when only the ground-plan of the Russian future had been laid out by Ivan the Great, such an outcome was inconceivable, and so it long remained. The first man formally to bear the title of 'Tsar of all the Russias' was his grandson Ivan IV, crowned in 1547; and the conferment of the title at his coronation was meant to say that the Grand Prince of Muscovy had become an emperor ruling many peoples. In spite of a ferocious vigour which earned him his nickname 'the Terrible', he played no significant role in European affairs. So little was Russia known even in the next century that a French king could write to a Tsar, not knowing that the prince whom he addressed had been dead for ten years. The shape of a future Russia was slowly determined, and almost unnoticed in the West. Even after Ivan the Great, Russia had remained territorially ill-defined and exposed. The Turks had pushed into south-east Europe. Between them and Muscovy lay the Ukraine, the lands of the Cossacks, peoples who fiercely protected their independence. So long as they had no powerful neighbours, they found it easy to do so. To the east of Russia, the Urals provided a theoretical though hardly a realistic frontier. Russia's rulers have always found it easy to feel isolated in the middle of hostile space. Almost instinctively, they have sought natural frontiers at its edges or a protective glacis of clients.

The first steps had to be the consolidation of the gains of Ivan the Great which constituted the Russian heartland. Then came penetration of the wilderness of the north. When Ivan the Terrible came to the throne, Russia had a small Baltic coast and a vast territory stretching up to the White Sea, thinly inhabited by scattered and primitive peoples, but providing a route to the west; in 1584 the port of Archangel was founded. Ivan could do little on the Baltic front but successfully turned on the Tatars after they burned Moscow yet again in 1571. He drove them from Kazan and Astrakhan and won control of the whole length of the Volga, carrying Muscovite power to the Caspian.

The other great thrust which began in his reign was across the Urals, into Siberia, and was to be less one of conquest than of settlement. Even today, most of the Russian republic is in Asia, and for nearly two centuries a world as well as a European power was ruled by the Tsars and their successors. The first steps towards this outcome were an ironic anticipation of what was to be a theme of the major Siberian frontier in later times: the first Russian settlers across the Urals seem to have been political refugees from Novgorod. Among those who followed were others fleeing from serfdom (there were no serfs in Siberia) and aggrieved Cossacks. By 1600 there were Russian settlements as much as six hundred miles beyond the Urals, closely supervised by a competent bureaucracy out to assure the state tribute in furs. The rivers were the keys to the region, more important even than those of the American frontier. Within fifty years a man and his goods could travel by river with only three portages from Tobolsk, three hundred miles east of the Urals, to the port of Okhotsk, three thousand miles away. There he would be only four hundred miles by sea from Sakhalin, the northernmost of the major islands of the chain which makes up Japan – a sea-passage about as long as that from Land's End to Antwerp. By 1700 there were 200,000 settlers east of the Urals: it had by then been possible already to agree the treaty of Nerchinsk with the Chinese and some Russians, we are told, talked of the conquest of China.

The movement eastward was not much affected by the upheavals and dangers of the 'Time of Troubles' which followed Ivan's death, though in the west there were moments when the outlet to the Baltic was lost and when even Moscow and Novgorod were occupied by Lithuanians or Poles. Russia was still not a serious European power in the early seventeenth century. The then rising strength of Sweden was thrown against her and it was not until the great war of 1654–67 that the Tsars finally regained Smolensk and Little Russia, not to be lost again (and then only briefly) until 1812. Maps and treaties now began to define Russia in the west in a way which had some reality. By 1700, though she still had no Black Sea coast, her south-western frontier ran on the western side of the Dnieper for most of its length, embracing the great historic city of Kiev and the Cossacks who lived on the east bank. They had appealed to the Tsar for protection from the Poles and were granted special, semi-autonomous governmental arrangements which survived until Soviet times. Most Russian gains had been at the expense of Poland, long preoccupied with fighting off Turk and Swede. But Russian armies joined the Poles in alliance against the Ottomans in 1687; this was a historic moment, too, for it was the beginning of the classical Eastern Question which was to trouble European statesmen until 1918, when they found that the problem of deciding what limit, if any, should be placed upon Russian encroachment on the Ottoman empire had at last disappeared with the protagonists themselves.

The making of Russia was overwhelmingly a political act. The monarchy was its centre and motor; the country had no racial unity to pre-ordain its existence and precious little geographical definition to impose a shape. If it was united by Orthodoxy, other Slavs were Orthodox, too. The growth of the personal domain and power of the Tsars was the key to the building of the nation. Ivan the Terrible was an administrative reformer. Under him appeared the beginnings of a nobility owing military service in return for their estates, a development of a system employed by the princes of Muscovy to obtain levies to fight the Tatars. It made possible the raising of an army which led the King of Poland to warn the English queen, Elizabeth I, that if they got hold of western technical skills the Russians would be unbeatable; the danger was remote, but this was prescient.

From time to time there were set-backs, though the survival of the state does not seem in retrospect to have been at stake. The last Tsar of the house of Rurik

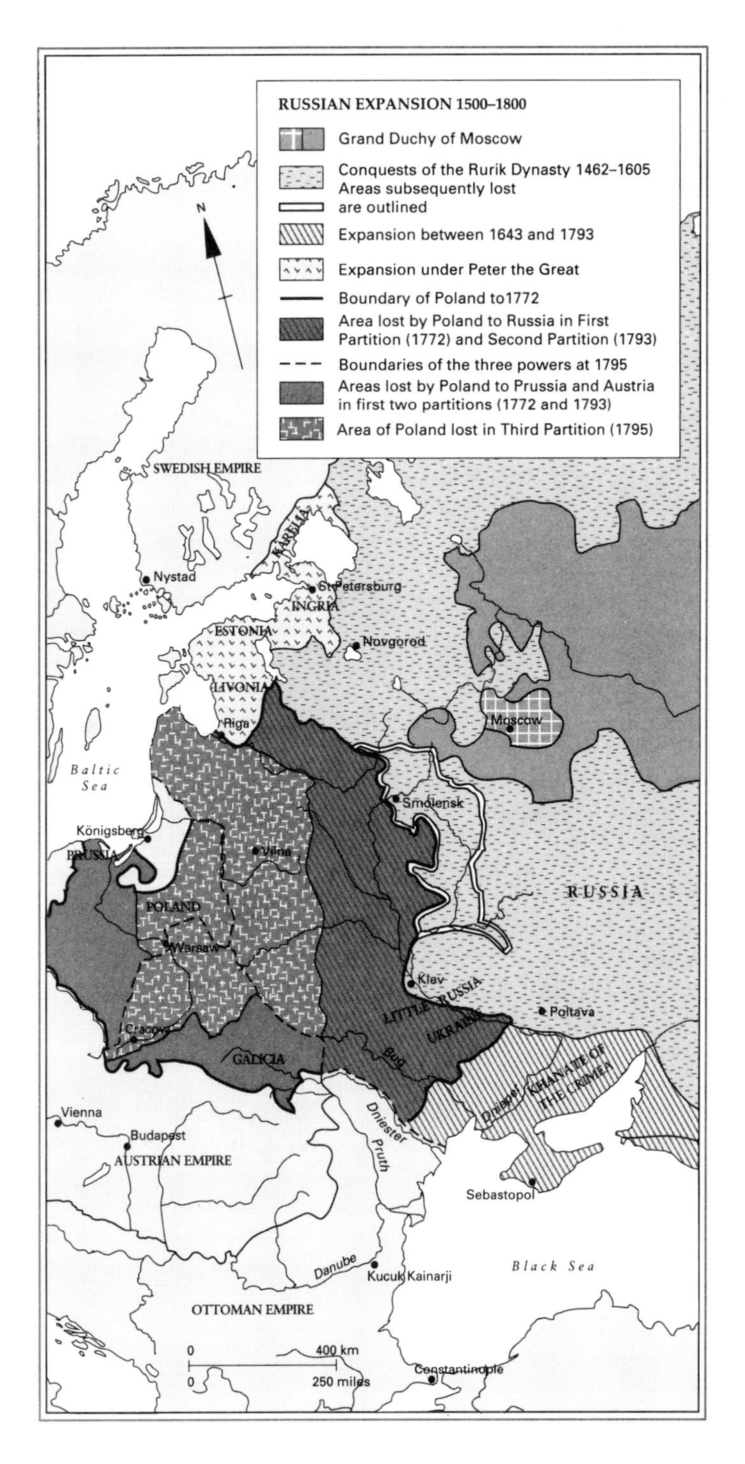

died in 1598. Usurpation and the disputing of the throne between noble families and Polish interventionists went on until 1613, when the first Tsar of a new house, Michael Romanov, emerged. Though a weak ruler who lived in the shadow of his dominating father, he founded a dynasty which was to rule Russia for three hundred years, until the tsarist state itself collapsed. His immediate successors fought off rival nobles and humbled the great ones among them, the boyars, who had attempted to revive a power curbed by Ivan the Terrible. Beyond their ranks the only potential internal rival was the Church. In the seventeenth century it was weakened by schism and in 1667 a great step in Russian history was taken when the patriarch was deprived after a quarrel with the Tsar. There was to be no Investiture Contest in Russia. After this time the Russian Church was structurally and legally subordinated to a lay official. Among believers there would emerge plenty of spontaneous doctrinal and moral opposition to current Orthodoxy, and there began the long-lived and culturally very important movement of underground religious dissent called the *raskol* which would eventually feed political opposition. But Russia was never to know the conflict of Church and state which was so creative a force in western Europe, any more than she was to know the stimulus of the Reformation.

The outcome was the final evolution of the enduring Russian governmental form, tsarist autocracy. It was characterized by the personification in the ruler of a semi-sacrosanct authority unlimited by clear legal checks, by an emphasis on the service owed to him by all subjects, by the linking of landholding to this idea, by the idea that all institutions within the state except the Church derived from it and had no independent standing of their own, by the lack of a distinction of powers and the development of a huge bureaucracy, and by the paramountcy of military needs. These qualities, as the scholar who listed them pointed out, were not all present at the start, nor were all of them equally operative and obvious at all times. But they mark tsardom off from monarchy in western Christendom where, far back in the Middle Ages, towns, estates of the realm, guilds and many other bodies had established the privileges and liberties on which later constitutionalism was to be built. In old Muscovy, the highest official had a title which meant 'slave' or 'servant' at a time when, in neighbouring Poland-Lithuania, his opposite number was designated 'citizen'. Even Louis XIV, though he might believe in Divine Right and aspire to unrivalled power, always conceived it to be a power explicitly restricted by rights, by religion, by divinely ordained law. Though his subjects knew he was an absolute monarch, they were sure he was not a despot. In England an even more startingly different monarchy was developing, one under the control of Parliament. Divergent from one another though English and French monarchical practice might be, they both accepted practical and theoretical limitations inconceivable to tsardom; they bore the stamp of a western tradition Russia had never known. For the whole of its existence the Russian autocracy was to be in the West a byword for despotism.

Yet it suited Russia. Moreover, the attitudes which underlay it seem in some measure to suit Russia still. Eighteenth-century sociologists used to suggest that big, flat countries favoured despotism. This was over-simple, but there were always latent centrifugal tendencies in a country so big as Russia, embracing so many natural regions and so many different peoples. The Tsars' title, significantly, was 'Tsar of all the Russias', and to this day events have reflected this diversity. Russia had always to be held together by a strong pull towards the centre if the divergences within it were not to be exploited by the enemies on the borders.

The humbling of the boyars left the ruling family isolated in its eminence. The Russian nobility was gradually brought to depend on the state on the grounds that nobility derived from service, which was indeed often rewarded in the seventeenth century with land and later with the grant of serfs. All land came to be held on the condition of service to the autocracy as defined in a Table of Ranks in 1722. This effectively amalgamated all categories of nobility into a single class. The obligations laid on noblemen by it were very large, often extending to a man's lifetime, though in the eighteenth century they came to be progressively diminished and were finally removed altogether. Nevertheless, service still continued to be the route to an automatic ennoblement, and Russian nobles never acquired quite such independence of their monarch as those of other countries. New privileges were conferred upon them but no closed caste emerged. Instead nobility grew hugely by new accessions and by natural increase. Some of its members were very poor, because there was neither primogeniture nor entail in Russia and property could be much subdivided in three or four generations. Towards the end of the eighteenth century most nobles owned fewer than a hundred serfs.

Of all imperial Russia's rulers the one who made the most memorable use of the autocracy and most deeply shaped its character was Peter the Great. He came to the throne as a ten-year-old child and when he died something had been done to Russia which could never be quite eradicated. In one way he resembled twentieth-century strong men who have striven ruthlessly to drag traditional societies into modernity, but he was very much a monarch of his own day, his attention focused on victory in war – Russia was only at peace for one year in his entire reign – and he accepted that the road to that goal ran through westernizing and modernizing. His ambition to win a Russian Baltic coast supplied the driving force behind the reforms which would open the way to it. That he should be sympathetic to such a course may owe something to his childhood, growing up as he did in the 'German' quarter of Moscow where foreign merchants and their retinues lived. A celebrated pilgrimage he made to western Europe in 1697–8 showed that his interest in technology was real. Probably in his own mind he did not distinguish the urge to modernize his countrymen from the urge to free them forever from the fear of their neighbours. Whatever the exact balance of his motives, his reforms have ever since served as something of an ideological touchstone; generation after generation of Russians were to look back with awe and ponder what he had done and its meaning for Russia. As one of them wrote in the nineteenth century, 'Peter the Great found only a blank page ... he wrote on it the words Europe and Occident.'

The 'bronze horseman': a nineteenth-century drawing of the monument to Peter the Great in St Petersburg.

His territorial achievement is the easiest to assess. Though he sent expeditions off to Kamchatka and the oases of Bokhara and ceased to pay to the Tatars a tribute levied on his predecessors, his driving ambition was to reach the sea to the west. For a while he had a Black Sea fleet and annexed Azov, but had to abandon it later because of distractions elsewhere, from the Poles and, above all, the Swedes. The wars with Sweden for the Baltic outlet were a struggle to the death. The Great Northern War, as contemporaries termed the last of them, began in 1700 and lasted until 1721. The world recognized that something decisive had happened when in 1709 the Swedish king's army, the best in the world, was destroyed far away from home at Poltava, in the middle of the Ukraine where its leader had sought to find allies among the Cossacks. The rest of Peter's reign drove home the point and at the peace Russia was established firmly on the Baltic coast, in Livonia, Estonia and the Karelian isthmus. Sweden's days as a great power were over; she had been the first victim of a new one.

A few years before this, the French *Almanach Royale* for the first time listed the Romanovs as one of the reigning families of Europe. Victory had opened the way to further contact with the West, and Peter had already anticipated the peace by beginning in 1703 to build, on territory captured from the Swedes, St Petersburg, the beautiful new city which was to be for two centuries the capital of Russia. The political and cultural centre of gravity thus passed from the isolation of Muscovy to the edge of Russia nearest the developed societies of the West. Now the westernizing of Russia could go ahead more easily. It was a deliberate break with the past.

Even Muscovy, of course, had never been completely isolated from Europe. A pope had helped to arrange Ivan the Great's marriage, hoping he would turn to the western Church. There was always intercourse with the neighbours, the Roman Catholic Poles, and English merchants had made their way to Moscow under Elizabeth I, where to this day they are commemorated in the Kremlin by the presence of a magnificent collections of the work of English silversmiths. Trade continued, and there also came to Russia the occasional foreign expert from the West. In the seventeenth century the first permanent embassies from European monarchs were established. But there was always a tentative and suspicious response among Russians; as in later times, efforts were made to segregate foreign residents.

Peter threw this tradition aside. He wanted experts – shipwrights, gun-founders, teachers, clerks, soldiers – and he gave them privileges accordingly. In administration he broke with the old assumption of inherited family office and tried to institute a bureaucracy selected on grounds of merit. He set up schools to teach technical skills and founded an Academy of Sciences, thus introducing the idea of science to Russia, where all learning had hitherto been clerical. Like many other great reformers he also put much energy into what might be thought superficialities. Courtiers were ordered to wear European clothes; the old long beards were cut back and women were told to appear in public in German fashions. Such psychological shocks were indispensable in so backward a country. Peter was virtually without allies in what he was trying to do and in the end such things as he achieved had to be driven through. They rested on his autocratic power and little else. The old Duma of the boyars was abolished and a new senate of appointed men took its place. Those who resisted were ruthlessly broken, but it was less easy for Peter to dispose of a conservative cost of mind; he had at his disposal only an administrative machine and communications that would seem inconceivably inadequate to any modern government.

The most striking sign of successful modernization was Russia's new military power. More complicated tests are harder to come by. The vast majority of Russians were untouched by Peter's educational reforms, which only obviously affected technicians and a few among the upper class. The result was a fairly westernized higher nobility, focused at St Petersburg; by 1800 its members were largely French-speaking and in touch with the currents of thought which arose in western Europe. But they were often resented by the provincial gentry and formed a cultural island in a backward nation. The mass of the nobility for a long time did not benefit from the new schools and academies. Further down the social scale, the Russian masses remained illiterate; those who learnt to read did so for the most part at the rudimentary level offered by the teaching of the village priest, often only one generation removed from illiteracy himself. A literate Russia had to wait for the twentieth century.

Her social structure, too, tended more and more to mark off Russia. She was to be the last country in Europe to abolish serfdom; among Christian countries only Ethiopia and the United States kept bonded labour for longer. While the eighteenth century saw the institution weakening almost everywhere, in Russia it spread. This was largely because labour was always scarcer than land; significantly, the value of a Russian estate was usually assessed in the number of 'souls' – that is, serfs – tied to it, not its extent. The number of serfs had begun to go up in the seventeenth century, when the Tsars found it prudent to gratify nobles by giving them land, some of which already had free peasants settled on it. Debt tied them to their landlords and many of them entered into bondage to the estate to work it off. Meanwhile, the law imposed more and more restrictions on the serf and rooted the structure of the state more and more in the economy. Legal powers to recapture and restrain serfs were steadily increased and landlords had been given a special interest in using such powers when Peter had made them responsible for the collection of the poll-tax and for military conscription. Thus, economy and administration were bound together in Russia more completely than in any western country. Russia's aristocrats tended to become hereditary civil servants, carrying out tasks for the Tsar.

Formally, by the end of the eighteenth century, there was little that a lord could not do to his serfs short of inflicting death on them. If they were not obliged to carry out heavy labour services, money dues were levied upon them almost arbitrarily. There was a high rate of desertion, serfs making for Siberia or even volunteering for the galleys. About a half of the Russian people were in bondage to their lords in 1800, a large number of the rest owing almost the same services to the Crown and always in danger of being granted away to nobles by it.

As new lands were annexed, their populations, too, passed into serfdom even if they had not known it before. The result was a huge inertia and a great rigidifying of society. By the end of the century, Russia's greatest problem for the next hundred years was already there: what to do with so huge a population when both economic and political demands made serfdom increasingly intolerable, but when its scale presented colossal problems of reform. It was like the man riding an elephant; it is all right so long as he keeps going but there are problems when he wants to get off.

Servile labour had become the backbone of the economy. Except in the famous Black Earth zone, only beginning to be opened up in the eighteenth century, Russian soil is by no means rich, and even on the best land farming methods were poor. It seems unlikely that production ever kept pace with population until the twentieth century though periodic famine and epidemics were the natural restoratives of balance. Population nearly doubled in the eighteenth century, about seven million of the thirty-six million or so at which it stood at the end having been acquired with new territories, the rest having accumulated by natural increase. This was a faster rate of growth than any other European country. Of this population, only about one in twenty-five at most were townsmen.

Yet the Russian economy made striking progress during the century and was unique in utilizing serfdom to industrialize. Here, it may be thought, was one of Peter's unequivocal successes; though there had been beginnings under the first two Romanovs, it was he who launched Russian industrialization as a guided movement. True, the effect was not quickly apparent. Russia's starting level was very low, and no eighteenth-century European economy was capable of rapid growth. Though grain production went up and the export of Russian cereals (later a staple of Russian foreign trade) began in the eighteenth century, it was done by the old method of bringing more land under cultivation and perhaps by the more successful appropriation of the surplus by the landlord and tax-collector. The peasant's consumption declined. This was to be the story throughout most of the imperial era and sometimes the load was crushing: it has been estimated that taxes took 60 per cent of the peasant's crop under Peter the Great. The techniques were not there to increase productivity and the growing rigidity of the system held it down more and more firmly. Even in the second half of the nineteenth century the typical Russian peasant wasted what little time was left to him after work for his lord by trudging around the collection of scattered strips which made up his holding. Often he had no plough, and crops had to be raised from the shallow scratching of the soil which was all that was possible.

None the less, this agricultural base somehow supported both the military effort which made Russia a great power, and the first phase of her industrialization. By 1800 Russia produced more pig-iron and exported more iron ore than any other country in the world. Peter, more than any other man, was responsible for this. He grasped the importance of Russia's mineral resources and built the administrative apparatus to grapple with them. He initiated surveys and imported the miners to exploit them. By way of incentive, the death penalty was prescribed for landlords who concealed mineral deposits on their estates or tried to prevent their use. Communications were developed to allow access to these resources and slowly the centre of Russian industry shifted towards the Urals. The rivers were crucial. Only a few years after Peter's death the Baltic was linked by water to the Caspian.

Manufacturing grew up around the core of extractive mineral and lumber industry which ensured Russia a favourable balance of trade for the whole century. Less than a hundred factories in Peter's reign became more than three thousand by 1800. After 1754, when internal customs barriers were abolished, Russia was the largest free-trade area in the world. In this, as in the granting of serf labour or of monopolies, the state continued to shape the Russian economy; Russian industry did not emerge from free enterprise, but from regulation, and this had to be, for industrialization ran against the grain of Russian social fact. There might be no internal customs barriers, but nor was there much long-distance internal trade. Most Russians lived in 1800 as they had done in 1700, within self-sufficient local communities depending on their artisans for a small supply of manufactures and hardly emerging into a money economy. Such 'factories' as there were seem sometimes to have been little more than agglomerations of artisans. Over huge areas labour service, not rent, was the basis of tenure. Foreign trade was still mainly in the hands of foreign merchants. Moreover, though state grants to exploit their resources and allocations of serfs encouraged mine-owners, the need of such encouragement shows that the stimuli for maintained growth which were effective elsewhere were lacking in Russia.

After Peter, in any case, there was a notable flagging of state innovation. The impetus could not be maintained; there were not enough educated men to allow the bureaucracy to keep up the pressure once his driving power had gone. Peter had not named a successor (he had his own son tortured to death). Those who followed him faced a renewed threat of hostility from the great noble families without his force of character and the terror he had inspired. The direct line was broken in 1730 when Peter's grandson died. Yet factional quarrels could be exploited by monarchs, and his replacement by his niece, Anna, was something of a recovery for the Crown. Though put on the throne by the nobles who had dominated her predecessor, she quickly curbed them. Symbolically, the court returned to St Petersburg from Moscow, to which (to the delight of the conservatives) it had gone after Peter's death. Anna turned to foreign-born ministers for help and this worked well enough until her death in 1740. Her successor and infant grand-nephew was within a year set aside (to be kept in prison until murdered more than twenty years later) in favour of Elizabeth, daughter of Peter the Great, who relied on the support of the Guards regiments and Russians irritated by foreigners. She was succeeded in 1762 by a nephew who reigned barely six months before he was forced to abdicate. The mistress of the overmighty subject who subsequently murdered the deposed Tsar was the new Tsarina and widow of the deposed victim, a German princess who became Catherine II and known, like Peter, as 'The Great'.

The glitter with which Catherine subsequently surrounded herself masked a great deal and took in many of her contemporaries. Among the things it almost hid was the bloody and dubious route by which she came to the throne. It is probably true, though, that she rather than her husband might have been the victim if she had not struck first. In any case, the circumstances of her accession and of those of her predecessors showed the weakening the autocracy had undergone since Peter. The first part of her reign was a ticklish business; powerful interests existed to exploit her mistakes and for all her identification with her new country (she had renounced her Lutheran religion to

The 'poor girl with three or four dresses' of thirty years earlier: Catherine II, empress of Russia, painted at the end of her life, after she had rewarded her adopted country by winning more new territory for it than any ruler since Peter the Great.

become Orthodox) she was a foreigner. 'I shall perish or reign,' she once said, and reign she did, to great effect.

Though Catherine's reign was more spectacular than that of Peter the Great, its innovatory force was less. She, too, founded schools and patronized the arts and sciences. The difference was that Peter was concerned with practical effect; Catherine rather to associate the prestige of enlightened thinkers with her court and legislation. The forms were often forward-looking while the reality was reactionary. Close observers were not taken in by legislative rhetoric, the reality was shown by the exile of the young Radischev, who had dared to criticize the regime and has been seen as Russian's first dissentient intellectual. Such reforming impulses as Catherine showed perceptibly weakened as the reign went on and foreign considerations distracted her.

Her essential caution was well shown by her refusal to tamper with the powers and privileges of the nobility. She was the Tsarina of the landlords, giving them greater power over the local administration of justice and taking away from their serfs the right to petition against their masters. Only twenty times in Catherine's thirty-four year reign did the government act to restrain landlords abusing their powers over their serfs. Most significant of all, the obligation to service was abolished in 1762 and a charter of rights was later given to the nobility which sealed a half-century of retreat from Peter's policies towards them. The gentry were exempted from personal taxation, corporal punishment and billeting, could be tried (and be deprived of their rank) only by their peers, and were given the exclusive right to set up factories and mines. The landowner was in a sense taken into partnership by the autocracy.

In the long run this was pernicious. Under Catherine, Russia began to truss herself more and more tightly in the corset of her social structure at a time when other countries were beginning to loosen theirs. This would increasingly unfit Russia to meet the challenges and changes of the next half-century. One sign of trouble was the scale of serf revolt. This had begun in the seventeenth century, but the most frightening and dangerous crisis came in 1773, the rebellion of Pugachev, the worst of the great regional uprisings which studded Russian agrarian history before the nineteenth century. Later, better policing would mean that revolt was usually local and containable, but it continued through almost the whole of the imperial era. Its recurrence is hardly surprising. The load of labour services piled on the peasant rose sharply in the Black Earth zone during Catherine's reign. Soon critics would appear among the literate class and the condition of the peasant would be one of their favourite themes, thus providing an early demonstration of a paradox evident in many developing countries in the next two centuries. It was becoming clear that modernization was more than a matter of technology; if you borrowed western ideas, they could not be confined in their effect. The first critics of Orthodoxy and autocracy were beginning to appear. Eventually the need to preserve an ossifying social system would virtually bring to a halt the changes which Russia needed to retain the place that courageous and unscrupulous leadership and seemingly inexhaustible military manpower had given her.

By 1796, when Catherine died, this place was indeed impressive. The most solid ground of her prestige was her armies and diplomacy. She had given Russia seven million new subjects. She said she had been well treated by Russia, to which she had come 'a poor girl with three or four dresses', but that she had paid her debts to it with Azov, the Crimea and the Ukraine. This was in the line of her predecessors. Even when the monarchy was weak, the momentum of Peter's reign carried the foreign policy of Russia forward along two traditional lines of thrust, into Poland and towards Turkey. It helped that Russia's likely opponents laboured under growing difficulties for most of the eighteenth century. Once Sweden was out of the running, only Prussia or the Habsburg empire could provide a counterweight, and since these two were often at loggerheads Russia could usually have her own way both over an ailing Poland and a crumbling Ottoman empire.

In 1701 the Elector of Brandenburg, with the consent of the emperor, became a king; his kingdom, Prussia, was to last until 1918. The Hohenzollern dynasty had provided a continuous line of electors since 1415, steadily adding to their ancestral domains, and Prussia, then a duchy, had been united to Brandenburg in the sixteenth century, after a Polish king had ousted the Teutonic Knights who ruled it. Religious toleration had been Hohenzollern policy after an elector was converted to Calvinism in 1613, while his subjects remained Lutheran. One problem facing the Hohenzollerns was the spread and variety of their lands, which stretched from East Prussia to the west bank of the Rhine. The Swedes provided infilling for this scatter of territories in the second half of the seventeenth century, though there were setbacks even for the 'Great Elector', Frederick William, the creator of the Prussian standing army and winner of the victories against the

Swedes which were the basis of the most enduring military tradition in modern European history. Arms and diplomacy continued to carry forward his successor to the kingly crown he coveted and to participation in the Grand Alliance against Louis XIV. Prussia was by that fact alone clearly a power. This imposed a heavy cost but careful housekeeping had again built up the best army and one of the best-filled treasuries in Europe by 1740, when Frederick II came to the throne.

He was to be known as 'The Great' because of the use he made of them, largely at the expense of the Habsburgs and the kingdom of Poland, though also at the expense of his own people whom he subjected to heavy taxation and exposed to foreign invasion. It is difficult to decide whether he was more or less attractive than his brutal father (whom he hated), for he was malicious, vindictive and completely without scruple. But he was highly intelligent and much more cultivated, playing and composing for the flute, and enjoying the conversation of clever men. He was also utterly devoted to the interests of his dynasty, which he saw as the extension of its territories and the magnification of its prestige.

Frederick gave up some possessions too remote to be truly incorporated in the state, but added to Prussia more valuable territories. The opportunity for the conquest of Silesia came when the emperor died in 1740 leaving a daughter, whose succession he had sought to assure but whose prospects were uncertain. This was Maria Theresia. She remained Frederick's most unforgiving opponent until her death in 1780 and her intense personal dislike for him was fully reciprocated. A general European war 'of the Austrian Succession' left Prussia holding Silesia. It was not to be lost in later wars and in the last year of his reign Frederick formed a League of German Princes to thwart the attempts of Maria Theresia's son and successor, Joseph II, to negotiate the acquisition of Bavaria as a recompense for the Habsburg inheritance.

This episode matters more to European history as a whole than might be expected of a contest for a province, however rich, and for the leadership of the princes of Germany. At first sight a reminder of how alive still in the eighteenth century were the dynastic preoccupations of the past, it is also, and more importantly, the opening of a theme with a century of life to it, and consequences great for Europe. Frederick launched a struggle between Habsburg and Hohenzollern for the mastery of Germany which was only to be settled in 1866. That is further ahead than may be usefully considered at present; but this context gives perspective to the Hohenzollern appeal to German patriotic sentiment against the emperor, many of whose essential interests were non-German. There would be periods of good relations, but in the long struggle which began in 1740 Austria's great handicap would always be that she was both more and less than a purely German state.

The disadvantages of the spread of her interests were made very obvious during the reign of Maria Theresia. The Austrian Netherlands were an administrative nuisance rather than a strategic advantage, but it was in the east that the worst distractions from German problems arose, and they became increasingly pressing as the second half of the century brought more and more clearly into view the likelihood of a long and continuing confrontation with Russia over the fate of the Ottoman empire. For thirty years or so Russo-Turkish relations had been allowed to slumber with only occasional minor eruptions over the building of a fort or the raids of the Crimean Tatars, one of the peoples originating in a fragment of the Golden Horde and under Turkish suzerainty. Then, between 1768 and 1774, Catherine fought her most successful war. A peace treaty with the Ottomans signed in an obscure Bulgarian village called Kutchuk Kainarji was one of the most important of the whole century. The Turks gave up their suzerainty over the Crimean Tatars (an important loss both materially, because of their military manpower, and morally, because this was the first Islamic people over which the Ottoman empire ceded control), and Russia took the territory between the Bug and Dnieper, together with an indemnity, and the right of free navigation on the Black Sea and through the straits. In some ways the most pregnant with future opportunity of the terms, was a right to take up with the Turks the interests of 'the church to be built in Constantinople and those who serve it'. This meant that the Russian government was recognized as the guarantor and

The first Partition of Poland: a contemporary print shows the east European monarchs dividing the spoils.

protector of new rights granted to the Greek – that is, Christian – subjects of the Sultan. It was to prove a blank cheque for Russian interference in Turkish affairs.

This was a beginning, not an end. In 1783 Catherine annexed the Crimea. Another war with the Turks carried her frontier up to the line of the Dniester. The next obvious boundary ahead was the Pruth, which meets the Danube a hundred miles or so from the Black Sea. The possibility of Russia's installation at the mouth of the Danube was to remain an Austrian nightmare, but the danger which appeared in the east before this was that Russia would swallow Poland. With the eclipse of Sweden, Russia had effectively had her own way at Warsaw. She left her interests to be secured through a compliant Polish king. The factions of the magnates and their quarrels blocked the road to reform and without reform Polish independence would be a fiction because effective resistance to Russia was impossible. When there seemed to be for a moment a slight chance of reforms these were checkmated by skilful Russian exploitation of religious divisions to produce confederations which speedily reduced Poland to civil war.

The last phase of Poland's independent history had opened when the Turks declared war on Russia in 1768, with the excuse that they wished to defend Polish liberties. Four years later, in 1772, came the first 'Partition' of Poland, in which Russia, Prussia and Austria shared between them about one-third of Poland's territory and one-half of her inhabitants. The old international system which had somewhat artificially preserved Poland had now disappeared. After two more partitions Russia had done best on the map, absorbing something like 180,000 square miles of territory (though in the next century it would be clear that a population of dissident Poles by no means an unambiguous gain) and Prussia also did well, emerging from the division of booty with more Slav than German subjects. The transformation of eastern Europe since 1500 was complete and the stage was set for the nineteenth century, when there would be no booty left to divert Austria and Russia from the Ottoman succession problem. Meanwhile, independent Poland disappeared for a century and a quarter.

Catherine rightly claimed to have done much for Russia, but she had only deployed a strength already apparent. Even in the 1730s, a Russian army had been as far west as the Neckar; in 1760 another marched into Berlin. In the 1770s there was a Russian fleet in the Mediterrancan. A few years later a Russian army was campaigning in Switzerland and, after twenty years, another entered Paris. The paradox at the heart of such evidences of strength was that this military power was based on a backward social and economic structure. Perhaps this was inherent in what Peter had done. The Russian state rested on a society with which it was fundamentally incompatible, and later Russian critics would make much of this theme. Of course, this did not mean that the clock could be put back. The Ottoman empire was for ever gone as a serious competitor for power while Prussia's emergence announced a new age as much as did Russia's. The future international weight of the United Provinces and Sweden had been unimaginable in 1500, but their importance, too, had come and gone by 1800; they were then still important nations, but of the second rank. France was still to be a front-rank power in an age of national states as she had been in the days of sixteenth-century dynastic rivalry; indeed, her power was relatively greater and the peak of her dominance in western Europe was still to come. But she faced a new challenger, too, and one which had already defeated her. From the little English kingdom of 1500, cooped up in an island off the coast of Europe under an upstart dynasty, had emerged the world power of Great Britain.

This was a transformation almost as surprising and sudden as Russia's. It transcended the old categories of European diplomacy quite as dramatically. In three hundred years, the major zones of European conflict and dispute had migrated from the old battlegrounds of Italy, the Rhine and the Netherlands, moving from them to central and eastern Germany, the Danube valley, Poland and Carpathia, and the Baltic, but also, greatest change of all, across the oceans. A new age had indeed opened, signalled not only by the remaking of Eastern Europe, but in the wars of Louis XIV, the first world wars of the modern era, imperial and oceanic in their scope.

4
Europe's Assault on the World

The change which came about in world history after 1500 is quite without precedent. Never before had one culture spread over the whole globe. From the earliest observable stage of prehistory, the tendency had always been towards differentiation. Now the cultural tide was turning. The essentials of what was happening were evident even by the end of the eighteenth century. Including Russia, European nations by then already laid claim to more than half the world's land surface and, in varying degree, actually controlled about a third of it. To the western hemisphere they had by then already transplanted settler populations large enough to constitute new centres of civilization; a new nation had emerged from former British territory in North America, and to the south the Spaniards had destroyed two mature civilizations to implant their own. To the east, the story was different, but equally impressive. Once past the Cape of Good Hope (where something like 20,000 Dutch lived), an Englishman travelling on an East Indiaman in 1800 would not touch at European colonial communities like those of the Americas unless he wandered as far off course as Australia, just beginning to receive its settlers. But in East Africa, Persia, India, Indonesia he would find Europeans come to do business and then, in the long or short run, planning to return home to enjoy the profits. They could even be found in Canton, or, in very small numbers, in the closed island kingdom of Japan. Only the interior of Africa, still protected by disease and climate, seemed impenetrable.

The remarkable transformation thus begun (and to go much further) was almost entirely a one-way process. Europeans went out to the world, it did not come to them. Few non-Europeans other than Turks entered Europe except as exotic imports or slaves. The Arabs and Chinese were by no means unskilful sailors. They had made oceanic voyages and knew about the compass, while the island peoples of the Pacific made long sea crossings on their mysterious errands, but the ships which came round the Horn or the tip of Africa to Atlantic ports were European and homeward bound, not Asiatic ones.

This was a great transformation of world relationships and it was the work of Europeans. Underpinning it lay layer upon layer of exploration, enterprise, technical advantage and governmental patronage. The trend seemed irreversible by the end of the eighteenth century and, in a sense, so it was to prove, even if direct European rule was to dissolve more quickly than it was built up. No civilization had been more rapidly and dramatically successful, so untroubled in its expansion by any but temporary and occasional setbacks.

One advantage possessed by Europeans had been the powerful motives they had to succeed. The major thrust behind the Age of Reconnaissance had been their wish to get into easier and more direct contact with the Far East, the source of things badly wanted in Europe, at a time when the Far East wanted virtually nothing Europe could offer in exchange. When Vasco da Gama showed what he brought to give to a king, the inhabitants of Calicut laughed at him; he had nothing to offer which could compare with what Arab traders already brought to India from other parts of Asia. It was indeed just the known superiority of so much of the civilization of the Orient that spurred Europeans on to try to reach it on some more regular and assured basis than the occasional trip of a Marco Polo. Coincidentally, China, India and Japan were at something like a cultural peak in the sixteenth and seventeenth centuries. The land blockade of Europe by the Turk made them even more attractive to Europeans than they had been before. There were huge profits to be made and great efforts could be justified.

If the expectation of reward is a good recipe for high morale, so is the expectation of success. By 1500 enough had been done for the business of exploration and new enterprise to be attacked confidently; there was a cumulative factor at work, as each successful voyage added both to knowledge and to the certainty that more could be done. As time went by, there would also be profits for the financing of future expansion. Then there was the psychological asset of Christianity. Soon after the establishment of settlement this found a vent in missionary enterprises, but it was always present as a cultural fact, assuring the European of his superiority to the peoples with which he began to come into touch for the first time. In the next four centuries, it was often to have disastrous effects. Confident in the possession of the true religion, Europeans were impatient and contemptuous of the values and achievements of the peoples and civilizations they disturbed. The result was always uncomfortable and often brutal. It is also true that religious zeal could blur easily into less avowable motives. As the greatest Spanish historian of the American conquests put it when describing why he and his colleagues had gone to the Indies, they thought 'to serve God and his Majesty, to give light to those who sat in darkness and to grow rich as all men desire to do'.

Greed quickly led to the abuse of power, to domination and exploitation by force. In the end this led to great crimes - though they were often committed unconciously. It sometimes brought about the destruction of whole societies, but this was only the worst aspect of a readiness to dominate present from the beginning in European enterprise. The adventurers who first reached the coasts of India were soon boarding Asian merchantmen, torturing and slaughtering their crews and passengers, looting their cargoes and burning the ravaged hulks. Europeans could usually exact what they wanted in the end because of a technical superiority which exaggerated the power of their tiny numbers and for a few centuries turned the balance against the great historic agglomerations of population.

The next Portuguese captain after da Gama to go there provided a fitting symbol of this by bombarding Calicut. A little later, when in 1517 the Portuguese reached Canton, they fired a salute as a gesture of friendship and respect, but the noise of their guns horrified the Chinese (who at first called them *folangki* – a remote corruption of 'Franks'). These weapons were much more powerful than anything China had. There had long been guns in Asia, and the Chinese had known about gunpowder centuries before Europe, but the technology of artillery had stood still there. European craftsmanship and metallurgy had in the fifteenth century made great strides, producing weapons better than any available elsewhere in the world. There were still more dramatic improvements to come, so that the comparative advantage of Europeans was to increase, right down to the twentieth century. This progress had been and was to be, again, paralleled in other fields, notably by the developments in shipbuilding and handling which have already been touched upon. When combined, such advances produced the remarkable weapon with which Europe opened up the world, the sailing-ship which was a gun-carrier. Again, evolution was far from complete in 1517, but already the Portuguese had been able to fight off the fleets organized by the Turks to keep them out of the Indian Ocean. (The Turks had more success in keeping control of the Red Sea, because in those narrower waters the oar-propelled galley which closed with its enemies to grapple and board retained more of its usefulness. Even there, though, the Portuguese were able to penetrate as far as the Suez isthmus.) The Chinese war-junk would do no better than the rowed galley. The abandonment of the oar for propulsion and the mounting, broadside, of large numbers of guns, enormously multiplied the value of Europe's scanty manpower.

This advantage was clear to contemporaries. As early an 1481 the pope forbade the sale of arms to Africans. The Dutch in the seventeenth century were very anxious to keep to themselves the secrets of gun-founding and not to allow them to pass into the hands of Asiatics. Yet pass they did. There had been Turkish gunners in India in the fifteenth century and before they reached China the Portuguese were supplying the Persians with cannon and teaching them how to cast more in order to embarrass the Turks.

The menacing European: a Benin ivory salt-cellar of the seventeenth century, a ship surmounting it, and Portuguese soldiers gathered about its base.

In the seventeenth century their knowledge of gun-founding and gunnery was one of the attractions which kept the Jesuit Fathers in favour with the Chinese authorities.

Yet although, as the Dutch feared, the knowledge of up-to-date gun-founding penetrated oriental societies it did not offset the European advantage. Chinese artillery remained inferior in spite of the Jesuits' training. There was more to the technological disparity of Europe and the world than mere know-how. One of the assets Europe enjoyed at the beginning of her era was not only knowledge, but an attitude to knowledge different from that of other cultures. There was a readiness to bring it to bear upon practical problems, a technological instinct for the useful. In it lay the roots of another psychological characteristic of Europeans, their growing confidence in the power to change things. Here, perhaps, was the most fundamental difference of all between them and the rest of the world. Europe was open to the future and its possibilities in a way that other cultures were not. On this confidence would long rest a psychological advantage of the greatest importance. Even in 1500 some Europeans had seen the future – and it worked.

Africa and Asia were the first targets against which Europeans' advantages were deployed. In these continents, the Portuguese led for a century and more. They figured so largely and were so successful in the opening of routes to the East that their king took the title (confirmed by the pope) 'Lord of the conquest, navigation and commerce of India, Ethiopia, Arabia and Persia', which sufficiently indicates both the scope and the eastern bias of Portuguese enterprise, but is slightly misleading in its reference to Ethiopia, with which Portuguese contacts were small. Penetration of Africa was impossible on any more than a tiny and hazardous basis. The Portuguese suggested that God had especially set a barrier about the African interior in its mysterious and noxious diseases (which were to hold Europeans at bay until the end of the nineteenth century). Even the coastal stations of West Africa were unhealthy and could only be tolerated because of their importance in the slave trade and the substructure of long-range commerce. The East African stations were less unhealthy, but they, too, were of interest not as jumping-off points for the interior, but because they were part of a commercial network created by Arabs, whom the Portuguese deliberately harried so as to send up the cost of the spices purveyed by way of the Red Sea and the Middle East to the Venetian merchants of the eastern Mediterranean. The successors of the Portuguese were to leave the interior of Africa alone as they had done, and the history of that continent for another two centuries was still to move largely to its own rhythms in the obscure fastnesses of its forests and savannahs, its inhabitants only coming into corrosive and stimulating contact with Europeans at its fringes. It is also true, though, that the opening of the European age in Asia showed that none of the powers concerned was in the first place interested in the subjugation or settlement of large areas. The period down to the middle of the eighteenth century was marked by the multiplication of trading posts, concessions in port facilities, protective forts and bases on the coast, for these by themselves would assure the only thing early imperialism sought in Asia, secure and profitable trade.

The Portuguese dominated this trade in the sixteenth century; their fire-power swept all before them and they rapidly built up a chain of bases and trading posts. Twelve years after Vasco da Gama arrived at Calicut the Portuguese established their main Indian Ocean trading station some three hundred miles further up the western Indian coast, at Goa. It was to become a missionary as well as a commercial centre; once established, the Portuguese empire strongly supported the propagation of the faith, and the Franciscans played a large part in this. In 1513 the first Portuguese ships reached the Moluccas, the legendary spice islands, and the incorporation of Indonesia, south-east Asia, and islands as far south as Timor within the European horizon began. Four years later the first Portuguese ships reached China and opened direct European trade with that empire. Ten years later they were allowed to use Macao; in 1557 they obtained a permanent settlement there. When Charles V gave up to them the rights which Spain had claimed as a result of exploration in the Moluccas, keeping only the Philippines in the Far East, and renouncing any interest in the Indian Ocean area, the Portuguese were in possession of a monopoly of eastern empire for the next half-century.

It was a trading monopoly, and not only one of trade with Europe; there was much business to be done as carriers between Asian countries; Persian carpets went to India, cloves from the Moluccas to China, copper and silver from Japan to China, Indian cloth to Siam. Both the Portuguese and their successors found this a profitable source of income to offset some of the costs of Europe's unfavourable balance of trade with Asia. For a long time Asia wanted little from Europe except silver. The only serious competitors at sea were the Arabs and they were controlled effectively by Portuguese squadrons operating from the East African bases, from Socotra, at the mouth of the Red Sea, where they had established themselves in 1507, from Ormuz, on the northern side of the entrance to the Persian Gulf, and from Goa. From these places the Portuguese expanded their commerce further and eventually pushed into the Red Sea as far as Massawa and up to the head of the Persian Gulf, where they established a factory at Basra. They had also secured trading privileges in Burma and Siam and in the 1540s were the first Europeans to land in Japan. This network of stations and privileges was supported by a diplomacy of agreements with local rulers and the superiority of Portuguese fire-power at sea. Even if they had wished to do so, they could not have developed this power on land because they lacked men, so that a commercial empire was not only economic sense but was all that could be created with the means available.

Portugal's supremacy disguised fundamental weaknesses, a lack of manpower and a shaky financial base. It lasted until the end of the century and was then replaced by that of the Dutch, who carried the technique and institutions of commercial empire to their furthest development. The Dutch were the trading imperialists *par excellence*, though in the end they also carried out some settlement in Indonesia. Their opportunity arose when Portugal was united with Spain in 1580. This change provided a stimulus to Dutch seamen now excluded from the profitable re-export trade of oriental goods from Lisbon to northern Europe which had been mainly in their hands. The background of the Eighty Years' War with Spain was an additional incentive for the Dutch to enter areas where they might make profits at the expense of the Spanish. With less than two million people, their survival depended on a narrow base; commercial wealth was therefore vitally important to them. They had large advantages in the pool of naval manpower, ships, wealth and experience built up by their ascendancy in fishing and carrying in northern waters, and commercial expertise at home made it easy to mobilize resources for new enterprises. The Dutch were assisted, too, by the simultaneous recovery of the Arabs, who took back the East African stations north of Zanzibar as Portuguese power wavered in the aftermath of the Spanish union.

The first decades of the seventeenth century brought the collapse of much of the Portuguese empire in the East and its replacement by the Dutch. The main objective of the Dutch was the Moluccas. A brief period of individual voyages (sixty-five in seven years, some round the Straits of Magellan, some round Africa) ended when in 1602, at the initiative of the States General, the government of the United Provinces, there was set up the Dutch United East India Company, the organization which was to prove the decisive instrument of Dutch commercial supremacy in the East. Like the Portuguese before them, the company's servants worked through diplomacy with native rulers to exclude competitors, and through a system of trading-stations. How unpleasant the Dutch could be to rivals was shown in 1623, when ten Englishmen were murdered at Amboyna; this ended any English attempt to intervene directly in the spice trade. Amboyna had been one of the first Portuguese bases to be seized in a rapid sweeping-up of Portuguese interests, but it was not until 1609, when a resident governor-general was sent to the East, that the reduction of the major Portuguese forts could begin. The centre of these operations was the establishment of the Dutch headquarters at Jakarta (renamed Batavia) in Java, where it was to remain until the end of Dutch colonial rule. It became the centre of an area of settlement, where Dutch planters could rely upon the company to back them up in a ruthless control of their labour force. The early history of the Dutch colonies is a grim one of insurrection, deportation, enslavement and extermination. The trade of local shippers – and of the Chinese junks – was deliberately destroyed in order to concentrate all sources of profit in the hands of the Dutch.

The spice trade to Europe was the centre of Dutch attention and was a huge prize. It accounted during most of the century for over two-thirds of the values of the cargoes sent back to Amsterdam. But the Dutch also set about replacing the Portuguese in the valuable East Asian trade. They could not expel the Portuguese from Macao, though they sent expeditions against it, but succeeded in setting themselves up in Formosa, from which they built up an indirect trade with the mainland of China. In 1638 the Portuguese were expelled from Japan and the Dutch succeeded them there. In the next two decades, the Portuguese were replaced by the Dutch in Ceylon, too. Their successful negotiation of a monopoly of trade to Siam, on the other hand, was overtaken by another power, France. This country's connexion with the area was opened by accident in 1660 when circumstance took three French missionaries to the Siamese capital. Thanks to their establishment of a mission centre and the presence of a Greek adviser at the Siamese court there followed a French diplomatic and military mission in 1685. But these promising beginnings ended in civil war and failure and Siam moved again out of the sphere of European influence for another two centuries.

In the early eighteenth century there thus existed a Dutch supremacy in the Indian Ocean and Indonesia, and an important Dutch interest in the China seas. To a remarkable degree this reproduced the earlier Portuguese pattern, although there survived Portuguese stations such as Goa and Macao. The heart of Dutch power was the Malacca Strait, from which it radiated through Malaysia and Indonesia, to Formosa and the trading links with China and Japan, and down to the south-east to the crucial Moluccas. This area was by now enjoying an internal trade so considerable that it was beginning to be self-financing, bullion from Japan and China providing its flow of currency rather than bullion from Europe as in the early days. Further west, the Dutch were also established at Calicut, in Ceylon and at the Cape of Good Hope, and had set up factories in Persia. Although Batavia was a big town and the Dutch were running plantations to grow the goods they needed, this was still a littoral or insular commercial empire, not one of internal dominion over the mainland. In the last resort it rested on naval power and it was to succumb, though not to disappear, as Dutch naval power was surpassed.

This was clearly beginning to happen in the last decades of the seventeenth century. The unlikely challenger for Indian Ocean supremacy was England. At an early date the English had sought to enter the spice trade. There had been an East Indian Company under James I, but its factors had got bloody noses for their pains, both when they tried to cooperate with the Dutch and when they fought them. The upshot of this was that by 1700 the English had in effect drawn a line under their accounts east of the Malacca Strait. Like the Dutch in 1580, they were faced with a need to change course and did so. The upshot was the most momentous event in British history between the Protestant Reformation and the onset of industrialization, the acquisition of supremacy in India.

In India the main rivals of the English were not the Dutch or Portuguese, but the French. What was at stake did not emerge for a long time. The rise of British power in India was very gradual. After the establishment of Fort St George at Madras and the acquisition of Bombay from the Portuguese as a part of the dowry of Charles II's queen, there was no further English penetration of India until the end of the century. From their early footholds (Bombay was the only territory they held in full sovereignty) Englishmen conducted a trade in coffee and textiles less glamorous than the Dutch spice trade, but one which grew in value and importance. It also changed their national habits, and therefore society, as the establishment of coffee-houses in London showed. Soon, ships began to be sent from India to China for tea; by 1700 Englishmen had found their national beverage and a poet would soon commemorate what he termed 'cups that cheer but not inebriate'.

As a defeat of the East India Company's forces in 1689 showed, military domination in India was unlikely to prove easy. Moreover, it was not necessary to prosperity and the company did not wish to fight if it could avoid it. Though at the end of the century a momentous acquisition was made when the company was allowed to occupy Fort William, which it had built at Calcutta, the directors in 1700 rejected the idea of acquiring fresh territory or planting colonies in India as quite unrealistic. Yet all preconceptions were to be changed by the collapse of the Moghul empire after the death of Aurungzebe in 1707. The consequences emerged slowly, but their total effect was that India dissolved into a collection of autonomous states with no paramount power.

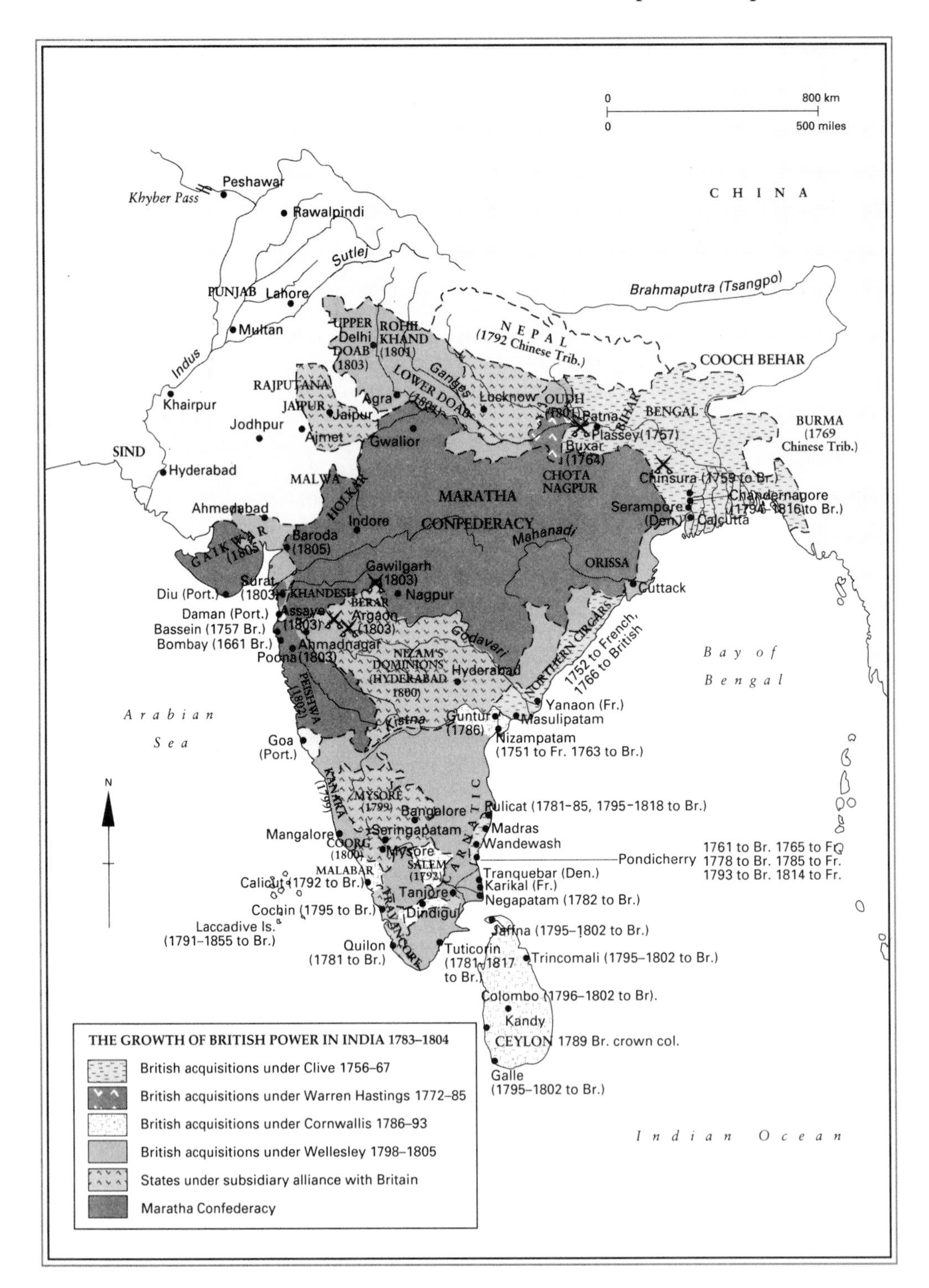

The Moghul empire had already before 1707 been troubled by the Marathas. The centrifugal tendencies of the empire had always favoured the *nawabs*, or provincial governors, too, and power was divided between them and the Marathas with increasing obviousness. The Sikhs provided a third focus of power. Originally appearing as a Hindu sect in the sixteenth century, they had turned against the Moghuls but had also drawn away from orthodox Hinduism to become virtually a third religion with it and Islam. The Sikhs formed a military brotherhood, had no castes, and were well able to look after their own interests in a period of disunion. Eventually a Sikh empire appeared in north-west India which was to endure until 1849. Meanwhile, there were signs in the eighteenth century of an increasing polarity between Hindu and Moslem. The Hindus withdrew more into their own communities, hardening the ritual practices which publicly distinguished them. The Moslems reciprocated. On this growing dislocation, presided over by a Moghul military and civil administration which was conservative and unprogressive, there fell also a Persian invasion in the 1730s and consequent losses of territory.

There were great temptations to foreign intervention in this situation. In retrospect it seems remarkable that both British and French took so long to take advantage; even in the 1740s the British East India Company was still less wealthy and powerful than the Dutch. This delay is a testimony to the importance still attached to trade as their main purpose. When they did begin to intervene, largely moved by hostility to the French and fear of what they might do, the British had several important advantages. The possession of a station at Calcutta placed them at the door to that part of India which was potentially the richest prize, Bengal and the lower Ganges valley. They had assured sea communications with Europe, thanks to British naval power, and ministers listened to the East India merchants in London as they did not listen to French merchants at Versailles. The French were the most dangerous potential competitors but their government was always likely to be distracted by its European continental commitments. Finally, the British lacked missionary zeal; this was true in the narrow sense that Protestant interest in missions in Asia quickened later than Catholic, and also, more generally, in that they had no wish to interfere with native custom or institution but only – somewhat like the Moghuls – to provide a neutral structure of power within which Indians could carry on their lives as they wished while the commerce from which the company profited prospered in peace.

The way into an imperial future led through Indian politics. Support for rival Indian princes was the first, indirect, form of conflict between French and British. In 1744 this led for the first time to armed struggle between British and French forces in the Carnatic, the south-eastern coastal region. India had been irresistibly sucked into the world-wide conflict between British and French power. The Seven Years' War (1756–63) was decisive. Before its outbreak, there had in fact been no remission of fighting in India, even while France and Great Britain were officially at peace after 1748. The French cause had prospered under a brilliant French governor in the Carnatic, Dupleix, who caused great alarm to the British by his extension of French power among native princes by force and diplomacy. But he was recalled to France and the French Indian company was not to enjoy the wholehearted support of the metropolitan government which it needed to emerge as the new paramount power. When war broke out again, in 1756, the *nawab* of Bengal attacked and captured Calcutta. His treatment of his English prisoners, many of whom were suffocated in the soon legendary 'Black Hole', gave additional offence. The East India Company's army, commanded by its employee, Robert Clive, retook the city from him, seized the French station at Chandernagore and then on 22 June 1757 won a battle over the *nawab*'s much larger armies at Plassey, about a hundred miles up the Hooghly from Calcutta.

It was not very bloody (the *nawab*'s army was suborned) but it was one of the decisive battles of world history. It opened to the British the road to the control of Bengal and its revenues. On these was based the destruction of French power in the Carnatic; that opened the way to further acquisitions which led, inexorably, to a future British monopoly of India. Nobody planned this. The British government, it is true, had begun to grasp what was immediately at stake in terms of a threat to trade and sent out a battalion of regular troops to help the company; the gesture is doubly revealing, both because it recognized that a national interest was involved, but also because of the tiny scale of this military effort. A very small number of European troops with European field artillery could be decisive. The fate of India turned on the company's handful of European and European-trained soldiers, and on the diplomatic skills and acumen of its agents on the spot. Upon this narrow base and the need for government in a disintegrating India was built the British Raj.

In 1764 the East India Company became the formal ruler of Bengal. This had by no means been the intention of the Company's directors who sought not to govern but to trade. However, if Bengal could pay for its own government, then the burden could be undertaken. There were now only a few scattered French bases; the peace of 1763 left five trading posts on condition that they were not fortified. In 1769 the French *Compagnie des Indes* was dissolved. Soon after, the British took Ceylon from the Dutch and the stage was cleared for a unique example of imperialism.

The road would be a long one and was for a long time followed reluctantly, but the East India Company was gradually drawn on by its revenue problems and by the disorder of native administrations in contiguous territories to extend its own governmental aegis. The obscuring of the company's primary commercial role was not good for business. It also gave its employees even greater opportunities to feather their own nests. This drew the interest of British politicians, who first cut into the powers of the directors of the company and then brought it firmly under the control of the Crown, setting up in 1784 a system of 'dual control' in India which was to last until 1858. In the same Act were provisions against further interference in native affairs; the British government hoped as fervently as the Company to avoid being dragged any further into the role of imperial power in India. But this was what happened in the next half-century, as many more acquisitions followed. The road was open which was to lead eventually to the enlightened despotism of the nineteenth-century Raj. India was quite unlike any other dependency so far acquired by a European state in that hundreds of millions of subjects were to be added to the empire without any conversion or assimilation of them being envisaged. The character of the British imperial structure would be profoundly transformed by this, and

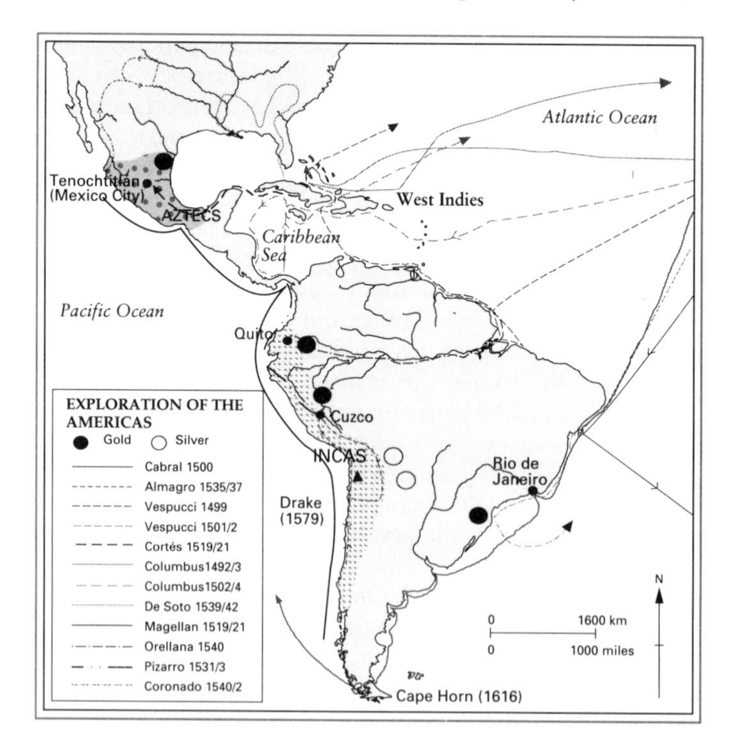

so, eventually, would be British strategy, diplomacy, external trade patterns and, even, outlook.

Except in India and Dutch Indonesia, no territorial acquisitions in the East in these centuries could be compared to the vast seizures of lands by Europeans in the Americas. Columbus' landing had been followed by a fairly rapid and complete exploration of the major 'West Indian' islands. It was soon clear that the conquest of American lands was attractively easy by comparison with the struggles to win north Africa from the Moors which had immediately followed the fall of Granada and the completion of the Reconquest on the Spanish mainland. Settlement rapidly made headway, particularly in Hispaniola and Cuba. The cornerstone of the first cathedral in the Americas was laid in 1523; the Spaniards, as their city-building was intended to show, had come to stay. Their first university (in the same city, Santo Domingo) was founded in 1538.

The Spanish settlers looked for land, as agriculturalists, and gold, as speculators. They had no competitors and, indeed, with the exception of Brazil, the story of the opening up of Central and South America remains Spanish until the end of the sixteenth century. The first Spaniards in the islands were often Castilian gentry, poor, tough and ambitious. When they went to the mainland they were out for booty, though they spoke as well of the message of the Cross and the greater glory of the crown of Castile. The first penetration of the mainland had come in Venezuela in 1499. Then, in 1513, Balboa crossed the isthmus of Panama and Europeans for the first time saw the Pacific. His expedition built houses and sowed crops; the age of the *conquistadores* had begun. One among them whose adventures captured and held the imagination of posterity was Hernán Cortés. Late in 1518 he left Cuba with a few hundred followers. He was deliberately flouting the authority of its governor and subsequently justified his acts by the spoils he brought to the Crown. After landing on the coast of Vera Cruz in February 1519, he burnt the ships his men had come in to ensure that they could not go back and then began the march to the high central plateau of Mexico which was to provide one of the most dramatic stories of the whole history of imperialism. When they reached the city of Mexico itself, they were astounded by the civilization they found there. Besides its wealth of gold and precious stones, it was situated in a land suitable for the kind of estate cultivation familiar to Castilians at home.

Though Cortés' followers were few and their conquest of the Aztec empire which dominated the central plateau heroic, they had great advantages and a lot of luck. The people upon whom they advanced were technologically primitive, easily impressed by the gunpowder, steel and horses the *conquistadores* brought with them. Aztec resistance was hampered by an uneasy feeling that Cortés might be an incarnation of their god, whose return to them they one day expected. The Aztecs were very susceptible to imported diseases, too. Furthermore, they were themselves an exploiting race and a cruel one; their Indian subjects were happy to welcome the new conquerors as liberators or at least as a change of masters. Circumstances thus favoured the Spaniards. Nevertheless, in the end their own toughness, courage and ruthlessness were the decisive factors.

In 1531 Pizarro set out upon a similar conquest of Peru. This was an even more remarkable achievement than the conquest of Mexico and, if possible, displayed even more dreadfully the rapacity and ruthlessness of the *conquistadores*. Settlement of the new empire began in the 1540s and almost at once there was made one of the most important mineral discoveries of historical times, that of a mountain of silver at Potosí, which was to be Europe's main source of bullion for the next three centuries.

By 1700, the Spanish empire in the Americas nominally covered a huge area from the modern New Mexico to the River Plate. By way of Panama and Acapulco it was linked by sea to the Spanish in the Philippines. Yet this huge extent on the map was misleading. The Californian, Texan and New Mexican lands north of the Rio Grande were very thinly inhabited; for the most part occupancy meant a few forts and trading posts and a large number of missions. Nor, to the south, was Chile well settled. The most important and most densely populated regions were three: New Spain (as Mexico was called), which quickly became the most developed part of Spanish America, Peru, which was important for its mines and intensively occupied, and some of the larger and long-settled Caribbean islands. Areas unsuitable for settlement by Spaniards were long neglected by the administration.

The Indies were governed by viceroys at Mexico and Lima as sister kingdoms of Castile and Aragon, dependent upon the Crown of Castile. They had a royal council of their own through which the king exercised direct authority. This imposed a high degree of centralization in theory; in practice, geography and topography made nonsense of such a pretence. It was impossible to control New Spain or Peru closely from Spain with the communications available. The viceroys and captains-general under them enjoyed an important and real independence in their day-to-day business. But the colonies could be run by Madrid for fiscal advantage and, indeed, the Spaniards and Portuguese were the only powers colonizing in the western hemisphere for over a century which managed to make their American possessions not only pay for themselves but return a net profit for the metropolis. This was largely because of the flow of precious metals. After 1540 silver flooded across the Atlantic, to be dissipated, unfortunately for Spain, in the wars of Charles V and Philip II. By 1650, 16,000 tons of silver had come to Europe, to say nothing of 180 tons of gold objects.

Whether Spain got other economic benefits is harder to say. She shared with other colonizing powers of the age the belief that there was only a limited amount of trade to go round; it followed that trade with her colonies should be

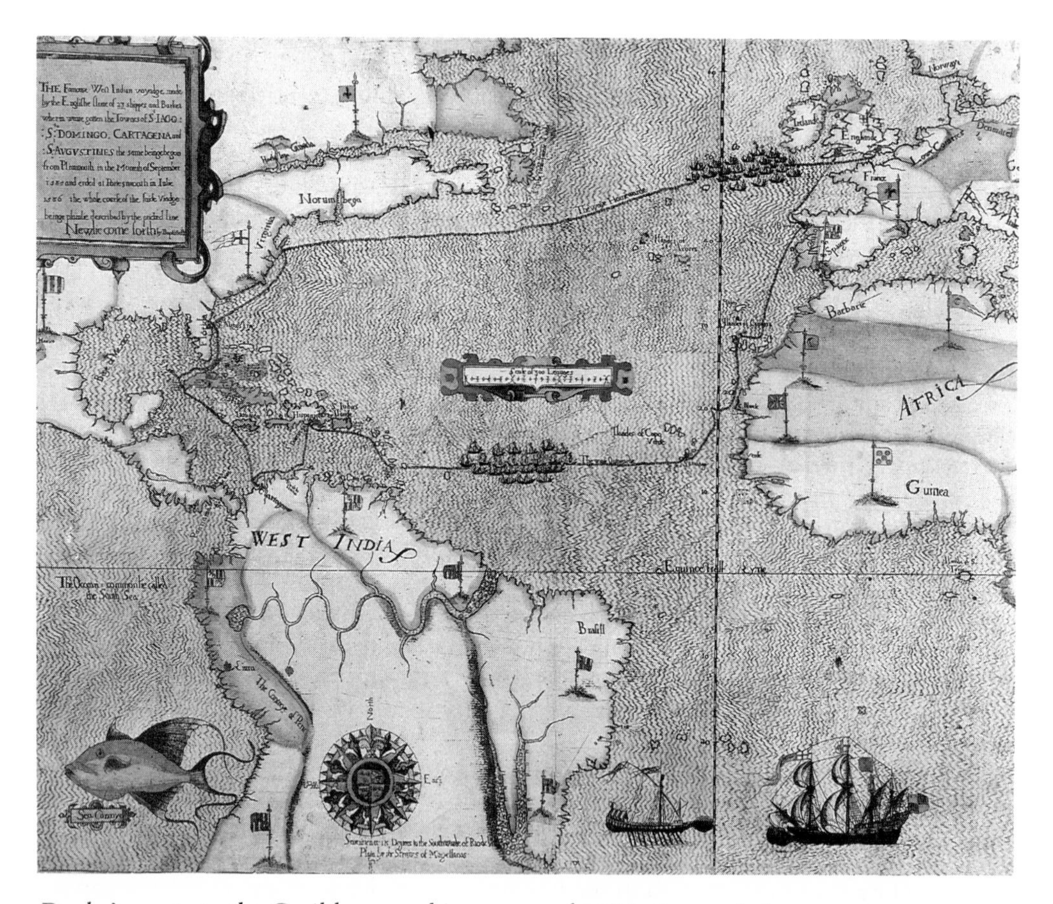

Drake's route to the Caribbean on his voyage of 1584–6 recorded on a contemporary map. This ferocious raid was one of the straws which finally broke the patience of Phillip II of Spain and provoked the launching of the Armada. The story that this was the voyage which introduced both tobacco and the potato to Europe remains, at best, plausible.

reserved to her by regulation and force of arms. Furthermore, she endorsed another commonplace of early colonial economic theory, the view that colonies should not be allowed to develop industries which might reduce the opportunities available to the home country in their markets. Unfortunately, Spain was less successful than other countries in drawing advantage from this. Though they successfully prevented the development of any but extractive or handicraft industry in America, the Spanish authorities were increasingly unable to keep out foreign traders (interlopers as they came to be called) from their territories. Spanish planters soon wanted what metropolitan Spain could not supply – slaves, especially. Apart from mining, the economy of the islands and New Spain rested on agriculture. The islands soon came to depend on black slavery; in the mainland colonies, a Spanish government unwilling to countenance the enslavement of the conquered populations evolved other devices to assure the supply of labour. The first, started in the islands and extended to Mexico, was a kind of feudal lordship: a Spaniard would be given an *encomienda*, a group of villages over which he extended protection in return for a share of their labour. The general effect was not always easily distinguishable from serfdom, or even from slavery.

The presence from the start of large pre-colonial native populations to provide labour did as much as the nature of the occupying power to differentiate the colonialism of Central and South America from that of the north. Centuries of Moorish occupation had accustomed the Spanish and Portuguese to the idea of living in a multi-racial society. There soon emerged in Latin America a population of mixed blood. In Brazil, which the Portuguese finally secured from the Dutch after thirty years' fighting, there was much interbreeding with the growing Negro population whose origins lay in imported slaves. (In Africa, too, the Portuguese showed no concern at racial interbreeding and its lack of a colour bar was always a palliative feature of Portuguese imperialism.)

None the less, though the establishment of racially mixed societies over huge areas was one of the enduring legacies of the Spanish and Portuguese empires, these societies were stratified along racial lines. The governing classes were always the Iberian-born and the *creoles*, persons of European blood born in the colonies. As time passed, the latter came to feel that the former, called *peninsulares*, excluded them from key posts and were antagonistic towards them. From the *creoles* there led downwards a blurred incline of increasing gradations of blood to the poorest and most oppressed, and these were always the pure Indians. Though their languages survived often,and thanks to the efforts of the Spanish missionaries, the dominant languages of the continent became, of course, those of the conquerors. This was the greatest single formative influence making for the cultural unification of the continent.

Another influence of comparable importance was the effective domination of a whole continent by Roman Catholicism. The Church played an enormous part in the opening of Spanish (and Portuguese) America. The lead was taken from the earliest years by the missionaries of the mendicant orders – Franciscans, in particular – but for three centuries their successors worked away at the civilization of native Americans. They took Indians from their tribes and villages, taught them Christianity and Latin (the early friars often kept them from learning Spanish, to protect them from corruption by the settlers), put them in trousers and sent them back to spread the light among their compatriots. The mission stations of the frontier determined the shapes of countries which would only come into existence centuries later.

For good and ill the Church saw itself from the start as the protector of the Indian subjects of the Crown. The eventual effect of this would only be felt after centuries had brought important changes in the demographic centre of gravity within the Roman communion, but it had many implications visible much earlier than this. It was in 1511 that the first sermon against the way the Spanish treated their new subjects was preached (by a Dominican) at Santo Domingo. From the start, the Spanish monarchy believed it had a moral and Christian mission in the New World. Laws were passed to protect the Indians and the advice of churchmen was sought about their rights and what could be done to secure them. But America was far away, and enforcement of laws difficult. It was all the harder to protect the native population when a catastrophic drop in its numbers created a labour shortage. The early settlers had brought smallpox to the Caribbean (its original source seems to have been Africa) and one of Cortés' men took it to the mainland; this was probably the main cause of the demographic disaster of the first century of Spanish empire in America.

The Church, meanwhile, was almost continuously seeking to convert the natives (two Franciscans baptised 15,000 Indians in a single day at Xocomilcho) and then to throw around them the protection of the mission and the parish. Others did not cease to make representations to the Crown. The name of one, Bartolomé de Las Casas, a Dominican, cannot be ignored. He had come out as a settler, only to become the first priest ordained in the Americas and thereafter, as theologian and bishop, he spent his life trying to influence Charles' government, and not without success. He brought to bear the discipline of refusing absolution even in the last rites to those whose confessions left him unsatisfied over their treatment of Indians, and argued against opponents on a thoroughly medieval basis. He assumed, with Aristotle, that some men were 'by nature' slaves (he had black slaves of his own) but denied that the Indians were among them. He was to pass into historical memory, anachronistically, as one of the first critics of colonialism, largely because of the use made of his writings two hundred years later by a publicist of the Enlightenment.

For centuries, the preaching and rituals of the Church were virtually the only access to European culture for the peasant, who found some of Catholicism's features sympathetic and comprehensible (the cult of the Virgin Mary, in particular, was very successful, perhaps because it was not difficult to assimilate, at least in Mexico, with indigenous tradition). To European education, only a few had access; Mexico had no native bishop until the seventeenth century, and education, except for the priesthood, did not take a peasant much further than the catechism. The Church tended, in fact, for all the devoted work of many of its clergy, to remain an imported, colonial church. Ironically, even the attempts of churchmen to protect the native Christians had the effect of isolating them (by, for instance, not teaching them Spanish) from the routes to integration with the possessors of power in their societies.

Perhaps this was inevitable. The Catholic monopoly in Spanish and Portuguese America was bound to mean a large measure of identification of the Church with the political structure: it was an important reinforcement for a thinly spread administrative apparatus and it was not only crusading zeal which made the Spanish enthusiastic proselytizers. The Inquisition was soon set up in New Spain and it was the Church of the Counter-Reformation which shaped American Catholicism south of the Rio Grande. This had important consequences much later; although some priests were to play important parts in the revolutionary and independence movements of South America, the Church as an organization never found it easy to adopt a progressive stance. In the long run, this meant that clericalism would become an issue in the politics of independent Latin America, where liberalism took on the associations of anti-clericalism it was to have in Catholic Europe. This was all in marked contrast to the religiously pluralistic society which appeared contemporaneously in British North America.

For all the spectacular inflow of bullion from the mainland colonies, the New World was probably of the greatest economic importance to Europe throughout most of the early modern period because of the Caribbean islands. This importance rested on their agricultural produce, above all on sugar, introduced first by the Arabs to Europe, in Sicily and Spain, and then carried by Europeans first to Madeira and the Canaries, and then to the New World. Both the Caribbean and Brazil were transformed economically by this crop. Medieval man had sweetened his food with honey; by 1700 sugar, though still expensive, was a European necessity and was, with tobacco, hardwood and coffee, the main product of the islands. Together, these exports gave the planters great importance in the affairs of their metropolitan countries.

The story of large-scale Caribbean agriculture began with the Spanish settlers, who quickly started growing fruit (which they had brought from Europe) and raising cattle. When they introduced rice and sugar, production was for a long time held back by a shortage of labour, as the native populations of the islands succumbed to European ill-treatment and disease. The next economic phase was the establishment by later arrivals of parasitic industries: piracy and smuggling. The Spanish occupation of the larger Caribbean islands – the Greater Antilles – still left hundreds of smaller islands unoccupied, most of them on the Atlantic fringe. These attracted the attention of English, French and Dutch captains who found them useful as bases from which to prey on Spanish ships going home from New Spain, and for contraband trade with the Spanish colonists who wanted their goods. European settlements appeared, too, on the Venezuelan coast where there was salt to be had for preserving meat. Where individuals led, governmental enterprises in the form of English royal concessions and the Dutch West India Company followed in the seventeenth century.

By then, the English had for decades been looking for suitable places for what contemporaries called 'plantations' – that is, settler colonies – in the New World. They tried the North American mainland first. Then, in the 1620s they established their first two successful West Indian colonies, on St Christopher in the Leeward Isles, and Barbados. Both prospered; by 1630 St Christopher had about three thousand inhabitants and Barbados about two thousand. This success was based on tobacco, the drug which, with syphilis (first reported in Europe at Cadiz in 1493) and the cheap automobile, was to be the New World's revenge for its disruption by the old. These tobacco colonies rapidly became of great importance to England not only because of the customs revenue they supplied, but also because the new growth of population in the Caribbean stimulated demand for exports and provided fresh opportunities for interloping in the trade of the Spanish empire. Soon the English were joined by the French in this lucrative business, the French occupying the Windward Isles, the English the rest of the Leewards. In the 1640s there were about seven thousand French in the West Indies, and over fifty thousand English.

After this time the tide of English emigration to the New World was diverted to North America and the West Indies were not again to reach such high figures of white settlement. This was partly because sugar joined tobacco as a staple crop. Tobacco can be produced economically in small quantities; it had therefore suited the multiplication of small holdings and the building up of a large immigrant population of whites. Sugar was economic only if cultivated in large units; it suited the big plantation, worked by large numbers, and these were likely to be black slaves, given the decline of local population in the sixteenth century. The Dutch supplied the slaves and aspired to the sort of general commercial monopoly in the western hemisphere which they were winning in the Far East, working out of a base at the mouth of the Hudson river, New Amsterdam. This was the beginning of a great demographic change in the Caribbean. In 1643 Barbados had thirty-seven thousand white inhabitants and only six thousand Negro slaves; by 1660 there were over fifty thousand of the latter.

With the appearance of sugar, the French colonies of Guadeloupe and Martinique took on a new importance and they, too, wanted slaves. A complex process of growth was under way. The huge and growing Caribbean market for slaves and imported European goods was added to that already offered by a Spanish empire increasingly unable to defend its economic monopoly. This set the role of the West Indies in the relationships of the powers for the next century. They were long a prey to disorder, the Caribbean an area where colonial frontiers met and policing was poor and there were great prizes to be won (in one year a Dutch captain took the great *flota* bearing home the year's treasure from the Indies to Spain). Not surprisingly, they became the classical and, indeed, legendary hunting-ground of pirates, whose heyday was the last quarter of the seventeenth century. Gradually, the great powers fought out their disputes until they arrived at acceptable agreements, but this was to take a long time. Meanwhile, through the eighteenth century the West Indies provided the great market for slaves and sustained most of that trade. As time passed, too, it became involved in another economy besides those of Europe, Africa and New Spain: that of a new North America.

Travellers' tales have always been popular. Captain John Smith (depicted here) was among the founders of the first successful English colony in North America and one of the first English authors to cater for a new demand for tales from the Americas.

For a long time, by all the standards of classical colonial theory, settlement in North America was a poor second in attractiveness to Latin America or the Caribbean. Precious metals were not discovered there and though there were furs in the north, there seemed to be little else that Europe wanted from that region. Yet there was nowhere else to go, given the Spanish monopoly to the south, and a great many nations tried it. The Spanish expansion north of the Rio Grande need not concern us, for it was hardly an occupation, but rather a missionary exercise, while Spanish Florida's importance was strategical, for it gave some protection to Spanish communications with Europe by the northern outlet from the Caribbean. It was the settlement of the Atlantic coast which drew other Europeans. There was even briefly a New Sweden, taking its place beside New Netherland, New England and New France.

The motives for settling North America were often those which operated elsewhere, though the crusading, missionary zeal of the Reconquest mentality was almost entirely missing further north. For most of the sixteenth century the Englishmen who were the most frequent explorers of North American possibilities thought there might be mines there to rival those of the Spanish Indies. Others believed that population pressure made emigration desirable and increasing knowledge revealed ample land in temperate climates with, unlike Mexico, very few native inhabitants. There was also a constant pull in the lure of finding a north-west passage to Asia.

By 1600, these impulses had produced much exploration, but only one (unsuccessful) settlement north of Florida, at Roanoke, Virginia. The English were too weak, the French too distracted, to achieve more. With the seventeenth century there came more strenuous, better-organized and better-financed efforts, the discovery of the possibility of growing some important staples on the mainland, a set of political changes in England which favoured emigration, and the emergence of England as a great naval power. Between them, these facts brought about a revolutionary transformation of the Atlantic littoral. The wilderness of 1600, inhabited by a few Indians, was a hundred years later an important centre of civilization. In many places settlers had pushed as far inland as the mountain barrier of the Alleghenies. Meanwhile the French had established a line of posts along the valley of the St Lawrence and the Great Lakes. In this huge right-angle of settlement lived about a half-million white people, mainly of British and French stock.

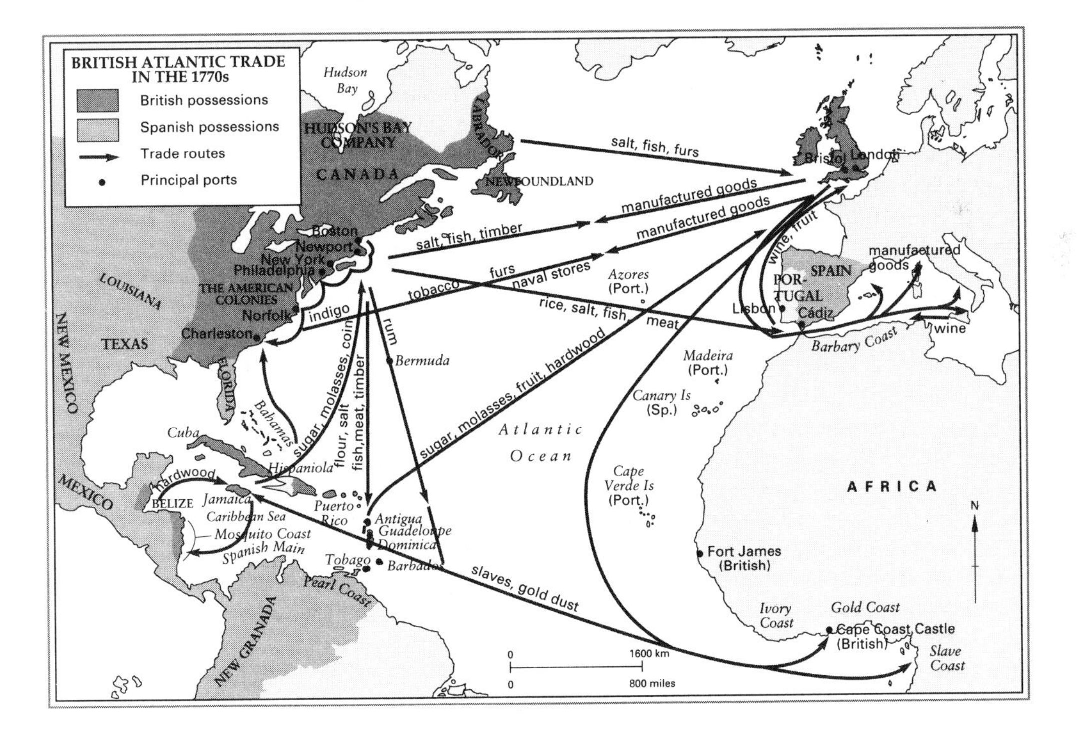

Spain claimed all North America, but this had long been contested by the English on the ground that 'prescription without possession availeth nothing'. The Elizabethan adventurers had explored much of the coast and gave the name 'Virginia', in honour of their queen, to all the territory north of thirty degrees of latitude. In 1606 James I granted a charter to a Virginia company to establish colonies. This was only formally the beginning; the company's affairs soon required revision of its structure and there were unprofitable initiatives in plenty, but in 1607 there was already established the first English settlement in America which was to survive, at Jamestown, in modern Virginia. It only just came through its early trials but by 1620 its 'starving time' was far behind it and it prospered. In 1608, the year after Jamestown's foundation, the French explorer, Samuel de Champlain, built a small fort at Quebec. For the immediate future the French colony was so insecure that its food had to be brought from France, but it was the beginning of settlement in Canada. Finally, in 1609, the Dutch sent an English explorer, Henry Hudson, to find a north-east passage to Asia. When he was unsuccessful he turned completely round and sailed across the Atlantic to look for a north-western one. Instead, he discovered the river that bears his name and established a preliminary Dutch claim by doing so. Within a few years there were Dutch settlements along the river, on Manhattan and on Long Island.

The English were in the lead and remained so. They prospered because of two new facts. One was the technique, of which they were the first and most successful exponents, of transporting whole communities, men, women and children. These set up agricultural colonies which worked the land with their own labour and soon became independent of the mother country for their livelihood. The second was the discovery of tobacco, which became a staple first for Virginia and then for Maryland, a colony whose settlement began in 1634. Further north, the availability of land which could be cultivated on European lines assured the survival of the colonies; although interest in the area had originally been awoken by the prospects of fur-trading and fishery there was soon a small surplus of grain for export. This was an attractive prospect for the land-hungry Englishman in a country widely believed in the early seventeenth century to be over-populated. Something like twenty thousand went to New England in the 1630s.

Another distinctive feature of the New England colonies was their association with religious dissent and Calvinistic Protestantism. They would not have been what they were without the Reformation. Although the usual economic motives were at work in the settlements, the leadership among immigrants to Massachusetts in the 1630s of men associated with the Puritan wing of English Protestantism bore fruit in a group of colonies whose constitutions varied from theocratic oligarchy to democracy. Though often the result of initiatives from within the English gentry and led by its members, they shed more rapidly than the southern colonies their inhibitions about radical departures from English social and political practice, and their religious nonconformity did as much as the conditions in which they had to survive to bring this about. At some moments during the English constitutional troubles of the mid-century it even seemed that the colonies of New England might escape from the control of the Crown altogether, but as yet this did not happen.

After the Dutch settlements of what was subsequently New York State had been swallowed up by the English, the North American littoral in 1700 from Florida north to the Kennebec river was organized as twelve colonies (a thirteenth, Georgia, appeared in 1732). In them lived some four hundred thousand whites and perhaps a tenth as many black slaves. Further north lay still disputed territory and then lands that were indisputably French. In these, colonists were much thinner on the ground than in the English settlements. French North Americans numbered perhaps fifteen thousand in all and had benefited from no such large migrations of communities as had the English colonies. Many of them were hunters and trappers, missionaries and explorers, strung out over the length of the St Lawrence and dotted about in the Great Lakes region and even beyond. New France was a huge area on the map, but outside the St Lawrence valley and Quebec it was only a scatter of strategically and commercially important forts and trading posts. Nor was density of settlement the only difference between the French and English colonial zones. New France was closely supervized from home; after 1663 a company structure had been abandoned in favour of direct royal rule and Canada was governed by a French governor with the advice of the *intendant* much as a French province was governed at home. There was no religious liberty; the Church in Canada was monopolistic and missionary. Its history is full of glorious examples of bravery and martyrdom, and also of bitter intransigence. The farms of the settled area were grouped in *seigneuries*, a device which had some value in decentralizing administrative responsibility. Social forms therefore reproduced those of the Old World much more than those in the English settlements, even to the extent of throwing up a nobility with Canadian titles.

The English colonies from which were to appear the future United States of America were very diverse. Strung out as they were over almost the whole Atlantic seaboard, they contained a great variety of climate, economy and terrain. Their origins reflected a wide range of motives and methods of foundation. They were soon somewhat changed ethnically, for after 1688 Scotch, Irish, German, Huguenot and Swiss emigrants had begun to arrive in appreciable numbers, though for a very long time the predominance of the English language and the relatively small numbers of non-English-speaking immigrants would maintain a culture overwhelmingly Anglo-Saxon. There was even by 1700 a large measure of effective religious toleration, though some of the colonies had close association with specific religious denominations. Religious pluralism increased the colonies' difficulty in seeing themselves as one society. They had no American centre; the Crown and the home country were the foci of the colonies' collective life, as English culture was still their background. None the less, it was already obvious that the American colonies were different: they offered to individuals opportunities for advancement unavailable either in the more strictly and closely regulated society of Canada or at home in Europe.

By 1700, some colonies had already shown a tendency to grasp whatever independence of royal control was available to them. It is tempting to look back a long way for evidence of the spirit of independence which was later to play so big a part in popular tradition. In fact, it would be a misconception to read the prehistory of the United States in these terms. The 'Pilgrim Fathers' who landed at cape Cod in 1620 were not rediscovered or inserted in their prominent place in the national mythology until the end of the eighteenth century. What can be seen much earlier than the idea of independence is the emergence of facts which would in the future make it easier to think in terms of independence and unity. One was the slow strengthening of a representative tradition in the first century of settlement. For all their initial diversity, in the early eighteenth century each colony settled down to work through some sort of representative assembly which spoke for its inhabitants to a royal governor appointed in London. Some of the settlements had needed to cooperate with one another against the Indians at an early date, and in the French wars this had become even more important. When the French loosed their Huron allies against the British colonists, it helped to create a sense of common interest among the individual colonies (as well as spurring on the British to enlist on their side the Iroquois who were hereditary foes of the Huron). From economic diversity, too, a measure of economic inter-relatedness was emerging. The middle and southern colonies produced the plantation crops of rice, tobacco, indigo and timber; New England built ships, refined and distilled molasses and grain spirits, grew corn and fished. There was a growing and apparent logic in thinking that the Americans might perhaps be able to run their affairs in their own interest – including that of the West Indian colonies – better than in that of the mother country. Economic growth changed the attitudes of some Englishmen, too. The northern mainland colonies of New England were on the whole underprized and even disliked in the mother country. They competed with the English in shipbuilding and, though illegally, in the Caribbean trade; unlike plantation colonies, they produced nothing that the mother country wanted. Besides, they were religious dissenters.

In the eighteenth century British America made great progress in wealth and civilization. The total colonial population had continued to grow and was well over a million by halfway through the century. As was being pointed out in the 1760s, the mainland colonies were going to be worth much more to Great Britain than the West Indies had been. By 1763, Philadelphia could rival many European cities in stylishness and cultivation. A great uncertainty had been removed in 1763, too, for Canada had been conquered and was by the peace treaty of that year to remain British. This changed

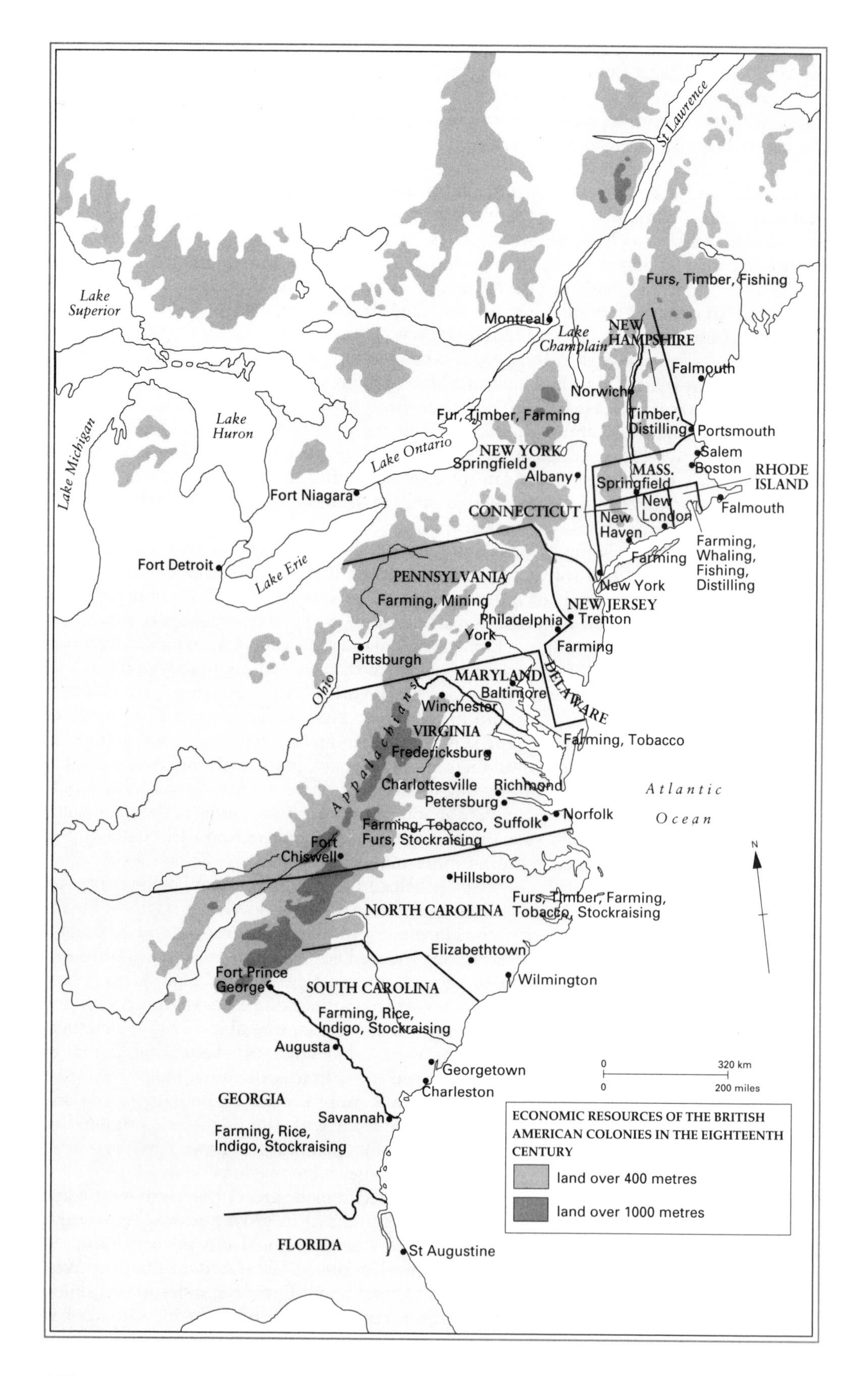
St Lawrence
Lake Superior
Montreal
Lake Champlain
NEW HAMPSHIRE
Furs, Timber, Fishing
Falmouth
Norwich
Lake Huron
Lake Michigan
Fur, Timber, Farming
Timber, Distilling
Portsmouth
Lake Ontario
NEW YORK
Springfield
Albany
Salem
Boston
MASS.
Springfield
RHODE ISLAND
Fort Niagara
CONNECTICUT
New Haven
New London
Falmouth
Fort Detroit
Lake Erie
PENNSYLVANIA
Farming
Farming, Whaling, Fishing, Distilling
New York
Farming, Mining
NEW JERSEY
Philadelphia
Trenton
York
Farming
Pittsburgh
Ohio
MARYLAND
Baltimore
DELAWARE
Winchester
Appalachians
VIRGINIA
Farming, Tobacco
Fredericksburg
Charlottesville
Richmond
Atlantic Ocean
Petersburg
Norfolk
Suffolk
Farming, Tobacco, Furs, Stockraising
Fort Chiswell
N
Hillsboro
Furs, Timber, Farming, Tobacco, Stockraising
NORTH CAROLINA
Elizabethtown
Wilmington
Fort Prince George
SOUTH CAROLINA
Farming, Rice, Indigo, Stockraising
Augusta
Georgetown
Charleston
0
320 km
0
200 miles
GEORGIA
Savannah
Farming, Rice, Indigo, Stockraising
ECONOMIC RESOURCES OF THE BRITISH AMERICAN COLONIES IN THE EIGHTEENTH CENTURY
land over 400 metres
land over 1000 metres
FLORIDA
St Augustine

the outlook of many Americans both towards the value of the protection afforded by the imperial government and towards the question of further expansion to the west. As farming settlers tended to fill up the coastal plain they came to press through the mountain barrier and down the river valleys beyond, eventually to the upper Ohio and the north-west. The danger of conflict with the French as a result was now removed, but this was not the only consideration which faced the British government in handling this movement after 1763. There were the rights and the likely reactions of the Indians to take into account. To antagonize them would be to court danger, but if Indian wars were to be avoided by holding the colonists back, then the frontier would have to be policed by British troops for that purpose, too. The result was a decision of government in London to impose a western land policy which would limit expansion, to raise taxes in the colonies to pay for the costs of defending forces, and to tighten up the commercial system and cease to wink at infringements in its working. It was unfortunate that all this was coming to a head in the last years in which the old assumptions about the economics of colonial dependencies and their relationship to the mother country were accepted without demur by the makers of colonial policy in London.

By then about two and a half centuries had gone by since European settlement in the New World began. The overall effect of expansion in the Americas upon European history had been immense, but is far from easy to define. Eventually, it is clear, all the colonial powers had, by the eighteenth century, been able to extract some economic profit from their colonies, though they did so in different ways. The flow of silver to Spain was the most obvious, but this had, of course, implications for the European economy as a whole and even for Asia. Growing colonial populations also helped to stimulate European exports and manufactures. In this respect the English colonies were of the greatest importance, pointing the way to a growing flow of emigrants from Europe which was to culminate in the last of that continent's major folk-migrations in the nineteenth and early twentieth centuries. To colonial expansion, too, must be linked the enormous growth of European shipping and shipbuilding. Whether engaged in slaving, contraband trading, legal import and export between metropolis and colony or fishing to supply new consumer markets, shipbuilders, shipowners and captains benefited. There was an incremental and incalculable effect at work. It is thus very hard to sum up the total effect of the possession of American colonies on the imperialist powers in the first age of imperialism.

Of the political and cultural importance of the fact it is easier to speak with confidence. Overall, the fate of the western hemisphere was to be culturally European and that meant, politically, that from Tierra del Fuego to Hudson's Bay it would eventually be organized in a series of sovereign states based on European legal and administrative principles where it did not remain directly dependent on a colonial power. It would also be Christian; when Hinduism or Islam came to the western hemisphere it would be as the possession of small minorities, not as rivals to a basically Christian culture.

More specifically, the greatest political importance of the era lay in the differentiation of America, north and south. It is not fanciful to see at the outset an ancient parallel. The ancient Greek city colonies were offshoots of parent communities not unlike the British colonies of the North American littoral. Once established, they tended to evolve towards a self-conscious identity of their own. The Spanish empire displayed the deployment of a regular pattern of institutions essentially metropolitan and imperial, rather as had done the provinces of imperial Rome. It took time for it to be clear that the basic forms already given to the evolution of British North America were to shape the kernel of a future world power. That evolution was therefore to prove a shaper of world as well as American history. Two great transforming factors had still to operate before the North American future was fixed in its main lines; the differing environments revealed as the continent filled up by movement to the west, and a much greater flow of non-Anglo-Saxon immigration. But these forces would flow into and around moulds set by the English inheritance, which would leave its mark on the future United States as Byzantium left its own on Russia. Nations do not shake off their origins, they only learn to see them in different ways. Sometimes outsiders can see this best. It was a great German statesman who remarked towards the end of the nineteenth century that its most important international fact had been that Great Britain and the United States spoke the same language.

5
World History's New Shape

In 1776 there began in America the first of a series of colonial revolts which were to take several decades to work themselves out. Besides marking an epoch in the history of the American continents these upheavals also provide a convenient vantage-ground from which to consider the first phase of European hegemony as a whole. In other parts of the world, too, something of a change of rhythm was marked by such facts as the elimination of serious French competition to the British in India, and the opening of Australasia, the last discovered and habitable continent, to settlement. At the end of the eighteenth century there is a sense of completing one era and opening another; it is a good point for assessment of the difference made by the previous three centuries to the history of the globe.

During them, outright conquest and occupation were the main form of European hegemony. They provided wealth Europe could use to increase still further its relative superiority over other civilizations and they set up political structures which diffused other forms of European influence. They were the work of a handful of European states which were the first world powers in the geographical range of their interests, even if not in their strength: the Atlantic nations to which the age of discoveries had given opportunities and historical destinies distinct from those of other European states.

The first to seize these opportunities had been Spain and Portugal, the only great colonial powers of the sixteenth century. They had long passed their zenith by 1763, when the Peace of Paris which ended the Seven Years' War was signed. This treaty is a convenient marker of a new world order which had already replaced that dominated by Spain and Portugal. It registered the ascendancy of Great Britain in the rivalry with France overseas which had preoccupied her for nearly three-quarters of a century. The duel was not over, and Frenchmen could still be hopeful that they would recover lost ground. Great Britain, none the less, was the great imperial power of the future. These two nations had eclipsed the Dutch, whose empire had been built, like theirs, in the seventeenth century, in the era of declining Portuguese and Spanish power. But Spain, Portugal and the United Provinces all still held important colonial territories and had left enduring marks on the world map.

These five nations had by the eighteenth century been differentiated by their oceanic history from the landlocked states of central Europe or those of the Mediterranean so important in earlier centuries. Their special colonial and overseas trade interests had given their diplomats new causes and places over which to compete. Most other states had been slower to recognize how important issues outside Europe might be, and so, indeed, had even some of these five at times. Spain had fought grimly enough (first for the Habsburgs in Italy, then against the Ottomans, and finally for European supremacy in the Thirty Years' War) to waste the treasure of the Indies in the process. In their long duel with the British, the French were always more liable than their rivals to distraction and the diversion of their resources to continental ends. At the outset, the discernment that extra-European issues might be intrinsically tangled with European interests in diplomacy had, after all, barely existed. Once the Spanish and Portuguese had demarcated their interests to their own satisfaction there was little to concern other European nations. The fate of a French Huguenot settlement in Florida or the flouting of the vague Spanish claims which was implicit in the Roanoke voyages hardly troubled the minds of European diplomats, let alone shaped their negotiations. This situation began to change when English pirates and adventurers countenanced by Elizabeth I began to inflict real damage on the Spanish fleets and colonies. They were soon joined by the Dutch and from this time one of the great themes of the diplomacy of the next century was apparent; as a French minister wrote under Louis XIV, 'Trade is the cause of a perpetual combat in war and in peace between the nations of Europe'. So much had things changed in two hundred years.

Rulers had, of course, always been concerned with wealth and the opportunity of increasing it. Venice had long defended her commerce by diplomatic means and the English had often safeguarded their cloth exports to Flanders by treaty. It was widely accepted that there was only so much profit to go round and that one country could therefore only gain at the expense of others. But it was a long time before diplomacy had to take account of the pursuit of wealth outside Europe. There was even an attempt to segregate such matters; in 1559 the French and Spanish agreed that what their captains did to one another 'beyond the line' (which meant at that time west of the Azores, and south of the Tropic of Cancer) should not be taken as a reason for hostility between the two states in Europe.

The change to a new set of diplomatic assumptions, if that is the way to put it, began in conflicts over trade with the Spanish empire. Contemporary thinking took it for granted that in the colonial relationship the interests of the metropolitan power were always paramount. In so far as those interests were economic, settlement colonies were intended to produce, either by exploiting their mineral and natural resources, or by their balance of trade with the mother country, a net advantage to the latter and, if possible, self-sufficiency, while her trading bases gave her the domination of certain areas of international traffic. By 1600 it was clear that claims would be settled by sea-power, and since the defeat of the Armada Spanish sea-power no longer commanded the respect it had done. Essentially, Philip was caught in a dilemma: the dispersal of his effort and interest between Europe – where the struggle with the Valois and Elizabeth, the Dutch Revolt, and the Counter-Reformation all claimed his resources – and the Indies, where safety could have lain only in sea-power and the organization of effective Spanish supply of the colonists' needs. The choice was to try to keep the empire, but to use it to pay for European policies. This was to underrate the difficulties of controlling so huge an empire through sixteenth-century bureaucracy and communications. Nevertheless a huge and complicated system of regular sailings in convoy, the concentration of colonial trade in a few authorized ports and policing by coastguard squadrons were ways in which the Spanish tried to keep the wealth of the Indies to themselves.

It was the Dutch who first made it clear that they were prepared to fight for a share of such prizes and therefore first forced diplomats to turn their attention and skills to regulating matters outside Europe. For the Dutch, predominance in trade overrode other considerations. What they would do for it was made clear from the start of the seventeenth century, in the East Indies, the Caribbean, and Brazil, where they engaged great fleets against the Spanish-Portuguese defence of the world's chief producer of sugar. The last provided their only serious setback, for in 1654 the Portuguese were able to evict the Dutch garrisons and resume control without subsequent challenge.

This quest for commercial wealth cut across the wishes of the most Protestant of English seventeenth-century governments; England had been an ally of the Dutch rebels in the previous century and Cromwell would have liked nothing better than the leadership of a Protestant alliance against Catholic Spain. Instead he found himself fighting the first of three Anglo-Dutch wars. The first (1652–4) was essentially a trade war. What was at issue was the English decision to restrict imports to England to goods travelling in English ships or those of the country producing the goods. This was a deliberate attempt to encourage English shipping and put it in a position to catch up with the Dutch. It struck at the heart of Dutch prosperity, its European carrying trade and, in particular, that in Baltic goods. The Commonwealth had a good navy and won. The second round came in 1665, after the Dutch had been further provoked by the English

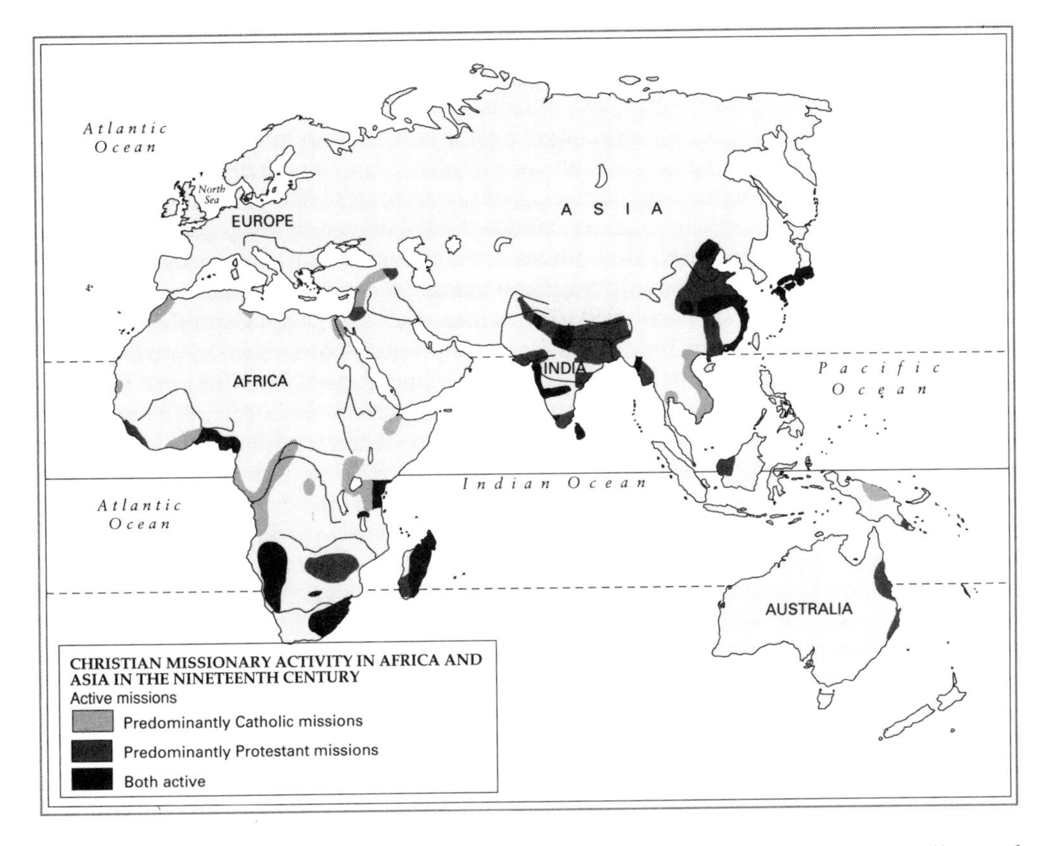

seizure of New Amsterdam. In this war the Dutch had the French and Danes as allies and also had the best of it at sea. At the peace they were therefore able to win an easing of the English restrictions on imports though they left New Amsterdam to the English in exchange for an offshoot of Barbados at Surinam. This was decided by the Treaty of Breda (1667), the first multilateral European peace settlement to say as much about the regulation of extra-European affairs as European. By it France surrendered West Indian islands to England and received in return recognition of her possession of the uninhabited and uninviting but strategically important territory of Acadia. The English had done well; the new Caribbean acquisitions followed in a tradition established under the Commonwealth, when Jamaica had been taken from Spain to be added to the existing plantation colonies.

Cromwell's policies have been seen as a decisive turn towards conscious imperial policy. This may be attributing too much to his vision. The returned Stuarts indeed kept intact most of the 'Navigation' system for the protection of shipping and colonial trade, as well as hanging on to Jamaica and continuing to recognize the new importance of the West Indies. Charles II gave a charter to a new company, named after Hudson's Bay, to contest with the French the fur trade of the north and west. He and his in other ways inadequate successor, James II, at least maintained (even if with some setbacks) English naval strength so that it was available to William of Orange in his wars with Louis XIV.

It would be tedious to trace the detailed changes of the next century during which the new imperial emphasis first of English and then of British diplomacy came to maturity. A brief third Anglo-Dutch war (it had virtually no important consequences) does not really belong to this epoch which is dominated by the long rivalry of England and France. The War of the League of Augsburg (or King William's War, as it was called in America) brought much colonial fighting, but no great changes. The War of the Spanish Succession was very different. It was a world war, the first of the modern era, about the fate of the Spanish empire as well as about French power. At its close, the British not only won Acadia (henceforth Nova Scotia) and other acquisitions in the western hemisphere from the French, but also the right to supply slaves to the Spanish colonies and to send one ship a year with merchandise to trade with them.

Overseas matters loomed larger and larger in British foreign policy after this. European considerations mattered less, in spite of the change of dynasty in 1714, when the Elector of Hanover became the first king of Great Britain. Though there were some embarrassing moments, British policy remained remarkably consistent, always swinging back to the goals of promoting, sustaining and extending British commerce. Often this was best done by seeking to maintain a general peace, sometimes by diplomatic pressure (as when the Habsburgs were persuaded to withdraw a scheme for an Ostend company to trade with Asia), sometimes by fighting to maintain privileges or strategical advantage.

The importance of war became clearer and clearer. The first time that two European powers ever went to war on a purely non-European issue came in 1739 when the British government began hostilities with Spain over, in essence, the Spanish right of search in the Caribbean – or, as the Spanish might have put it, over the steps they properly took to secure their empire against abuse of the trading privileges granted in 1713. This was to be remembered as the 'War of Jenkins' Ear' – the organ produced in pickle by its owner in the House of Commons, whose sensitive patriotism was inflamed and outraged to hear of the alleged mutilation by a Spanish coastguard. The conflict was soon caught up with the War of the Austrian Succession, and therefore became an Anglo-French struggle. The peace of 1748 did not much change the respective territorial position of the two rivals, nor did it end fighting in North America, where the French appeared to be about to cut off the British settlements for ever from the American west by a chain of forts. The British government sent its first regular contingents to America to meet this danger, but unsuccessfully; only in the Seven Years' War did a British minister grasp that the chance of a final decision in the long duel existed because of France's commitment to her ally Austria in Europe. Once British resources were allocated in accordance with this, sweeping victories in North America and India were followed by others in the Caribbean, some at the expense of Spain. A British force even seized the Philippines. It was global war.

The peace of 1763 did not in fact go so far in crippling France and Spain as many Englishmen had wanted. But it virtually eliminated French competition in North America and India. When it was a question of retaining Canada or Guadeloupe, a sugar-producing island, one consideration in favour of keeping Canada was that competition from increased sugar production within the empire was feared by Caribbean planters already under the British flag. The result was a huge British empire. By 1763, the whole of eastern North America and the Gulf Coast as far west as the mouth of the Mississippi was British. The elimination of French Canada had blown away the threat – or, from the French point of view, the hope – of a French empire of the Mississippi valley, stretching from the St Lawrence to New Orleans, which had been created by the great French explorers of the seventeenth century. Off the continental coast the Bahamas were the northern link of an island chain that ran down through the lesser Antilles to Tobago, and all but enclosed the Caribbean. Within it, Jamaica, Honduras and the Belize coast were British. In the Peace of 1713, the British had exacted a limited legal right to trade in slaves with the Spanish empire which they quickly pressed far beyond its intended limits. In Africa there were only a few British posts on the Gold Coast but these were the bases of the huge African slave trade. In Asia the direct government of Bengal was about to provide a start to the territorial phase of British expansion in India.

British imperial supremacy was based on sea-power. Its ultimate origins could be sought in the ships built by Henry VIII, among the greatest warships of the age (the *Harry Grâce à Dieu* carried 186 guns), but this early start was not followed up until the reign of Elizabeth I. Her captains, with little financing available either from Crown or commercial investors, built both a fighting tradition and better ships from the profits of operations against the Spanish. Again, there was an ebbing of interest and effort under the early Stuart kings. The royal administration could not afford ships (and paying for new ones was, indeed, one of the causes of the royal taxes Parliament had raged over). It was only under the Commonwealth, ironically, that the serious and continuing interest in naval power which sustained the Royal Navy of the future began. By that time, the connection between Dutch superiority in merchant shipping and their naval strength had been taken to heart and the upshot was the Navigation Act which provoked the first Anglo-Dutch war. A strong merchant marine provided a nursery of seamen for fighting vessels and the flow of trade whose taxation by customs dues would finance the upkeep of specialized warships. A strong merchant marine could only be built upon carrying the goods of other nations: hence the importance of competing, if necessary by gunfire, and of breaking into such reserved areas as the Spanish American trade.

The machines which were evolved to do the fighting in this competition underwent steady improvement and specialization, but no revolutionary change, between the fifteenth and nineteenth centuries. Once square-rigging and broadside firing had been adopted, the essential shape of vessels was determined, though individual design could still do much to give sailing superiority and the French usually built better ships than Great Britain during the eighteenth-century duel between the two countries. In the sixteenth century, under English influence, ships grew longer in proportion to their beam. The relative height of the forecastle and poop above the deck gradually came down, too, over the whole period. Bronze guns reached a high level of development even in the early seventeenth century; thereafter gunnery changed by improvement in design, accuracy and weight of shot. There were two significant eighteenth-century innovations, the short-range but large-calibre and heavy-shotted iron carronade, which greatly increased the power of even small vessels, and a firing mechanism incorporating a flintlock which made possible more precise control of the guns.

Specialization of function and design between warships and merchant vessels was accepted by the middle of the seventeenth century, though the line was still somewhat blurred by the existence of older vessels and the practice of privateering. This was a way of obtaining naval power on the cheap. In time of war, governments authorized individual private captains or their employers to prey upon enemy shipping, taking profits from the prizes they made. It was a form of regularized piracy and English, Dutch and French privateers all operated at various times with great success against one another's traders. The first great privateering war was that fought unsuccessfully against the English and Dutch under King William by the French.

Other seventeenth-century innovations were tactical and administrative. Signalling became formalized and the first Fighting Instructions were issued to the Royal Navy. Recruitment became more important; the press-gang appeared in England (the French used naval conscription in the maritime provinces). In this way large fleets were manned and it became clear that, given equality of skill and the limited damage which could be done even by heavy guns, numbers were always likely to be decisive in the end.

From the seminal period of development in the seventeenth century there emerged a naval supremacy which was to last over two centuries and underpin a world-wide *pax Britannica*. Dutch competition dropped away as the Republic bent under the strain of defending its independence on land against the French. The important maritime rival of the English was France and here it is possible to see that a decisive point had been passed by the end of King William's reign. By then, the dilemma of being great on land or sea had been decided by the French in favour of the land. From that time, the promise of a French naval supremacy was never to be revived, though French shipbuilders and captains would still win victories by their skill and courage. The English were not so distracted from oceanic power; they had only to keep their continental allies in the field, not to keep up great armies themselves. But there was a little more to it than a simple concentration of resources. British maritime strategy also evolved in a way very different from that of other sea-powers. Here, the French loss of interest in the navy of Louis XIV is in point, for it came after the English had inflicted a resounding defeat in a fleet action in 1692 which discredited the French admirals. It was the first of many such victories which demonstrated an appreciation of the strategical reality that sea-power was in the end a matter of commanding the surface of the sea so that friendly ships could move on it in safety while those of the enemy could not. The key to this desirable end was the neutralization of the enemy's fleet. So long as it was there, a danger existed. The early defeat of the enemy's fleet in battle therefore became the supreme aim of British naval commanders for a century during which it gave the Royal Navy almost uninterrupted command of the seas and a formidable offensive tradition.

Samuel Pepys, an outstanding civil servant, guardian of the interests of the Royal Navy of Charles II and celebrated diarist. Here he holds a song of his own composition; he was a keen amateur musician.

Naval strategy fed imperial enterprise indirectly as well as directly because it made more and more necessary the acquisition of bases from which squadrons could operate. This was particularly important in building the British empire. In the late eighteenth century, too, that empire was about to undergo the loss of much of its settled territory and this would bring further into relief the way in which European hegemony was, outside the New World, still in 1800 a matter of trading stations, island plantations and bases, and the control of carrying trade, rather than of occupation of large areas.

Less than three centuries of even this limited form of imperialism revolutionized the world economy. Before 1500, there had been hundreds of more or less self-supporting and self-contained economies, some of them linked by trade. The Americas and Africa were almost, Australasia entirely, unknown to Europe, communication within them was tiny in proportion to their huge extent, and there was a thin flow of luxury trade from Asia to Europe. By 1800, a world-wide network of exchange had appeared. Even Japan was a part of it and central Africa, though still mysterious and unknown, was linked to it through slaving and the Arabs. Its first two striking adumbrations had been the diversion of Asian trade with Europe to the sea routes dominated by the Portuguese and the flow of bullion from America to Europe. Without that stream, above all of silver, there could hardly have been a trade with Asia for there was almost nothing produced in Europe that Asia wanted. This may have been the main importance of the bullion from the Americas, whose flow reached its peak at the end of the sixteenth century and in the early decades of the next.

Although a new abundance of precious metals was the first and most dramatically obvious economic effect of Europe's new interplay with Asia and America, it was less important than the general growth of trade, of which slaves from Africa for the Caribbean and Brazil formed a part. The slave-ships usually made their voyage back to Europe from the Americas loaded with the colonial produce which more and more became a necessity to Europe. In Europe, first Amsterdam and then London surpassed Antwerp as international ports, in large measure because of the huge growth of the re-export trade in colonial goods which were carried by Dutch and English ships. Around these central flows of trade there proliferated branches and sub-branches which led to further specializations and ramifications. Shipbuilding, textiles and, later, financial services such as insurance all prospered together, sharing in the consequences of a huge expansion in sheer volume. Eastern trade in the second half of the eighteenth century made up a quarter of the whole of Dutch external commerce and during that century the number of ships sent out by the East India Company from London went up threefold. These ships, moreover, improved in design, carried more and were worked by fewer men than those of earlier times.

The material consequences of Europe's new involvement with the world are much easier to measure than some of the others. European diet remains one of the most varied in the world and this came about in the early modern age. The coming of tobacco, coffee, tea and sugar alone brought about a revolution in taste, habit and housekeeping. The potato was to change the lives of many countries by sustaining much larger populations than its predecessors. Scores of drugs were added to the European pharmacopoeia, mainly from Asia.

Beyond such material effects it is harder to proceed. The interplay of new knowledge of the world with European mentality is especially hard to pin down. Men's minds were changing, as the great increase in the numbers of books about discoveries and voyages in both East and West showed as early as the sixteenth century. Oriental studies may be said to have been founded as a science in the seventeenth century, though Europeans only begin to show the impact of knowledge of the anthropologies of other people towards its close. Such developments were intensified in the unrolling of their effects by the fact that they took place in an age of printing, too, and this makes the novelty of interest in the world outside Europe hard to evaluate. By the early eighteenth century, though, there were signs of an important intellectual impact at a deep level. Idyllic descriptions of savages who lived moral lives without the help of Christianity provoked reflexion; an English philosopher, John Locke, used the evidence of other continents to show that men did not share any God-given innate ideas. In particular, China furnished examples for speculation on the relativity of social institutions, while the penetration of Chinese literature (much aided by the studies of the Jesuits) revealed a chronology whose length made nonsense of traditional calculations of the date of the Flood described in the Bible as the second beginning of all men.

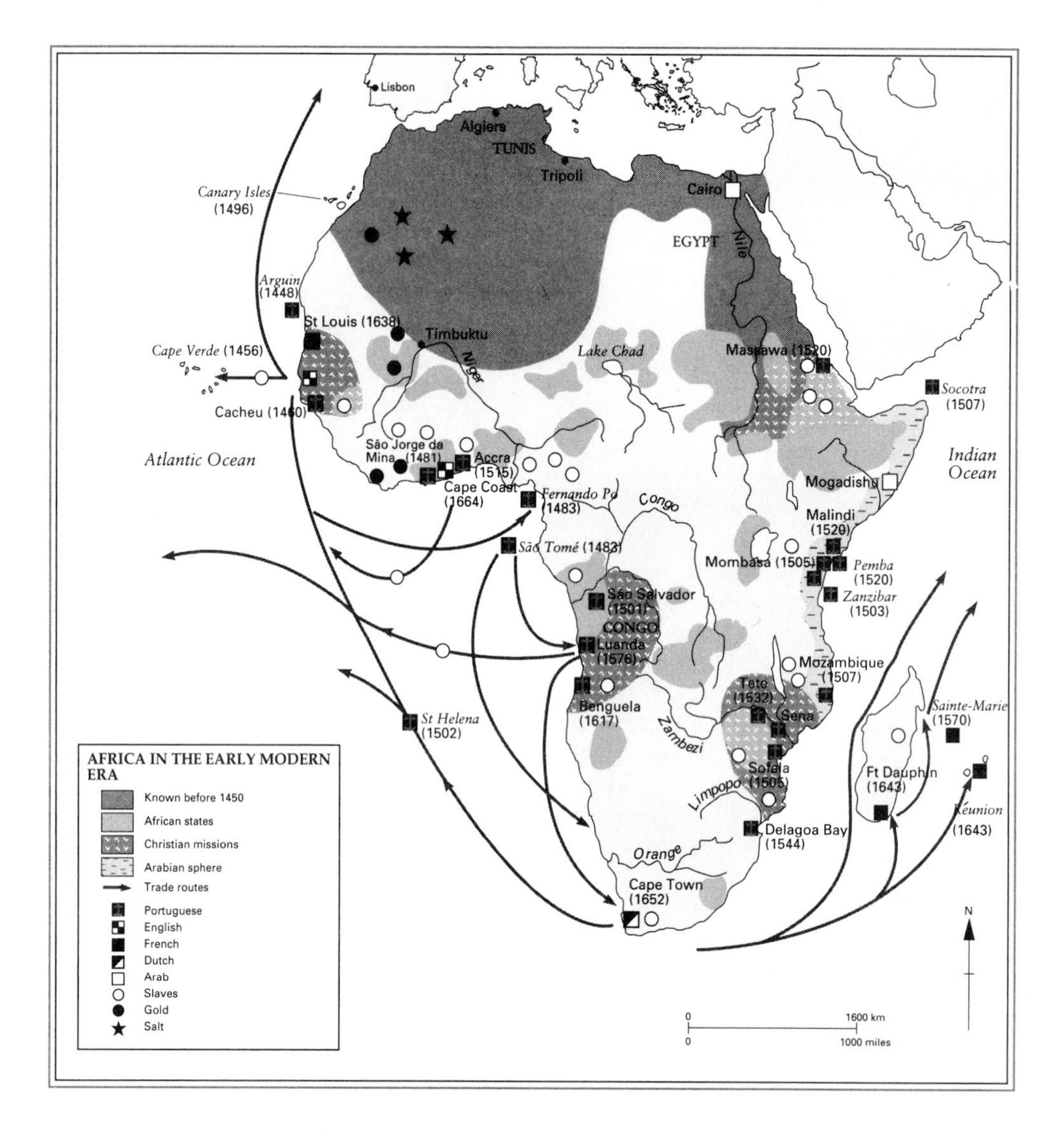

As its products became more easily available, China also provoked in Europe an eighteenth-century craze for oriental styles in furniture, porcelain and dress. As an artistic and intellectual influence this has remained more obvious than the deeper perspective given to the observation of European life by an awareness of different civilizations with different standards elsewhere. But while such comparisons may have had some disquieting aspects, revealing that Europe had, perhaps, less to be proud of in its attitude to other religions than China, there were still others suggested by exploits such as those of the *conquistadores* which fed Europeans' notions of their superiority.

The impact of Europe on the world is no easier to encapsulate in a few simple formulae than that of the world upon Europe, but it is, in some of its manifestations at least, at times more dramatically obvious. It is an appalling fact that almost nowhere in the world can the inhabitants of non-European countries be shown to have benefited materially from the first phase of Europe's expansion; far from it, they often suffered terribly. Yet this was not always something for which blame attaches to the Europeans – unless they should be blamed for being there at all. In an age with no knowledge of infectious disease beyond the most elementary, the devastating impact of smallpox or other diseases brought from Europe to the Americas could not have been anticipated. But it was disastrous. It has been calculated that the population of Mexico fell by three-quarters in the sixteenth century; that of some Caribbean islands was wiped out altogether.

The gravestone of a slave in a an English churchyard.

Such facts as the ruthless exploitation of those who survived, on the other hand, whose labour was so much more valuable after this demographic collapse, are a different matter. Here is expressed that *leitmotiv* of subjection and domination which runs through well-nigh every instance of Europe's early impact on the rest of the world. Different colonial environments and different European traditions present gradations of oppression and exploitation. Not all colonial societies were based on the same extremes of brutality and horror. But all were tainted. The wealth of the United Provinces and its magnificent seventeenth-century civilization were fed by roots which, at least in the spice islands and Indonesia, lay in bloody ground. Long before expansion in North America went west of the Alleghenies, the brief good relations of the first English settlers of Virginia with the Red Indian had soured and extermination and eviction had begun. Though the populations of Spanish America had been in some measure protected by the state from the worst abuses of the *encomienda* system, they had for the most part been reduced to peonage, while determined efforts were made (for the highest motives) to destroy their culture. In South Africa the fate of the Hottentot and in Australia that of the Aborigine would repeat the lesson that European culture could devastate those whom it touched, unless they had the protection of old and advanced civilizations such as those of India or China. Even in those great countries, much damage would be done, nor would they be able to resist the European once he decided to bring sufficient force to bear. But it was the settled colonies that showed most clearly the pattern of domination.

The prosperity of many of them long depended on the African slave trade, whose economic importance has already been touched upon. Since the eighteenth century it has obsessed critics who have seen in it the most brutal example of the inhumanity of man to man, whether that of white to black, of European to non-European, or of capitalist to labourer. It has properly dominated much of the historiography of Europe's expansion and American civilization, for it was a major fact in both. Less usefully, it has, because of its importance in shaping so much of the New World, diverted attention from other forms of slavery at other times – or even alternative fates to slavery, such as the extermination, intentional or unintentional, which overtook other peoples.

Outlets in the New World settler colonies were to dominate the direction of the slave trade until its abolition in the nineteenth century. First in the Caribbean islands and then on the American mainland, the slavers found their most reliable customers. The Portuguese who had first dominated the trade were soon elbowed out by the Dutch and then by Elizabeth I's 'sea-dogs'. Meanwhile, Portuguese captains turned to importing slaves to Brazil instead as the sixteenth century went on. Early in the seventeenth century the Dutch founded a company to ensure a regular supply of slaves to the West Indies, but by 1700 their lead had been overtaken by French and English slavers who had established their own posts on the 'slave coast' of Africa. Altogether, their efforts sent between nine and ten millions of black slaves to the western hemisphere, 80 per cent of them after 1700. The eighteenth century saw the greatest prosperity of the trade; some six million slaves were shipped then. European ports like Bristol and Nantes built a new age of commercial wealth on slaving. New lands were opened as black slave labour made it possible to work them. Larger-scale production of new crops brought, in turn, great changes in European demand, manufacturing and trading patterns. Racially, too, we still live with the results.

What has disappeared and can now never be measured is the human misery involved, not merely in physical hardship (a black might live only a few years on a West Indian plantation even if he survived the horrible conditions of the voyage) but in the psychological and emotional tragedies of this huge migration. Historians still debate whether slavery 'civilized' blacks in the Americas by bringing them into contact, willy-nilly, with higher civilizations, or whether it retarded them in quasi-infantile dependence. The question seems as insoluble as the degree of cruelty involved is incalculable; on the one hand is the evidence of the fetters and the whipping-block, on the other the reflexion that these were commonplaces of European life too, and that, *a priori*, self-interest should have prompted the planters to care for their investment. That it did not always do so, slave rebellions showed. Such resistance, though, was infrequent, a fact which also bears reflexion. It is unlikely that the debate will end.

Estimates of the almost unrecorded damage done in Africa are even harder to arrive at, for the evidence is even more subject to conjecture. The obvious demographic loss may (some have hazarded) be balanced against the introduction to Africa of new foodstuffs from America. Conceivably, such by-products of a European contact determined by the hunt for slaves actually led to population growth, but the hypothesis can hardly be weighed against the equally immeasurable effects of imported disease.

It is notable that the African slave trade for a long time awoke no misgivings such as those which had been shown by Spanish churchmen in defence of the American Indians, and the arguments with which some Christians actually resisted the restriction of this traffic still retain a certain gruesome fascination. Feelings of responsibility and guilt began to emerge widely only in the eighteenth century and mainly in France and England. One expression of it was the British acquisition of another dependency in 1787, Sierra Leone, soon adopted by philanthropists as a refuge for African slaves freed in England. Once combined with a favourable political and economic conjuncture, the current of public feeling educated by humanitarian thought would destroy the slave trade and, in the European world, slavery. But that is part of a different story. In the unfolding of

European world power, slavery was a huge social and economic fact. It was to become a great mythical one, symbolizing at its harshest the triumph of force and cupidity over humanity. Sadly, it was also only the outstanding expression of a general dominance by force of advanced societies over weaker ones.

Some Europeans recognized this but none the less believed that any evil was outweighed by what they offered to the rest of the world, above all, by the bringing of Christianity. It was a bull of Paul III, the pope who summoned the Council of Trent, which proclaimed that 'the Indians are truly men and ... are not only capable of understanding the Catholic faith but according to our information, they desire exceedingly to receive it.' Such optimism was not merely an expression of the Counter-Reformation spirit, for the missionary impulse had been there from the start in the Spanish and Portuguese possessions. Jesuit missionary work began in Goa in 1542 and radiated from there all over the Indian Ocean and south-east Asia and even reached Japan. Like the other Catholic powers, the French, too, emphasized missionary work, even in areas where France was not herself economically or politically involved. A new vigour was none the less given to missionary enterprise in the sixteenth and seventeenth centuries and may be acknowledged as one invigorating effect of the Counter-Reformation. Formally at least, Roman Christianity took in more converts and greater tracts of territory in the sixteenth century than in any earlier. What this really meant is harder to assess, but what little protection the native had was provided by the Roman Catholic Church, whose theologians kept alive, however dimly at times, the only notion of trusteeship towards subject peoples which existed in early imperial theory.

The quaker William Penn founded the colony of Pennsylvania. His unusually scrupulous relations with Indians became near-legendary and are commemorated in this picture showing an agreement of a treaty with them.

Protestantism lagged far behind in concern about the natives of settlement colonies, as it did in missionary work. The Dutch hardly did anything and the English American colonists not only failed to convert, but actually enslaved some of their Red

Indian neighbours (the Quakers of Pennsylvania were laudable exceptions). The origins of the great Anglo-Saxon overseas missionary movements are not to be detected until the end of the seventeenth century. Furthermore, even in the gift of the Gospel to the world when it came there lay a tragic ambiguity. It, too, was a European export of enormously corrosive potential, challenging and undermining traditional structures and ideas, threatening social authority, legal and moral institutions, family and marriage patterns. The missionaries, often in spite of themselves, became instruments of the process of domination and subjugation which runs through the story of Europe's intercourse with the rest of the globe.

Perhaps there was nothing Europeans brought with them which would not in the end turn out to be a threat, or at least double-edged. The food plants which the Portuguese carried from America to Africa in the sixteenth century – cassava, sweet potatoes, maize – may have improved African diet, but (it has been argued) may also have provoked population growth which led to social disruption and upheaval. Plants taken to the Americas, on the other hand, founded new industries which then created a demand for slaves; coffee and sugar were commodities of this sort. Further north, wheat-growing by British settlers did not require slaves, but intensified the demand for land and added to the pressures driving the colonists into the ancestral hunting-grounds of the Indians, whom they ruthlessly pushed out of the way.

The lives of generations unborn when such transplants were first made were to be shaped by them, and a longer perspective than one confined by 1800 is helpful here. Wheat was, after all, ultimately to make the western hemisphere the granary of European cities; in our own century even Russia and Asian countries have drawn on it. A still-flourishing wine industry was implanted by the Spanish in the Madeiras and America as early as the sixteenth century. When bananas were established in Jamaica, coffee in Java and tea in Ceylon, the groundwork was laid of much future politics. All such changes, moreover, were in the nineteenth century complicated by variations in demand, as industrialization increased the demand for old staples such as cotton (in 1760 England imported two and a half million pounds of raw cotton - in 1837 the figure was three hundred and sixty million) and sometimes created new ones; it was a consequence of this that rubber was successfully transplanted from South America to Malaya, a change fraught with great strategic significance for the future.

Captain Cook's crew improve their acquaintance with a threatened species.

The scope of such implications for the future in the early centuries of European hegemony will appear sufficiently in what follows. Here it is only important to note one more, often-repeated, characteristic of this pattern, its unplanned, casual nature. It was the amalgam of many individual decisions by comparatively few men. Even their most innocent innovations could have explosive consequences. It is worth recalling that it was the importation of a couple of dozen rabbits in 1859 which led to the devastation of much of rural Australia by millions of them within a few decades. Similarly, but on a smaller scale, Bermuda was to be plagued with English toads.

Conscious animal importations, though, were even more important (the first response to the Australian rabbit scourge was to send for English stoats and weasels; a better answer had to wait for myxomatosis). Almost the entire menagerie of European domesticated animals was settled in the Americas by 1800. The most important were cattle and horses. Between them they would revolutionize the life of the Plains Indians; later, after the coming of refrigerated ships, they were to make South America a great meat exporter just as Australasia was to be made one by the introduction of sheep the English had themselves imported originally from Spain. And, of course, the Europeans brought human blood-stock, too. Like the British in America, the Dutch for a long time did not encourage the mixing of races. Yet in Latin America, Goa and Portuguese Africa the effects were profound. So, in an entirely different and negative way, were they in British North America, where racial intermarriage was not significant and the near-exact coincidence of colour and legally servile status bequeathed an enormous legacy of political, economic, social and cultural problems to the future.

The creation of large colonial populations shaped the future map, but also presented problems of government. The British colonies nearly always had some form of representative institution which reflected parliamentary tradition and practice while France, Portugal and Spain all followed a straightforward authoritarian and monarchical institutional system. None of them envisaged any sort of independence for their colonists, nor any need to safeguard their interests against those of the mother country, whether these were conceived as paramount or complementary. This would in the end cause trouble and by 1763 there were signs at least in the British North American colonies that it might be on lines reminiscent of seventeenth-century England's struggles between Crown and Parliament. And in their struggles with other nations, even when their governments were not formally at war with them, the colonists always showed a lively sense of their own interests. Even when Dutch and English were formally allied against France their sailors and traders would fight one another 'beyond the line'.

Problems of imperial government in the eighteenth century were, though, largely a matter of the western hemisphere. That was where the settlers had gone. Elsewhere in the world in 1800, even in India, trade still mattered more than possession and many important areas had still to feel the full impact of Europe. As late as 1789 the East India Company was sending only twenty-one ships in the year to Canton; the Dutch were allowed two a year to Japan. Central Asia was at that date still only approachable by the long land routes used in the days of Chinghis Khan and the Russians were still far from exercising effective influence over the hinterland. Africa was protected by climate and disease. Discovery and exploration still had to complete that continent's map before European hegemony could become a reality.

In the Pacific and 'South Seas', things were moving faster. Dampier's voyage of 1699 had opened a century of exploration which added Australasia, the last unknown continent, to the map. In the north, the existence of the Bering Straits had been demonstrated by 1730. The voyages of Bougainville and Cook, in the 1760s and 1770s, added Tahiti, Samoa, eastern Australia, Hawaii and New Zealand to the last New World to be opened. Cook even penetrated the Antarctic Circle. In 1788 the first cargo of convicts, 717 of them, was landed in New South Wales. British judges were calling into existence a new penal world to redress the balance of the old, since the American colonies were now unavailable for dumping English undesirables, and were incidentally founding another new nation. More important still, a few years later the first sheep arrived and so was founded the industry to ensure that nation's future. Along with animals, adventurers and ne'er-do-wells there came to the South Pacific, also, the Gospel. In 1797 the first missionaries arrived in Tahiti. With them, the blessings of European civilization may be reckoned at last to have appeared, at least in embryonic form, in every part of the habitable world.

6
Ideas Old and New

The essence of the civilization Europe was exporting to the rest of the globe lay in ideas. The limits they imposed and the possibilities they offered shaped the way in which that civilization operated. What is more, although the twentieth century has done great damage to them, the leading ideas adumbrated by Europeans between 1500 and 1800 still provide most of the guide-posts by which we make our way. European culture was then given a secular foundation; it was then, too, that there took hold a progressive notion of historical development as movement towards an apex at which Europeans felt themselves to stand. Finally it was then that there grew up a confidence that scientific knowledge used in accordance with utilitarian criteria would make possible limitless progress. In short, the civilization of the Middle Ages at last came to an end in the minds of thinking men.

For all that, things rarely happen cleanly and neatly in history, and only a few Europeans would have been aware of this change by 1800. The traditional institutions of monarchy, hereditary status society and religion still held sway over most of the continent in that year. Only a hundred years before there had been no civil marriage anywhere in Europe and there was still none over most of it. Barely twenty years before 1800 the last heretic had been burned in Poland and even in England an eighteenth-century monarch had, like medieval kings, touched for the King's Evil. The seventeenth century, indeed, had in one or two respects even shown regression. In both Europe and North America there was an epidemic of witch-hunting far more widespread than anything in the Middle Ages (Charlemagne had condemned witch-burners to death and canon law had forbidden belief in the night flights and other supposed pranks of witches as pagan). Nor was this the end of superstition. The last English wizard was harried to his death by his neighbours well after 1700 and a Protestant Swiss was legally executed by his countrymen for witchcraft in 1782. The Neapolitan cult of St Januarius was still of political importance in the era of the French Revolution because the successful or unsuccessful liquefaction of the saint's blood was believed to indicate divine pleasure or displeasure at what the government was doing. Penology was still barbarous; some crimes were thought so atrocious as to merit punishment of exceptional ferocity and it was as parricides that the assassin of Henry IV of France and the attempted assassin of Louis XV suffered their abominable torments. The second died under them in 1757, only a few years before the publication of the most influential advocacy of penal reform that has ever been written. The glitter of modernity in the eighteenth century can easily deceive us; in societies which produced art of exquisite refinement and outstanding examples of chivalry and honour, popular amusements focused on the pleasures of bear-baiting, cock-fighting or pulling the heads off geese.

The culture of the people is often an aspect of society which shows most obviously the weight of the past, but until almost the end of these three centuries much of the formal and institutional apparatus which upheld the past also remained intact over most of Europe. The most striking example to modern eyes would be the primacy still enjoyed almost everywhere in the eighteenth century by organized religion. In every country, Catholic, Protestant and Orthodox alike, even ecclesiastical reformers took it for granted that religion should be upheld and protected by the law and the coercive apparatus of the state. Only a very few advanced thinkers questioned this. In much of Europe there was still no toleration for views other than those of the established Church. The coronation oath taken by a French king imposed on him the obligation to stamp out heresy and only in 1787 did non-Catholic Frenchmen gain any recognized civic status and therefore the right to legitimize their children by contracting legal marriage. In Catholic countries the censorship, though often far from effective, was still supposed and sometimes strove to prevent the dissemination of writings inimical to Christian belief and the authority of the Church. Although the Counter-Reformation spirit had ebbed and the Jesuits were dissolved, the Index of prohibited books and the Inquisition which had first compiled it were maintained. The universities everywhere were in clerical hands; even in England, Oxford and Cambridge were closed to nonconformist dissenters and Roman Catholics. Religion also largely determined the content of their teaching and the definition of the studies they pursued.

The Royal Observatory was built at Greenwich in 1675 and its primacy is commemorated in the phrase 'Greenwich Mean Time'. A portrait of its royal patron, Charles II, looks down from the wall of the room shown in this engraving.

The institutional fabric of society, it is true, showed also the onset of innovation. One of the reasons why universities lost importance in these centuries was that they no longer monopolized the intellectual life of Europe. From the middle of the seventeenth century there appeared in many countries, and often under the highest patronage, academies and learned societies such as the English Royal Society, which was given a charter in 1662, or the French *Acadèmie des Sciences*, founded four years later. In the eighteenth century such associations greatly multiplied; they were diffused through smaller towns and founded with more limited and special aims, such as the promotion of agriculture. A great movement of voluntary socialization was apparent; though most obvious in England and France, it left few countries in western Europe untouched. Clubs and societies of all sorts were a characteristic of an age no longer satisfied to exhaust its potential in the social institutions of the past, and they sometimes attracted the attention of government. Some of them made no pretension to have as their sole end literary, scientific or agricultural activity, but provided gatherings and meeting-places at which general ideas were debated, discussed or merely chatted about. In this way they assisted the circulation of new ideas. Among such associations the most remarkable was the international brotherhood of freemasons. It was introduced from England to continental Europe in the 1720s and within a half-century spread widely; there may have been more than a quarter-million masons by 1789. They were later to be the object of much calumny; the myth was propagated that they had long had revolutionary and subversive aims. This was not true of the craft as a body, however true it may have been of a few individual masons, but it is easy to believe that so far as masonic lodges, like other gatherings, helped in the publicity and discussion of new ideas, they contributed to the breaking up of the ice of tradition and convention.

The increased circulation of ideas and information did not, of course, rest primarily on such meetings, but on the diffusion of the written word through print. One of the crucial transformations of Europe after 1500 was that it became more literate; some have summed it up as the change from a culture focused on the image to one focused on the word. Reading and writing (and especially the former), though not universally diffused, had, nevertheless, become widespread and in some places common. They were no longer the privileged and arcane knowledge of a small èlite, nor were they any longer mysterious in being intimately and specially connected with religious rites.

In assessing this change we can emerge a little way from the realm of imponderables and enter that of measurable data which shows that somehow, for all the large pools of illiteracy which still existed in 1800, Europe was by then a literate society as it was not in 1500. That is, of course, not a very helpful statement as it stands. There are many degrees of attainment in both reading and writing. Nevertheless, however we define our terms, Europe and its dependencies in 1800 probably contained most of the literate people in the world. It therefore had a higher proportion of literates than other cultures. This was a critical historical change. By then, Europe was well into the age of the predominance of print, which eventually superseded, for most educated people, the spoken word and images as the primary means of instruction and direction, and lasted until the twentieth century restored oral and visual supremacy by means of radio, cinema and television.

The sources for assessing literacy are not good until the middle of the nineteenth century – when, it appears, somewhere about half of all Europeans still could neither read nor write – but they all suggest that the improvement from about 1500 was cumulative but uneven. There were important differences between countries, between the same countries at different periods, between town and country, between the sexes, and between occupations. All this is still true, though in diminished degree, and it greatly simplifies the problem of making general statements: none but the vaguest are possible until recent times. But specific facts are suggestive about trends.

The first signs of the educational effort underlying the increase of literacy can be seen before the invention of printing. They appear to be another part of that revival and invigoration of urban life between the twelfth and thirteenth centuries whose importance has already been noted. Some of the earliest evidence of the commissioning of schoolmasters and provision of school places comes from the Italian cities which were then the vanguard of European civilization. In them there soon appeared a new appreciation, that literacy is an essential qualification for certain kinds of office. We find, for example, provisions that judges should be able to read, a fact with interesting implications for the history of earlier times.

The early lead of the Italian cities had given way by the seventeenth century to that of England and the Netherlands (both countries with, for the age, a high level of urbanization). These have been thought to be the European countries with the highest levels of literacy in about 1700; the transfer of leadership to them illustrates the way in which the history of rising literacy is geographically an uneven business. Yet French was to be the international language of eighteenth-century publication and the bedrock of the public which sustained this must surely have been found in France. It would not be surprising if levels of literacy were higher in England and the United Provinces, but the numbers of the literate may well have been larger in France, where the total population was so much bigger.

An outstanding place in the overall trend to literacy must surely be given to the spread of printing. By the seventeenth century there was in existence a corpus of truly popular publishing, represented in fairy-stories, tales of true and unrequited love, almanacs and books of astrology, hagiographies. The existence of such material is evidence of demand. Printing had given a new point to being literate, too, for the consultation of manuscripts had necessarily been difficult and time-consuming, because of their relative inaccessibility. Technical knowledge could now be made available in print very quickly and this meant that it was in the interest of the specialist to read in order to maintain his skill in his craft.

Another force making for literacy was the Protestant Reformation. Almost universally, the reformers themselves stressed the importance of teaching believers how to read; it is no coincidence that by the nineteenth century Germany and Scandinavia both reached higher levels of literacy than many Catholic countries. The Reformation made it important to read the Bible and it had rapidly become available in print in the vernaculars which were thus strengthened and disciplined by the diffusion and standardization which print brought with it. Bibliolatry, for all its more obviously unfortunate manifestations, was a great force for enlightenment; it was both a stimulus to reading and a focus for intellectual activity. In England and Germany its importance in the making of a common culture can hardly be exaggerated, and in each country it produced a translation of the Bible which was a masterpiece.

As the instance of the reformers shows, authority was often in favour of greater literacy, but this was not confined to the Protestant countries. In particular, the legislators of innovating monarchies in the eighteenth century often strove to promote education – which meant in large measure primary education. Austria and Prussia were notable in this respect. Across the Atlantic the puritan tradition had from the start imposed in the New England communities the obligation to provide schooling. In other countries education was left to the informal and unregulated operation of private enterprise and charity (as in England), or to the Church. From the sixteenth century begins the great age of particular religious orders devoted to teaching (as in France).

An important consequence, promoter and concomitant of increased literacy was the rise of the periodical press. From broadsheets and occasional printed newsletters there evolved by the eighteenth century journals of regular publication. They met various needs. Newspapers began in seventeenth-century Germany, a daily coming out in London in 1702, and by the middle of the century there was an important provincial press and millions of newspapers were being printed each year. Magazines and weekly journals began to appear in England in the first half of the eighteenth century and the most important of them, the *Spectator*, set a model for journalism by its conscious effort to shape taste and behaviour. Here was something new. Only in the United Provinces did journalism have such success as in England; probably this was because all other European countries enjoyed censorships of varying degrees of efficacy as well as different levels of literacy. Learned and literary journals appeared in increasing numbers, but political reporting and comment were rarely available. Even in eighteenth-century France it was normal for the authors of works embodying advanced ideas to circulate them only in manuscript; in this stronghold of critical thought there was still a censorship, though one arbitrary and unpredictable and, as the century wore on, less effective in its operation.

It may have been a growing awareness of the subversive potential of easily accessible journalism which led to a change of wind in official attitudes to education. Until the eighteenth century there was no very widespread feeling that education and literacy might be dangerous and should not be widely extended. Though formal censorship had always been a recognition of the potential dangers brought by literacy, there was a tendency to see this in predominantly religious terms; one duty of the Inquisition was to maintain the effectiveness of the Index. In retrospect it may well seem that the greater opportunity which literacy and printing gave for the criticism and questioning of authority in general was a more important effect than their subversion of religion. Yet this was not their only importance. The diffusion of technical knowledge also accelerated other kinds of social change. Industrialization would hardly have been possible without greater literacy and a part of what has been called a 'scientific revolution' in the seventeenth century must be attributed to the simple cumulative effect of more rapidly and widely circulated information.

The fundamental sources of this 'revolution' none the less lie deeper than this, in changed intellectual attitudes. Their core was a changed view of Man's relation to nature. From a natural world observed with bemused awe as evidence of God's mysterious ways, men somehow made the great step to a conscious search for means to achieve its manipulation. Although the work of medieval scientists had been by no means as primitive and uncreative as it was once the fashion to believe, it suffered from two critical limitations. One was that it could provide very little knowledge that was of practical use. This inhibited attention to it. The second was its theoretical weakness; it had to be surpassed at a conceptual as well as a technical level. In spite of its beneficial irrigation by ideas from Arab sources and a healthy emphasis on definition and diagnosis in some of its branches, medieval science rested on assumptions which were untested, in part because the means of testing them could not be grasped, in part because the wish to test them did not exist. The dogmatic assertion of the theory that the four elements, Fire, Air, Earth and Water, were the constituents of all things, for example, went unrefuted by experiment. Although experimental work of a sort went on within the alchemical and hermetic traditions, and with Paracelsus came to be directed towards other ends than a search for gold, it was still directed by mythical, intuitive conceptions.

This remained broadly true until the seventeenth century. The Renaissance had its scientific manifestations but they found expression usually in descriptive studies (an outstanding example was that of Vesalius' human anatomy of 1543) and in the solution of practical problems in the arts (such as those of perspective) and mechanical crafts. One branch of this descriptive and classificatory work was particularly impressive, that addressed to making sense of the new geographical knowledge revealed by the discoverers and cosmographers. In geography, said a French physician of the early sixteenth century, 'and in what pertains to astronomy, Plato, Aristotle, and the old philosophers made progress, and Ptolemy added a great deal more. Yet, were one of them to return today, he would find geography changed past recognition.' Here was one of the stimuli for a new intellectual approach to the world of nature.

It was not a stimulus quick to operate. A tiny minority of educated men, it is true, would already in 1600 not have found it easy to accept the conventional world picture based on the great medieval synthesis of Aristotle and the Bible. Some of them felt an uneasy loss of coherence, a sudden lack of bearings, an alarming uncertainty. But for most men who considered the matter at all, the old picture still held true, the whole universe still centred on the earth, and the life of the earth upon man, its only rational inhabitant. The greatest intellectual achievement of the next century was to make it impossible for educated men to think like this. It was so important that it has been seen as the essential change to the modern from the medieval world.

Early in the seventeenth century something new is already apparent in science. The changes which then manifested themselves meant that an intellectual barrier was crossed and the nature of civilization was altered for ever. There appeared in Europe a new attitude, deeply utilitarian, encouraging men to invest time, energy and resources to master nature by systematic experiment. When a later age came to look back for its precursors in this attitude they found the outstanding one to have been Francis Bacon, sometime Lord Chancellor of England, fondly supposed by some later admirers to be the author of the plays of Shakespeare, a man of outstanding intellectual energy and many unlikeable personal traits. His works seem to have had little or no contemporary effect but

The title-page of the great work on human anatomy by the Flemish teacher of surgery at Padua, Andreas Vesalius. The word 'physiology' was coined in the year of its publication.

they attracted posterity's attention for what seemed a prophetic rejection of the authority of the past. Bacon advocated a study of nature based upon observation and induction and directed towards harnessing it for human purposes. 'The true and lawful end of the sciences', he wrote, 'is that human life be enriched by new discoveries and powers.' Through them could be achieved a 'restitution and reinvigorating [in great part] of man to the sovereignty and power... which he had in his first state of creation'. This was ambitious indeed - nothing less than the redemption of mankind from the consequences of Adam's Fall – but Bacon was sure it was possible if scientific research was effectively organized; in this, too, he was a prophetic figure, precursor of later scientific societies and institutions.

The modernity of Bacon was later exaggerated and other men – notably his contemporaries Kepler and Galileo – had much more to say which was of importance in the advance of science. Nor did his successors adhere so closely as he would have wished to a programme of practical discovery of 'new arts, endowments, and commodities for the bettering of man's life' (that is, to a science dominated by technology). Nevertheless, he rightly acquired something of the status of a mythological figure because he went to the heart of the matter in his advocacy of observation and experiment instead of deduction from *a priori* principles. Appropriately, he is said even to have achieved scientific martyrdom, having caught cold while stuffing a fowl with snow one freezing March day, in order to observe the effects of refrigeration upon the flesh. Forty years later, his central ideas were the commonplace of scientific discourse. 'The management of this great machine of the world', said an English scientist in the 1660s, 'can be explained only by the experimental and mechanical philosophers.' Here were ideas which Bacon would have understood and approved and which are central to the world which we still inhabit. Ever since the seventeenth century it has been a characteristic of the scientist that he answers questions by means of experiment and for a long time it was to lead to new attempts to understand what was revealed by these experiments by constructing systems.

This led at first to concentration on the physical phenomena which could best be observed and measured by the techniques available. Technological innovation had arisen from the slow accretion of skills by European workmen over centuries; these skills could now be directed to the solution of problems which would in turn permit the solution of other, intellectual problems. The invention of logarithms and calculus was a part of an instrumentation which had among other components the building of better clocks and optical instruments. When the clockmaker's art took a great stride forward with the seventeenth-century introduction of the pendulum as a controlling device it in turn made the measurement of time by precision instruments, and therefore astronomy, much easier. With the telescope came new opportunities to scrutinize the heavens; Harvey discovered the circulation of the blood as the result of a theoretical investigation by experiment, but *how* circulation took place was only made comprehensible when the microscope made it possible to see the tiny vessels through which blood flowed. Telescopic and microscopic observation were not only central to the discoveries of the scientific revolution, moreover, but made visible to laymen something of what was implied in a new world outlook.

What was not achieved for a long time was the line of demarcation between the scientist and philosopher which we now recognize. Yet a new world of scientists had come into being, a true scientific community and an international one, too. Here we come back to printing. The rapid diffusion of new knowledge was very important. The publication of scientific books was not its only form; the *Philosophical Transactions* of the Royal Society were published and so were, increasingly, the memoirs and proceedings of other learned bodies. Scientists moreover kept up voluminous private correspondences with one another, and much of the material they recorded in them has provided some of the most valuable evidence for the way in which scientific revolution actually occurred. Some of these correspondences were published; they were more widely intelligible and read than would be the exchanges of leading scientists today.

One feature of the scientific revolution remarkable to the modern eye is that it was something in which amateurs and part-time enthusiasts played a big part. It has been suggested that one of the most important facts explaining why science progressed in

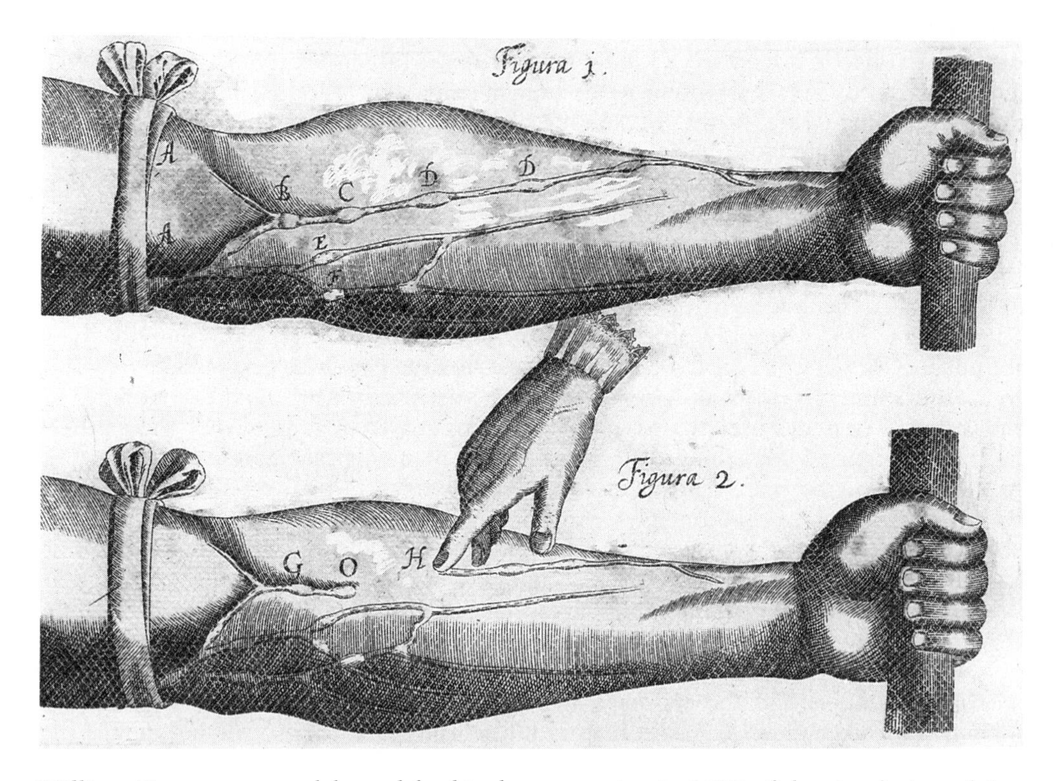

William Harvey, most celebrated for his demonstration in 1628 of the circulation of the blood, was also the author of a treatise on human anatomy from which this illustration is taken.

Europe while stagnation overtook even outstanding technical achievement in China, was the association with it in Europe of the social prestige of the amateur and the gentleman. The membership of the learned societies which began to appear more widely at about the mid-century was full of gentiemanly dabblers who could not by any stretch of imagination have been called professional scientists but who lent to these bodies the indefinable but important weight of their standing and respectability whether or not they got their hands dirty in experimental work.

By 1700 specialization between the major different branches of science already existed though it was by no means as important as it was to become. Nor was science in those days relentlessly demanding on time; scientists could still make major contributions to their study while writing books on theology or holding administrative office. This suggests some of the limitations of the seventeenth-century revolution; it could not transcend the limits of the techniques available and while they permitted great advances in some fields, they tended to inhibit attention to others. Chemistry, for example, made relatively small progress (though few still accepted the Aristotelian scheme of four elements which had still dominated thinking about the constituents of matter in 1600), while physics and cosmology went ahead rapidly and indeed arrived at something of a plateau of consolidation which resulted in less spectacular but steady advance well into the nineteenth century, when new theoretical approaches reinvigorated them.

Altogether, the seventeenth-century scientific achievement was a huge one. First and foremost, it replaced a theory of the universe which saw phenomena as the direct and often unpredictable operation of divine power by a conception of it as a mechanism, in which change proceeded regularly from the uniform and universal working of laws of motion. This was still quite compatible with belief in God. His majesty was not perhaps shown in daily direct intervention but in His creation of a great machine; in the most celebrated analogy God was the great watch-maker. Neither the typical student of science, nor the scientific world view of the seventeenth century was anti-religious or anti-theocentric. Though it was indubitably important that new views on astronomy, by displacing man from the centre of the universe, implicitly challenged his uniqueness (it was in 1686 that a book appeared arguing that there might be more than one inhabited world), this was not what preoccupied the men who made the cosmological revolution. For them it was only an accident that the authority of the Church became entangled with the proposition that the sun went round the earth. The new views they put forward merely emphasized the greatness and mysteriousness of God's ways. They took for granted the possibility of christening the new knowledge as Aristotle had been christened by the Middle Ages.

The great roll of the cosmological revolution has always been headed by the name of Copernicus, a Polish cleric whose book *On the Revolutions of the Celestial Orbs* was published in 1543. This was the same year as Vesalius' great work on anatomy (and, curiously, of the first edition of the works of Archimedes); Copernicus was a Renaissance humanist rather than a scientist – not surprisingly, considering when he lived. In part for philosophic and aesthetic reasons he hit upon the idea of a universe of planets moving round the sun, explaining their motion as a system of cycles and epicycles. It was (so to speak) a brilliant guess, for he had no means of testing the hypothesis and most common-sense evidence told against it.

The first true scientific data in support of heliocentricity was in fact provided by a man who did not accept it, the Dane Tycho Brahe. Besides possessing the somewhat striking distinction of an artificial nose, Brahe began recording the movements of planets, first with rudimentary instruments and then, thanks to a munificent king, from the best-equipped observatory of his age. The result was the first systematic collection of astronomical data to be made within the orbit of the western tradition since the Alexandrian era. Kepler, the first great Protestant scientist, who was invited by Brahe to assist him, went on to make even more careful observations of his own and provide a second major theoretical step forward. He showed that the movements of planets could be explained as regular if their courses followed ellipses at irregular speeds. This broke at last with the Ptolemaic framework within which cosmology had been more and more cramped and provided the basis of planetary explanation until the twentieth century. Then came Galileo Galilei, who eagerly seized upon the telescope, an instrument seemingly discovered about 1600, possibly by chance. Galileo was an academic, professor at Padua of two subjects characteristically linked in early science: physics and military engineering. His use of the telescope finally shattered the Aristotelian scheme; Copernican astronomy was made visible and the next two centuries were to apply to the stars what was known of the nature of the planets.

Galileo's major work, nevertheless, was not in observation but theory. He first described the physics which made a Copernican universe possible by providing a mathematical treatment of the movement of bodies. With his work, mechanics left the world of the craftman's knowhow, and entered that of science. What is more, Galileo came to his conclusions as a result of systematic experiment. On this rested what Galileo called 'two new sciences', statics and dynamics. The published result was the book in which has been seen the first statement of the revolution in scientific thought, Galileo's *Dialogue on the Two Great Systems of the World* (that of Ptolemy and that of Copernicus) of 1632. Less remarkable than its contents, but still interesting, are the facts that it was written not in Latin but the vernacular Italian, and dedicated to the pope; Galileo was undoubtedly a good Catholic. Yet the book provoked an uproar, rightly, for it meant the end of the Christian-Aristotelian world view which was the great cultural triumph of the medieval Church. Galileo's trial followed. He was condemned and recanted, but this did not diminish the effect of his work. Copernican views henceforth dominated scientific thinking.

In the year that Galileo died, Newton was born. It was his achievement to provide the physical explanation of the Copernican universe; he showed that the same mechanical laws explained both what Kepler and what Galileo had said, and finally brought together terrestial and celestial knowledge. He employed a new mathematics, the 'method of fluxions' or, in later terminology, the infinitesimal calculus. Newton did not invent this; he applied it to physical phenomena. It provided a way of calculating the positions of bodies in motion. His conclusions were set out in a discussion of the movements of the planets contained in a book which was the most important and influential scientific work since that of Euclid. The *Principia*, as it is called for short (or, Anglicized, *The Mathematical Principles of Natural Philosophy*), demonstrated how gravity sustained the physical universe. The general cultural consequences of this discovery were comparable with those within science. We have no proper standard of measurement, but perhaps they were even greater. That a single law, discovered by observation and calculation, could explain so much was an astonishing revelation of what the new scientific thinking could achieve. Pope has been quoted to excess, but his epigram still best summarizes the impact of Newton's work on educated men:

Nature and Nature's laws lay hid in night:
God said, 'Let Newton be!' and all was light.

Newton thus in due time became, with Bacon, the second of the canonized saints of a New Learning. There was little exaggeration in this in Newton's case. He was a man of almost universal scientific interests and, as the phrase has it, touched little that he did not adorn. Yet the significance of much of what Newton did is bound to elude the non-scientist. Manifestly, he completed the revolution begun with Copernicus. A dynamic conception of the universe had replaced a static one. His achievement was great enough to provide the physics of the next two centuries and to underpin all the other sciences with a new cosmology.

What was not anticipated by Newton and his predecessors was that this might presage an insoluble conflict of science and religion. Newton, indeed, seems even to have been pleased to observe that the law of gravity did not adequately sustain the view that the universe was a self-regulated system, self-contained once created; if it was not just a watch, its creator could do more than invent it, build it, wind it up and then stand back. He welcomed the logical gap which he could fill by postulating divine intervention, for he was a passionate Protestant apologist. Churchmen, especially Catholic, nevertheless did not find it easy to come to terms with the new science. In the Middle Ages clerics had made important contributions to science, but from the seventeenth to the mid-nineteenth century, very little first-rank scientific work was done by churchmen. This was truer, certainly, of the countries where the Counter-Reformation had triumphed than those where it had not. In the seventeenth century there opened that split between organized religion and science which has haunted European intellectual history ever since, whatever efforts have from time to time been made to patch it up. The symbolic crisis was that of the Neapolitan Bruno. He was not a scientist but a speculator, formerly a Dominican monk who broke with his order and wandered about Europe publishing controversial works, dabbling in a magical 'secret science' supposedly derived from ancient Egypt. In the end the Inquisition took him and after eight years in its hands he was burned at Rome for heresy. His execution became one of the foundations of the later historical mythology of the development of 'free thought', of the struggle between progress and religion as it was to come to be seen.

In the seventeenth century such an antithesis was not much felt by scientists and philosophers. Newton, who wrote copiously on biblical and theological topics and believed his work on the prophetical books to be as flawless as the *Principia*, seems to have held that Moses knew about the heliocentric theory and recommended his readers to 'beware of Philosophy and vain deceit and oppositions of science falsely so called' and to have recourse to the Old Testament. Napier, the inventor of logarithms, was delighted to have in them a new tool to deploy in deciphering the mysterious references in the Book of Revelation to the Number of the Beast. The French philosopher Descartes formulated what he found to be satisfactory philosophical defences of religious belief and Christian

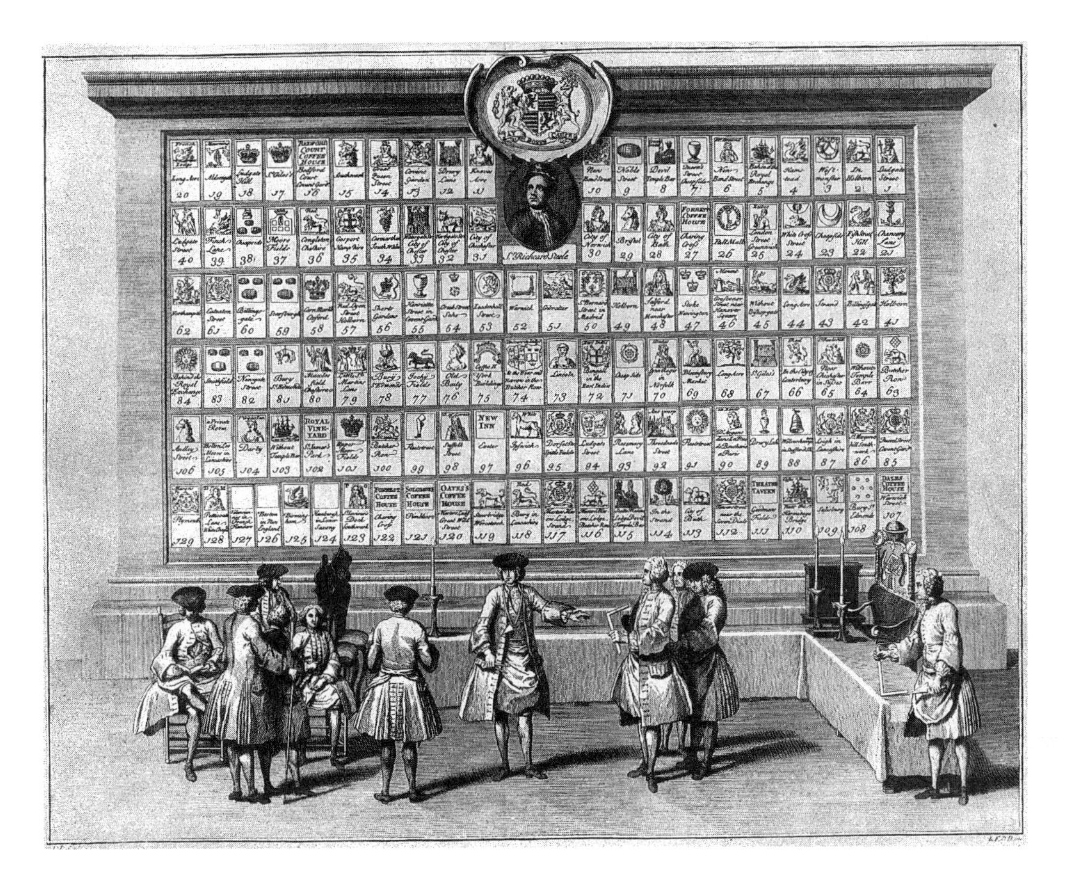

An English masonic print of the eighteenth century, when English lodges still met usually in taverns – hence the inn-signs displayed behind the freemasons, who wear masonic aprons and carry symbols of the craft.

truth coherent with his technically sceptical approach to his subject. This did not prevent him (or the philosophical movement which took its name from him, Cartesianism) from attracting the hostility of the Church. The traditional defenders of religious belief correctly recognized that what was at stake was not only the conclusions people arrived at, but the way that they arrived at them. A rationally argued acceptance of religious belief which started from principles of doubt and demonstrated they could satisfactorily be overcome was a poor ally for a Church which taught that truth was declared by Authority. The Church was quite right in setting aside as irrelevant Descartes' own devotion and Christianity and correctly (from its own point of view) put all his works on the Index.

The argument from authority was taken up by a French Protestant clergyman of the later seventeenth century, Pierre Bayle, who pointed out that it had an unsatisfactory open-endedness. What authority prescribed the authority? In the end it seemed to be a matter of opinion. Every dogma of traditional Christianity, he suggested, might be refuted if not in accordance with natural reason. With such ideas a new phase in the history of European thought announced itself; it has been called the Enlightenment.

This word and similar ones were used in the eighteenth century in most European languages to characterize the thinking which men felt distinguished their own age and cut it off from what had gone before. The key image is of the letting in of light upon what was dark, but when the German philosopher Kant asked the question 'What is enlightenment?' in a famous essay he gave a different answer: liberation from self-imposed tutelage. At its heart lay a questioning of authority. The great heritage to be left behind by the Enlightenment was the generalizing of the critical attitude. From this time, everything was exposed to scrutiny. Some felt – and it came in the very long run to be true – that nothing was sacred, but this is somewhat misleading. Enlightenment had its own authority and dogmas; the critical stance itself long went unexamined. Furthermore, Enlightenment was as much a bundle of attitudes as a collection of ideas and here lies another difficulty in coming to terms with it. Many streams flowed into this result but by no means did they all follow the same course. The roots of Enlightenment are as confused as its development which always resembled a continuing debate – sometimes a civil war – much more than the advance of a united army of the enlightened.

Descartes had taught men that systematic doubt was the beginning of firm knowledge. Fifty years later, the English philosopher John Locke provided an account of the psychology of knowledge which reduced its primary constituents to the impressions conveyed by the senses to the mind; there were not, he argued against Descartes, ideas innate in man's nature. The mind contained only sense-data and the connexions it made between them. This was, of course, to imply that mankind had no fixed ideas of right and wrong; moral values, Locke taught, arose as the mind experienced pain and pleasure. There was to be an enormous future for the development of such ideas; from them would flow ideas about education, about society's duty to regulate material conditions and about many other derivations from environmentalism. There was also a huge past behind them: the dualism which Descartes and Locke both expressed in their distinctions of body and mind, physical and moral, have their roots in Plato and Christian metaphysics. Yet what is perhaps most striking at this point is that his ideas could still be associated by Locke with the traditional framework of Christian belief.

Such incoherences were always to run through the Enlightenment, but its general trend is clear. The new prestige of science, too, seemed to promise that the observations of the senses were, indeed, the way forward to knowledge, and knowledge whose value was proved by its utilitarian efficacy. It could make possible the improvement of the world in which men lived. Its techniques could unlock the mysteries of nature and reveal their logical, rational foundations in the laws of physics and chemistry.

All this was long an optimistic creed (the word *optimiste* entered the French language in the seventeenth century). The world was getting better and would continue to do so. In 1600 things had been very different. Then, the Renaissance worship of the classical past had combined with the upheavals of war and the always latent feeling of religious men that the end of the world could not long be delayed, to produce a pessimistic mood and a sense of decline from a great past. In a great literary debate over whether the achievements of the ancients excelled those of modern times the writers of the late seventeenth century crystallized the idea of progress which emerged from the Enlightenment. It was also a non-specialists' creed. In the eighteenth century it was still possible for an educated man to tie together in a manner satisfactory at least for himself the logic and implications of many different studies. Voltaire was famous as a poet and playwright, but wrote at length on history (he was for a time the French historiographer royal) and expounded Newtonian physics to his countrymen. Adam Smith was renowned as a moral philosopher before he dazzled the world with his *Wealth of Nations*, a book which may reasonably be said to have founded the modern science of economics.

In such eclecticism religion, too, found a place, yet (as Gibbon put it) 'in modern times, a latent, and even involuntary, scepticism adheres to the most pious disposition.' In 'enlightened' thought there seemed to be small room for the divine and the theological. It was not just that men no longer felt hell gaping about them and that the world became less mysterious; it also promised to be less tragic. More and more troubles seemed not inseparable from being, but man-made. Awkward problems, it was true, might still be presented by appalling natural disasters such as earthquakes, but if the relief of most ills was possible, if, as one thinker put it, 'Man's proper business is to seek happiness and avoid misery', what was the relevance of the dogmas of Salvation and Damnation? God could still be included in a perfunctory way in the philosopher's account of the universe, as the First Cause that had started the whole thing going and the Great Mechanic who prescribed the rules on which it ran, but was there any place for His subsequent intervention in its working, either directly by incarnation or indirectly through His Church and the sacraments it conveyed? Inevitably, the Enlightenment brought revolt against the Church, the supreme claimant to intellectual and moral authority.

For all that had been done by Leonardo and the great scientific engravers, science was slow to become part of the subject-matter of fine art. Joseph Wright of Derby was one of the first, and has remained one of the few English painters to choose subjects from science and technology. This picture is entitled 'The air-pump'.

Here was a fundamental conflict. The rejection of authority by thinking men in the seventeenth and eighteenth centuries was only rarely complete, in the sense that new authority was sought and discovered in what were believed to be the teachings of science and reason. Yet increasingly and more and more emphatically the authority of the past was rejected. As the literary argument over ancient and modern culture advanced the destruction of the authority of classical teaching, so had the Protestant Reformation exploded the authority of the Catholic Church, the other pillar of traditional European culture. When the Protestant reformers had replaced old priest by new presbyter (or by the Old Testament) they could not undo the work of undermining religious authority which they had begun and which the men of the Enlightenment were to carry much further.

Such implications took some time to emerge, whatever the quickly formulated and justified misgivings of churchmen. The characteristics of advanced thought in the eighteenth century tended to express themselves in fairly practical and everyday recommendations which in a measure masked their tendency. They are probably best summarized in terms of the fundamental beliefs which underlay them and of which they were consequences. At the basis of all others was a new confidence in the power of mind; this was one reason why the Enlightened so much admired Bacon, who shared this with them, yet even the creative giants of the Renaissance did not do so much to give European man a conviction of intellectual power as did the eighteenth century. On this rested the assurance that almost indefinite improvement was possible. Most thinkers of the age were optimists who saw it as the apex of history. Confidently they looked forward to the improvement of the lot of mankind by the manipulation of nature and the unfolding to man of the truths which reason had written in his heart. Innate ideas bundled out by the front door crept in again by the back stairs. Optimism was qualified only by the realization that there were big practical obstacles to be overcome. The first of these was simply ignorance. Perhaps a knowledge of Final Causes was impossible (and certainly science seemed to suggest this as it revealed more and more complexity in nature) but this was not the sort of ignorance which worried the Enlightened. They had a more everyday level of experience in mind and the combination of reason and knowledge gave confidence that ignorance could be dispersed. The greatest literary embodiment of Enlightenment had precisely this aim. The great *Encyclopédie* of Diderot and D'Alembert was a huge compilation of information and propaganda in twenty-one volumes published between 1751 and 1765. As some of its articles made clear, another great obstacle to enlightenment was intolerance – especially when it interfered with freedom of publication and debate. Parochialism was yet another barrier to happiness. The values of the Enlightenment, it was assumed, were those of all civilized men. They were universal. Never, except perhaps in the Middle Ages, has the European intellectual élite been more cosmopolitan or shared more of a common language. Its cosmopolitanism was increased by knowledge of other societies, for which the Enlightenment showed an extraordinary appetite. In part this was because of genuine curiosity; travel and discovery brought to men's notice new ideas and institutions and thus made them more aware of social and ethical relativity and provided new grounds for criticism. Above all, a supposedly humane and enlightened China captured the imagination of eighteenth-century Europeans, a fact which perhaps suggests how superficial was their acquaintance with its realities.

Once ignorance, intolerance and parochialism were removed, it was assumed that the unimpeded operation of the laws of nature, uncovered by reason, would promote the reform of society in everyone's interest except that of those wedded to the past by their blindness or their enjoyment of indefensible privilege. The *Lettres persanes* of the French author Montesquieu began the tradition of suggesting that the institutions of existing societies – in his case the laws of France – could be improved by comparison with the laws of nature. In articulating such a programme, the men of the Enlightenment were appointing themselves as the priesthood of a new social order. In their vision of their role as critics and reformers there emerged for the first time a social ideal which has been with us ever since, that of the intellectual. Moralists, philosophers, scholars, scientists already existed; their defining characteristic was specialized competence. What the Enlightenment invented was the ideal of the generalized critical intellect. Autonomous, rational, continuous and universal criticism was institutionalized as never before and the modern intellectual is the outcome.

The eighteenth century did not use this term. It had the type, but called its exemplars simply 'philosophers'. This was an interesting adaptation and broadening of a word already familiar; it came to connote not the specialized mental pursuit of philosophical studies but the acceptance of a common outlook and critical stance. It was a term with moral and evaluative tones, used familiarly by enemies as well as friends to indicate also a zeal to propagate the truths revealed by critical insight to a large and lay public. The archetypes were a group of French writers soon lumped together in spite of their differences and referred to as *philosophes*. Their numbers and celebrity correctly suggest the preponderance of France in the central period of Enlightenment thought. Other countries neither produced so many and such conspicuous figures within this tradition, nor did they usually confer such prestige and eminence on those they had. Yet the presiding deities of the early Enlightenment were the English Newton and

Locke; it could be reasonably claimed too that the philosopher who expressed the most extreme development of Enlightenment ideals and methods was Bentham, and that its greatest historiographical monument is Gibbon's work. Further north, Scotland had a great eighteenth-century cultural efflorescence and produced in Hume one of the most engaging as well as the most acute of the Enlightenment's technical philosophers, who combined extreme intellectual scepticism with good nature and social conservatism, and in Adam Smith the author of one of the great creative books of modern times. Among Latin countries, Italy was, outside France, most prolific in its contribution to the Enlightenment in spite of the predominance there of the Roman Church. The Italian Enlightenment would be assured of a remembrance even if it had thrown up only Beccaria, the author of a book which founded penal reform and the criticism of penology and gave currency to one of the great slogans of history, 'the greatest happiness of the greatest number'. The German Enlightenment was slower to unroll and less productive of figures who won universal acclaim (possibly for linguistic reasons) but produced in Kant a thinker who, if he consciously sought to go beyond the Enlightenment, nevertheless embodied in his moral recommendations much of what it stood for. Only Spain seemed to lag conspicuously. It was not an unfair impression even allowing for the work of one or two enlightened statesmen; Spanish universities in the eighteenth century were still rejecting Newton.

Important though the work of other nations was for the history of civilization, that of the French struck contemporaries the most forcefully. There were many reasons: a simple one is that men are always fascinated by power and France under Louis XIV had acquired a prestige long to endure. Another reason is the magnificent instrument for the diffusion of French culture which lay to hand in the French language. It was in the eighteenth century the *lingua franca* of Europe's intellectuals and its people of fashion alike; Maria Theresia and her children used it for their family correspondence and Frederick II wrote (bad) verses in it. A European audience was assured for any book written in French and it seems likely that the success of that language actually held back cultural advance in the German tongue.

A common language made possible propaganda, discussion and critical comment, but what would actually be achieved by way of practical reform in the short term was bound to depend on political circumstance. Some statesmen attempted to put 'enlightened' ideas into practice, because there were coincidences between the interests of states and the aims of philosophers. This was especially apparent when 'enlightened despotisms' found themselves running into opposition from vested interest and conservatism. Such conflicts were obvious in the enforcement of educational reform at the expense of the Church inside the Habsburg dominions, or in Voltaire's attacks, written to the brief of a royal minister, on the *parlement* of Paris when it stood in the way of fiscal innovation. Some rulers, like Catherine the Great of Russia, ostentatiously paraded the influence of Enlightenment ideas on their legislation. Perhaps the most important and influential impact of such ideas, apart from those of utilitarian reform which were deployed against the Church, was always in educational and economic matters. In France, at least, the economic recommendations of enlightened thinkers made their mark on administration.

Religious questions drew the attention of the *philosophes* with unique power. The Church and the effects of its teaching were, of course, still inseparable from every side of Europe's life. It was not just that the Church claimed authority in so much, but also that it was physically omnipresent as a great corporate interest, both social and economic; it was involved in some measure in every aspect of society to which the attention of reformers might be drawn. Whether it was because the abuse of sanctuary or clerical privilege stood in the way of judicial reform, or mortmain impeded economic improvement, or a clerical monopoly of education encumbered the training of administrators, or dogma prevented the equal treatment of loyal and valued subjects, the Church seemed to find itself always opposing improvement. But this was not all that drew the fire of the *philosophes*. Religion could also lead, they thought, to crime. One of the last great scandals of the era of religious persecution was the execution of a Protestant at Toulouse in 1762 on the charge of converting Catholics to heresy. For this he was tortured, tried, convicted and executed. Voltaire made this a *cause célèbre*. His efforts did not change the law, but for all the violence of feeling which continued to divide Catholic and Protestant in southern France, they made it impossible for such a judicial murder ever to be repeated there – or, probably, in France as a whole. Yet France did not give even a limited legal toleration to Protestants until 1787 and then did not extend it to Jews. By that time Joseph II had already introduced religious toleration into his Catholic territories.

This suggests an important limit to the practical success of enlightenment. For all its revolutionary power, it had to operate within the still very restrictive institutional and moral framework of the ancien régime. Its relationship with despotism was ambiguous: it might struggle against the imposition of censorship or the practice

The patriarch of the Enlightenment, Voltaire, gets up, dictating to a secretary while he dresses.

Self-appointed advocate of the pure heart and uncorrupted feeling, the tormented Jean-Jacques Rousseau.

of religious intolerance in a theocratic monarchy, but could also depend on despotic power to carry out reform. Nor, it must be remembered, were enlightened ideas the only stimulus to improvement. The English institutions Voltaire admired did not stem from enlightenment and many changes in eighteenth-century England owed more to religion than to 'philosophy'.

The greatest political importance of the Enlightenment lay in its legacies to the future. It clarified and formulated many of the key demands of liberalism, though here, too, its legacy is ambiguous, for the men of the Enlightenment sought not freedom for its own sake but freedom for the consequences it would bring. The possibility of contriving that mankind should be happy on earth was the key invention of the eighteenth century; it may be said, indeed, not merely to have invented earthly happiness as a feasible goal but also the thought that it could be measured (Bentham wrote of a 'felicific calculus') and that it could be promoted through the exercise of reason. Those ideas all had profound political implications.

Apart from this, the age made its best-known contribution to the future European liberal tradition in a more specific and negative form; the Enlightenment created classical anti-clericalism. Criticism of what the Church had done led to support for attacks by the state upon ecclesiastical organizations and authority. The struggles of Church and State had many roots other than philosophical, but could always be presented as a part of a continuing war of Enlightenment and rationality against superstition and bigotry. In particular, the Papacy attracted criticism – or contempt; Voltaire seems to have once believed that it would in fact disappear before the end of the century. The greatest success of the *philosophes* in the eyes of their enemies and of many of their supporters was the papal dissolution of the Society of Jesus.

A few *philosophes* carried their attacks on the Church beyond institutions to an attack on religion itself. Out-and-out atheism (together with deterministic materialism) had its first serious expression in the eighteenth century, but this was unusual. Most of those during the Enlightenment era who thought about these things were probably sceptical about the dogmas of the Church, but kept up a vague theism. Certainly, too, they believed in the importance of religion as a social force. As Voltaire said, 'one must have religion for the sake of the people'. He, in any case, continued throughout his life to assert, with Newton, the existence of God and died formally at peace with the Church.

Here is a hint of something always in danger of being lost to sight in the Enlightenment, the importance of the non-intellectual and non-rational side of human nature. The most prophetic figure of the century in this respect and one who quarrelled bitterly with many of the leading figures among the 'enlightened' and the *philosophes* was the Genevan Rousseau. His importance in the history of thought lies in his impassioned pleas that due weight be given to the feelings and the moral sense, both in danger of eclipse by rationality. Because of this, he thought, the men of his day were stunted creatures, partial and corrupt beings, deformed by the influence of a society which encouraged this eclipse.

European culture owes an enormous amount to Rousseau's vision, much of it to prove pernicious in its effect. He planted (it has been well said) a new torment in every soul. There can be found in his writings a new attitude to religion (which was to revivify it), a new psychological obsession with the individual which was to flood into art and literature, the invention of the sentimental approach to nature and natural beauty, the origins of the modern doctrine of nationalism, a new child-centredness in educational theory, a secularized puritanism (rooted in a mythical view of ancient Sparta), and much else besides. All these things had both good and bad consequences; Rousseau was, in short, the key figure in the making of what has been called Romanticism. In much he was an innovator, and often one of genius. Much, too, he shared with other men. His distaste for the Enlightenment erosion of community, his sense that men were brothers and members of a social and moral whole was, for example, expressed just as eloquently by the Irish author Edmund Burke, who nevertheless drew from it very different conclusions. Rousseau was in some measure voicing views beginning to be held by others as the age of Enlightenment passed its zenith. Yet of Rousseau's central and special importance to Romanticism, there can be no doubt.

Romanticism is a much used and much misused term. It can be properly applied to things which seem diametrically opposed. Soon after 1800, for example, some men would deny any value to the past and seek to overthrow its legacies just as violently as men of the Enlightenment had done, while at the same time others tenaciously defended historic institutions. Both can be (and have been) called Romantics, because in each of them moral passion counted for more than intellectual analysis. The clearest link between such antitheses lay in the new emphasis of romantic Europe on feeling, intuition, and, above all, the natural. Romanticism, whose expressions were to be so manifold, started almost always from some objection to enlightened thought, whether from disbelief that science could provide an answer to all questions, or from a revulsion against rational self-interest. But its positive roots lay deeper than this, in the Reformation's displacement of so many traditional values by the one supreme value of sincerity; it was not entirely wrong to see Romanticism as some Catholic critics saw it, as a secularized Protestantism, for above all it sought authenticity, self-realization, honesty, moral exaltation. Unhappily it did so all too often without regard to cost. The great effects were to reverberate through the nineteenth century, usually with painful results, and in the twentieth century would affect many other parts of the world as one of the last manifestations of the vigour of European culture.

The Queen-Empress. Victoria R.I., ruler of an empire larger in extent and more numerous than any in history, looks out over the maidan of Calcutta, long the capital of British India.

BOOK VI

THE GREAT ACCELERATION

In the middle of the eighteenth century most people in the world (and perhaps most Europeans, too) could still believe that history would go on much as it seemed always to have done. The weight of the past was everywhere enormous and often it was immovable: some of the European efforts to shake it off have been touched upon, but nowhere outside Europe was even the possibility of doing so grasped. Though in many parts of the world a few people's lives had begun to be revolutionized by contact with Europeans, most of it was unaffected and much of it was untouched by such contamination of traditional ways.

In the next century and a half change was to come thick and fast almost everywhere and to ignore the fact was to be much harder if not impossible. By 1900 it was obvious that in Europe and the European world of settlement it had irreversibly cut off much of the traditional past. Just as important, impulses from northern Europe and the Atlantic world have also radiated outwards to transform both Europe's relations with the other continents and the very foundations of their lives for many of their peoples, however much some of them regretted and resisted it. By the end of the nineteenth century (and this is only an approximate and convenient marker) a world once regulated by tradition was on a new course. Its destiny was now to be continuing and accelerating transformation and the second adjective was as important as the first. A man born in 1800 who lived out the psalmist's span of three-score years and ten could have seen the world more changed in his lifetime than it had been in the previous thousand years.

The consolidation of the European world hegemony was central to these changes and one of the great motors propelling them. By 1900 European civilization had shown itself to be the most successful which had ever existed. Men might not always agree on what was most important about it but no one could deny that it had produced wealth on an unprecedented scale and that it dominated the rest of the globe by power and influence as no previous civilization had ever done. Europeans (or their descendants) ran the world. Much of their domination was political, a matter of direct rule. Large areas of the world had been peopled by European stocks. As for the non-European countries still formally and politically independent of Europe, most of them had in practice to defer to European wishes and accept European interference in their affairs. Few indigenous peoples could resist, and if they did Europe often won its subtlest victory of all, for successful resistance required the adoption of European practices and, therefore, Europeanization in another form.

1
Long-Term Change

In 1798 Thomas Malthus, an English clergyman, published an *Essay on Population* which was to prove the most influential book ever written on the subject. He described what appeared to be the laws of population growth but his book's importance transcended this apparently limited scientific task. Its impact on, for example, economic theory and biological science was to be just as important as the contribution it made to demographic studies. Here, though, such important consequences matter less than the book's status as a symptom of a change in thinking about population. Roughly speaking, for two centuries or so European statesmen and economists had agreed that a rising population was a sign of prosperity. Kings should seek to increase the number of their subjects, it was thought, not merely because this would provide more taxpayers and soldiers but because a bigger population both quickened economic life and was an indication that it had done so. Obviously, larger numbers showed that the economy was providing a living for more people. This view was in its essentials endorsed by no less an authority than the great Adam Smith himself, whose *Wealth of Nations*, a book of huge influence, had agreed as recently as 1776 that an increase in population was a good rough test of economic prosperity.

Malthus doused this view with very cold water. Whatever the consequences for society as a whole might be judged to be, he concluded that a rising population sooner or later spelt disaster and suffering for most of its members, the poor. In a famous demonstration he argued that the produce of the earth had finite limits, set by the amount of land available to grow food. This in turn set a limit to population. Yet population always tended to grow in the short run. As it grew, it would press increasingly upon a narrowing margin of subsistence. When this margin was exhausted, famine must follow. The population would then fall until it could be maintained with the food available. This mechanism could only be kept from operating if men and women abstained from having children (and prudence, as they regarded the consequences, might help them by encouraging late marriage) or by such horrors as the natural checks imposed by disease or war.

Much more could be said about the complexity and refinement of this gloomy thesis. It aroused huge argument and counter-argument, and whether true or false, a theory attracting such attention must tell us much about the age. Somehow, the growth of population had begun to worry people so that even prose so unattractive as that of Malthus had great success. People had become aware of population growth as they had not been aware of it before and had done so just as it was to become faster than ever. In the nineteenth century, in spite of what Malthus had said, the numbers of some divisions of the human race went up with a rapidity and to levels hitherto inconceivable.

A long view is best for measuring such a change; there is nothing to be gained and much to be lost by worrying about precise dates and the overall trends run on well into the twentieth century. If we include Russia (whose population has until very recent times to be estimated from very poor statistics) then a European population of about one hundred and ninety million in 1800 rose to about four hundred and twenty million a century later. As the rest of the world seems to have grown rather more slowly, this represented a rise in Europe's share of the total population of the world from about

The problems of paternity: a Mr Quiverful grapples with the problems of the new census return in an English cartoon of the early nineteenth century.

one-fifth to one-quarter; for a little while, her disadvantage in numbers by comparison with the great Asiatic centres of population was reduced (while she continued to enjoy her technical and psychological superiority). Moreover, at the same time, Europe was sustaining a huge emigration of her stocks. In the 1830s European emigration overseas first passed the figure of a hundred thousand a year; in 1913 it was over a million and a half. Taking an even longer view, perhaps fifty million people left Europe to go overseas between 1840 and 1930, most of them to the western hemisphere. All these people *and their descendants* ought to be added to the totals in order to grasp how much European population growth accelerated in these years.

This growth was not shared evenly within Europe and this made important differences to the standing of great powers. Their strength was usually reckoned in terms of military manpower and it was a crucial change that Germany replaced France as the largest mass of population under one government west of Russia in the second half of the nineteenth century. A dramatic increase had been shown earlier in the United Kingdom, whose population grew from about 8 million when Malthus wrote to 22 million by mid-century. It was to reach 36 million by 1914. Another way of looking at such changes would be to compare the respective shares of Europe's population enjoyed by the major military powers at different dates. Between 1800 and 1900, for example, that of Russia grew from 21 to 24 per cent of the total, Germany's from 13 to 14, while France's fell from 15 to 10 per cent, and that of Austria slightly less, from 15 to 12.

Yet population grew everywhere, though at different rates at different times. The poorest agrarian regions of eastern Europe, for example, experienced their highest growth rates only in the 1920s and 1930s. This is because the basic mechanism of population increase in this period, underlying change everywhere, was a fall in mortality. Never in history has there been so spectacular a fall in death-rates as in the last hundred years, and it showed first in the advanced countries of Europe in the nineteenth century. Roughly speaking, before 1850 most European countries had birth-rates which slightly exceeded death-rates and both were about the same in all countries. They showed, that is to say, how little impact had been made by that date upon the fundamental determinants of human life in a still overwhelmingly rural society. After 1880 this changed rapidly. The death-rate in advanced European countries fell pretty steadily, from about 35 per thousand inhabitants per year to about 28 by 1900; it would be about 18 fifty years later. Less advanced countries still maintained rates of 38 per thousand between 1850 and 1900, and 32 down to 1950. This produced a striking inequality between two Europes. In the richer, expectation of life was much higher. Since, in large measure, advanced European countries lay in the west, this was (leaving out Spain, a poor country with high mortality) a fresh intensification of older divisions between east and west, a new accentuation of the imaginary line from the Baltic to the Adriatic.

Other factors besides lower mortality helped. Earlier marriage and a rising birth-rate had showed themselves in the first phase of expansion, as economic opportunity increased, but now they mattered much more, since from the nineteenth century onwards, the children of earlier marriages were much more likely to survive, thanks to greater humanitarian concern, cheaper food and medical and engineering progress. Of these, medical science and the provision of medical services were the last to influence population trends. Doctors only came to grips with the great killing diseases from about 1870 onwards; these were the child-killers: diphtheria, scarlet fever, whooping-cough, typhoid. Infant mortality was thus dramatically reduced and expectation of life at birth greatly increased. But earlier than this, social reformers and engineers had already done much to reduce the incidence of these and other diseases (though not their fatality) by building better drains and devising better cleaning arrangements for the growing cities. Cholera was eliminated in industrial countries by 1900, though it had devastated London and Paris in the 1830s and 1840s. No western European country had an important plague outbreak after 1899. As such changes affected more and more countries, their general tendency was everywhere to raise the average age of death with, in the long run, dramatic results. By the second quarter of the twentieth century, men and women in North America, the United Kingdom, Scandinavia and industrial Europe could expect to live two or three times as long as their medieval ancestors. Immense consequences flowed from this.

Just as accelerated population increase first announced itself in those countries which were economically the most advanced, so did the slowing down of growth which was the next discernible demographic trend. This was produced by a declining number of births, though it was for a long time masked because the fall in the death-rate was even faster. In every society this showed itself first among the better-off; to this day, it remains a good rough working rule that fecundity varies inversely with income (celebrated exceptions among wealthy American political dynasties notwithstanding). In some societies (and in western rather than eastern Europe) this was because marriage tended to be put off longer so that women were married for less of their fertile lives; in some it was because couples chose to have fewer children – and could now do so with confidence, thanks to effective contraceptive techniques. Possibly there had long been some knowledge of such techniques in some European countries; it is at least certain that the nineteenth century brought both improvements in them (some made possible by scientific and technical advance in manufacturing the necessary devices) and propaganda which spread knowledge of them. Once more, a social change touches upon a huge ramification of influences, because it is difficult not to connect such spreading knowledge with, for example, greater literacy, and with rising expectations. Although people were beginning to be wealthier than their ancestors, they were all the time adjusting their notion of what was a tolerable life – and therefore a tolerable size of family. Whether they followed the calculation by putting off the date of marriage (as French and Irish peasants did) or by adopting contraceptive techniques (as the English and French middle classes seem to have done) was shaped by other cultural factors.

Changes in the ways men and women died and lived in their families transformed the structures of society. On the one hand, the western countries in the nineteenth and twentieth centuries had absolutely more young people about and, for a time, also had them about in a greater proportion than ever before. It is difficult not to think that the expansiveness, buoyancy and vigour of nineteenth-century Europe owed much to this. On the other hand, advanced societies gradually found a much higher percentage of their members surviving into old age than ever before. This increasingly strained the social mechanism which had in earlier centuries maintained the old and those incapable of work; the problem grew worse as competition for industrial employment became more intense. By 1914, in almost every European or North American country much thought had been given to ways of confronting the problems of poverty and dependence, however great the differences in scale and success of efforts to cope with them.

The soup-kitchen of the Victorian slum replaced the monastery's distribution of its broken meats in the Middle Ages. Once destitution was identified as a responsibility for society as a whole, the first relief measures sought to improve, rationalize and enlarge the traditional devices of charity.

Such trends would not begin to show in eastern Europe until after 1918, when their general pattern was already well established in the advanced western countries. Death-rates long continued to fall more sharply than did the birth-rate, even in advanced countries, so that down to the present the population of Europe and the European world has continued to rise. It is one of the most important themes in the history of the era, linked to almost every other. Its material consequences can be seen in unprecedented urbanization and the rise of huge consumer markets for manufacturing industry. The social consequences range from strife and unrest to changing institutions to grapple with them. There were international repercussions as statesmen took into account population figures in deciding what risks they could (and which they had to) take, or as people became more and more alarmed about the consequences of overcrowding. Worries in the nineteenth-century United Kingdom over the prospect of too many poor and unemployed led to the encouragement of emigration which, in its turn, shaped people's thinking and feelings about empire. Later, the Germans discouraged emigration because they feared the loss of military potential, while the French and Belgians pioneered the award of children's allowances for the same reason.

Some of these measures suggest, correctly, that the gloomy prophecies of Malthus tended to be forgotten as the years went by and the disasters he feared did not take place. The nineteenth century still brought demographic calamities to Europe; Ireland and Russia had spectacular famines and near-famine conditions occurred in many places. But such disaster grew rarer. As famine and dearth were eliminated from advanced countries, this in turn helped to make disease demographically less damaging. Meanwhile Europe north of the Balkans enjoyed two long periods of virtually undisturbed peace from 1815 to 1848 and from 1871 to 1914; war, another of Malthus' checks, also seemed to be less of a scourge. Finally, his diagnosis actually seemed to be disproved when a rise in population was accompanied by higher standards of living – as rises in the average age

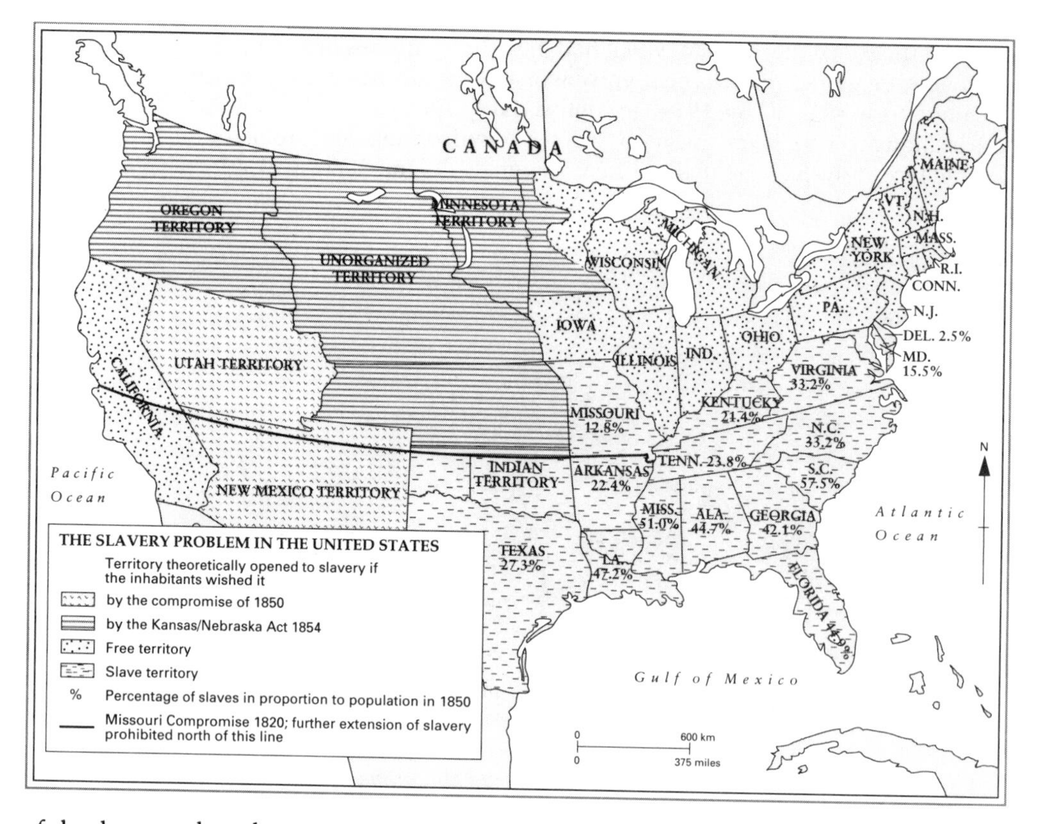

of death seemed to show. Pessimists could only reply (reasonably) that Malthus had not been answered; all that had happened was that there had turned out to be much more food available than had been feared. It did not follow that supplies were limitless.

In fact, there was occurring another of those few great changes of history which have decisively transformed the basic conditions of human life. It can reasonably be called a food-producing revolution. Its beginnings have already been traced. In the eighteenth century European agriculture was already capable of obtaining about two and a half times the yield on its seed normal in the Middle Ages. Now even greater agricultural improvement was at hand. Yields would go up to still more spectacular levels. From about 1800, it has been calculated, Europe's agricultural productivity grew at a rate of about one per cent a year, dwarfing all previous advance. More important still, as time passed European industry and commerce would make it possible to tap huge larders in other parts of the world. Both of these changes were aspects of a single process, the accelerating investment in productive capacity which made Europe and North America by 1870 clearly the greatest concentration of wealth on the face of the globe. Agriculture was fundamental to it. People have spoken of an 'agricultural revolution' and provided this is not thought of as implying rapid change, it is an acceptable term; nothing much less strong will describe the huge surge in world output achieved between 1750 and 1870 (and, later, even surpassed). But it was a process of great complexity, drawing on many different sources and linked to the other sectors of the economy in indispensable ways. It was only one aspect of a worldwide economic change which involved in the end not merely continental Europe, but the Americas and Australasia as well.

These important qualifications once made, it is possible to particularize. By 1750 the best English agriculture was the best in the world. The most advanced techniques were practised and the integration of agriculture with a commercial market economy had gone furthest in England, whose lead was to be maintained for another century or so. European farmers went there to observe methods, buy stock and machinery, seek advice. Meanwhile, the English farmer, benefiting from peace at home (that there were no large-scale and continuous military operations on British soil after 1650 was of literally incalculable benefit to the economy) and a rising population to buy his produce, generated profits which provided capital for further improvement. His willingness to invest them in this way was, in the short run, an optimistic response to the likely commercial prospects but also says something deeper about the nature of English society. The benefits of better farming went in England to individuals who owned their own land or held it securely as leaseholding tenants on terms shaped by market realities. English agriculture was part of a capitalist market economy in which land was even by the eighteenth century treated almost as a commodity like any other. Restraints on its use familiar in European countries had disappeared faster and faster ever since Henry VIII's sequestration of ecclesiastical property. After 1750, the last great stage of this came with the spate of Enclosure Acts at the turn of the century (significantly coincident with high prices for grain) which mobilized for private profit the English peasant's traditional rights to pasture, fuel or other economic benefits. One of the most striking contrasts between English and European agriculture in the early nineteenth century was that the traditional peasant all but disappeared in England. England had wage labourers and smallholders, but the huge European rural populations of individuals with some, if minuscule, legal rights linking them to the soil through communal usages and a mass of tiny holdings, did not exist.

Inside the framework provided by prosperity and English social institutions, technical progress was continuous. For a long time, much of this was hit-and-miss. Early breeders of better animals succeeded not because of a knowledge of chemistry, which was in its infancy, or of genetics, which did not exist, but because they backed hunches. Even so, the results were remarkable. The appearance of the livestock inhabiting the landscape changed; the scraggy medieval sheep whose backs resembled, in section, the Gothic arches of the monasteries which bred them, gave way to the fat, square, contented-looking animal familiar today. 'Symmetry, well-covered' was an eighteenth-century farmer's toast. The appearance of farms changed as draining and hedging progressed and big, open medieval fields with their narrow strips, each cultivated by a different peasant, gave way to enclosed fields worked in rotation and made a huge patchwork of the English countryside. In some

Early Sydney; a scatter of buildings interspersed with farms and still preserving styles familiar in Old Europe.

of these fields machinery was at work even by 1750. Much thought was given to its use and improvement in the eighteenth century, but it does not seem that it really made much of a contribution to output until after 1800, when more and more large fields became available, and it became more productive in relation to cost. It was not long before steam engines were driving threshers; with their appearance in English fields, the way was open which would lead eventually to an almost complete replacement of muscular by machine power on the twentieth-century farm.

Such improvements and changes spread, *mutatis mutandis* and with a lag in time, to continental Europe. Except by comparison with earlier centuries of quasi-immobility, progress was not always rapid. In Calabria or Andalusia, it might be imperceptible over a century. Nevertheless, rural Europe changed, and the changes came by many routes. The struggle against the inelasticities of food-supply was in the end successful, but it was the outcome of hundreds of particular victories over fixed crop rotations, out-dated fiscal arrangements, poor standards of tillage and husbandry, and sheer ignorance. The gains were better stock, more effective control of plant blight and animal disease, the introduction of altogether new species, and much else. Change on so comprehensive a basis often had to work against the social and political grain, too. The French formally abolished serfdom in 1789; this probably did not mean much, for there were few serfs in France at that date. The abolition of the 'feudal system' in the same year was a much more important matter. What was meant by this vague term was the destruction of a mass of traditional and legal usages and rights which stood in the way of the exploitation of land by individuals as an investment like any other. Almost at once, many of the peasants who had thought they wanted this discovered that they did not altogether like it in practice; they discriminated. They were happy to abolish the customary dues paid to the lord of the manor, but did not welcome the loss of customary rights to common land. The whole change was made still more confusing and difficult to measure by the fact that there took place at the same time a big redistribution of property. Much land previously belonging to the Church was sold within a few years to private individuals. The consequent increase in the number of people owning land outright and growth in the average size of properties should, on the English analogy, have led to a period of great agricultural advance for France, but it did not. There was very slow progress and little consolidation of properties on the English pattern.

This suggests, rightly, that generalizations about the pace and uniformity of what was happening should be cautious and qualified. For all the enthusiasm Germans were showing for travelling exhibitions of agricultural machinery in the 1840s, theirs was a huge country and one of those (France was the other) of which a great economic historian commented that 'broadly speaking, no general and thorough-going improvement can be registered in peasant life before the railway age'. Yet the dismantling of medieval institutions standing in the way of agricultural improvement did go on steadily before that and prepared the way for it. It was accelerated in some places by the arrival during the Napoleonic period of French armies of occupation which introduced French law, and after this by other forces, so that by 1850 peasants tied to the soil and obligatory labour had disappeared from most of Europe. This did not mean, of course, that attitudes from the *ancien régime* did not linger after its institutions had disappeared. Prussian, Magyar and Polish landlords seem, for good and ill, to have maintained much of their more or less patriarchal authority in the manor even after its legal supports had vanished, and did so as late as 1914. This was important in assuring a continuity of conservative aristocratic values in a much more intense and concentrated way in these areas than in western Europe. The Junker often accepted the implications of the market in planning his own estate management, but not in his relations with his tenants.

The longest resistance to change in traditional legal forms in agriculture came in Russia. There, serfdom itself persisted until abolished in 1861. This act did not at once bring Russian agriculture entirely under the operation of individualist and market economy principles, but with it an era of European history had closed. From the Urals to Corunna there no longer survived in law any substantial working of land on the basis of serfdom, nor were peasants any longer bound to landlords whom they could not leave. It was the end of a system which had been passed from antiquity to western Christendom in the era of the barbarian invasions and had been the basis of European civilization for centuries. After 1861, Europe's rural proletariat everywhere worked for wages or keep; the pattern which had begun to spread in England and France with the fourteenth-century agricultural crisis had become universal.

Formally, the medieval usage of bond labour lasted longest in some of the American countries forming part of the European world. Obligatory labour in its most unqualified form, slavery, was legal in some of the United States, the most important of these countries, until the end of a great civil war in 1865, when its abolition (though promulgated by the victorious government two years before) became effective throughout the whole republic. The war which had made this possible had been in some measure a distraction from the already rapid development of the country, now to be resumed and to become of vital significance to Europe. Even before the war, cotton-growing, the very agricultural operation which had been the centre of debates over slavery, had already shown how the New World might supplement European agriculture on such a scale as to become almost indispensable. After the war the way was open for the supply to Europe not merely of products such as cotton, which she could not easily grow, but also of food.

The mechanical reaper was one of the major developments of nineteenth-century agricultural technology. By the end of the century giant machines like this were at work in the vast cornlands of the American and Canadian prairies.

The United States – and Canada, Australia and New Zealand, the Argentine and Uruguay – were soon to show they could offer food at much cheaper prices than Europe herself. Two things made this possible. One was the immense natural resources of these new lands. The American plains, the huge stretches of pasture in the South American pampas and the temperate regions of Australasia provided vast areas for the growing of grain and the raising of livestock. The second was a revolution in transport which made them exploitable for the first time. Steam-driven railways and ships came into service in increasing numbers from the 1860s. These quickly brought down transport costs and did so all the faster as lower prices bred growing demand. Thus further profits were generated to be put into more capital investment on the ranges and prairies of the New World. On a smaller scale the same phenomenon was at work inside Europe, too. From the 1870s the eastern European and German farmers began to see that they had a competitor in Russian grain, able to reach the growing cities much more cheaply once railways were built in Poland and western Russia and steamships could bring it from Black Sea ports. By 1900 the context in which European farmers worked, whether they knew it or not, was the whole world; the price of Chilean guano or New Zealand lamb could already settle what went on in their local markets.

Even in such a sketch the story of agricultural expansion bursts its banks; after first creating civilization and then setting a limit to its advance for thousands of years, agriculture suddenly became its propellant; within a century or so it suddenly demonstrated that it could feed many more people than ever before. The demand of the growing cities, the coming of railways, the availability of capital, all point to its inseparable interconnexion with other sides of a growing trans-oceanic economy between 1750 and 1870. For all its chronological primacy and its huge importance as a generator of investment capital, the story of agriculture in this period should only for convenience be separated from that of overall growth registered in the most obvious and spectacular way by the appearance of a whole new society, one based on large-scale industrialization.

This is a colossal subject. It is not even easy to see just how big it is. It produced the most striking change in European history since the barbarian invasions, but it has been seen as even more important, as the biggest change in human history since the coming of agriculture, iron or the wheel. Within a fairly short time – a century and a half or so – societies of peasants and craftsmen turned into societies of machine-tenders and bookkeepers. Ironically, it ended the ancient primacy of agriculture from which it had sprung. It was one of the major facts turning human experience back from the differentiation produced by millennia of cultural evolution to common experiences which would tend once more towards cultural convergence.

Even to define it is by no means easy, though the processes which lie at its heart are obvious around us. One is the replacement of human or animal labour by machines driven by power from other, usually mineral, sources. Another is the organization of production in much larger units. Another is the increasing specialization of manufacturing. But all these things have implications and ramifications which quickly take us far beyond them. Although it embodied countless conscious decisions by countless *entrepreneurs* and customers, industrialization also looks like a blind force sweeping across social life with transforming power, one of the 'senseless agencies' a philosopher once detected as half the story of revolutionary change. Industrialization meant new sorts of towns, needed new schools and new forms of higher learning, even new patterns of daily existence and living together.

The roots which made such a change possible go back far beyond the early modern age. Capital for investment had been accumulated slowly over many centuries of agricultural and commercial innovation. Knowledge had been built up, too. Canals were to provide the first network of communication for bulk transport once industrialization got under way, and from the eighteenth century they began to be built as never before in Europe (in China, of course, the story was different). Yet even Charlemagne's men had known how to build them. Even the most startling technical innovations had roots deep in the past. The men of the 'Industrial Revolution' (as a Frenchman of the early nineteenth century named the great upheaval of his era) stood on the shoulders of innumerable craftsmen and artificers of pre-industrial times who had slowly built up skills and experience for the future. Fourteenth-century Rhinelanders, for example, learnt to make cast iron; by 1600 the gradual spread of blast furnaces had begun to remove the limits hitherto set to the use of iron by its high cost and in the eighteenth century came the inventions making it possible to use coal instead of wood as fuel for some processes. Cheap iron, even in what were by later standards small quantities, led to experiment with new ways of using it; further changes would then follow. New demand meant that areas where ore was easily to be found became important. When new techniques of smelting permitted the use of mineral rather than vegetable fuel, the location of supplies of coal and iron began to fashion the later industrial geography of Europe and North America. In the northern hemisphere lies much of the discovered coal supply of the world, in a great belt running from the basin of the Don, through Silesia, the Ruhr, Lorraine, the north of England and Wales, to Pennsylvania and West Virginia.

Coalbrookdale, Shropshire, in 1758. Fifty years after the first smelting of iron with coke instead of charcoal, the appearance of this early industrial centre was still remarkably rural.

Better metal and richer fuel made their decisive contribution to early industrialization with the invention of a new source of energy, the steam-engine. Again, the roots are very deep. That the power of steam could be used to produce movement was known in Hellenistic Alexandria. Even if (as some believe) there existed the technology to develop this knowledge, contemporary economic life did not make it worthwhile to strain to do so. The eighteenth century brought a series of refinements so important that they can be considered as fundamental changes, and did so when there was money to invest in them. The result was a source of power rapidly recognized as of revolutionary importance. The new steam-engines were not only the product of coal and iron, they also consumed them, directly both as fuel and as materials used in their own construction. Indirectly they stimulated production by making possible other processes which led to increased demand for them. The most obvious and spectacular was railway-building. It required huge quantities of first iron and then steel for rails and rolling-stock. But it also made possible the movement of objects at much lower cost. What the new trains moved might well again be coal, or ore, thus allowing these materials to be used cheaply far from where they were easily found and dug. New industrial areas followed the lines, from which the railway could carry away goods to distant markets.

The railway was not the only change steam made to transport and communications. The first steamship went to sea in 1809. By 1870, though there were still many sailing-ships and navies were still building battleships with a full spread of sail, regular ocean sailings by 'steamers' were commonplace. The economic effect was dramatic. Oceanic transport's real cost in 1900 was a seventh of what it had been a hundred years earlier. The shrinking of costs, of time spent in transit, and of space, which steamships and railways produced, overturned men's ideas of the possible. Since the domestication of the horse and the invention of the wheel, people and goods had been conveyed at speeds which certainly varied according to the local roads available, but probably only within limits of no more than one and five miles per hour over any considerable distance. Faster travel was possible on water and this had perhaps increased somewhat over the millennia in which ships underwent quite considerable modification. But all such slow improvement was dwarfed when in a man's lifetime he could witness the difference between travel on horseback and in a train capable of forty or even fifty miles an hour for long periods.

A new scale of industrial organization: the big iron looms of an English textile mill in the 1840s, centrally powered from one engine-house.

We have now lost one of the most pleasant of industrial sights, the long, streaming plume of steam from the funnel of a locomotive at speed, hanging for a few seconds behind it against a green landscape before disappearing. It greatly struck those who first saw it and so, less agreeably, did other visual aspects of the industrial transformation. One of the most terrifying was the industrial town, dominated by a factory with smoking chimneys, as the pre-industrial town had been by the spire of church or cathedral. So dramatic and novel was the factory, indeed, that it has often gone unremarked that it was an unusual expression of the early stages of industrialization, not a typical one. Even in the middle of the nineteenth century most English industrial workers worked in manufacturing enterprises employing fewer than fifty. For a long time great agglomerations of labour were to be found only in textiles; the huge Lancashire cotton mills which first gave that area a visual and urban character distinct from earlier manufacturing towns were startling because they were unique. Yet by 1850 it was apparent that in more and more manufacturing processes the trend was towards the centralization under one roof made attractive by economies of transport, specialization of function, the use of more powerful machinery and the imposition of effective work discipline.

In the middle of the nineteenth century the changes of which these were the most striking had only created a mature industrial society in one country, Great Britain. Long and unconscious preparation lay behind this. Domestic peace and less rapacious government than on the continent had bred confidence for investment. Agriculture had provided its new surpluses first in England. Mineral supplies were available to exploit the new technological apparatus resulting from two or three generations of remarkable invention. An expanding overseas commerce generated further profits for investment and the basic machinery of finance and banking was already in being before industrialization needed to call on it and seemed to have readied society psychologically for change; observers detected an exceptional sensitivity to pecuniary and commercial opportunity in eighteenth-century England. Finally, an increasing population was beginning to offer both labour and a rising demand for manufactured goods. All these forces flowed together and the result was unprecedented and continuing industrial growth, first apparent as something totally new and irreversible in the second quarter of the nineteenth century. By 1870 Germany, France, Switzerland, Belgium and the United States had joined Great Britain in showing the capacity for self-sustained economic growth but she was still first among them both in the scale of her industrial plant and in her historic primacy. The inhabitants of 'the workshop of the world', as the British liked to think of themselves, were fond of running over the figures which showed how wealth and power had followed upon industrialization. In 1850 the United Kingdom owned half the world's ocean-going ships and contained half the world's railway track. On those railways trains ran with a precision and regularity and even a speed not much improved upon for a hundred years after. They were regulated by 'time-tables' which were the first examples of their kind (and occasioned the first use of the word) and their operation relied on the electric telegraph. They were ridden in by men and women who had a few years before only ridden in stage-coaches or carters' wagons. In 1851, a year when a great international exhibition at London advertised her new supremacy, Great Britain smelted two and a half million tons of iron. It does not sound much, but it was five times as much as the United States of America and ten times as much as Germany. At that moment, British steam-engines could produce more than 1.2 million horsepower, more than half that of all Europe together.

By 1870 a change had already started to appear in relative positions. Great Britain was still in most ways in the lead, but less decisively, and was not long to remain there. She still had more steam horsepower than any other European country, but the United States (which had already had more in 1850) was ahead of her and Germany was coming up fast. In the 1850s both Germany and France had made the important transition already made in Great Britain from smelting most of their iron by charcoal to smelting with mineral fuels. British superiority in manufacturing iron was still there and her pig-iron output had gone on rising, but now it was only three and a half times that of the United States and four times that of Germany. These were still huge superiorities, none the less, and the age of British industrial dominance had not yet closed.

The industrial countries of which Great Britain was the first were puny creatures in comparison with what they were to become. Among them only Great Britain and Belgium had a large majority of their population living in urban districts in the middle of the nineteenth century and the census of 1851 showed that agriculture was still the biggest single employer of labour among British industries (it was rivalled only by domestic service). But in these countries the growing numbers engaged in manufacturing industries, the rise of new concentrations of economic wealth and a new scale of urbanization all made very visible the process of change which was going forward.

This was especially true of the towns. They grew at a spectacular rate in the nineteenth century, particularly in its second half, when the appearance of big centres that would be the nuclei of what a later age would call 'conurbations' was especially marked. For the first time, some European cities ceased to depend on rural immigration for their growth. There are difficulties in reckoning indices of urbanization, largely because in different countries urban areas were defined in different ways, but this does not obscure the main lines of what was happening. In 1800 London, Paris and Berlin had, respectively, about 900,000, 600,000 and 170,000 inhabitants. In 1900 the corresponding figures were 4.7 million, 3.6 million and 2.7 million. In that year, too, Glasgow, Moscow, St Petersburg and Vienna also had more than a million inhabitants each. These were the

Isambard Kingdom Brunel, cigar in mouth, greatest of Victorian engineers, was a Portsmouth boy. After attending a famous Parisian lyceé, he made his reputation as a bridge-builder and railway engineer. England is still studded with his achievements. Here, he was viewing his last ship, the Great Eastern, whose trial voyage began only a few days before his death at the age of 53.

giants; just behind them were sixteen more European cities with over 500,000, a figure only passed by London and Paris in 1800. These great cities and the smaller ones which were still immeasurably bigger than the old ones they overshadowed were still attracting immigrants in large numbers from the countryside, notably in Great Britain and Germany. This reflected the tendency for urbanization to be marked in the relatively few countries where industrialization first made headway, because it was the wealth and employment generated by industry which to begin with drew men to them. Of the twenty-three cities of more than a half-million inhabitants in 1900, thirteen were in four countries, the United Kingdom (6), Germany (3), France (3) and Belgium (1).

Opinion about cities has undergone many changes. As the eighteenth century ended, something like a sentimental discovery of rural life was in full swing. This coincided with the first phase of industrialization and the nineteenth century opened with the tide of aesthetic and moral comment on the turn against a city life which was indeed about to reveal a new and often unpleasant face. That urbanization seemed an unwelcome, even unhealthy, change to many people, was a tribute to the revolutionary force of what was going on. Conservatives distrusted and feared cities. Long after European governments had demonstrated the ease with which they could control urban unrest, the cities were regarded suspiciously as likely nests of revolution. This is hardly surprising; conditions in many of the new metropolitan centres were often harsh and terrible for the poor. The East End of London could present appalling evidence of poverty, filth, disease and deprivation to anyone who chose to penetrate its slums. A young German businessman, Friedrich Engels, wrote in 1844 one of the most influential books of the century *The Condition of the English Working-Class in England* to expose the appalling conditions in which lived the poor of Manchester, but many other writers took up to similar themes. In France the phenomenon of the 'dangerous classes' (as the Parisian poor were called) preoccupied governments for the first half of the century, and misery fired a succession of revolutionary outbreaks between 1789 and 1871. Clearly, it was not unreasonable to fear that the growing cities could breed resentment and hatred of society's rulers and beneficiaries, and that this was a potentially revolutionary force.

It was also reasonable to predicate that the city made for ideological subversion. It was the great destroyer of traditional patterns of behaviour in nineteenth-century Europe and a crucible of new social forms and ideas, a huge and anonymous thicket in which men and women easily escaped the scrutiny of priest, squire and neighbours which had been the regulator of rural communities. In it (and this was especially true as literacy slowly spread downwards) new ideas were brought to bear upon long-unchallenged assumptions. Upper-class nineteenth-century Europeans were particularly struck by the seeming tendency of city life to atheism and irreligion. More was at stake, it was felt, than religious truth and sound doctrine (about which the upper classes themselves had long comfortably tolerated disagreement). Religion was the great sustainer of morals and the support of the established social order. A revolutionary writer sneered that religion was 'the opium of the people'; the possessing classes would hardly have put it in the same terms, but they acknowledged the importance of religion as social cement. One result was a long-continued series of attempts both in Catholic and Protestant countries to find a way of recapturing the towns for Christianity. The effort was misconceived in so far as it presumed that the Churches had ever had any footing in the urban areas which had long since swamped the traditional parish structures and religious institutions of the old towns and villages at their hearts. But it had a variety of expressions, from the building of new churches in industrial suburbs to the creation of missions combining evangelism and social service which taught churchmen the facts of modern city life. By the end of the century the religious-minded were at least well aware of the challenge they faced, even if their predecessors had not been. One great English evangelist used in the title of one of his books words precisely calculated to emphasize the parallel with missionary work in pagan lands overseas: *Darkest England*. His answer was to found a quite new instrument of religious propaganda, designed to appeal specifically to a new kind of population and to combat specifically the ills of urban society, the Salvation Army.

Here once more the revolution brought by industrialization has an impact far beyond material life. It is an immensely complicated problem to distinguish how modern civilization, the first, so far as we know, which does not have some formal structure of religious belief at its heart, came into being. Perhaps we cannot separate the role of the city in breaking down traditional religious observance from, say, that of science and philosophy in corrupting the belief of the educated. Yet a new future was visible already in the European industrial population of 1870, much of it literate, alienated from traditional authority, secular-minded and beginning to be conscious of itself as an entity. This was a different basis for civilization from anything yet seen.

Home: a Russian working-class family in the corner which they rented in a St Petersburg room in the 1890s.

This is to anticipate, but legitimately, for it suggests once again how rapid and deep was the impact of industrialization on every side of life. Even the rhythm of life changed. For the whole of earlier history, the economic behaviour of most of mankind had been regulated fundamentally by the rhythms of nature. In an agricultural or pastoral economy they imposed a pattern on the year which dictated both the kind of work which had to be undertaken and the kind which could be. Operating within the framework set by the seasons were the subordinate divisions of light and darkness, fair weather and foul. Men lived in great intimacy with their tools, their animals and the fields in which they won their bread. Even the relatively few townsmen lived, in large measure, lives shaped by the forces of nature; in Great Britain and France a bad harvest could still blight the whole economy well after 1850. Yet by then many people were already living lives whose rhythms were dictated by quite different pacemakers. Above all they were set by the means of production and their demands – by the need to keep machines economically employed, by the cheapness or dearness of investment capital, by the availability of labour. The symbol of this was the factory whose machinery set a pattern of work in which accurate time-keeping was essential. Men began to think in a quite new way about time as a consequence of their industrial work.

As well as imposing new rhythms, industrialism also related the labourer to his work in new ways. It is difficult, but important, to avoid sentimentalizing the past in assessing this. At first sight the disenchantment of the factory worker with his monotonous routine, with its exclusion of personal involvement and its background of the sense of working for another's profit justifies the rhetoric it has inspired, whether this takes the form of regret for a craftsman's world that has vanished or analysis of what has been identified as the alienation of the worker from his product. But the life of the medieval peasant was monotonous, too, and much of it was spent working for another's profit. Nor is an iron routine necessarily less painful because it is set by sunset and sunrise instead of an employer, or more agreeably varied by drought and tempest than by commercial slump and boom. Yet the new disciplines involved a revolutionary transformation of the ways men won their livelihood, however we may evaluate the results by comparison with what had gone before.

A clear example can be found in what soon became notorious as one of the persistent evils of early industrialism, its abuse of child labour. A generation of Englishmen morally braced by the abolition of slavery and by the exaltation that accompanied it was also one intensely aware of the importance of religious training – and therefore of anything which might stand between it and the young – and one disposed to be sentimental about children in a way earlier generations had not been. All this helped to create an awareness of this problem (first, in the United Kingdom) which perhaps distracted attention from the fact that the brutal exploitation of children in factories was only one part of a total transformation of patterns of employment. About the use of children's labour in itself there was nothing new. Children had for centuries provided swineherds, birdscarers, gleaners, maids-of-all-work, crossing-sweepers, prostitutes and casual drudges in Europe (and still do in most non-European societies). The terrible picture of the lot of unprotected children in Hugo's great novel *Les Misérables* (1862) is a picture of their life in a *pre*-industrial society. The difference made by industrialism was that their exploitation was regularized and given a quite new harshness by the institutional forms of the factory. Whereas the work of children in an agricultural society had perforce been clearly differentiated from that of adults by their inferior strength, there existed in the tending of machines a whole range of activity in which children's labour competed directly with that of adults. In a labour market normally over-supplied, this meant that there were irresistible pressures upon the parent to send the child into the factory to earn a contribution to the family income as soon as possible, sometimes at the age of five or six. The consequences were not only often terrible for the victims, but also revolutionary in that the relation of child to society and the structure of the family were blighted. This was one of the 'senseless agencies' of history at its most dreadful.

The problems created by such forces were too pressing to remain without attention and men quickly made a start in taming the most obvious evils of industrialism. By 1850, the law of England had already begun to intervene to protect, for example, women and children in mines and factories; in all the millennia of the history of agriculturally-based economies, it had still been impossible by that date to eradicate slavery even in the Atlantic world. Given the unprecedented scale and speed with which social transformation was upon them, the men of early industrial Europe need not be blamed without qualification for not acting more quickly to remedy ills whose outlines they could only dimly grasp. Even in the early stage of English industrialism, when, perhaps, the social cost was most heavy, it was difficult to cast off the belief that the liberation of the economy from legal interference was essential to the enormous generation of new wealth which was going on.

True, it is almost impossible to find economic theorists and publicists of the early industrial period who advocated absolute non-interference with the economy. Yet there was a broad, sustaining current which favoured the view that much good would result if the market economy was left to operate without the help or hindrance of politicians and civil servants. One force working this way was the teaching often summed up in a phrase made famous by a group of Frenchmen: *laissez-faire*. Broadly speaking, economists after Adam Smith had said with growing consensus that the production of wealth would be accelerated, and therefore the general well-being would increase, if the use of economic resources followed the 'natural' demands of the market. Another reinforcing trend was individualism, embodied in both the assumption that individuals knew their own business best and the increasing organization of society around the rights and interests of individuals.

From the inside of 'the Free-Trade hat', a graphic representation of the creed of Free Trade which for some Victorians came to be revered almost as much as the teachings of Holy Scripture themselves.

These were the sources of the long-enduring association between industrialism and liberalism; they were deplored by conservatives who regretted a hierarchical, agricultural order of mutual obligations and duties, settled ideas, and religious values. Yet liberals who welcomed the new age were by no means taking their stand on a simply negative and selfish base. The creed of 'Manchester', as it was called because of the symbolic importance of that city in English industrial and commercial development, was for its leaders much more than a matter of mere self-enrichment. A great political battle which for years preoccupied Englishmen in the early nineteenth century made this clear. Its focus was a campaign for the repeal of what were called the 'Corn Laws', a tariff system originally imposed to provide protection for the British farmer from imports of cheaper foreign grain. The 'repealers', whose ideological and political leader was a none-too-successful businessman, Richard Cobden, argued that much was at stake. To begin with, retention of the duties on grain demonstrated the grip upon the legislative machinery of the agricultural interest, the traditional ruling class, who ought not to be allowed a monopoly of power. Opposed to it were the dynamic forces of the future which sought to liberate the national economy from such distortions in the interest of particular groups. Back came the reply of the anti-repealers: the manufacturers were themselves a particular interest who only wanted cheap food imports in order to be able to pay lower wages; if they wanted to help the poor, what about some regulation of the conditions under which they employed women and children in factories? There, the inhumanity of the production process showed a callous disregard for the obligations of privilege which would never have been tolerated in rural England. To this, the repealers responded that cheap food would mean cheaper goods for export. And in this, for someone like Cobden, much more than profit was involved. A worldwide expansion of Free Trade untrammelled by the interference of mercantilist governments would lead to international progress both material and spiritual, he thought; trade brought peoples together, exchanged and multiplied the blessings of civilization and increased the power in each country of its progressive forces. On one occasion he committed himself to the view that Free Trade was the expression of the Divine Will (though even this was not to go as far as the British consul at Canton who had proclaimed that 'Jesus Christ is Free Trade, and Free Trade is Jesus Christ').

There was much more to the Free Trade issue in Great Britain (of which the Corn Law debate was the focus) than a brief summary can resume. The more it is expounded, the more it becomes clear that industrialism involved creative, positive ideologies which implied intellectual, social and political challenge to the past. This is why it should not be the subject of simple moral judgements, though both conservatives and liberals thought it could be at the time. The same man might resist legislation to protect the workman against long hours while proving himself a model employer, actively supporting educational and political reform and fighting the corruption of public interest by privileged birth. His opponent might struggle to protect children working in factories and act as a model squire, a benevolent patriarch to his tenants, while bitterly resisting the extension of the franchise to those not members of the established Church or any reduction of the political influence of landlords. It was all very muddled. In the specific issue of the Corn Laws the outcome was paradoxical, too, for a conservative Prime Minister was in the end convinced by the arguments of the repealers. When he had the opportunity to do so without too obvious an inconsistency he persuaded parliament to make the change in 1846. His party contained men who never forgave him and this great climax of Sir Robert Peel's political career, for which he was to be revered by his liberal opponents once he was safely out of the way, came shortly before he was dismissed from power by his own followers.

Only in England was the issue fought out so explicitly and to so clear-cut a conclusion. In other countries, paradoxically, the protectionists soon turned out to have the best of it. Only in the middle of the century, a period of expansion and prosperity, especially for the British economy, did Free Trade ideas get much support outside the United Kingdom, whose prosperity was regarded by believers as evidence of the correctness of their views and even mollified their opponents; Free Trade became a British political dogma, untouchable until well into the twentieth century. The prestige of British economic leadership helped to give it a brief popularity elsewhere, too. The prosperity of the era in fact owed as much to other influences as to this ideological triumph, but the belief added to the optimism of economic liberals. Their creed was the culmination of the progressive view of Man's potential as an individual, whose roots lay in Enlightenment ideas.

The solid grounds for this optimism can nowadays be too easily overlooked. In assessing the impact of industrialism we labour under the handicap of not having before us the squalor of the past it left behind. For all the poverty and the slums (and the very worst was over by then), the people who lived in the great cities of 1900 consumed more and lived longer than their ancestors. This did not, of course, mean they were either tolerably off, by later standards, or contented. But they were often, and probably for the most part, materially better off than their predecessors or most of their contemporaries in the non-European world. Amazing as it may seem, they were part of the privileged minority of mankind. Their longer lives were the best evidence of it.

2
Political Change in an Age of Revolution

In the eighteenth century the word 'revolution' came to have a new meaning. Traditionally it meant only a change in the composition of government and not necessarily a violent one (though one reason why the English 'Glorious Revolution' of 1688 was thought glorious was that it had been non-violent, Englishmen learnt to believe). Men could speak of a 'revolution' occurring at a particular court when one minister replaced another. After 1789 this changed. Men came to see that year as the beginning of a new sort of revolution, a real rupture with the past, characterized by violence, by limitless possibilities for fundamental change, social, political and economic, and began to think, too, that this new phenomenon might transcend national boundaries and have something universal and general about it. Those who disagreed very much about the desirability of such a revolution could none the less agree that this new sort of revolution existed and that it was fundamental to the politics of their age.

It would be misleading to seek to group all the political changes of this period under the rubric of 'revolution' conceived in such terms as these. But we can usefully speak of an 'age of revolution' for two reasons. One is that there were indeed within a century or so many more political upheavals than hitherto which could be called revolutions in this extreme sense, even though many of them failed and others brought results far different from those they had led people to expect. In the second place, if we give the term a little more elasticity, and allow it to cover examples of greatly accelerated and fundamental political change which certainly go beyond the replacement of one set of governors by another, then there are many less dramatic political changes in these years which are distinctly revolutionary in their effect. The first and most obvious was the dissolution of the first British empire, whose central episode later became known as the American Revolution.

In 1763 British imperial power in North America was at its height. Canada had been taken from the French; the old fear of a Mississippi valley cordon of French forts enclosing the thirteen colonies had been blown away. This might seem to dispose of any grounds for future misgiving, yet some prophets had already suggested, even before the French defeat, that their removal might not strengthen but weaken the British grasp on North America. In the British colonies, after all, there were already more colonists then there were subjects in many sovereign states of Europe. Many were neither of English descent nor native English-speakers. They had economic interests not necessarily congruent with those of the imperial power. Yet the grip of the British government on them was bound to be slack, simply because of the huge distances which separated London from the colonies. Once the threat from the French (and from the Indians whom the French had egged on) was gone, the ties of empire might have to be allowed to grow slacker still.

Difficulties appeared almost at once in the wake of the French expulsion. How was the West to be organized? What relation was it to have to the existing colonies? How were the new Canadian subjects of the Crown to be treated? These problems were given urgency by 'Pontiac's rebellion', an Indian revolt in the Ohio valley in 1763 in response to pressure by the colonists who saw the West as their proper domain for settlement and trade. The imperial government immediately proclaimed the area west of the Alleghenies closed to settlement. This, as a start, offended many colonials who had looked forward to the exploitation of these regions, and it was followed by still more irritation as British administrators negotiated treaties with Indians and worked out arrangements for a garrisoned frontier to protect the colonists and Indians from one another.

Ten years followed during which the dormant potential for American independence matured and came to a head. Grumbles about grievances turned first into resistance, then rebellion. Time after time, colonial politicians used provocative British legislation to radicalize American politics by making the colonists believe that the practical liberty they already enjoyed was in danger. The pace throughout was set by British initiatives. Paradoxically, Great Britain was ruled at this time by a succession of ministers anxious to carry out reforms in colonial affairs; their excellent intentions helped to destroy a *status quo* which had previously proved workable. They thus provide one of

The famous Boston Massacre, a godsend to radical propagandists though the soldiers (who were tried by a local court) were acquitted of the charge of murder.

the first examples of what was to be a frequent phenomenon of the next few decades, the goading of vested interests into rebellion by well-meant but politically ill-judged reform.

One principle firmly grasped in London was that the Americans ought to pay a proper share of the taxes which contributed to their defence and the common good of the empire. There were two distinct attempts to assure this. The first, in 1764–5, took the form of imposing duties on sugar imported to the colonies and a Stamp Act which was to raise money from revenue stamps to be put on various classes of legal documents. The important thing about these was not the amounts they proposed to raise nor even the novelty of taxing the internal transactions of the colonies (which was much discussed) but rather that these were, as both English politicians and American taxpayers saw, unilateral acts of legislation by the imperial parliament. The usual way in which colonial affairs were handled and revenue raised hitherto had been by haggling with their own assemblies. What was now brought into question was something so far hardly even formulated as a question: whether the undoubted legislative sovereignty of the parliament of the United Kingdom also extended to its colonies. Riots, non-importation agreements and angry protest followed. The unhappy officials who held the stamps were given a bad time. Ominously, representatives of nine colonies attended a Stamp Act Congress to protest. The Stamp Act was withdrawn.

The London government then took a different tack in its second fiscal initiative. It turned to external duties on paint, paper, glass and tea. As these were not internal taxes and the imperial government had always regulated trade, they seemed more promising. But it proved an illusion. Americans were by now being told by their radical politicians that no taxation at all should be levied on them by a legislature in which they were not represented. As George III saw, it was not the Crown but parliament whose power was under attack. There were more riots and boycotts and one of the first of those influential scuffles which make up so much of the history of decolonization, when the death of possibly five rioters in 1770 was mythologized into a 'Boston Massacre'.

Once more, the British government retreated. Three of the duties were withdrawn: that on tea remained. Unfortunately, the issue was by now out of hand; it transcended taxation, as the British government saw, and had become one of whether or not the imperial parliament could make laws enforceable in the colonies. As George III put it a little later: 'We must either master them, or totally leave them to themselves.' The issue was focused in one place, though it manifested itself throughout the colonies. By 1773, after the destruction of a cargo of tea by radicals (the 'Boston Tea Party'), the crucial question for the British government was: could Massachusetts be governed?

There were to be no more retreats: George III, his ministers and the majority of the House of Commons were agreed on this. A number of coercive acts were passed to bring Boston to heel. The New England radicals were heard all the more sympathetically in the other colonies at this juncture because a humane and sensible measure providing for the future of Canada, the Quebec Act of 1774, stirred up wide feeling. Some disliked the privileged position it gave to Roman Catholicism (it was intended to leave French Canadians as undisturbed as possible in their ways by their change of rulers), while others saw its extension of Canadian boundaries south to the Ohio as another block to expansion in the west. In September the same year there met a Continental Congress of delegates from the colonies at Philadelphia. It severed commercial relations with the United Kingdom and demanded the repeal of much existing legislation, including the Quebec Act. By this time the recourse to force was probably inevitable. The radical colonial politicians had brought out into the open the practical sense of independence already felt by many Americans. But it was inconceivable that any eighteenth-century imperial government could have grasped this. The British government was in fact remarkably reluctant to act on its convictions by relying simply on force until disorder and intimidation of the law-abiding and moderate colonials had already gone very far. At the same time, it made it clear that it would not willingly bend on the principles of sovereignty.

Arms were gathered in Massachusetts. In April 1775 a detachment of British soldiers sent to Lexington to seize some of them fought the first action of the American

Revolution. It was not quite the end of the beginning. It took a year more for the feelings of the colonists' leaders to harden into the conviction that only complete independence from Great Britain would rally an effective resistance. The result was the Declaration of Independence of July 1776, and the debate was transferred to the battlefield.

The British lost the war which followed because of the difficulties imposed by geography, because American generalship succeeded in avoiding superior forces long enough to preserve an army which could impose its will on them at Saratoga in 1777, because the French entered the war soon afterwards to win a return match for the defeat of 1763, and because the Spanish followed them and thus tipped the balance of naval power. The British had a further handicap; they dared not fight the kind of war which might win military victory by terrorizing the American population and thus encouraging those who wished to remain under the British flag to cut off the supplies and freedom of movement which General Washington's army enjoyed. They could not do this because their over-riding aim had to be to keep open the way to a conciliatory peace with colonists willing again to accept British rule. In these circumstances, the Bourbon coalition was fatal. The military decision came in 1781, when a British army found itself trapped at Yorktown between the Americans on land and a French squadron at sea. Only seven thousand or so men were involved, but their surrender was the worst humiliation yet undergone by British arms and the end of an era of imperial rule. Peace negotiations soon began and two years later, at Paris, a treaty was signed in which Great Britain recognized the independence of the United States of America, whose territory the British negotiators had already conceded should run to the Mississippi. This was a crucial decision in the shaping of a new nation; the French, who had envisaged making a recovery in the Mississippi valley, were disappointed. The northern continent was to be shared by the rebels only with Spain and Great Britain, it appeared.

For all the loose ends which would need to be tied up, and a number of boundary disputes which dragged on for decades to come, the appearance of a new state of great potential resources in the western hemisphere was by any standard certainly a revolutionary change. If it was at first often seen as something less than this by foreign observers, that was because the weaknesses of the new nation were at the time more apparent than its potential. Indeed, it was far from clear that it was a nation at all; the colonies were weak and divided and many expected them to fall to quarrelling and disunion. Their great and inestimable advantage was their remoteness. They could work out their problems virtually untroubled by foreign intervention, a blessing crucial to much that was to follow.

Victory in war was followed by a half-dozen critical years during which a handful of American politicians took decisions which were to shape much of the future history of the world. As in all civil wars and wars of independence, deep divisions had been created which accentuated political weakness. Among these, those which divided loyalists from rebels were, for all their bitterness, perhaps the least important. That problem had been solved, brutally, by emigration of the defeated; something like 80,000 loyalists left the rebel colonies, for a variety of motives running from dislike of intimidation and terror to simple loyalty to the Crown. Other divisions were likely to cause more trouble in the future. Class and economic interests separated farmers, merchants and plantation-owners. There were important differences between the new states which had replaced the former colonies and between the regions or sections of a rapidly developing country; one of these, that imposed by the economic importance of black slavery to the southern states, was to take decades to work out. On the other hand, the Americans also had great advantages as they set about nation-building. They faced the future without the terrible incubus of a huge illiterate and backward peasant population such as stood in the way of evolving a democratic system in many other countries. They had ample territory and great economic resources even in their existing areas of occupation (though the extent of these could not yet be known). Finally, they had European civilization to draw upon, subject only to the modifications its legacies might undergo in transplantation to a virgin – or near-virgin – continent.

IN CONGRESS, JULY 4, 1776

The unanimous Declaration of the thirteen united States of America,

The Declaration of Independence, the central document of the American Revolution and still a source of inspiration and renewal in the public life of the United States. Perhaps that is because the man who drafted it, Thomas Jefferson, resolutely set out to write a profoundly conservative document: it looked back to accepted truths rather than strove to startle or excite men with new discoveries. 'It was', he said, 'intended to be an expression of the American mind.' His purpose was 'not to find out new principles, or new arguments never before thought of, not merely to say things which had never been said before, but to place before mankind the common sense of the subject'. Therein, no doubt, lies the genius of the noble prose which was the result.

The war against the British had imposed a certain discipline. Articles of Confederation had been agreed between the former colonies and came into force in 1781. In them appeared the name of the new nation, the United States of America.

The peace brought a growing sense that these arrangements were unsatisfactory. There were two areas of particular concern. One was disturbance arising fundamentally from disagreement about what the Revolution ought to have meant in domestic affairs. The central government came to many Americans to appear to be far too weak to deal with disaffection and disorder. The other arose from a post-war economic depression, particularly affecting external trade and linked to currency problems arising from the independence of individual states. To deal with these, too, the central government seemed ill-equipped. It was accused of neglecting American economic interests in its conduct of relations with other countries. Whether true or not, this was widely believed. The outcome was a meeting of delegates from the states in a constitutional convention at Philadelphia in 1787. After four months' work they signed a draft constitution which was then submitted to the individual states for ratification. When nine states had ratified it the constitution came into effect in the summer of 1788. In April 1789 George Washington, the former commander of the American forces in the war against the British, took the oath of office as the first president of the new republic, thus inaugurating a series of presidencies which has continued unbroken to this day.

Much was said about the need for simple institutions and principles clear in their intention, yet the new constitution was still to be revealing its potential for development two hundred years later. For all the determination of its drafters to provide a document which would unambiguously resist reinterpretation, they were (fortunately) unsuccessful. The United States' constitution was to prove capable of spanning a historical epoch which turned a scatter of largely agricultural societies into a giant and industrial world power. In part this was because of the provision for conscious amendment, but in larger measure it was due to the evolving interpretation of the doctrines it embodied. But also much remained unchanged; though often formal, these features of the constitution are very important. Besides them, too, there were fundamental principles which were to endure, even if there was much argument about what they might mean.

To begin with the most obvious fact: the constitution was republican. This was by no means normal in the eighteenth century and should not be taken for granted. Some Americans felt that republicanism was so important and so insecure that they even disapproved of the constitution because they thought it (and particularly its installation of a president as the head of the executive) 'squinted towards monarchy', as one of them put it. The ancient republics were as familiar to classically educated Europeans for their tendency to decay and faction as for their legendarily admirable morals. The history of the Italian republics was unpromising, too, and much more unedifying than that of Athens and Rome. Republics in eighteenth-century Europe were few and apparently unflourishing. They seemed to persist only in small states, though it was conceded that the remoteness of the United States might protect republican forms which would elsewhere ensure the collapse of a large state. Still, observers were not sanguine about the new nation. The later success of the United States was therefore to be of incalculable importance in reversing opinion about republicanism. Very soon, its capacity to survive, its cheapness and a liberalism wrongly thought to be inseparable from it focused the attention of critics of traditional governments all over the civilized world. European advocates of political change soon began to look to America for inspiration; soon, too, the influence of republican example was to spread from the northern to the southern American continent.

The second characteristic of the new constitution which was of fundamental importance was that its roots lay largely in British political experience. Besides the law of England, whose Common Law principles passed into the jurisprudence of the new state, this was true also of the actual arrangement of government. The founding fathers had all grown up in the British colonial system in which elected assemblies had debated the public interest with monarchical governors. They instituted a bicameral legislature (although they excluded any hereditary element in its composition) on the English model, to offset a president. They thus followed English constitutional theory in putting a monarch, albeit an elected one, at the head of the executive machinery of government. This was not how the British constitution of the eighteenth century actually worked, but it was a good approximation to its theory. What they did, in fact, was to take the best constitution they knew, purge it of its corruptions (as they saw them) and add to it the modifications appropriate to American political and social circumstance. What they did not do was to emulate the alternative principle of government available in contemporary Europe, monarchical absolutism, even in its enlightened form. The Americans wrote a constitution for free men because they believed that the British already lived under one. They thought it had failed only in so far as it had been corrupted, and that it had been improperly employed to deprive Americans of the rights they too ought to have possessed under it. Because of this, the same principles of government (albeit in much evolved forms) would one day be propagated and patronized in areas which shared none of the cultural assumptions of the Anglo-Saxon world on which they rested.

One way in which the United States differed radically from most other existing states and diverged consciously from the British constitutional model was in adherence to the principle of federalism. This was indeed fundamental to it, since only large concessions to the independence of individual states made it possible for the new union to come into existence at all. The former colonies had no wish to set up a new central government which would bully them as they believed the government of King George had done. The federal structure provided an answer to the problem of diversity – *e pluribus unum*. It also dictated much of the form and content of American politics for the next eighty years. Question after question whose substance was economic or social or ideological would find itself pressed into the channels of a continuing debate about what were the proper relations of the central government and the individual states. It was a debate which would in the end come within an ace of destroying the Union. Federalism would also promote a major readjustment within the constitution, the rise of the Supreme Court as an instrument of judicial review. Outside the Union, the nineteenth century would show the appeal of federalism to many other countries, impressed by what appeared to have been achieved by the Americans. Federalism was to be seen by European liberals as a crucial device for reconciling unity with freedom and British governments found it a great standby in their handling of colonial problems.

Finally, in any summary, however brief, of the historic significance of the constitution of the United States, attention must be given to its opening words: 'We the People' (even though they seem to have been included almost casually). The actual political arrangements in several of the states of 1789 were by no means democratic, but the principle of popular sovereignty was enunciated clearly from the start. In whatever form the mythology of a particular historical epoch might cloak it, the popular will was to remain the ultimate court of appeal in politics for Americans. Here was a fundamental departure from British constitutional practice, and it owed something to the way in which seventeenth-century colonists had sometimes given themselves constitutions. Yet this had been within the overall framework of kingship. British constitutionalism was prescriptive; the sovereignty of king in parliament was not there because the people had once decided it should be, but because it was there and was unquestioned. As the great English constitutional historian Maitland once put it, Englishmen had taken the authority of the Crown as a substitute for the theory of the state. The new constitution broke with this and with every other prescriptive theory (though this was not quite a clean break with British political thinking, for Locke had said in the 1680s that governments held their powers on trust and that the people could upset governments which abused that trust and on this ground, among others, some Englishmen had justified the Glorious Revolution).

The American adoption of a democratic theory that all governments derive their just powers from the consent of the governed, as it had been put as early as the Declaration of Independence, was epoch-marking. But it by no means solved the problems of political authority at a stroke. Many Americans feared what a democracy might do and sought to restrict the popular element in the political system right from the start. Another problem was suggested by the fundamental rights set out in the first ten amendments to the constitution at the end of 1789. These were presumably as much open to reamendment at the hands of popular sovereignty as any other part of the constitution. Here was an important source of disagreement for the future: Americans have always found it easy to be somewhat confused (especially in the affairs of other countries but even in their own at times) about whether democratic principles consist in following the wishes of the majority or in upholding certain fundamental rights. Nevertheless, the adoption of the democratic principle in 1787 was immensely important and justifies the consideration of the constitution as a landmark in world history. For generations to come the new United States would become the focus of the aspirations of men longing to be free the world over – 'the world's last, best hope', as one American said. Even today, when America has become a great conservative nation, the democratic ideal of which for so long it was the custodian and exemplar retains its power, and the institutions it fertilized are still working.

Paris was the centre of social and political discussion in Europe. To it returned some of the French soldiers who had helped to bring to birth the young American Republic. It is hardly surprising, then, that although most European nations in some measure responded to the transatlantic revolution, Frenchmen were especially aware of it. American example and the hopes it raised were a contribution, though a subsidiary one, to the huge release of forces which is still, in spite of later upheavals, called *the* French Revolution. Unfortunately, this all too familiar and simple term puts obstacles in the way of understanding. Politicians and scholars have offered many different interpretations of what the essence of the Revolution was, have disagreed about how long it went on and what were its results, and even about when it began. They agree about little except that what began in 1789 was very important. Within a very short time, indeed, it changed the whole concept of revolution, though there was much in it that looked to the past rather than the future. It was a great boiling-over of the pot of French society and the pot's contents were a jumbled mixture of conservative and innovating elements much like those of the 1640s in England, and equally confused in their mixture of consciousness and unconsciousness of direction and purposes, too.

This confusion was the symptom of big dislocations and maladjustments in the material life and government of France. She was the greatest of European powers and her rulers neither could nor wished to relinquish her international role. The first way in which the American Revolution had affected her was by providing an opportunity for revenge; Yorktown was the retaliation for defeat at the hands of the British in the Seven Years' War, and to deprive them of the Thirteen Colonies was some compensation for the

The ancient Estates General of France meets under the presidency of its king in May 1789 at Versailles.

French loss of India and Canada. Yet the successful effort was costly. The second great consequence was that for no considerable gain beyond the humiliation of a rival, France added yet another layer to the huge and accumulating debt piled up by her efforts since the 1630s to build and maintain her European supremacy.

Attempts to liquidate this debt and cut the monarchy free from the cramping burden it imposed (and it was becoming clear after 1783 that France's real independence in foreign affairs was narrowing sharply because of it) were made by a succession of ministers under Louis XVI, the young, somewhat stupid, but high-principled and well-meaning king who came to the throne in 1774. None of them succeeded in even arresting the growth of the debt, let alone in reducing it. What was worse, their effects only advertised the facts of failure. The deficit could be measured and the figures published as would never have been possible under Louis XIV. If there was a spectre haunting France in the 1780s, it was not that of revolution but of state bankruptcy.

The essential problem was that the whole social and political structure of France stood in the way of tapping the wealth of the better-off, the only sure way of emerging from the financial *impasse.* Ever since the days of Louis XIV himself, it had proved impossible to levy a due weight of taxation on the wealthy without resorting to force, for the whole legal and social structure of France, its privileges, special immunities and prescriptive rights, blocked the way ahead. The paradox of eighteenth-century government was at its most evident in France; a theoretically absolute monarchy could not infringe the mass of liberties and rights which made up the essentially medieval constitution of the country without threatening its own foundations. Monarchy itself rested on prescription.

To more and more Frenchmen it appeared that France needed a transformation of her governmental and constitutional structure if she was to emerge from her difficulties. But some went further. They saw in the inability of government to share fiscal burdens equitably between classes the extreme example of a whole range of abuses which needed reform. The issue was more and more exaggerated in terms of polarities: of reason and superstition, of freedom and slavery, of humanitarianism and greed. Above all, it tended to concentrate on the symbolic question of legal privilege. The class which focused the anger this aroused was the nobility, an immensely diverse and very large body (there seem to have been between 200,000 and 250,000 noble males in France in 1789) about which cultural, economic or social generalization is impossible, but whose members all shared a legal status which is some degree conferred privilege at law.

While the logic of financial extremity pushed the governments of France more and more towards conflict with the privileged, there was a natural unwillingness on the part of many of the royal advisers, themselves usually noblemen, and of the king himself to proceed except by agreement. When in 1788 a series of failures nerved the government to accept that conflict was inevitable, it still sought to confine the conflict to legal channels, and, like Englishmen in 1640, turned to historic institutions for means to do so. Not having Parliament to hand, they trundled out from the attic of French constitutionalism the nearest thing to a national representative body that France had ever possessed, the Estates General. This body of representatives of nobles, clergy and commoners had not met since 1614. It was hoped that it would provide sufficient moral authority to squeeze agreement from the fiscally privileged for the payment of higher taxes. It was an unimpeachably constitutional step, but as a solution had the disadvantage that great expectations were aroused while what the Estates General could legally do was obscure. More than one answer was given. Some already said the Estates General could legislate for the nation, even if historic and undoubted legal privileges were at issue.

This very complicated political crisis was coming to a head at the end of a period in which France was also under other strains. Underlying everything was an increasing population. Since the second quarter of the century this had risen at what a later age would think a slow rate, but still fast enough to outstrip growth in the production of food. This sustained a long-run inflation of food prices which bore most painfully upon the poor, the vast majority of Frenchmen who were peasants with little or no land. Given the coincidence of the fiscal demands of governments – which for a long time staved off the financial crisis by borrowing or by putting up the direct and indirect taxes which fell most heavily on the poor – and the efforts of landlords to protect themselves in inflationary times by holding down wages and putting up rents and dues, the life of the poor Frenchman was growing harsher and more miserable for nearly all the century. To this general impoverishment should be added the special troubles which from time to time afflicted particular regions or classes, but which, coincidentally, underwent something of an intensification in the second half of the 1780s. Bad harvests, cattle disease, and recession which badly affected the areas where peasants' families produced textiles as a supplement to their income, all sapped the precarious health of the economy in the 1780s. The sum effect was that the elections to the Estates General in 1789 took place in a very excited and embittered atmosphere. Millions of Frenchmen were desperately seeking some way out of their troubles, were eager to seek and blame scapegoats, and had quite unrealistic and inflated notions of what good the king, whom they trusted, could do for them.

Thus a complex interplay of governmental impotence, social injustice, economic hardship and reforming aspiration brought about the French Revolution. But before this complexity is lost to sight in the subsequent political battles and the simplifying slogans they generated, it is important to emphasize that almost no one either anticipated this outcome or desired it. There was much social injustice in France, but no more than many other eighteenth-century states found it possible to live with. There was a welter of expectant and hopeful advocates of particular reforms ranging from the abolition of the censorship to the prohibition of immoral and irreligious literature, but no one doubted that such changes could easily be carried out by the king, once he was informed of his people's wishes and needs. What did not exist was a party of revolution clearly confronting a party of reaction.

Parties only came into existence when the Estates General had met. This is one reason why the day on which they did so, 5 May 1789 (a week after George Washington's inauguration), is a date in world history, because it opened an era in which to be for or against the Revolution became the central political question in most continental countries, and even tainted the very different politics of Great Britain and the United States. What happened in France was bound to matter elsewhere. At the simplest level this was because she was the greatest European power; the Estates General would either paralyse her (as many foreign diplomats hoped) or liberate her from her difficulties to play again a forceful role. Beyond this, France was also the cultural leader of Europe. What her writers and politicians said and did was immediately accessible to people elsewhere because of the universality of the French language, and it was bound to be given respectful attention because people were in the habit of looking to Paris for intellectual guidance.

In the summer of 1789 the Estates General turned itself into a national assembly claiming sovereignty. Breaking with the assumption that it represented the great medieval divisions of society, the majority of its members claimed to represent all Frenchmen without distinction. It was able to take this revolutionary step because the turbulence of France frightened the government and those deputies to the assembly who opposed such a change. Rural revolt and Parisian riot alarmed ministers no longer sure that they could rely upon the army. This led first to the monarchy's abandonment of the privileged classes, and then its concession, unwillingly and uneasily, of many other things asked for by the politicians who led the new National Assembly. At the same time these concessions created a fairly clear-cut division between those who were for the Revolution and those who were against it; in language to go round the world they were soon called Left and Right (because of the places in which they sat in the assembly).

The main task which the National Assembly set itself was the writing of a constitution, but in the process it transformed the whole institutional structure of France. Before 1791, when it dispersed, it had nationalized the lands of the Church, abolished what it termed 'the feudal system', ended censorship, created a system of centralized representative government, obliterated the old provincial and local divisions and replaced them with the 'departments' under which Frenchmen still live, instituted equality before the law, and separated the executive from the legislative power. These were only the most remarkable things done by one of the most remarkable parliamentary bodies the world has ever seen. Its failures tend to mask this huge achievement; they should not be allowed to do so. Broadly speaking, they removed the legal and institutional checks on the modernization of France. Popular sovereignty, administrative centralization, and individual legal equality were from this time poles towards which her institutional life always returned.

Many Frenchmen did not like all this; some liked none of it. By 1791 the king had given clear evidence of his own misgivings, the goodwill which had supported him in the early Revolution was gone and he was suspected as an anti-revolutionary. Some noblemen had already disliked enough of what was going on to emigrate; they were led by two of the king's brothers, which did not improve the outlook for royalty. But most important of all, many Frenchmen turned against the Revolution when, because of papal policy, the National Assembly's settlement of Church affairs was called in question. Much in it had appealed deeply to many Frenchmen, churchmen among them, but the pope rejected it and this raised the ultimate question of authority. French Catholics had to decide whether the authority of the pope or that of the French constitution was supreme for them. This created the most important division which came to embitter revolutionary politics.

In 1792 a great crisis occurred. France went to war with Austria at the beginning of the year and with Prussia soon after. The issue was complicated but many Frenchmen believed that foreign powers wished to intervene to bring the Revolution to an end and put the clock back to 1788. By the summer, as things went badly and shortages and suspicion mounted at home, the king was discredited. A Parisian insurrection overthrew the monarchy and led to the summoning of a new assembly to draw up a new and, this time, republican constitution. This body, called the Convention, was the centre of French government until 1796. Through civil and foreign war and economic and ideological crisis it achieved the survival of the Revolution. Most of its members were politically not much more advanced in their views than their predecessors. They believed in the individual and the sanctity of property (they prescribed the death penalty for anyone proposing a law to introduce agrarian communism) and that the poor are always with us, though they allowed some of them a small say in affairs by supporting direct universal adult male suffrage. What distinguished them from their predecessors was that they were willing to go rather further to meet emergencies than earlier French assemblies (especially when frightened by the possibility of defeat); they sat in a capital city which was for a long time manipulated by more extreme politicians to push them into measures more radical than they really wanted, and into using very democratic language. Consequently, they frightened Europe much more than their predecessors had done.

Their symbolic break with the past came when the Convention voted for the execution of the king in January 1793. The judicial murder of kings had hitherto been believed to be an English aberration; now the English were as shocked as the rest of Europe. They, too, now went to war with France, because they feared the strategical and commercial result of French success against the Austrians in the Netherlands. But the war looked more and more like an ideological struggle and to win it the French government appeared increasingly bloodthirsty at home. A new instrument for humane execution, the guillotine (a characteristic product of pre-revolutionary enlightenment, combining as it did technical efficiency and benevolence in the swift, sure death it afforded its victims), became the symbol of the Terror, the name soon given to a period during which the Convention strove by intimidation of its enemies at home to assure survival to the Revolution. There was much that was misleading in this symbolism. Some of the Terror was only rhetoric, the hot air of politicians trying to keep up their own spirits and frighten their opponents. In practice it often reflected a jumble of patriotism, practical necessity, muddled idealism, self-interest and petty vengefulness, as old scores were settled in the name of the republic. Many people died, of course – something over 35 000, perhaps – and many emigrated to avoid danger, yet the guillotine killed only a minority of the victims, most of whom died in the provinces, often in conditions of civil war and sometimes with arms in their hands. In eighteen months or so the Frenchmen whom contemporaries regarded as monsters killed about as many of their countrymen as died in ten days of street-fighting and firing-squads in Paris in 1871. To take a different but equally revealing measure, the numbers of those who died in this year and a half are roughly twice those of the British soldiers who died on the first day of the battle of the Somme in 1916. Such bloodshed drove divisions even deeper between Frenchmen, but their extent should not be exaggerated. All noblemen, perhaps, had lost something in the Revolution, but only a minority of them found it necessary to emigrate. Probably the clergy suffered more, man for man, than the nobility, and many priests fled abroad. Yet there were fewer emigrants from France during the Revolution than from the American colonies after 1783. A much larger proportion of Americans felt too intimidated or disgusted with their Revolution to live in the United States after independence than the proportion of Frenchmen who could not live in France after the Terror.

The Convention won victories and put down insurrection at home. By 1797, only Great Britain had not made peace with France, the Terror had been left behind, and the republic was ruled by something much more like a parliamentary regime under the constitution whose adoption closed the Convention era in 1796. The Revolution was safer than ever. But it did not seem so. Abroad, the royalists strove to get allies with whom to return and also intrigued with malcontents inside France. The return of the old order was a prospect which few Frenchmen would welcome, though. On the other hand, there were those who argued that the logic of democracy should be pressed further, that there were still great divisions between rich and poor which were as offensive as had been the old distinctions of legally privileged and unprivileged, and that the Parisian radicals should have a greater say in affairs. This was almost as alarming as fears of a restoration to those who had benefited from the Revolution or simply wanted to avoid further bloodshed. Thus pressed from Right and Left, the Directory (as the new regime was called) was in a way in a good position, though it made enemies who found the (somewhat zigzag) *via media* it followed unacceptable. In the end it was destroyed from within when a group of politicians intrigued with soldiers to bring about a *coup d'état* which instituted a new regime in 1799.

European government was changed much more during the nineteenth century by such blind forces as those of population growth, urbanization, railways and mass literacy than by revolution. Yet the idea of revolution went on fascinating and terrifying men even when it almost always failed in practice. France, understandably, was the supreme example of a country where the myth of revolution bemused politicians and artists alike. Delacroix's celebrated painting of Liberty leading the people across the barricades, symbolic Phrygian cap on her head and tricolour in her hand, is still one of the most moving evocations of the idealization of what a priest once called la sainte canaille *– the mob whose brutality is seen as purged of evil in its commitment to the cause of freedom for all men.*

At that moment, ten years after the meeting of the Estates General, it was at least clear to most observers that France had for ever broken with the medieval past. In law this happened very rapidly. Nearly all the great reforms underlying it were legislated at least in principle in 1789. The formal abolition of feudalism, legal privilege and theocratic absolutism and the organization of society on individualist and secular foundations were the heart of the 'principles of '89' then distilled in the Declaration of the Rights of Man and the Citizen which prefaced the constitution of 1791. Legal equality and the legal protection of individual rights, the separation of Church and State and religious toleration were their expressions. The derivation of authority from popular sovereignty acting through a unified National Assembly, before whose legislation no privilege of locality or group could stand, was the basis of the jurisprudence which underlay them. It showed both that it could ride out financial storms far worse than those the old monarchy, had failed to master (national bankruptcy and the collapse of the currency among them), and that it could carry out administrative change of which enlightened despotism had only dreamed. Other Europeans watched aghast or at least amazed as this powerful legislative engine was employed to overturn and rebuild institutions at every level of French life. Legislative sovereignty was a great instrument of reform, as the enlightened despots had known. Judicial torture came to an end, and so did titular nobility, juridical inequality and the old corporate guilds of French workmen. Incipient trades unionism was scotched in the egg by legislation forbidding association by workers or employers for collective economic ends. In retrospect, the signposts to market society seem pretty plain. Even the old currency based on units in the Carolingian ratios of 1:20:12 (*livres, sous,* and *deniers*) gave way to a decimal system of *francs* and *centimes*, just as the chaos of old-fashioned weights and measures was (in theory) replaced by the metric system later to become almost universal.

Such great changes were bound to be divisive, the more so because men's minds change more slowly than their laws. Peasants who eagerly welcomed the abolition of feudal dues were much less happy about the disappearance of the communal usages from which they benefited and which were also part of the 'feudal' order. Such conservatism was especially hard to interpret in religious affairs, yet was very important. The holy vessel kept at Rheims from which the kings of France had been anointed since the Middle Ages was publicly destroyed by the authorities during the Terror, an altar to Reason replaced the Christian one in the cathedral of Notre Dame and many priests underwent fierce personal persecution. Clearly, the France which did this was no longer Christian in the traditional sense, and the theocratic monarchy went unmourned by most Frenchmen. Yet the treatment of the Church aroused popular opposition to the Revolution as nothing else had done, the cults of quasi-divinities such as Reason and the Supreme Being which some revolutionaries promoted were a flop, and many Frenchmen (and perhaps most Frenchwomen) would happily welcome the official restoration of the Catholic Church to French life when it eventually came. By then, is had long been restored *de facto* in the parishes by the spontaneous actions of French men and women.

The divisions aroused by revolutionary change in France could no more be confined within its borders than could the principles of '89. These had as first commanded much admiration and not much explicit condemnation or distrust in other countries, though this soon changed, in particular when French governments began to export their principles by propaganda and war. Change in France soon generated debate about what should happen in other countries. Such debate was bound to reflect the terminology and circumstances in which it arose. In this way France gave her politics to Europe and this is the second great fact about the revolutionary decade. That is when modern European politics began, and the terms Right and Left have been with us ever since. Liberals and conservatives (though it was to be a decade or so before the terms were used) came into political existence when the French Revolution provided what appeared to be a touchstone or litmus paper for political standpoints. On one side were republicanism, a wide suffrage, individual rights, free speech and free publication; on the other were order, discipline, and emphasis on duties rather than rights, the recognition of the social function of hierarchy and a wish to temper market forces by morality.

Some Frenchmen had always believed that the French Revolution had universal significance. In the language of enlightened thought they advocated the acceptance by other nations of the recipes they employed for the settlement of French problems. This was not entirely arrogant. Societies in pre-industrial and traditional Europe still had many features in common; all could learn something from France. In this way the forces making for French influence were reinforced by conscious propaganda and missionary effort. This was another route by which events in France entered universal history.

That the Revolution was of universal, unprecedented significance was not an idea confined to its admirers and supporters. It also lay at the roots of European conservatism as a self-conscious force. Well before 1789, it is true, many of the constituent elements of modern conservative thought were lying about in such phenomena as irritation over the reforming measures of enlightened despotism, clerical resentment of the prestige and effect of 'advanced' ideas, and the emotional reaction from the fashionable and consciously rational which lay at the heart of romanticism. Such forces were especially prevalent in Germany, but it was in England that there appeared the first and in many ways the greatest statement of the conservative, anti-revolutionary argument. This was the *Reflections on the Revolution in France,* published in 1790 by Edmund Burke. As might easily be inferred from his former role as defender of the rights of the American colonists, this book was far from a mindless defence of privilege. In it a conservative stance shook itself clear of the legalistic defence of institutions and expressed itself in a theory of society as the creation of more than will and reason and the embodiment of morality. The Revolution, by contrast, was condemned as the expression of the arrogance of the intellect, of arid rationalism, and of pride – deadlines and of pride – deadliest of all the sins.

The new polarization which the Revolution brought to Europe's politics promoted also the new idea of revolution itself, and that was to have great consequences. The old idea that a political revolution was merely a circumstantial break in an essential continuity was replaced by one which took it as radical, comprehensive upheaval, leaving untouched no institution and limitless in principle, tending, perhaps, even to the subversion of such basic institutions as the family and property. According to whether people felt heartened or dismayed by this prospect, they sympathized with or deplored revolution wherever it occurred as a manifestation of a universal phenomenon. In the nineteenth century they came even to speak of *the* Revolution as a universally, eternally present force. This idea was the extreme expression of an ideological form of politics which is by no means yet dead. There are still those who, broadly speaking, feel that all insurrectionary and subversive movements should, in principle, be approved or condemned without regard to the particular circumstances of cases. This mythology has produced much misery, but first Europe and then the world which Europe transformed have had to live with those who respond emotionally to it, just as earlier generations had to live with the follies of religious divisions. Its survival, unhappily, is testimony still of the impact of the French Revolution.

Many dates can be chosen as the 'beginning' of the French Revolution; a specific date to 'end' it would be meaningless. The year 1799 none the less was an important punctuation mark in its course. The *coup d'état* which then swept the Directory away brought to power a man who quickly inaugurated a dictatorship which was to last until 1814 and turn international relations upside-down. This was Napoleon Bonaparte, formerly general of the republic, now First Consul of the new regime and soon to be the first Emperor of France. Like most of the leading figures of his age, he was still a young man when he came to power. He had already shown exceptional brilliance and ruthlessness as a soldier. His victories combined with a shrewd political sense and a readiness to act in an insubordinate manner to win him a glamorous reputation; in many ways he now seems the greatest example of the eighteenth-century type of 'the adventurer'. In 1799 he had a great personal prestige and popularity. No one except the defeated politicians much regretted it when he shouldered them aside and assumed power. Immediately he justified himself by defeating the Austrians (who had joined again in a war against France) and making a victorious peace for France (as he had done once already). This removed the threat to the Revolution; no one doubted Bonaparte's own commitment to its principles. His consolidation of them was his most positive achievement.

Although Napoleon (as he was called officially after 1804, when he proclaimed his empire) reinstituted monarchy in France, it was in no sense a restoration. Indeed, he took care so to affront the exiled Bourbon family that any reconciliation with

Bonaparte, 'calm on a fiery horse', as he asked the painter David to depict him, crossing the St Bernard pass in 1800. In 1802 a plebiscite confirmed him as First Consul for life and in 1804 another declared him Emperor of France.

it was inconceivable. He sought popular approval for the empire in a plebiscite and got it. This was a monarchy Frenchmen had voted for; it rested on popular sovereignty, that is, the Revolution. It assumed the consolidation of the Revolution which the Consulate had already begun. All the great institutional reforms of the 1790s were confirmed or at least left intact; there was no disturbance of the land sales which had followed the confiscation of Church property, no resurrection of the old corporations, no questioning of the principle of equality before the law. Some measures were even taken further, notably when each department was given an administrative head, the prefect, who was in his powers something like one of the emergency emissaries of the Terror (many former revolutionaries became prefects). Such further centralization of the administrative structure would, of course, have been approved also by the enlightened despots. In the actual working of government, it is true, the principles of the Revolution were often infringed in practice. Like all his predecessors in power since 1793, Napoleon controlled the press by a punitive censorship, locked up people without trial and in general gave short shrift to the Rights of Man so far as civil liberties were concerned. Representative bodies existed under consulate and empire, but not much attention was paid to them. Yet it seems that this was what Frenchmen wanted, as they had wanted Napoleon's shrewd recognition of reality, a concordat with the pope which reconciled Catholics to the regime by giving legal recognition to what had already happened to the Church in France.

All in all, this amounted to a great consolidation of the Revolution and one guaranteed at home by firm government and by military and diplomatic strength abroad. Both were eventually to be eroded by Napoleon's huge military efforts. These for a time gave France the dominance of Europe; her armies fought their way to Moscow in the east and Portugal in the west and garrisoned the Atlantic and northern coast from Corunna to Stettin. Nevertheless, the cost of this was too great; even ruthless exploitation of occupied countries was not enough for France to sustain this hegemony indefinitely against the coalition of all the other European countries which Napoleon's arrogant assertion of his power aroused. When he invaded Russia in 1812 and the greatest army he ever led crumbled into ruins in the snows of the winter, he was doomed unless his enemies should fall out with one another. This time they did not. Napoleon himself blamed the British, who had been at war with him (and, before him, with the Revolution) with only one short break since 1792. There is something in this; the Anglo-French war was the last and most important round in a century of rivalry, as well as a war of constitutional monarchy against military dictatorship. It was the Royal Navy at Aboukir in 1798 and Trafalgar in 1805 which confined Napoleon to Europe, British money which financed the allies when they were ready to come forward, and a British army in the Iberian peninsula which kept alive there from 1809 onwards a front which drained French resources and gave hope to other Europeans.

By the beginning of 1814, Napoleon could defend only France. Although he did so at his most brilliant, the resources were not available to fight off Russian, Prussian and Austrian armies in the east, and a British invasion in the south-west. At last his generals and ministers were able to set him aside and make peace without a popular outcry, even though this meant the return of the Bourbons. But it could not by then mean the return of anything else of significance from the years before 1789. The Concordat remained, the departmental system remained, equality before the law remained, a representative system remained; the Revolution, in fact, had become part of the established order in France. Napoleon had provided the time, the social peace and the institutions for that to happen. Nothing survived of the Revolution except what he had confirmed.

This makes him very different from a monarch of the traditional stamp, even the most modernizing – and, in fact, he was often very conservative in his policies, distrusting innovation. In the end he was a democratic despot, whose authority came from the people, both in the formal sense of the plebiscites, and in the more general one that he had needed (and won) their goodwill to keep his armies in the field. He is thus nearer in style to twentieth-century rulers than to Louis XIV. Yet he shares with that monarch the credit for carrying French international power to an unprecedented height and because of this both of them have retained the admiration of their countrymen. But again there is an important, and twofold, difference: Napoleon not only dominated Europe as Louis XIV never did, but because the Revolution had taken place his hegemony represented more than mere national supremacy, though this fact should not be sentimentalized. The Napoleon who was supposed to be a liberator and a great European was the creation of later legend. The most obvious impact he had on Europe between 1800 and 1814 was the bloodshed and upheaval he brought to every corner of it, often as a consequence of megalomania and personal vanity. But there were also important side-effects, some intentional, some not. They all added up to the further spread and effectiveness of the principles of the French Revolution.

Their most obvious expression was on the map. The patchwork quilt of the European state system of 1789 had undergone some revision already before Napoleon took power, when French armies in Italy, Switzerland and the United Provinces had created new satellite republics. But these had proved incapable of survival once French support was withdrawn and it was not until French hegemony was re-established under the Consulate that there appeared a new organization which would have enduring consequences in some parts of Europe. The most important of these were in west Germany where political structure was revolutionized and medieval foundations swept away. German territories on the left bank of the Rhine were annexed to France for the whole of the period from 1801 to 1814, and this began a period of destroying historic German polities. Beyond the river, France provided the plan of a reorganization which secularized the ecclesiastical states, abolished nearly all the Free Cities, gave extra territory to Prussia, Hanover, Bavaria and Baden to compensate them for losses elsewhere, and abolished the old independent imperial nobility. The practical effect was to diminish the

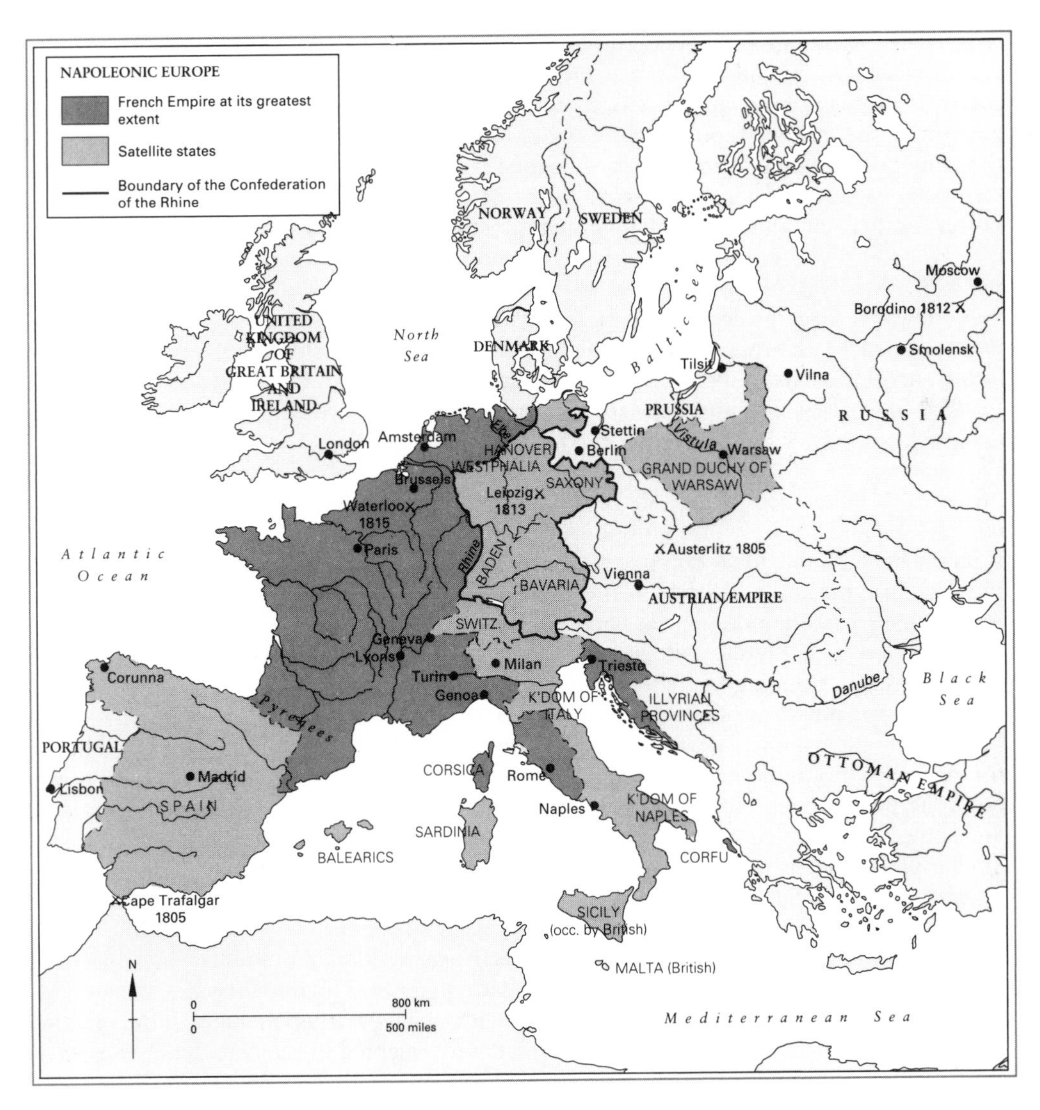

Catholic and Habsburg influence in Germany while strengthening the influence of its larger princely states (especially Prussia). The constitution of the Holy Roman Empire was revised, too, to take account of these changes. In its new form it lasted only until 1806, when another defeat of the Austrians led to more changes in Germany and its abolition. So came to an end the institutional structure which, however inadequately, had given Germany such political coherence as it had possessed since Ottonian times. A Confederation of the Rhine was now set up which provided a third force balancing that of Prussia and Austria. Thus were triumphantly asserted the national interests of France in a great work of destruction. Richelieu and Louis XIV would have enjoyed the contemplation of a French frontier on the Rhine with, beyond it, a Germany divided into interests likely to hold one another in check. But there was another side to it; the old structure, after all, had been a hindrance to German consolidation. No future rearrangement would ever contemplate the resurrection of the old structures. When, finally, the allies came to settle post-Napoleonic Europe, they too provided for a German Confederation. It was different from Napoleon's. Prussia and Austria were members of it in so far as their territories were German, but there was no going back on the fact of consolidation. A complicated structure of over three hundred political units with different principles of organization in 1789 was reduced to thirty-eight states in 1815.

Reorganization was less dramatic in Italy and its effects less revolutionary. The Napoleonic system provided in the north and south of the peninsula two large units which were nominally independent, while a large part of it (including the papal states) was formally incorporated in France and organized in departments. None of this structure survived in 1815, but neither was there a complete restoration of the old regime. In particular, the ancient republics of Genoa and Venice were left in the tombs to which the armies of the Directory had first consigned them. They were absorbed by bigger states, Genoa by Sardinia, Venice by Austria. Elsewhere in Europe, at the height of Napoleonic power, France had annexed and governed directly a huge block of territory whose coasts ran from the Pyrenees to Denmark in the north and from Catalonia almost without interruption to the boundary between Rome and Naples in the south. Lying detached from it was a large piece of modern Yugoslavia. Satellite states and vassals of varying degrees of real independence, some of them ruled over by members of Napoleon's own family, divided between them the rest of Italy, Switzerland and Germany west of the Elbe. Isolated in the east was another satellite, the 'grand duchy' of Warsaw, which had been created from former Russian territory.

In most of these countries common administrative practice and institutions provided a large measure of shared experience. That experience, of course, was of institutions and ideas which embodied the principles of the Revolution. They hardly reached beyond the Elbe except in the brief Polish experiment and thus the French Revolution came to be another of those great shaping influences which again and again have helped to differentiate eastern and western Europe. Within the French empire, Germans, Italians, Illyrians, Belgians and Dutch were all governed by the Napoleonic legal codes; the bringing of these to fruition was the result of Napoleon's own initiative and insistence, but the work was essentially that of revolutionary legislators who had never been able in the troubled 1790s to draw up the new codes so many Frenchmen had hoped for in 1789. With the codes went concepts of family, property, the individual, public power and others which were thus generally spread through Europe. They sometimes replaced and sometimes supplemented a chaos of local, customary, Roman and ecclesiastical law. Similarly, the departmental system of the empire imposed a common administrative practice, service in the French armies imposed a common discipline and military regulation and French weights and measures, based on the decimal system, replaced many local scales. Such innovations exercised an influence beyond the actual limits of French rule, providing models and inspiration to modernizers in other countries. The models were all the more easily assimilated because French officials and technicians worked in many of the satellites while many nationalities other than French were represented in the Napoleonic service.

Such changes took time to produce their full effect, but it was a deep one and was revolutionary. It was by no means necessarily liberal; even if the Rights of Man formally followed the Tricolour of the French armies, so did Napoleon's secret police, quartermasters and customs officers. A more subtle revolution deriving from the Napoleonic impact lay in the reaction and resistance it provoked. In spreading revolutionary principles the French were often putting a rod in pickle for their own backs. Popular sovereignty lay at the heart of the Revolution and it is an ideal closely linked to that of nationalism. French principles said that peoples ought to govern themselves and that the proper unit in which they should do so was the nation: the revolutionaries had proclaimed their own republic 'one and indivisible' for this reason. Many of their foreign admirers applied this principle to their own countries; manifestly, Italians and Germans did not live in national states, and perhaps they should. But this was only one side of the coin. French Europe was run for the benefit of France, and it thus denied the national rights of other Europeans. They saw their agriculture and commerce sacrificed to French economic policy, found they had to serve in the French armies, or to receive at the hands of Napoleon French (or Quisling) rulers and viceroys. When even those who had welcomed the principles of the Revolution felt such things as grievances, it is hardly surprising that those who had never welcomed them at all should begin to think in terms of national resistance, too. Nationalism in Europe was given an immense fillip by the Napoleonic era, even if governments distrusted it and felt uneasy about employing it. Germans began to think of themselves as more than Westphalians and Bavarians, and Italians began to believe they were more than Romans or Milanese, because they discerned a common interest against France. In Spain and Russia the identification of patriotic resistance with resistance to the Revolution was virtually complete.

In the end, then, though the dynasty Napoleon hoped to found and the empire he set up both proved fragile, his work was of great importance. He unlocked huge reserves of energy in other countries just as the Revolution had unlocked them in France, and afterwards they could never be quite shut up again. He ensured the legacy of the Revolution its maximum effect and this was his greatest achievement, whether he desired it or not.

His unconditional abdication in 1814 was not quite the end of the story. Just under a year later the emperor returned to France from Elba where he had lived in a pensioned exile, and the restored Bourbon regime crumbled at a touch. The allies none the less determined to overthrow him, for he had frightened them too much in the past. Napoleon's attempt to anticipate the gathering of overwhelming forces against him came to an end at Waterloo, on 18 June 1815, when the threat of a revived French empire was destroyed by the Anglo-Belgian and Prussian armies. This time the victors sent him to St Helena, thousands of miles away in the South Atlantic, where he died in 1821. The final alarm that he had given them strengthened their determination to make a peace that would avoid any danger of a repetition of the quarter-century of almost continuous war which Europe had undergone in the wake of the Revolution. Thus Napoleon still shaped the map of Europe, not only by the changes he had made in it, but also by the fear France had inspired under his leadership.

3
Political Change: A New Europe

Whatever conservative statesmen hoped in 1815, an uncomfortable and turbulent era had only just begun. This can be seen most easily in the way the map of Europe changed in the next sixty years. By 1871, when a newly-united Germany took its place among the great powers, most of Europe west of a line drawn from the Adriatic to the Baltic was organized in states based on the principle of nationality, even if some minorities still denied it. Even to the east of that line there were some states which were already identified with nations. By 1914 the triumph of nationalism was to go further still, and most of the Balkans would be organized as nation-states, too.

Nationalism, one aspect of a new kind of politics, had origins which went back a long way, to the examples set in Great Britain and some of Europe's smaller states in earlier times. Yet its great triumphs were to come after 1815, as part of the appearance of a new politics. At their heart lay an acceptance of a new framework of thought which recognized the existence of a public interest greater than that of individual rulers or privileged hierarchies. It also assumed that competition to define and protect that interest was legitimate. Such competition was thought increasingly to require special arenas and institutions; old juridical or courtly forms no longer seemed sufficient to settle political questions.

An institutional framework for this transformation of public life took longer to emerge in some countries than others. Even in the most advanced it cannot be identified with any single set of practices. It always tended, though, to be strongly linked with the recognition and promotion of certain principles. Nationalism was one of them which told most against older principles – that of dynasticism, for instance. It was more and more a commonplace of European political discourse, as the nineteenth century went on, that the interests of those recognized to be 'historic' nations should be protected and promoted by governments. This was, of course, wholly compatible with bitter and prolonged disagreement about which nations were historic, how their interests should be defined, and to what extent they could and should be given weight in statesmen's decisions.

There were also other principles in play besides nationalism. Terms like democracy and liberalism do not help very much in defining them, though they must be used in default of better ones. In most countries there was a general trend towards accepting representative institutions as a way of associating (even if only formally) more and more people with the government. Liberals and democrats almost always asked for more people to be given votes and for better electoral representation. More and more, too, the individual became the basis of political and social organization in economically advanced countries. The individual's membership of communal, religious, occupational and family units came to matter much less than his individual rights. Though this led in some ways to greater freedom, it sometimes led to less. The state became much more juridically powerful in relation to its subjects in the nineteenth century than ever before, and slowly, as its apparatus became technically more efficient, came to be able to coerce them more effectively.

The French Revolution had been of enormous importance in actually launching such changes but its continuing influence as example and a source of mythology mattered just as much. For all the hopes and fears that the Revolution was over by 1815, its full Europe-wide impact was then still to come. In many other countries institutions already swept away in France invited criticism and demolition. They were the more vulnerable because other forces of economic and social change were also at work. This gave revolutionary ideas and traditions new opportunities. There was a widespread sense that all Europe faced, for good or ill, potential revolution. This encouraged the upholders and would-be-destroyers of the existing order alike to sharpen political issues and fit them into the frameworks of the principles of 1789, nationalism and liberalism. By and large, these ideas dominated the history of Europe down to about 1870 and vitalised its politics. They did not achieve all their advocates hoped. Their realization in practice had many qualifications, they frequently and thwartingly got in one another's way, and they had many opponents. Yet they remain useful guiding threads in the rich and turbulent history of nineteenth-century Europe, already a political laboratory whose experiments, explosions and discoveries were changing the history of the rest of the world.

These influences could already be seen at work in the negotiations shaping the foundation deed of the nineteenth-century international order, the Treaty of Vienna of 1815, which closed the era of the French wars. Its central aim was to prevent their repetition. The peacemakers sought the containment of France and the avoidance of revolution, using as their materials the principle of legitimacy which was the ideological core of conservative Europe and certain practical territorial arrangements against future French aggression. Thus Prussia was given large acquisitions on the Rhine, a new northern state appeared under a Dutch king ruling both Belgium and the Netherlands, the kingdom of Sardinia was given Genoa, and Austria not only recovered her former Italian possessions, but kept Venice and was allowed a virtually free hand in keeping the other Italian states in order. In most of these cases legitimacy bowed to expediency; those despoiled by the revolutionaries or Napoleon did not all obtain restoration. But the powers talked legitimacy all the same, and (once the arrangements were complete) did so with some success. For nearly forty years the Vienna settlement provided a framework within which disputes were settled without war. Most of the regimes installed in 1815 were still there, even if some of them were somewhat shaken, forty years later.

This owed much to the salutary fear of revolution. In all the major continental states the restoration era (as the years after 1815 have been termed) was a great period for policemen and plotters alike. Secret societies proliferated, undiscouraged by failure after failure. This record showed, though, that there was no subversive threat that could not be handled easily enough. Austrian troops dealt with attempted coups in Piedmont and Naples, French soldiers restored the power of a reactionary Spanish king hampered by a liberal constitution, the Russian empire survived a military conspiracy and a Polish revolt. The Austrian predominance in Germany was not threatened at all and it is difficult in retrospect to discern any very real danger to any part of the Habsburg monarchy before 1848. Russian and Austrian power, the first in reserve, the second the main force in central Europe and Italy from 1815 to 1848, were the two rocks on which the Vienna system rested.

Mistakenly, liberalism and nationalism were usually supposed to be inseparable; this was to prove terribly untrue in later times, but in so far as a few people did seek to change Europe by revolution before 1848, it is broadly true that they wanted to do so by advancing both the political principles of the French Revolution – representative government, popular sovereignty, freedom of the individual and the press – and those of nationality. Many confused the two; the most famous and admired of those who did so was Mazzini, a young Italian. By advocating an Italian unity most of his countrymen did not want and conspiring unsuccessfully to bring it about, he became an inspiration and model for other nationalists and democrats in every continent for over a century and one of the first idols of radical chic. The age of the ideas he represented had not yet come.

For forty years, Giuseppe Mazzini was the advocate of Italian unification and revolution. More generally, he is rightly remembered as one of the purest embodiments of the belief in nationalism. Yet he died in exile, deeply disappointed and embittered by the unity Italy achieved in his lifetime.

To the west of the Rhine, where the writ of the Holy Alliance (as was termed the group of three conservative powers, Russia, Austria and Prussia) did not run, the story was different; there, legitimism was not to last long. The very restoration of the Bourbon dynasty in 1814 had itself been a compromise with the principle of legitimacy. Louis XVIII was supposed to have reigned like any other king of France since the death of his predecessor, Louis XVII, in a Paris prison in 1795. In fact, as everyone knew but legitimists tried to conceal, he came back in the baggage train of the Allied armies which had defeated Napoleon and he only did so on terms acceptable to the French political and military élites of the Napoleonic period and, presumably, tolerable by the mass of Frenchmen. The restored regime was regulated by a charter which created a constitutional monarchy, albeit with a limited suffrage. The rights of individuals were guaranteed and the land settlement resulting from revolutionary confiscations and sales was unquestioned; there was to be no return to 1789.

Nevertheless, there was some uncertainty about the future; battle between Right and Left began with arguments about the charter itself – was it a contract between king and people, or a simple emanation of the royal benevolence which might therefore be withdrawn as easily as it had been granted? – and went on over a whole range of issues which raised questions of principle (or were thought to do so) about ground won for liberty and the possessing classes in the Revolution.

What was implicitly at stake was what the Revolution had actually achieved. One way of describing that would be to say that those who had struggled to be recognized as having a voice in ruling France under the *ancien régime* had won; the political weight of the 'notables', as they were sometimes called, was assured and they, whether drawn from the old nobility of France, those who had done well out of the Revolution, Napoleon's lackeys, or simply substantial landowners and businessmen, were now the real rulers of France. Another change had been the nation-making brought about in French institutions; no person or corporation could now claim to stand outside the operative sphere of the national government of France. Finally and crucially, the Revolution had changed Frenchmen's thinking. Among other things, the terms in which French public affairs would be discussed and debated had been transformed. Wherever the line was to be drawn between Right and Left, conservatives or liberals, it was on that line that political battle now had to be centred, not over the privilege of counselling a monarch by Divine Right. This was just what the last king of the direct Bourbon line, Charles X, failed to see. He foolishly attempted to break out of the constitutional limitations which bound him, by what was virtually a *coup d'état*. Paris rose against him in the 'July Revolution' of 1830, liberal politicians hastily put themselves at its head, and to the chagrin of republicans, ensured that a new king replaced Charles.

Louis Philippe was head of the junior branch of the French royal house, the Orléans family, but to many conservative eyes he was the Revolution incarnate. His father had voted for the execution of Louis XVI (and went to the scaffold himself soon after) while the new king had fought as an officer in the republican armies. He had even been a member of the notorious Jacobin club which was widely believed to have been a deep-rooted conspiracy, and certainly had been a forcing-house for some of the Revolution's most prominent leaders. To liberals Louis Philippe was attractive for much the same reasons; he reconciled the Revolution with the stability provided by monarchy, though the radicals were disappointed. The regime over which he was to preside for eighteen years proved unimpeachably constitutional and preserved essential political freedoms, but protected the interests of the well-to-do. It vigorously suppressed urban disorder (of which poverty produced plenty in the 1830s) and this made it unpopular with the Left. One prominent politician told his fellow-countrymen to enrich themselves – a recommendation much ridiculed and misunderstood, though all he was trying to do was to tell them that the way to obtain the vote was through the qualification which a high income conferred (in 1830 only about a third as many Frenchmen as Englishmen had a vote for their national representatives, while the population of France was about twice that of England). Nevertheless, in theory, the July Monarchy rested on popular sovereignty, the

In 1848 monarchs everywhere trembled. In this Punch *cartoon, the uncrowned figure in the boat is Louis-Philippe, already turned off the French throne when it was published.*

revolutionary principle of 1789.

This gave it a certain special international standing in a Europe divided by ideology. In the 1830s there were sharply evident differences between a Europe of constitutional states – England, France, Spain and Portugal – and that of the legitimist, dynastic states of the East, with their Italian and German satellites. Conservative governments had not liked the July revolution. They were alarmed when the Belgians rebelled against their Dutch king in 1830, but could not support him because the British and French favoured the Belgians and Russia had a Polish rebellion on her hands. It took until 1839 to secure the establishment of an independent Belgium, and this was until 1848 the only important change in the state system created by the Vienna settlement, though the internal troubles of Spain and Portugal caused ripples which troubled European diplomacy.

Elsewhere, in south-east Europe, the pace of change was quickening. A new revolutionary era was opening there just as that of western Europe moved to its climax. In 1804 a well-to-do Serbian pork dealer had led a revolt by his countrymen against the undisciplined Turkish garrison of Belgrade. At that moment, the Ottoman regime was willing to countenance his actions in order to bridle its own mutinous soldiers. But the eventual cost to the empire was the establishment of an autonomous Serbian princedom in 1817. By then the Turks had also ceded Bessarabia to Russia, and had been forced to recognize that their hold on much of Greece and Albania was little more than formal, real power being in the hands of the local Pashas.

This was, though hardly yet visibly so, the opening of the Eastern Question of the nineteenth century: who or what was to inherit the fragments of the crumbling Ottoman empire? It was to take more than a century and a world war to solve the question in Europe; in what were once Asian provinces of the empire, the wars of the Ottoman Succession are still going on today. Racial, religious, ideological, and diplomatic issues were entangled from the start. The Vienna settlement did not include Ottoman territories among those covered by guarantees from the great powers. When what was soon represented as a 'revolution' of 'Greeks' (that is, Orthodox Christian subjects of the

Sultan, many of whom were bandits and pirates) began against the Turks in 1821, Russia favoured the rebels; this cut across conservative principles, but religion and the old pull of Russian imperialist expansion to the south-east made it impossible for the Holy Alliance to support the sultan as it supported other rulers. In the end, the Russians went to war with the Ottomans and defeated them. It was now evident that the nineteenth-century Eastern Question was going to be further complicated by nationalism, for the new kingdom of Greece which emerged in 1832 was bound to give ideas to other peoples in the Balkans.

In 1848 came a new revolutionary explosion. Briefly, it seemed that the whole 1815 settlement was in jeopardy. The 1840s had been years of economic hardship, food shortages and distress in many places, particularly in Ireland where, in 1846, there was a great famine and then in central Europe and France in 1847, where a commercial slump starved the cities. Unemployment was widespread. This bred violence which gave new edge to radical movements everywhere. One disturbance inspired another; example was contagious and weakened the capacity of the international security system to deal with further outbreaks. The symbolic start came in February, in Paris, where Louis Philippe abdicated after discovering the middle classes would no longer support his continued opposition to the extension of the suffrage. By the middle of the year, government had been swept aside or was at best on the defensive in every major European capital except London and St Petersburg. When a republic appeared in France after the February Revolution every revolutionary and political exile in Europe had taken heart. The dreams of thirty years' conspiracies seemed realizable. The *Grande Nation* would be on the move again and the armies of the Great Revolution might march once more to spread its principles. What happened, though, was very different. France made a diplomatic genuflexion in the direction of martyred Poland, the classical focus of liberal sympathies, but the only military operations it undertook were in defence of the pope, an unimpeachably conservative cause.

This was symptomatic. The revolutionaries of 1848 were provoked by very different situations, and many different aims, and followed divergent and confusing paths. In most of Italy and central Europe they rebelled against governments which they thought oppressive because they were illiberal; there, the great symbolic demand was for constitutions to guarantee essential freedoms. When such a revolution occurred in Vienna itself, the chancellor Metternich, architect of the conservative order of 1815, fled into exile. Successful revolution at Vienna meant the paralysis and therefore the dislocation of the whole of central Europe. Germans were now free to have their revolutions without fear of Austrian intervention in support of the *ancien régime in* the smaller states. So were other peoples within the Austrian dominions; Italians (led by an ambitious but apprehensive conservative king of Sardinia) turned on the Austrian armies in Lombardy and Venetia, Hungarians revolted at Budapest, and Czechs at Prague. This much complicated things. Many of these revolutionaries wanted national independence rather than constitutionalism, though constitutionalism seemed for a time the way to independence because it attacked dynastic autocracy.

If the liberals were successful in getting constitutional governments installed in all the capitals of central Europe and Italy, then it followed there would actually come into existence nations hitherto without state structures of their own, or at least without them for a very long time. If Slavs achieved their own national liberation then states previously thought of as German would be shorn of huge tracts of their territory, notably in Poland and Bohemia. It took some time for this to sink in. The German liberals suddenly fell over this problem in 1848 and quickly drew their conclusions; they chose nationalism. (The Italians were still to be coming to terms with their own version of the dilemma in the South Tyrol a hundred years later.) The German revolutions of 1848 failed, essentially, because the German liberals decided that German nationalism required the preservation of German lands in the east. Hence, they needed a strong Prussia and must accept its terms for the future of Germany. There were other signs, too, that the tide had turned before the end of 1848. The Austrian army had mastered the Italians. In Paris a rising aiming to give the Revolution a further shove in the direction of democracy was crushed with great bloodshed in June. The republic was, after all, to be a conservative one. In 1849 came the end. The Austrians overthrew the Sardinian army which was the only shield of the Italian revolutions, and monarchs all over the peninsula then began to withdraw the constitutional concessions they had made while Austrian power was in abeyance. German rulers did the same, led by Prussia. The pressure was kept up on the Habsburgs by the Croats and Hungarians, but then the Russian army came to its ally's help.

Liberals saw 1848 as a 'springtime of the nations'. If it was one, the shoots had not lived long before they withered. By the end of 1849 the formal structure of Europe was once again much as it had been in 1847, in spite of important changes within some countries. Nationalism had certainly been a popular cause in 1848, but it had been shown that it was neither strong enough to sustain revolutionary governments nor necessarily an enlightened force. Its failure shows the charge that the statesmen of 1815 'neglected' to give it due attention is false; no new nation emerged from 1848 for none was ready to do so. The basic reason for this was that although nationalities might exist, over most of Europe nationalism was still an abstraction for the masses; only relatively few and well-educated, or at least half-educated, people much cared about it. Where national differences also embodied social issues there was sometimes effective action by people who felt they had an identity given them by language, tradition or religion, but it did not lead to the setting up of new nations. The Ruthene peasants of Galicia in 1847 had happily murdered their Polish landlords when the Habsburg administration allowed them to do so. Having thus satisfied themselves they remained loyal to the Habsburgs in 1848.

There were some genuinely popular risings in 1848. In Italy they were usually revolts of townsmen rather than peasants; the Lombard peasants, indeed, cheered the Austrian army when it returned, because they saw no good for them in a revolution led by the aristocrats who were their landlords. In parts of Germany, over much of which the traditional structures of landed rural society remained intact, the peasants behaved as their predecessors had done in France in 1789, burning their landlords' houses, not merely through personal animus but in order to destroy the hated and feared records of rents, dues and labour services. Such outbreaks frightened urban liberals as much as the Parisian outbreak of despair and unemployment in the June Days frightened the middle classes in France. There, because the peasant was since 1789 (speaking broadly) a conservative, the government was assured the support of the provinces in crushing the Parisian poor who had given radicalism its brief success. But conservatism could be found within revolutionary movements, too. German working-class turbulence alarmed the better-off, but this was because the leaders of German workers talked of 'socialism' while actually seeking a return to the past. They had the safe world of guilds and apprenticeships in mind, and feared machinery in factories, steamboats on the Rhine which put boatmen out of work, the opening of unrestricted entry to trades – in short, the all-too-evident signs of the onset of market society. Almost always, liberalism's lack of appeal to the masses was shown up in 1848 by popular revolution.

Altogether, the social importance of 1848 is as complex and escapes easy generalization as much as its political content. It was probably in the countryside of eastern and central Europe that the revolutions changed society most. There, liberal principles and the fear of popular revolt went hand in hand to impose change on the landlords. Wherever outside Russia obligatory peasant labour and bondage to the soil survived, it was abolished as a result of 1848. That year carried the rural social revolution launched sixty years earlier in France to its conclusion in central and most of eastern Europe. The way was now open for the reconstruction of agricultural life in Germany and the Danube valley on individualist and market lines. Though many of its practices and habits of mind were still to linger, feudal society had in effect now come to an end all over Europe. The political components of French revolutionary principles, though, would have to wait longer for their expression.

In the case of nationalism this was not very long. A dispute over Russian influence in the Near East, where Turkish power was visibly declining, in 1854 ended the long peace between great powers which had lasted since 1815. The Crimean War, in which the French and British fought as allies of the Turks against the Russians, was in many ways a notable struggle. Fighting took place in the Baltic, in southern Russia, and in the Crimea, the last theatre attracting most attention. There, the allies had set themselves to capture Sebastopol, the naval base which was the key to Russian power in the Black Sea. Some of the results were surprising. The British army fought gallantly, as did its opponents and allies, but was especially distinguished by the inadequacy of its administrative arrangements; the scandal these caused launched an important wave of radical reform at home. Incidentally the war also helped to found the prestige of a new profession for women, that of nursing, for the collapse of British medical services had been particularly striking. Florence Nightingale's work launched the first major extension of the occupational opportunities available to respectable women since the creation of female religious communities in the Dark Ages. The conduct of the war is also noteworthy in another way as an index of modernity: it was the first between major powers in which steamships and a railway were employed.

Yet these things, however portentous, mattered less in the short run than what the war did to international relations. Russia was defeated and her long enjoyment of a power to intimidate Turkey was bridled for a time. A step was taken towards the establishment of another new Christian nation, Romania, which was finally brought about in 1862. Once more, nationality triumphed in former Turkish lands. But the crucial effect of the war was that the Holy Alliance had disappeared. The old rivalry of the eighteenth century between Austria and Russia over what would happen to the Turkish inheritance in the Balkans had broken out again when Austria warned Russia not to occupy the Danube principalities (as the future Romania was termed) during the war and then occupied them herself. This was five years after Russia had intervened to restore Habsburg power by crushing the Hungarian revolution. It was the end of friendship between the two powers. The next time Austria faced a threat she would have to do so without the Russian policeman of conservative Europe at her side.

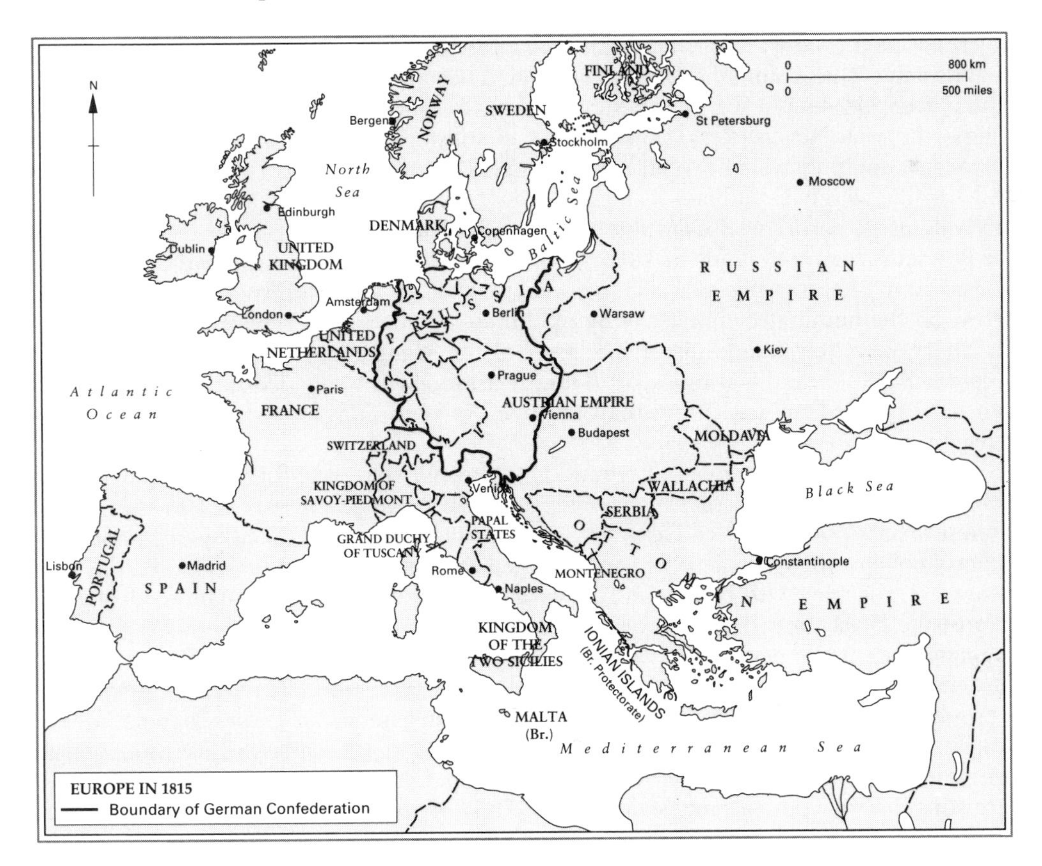

In 1856, when peace was made, few people can have anticipated how quickly that time would come. Within ten years Austria lost in two short, sharp wars her hegemony both in Italy and in Germany, and those countries were united in new national states. Nationalism had indeed triumphed, and at the cost of the Habsburgs, as had been prophesied by enthusiasts in 1848, but in a totally unexpected way. Not revolution, but the ambitions of two traditionally expansive monarchical states, Sardinia and Prussia, had led each to set about improving its position at the expense of Austria, whose isolation was at that moment complete. Not only had she sacrificed the Russian alliance, but after 1852 France was ruled by an emperor who again bore the name Napoleon (he was the nephew of the first to do so). He had been elected president of the Second Republic, whose constitution he then set aside by *coup d'état*. The name Napoleon was itself terrifying. It suggested a programme of international reconstruction – or revolution. Napoleon III (the second was a legal fiction, a son of Napoleon I who had never ruled) stood for the destruction of the anti-French settlement of 1815 and therefore of the Austrian predominance which propped it up in Italy and Germany. He talked the language of nationalism with less inhibition than most rulers and seems to have believed in it. With arms and diplomacy he forwarded the work of two great diplomatic technicians, Cavour and Bismarck, the prime ministers respectively of Sardinia and of Prussia.

In 1859 Sardinia and France fought Austria; after a brief war the Austrians were left with only Venetia in Italy. Cavour now set to work to incorporate other Italian states into Sardinia, a part of the price being that Sardinian Savoy had to be given to France. Cavour died in 1861, and debate still continues over what was the extent of his real aims, but by 1871 his successors had produced a united Italy under the former King of Sardinia, who was thus recompensed for the loss of Savoy, the ancestral duchy of his house. In that year Germany was united, too. Bismarck had begun by rallying German liberal sentiment to the prussian cause once again in a nasty little war against Denmark in 1864. Two years later Prussia defeated Austria in a lightning campaign in Bohemia, thus at last ending the Hohenzollern-Habsburg duel for supremacy in Germany begun in 1740 by Frederick II. The war which did this was rather a registration of an accomplished fact than its achievement, for since 1848 Austria had been much weakened in German affairs. In that year, German liberals had offered a German crown, not to the emperor, but to the King of Prussia. Nevertheless, some states had still looked to Vienna for leadership and patronage, and they were now left alone to meet Prussian bullying. The Habsburg empire now became wholly Danubian, its foreign policy preoccupied with south-east Europe and the Balkans. It had retired from the Netherlands in 1815, Venetia had been exacted by the Prussians for the Italians in 1866, and now it left Germany to its own devices, too. Immediately after the peace the Hungarians seized the opportunity to inflict a further defeat on the humiliated monarchy by obtaining a virtual autonomy for the half of the Habsburg monarchy made up of the lands of the Hungarian crown. The empire thus became, in 1867, the Dual or Austro-Hungarian monarchy, divided rather untidily into two halves linked by little more than the dynasty itself and the conduct of a common foreign policy.

German unification required one further step. It had gradually dawned on France that the assertion of Prussian power beyond the Rhine was not in the French interest; instead of a disputed Germany, she now faced one dominated by one important military power. The Richelieu era had crumbled away unnoticed. Bismarck used this new awareness, together with Napoleon III's weaknesses at home and international isolation, to provoke France into a foolish declaration of war in 1870. Victory in this war set the coping-stone on the new edifice of German nationality, for Prussia had taken the lead in 'defending' Germany against France – and there were still Germans alive who could remember what French armies had done in Germany under an earlier Napoleon. The Prussian army destroyed the Second Empire in France (it was to be the last monarchical regime in that country) and created the German empire, the Second Reich, as it was called, to distinguish it from the medieval empire. In practice, it was a Prussian domination cloaked in federal forms, but as a German national state it satisfied many German liberals.

It was dramatically and appropriately founded in 1871 when the king of Prussia accepted the crown of united Germany (which his predecessor had refused to take from German liberals in 1848) from his fellow-princes in the palace of Louis XIV at Versailles.

The Unification Process in Nineteenth-Century Germany: Some Milestones

1806	Organization of *The Confederation of the Rhine* (12 July) of sixteen princes under French domination.
	Francis II gives up title of *Holy Roman Emperor* (6 August).
1815	Creation by the Treaty of Vienna of the *Germanic Confederation* of thirty-eight sovereign powers.
1819	Prussia establishes a uniform tariff for all her territories and signs a tariff agreement with Schwarzburg-Sondershausen (October).
	The beginning of the *Zollverein* (Customs Union) which by
1844	includes most states of the Germanic Confederation but not Austria.
1848	In wake of revolutionary 'March days' the King of Prussia promises that Prussia will be 'merged into Germany'.
	Frankfurt National Assembly meets May 18, the first all-Germany parliament, elected by democratic suffrage. It adopts a constitution in
1849	and elects the King of Prussia 'Emperor of the Germans' (March 28), but he refuses the crown.
1850	The old *Germanic Confederation* re-established.
1853	Hanover, Brunswick, Mecklenburg join *Zollverein* (all non-Austrian Germany now members).
1863–5	A crisis over Danish claims to Schleswig-Holstein leaves Prussia, not Austria, as focus of German nationalist hopes.
1866	Prussia engineers war with Austria and declares the *Germanic Confederation* at an end. After victory at Sadowa (Königgrätz, 3 July), peace is made (Treaty of Prague, 23 August).
1867	Treaties between Prussia and German states north of the river Main lead to *The North German Confederation*, a federal structure dominated by Prussia.
	Zollparlament set up (July 8) including representation of four South German states as well as the North German *Reichstag* to deal with customs question.
1870	France declares war on Prussia (July 19), which receives support of South German states. Decisive French defeat at Sedan (1 September).
1871	William I of Prussia proclaimed German Emperor at Versailles after negotiation with all German states, North and South.
	The new, federal, Empire consists of twenty-five states, with a common *Reichstag*.

There had thus been in fifty years a revolution in international affairs and is would have consequences for world, as well as European, history. Germany had replaced France as the dominant land-power in Europe as France had replaced Spain in the seventeenth century. This fact was to overshadow Europe's international relations until they ceased to be determined by forces originating within her. It owed just a little to revolutionary politics in the narrow and strict sense. The conscious revolutionaries of the nineteenth century had achieved nothing comparable with the work of Cavour, Bismarck and, half in spite of himself, Napoleon III. This is very odd, given the hopes entertained of revolution in this period, and the fears felt for it. Revolution had achieved little except at the fringes of Europe and had even begun to show signs of flagging. Down to 1848 there had been plenty of revolutions, to say nothing of plots, conspiracies and *pronunciamientos* which did not justify the name. After 1848 there were very few. Another Polish revolution took place in 1863, but this was the only outbreak of note in the lands of the great powers until 1871.

An ebbing of revolutionary effort by then is understandable. Revolutions seemed to have achieved little outside France and had there brought disillusion and dictatorship. Some of their goals were being achieved in other ways. Cavour and his followers had created a united Italy, after all, greatly to the chagrin of Mazzini, since it was not one of which that revolutionary could approve, and Bismarck had done what many of the German liberals of 1848 had hoped for by providing a Germany which was indisputably a great power. Other ends were being achieved by economic progress; for all the horrors of the poverty which it contained, nineteenth-century Europe was getting richer and was giving more and more of its peoples a larger share of its wealth. Even quite short-term factors helped here. The year 1848 was soon followed by the great gold discoveries of California which provided a flow of bullion to stimulate the world economy in the 1850s and 1860s; confidence grew and unemployment fell in these decades and this was good for social peace.

A more fundamental reason why revolutions were less frequent was, perhaps, that they became more difficult to carry out. Governments were finding it steadily easier to grapple with them, largely for technical reasons. The nineteenth century created modern police forces. Better communications by rail and telegraph gave new power to central government in dealing with distant revolt. Above all, armies had a growing technical superiority to rebellion. As early as 1795 a French government showed that once it had control of the regular armed forces, and was prepared to use them, it could master Paris. During the long peace from 1815 to 1848 many European armies in fact became much more instruments of security, directed potentially against their own populations, than means of international competition, directed against foreign enemies. It was only the defection of important sections of the armed forces which permitted successful revolution in Paris in 1830 and 1848; once such forces were available to the government, battles like that of the June Days of 1848 (which one observer called the greatest slave-war in history) could only end with the defeat of the rebels. From that year, indeed, no popular revolution was ever to succeed in a major European country against a government whose control of its armed forces was unshaken by defeat in war or by subversion, and which was determined to use its power.

This was vividly and bloodily demonstrated in 1871, when a rebellious Paris was once again crushed by a French government in little more than a week with a toll of dead as great as that exacted by the Terror of 1793-4. A popular regime which drew to itself a wide range of radicals and reformers set itself up in the capital as the 'Commune' of Paris, a name evocative of traditions of municipal independence going back to the Middle Ages and, more important, to 1793, when the Commune (or city council) of Paris had been the centre of revolutionary fervour. The Commune of 1871 was able to take power because in the aftermath of defeat by the Germans the government could not disarm the capital of the weapons with which it had successfully withstood a siege, and because the same defeat had inflamed many Parisians against the government they believed to have let them down. During its brief life (there were a few weeks of quiet while the government prepared its riposte) the Commune did very little, but it produced a lot of left-wing rhetoric and was soon seen as the embodiment of social revolution. This gave additional bitterness to the efforts to suppress it. They came when the government had reassembled its forces from returning prisoners of war to reconquer Paris, which became the scene of brief but bloody street-fighting. Once again, regularly constituted armed forces overcame workmen and shopkeepers manning hastily improvised barricades.

If anything could do so, the ghastly failure of the Paris Commune should have killed the revolutionary myth, both in its power to terrify and its power to inspire. Yet it did not. If anything, it strengthened it. Conservatives found it a great standby to have the Commune example to hand in evoking the dangers lurking always ready to burst out from under the surface of society. Revolutionaries had a new episode of heroism and martyrdom to add to an apostolic succession of revolutionaries running already from 1789 to 1848. But the Commune also revivified the revolutionary mythology because of a new factor whose importance had already struck both Left and Right. This was socialism.

This word (like its relative, 'socialist') has come to cover a great many different things, and did so almost from the start. Both words were first commonly used in France round about 1830 to describe theories and men opposed to a society run on market principles and to an economy operated on *laissez-faire* lines, of which the main beneficiaries (they thought) were the wealthy. Economic and social egalitarianism is fundamental to the socialist idea. Most socialists have been able to agree on that. They have usually believed that in a good society there would be no classes oppressing one another through the advantages given to one by the ownership of wealth. All socialists, too, could agree that there was nothing sacred about property, whose rights buttressed injustice; some sought its complete abolition and were called communists. 'Property is theft' was one very successful slogan.

Mr and Mrs Karl Marx.

Such ideas might be frightening, but were not very novel. Egalitarian ideas have fascinated men throughout history and the Christian rulers of Europe had managed without difficulty to reconcile social arrangements resting on sharp contrasts of wealth with the practice of a religion one of whose greatest hymns praised God for filling the hungry with good things and sending the rich empty away. What happened in the early nineteenth century was that such ideas seemed to become at once more dangerous and more widespread. There was also a need for new terms because of other developments. One was that the success of liberal political reform appeared to show that legal equality was not enough, if it was deprived of content by dependence on other men's economic power, or denatured by poverty and attendant ignorance. Another was that already in the eighteenth century a few thinkers had seen big discrepancies of wealth as irrationalities in a world which could and should (they thought) be regulated to produce the greatest good of the greatest number. In the French Revolution some thinkers and agitators already pressed forward demands in which later generations would see socialist ideas. Egalitarian ideas none the less only became socialism in a modern sense when they began to grapple with the problems of the new epoch of economic and social change, above all with those presented by industrialization.

This often required a great perspicacity, for these changes were very slow in making their impact outside Great Britain and Belgium, the first continental country to be industrialized in the same degree. Yet perhaps because the contrast they presented with traditional society was so stark, even the small beginnings of concentration in capitalist finance and manufacturing were remarked. One of the first men to grasp their potentially very great implications for social organization was a French nobleman, Claude Saint-Simon. His seminal contribution to socialist thought was to consider the impact on society of technological and scientific advance. Saint-Simon thought that they not only made planned organization of the economy imperative, but implied (indeed, demanded) the replacement of the traditional ruling classes, aristocratic and rural in their outlook, by élites representing new economic and intellectual forces. Such ideas influenced many thinkers (most of them French) who in the 1830s advocated greater egalitarianism; they seemed to show that on rational as well as ethical grounds such change was desirable. Their doctrines made enough impact and considerations were enough talked about to terrify the French possessing classes in 1848, who thought they saw in the June Days a 'socialist' revolution. Socialists willingly identified themselves for the most part with the tradition of the French Revolution, picturing the realization of their ideals as its next phase, so the misinterpretation is understandable.

In 1848, at this juncture, there appeared a pamphlet which is the most important document in the history of socialism. It is always known as *The Communist Manifesto* (though this was not the title under which it appeared). It was largely the work of a young German Jew, Karl Marx, and with it the point is reached at which the prehistory of socialism can be separated from its history. Marx proclaimed a complete break with what he called the 'utopian socialism' of his predecessors. Utopian socialists attacked industrial capitalism because they thought it was unjust; Marx thought this beside the point. Nothing, according to Marx, could be hoped for from arguments to persuade people that change was morally desirable. Everything depended on the way history was going, towards the actual and inevitable creation of a new working class by industrial society, the rootless wage-earners of the new industrial cities which he termed the industrial proletariat. This class was bound, according to him, to act in a revolutionary way. History was working upon them so as to generate revolutionary capacity and mentality. It would present them with conditions to which revolution was the only logical outcome and that revolution would be, by those conditions, guaranteed success. What mattered was not that capitalism was morally wrong, but that it was already out-of-date and therefore historically doomed. Marx asserted that every society had a particular system of property rights and class relationships, and these accordingly shaped its particular political arrangements. Politics were bound to express economic forces. They would change as the particular organization of society changed under the influence of economic developments, and therefore, sooner or later (and Marx seems to have thought sooner), the revolution would sweep away capitalist society and its forms as capitalist society had already swept away feudal.

There was much more to Marx than this, but this was a striking and encouraging message which gave him domination of the international socialist movement which emerged in the next twenty years. The assurance that history was on their side was a great tonic to revolutionaries. They learnt with gratitude that the cause to which they were impelled anyway by motives ranging from a sense of injustice to the promptings of envy was predestined to triumph. This was essentially a religious faith. For all its intellectual possibilities as an analytical instrument, Marxism came to be above all a popular mythology, resting on a view of history which said that men were bound by necessity because their institutions were determined by the evolving methods of production, and on a faith that the working class were the Chosen People whose pilgrimage through a wicked world would end in the triumphal establishment of a just society in which necessity's iron law would cease to operate. Social revolutionaries could thus feel confident of scientifically irrefutable arguments for irresistible progress towards the socialist millennium while clinging to a revolutionary activism it seemed to make unnecessary. Marx himself seems to have followed his teaching more cautiously, applying it only to the broad, sweeping changes in history which individuals are powerless to resist and not to its detailed unfolding. Perhaps it is not surprising that, like many masters, he did not recognize all his pupils: he came later to protest that he was not a Marxist.

This new religion was an inspiration to working-class organization. Trades unions and cooperatives already existed in some countries, the first international organization of working men appeared in 1863. Though it included many who did not subscribe to Marx's views (anarchists, among others), his influence was paramount within it (he was its secretary). Its name frightened conservatives, some of whom blamed the Paris Commune on it. Whatever their justification, their instincts were right. What happened in the years after 1848 was that socialism captured the revolutionary tradition from the liberals, and a belief in the historical role of an industrial working class still barely visible outside England (let alone predominant in most countries) was tacked on to the tradition which held that, broadly speaking, revolution could not be wrong. Forms of thinking about politics evolved in the French Revolution were thus transferred to societies to which they would prove increasingly inappropriate. How easy such a transition could be was shown by the way Marx snapped up the drama and mythical exaltation of the Paris Commune for socialism. In a powerful tract he annexed it to his own theories, though it was, in fact, the product of many complicated and differing forces and expressed very little in the way of egalitarianism, let alone 'scientific' socialism. It emerged, moreover, in a city which, though huge, was not one of the great manufacturing centres in which he predicted proletarian revolution would mature. These remained, instead, stubbornly quiescent. The Commune was, in fact, the last and greatest example of revolutionary and traditional Parisian radicalism. It was a great failure (and socialism suffered from it, too, because of the repressive measures it provoked), yet Marx made it central to socialist mythology.

Russia seemed, except in her Polish lands, immune to the disturbances troubling other continental great powers. The French Revolution had been another of those experiences which, like feudalism, Renaissance or Reformation, decisively shaped western Europe and passed Russia by. Although Alexander I, the tsar under whom Russia met the 1812 invasion, had indulged himself with liberal ideas and had even thought of a constitution, nothing seemed to come of this. A formal liberalization of Russian institutions did not begin until the 1860s, and even then its source was not revolutionary contagion. It is true that liberalism and revolutionary ideologies did not quite leave Russia untouched before this. Alexander's reign had seen something of an opening of Pandora's box of ideas and it had thrown up a small group of critics of the regime who found their models in western Europe. Some of the Russian officers who went there with the armies which pursued Napoleon to Paris were led by what they saw and heard to make unfavourable comparisons with their homeland; this was the beginning of Russian political opposition. In an autocracy opposition was bound to mean conspiracy. Some of them took part in the organization of secret societies which attempted a *coup* amid the uncertainty caused by the death of Alexander in 1825; this was called the 'Decembrist' movement. It collapsed but only after giving a fright to Nicholas I, a tsar who decisively and negatively affected Russia's historical destiny at a crucial moment by ruthlessly turning on political liberalism and seeking to crush it. In part because of the immobility which he imposed upon her, Nicholas' reign influenced Russia's destiny more than any since that of Peter the Great. A dedicated believer in autocracy, he confirmed the Russian tradition of authoritarian bureaucracy, the management of cultural life, and the rule of the secret police just when the other great conservative powers were, however unwillingly, beginning to move in the opposite direction. There was, of course, much to build on already in the historical legacies which differentiated Russian autocracy from western European monarchy. But there were also great challenges to be met and Nicholas' reign was a response to these as well as a simple deployment of the old methods of despotism by a man determined to use them.

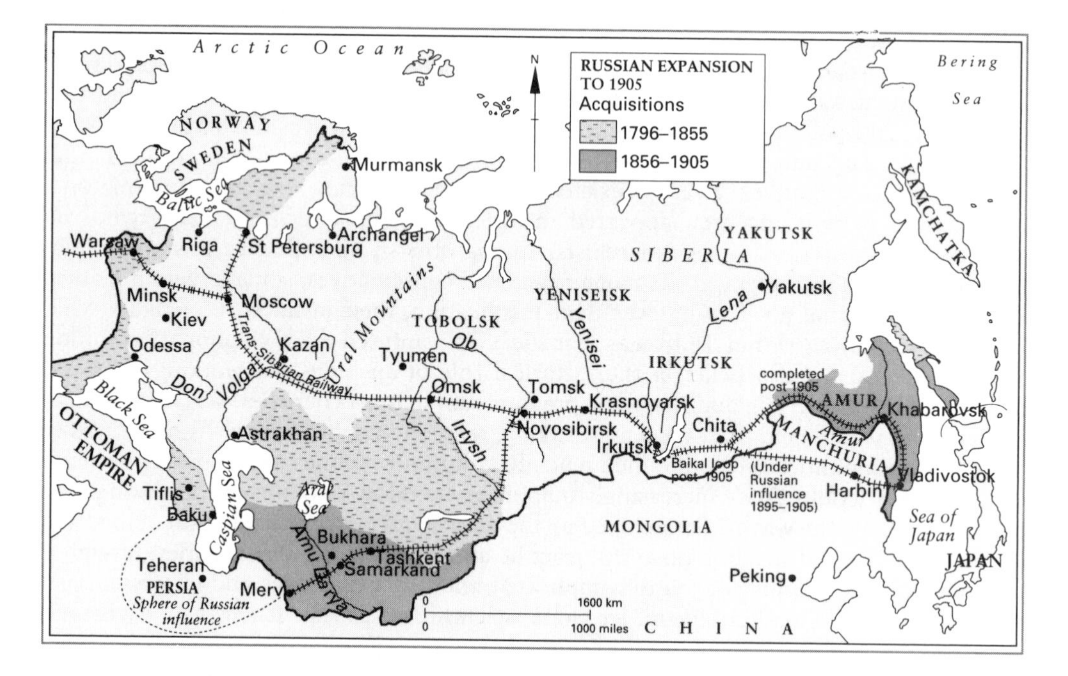

The ethnic, linguistic and geographical diversity of the empire had begun to pose problems far outrunning the capacity of Muscovite tradition to deal with them. The population of the empire itself more than doubled in the forty years after 1770. This ever-diversifying society none the less remained overwhelmingly backward; its few cities were hardly a part of the vast rural expanses in which they stood and often seemed insubstantial and impermanent, more like temporary though huge encampments than settled centres of civilization. The greatest expansion had been to the south and south-east; here new élites had to be incorporated in the imperial structure and to stress the religious ties between the Orthodox was one of the easiest ways to do this. As the conflict with Napoleon had compromised the old prestige of things French and the sceptical ideas of the Enlightenment associated with that country, a new emphasis was now given to religion in the evolution of a new ideological basis for the Russian empire under Nicholas. 'Official Nationality', as it was called, was Slavophile and religious in doctrine, bureaucratic in form and gave Russia an ideological unity it had lost since outgrowing its historic centre in Muscovy.

The importance of official ideology was from this time one of the great differences between Russia and western Europe. Until the last decade of the twentieth

The actual appearance of the battlefield began to be photographed in the Crimean War. A Russian battery after its capture by the French.

century Russian governments never gave up their belief in official ideology as a unifying force. Yet this did not mean that daily life in the middle of the century, either for the civilized classes or the mass of a backward population, was much different from that of other parts of eastern and central Europe. Yet Russian intellectuals argued about whether Russia was or was not a European country, and this is not surprising; Russia's roots were different from those of countries further west. What is more, a decisive turn was taken under Nicholas, from the beginning of whose reign possibilities of change which were at least being felt in other dynastic states in the first half of the nineteenth century were simply not allowed to appear in Russia. It was the land *par excellence* of censorship and police. In the long run this was bound to exclude certain possibilities of modernization (though other obstacles rooted in Russian society seem equally important), but in the short run it was highly successful. Russia passed through the whole nineteenth century without revolution; revolts in Russian Poland in 1830–1 and 1863–4 were ruthlessly suppressed, the more easily because Poles and Russians cherished traditions of mutual dislike.

The other side of the coin was the almost continuous violence and disorder of a savage and primitive rural society, and a mounting and more and more violent tradition of conspiracy which perhaps incapacitated Russia even further for normal politics and the shared assumptions they required. Unfriendly critics variously described Nicholas' reign as an ice age, a plague zone and a prison, but not for the last time in Russian history the preservation of a harsh and unyielding despotism at home was not incompatible with a strong international rôle. This rested upon Russia's huge military superiority. When armies contended with muzzle-loaders and no important technological differences distinguished one from another her vast numbers were decisive. On Russian military strength rested the anti-revolutionary international security system, as 1849 showed. But Russian foreign policy had other successes, too. Pressure was consistently kept up on the central Asian khanates and on China. The left bank of the Amur became

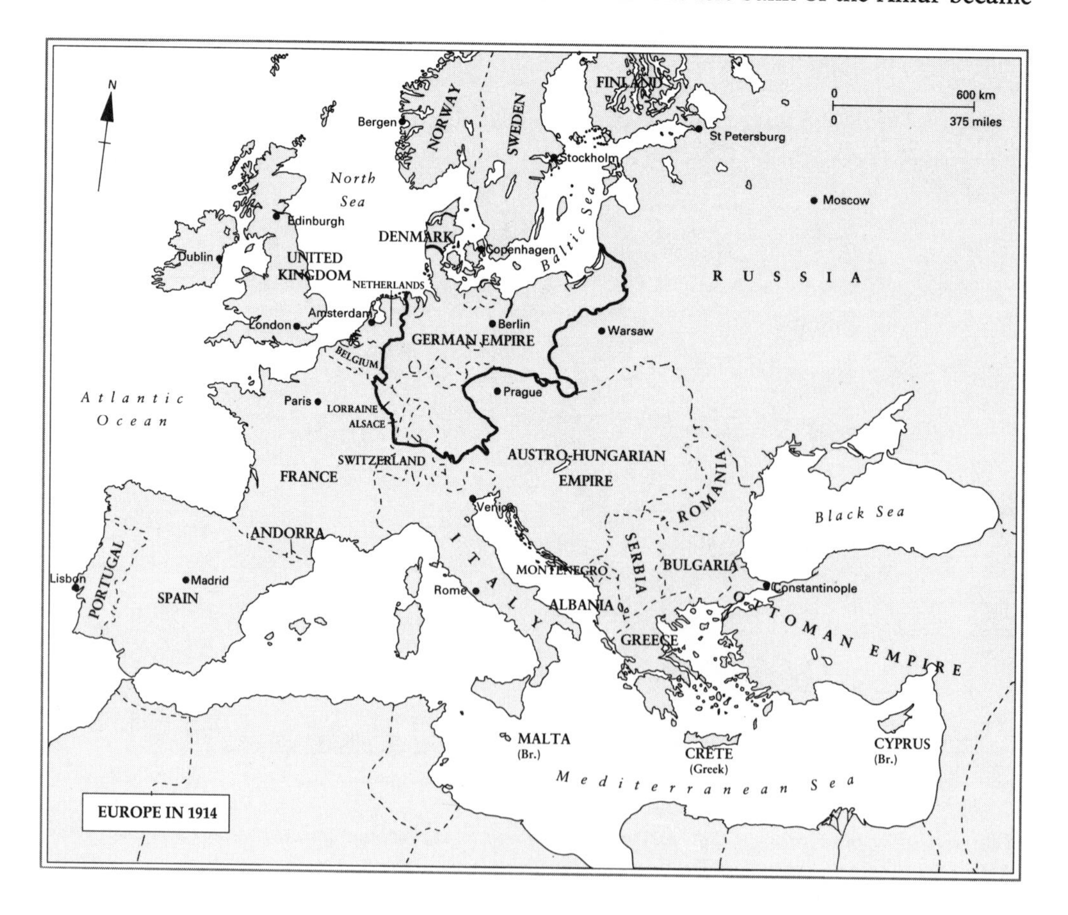

EUROPE IN 1914

Russian and in 1860 Vladivostok was founded. Great concessions were exacted from Persia and during the nineteenth century Russia absorbed Georgia and a part of Armenia. For a time there was even a determined effort to pursue Russian expansion in North America, where there were forts in Alaska and settlements in northern California until the 1840s.

The major effort of Russian foreign policy, nevertheless, was directed to the south-west, towards Ottoman Europe. Wars in 1806–12 and 1828 carried the Russian frontier across Bessarabia to the Pruth and the mouth of the Danube. It was by now clear that the partition of the Ottoman empire in Europe would be as central to nineteenth-century diplomacy as the partition of Poland had been to that of the eighteenth, but there was an important difference: the interests of more powers were involved this time and the complicating factor of national sentiment among the subject peoples of the Ottoman empire would make an agreed outcome much more difficult. As it happened, the Ottoman empire survived far longer than might have been expected, and an eastern question is still bothering statesmen.

Some of these complicating factors led to the Crimean War, which began with a Russian occupation of Ottoman provinces on the lower Danube. In Russia's internal affairs the war was more important than in those of any other country. It revealed that the military colossus of the 1815 restoration now no longer enjoyed an unquestioned superiority. She was defeated on her own territory and obliged to accept a peace which involved the renunciation for the foreseeable future of her traditional goals in the Black Sea area. Fortunately, in the middle of the war Nicholas I had died. This simplified the problems of his successor; defeat meant that change had to come. Some modernization of Russian institutions was unavoidable if Russia was again to generate a power commensurate with her vast potential, which had become unrealizable within her traditional framework. When the Crimean War broke out there was still no Russian railway south of Moscow. Russia's once important contribution to European industrial production had hardly grown since 1800 and was now far outstripped by others'. Her agriculture remained one of the least productive in the world and yet her population steadily rose, pressing harder upon its resources. It was in these circumstances that Russia at last underwent radical change. Though less dramatic than many upheavals in the rest of Europe it was in fact more of a revolution than much that went by that name elsewhere, for what was at last uprooted was the institution which lay at the very roots of Russian life, serfdom.

Its extension had been the leading characteristic of Russian social history since the seventeenth century. Even Nicholas had agreed that it was the central evil of Russian society. His reign had been marked by increasingly frequent serf insurrections, attacks on landlords, crop-burning and cattle-maiming. The refusal of dues was almost the least alarming form of popular resistance to it. Yet it was appallingly difficult for the rider to get off the elephant. The vast majority of Russians were serfs. They could not be transformed overnight by mere legislation into wage labourers or smallholders, nor could the state accept the administrative burden which would suddenly be thrown upon it if the services discharged by the manorial system should be withdrawn and nothing put in their place. Nicholas had not dared to proceed. Alexander II did. After years of study of the evidence and possible advantages and disadvantages of different forms of abolition, the tsar issued in 1861 the edict which marked an epoch in Russian history and won him the title of the 'Tsar Liberator'. The one card Russian government could play was the unquestioned authority of the autocrat and it was now put to good use.

The edict gave the serfs personal freedom and ended bond labour. It also gave them allotments of land. But these were to be paid for by redemption charges whose purpose was to make the change acceptable to the landowners. To secure the repayments and offset the dangers of suddenly introducing a free labour market, the peasant remained to a considerable degree subject to the authority of his village community, which was given the charge of distributing the land allotments on a family basis.

It would not be long before a great deal would be heard about the shortcomings of this settlement. Yet there is much to be said for it and in retrospect it seems a massive achievement. A few years later the United States would emancipate its Negro slaves. There were far fewer of them than there were Russian peasants and they lived in a country of much greater economic opportunity, yet the effect of throwing them on the labour market, exposed to the pure theory of *laissez-faire* economic liberalism, was to exacerbate a problem with whose ultimate consequences the United States is still grappling. In Russia the largest measure of social engineering in recorded history down to this time was carried out without comparable dislocation and it opened the way to modernization for what was potentially one of the strongest powers on earth. It was the indispensable first step towards making the peasant look beyond the estate for available industrial employment.

More immediately, liberation opened an era of reform; there followed other measures which by 1870 gave Russia a representative system of local government and a reformed judiciary. When, in 1871, the Russians took advantage of the Franco-Prussian War to denounce some of the restrictions placed on their freedom in the Black Sea in 1856, there was almost a symbolic warning to Europe in what they did. After tackling her greatest problem and beginning to modernize her institutions Russia was again announcing that she would after all be master in her own house. The resumption of the most consistently and long-pursued policies of expansion in modern history was only a matter of time.

4

Political Change: The Anglo-Saxon World

No European nation has so successfully seeded the globe with its own stocks as the United Kingdom. By the end of the nineteenth century it had created an Anglo-Saxon world which was an identifiable sub-unit within the ambit of European civilization, with an historical destiny diverging from that of the European continent. Its components included growing British communities in Canada, Australia, New Zealand and South Africa (the first and last containing other important national elements, too). At the heart of this world were the two great Atlantic nations, one the greatest world power of the nineteenth century, one that of the next.

They were at first sight so different – and so many people found it profitable to keep on pointing it out – that it is easy to overlook how much the United Kingdom and the United States of America had in common for much of the nineteenth century. Though one was a monarchy and the other a republic, both countries escaped first the absolutist and then the revolutionary currents of continental Europe. Anglo-Saxon politics, of course, changed quite as radically as those of any other countries in the nineteenth century. But they were not transformed by the same political forces as those of continental states nor in the same way.

Their similarity arose in part because the two countries shared more than they usually admitted. One aspect of their curious relations was that the Americans could still without a sense of paradox call England the mother country. The heritage of English culture and language was for a long time paramount in the United States; immigration from other European countries only became overwhelming in the second half of the nineteenth century. Though by the middle of the century many Americans – perhaps most – already had the blood of other European nations in their veins, the tone of society was long set by those of British stock. It was not until 1837 that there was a president who did not have an English, Scotch, or Irish surname (the next would not be until 1901, and there have been only four such down to the present day).

Post-colonial problems made, as they did in far later times, for emotional, sometimes violent, and always complex relations between the United States and the United Kingdom. But they were also much more than this. They were, for example, shot through with economic connexions. Far from dwindling (as had been feared) after independence, commerce between the two countries had gone on from strength to strength. English capitalists found the United States an attractive place for investment even after repeated and unhappy experiences with the bonds of defaulting states. British money was heavily invested in American railroads, banking, and insurance. Meanwhile the ruling élites of the two countries were at once fascinated and repelled by each other. Englishmen commented acidly on the vulgarity and rawness of American life but warmed as if by instinct to its energy, optimism and opportunity; Americans found it hard to come to terms with monarchy and hereditary titles but sought to penetrate the fascinating mysteries of English culture and society no less eagerly for that.

More striking than the huge differences between them was what the United Kingdom and the United States had in common when considered from the standpoint of continental Europe. Above all, both were able to combine liberal and democratic politics with spectacular advances in wealth and power. They did this in very different circumstances, but at least one was common to both, the fact of isolation: Great Britain had the Channel between herself and Europe, the United States had the Atlantic Ocean. This physical remoteness long masked from Europeans the international potential of the young republic and the huge opportunities facing it in the West. At the peace of 1783 the British had defended the Americans' frontier interests in such a way that there inevitably lay ahead a period of expansion for the United States; what was not clear was how far it might carry nor what other powers it might involve. This was in part a matter of geographical ignorance. No one knew for certain what the western half of the continent might contain. For decades the huge spaces just across the eastern mountain ranges would provide a big enough field of expansion. In 1800 the United States was still psychologically and actually very much a matter of the Atlantic seaboard and the Ohio valley.

On 7 November 1861 a British steamer homeward bound from Havana was stopped by a United States sloop, boarded and forced to surrender two passengers, representatives of the rebellious Confederate government of the South. The result was the last crisis in which there seemed a real danger of war between the two great Anglo-Saxon powers. Popular feeling was greatly excited on both sides of the Atlantic. During the weeks of uncertainty Punch *published these two cartoons. In the end, the United States government released its prisoners. There was never again to be so great a danger of collision, and even at that date no British government could contemplate a war with the United States with anything but dismay.*

Its political frontiers were then ill-defined, but imposed relations with France, Spain and the United Kingdom. None the less, if frontier questions could be settled, then a practical isolation might be attained, for the only other interests which might involve Americans in the affairs of other countries were, on the one hand, trade and the protection of her nationals, and, on the other, foreign intervention in the affairs of the United States. The French Revolution appeared briefly to pose the chance of the latter, and caused a quarrel, but for the most part it was frontiers and trade which preoccupied American diplomacy under the young republic. Both also aroused powerful and often divisive or potentially divisive forces in domestic politics.

The American aspiration to non-involvement with the outside world was already clear in 1793, when the troubles of the French revolutionary war led to a Neutrality Proclamation rendering American citizens liable to prosecution in American courts if they took any part in the Anglo-French war. The bias of American policy already expressed in this received its classical formulation in 1796. In the course of Washington's Farewell Address to his 'Friends and Fellow-Citizens' as his second term as president drew to a close, he chose to comment on the objectives and methods which a successful republican foreign policy should embody, in language to be deeply influential both on later American statesmen and on the national psychology. In retrospect, what is now especially striking about Washington's thoughts is their predominantly negative and passive tone. 'The great rule of conduct for us', he began, 'in regard to foreign nations is, in extending our commercial relations, to have with them as little political connection as possible.' 'Europe has a set of primary interests', he continued, 'which to us have none, or a very remote relation ... Our detached and distant situation invites and enables us to pursue a different course ... It is our true policy to steer clear of permanent alliances with any portion of the foreign world.' Moreover, Washington also warned his countrymen against assumptions of permanent or special hostility or friendship with any other nation. In all this there was no hint of America's future destiny as a world power (Washington did not even consider other than European relations; America's future Pacific and Asian role was inconceivable in 1796).

By and large, a pragmatic approach, case by case, to the foreign relations of the young republic was indeed the policy pursued by Washington's successors in the presidency. There was only one war with another great power, that between the United States and Great Britain in 1812. Besides contributing to the growth of nationalist feeling in the young republic, the struggle led both to the appearance of Uncle Sam as the caricature embodiment of the nation and to the composition of the 'Star-spangled Banner'. More importantly, it marked an important stage in the evolving relations of the two countries. Officially, British interference with trade during the struggle with the Napoleonic blockade had caused the American declaration of war, but more important had been the hopes of some Americans that the conquest of Canada would follow. It did not, and the failure of military expansion did much to determine that the future negotiation of the boundary problems with the British should be by peaceful negotiation. Though Anglophobia had been aroused again in the United States by the war, the fighting (which had its humiliations for both sides) cleared the air. In future boundary disputes it was tacitly understood that neither American nor British governments were willing to consider war except under extreme provocation. In this setting the northern boundary of the United States was soon agreed as far west as the 'Stony Mountains' (as the Rockies were then called); in 1845 it was carried further west to the sea and by then the disputed Maine boundary, too, had been agreed.

The greatest change in American territorial definition was brought about by the Louisiana Purchase. Roughly speaking, 'Louisiana' was the area between the Mississippi and the Rockies. In 1803 it belonged, if somewhat theoretically, to the French, the Spanish having ceded it to them in 1800. This change had provoked American interest; if Napoleonic France envisaged a revival of French American empire, New Orleans, which controlled the mouth of the river down which so much American commerce already passed, was of vital importance. It was to buy freedom of navigation on the Mississippi that the United States entered a negotiation which ended with the purchase of an area larger than the then total area of the republic. On the modern map it includes Louisiana, Arkansas, Iowa, Nebraska, both the Dakotas, Minnesota west of the Mississippi, most of Kansas, Oklahoma, Montana, Wyoming and a big piece of Colorado. The price was $11,250,000.

This was the largest sale of land of all time and its consequences were appropriately huge. It transformed American domestic history. The opening of the way to the trans-Mississippi West was to lead to a shift in demographic and political balance of revolutionary import for the politics of the young republic. This shift was already showing itself in the second decade of the century when the population living west of the Alleghenies more than doubled. When the Purchase was rounded off by the acquisition of the Floridas from Spain, the United States had by 1819 legal sovereignty over territory bounded by the Atlantic and Gulf coasts from Maine to the Sabine river, the Red and Arkansas rivers, the Continental Divide and the line of the 49th Parallel agreed with the British.

The United States was already the most important state in the Americas. Though there were still some European colonial possessions there, a major effort would be required to contest this fact, as the British had discovered in war. None the less, alarm about a possible European intervention in Latin America, together with Russian activity in the Pacific north-west, led to a clear American statement of the republic's determination to rule the roost in the western hemisphere. This was the 'Monroe doctrine', enunciated in 1823, which said that no future European colonization in the hemisphere could be considered and that intervention by European powers in its affairs would be seen as unfriendly to the United States. As this suited British interests, the Monroe doctrine was easily maintained. It had the tacit underwriting of the Royal Navy and no European power could conceivably mount an American operation if British seapower was used against it.

The Monroe doctrine remains the bedrock of American hemisphere diplomacy to this day. One of its consequences was that other American nations would not be able to draw upon European support in defending their own independence against the United States. The main sufferer before 1860 was Mexico. American settlers within its borders rebelled and set up an independent Texan republic which was subsequently annexed by the United States. In the war that followed Mexico did very badly. The peace of 1848 stripped her, in consequence, of what would one day become Utah, Nevada, California and most of Arizona, an acquisition of territory which left only a small purchase of other Mexican land to be made to round off the mainland territory of the modern United States by 1853.

In the seventy years after the Peace of Paris the republic thus expanded to fill half a continent. Nearly four million people in 1790 had become nearly twenty-four million by 1850. Most of these still lived east of the Mississippi, it was true, and the only cities with more than 100,000 inhabitants were the three great Atlantic ports of Boston, New York and Philadelphia: none the less, the centre of gravity of the nation was moving westward. For a long time the political, commercial and cultural élites of the eastern seaboard would continue to dominate American society. But from the moment that the Ohio valley had been settled a western interest had been in existence; Washington's Farewell Address had already recognized its importance. The West was an increasingly decisive contributor to the politics of the next seventy years, until there came to a head the greatest crisis in the history of the United States and one which settled her destiny as a world power.

Expansion, both territorial and economic shaped American history as profoundly as the democratic bias of her political institutions. Its influence on those institutions, too, was very great and sometimes glaring; sometimes they were transformed. Slavery is the outstanding example. When Washington began his presidency there were a little under 700,000 black slaves within the territories of the Union. This was a large number, but the framers of the constitution paid no special attention to them, except in so far as questions of political balance between the different states were involved. In the end it had been decided that a slave should count as three-fifths of a free man in deciding how many representatives each state should have.

Within the next half-century three things revolutionized this state of affairs. The first was an enormous extension of slavery. It was driven by a rapid increase in the world's consumption of cotton (above all in its consumption by the mills of England). This led to a doubling of the American crop in the 1820s and then its doubling again in the 1830s: by 1860, cotton provided two-thirds of the value of the total exports of the United States. This huge increase was obtained largely by cropping new land, and new plantations meant more labour. By 1820 there were already a million and a half

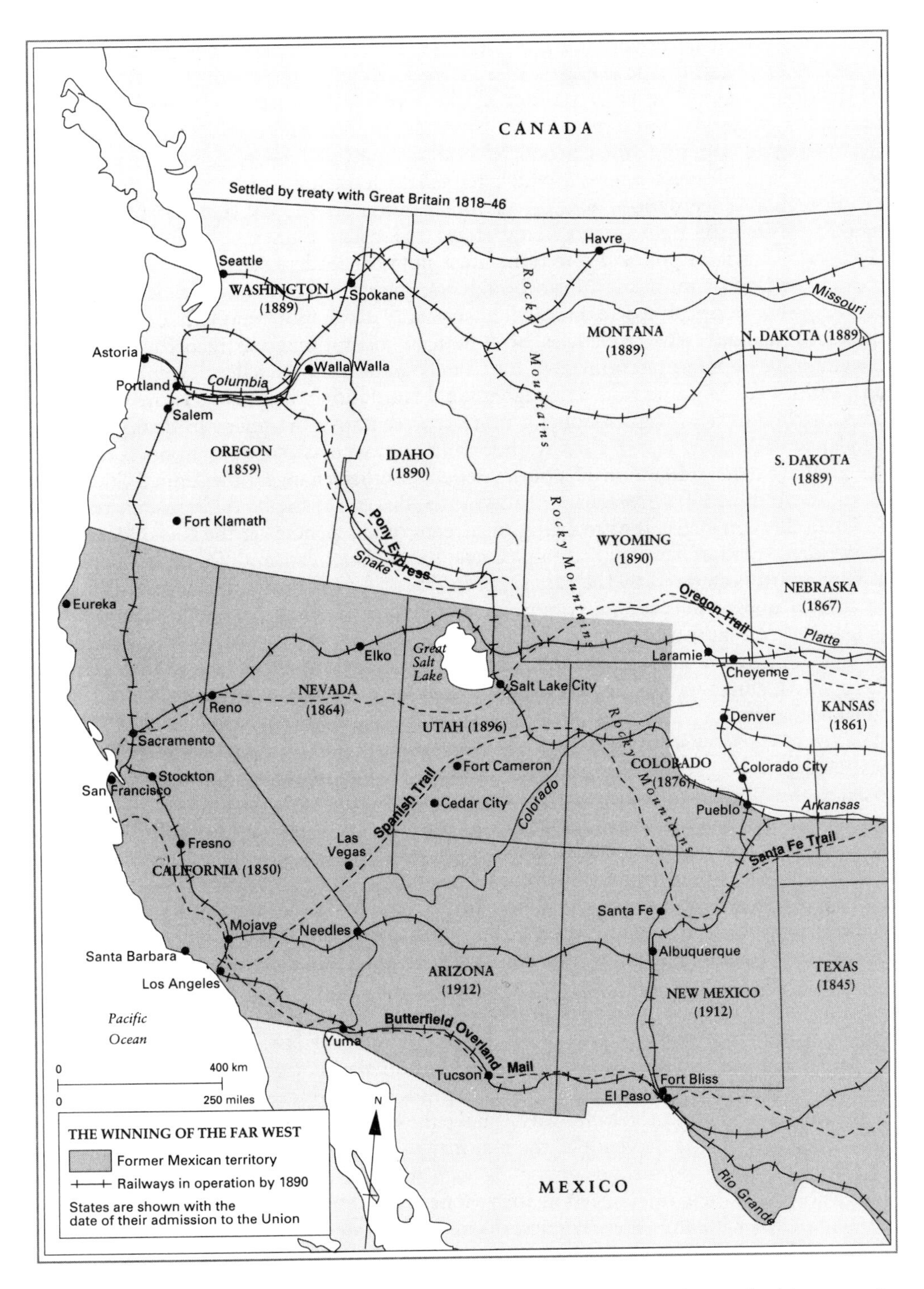

black slaves, by 1860 about four million. In the southern states slavery had become the foundation of the economic system. Because of this, Southern society became even more distinctive; it had always been aware of the ways it differed from the more mercantile and urban northern states, but now its 'peculiar institution', as slavery was called, came to be regarded by Southerners as the essential core of a particular civilization. By 1860 many of them thought of themselves as a nation, with a way of life they idealized and believed to be threatened by tyrannous interference from the outside. The expression and symbol of this interference was, in their view, the growing hostility of Congress to slavery.

That slavery became a political issue was the second development changing its rôle in American life. It was part of a general evolution in American politics evident also in other ways. The early politics of the republic had reflected what were to be later called 'sectional' interests and the Farewell Address itself had drawn attention to them. Roughly speaking, they produced political parties reflecting, on the one hand, mercantile and business interests which tended to look for strong federal government and protectionist legislation, and, on the other, agrarian and consumer interests which tended to assert the right of individual states and cheap money policies. At that stage slavery was hardly a political question; most political leaders seem to have thought of it as an evil which must succumb (though no one quite knew how) with the passage of time. This political world gradually gave way to a more modern one, partly as a result of the inherent tendencies of American institutions, partly because of social change. Judicial interpretation gave a strongly national and federal emphasis to the constitution. At the same time as congressional legislation was thus given new potential force, the law-makers were becoming more representative of American democracy; the presidency of Andrew Jackson has traditionally been seen as especially important in this. The growing democratization of politics reflected other changes; the United States was not to be troubled by a proletariat of townsmen driven off the land because in the West the possibility long existed of realizing the dream of independence; the social ideal of the independent smallholder could remain central to the American tradition. The opening up of the western hinterland by the Louisiana Purchase was as important in revolutionizing the wealth and population distributions which shaped American politics as was the commercial and industrial growth of the North.

Above all, the opening of the West transformed the question of slavery. There was great scope for dispute about the terms on which new territories should be joined to the Union. As the organization first of the Louisiana Purchase and then territory taken from Mexico had to be settled, the inflammatory question was bound to be raised: was slavery to be permitted in the new territories? A fierce anti-slavery movement had arisen in the North which dragged the slavery issue to the forefront of American politics and kept it there until it overshadowed every other question. Its campaign for the ending of the slave trade and the eventual emancipation of the black stemmed from much the same forces which had produced similar demands in other countries towards the end of the eighteenth century. But the American movement was importantly different, too. In the first place it was confronted with a *growth* of slavery at a time when it was disappearing elsewhere in the European world, so that the universal trend seemed to be at least checked, if not reversed, in the United States. Secondly, it involved a tangle of constitutional questions because of argument about the extent to which private property could be interfered with in individual states where local laws upheld it, or even in territories that were not yet states. Moreover the anti-slavery politicians brought forward a question which lay at the heart of the constitution, and, indeed, of the political life of every European country, too: who was to have the last word? The people were sovereign, that was clear: but was the 'people' the majority of its representatives in Congress, or the populations of individual states acting through their state legislatures and asserting the indefeasibility of their rights even against Congress? Thus slavery came by mid-century to be entangled with almost every question raised by American politics.

These great issues were just contained so long as the balance of power between the Southern and Northern states remained roughly the same. Although the North had a slight preponderance of numbers, the crucial equality in the Senate (where each state had two senators, regardless of its population or size) was maintained. Down to 1819, new states were admitted to the Union on an alternating system, one slave, one free; there were then eleven of each. Then came the first crisis, over the admission of the state of Missouri. In the days before the Louisiana Purchase French and Spanish law permitted slavery there and its settlers expected this to continue. They were indignant, and so were representatives of the Southern states, when a Northern congressman proposed restrictions upon slavery in the new state's constitution. There was great public stir and debate about sectional advantage; there was even talk of secession from the Union, so strongly did some Southerners feel. Yet the moral issue was muted. It was still possible to reach a political answer to a political question by the 'Missouri Compromise' which admitted Missouri as a slave state, but balanced her by admitting Maine at the same time, and prohibiting any further extension of slavery in United States territory north of a line of latitude 36° 30′. This confirmed the doctrine that Congress had the right to keep slavery out of new territories if it chose to exercise it, but there was no reason to believe that the question would again arise for a long time. Indeed, so it proved until a generation had passed. But already some had seen the future it contained: Thomas Jefferson, a former president and the man who drafted the Declaration of Independence, wrote that he 'considered it at once as the knell of the Union', and another (future) president wrote in his diary that the Missouri question was 'a mere preamble – a title-page to a great, tragic volume'.

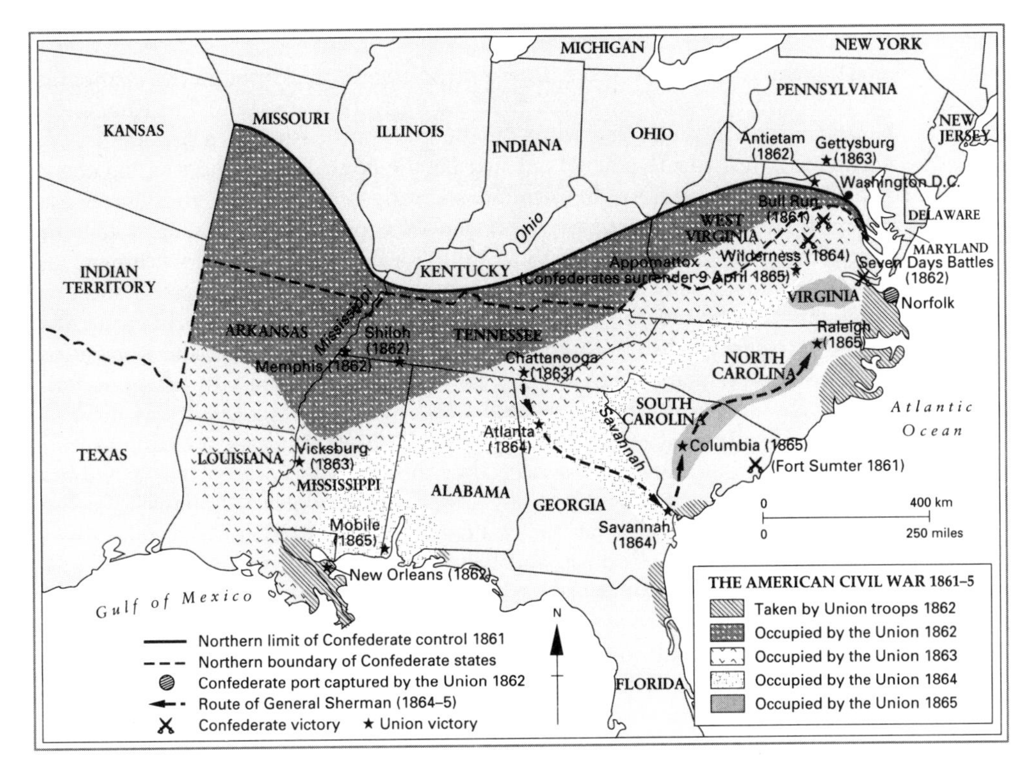

Yet the tragedy did not come to a head for forty years. In part this was because Americans had much else to think about – territorial expansion above all – and in part because no question arose of incorporating territories suitable for cotton-growing, and therefore requiring slave labour, until the 1840s. But there were soon forces at work to agitate public opinion and they would be effective when the public was ready to listen. It was in 1831 that a newspaper was established in Boston to advocate the unconditional emancipation of Negro slaves. This was the beginning of the 'abolitionist' campaign of increasingly embittered propaganda, electoral pressure upon politicians in the North, assistance to runaway slaves and opposition to their return to their owners after recapture, even when the law courts said they must be sent back. Against the background abolitionists provided, a struggle raged in the 1840s over the terms on which territory won from Mexico should be admitted. It ended in 1850 in a new Compromise, but one not to last long. From this time, politics were strained by increasing feelings of persecution and victimization among the Southern leaders and a growing arrogance on their part in the defence of their states' way of life. National party allegiances were already affected by the slavery issue; the Democrats took their stand on the finality of the 1850 settlement.

The 1850s brought the descent to disaster. The need to organize Kansas blew up the truce which rested on the 1850 Compromise and brought about the first bloodshed as abolitionists strove to bully pro-slavery Kansas into accepting their views. There emerged a Republican party in protest against the proposal that the people living in the territory should decide whether Kansas should be slave or free: Kansas was north of the 36° 30′ line. The anger of abolitionists now mounted, too, whenever the law supported the slave-owner, as it did in a notable Supreme Court decision in 1857 (in the 'Dred Scott' case) which returned a slave to his master. In the South, on the other hand, such outcries were seen as incitements to disaffection among the blacks and a determination to use the electoral system against Southern liberties – a view which was, of course, justified, because the abolitionists, at least, were not men who would compromise, though they could not get the Republican party to support them. The Republican presidential candidate in the election of 1860 campaigned on a programme which in so far as it concerned slavery envisaged only the exclusion of slavery from all territories to be brought into the Union in the future.

This was already too much for some Southerners. Although the Democrats were divided, the country voted on strictly sectional grounds in 1860; the Republican candidate Abraham Lincoln, who was to prove the greatest of American presidents, was elected by northern states, together with the two Pacific coast ones. This was the end of the line for many Southerners. South Carolina formally seceded from the Union as a protest against the election. In February 1861 she was joined by six other states, and the Confederate States of America which they set up had its provisional government and president installed a month before President Lincoln was inaugurated in Washington.

Each side accused the other of revolutionary designs and behaviour. It is very difficult not to agree with both of them. The heart of the Northern position, as Lincoln saw, was that democracy should prevail, a claim assuredly of potentially limitless revolutionary implication. In the end, what the North achieved was indeed a social revolution in the South. On the other side, what the South was asserting in 1861 (and three more states joined the Confederacy after the first shots were fired) was that it had the same right to organize its life as had, say, revolutionary Poles or Italians in Europe. It is unfortunate, but generally true, that the coincidence of nationalist claims with liberal institutions is rarely exact, or even close, and never complete, but the defence of slavery was also a defence of self-determination. At the same time, though such great issues of principle were certainly at stake, they presented themselves in concrete, personal and local terms which make it very difficult to state clearly the actual lines along which the Republic divided for the great crisis of its history and identity. They ran through families, towns and villages, religions, and sometimes ran round groups of different colours. It is the tragedy of civil wars to be like that.

Once under way, war has a revolutionary potential of its own. Much of the particular impact of what one side called 'the Rebellion' and the other side 'the War between the States' grew out of the necessities of the struggle. It took four years for the Union forces to beat the Confederacy and in that time an important change had occurred in Lincoln's aims. At the beginning of the war he had spoken only of restoring the proper order of affairs: there were things happening in the Southern states, he told the people, 'too powerful to be suppressed by the ordinary course of judicial proceedings' and they would require military operations. This view broadened into a consistent reiteration that the war was fundamentally about preserving the Union; Lincoln's aim in fighting was to reunite the states which composed it. For a long time this meant that he failed to satisfy those who sought from the war the abolition of slavery. But in the end he came round to it. In 1862 he could still say in a public letter that

> *If I could save the Union without freeing any slave, I would do it; and if I could save it by freeing all the slaves I would do it; and if I could save it by freeing some and leaving others alone, I would also do that,*

but he did so at a moment when he had already decided that he must proclaim the emancipation of slaves in the rebel states. It became effective on New Year's Day 1863; thus the nightmare of Southern politicians was reality at last, though only because of the war they had courted. It transformed the nature of the struggle, though not very obviously. In 1865 the final step was taken in an amendment to the constitution which prohibited slavery anywhere in the United States. By that time the Confederacy was defeated, Lincoln had been murdered and the cause which he had imperishably summed up as 'government of the people, by the people, for the people' was safe.

In the aftermath of its military victory that cause could hardly appear as an unequivocally noble or righteous one to all Americans, but its triumph was pregnant with importance not only for America but for mankind. It was the only political event of the century whose implications were as far-reaching as, say, the Industrial Revolution. The war settled the future of the continent; one great power would continue to dominate the Americas and exploit the resources of the richest untapped domain yet known to be open to man. That fact in due course settled the outcome of two world wars and therefore the history of the world. The Union armies also decided that the system which would prevail in American politics would be the democratic one; this was not, perhaps, always true in the sense of Lincoln's words but the political institutions which in principle provided for the rule of the majority were henceforth secure from direct challenge. This was to have the incidental effect of linking democracy and material well-being closely in the minds of Americans; industrial capitalism in the United States would have a great pool of ideological commitment to draw upon when it faced its later critics.

There were other domestic consequences, too. The most obvious was the creation of a new colour problem. In a sense there had been no colour problem while slavery existed. Servile status was the barrier separating the overwhelming majority of

Those for and about whom the United States divided: black field hands on a Tennessee plantation after the War between the States.

blacks (there had always been a few free among them) from whites, and it was upheld by legal sanctions. Emancipation swept away the framework of legal inferiority and replaced this with the framework, or myth, of democratic equality when very few Americans were ready to give this social reality. Millions of blacks in the South were suddenly free. They were also for the most part uneducated, largely untrained except for field labour, and virtually without leadership of their own race. For a little while in the Southern states they leant for support on the occupying armies of the Union; when this prop was removed blacks disappeared from the legislatures and public offices of the Southern states to which they had briefly aspired. In some areas they disappeared from the polling-booths, too. Legal disabilities were replaced by a social and physical coercion which was sometimes harsher than had been the old regime of slavery. The slave at least had the value to his master of being an investment of capital; he was protected like other property and was usually ensured a minimum security and maintenance. Competition in a free labour market at a moment when the economy of large areas of the South was in ruins, with impoverished whites struggling for subsistence, was disastrous for the black. By the end of the century he had been driven by a poor white population bitterly resentful of defeat and emancipation into social subordination and economic deprivation. From this was to stem emigration to the North in the twentieth century and racial problems in our own day.

As another consequence of the war the United States retained a two-party system. Between them, Republicans and Democrats have continued to divide the presidency to this day, though sometimes threatened by third parties. There was nothing to make this probable before 1861. Many parties had come and gone, reflecting different movements in American society. But the war was to rivet upon the Democratic party a commitment to the Southern cause which at first was a grave handicap because it carried the stigma of disloyalty (no Democrat was president until 1885). Correspondingly, it won for the Republicans the loyalty of Northern states and the hopes of radicals who saw in them the saviours of the Union and democracy, and the liberators of the slave. Before the inadequacy of these stereotypes was clear, the parties were so deeply rooted in certain states that their predominance in them, let alone survival, was unchallengeable. Twentieth-century American politics would proceed by internal transformation of the two great parties which long reflected their primitive origins.

For the moment the Republicans of 1865 had it all their own way. Perhaps they would have found a way to reconcile the South if Lincoln had lived. As it was, the impact of their policies upon a defeated and devastated South made the 'Reconstruction' years bitter ones. Many Republicans strove honestly to use the power they had to ensure democratic rights for the blacks; thus they ensured the future hegemony in the South of the Democrats. But they did not do too badly. Soon, too, the economic tide was with them as the great expansion interrupted briefly by the war was resumed.

It had been going on for seventy years and was already prodigious. Its most striking manifestation had been territorial; it was about to become economic. The phase of America's advance to the point at which her citizens would have the highest *per capita* income in the world was just opening in the 1870s. In the euphoria of this huge blossoming of confidence and expectation, all political problems seemed for a while to have been solved. Under Republican administrations America turned, not for the last time, to the assurance that the business of America was not political debate but economic advance. The South remained largely untouched by the new prosperity and slipped even further behind the North; it had no political leverage until an issue capable of bringing support to the Democrats in other sections turned up.

Meanwhile, the North and West could look back with confidence that the astonishing changes of the previous seventy years promised even better times ahead. Foreigners could feel this, too; that is why they were coming to the United States in growing numbers – two and a half million in the 1850s alone. They fed a population which had grown from just over five and a quarter millions in 1800 to nearly forty millions in 1870. About half of them by then lived west of the Alleghenies and the vast majority of them in rural areas. The building of railroads was opening the Great Plains to settlement and exploitation which had not yet really begun. In 1869 the golden spike was driven which marked the completion of the first trans-continental railroad link. In the new West the United States would find its greatest agricultural expansion; already, thanks to the shortage of labour experienced in the war years, machines were being used in numbers which pointed to a quite new scale of farming, the way to a new phase of the world's agricultural revolution which would make North America the granary of Europe. There were a quarter of a million mechanical reapers alone at work by the end of the war. Industrially, too, great years lay ahead; the United States was not yet an industrial power to compare with Great Britain (in 1870 there were still less than two million Americans employed in manufacturing), but the groundwork was done. With a large, increasingly well-off domestic market the prospects for American industry were bright.

Poised on the brink of their most confident and successful era, Americans were not being hypocritical in forgetting the losers. They understandably found it easy to do so in the general sense that the American system worked well. The blacks and the poor whites of the South had now joined the Indian, who had been a loser steadily for two centuries and a half, as the forgotten failures. The new poor of the growing northern cities should probably not be regarded, comparatively, as losers; they were at least as well off, and probably better, than the poor of Manchester or Naples. Their willingness to come to the United States showed that she was already a magnet of great power. Nor was that power only material. Besides the 'wretched refuse', there were the 'huddled masses yearning to breathe free'. The United States was in 1870 still a political inspiration to political radicals elsewhere, though perhaps her political practice and forms had more impact in Great Britain – where people linked (both approvingly and disapprovingly) democracy with the 'Americanization' of British politics – than in continental Europe.

Such transatlantic influences and connexions were aspects of the curious, fitful, but tenacious relations between the two Anglo-Saxon countries. They both underwent similar revolutionary change and here, perhaps, the achievement of Great Britain in the early nineteenth century is even more remarkable than the transformation of the United States. At a time of unprecedented and potentially dislocating social upheaval, which turned her within a single lifetime into the first industrialized and urbanized society of modern times, Great Britain managed to maintain an astonishing constitutional and political continuity. At the same time, too, she was acting as a world and European power as the United States never had to, and ruled a great empire. In this setting she began the democratization of her institutions while retaining most of her buttresses of individual liberty.

Socially the United Kingdom was far less democratic than the United States in 1870 (if the blacks are set aside as a special case). Social hierarchy (conferred by birth and land if possible, but if not, money would sometimes do) stratified the United Kingdom; every observer was struck by the assured confidence of the English ruling classes that they were meant to rule. There was no American West to offset the deep swell of deference with the breeze of frontier democracy; Canada and Australia attracted restless emigrants, but in so doing removed the possibility of their changing the tone of English society. Political democracy developed faster than social, on the other hand, even if the universal male suffrage already long-established in the United States would not be introduced until 1918; the democratization of English politics was already past the point of reversibility by 1870.

This great change had come about within a few decades. Though it had deeply libertarian institutions – equality at law, effective personal liberty, a representative system – the English constitution of 1800 had not rested on democratic principles. Its basis was the representation of certain individual and historic rights and the sovereignty of the Crown in parliament. The accidents of the past produced from these elements an electorate large by contemporary European standards, but as late as 1832, the word 'democratic' was a pejorative one and few thought it indicated a desirable goal. To most Englishmen, democracy meant the French Revolution and military despotism. Yet the most important step towards democracy in the English political history of the century was taken in 1832.

In the 1870s the landlord's grip on England was still very strong. These labourers and their families are standing outside the cottage from which they have just been evicted for membership of a trades union.

This was the passing of a Reform Act which was not itself democratic and was, indeed, intended by many of those who supported it to act as a barrier to democracy. It carried out a great revision of the representative system, removing anomalies (such as the tiny constituencies which had been effectively controlled by patrons), to provide parliamentary constituencies which better (though still far from perfectly) reflected the needs of a country of growing industrial cities, and above all to change and make more orderly the franchise. It had been based on a jumble of different principles in different places; now, the main categories of persons given the vote were freeholders in the rural areas, and householders who owned or paid rent for their house at a middle-class level in the towns. The model elector was the man with a stake in the country, although dispute about the precise terms of the franchise still left some oddities. The immediate result was an electorate of about 650,000 and a House of Commons which did not look very different from its predecessor. None the less, dominated by the aristocracy as it still was, it marked the beginning of nearly a century during which British politics were to be completely democratized, because once the constitution had been changed in this way, then it could be changed again and the House of Commons more and more claimed the right to say what should be done. In 1867, another Act produced an electorate of about two millions and in 1872 the decision that voting should take place by secret ballot followed: a great step.

This process would not be completed before the twentieth century, but it soon brought other changes in the nature of British politics. Slowly, and somewhat grudgingly, the traditional political class began to take account of the need to organize parties which were something more than family connexions or personal cliques of members of parliament. This was much more obvious after the emergence of a really big electorate in 1867. But the implication – that there was a public opinion to be courted which was more than that of the old landed class – was grasped sooner than this. All the greatest of English parliamentary leaders in the nineteenth century were men whose success rested on their ability to catch not only the ear of the House of Commons, but that of important sections of society outside it. The first and possibly most significant example was Sir Robert Peel, who created English conservatism. By accepting the verdicts of public opinion he gave conservatism a pliability which always saved it from the intransigence into which the right was tempted in so many European countries.

The great political crisis of Corn Law Repeal demonstrated this. It was not only about economic policy; it was also about who should govern the country, and was in some ways a complementary struggle to that for parliamentary reform before 1832. By the middle of the 1830s the conservatives had been brought by Peel to accept the consequences of 1832, and in 1846 he was just able to make them do the same over the protective Corn Laws, whose disappearance showed that landed society no longer had the last word. Vengefully, his party, the stronghold of the country gentlemen who considered the agricultural interests the embodiment of England and themselves the champions of the agricultural interest, turned on Peel soon afterwards and rejected him. They were right in sensing that the whole tendency of his policy had been directed to the triumph of the Free Trade principles which they associated with the middle-class manufacturers. Their decision divided their party and condemned it to paralysis for twenty years, but Peel had in fact rid them of an incubus. He left it free when reunited to compete for the electorate's goodwill untrammelled by commitment to the defence of only one among several economic interests.

The redirection of British tariff and fiscal policies towards Free Trade was one side, though in some ways the most spectacular, of a general alignment of British politics towards reform and liberalization in the central third of the century. During this time a beginning was made with local government reform (significantly, in the towns, not in the countryside where the landed interest was still the master), a new Poor Law was introduced, factory and mining legislation was passed and began to be effectively policed by inspection, the judicial system was reconstructed, disabilities on protestant nonconformists, Roman Catholics and Jews were removed, the ecclesiastical monopoly of matrimonial law which went back to Anglo-Saxon times was ended, a postal system was set up which became the model on which other nations would shape theirs, and a beginning was even made with tackling the scandalous neglect of public education. All this was accompanied by unprecedented growth in wealth, whose confident symbol was the holding in 1851 of a Great Exhibition of the world's wares in London under the patronage of the queen herself and the direction of her consort. If Englishmen were inclined to bumptiousness, as they seem to have been in the central decades of Victoria's reign, then it may be said that they had grounds. Their institutions and economy had never looked healthier.

Not that everyone was pleased. Some moaned about a loss of economic privilege: in fact, the United Kingdom continued to display extremes of wealth and poverty as great as any other country's. There was somewhat more substance to the fear of creeping centralization. Parliamentary legislative sovereignty led to bureaucracy invading more and more areas which had previously been immune to government intervention in practice. England in the nineteenth century was very far from concentrating power in her state apparatus to the degree which has now become usual in all countries. Yet some people felt worried that she might be going the way of France, a country whose highly centralized administration was taken to be sufficient explanation of the failure to achieve liberty which had accompanied the French success in establishing equality. In offsetting such a tendency, the Victorian reforms of local government, some of which came only in the last two decades of the century, were crucial.

Some foreigners admired. Most wondered how, in spite of the appalling conditions of her factory towns, the United Kingdom had somehow navigated the rapids of popular unrest which had proved fatal to orderly government in other states. She had deliberately undertaken huge reconstructions of her institutions at a time when the dangers of revolution were clearly apparent elsewhere, and had emerged unscathed, her power and

The 'machinery corner' inside the Crystal Palace, the site of the Great Exhibition of 1851.

wealth enhanced and the principles of liberalism even more apparent in her politics. Her statesmen and historians gloried in reiterating that the essence of the nation's life was freedom, in a famous phrase, 'broadening down from precedent to precedent'. Englishmen seemed fervently to believe this, yet is did not lead to license. The country did not have the advantages of geographical remoteness and almost limitless land which were enjoyed by the United States – and even the United States had fought one of the bloodiest wars in human history to contain a revolution. How, then, had Great Britain done it?

This was a leading question, though one historians still sometimes ask without thinking about its implications that there exist certain conditions which make revolution likely and that British society seems to have fulfilled them. It may be, rather, that no such propositions need to be conceded. Perhaps there never was a potentially revolutionary threat in this rapidly changing society. Many of the basic changes which the French Revolution brought to Europe had already existed in Great Britain for centuries, after all. The fundamental institutions, however rusty or encrusted with inconvenient historic accretion they might be, offered large possibilities. Even in unreformed days, the House of Commons and House of Lords were not the closed corporate institutions which were all that was available in many European states. Already before 1832, they had shown their capacity to meet new needs, even if slowly and belatedly; the first Factory Act (not, admittedly, a very effective one) had been passed as early as 1801. Once 1832 was past, then there were good grounds for thinking that if parliament were only pressed hard enough from the outside, it would carry out any reforms that were required. There was no legal restraint on its power to do so. Even the oppressed and angry seem to have seen this. There were many outbreaks of desperate violence and many revolutionaries about in the 1830s and 1840s (which were especially hard times for the poor) but it is striking that the most important popular movement of the day, the great spectrum of protest gathered together in what was called 'Chartism', asked in the People's Charter which was its programme for measures which would make parliament more responsive to popular needs, not for its abolition.

Yet it is not likely that parliament would have been called upon to provide reform unless other factors had operated. Here it is perhaps significant that the great reforms of Victorian England were all ones which interested the middle classes as much as the masses, with the possible exception of factory legislation. The English middle class came to an early share in political power as its continental counterparts had not and could therefore use it to obtain change; it was not tempted to ally with revolution, the recourse of desperate men to whom other avenues were closed. But in any case it does not seem that the English masses were themselves very revolutionary. At any rate, their failure to act in a revolutionary way has caused much distress to later left-wing historians. Whether this is because their sufferings were too great, not great enough or whether simply there were too many differences between different sections of the working class has been much disputed. But it is at least worth noticing, as did contemporary visitors, that in England traditional patterns of behaviour died hard; it was long to remain a country with habits of deference to social superiors which much struck foreigners – especially Americans. Moreover, there were working-class organizations which provided alternatives to revolution. They were often 'Victorian' in their admirable emphasis on self-help, caution, prudence, sobriety. Of the elements making up the great English Labour movement, only the political party which bears that name was not in existence already before 1840; the others were mature by the 1860s. The 'friendly societies' for insurance against misfortune, the cooperative associations and, above all, the trades unions all provided effective channels for personal participation in the improvement of working-class life, even if at first only to a few and slowly. This early maturity was to underlie the paradox of English socialism, its later dependence on a very conservative and unrevolutionary trade-union movement, long the largest in the world.

Once the 1840s were over, economic trends may have helped to allay discontent. At any rate working-class leaders often said so, almost regretfully; they, at least, thought that betterment told against a revolutionary danger in England. As the

THE BRITISH LION IN 1850;

OR, THE EFFECTS OF FREE TRADE.

A certain complacency appears in British public life during the prosperous mid-century.

international economy picked up in the 1850s good times came to the industrial cities of a country which was the workshop of the world and its merchant, banker, and insurer, too. As employment and wages rose, the support which the Chartists had mustered crumbled away and they were soon only a reminiscence.

The symbols of the unchanging form which contained so much change were the central institutions of the kingdom: parliament and the Crown. When the Palace of Westminster was burned down and a new one was built, a mock-medieval design was chosen to emphasize the antiquity of the Mother of Parliaments. The violent changes of the most revolutionary era of British history thus continued to be masked by the robes of custom and tradition. Above all, the monarchy continued. Already when Victoria ascended the throne, it was second only to the Papacy in antiquity among the political institutions of Europe; yet it was much changed in reality, for all that. It had been brought very low in public esteem by George III's successor, the worst of English kings, and not much enhanced by his heir. Victoria and her husband were to make it again unquestioned. In part this was against the grain for the queen herself; she did not pretend to like the political neutrality appropriate to a constitutional monarch when the Crown had withdrawn above the political battle. None the less, it was in her reign that this withdrawal was seen to be made. She also domesticated the monarchy; for the first time since the days of the young George III the phrase 'the Royal Family' was a reality and could be seen to be such. It was one of many ways in which her German husband, Prince Albert, helped her, though he got little thanks for it from an ungrateful English public.

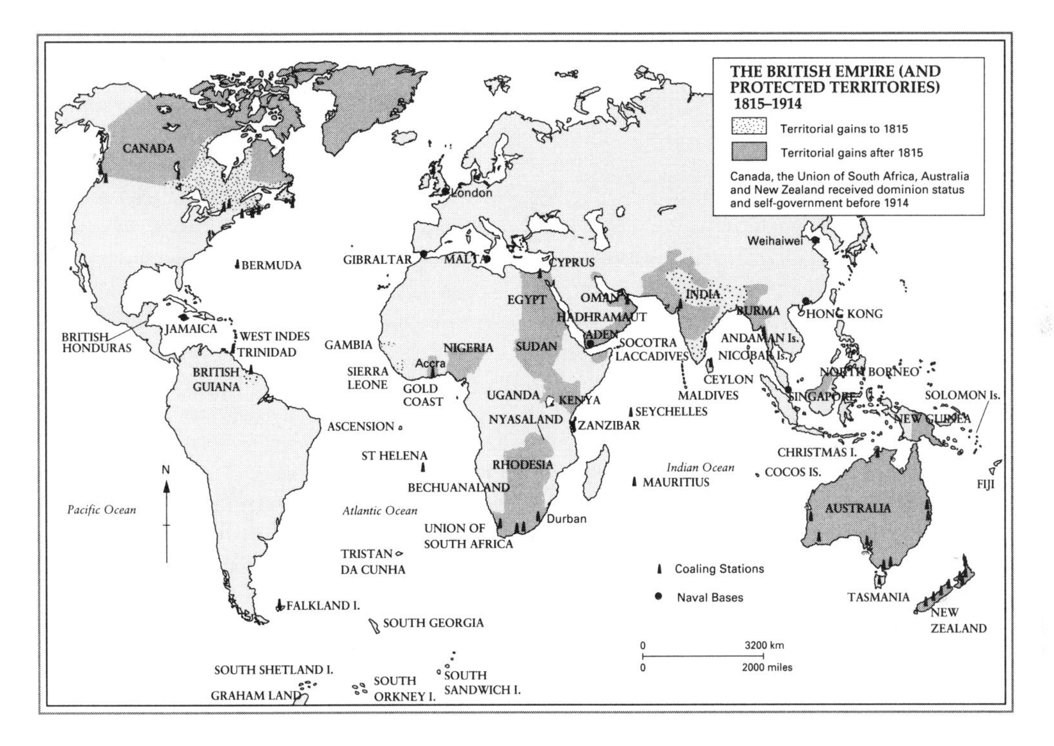

Only in Ireland did their capacity for imaginative change seem always to fail the British people. They had faced a real revolutionary danger and had had to put down a rebellion there in 1798. In the 1850s and 1860s things were quiet. But the reason was in large measure an appalling disaster which overtook Ireland in the middle of the 1840s when the failure of the potato crop was followed by famine, disease and thus, brutally, a Malthusian solution of Ireland's over-population. For the moment, the demand for the repeal of the Act of Union which had joined her to Great Britain in 1801 was muted, the dislike of her predominantly Catholic population for an alien and established Protestant Church was in abeyance and there was no serious disturbance among a peasant population feeling no loyalty to absentee English landlords. Problems none the less remained and the Liberal government which took office in 1868 addressed itself above all else to them. All that emerged was a new Irish nationalism movement, demanding 'Home Rule'. This demand was to haunt British politics and overturn their combinations and settlements for the rest of the century and beyond. Through the capacity Home Rule was to show to wreck British Liberalism, Ireland again became a force in world history after over a thousand years. She was also to make another important, if less direct, impact upon it at about the same time, through the Irish emigration to the United States.

5
The European World Hegemony

By 1900 the peoples of Europe and European stocks overseas dominated the globe. They did so in many different ways, some explicit and some implicit, but the qualifications matter less than the general fact. For the most part, the world responded to European initiatives and marched increasingly to European tunes. It is easy to overlook the fact that this was a sign of a unique development in world history. For the first time, one civilization established itself as a leader right round the globe. One minor consequence is that the remainder of this book will be increasingly concerned with a single, global, history.

It is important not to think only of the direct formal rule of the majority of the world's land-surface by European states (some people would prefer the term 'Western' but this is unnecessarily finicky – the Americas and Antipodes are dominated by culture of European origin, not of Asian or African – and is also liable to suggest too much, because of the use of that word recently in a narrow political sense). There is economic and cultural hegemony to be considered, and European ascendancy was often expressed in influence as well as in overt control. The important distinction is between European forces which are aggressive, shaping, manipulative, and indigenous cultures and peoples which are the objects of those forces, and not often able to resist them effectively. It was by no means always to the disadvantage of non-Europeans that this was so, but they tended almost always to be the underdogs, those who adapted to the Europeans' world. At times they did so willingly, when they succumbed to the attractive force of Europe's progressive ideals or, most subtly of all, to new sets of expectations aroused by European teaching and example.

One way of envisaging the Europeans' world of 1900 is as a succession of concentric circles. The innermost was old Europe itself, which had grown in wealth and population for three centuries thanks to an increasing mastery first of its own and then of the world's resources. Europeans distinguished themselves more and more from other human beings by taking and consuming a growing share of the world's goods and by the energy and skill they showed in manipulating their environment. Their civilization was already rich in the nineteenth century and was all the time getting richer. Industrialization had confirmed its self-feeding capacity to open up and create new resources; furthermore, the power generated by new wealth made possible the appropriation of the wealth of other parts of the world. The profits of Congo rubber, Burmese teak or Persian oil would not for a long time be reinvested in those countries. The poor European and American benefited from low prices for raw materials, and improving mortality rates tell the story of an industrial civilization finding it possible to give its peoples a richer life. Even the European peasant could buy cheap manufactured clothes and tools while his contemporaries in Africa and India still lived in the Stone Age.

This wealth was shared by the second circle of European hegemony, that of the European cultures transplanted overseas. The United States is the greatest example; Canada, Australia, New Zealand, South Africa and the countries of South America make up the list. They did not all stand on the same footing towards the Old World, but together with Europe proper they were what is sometimes called the 'Western world', an unhelpful expression, since they are scattered all round the globe, but one which seeks to express the important fact of the similarity of the ideas and institutions from which they were sprung.

Of course, these were not all that had shaped them. They all had their distinctive frontiers, they all had faced special environmental challenges and unique historical circumstances. But what they had in common were ways of dealing with these challenges, institutions which different frontiers would reshape in different ways. They were all formally Christian – no one ever settled new lands in the name of atheism until the twentieth century – all regulated their affairs by European systems of law, and all had access to the great cultures of Europe with which they shared their languages.

In 1900 this world was sometimes called the 'civilized world'. It was called that just because it was a world of shared standards; the confident people who used the phrase could not easily see that there was much else deserving of the name of civilization in the world. When they looked for it, they tended to see only heathen, backward, benighted people or a few striving to join the civilized. Such an attitude was an important part of the story of European success; what were taken to be demonstrations of the inherent superiority of European ideas and values nerved men to fresh assaults on the world and inspired fresh incomprehension of it. The progressive values of the eighteenth century provided new arguments for superiority to reinforce those originally stemming from religion. By 1800, Europeans had lost almost all of their former respect for other civilizations. Their own social practice seemed obviously superior to the unintelligible barbarities found elsewhere. The advocacy of individual rights, a free press, universal suffrage, the protection of women, children (and even animals) from exploitation, have been ideals pursued right down to our own day in other lands by Europeans and Americans, often wholly unconscious that they might be inappropriate. Philanthropists and progressives long continued to be confident that the values of European civilization should be universalized as were its medicine and sanitation, even when deploring other assertions of European superiority. Science, too, has often seemed to point in the same direction, to the destruction of superstition and the bringing of the blessings of a rational exploitation of resources, the provision of formal education and the suppression of backward social customs. There was a well-nigh universal assumption that the values of European civilization were better than indigenous ones (obviously, too, they often were) and a large obliviousness to any disruptive effects they might have.

Fortunately, it was thought, for the peoples of some of the lands over which 'thick darkness' (as the Victorian hymn put it) brooded, they were by 1900 often ruled directly by Europeans or European stocks: subject peoples formed the third concentric circle of European civilization. In many colonies enlightened administrators toiled to bring the blessings of railways, Western education, hospitals, law and order to peoples whose own institutions had clearly failed (it was taken as evidence of their inadequacy that they had failed to stand up to the challenge and competition of a superior civilization). Even when native institutions were protected and preserved, it was from a position which assumed the superiority of the culture of the colonial power.

Such conscious superiority is no longer admired or admissible. In one respect, nevertheless, it achieved an end which the most scrupulous critics of colonialism still accept as a good one, even when suspecting the motives behind it. This was the abolition of slavery in the European world and the deployment of force and diplomacy to combat it even in countries Europeans did not control. The crucial steps were taken in 1807 and 1834, when the British parliament abolished first the trade in slaves and then slavery itself within the British Empire. This action by the major naval, imperial and commercial power was decisive; similar measures were soon enforced by other European nations, and slavery finished in the United States in 1865. The end of the process may be reckoned to be the emancipation of slaves in Brazil in 1888, at which date colonial governments and the Royal Navy were pressing hard on the operations of Arab slavetraders in the African continent and the Indian Ocean. Many forces, intellectual, religious, economic and political, went into this great achievement, and debate about their precise individual significance continues. It is perhaps worth pointing out here that though it was only after three hundred years and more of large-scale slave-trading that abolition came, Europe's is also the only civilization which has ever eradicated slavery for itself.

Though in the present century slavery briefly returned to Europe, it could not be sustained except by force, nor was it avowable as slavery. It cannot have been much consolation to their unhappy occupants, but the forced-labour camps of our own century were run by men who had to pay the tribute of hypocrisy to virtue by disguising their slaves as the subjects of re-education or judicial punishment.

Beyond this outermost circle of directly-ruled territories lay the rest of the world. Its peoples were shaped by Europe, too. Sometimes their values and institutions were corroded by contact with it – as was the case in the Chinese and Ottoman empires – and this might lead to indirect European political interference as well as the weakening of traditional authority. Sometimes they were stimulated by such contacts and exploited them: Japan is the only example of an important nation doing this with success. What was virtually impossible was to remain untouched by Europe. The busy, bustling energy of the

At the beginning of this century Christian missionaries were still the pioneers and spearhead of European influence in many parts of the world. When photographed, these indomitable ladies were about to set off on a West African journey.

European trader would alone have seen to that. In fact, it is the areas which were not directly ruled by Europeans which make the point of European supremacy most forcibly of all. European values were transferred on the powerful wings of aspiration and envy. Geographical remoteness was almost the only security (but even Tibet was invaded by the British in 1904). Ethiopia is virtually the solitary example of successful independence; it survived British and Italian invasion in the nineteenth century, but, of course, had the important moral advantage of claiming to have been a Christian country, albeit not a Western one and only intermittently, for some fourteen centuries.

Whoever opened the door, a whole civilization was likely to try to follow them through it, but one of the most important agencies bringing European civilization to the rest of the world had always been Christianity, because of its virtually limitless interest in all sides of human behaviour. The territorial spread of the organized churches and the growth in their numbers of official adherents in the nineteenth century made this the greatest age of Christian expansion since apostolic times. Much of this was the result of a renewed wave of missionary activity; new orders were set up by Catholics, new societies for the support of overseas missions appeared in Protestant countries. Yet the paradoxical effect was the intensifying of the European flavour of what was supposedly a creed for all sorts and conditions of men. In most of the receiving countries, Christianity was long seen as just one more aspect of European civilization, rather than as a spiritual message which might use a local idiom. An interesting if trivial example was the concern missionaries often showed over dress. Whereas the Jesuits in seventeenth-century China had discreetly adopted the costume of their hosts, their nineteenth-century successors set to work with zeal to put Bantus and Solomon Islanders into European garments which were often of almost freakish unsuitability. This was one way in which Christian missionaries diffused more than a religious message. Often, too, they brought important material and technical benefits: food in time of famine, agricultural techniques, hospitals and schools, some of which could be disruptive of the societies which received them. Through them filtered the assumptions of a progressive civilization.

The ideological confidence of Europeans, missionaries and non-missionaries alike, could rest in the last resort on the knowledge that they could not be kept away even from countries which were not colonized. There appeared to be no part of the world where Europeans could not, if they wished, impose themselves by armed strength. The development of weapons in the nineteenth century gave Europeans an even greater relative advantage than they had enjoyed when the first Portuguese broadside was fired at Calicut. Even when advanced devices were available to other peoples, they could rarely deploy them effectively. At the battle of Omdurman in the Sudan in 1898 a British regiment opened fire on its opponents at 2,000 yards' range with the ordinary magazine rifle of the British army of the day. Soon afterwards, shrapnel shell and machine-guns were shredding to pieces the masses of the Mahdist army, who never reached the British line. By the end of the battle 10,000 of them had been killed for a loss of 48 British and Egyptian soldiers. It was not, as an Englishman put it soon afterwards, simply the case that

Whatever happens, we have got
The Maxim gun, and they have not,

for the Khalifa had machine-guns in his armoury at Omdurman, too. He also had telegraph apparatus to communicate with his forces and electric mines to blow up the British gunboats on the Nile. But none of these things was properly employed; not only a technical, but a mental transformation was required before non-European cultures could turn the instrumentation of the Europeans against them.

There was also one other sense, more benevolent and less disagreeable, in which European civilization rested upon force. This was because of the *Pax Britannica* which throughout the whole nineteenth century stood in the way of European nations fighting each other for mastery of the non-European world. There was no repeat performance of the colonial wars of the seventeenth and eighteenth centuries in the nineteenth, though the greatest extension of direct colonial rule in modern times was then going on. Traders of all nations could move without let or hindrance on the surface of the seas. British naval supremacy was a precondition of the informal expansion of European civilization.

The water-cooled Maxim gun, fully automatic, self-loading, and the model for most machine-guns in service down to the Great War.

It guaranteed, above all, the international framework of trade whose centre, by 1900, was Europe. The old peripheral exchanges by a few merchants and enterprising captains had, from the seventeenth century onwards, been replaced gradually by integrated relationships of interdependence based on a broad distinction of role between industrial and non-industrial countries; the second tended to be primary producers meeting the needs of the increasingly urbanized populations of the first. But this crude distinction needs much qualification. Individual countries often do not fit it; the United States, for example, was both a great primary producer and the world's leading manufacturing power in 1914, with an output as great as those of Great Britain, France and Germany together. Nor was this distinction one which ran exactly between nations of European and non-European culture. Japan and Russia were both industrializing faster than China or India in 1914, but Russia, though European, Christian and imperialist, could certainly not be regarded as a developed nation, and most Japanese (like most Russians) were still peasants. Nor could a developed economy be found in Balkan Europe. All that can be asserted is that in 1914 a nucleus of advanced countries existed with social and economic structures quite different from those of traditional society, and that these were the core of an Atlantic group of nations which was increasingly the world's main producer and consumer.

The world economy came to a sharp focus in London, where the financial services which sustained the flow of world trade were centred. A huge amount of the world's business was transacted by means of the sterling bill of exchange; it rested in turn upon the international gold standard which sustained confidence by ensuring that the main currencies remained in fairly steady relationships with one another. All major countries had gold currencies and travel any where in the world was possible with a bag of gold sovereigns, five-dollar pieces, gold francs or any other major medium of exchange without any doubts about their acceptability.

London was also in another sense the centre of the world economy because although the United Kingdom's gross output was by 1914 overtaken in important respects by that of the United States and Germany, she was the greatest of trading nations. The bulk of the world's shipping and carrying trade was in British hands. She was the main importing and exporting nation and the only one which sent more of its manufactures to non-European nations than to European. Great Britain was also the biggest exporter of capital and drew a huge income from her overseas investments, notably those in the United States and South America. Her special rôle imposed a roughly triangular system of international exchange. The British bought goods, manufactured and otherwise, from Europe and paid for them with their own manufactures, cash and overseas produce. To the rest of the world they exported manufactures, capital and services, taking in return food and raw materials and cash. This complex system illustrates how little the European relationship with the rest of the world was a simple one of exchanging manufactures for raw materials. And there was, of course, always the unique instance of the United States, little involved in export, but gradually commanding a greater and greater share of its own domestic market for manufactured goods, and still a capital importer.

Most British economists believed in 1914 that the prosperity which this system enjoyed and the increasing wealth which it made possible showed the truth of Free Trade doctrine. Their own country's prosperity had grown most rapidly in the heyday of such ideas. Adam Smith had predicted that prosperity would continue if a closed imperial system reserving trade to the mother country were abandoned and so, in the case of America, it had soon proved, for a big expansion had come to the Anglo-American trade within a few years of the peace of 1783. By 1800 a majority of British exports were already going outside Europe and there then still lay ahead the greatest period of expansion of trade in India and East Asia. Understandably, British imperial policy was directed not to the potentially embarrassing acquisition of new colonies, but to the opening of areas closed to trade, for that was where prosperity was deemed to lie. One outstanding example was the Opium War of 1839–42. The outcome was the opening of five Chinese ports to European trade and the *de facto cession* to Great Britain of Hong Kong as a base for the exercise of a jurisdiction inseparable from the management of commerce.

In the middle of the nineteenth century there had been for a couple of decades a high tide of Free Trade ideas when more governments seemed willing to act upon them than ever before or after. In this phase, tariff barriers were demolished and the comparative advantage of the British, first among trading and manufacturing nations, had continued. But this era passed in the 1870s and 1880s. The onset of a world-wide recession of economic activity and falling prices meant that by 1900 Great Britain was again the only major nation without tariffs for protection and even in that country questioning of the old Free Trade dogmas was beginning to be heard as competition from Germany grew fiercer and more alarming.

Nevertheless, by comparison with that of today the economic world of 1914 must still seem to be one of astonishing economic freedom and confidence. A long European peace provided the soil in which trading connexions could mature. Stable currencies assured great flexibility to a world price system; exchange control existed nowhere in the world and Russia and China were by then as completely integrated into this market as other countries. Freight and insurance rates had grown cheaper and cheaper, food prices had shown a long-term decline and wages had shown a long-term rise. Interest rates and taxation were low. It seemed as if a capitalist paradise might be achievable.

As this system had grown to incorporate Asia and Africa, it, too, came to be instrumental in a diffusion of ideas and techniques originally European, but soon acclimatized in other lands. Joint stock companies, banks, commodity and stock exchanges spread round the world by intrusion and imitation; they began to displace traditional structures of commerce. The building of docks and railways, the infrastructures of world trade, together with the beginnings of industrial employment, began in some places to turn peasants into an industrial proletariat. Sometimes the effects on local economies could be bad; the cultivation of indigo in India, for example, more or less collapsed when synthetic dyes became available in Germany and Great Britain. Isolation first disturbed by explorers, missionaries and soldiers was destroyed by the arrival of the telegraph and the railway; in the twentieth century the motor car would take this further. Deeper relationships were being transformed, too; the canal opened at Suez in 1869 not only shaped British commerce and strategy, but gave the Mediterranean new importance, not this time as a centre of a special civilization, but as a route.

A European magnate in his chauffeur-driven car; an image from the Belgian Congo in the early years of this century.

Economic integration and institutional change were inseparable from cultural contamination. The formal instruments of missionary religion, educational institutions and government policy are only a tiny part of this story. The European languages which were used officially, for example, took with them European concepts and opened to educated élites in non-European countries the heritage not only of Christian civilization, but of secular and 'enlightened' European culture, too. Missionaries spread more than dogma or medical and educational services; they also provoked the criticism of the colonial regime itself, because of the gap between its performance and the pretensions of the culture it imposed.

In the perspectives of the twentieth century, much of what is most durable and important in the impact of Europe on the world can be traced to such unintended, ambiguous effects as these. Above all, there was the simple urge to imitate, whether expressed ludicrously in the adoption of European dress or, much more importantly, in the conclusion drawn by many who sought to resist European hegemony that to do so it was necessary to adopt European ways. Almost everywhere, radicals and reformers advocated Europeanization. The ideas of 1776, 1789 and 1848 are still at work in Asia and Africa and the world still debates its future in European terms.

This extraordinary outcome is too often overlooked. In its unravelling, 1900 is only a vantage point, not the end of the story. The Japanese are a gifted people who have inherited exquisite artistic traditions, yet they have adopted not only western industrialism (which is understandable enough) but western art forms, western dress and even western drink in preference to their own. Though the Japanese find whisky and claret fashionable, *sake* has not yet swept Europe or the United States. The Chinese officially revere Marx, a German philosopher who articulated a system of thought rooted in nineteenth-century German idealism and English social and economic facts, who rarely spoke of Asia except with contempt and never went east of Prussia in his life. This suggests another curious fact: the balance-sheet of cultural influence is overwhelmingly one-sided. The world gave back to Europe occasional fashions, but no idea or institution of comparable effect to those Europe gave to the world. The teaching of Marx was long a force throughout twentieth-century Asia; the last non-European whose words had any comparable authority in Europe was Jesus Christ.

One physical transmission of culture was achieved by the movement of Europeans to other continents. Outside the United States, the two most numerous groups of European communities overseas were (as they still are) in South America and the former British colonies of white settlement which, though formally subject to London's direct rule for much of the nineteenth century, were in fact long oddly hybrid, not quite independent nations, but not really colonies either. Both groups were fed during the nineteenth century, like the United States, by the great diaspora of Europeans whose numbers justify one name which has been given to this era of European demography: the Great Resettlement. Before 1800, there was little European emigration except from the British Isles. After that date, something like sixty million Europeans went overseas, and this tide began to flow strongly in the 1830s. In the nineteenth century most of it went to North America, and then to Latin America (especially Argentina and Brazil), to Australia and South Africa. At the same time a concealed European emigration was also occurring across land within the Russian empire, which occupied one-sixth of the world's land surface and which had vast spaces to draw migrants in Siberia. The peak of European emigration overseas actually came on the eve of the First World War, in 1913, when over a million and a half left Europe; over a third of these were Italians, nearly four hundred thousand were British and two hundred thousand Spanish. Fifty years earlier, Italians figured only to a minor degree, Germans and Scandinavians loomed much larger. All the time, the British Isles contributed a steady flow; between 1880 and 1910 eight and a half million Britons went overseas (the Italian figure for this period was just over six million).

The greatest number of British emigrants went to the United States (about 65 per cent of them between 1815 and 1900), but large numbers went also to the self-governing colonies; this ratio changed after 1900 and by 1914 a majority of British emigrants was going to the latter. Italians and Spaniards also went to South America in large numbers, though Italians preferred the United States. That country remained the greatest of the receivers for all other nationalities; between 1820 and 1950 the United States benefited by the arrival of over thirty-three million Europeans.

Explanations of this striking demographic evolution are not far to seek. Politics sometimes contributed to the flow, as it did after 1848. Rising populations in Europe always pressed upon economic possibilities as the discovery of the phenomenon of 'unemployment' shows. In the last decades of the nineteenth century, too, when emigration was rising fastest, European farmers were pressed by overseas competition. Above all, it mattered that for the first time in human history there were obvious opportunities in other lands, where labour was needed, at a moment when there were suddenly easier and cheaper means of getting there. The steamship and railroad greatly changed demographic history and they both began to produce their greatest effect after 1880. They permitted much greater local mobility, so that temporary migrations of labour and movements within continents became much easier. Great Britain exported Irish peasants, Welsh miners and steelworkers and English farmers; she took in at the end of the nineteenth century an influx of Jewish communities from eastern Europe which remained a distinguishable element in British society. To the seasonal migration of labour which had always characterized such border districts as southern France were now added longer-term movements as Poles came to France to work in coal-mines and Italian waiters and ice-cream men became part of British folklore. When political changes made the North African shore accessible, it, too, was changed by short-range migration from Europe; Italians, Spaniards and Frenchmen were drawn there to settle or trade in the coastal cities and thus created a new society with interests distinct both from those of the societies from which the migrants had come and from those of the native societies beside which they had settled.

Easier travel did not only ease European migration. Chinese and Japanese settlement on the Pacific coasts of North America was already important by 1900. Chinese migrants also moved down into south-east Asia, Japanese to Latin America; the spectacle frightened Australians, who sought to preserve a 'White Australia' by limiting immigration by racial criteria. The British Empire provided a huge framework within which Indian communities spread round the world. But these movements, though important, were subordinate to the major phenomenon of the nineteenth century, the last great *Völker wanderung* of the European peoples, and one as decisive for the future as the barbarian invasions had been.

In 'Latin America' (the term was invented in the middle of the nineteenth century), which attracted in the main Italians and Spaniards, southern Europeans could find much that was familiar. There was the framework to cultural and social life provided by Catholicism; there were Latin languages and social customs. The political and legal framework also reflected the imperial past, some of whose institutions had persisted through an era of political upheaval at the beginning of the nineteenth century which virtually ended Spanish and Portuguese colonial rule on the mainland. This happened because events in Europe had led to a crisis in which weaknesses in the old empires proved fatal.

This was not for want of effort, at least on the part of the Spanish. In contrast to the British in the north, the metropolitan government had carried out sweeping reforms in the eighteenth century. When the Bourbons replaced the last Habsburg on the Spanish throne in 1701 a new era of Spanish imperial development had begun, though it took some decades to become apparent. When changes came they led first to reorganization and then to 'enlightened' reform. The two viceroyalties of 1700 became four, two more appearing in New Granada (Panama and the area covered by Ecuador, Colombia and Venezuela), and La Plata, which ran from the mouth of the river across the continent to the border of Peru. This structural rationalization was followed by relaxations of the closed commercial system, at first unwillingly conceded and then consciously promoted as a means to prosperity. These stimulated the economy both of the colonies and of those parts of Spain (notably the Mediterranean littoral) which benefited from the ending of the monopoly of colonial trade hitherto confined to the port of Seville.

None the less, these healthy tendencies were offset by grave weaknesses which they did not touch. A series of insurrections revealed deep-seated ills. In Paraguay (1721–35), Colombia (1781) and, above all, Peru (1780) there were real threats to colonial government which could only be contained by great military efforts. Among others, these required levies of colonial militia, a double-edged expedient, for it provided the *creoles* with military training which they might turn against Spain. The deepest division in Spanish colonial society was between the Indians and the colonists of Spanish descent, but that between the *creoles* and *peninsulares* was to have more immediate political importance. It had widened with the passage of time. Resentful of their exclusion from high office, the *creoles* noted the success of the British colonists of North America in shaking off imperial rule. The French Revolution, too, at first suggested possibilities rather than dangers.

As these events unrolled, the Spanish government was embarrassed in other ways. In 1790 a quarrel with Great Britain led at last to surrender of the remnants of the old Spanish claim to sovereignty throughout the Americas, when it conceded that the right to prohibit trade or settlement in North America only extended within an area of thirty

miles around a Spanish settlement. Then came wars, first with France, then with Great Britain (twice), and finally with France again, during the Napoleonic invasion. These wars not only cost Spain Santo Domingo, Trinidad and Louisiana, but also its dynasty, which was forced by Napoleon to abdicate in 1808. The end of Spanish sea-power had already come at Trafalgar. In this state of disorder and weakness, when, finally, Spain itself was engulfed by French invasion, the *creoles* decided to break loose and in 1810 the Wars of Independence began with risings in New Granada, La Plata and New Spain. The *creoles* were not at first successful and in Mexico they found that they had a racial war on their hands when the Indians took the opportunity to turn on all whites. But the Spanish government was not able to win them over nor to muster sufficient strength to crush further waves of rebellion. British sea-power guaranteed that no conservative European power could step in to help the Spanish and thus practically sustained the Monroe doctrine. So, there emerged from the fragments of the former Spanish empire a collection of republics, most of them ruled by soldiers.

In Portuguese Brazil the story had gone differently, for though a French invasion of Portugal had in 1807 provoked a new departure, it was different from that of the Spanish empire. The Prince Regent of Portugal had himself removed from Lisbon to Rio de Janeiro which thus became the effective capital of the Portuguese empire. Though he went back to Portugal as king in 1820 he left behind his son, who took the lead in resisting an attempt by the Portuguese government to reassert its control of Brazil and, with relatively little fighting, became the emperor of an independent Brazil in 1822.

A glance at the map of contemporary South America reveals the most obvious of many great differences between the revolution of 1776 and those of 1810 onwards: no United States of South America emerged from the Wars of Independence. Although the great hero and leader of independence, Bolivar, hoped for much from a Congress of the new states at Panama in 1826, nothing came of it. It is not difficult to understand why. For all the variety of the thirteen British colonies and difficulties facing them, they had after their victory relatively easy intercommunication by sea and few insurmountable obstacles of terrain. They also had some experience of cooperation and a measure of direction of their own affairs even while under imperial rule. With these advantages, their divisions still remained important enough to impose a constitution which gave very limited powers to the national government. It is hardly surprising that the southern republics could not achieve continental unity for all the advantages of the common background of Spanish rule which most of them shared.

The absence of unifying factors may not have been disadvantageous, for the Latin Americans of the early nineteenth century faced no danger or opportunity which made unity desirable. Against the outside world they were protected by Great Britain and the United States. At home the problems of post-colonial evolution were far greater than had been anticipated and were unlikely to be tackled more successfully by the creation of an artificial unity. Indeed, as in Africa a century and a half later, the removal of colonial rule revealed that geography and community did not always suit political units which corresponded to the old administrative divisions. The huge, thinly populated states which emerged from the Wars of Independence were constantly in danger of falling apart into small units as the urban minorities who had guided the independence movement found it impossible to control their followers. Some did break up. There were racial problems, too; the social inequalities they gave rise to were not removed by independence. Not every country experienced them in the same way. In Argentina, for example, the relatively small Indian population underwent near-extermination at the hands of the army. That country was celebrated by the end of the nineteenth century for the extent to which it resembled Europe in the domination of European strains in its population. At the other extreme, Brazil had a population the majority of which was black and, at the time of independence, much of it still in slavery. Intermarriage was not frowned upon, and the result is a population which may well be the most successfully integrated ethnic mix in the world today.

The new Latin American states could not draw upon any tradition of self-government in facing their many problems, for the colonial administrations had been absolutist and had not thrown up representative institutions. For the political principles they sought to apply, the leaders of the republics looked in the main to the French Revolution, but these were advanced ideas for states whose tiny élites did not even share among themselves agreement about accepted practice; they could hardly produce a framework of mutual tolerance. Worse still, revolutionary principles quickly brought the Church into politics, a development which was perhaps in the long run inevitable, given its huge power as a landowner and popular influence, but unfortunate in adding anti-clericalism to the woes of the continent. In these circumstances, it was hardly surprising that during most of the century each republic found that its affairs tended to drift into the hands of *caudillos*, military adventurers and their cliques who controlled armed forces sufficient to give them sway until more powerful rivals came along.

The cross-currents of civil war and wars between the new states – some very bloody – led by 1900 to a map which is still much the same today. Mexico, the most northern of the former Spanish colonies, had lost land in the north to the United States. Four mainland central American republics had appeared and two island states, the Dominican Republic and Haiti. Cuba was on the point of achieving independence. To the south were the ten states of South America. All of these countries were republican; Brazil had given up its monarchy in 1889. Though all had been through grave civic disorders, they represented very different degrees of stability and constitutional propriety. In Mexico, an Indian had indeed become president, to great effect, in the 1850s, but everywhere there remained the social divisions between Indians and the *mestizos* (those of mixed blood) on the one hand, and those of European blood (much reinforced in numbers when immigration became more rapid after 1870) on the other. The Latin American countries had contained about 19 million people in 1800; a century later they had 63 million.

This argues a certain increase in wealth. Most of the Latin American countries had important natural resources in one form or another. Sometimes, they fought over them for such advantages became even more valuable as Europe and the United States became more industrialized. Argentina had space and some of the finest

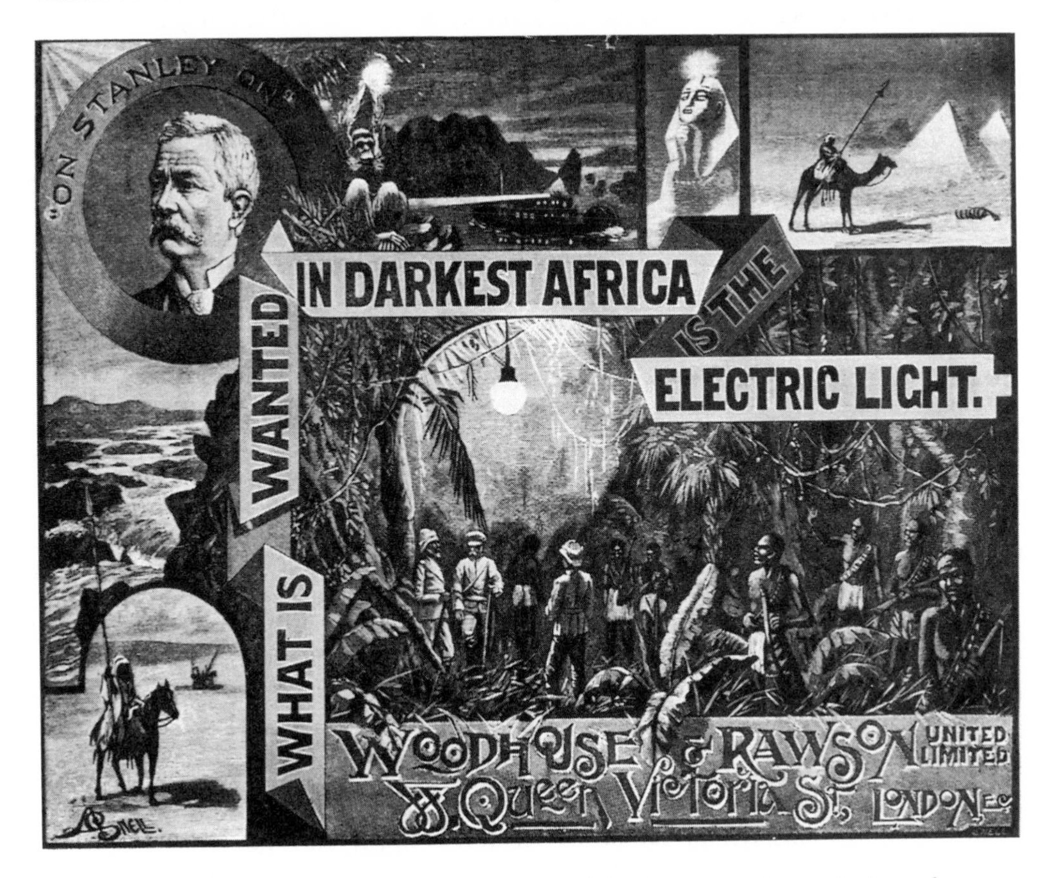

An advertisement of the 1890s and a true sign of the times, tapping, as it does, the glamour of exploration, technological pride and a conviction of racial superiority.

pasture in the world: the invention of refrigerator ships in the 1880s made it England's butcher and later grain grower as well. At the end of the nineteenth century, she was the richest of the Latin American countries. Chile had nitrates (taken from Bolivia and Peru in the 'War of the Pacific' of 1879–83) and Venezuela had oil; both grew more important with the twentieth century. Mexico had oil, too. Brazil had practically everything except oil, coffee and sugar above all. The list could be continued but would confirm that the growing wealth of Latin America came above all from primary produce. The capital to exploit this came from Europe and the United States and this produced new ties between these European nations overseas and Europe itself.

This increase in wealth nevertheless was connected with two related drawbacks. One was that it did little to reduce the disparities of wealth to be found in these countries; indeed, they may have increased. In consequence, social, like racial, tensions remained largely unresolved. An apparently Europeanized urban élite lived lives wholly unlike those of the Indian and *mestizo* masses. This was accentuated by the dependence of Latin America on foreign capital. Not unreasonably, foreign investors sought security. They by no means always got it, but it tended to lead them to support of the existing social and political authorities, who thus enhanced still further their own wealth. It would take only a few years of the twentieth century for conditions resulting from this sort of thing to produce social revolution in Mexico.

The irritation and disappointment of foreign investors who could not collect the debts due to them led sometimes to diplomatic conflicts and even armed intervention. The collection of debt was, after all, not seen as a revival of colonialism and European governments sent stiff messages and backed them up with force on several occasions during the century. When in 1902 Great Britain, Germany and Italy together instituted a naval blockade of Venezuela in order to collect debts due to their subjects who had suffered in revolutionary troubles, this provoked the United States to go further than the Monroe doctrine.

From the days of the Texan republic onwards, the relations of the United States with its neighbours had never been easy: nor are they today. Too many complicating factors were at work. The Monroe doctrine expressed the basic interest of the United States in keeping the hemisphere uninvolved with Europe and the first Pan-American Congress was another step in this direction when the United States organized it in 1889. But this could no more prevent the growth of economic links with Europe than could the Revolution sever those of the United States with Great Britain (and North Americans were among the investors in South American countries and soon had their special pleas to make to their government). Moreover, as the century came to an end, it was clear that the strategic situation which was the background to the Monroe doctrine had changed. Steamships and the rise of American interest in the Far East and the Pacific were the main causes of change. This made the United States much more sensitive, in particular, to developments in central America and the Caribbean, where an isthmian canal was more and more likely to be built.

The outcome was greater heavy-handedness and even arrogance in United States policy towards its neighbours in the early twentieth century. When, after a brief war with Spain, the United States won Cuba its independence (and took Puerto Rico from Spain for herself), special restraints were incorporated in the new Cuban constitution to ensure she would remain a satellite. The territory of the Panama Canal was obtained by intervention in the affairs of Colombia. The Venezuelan debt affair was followed by an even more remarkable assertion of American strength – a 'corollary' to the Monroe doctrine. This was the announcement (almost at once given practical expression in Cuba and the Dominican Republic) that the United States would exercise a right of intervention in the affairs of any state in the western hemisphere whose internal affairs were in such disorder that they might tempt European intervention. Later, one American president sent marines to Nicaragua in 1912 on this ground and another occupied the Mexican port of Vera Cruz in 1914 as a way of coercing a Mexican government. In 1915 a protectorate was established by treaty over Haiti which was to last forty years.

This was not the end of an unhappy story of relations between the United States and her neighbours, though far enough to take it for the moment. Their importance here, in any case, is only symptomatic of the ambiguous standing of the Latin American States in relation to Europe. Rooted in its culture, tied to it by economics, they none the less were constrained politically to avoid entanglement with it. This did not, of course, mean that they did not stand, so far as Europeans were concerned, on the white man's side of the great distinction more and more drawn between those within the pale of European civilization and those outside it. When the European thought of 'Latin Americans' he thought of those of European descent, the urban, literate, privileged minority, not the Indian and black masses.

The crumbling of the Spanish empire so soon after the defection of the thirteen colonies long led many people to expect that the other settler colonies of the British Empire would soon throw off the rule of London, too. In a way, this happened, but hardly as had been anticipated. At the end of the nineteenth century, the British magazine *Punch* printed a patriotic cartoon in which the British Lion looked approvingly at rows of little lion-cubs, armed and uniformed, who represented the colonies overseas. They were appropriately dressed as soldiers, for the volunteer contingents sent from other parts of the empire to fight for the British in the war they were then engaged upon in South Africa were of major importance. A century earlier, no one could have anticipated that a single colonial soldier would be available to the mother country. The year 1783 had burnt deep into the consciousness of British statesmen. Colonies, they thought they had learnt, were tricky things, costing money, conferring few benefits, engaging the metropolitan country in fruitless strife with other powers and native peoples and in the end usually turning round to bite the hand that fed them. The distrust of colonial entanglements which such views engendered helped to swing British imperial interest towards the East at the end of the eighteenth century, towards the possibilities of Asian trade. It seemed that in the Far East there would be no complications caused by European settlers and in Eastern seas no need for expensive forces which could not easily be met by the Royal Navy.

Broadly speaking, this was to be the prevailing attitude in British official circles during the whole nineteenth century. It led them to tackle the complicated affairs of each colony in ways which sought, above all else, economy and the avoidance of trouble. In the huge spaces of Canada and Australia this led, stormily, to the eventual uniting of the individual colonies in federal structures with responsibility for their own government. In 1867 the Dominion of Canada came into existence, and in 1901 there followed the Commonwealth of Australia. In each case, union had been preceded by the granting of responsible government to the original colonies and in each case there had been special difficulties. In Canada the outstanding one was the existence of a French Canadian community in the province of Quebec, in Australia the clashes of interest between settlers and convicts – of whom the last consignment was delivered in 1867. Each, too, was a huge, thinly populated country which could only gradually be pulled together to generate a sense of nationality. In each case the process was slow: it was not until 1885 that the last spike was driven on the transcontinental line of the Canadian Pacific Railway, and transcontinental railways in Australia were delayed for a long time by the adoption of different gauges in different states. In the end, nationalism was assisted by the growth of awareness of potential external threats – United States economic strength and Asian immigration – and, of course, by bickering with the British.

New Zealand also achieved responsible government, but one less decentralized, as befitted a much smaller country. Europeans had arrived there from the 1790s onwards and they found a native people, the Maori, with an advanced and complex culture, whom the visitors set about corrupting. Missionaries followed, and did their best to keep out settlers and traders. But they arrived just the same. When it seemed that a French *entrepreneur* was likely to establish a French interest, the British government at last reluctantly gave way to the pressure brought upon it by missionaries and some of the settlers and proclaimed British sovereignty in 1840. In 1856 the colony was given responsible government and only wars with the Maoris delayed the withdrawal of British soldiers until 1870. Soon afterwards, the old provinces lost their remaining legislative powers. In the later years of the century, New Zealand governments showed remarkable independence and vigour in the pursuit of advanced social welfare policies and achieved full self-government in 1907.

That was the year after a Colonial Conference in London had decided that the name 'Dominion' should in future be used for all the self-governing dependencies, which meant, in effect, the colonies of white settlement. One more remained to be given this status before 1914, the Union of South Africa, which came into existence in 1910. This was the end of a long and unhappy chapter – the unhappiest in the history of the British Empire and one which closed only to open another in the history of Africa which within a few decades looked even more bleak.

No British colonists had settled in South Africa until after 1814, when Great Britain for strategical reasons retained the former Dutch colony at the Cape of Good Hope. This was called 'Cape Colony' and there soon arrived some thousands of British settlers who, though outnumbered by the Dutch, had the backing of the British government in introducing British assumptions and law. This opened a period of whittling away of the privileges of the Boers, as the Dutch farmers were called. In particular, they were excited and irked by any limitation of their freedom to deal with the native African as they wished. Their especial indignation was aroused when, as a result of the general abolition of slavery in British territory, some 35,000 of their blacks were freed with, it was said, inadequate compensation. Convinced that the British would not abandon a policy favourable to the native African – and, given the pressures upon British governments, this was a reasonable view – a great exodus of Boers took place in 1835. This Great Trek north across the Orange River was of radical importance in making the Afrikaner consciousness. It was the beginning of a long period during which Anglo-Saxon and Boer struggled to live sometimes apart, sometimes together, but always uncomfortably, their decisions as they did so dragging in their train others about the fate of the black African.

A Boer republic in Natal was soon made a British colony, in order to protect the Africans from exploitation, and prevent the establishment of a Dutch port which might some day be used by a hostile power to threaten British communications with the Far East. Another exodus of Boers followed, this time north of the Vaal River. This was the first extension of British territory in South Africa but set a pattern which was to be repeated. Besides humanitarianism, the British government and the British colonists on the spot were stirred by the need to establish good relations with African peoples which would otherwise (as the Zulus had already shown against the Boers) present a continuing security problem (not unlike that posed by Indians in the American colonies in the previous century). By mid-century, there existed two Boer republics in the north (the Orange Free State and the Transvaal), while Cape Colony and Natal were under the British flag, with elected assemblies for which the few black men who met the required economic tests could vote. There were also native states under British protection. In one of these, Basutoland, this actually placed Boers under black jurisdiction, an especially galling subjection for them.

Happy relations were unlikely in these circumstances and, in any case, British governments were often in disagreement with the colonists at the Cape, who, after 1872, had responsible government of their own. New facts arose, too. The discovery of diamonds led to the British annexation of another piece of territory which, since it lay north of the Orange, angered the Boers. British support for the Basutos, whom the Boers had defeated, was a further irritant. Finally, the governor of Cape Colony committed an act of folly by annexing the Transvaal republic. After a successful Boer rising and a nasty defeat of a British force, the British government had the sense not to persist and restored independence to the republic in 1881, but from this moment Boer distrust of British policy in South Africa was probably insurmountable.

Within twenty years this led to war, largely because of two further unanticipated changes. One was a small-scale industrial revolution in the Transvaal republic, where gold was found in 1886. The result was a huge influx of miners and speculators, the involvement of outside financial interests in the affairs of South Africa, and the possibility that the Afrikaaner State might have the financial resources to escape from the British suzerainty it unwillingly accepted. The index of what had happened was Johannesburg, which grew in a few years to a city of 100,000 – the only one in Africa south of the Zambezi. The second change was that other parts of Africa were being swallowed in the 1880s and 1890s by other European powers and the British government was reacting by stiffening in its determination that nothing must shake the British presence at the Cape, deemed essential to the control of sea-routes to the East and increasingly dependent on traffic to and from the Transvaal for its revenues. The general effect was to make British governments view with concern any possibility of the Transvaal obtaining independent access to the Indian Ocean. This concern made them vulnerable to the pressure of an oddly-assorted crew of idealistic imperialists, Cape politicians, English demagogues and shady financiers who provoked a confrontation with the Boers in 1899 which ended in an ultimatum from the Transvaal's president, Paul Kruger, and the outbreak of the Boer War. Kruger had a deep dislike of the British; as a boy he had gone north on the Great Trek.

The well-known traditions of the British army of Victorian times were amply sustained in the last war of the reign, both in the level of ineptness and incompetence shown by some higher commanders and administrative services and in the gallantry shown by regimental officers and their men in the face of a brave and well-armed enemy whom their training did not prepare them to defeat. But of the outcome there could be no doubt; as the Queen herself remarked, with better strategical judgement than some of her subjects, the possibilities of defeat did not exist. South Africa was a theatre isolated by British sea-power; no other European nation could help the Boers and it was only a matter of time before greatly superior numbers and resources were brought to bear upon them. This cost a great deal – over a quarter of a million soldiers were sent to South Africa – and aroused much bitterness in British domestic politics; further, it did not present a very favourable picture to the outside world. The Boers were regarded as an oppressed nationality; so they were, but the nineteenth-century liberal obsession with nationality in this case as in others blinded observers to some of the shadows it cast. Fortunately, British statesmanship recovered itself sufficiently to make a generous treaty to end the war in 1902 when the Boers had been beaten in the field. This was the end of the Boer republics. But concession rapidly followed; by 1906 the Transvaal had a responsible government of its own which in spite of the large non-Boer population brought there by mining, the Boers controlled after an electoral victory the following year. Almost at once they began to legislate against Asian immigrants, mainly Indian. (One young Indian lawyer, Mohandas Gandhi, now entered politics as the champion of his community.) When, in 1909, a draft constitution for a South African Union was agreed, it was on terms of equality for the Dutch and English languages and, all-important, it provided for government by an elected assembly to be formed according to the electoral regulations decided in each province. In the Boer provinces the franchise was confined to white men.

At the time, there was much to be said for the settlement. When people then spoke of a 'racial problem' in South Africa they meant the problem of relations between the British and Boers whose conciliation seemed the most urgent need. The defects of the settlement would take some time to appear. When they did it would be not only because the historical sense of the Afrikaner proved to be tougher than people had hoped, but also because the transformation of South African society which had begun by the industrialization of the Rand could not be halted and would give irresistible momentum to the issue of the black Africans.

In this respect, South Africa's future had been just as decisively influenced as had those of all the other British white dominions by being caught up in the trends of the whole world economy. Canada, like the United States, had become with the building of the railroads on her plains one of the great granaries of Europe. Australia and New Zealand first exploited their huge pastures to produce the wool for which European factories were increasingly in the market; then, with the invention of refrigeration, they used them for meat and, in the case of New Zealand, dairy produce. In this way these new nations found staples able to sustain economies much greater than those permitted by the tobacco and indigo of the seventeenth-century plantations. The case of South Africa was to be different in that she was to reveal herself only gradually (as much later would Australia) as a producer of minerals. The beginning of this was the diamond industry, but the great step forward was the Rand gold discovery of the 1880s. The exploitation of this sucked in capital and expertise to make possible the eventual exploitation of other minerals. The return which South Africa provided was not merely in the profits of European companies and shareholders, but also an augmentation of the world's gold supply which stimulated European commerce much as had done the California discoveries of 1849.

Freedom fighters of 1899: three generations of Boer farmers in dress typical of that in which most Boers took the field against the British.

The growth of humanitarian and missionary sentiment in England and the well-founded Colonial Office tradition of distrust of settler demands made it harder to forget the native populations of the white dominions than it had been for Americans to sweep aside the Plains Indians. Yet in the British colonies, modernity made its impact not upon ancient civilizations but on primitive societies, some of which were at a very low stage of achievement indeed, Neolithic if not Palaeolithic, and correspondingly vulnerable. The Canadian Indians and Eskimos were relatively few and presented no such important obstacle to the exploitation of the west and north-west as had done the Plains Indians' struggle to keep their hunting-grounds. The story in Australia was far bloodier. The hunting and gathering society of the aborigine was disrupted by settlement, tribes were antagonized and provoked into violence by the uncomprehending brutality of the white Australians, and new diseases cut fast into their numbers. The early decades of each Australian colony are stained by the blood of massacred aborigines; the later years are notorious for the neglect, bullying and exploitation of the survivors. There is perhaps no other population inside former British territory which underwent a fate so like that of the American Indian. In New Zealand, the arrival of the first white men brought guns to the Maori, who employed them first on one another, with disruptive effects upon their societies. Later came the wars with the government, whose essential origin lay in the settlers' displacement of the Maori from their lands. At their conclusion, the government took steps to safeguard these tribal lands from further expropriation, but the introduction of English notions of individual ownership led to the disintegration of the tribal holdings and the virtual loss of their lands by the end of the century. The Maoris, too, declined in numbers, but not so violently or irreversibly as did the Australian aborigines. There are now many more Maoris than in 1900 and their numbers grow faster than those of New Zealanders of European stock.

As for South Africa, the story is a mixed one. British protection enabled some of its native peoples to survive into the twentieth century on their ancestral lands living lives which changed only slowly. Others were driven off or exterminated. In all cases, though, the crux of the situation was that in South Africa, as elsewhere, the fate of the native inhabitants was never in their own hands. They depended for their survival upon the local balance of governmental interest and benevolence, settler needs and traditions, economic opportunities and exigencies. Although in the short run they could sometimes present formidable military problems (as did the Zulus of Cetewayo, or the guerrilla warfare of the Maoris) they could not in the end generate from their own resources the means of effective resistance any more than had the Aztecs been able successfully to resist Cortés. For non-European peoples to do that, they would have to Europeanize. The price of establishing the new European nations beyond the seas always turned out to be paid by the native inhabitant, often to the limit of his ability.

This should not be quite the last word. There remains the problem of self-justification: Europeans witnessed these things happening and did not stop them. It is too simple to explain this by saying they were all bad, greedy men (and, in any case, the work of the humanitarians among them makes the blackest judgement untenable). The answer must lie somewhere in mentality. Many Europeans who could recognize that the native populations were damaged, even when the white contact with them was benevolent in intention, could not be expected to understand the corrosive effect of this culture on established structures. This requires an anthropological knowledge and insight Europe had still to achieve. It was all the more difficult when, clearly, a lot of native culture was simple savagery and the European's missionary confidence was strong. He *knew* he was on the side of Progress and Improvement, and might well still see himself as on the side of the Cross, too. This was a confidence which ran through every side of European expansion, the white settler colonies, directly-ruled possessions or the arrangements made with dependent societies. The confidence in belonging to a higher civilization was not only a licence for predatory habits as Christianity had earlier been, but the nerve of an attitude akin, in many cases, to that of crusaders. It was their sureness that they brought something better that blinded men all too often to the actual and material results of substituting individual freehold for tribal rights, of turning the hunters and gatherers, whose possessions were what they could carry, into wage-earners.

6
European Imperialism and Imperial Rule

The ruling of alien peoples and other lands by Europeans was the most striking evidence that they ran the world. In spite of enormous argument about what imperialism was and is, it seems helpful to confine the notion of it in the first place to such direct and formal overlordship, blurred though its boundaries with other forms of power over the non-European world may be. This does not raise or answer questions about causes or motives, on which much time, ink and thought have been spent. From the outset different and changing causes were at work, and not all the motives involved were unavowable or self-deceiving. Nor was imperialism the manifestation of only one age, for it has gone on all through history; nor was it peculiar to Europe's relations with non-Europeans overseas, for imperial rule had advanced overland as well as across the seas and some Europeans have ruled others and some non-Europeans Europeans. None the less, in the nineteenth and twentieth centuries the word came to be particularly associated with European expansion and the direct domination of Europeans over the rest of the world had by then become much more obvious than ever before. Although events in the Americas early in the nineteenth century suggested that the European empires built up over the preceding three centuries were in decline, in the next hundred years European imperialism was carried further and became more effective than ever before. This happened in two distinguishable phases, and one running down to about 1870 can conveniently be considered first. Some of the old imperial powers then continued to enlarge their empires impressively; such were Russia, France and Great Britain. Others stood still or found theirs reduced; these were the Dutch, Spanish and Portuguese.

The Russian expansion has at first sight something in common both with the American experience of filling up the North American continent and dominating its weaker neighbours, and something with that of the British in India, but was in fact a very special case. To the west Russia faced matured, established European states where there was little hope of successful territorial gain. The same was only slightly less true of expansion into the Turkish territories of the Danubian regions, for here the interests of other powers were always likely to come into play against Russia and check her in the end. Her main freedom of action lay southwards and eastwards; in both directions the first three-quarters of the nineteenth century brought great acquisitions. A successful war against Persia (1826–8) led to the establishment of Russian naval power on the Caspian as well as gains of territory in Armenia. In central Asia an almost continuous advance into Turkestan and towards the central oases of Bokhara and Khiva culminated in the annexation of the whole of Transcaspia in 1881. In Siberia, aggressive expansion was followed by the exaction of the left bank of the Amur down to the sea from China and the founding in 1860 of Vladivostok, the Far Eastern capital. Soon after, Russia liquidated her commitments in America by selling Alaska to the United States; this seemed to show she sought to be an Asian and Pacific, but not an American, power.

The other two dynamic imperial states of this era, France and Great Britain, expanded overseas. Yet many of the Britain gains were made at the expense of France; the Revolutionary and Napoleonic Wars were in this respect the final round of the great colonial Anglo-French contest of the eighteenth century. As in 1714 and 1763, many of Great Britain's acquisitions at a victorious peace in 1815 were intended to reinforce her maritime strength. Malta, St Lucia, the Ionian islands, the Cape of Good Hope, Mauritius and Trincomalee were all kept for this reason. As steamships took their station in the Royal Navy, the situation of bases had to take coaling into account; this could lead to further acquisitions. In 1839, an internal upheaval in the Ottoman empire gave the British the opportunity to take Aden, a base of strategic importance on the route to India, and others were to follow. No power could successfully challenge such action after Trafalgar. It was not that resources did not exist elsewhere which, had they been assembled, could have wrested naval supremacy form Great Britain. But to do so would have demanded a huge effort. No other nation operated either the number of ships or possessed the bases which could make it worth while to challenge this thalassocracy. There were advantages to other nations in having the world's greatest commercial power undertake a policing of the seas from which all could benefit.

Gibraltar was one of the most important links in the chain of British naval supremacy. A Highland regiment is here photographed posting sentries on one of the batteries, in the 1860s.

Naval supremacy guarded the trade which gave the British colonies participation in the fastest-growing commercial system of the age. Already before the American Revolution British policy had been more encouraging to commercial enterprise than the Spanish or French. Thus the old colonies themselves had grown in wealth and prosperity and the later Dominions were to benefit. On the other hand, settlement colonies went out of fashion in London after the American Revolution; they were seen mainly as sources of trouble and expense. Yet Great Britain was the only European country sending out new settlers to existing colonies in the early nineteenth century, and those colonies sometimes drew the mother country into yet further extensions of territory rule over alien lands.

In some acquisitions (notably in South Africa) there can be seen at work a new concern about imperial communication with Asia. This is a complicated business. No doubt American independence and the Monroe doctrine diminished the attractiveness of the western hemisphere as a region of imperial expansion, but the origins of a shift of British interest to the East can be seen before 1783, in the opening up of the South Pacific and in a growing Asian trade. War with the Netherlands, when it was a French satellite, subsequently led to new British enterprise in Malaya and Indonesia. Above all, there was the steadily-deepening British involvement in India. By 1800 the importance of the Indian trade was already a central axiom of British commercial and colonial thinking. By 1850, it has been urged, much of the rest of the empire had only been acquired because of the strategical needs of India. By then, too, the extension of British control inside the subcontinent itself was virtually complete. It was and remained the centre-piece of British imperialism.

This had hardly been expected or even foreseeable. In 1784 the institution of 'Dual Control' had been accompanied by decisions to resist further acquisition of Indian territory; the experience of American rebellion had reinforced the view that new commitments were to be avoided. Yet there was a continuing problem, for through its revenue management the Company's affairs inevitably became entangled in native administration and politics. This made it more important than ever to prevent excesses by its individual officers such as had been tolerable in the early days of private trading; slowly, agreement emerged that the government of India was not only of interest to Parliament because it might be a great source of patronage, but also because the government in London had a responsibility for the good government of Indians. There began to be articulated a notion of trusteeship. It is perhaps hardly surprising that, during a century in which the idea that government should be for the benefit of the governed was gaining ground in Europe, the same principle should be applied, sooner or later, to rule over colonial peoples. Since the days of Las Casas, exploitation of indigenous peoples had its vociferous critics. In the mid-eighteenth century, a bestselling book by the *abbé* Raynal (it went through thirty editions and many translations in twenty years) had put the criticisms of the churchmen into the secular terms of enlightenment humanitarianism. Against this deep background Edmund Burke in 1783 put it to the House of Commons in a debate on India that 'all political power which is set over men ... ought to be some way or other exercised ultimately for their benefit'.

The background against which Indian affairs were considered was steadily changing. Across two centuries, the awe and amazement inspired by the Moghul court in the first merchants to reach it had given way rapidly to contempt for what was seen on closer acquaintance as backwardness, superstition and inferiority. But now there were signs of another change. While Clive, the victor of Plassey, never learnt to speak with readiness in any Indian tongue, Warren Hastings, the first governor general of India, strove to get a chair of Persian set up at Oxford, and encouraged the introduction of the first printing-press to India and the making of the first fount of a vernacular (Bengali) type. There was greater appreciation of the complexity and variety of Indian culture. In 1789 there began to be published in Calcutta the first journal of oriental studies, *Asiatick Researches*. Meanwhile, at the more practical level of government, company judges were already enjoined to follow Islamic law in family cases involving Moslems, while the revenue authority of Madras both regulated and funded Hindu temples and festivals. From 1806 Indian languages were taught at the Company's college, Haileybury.

The periodic renewals of the Company's charter took place, therefore, in the light of changing influences and assumptions about Anglo-Indian relationships. Meanwhile, government's responsibilities grew. In 1813 renewal strengthened London's control further, and abolished the Company's monopoly of trade with India. By then, the wars with France had already led to the extension of British power over south India through annexation and the negotiation of treaties with native rulers which secured control of their foreign policy. By 1833, when the charter was again renewed, the only important block of territory not ruled directly or indirectly by the Company was in the north-west. The annexation of the Punjab and Sind followed in the 1840s and with their paramountcy established in Kashmir, the British held sway over virtually the whole subcontinent.

The Company had ceased to be a commercial organization and had become a government. The 1833 charter took away its trading functions (not only those with India but the monopoly of trade with China); in sympathy with current thinking, Asian trade was henceforth to be conducted on a free trade basis, and the Company confined to an administrative role. The way was open to the consummation of many real and symbolic breaks with India's past and the final incorporation of the subcontinent in a modernizing world. Symbolically, the name of the Moghul emperor was removed from the coinage, but it was more than a symbol that Persian ceased to be the legal language of record and justice. This step not only marked the advance of English as the official language (and therefore of English education), but also disturbed the balance of forces between Indian communities. Anglicized Hindus would prove to do better than less enterprising Moslems. In a subcontinent so divided in so many ways, the adoption of English as the language of administration was complemented by the important decision to provide primary education through instruction given in English.

At the same time an enlightened despotism exercised by successive governors-general began to impose material and institutional improvement. Roads and canals were built and the first railway followed in 1853. Legal codes were introduced. English officials for the Company's service began to be trained specially in the college established for this purpose. The first three universities in India were founded in 1857 and were based on institutions already founded well before this. There were other educational structures, too; as far back as 1791 a Scotchman had founded a Sanskrit college at Benares, the Lourdes of Hinduism. Much of the transformation which India was gradually undergoing arose not from the direct work of government but from the increasing freedom with which these and other agencies were allowed to operate. From 1813 the arrival of missionaries (the Company had hitherto kept them out) gradually built up another constituency at home with a stake in what happened in India – often to the embarrassment of official India. Two philosophies, in effect, were competing to make government act positively. Utilitarianism looked for the promotion of happiness, evangelical Christianity to the salvation of souls. Both were sure they knew what was best for India and both were equally arrogant in their judgements of it. Both subtly changed British attitudes as time passed.

The coming of the steamship had its influence, too. It brought India nearer. More Englishmen and Scotchmen began to live in India and make their careers there. This gradually transformed the nature of the British presence. The comparatively few officers of the eighteenth-century Company had been content to live the lives of exiles, seeking rewards in their commercial opportunities and relaxation in a social life sometimes closely integrated with that of the Indians. They often lived much in the style of Indian gentlemen, some of them taking to Indian dress and food, some to Indian wives and concubines. Reform-minded officials, intent on the eradication of the backward and barbaric in native practice – and in such practices as female infanticide and suttee they had good cause for concern – missionaries with a creed to preach which was corrosive of the whole structure of Hindu or Moslem society and, above all, the Englishwomen who arrived to make homes in India for two or three years at a time while their husbands worked there, all changed the temper of the British community. The British in India lived more and more apart from the natives, more and more convinced of their superiority and that it sanctioned the ruling of Indians who were (they felt) culturally and morally inferior. They grew consciously more alien from those they ruled. One of them spoke approvingly of his countrymen as representatives of a 'belligerent civilization' and defined their task as 'the introduction of the essential parts of European civilization into a country densely peopled, grossly ignorant, steeped in idolatrous superstition, unenergetic, fatalistic, indifferent to most of what we regard as the evils of life and preferring the response of submitting to them to the trouble of encountering and trying to remove them'. This robust creed was far from that of the Englishmen of the previous century. They had innocently sought to do no more in India than make money. Now, while new laws antagonized powerful native interests, those who shared this creed had less and less social contact with Indians; more and more they confined the educated Indian to the lower ranks of the administration and withdrew into an enclosed, but conspicuously privileged, life of their own. Earlier conquerors had been absorbed by Indian society in greater or lesser measure; the Victorian British, thanks to a modern technology which continuously renewed their contacts with the homeland and their confidence in their intellectual and religious superiority, remained immune, increasingly aloof, as no earlier conqueror had been. They could not be untouched by India, as many legacies to the English language and the English breakfast and dinner-table still testify, but they created a civilization that was not Indian, even if it was not wholly English; 'Anglo-Indian' in the nineteenth century was a word applied not to a person of mixed blood, but to an Englishman who made his career in India, and it indicated a cultural and social distinctiveness.

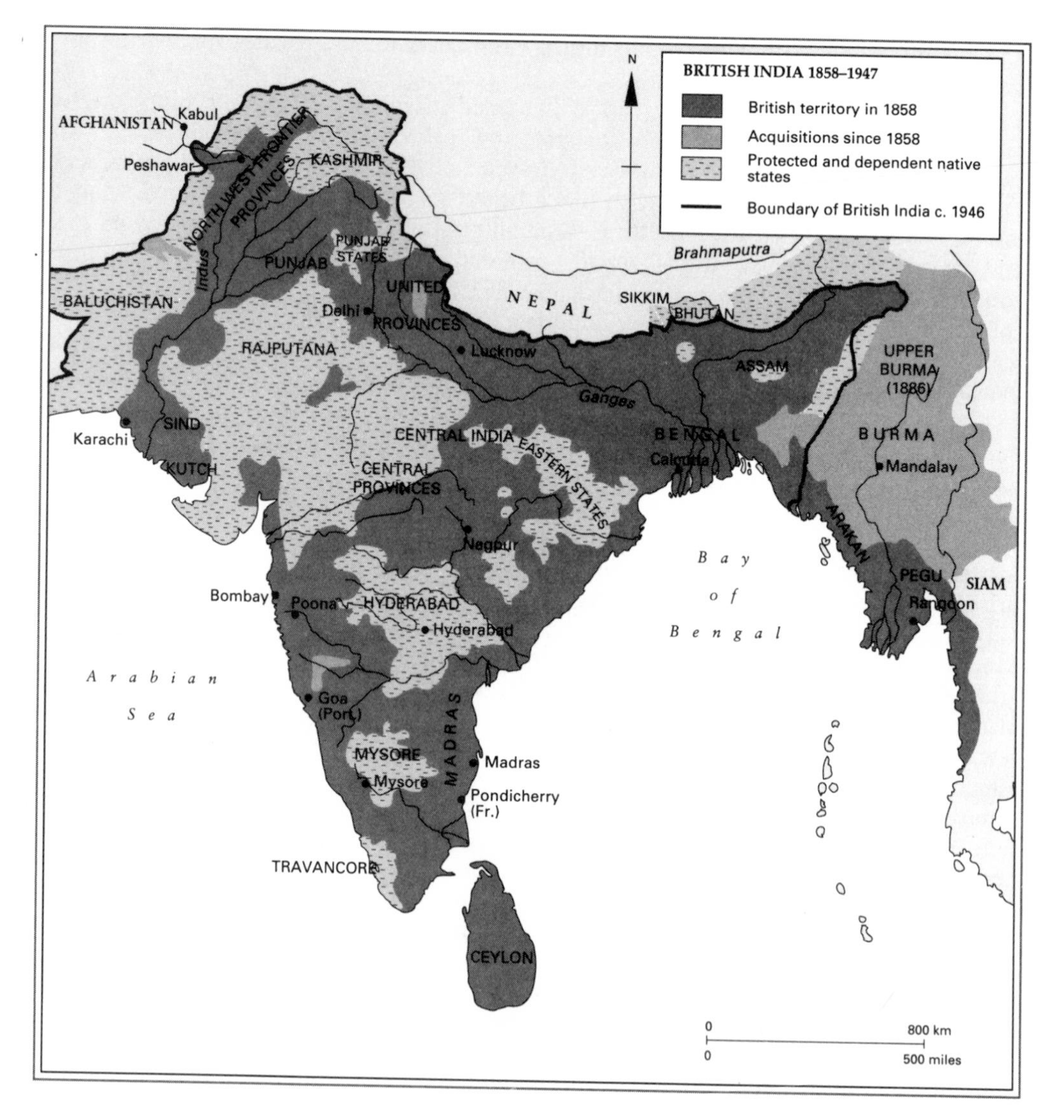

The separateness of Anglo-Indian society from India was made virtually absolute by the appalling damage done to British confidence by the rebellions of 1857 called the Indian Mutiny. Essentially, this was a chain reaction of outbreaks, initiated by a mutiny of Hindu soldiers who feared the polluting effect of using a new type of cartridge, greased with animal fat. This detail is significant. Much of the rebellion was the spontaneous and reactionary response of traditional society to innovation and modernization. By way of reinforcement there were the irritations of native rulers, both

Moslem and Hindu, who regretted the loss of their privileges and thought that the chance might have come to recover their independence; the British were after all very few. The response of those few was prompt and ruthless. With the help of loyal Indian soldiers the rebellions were crushed, though not before there had been some massacres of British captives and a British force had been under siege at Lucknow, in rebel territory, for some months.

The Mutiny and its suppression were disasters for British India, though not quite unmitigated. It did not much matter that the Moghul empire was at last formally brought to an end by the British (the Delhi mutineers had proclaimed the last emperor their leader). Nor was there, as later Indian nationalists were to suggest, a crushing of a national liberation movement whose end was a tragedy for India. Like many episodes important in the making of nations, the Mutiny was to be important as a myth and an inspiration; what it was later believed to have been was more important than what it actually was, a jumble of essentially reactionary protests. Its really disastrous and important effect was the wound it gave to British goodwill and confidence. Whatever the expressed intentions of British policy, the consciousness of the British in India was from this time suffused by the memory that Indians had once proved almost fatally untrustworthy. Among Anglo-Indians as well as Indians the mythical importance of the Mutiny grew with time. The atrocities actually committed were bad enough, but unspeakable ones which had never occurred were also alleged as grounds for a policy of repression and social exclusiveness. Immediately and institutionally, the Mutiny also marked an epoch because it ended the government of the Company. The governor-general now became the queen's viceroy, responsible to a British cabinet minister. This structure provided the framework of the British Raj for the whole of its life.

The Mutiny thus changed Indian history only by thrusting it more firmly in a direction to which it already tended. Another fact which was equally revolutionary for India was much more gradual in its effects. This was the nineteenth-century flowering of the economic connexion with Great Britain. Commerce was the root of the British presence in the subcontinent and it continued to shape its destiny. The first major development came when India became the essential base for the China trade. Its greatest expansion came in the 1830s and 1840s when, for a number of reasons, access to China became much easier. It was at about the same time that there took place the first rapid rise in British exports to India, notably of textiles, so that, by the time of the Mutiny, a big Indian commercial interest existed which involved many more Englishmen and English commercial houses than the old Company had ever done.

The story of Anglo-Indian trade was now locked into that of the general expansion of British manufacturing supremacy and world commerce. The Suez Canal brought down the costs of shipping goods to Asia by a huge factor. By the end of the century the volume of British trade with India had more than quadrupled. The effects were felt in both countries, but were decisive in India, where a check was imposed on an industrialization which might have gone ahead more rapidly without British competition. Paradoxically, the growth of trade thus delayed India's modernization and alienation from its own past. But there were other forces at work, too. By the end of the century the framework provided by the Raj and the stimulus of the cultural influences it permitted had already made impossible the survival of wholly unmodernized India.

No other nation in the early nineteenth century so extended its imperial possessions as Great Britain, but the French, too, had made substantial additions to the empire with which they had been left in 1815. In the next half-century France's interests elsewhere (in West Africa and the South Pacific, for example) were not lost to sight but the first clear sign of a reviving French imperialism came in Algeria. The whole of North Africa was open to imperial expansion by European predators because of the decay of the formal overlordship of the Ottoman sultan. Right round the southern and eastern Mediterranean coasts the issue was posed of a possible Turkish partition. French interest in the area was natural; it went back to a great extension of the country's Levant trade in the eighteenth century and an expedition to Egypt under Bonaparte in 1798.

Algeria's conquest began uncertainly in 1830. A series of wars not only with its native inhabitants but with the sultan of Morocco followed until, by 1870, most of the country had been subdued. This was, in fact, to open a new phase of expansion, for the French then turned their attention to Tunis, which accepted a French protectorate in 1881. To both these sometime Ottoman dependencies there now began a steady flow of European immigrants, not only from France, but from Italy and, later, Spain. In a few cities this built up substantial settler populations which were to complicate the story of French rule. The day was past when the African Algerian might have been exterminated or all but exterminated, like the Aztec, American Indian or Australian aborigine. His society, in any case, was more resistant, formed in the crucible of an Islamic civilization which had once contested Europe so successfully. None the less, he suffered, notably from the introduction of land law which broke up his traditional usages and impoverished the peasant by exposing him to the full blast of market economics.

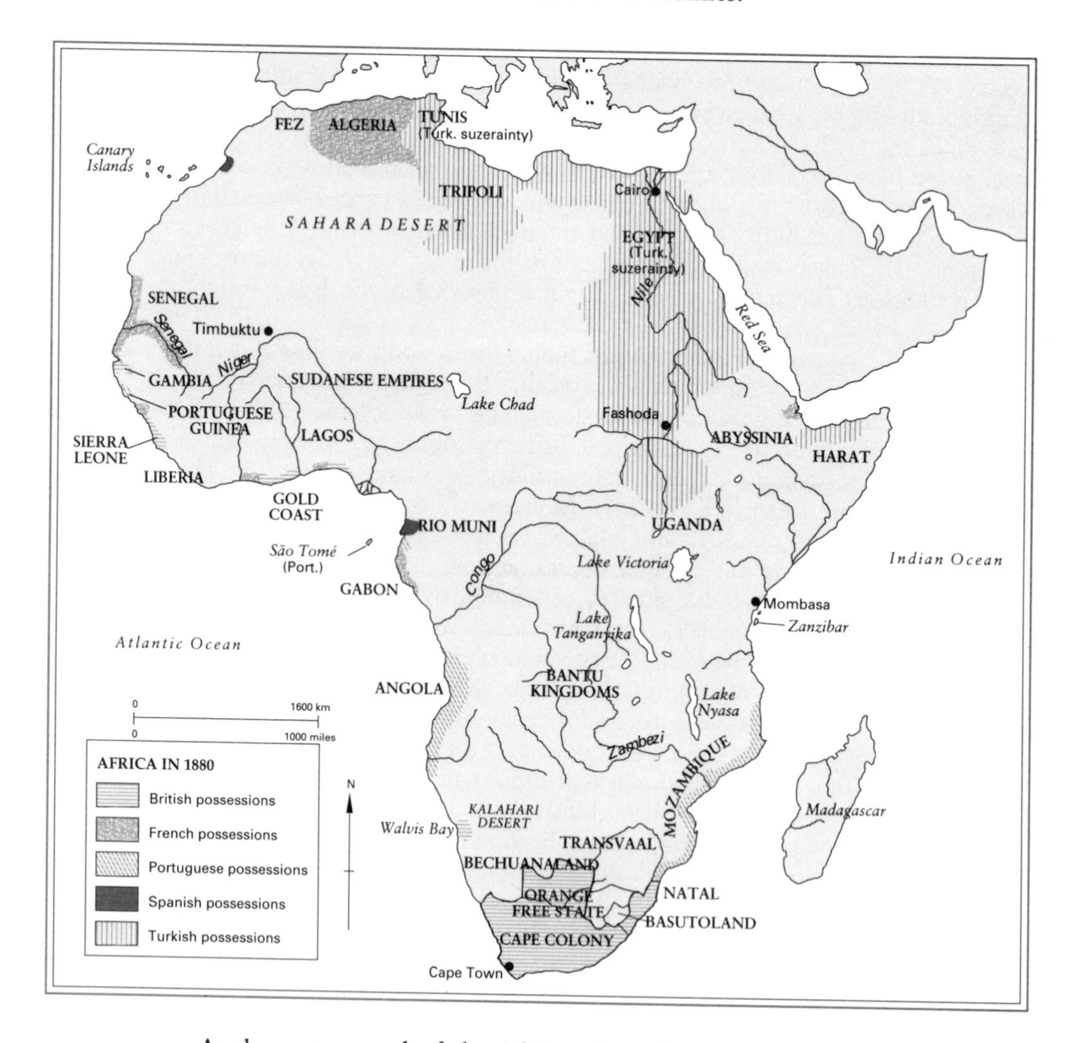

At the eastern end of the African littoral a national awakening in Egypt led to the emergence there of the first great modernizing nationalist leader outside the European world, Mehemet Ali, pasha of Egypt. Admiring Europe, he sought to borrow its ideas and techniques while asserting his independence of the sultan. When he was later called upon for help by the sultan against the Greek revolution, Mehemet Ali went on to attempt to seize Syria as his reward. This threat to the Ottoman empire provoked an international crisis in the 1830s in which the French took his side. They were not successful, but thereafter French policy continued to interest itself in the Levant and

Syria, too, an interest which was eventually to bear fruit in the brief establishment in the twentieth century of a French presence in the area.

The feeling that Great Britain and France had made good use of their opportunities in the early part of the nineteenth century was no doubt one reason why other powers tried to follow them from 1870 onwards. But envious emulation does not go far as an explanation of the extraordinary suddenness and vigour of what has sometimes been called the 'imperialist wave' of the late nineteenth century. Outside Antarctica and the Arctic less than a fifth of the world's land surface was not under a European flag or that of a country of European settlement by 1914; and of this small fraction only Japan, Ethiopia and Siam enjoyed real autonomy.

Why this happened has been much debated. Clearly one part of the story is that of the sheer momentum of accumulated forces. The European hegemony became more and more irresistible as it built upon its own strength. The theory and ideology of imperialism were up to a point mere rationalizations of the huge power the European world suddenly found itself to possess. Practically, for example, as medicine began to master tropical infection and steam provided quicker transport, it became easier to establish permanent bases in Africa and to penetrate its interior; the Dark Continent had long been of interest but its exploitation began to be feasible for the first time in the 1870s. Such technical developments made possible and attractive a spreading of European rule which could promote and protect trade and investment. The hopes such possibilities aroused were often ill-founded and usually disappointed. Whatever the appeal of 'undeveloped estates' in Africa (as one British statesman liked to put it), or the supposedly vast market for consumer goods constituted by the penniless millions of China, industrial countries still found other industrial countries their best customers and trading partners and it was former or existing colonies rather than new ones which attracted most overseas capital investment. By far the greatest part of British money invested abroad went to the United States and South America; French investors preferred Russia to Africa, and German money went to Turkey.

On the other hand, economic expectation excited many individuals. Because of them, imperial expansion always had a random factor in it which makes it hard to generalize about. Explorers, traders and adventurers on many occasions took steps which led governments, willingly or not, to seize more territory. They were often popular heroes, for this most active phase of European imperialism coincided with a great growth of popular participation in public affairs. By buying newspapers, voting, or cheering in the streets, the masses were more and more involved in politics which (among other things) emphasized imperial competition as a form of national rivalry. The new cheap press often pandered to this by dramatizing exploration and colonial warfare. Some thought, too, that social dissatisfactions might be soothed by the contemplation of the extension of the rule of the national flag over new areas even when the experts knew that nothing was likely to be forthcoming except expense.

But cynicism is no more the whole story than is the profit motive. The idealism which inspired some imperialists certainly salved the conscience of many more. Men who believed that they possessed true civilization were bound to see the ruling of others for their good as a duty. Kipling's famous poem did not urge Americans to take up the White Man's Booty, but his Burden.

All these elements were tangled together more than ever after 1870. They all had to find a place in a context of changing international relationships which imposed its own logic on colonial affairs. The story need not be explained in detail, but two continuing themes stand out in it. One is that, because she was the only truly world-wide imperial power, Great Britain quarrelled more than anyone else with other states over colonies. She had possessions everywhere. The centre of her concerns was more than ever India; the acquisition of African territory to safeguard the Cape route and the new one via Suez, and frequent alarms over dangers to the lands which were India's glacis both showed this. Between 1870 and 1914 the only crises over non-European issues which made war between Great Britain and another great power seem possible arose over

Russian dabblings in Afghanistan and a French attempt to establish themselves on the Upper Nile. British officials were also much concerned about French penetration of West Africa and Indo-China, and Russian influence in Persia.

These facts indicate the second continuing theme. Though European nations quarrelled about what happened overseas for forty years or so, and though the United States went to war with one of them (Spain), the partition of the non-European world was amazingly peaceful. When a Great War at last broke out in 1914, Great Britain, Russia and France, the three nations which had quarrelled with one another most over imperial difficulties, would be on the same side; it was not overseas colonial rivalry which caused the conflict. Only once after 1900, in Morocco, did a real danger of war occasioned by a quarrel over non-European lands arise between two European great powers and here the issue was not really one of colonial rivalry, but of whether Germany could bully France without fear of her being supported by others. Quarrels over non-European affairs before 1914 seem in fact to have been a positive distraction from the more dangerous rivalries of Europe itself; they may even have helped to preserve European peace.

Imperial rivalry had its own momentum. When a power got a new concession or a colony, it almost always spurred on others to go one better. The imperialist wave was in this way self-feeding. By 1914 the most striking results were to be seen in Africa. The activities of explorers, missionaries, and the campaigners against slavery early in the nineteenth century had encouraged the belief that extension of European rule in the 'Dark Continent' was a matter of spreading enlightenment and humanitarianism – the blessings of civilization, in fact. On the African coasts, centuries of trade had shown that desirable products were available in the interior. The whites at the Cape were already pushing further inland (often because of Boer resentment of British rule). Such facts made up an explosive mixture, which was set off in 1881 when a British force was sent to Egypt to secure that country's government against a nationalist revolution whose success (it was feared) might threaten the safety of the Suez Canal. The corrosive power of European culture – the source of the ideas of the Egyptian nationalists – thus both touched off another stage in the decline of the Ottoman empire of which Egypt was still formally a part and launched what was called the 'Scramble for Africa'.

The British had hoped to withdraw their soldiers from Egypt quickly; in 1914 they were still there. British officials were by then virtually running the administration of the country while, to the south, Anglo-Egyptian rule had been pushed deep into the Sudan. Meanwhile, Turkey's western provinces in Libya and Tripolitania had been taken by the Italians (who felt unjustly kept out of Tunisia by the French protectorate there), Algeria was French, and the French enjoyed a fairly free hand in Morocco, except where the Spanish were installed. Southwards from Morocco to the Cape of Good Hope, the coastline was entirely divided between the British, French, Germans, Spanish, Portuguese and Belgians, with the exception of the isolated black republic of Liberia. The empty wastes of the Sahara were French; so was the basin of the Senegal and much of the northern side of that of the Congo. The Belgians were installed in the rest of it on what was soon to prove some of the richest mineral-bearing land in Africa. Further east, British territories ran from the Cape up to Rhodesia and the Congo border. On the east coast they were cut off from the sea by Tanganyika (which was German) and Portuguese East Africa. From Mombasa, Kenya's port, a belt of British territory stretched through Uganda to the borders of the Sudan and the headwaters of the Nile. Somalia and Eritrea (in British, Italian and French hands) isolated Ethiopia, the only African country other than Liberia still independent of European rule. The ruler of this ancient but Christian polity was the only non-European ruler of the nineteenth century to avert the threat of colonization by a military success, the annihilation of an Italian army at Adowa in 1896. Other Africans did not have the power to resist successfully, as the French suppression of Algerian revolt in 1871, the Portuguese mastery (with some difficulty) of insurrection in Angola in 1902 and again in 1907, the British destruction of the Zulu and Matabele, and, worst of all, the German massacre of the Herrero of South-West Africa in 1907, all showed.

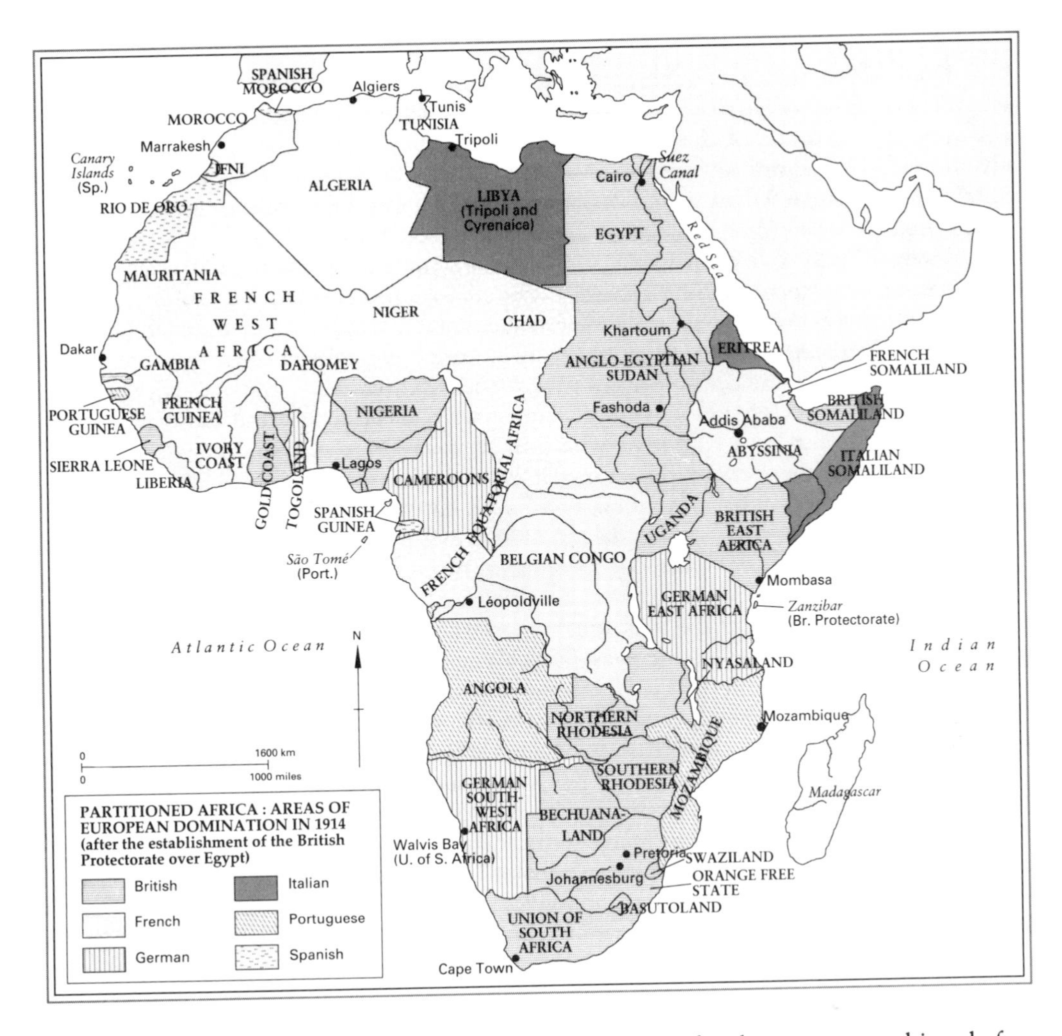

This colossal extension of European power, for the most part achieved after 1881, transformed African history. The bargains of European negotiators, the accidents of discovery and the convenience of colonial administrations in the end settled the ways in which modernization came to Africa. The suppression of inter-tribal warfare and the introduction of even elementary medical services released population growth in some areas. As in America centuries earlier, the introduction of new crops made it possible to feed more people. Different colonial regimes had different cultural and economic impact, however. Long after the colonialists had gone, there would be big distinctions between countries where, say, French administration or British judicial practice had taken root. All over the continent Africans found new patterns of employment, learnt something of European ways through European schools or service in colonial regiments, saw different things to admire or hate in the white man's ways which now came to regulate their lives. Even when, as in some British possessions, great emphasis was placed on rule through traditional native institutions, they had henceforth to work in a new context. Tribal and local unities would go on asserting themselves but more and more did so against the grain of new structures created by colonialism and left as a legacy of independent Africa. Languages, Christian monogamy, entrepreneurial attitudes, new knowledge, all contributed finally to a new self-consciousness and greater individualism. From such influences would emerge the new Africans of the twentieth century. Without imperialism, for good or ill, those influences could never have been so effective so fast.

Europe, by contrast, was hardly changed by the African adventure, and it is easy to overestimate its brief effect there. Clearly, it was important that Europeans could lay their hands on yet more easily exploitable wealth, yet probably only Belgium drew from Africa resources making a real difference to its national future. Sometimes, too, the exploiting of Africa aroused political opposition in European countries; there was more than a touch of the *conquistadores* about some of the late nineteenth-century adventurers. The administration of the Congo by the Belgian King Leopold and forced labour in Portuguese Africa were notorious examples, but there were other places where Africa's natural resources – human and material – were ruthlessly exploited or despoiled in the interests of profit by Europeans with the connivance of imperial authorities, and this soon created an anti-colonial movement. Some nations recruited African soldiers, though not for service in Europe, where only the French hoped to employ them to offset the weight of German numbers. Some countries hoped for outlets for emigration which would ease social problems, but the opportunities presented by Africa for European residence were very mixed. The two large blocks of settlement were in the south, and the British would later open up Kenya and Rhodesia where there were lands suitable for white farmers. Apart from this, there were the cities of French North Africa, and also a growing community of Portuguese in Angola. The hopes entertained of Africa as an outlet for Italians, on the other hand, were disappointed, and German emigration was tiny and almost entirely temporary. Some European countries – Russia, Austria, Hungary and the Scandinavian nations – sent virtually no settlers to Africa at all.

Of course, there was much more than Africa to the story of nineteenth century imperialism. The Pacific was partitioned less dramatically but in the end no independent political unit survived among its island peoples. There was also a big expansion of British, French and Russian territory in Asia. The French established themselves in Indo-China, the British in Malaya and Burma, which they took to safeguard the approaches to India. Siam retained her independence because it suited both powers to have a buffer between them. The British also asserted their superiority by an expedition to Tibet, with the same considerations of Indian security in mind. Most of these areas, like much of the zone of Russian overland expansion, were formally under Chinese suzerainty. Their story is part of that of the crumbling Chinese empire, a story which paralleled the corrosion of other empires, such as the Ottoman, Moroccan and Persian, by European influence, though it has even greater importance for world history. At one moment it looked as if a Scramble for China might follow the partition of Africa. That story is better considered elsewhere. Here it is convenient to notice that the imperialist wave in the Chinese sphere as in the Pacific was importantly different from that in Africa because the United States of America took part.

Americans had always been uneasy and distrustful over imperial ventures outside the continent they long regarded as God-given to them. Even at its most arrogant, imperialism had to be masked, muffled and muted in the republic in a way unnecessary in Europe. The very creation of the United States had been by successful rebellion against an imperialist power. The constitution contained no provision for the ruling of colonial possessions and it was always very difficult to see what could be the position under it of territories which could not be envisaged as eventually moving towards full statehood, let alone that of non-Americans who stayed under American rule. On the other hand, there was much that was barely distinguishable from imperialism in the nineteenth-century territorial expansion of the United States, though Americans might not recognize it when it was packaged as a 'Manifest Destiny'. The most blatant examples were the war of 1812 against the British and the treatment of Mexico in the middle of the century. But there was also the dispossession of the Indians to consider and the dominating implications of the Monroe doctrine.

In the 1890s the overland expansion of the United States was complete. It was announced that the continuous frontier of domestic settlement no longer existed. At this moment, economic growth had given great importance to the influence of business interests in American government, sometimes expressed in terms of economic nationalism and high tariff protection. Some of these interests directed the attention of American public opinion abroad, notably to Asia. The United States was thought by some to be in danger of

An explosion which detonated a war. The USS Maine blows up.

exclusion from trade there by the European powers. There was an old connexion at stake (the first American Far Eastern squadron had been sent out in the 1820s) as a new era of Pacific awareness dawned with California's rapid growth in population. A half-century's talk of a canal across Central America also came to a head at the end of the century; it stimulated interest in the doctrines of strategists who suggested that the United States might need an oceanic glacis in the Pacific to maintain the Monroe doctrine.

All these currents flowed into a burst of expansion which has remained to this day a unique example of American overseas imperialism because, for a time, it set aside traditional restraint on the acquisition of new territory overseas. The beginnings must be traced back to the increased opening of China and Japan to American commerce in the 1850s and 1860s and to participation with the British and the Germans in the administration of Samoa (where a naval base obtained in 1878 has remained a United States possession). This was followed by two decades of growing intervention in the affairs of the kingdom of Hawaii, to which the protection of the United States had been extended since the 1840s. American traders and missionaries had established themselves there in large numbers. Benevolent patronage of the Hawaiians gave way to attempts to achieve annexation to the United States in the 1890s. Washington already had the use of Pearl Harbor as a naval base but was led to land marines in Hawaii when a revolution occurred there. In the end, the government had to give way to the forces set in motion by the settlers and a short-lived Hawaiian Republic was annexed as a United States Territory in 1898.

In that year, a mysterious explosion destroyed an American cruiser, the USS *Maine*, in Havana harbour. This led to a war with Spain. In the background was both the long Spanish failure to master revolt in Cuba, where American business interests were prominent and American sentiment was aroused, and the growing awareness of the importance of the Caribbean approaches to a future canal across the isthmus. In Asia, American help was given to another rebel movement against the Spaniards in the

Philippines. When American rule replaced Spanish in Manila, the rebels turned against their former allies and a guerrilla war began. This was the first phase of a long and difficult process of disentangling the United States from her first Asian colony. At that moment, given the likelihood of the collapse of the Chinese empire, it seemed best in Washington not to withdraw. In the Caribbean, the long history of Spanish empire in the Americas at last came to an end. Puerto Rico passed to the United States and Cuba obtained its independence on terms which guaranteed its domination by the United States. American forces went back to occupy the island under these terms from 1906 to 1909, and again in 1917.

Two-ocean power. A United States battleship passes through the Panama Canal soon after its opening.

This was the prelude to the last major development in this wave of American imperialism. The building of an isthmian canal had been canvassed since the middle of the nineteenth century and the completion of Suez gave it new plausibility. American diplomacy negotiated a way round the obstacle of possible British participation; all seemed plain sailing but a snag arose in 1903 when a treaty providing for the acquisition of a canal zone from Colombia was rejected by the Colombians. A revolution was engineered in Panama, where the canal was to run. The United States prevented its suppression by the Colombian government and a new Panamanian republic emerged which gratefully bestowed upon the United States the necessary territory together with the right to intervene in its affairs to maintain order. The work could now begin and the canal was opened in 1914. The possibility of transferring ships rapidly from one ocean to another made a great difference to American naval strategy. It was also the background to the 'corollary' to the Monroe doctrine proposed by President Theodore Roosevelt; when the Canal Zone became the key to the naval defence of the hemisphere, it was more important than ever to assure its protection by stable government and United States predominance in the Caribbean states. A new vigour in American intervention in them was soon evident.

Though its motives and techniques were different – for one thing, there was virtually no permanent American settlement in the new possessions – the actions of the United States were part of the last great seizure of territories carried out by the European peoples. Almost all of them had taken part except the South Americans; even the Queenslanders had tried to annexe New Guinea. By 1914 a third of the world's surface was under two flags, those of the United Kingdom and Russia (though how much Russian territory should be regarded as colonial is, of course, debatable). To take a measure which excludes Russia, in 1914 the United Kingdom ruled four hundred million subjects outside its own borders, France over fifty million and Germany and Italy about fourteen million each; this was an unprecedented aggregation of formal authority.

At the date, though, there were already signs that imperialism overseas had run out of steam. China had proved a disappointment and there was little left to divide even if Germany and Great Britain discussed the possibility of partitioning the Portuguese empire, which seemed to be about to follow the Spanish. The most likely area left for further European imperialism was the decaying Ottoman empire, whose dissolution seemed at last to be imminent when the Italians seized Tripoli in 1912 and a Balkan coalition formed against Turkey took away almost all that was left of her European territories in the following year. Such a prospect did not seem likely to be so free from conflict between great powers as had been the partition of Africa; much more crucial issues would be at stake for them so near to home.

7
Asia's Response to a Europeanizing World

A perceptive Chinese observer might have found something revealing in the disgrace which in the end overtook the Jesuits at K'ang-hsi's court. For more than a century these able men had judiciously and discreetly sought to ingratiate themselves with their hosts. To begin with they had not even spoken of religion, but had contented themselves with studying the language of China. They had even worn Chinese dress, which, we are told, created a very good impression. Great successes had followed. Yet the effectiveness of their mission was suddenly paralysed; their acceptance of Chinese rites and beliefs and their sinicizing of Christian teaching led to the sending of two papal emissaries to China to check such improper flexibility. This was striking: evidently Europeans, unlike all other conquerors of China, would not easily succumb to its cultural pull. There was a message for all Asia in this revelation of the intransigence of European culture. It was more important to what was going to happen in Asia – and was already under way there – than even the technology of the newcomers. It was certainly more decisive than any temporary or special weaknesses of the eastern empires, as China's own history was to show. Under the immediate successors of K'ang-hsi, the Manchu empire was already past its peak; its slow and eventually fatal decline would not have in itself been surprising given the cyclical pattern of dynastic rise and fall in the past. What made the fate of the Ch'ing dynasty different from that of its predecessors was that it survived long enough to preside over the country while it faced a quite new threat, one from a culture stronger than that of traditional China. For the first time in nearly two thousand years, Chinese society would have to change, not the imported culture of a new wave of barbarian conquerors. The Chinese Revolution was about to begin.

In the eighteenth century, no Chinese official could have been expected to discern this. When Lord Macartney arrived in 1793 to ask for equality of diplomatic representation and free trade the confidence of centuries was intact. The first western advances and encroachments had been successfully rebuffed or contained. The representative of George III could only take back polite but unyielding messages of refusal to what the Chinese emperor was pleased to call 'the lonely remoteness of your island, cut off from the world by intervening wastes of sea'. It can hardly have improved matters that George was also patted on the back for his 'submissive loyalty in sending this tribute mission' and encouraged to 'show even greater devotion and loyalty in future'.

The assumption of their own cultural and moral superiority was then as unquestioned a part of the mental world of the educated Chinese as it was of that of the European and American missionaries and philanthropists of the next century who unconsciously patronized the people they came to serve. Such language embodied the Chinese world view, in which all nations paid tribute to the emperor, possessed of the Mandate of Heaven, and in which China already had all the materials and skills for the highest civilization and would only waste her time and energy in indulging relations with Europe going beyond the limited trade tolerated at Canton (where by 1800 there were perhaps a thousand Europeans). Nor was this nonsense. Nearly three centuries of trade with China had failed to reveal any manufactured goods from Europe which the Chinese wanted except the mechanical toys and clocks which they found amusing. European trade with China rested on the export to her of silver or other Asian products.

As a British merchant concisely put it in the middle of the eighteenth century, the 'East India trade ... exports our bullion, spends little of our product or manufactures and brings in commodities perfectly manufactured which hinder the consumption of our own'.

Lord Macartney's reception by the emperor – as seen by an English caricaturist who reveals more of his countrymen's attitude than those of the Chinese.

At that time, official China still felt confident about her internal regime and cultural superiority, but in retrospect signs of future difficulties can be discerned. The secret societies and cults which kept alive a smouldering national resentment against a foreign dynasty and the central power, still survived and even prospered. They found fresh support as the surge of population became uncontainable; in the century before 1850 numbers seem to have more than doubled to reach about four-hundred and thirty million by 1850. Pressure on cultivated land became much more acute because the area worked could be increased only by a tiny margin; times grew steadily harder and the lot of the peasantry more and more miserable. There had been warning signs in the 1770s and 1780s when a century's internal peace was broken by great revolts such as those which had so often in the past been the sign of dynastic decline. Early in the next century they became more frequent and destructive. To make matters worse, they were accompanied by another economic deterioration, inflation in the price of the silver in which taxes had to be paid. Most daily transactions (including the payment of wages) were carried out in copper, so this added to the crushing burdens already suffered by the poor. Yet none of this seemed likely to be fatal, except, possibly, to the dynasty. It could all be fitted into the traditional pattern of the historic cycle. All that was required was that the service gentry should remain loyal, and even if they did not, then, though a collapse of government might follow, there was no reason to believe that in due course another dynasty would not emerge to preserve the imperial framework of an unchanging China.

It was not, in the end, to turn out like this because of the drive and power of the nineteenth-century barbarian challenge. The inflation itself was a result of a change of China's relations with the outside world which within a few decades made nonsense of the reception given to Macartney. Before 1800 the West could offer China little that she wanted except silver, but within the next three decades of the nineteenth century this ceased to be so, largely because British traders at last found a commodity the Chinese wanted and India could supply: opium. Naval expeditions forced the Chinese to open their country to sale (albeit at first under certain restrictions) of this drug, but the 'Opium War' which began in 1839 ended in 1842 with a treaty which registered a fundamental change in China's relations with the West. The Canton monopoly and the tributary status of the foreigner came to an end together. Once the British had kicked ajar the door to western trade, others were to follow them through it.

Unwittingly, the government of Queen Victoria thus launched the Chinese Revolution. The 1840s were the beginning of a period of upheaval which took over a century to come to completion. The revolution would slowly reveal itself as a double repudiation, both of the foreigner and of much of the Chinese past. The first would increasingly express itself in the nationalist modes and idioms of the progressive European world. Because such ideological forces could not be contained within the traditional framework, they would in the end prove fatal to it, when the Chinese sought to remove the obstacles to modernization and national power. More than a century after the Opium War the Chinese Revolution at last finally shattered for good a social system which had been the foundation of Chinese life for thousands of years. By that time, though, much of old China would already have vanished. By the time, too, it would appear that China's troubles had also been a part of a much wider upheaval, a Hundred Years' War of Asia and the West whose turning-point came in the early twentieth century.

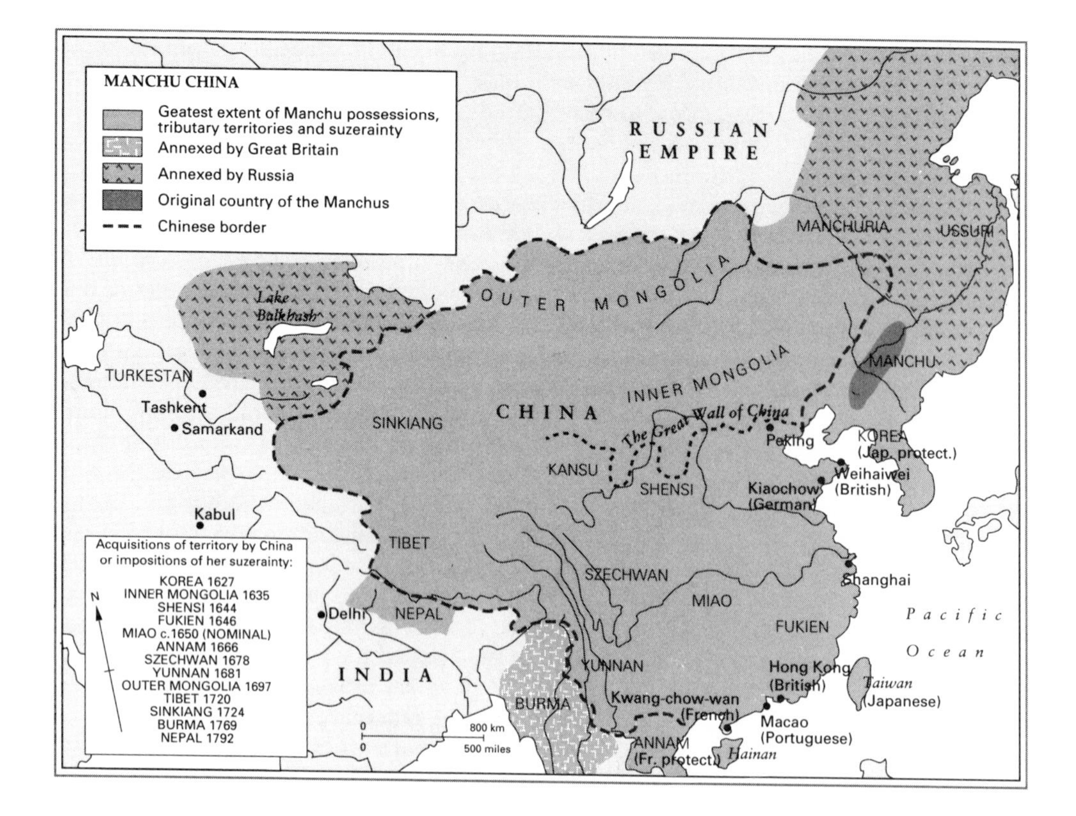

These implications matured only slowly. In the beginning, western encroachments in China usually produced only a simple, xenophobic hostility and even this was not universal. After all, for a long time very few Chinese were directly or obviously much concerned with the coming of the foreigners. A few (notably Canton merchants involved in the foreign trade) even sought accommodation with them. Hostility was a matter of anti-British mobs in the towns and of the rural gentry. At first many officials saw the problem as a limited one: that of the addiction of the subjects of the empire to a dangerous drug. They were humiliated, in particular, by the weaknesses which this revealed in their own people and administration; there was much connivance and corruption involved in the opium trade. They do not at first seem to have seen the deeper issue of the future, that of the questioning of an entire order, or sense a cultural threat; China had suffered defeats in the past and its culture had survived unscathed.

The first portent of a deeper danger came when, in the 1840s, the imperial government had to concede that missionary activity was legal. Though still limited, this was obviously corrosive of tradition. Officials in the Confucian mould who felt its danger stirred up popular feeling against missionaries – whose efforts made them easy targets – and there were scores of riots in the 1850s and 1860s. Such demonstrations often made things worse. Sometimes foreign consuls would be drawn in; exceptionally a gunboat would be sent. The Chinese government's prestige would suffer in the ensuing exchange of apologies and punishment of culprits. Meanwhile, the activity of the missionaries was steadily undermining the traditional society in more direct and didactic ways by preaching an individualism and egalitarianism alien to it and by acting as a magnet to converts to whom it offered economic and social advantages.

The process of undermining China also went forward directly by military and naval means; there were further impositions of concessions by force. But there was a growing ambiguity in the Chinese response. The authorities did not always resist the arrival of the foreigners. First the gentry of the areas immediately concerned and then the Peking government came round to feeling that foreign soldiers might not be without their value for the regime. Social disorder was growing. It could not be canalized solely against the foreigners and was threatening the establishment; China was beginning to undergo a cycle of peasant revolts which were to be the greatest in the whole of human history. In the middle decades of the century the familiar symptoms multiplied: banditry, secret societies. In the 1850s the 'Red Turbans' were suppressed only at great cost. Such troubles frightened the establishment, and threw it on to the defensive, leaving it with little spare capacity to resist the steady gnawing of the West. These great rebellions were fundamentally caused by hunger for land and the most important and distinctive of them was the Taiping rebellion or, as it may more appropriately be called, revolution, which lasted from 1850 to 1864.

The heart of this great convulsion, which cost the lives of more people than died the world over in the First World War, was a traditional peasant revolt. Hard times and a succession of natural disasters had helped to provoke it. It drew on a compound of land hunger, hatred of tax-gatherers, social envy and national resentment against the Manchus (though it is hard to see exactly what this meant in practice, for most of the officials who actually administered the empire were, of course, themselves Chinese). It was also a regional outbreak, originating in the south and even there promoted by an isolated minority of settlers from the north. The new feature behind the revolt and one which made it ambiguous in the eyes of both Chinese and Europeans was that its leader, Hung Hsiu-ch'uan, had a superficial but impressive acquaintance with the Christian religion in the form of American protestantism and this led him to denounce the worship of other gods, destroy idols – which included the Confucian ancestor tablets – and to talk of establishing the kingdom of God on earth. He had been rejected by his own culture, for he had been an unsuccessful candidate in the examinations which conferred status on low-born Chinese. Within the familiar framework of one of the periodic peasant upheavals of old China, that is to say, a new ideology was at work and showing itself subversive of the traditional culture and state. Some of its opponents at once grasped this and saw the Taiping movement as an ideological as well as a social challenge. Thus the revolution can be seen as an epoch in the Western disruption of China.

Soldiers of the 'ever-victorious' army raised and trained to fight the Taipings by the British general, Gordon, on behalf of the imperial government. Later, in 1883 at Khartoum in the Sudan, Gordon was to die what many of his countrymen regarded as a martyr's death.

The Taiping army at first had a series of spectacular successes. By 1853 they had captured Nanking and established there the court of Hung Hsiuch'uan, now proclaimed the 'Heavenly King'. In spite of alarm further north, this was as far as they went. After 1856 the revolution was on the defensive. Nevertheless, it announced important social changes (which were to make it a source of inspiration for later Chinese reformers) and although it is by no means clear how widely these were effective or even appealing, their disruptive ideological effect was considerable. The basis of Taiping society was communism: there was no private property but communal provision for general needs. The land was in theory distributed for working in plots graded by quality to provide just shares. Even more revolutionary was the extension of social and educational equality to women. The traditional binding of their feet was forbidden and a measure of sexual austerity marked the movement's aspirations (though not the conduct of the 'Heavenly King' himself). These things reflected the mixture of religious and social elements which lay at the root of the Taiping cult and the danger it presented to the traditional order.

The movement benefited at first from the demoralization brought about in the Manchu forces by their defeats at the hands of the Europeans and from the usual weaknesses shown by central government in China in a relatively distant and distinct region. As time passed and the Manchu forces were given abler (sometimes European) commanders, the bows and spears of the Taipings were proved insufficient. The foreigners, too, came to see the movement as a threat but kept up their pressure while the imperial government grappled with the Taipings. Treaties with France and the United States which followed that with Great Britain guaranteed the toleration of Christian missionaries and began the process of reserving jurisdiction over foreigners to consular and mixed courts.

The danger presented by the Taiping revolution brought yet more concessions: new fighting resulted in the opening of more Chinese ports to foreign trade, the introduction to the Chinese customs administration of foreign superiors, the legalization of the sale of opium and the cession to the Russians of the province in which Vladivostok was to be built. It is hardly surprising that in 1861 the Chinese decided for the first time to set up a new department to deal with foreign affairs. The old myth that all the world recognized the Mandate of Heaven and owed tribute to the imperial court was dead.

In the end, the foreigners joined in against the Taipings. Whether their help was needed to end it is hard to say; certainly the Taiping movement was already failing. In 1864 Hung died and shortly afterwards Nanking fell to the Manchu. This was a victory for traditional China: the rule of the bureaucratic gentry had survived one more threat from below. None the less, an important turning-point had been reached. A rebellion had offered a revolutionary programme announcing a new danger, that the old challenge of peasant rebellion might be reinforced by an ideology from outside deeply corrosive to Confucian China. Nor did the end of the Taiping rebellion mean internal peace; from the middle of the 1850s until well into the 1870s there were great Moslem risings in the north-west and south-west as well as other rebellions.

Immediately, China showed even greater weakness in the face of the Western barbarians. Large areas had been devastated in the fighting; in many of them the soldiers were powerful and threatened the control of the bureaucracy. If the enormous loss of life did something to reduce pressure on land, this was probably balanced by a decline in the prestige and authority of the dynasty. Concessions had already had to be made to the Western powers under and because of these disadvantaged conditions; between 1856 and 1860 British and France forces were engaged every year against the Chinese. A treaty in 1861 brought to nineteen the number of 'treaty ports' open to Western merchants and provided for a permanent British ambassador at Peking. Meanwhile, the Russians exploited the Anglo-French successes to secure the opening of their entire border with china to trade. Further concessions would follow. It was evident that methods which had drawn the sting of nomadic invaders were not likely to work with confident Europeans whose ideological assurance and increasing technical superiority protected them from the seduction of Chinese civilization. When Roman Catholic missionaries were given the right to buy land and put up buildings Christianity was linked to economic penetration, soon, trying to protect converts meant involvement in the internal affairs of public order and police. It was impossible to contain the slow but continuous erosion of Chinese sovereignty. Never formally a colony, China was beginning none the less to undergo a measure of colonization.

Then there were territorial losses as the century wore on. In the 1870s the Russians seized the Ili valley (though they later handed much of it back) and in the next decade the French established a protectorate in Annam. Loosely asserted but ancient Chinese suzerainty was being swept away; the French began to absorb Indo-China and the British annexed Burma in 1886. The worst blow came from another Asiatic state; in the war of 1894–5 the Japanese took Formosa and the Pescadores, while China had to recognize the independence of Korea, from which they had received tribute since the seventeenth century. Following the Japanese success came further encroachments by other powers; provoked by the Russians, who established themselves in Port Arthur, England, France and Germany all extracted long leases of ports at the end of the century. Before this, the Portuguese, who had been in China longer than any other Europeans, converted their tenure of Macao into outright ownership. Even the Italians were in the market, though they did not actually get anything. And long before this, concessions, loans and agreements had been exacted by Western powers to protect and foster their own economic and financial interests. It is hardly surprising that when a British prime minister spoke at the end of the century of two classes of nation, the 'living and the dying', China was regarded as an outstanding example of the second. Statesmen began to envisage her partition.

Before the end of the nineteenth century it was evident to many Chinese intellectuals and civil servants that the traditional order would not generate the energy

Pu-Yi, last emperor of China. In 1912, a year after this picture was taken, he abdicated and twenty-five centuries of imperial rule came to an end. Pu-Yi was once more restored to the Chinese throne for a few days in 1917 and then again reigned - nominally – as a puppet emperor of Manchukuo on behalf of the Japanese from 1934 to 1945. Five years in Russian hands were then followed by political 're-education' by the Chinese communists, from which the former emperor emerged to take a job as an archivist in 1959. In 1967 he died aged sixty-one.

necessary to resist the new barbarians. Attempts along the old lines had failed. New tendencies began to appear. A 'society for the study of self-strengthening' was founded to consider Western ideas and inventions which might be helpful. Its leaders cited the achievements of Peter the Great and, more significantly, those of contemporary reformers in Japan, an example all the more telling because of the superiority shown by the Japanese over China in war in 1895. Yet the would-be reformers still hoped that they would be able to root change in the Confucian tradition, albeit one purified and invigorated. They were members of the gentry and they succeeded in obtaining the ear of the emperor; they were thus working within the traditional framework and machinery of power to obtain administrative and technological reform without compromising the fundamentals of Chinese culture and ideology.

Unfortunately this meant that the Hundred Days of Reform of 1898 (as it came to be known) was almost at once tangled up in the court politics of the rivalry between the emperor and the dowager empress, to say nothing of Chinese-Manchu antagonism. Though a stream of reform edicts was published, they were swiftly overtaken by a *coup d'état* by the empress, who locked up the emperor. The basic cause of the reformers' failure was the provocation offered by their inept political behaviour. Yet although the movement had failed, it was important that it had existed at all. It was to be a great stimulus to wider and deeper thinking about China's future.

For the moment, China seemed to have turned back to older methods of confronting the threat from outside, as a dramatic episode, the 'Boxer movement', showed. Exploited by the empress, this was essentially a backward-looking and xenophobic upheaval which was given official encouragement. Missionaries and converts were murdered, a German minister killed and the foreign legations at Peking besieged, but the Boxers were unable to stand up to Western fighting-power. They once more revealed the hatred of foreigners which was waiting to be tapped, but also showed how little could be hoped for from the old structure, for its most conservative forces had dominated the movement, not the few reformers who became involved in it. It was suppressed by a military intervention which provides the only example in history of the armed forces of all the great powers operating under the same commander (a German, as it happened). The sequel was yet another diplomatic humiliation for China; an enormous indemnity was settled on customs henceforth under foreign direction.

The ending of the Boxer movement left China still more unstable. Reform had failed in 1898; so now had reaction. Perhaps only revolution lay ahead. Officers in the parts of the army which had undergone reorganization and training on Western lines began to think about it. Students in exile had already begun to meet and discuss their country's future, above all in Tokyo. The Japanese were happy to encourage subversive movements which might weaken their neighbour; in 1898 they had set up an 'East Asian Cultural Union' from which emerged the slogan 'Asia for the Asians'. The Japanese had great prestige in the eyes of the young Chinese radicals as Asians who were escaping from the trap of traditional backwardness which had been fatal to India and seemed to be about to engulf China. Japan could confront the West on terms of equality. Other students looked elsewhere for support, some to the long-enduring secret societies. One of them was a young man called Sun Yat-sen. His achievement has often been exaggerated, but, nevertheless, he attempted revolution ten times altogether. In the 1890s, he and others were asking only for a constitutional monarchy, but it was a very radical demand at that time.

Discontented exiles drew on support from Chinese businessmen abroad, of whom there were many, for the Chinese had always been great traders. They helped Sun Yat-sen to form in 1905 in Japan a revolutionary alliance aiming at the expulsion of the Manchus and the initiation of Chinese rule, a republican constitution, land reform. It sought to conciliate the foreigners, at this stage a wise tactical move, and the new programme showed the influence of Western thinkers (notably that of the English radical John Stuart Mill and the American economic reformer Henry George); once again the West provided the stimulus and some of the ideological baggage of a Chinese reform movement. This was the beginning of the party eventually to emerge as the dominant clique in the Chinese Republic.

Its formation, though, may well be thought less significant than another event of the same year, the abolition of the traditional examination system. More than any other institution, the examination system had continued to hold Chinese civilization

THE SAME OLD BEAR.

Russian Bear (to British Lion). "YOU 'VE GOT SO MUCH TO DO ELSEWHERE, *I'LL* TACKLE THIS OBSTREPEROUS PARTY."
British Lion. "OH, THANKS! BUT I WOULDN'T LEAVE YOU ALONE WITH HIM FOR WORLDS!"

Every intervention by a foreign power in China had implications for other powers who might find the balance of their own interests threatened. The anti-Boxer campaign attracted wide support for just this reason, as Punch *noted.*

together by providing the bureaucracy it recruited with its internal homogeneity and cohesion. These would not quickly wane, but the distinction between the mass of Chinese subjects and the privileged ruling class was now gone. Meanwhile, returning students from abroad, dissatisfied with what they found and no longer under the necessity of accommodating themselves to it by going through the examination procedure if they wished to enter government service, exercised a profoundly disturbing influence. They much increased the rate at which Chinese society began to be irradiated by Western ideas. Together with the soldiers in a modernized army, more and more of them looked to revolution for a way ahead.

There were a number of rebellions (some directed by Sun Yat-sen from Indo-China with French connivance) before the empress and her puppet emperor died on successive days in 1908. The event raised new hopes but the Manchu government continued to drag its feet over reform. On the one hand it made important concessions of principle and promoted the flow of students abroad; on the other it showed that it could not achieve a decisive break with the past or surrender any of the imperial privileges of the Manchus. Perhaps more could not have been asked for. By 1911, the situation had deteriorated badly. The gentry class showed signs of losing its cohesion: it was no longer to back the dynasty in the face of subversion as it had done in the past. Governmentally, there existed a near-stalemate of internal power, the dynasty effectively controlling only a part of China. In October a revolutionary headquarters was discovered at Hankow. There had already been revolts which had been more or less contained earlier in the year. This precipitated one which at last turned into a successful revolution. Sun Yat-Sen, whose name was used by the early rebels, was in the United States at the time and was taken by surprise.

Yuan Shih-k'ai at his inauguration as president in 1912.

The course of the revolution was decided by the defection from the regime of its military commanders. The most important of these was Yuan Shih-k'ai; when he turned on the Manchus, the dynasty was lost. The Mandate of Heaven had been withdrawn until on 12 February 1912 the last and six-year-old Manchu emperor abdicated. A republic had already been proclaimed, with Sun Yat-sen its president, and a new nationalist party soon appeared behind him. In March he resigned the presidency to Yüan Shihk'ai; thus acknowledging where power really lay in the new Republic and inaugurating a new phase of Chinese government, in which an ineffective constitutional regime at Peking disputed the practical government of China by warlords. This alone meant that China had still a long way to travel before she would be a modern nation-state. None the less, she had begun the half-century's march which would recover for her an independence lost in the nineteenth century to foreigners.

At the beginning of the nineteenth century, there was little to show a superficial observer that Japan might adapt more successfully than China to challenges from the West. She was to all appearances deeply conservative. Yet much had already changed since the establishment of the shogunate and there were signs that the changes would cut deeper and faster as the years went by. One paradox is that this is in part attributable to the success of the Tokugawa era itself. It had brought peace. An obvious result was that Japan's military system became old-fashioned and inefficient. The samurai themselves were evidently a parasitic class; warriors, there was nothing for them to do except to cluster in the castle-towns of their lords, consumers without employment, a social and economic problem. The prolonged peace also led to the surge of growth which was the most profound consequence of the Tokujawa era. Japan was already a semi-developed, diversifying society, with a money economy, the beginnings of a quasi-capitalist structure in agriculture which eroded the old feudal relationships, and a growing urban population. Osaka, the greatest mercantile centre, had between three and four hundred thousand inhabitants in the last years of the shogunate. Edo may have had a million. These great centres of consumption were sustained by financial and mercantile arrangements which had grown enormously in scale and complication since the seventeenth century. They made a mockery of the old notion of the inferiority of the merchant order. Even their techniques of salesmanship were modern; the eighteenth-century house of Mitsui (still today one of the great pillars of Japanese capitalism) gave free umbrellas decorated with their trademark to customers caught in their shops by the rain.

Many of these changes registered the creation of new wealth from which the shogunate had not itself benefited, largely because it was unable to tap it at a rate which kept pace with its own growing needs. The main revenue was the rice tax which flowed through the lords, and the rate at which the tax was levied remained fixed at the level of a seventeenth-century assessment. Taxation therefore did not take away the new wealth arising from better cultivation and land reclamation and, because this remained in the hands of the better-off peasants and village leaders, this led to sharpening contrasts in the countryside. The poorer peasantry was often driven to the labour markets of the towns. This was another sign of disintegration in the feudal society. In the towns, which suffered from an inflation made worse by the shogunate's debasement of the coinage, only the merchants seemed to prosper. A last effort of economic reform failed in the 1840s. The lords grew poorer and their retainers lost confidence; before the end of the Tokugawa, some samurai were beginning to dabble in trade. Their share of their lord's tax yield was still only that of their seventeenth-century predecessors; everywhere could be found impoverished, politically discontented swordsmen – and some aggrieved families of great lords who recalled the days when their race had stood on equal terms with the Tokugawa.

The obvious danger of this potential instability was all the greater because insulation against Western ideas had long since ceased to be complete. A few learned men had interested themselves in books which entered Japan through the narrow aperture of the Dutch trade. Japan was very different from China in its technical receptivity. 'The Japanese are sharp-witted and quickly learn anything they see', said a sixteenth-century Dutchman. They had soon grasped and exploited as the Chinese never did the advantages of European firearms, and began to make them in quantity. They copied the European clocks, which the Chinese treated as toys. They were eager to learn from Europeans, as unhampered by their traditions as the Chinese seemed bogged down in theirs. On the great fiefs there were notable schools or research centres of 'Dutch studies'. The shogunate itself had authorized the translation of foreign books, an important step in so literate a society, for education in Tokugawa Japan had been almost too successful: even young samurai were beginning to enquire about Western ideas. The islands were relatively small and communications good, so that new ideas got about easily. Thus, Japan's posture when she suddenly had to face a new and unprecedented challenge from the West was less disadvantageous than that of China.

The first period of Western contact with Japan had ended in the seventeenth century, with the exclusion of all but a few Dutchmen allowed to conduct trade from an island at Nagasaki. Europeans had not then been able to challenge this outcome. That this was not likely to continue to be the case was shown in the 1840s by what happened to China, whose fate some of Japan's rulers observed with increasing alarm. The Europeans and North Americans seemed to have both a new interest in breaking into Asian trade and new and irresistible strength to do it. The King of the Netherlands warned the shogun that exclusion was no longer a realistic policy. But there was no agreement among Japan's rulers about whether resistance or concession was the better. Finally, in 1851 the President of the United States commissioned a naval officer, Commodore Perry, to open relations with Japan. Under him, the first foreign squadron to sail into Japanese waters entered Edo Bay in 1853. In the following year it returned and the first of a series of treaties with foreign powers was made by the shogunate.

Perry's arrival could be seen in Confucian terms as an omen that the end of the shogunate was near. No doubt some Japanese saw it in that way. Yet this did not at once follow and there were a few years of somewhat muddled response to the barbarian threat. Japan's rulers did not straightway come round to a wholehearted policy of concession (there was one further attempt to expel foreigners by force) and Japan's future course was not set until well into the 1860s. Within a few years the success of the West was none the less embodied in and symbolized by a series of so-called 'unequal

The Japanese capture of Tientsin in the Sino-Japanese war of 1894 celebrated in a contemporary Japanese print.

treaties'. Commercial privileges, extra-territoriality for Western residents, the presence of diplomatic representatives and restrictions on the Japanese export of opium were the main concessions won by the United States, Great Britain, France, Russia and the Netherlands. Soon afterwards the shogunate came to an end; its inability to resist the foreigner was one contributing factor and another was the threat from two great aggregations of feudal power which had already begun to adopt Western military techniques in order to replace the Tokugawa by a more effective and centralized system under their control. There was fighting between the Tokugawa and their opponents, but it was followed not by a relapse into disorder and anarchy but by a resumption of power by the imperial court and administration in 1868 in the so-called 'Meiji Restoration'.

The re-emergence of the emperor and the wide-spread acceptance of the revolutionary renewal which followed was attributable above all to the passionate desire of most literate Japanese to escape from a 'shameful inferiority' to the West which might have led them to share the fate of the Chinese and Indians. In the 1860s both the *bakufu* and some individual clans had already sent several missions to Europe. Anti-foreign agitation was dropped in order to learn from the West the secrets of its strength. There was a paradox in this. As in some European countries, a nationalism rooted in a conservative view of society was to dissolve much of the tradition it was developed to defend.

The transference of the court to Edo was the symbolic opening of the Meiji 'Restoration' and the regeneration of Japan; its indispensable first stage was the abolition of feudalism. What might have been a difficult and bloody business was made simple by the voluntary surrender to the emperor of their lands by the four greatest clans, who set out their motives in a memorial they addressed to the emperor. They were returning to the emperor what had originally been his, they said, 'so that a uniform rule may prevail throughout the empire. Thus the country will be able to rank equally with the other nations of the world.' This was a concise expression of the patriotic ethic which was to inspire Japan's leaders for the next half-century and was widely spread in a country with a large degree of literacy where local leaders could make possible the acceptance of national goals to a degree impossible elsewhere. True, such expressions were not uncommon in other countries. What was peculiar to Japan was the urgency which observation of the fate of China lent to the programme, the emotional support given to the idea by Japanese social and moral tradition, and the fact that in the imperial throne there was available within the established structure a source of moral authority not committed merely to maintaining the past. These conditions made possible a Japanese 1688: a conservative revolution opening the way to radical change.

The first railway in Japan ran from Tokyo to Yokohama. One of its very European-looking stations appears in this print.

Rapidly, Japan adopted many of the institutions of Western government and Western society. A prefectorial system of administration, posts, a daily newspaper, a ministry of education, military conscription, the first railway, religious toleration and the Gregorian calendar all arrived within the first five years. A representative system of local government was inaugurated in 1879 and ten years later a new constitution set up a bi-cameral Parliament (a peerage had already been created in preparation for the organization of the upper house). In fact, this was somewhat less revolutionary than might appear, given the strong authoritarian strain in the document. At about the same time, too, the innovatory passion was beginning to show signs of flagging; the period when things Western were a craze was over; no such enthusiasm was to be seen again until the second half of the twentieth century. In 1890 an imperial Rescript on Education, subsequently to be read on great days to generations of Japanese school-children, enjoined the observation

RELIQUARY *Byzantine art was almost purely religious and ecclesiastical in mode. Reliquaries were often, like this one now in the cathedral treasury of Limburg, decorated with extraordinary richness. They evoked a theme almost literally at the very roots of Byzantine civilization, for when Constantine's city was dedicated in 330 there were buried at the site not only relics of saints, but, allegedly, crumbs from the bread with which Christ fed the five thousand, the crosses of the thieves who died with Him, the adze with which Noah built the Ark and even the palladium which Aeneas took from Troy to Italy to found Rome.*

MOZARABIC ART *The eighth-century* Commentary on the Apocalypse *by the monk Beatus of Liebana became one of the most popular works of scholarship in Spain and was much copied and re-copied. It became the vehicle for some of the finest expressions of the art which came to be called 'Mozarabic' – over twenty illuminated manuscripts of the work survive. They were written and illustrated in a curiously attractive and eclectic style; this eleventh-century example, a depiction of Babylon in flames, brings together Moorish arches and Visigothic ornament in its image of the city.*

of the traditional Confucian duties of filial piety and obedience and the sacrifice of self to the state if need be.

Much – perhaps the most important part – of old Japan was to survive the Meiji revolution and was to do so very obviously; this is in part the secret of modern Japan. But much, too, had gone. Feudalism could never be restored, generously compensated with government stock though the lords might be. Another striking expression of the new direction was the abolition of the old ordered class system. Care was shown in removing the privileges of the *samurai*; some of them could find compensation in the opportunities offered to them by the new bureaucracy, in business – no longer to be a demeaning activity – and in the modernized army and navy. For these foreign instruction was sought, because the Japanese sought proven excellence. Gradually they dropped their French military advisers and took to employing Germans after the Franco-Prussian War; the British provided instructors for the navy. Young Japanese were sent abroad to learn at first hand other secrets of the wonderful and threatening puissance of the West. It is still hard not to be moved by the ardour of many of these young men and of their elders and impossible not to be impressed by their achievement, which went far beyond Japan and their own time. The *shishi* (as some of the most passionate and dedicated activists of reform were called) later inspired national leaders right across Asia, from India to China. Their spirit was still at work in the young officers of the 1930s who were to launch the last and most destructive wave of Japanese imperialism.

The crudest indexes of the success of the reformers are the economic, but they are very striking. They built on the economic effects of the Tokugawa peace. It was not only the borrowing of Western technology and expertise which ensured the release in Japan of a current of growth achieved by no other non-Western state. The country was lucky in being already well-supplied with entrepreneurs who took for granted the profit motive and it was undoubtedly richer than, say, China. Some of the explanation of the great leap forward by Japan lay also in the overcoming of inflation and the liquidation of feudal restraints which had made it hard to tap Japan's full potential. The first sign of change was a further increase in agricultural production, little though the peasants, who made up four-fifths of the population in 1868, benefited from it. Japan managed to feed a growing population in the nineteenth century by bringing more land under cultivation for rice and by cultivating existing fields more intensively. Though the dependence on the land tax lessened as a bigger portion of the revenue could be found from other sources, it was still upon the peasant that the cost of the new Japan fell most heavily. As late as 1941, Japanese farmers reaped few of the gains from modernization. Relatively they had fallen behind; their ancestors a century earlier had a life expectancy and income approximating to that of their British equivalents, and even by 1900 this was far from true of their successors. There were few non-agricultural resources. It was the increasingly productive tax on land which paid for investment. Consumption remained low, though there was not the suffering of, say, the later industrialization process of Stalin's Russia. A high rate of saving (12 per cent in 1900) spared Japan dependence on foreign loans but, again, restricted consumption. This was the other side of the balance sheet of expansion whose credit entries were clear enough: the infrastructure of a modern state, an indigenous arms industry, a usually high credit rating in the eyes of foreign investors and a big expansion of cotton-spinning and other textile manufacturing, by 1914.

In the end a heavy spiritual cost had to be paid for these successes. Even while seeking to learn from the West, Japan turned inward. The 'foreign' religious influences of Confucianism and even, at first, Buddhism, were attacked by the upholders of the state Shintoist cult, which, already under the shogunate had begun to stress and enhance the role of the emperor as the embodiment of the divine. The demands of loyalty to the emperor as the focus of the nation came to over-ride the principles embodied in the new constitution which might have been developed in liberal directions in a different cultural setting. The quality of the regime did not express itself in such institutions, but in the repressive actions of the imperial police. Yet it is difficult to see how an authoritarian emphasis could, in fact, have been avoided, given the two great tasks facing the statesmen of the Meiji Restoration. The modernization of the economy meant not planning in the modern sense but a strong governmental initiative and harsh fiscal policies. The other problem was order. The imperial power had once before gone into eclipse because of its failure to meet the threat on this front and now there were new dangers, because not all conservatives could be reconciled to the new model Japan. Discontented *ronin* – rootless *samurai* without masters – were one source of trouble. Another was peasant misery; there were scores of agrarian revolts n the first decade of the Meiji era. The Satsuma rebellion of 1877 showed that the government's new conscript forces could handle popular disorder. It was the last of several rebellions against the Restoration and the last great challenge from conservatism.

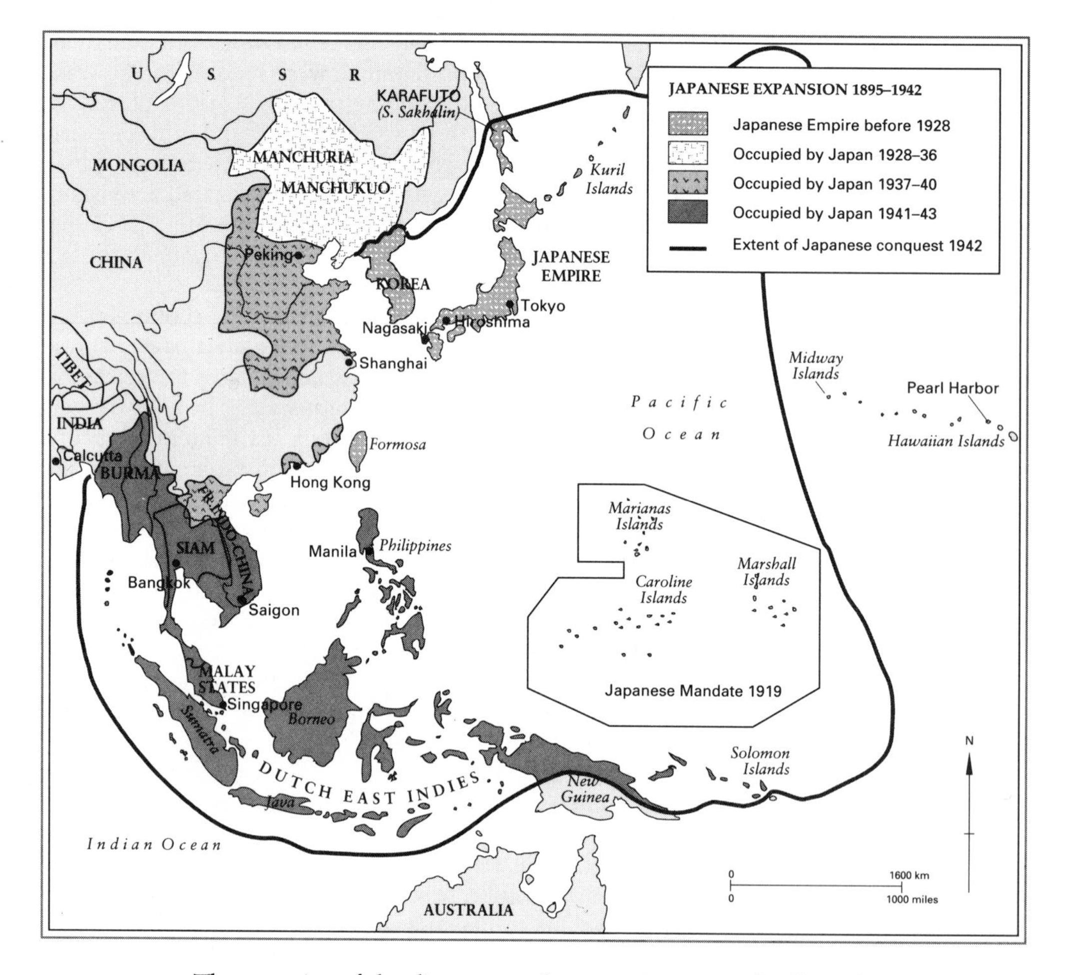

The energies of the discontented *samurai* were gradually to be siphoned off into the service of the new state, but this did not mean that the implications for Japan were all beneficial. They intensified in certain key sectors of the national life an assertive nationalism which was to lead eventually to aggression abroad. Immediately, this was likely to find expression not only in resentment of the West but also in imperial ambitions directed towards the nearby Asian mainland. Modernization at home and adventure abroad were often in tension in Japan after the Meiji Restoration, but in the long run they pulled in the same direction. The popular and democratic movements especially felt the tug of imperialism.

China was the predestined victim and was to be served as harshly by her fellow-Asians as by any of the Western states. At first she was threatened only indirectly. Just as China's supremacy over the dependencies on her borders was challenged in Tibet, Indo-China and Manchuria by Europeans, so the Japanese threatened it in the ancient empire of Korea, long a tributary of Peking. Japanese interests there went back a long way. In part this was strategic; the Tsushima straits were the place where the mainland was nearest. In 1876 an overt move was made; under the threat of military and naval action which was very much like that used on China by the Europeans – and on Japan by Perry – the Koreans agreed to open three of their ports to the Japanese and to exchange diplomatic representation. Some Japanese already wanted more. They remembered earlier Japanese invasions and successful piracy on the Korean coast and pointed to the mineral and natural wealth of Korea. The statesmen of the Restoration did not at once give way to such pressure, but in a sense they were only making haste slowly. In the 1890s another step forward was taken which led Japan into her first major war since the Restoration, and it was against China. It was sweepingly successful but was followed by an appalling national humiliation, when in 1895 a group of Western powers forced Japan to accept a peace treaty much less advantageous than the one she had imposed on the Chinese (which had included a declaration of Korea's independence).

At this point resentment of the West fused with enthusiasm for expansion in Asia. Popular dislike of the 'unequal treaties' had been running high and the 1895 disappointment brought it to a head. The Japanese government had its own interests in backing Chinese revolutionary movements and now it had a slogan to offer them: 'Asia for the Asians'. It was becoming clear, too, to the western powers that dealing with Japan was a very different matter from bullying China. Japan was increasingly recognized to be a 'civilized' state, not to be treated like other non-European nations. One symbol of the change was the ending in 1899 of one humiliating sign of European predominance, extra-territoriality. Then, in 1902, came the clearest acknowledgment of Japan's acceptance as an equal by the West, an Anglo-Japanese alliance. Japan, it was said, had joined Europe.

A parliament house of what had become the standard European layout was built for the new Japanese parliament, here being opened in 1891 by the emperor, seated at the right. Not only have the members been made to look very Europeanized in this print, but the ladies in the gallery, too, wear European dress.

Russia was at that moment the leading European power in the Far East. In 1895 her role had been decisive; her subsequent advance made it clear to the Japanese that the longed-for prize of Korea might elude them if they delayed. Railway-building, the development of Vladivostock, and Russian commercial activity in Korea – where politics was little more than the struggle of pro-Russian and pro-Japanese factions –

were alarming. Most serious of all, the Russians had leased the naval base of Port Arthur from the enfeebled Chinese. In 1904 the Japanese struck. The result, after a year of war in Manchuria, was a humiliating defeat for the Russians. It was the end of tsarist pretensions in Korea and South Manchuria, where Japanese influence was henceforth dominant, and other territories passed into Japanese possession to remain there until 1945. But there was more to the Japanese victory than that. For the first time since the Middle Ages, non-Europeans had defeated a European power in a major war. The reverberations and repercussions were colossal.

The formal annexation of Korea by Japan in 1910, together with the Chinese Revolution of the following year and the end of Manchu rule, can now be seen as a milestone, the end of the first phase of Asia's response to the West, and as a turning-point. Asians had shown very differing reactions to Western challenges. One of the two states which were to be the great Asian powers of the second half of the century was Japan, and she had inoculated herself against the threat from the West by accepting the virus of modernization. The other, China, had long striven not to do so.

In each case, the West provided both direct and indirect stimulus to upheaval, though in one case it was successfully contained and in the other it was not. In each case, too, the fate of the Asian power was shaped not only by its own response, but by the relations of the Western powers among themselves. Their rivalries had generated the scramble in China which had so alarmed and tempted the Japanese. The Anglo-Japanese alliance assured them that they could strike at their great enemy, Russia, and find her unsupported. A few years more and Japan and China would both be participants as formal equals with other powers in the First World War. Meanwhile, Japan's example and, above all, its victory over Russia, were an inspiration to other Asians, the greatest single reason for them to ponder whether European rule was bound to be their lot. In 1905 an American scholar could already speak of the Japanese as the 'peers of western peoples'; what they had done, by turning Europe's skills and ideas against her, might not other Asians do in their turn?

Everywhere in Asia European agencies launched or helped to launch changes which speeded up the crumbling of Europe's political hegemony. They had brought with them ideas about nationalism and humanitarianism, the Christian missionary's dislocation of local society and belief, and a new exploitation not sanctioned by tradition; all of which helped to ignite political, economic and social change. Primitive, almost blind, responses like the Indian Mutiny or Boxer rebellion were the first and obvious outcome, but there were others which had a much more important future ahead. In particular, this was true in India, the biggest and most important of all colonial territories.

In 1877 Parliament had bestowed the title of 'Empress of India' upon Queen Victoria; some Englishmen laughed and a few disapproved, but it does not seem that there were many who did either. Most took the British supremacy there to be permanent or near-permanent and were not much concerned about names. They would have agreed with their compatriot who said 'we are not in India to be pleasant' and held that only a severe and firm government could be sure to prevent another Mutiny. Others would also have agreed with the British viceroy who declared as the twentieth century began that 'As long as we rule India, we are the greatest power in the world. If we lose it, we shall drop straightaway to a third-rate power.' Two important truths underlay this assertion. One was that the Indian taxpayer paid for the defence of much or most of the British empire; Indian troops had been used to sustain it from Malta to China and in the subcontinent there was always a strategical reserve. The second was that Indian tariff policy was subordinated to British commercial and industrial realities.

These were the harsh facts, whose weight was harder and harder to ignore. Yet they were not the whole story of the Raj. There was more to it than just fear, greed, cynicism or the love of power. Human beings do not find it easy to pursue collective purposes without some sort of myth to justify them; nor did the British in India. Some of them saw themselves as the heirs of the Romans whom a classical education taught them to admire, stoically bearing the burden of a lonely life in an alien land to bring peace to the warring and law to peoples without it. Others saw in Christianity a precious gift with which they must destroy idols and cleanse evil custom. Some never formulated such clear views but were simply convinced that what they brought was better than what they found and therefore what they were doing was good. At the basis of all these views there was a conviction of superiority and there was nothing surprising about this; it had always animated some imperialists. But in the later nineteenth century it was especially reinforced by fashionable racialist ideas and a muddled reflexion of what was thought to be taught by current biological science about the survival of the fittest. Such ideas provided another rationale for the much greater social separation of the British in India from native Indians after the shock of the Mutiny. Although there was a modest intake of nominated Indian landlords and native rulers into the legislative branch of government it was not until the very end of the century that these were joined by elected Indians. Moreover, though Indians could compete to enter the civil service, there were important practical obstacles in the way of their entry to the ranks of the decision-makers. In the army, too, Indians were kept out of the senior commissioned ranks.

The largest single part of the British army was always stationed in India, where its reliability and monopoly of artillery combined with the officering of the Indian regiments by Europeans of ensure that there would be no repetition of the Mutiny. The coming of railways, telegraphs and more advanced weapons in any case told in favour of government in India as much as in any European country. But armed force was not the explanation of the self-assuredness of British rule, any more than was a conviction of racial superiority. The Census Report of 1901 recorded that there were just under three-hundred million Indians. These were governed by about 900 white civil servants. Usually there was about one British soldier for every four thousand Indians. As an Englishman once put it, picturesquely, had all the Indians chosen to spit at the same moment, his countrymen would have been drowned.

The Raj rested also on carefully administered policies. One assumption underlying them after the Mutiny was that Indian society should be interfered with as little as possible. Female infanticide, since it was murder, was forbidden, but there was to be no attempt to prohibit polygamy or child marriage (though after 1891 it was not legal for a marriage to be consummated until the wife was twelve years old). The line of the law was to run outside what was sanctioned by Hindu religion. This conservatism was reflected in a new attitude towards the native Indian rulers. The Mutiny had shown that they were usually loyal; those who turned against the government had been provoked by resentment against British annexation of their lands. Their rights were therefore scrupulously respected after the Mutiny; the princes ruled their own states independently and virtually irresponsibly, checked only by their awe of the British political officers resident at their courts. The native states included over a fifth of the population. Elsewhere, the British cultivated the native aristocracy and the landlords. This was part of a search for support from key groups of Indians, but often led the British to lean on those whose own leadership powers were already being undermined by social change. Enlightened despotism at their expense, but in the interests of the peasantry (such as had been shown earlier in the century) none the less now disappeared. These were all some of the unhappy consequences of the Mutiny.

Yet no more than any other imperial government was the Raj able permanently to ensure itself against change. Its very success told against it. The suppression of warfare favoured the growth of population – and one consequence was more frequent famine. But the provision of ways of earning a living other than by agriculture (which was a possible outlet from the problem of an over-populated countryside) was made very difficult by the obstacles in the way of Indian industrialization. These arose in large measure from a tariff policy in the interest of British manufactures. A slowly emerging class of Indian industrialists did not, therefore, feel warmly towards government, but were antagonized by it. Their ranks also included many of the growing number of Indians who had received an education along English lines and had subsequently been irritated to compare its precepts with the practice of the British community in India. Some who had gone to England to study at Oxford, Cambridge or the Inns of Court found the contrast especially galling: in England there were even Indian members of parliament in the nineteenth century, while an Indian graduate in India might be slighted by a British private soldier, and there was an outcry from British residents when, in the 1880s, a viceroy wished to remove the 'invidious distinction' which prevented a European from being brought before an Indian magistrate. Some, too, had pondered what they read in their studies; John Stuart Mill and Mazzini were thus to have a huge influence in India, and, through its leaders, in the rest of Asia.

Resentment was especially felt among the Hindus of Bengal, the historic centre of British power: Calcutta was the capital of India. In 1905 this province was divided in two. This partition was an important landmark, for it for the first time brought the Raj into serious conflict with something which had not existed in 1857, the Indian nationalist movement.

At every stage this movement had been fed and stimulated by non-Indian forces. British orientalists, at the beginning of the nineteenth century, had begun the rediscovery of classical Indian culture which was essential both to the self-respect of Hindu nationalism and the overcoming of the subcontinent's huge divisions. Indian scholars began to bring to light, under European guidance, the culture and religion embedded in the neglected Sanskrit scriptures; through these they could formulate a conception of a Hinduism far removed from the rich and fantastic, but also superstitious, accretions of its popular form. By the end of the nineteenth century this recovery of the Aryan and Vedic past – Islamic India was virtually disregarded – had gone far enough for Hindus to meet with confidence the reproaches of Christian missionaries and offer a cultural counter-attack; a Hindu emissary to a 'Parliament of Religions' in Chicago in 1893 not only awoke great personal esteem and obtained serious attention for his assertion that Hinduism was a great religion capable of revivifying the spiritual life of other cultures, but actually made converts.

The Hindu national consciousness, like the political activity it was to reinforce, was for a long time confined to a few. The proposal that Hindi should be India's language seemed wildly unrealistic when the hundreds of languages and dialects which fragmented Indian society were considered and could only appeal to an élite seeking to strengthen its links across a subcontinent. The definition of its membership was education rather than wealth: its backbone was provided by those Hindus, often Bengali, who felt especially disappointed at the failure of their educational attainments to win them an appropriate share in the running of India. The Raj seemed determined to maintain the racial predominance of Europeans and to rely upon such conservative interests as the princes and landlords, to the exclusion and, possibly even more important, the humiliation of the *babu*, the educated, middle-class, urban Hindu.

A new cultural self-respect and a growing sense of grievance over rewards and slights were the background to the formation of the Indian National Congress. The immediate prelude was a flurry of excitement over government proposals, subsequently modified because of the outcry of European residents, to equalize the treatment of Indians and Europeans in the courts. Disappointment caused an Englishman, a former civil servant, to take the steps which led to the first conference of the Indian National Congress in Bombay in December 1885. Vice-regal initiatives, too, had played a part in this, and Europeans were long to be prominent in the administration of Congress. Even longer would they patronize it with protection and advice in London. It was an appropriate symbol of the complexity of the European impact on India that some Indian delegates attended in European dress, improbably attired in morning-suits and top-hats of comical unsuitability to the climate of their country, but the formal attire of its rulers.

Congress was soon committed by its declaration of principles to national unity and regeneration: as in Japan already and China and many other countries later, this was the classical product of the impact of European ideas. But it did not at first aspire to self-government. Congress sought, rather, to provide a means of communicating Indian views to the viceroy and proclaimed its 'unswerving loyalty' to the British Crown. Only after twenty years, in which much more extreme nationalist views had won adherents among Hindus, did it begin to discuss the possibility of independence. During this time its attitude had been soured and stiffened by the vilification it received from British residents who declared it unrepresentative, and the unresponsiveness of an administration which endorsed this view and preferred to work through more traditional and conservative social forces. Extremists became more insistent. In 1904 came the inspiring victories of Japan over Russia. The issue for a clash was provided in 1905 by the administrative division of Bengal.

Indian India: a young prince and his advisers. A British political officer was stationed at each of the courts of the native Indian states, whose internal administration was left in their own hands – unless intervention became necessary because of intolerable inadequacy or potential danger to the Raj.

The purpose of the partition was twofold: it was administratively convenient and it would possibly weaken Bengal's nationalism. It produced a West Bengal where there was a Hindu majority, and an East Bengal with a Moslem majority. This detonated a mass of explosive materials long accumulating. Immediately, there was a struggle for power in Congress. At first a split was avoided by agreement on the aim of *swaraj*, which in practice might mean independent self-government such as that enjoyed by the white dominions: their example was suggestive. The extremists were heartened by anti-partition riots. A new weapon was deployed against the British, a boycott of goods, which, it was hoped, might be extended to other forms of passive resistance such as non-payment of taxes and the refusal of soldiers to obey orders. By 1908 the extremists were excluded from Congress. By this time, a second consequence was apparent: extremism was producing terrorism. Again, foreign models were important. Russian revolutionary terrorism now joined the works of Mazzini and the biography of Garibaldi, the guerrilla leader hero of Italian independence, as formative influences on an emerging India. The extremists argued that political murder was not ordinary murder. Assassination and bombing were met with special repressive measures.

The third consequence of partition was perhaps the most momentous. It brought out into the open the division of Moslem and Hindu.

For reasons which went back to the percolation of Moslem India before the Mutiny by an Islamic reform movement, the Arabian Wahhabi sect, Indian Moslems had for a century felt themselves more and more distinct from Hindus. Distrusted by the British because of attempts to revivify the Moghul empire in 1857, they had little success in winning posts in government or on the judicial bench. Hindus had responded more eagerly than Moslems to the educational opportunities offered by the Raj; they were of more commercial weight and had more influence on government. But Moslems too, had found their British helpers, who had established a new Islamic college providing the English education they needed to compete with Hindus, and had helped to set up Moslem political organizations. Some English civil servants began to grasp the potential for balancing Hindu pressure which this could give the Raj. Intensification of Hindu ritual practice such as a cow protection movement was not likely to do anything but increase the separation of the two communities.

Medieval English kings held 'crown-wearings' which asserted their title to their office and their dignity. In 1911, in India, a similar ceremony, but blended with the monarchic rituals of Asia, was held. It was the only occasion in the whole history of British India when the reigning monarch went there, to attend a great Coronation Durbar at Delhi. George V, the King-Emperor, and his consort, Queen Mary, wore the crowns and robes of English majesty, and were accompanied by heralds in the tabards of the Middle Ages, but were paged by the sons of Indian princes who had come to do homage as they would have done to a Moghul emperor.

Nevertheless, it was only in 1905 that the split became, as it remained, one of the fundamentals of the subcontinent's politics. The anti-partitionists campaigned with a strident display of Hindu symbols and slogans. The British governor of eastern Bengal favoured Moslems against Hindus and strove to give them a vested interest in the new province. He was dismissed, but his inoculation had taken: Bengal Moslems deplored his removal. An Anglo-Moslem *entente* was in the making. This further inflamed Hindu terrorists. To make things worse, all this was taking place during five years (from 1906 to 1910) in which prices rose faster than at any time since the Mutiny.

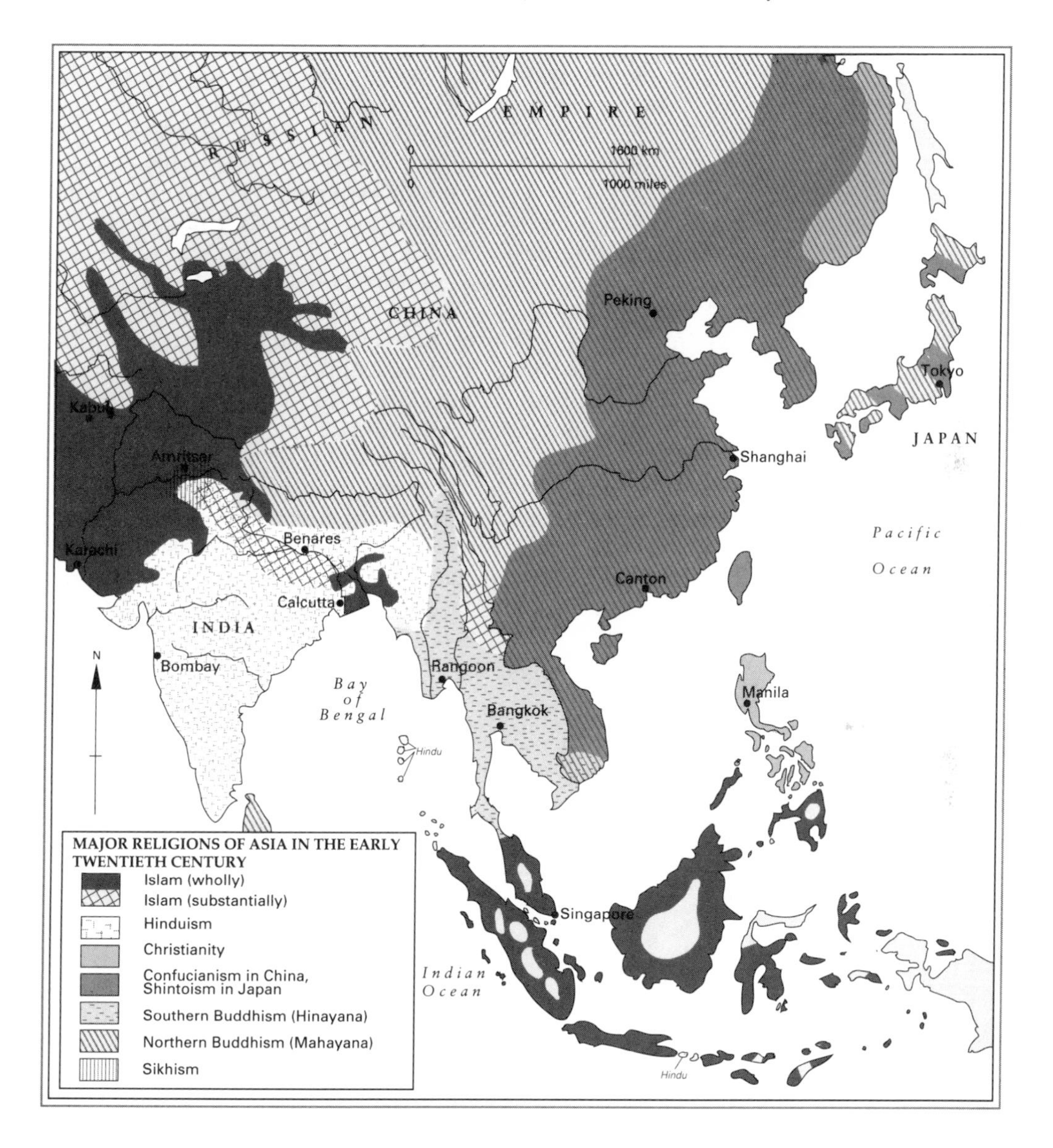

An important set of political reforms conceded in 1909 did not do more than change somewhat the forms with which to operate the political forces which were henceforth to dominate the history of India until the Raj came to an end nearly forty years later. Indians were for the first time appointed to the council which advised the British minister responsible for India and, more important, further elected places were provided for Indians in the legislative councils. But the elections were to be made by electorates which had a communal basis; the division of Hindu and Moslem India, that is to say, was institutionalized.

In 1911, for the first and only time, a reigning British monarch visited India. A great imperial durbar was held at Delhi, the old centre of Moghul rule, to which the capital of British India was now transferred from Calcutta. The princes of India came to do homage; Congress did not question its duty to the throne. The accession of George V that year had been marked by the conferring of real and symbolic benefits, of which the most notable and politically significant was the reuniting of Bengal. If there was a moment at which the Raj was at its apogee, this was it.

Yet India was far from settled. Terrorism and seditious crime continued. The policy of favouring the Moslems had made Hindus more resentful while Moslems now felt that the government had gone back on its understandings with them in withdrawing the partition of Bengal. They feared the resumption of a Hindu ascendancy in the province. Hindus, on the other hand, took the concession as evidence that resistance had paid and began to press for the abolition of the communal electoral arrangements which the Moslems prized. The British had therefore done much to alienate Moslem support when a further strain appeared. The Indian Moslem élites which had favoured cooperation with the British were increasingly under pressure from more middle-class Moslems susceptible to the violent appeal of a pan-Islamic movement. The pan-Islamists could point to the fact that the British had let the Moslems down in Bengal, but also noted that in Tripoli (which the Italians attacked in 1911) and the Balkans in 1912 and 1913, Christian powers were attacking Turkey, the seat of the Caliphate, the institutional embodiment of the spiritual leadership of Islam, and Great Britain was, indisputably, a Christian power. The intense susceptibilities of lower-class Indian Moslems were excited to the point at which even the involvement of a mosque in the replanning of a street could be presented as a part of a deliberate plot to harry Islam. When in 1914 Turkey decided to go to war with Great Britain, though the Moslem League remained loyal, some Indian Moslems accepted the logical consequence of the Caliphate's supremacy, and began to prepare revolution against the Raj. They were few. What was more important for the future was that by that year not two but three forces were making the running in Indian politics, the British, Hindu and Moslem. Here was the origin of the future partition of the only complete political unity the subcontinent had ever known, and, like that unity, it was as much the result of the play of non-Indian as of Indian forces.

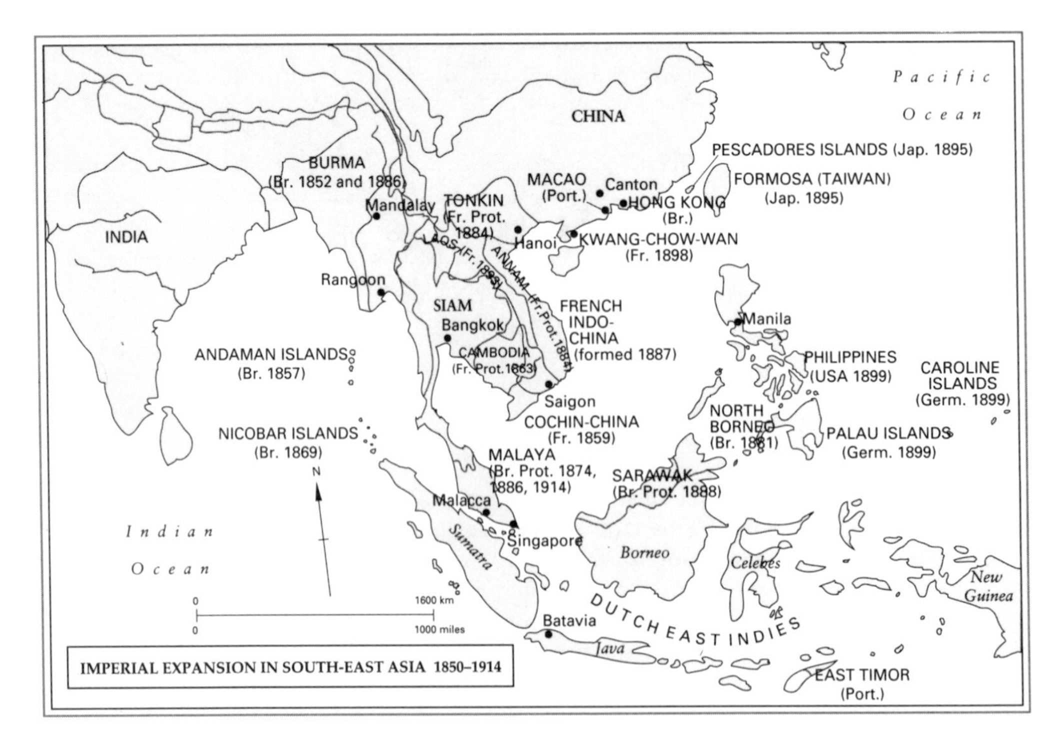

IMPERIAL EXPANSION IN SOUTH-EAST ASIA 1850–1914

India was the largest single mass of non-European population and territory under European rule in Asia, but to the south-east and in Indonesia, both part of the Indian cultural sphere, lay further imperial possessions. Few generalizations are possible about so huge an area and so many peoples and religions. One negative fact was observable: in no other European possession in Asia was there such transformation before 1914 as in India, though in all of them modernization had begun the corrosion of local tradition. The forces which produced this were those which have already been noted at work elsewhere: European aggression, the example of Japan, and the diffusion of European culture. But the first and last of these forces operated in the region for a shorter time before 1914 than in China and India. In 1880 most of mainland South-East Asia was still ruled by native princes who were independent rulers, even if they had to make concessions in 'unequal treaties' to European power. In the following decade this was rapidly changed by the British annexation of Burma and continuing French expansion in Indo-China. The sultans of Malaya acquired British residents at their courts who directed policy through the native administration, while the 'Straits settlements' were ruled directly as a colony. By 1900 among the kingdoms of this region only Siam was still independent.

Most of them had been shaped by cultural influences which were Indian in origin. The only one which was culturally more closely linked to China was Vietnam, that part of Indo-China known in the early stages of French Asian expansion as Annam. Vietnam had the longest tradition of national identity and a history of national revolt long before the European imperial era. It is not therefore surprising to find that it was here that Indo-Chinese resistance to Europeanization was most marked.

French interest in the area went back to Christian missions in the seventeenth century. In the 1850s the persecution of Christianity led the French (briefly assisted by the Spanish) to intervene in south Vietnam, then known as Cochin China. This brought about diplomatic conflict with China, which claimed sovereignty over the country. In 1863 the emperor of Annam ceded part of Cochin China under duress to the French and Cambodia accepted a French protectorate. This was followed by further French advance and the arousing of Indo-Chinese resistance. In the 1870s the French occupied the Red River delta; soon, other quarrels led to a war with China, the paramount power, which confirmed the French grip on India-China. In 1887 they set up a *Union Indo-Chinoise* which disguised a centralized regime behind a system of protectorates. Though this meant the preservation of native rulers (the emperor of Annam and the kings of Cambodia and Laos), the aim of French colonial policy was always assimilation. French culture was to be brought to new French subjects whose élites were to be gallicized because this was believed to be the best way to modernization and civilization.

The centralizing tendencies of French administration soon made it clear that the formal structure of native government was a sham. Unwittingly, the French thus sapped local institutions without replacing them with others enjoying the loyalty of the people. This was a dangerous course. There were also other important by-products of the French presence. It brought with it, for example, French tariff policy, which was to slow down industrialization. This eventually led Indo-Chinese businessmen, like their Indian equivalents, to wonder in whose interests their country was run. Moreover, the conception of an Indo-China which was integrally a part of France, whose inhabitants should be turned into Frenchmen, also brought problems. The French administration had to grapple with the paradox that access to French education could lead to reflexion on the inspiring motto to be found on official buildings and documents of the Third Republic: 'liberty, equality and fraternity'. Finally, French law and notions of property broke down the structure of village landholding and threw power into the hands of money-lenders and landlords. With a growing population in the rice-growing areas, this was to build up a revolutionary potential for the future.

Japan and China provided catalysts for Indo-Chinese grievances embodied in these facts and the legacy of traditional Vietnamese nationalism soon made itself felt. The Japanese victory over Russia led several young Vietnamese to go to Tokyo, where they met Sun Yat-sen and the Japanese sponsors of 'Asia for the Asians'. After the Chinese revolution of 1911, one of them organized a society for a Vietnamese republic. None of this much troubled the French who were well able to contain such opposition before 1914, but it curiously paralleled conservative opposition to them among the Vietnamese Confucian scholar class. Though they opened a university in 1907, the French had to close it almost at once and it remained closed until 1918 because of fears of unrest among the intellectuals. This important section of Vietnamese opinion was already deeply alienated by French rule within a couple of decades of its establishment.

Further south, too, French history had already had an indirect impact in Indonesia. By the end of the nineteenth century there were some sixty million Indonesians; population pressure had not yet produced there the strains that were to come, but it was the largest group of non-Europeans ruled by a European state outside India. Their ancestors had nearly two centuries of sometimes bitter experience of Dutch rule before the French Revolution led to the invasion of the United Provinces, the setting up of a new revolutionary republic there in 1795 and the dissolving of the Dutch East India Company. Equally important, there soon followed a British occupation of Java. The British troubled the waters by important changes in the revenue system, but there were also other influences now at work to stir up Indonesia. Though originally an outcropping of the Hindu civilization of India, it was also part of the Islamic world, with large numbers of at least nominal Moslems among its peoples, and commercial ties with Arabia. In the early years of the nineteenth century this had new importance. Indonesian pilgrims, some of them of birth and rank, went to Mecca and then sometimes went on to Egypt and Turkey. There they found themselves directly in touch with reforming ideas from further west.

The instability of the situation was revealed when the Dutch returned and had, in 1825, to fight a 'Java War' against a dissident prince which lasted five years. It

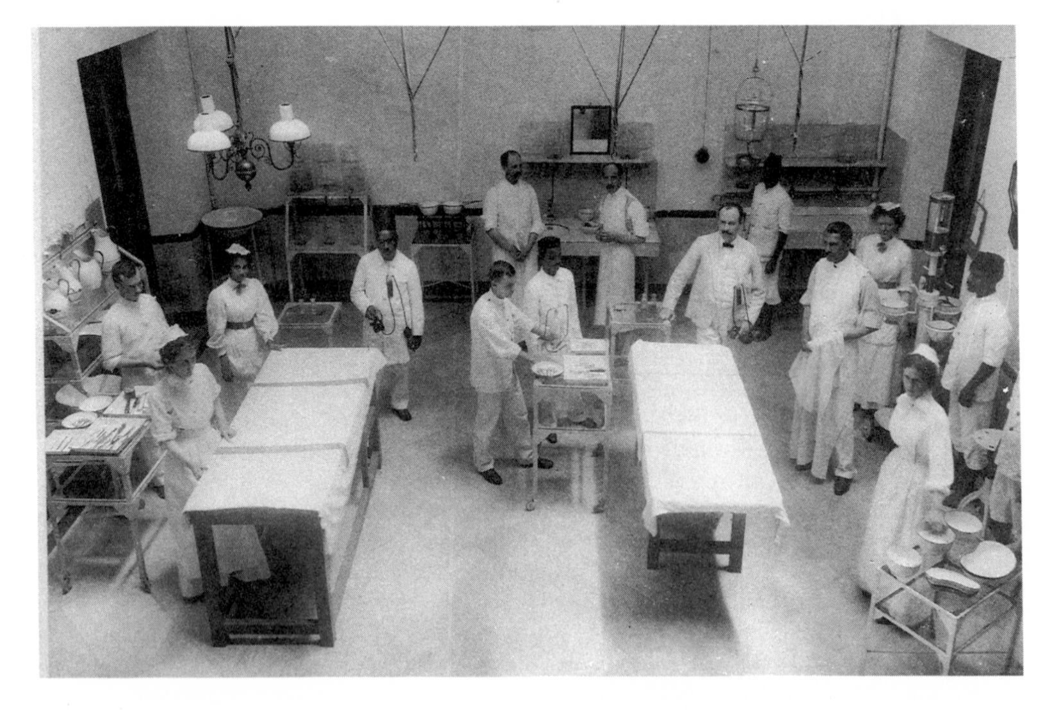

Universities, colleges and special schools were the main channel through which European ideas and standards reached Indians. By the beginning of this century, Indian staff, as well as orderlies and assistants, were available in increasing numbers, as this picture of the operating theatre of a Bombay medical college shows. Most spectacular is the Indian - or, more probably, Anglo-Indian – girl in nurse's uniform; one of the earliest signs of the liberation of women in India from traditional roles.

damaged the island's finances so that the Dutch were constrained to introduce further changes. The result was an agricultural system which enforced the cultivation of crops for the government. The workings of this system led to grave exploitation of the peasant which began in the later nineteenth century to awaken among Dutchmen an uneasiness about the conduct of their colonial government. This culminated in a great change of attitude; in 1901 a new 'Ethical Policy' was announced which was expressed in decentralization and a campaign to achieve improvement through village administration. But this programme often proved so paternalistic and interventionist in action that it, too, sometimes stimulated hostility. This was utilized by the first Indonesian nationalists, some of them inspired by Indians. In 1908 they formed an organization to promote national education. Three years later an Islamic association appeared whose early activities were directed as much against Chinese traders as against the Dutch. By 1916 it had gone so far as to ask for self-government while remaining in union with the Netherlands. Before this, however, a true independence party had been founded in 1912. It opposed Dutch authority in the name of native-born Indonesians, of any race; a Dutchman was among its three founders and others followed him. In 1916 the Dutch took the first step towards meeting the demands of these groups by authorizing a parliament with limited powers for Indonesia.

Though European ideas of nationalism were by the early years of the twentieth century at work in almost all Asian countries, they took their different expressions from different possibilities. Not all colonial regimes behaved in the same way. The British encouraged nationalists in Burma, while the Americans doggedly pursued a benevolent paternalism in the Philippines after suppressing insurrection originally directed against their Spanish predecessors. Those same Spanish, like the Portuguese elsewhere in Asia, had vigorously promoted Christian conversion, while the British Raj was very cautious about interference with the native religions. History also shaped the futures of colonial Asia, because of the different legacies different European regimes played there. Above all, the forces of historical possibilities and historical inertia showed themselves in Japan and China, where direct European influence was just as dramatic in its effects as in India or Vietnam. In every instance, the context in which that influence operated was decisive in shaping the future; at the end of a Asia much (perhaps most) of that context remained intact. A huge residue of customary thought and practice remained undisturbed. Too much history was present for European expansion alone to explain twentieth-century Asia. The catalytic and liberating power of that expansion, none the less, brought Asia into the modern era.

The reward of the Red Army: the Russian flag over the Reichstag building in Berlin, May 1945.

BOOK VII

THE END OF THE EUROPEANS' WORLD

In 1900 Europeans could look back on two, perhaps three, centuries of astonishing growth. Most of them would have said that it was growth for the better – that is, progress. Their history since the Middle Ages looked very much like a continuing advance to evidently worthwhile goals questioned by few. Whether the criteria were intellectual and scientific, or material and economic (even if they were moral and aesthetic, some said, so persuasive was the gospel of progress), a look at their own past assured them that they were set on a progressive course – which meant that the world was set on a progressive course, for their civilization was spread world-wide. What was more, limitless advance seemed to lie ahead. Europeans showed in 1900 much the same confidence in the continuing success of their culture as the Chinese élite had shown in theirs a century earlier. The past, they were sure, proved them right.

Even so, a few did not feel so confident. They felt that the evidence could equally well imply a pessimistic conclusion. Though there were far fewer pessimists than optimists, they numbered in their ranks men of acknowledged standing and powerful minds. Some of them argued that the civilization in which they lived had yet to reveal its full self-destructive potential and sensed that the time when it would do so might not be far away. Some of them saw a civilization more and more obviously drifting away from its moorings in religion and moral absolutes, carried along by the tides of materialism and barbarity – probably to complete disaster.

As it turned out, neither optimists nor pessimists were wholly right, perhaps because their eyes were glued too firmly to what they thought were the characteristics of European civilization. They looked to its own inherent powers, tendencies, or weaknesses for guidance about the future; not many of them paid much attention to the way Europe was changing the world in which her own ascendancy had been built and was thus to alter once again the balance between the major centres of civilization. Few looked further than Europe and Europe beyond the seas except the unbalanced cranks who fussed about the 'Yellow Peril', though Napoleon had a century earlier warned that China was a sleeping giant best left undisturbed.

It is tempting to say in retrospect that the pessimists have had the best of the argument; it may even be true. But hindsight is sometimes a disadvantage to the historian; in this instance it makes it difficult to see how the optimists could once have felt so sure of themselves. Yet we should try to do so. For one thing, there were men of vision and insight among them; for another, optimism was for so long an obstacle to the solution of certain problems in this century that it deserves to be understood as a historical force in its own right. And much of what the pessimists said was wrong too. Appalling though the disasters of the twentieth century have been, they fell on societies more resilient than those shattered by lesser troubles in earlier times, and they have not always been those feared nearly a century ago. In 1900, optimists and pessimists alike had to work with data which could be read in more than one way. It is not reprehensible, merely tragic, that they found it so hard to judge exactly what lay ahead. With better information available to us, we have not been so successful in shorter-term prediction that we are in a position to condemn them.

1
Strains in the System

One historical trend very obvious as the twentieth century opened was the continuing increase of population in the European world. In 1900 Europe had about four-hundred million inhabitants – a quarter of them Russians – the United States about seventy-six million and the British overseas Dominions about 15 million between them. This kept the dominant civilization's share of world population high. On the other hand, growth was already beginning to slow down in some countries in the first decade of this century. This was most obvious in the advanced nations which were the heart of western Europe, where growth depended more and more on falling death-rates. In them there was evidence that keeping your family small was a practice now spreading downwards through society. Traditional contraceptive knowledge of a sort had long been available, but the nineteenth century had brought to the better-off more effective techniques. When these were taken up more widely (and there were soon signs that they were), their impact on population structure would be very great.

In eastern and Mediterranean Europe, on the other hand, such effects were far away. There, rapid growth was only just beginning to produce grave strains. The growing availability of outlets though emigration in the nineteenth century had made it possible to overcome them; there might be trouble to come if those outlets ceased to be so easily available. Further afield, even more pessimistic reflexions might be prompted by considering what would happen when the agencies at work to reduce the death-rate in Europe came to spread to Asia and Africa. In the world civilization the nineteenth century had created, this could not be prevented. In that case, Europe's success in imposing itself would have guaranteed the eventual loss of the demographic advantage recently added to her technical superiority. Worst still, the Malthusian crisis once feared (but lost to sight as the nineteenth-century economic miracle removed the fear of over-population) might at last become a reality.

It had been possible to set aside Malthus' warnings because the nineteenth century brought about the greatest expansion of wealth the world had ever known. Its sources lay in the industrialization of Europe, and the techniques for assuring the continuance of this growth were by no means exhausted or compromised in 1900. There had not only been a vast and accelerating flow of commodities available only in (relatively) tiny quantities a century before, but whole new ranges of goods had come into existence. Oil and electricity had joined coal, wood, wind and water as sources of energy. A chemical industry existed which could not have been envisaged in 1800. Growing power and wealth had been used to tap seemingly inexhaustible natural resources, both agricultural and mineral. Railways, electric trams, steamships, motor cars and bicycles gave to millions a new control over their environment; they quickened travel from place to place and eased transport for the first time since animals had been harnessed to carts thousands of years before. The overall result of such changes had been that in many countries a growing population had been easily carried on an even faster growing-production of wealth; between 1870 and 1900, for example, Germany's output of pig-iron increased sixfold, but her population rose only by about a third. In terms of consumption, or of the services to which they had access, or in the enjoyment of better health, even the mass of the population in developed countries were much better off in 1900 than their predecessors a hundred years before. This still left out people like the Russian or Andalusian peasants

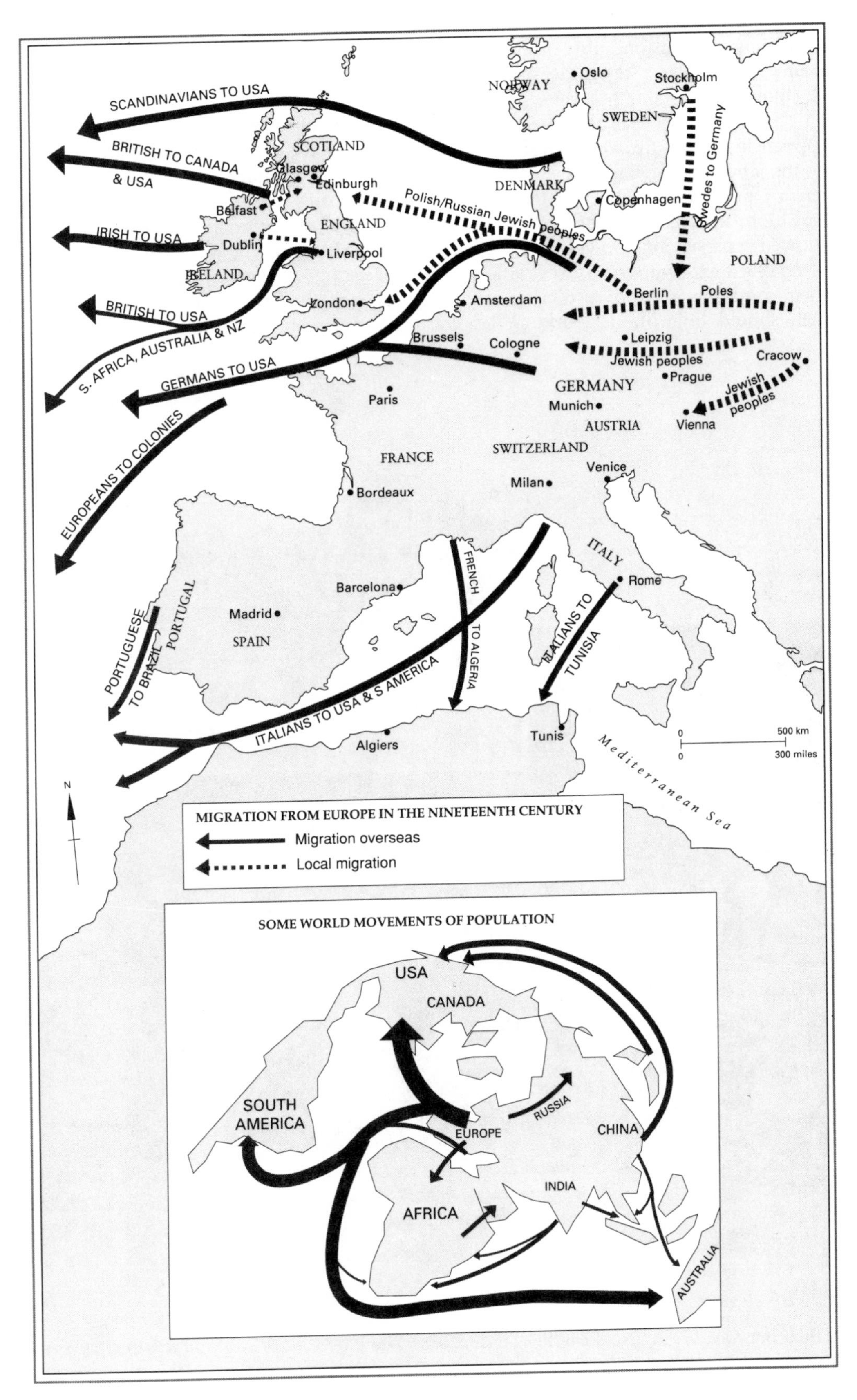

SCANDINAVIANS TO USA
BRITISH TO CANADA & USA
IRISH TO USA
BRITISH TO USA
S. AFRICA, AUSTRALIA & NZ
GERMANS TO USA
EUROPEANS TO COLONIES
Polish/Russian Jewish peoples
Swedes to Germany
Poles
Jewish peoples
Jewish peoples
PORTUGUESE TO BRAZIL
FRENCH TO ALGERIA
ITALIANS TO TUNISIA
ITALIANS TO USA & S AMERICA
NORWAY
Oslo
Stockholm
SWEDEN
SCOTLAND
Glasgow
Edinburgh
DENMARK
Copenhagen
Belfast
ENGLAND
Dublin
Liverpool
IRELAND
POLAND
Berlin
London
Amsterdam
Leipzig
Brussels
Cologne
Cracow
Prague
GERMANY
Paris
Munich
AUSTRIA
Vienna
SWITZERLAND
FRANCE
Venice
Milan
Bordeaux
ITALY
Barcelona
Rome
Madrid
PORTUGAL
SPAIN
Algiers
Tunis
Mediterranean Sea
500 km
300 miles
N
MIGRATION FROM EUROPE IN THE NINETEENTH CENTURY
Migration overseas
Local migration
SOME WORLD MOVEMENTS OF POPULATION
USA
CANADA
SOUTH AMERICA
EUROPE
RUSSIA
CHINA
AFRICA
INDIA
AUSTRALIA

(though an assessment of their condition is by no means easy to make nor the result a foregone conclusion). But, none the less, the way ahead looked promising even for them, inasmuch as a key to prosperity had been found which could be made available to all countries.

In spite of this cheerful picture, doubts could break in. Even if what might happen in the future were ignored, contemplation of the cost of the new wealth and doubts about the social justice of its distribution were troubling. Most people were still terribly poor, whether or not they lived in rich countries where the incongruity of this was more striking than in earlier times. Poverty was all the more afflicting when society showed such obvious power to create new wealth. Here was the beginning of a change of revolutionary import in expectations. Another change in the way men thought about their condition arose over their power to get a livelihood at all. It was not new that men should be without work. What was new was that situations should suddenly

English elegance on display; the start of a balloon race at Ranelagh, 1907.

The other face of English society: children from the East End of London waiting for food during a dock strike, 1912.

arise in which the operation of blind forces of boom and slump produced millions of men without work concentrated in great towns. This was 'unemployment', the new phenomenon for which a new word had been needed. Some economists thought that it might be an inevitable concomitant of capitalism. Nor were the cities themselves yet rid of all the evils which had so struck the first observers of industrial society. By 1900 the majority of western Europeans were town-dwellers. By 1914 there were more than 140 cities of over a hundred thousand inhabitants. In some of them, millions of people were cramped, ill-housed, under-provided with schools and fresh air, let alone amusement other than that of the streets, and this often in sight of the wealth their society helped to produce. 'Slums' was another word invented by the nineteenth century. Two converging conclusions were often drawn from contemplating them. One was that of fear; many sober statesmen at the end of the nineteenth century still distrusted the cities as centres of revolutionary danger, crime and wickedness. The other was hopeful: the condition of the cities gave grounds for assurance that revolution against the injustice of the social and economic order was inevitable. What both these responses neglected, of course, was the accumulating evidence of experience that revolution in western Europe was in fact less and less likely.

The fear of revolution was fed also by disorder, even if its nature was misinterpreted and exaggerated. In Russia, a country which was clearly a part of Europe if it is contrasted with the rest of the world, but one which had not moved forward rapidly along the lines of economic and social progress, reform had not modified autocracy sufficiently and there was a continuing revolutionary movement. It broke out in terrorism – one of whose victims was a tsar – and was supplemented by continuing and spontaneous agrarian unrest. Peasant attacks on landlords and their bailiffs reached a peak in the early years of this century. When they were followed by defeat in war at the hands of the Japanese and the momentary shaking of the regime's confidence, the result was a revolution in 1905. Russia might be, and no doubt was, a special case, but Italy, too, had something that some observers thought of as barely contained revolution in 1898 and again in 1914, while one of the great cities of Spain, Barcelona, exploded into bloody street-fighting in 1909. Strikes and demonstrations could become violent in industrialized countries without revolutionary traditions, as the United States amply showed in the 1890s; even in Great Britain deaths sometimes resulted from them. This was the sort of data which, when combined with the sporadic activities of anarchists, kept policemen and respectable citizens on their toes. The anarchists especially succeeded in pressing themselves on the public imagination. Their acts of terrorism and assassinations during the 1890s received wide publicity; the importance of such acts transcended success or failure because the growth of the press had meant that great publicity value could be extracted from a bomb or a dagger-stroke. In using such methods, not all anarchists shared the same aims, but they were children of their epoch: they protested not only against the state in its governmental aspects, but also against a whole society which they judged unjust.

Socialists contributed most to the rhetoric which sustained the fear of revolution. By 1900 socialism almost everywhere meant Marxism. An important alternative tradition and mythology existed only in England, where the early growth of a numerous trade-union movement and the possibility of working through one of the major established political parties produced a non-revolutionary radicalism. The supremacy of Marxism among continental socialists, by contrast, was formally expressed in 1896, when the 'Second International', an international working-class movement set up seven years before to coordinate socialist action in all countries, expelled the anarchists who had until then belonged to it. There was a well-established tradition of hostility between anarchism and the Marxism which increasingly dominated socialist organizations. Four years later, the International opened a permanent office in Brussels. Within this movement, numbers, wealth and theoretical contributions made the German Social Democratic Party preponderant. This party had prospered in spite of police persecution thanks to Germany's rapid industrialization, and by 1900 was an established fact of German politics, their first truly mass organization. Its numbers and wealth alone would have made it likely that Marxism, the official creed of the German party, would be that of the international socialist movement, even had Marxism not had its own intellectual and emotional appeal. This appeal lay above all in its assurance that the world was going the way socialists hoped, and the emotional satisfaction it provided of participating in a struggle of classes, which, Marxists insisted, must end in violent revolution.

Though such a mythology confirmed the fears of the established order, some intelligent Marxists had noticed that after 1880 or so the facts by no means obviously supported it. Manifestly, great numbers of people had been able to obtain a higher standard of living within the capitalist system. The unfolding of that system in all its complexity was not simplifying and sharpening class conflict in the way Marx had predicted. Moreover, capitalist political institutions had been able to serve the working class. This was very important; in Germany, above all, but also in England, important advantages had been won by socialists using the opportunities provided by parliaments. The vote was available as a weapon and they were not disposed to ignore it while waiting for the revolution. This led some socialists to attempt to restate official Marxism so as to take account of such trends; they were called 'Revisionists' and, broadly speaking, they advocated a peaceful advance towards the transformation of society by socialism. If people liked to call that transformation, when it came, a revolution, then only an argument about usage was involved. Inside this theoretical position and the conflict it provoked was a practical issue which came to a head at the end of the century: whether socialists should or should not sit as ministers in capitalist governments.

The debate which this aroused took years to settle. What emerged in the end was explicit condemnation of revisionism by the Second International while national parties, notably the Germans, continued to act on it in practice, doing deals with the existing system as suited them. Their rhetoric continued to be about revolution. Many socialists even hoped that this might be made a reality by refusing to fight as conscripts if their governments tried to make them go to war. One socialist group, the majority in the Russian party, continued vigorously to denounce revisionism and advocate violence; no doubt this recognized the peculiarity of the Russian situation, where there was little opportunity for effective parliamentary activity and a deep tradition of revolution and terrorism. This group was called Bolshevik, from the Russian word meaning a majority, and more was to be heard of it.

Socialists claimed to speak for the masses. Whether they did so or did not, by 1900 many conservatives worried that the advances gained by liberalism and democracy in the nineteenth century might well prove irresistible except by force. A few of them were still living in a mental world which was pre-nineteenth rather than pre-twentieth century. In much of eastern Europe quasi-patriarchal relationships and the traditional authority of the landowner over his estates were still intact. Such societies could still produce aristocratic conservatives who were opposed in spirit not merely to encroachments upon their material privilege, but also to the values and assumptions of what was to be called 'market society'. But this line was more and more blurred and, for the most part, conservative thinking tended to fall back upon the defence of capital, a position which, of course, would in many places half a century earlier have been regarded as radically liberal, because individualist. The new form of capitalist, industrial conservatism opposed itself more and more vigorously to the state's interference with its wealth, an interference which had grown steadily with the state's acceptance of a larger and larger role in the regulation of society. There was a crisis in England on the issue which led to a revolutionary transformation of what was left of the 1688 constitution in 1911 by the crippling of the power of the House of Lords to restrain an elected House of Commons. In the background were many issues, among them higher taxation of the rich to pay for social services. Even France had by 1914 accepted the principle of an income tax.

Such changes registered the logic of the democratizing of politics in advanced societies. By 1914, universal adult male suffrage existed in France, Germany and several smaller European countries; Great Britain and Italy had electorates big enough to come near to meeting this criterion. This brought forward another disruptive question: if men had, should not women have the vote in national politics? The issue was already causing uproar in English politics. But in Europe only Finland and Norway had women in their parliamentary electorates by 1914, though, further afield, New Zealand, two Australian states and some in the United States had given women the franchise by then. The issue was to remain open in many countries for another thirty years.

Political rights were only one side of the whole question of women's rights in a society whose overall bias, like that of every other great civilization which had preceded it, was towards the interests and values of men. Yet discussion of women's role in society had begun in the eighteenth century and it was not long before cracks appeared in the structure of assumptions which had so long enclosed it. Women's rights to education, to employment, to control of their own property, to moral independence, even to wear more comfortable clothes, had increasingly been debated in the nineteenth century. Ibsen's play *A Doll's House* was interpreted as a trumpet-call for the liberation of women instead of, as the author intended, a plea for the individual. The bringing forward of such issues implied a real revolution. The claims of women in Europe and North America threatened assumptions and attitudes which had not merely centuries, but even millennia, of institutionalization behind them. They awoke complex emotions, for they were linked to deep-seated notions about the family and sexuality. In these ways, they troubled some people – men and women alike – more deeply than the threat of social revolution or political democracy. People were right to see the question in this dimension. In the early European feminist movement was the seed of something whose explosive content would be even greater when transferred (as it soon was) to other cultures and civilizations assaulted by western values.

The politicization of women and political attacks on the legal and institutional structures which were felt by them to be oppressive probably did less for women than did some other changes. Three of these were of slowly growing but, eventually, gigantic importance in undermining tradition. The first was the growth of the advanced capitalist economy. By 1914 this already meant great numbers of new jobs – as typists, secretaries, telephone operators, factory hands, department store assistants and teachers – for women in some countries. Almost none of these had existed a century earlier. They brought a huge practical shift of economic power to women: if they could earn their own living, they were at the beginning of a road which would eventually transform family structures. Soon, too, the demands of warfare in the industrial societies would accelerate this advance as the need for labour opened an even wider range of occupations to them. Meanwhile, for growing numbers of girls even by 1900, a job in industry or commerce at once meant a chance of liberation from parental regulation and the trap of married drudgery. Most women did not by 1914 so benefit, but an accelerating process was at work, because such developments would stimulate other demands, for example, for education and professional training.

The second great transforming force was even further from showing its full potential to change women's lives by 1914. This was contraception. It had already decisively affected demography. What lay ahead was a revolution in power and status as more women absorbed the idea that they might control the demands of childbearing and rearing which hitherto had throughout history dominated most women's lives; beyond that lay an even deeper change, only beginning to be discerned in 1914, as women came to see that they could pursue sexual satisfaction without necessarily entering the obligation of lifelong marriage.

To the third great tendency moving women imperceptibly but irresistibly towards liberation from ancient ways and assumptions it is much harder to give an identifying single name, but if it has a governing principle, it is technology. It was a process made up of a vast number of innovations, some of them slowly accumulating already for decades before 1900 and all tending to cut into the iron time-tables of domestic routine and drudgery, however marginally at first. The coming of piped water, or of gas for heating and lighting, are among the first examples; electricity's cleanliness and flexibility

The campaign for women's political rights in the United Kingdom provoked extraordinary outbursts of feeling on both sides – perhaps because the contenders were not themselves conscious of the depth and complexity of the emotions awoken by feminist demands. This lady attempted to chain herself to the railings of Buckingham Palace in 1914.

was later to have even more obvious effects. Better shops were the front line of big changes in retail distribution which not only gave a notion of luxury to people other than the rich, but also made it easier to meet household needs. Imported food, with its better processing and preserving, slowly changed habits of family catering once based – as they are still often based in India or Africa – on daily or twice daily visits to the market. The world of detergents and easily cleaned artificial fibres still lay in the future in 1900, but already soap and washing soda were far more easily and cheaply available than a century before, while the first domestic machines – gas cookers, vacuum cleaners, washing machines – began to make their appearance at least in the homes of the rich early in this century.

Historians who would recognize at once the importance of the introduction of the stirrup or the lathe in earlier times have none the less strangely neglected the cumulative force of such humble commodities and instruments as these. Yet they implied a revolution for half the world. It is more understandable that their long-term implications interested fewer people at the beginning of this century than the antics of the 'suffragettes', as women who sought the vote were called in England. The immediate stimulus to their activity was the evident liberalization and democratization of political institutions in the case of men. This was the background which their campaign presupposed. Logically, there were grounds for pursuing democracy across the boundaries of sex even if this meant doubling the size of electorates.

But the formal and legal structure of politics was not the whole story of their tendency to show more and more of a 'mass' quality. The masses had to be organized. By 1900 there had appeared to meet this need the modern political party, with its simplifications of issues in order to present them as clear choices, its apparatus for the spread of political awareness, and its cultivation of special interests. From Europe and the United States it spread round the world. Old-fashioned politicians deplored the new model of party and by no means always did so insincerely, because it was another sign of the coming of mass society, the corruption of public debate and the need for traditional élites to adapt their politics to the ways of the man in the street.

The importance of public opinion had begun to be noticed in England early in the nineteenth century. It had been thought decisive in the struggles over the Corn Laws. By 1870, the French emperor felt he could not resist the popular clamour for a war which he feared and was to lose. Bismarck, the quintessential conservative statesman, felt he must give way to public opinion and promote Germany's colonial interests. The manipulation of public opinion, too, seemed to have become possible (or so, at least, many newspaper owners and statesmen believed). Growing literacy had two sides to it. It had been believed on the one hand that investment in mass education was necessary in order to civilize the masses for the proper use of the vote. What seemed to be the consequence of rising literacy, however, was that a market was created for a new cheap press which often pandered to emotionalism and sensationalism, and for the sellers and devisers of advertising campaigns, another invention of the nineteenth century.

The political principle which undoubtedly still had the most mass appeal was nationalism. Moreover, it kept its revolutionary potential. This was clear in a number of places. In Turkish Europe, from the Crimean War onwards, the successes of nationalists in fighting Ottoman rule and creating new nations had never flagged. Serbia, Greece and Romania were solidly established by 1870. By the end of the century they had been joined by Bulgaria and Montenegro. In 1913, in the last wars of the Balkan states against Turkey before a European conflict swallowed the Turkish question, there appeared Albania, and by then an autonomous Crete already had a Greek governor. These nationalist movements had at several times dragged greater states into their affairs and always presented a potential danger to peace. This was not so true of those within the Russian empire, where Poles, Jews, Ukrainians and Lithuanians felt themselves oppressed by the Russians. War, though, seemed a more likely outcome of strains in the Austro-Hungarian empire, where nationalism presented a real revolutionary danger in the lands within the Hungarian half of the monarchy. Slav majorities there looked across the border to Serbia for help against Magyar oppressors. Elsewhere in the empire – in Bohemia and Slovakia, for example – feeling was less high, but nationalism was no less the dominant question. Great Britain faced no such dangers as these, but even she had a nationalist problem, in Ireland. Indeed, she had two. That of the Catholic Irish was for most of the nineteenth century the more obvious. Important reforms and concessions had been granted, though they fell short of the autonomous state of 'Home Rule' to which the British Liberal Party was committed. By 1900, however, agricultural reform and better economic conditions had drawn much of the venom from this Irish question. It was reinserted by the appearance of another Irish nationalism, that of the Protestant majority of the province of Ulster, which was excited to threaten revolution if the government in London gave Home Rule to the Roman Catholic

Irish nationalists. This was much more than merely embarrassing. When the machinery of English democracy did finally deliver Home Rule legislation in 1914, some foreign observers were misled into thinking that British policy would be fatally inhibited from intervention in European affairs by revolution at home.

All those who supported such expressions of nationalism believed themselves with greater or less justification to do so on behalf of the oppressed. But the nationalism of the great powers was also a disruptive force. France and Germany were psychologically deeply sundered by the transfer of two provinces, Alsace and Lorraine, to Germany in 1871. French politicians whom it suited to do so, long and assiduously cultivated the theme of *revanche*. Nationalism in France gave especial bitterness to political quarrels because they seemed to raise questions of loyalty to great national institutions. Even the supposedly sober British from time to time grew excited about national symbols. There was a brief but deep enthusiasm for imperialism and always great sensitivity over the preservation of British naval supremacy. More and more this appeared to be threatened by Germany, a power whose obvious economic dynamism caused alarm by the danger it presented to British supremacy in world commerce. It did not matter that the two countries were one another's best customers; what was more important was that they appeared to have interests opposed in many specific ways. Additional colour was given to this by the stridency of German nationalism under the reign of the emperor, Wilhelm II. Conscious of Germany's potential, he sought to give it not only real but symbolic expression. One effect was his enthusiasm for building a great navy; this especially annoyed the British who could not see that it could be intended for use against anyone but them. But there was a generally growing impression in Europe, far from unjustified, that the Germans were prone to throw their weight about unreasonably in international affairs. National stereotypes cannot be summarized in a phrase, but because they helped to impose terrible simplifications upon public reactions they are part of the story of the disruptive power of nationalist feeling at the beginning of this century.

'One empire, one people, one God' says the inscription on this stamp. Germany had been united less than forty years when it was issued.

Those who felt confident could point to the diminution of international violence in the nineteenth century; there had been no war between European great powers since 1876 (when Russia and Turkey had come to blows) and, unhappily, European soldiers and statesmen failed to understand the portents of the American civil war, the first in which one commander could control over a million men, thanks to railway and telegraph, and the first to show the power of modern mass-produced weapons to inflict huge casualties. While such facts were overlooked, the summoning of congresses in 1899 and 1907 to halt competition in armaments could be viewed optimistically, though they failed in their aim. Certainly acceptance of the practice of international arbitration had grown and some restrictions on the earlier brutality of warfare were visible. A significant phrase was used by the German emperor when he sent off his contingent to the international force fielded against the Chinese Boxers. Stirred to anger by reports of atrocities against Europeans by Chinese, he urged his soldiers to behave 'like Huns'. The phrase stuck in people's memories. Though thought to be excessive even at the time, its real interest lies in the fact that he should have believed such a recommendation was needed. Nobody would have had to tell a seventeenth-century army to behave like Huns, because it was in large measure then taken for granted that they would. By 1900, European troops were not expected to behave in this way and had therefore to be told to do so. So far had the humanizing of war come. 'Civilized warfare' was a nineteenth-century concept and far from a contradiction in terms. In 1899 it had been agreed to forbid, albeit for a limited period, the use of poison gas, dum-dum bullets and even the dropping of bombs from the air.

The restraint exercised on European rulers by the consciousness of any unity other than that of a common resistance to revolution had, of course, long since collapsed together with the idea of Christendom. Nineteenth-century religion was only a palliative or mitigation of international conflict, a minor and indirect force, trickling through to a humanitarianism and pacifism fed from other sources. Christianity had proved as feeble a check to violence as would the hopes of socialists that the workers of the world would refuse to fight one another in the interests of their masters. Whether this was a result of a general loss of power by organized religion is not clear. Certainly much misgiving was felt by 1900 about its declining force as an agent regulating behaviour. This was not because a new religion of traditional form challenged the old Christian Churches. There had been, rather, a continuing development of trends observable in the eighteenth century and much more marked since the French Revolution. Almost all the Christian communions seemed more and more touched by the blight of one or other of the characteristic intellectual and social advances of the age. Nor did they seem able to exploit new devices – the late nineteenth-century appearance of mass-circulation newspapers, for instance – which might have helped them. Indeed, some of them, above all the Roman Catholic Church, positively distrusted such developments.

Though they all felt a hostile current, the Roman Catholic Church was the most obvious victim, the Papacy having especially suffered both in its prestige and power. It had openly proclaimed its hostility to progress, rationality and liberalism in statements which became part of the dogmas of the Church. Politically, Rome had begun to suffer from the whittling away of the Temporal Power in the 1790s, when the French revolutionary armies brought revolutionary principles and territorial change to Italy. Often, later infringements of the Papacy's rights were to be justified in terms of the master-ideas of the age: democracy, liberalism, nationalism. Finally, in 1870, the last territory of the old papal states still outside the Vatican itself was taken by the new kingdom of Italy and the Papacy became almost entirely a purely spiritual and ecclesiastical authority. This was the end of an era of temporal authority stretching back to Merovingian times and some felt it to be an inglorious one for an institution long the centre of European civilization and history.

In fact, it was in some ways to prove advantageous. Nevertheless, at the time the spoliation confirmed both the hostility to the forces of the century which the Papacy had already expressed and the derision in which it was held by many progressive thinkers. Feeling on both sides reached new heights when in 1870 it became a part of the dogma of the Church that the pope, when he spoke *ex cathedra* on faith and morals, did so with infallible authority. There followed two decades in which anti-clericalism and priest-baiting were more important in the politics of Germany, France, Italy and Spain than ever before. Governments took advantage of anti-papal prejudice to advance their own legal powers over the Church, but they were also increasingly pushing into areas where the Church had previously been paramount – above all, elementary and secondary education.

Persecution bred intransigence. In conflict, it emerged that whatever view might be taken on the abstract status of the teachings of the Roman Church, it could still draw on vast reservoirs of loyalty among the faithful. Moreover, these were still being recruited by conversion in the mission field overseas and would soon be added to in still greater numbers by demographic trends. Though the Church might not make much progress among the new city-dwellers of Europe, untouched by an inadequate ecclesiastical machine and paganized by the slow stain of the secular culture in which they were immersed, it was far from dying, let alone dead, as a political and social force. Indeed, the liberation of the Papacy from its temporal role made it easier for Catholics to feel uncompromised loyalty towards it.

The Roman Catholic Church had been the most demanding of the Christian denominations and was in the forefront of the battle of religion with the age, but the claims of revelation and the authority of priest and clergyman were everywhere questioned. This was one of the most striking features of the nineteenth century, all the more so because so many Europeans and Americans still retained simple and literal beliefs in the dogmas of their churches and the story contained in the Bible. They felt great anxiety when such beliefs were threatened, yet this was happening increasingly and in all countries. Traditional religious belief was at first threatened only among an intellectual élite, often consciously holding ideas drawn from Enlightenment sources: 'Voltairean' was a favourite nineteenth-century adjective to indicate anti-religious and sceptical views. As the nineteenth century proceeded, such ideas were reinforced by two other intellectual currents, both also at first a concern of élites, but increasingly with a wider effect in an age of growing mass literacy and cheap printing.

The first of these influential groups of ideas came from biblical scholars, the most important of them German, who from the 1840s onwards not only demolished many assumptions about the value of the Bible as historical evidence, but also, and perhaps more fundamentally, brought about something of a psychological change in the whole attitude to the scriptural text. In essence this change made it possible henceforth simply to regard the Bible as a historic text like any other, to be approached critically. An immensely successful (and scandal-provoking) *Life of Jesus*, published in 1863 by a French scholar, Ernest Renan, brought such an attitude before a wider public than ever before. The book which had been the central text of European civilization since its emergence in the Dark Ages would never recover its position.

The second source of ideas damaging to traditional Christian faith-and therefore to the morality, politics, and economics for so long anchored in Christian assumptions – was to be found in science. Enlightenment attacks on internal and logical inconsistency in the teaching of the Church became much more alarming when science began to produce empirical evidence that things said in the Bible (and therefore based on the same authority as everything else in it) plainly did not fit observable fact. The starting-point was geology; ideas which had been about since the end of the eighteenth century were given a much wider public in the 1830s by the publication of *Principles of Geology* by Scotch scientist, Charles Lyell. This book explained landscape and geological structure in terms of forces still at work, that is, not as the result of a single act of creation, but of wind, rain and so on. Moreover, Lyell pointed out that if this were correct, then the presence of fossils of different forms of life in different geological strata implied that the creation of new animals had been repeated in each geological age. If this were so, the biblical account of creation was clearly in difficulties. That biblical chronology was simply untrue in relation to man was increasingly suggested by discoveries of stone tools in British caves along with the fossilized bones of extinct animals. The argument that man was much older than the biblical account allowed may perhaps be regarded as officially conceded when, in 1859, British learned societies heard and published papers establishing 'that in a period of antiquity remote beyond any of which we have hitherto found traces' men had lived in Palaeolithic societies in the Somme valley.

It is an over-simplification, but not grossly distorting, to say that the same year brought many of these questions to a head by an approach along a different line, the biological, when an English scientist, Charles Darwin, published one of the seminal books of modern civilization, the book called, for short, *The Origin of Species*. Much in it he owed without acknowledgement to others. Its publication came at a moment and in a country where it was especially likely to cause a stir; the issue of the rightfulness of the traditional dominance of religion (for example, in education) was in the air. The word 'evolution' with which Darwin's name came especially to be connected was by then already familiar, though he tried to avoid using it and did not let it appear in *The Origin of Species* until its fifth edition, ten years after the first. Nevertheless, his book was the greatest single statement of the evolutionary hypothesis – namely, that living things were what they were because their forms had undergone long evolution from simpler ones. This, of course, included man, as he made explicit in another book, *The Descent of Man*, of 1871. Different views were held about how this evolution had occurred. Darwin,

The idea of biological evolution by natural selection was the most influential scientific idea of the nineteenth century. Although at least one other scientist arrived at the idea independently of him, and although he himself attributed crucial importance to the influence of Malthus, it is associated above all with the name of Darwin.

impressed by Malthus' vision of the murderous competition of mankind for food, took the view that the qualities which made success likely in hostile environments ensured the 'natural selection' of those creatures embodying them: this was a view to be vulgarized (and terribly misrepresented) by the use as a slogan of the phrase 'survival of the fittest'. But important though many aspects of his work were to be in inspiring fresh thought, here it is important rather to see that Darwin dealt a blow against the biblical account of creation (as well as against the assumption of the unique status of Man) which had wider publicity than any earlier one. In combination with biblical criticism and geology, his book made it impossible for any conscientious and thoughtful man to accept – as he had still been able to do in 1800 – the Bible as literally true.

The undermining of the authority of scripture remains the most obvious single way in which science affected formulated beliefs. Yet just as important, if not more so, was a new, vague but growing, prestige which science was coming to have among a public more broadly-based than ever before. This was because of its new status as the supreme instrument for the manipulation of nature, which was seen as increasingly powerless to resist. Here was the beginning of what was to grow into a mythology of science. Its essence lay in the fact that while the great achievements of seventeenth-century science had not often resulted in changes in the lives of ordinary men and women, those of the nineteenth century increasingly did. Men who understood not a word of what might be written by Joseph Lister, who established the need for (and technique of using) antiseptics in surgery, or by Michael Faraday, who more than any other man made possible the generation of electricity, knew none the less that the medicine of 1900 was different from that of their grandfathers and often saw electricity about them in their work and homes. By 1914, radio messages could be sent across the Atlantic, flying-machines which did not rely upon support by bags of gas of lower density than air were common, aspirins were easily available and an American manufacturer was selling the first cheap mass-produced automobile. The objective achievements of science were by no means adequately represented by such facts, but material advance of this sort impressed the average man and led him to worship at a new shrine.

His awareness of science came through technology because for a long time this was almost the only way in which science had a positive impact on the lives of most people. Respect for it therefore usually grew in proportion to spectacular results in engineering or manufacture and even now, though science makes its impact in other ways, it still makes it very obviously through industrial processes. But though deeply entwined in this way with the dominant world civilization and so interwoven with society, the growth of science meant much more than just a growth of sheer power. In the years down to 1914 the foundations were laid for what would be evident in the second half of the twentieth century, a science which was as much as anything the mainspring of the dominant world culture. So rapid has been the advance to this state of affairs that science has already affected every part of human life while people are still trying to grapple with some of its most elementary philosophical implications.

The easiest observations of this change which can be made (and the easiest to take as a starting-point) are those which display the status of science as a social and material phenomenon in its own right. From the moment when the first great advances in physics were made, in the seventeenth century, science was already a social fact. Institutions were then created in which men came together to study nature in a way which a later age could recognize as scientific, and scientists even then were sometimes employed by rulers to bring to bear their expertise on specific problems. It was noticeable, too, that in the useful arts – and they were more usually called arts than sciences – such as navigation or agriculture, experiment by those who were not themselves practising technicians could make valuable contributions. But a terminological point helps to set this age in perspective and establish its remoteness from the nineteenth century and after: at this time scientists were still called 'natural philosophers'. The word 'scientist' was not invented until about a third of the way through the nineteenth century, when men felt that there was need to distinguish a rigorous experimental and observational investigation of nature from speculation on it by unchecked reason. Even then, though, there was little distinction in most men's minds between the man who carried out such an investigation and the applied scientist or technologist who was the much more conspicuous representative of science in an age of engineering, mining and manufacturing on an unprecedented scale.

The nineteenth century was none the less the first in which science was taken for granted by educated men as a specialized field of study, whose investigators had professional standing. Its new status was marked by the much larger place given to science in education, both by the creation of new departments at existing universities, and by the setting up in some countries, notably France and Germany, of special scientific and technical institutions. Professional studies, too, incorporated larger scientific components. Such developments accelerated as the effects of science on social and economic life became increasingly obvious. The sum effect was to carry much further an already long-established trend. Since about 1700 there has been a steady and exponential increase in the world population of scientists: their numbers have doubled roughly every fifteen years (which explains the striking fact that ever since then there have always been, at any moment, more scientists alive than dead). For the nineteenth century, other measurements of the growth of science can be used (the establishment of astronomical observatories, for example) and these, too, provide exponential curves.

This social phenomenon underlay the growing control of his environment and the improvement of his life which were so easily grasped by the layman. This was what made the nineteenth century the first in which science truly became an object of religion – perhaps of idolatry. By 1914, educated Europeans and Americans took for granted anaesthetics, the motor car, the steam turbine, harder and specialized steels, the aeroplane, the telephone, the wireless and many more marvels which had not existed a century previously; their effects were already very great. Perhaps the most widely apparent were those stemming from the availability of cheap electrical power; it was already shaping cities by making electric trams and trains available to suburban householders, work in factories through electric motors, and domestic life through the electric light. Even animal populations were affected: the 36,000 horses pulling trams in Great Britain in 1900 had only 900 successors in 1914. Of course, the practical application of science was by no means new. There has never been a time since the seventeenth century when there has not been some obvious technological fall-out from scientific activity though, to begin with, it was largely confined to ballistics, navigation and map-making, agriculture and a few elementary industrial processes. Only in the nineteenth century did science really begin to play an important role in sustaining and changing society which went beyond a few obviously striking and spectacular accomplishments. The chemistry of dyeing, for example, was a vast field in which nineteenth-century research led to sweeping innovations which flooded through the manufacture of drugs, explosives, antiseptics – to mention only a few. These had human and social, as well as economic, repercussions. The new fast dyes themselves affected millions of people; the unhappy Indian grower of indigo found that his market dried up on him, and the industrial working classes of the West found they could buy marginally less drab clothes and thus began to move slowly forward along the road at the end of which mass-production methods and man-made fibres all but obliterated visible difference between the clothes of different classes.

This already takes us across the boundary between sustaining life and changing it. Fundamental science was to go on changing society, though some of what was done before 1914 – in physics, for example – is better left for discussion at a later point. One area in which effects are easier to measure was medicine. By 1914, advances had been made which were huge by comparison with a century earlier. A skill had become a science. Great bridgeheads had been driven into the theory and control of infection; antiseptics, having been introduced by Lister only in the 1860s, were taken for granted a couple of decades later, and he and his friend Louis Pasteur, the most famous and greatest of French chemists, laid the foundations of bacteriology. Queen Victoria herself had been a pioneer in the publicizing of new medical methods; the use of anaesthetics during the birth of a prince or princess was important in winning quick social acceptance for techniques only in their infancy in the 1840s. Fewer people, perhaps, would have been aware of the importance of such achievements as the discovery in 1909 of Salvarsan, a landmark in the development of selective treatment of infection, or the identification of the carrier of malaria, or the discovery of 'X-rays'. Yet all these advances, though of great importance, were to be far surpassed in the next fifty years.

Enough impact was made by science even before 1914 to justify the conclusion that it generated its own mythology. In this context, 'mythology' implies no connotations of fiction or falsity. It is simply a convenient way of calling attention to the fact that science, the vast bulk of its conclusions no doubt validated by experiment and therefore 'true', has also come to act as an influence shaping the way men look at the world, just as great religions have done in the past. It has, that is to say, come to be important as more than a method for exploring and manipulating nature. It has been thought also to provide guidance about metaphysical questions, the aims men ought to pursue, the standards they should employ to regulate behaviour. Above all it has been a pervasive influence in shaping popular attitudes. All this, of course, has no intrinsic or necessary connexion with science as the pursuit of scientists. But the upshot in the end was a civilization whose élites had, except vestigially, no dominant religious belief or transcendent ideals. It is a civilization whose core, whether or not this is often articulated, lies in belief in the promise of what can be done by manipulating nature. In principle, it believes that there is no problem which need be regarded as insoluble, given sufficient resources of intellect and money; it has room for the obscure, but not for the essentially mysterious. Many scientists have drawn back from this conclusion. The implications of it are still far from being grasped. But it is the assumption on which a dominant world view now rests and it was already formed in essentials before 1914

Confidence in science in its crudest from has been called 'scientism', but probably very few people held it with complete explicitness and lack of qualification, even in the late nineteenth century, its heyday. Equally good evidence of the prestige of the scientific method, though, is provided by the wish shown by intellectuals to extend it beyond the area of the natural sciences. One of the earliest examples can be detected in the wish to found 'social sciences' which can be seen in the utilitarian followers of the English reformer and intellectual Jeremy Bentham, who hoped to base the management of society upon calculated use of the principles that men responded to pleasure and pain, and that pleasure should be maximized and pain minimized, it being understood that what was to be taken into account were the sensations of the greatest number and their intensity. Later in the nineteenth century, Marx (who was greatly impressed by the work of Darwin) also exemplified the wish to found a science of society. A name for one was provided by the French philosopher Auguste Comte – sociology. These (and many other) attempts to emulate the natural sciences proceeded on a basis of a search for general quasi-mechanical laws; that the natural sciences were at that moment abandoning the search for such laws does not signify here, the search itself still testifying to the scientific model's prestige.

Paradoxically, science too was thus contributing by 1914 to an ill-defined sense of strain in European civilization. This showed most obviously in the problems posed to traditional religion, without doubt, but it also operated in a more subtle way; in determinisms such as those many men drew from thinking about Darwin, or through a relativism suggested by anthropology or the study of the human mind, science itself sapped the confidence in the values of objectivity and rationality which had been so important to it since the eighteenth century. By 1914 there were signs that liberal, rational, enlightened Europe was under strain just as much as traditional, religious and conservative Europe.

Doubt must not loom too large. The most obvious fact about early twentieth-century Europe is that although many Europeans might be sceptical or fearful about its future, it was almost never suggested that it would not continue to be the centre of the world's affairs, the greatest concentration of political power in the globe and the real maker of the world's destinies. Diplomatically and politically, European statesmen could ignore the rest of the world for all important matters, except in the western hemisphere, where another nation of European origins, the United States, was paramount, and the Far East, where Japan was increasingly important and the Americans had interests which they might require others to respect. It was their relationships with one another, nevertheless, that fascinated most European statesmen in 1900; for most of them there was nothing else so important to worry about at this time.

2
The Era of the First World War

Against the one clear favourable fact that great wars had been successfully averted by European states ever since 1870, could be set some political evidence that the international situation was none the less growing dangerously unstable by 1900. Some major states had grave internal problems, for example, which might imply external repercussions. For all the huge difference between them, United Germany and United Italy were new states; they had not existed forty years earlier and this made their rulers especially sensitive to internal divisive forces and consequently willing to court chauvinistic feeling. Some of Italy's leaders went in for disastrous colonial ventures, keeping alive suspicion and unfriendliness towards Austria-Hungary (formally Italy's ally, but the ruler of territories still regarded by Italians as 'unredeemed') and finally plunged their country into war with Turkey in 1911. Germany had the advantages of huge industrial and economic success to help her, yet after the cautious Bismarck had been sent into retirement her foreign policy was conducted more and more with an eye to winning the impalpable and slippery prizes of respect and prestige – a 'place in the sun', as some Germans summed it up. Germany had also to face the consequences of industrialization. The new economic and social forces it spawned were increasingly difficult to reconcile with the conservative character of her constitution which gave so much weight in imperial government to a semi-feudal, agrarian aristocracy.

Nor was internal tension confined to new states. The two great dynastic empires of Russia and Austria-Hungary each faced grave internal problems; more than any other states they still fitted the assumption of the Holy Alliance era that governments were the natural opponents of their subjects. Yet both had undergone great change in spite of apparent continuity. The Habsburg monarchy in its new, hyphenated form was itself the creation of a successful nationalism, that of the Magyars. In the early years of the twentieth century there were signs that it was going to be more and more difficult to keep the two halves of the monarchy together without provoking other nations inside it beyond endurance. Moreover, here, too, industrialization (in Bohemia and Austria) was beginning to add new tensions to old. Russia, as has been indicated, actually exploded in political revolution in 1905, and was also changing more deeply. Autocracy and terrorism between them destroyed the liberal promise of the reforms of Alexander II, but they did not prevent the start of faster industrial growth by the end of the century. This was the beginning of an economic revolution to which the great emancipation had been the essential preliminary. Policies designed to exact grain from the peasant provided a commodity for export to pay for the service of foreign loans. With the twentieth century, Russia began to show at last a formidable rate of economic advance. The quantities were still small – in 1910 Russia produced less than a third as much pig-iron as the United Kingdom and only about a quarter as much steel as Germany. But these quantities had been achieved at a very high rate of growth. Probably more important, there were signs that by 1914 Russian agriculture might at last have turned the corner and be capable of producing a grain harvest which would grow faster than population. A determined effort was made by one minister to provide Russia with a class of prosperous independent farmers whose self-interest was linked to raising productivity, by removing the last of the restraints on individualism imposed by the terms of serfdom's abolition. Yet there was still

The Black Hundreds. A demonstration in Odessa by the reactionary, anti-semitic movement used by the Tsarist authorities against revolutionary forces. They carry a picture of the Tsar.

much backwardness to overcome. Even in 1914 less than 10 per cent of Russians lived in towns and only about three million out of a total population of more than one-hundred and fifty million worked in industry. The debit side still loomed large in of Russia's progress. She might be a potential giant, but was one entangled with grievous handicaps. The autocracy governed badly, reformed unwillingly and opposed all change (though it was to make constitutional concessions in 1905). The general level of culture was low and unpromising; industrialization would demand better education and that would cause new strains. Liberal traditions were weak; terrorist and autocratic traditions were strong. Russia was still dependent on foreign suppliers for the capital she needed, too.

Most of this came from her ally, France. With the United Kingdom and Italy, the Third Republic represented liberal and constitutional principles among Europe's great powers. Socially conservative, France was, in spite of her intellectual vitality, uneasy and conscious of weakness. In part, a superficial instability was a matter of bitter exchanges between politicians; in part it was because of the efforts of some who strove to keep alive the revolutionary tradition and rhetoric. Yet the working-class movement was weak. France moved only slowly towards industrialization and, in fact, the Republic was probably as stable as any other regime in Europe, but slow industrial development indicated another handicap of which Frenchmen were very aware, their military inferiority. The year 1870 had shown that the French could not on their own beat the German army. Since then, the disparity of the two countries' positions had grown ever greater. In manpower, France had fallen further still behind and in economic development, too, she had been dwarfed by her neighbour. Just before 1914, France was raising about one-sixth as much coal as Germany, made less than a third as much pig-iron and a quarter as much steel. If there was ever to be a return match for 1870, Frenchmen knew they needed allies.

An ally was not, in 1900, to be looked for across the Channel. This was mainly because of colonial issues; France (like Russia) came into irritating conflict with the United Kingdom in a great many places around the globe where British interests lay. For a long time, the United Kingdom found she could remain clear of European entanglements; this was an advantage, but she, too, had troubles at home. The first industrial nation was also one of the most troubled by working-class agitation and, increasingly, by uncertainty about her relative strength. By 1900 some British businessmen were clear that Germany was a major rival; there were plenty of signs that in technology and method German industry was greatly superior to British. The old certainties began to give way; Free Trade itself was called in question. There were even signs, in the violence of Ulstermen and suffragettes and the embittered struggles over social legislation with a House of Lords determined to safeguard the interests of wealth, that parliamentarianism itself might be threatened. There was no longer a sense of the sustaining consensus of mid-Victorian politics. Yet there was also a huge solidity about British institutions and political habits. Parliamentary monarchy had proved able to carry through vast changes since 1832 and there was little reason for fundamental doubt that it could continue to do so.

Only a perspective which Englishmen of the day found hard to recognize reveals the fundamental change which had come about in the international position of the United Kingdom within the preceding half-century or so. This is provided by a view from Japan or the United States, the two great extra-European powers. The Japanese portent was the more easily discerned of the two, perhaps, because of the military victory over Russia, yet there were signs for those who could interpret them that the United States would shortly emerge as a power capable of dwarfing Europe and as the most powerful nation in the world. Her nineteenth-century expansion had come to a climax with the establishment of her supremacy on an unquestionable footing of power in her own hemisphere. The war with Spain and the building of the Panama canal rounded off the process. American domestic, social and economic circumstances were such that the political system proved easily able to handle the problems it faced once the great mid-century crisis was surmounted. Amongst these, some of the gravest resulted from industrialization. The confidence that all would go well if the economically strongest were simply allowed to drive all others to the wall first began to be questioned towards the end of the nineteenth century. But this was after an industrial machine of immense scale had already matured. It would be the bedrock of future American power. By 1914 American production of pig-iron and steel was more than twice that of Great Britain and Germany together; the United States mined almost enough coal to outpace them, too. At the same time the standard of living of her citizens continued to act as a magnet to immigration; in her natural resources and a stream of cheap, highly motivated labour lay two of the sources of America's economic might. The other was foreign capital. She was the greatest of debtor nations.

Though her political constitution was older in 1914 than that of any major European state except Great Britain or Russia, the arrival of new Americans long helped to give the United States the characteristics and psychology of a new nation. A need to integrate her new citizens often led to the expression of strong nationalist feeling. But because of geography, a tradition of rejecting Europe, and the continuing domination of American government and business by élites formed in the Anglo-Saxon tradition, this did not take violent forms outside the western hemisphere. The United States in 1914 was still a young giant waiting in the wings of history, whose full importance would only become manifest when Europe needed to involve America in its quarrels.

In that year a war began as a result of those quarrels. Though it was not the bloodiest nor most prolonged war in history, nor strictly, as it was later termed, the 'first' World War, it was the most intensely fought struggle and the greatest in geographical extent to have occurred down to that time. Nations in every continent took part. It was also costlier than any earlier war and made unprecedented demands upon resources. Whole societies were mobilized to fight it, in part because it was also the first war in which machines played an overwhelmingly important part; war was for the first time transformed by science. The best name to give it remains the simple one used by those who fought in it: the *Great* War. This is justified by its unprecedented psychological effect.

It was also the first of two wars whose central issue was the control of German power. The damage they did ended Europe's political, economic and military supremacy. Each of these conflicts began in essentially European issues and the war always had a predominantly European flavour; like the next great struggle detonated by Germany though, it sucked into it other conflicts and jumbled together a whole anthology of issues. But Europe was the heart of the matter and self-inflicted damage in the end deprived her of world hegemony. This did not happen by 1918, when the Great War ended (though irreparable damage had already been done, even by then), but it was obvious in 1945, at the end of a 'Second World War'. That left behind a continent whose pre-1914 structure had vanished. It has led some historians to speak of the whole era from 1914 to 1945 as an entity, as a European 'civil war' – not a bad metaphor, provided it is borne in mind that it is a metaphor. Europe had never been free from wars for long and the containment of internal disorder is the fundamental presupposition of a state: Europe had never been united and could not therefore have a true civil war. But it was the source and seat of a civilization which was a unity; Europeans saw themselves as having more in common with other Europeans than with black, brown or yellow men. Furthermore, it was a system of power which in 1914 was an economic unity and had just experienced its longest period of internal peace. These facts, all of which were to vanish by 1945, make the metaphor of civil war vivid and acceptable; it signifies the self-destructive madness of a civilization.

A European balance had kept the peace between great states for over forty years. By 1914 this was dangerously disturbed. Too many people had come to feel that the chances of war might offer them more than a continued peace. This was especially so in the ruling circles of Germany, Austria-Hungary and Russia. By the time that they had come to feel this, there existed a complicated set of ties, obligations and interests between states which so involved them with one another that it was unlikely that a conflict could be limited to two, or even to a few of them. Another force making for instability was the existence of small countries enjoying special relations with larger ones; some of them were in a position to take the effective power to make decisions from the hands of those who would have to fight a major war.

This delicate situation was made all the more dangerous by the psychological atmosphere in which statesmen had to work by 1914. It was an age when mass emotions were easily aroused, in particular by nationalist and patriotic stimuli. There was widespread ignorance of the dangers of war, because nobody except a tiny minority foresaw a war which would be different from that of 1870; they remembered the France of that year, and forgot how, in Virginia and Tennessee only a few years earlier, modern war had first shown its face in prolonged slaughter and huge costs (more Americans died in the Civil War than have died in all the other wars in which the United States has taken part, even to the present

DH-4 day bombers of the Royal Flying Corps, France, early 1918. The design was one of the most successful of the Great War, in spite of the near impossibility of pilot and observer communicating in flight, given the distance between them, and a tendency to burst into flames in any crash landing.

day. Everyone knew that wars could be destructive and violent, certainly, but also believed that in the twentieth century they would be swiftly over. The very cost of armaments made it inconceivable that civilized states could sustain a prolonged struggle such as that with Napoleonic France; the complex world economy and the taxpayer, it was said, could not survive one. This perhaps diminished misgivings about courting danger. There are even signs, too, that many Europeans were bored by their lives in 1914 and saw in war an emotional release purging away a sense of decadence and sterility. Revolutionaries, of course, welcomed international conflict because of the opportunities they thought it might bring. Finally, it is worth remembering that the long success of diplomats in negotiating grave crises without war was itself a danger. Their machinery had worked so many times that when it was presented with facts more than ordinarily recalcitrant in July 1914, their significance for a time seemed to escape many of those who had to deal with them. On the very eve of conflict, statesmen were still finding it difficult to see why another conference of ambassadors or even a European congress should not extricate them from their problems.

One of the conflicts which came to a head in 1914 went back a very long way. This was the old rivalry of Austria-Hungary and Russia in south-eastern Europe. Its roots lay deep in the eighteenth century, but its last phase was dominated by the accelerated collapse of the Ottoman empire in Europe from the Crimean War onwards. For this reason the First World War is from one point of view to be seen as another war of the Ottoman succession. After the Congress of Berlin in 1878 had pulled Europe through one dangerous moment, Habsburg and Romanov policy had settled down to a sort of understanding by the 1890s. This lasted until Russian interest in the Danube valley revived after the checking of Russian imperial ambition in the Far East by the Japanese.

At that moment, events outside the Habsburg and Turkish empires were giving a new aggressiveness to Austro-Hungarian policy, too.

At the root of this was revolutionary nationalism. A reform movement looked for a while as if it might put the Ottoman empire together again and this provoked the Balkan nations to try to undo the *status quo* established by the great powers and the Austrians to look to their own interests in a situation once again fluid. They offended the Russians by a mismanaged annexation of the Turkish province of Bosnia in 1909; the Russians had not been given a corresponding and compensating gain. Another consequence of Bosnia's annexation was that the Dual Monarchy acquired more Slav subjects. There was already discontent among the monarchy's subject peoples, in particular, the Slavs who lived under Magyar rule. More and more under the pressure of Magyar interests, the government in Vienna had shown itself hostile to Serbia, a nation to which these Slav subjects might look for support. Some of them saw Serbia as the nucleus of a future state embracing all South Slavs, and its rulers seemed unable (and perhaps unwilling) to restrain South Slav revolutionaries who used Belgrade as a base for terrorism and subversion in Bosnia. Lessons from history are often unfortunate; the Vienna government was only too ready to conclude that Serbia might play in the Danube valley the role that Sardinia had played in uniting Italy. Unless the serpent were scotched in the egg, another loss of Habsburg territory would then follow. Having been excluded from Germany by Prussia and from Italy by Sardinia, a potential Yugoslavia now seemed to some Habsburg counsellors to threaten the empire with exclusion from the lower Danube valley. This would mean its end as a great power and an end, too, of Magyar supremacy in Hungary, for fairer treatment of Slavs who remained in Hungarian territory would be insisted upon by south Slavdom. The continuing subsidence of the Ottoman empire could then only benefit Russia, the power which stood behind Serbia, determined there should not be another 1909.

Into this complicated situation, the other powers were pulled by interest, choice, sentiment and formal diplomacy. Of these, the last was perhaps less important than was once thought. Bismarck's efforts in the 1870s and 1880s to ensure the isolation of France and the supremacy of Germany had spawned a system of alliances unique in peacetime. Their common characteristic was that they defined conditions on which countries would go to war to support one another, and this seemed to cramp diplomacy. But in the end they did not operate as planned. This does not mean that they did not contribute to the coming of war, only that formal arrangements can only be effective if people want them to be, and other factors decided that in 1914.

At the root of the whole business was the German seizure of Alsace and Lorraine from France in 1871, and the consequent French restlessness for revenge. Bismarck guarded against this first by drawing together Germany, Russia and Austria-Hungary on the common ground of dynastic resistance to revolutionary and subversive dangers which France, the only republic among the major states, was still supposed to represent; there were still alive in 1871 people born before 1789 and many others who could remember the comments of those who had lived through the years of the great Revolution, while the upheaval of the Paris Commune revived all the old fears of international subversion. The conservative alliance none the less lapsed in the 1880s, essentially because Bismarck felt he must in the last resort back Austria-Hungary if a conflict between her and Russia proved unavoidable. To Germany and the Dual Monarchy was then added Italy; thus was formed in 1882 the Triple Alliance. But Bismarck still kept a separate 'Reinsurance' treaty with Russia, though he seems to have felt uneasy about the prospect of keeping Russia and Austria-Hungary at peace in this way.

Yet a conflict between them did not again look likely until after 1909. By then, Bismarck's successors had allowed his Reinsurance treaty to lapse and Russia had become in 1892 an ally of France. From that date the road led away from Bismarck's Europe, where everyone else had been kept in equilibrium by Germany's central rôle, to a Europe divided into two camps. This was made worse by German policy. In a series of crises, she showed that she wanted to frighten other nations with her displeasure and make herself esteemed. In particular, in 1905 and 1911 her irritation was directed against France, and commercial and colonial issues were used as excuses to show by displays of force that France had not won the right to disregard German wishes by making an ally of Russia. German military planning had already by 1900 accepted the need to fight a two-front war if necessary, and made preparations to do so by a quick overthrow of France while the resources of Russia were slowly mobilized.

As the twentieth century opened, it had therefore become highly probable that if an Austro-Russian war broke out, Germany and France would join in. Moreover, Germans had within a few years made this more likely by patronizing the Turks. This was much more alarming to the Russians than it would have been earlier, because a growing export trade in grain from Russia's Black Sea ports had to pass through the Straits. The Russians began to improve their fighting-power. One essential step in this was the completion of a railway network which would make possible the mobilization and delivery to the battlefields of eastern Europe of Russia's vast armies.

In all this, there was no obvious need for Great Britain to be concerned, had not German policy perversely antagonized her. At the end of the nineteenth century Great Britain's quarrels were almost all with France and Russia. They arose where imperial ambitions clashed, in Africa and central and south-eastern Asia. Anglo-German relations were more easily managed, if occasionally prickly. As Great Britain entered the new century she was still preoccupied with empire, not with Europe. The first peacetime alliance she had made since the eighteenth century was with Japan, to safeguard her interests in the Far East. Then came a settlement of long outstanding disputes with France in 1904; this was in essence an agreement about Africa, where France was to be given a free hand in Morocco in return for Great Britain having one in Egypt – a way of settling another bit of the Ottoman succession – but it rounded up other colonial quarrels the world over, some going back as far as the peace of Utrecht. A few years later, Great Britain made a similar (though less successful) agreement with Russia about spheres of interest in Persia. But the Anglo-French settlement grew into much more than a clearing away of grounds for dispute. It became what was called an *entente*, or special relationship.

This was Germany's doing. Irritated by the Anglo-French agreement, the German government decided to show France that it would have to have its say in Morocco's affairs at an international conference. They got it, but bullying France solidified the *entente*; the British began to realize that they would have to concern themselves for the first time in decades with the continental balance of power. If they did not, Germany would dominate it. At the end of this road would be their acceptance of a rôle as a great military power on land, a change of the assumptions which British strategy had followed since the days of Louis XIV and Marlborough, the last age in which the country had put its major weight into prolonged effort on the Continent. Secret military talks with the French explored what might be done to help their army against a German invasion through Belgium. This was not going far, but Germany then threw away the chance to reassure British public opinion, by pressing forward with plans to build a great navy. It was inconceivable that such a step could be directed against any power except Great Britain. The result was a naval race which most British were determined to win (if they could not end it) and therefore the further inflammation of popular feeling. In 1911, when the gap between the two countries' fleets was narrowest and most felt in Great Britain, German diplomacy provoked another crisis over Morocco. This time, a British minister said publicly something that sounded very much like an assertion that Great Britain would go to war to protect France.

Yet when war came, it was in the South Slav lands. Serbia did well in the Balkan Wars of 1912–13, in which the young Balkan nations first despoiled Turkey of most that was left of her European territory and then fell out over the spoils. But Serbia might have got more had the Austrians not objected. Behind Serbia stood Russia, launched on the programme of rebuilding and expanding her forces which would take three or four years to bring to fruition. If South Slavs were to be shown that the Dual Monarchy could humiliate Serbia so that they could not hope for her support, then the sooner the better.

The Western Front, 1916. A British battalion moves to the rear with its wounded past a supply column moving up to the front.

Given that Germany was the Dual Monarch's ally she, in turn, was unlikely to seek to avoid fighting Russia while there was still time to be sure of winning.

The crisis came when an Austrian archduke was assassinated by a Bosnian terrorist at Sarajevo in June 1914. The Austrians believed that the Serbians were behind it. They decided that the moment had come to teach Serbia her lesson and kill for ever the pan-Slav agitation. The Germans supported them. The Austrians declared war on Serbia on 28 July. A week later all the great powers were at war (though, ironically, the Austro-Hungarians and Russians were still at peace with one another; it was only on 6 August that the Dual Monarchy at last declared war on its old rival). Before that date German military planning had dictated the timetable of events. The key decision to attack France before Russia had been made years before; German planning required such an attack to be made through Belgium, whose neutrality the British among others had guaranteed. Thereafter the sequence of events fell almost automatically into place. When Russia mobilized to bring pressure on Austria-Hungary for Serbia's protection, the Germans declared war on Russia. Having done that, they had to attack the French and, finding a pretext, formally declared war on them. Thus, the Franco-Russian alliance never actually operated. By Germany's violation of Belgian neutrality, the British government, uneasy about a German attack on France, but not seeing clearly on what grounds they could justify intervention to prevent it, was given an issue to unite the country and take it into war against Germany on 4 August.

Just as the duration and intensity of the war were to outrun all expectations, so did its geographical spread. Japan and the Ottoman empire joined in soon after the outbreak; the former on the side of the Allies (as France, Great Britain and Russia were called) and Turkey on that of the Central Powers (Germany and Austria-Hungary). Italy joined the Allies in 1915, in return for promises of Austrian territory. Other efforts were made to pick up new supporters by offering cheques to be cashed after a victorious peace; Bulgaria joined the Central Powers in September 1915 and Romania the Allies in the following year. Greece became another Allied state in 1917. Portugal's government had tried to enter the war in 1914, but though unable to do so because of internal troubles was finally faced with a German declaration of war in 1916. Thus, by the end of that year, the original issues of Franco-German and Austro-Russian rivalry had been thoroughly confused by other struggles. The Balkan states were fighting a third Balkan war (the war of the Ottoman succession in its European theatre), the British a war against German naval and commercial power, the Italians the last war of the *Risorgimento*, while outside Europe the British, Russians and Arabs had begun a war of Ottoman partition in Asia and the Japanese were pursuing another cheap and highly profitable episode in the assertion of their hegemony in the Far East.

Indian soldiers were for a time deployed by the British in France in the Great War though, mainly for climatic reasons, they were later withdrawn from this theatre.

One reason why there was a search for allies in 1915 and 1916 was that the war then showed every sign of bogging down in a stalemate no one had expected. The nature of the fighting had surprised almost everyone. It had opened with a German sweep into northern France. This did not achieve the lightning victory which was its aim but gave the Germans possession of all but a tiny scrap of Belgium and much French territory, too. In the east, Russian offensives had been stopped by the Germans and Austrians. Thereafter, though more noticeably in the west than the east, the battlefields settled down to siege warfare on an unprecedented scale. This was because of two things. One was the huge killing-power of modern weapons. Magazine rifles, machine-guns and barbed wire could stop any infantry attack not preceded by pulverizing bombardment. Demonstrations of this truth were provided by the huge casualty lists. By the end of 1915 the French army alone had lost 300,000 dead; that was bad enough, but in 1916 one seven-month battle before Verdun added another 315,000 to this total. In the same battle 280,000 Germans died. While it was going on, another struggle further north, on the Somme, cost the British 420,000 casualties and the Germans about the same. The first day of that battle, 1 July, remains the blackest in the history of the British army, when it suffered 60,000 casualties, of whom more than a third died.

Such figures made nonsense of the confident predictions that the cost of modern war would be bound to make the struggle a short one. This was a reflection of the second surprise, the revelation of the enormous war-making power of industrial societies. Plenty of people were weary by the end of 1916, but by then the warring states had already amply demonstrated a capacity greater than had been imagined to conscript and organize their peoples as never before in history to produce unprecedented quantities of material and furnish the recruits for new armies. Whole societies were engaged against one another; the international solidarity of the working class might never have been thought of for all the resistance it opposed to this, nor the international interests of ruling classes against subversion.

Inability to batter one another into submission on the battlefields accelerated the strategic and technical expansion of the struggle. This was why diplomats had sought new allies and generals new fronts. The Allies in 1915 mounted an attack on Turkey at the Dardanelles in the hope, not to be realized, of knocking her out of the war and opening up communication with Russia through the Black Sea. The same search for a way round the French deadlock later produced a new Balkan front at Salonika; it replaced the one which had collapsed when Serbia was overrun. Colonial possessions, too, had ensured from the first that there would be fighting all round the globe, even if on a small scale. The German colonies could be picked up fairly easily, thanks to the British command of the seas, though the African ones provoked some lengthy campaigning. The most important and considerable extra-European operations, though, were in the eastern and southern parts of the Turkish empire. A British and Indian army entered Mesopotamia. Another force advanced from the Suez canal towards Palestine. In the Arabian desert, an Arab revolt against the Turks provided some of the few romantic episodes to relieve the brutal squalor of industrial war.

The technical expansion of the war was most noticeable in its industrial effects and in the degeneration of standards of behaviour. The American civil war a half-century before had prefigured the first of these, too, in revealing the economic demands of mass war in a democratic age. The mills, factories, mines and furnaces of Europe now worked as never before. So did those of the United States and Japan, the former accessible to the Allies but not to the Central Powers because of the British naval supremacy. The maintenance of millions of men in the field required not only arms and ammunition, but food, clothing, medical equipment, and machines in huge quantities. Though a war in which millions of animals were needed, it was also the first war of the internal-combustion engine; trucks and tractors swallowed petrol as avidly as horses and mules ate their fodder. Many statistics illustrate the new scale of war but one must suffice; in 1914 the whole British empire had 18,000 hospital beds and four years later it had 630,000.

The repercussions of this vast increase in demand rolled outwards through society, leading in all countries in varying measure to the governments' control of the economy, conscription of labour, the revolutionizing of women's employment, the introduction of new health and welfare services. They also rolled overseas. The United States ceased to be a debtor nation; the Allies liquidated their investments there to pay for what they needed and became debtors in their turn. Indian industry received the fillip it had long required. Boom days came to the ranchers and farmers of the Argentine and the British white Dominions. The latter also shared the military burden, sending soldiers to Europe and fighting the Germans in their colonies.

Technical expansion also made the war more frightful. This was not only because machine-guns and high explosive made possible such terrible slaughter. It was not even because of new weapons such as poison gas, flame-throwers or tanks, all of which made their appearance as soldiers strove to find a way out of the deadlock of the battlefield. It was also because the fact that whole societies were engaged in warfare brought with it the realization that whole societies could be targets for war-like operations. Attacks on the morale, health and efficiency of civilian workers and voters became desirable. When such attacks were denounced, the denunciations were themselves blows in another sort of campaign, that of propaganda. The possibilities of mass literacy and the recently created cinema industry supplemented and overtook such old standbys as pulpit and school in this kind of warfare. To British charges that the Germans who carried out primitive bombing raids on London by airship were 'baby-killers', Germans retorted that the same could be said of the sailors who sustained the British blockade. The rising figures of German infant mortality bore them out.

'Tank' was a word adopted by the British to keep secret a new invention they were preparing – a self-propelled fortress which could cross all but very large obstacles, impervious to the barbed-wire and machine-gun fire which brought infantry to a bloody halt so many times. By the end of the war, tanks had been developed for specialized roles. This French example was a light, two-man machine, and comparatively fast.

In part because of the slow but apparently irresistible success of the British blockade and because of its unwillingness to risk the fleet whose building had done so much to poison pre-war feeling between the two countries, the German High Command devised a new use for a weapon whose power had been underrated in 1914, the submarine. It was launched at Allied shipping and the ships of neutrals who were supplying the Allies, attacks often being made without warning and on unarmed vessels. This was first done early in 1915, though few submarines were then available and they did not do much damage. There was an outcry when a great British liner was torpedoed that year, with the loss of 1200 lives, many of them Americans, and the unrestricted sinking of shipping was called off by the Germans. It was resumed at the beginning of 1917. By then it was clear that if Germany did not starve Great Britain first, she herself would be choked by British blockade. During that winter there was famine in Balkan countries and people were starving in the suburbs of Vienna. The French had by then suffered 3,350,000 casualties and the British over a million, the Germans had lost nearly two and a half million and were still fighting a war on two fronts. Food riots and strikes were becoming more frequent; infant mortality was rising towards a level 50 per cent higher than that of 1915. There was no reason to suppose that the German army, divided between east and west, would be any more likely to achieve a knockout than had been the British and French and it was in any case more favourably placed to fight on the defensive. In these circumstances the German general staff chose to resume unrestricted submarine warfare, the decision which brought about the first great transformation of the war in 1917, the entry into it of the United States. The Germans knew this would happen, but gambled on bringing Great Britain to her knees – and thus France – before American weight could be decisive.

American opinion, by no means favourable to one side or the other in 1914, had come a long way since then. Allied propaganda and purchases had helped; so had the first German submarine campaign. When the Allied governments began to talk about war aims which included the reconstruction of Europe on the basis of safeguarding the interests of nationalities it had an appeal to 'hyphenated' Americans. The resumption of unrestricted submarine warfare was decisive; it was a direct threat to American interests and the safety of her citizens. When it was also revealed to the American government that Germany hoped to negotiate an alliance with Mexico and Japan against the United States, the hostility aroused by the submarines was confirmed. Soon, an American ship was sunk without warning and the United States declared war shortly afterwards.

The impossibility of breaking the European deadlock by means short of total war had thus sucked the New World into the quarrels of the Old, almost against its will. The Allies were delighted; victory was now assured. Immediately, though, they faced a gloomy year. The year 1917 was even blacker for Great Britain and France than 1916. Not only did the submarine take months to master but a terrible series of battles in France (usually lumped under one name, Passchendaele) inflicted an ineffaceable scar upon the British national consciousness and cost another 400,000 men to gain five miles of mud. Worn out by heroic efforts in 1916, the French army underwent a series of mutinies. Worst of all for the Allies, the Russian empire collapsed and Russia ceased, by the end of the year, to be a great power for the foreseeable future.

The Russian state was destroyed by the war. This was the beginning of the revolutionary transformation of central and eastern Europe, too. The makers of what was called a 'revolution' in Russia in February 1917 were the German armies which had in the end broken the hearts of even the long-enduring Russian soldiers, who had behind them cities starving because of the breakdown of the transport system and a government of incompetent and corrupt men who feared constitutionalism and liberalism as much as defeat. At the beginning of 1917 the security forces themselves could no longer be depended upon. Food riots were followed by mutiny and the autocracy was suddenly seen to be powerless. A provisional government of liberals and socialists was formed and the Tsar abdicated. The new government then itself failed, in the main because it attempted the impossible, the continuation of the war; the Russians wanted peace and bread, as Lenin, the leader of the Bolsheviks, saw. His determination to take power from the moderate provisional government was the second reason for their failure. Presiding over a disintegrating country, administration and army, still facing the unsolved problems of privation in the cities, the provisional government was itself swept away in a second change, the *coup* called the October Revolution which, together with the American entry into the war, marks 1917 as a break between two eras of European history. Previously, Europe had settled its own affairs; now the United States would be bound to have a

Lenin in Red Square, Moscow, May Day 1919, with some evidence of bourgeois support in the background.

large say in its future and there had come into being a state which was committed by the beliefs of its founders to the destruction of the whole pre-war European order, a truly and consciously revolutionary centre for world politics.

The immediate and obvious consequence of the establishment of the Union of Soviet Socialist Republics (USSR), as Russia was now called, after the workers' and soldiers' councils which were her basic political institution after the revolution, was a new strategic situation. The Bolsheviks consolidated their *coup d'état* by dissolving (since they did not control it) the only freely-elected representative body based on universal suffrage Russia ever had and by trying to secure the peasant's loyalties by promises of land and peace. This was essential if they were to survive; the backbone of the party which now strove to assert its authority over Russia was the very small industrial working class of a few cities. Only peace could provide a safer and broader foundation. At first the terms demanded by the Germans were thought so objectionable that the Russians stopped negotiation; they then had to accept a much more punitive outcome, the Treaty of Brest-Litovsk, in March 1918. It imposed severe losses of territory, but gave the new order the peace and time it desperately needed to tackle its internal troubles.

The Allies were furious. They saw the Bolsheviks' action as a treacherous defection. Nor was their attitude towards the new regime softened by the intransigent revolutionary propaganda it directed against their citizens. The Russian leaders expected a revolution of the working class in all the advanced capitalist countries. This gave an extra dimension to a series of military interventions in the affairs of Russia by the Allies. Their original purpose was strategic, in that they hoped to stop the Germans exploiting the advantage of being able to close down their eastern front, but they were quickly interpreted by many people in the capitalist countries and by all Bolsheviks as anti-communist crusades. Worse still, they became entangled in a civil war which seemed likely to destroy the new regime. Even without the doctrinal filter of Marxist theory through which Lenin and his colleagues saw the world, these episodes would have been likely to sour relations between Russia and the capitalist countries for a long time; once translated into Marxist terms they seemed a confirmation of an essential and ineradicable hostility. Their recollection has dogged Russian attitudes ever since. They also helped to justify the Russian revolution's turn downwards into authoritarian government. Fear of the invader as a restorer of the old order and patron of landlords combined with Russian traditions of autocracy and police terrorism to stifle any liberalization of the regime.

The Russian communists' conviction that revolution was about to occur in central and western Europe was in one sense correct, yet crucially wrong. In its last year, war's revolutionary potential indeed became plain, but in national, not class, forms. The Allies were provoked (in part by the Bolsheviks) to a revolution strategy of their own. The military situation looked bleak for them at the end of 1917. It was obvious that they would face a German attack in France in the spring without the advantage of a Russian army to draw off their enemies and that it would be a long time before American troops arrived in large numbers to help them in France. But they could adopt a revolutionary weapon. They could appeal to the nationalities of the Austro-Hungarian empire. This had the additional advantage of emphasizing in American eyes the ideological purity of the Allied cause now that it was no longer tied to tsardom. Accordingly, in 1918, subversive propaganda was directed at the Austro-Hungarian armies and encouragement was given to Czechs and South Slavs in exile. Before Germany gave in, the Dual Monarchy was already dissolving under the combined effects of reawakened national sentiment and a Balkan campaign which at last began to provide victories. This was the second great blow to old Europe. The political structure of the whole area bounded by the Urals, the Baltic and the Danube valley was now in question as it had not been for centuries. There was even a Polish army again in existence. It was patronized by the Germans as a weapon against Russia, while the American president announced that an independent Poland was an essential of Allied peacemaking. All the certainties of the past century seemed to be in the melting-pot.

The crucial battles were fought against this increasingly revolutionary background. By the summer, the Allies had managed to halt the last great German offensive. It had made huge gains, but not enough. When the Allied armies began to move forward victoriously in their turn, the German leaders sought an end: they, too, thought they saw signs of revolutionary collapse at home. When the Kaiser abdicated, the third of the dynastic empires had fallen; the Habsburgs had already gone, so that the Hohenzollerns just outlasted their old rivals. A new German government requested an armistice and the fighting came to an end.

The cost of this huge conflict has never been adequately computed. One figure which is approximate indicates its scale: about ten million men had died as a result of direct military action. Yet typhus probably killed another million in the Balkans alone. Even such horrible figures as these do not indicate the physical cost in maiming, blinding,

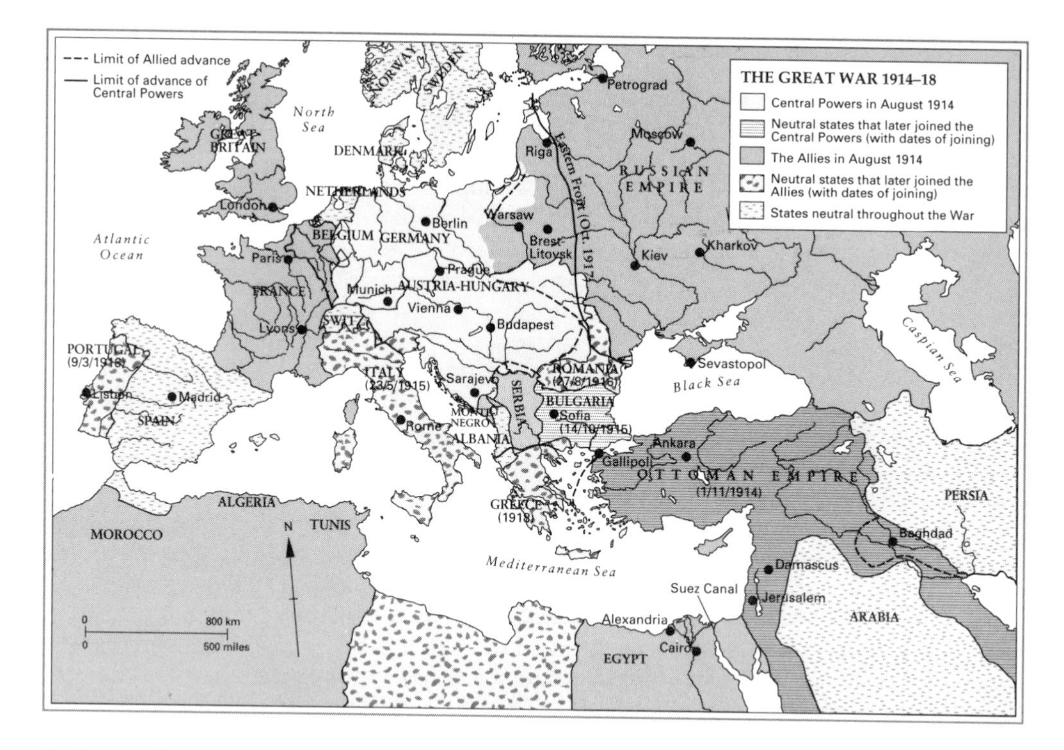

the loss to families of fathers and husbands, the spiritual havoc in the destruction of ideals, confidence and goodwill. Europeans looked at their huge cemeteries and were appalled at what they had done. The economic damage was immense, too. Over much of Europe people starved. A year after the war manufacturing output was still nearly a quarter below that of 1914; Russia's was only 20 per cent of what it had then been. Transport was in some countries almost impossible to procure. Moreover all the complicated, fragile machinery of international exchange was smashed and some of it could never be replaced. At the centre of this chaos lay, exhausted, a Germany which had been the economic dynamo of central Europe. 'We are at the dead season of our fortunes', wrote J.M. Keynes, a young British economist at the peace conference. 'Our power of feeling or caring beyond the immediate questions of our own material well-being is temporarily eclipsed ... We have been moved beyond endurance, and need rest. Never in the lifetime of men now living has the universal element in the soul of man burnt so dimly.'

Delegates to a peace conference began to assemble at the end of 1918. It was once the fashion to emphasize their failures, but perspective and the recognition of the magnitude of their tasks imposes a certain respect for what they did. It was the greatest settlement since 1815 and its authors had to reconcile great expectations with stubborn facts. The power to make the crucial decisions was remarkably concentrated: the British and French prime ministers and the American president Woodrow Wilson, dominated the negotiations. These took place between the victors; the defeated Germans were subsequently presented with their terms. In the diverging interests of France, aware above all of the appalling danger of any third repetition of German aggression, and of the Anglo-Saxon nations, conscious of standing in no such peril, lay the central problem of European security, but many others surrounded and obscured it. The peace settlement had to be a world settlement. It not only dealt with territories outside Europe – as earlier great settlements had done – but many non-European voices were heard in its making. Of twenty-seven states whose representatives signed the main treaty, a majority, seventeen, lay in other continents. The United States was the greatest of these; with Japan, Great Britain, France and Italy she formed the group described as the 'principal' victorious powers. For a world settlement, nevertheless, it was ominous that no representative attended from Russia, the only great power with both European and Asian frontiers.

Technically, the peace settlement consisted of a group of distinct treaties made not only with Germany, but Bulgaria, Turkey and the 'succession states' which claimed the divided Dual Monarchy. Of these a resurrected Poland, an enlarged Serbia called Yugoslavia and an entirely new Czechoslovakia were present at the conference as allies, while a much reduced Hungary and the Germanic heart of old Austria were treated as defeated enemies with whom peace had to be made. All of this posed difficult problems. But the main concern of the Peace Conference was the settlement with Germany embodied in the Treaty of Versailles signed in June 1919.

This was a punitive settlement and explicitly stated that the Germans were responsible for the outbreak of war. But most of the harshest terms arose not from this moral guilt but from the French wish, if possible, so to tie Germany down that any third German war was inconceivable. This was the purpose of economic reparations, which were the most unsatisfactory part of the settlement. They angered Germans and made acceptance of defeat even harder. Moreover they were economic nonsense. Nor was the penalizing of Germany supported by arrangements to ensure that Germany might not one day try to reverse the decision by force of arms, and this angered the French. Germany's territorial losses, it went without saying, included Alsace and Lorraine, but were otherwise greatest in the east, to Poland. In the west the French did not get much more reassurance than an undertaking that the German bank of the Rhine should be 'demilitarized'.

The second leading characteristic of the peace was its attempt where possible

The appalling bloodshed of the Great War led some to think that Europe's main handicap in setting her house in order after 1918 was the absence of a generation slaughtered on the battlefields. Some statesmen, certainly, for all their goodwill and ability, still leave an impression of wrestling with issues born in a world they did not understand and perhaps this was more marked in the great democracies than elsewhere. Two British prime ministers of these years, Stanley Baldwin and Ramsay MacDonald, at a meeting of the League of Nations.

to follow the principles of self-determination and nationality. In many places this merely meant recognizing existing facts; Poland and Czechoslovakia were already in existence as states before the peace conference met, and Yugoslavia was built round the core of the former Serbia. By the end of 1918, therefore, these principles had already triumphed over much of the area occupied by the old Dual Monarchy (and were soon to do so also in the former Baltic provinces of Russia). After outlasting even the Holy Roman Empire, the Habsburgs were gone at last and in their place appeared states which, though not uninterruptedly, were to survive most of the rest of the century. The principle of self-determination was also followed in providing that certain frontier zones should have their destiny settled by plebiscite.

Unfortunately, the principle of nationality could not always be applied. Geographical, historical, cultural and economic realities cut across it. When it prevailed over them – as in the destruction of the Danube's economic unity – the results could be bad; when it did not they could be just as bad because of the aggrieved feelings left behind. Eastern and central Europe was studded with national minorities embedded resentfully in nations to which they felt no allegiance. A third of Poland's population did not speak Polish; more than a third of Czechoslovakia's consisted of minorities of Poles, Russians, Germans, Magyars and Ruthenes; an enlarged Romania now contained over a million Magyars. In some places, the infringement of the principle was felt with especial acuteness as an injustice. Germans resented the existence of a 'corridor' connecting Poland with the sea across German lands, Italy was disappointed of Adriatic spoils held out to her by her allies when they had needed her help, and the Irish had still not got Home Rule after all.

The most obvious non-European question concerned the disposition of the German colonies. Here there was an important innovation. Old-fashioned colonial greed was not acceptable to the United States; instead, tutelage for non-European peoples formerly under German or Turkish rule was provided by the device of trusteeship. 'Mandates' were given to the victorious powers (though the United States declined any) by a new 'League of Nations' to administer these territories while they were prepared for self-government; it was the most imaginative idea to emerge from the settlement, even though it was used to drape with respectability the last major conquests of European imperialism.

The League of Nations owed much to the enthusiasm of the American president, Woodrow Wilson who ensured its Covenant – its constitution – pride of place as the first part of the Peace Treaty. It was the one instance in which the settlement transcended the idea of nationalism (even the British empire had been represented as individual units, one of which, significantly, was India). It also transcended that of Europe; it is another sign of the new era that twenty-six of the original forty-two members of the League were countries outside Europe. Unfortunately, because of domestic politics Wilson had not taken into account, the United States was not among them. This was the most fatal of several grave weaknesses which made it impossible for the League to satisfy the expectations it had aroused. Perhaps these were all unrealizable in principle, given the actual state of world political forces. None the less, the League was to have its successes in handling matters which might without its intervention have proved dangerous. If exaggerated hopes had been entertained that it might do more, it does not mean it was not a practical as well as a great and imaginative idea.

Russia was absent from the League just as she was from the peace conference. Probably the latter was the more important. The political arrangements to shape the next stage of European history were entered into without consulting her, though in eastern Europe this meant the drawing of boundaries in which any Russian government was bound to be vitally interested. It was true that the Bolshevik leaders did all they could to provide excuses for excluding them. They envenomed relations with the major powers by revolutionary propaganda, for they were convinced that the capitalist countries were determined to overthrow them. The British prime minister, Lloyd George, and Wilson were in fact more flexible – even sympathetic – than many of their colleagues and electors in dealing with Russia. Their French colleague, Clemenceau, on the other

EAST INDIA COMPANY OFFICIAL *Surrounded by the usual dignified appurtenances of Mughal India, by then in decline, an official of the eighteenth century East India Company, harbinger of a new order, goes escorted about his business. Bearers and guards with ceremonial staffs precede him, the elephant (conveniently supplied with ladder for ease of mounting and dismounting) is richly decorated and caparisoned, and a mysterious enclosed palanquin accompanies him, probably conveying his Indian lady.*

BENIN *The art of the West African kingdom of Benin was royal art, focussed upon the glorifying of the Oba or 'divine king', his predecessors, famous ancestors and great events in the kingdom's history. It was as firmly directed to that end as the self-glorifying paintings and sculptures of European princes, but was more limited in form, to bronze, cast brass and ivory. At the end of the nineteenth century, a huge amount of this art became available to European collectors and museums, thanks to a British punitive expedition in 1897 which deposed the Oba, destroyed his palace and carried off thousands of specimens of Benin art. Thriftily, the British government sold them off to pay for the expedition, among them this male head, now in the British Museum.*

hand, was passionately anti-Bolshevik and had the support of many French ex-soldiers and investors in being so; Versailles was the first great European peace to be made by powers all the time aware of the dangers of disappointing democratic electorates. But however the responsibility is allocated, the outcome was that Russia, the European power which had, potentially, the greatest weight of all in the affairs of the continent, was not consulted in the making of a new Europe. Though for the time being virtually out of action, she was bound eventually to join the ranks of those who wished to revise the settlement or overthrow it. It only made it worse that her rulers detested the social system it was meant to protect.

Huge hopes had been entertained of this settlement. They were often unrealistic, yet in spite of its manifest failures, the peace has been over-condemned, for it had many good points. When it failed, it was for reasons which were for the most part beyond the control of the men who made it. In the first place, the days of a European world hegemony in the narrow political sense were over. The peace treaties of 1919 could do little to secure the future beyond Europe. The old imperial policemen were now too weakened to do their job inside Europe, let alone outside; some had disappeared altogether. In the end the United States had been needed to ensure Germany's defeat but now she plunged into a period of artificial isolation. Nor did Russia wish to be involved in the continent's stabilizing. The isolationism of the one power and the sterilization of the other by ideology left Europe to its own inadequate devices. When no revolution broke out in Europe, the Russians turned in on themselves; when Americans were given the chance by Wilson to be involved in Europe's peace-keeping, they refused it. Both decisions are comprehensible, but their combined effect was to preserve an illusion of European autonomy which was no longer a reality and could no longer be an adequate framework for handling its problems. Finally, the settlement's gravest immediate weakness lay in the economic fragility of the new structures it presupposed. Here its terms were more in question: self-determination often made nonsense of economics. But it is difficult to see on what grounds self-determination could have been set aside. Ireland's problems are still with us seventy years after an independent Irish Free State appeared in 1922.

The situation was all the more likely to prove unstable because many illusions persisted in Europe and many new ones arose. Allied victory and the rhetoric of peace-making made many think that there had been a great triumph of liberalism and democracy. Four autocratic anti-national illiberal empires had collapsed, after all, and to this day the peace settlement retains the distinction of being the only one in history made by great powers all of whom were democracies. Liberal optimism also drew strength from the ostentatious stance taken by Wilson during the war; he had done all he could to make it clear that he saw the participation of the United States as essentially different in kind from that of the other Allies, being governed (he reiterated) by high-minded ideals and a belief that the world could be made safe for democracy if other nations would give up their bad old ways. Some thought that he had been shown to be right; the new states, above all the new Germany, adopted liberal, parliamentary constitutions and often republican ones, too. Finally, there was the illusion of the League; the dream of a new international authority which was not an empire seemed at last a reality.

Yet all this was rooted in fallacy and false premise. Since the peacemakers had been obliged to do much more than enthrone liberal principles – they had also to pay debts, protect vested interests, and take account of intractable facts – those principles had been much muddied in practice. Above all, they had left much unsatisfied nationalism about and had created new and fierce nationalist resentments in Germany. Perhaps this could not be helped, but it was a soil in which things other than liberalism could grow. Further, the democratic institutions of the new states – and the old ones, too, for that matter – were being launched on a world whose economic structure was terribly damaged. Everywhere, poverty, hardship and unemployment exacerbated political struggle and in many places they were made worse by the special dislocations produced by respect for national sovereignty. The crumbling of old economic patterns of exchange in the war made it much more difficult to deal with problems like peasant poverty and unemployment, too;

Russia, once the granary of much of western Europe, was now inaccessible economically. This was a background which revolutionaries could exploit. The communists were happy and ready to do this, for they believed that history had cast them for this rôle, and soon their efforts were reinforced in some countries by another radical phenomenon, fascism.

Following upon war, revolution and civil war, famine brought immeasurable suffering to millions of Russians and Ukrainians in 1921.

Communism threatened the new Europe in two ways. Internally, each country soon had a revolutionary communist party. They effected little that was positive, but caused great alarm. They also did much to prevent the emergence of strong progressive parties. This was because of the circumstances of their birth. A 'Comintern', or Third International, was devised by the Russians in March 1919 to provide leadership for the international socialist movement which might otherwise, they feared, rally again to the old leaders whose lack of revolutionary zeal they blamed for a failure to exploit the opportunities of the war. The test of socialist movements for Lenin was adherence to the Comintern, whose principles were deliberately rigid, disciplined and uncompromising, in accordance with his view of the needs of an effective revolutionary party. In almost every country this divided socialists into two camps. Some adhered to the Comintern and took the name communist; others, though sometimes claiming still to be Marxists, remained in the rump national parties and movements. They competed for working-class support and fought one another bitterly.

The new revolutionary threat on the Left was all the more alarming to many Europeans because there were plenty of revolutionary possibilities for communists to exploit. The most conspicuous led to the installation of a Bolshevik government in Hungary, but more startling, perhaps, were attempted communist coups in Germany, some briefly successful. The German situation was especially ironical, for the government of the new republic which emerged there in the aftermath of defeat was dominated by socialists who were forced back to reliance upon conservative forces – notably the professional soldiers of the old army – in order to prevent revolution. This happened

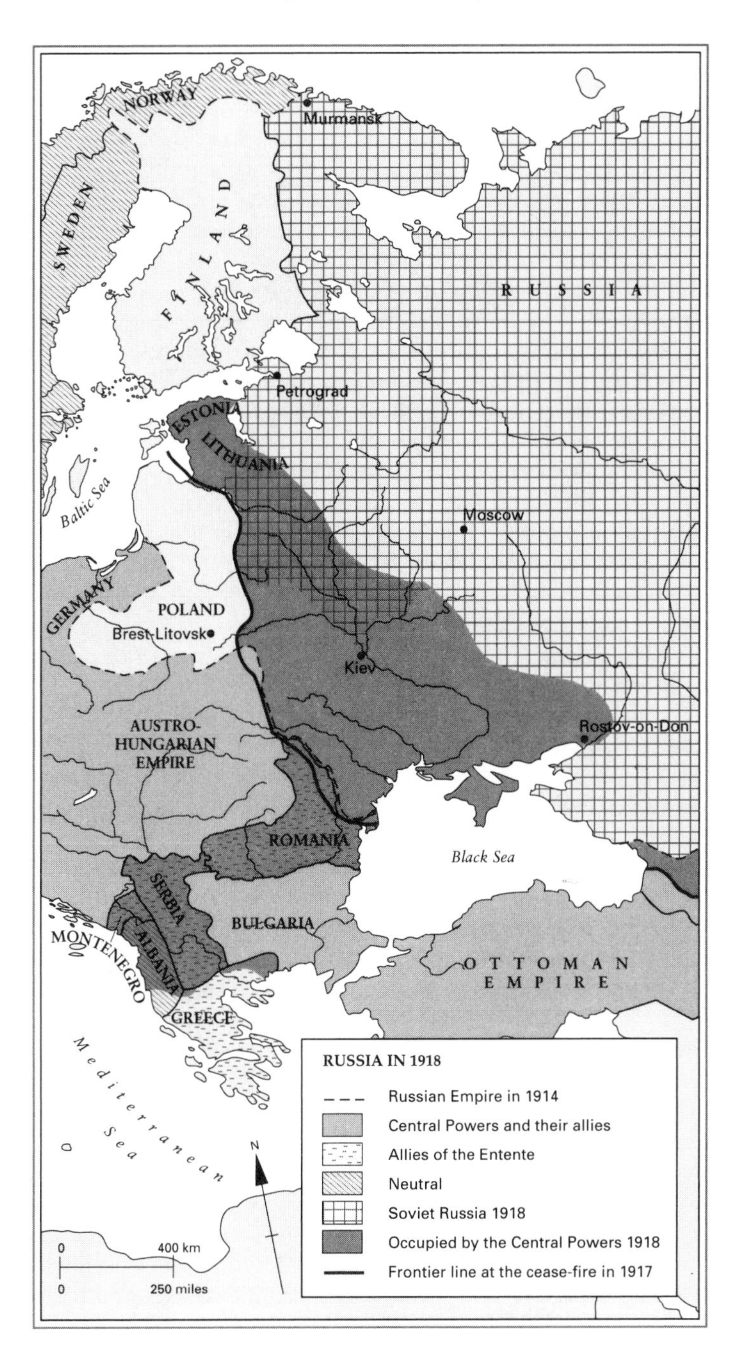

even before the founding of the Comintern and it gave a special bitterness to the divisions of the Left in Germany. But everywhere, communist policy made united resistance to conservatism more difficult, frightening moderates with revolutionary rhetoric and conspiracy.

In eastern Europe, the social threat was often seen also as a Russian threat. The Comintern was manipulated as an instrument of Soviet foreign policy by the Bolshevik leaders; this was justifiable, given their assumption that the future of world revolution depended upon the preservation of the first socialist state as a citadel of the international working class. In the early years of civil war and slow consolidation of Bolshevik power in Russia that belief led to the deliberate incitement of disaffection abroad in order to preoccupy capitalist governments. But in eastern and central Europe there was more to it than this, because the actual territorial settlement of that area was in doubt long after the Versailles treaty. The First World War did not end there until in March 1921 a peace treaty between Russia and the new Polish Republic provided frontiers lasting until 1939. Poland was the most anti-Russian by tradition, the most anti-Bolshevik by religion, as well as the largest and most ambitious of the new nations. But all of them felt threatened by a recovery of Russian power, especially now that it was tied up with the threat of social revolution. This connexion helped to turn many of these states before 1939 to dictatorial or military governments which would at least guarantee a strong anti-communist line.

Fear of communist revolution in eastern and central Europe was most evident in the immediate post-war years, when economic collapse and uncertainty about the outcome of the Polish-Russian war (which, at one time, appeared to threaten Warsaw itself) provided the background. In 1921, with peace at last and, symbolically, the establishment of orderly official relations between the USSR and Great Britain, there was a noticeable relaxation. This was connected with the Russian government's own sense of emerging from a period of acute danger in the civil war. It did not produce much in the way of better diplomatic manners, and revolutionary propaganda and denunciation of capitalist countries did not cease, but the Bolsheviks could now turn to the rebuilding of their own shattered land. In 1921 Russian pig-iron production was about one-fifth of its 1913 level, that of coal a tiny 3 per cent or so, while the railways had less than half as many locomotives in service as at the start of the war. Livestock had declined by over a quarter and cereal deliveries were less than two-fifths of those of 1916. On to this impoverished economy there fell in 1921 a drought in south Russia. More than two million died in the subsequent famine and even cannibalism was reported.

Liberalization of the economy brought about a turnround. By 1927 both industrial and agricultural production were nearly back to pre-war levels. The regime in these years was undergoing great uncertainty in its leadership. This had already been apparent before Lenin died in 1924, but the removal of a man whose acknowledged ascendancy had kept forces within it in balance opened a period of evolution and debate in the Bolshevik leadership. It was not about the centralized, autocratic nature of the regime which had emerged from the 1917 revolution, for none of the protagonists considered that political liberation was conceivable or that the use of secret police and the party's dictatorship could be dispensed with in a world of hostile capitalist states. But they could disagree about economic policy and tactics and personal rivalry sometimes gave extra edge to this.

Broadly speaking, two viewpoints emerged. One emphasized that the revolution depended on the goodwill of the mass of Russians, the peasants; they had first been allowed to take the land, then antagonized by attempts to feed the cities at their expense, then conciliated again by the liberalization of the economy and what was known as 'NEP', the New Economic Policy which Lenin had approved as an expedient. Under it, the peasants had been able to make profits for themselves and had begun to grow more food and to sell it to the cities. The other viewpoint showed the same facts in a longer perspective. To conciliate the peasants would slow down industrialization, which Russia needed to survive in a hostile world. The party's proper course, argued those who took this view, was to rely upon the revolutionary militants of the cities and to exploit the still non-Bolshevized peasants in their interest while pressing on with industrialization and the promotion of revolution abroad. The communist leader Trotsky took this view.

What happened was roughly that he was shouldered aside, but his view prevailed. From the intricate politics of the party there emerged eventually the ascendancy of a member of its bureaucracy, Joseph Stalin, a man far less attractive intellectually than either Lenin or Trotsky, equally ruthless, and of greater historical importance. Gradually arming himself with a power which he used against former colleagues and old Bolsheviks as willingly as against his enemies, he carried out the real Russian revolution to which the Bolshevik seizure of power had opened the way and created a new élite on which a new Russia was to be based. For him industrialization was paramount. The road to it lay through finding a way of forcing the peasant to pay for it by supplying the grain he would rather have eaten if not offered a good profit. Two 'Five Year Plans' carried out an industrialization programme from 1928 onwards, and their roots lay in the collectivization of agriculture. The Party now for the first time conquered the countryside. In a new civil war millions of peasants were killed or transported, and grain levies brought back famine. But the towns were fed, though the police apparatus kept consumption down to the minimum. There was a fall in real wages. But by 1937 80 per cent of Russian industrial output came from plant built since 1928. Russia was again a great power and the effects of this alone would assure Stalin a place in history.

The price in suffering was enormous. The enforcement of collectivization was only made possible by brutality on a scale far greater than anything seen under the tsars and it made Russia a totalitarian state far more effective than the old autocracy had been. Stalin, though himself a Georgian, looks a fully Russian figure, a despot whose ruthless use of power is anticipated by an Ivan the Terrible or a Peter the Great. He was also a somewhat paradoxical claimant to Marxist orthodoxy, which taught that the economic structure of society determined its politics. Stalin precisely inverted this; he demonstrated that if the will to use political power was there, the economic substructure could be revolutionized by force.

Critics of liberal capitalist society in other countries often held up Soviet Russia, of which they had a very rosy picture, as an example of the way in which a society might achieve progress and a revitalization of its cultural and ethical life. But this was not the only model offered to those who found the civilization of the West disappointing. In the 1920s in Italy a movement appeared called fascism. It was to lend its name to a number of other and only loosely related radical movements in other countries which had in common a rejection of liberalism and strong anti-Marxism. The Great War had badly strained constitutional Italy. Though poorer than other countries regarded in 1914 as great powers, her share of fighting had been disproportionately heavy and often unsuccessful and much of it had taken place on Italian territory. Inequalities had accentuated social divisions as the war went on. With peace came even faster inflation, too. The owners of property, whether agricultural or industrial, and those who could ask higher wages because of a labour shortage, were more insulated against it than the middle classes and those who lived on investment or fixed incomes. Yet these were on the whole the most convinced supporters of the unification completed in 1870. They had sustained a constitutional and liberal state while conservative Roman Catholics and revolutionary socialists had long opposed it. They had seen the war Italy entered in 1915 as an extension of the *Risorgimento*, the nineteenth-century struggle to unite Italy as a nation, a crusade to remove Austria from the last soil she ruled which was inhabited by those of Italian blood or speech. Like all nationalism, this was a muddled, unscientific notion, but it was powerful.

Peace brought to Italians disappointment and disillusion; many nationalist dreams were left unrealized. Moreover, as the immediate post-war economic crisis deepened, the socialists grew stronger in parliament and seemed more alarming now that a socialist revolutionary state existed in Russia. Disappointed and frightened, tired of socialist anti-nationalism, many Italians began to cast loose from liberal parliamentarianism and to look for a way out of Italy's disappointments. They were sympathetic to intransigent nationalism abroad for example, to an adventurer who seized the Adriatic port of Fiume which the Peace Conference had failed to give to Italy) and violent anti-Marxism at home. The second was bound to be attractive in a Roman Catholic country, but it was not only from the traditionally conservative Church that the new leadership against Marxism came.

In 1919 a journalist and ex-serviceman who had before the war been an extreme socialist, Benito Mussolini, formed a movement called the *fascio di combattimento*, which can be roughly translated as 'union for struggle'. It sought power by any means, among them violence by groups of young thugs, directed at first against socialists and working-class organizations, then against elected authorities. The movement prospered. Italy's constitutional politicians could neither control it nor tame it by cooperation. Soon the fascists (as they came to be called) often enjoyed official or quasi-official patronage and protection from local officials and police. Gangsterism was semi-institutionalized. By 1922 they had not only achieved important electoral success but had virtually made orderly government impossible in some places by terrorizing their political enemies, especially if they were communist or socialist. In that year, other politicians having failed to master the fascist challenge, the king called Mussolini to form a government; he did so, on a coalition basis, and the violence came to an end. This was what was called in later fascist mythology the 'March on Rome', but was not quite the end of constitutional Italy. Mussolini only slowly turned his position into a dictatorship. In 1926 government by decree began; elections were suspended. There was little opposition.

The new regime had terrorism in plenty at its roots, and it explicitly denounced liberal ideals, yet Mussolini's rule was far short of totalitarian and was much less brutal than the Russian (of which he sometimes spoke admiringly). He undoubtedly had aspirations to revolutionary change, and many of his followers much stronger ones, but revolution turned out in practice to be largely a propaganda claim; Mussolini's own temperamental impatience with an established society from which he felt excluded lay behind it, as much as real radical pressure in his movement. Italian fascism in practice and theory rarely achieved coherence; instead, it reflected more and more the power of established Italy. Its greatest domestic step was a diplomatic agreement with the Papacy, which in return for substantial concessions to the authority of the Church in Italian life (which persist to this day) recognized the Italian state officially for the first time. For all fascism's revolutionary rhetoric, the Lateran treaties of 1929 which embodied this agreement were a concession to the greatest conservative force in Italy. 'We have given back God to Italy and Italy to God,' said the pope. Just as unrevolutionary were the results of fascist criticism of free enterprise. The subordination of individual interest to the state boiled down to depriving trades unions of their power to protect their members' interests. Few checks were placed on the freedom of employers and fascist economic planning was a mockery. Only agricultural production notably improved.

The same divergence between style and aspiration on the one hand and achievement on the other was also to be marked in movements elsewhere which have been called fascist. Though indeed reflecting something new and post-liberal – they were inconceivable except as expressions of mass society – such movements almost always in practice made compromising concessions to conservative influences. This makes it difficult to speak of the phenomenon 'fascism' at all precisely; in many countries regimes appeared which were authoritarian – even totalitarian in aspiration – intensely nationalist, and anti-Marxist. But fascism was not the only possible source of such ideas. Such governments as those which emerged in Portugal and Spain, for example, drew upon traditional and conservative forces rather than upon those which arose from the new phenomenon of mass politics. In them, true radicals who were fascists often felt discontented at concessions made to the existing social order. Only in Germany, in the end, did a movement some termed 'fascist' succeed in a revolution which mastered historical conservatism. For such reasons, the label of fascism sometimes confuses as much as it clarifies.

Perhaps it is best merely to distinguish two separable phenomena of the twenty years after 1918. One is the appearance (even in stable democracies such as Great Britain and France) of ideologists and activists who spoke the language of a new, radical politics, emphasized idealism, will-power and sacrifice, and looked forward to rebuilding society and the state on new lines without respect to vested interests or concessions to materialism. This was a phenomenon which, though widespread, triumphed in only two major states, Italy and Germany. In each of these, economic collapse, outraged nationalism and anti-Marxism were the sources of success, though that in Germany did not come until 1933. If one word is wanted for this, let it be fascism. In other countries, usually underdeveloped economically, it might be better to speak of authoritarian, rather than fascist, regimes, especially in eastern Europe. There, large agricultural populations presented problems aggravated by the peace settlement. Sometimes alien national minorities appeared to threaten the state. Liberal institutions were only superficially implanted in many of the new countries and traditional conservative social and religious forces were strong. As in Latin America, where similar economic conditions could be found, their apparent constitutionalism tended to give way sooner or later to the rule of strong men and soldiers. This proved the case before 1939 in the new Baltic states, Poland and all the successor states of the Dual Monarchy except Czechoslovakia, the one effective democracy in central Europe or the Balkans. The need of these states to fall back on such regimes demonstrated both the unreality of the hopes entertained of their political maturity in 1918 and the new fear of Marxist communism, especially acute on Russia's borders. Such pressure operated also – though less acutely – in Spain and Portugal, where the influence of traditional conservatism was even stronger and Catholic social thinking counted for more than fascism.

The failures of democracy between the wars did not proceed at an even pace; in the 1920s a bad economic start was followed by a gradual recovery of prosperity in which most of Europe outside Russia shared, and the years from 1925 to 1929 were on the whole good ones. This permitted optimism about the political future of the new democratic nations. Currencies emerged from appalling inflation in the first half of the decade and were once more stable; the resumption by many countries of the gold standard was a sign of confidence that the old pre-1914 days were returned. In 1925 the production of food and raw materials in Europe for the first time passed the 1913 figure and a recovery of manufacturing was also under way. With the help of a world-wide recovery of trade and huge investment from the United States, now an exporter of capital, Europe reached in 1929 a level of trade not to be touched again until 1954.

Yet collapse followed. Economic recovery had been built on insecure foundations. When faced with a sudden crisis, the new prosperity crumbled rapidly. There followed not merely a European but a world economic crisis which was the single most important event between two world wars.

The complex but remarkably efficient economic system of 1914 had in fact been irreparably damaged. International exchange was hampered by a huge increase of restrictions immediately after the war as new nations strove to protect their infant economies with tariffs and exchange control, and bigger and older nations tried to repair their enfeebled ones. The Versailles treaty made things worse by saddling Germany, the most important of all the European industrial states, with an indefinite burden of reparation in kind and in cash. This not only distorted her economy and delayed its recovery for years, but also took away much of the incentive to make it work. To the east, Germany's greatest potential market, Russia, was almost entirely cut off behind an economic frontier which little trade could penetrate; the Danube valley and the Balkans, another great area of German enterprise, was divided and impoverished. Temporarily, these difficulties were gradually overcome by the availability of American money, which Americans were willing to supply (though they would not take European goods and retired behind their tariff walls). But this brought about a dangerous dependence on the continued prosperity of the United States.

In the 1920s the United States produced nearly 40 per cent of the world's coal and over half the world's manufactures. This abundance, enhanced by the demands of war, had transformed the life of many Americans, the first people in the world to be able to take for granted the possession of family automobiles. Unfortunately, American domestic prosperity carried the world. On it depended the confidence which provided American capital for export. Because of this, a swing in the business cycle turned into a world economic disaster. In 1928 short-term money began to be harder to get in the United States. There were also signs that the end of the long boom might be approaching. These two factors led to the calling back of American loans from Europe. Soon some European borrowers were in difficulties. Meanwhile, demand was slackening in the United States as people began to think a severe slump might be on the way. Almost accidentally, this detonated a particularly sudden and spectacular stock market collapse in October 1929. It did not matter that there was thereafter a temporary rally and that great bankers bought to restore confidence. It was the end of American business confidence and of overseas investment. After a last brief rally in 1930 American money for investment abroad dried up. The world slump began.

Economic growth came to an end because of the collapse of investment, but another factor was soon operating to accelerate disaster. As the debtor nations tried to put their accounts in order, they cut imports. This caused a drop in world prices, so that countries producing primary goods could not afford to buy abroad. Meanwhile, at the centre of things, both the United States and Europe went into a financial crisis; as countries struggled, unsuccessfully, to keep the value of their currencies steady in relation to gold (an internationally acceptable means of exchange – hence the expression 'gold standard') they adopted deflationary policies to balance their books which again cut demand. By 1933 all the major currencies, except the French, were off gold. This was the symbolic expression of the tragedy, the dethronement of one old idol of liberal economics. Its reality was a level of unemployment which may have reached thirty million in the industrial world. In 1932 (the worst year for industrial countries) the index of industrial production for the United States and Germany was in each case only just above half of what it had been in 1929.

The effects of economic depression rolled outwards with a ghastly and irresistible logic. The social gains of the 1920s, when many people's standard of living had improved, were wiped out. No country had a solution to unemployment and though it was at its worst in the United States and Germany it existed in a concealed form all round the world in the villages and farmlands of the primary producers. The national income of the United States fell by 38 per cent between 1929 and 1932; this was exactly the figure by which the prices of manufactured goods fell, but at the same time raw material prices fell by 56 per cent and foodstuffs by 48 per cent respectively. Everywhere, therefore, the poorer nations and the poorer sectors of the mature economies suffered disproportionately. They may not always have seemed to do so, because they had less far to fall; an eastern European or an Argentinian peasant may not have been absolutely much worse off for he was always badly off, while an unemployed German clerk or factory hand certainly was worse off and knew it.

There was to be no world recovery before another great war. Nations cut themselves off more and more behind tariffs and strove in some cases to achieve economic self-sufficiency by an increasing state control of their economic life. Some did better than others, some very badly. The disaster was a promising setting for the communists and fascists who expected or advocated the collapse of liberal civilization and now began to flap expectantly about the enfeebled carcass. The end of the gold standard and the belief in non-interference with the economy mark the collapse of a world order in its economic dimension as strikingly as the rise of totalitarian regimes and the rise of nationalism to its destructive climax mark it in its political. Liberal civilization, frighteningly, had lost its power to control events. Many Europeans still found it hard to see this, though, and they continued to dream of the restoration of an age when that civilization enjoyed unquestioned supremacy. They forgot that its values had rested on a political and economic hegemony which, for a time, had worked but was already visibly in decay all round the world.

3
A New Asia in the Making

Europe's troubles could not be confined to one continent. They were bound soon to cramp her ability to dominate affairs elsewhere and the earliest signs of this came in Asia. European colonial power in Asia was, in the perspective of world history, only very briefly unchallengeable and unchallenged. By 1914 one European power, Great Britain, had made an ally of Japan in order to safeguard her interests in the Far East, rather than rely on her own resources. Another, Russia, had been beaten by Japan in war and had turned back towards Europe after twenty years of pressure towards the Yellow Sea. Even the bullying of China which seemed likely to prove fatal at the time of the Boxer rebellion was relaxed; she lost no more territory to European imperialists after that. As tensions in Europe mounted and the difficulty of frustrating Japanese ambitions indefinitely became clear, European statesmen realized that the time for acquiring new ports or dreaming of partitions of the Sick Man of the Far East was over. It would suit everyone better to turn to what was always, in effect, British policy, that of an Open Door through which all countries might seek their own commercial advantage. That advantage, too, showed signs of being much less spectacular than had been thought in the sanguine days of the 1890s and here was another reason to tread more softly in the Far East.

The peak of the European onslaught on Asia and its greatest successes were therefore past by 1914. The revolutionizing of Asia by colonialism, cultural interplay, and economic power had already produced defensive reflexes which had to be taken seriously.They had gone furthest in Japan and it was their indirect operation as catalysts of modernization, channelled through this local and Asian force, which set the pace of the next phase of the Hundred Years' War of East and West. Japanese dynamism dominates Asian history in the first forty years of this century; China's revolution had no similar impact until after 1945 when, together with new change-making forces from outside, that country again surpassed Japan in importance as a shaper of Asian affairs and closed the Western Age in Asia.

Japan's dynamism showed itself both in economic growth and territorial aggressiveness. For a long time the first was more obvious. It was part and parcel of an overall process of what was seen as 'westernizing' which could in the 1920s still sustain a mood of liberal hopefulness about Japan and helped to mask Japanese imperialism. In 1925 universal suffrage was introduced and in spite of much European evidence that this had no necessary connexion with liberalism or moderation, it seemed to confirm once again a pattern of steady constitutional progress begun in the nineteenth century.

This confidence, shared both by foreigners and by Japanese, was for a time helped by Japan's industrial growth, notably in the mood of expansive optimism awoken by the Great War, which gave her great opportunities: markets (especially in Asia) in which she had been faced by heavy western competition were abandoned to her when their former exploiters found they could not meet the demands of the war in their own countries; the Allied governments ordered great quantities of munitions from Japanese factories; a world shipping shortage gave her new shipyards the work they needed. The Japanese gross national product went up by forty per cent during the war years. Though interrupted in 1920 expansion was resumed later in the decade and in 1929 the Japanese had an industrial base which (though it still engaged less than one in five of the population) had in twenty years seen its steel production rise almost tenfold, its textile production tripled, and its coal output doubled. Her manufacturing sector was beginning to influence other Asian countries, too; she imported iron ore from China and Malaya, coal from Manchuria. Still small though her manufacturing industry was by comparison with that of the Western powers, and though it coexisted with an enduring small-scale and artisan sector, Japan's new industrial strength was beginning to shape both domestic politics and foreign relations in the 1920s. In particular, it affected her relations with mainland Asia.

Asia contemplates the instruments of recovery from western domination; Chinese officers examine some of the first products of the new Nanking arsenal, built at the end of the nineteenth century. In spite of such innovations, the imperial regime could not sufficiently liberate itself from the past to use China's resources adequately.

A contrast to the pre-eminent and dynamic role of Japan was provided there by the continuing eclipse of China, potentially the greatest of Asian and world powers. The 1911 revolution had been of enormous importance, but did not by itself end this eclipse. In principle, it marked an epoch far more fundamentally than the French or Russian revolutions: it was the end of more than two thousand years of history during which the Confucian state had held China together and Confucian ideals had dominated Chinese culture and society. Inseparably intertwined, Confucianism and the legal order fell together. The 1911 revolution proclaimed the shattering of the standards by which traditional China lived. On the other hand, the revolution was limited, in two ways especially. In the first place, it was destructive rather than constructive. The monarchy had held together a vast country, virtually a continent, of widely different regions. Its collapse meant that the centrifugal regionalism which so often expressed itself in Chinese history could again have full rein. Many of the revolutionaries were animated by a bitter envy and distrust of Peking. Secret societies, the gentry and military commanders were all ready and willing to step forward and take a grip of affairs in their own regions. These tendencies were somewhat masked while Yüan Shih-k'ai remained at the head of affairs (until 1916), but then burst out. The revolutionaries were split between a group round Sun Yat-sen called the Chinese National People's Party, or *Kuomintang* (KMT), and those who upheld the central government based on the parliamentary structure at Peking. Sun's support was drawn mainly from Canton businessmen and certain soldiers in the south. Against this background warlords thrived. They were soldiers who happened to have control of substantial forces and arms at a time when the central government was continuously weak. Between 1912 and 1928 there were some 1300 of them, often controlling important areas. Some of them carried out reforms. Some were simply bandits. Some had considerable status as plausible pretenders to government power. It was a little like the end of the Roman empire, though less drawn out. Nothing took the place of the old scholar-bureaucrats and the soldiers stepped forward to fill the void. Yüan Shih-k'ai himself can be regarded as the outstanding example of the type.

This reflected the second limitation of the revolution of 1911: it provided no basis of agreement for further progress. Sun Yat-sen had said that the solution of the national question would have to precede that of the social. But even about the shape of a nationalist future there was much disagreement, and the removal of the dynasty took away a common enemy that had delayed its emergence. Although eventually creative, the intellectual confusion marked among the revolutionaries in the first decade of the Chinese Revolution was deeply divisive and symptomatic of the huge task awaiting China's would-be renovators.

From 1916 a group of cultural reformers began to gather particularly at the university of Peking. The year before, one of them, Ch'en Tu-hsiu, had founded a journal called *New Youth* which was the focus of the debate they ignited. Ch'en preached to Chinese youth, in whose hands he believed the revolution's destiny to lie, a total rejection of the old Chinese cultural tradition. Like other intellectuals who talked of Huxley and Dewey and introduced to their bemused compatriots the works of Ibsen, Ch'en still thought the key lay in the West; in its Darwinian sense of struggle, its individualism and utilitarianism, it still seemed to offer a way ahead. But important though such leadership was and enthusiastic though its disciples might be, an emphasis on a western re-education for China was a handicap. Not only were many educated and patriotic Chinese sincerely attached to the traditional culture, but western ideas were only sure of a ready welcome among the most untypical elements of Chinese society, the seaboard city-dwelling merchants and their student offspring, often educated abroad. The mass of Chinese could hardly be touched by such ideas and appeals, and the demand of other reformers for a vernacular literature was one evidence of this fact.

In so far as they were touched by nationalist feeling the Chinese were likely to turn against the West and against the western-inspired capitalism which, for many of them, meant one more kind of exploitation and was the most obvious constituent of the civilization some modernizers urged them to adopt. But for the most part China's peasant masses seemed after 1911 relapsed in passivity, apparently unmoved by events and unaware of the agitation of angry and westernized young men. It is not easy to generalize about their economic state: China was too big and too varied. But it seems clear that while the population steadily increased, nothing was done to meet the peasants' hunger for land; instead, the number of the endebted and landless grew, their wretched lives frequently made even more intolerable by war, whether directly, or through its concomitants, famine and disease. The Chinese Revolution would only be assured success when it could activate these people, and the cultural emphasis of the reformers sometimes masked an unwillingness to envisage the practical political steps necessary for this.

China's weakness remained Japan's opportunity. A world war was the occasion to push forward again her nineteenth-century policies. The advantages offered by the Europeans' quarrels with one another could be exploited. Japan's allies could hardly object to her seizure of the German ports in China; even if they did, they could do nothing about it while they needed Japanese ships and manufactures. There was always the hope, too, that the Japanese might send their own army to Europe to fight, though nothing like this happened. Instead, the Japanese finessed by arousing fears that they might make a separate peace with the Germans and pressed ahead in China.

At the beginning of 1915 the Japanese government presented to the Chinese government a list of twenty-one demands, following them up with an ultimatum. In their entirety they amounted to a proposal for a Japanese protectorate over China. The United

Kingdom and United States did what diplomacy could do to have them reduced but, in the end, the Japanese got much of what they asked for. It included further confirmation of their special commercial and leasehold rights in Manchuria. Chinese patriots were enraged, but there was nothing they could do at a moment when their internal politics were in disorder. They were so confused, indeed, that Sun Yat-sen was himself at this moment seeking Japanese support.

The next intervention came in 1916, when Japanese pressure was brought to bear on the British to dissuade them from approving Yüan Shih-k'ai's attempt to restore stability by making himself emperor. In the following year came another treaty, this time extending the recognition of Japan's special interests as far as Inner Mongolia. In August 1917 the Chinese government went to war with Germany, partly in the hope of winning goodwill and support which would ensure her an independent voice at the peace, but only a few months later the United States formally recognized the special interests of Japan in China in return for endorsement of the principle of the 'Open Door' and a promise to maintain Chinese integrity and independence. All that the Chinese had got from the Allies was the ending of German and Austrian extra-territoriality and the concession that payment of Boxer indemnities to the Allies should be delayed. The Japanese, moreover, secured more concessions from China in secret agreements in 1917 and 1918.

Yet, when the peace came, it deeply disappointed both the great Asian powers of the future. Japan was now indisputably a world power; she had in 1918 the third largest navy in the world. It was true, too, that she won solid gains at the peace: she retained the former German rights in Shantung (promised to her by the British and French in 1917), and was granted a mandate over many of the former German Pacific islands and a permanent seat on the Council of the League of Nations. But the gain in 'face' implied in such recognition was offset in Japanese eyes by a failure to have a declaration in favour of racial equality written into the Covenant of the League. The Chinese had much more to feel aggrieved about, for in spite of widespread sympathy over the Twenty-One Demands (notably in the United States) they were unable to obtain a reversal of the Shantung decision. Disappointed of American diplomatic support and crippled by the divisions within their own delegation between the representatives of the Peking government and those of the Kuomintang at Canton, the Chinese refused to sign the treaty.

An almost immediate consequence was a movement in China to which some commentators have given an importance as great as that of the 1911 revolution itself. This was the 'May 4th Movement' of 1919. It stemmed from a student demonstration in Peking against the peace, which had been planned for 7 May, the anniversary of China's acceptance of the 1915 demands, but was brought forward to anticipate action by the authorities. It escalated, though at first only into a small riot and the resignation of the head of the university. This then led to a nation-wide student movement (one of the first political reflexions of the widely-spread establishment in China of new colleges and universities after 1911). This in turn spread to embrace others than students and to manifest itself in strikes and a boycott of Japanese goods. A movement which had begun with intellectuals and their pupils spread to include other city-dwellers, notably industrial workers and the new Chinese capitalists who had benefited from the war. It was the most important evidence yet seen of the mounting rejection of Europe by Asia.

For the first time, an industrial China entered the scene. China, like Japan, had enjoyed an economic boom during the war. Though a decline in European imports to China had been partly offset by increased Japanese and American sales, Chinese entrepreneurs in the ports had found it profitable to invest in production for the home market. The first important industrial areas outside Manchuria began to appear. They belonged to progressive capitalists who sympathized with revolutionary ideas all the more when the return of peace brought renewed western competition and evidence that China had not earned her liberation from tutelage to the foreigner. The workers, too, felt this resentment: their jobs were threatened. Many of them were first-generation town-dwellers, drawn into the new industrial areas from the countryside by the promise of employment. An uprooting from the tenacious soil of peasant tradition was even more important in China than in Europe a century before. Family and village ties were especially strong in China. The migrant to the town broke with patriarchal authority and the reciprocal obligations of the independent producing unit, the household: this was a further great weakening of the age-old structure which had survived the revolution and still tied China to the past. New material was thus made available for new ideological deployments.

The May 4th Movement first showed what could be made of such forces as these by creating the first broadly-based Chinese revolutionary coalition. Progressive western liberalism had not been enough; implicit in the movement's success was the disappointment of the hopes of many of the cultural reformers. Capitalist western democracy had been shown up by the Chinese government's helplessness in the face of Japan. Now, that government had another humiliation from its own subjects: the boycott and demonstration forced it to release the arrested students and dismiss its pro-Japanese ministers. But this was not the only important consequence of the May 4th Movement. For all their limited political influence, reformers had for the first time, thanks to the students, broken through into the world of social action. This aroused enormous optimism and greater popular political awareness than ever before. This is the case for saying that contemporary Chinese history begins positively in 1919 rather than 1911.

Yet ultimately the explosion had come because of an Asian force, Japanese ambition. That force, not in itself a new one in China's affairs, was by 1919 operating on a China whose cultural tradition was dissolving fast. The ending of the examination system, the return of the westernized exiles and the great literary and cultural debate of the war years had all pushed things too far for any return to the old stable state. The warlords could provide no new authority to identify and sustain orthodoxy. And now even the great rival of the Confucian past, western liberalism, was under attack because of its association with the exploiting foreigner. Western liberalism had never had mass appeal; now its charm for intellectuals was threatened just as another rival ideological force from the West had appeared on the scene. The Bolshevik Revolution gave Marxism a homeland to which its adherents abroad could look for inspiration, guidance, leadership and, sometimes, material support, a great new factor now introduced into an already-dissolving historical epoch, and bound to accelerate its end.

Both the February 1917 revolution and the Bolshevik victory had been warmly welcomed by one of the contributors to *New Youth*, Li Ta-chao, who was from 1918 a librarian at Peking University. Soon he saw in Marxism the motive force of world revolution and the means to vitalize the Chinese peasantry. At that moment of disillusion with the West, Russia was very popular among Chinese students. It seemed that the successors of the Tsar had driven out the old imperialist Adam, for one of the first acts of the Soviet government had been a formal renunciation of all extra-territorial rights and jurisdictions enjoyed by the Tsarist state. In the eyes of the nationalists, Russia, therefore, had clean hands. Moreover, her revolution – a revolution in a great peasant society – claimed to be built upon a doctrine whose applicability to China seemed especially plausible in the wake of the industrialization provoked by the war. In 1918 there had begun to meet at Peking University a Marxist study society. One of its members was an assistant in the university library, Mao Tse-tung, and others were prominent in the May 4th Movement. By 1920 there was an outlet for Marxist ideas in one of the student magazines which expressed the aspirations of that movement and the first attempts had been successfully made to deploy Marxist and Leninist principles by organizing strikes in support of it.

Yet Marxism opened divisions between the reformers. Ch'en Tu-hsiu himself turned to it as a solution for China's problems in 1920. He threw his energies into helping to organize the emerging Chinese Left around Marxism. The liberals were beginning to be left behind. The Comintern observed its opportunities and had sent its first man to China in 1919 to help Ch'en and Li Ta-chao. The effects were not entirely happy; there were quarrels. Nevertheless, in circumstances still obscure – we know precisely neither names nor dates – a Chinese communist party was formed in Shanghai in 1921 by delegates from different parts of China (Mao Tse-tung among them).

So began the last stage of the Chinese Revolution and the latest twist of that curious dialectic which has run through the relations of Europe with Asia. Once more an alien Western idea, Marxism, born and shaped in a society totally unlike the traditional

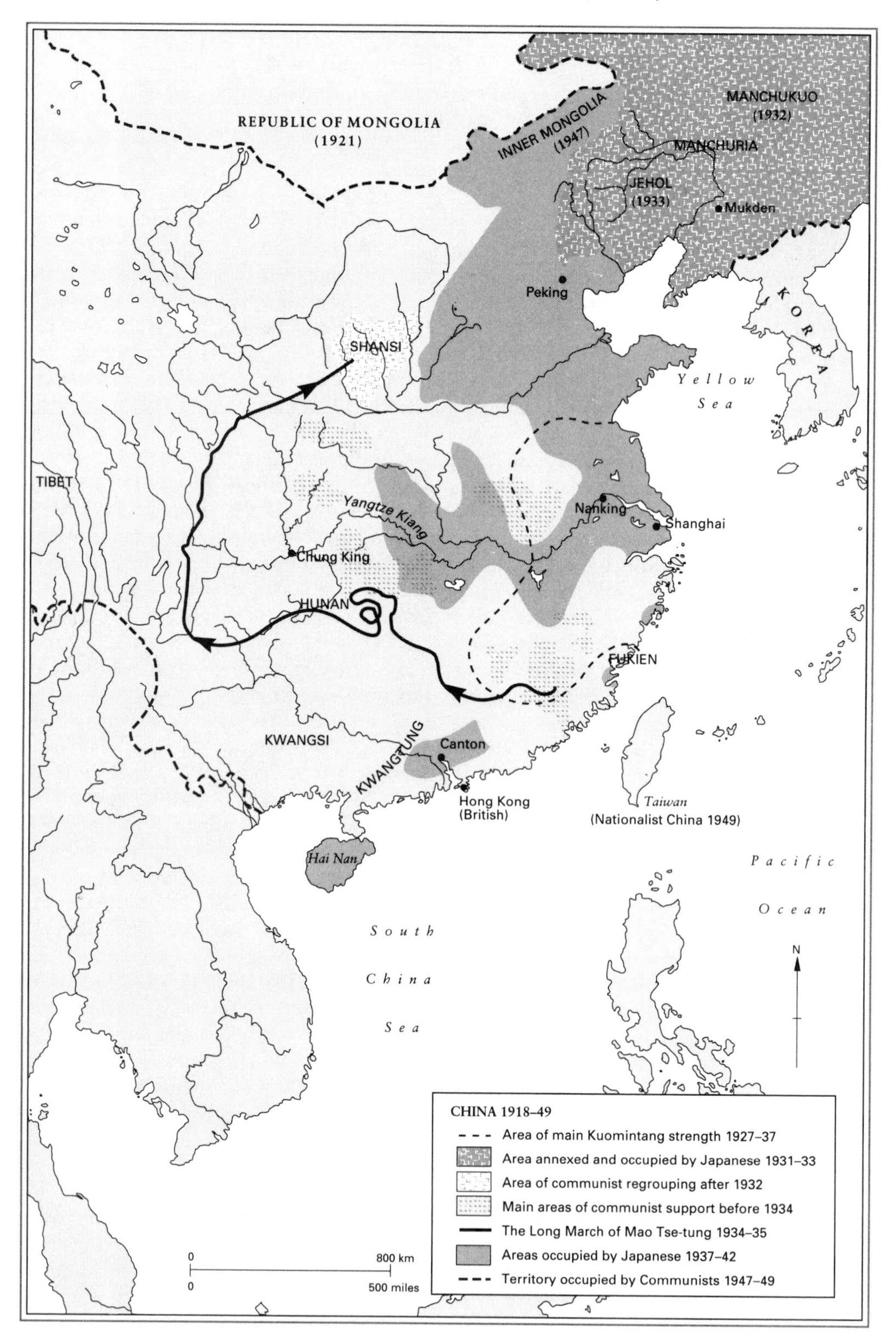

societies of the East, embodying a background of assumptions whose origins were rooted in Judaeo-Christian culture, was taken up by an Asian people and put to their use. It was to be deployed not merely against the traditional sources of inertia in China, in the name of the western goals of modernization, efficiency and universal human dignity and equality, but against the source from which it, too, came – the European world.

Communism benefited enormously in China from the fact that capitalism could easily be represented as the unifying, connecting principle behind foreign exploitation and aggression. In the 1920s, China's divisions were thought to make her of little account in international affairs, though nine powers with Asiatic interests were got to guarantee her territorial integrity and Japan agreed to hand back former German territories in China which she had taken in the Great War. This was part of a complicated set of agreements made at Washington whose core was the international limitation on naval strength (there was great uneasiness about the cost of armaments); these in the end left Japan relatively stronger. The four major powers guaranteed one another's possessions, too, and thus provided a decent burial for the Anglo-Japanese alliance, whose ending had long been sought by the Americans. But the guarantee to China, everyone knew, was worth no more than the preparedness of the Americans to fight to support it; the British had been obliged by the treaties *not* to build a naval base at Hong Kong. Meanwhile, foreigners continued to administer the customs and tax revenues on which the Peking government of an 'independent' China depended and foreign agents and businessmen dealt directly with the warlords when it suited them. Though American policy had further weakened the European position in Asia, this was not apparent in China.

The apparently-continuing grip of the foreign devils on China's life was one reason why Marxism's appeal to intellectuals went far beyond the boundaries of the formal structure of the Chinese communist party. Sun Yat-sen stressed his doctrinal disagreement with it but adopted views which helped to carry the KMT away from conventional liberalism and in the direction of Marxism. In his view of the world, Russia, Germany and Asia had a common interest as exploited powers against their oppressors and enemies, the four imperialist powers (Germany was well-regarded after she had undertaken in 1921 to place her relations with China on a completely equal footing). He coined a new expression, 'hypo-colony', for the state of affairs in which China was exploited without formal subordination as a dependency. His conclusion was collectivist: 'On no account must we give more liberty to the individual,' he wrote; 'let us secure liberty instead for the nation.' This was to give new endorsement to the absence of individual liberty which had always been present in the classical Chinese outlook and tradition. The claims of family, clan and state had always been paramount and Sun Yat-sen envisaged a period of one-party rule in order to make possible mass indoctrination to reconfirm an attitude which had been in danger of corruption by western ideas.

There was apparent, then, no grave obstacle to the cooperation of the Chinese Communist Party (CCP) and the KMT. The behaviour of the Western powers and of the warlords provided common enemies and the Russian government helped to bring them together. Cooperation with the anti-imperialist power with which China had her longest land frontier seemed at least prudent and potentially very advantageous. The policy of the Comintern, for its part, favoured cooperation with the KMT to safeguard Russian interests in Mongolia and as a step towards holding off Japan. Russia had been left out of the Washington conferences, though no power had greater territorial interests in the Far East. For her, cooperation with the likely winners in China was an obvious course even if Marxist doctrine had not also fitted such a policy. From 1924 onwards the CCP was working with the KMT under Russian patronage, in spite of some doubts among Chinese communists. As individuals, though not as a party, they could belong to the KMT. Sun Yat-sen's able young soldier, Chiang K'ai-shek, was sent to Moscow for training, and a military academy was founded in China to provide ideological as well as military instruction.

In 1925 Sun Yat-sen died; he had made communist cooperation with his followers easier, and the united front still endured. Sun Yat-sen's will (which Chinese

Chian K'ai-shek in the 1920s. The strong man of a new China.

schoolchildren learnt by heart) had said that the Revolution was not yet complete and while the communists made important advances in winning peasant support for the Revolution in certain provinces, the new revolutionary army led by idealistic young officers made headway against the warlords. By 1927 something of a semblance of unity had been restored to the country under the leadership of the KMT. Anti-imperialist feeling supported a successful boycott of British goods, which led the British government, alarmed by the evidence of growing Russian influence in China, to surrender its concessions at Hankow and Kiukiang. It had already promised to return Wei-hai-wei to China (1922), and the United States had renounced its share of the Boxer indemnity. Such successes added to signs that China was on the move at last.

One important aspect of this Revolution long went unremarked. Theoretical Marxism stressed the indispensable revolutionary role of the industrial proletariat. The Chinese communists were proud of the progress they had made in politicizing the new urban workers, but the mass of Chinese were peasants. Still trapped in the Malthusian vice of rising numbers and land shortage, their centuries of suffering were, if anything, intensified by the breakdown of central authority in the warlord years. Some Chinese communists saw in the peasants a revolutionary potential which, if not easy to reconcile with contemporary Marxist orthodoxy (as retailed by the Moscow theorists), none the less embodied Chinese reality. One of them was Mao Tse-tung. He and those who agreed with him turned their attention away from the cities to the countryside in the early 1920s and began an unprecedented effort to win over the rural masses to communism. Paradoxically, Mao seems to have continued to cooperate with the Kuomintang longer than other Chinese communists just because it was more sympathetic to the organization of the peasants than was his own party.

A great success followed. It was especially marked in Hunan, but altogether some ten million or so peasants and their families were by 1927 organized by the communists. 'In a few months', wrote Mao, 'the peasants have accomplished what Dr Sun Yat-sen wanted, but failed, to accomplish in the forty years he devoted to the national revolution.' Organization made possible the removal of many of the ills which beset the peasants. Landlords were not dispossessed, but their rents were often reduced. Usurious rates of interest were brought down to reasonable levels. Rural revolution had eluded all previous progressive movements in China and was identified by Mao as the failure of the 1911 revolution; the communist success in reaching this goal was based on the discovery that it could be brought about by using the revolutionary potential of the peasants themselves. This had enormous significance for the future, for it implied new possibilities of historical development through Asia. Mao grasped this and revalued urban revolution accordingly. 'If we allot ten points to the democratic revolution,' he wrote, 'then the achievements of the urban dwellers and the military units rate only three points, while the remaining seven points should go to the peasants in their rural revolution.' In an image twice-repeated in a report on the Hunan movement he compared the peasants to an elemental force; 'the attack is just like a tempest or hurricane; those who submit to it survive, those who resist perish'. Even the image is significant; here was something rooted deeply in Chinese tradition and the long struggle against landlords and bandits. If the communists tried hard to set aside tradition by eradicating superstition and breaking family authority, they nevertheless drew upon it, too.

Communism's rural lodgement was the key to its survival in the crisis which overtook its relations with the KMT after Sun Yat-sen's death. Sun's removal permitted a rift to open in the KMT between a 'left' and a 'right' wing. The young Chiang, who had been seen as a progressive, now emerged as the military representative of the 'right', which reflected mainly the interests of capitalists and, indirectly, landlords. Differences within the KMT over strategy were resolved when Chiang, confident of his control of his troops, committed them to destroying the left factions and the communist party's organization in the cities. This was accomplished with much bloodshed in Shanghai and Nanking in 1927, under the eyes of contingents of European and American soldiers who had been sent to China to protect the concessions. The CCP was proscribed, but this was not quite the end of its cooperation with the KMT, which continued in a few areas for some months, largely because of Russian unwillingness to break with Chiang. Russian direction had already made easier the destruction of the city communists; the Comintern in China, as elsewhere, myopically pursued what were believed to be Russian interests refracted through the mirror of dogmatic Marxism. These interests were for Stalin in the first place domestic; in external affairs, he wanted someone in China who could stand up to the British, the greatest imperialist power, and the KMT seemed the best bet for that. Theory fitted these choices; the bourgeois revolution had to precede the proletarian, according to Marxist orthodoxy. Only after the triumph of the KMT was clear did the Russians withdraw their advisers from the CCP, which gave up open politics to become a subversive, underground organization. Chinese nationalism had in fact done well out of Russian help even if the CCP had not.

Nevertheless, the KMT was left with grave problems and a civil war on its hands at a time when the Revolution needed to satisfy mass demands if it was to survive. The split within the Revolution was a setback, making it impossible to dispose finally of the warlord problem and, more serious, weakening the anti-foreign front. Pressure from Japan had continued in the 1920s after the temporary relaxation and handing back of Kiao-chou. Its domestic background was changing in an important way. When the wartime economic boom finally ended in 1920, hard times and growing social strains followed, even before the onset of the world economic depression. By 1931, half Japan's factories were idle; the collapse of European colonial markets and the entrenchment of what remained of then behind new tariff barriers had a shattering effect as Japanese exports of manufactures went down by two-thirds. The importance of Japan's outlets on the Asian mainland was now crucial. Anything that seemed to threaten them provoked intense irritation. The position of the Japanese peasant deteriorated, too, millions being ruined or selling their daughters into prostitution in order to survive. Grave political consequences were soon manifest, though less in the intensification of class conflict than in the provocation of nationalist extremism. The forces which were to pour into this had for a long time been absorbed in the struggle against the 'unequal treaties'. With those out of the way, a new outlet was needed, and the harsh operation of industrial capitalism in times of depression provided anti-western feeling with fresh fuel.

China 1932. Under the gaze of the languid sirens of Westernized film posters, Japanese sailors and marines man a strongpoint. Identification markers are laid out on the ground for the benefit of the Japanese air force.

The circumstances seemed propitious for further Japanese aggression in Asia. The Western colonial powers were clearly on the defensive, if not in full retreat. The Dutch faced rebellions in Java and Sumatra in the 1920s, the French a Vietnamese revolt in 1930; in both places there was the sinister novelty of communist help to nationalist rebels. The British were not in quite such difficulties in India. Yet though some Englishmen were not yet reconciled to the idea that India must move towards self-government, it was the proclaimed aim of British policy. In China the British had already shown in the 1920s that they wanted only a quiet accommodation with a nationalist movement they found hard to assess, and not too grave a loss of face. Their Far Eastern policies looked even feebler after economic collapse, which also knocked the stuffing out of American opposition to Japan. Finally, Russian power, too, seemed in eclipse after its attempt to influence events in China. Chinese nationalism, on the contrary, had won notable successes, showed no sign of retreat and was considered to be beginning to threaten the long-established Japanese presence in Manchuria. All these factors were present in the calculations made by Japanese statesmen as the depression deepened.

Manchuria was the crucial theatre. The Japanese presence there went back to 1905. Heavy investment had followed. At first the Chinese acquiesced, but in the 1920s began to question it, with support from the Russians who foresaw danger from the Japanese pushing their influence towards Inner Mongolia. In 1929 the Chinese in fact came into conflict with the Russians over control of the railway which ran across Manchuria and was the most direct route to Vladivostok, but this can only have impressed the Japanese with the new vigour of Chinese power; the nationalist KMT was reasserting itself in the territories of the old empire. There had been armed conflict in 1928 when the Japanese had tried to prevent KMT soldiers from operating against warlords in north China whom they found it convenient to patronize. Finally, the Japanese government was by no means unambiguously in control on the spot. Effective power in Manchuria rested with the commanders of the Japanese forces there, and when in 1931 they organized an incident near Mukden which they used as an excuse for taking over the whole province, those in Tokyo who wished to restrain them could not do so.

THE DOORMAT.

A British comment by an Australian cartoonist on Japanese respect for the League of Nations just after the seizure of Manchuria. Later in the year Japan withdrew from the League after an adverse report on the Manchuria affair by League commissioners; many Japanese felt that their case had not been considered fairly at Geneva.

There followed the setting up of a new puppet state, Manchukuo (to be ruled by the last Manchu emperor), an outcry at the League of Nations against Japanese aggression, assassinations in Tokyo, the establishment there of a government much more under military influence, and the expansion of the quarrel with China. In 1932 the Japanese replied to a Chinese boycott of their goods by landing forces at Shanghai; in the following year they came south across the Great Wall to impose a peace which left Japan dominating a part of historic China itself and trying unsuccessfully to organize a secessionist north China. There matters stood until 1937.

The KMT government thus proved unable, after all, to resist imperialist aggression. Yet from its new capital, Nanking, it appeared to control successfully all save a few border areas. It continued to whittle away at the treaties of inferiority and was helped by the fact that as the western powers saw in it a means of opposing communism in Asia, they began to show themselves somewhat more accommodating.

These achievements, considerable though they were, none the less masked important weaknesses which compromised the KMT's domestic success. The crux was that though the political revolution might have continued, the social revolution had come to a stop. Intellectuals withdrew their moral support from a regime which had not provided reforms of which a need to do something about land was the most pressing. The peasants had never given the KMT their allegiance as some of them had given it to the communists. Unfortunately for the regime, Chiang fell back more and more at this juncture upon direct government through his officers and showed himself more and more conservative at a time when the traditional culture had decayed beyond repair. The regime was tainted with corruption in the public finances, often at the highest level. The foundations of the new China were therefore insecure. And there was once more a rival waiting in the wings.

The central leadership of the CCP for some time continued to hope for urban insurrection; in the provinces, none the less, individual communist leaders continued to work along the lines indicated by Mao in Hunan. They dispossessed absentee landlords and organized local soviets, a shrewd appreciation of the value of the traditional peasant hostility to central government. By 1930 they had done better than this, by organizing an army in Kiangsi, where a Chinese Soviet Republic ruled fifty million people, or claimed to. In 1932 the CCP leadership abandoned Shanghai to join Mao in this sanctuary. KMT efforts were directed towards destroying this army, but always without success. This meant fighting on a second front at a time when Japanese pressure was strongest. The last great KMT effort had a partial success, it is true, for it drove the communists out of their sanctuary, thus forcing on them the 'Long March' to Shensi which began in 1934, the epic of the Chinese Revolution and an inspiration ever since. Once there, the seven thousand survivors found local communist support, but were still hardly safe; only the demands of resistance to the Japanese prevented the KMT from doing more to harass them.

Consciousness of the external danger explains why there were tentative essays in cooperation between CCP and KMT again in the later 1930s. They owed something, too, to another change in the policies of the Comintern; it was an era of 'Popular Fronts' elsewhere which allied communists with other parties. The KMT was also obliged to mute its anti-Western line and this won it a certain amount of easy sympathy in England and, above all, the United States. But neither the cooperation of communists nor the sympathies of western liberals could prevent the nationalist regime from being forced on the defensive when the Japanese launched their attack in 1937.

The 'China incident', as the Japanese continued to call it, was to take eight years' fighting and inflict grave social and physical damage on China. It has been seen as the opening of the Second World War. At the end of 1937 the Chinese government removed itself for safety's sake to Chungking in the far west while the Japanese occupied all the important northern and coastal areas. League condemnation of Japan and Russian deliveries of aircraft seemed equally unable to stem the onslaught. The only bonus in the first black years was an unprecedented degree of patriotic unity in China; communists and nationalists alike saw that the national revolution was at stake. This was the view of the Japanese, too; significantly, in the area they occupied, they encouraged the re-establishment of Confucianism. Meanwhile, the western powers felt deplorably unable to intervene. Their protests, even on behalf of their own citizens, were brushed aside by the Japanese who by 1939 made it clear that they were prepared to blockade the foreign settlements if recognition of the Japanese new order in Asia was not forthcoming. For British and French weakness there was an obvious explanation: they had troubles enough elsewhere. American ineffectiveness had deeper roots; it went back to a long-established fact that however the United States might talk about mainland Asia, Americans would not fight for it, perhaps wisely. When the Japanese bombed and sank an American gunboat near Nanking the State Department huffed and puffed but eventually swallowed Japanese 'explanations'. It was all very different from what had happened to the USS *Maine* in Havana harbour forty years before, though the Americans did send supplies to Chiang K'ai-shek.

Shanghai, China's major port, was occupied by the Japanese in November 1937. Japanese infantry celebrate the successful storming of the North Station.

By 1941, China was all but cut off from the outside world, though on the eve of rescue. At the end of that year her struggle would at last be merged with a world war. By then, though, much damage had been done to her. In the long duel between the potential Asian rivals, Japan had been so far clearly the winner. On the debit side of Japan's account had to be placed the economic cost of the struggle to her and the increasing difficulty experienced by her occupying forces in China. On the other hand, her international position had never seemed stronger; she showed it by humiliating western residents in China and by forcing the British in 1940 to close the Burma Road by which supplies reached China, and the French to admit an occupying army to Indo-China. Here was a temptation to further adventure, and it was not likely to be resisted while the prestige of the military and their power in government remained as high as it had been since the mid-1930s.

Yet there was also a negative side to this. Aggression made it more and more imperative for Japan to seize the economic resources of south-east Asia and Indonesia. Yet it also slowly prepared the Americans psychologically for armed defence of their interest. It was clear by 1941 that the United States would have to decide before long whether it was to be an Asian power at all and what that might mean. In the background, though, lay something even more important. For all her aggression against China, it was with the window-dressing slogan of 'Asia for the Asians' that Japan advanced on the crumbling western position in Asia. Just as her defeat of Russia in 1905 marked an epoch in the psychological relations of Europe and Asia, so did the independence and power which Japan showed in 1938-41. When followed by conquest of the European empires, as it was to be, it would signal the beginning of the era of decolonialization; this was to be fittingly inaugurated by the one Asian power at that time successful in its 'westernization'.

4

The Ottoman Heritage and the Western Islamic Lands

During the nineteenth century the Ottoman empire all but disappeared in Europe and Africa. In each continent, the basic causes were the same: the disintegrating effect of nationalism and the predatory activities of European powers. The Serbian revolt of 1804 and Mehemet Ali's establishment of himself as the governor of Egypt in 1805 together opened the final, though drawn-out, era of Turkish decline. In Europe the next step was the Greek revolt; from that time the story of the Ottoman empire in Europe can be told in the dates of the establishment of new nations, until in 1914 Turkey was left with only eastern Thrace. In Islamic Africa the decline of Ottoman power had by then gone even further, and much of North Africa had already been virtually independent of the sultan's rule early in the nineteenth century.

One result was that nationalism in Islamic Africa tended to be directed more against Europeans than against the Ottomans. It was also a culturally revolutionary force. The story again begins with Mehemet Ali. Though he himself never went further west than his birthplace, Kavalla, he admired European civilization and thought Egypt could learn from it. He imported technical instructors, employed foreign consuls in the direction of health and sanitation measures, printed translations of European books and papers on technical subjects, and sent boys to study in France and England. Yet he was working against the grain. His practical achievements disappointed him, though he opened Egypt to European (especially French) influence as never before. Much of it flowed through educational and technical institutions and reflected an old French interest in the trade and affairs of the Ottoman empire. French was soon the second language of educated Egyptians and a large French community grew up in Alexandria, one of the great cosmopolitan cities of the Mediterranean.

Few modernizing statesmen in the non-European world have been able to confine their borrowings from the West to technical knowledge. Soon, young Egyptians began to pick up political ideas, too; there were plenty of them available in French. A compost was forming which would in the end help to transform Europe's relations with Egypt. Egyptians would draw the same lesson as Indians, Japanese and Chinese: the European disease had to be caught in order to generate the necessary antibodies against it. So, modernization and nationalism became inextricably intertwined. Here lay the origin of an enduring weakness in Middle Eastern nationalism. It was long to be the creed of advanced élites cut off from a society whose masses lived in an Islamic culture still largely uncorroded by western ideas. Paradoxically, the nationalists were usually the most Europeanized members of Egyptian, Syrian and Lebanese societies, and this was true until well into the twentieth century. Yet their ideas were to come to have wider resonance. It was among Christian Arabs of Syria that there seems first to have appeared the idea of pan-Arabian or Arab nationalism (as opposed to Egyptian, Syrian or some other kind), an assertion that all Arabs, wherever they were, constituted a nation. This was an idea distinct from that of the brotherhood of Islam which not only embraced millions of non-Arabs, but also excluded many non-Moslem Arabs. The potential complications of this for any attempt actually to realize an Arab nation in practice were, like other weaknesses of pan-Arabist ideas, not to appear until well into the twentieth century.

Another landmark in the history of the former Ottoman lands was the opening of the Suez Canal in 1869. This did more (though indirectly) than any other single fact to doom Egypt to intervention and therefore increasing irritation by the foreigner. Yet the Canal was not the immediate cause of the start of nineteenth-century interference by Europeans in Egypt's government. That came about because of the actions of Ismail (the first ruler of Egypt to obtain from the sultan the title of khedive, in recognition of his substantial *de facto* independence). Educated in France, Ismail liked Frenchmen and up-to-date ideas, and travelled much in Europe. He was very extravagant. When he became ruler, in 1863, the price of cotton, Egypt's main export, was high because of the American Civil War and Ismail's financial prospects therefore looked good. Unhappily, his financial management was less than orthodox. The results were to be seen in the rise in the Egyptian national debt; £7,000,000 at Ismail's accession, it stood at nearly £100,000,000 only thirteen years later. The interest charges amounted to £5,000,000 a year, in an age when such sums mattered. In 1876 the Egyptian government was bankrupt and ceased to pay its debts, so foreign managers were put in. Two controllers, one British, one French, were appointed to make sure that Egypt was governed by Ismail's son with the priority of keeping up revenue and paying off the debt. They were soon blamed by nationalists for the huge burdens of taxation laid upon the Egyptian poor in order to provide the revenue to pay debt interest as well as for economies, such as the reduction of government salaries. The European officials who worked in the name of the khedive were, in the nationalists' eyes, simply the agents of foreign imperialism. There was growing resentment of the privileged legal position of the many foreigners in Egypt and their special courts.

A new linking of Asia and Europe. A French frigate passes through the newly opened Suez Canal.

These grievances led to nationalist conspiracy and eventually to revolution. As well as the westernizing xenophobes a few now began to urge the reform of Islam, the unity of the Moslem world and a pan-Islamic movement adapted to modern life. Some took a more local view, antagonized by the preponderance of Turks in the khedive's entourage. But such divisions mattered less than British intervention in frustrating a revolution in 1882. This was not intervention for financial reasons. It took place because British policy, even under a Liberal prime minister who favoured nationalism in other parts of the Ottoman empire, could not accept the danger that the security of the Canal route to India might be jeopardized by an unfriendly government at Cairo. It was unthinkable at the time, but British soldiers were to remain in the Canal Zone until 1956, tied down by strategical dogma.

After 1882 the British became the prime targets of nationalist hatred in Egypt. They said they wanted to withdraw as soon as a dependable government was available, but could not do so for none acceptable to them was conceivable. Instead, British administrators took on more and more of the government of Egypt. This was not wholly deplorable; they reduced the debt, and mounted irrigation schemes which made it possible to feed a growing population (it doubled to about twelve million between 1880 and 1914). They antagonized Egyptians, though, by keeping them out of government service in the interests of economy, by imposing high taxes and by being foreign. After 1900 there was growing unrest and violence. The British and the puppet Egyptian government proceeded firmly against agitation, and also sought ways out through reform. At first administrative, this led in 1913 to a new constitution providing for more representative elections to a more powerful legislative assembly. Unfortunately, the assembly met only for a few months before it was suspended at the outbreak of war. The Egyptian government was pushed into war with Turkey, a khedive suspected of anti-British plotting was replaced, and at the end of the year the British proclaimed a protectorate. The khedive now took the title of sultan.

By then, the Ottoman government had lost Tripolitania to the Italians, who had invaded it in 1911 partly because of another manifestation of reforming nationalism, this time in Turkey itself. In 1907 a successful rebellion had been started there by the 'Young Turk' movement, which had a complicated history, but a simple purpose. As one Young Turk put it: 'we follow the path traced by Europe .... even in our refusal to accept foreign intervention'. The first part of this meant that they wished to end the despotic rule of Abdul Hamid and restore a liberal constitution granted in 1876 and subsequently withdrawn. But they wanted this less for its own sake than because they thought it would revive and reform the empire, making possible modernization and an end to the process of decay. Both this programme and the Young Turks' methods of conspiracy owed much to Europe; they used, for example, masonic lodges as cover and organized secret societies such as those which had flourished among European liberals in the days of the Holy Alliance. But they much resented the increasing interference in Ottoman internal affairs by Europeans, notably in the management of finance, for, as in Egypt, the securing of interest on money lent for internal development had been followed by loss of independence. European bullying had also resulted (they felt) in the Ottoman government's long and humiliating retreat from the Danube valley and the Balkans.

After a series of mutinies and revolts, the sultan gave way over the constitution in 1908. Liberals abroad smiled on constitutional Turkey; it seemed that misrule was at last to end. But an attempted counter-revolution led to a Young Turk *coup* which deposed Abdul Hamid and installed a virtual dictatorship. From 1909 to 1914 the revolutionaries ruled with increasingly dictatorial means from behind the façade of constitutional monarchy. Ominously, one of them announced that 'there are no longer Bulgars, Greeks, Romanians, Jews, Moslems .... we glory in being Ottoman'. This was something quite new: the announcement of the end of the old multinational regime.

With hindsight, the Young Turks seem more comprehensible than they did at the time. They faced problems like those of many modernizers in non-European countries and their violent methods have been emulated by many since from necessity or imagined necessity. They threw themselves into reform of every branch of government (importing many European advisers). To seek (for instance) to improve the education of girls was a significant gesture in an Islamic country. But they took power in the middle of a shattering succession of diplomatic humiliations which weakened their appeal and led them to rely on force. After the Habsburg annexation of Bosnia, the ruler of Bulgaria won an acknowledgement of Bulgarian independence, and the Cretans announced their union with Greece. A brief pause then was followed by the Italian attack on Tripoli, and the Balkan Wars.

One minor but real operational success achieved by British imperial power between the two great wars was the cheap and humane policing of large areas of the Arab Middle East by air-power and motor cars. Much of Iraq, Transjordan and Palestine (where this picture was taken) was controlled by relatively small numbers of men of the Royal Air Force, who operated armoured car patrols as well as aircraft.

Under such strain, it was soon apparent that the post-reform harmony among the peoples to which liberals had looked forward was a chimera. Religion, language, social custom and nationality still fragmented even what was left of the empire. The Young Turks were driven back more and more upon the assertion of one nationalism among many, that of the Ottomans. This, of course, led to resentment among other peoples. The result was once more massacre, tyranny and assassination, the time-honoured instruments of rule at Constantinople; from 1913 they were deployed by a triumvirate of Young Turks who ruled as a collective dictatorship until the outbreak of the Great War.

Though they had disappointed many of their admirers, these men had the future on their side. They represented the ideas which would one day remake the Ottoman heritage: nationalism and modernization. They had even – willy-nilly – done something towards this by losing most of the little that was left of the Ottoman empire in Europe, thus releasing themselves from a burden. But their heritage was still too encumbering in 1914. Before them lay no better alternative as a vehicle for reform than nationalism. How little pan-Islamic ideas would mean was to be shown by what happened after 1914 in the largest remaining block of Ottoman territory, the largely Moslem provinces of Asia.

In 1914 these covered a large and strategically very important area. From the Caucasus the frontiers with Persia ran down to the Gulf near Basra, at the mouth of the Tigris. On the southern shore of the Gulf Turkish rule ran round Kuwait (with an independent Sheik and under British protection) and then back to the coast as far south as Qatar. From here the coasts of Arabia right round to the entrance of the Red Sea were in one way or another under British influence, but the whole interior and Red Sea coast were Ottoman. Under British pressure the Sinai desert had been surrendered to Egypt a few years before, but the ancient lands of Palestine, Syria and Mesopotamia were still all Turkish. This was the heartland of historical Islam, and the sultan was still caliph, its spiritual leader.

This heritage was to crumble as the strategy and politics of world war played upon it. Even within the historic Islamic heartland, there had been signs before 1914 that the new nationalist forces were at work. In part, they stemmed from old-established European cultural influences, which operated in Syria and the Lebanon much more strongly than in Egypt. French influence had been joined in those countries by American missionary efforts and the foundation of schools and colleges to which there came Arab boys, both Moslem and Christian, from all over the Arab world. The Levant was culturally advanced and literate. On the eve of the world war over a hundred Arabic newspapers were published in the Ottoman empire outside Egypt.

An important crystallization had followed the triumph of the Young Turks and their Ottomanizing tendencies. Secret societies and open groups of dissidents were formed among Arab exiles, notably in Paris and Cairo. In the background was another uncertain factor: the rulers of the Arabian peninsula, whose allegiance to the sultan was shaky. The most important of them was Hussein, Sherif of Mecca, in whom by 1914 the Turkish government had no confidence. A year earlier there had also been the ominous sign of a meeting of Arabs in Persia to consider the independence of Iraq. Against this, the Turks could only hope that the divisiveness of the different interests represented among the Arabs would preserve the *status quo*.

Finally, although it did not present an immediate danger, the latest converts to the religion of nationalism were the Jews. Their history had taken a new turn when, in 1897, there appeared a Zionist Congress whose aim was the securing of a national home. Thus, in the long history of Jewry, assimilation, still barely achieved in many European countries after the liberating age of the French Revolution, was now replaced as an ideal by that of territorial nationalism. The desirable location had not at once been clear; Argentina, Uganda were suggested at different times, but by the end of the country Zionist opinion had come to rest finally on Palestine. Jewish immigration there had begun, though still on a small scale. The unrolling of the war was to change its significance.

Curious parallels existed between the Ottoman and Habsburg empires in 1914. Both sought war, seeing it, in part, as a solution to their problems. Yet both were bound to suffer from it, because too many people inside and outside their borders saw in war an opportunity to score at their expense. In the end, both empires were to be destroyed by it. Even at the outset, Russia, the historic enemy, seemed likely to benefit since Turkey's entry to the war at last evaporated the long resistance of the British and

14 October 1927: the inauguration of the first Iraq oil well. A symbol of a new historical era for the Middle East.

French to the establishment of Tsarist power at Constantinople. For their part, the French had their own fish to fry in the Middle Eastern pan. Though their irritation over a British presence in Egypt had subsided somewhat with the making of the entente and a free hand for France in Morocco, there was a tradition of a special French role in the Levant. The evocations of St Louis and the crusaders with which some enthusiasts made play did not have to be taken seriously, but, undeniably, French governments had for a hundred years exercised a special protection of Catholicism in the Ottoman empire, especially in Syria, to which Napoleon III had sent a French army in the 1860s. There was also the cultural predominance evinced by the wide use of the French language among the educated in the Levant, and much French capital was invested there. These were not forces which could be overlooked.

Nevertheless, in 1914 Turkey's main military antagonists outside Europe were likely to be Russia in the Caucasus, and Great Britain at Suez. The defence of the Canal was the foundation of British strategic thinking in the area, but it soon became clear that no great danger threatened it. Then occurred events announcing the appearance of new factors which would in the end turn the Middle and Near East upside-down. At the end of 1914 an Indian-British army landed at Basra to safeguard oil supplies from Persia. This was the beginning of the interplay of oil and politics in the historical destiny of this area, though it was not to show itself fully until well after the Ottoman empire had ceased to exist. On the other hand, an approach which the British governor of Egypt made to Hussein in October 1914 bore fruit very quickly. This was the first attempt to use the weapon of Arab nationalism.

The attraction of striking a blow against Germany's ally became all the greater as fighting went on bloodily but indecisively in Europe. An attempt in 1915 to force the Dardanelles by combined naval and land operations, in the hope of taking Constantinople, bogged down. By then Europe's civil war had already set in train forces one day to be turned against her. But there was a limit to what could be offered to Arab allies. Terms were not agreed with Hussein until the beginning of 1916. He had demanded independence for all the Arab lands south of a line running along the 37th degree of latitude – that was about eighty miles north of a line from Aleppo to Mosul and included, in effect, the whole of the Ottoman empire outside Turkey and Kurdistan. It was much more than the British could take at the gallop. The French had to be consulted, too, because of their special interest in Syria. When an agreement was made between the British and French on spheres of influence in a partitioned Ottoman empire it left many questions still unsettled for the future, including the status of Iraq, but an Arab nationalist political programme had became a reality.

The future of such undertakings was soon in doubt. The Arab revolt began in June 1916 with an attack on the Turkish garrison of Medina. The rising was never to be more than a distraction from the main theatres of war, but it prospered and became a legend. Soon the British felt they must take the Arabs more seriously; Hussein was recognized as king of the Hejaz. Their own troops pressed forwards in 1917 into Palestine, taking Jerusalem. In 1918 they were to enter Damascus together with the Arabs. Before this, though, two other events had further complicated the situation. One was the American entry into the war; in a statement of war aims President Wilson said he favoured 'an absolute unmolested opportunity of development' for the non-Turks of the Ottoman empire. The other was the Bolshevik publication of their predecessors' secret diplomacy; this revealed Anglo-French proposals for spheres of influence in the Middle East. One part of this agreement had been that Palestine should be administered internationally. Another irritant was added when it was announced that British policy favoured the establishment of a national home in Palestine for the Jewish people. The 'Balfour Declaration' can be accounted the greatest success of Zionism down to this time. It was not strictly incompatible with what had been said to the Arabs, and President Wilson had joined in the good work by introducing to it qualifications to protect Palestinians who were not Jews, but it is almost inconceivable that it could ever have operated unchallenged, especially when further British and French expressions of goodwill towards Arab aspirations followed in 1918. On the morrow of Turkish defeat, the outlook was thoroughly confused.

Hussein was at that moment recognized as King of the Arab peoples by Great Britain, but this did little for him. It was mainly the British and French, with the help of the League of Nations, who were to lay out the main lines of the map of the modern Arab world, not Arab nationalism. During a confused decade the British and French became embroiled with the Arabs whom they had themselves conjured on to the stage of world politics, while the Arab leaders quarrelled among themselves. Islamic unity once more disappeared but, mercifully, so did the Russian threat (even if only briefly), so only two great powers were involved in the Middle East. They distrusted one another, but could agree, roughly on the basis that if the British had their way in Iraq, the French could have theirs in Syria. This was legitimized subsequently by the League of Nations awarding three main mandates to them for Arab lands. Palestine, Transjordan and Iraq went to the British and Syria to the French, who governed high-handedly from the start, having to install themselves by force after a national congress had asked for independence or a British or American mandate. They evicted the king the Arabs had chosen, Hussein's son. Subsequently they had to face a full-scale insurrection. The French were still holding their own by force in the 1930s, though there were by then signs that they would concede some power to the nationalists. Unfortunately, the Syrian situation soon also showed the disintegrating power of nationalism when the Kurdish people of north Syria revolted against the prospect of submergence in an Arab state, so introducing to Western diplomats another Middle Eastern problem with a long life before it.

The Arabian peninsula was meanwhile racked by a struggle between Hussein and yet another king with whom the British had negotiated a treaty (his followers, to make things more difficult still, were members of a particularly puritanical Islamic sect who added religious to dynastic and tribal conflict). Hussein was displaced, and in 1932 the new kingdom of Saudi Arabia emerged in the place of the Hejaz. From this flowed other problems, for sons of Hussein were by this time kings of Iraq and Transjordan. After heavy fighting had shown the difficulties ahead, the British had moved as fast as they dared towards the ending of the mandate over Iraq, seeking only to secure British strategic interests by preserving a military and air force presence. In 1932, accordingly, Iraq entered the League as an independent and fully sovereign state. Earlier, Transjordan had been recognized as independent by the British in 1928, again with some retention of military and financial powers.

Palestine was much more difficult. From 1921, when there were anti-Jewish riots by Arabs alarmed over Jewish immigration and Jewish acquisition of Arab land, that unhappy country was never to be long at peace. More was at stake than merely religious or national feeling. Jewish immigration meant the irruption of a new Westernizing and modernizing force, its operation changing economic relationships and imposing new demands on a traditional society. The British mandatory power was caught between the outcry of the Arabs if it did not restrict Jewish immigration, and the outcry of the Jews if it did. But Arab governments now had to be taken into account, and they occupied lands which were economically and strategically important to British security. World opinion was becoming involved, too. The question became more inflamed than ever when in 1933 there came to power in Germany a regime which persecuted Jews and began to take away the legal and social gains they had been making since the French Revolution. By 1937 there were pitched battles between Jews and Arabs in Palestine. Soon a British army was trying to hold down an Arab insurrection.

The collapse of the paramount power in the Arab lands had often in the past been followed by a period of disorder. What was unclear this time was whether disorder would be followed – as earlier periods of anarchy had eventually been – by the establishment of a new imperial hegemony. The British did not want that rôle; after a brief spell of imperial intoxication in the aftermath of victory, they desired only to secure their own fundamental interests in the area, the protection of the Suez Canal and the swelling flow of oil from Iraq and Iran. Between 1918 and 1934 a great pipeline had been built from northern Iraq across Transjordania and Palestine to Haifa, thus giving yet another new twist to the future of these territories. The consumption of oil in Europe was not yet so large that there was any general dependence on it, nor had the great discoveries been made which would again change the political position in the 1950s. But a new factor was making itself felt; the Royal Navy had turned over to oil for its ships.

The British believed Suez to be best secured by keeping forces in Egypt, but this caused increasing trouble. The war had intensified Egyptian feeling. Armies of occupation are never popular; when the war sent up prices the foreigner was blamed. Egyptian nationalist leaders attempted in 1919 to put their case to the Paris Peace Conference but were prevented from doing so; there followed a rising against the British which was quickly put down. But the British were in retreat. The protectorate was ended in 1922 in the hope of getting ahead of nationalist feeling. Yet the new kingdom of Egypt had an electoral system which returned nationalist majority after nationalist majority, thus making it impossible for an Egyptian government to come to terms for safeguarding British interests which any British government would find acceptable. The result was a prolonged constitutional crisis and intermittent disorder until in 1936 the British finally agreed to be content with a right to garrison the Canal Zone for a limited number of years. An end was also announced to the jurisdictional privileges of foreigners.

This was part of a British retreat from empire which can be detected elsewhere after 1918; it was in part a reflexion of an overstretching of power and resources, as British foreign policy began to be preoccupied by other challenges. Changes in world relationships far from the Middle East thus helped to shape post-Ottoman developments in Islamic lands. Another novel factor was Marxist communism. During the whole of the years between the wars, Russian radio broadcasting to the Arab countries supported the first Arab communists. But for all the worry they caused, communism showed no sign of being able to displace the strongest revolutionary influence of the area, still that of Arab nationalism, whose focus had come by 1938 to be Palestine. In that year a congress was held in Syria to support the Palestinian Arab cause. Arab resentment of the brutality of the French in Syria was beginning to be evident, too, as well as an Arab response to the outcry of the Egyptian nationalists against the British. In pan-Arab feeling lay a force which some thought might in the end over-ride the divisions of the Hashemite Kingdoms.

Allied agreements during the war also complicated the history of the Ottoman homeland, Turkey (as it was soon to be renamed) itself. The British, French, Greeks and Italians had all agreed on their shares of the booty; the only simplification brought by the war had been the elimination of the Russian claim to Constantinople and the Straits. Faced with Greek and Italian invasion, the sultan signed a humiliating peace. Greece was given large concessions, Armenia was to be an independent state, while what was left of Turkey was divided into British, French and Italian spheres of influence. This was the most blatant imperialism. To drive home the point, European financial control was re-established.

The outcome was the first successful revision of any part of the peace settlement. It was largely the work of one man, a former Young Turk and an outstanding soldier, Mustafa Kemal, who drove out French and Greeks in turn after frightening away the Italians. With Bolshevik help he crushed the Armenians. The British decided to negotiate and so a second treaty was made with Turkey in 1923. It was a triumph of nationalism over the decisions at Paris, and it was the only part of the peace settlement which was negotiated between equals and not imposed on the defeated. It was also the only one in which Russian negotiators took part and it lasted better than any of the other peace treaties. The capitulations and financial controls disappeared. Turkey gave up her claims to the Arab lands and the islands of the Aegean, Cyprus, Rhodes and the Dodecanese. A big exchange of Greek and Turkish population followed and the hatred of these peoples for one another received fresh reinforcement. So the Ottoman empire outside Turkey was wound up after six centuries, and a new republic came into existence in 1923 as a national state under a dictator who proved rapidly to be one of the most effective of modernizers.

Appropriately, the caliphate followed the empire, being abolished in 1924. This was the end of Ottoman history; of Turkish history, it was a new beginning. The Anatolian Turks were now for the first time in five or six centuries the majority people of their state.

Kemal, as he tended to call himself (the name meant 'Perfection'), was something of a Peter the Great (though he was not interested in territorial expansion after the successful revision of the dictated peace) and something of a more enlightened despot. The law was secularized (on the model of the Napoleonic code), the Moslem calendar abandoned, and in 1928 the constitution was amended to remove the statement that Turkey was an Islamic state. Polygamy was forbidden. In 1935 the weekly day of rest, formerly Friday, the Islamic holy day, became Sunday and a new word entered the language: *vikend* (the period from 1.00 p.m. Saturday to midnight Sunday). Schools ceased to give religious instruction. The fez was forbidden; although it had come from Europe it was considered Moslem. Kemal was conscious of the radical nature of the modernization he wished to achieve and such symbols mattered to him. They were signs, but signs of something very important, the replacement of traditional Islamic society by a European one. One Islamic ideologist urged his fellow-Turks to 'belong to the Turkish nation, the Moslem religion and European civilization' and did not appear to see difficulties in achieving that. The alphabet was latinized and this had great importance for education, henceforth obligatory at the primary level. A national past was rewritten in the school-books; it was said that Adam had been a Turk.

Kemal Ataturk undertakes a model lesson to publicize the latinization of Turkish script.

Kemal – on whom the National Assembly conferred the name of Ataturk, or 'Father of the Turks' – is an immensely significant figure. He is what Mehemet Ali perhaps wanted to be, the first transformer of an Islamic state by modernization. He remains strikingly interesting; until his death in 1938 he seemed determined not to let his revolution congeal. The result was the creation of a state in some ways among the most advanced in the world at that date. In Turkey, a much greater break with the past was involved in giving a new role to women than in Europe, but in 1934 Turkish women received the vote and they were encouraged to enter the professions.

The most important Islamic country neither under direct imperial rule by Europeans nor Ottomans before 1914 was Persia. The British and Russians had both interfered in her affairs after agreeing over spheres of influence in 1907, but Russian power had lapsed with the Bolshevik Revolution. British forces continued to operate on Persian territory until the end of the war. Resentment against the British was excited when a Persian delegation, too, was not allowed to state its case to the Peace Conference. There was a confused period during which the British struggled to find means of maintaining resistance to the Bolsheviks after withdrawal of their forces. There could be no question of retaining Persia by force, given the over-taxing of British strength. Almost by accident, a British general had already discovered the man who was to do this, though hardly in the way anticipated.

This was Reza Khan, an officer who carried out a *coup d'état* in 1921 and at once used the Bolshevik fear of the British to get a treaty conceding all Russian rights and property in Persia and the withdrawal of Russian forces. Reza Khan then went on to defeat separatists who had British support. In 1925 he was given dictatorial powers by the national assembly and a few months later was proclaimed 'Shah of Shahs'. He was to rule until 1941 (when the Russians and the British together turned him off the throne), somewhat in the style of an Iranian Kemal. The abolition of the veil and religious schools showed secularist aims, though they were not pressed so far as in Turkey. In 1928 the capitulations were abolished, an important symbolic step; meanwhile industrialization and the improvement of communications were pressed forward. A close association with Turkey was cultivated. Finally, the Persian strong man won in 1933 the first notable success in a new art, the diplomacy of oil, when the concession held by the Anglo-Persian Oil Company was cancelled. When the British government took the question to the League of Nations, another and more favourable concession was Reza Shah's greatest victory and the best evidence of the independence of Persia. A new era had opened in the Gulf, fittingly marked in 1935 by an official change of the name of the state: Persia became Iran.

5
The Second World War

The demonstration that the European age was at last over was made in another world war. It began (in 1939) like its predecessor, as a European struggle, and like it became a combination of wars. Like it, too, but to a far greater degree, it made unprecedented demands; this time they were on a scale which left nothing untouched, unmobilized, undisturbed. It was realistically termed 'total' war.

By 1939, there were already many signs for those with eyes to see that an historical era was ending. Though 1919 had brought a few last extensions of territorial control by colonial powers, the behaviour of the greatest of them, Great Britain, showed that imperialism was on the defensive, if not already in retreat. The vigour of Japan meant that Europe was no longer the only focus of the international power system; a prescient South African statesman said as early as 1921 that 'the scene has shifted away from Europe to the Far East and to the Pacific'. His prediction now seems more than ever justified and it was made when the likelihood that China might soon again exercise her due weight was far from obvious. Ten years after he spoke, the economic foundations of western preponderance had been shaken even more plainly than the political; the United States, greatest of industrial powers, had still ten million unemployed. Though none of the European industrial countries was by then in quite such straits, the confidence which took for granted the health of the basic foundations of the economic system had evaporated for ever. Industry might be picking up in some countries – largely because rearmament was stimulating it – but attempts to find recovery by international cooperation came to an end when a World Economic Conference broke down in 1933. After that, each nation had gone its own way; even the United Kingdom at last abandoned Free Trade. *Laissez-faire* was dead, even if people still talked about it. Governments were by 1939 deliberately interfering with the economy as they had not done since the heyday of mercantilism.

If the political and economic assumptions of the nineteenth century had gone, so had many others. It is more difficult to speak of intellectual and spiritual trends than of political and economic, but though many people still clung to old shibboleths, for the élite which led thought and opinion the old foundations were no longer firm. Many people still attended religious services – though only a minority, even in Roman Catholic countries – but the masses of the industrial cities lived in a post-Christian world in which the physical removal of the institutions and symbols of religion would have made little difference to their daily lives. So did intellectuals; they perhaps faced an even greater problem than that of loss of religious belief, because many of the liberal ideas which had helped to displace Christianity from the eighteenth century were by now being displaced in their turn. In the 1920s and 1930s, the liberal certainties of the autonomy of the individual, objective moral criteria, rationality, the authority of parents, and an explicable mechanical universe all seemed to be going under along with the belief in Free Trade.

Change was most obvious in the arts. For three or four centuries, since the age of humanism, Europeans had believed that the arts expressed aspirations, insights and pleasures accessible in principle to ordinary men, even though they might be raised to an exceptional degree of fineness in execution, or be especially concentrated in form so that not all individual men would always enjoy them. At any rate, it was possible for the whole of that time to retain the notion of the cultivated man who, given time and study, could discriminate with taste among the arts of his time because they were expressions of a shared culture with shared standards. This idea was somewhat weakened when the nineteenth century, in the wake of the romantic movement, came to idealize the artist as genius – Beethoven was one of the first examples – and formulated the notion of the *avant-garde*. By the first decade of the twentieth century, though, it was already very difficult for even trained eyes and ears to recognize art in much of what was done by contemporary artists. The best symbol of this was the dislocation of the image in painting. Here, the flight from the representational still kept a tenuous link with tradition as late as Cubism, but by then it had long ceased to be apparent to the average 'cultivated man' – if he still existed. Artists retired into a less and less accessible chaos of private visions, whose centre was reached in the world of Dada and Surrealism. The years after 1918 are of the greatest interest as something of a culmination of disintegration; in Surrealism even the notion of the objective disappeared, let alone its representation. As one Surrealist put it, the movement meant 'thought dictated in the absence of all control exerted by reason, and outside all aesthetic or moral preoccupations'. Through chance, symbolism, shock, suggestion and violence the Surrealists sought to go beyond consciousness itself. In so doing, they were only doing what many writers and musicians were trying to do, too.

Such phenomena provide evidence in widely different forms of the decay of liberal culture which was the final outcome of the high civilization of the European age. It is significant that such disintegratory movements were often prompted by a sense that the traditional culture was too limited because of its exclusion of the resources of emotion and experience which lay in the unconscious. Probably few of the artists who would have agreed with this would have read the work of the man who, more than any other, gave the twentieth century a language in which to explore this area and the confidence that it was there that the secrets of life lay.

This was Sigmund Freud, the founder of psychoanalysis. He deserves a place in the history of culture beside Newton or Darwin, for he changed the way educated men thought of themselves. He introduced several new ideas into ordinary discourse: the special meanings we now give to the words 'complex', 'unconscious' and 'obsession', and the appearance of the familiar terms 'Freudian slip' and 'libido' are monuments to the power of his teaching. His influence quickly spread into literature, personal relations, education, politics. Like that of many prophets, his message was often distorted. What he was believed to have said was much more important than the specific clinical studies which were his contribution to science. Again like Newton and Darwin, Freud's importance beyond science – where his influence was more complex – lay in providing a new mythology. It was to prove highly corrosive.

The message men took from Freud suggested that the unconscious was the real source of most significant behaviour, that moral values and attitudes were projections of the influences which had moulded this unconscious, that, therefore, the idea of responsibility was at best a myth and probably a dangerous one, and that perhaps rationality itself was an illusion. It did not matter much that Freud's own assertions would have been nonsense had this been true or that this left out the subtlety and science of his work. This was what many people believed he had proved – and still believe. Such a bundle of ideas called in question the very foundation of liberal civilization itself, the idea of the rational, responsible, consciously motivated individual, and this was its general importance.

Freud's teaching was not the only intellectual force contributing to the loss of certainty and the sense that men had little firm ground beneath their feet. But it was the most apparent in the intellectual life of the interwar period. From grappling with the insights he brought, or with the chaos of the arts, or with the incomprehensibility of a world of science which seemed suddenly to have abandoned Laplace and Newton, men plunged worriedly into the search for new mythologies and standards to give them bearings. Politically, this led to fascism, Marxism, and the more irrational of the old certainties, extreme nationalism, for example. People did not feel inspired or excited by tolerance, democracy, and the old individual freedoms.

DIE

TRAUMDEUTUNG

VON

DR. SIGM. FREUD.

»FLECTERE SI NEQUEO SUPEROS, ACHERONTA MOVEBO.«

LEIPZIG UND WIEN.

FRANZ DEUTICKE.

1900.

Freud's Interpretation of Dreams *(title of the English edition of 1913), the book through which he first began to reach a wide lay public. While none of his 'discoveries' has gone unchallenged and debate will long continue about the scientific value of his work, there is no one whose work in this century can more truly be said to have so influenced every aspect of its culture.*

Such influences made it all the more difficult to deal with the deepening uncertainty and foreboding clouding international relations in the 1930s. The heart of this lay in Europe, in the German problem which threatened a greater upheaval than could Japan. Germany had not been destroyed in 1918; it was a logical consequence, therefore, that she would one day again exercise her due weight. Geography, population and industrial power all meant that in one way or another a united Germany must dominate central Europe and overshadow France. What was at issue at bottom was whether this could be faced without war; only a few cranks thought it might be disposed of by dividing again the Germany united in 1871.

Germans soon began to demand the revision of the settlement of Versailles. This demand eventually became unmanageable, though in the 1920s it was tackled in a hopeful spirit. The real burden of reparations was gradually whittled away and the Treaty of Locarno was seen as a great landmark because by it Germany gave her consent to the Versailles territorial settlement in the west. But it left open the question of revision in the east and behind this loomed the larger question: how could a country potentially so powerful as Germany be related to its neighbours in a balanced, peaceful way, given the particular historical and cultural experience of the Germans?

Most people hoped this had been settled by the creation of a democratic German republic whose institutions would gently and benevolently reconstruct German society and civilization. It was true that the constitution of the Weimar Republic (as it was called from the place where its constituent assembly met) was very liberal, but too many Germans were out of sympathy with it from the start. That Weimar had solved the German problem was revealed as an illusion when economic depression shattered the narrow base on which the German Republic rested and set loose the destructive nationalist and social forces it had masked.

When this happened, the containment of Germany again became an international problem. But for a number of reasons, the 1930s were a very unpromising decade for containment to be easy. To begin with, some of the worst effects of the world economic crisis were felt in the relatively weak and agricultural economies of the new eastern countries. France had always looked for allies against a German revival there, but such allies were now gravely weakened. Furthermore, their very existence made it doubly difficult to involve Russia, again an indisputable (if mysterious) great power, in the containment of Germany. Her ideological distinction presented barriers enough to cooperation with the United Kingdom and France, but there was also her strategic remoteness. No Russian force could reach central Europe without crossing one or more of the east European states whose short lives were haunted by fear of Russia and communism: Romania, Poland and the Baltic states, after all, were built from, among other things, former Russian lands.

Nor were the Americans of any help. The whole trend of American policy since Wilson failed to get his countrymen to join the League had been back towards a self-absorbed isolation which was, of course, suited to traditional ideas. Americans who had gone to Europe as soldiers did not want to repeat the experience. Justified apparently by boom in the 1920s, isolation was paradoxically confirmed by slump in the 1930s. When Americans did not confusedly blame Europe for their troubles – the question of debts from the war years had great psychological impact because it was believed to be tied up with international financial problems (as indeed it was, though not quite as Americans thought) – they felt distrustful of further entanglement. Anyway, the depression left them with enough on their plate. With the election of a Democratic president in 1932 they were, in fact, at the beginning of an era of important change which would in the end sweep away this mood, too, but this could not be foreseen.

The next phase of American history was to be presided over by Democrats for five successive presidential terms, the first four of them after elections won by the same man, Franklin Roosevelt. To stand four successive times as presidential candidate was almost unprecedented (only the unsuccessful socialist, Eugene Debs, also did so); to win, astonishing. To do so with (on each occasion) an absolute majority of the popular vote was something like a revolution. No earlier Democratic candidate since the Civil War had ever had one at all (and no other was to have one until 1964). Moreover, Roosevelt was a rich, patrician figure. It is all the more surprising, therefore, that he should have emerged as one of the greatest leaders of the early twentieth century. He did so in an electoral contest which was basically one of hope versus despair. He offered confidence and the promise of action to shake off the blight of economic depression. A political transformation followed his victory, the building of a Democratic hegemony on a coalition of neglected constituencies – the South, the poor, the farmer, the Negro, the progressive liberal intellectual – which then attracted further support as it seemed to deliver results.

There was some degree of illusion in this. The 'New Deal' on which the Roosevelt administration embarked was still not grappling satisfactorily with the economy by 1939. None the less it changed the emphasis of the working of American capitalism and

Franklin D. Roosevelt, a cripple of magnetic personal quality whose greatest asset when he became President in the depths of the economic slump may have been the contagion of his jaunty confidence that America's problems were manageable if only tackled with courage.

its relations with government. A huge programme of unemployment relief with insurance was started, millions were poured into public works, new regulation of finance was introduced, and a great experiment in public ownership was launched in a hydro-electric scheme in the Tennessee valley. This in the end gave capitalism a new lease of life, in a new governmental setting. The New Deal brought the most important extension of the power of the Federal authorities over American society and the states that had ever occurred in peacetime and it has proved irreversible. Thus American politics reflected the same pressures towards collectivism which affected other countries in the twentieth century. In this sense, too, the Roosevelt era was historically decisive. It changed the course of American constitutional and political history as nothing had done since the Civil War and incidentally offered to the world a democratic alternative to fascism and communism by providing a liberal version of large-scale governmental intervention in the economy. This achievement is all the more impressive in that it rested almost entirely on the interested choices of politicians committed to the democratic process and not on the arguments of economists, some of whom were already advocating greater central management of the economy in capitalist nations. It was a remarkable demonstration of the ability of the American political system to deliver what people felt they wanted.

The same machinery, of course, could also only deliver as a foreign policy what most Americans would tolerate. Roosevelt was much more aware than the majority of his fellow-citizens of the dangers of persistent American isolation from Europe's problems. But he could reveal his own views only slowly. With Russia and the United States unavailable, only the western European great powers remained to confront Germany if she revived. Great Britain and France were badly placed to act as the policemen of Europe. They had memories of their difficulties in dealing with Germany even when Russia had been at their side. Furthermore, they had been much at odds with one another since 1918. They were also militarily weak. France, conscious of her inferiority in manpower should Germany ever rearm, had invested in a programme of strategic defence by fortification which looked impressive but effectively deprived her of the power to act offensively. The Royal Navy was no longer without a rival, nor, as in 1914, safe in concentrating its resources in European waters. British governments long pursued the reduction of expenditure on armaments at a time when world-wide commitments were a growing strain on her forces. Economic depression reinforced this tendency; it was feared that the costs of rearmament would cripple recovery by causing inflation. Many British voters, too, believed that Germany's grievances were just. They were disposed to make concessions in the name of German nationalism and self-determination, even by handing back German colonies. Both Great Britain and France were also troubled by a joker in the European pack, Italy. Under Mussolini, hopes that she might be enlisted against Germany had disappeared by 1938.

The cause lay in a belated attempt by Italy to participate in the Scramble for Africa. In 1935, her forces invaded Ethiopia. Such action posed the question of what should be done by the League of Nations; it was clearly a breach of its Covenant that one of its members should attack another. France and Great Britain were in an awkward position. As great powers, Mediterranean powers and African colonial powers, they were bound to take the lead against Italy at the League. But they did so feebly and half-heartedly, for they did not want to alienate an Italy they would like to have with them against Germany. The result was the worst possible one. The League failed to check aggression and Italy was alienated. Ethiopia lost its independence, though, it later proved, only for six years.

This was one of several moments at which it later looked as if a fatal error was committed. But it is impossible to say in retrospect at what stage the situation which developed from these facts became unmanageable. Certainly the emergence of a much more radical and ferociously opportunist regime in Germany was the major turning-point. But the depression preceded this and made it possible. Economic collapse also had another important effect. It made plausible an ideological interpretation of events in the 1930s and thus further embittered them. Because of the intensification of class conflict which economic collapse brought with it, interested politicians sometimes interpreted the development of international relations in terms of fascism versus communism, and even of Right versus Left or Democracy versus Dictatorship. This was easier after Mussolini, angered by British and French reactions to his invasion of Ethiopia, came to ally Italy to Germany and talked of an anti-communist crusade. But this was misleading, too. All ideological interpretations of international affairs in the 1930s tended to obscure the central nature of the German problem – and, therefore, to make it harder to tackle.

Here Russian propaganda was important, too. During the 1930s her internal situation was precarious. The industrialization programme was imposing grave strains and sacrifices. These were mastered – though perhaps also exaggerated – by a savage intensification of dictatorship which expressed itself not only in the collectivization struggle against the peasants, but in the turning of terror against the cadres of the regime itself from 1934 onwards. In five years millions of Russians were executed, imprisoned or exiled, often to forced labour. The world looked on amazed as batches of defendants grovelled with grotesque 'confessions' before Soviet courts. Nine out of ten generals in the army went, and, it has been estimated, half the officer corps. A new communist élite replaced the old one in these years; by 1939 over half the delegates who had attended the Party Congress of 1934 had been arrested. It was very difficult for outsiders to be sure what was happening, but it was clear to them that Russia was by no means either a civilized, liberal state nor necessarily a very strong potential ally.

More directly, this affected the international situation because of the propaganda which accompanied it. Much of this, no doubt, arose from the deliberate provocation inside Russia of a siege mentality; far from being relaxed, the habit of thinking of the world in terms of Us versus Them which had been born in Marxist dogma and the interventions of 1918–22 was encouraged in the 1930s. As this notion took hold, so, outside, did the preaching of the doctrine of international class-struggle by the Comintern. The reciprocal effect was predictable. The fears of conservatives everywhere were intensified. It became easy to think of any concession to Left-wing or even mildly progressive forces as a victory for the Bolsheviks. As attitudes thus hardened on the Right, so communists were given new evidence for the thesis of inevitable class-conflict and revolution.

But there was not one successful left-wing revolution. The revolutionary danger had subsided rapidly after the immediate post-war years. Labour governments peacefully and undramatically ruled Great Britain for part of the 1920s. The second ended in financial collapse in 1931, to be replaced by conservative coalitions which had overwhelming electoral support and proceeded to govern with remarkable fidelity to the tradition of progressive and piecemeal social and administrative reform which had marked Great Britain's advance into the 'Welfare State'. This direction had been followed even further in the Scandinavian countries, often held up for admiration for their combination of political democracy and practical socialism, and as a contrast to communism. Even in France, where there was a large and active communist party, there was no sign that its aims were acceptable to the majority of the electorate even after the depression. In Germany the communist party before 1933 had been able to get more votes, but it was never able to displace the Social Democrats in control of the working-class movement. In less advanced countries than these, communism's revolutionary success was even smaller. In Spain it had to compete with socialists and anarchists; Spanish conservatives certainly feared it and may have been right to fear also the tendency to slide towards social revolution felt under the republic which was established in 1931, but they had better grounds for seeing elsewhere than in Spanish communism the real danger facing them.

Yet the ideological interpretation had great appeal, even to many who were not communists. It was much strengthened by the accession to power of a new ruler in Germany, Adolf Hitler, whose success makes it very difficult to deny him political genius despite the pursuit of goals which make it difficult to believe him wholly sane. In the early 1920s he was only a disappointed agitator, who had failed in an attempt to overthrow a government (the Bavarian) and who poured out his obsessive nationalism and anti-semitism not only in hypnotically effective speeches but in a long, shapeless, semi-autobiographical book which few people read. In 1933, the National Socialist German Workers Party which he led ('Nazi' for short) was strong enough for him to be appointed Chancellor of the German republic. Politically, this may have been the most momentous single decision of the century, for it meant the revolutionizing of Germany, its redirection upon a course of aggression which ended by destroying the old Europe and Germany too, and that meant a new world.

Though Hitler's messages were simple, his appeal was complex. He preached that Germany's troubles had identifiable sources. The Treaty of Versailles was one. The international capitalists were another. The supposedly anti-national activities of German Marxists and Jews were others. He also said that the righting of Germany's political wrongs must be combined with the renovation of German society and culture, and that this was a matter of purifying the biological stock of the German people, by excising its non-Aryan components.

In 1922 such a message took him very little way. In 1930 it won Hitler 107 seats in the German parliament – more than the communists, who had 77. The Nazis were already the beneficiaries of economic collapse, and it was to get worse. There are several reasons why the Nazis reaped its political harvest, but one of the most important was that the communists spent as much energy fighting the socialists as their other opponents. This had fatally handicapped the Left in Germany all through the 1920s. Another reason was that under the democratic republic anti-semitic feeling had grown. It, too, was exacerbated by economic collapse. Anti-semitism, like nationalism, had an appeal which cut across classes as an explanation of Germany's troubles, unlike the equally simple Marxist explanation in terms of class war which, naturally, antagonized some as well as (it was hoped) attracting others.

By 1930 the Nazis showed they were a power in the land. They attracted more support, and won backers from those who saw in their street-fighting gangs an anti-communist insurance, from nationalists who sought rearmament and revision of the Versailles peace settlement and from conservative politicians who thought that Hitler was a party leader like any other who might now be valuable in their own game. The manoeuvres were complicated, but in 1932 the Nazis became the biggest party in the German parliament, though without a majority. In January 1933 Hitler was called to office legally and constitutionally by the head of the republic. There followed new elections, in which the regime's monopoly of the radio and use of intimidation still did not secure the Nazis a majority of seats; none the less, they had one when supported by some Right-wing members of parliament who joined them to vote special enabling powers to the government. The most important was that of governing by emergency decree. This was the end of parliament and parliamentary sovereignty. Armed with these powers, the Nazis proceeded to carry out a revolutionary destruction of democratic institutions. By 1939, there was virtually no sector of German society not controlled or intimidated by them. The conservatives, too, had lost. They soon found that Nazi interference with the independence of traditional authorities was likely to go very far.

Like Stalin's Russia, the Nazi regime rested in large measure on terror used mercilessly against its enemies. It was also unleashed against the Jews and an astonished Europe found itself witnessing revivals in one of its most advanced societies of the pogroms of medieval Europe or tsarist Russia. This was indeed so amazing that many people outside Germany found it difficult to believe that it was happening. Confusion over the nature of the regime made it even more difficult to deal with. Some saw Hitler simply as a nationalist leader bent, like an Ataturk, upon the regeneration of his country and the assertion of its rightful claims. Others saw him as a crusader against Bolshevism. Even when people only thought he might be a useful barrier against it, that increased the likelihood that men of the Left would see him as a tool of capitalism. But no simple formula will contain Hitler or his aims – and there is still great disagreement about what these were – and probably a reasonable approximation to the truth is simply to recognize that he expressed the resentments and exasperations of German society in their most negative and destructive

With one hand on a standard of the S.A. – the street-fighting units which carried Nazis to power – Hitler renews the ties of loyalty of the party rank-and-file at the Nuremburg rally of 1934.

forms and embodied them to a monstrous degree. When his personality was given scope by economic disaster, political cynicism and a favourable arrangement of international forces, he could release these negative qualities at the expense of, in the long run, all Europeans, his own countrymen included.

The path by which Germany came to be at war again in 1939 is complicated. There is still much argument about when, if ever, there was a chance of avoiding the final outcome. One important moment, clearly, was when Mussolini, formerly wary of German ambitions in central Europe, became Hitler's ally. After he had been alienated by British and French policy over his Ethiopian adventure, a civil war broke out in Spain when a group of generals mutinied against the left-wing republic. Hitler and Mussolini both sent contingents to support the man who emerged as the rebel leader, General Franco. This, more than any other single fact, gave an ideological colour to Europe's divisions. Hitler, Mussolini and Franco were all now identified as 'fascist' and Russian foreign policy began to coordinate support for Spain within western countries by letting local communists abandon their attacks on other left-wing parties and encouraging 'Popular Fronts'. Thus Spain came to be seen as a conflict between Right and Left in its purest form; this was a distortion, but it helped to accustom people to think of Europe as divided into two camps.

British and French governments were by this time well aware of the difficulties of dealing with Germany. Hitler had already in 1935 announced that her rearmament (forbidden at Versailles) had begun. Until their own rearmament was completed, they remained very weak. The first consequence of this was shown to the world when German troops re-entered the 'demilitarized' zone of the Rhineland from which they had been excluded by the Treaty of Versailles. No attempt was made to resist this move. After the civil war in Spain had thrown opinion in Great Britain and France into further disarray, Hitler then seized Austria. The terms of Versailles which forbade the fusion of Germany and Austria seemed hard to uphold; to the French and British electorates this could be presented as a matter of legitimately aggrieved nationalism. The Austrian republic had also long had internal troubles. The *Anschluss* (as union with Germany was called) took place in 1938. In the autumn came the next German aggression, the seizure of part of Czechoslovakia. Again, this was justified by the specious claims of self-determination; the areas involved were so important that their loss crippled the prospect of future Czechoslovak self-defence, but they were areas with many German inhabitants. Memel would follow, on the same grounds, the next year. Hitler was gradually fulfilling the old dream which had been lost when Prussia beat Austria – the dream of a united Great Germany, defined as all lands of those of German blood.

The dismemberment of Czechoslovakia was something of a turning-point. It was achieved by a series of agreements at Munich in September 1938 in which Great Britain and Germany took the main parts. This was the last great initiative of British foreign policy to try to satisfy Hitler. The British prime minister was still too unsure of rearmament to resist, but hoped also that the transference of the last substantial group of Germans from alien rule to that of their homeland might deprive Hitler of the motive for further revision of Versailles – a settlement which was now somewhat tattered in any case.

He was wrong, because Hitler went on to inaugurate a programme of expansion into Slav lands. The first step was the absorption of what was left of Czechoslovakia, in March 1939. This brought forward the question of the Polish settlement of 1919. Hitler resented the 'Polish Corridor' which separated East Prussia from Germany and contained Danzig, an old German city given an internationalized status in 1919. At this point the British government, though hesitatingly, changed tack and offered a guarantee to Poland and other east European countries against aggression. It also began a wary negotiation with Russia.

Russian policy remains hard to interpret. It seems that Stalin kept the Spanish civil war going with support to the republic as long as it seemed likely to tie up German attention, but then looked for other ways of buying time against the attack

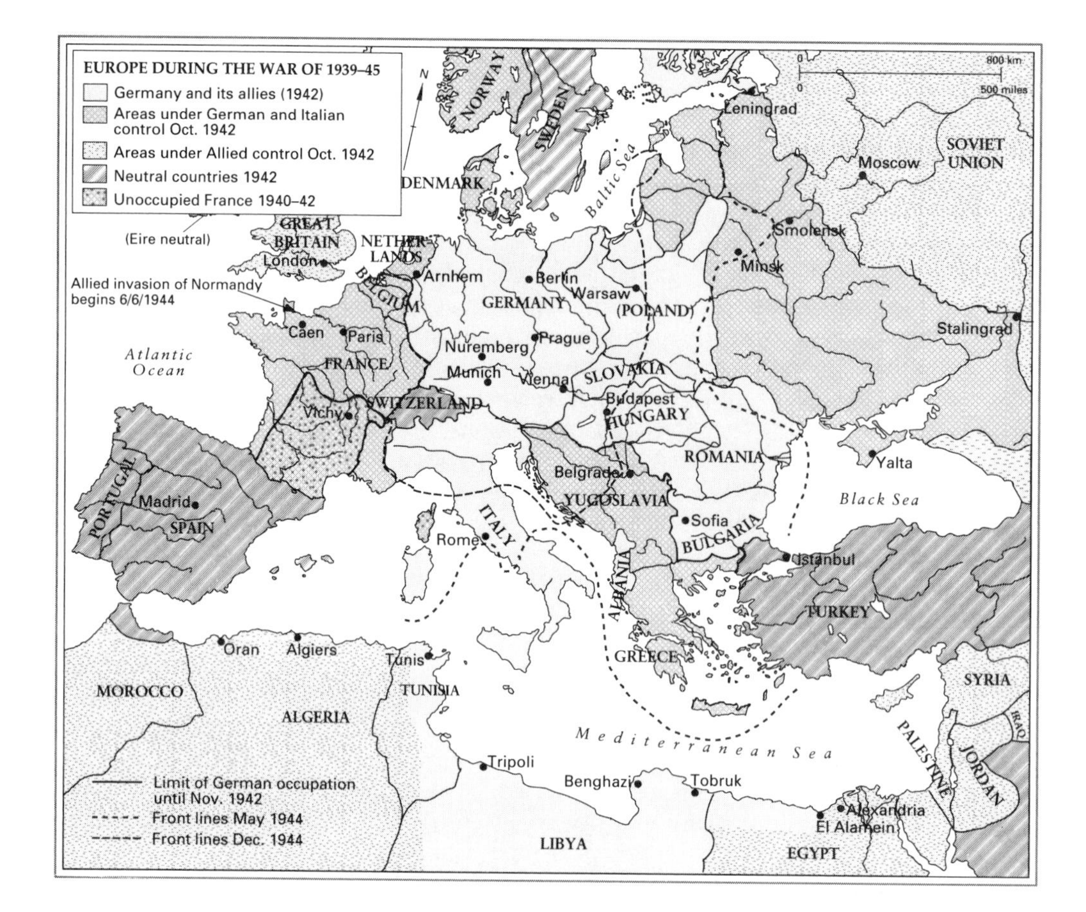

from the west which he always feared. To him, it seemed likely that a German attack on Russia might be encouraged by Great Britain and France who would see with relief the trouble they had so long faced turning on the workers' state. No doubt they would have done. There was little possibility of working with the British or French to oppose Hitler, however, even if they were willing to do so, because no Russian army could reach Germany except through Poland – and this the Poles would never permit. Accordingly, as a Russian diplomat remarked to a French colleague on hearing of the Munich decisions, there was now nothing for it but a fourth partition of Poland. This was arranged in the summer of 1939. After all the propaganda each had directed against Bolshevik-Slav barbarism and fascist-capitalist exploitation, Germany and Russia made an agreement in August which provided for the division of Poland between them; authoritarian states enjoy great flexibility in the conduct of diplomacy. Thus insured, Hitler went on to attack Poland. He thus began the Second World War on 1 September 1939. Two days later the British and French honoured their guarantee to Poland and declared war on Germany.

Their governments were not very keen on doing so, for it was obvious that they could not help Poland. That unhappy nation disappeared once more, divided by Russian and German forces about a month after the outbreak of war. But not to have intervened would have meant acquiescing to the German domination of Europe, for no other nation would have thought British or French support worth having. So, uneasily and without the excitement of 1914, the only two constitutional great powers of Europe found themselves facing a totalitarian regime. Neither their peoples nor governments had much enthusiasm for this role, and the decline of liberal and democratic forces since 1918 put them in a position much inferior to that of 1914, but exasperation with Hitler's long series of aggressions and broken promises made it hard to see what sort of peace could be made which would reassure them. The basic cause of the war was, as in 1914, German nationalism. But whereas then Germany had gone to war because *she* felt threatened, now Great Britain and France were responding to the danger presented by her expansion. *They* felt threatened this time.

To the surprise of many observers, and the relief of some, the first six months of the war were almost uneventful once the short Polish campaign was over. It was quickly plain that mechanized forces and airpower were to play a much more important part than between 1914 and 1918. The memory of the slaughter of the Somme and Verdun was too vivid for the British and French to plan anything but an economic offensive; the weapon of blockade, they hoped, would be effective. Hitler was unwilling to disturb them, because he was anxious to make peace. This deadlock was only broken when the British sought to intensify the blockade in Scandinavian waters. This coincided, remarkably, with a German offensive to secure ore supplies which conquered Norway and Denmark. Its launching on 9 April 1940 opened an astonishing period of fighting. Only a month later there began a brilliant German invasion first of the Low Countries and then of France. A powerful armoured attack through the Ardennes opened the way to the division of the Allied armies and the capture of Paris. On 22 June France signed an armistice with the Germans. By the end of the month, the whole European coast from the Pyrenees to the North Cape was in German hands. Italy had joined in on the German side ten days before the French surrender. A new French government at Vichy broke off relations with Great Britain after the British had seized or destroyed French warships they felt might fall into German hands. The Third Republic effectively came to an end with the installation of a French marshal, a hero of the First World War, as Head of State. With no ally left on the continent, Great Britain faced a worse strategical situation by far than that in which she had struggled against Napoleon.

After this huge change in the nature of the war, Great Britain was not quite alone. There were the Dominions, all of which had entered the war on her side, and a number of governments in exile from the overrun continent. Some of these commanded forces of their own and Norwegians, Danes, Dutchmen, Belgians, Czechs and Poles were to fight gallantly, often with decisive effect, in the years ahead. The most important exiled contingents were those of the French, but at this stage they represented a faction within France, not its legal government. A general who had left France before the armistice and was condemned to death *in absentia* was their leader: Charles de Gaulle. He was recognized by the British as 'leader of the Free French'. He saw himself as constitutional legatee of the Third Republic and the custodian of France's interests and honour. He soon began to show an independence which was in the end to make him the greatest servant of France since Clemenceau.

De Gaulle's position was important because of uncertainties about what might happen to parts of the French empire where he hoped to find sympathizers who wished to join him to continue the fight. This was one way in which the war was now extended geographically. Another resulted from Italy's entry into the war, since her African possessions and the Mediterranean sea-lanes then became operational areas. Finally, the availability of Atlantic and Scandinavian ports meant that what was later called the 'Battle of the Atlantic', the German struggle to sever British sea communications by submarine, surface, and air attack, was now bound to become much fiercer.

Immediately, the British Isles faced direct attack. The hour had already found the man to brace the nation against such a challenge. Winston Churchill, after a long and chequered political career, had become Prime Minister when the Norwegian campaign collapsed, because no other man commanded support in all parties in the House of Commons. To the coalition government which he immediately formed he gave vigorous leadership, something hitherto felt to be lacking. More important than this, he called forth in his people, whom he addressed frequently by radio, qualities they had forgotten they possessed. It was soon clear that only defeat after direct assault was going to get the British out of the war.

This was even more certain after a great air battle over southern England in August and September had been won by British science and the Royal Air Force. For a moment, Englishmen knew the pride and relief of the Greeks after Marathon. It was true, as Churchill said in a much-quoted speech, that 'never in the field of human conflict was so much owed by so many to so few'. This victory made a German seaborne invasion impossible (though a successful one was always unlikely). It also established that Great Britain could not be defeated by air bombardment alone. The islands had a bleak outlook ahead, but this victory changed the direction of the war, for it was the beginning of a period in which a variety of influences turned German attention in another direction. In December 1940 planning began for a German invasion of Russia.

By that winter, Russia had made further gains in the west, apparently with an eye to securing a glacis against a future German attack. A war against Finland gave her important strategic areas. The Baltic republics of Latvia Lithuania and Estonia were swallowed in 1940. Bessarabia, which Romania had taken from Russia in 1918, was now taken back, together with the northern Bukovina. In the last case, Stalin was going beyond tsarist boundaries. The German decision to attack Russia arose in part because of disagreements about the future direction of Russian expansion: Germany sought to keep Russia away from the Balkans and the Straits. It was also aimed at demonstrating, by a quick overthrow of Russia, that further British war-making was pointless. But there was also a deep personal element in the decision. Hitler had always sincerely and fanatically detested Bolshevism and maintained that the Slavs, a racially inferior group, should provide Germans with living-space and raw materials in the east. This was a last, perverted vision of the old struggle of the Teuton to impose western civilization on the Slav east. Many Germans responded to such a theme. It was to justify more appalling atrocities than any earlier crusading myth.

In a brief spring campaign, which provided an overture to the holocaust ahead, the Germans overran Yugoslavia and Greece, with which Italian forces had been unhappily engaged since October 1940. Once again British arms were driven from the mainland of Europe. Crete, too, was taken by a spectacular German airborne assault. Now all was ready for 'Barbarossa', as the great onslaught on Russia was named, after a crusading German emperor of the Middle Ages.

The attack was launched on 22 June 1941 and had huge early successes. Vast numbers of prisoners were taken and the Russian armies fell back hundreds of miles. The German advance guard came within a narrow margin of entering Moscow. But that margin was not quite eliminated and by Christmas the first successful Russian counter-attacks had announced that in fact Germany was pinned down. German strategy had lost the initiative. If the British and Russians could hold on and if they could keep in alliance with one another then, failing a radical technical modification of the war by the discovery of new weapons of great power, their access to American production would inexorably increase their strength. This did not, of course, mean that they would inevitably defeat Germany, only that they might bring her to negotiate terms.

The American president had believed since 1940 that in the interests of the United States Great Britain had to be supported up to the limits permitted by his own public and the law of neutrality. In fact, he went well beyond both at times. By the summer of 1941, Hitler knew that to all intents and purposes the United States was an undeclared enemy. A crucial step had been the American Lend-Lease Act of March that year which provided production and services to the Allies without payment. Soon afterwards, the American government extended naval patrols and the protection of its shipping further eastward into the Atlantic. After the invasion of Russia came a meeting between Churchill and Roosevelt which resulted in a statement of shared principles – the Atlantic Charter – in which a nation at war and another formally at peace spoke of the needs of a post-war world 'after the final destruction of the Nazi tyranny'. This was a long way from isolationism and was the background to Hitler's second fateful but foolish decision of 1941, a declaration of war on the United States on 11 December, after a Japanese attack on British and American territories four days earlier. Hitler had earlier promised the Japanese to do this. The war thus became global. The British and American declarations of war on Japan might have left two separate wars to rage, with only Great Britain engaged in both; Hitler's action threw away the chance that American power might have been kept out of Europe and deployed only in the Pacific. Few single acts have so marked the end of an epoch, for this announced the eclipse of European affairs. Europe's future would now be settled not by her own efforts by the two great powers on her flanks, the United States and Soviet Russia.

The Japanese decision was also a rash one, though the logic of Japanese policy had long pointed towards conflict with the United States. Japan's ties with Germany and Italy, though they had some propaganda value for both sides, did not amount to much in practice. What mattered in the timing of Japanese policy was the resolution of debates in Tokyo about the danger, or lack of it, of a challenge to the United States which must involve war. The crux of the matter was that Japan's needs for a successful conclusion of the war in China included oil which she could only obtain with the tacit consent of the United States that Japan was to destroy China. This no American government could have given. Instead, in October 1941 the American government imposed an embargo on all trade by United States citizens with Japan.

There followed the last stages of a process which had its origins in the ascendancy established in Japan by reactionary and militant forces in the 1930s. The question had by this time become for the Japanese military planners purely strategic and technical; since they would have to take the resources in south-east Asia which they needed by force, all that had to be settled was the nature of the war against the United States and its timing. Such a decision was fundamentally irrational, for the chances of ultimate success were very small; once arguments of national honour had won, though, the final calculations about the best point and moment of attack were carefully made. The choice was made to strike as hard a blow as possible against American sea-power at the outset in order to gain the maximum freedom of movement in the Pacific and South China

Kerch, 1944. Bodies of those murdered by the retreating German army discovered by survivors.

Sea. The result was the onslaught of 7 December, whose centre-piece was an air attack on the American fleet at Pearl Harbor which was one of the most brilliantly conceived and executed operations in the history of warfare. By mischance it fell just short of complete success, for it did not destroy American naval air power, though it gave the Japanese for months the strategical freedom they sought. This failure was fatal; after it the Japanese faced a prolonged war they were bound to lose in the end. Pearl Harbor united Americans as little else could have done. Isolationism could be virtually ignored after 8 December; Roosevelt had a nation behind him as Wilson never had.

When a few Japanese bombs were reported to have fallen on the American mainland, it was obvious that this was much more truly a world war than the first had been. The German operations in the Balkans had by the time of Pearl Harbor left Europe with only four neutral countries – Spain, Portugal, Sweden and Switzerland. The war in North Africa raged back and forth between Libya and Egypt. It was extended to Syria by the arrival there of a German mission and to Iraq when a nationalist government supported by German aircraft was removed by a British force. Iran had been occupied by the British and Russians in 1941. In Africa, Ethiopia was liberated and the Italian colonial empire destroyed.

With the opening of the Far Eastern war the Japanese wrought destruction on the colonial empires there, too. Within a few months they took Indonesia, Indo-China, Malaya, the Philippines. They pressed through Burma towards the Indian border and were soon bombing the north Australian port of Darwin from New Guinea. Meanwhile, the naval war was fought by German submarine forces, aircraft and surface raiders all over the Atlantic, Arctic, Mediterranean and Indian Ocean. Only a tiny minority of countries were left outside this struggle. Its demands were colossal and carried much further the

6 June 1944: a South Lancashire battalion lands in Normandy as part of the greatest amphibian operation in history.

mobilization of whole societies than had the First World War. The role of the United States was decisive. Her huge manufacturing power made the material preponderance of the 'United Nations' (as the coalition of states fighting the Germans, the Italians and Japanese was called from the beginning of 1942) incontestable.

None the less, the way ahead was still a hard one. The first part of 1942 was still very bleak for the United Nations. Then came the turning-point, in four great and very different battles. In June a Japanese fleet attacking Midway Island was broken in a battle fought largely by aircraft. Japanese losses in carriers and aircrews were such that she never regained the strategical initiative and the long American counter-attack in the Pacific now began to unroll. Then, at the beginning of November, the British army in Egypt decisively defeated the Germans and Italians and began a march west which was to end with the eviction of the enemy from all North Africa. The battle of El Alamein had coincided with landings by Anglo-American forces in French North Africa. They subsequently moved eastwards and by May 1942 German and Italian resistance on the continent had ceased. Six months earlier, at the end of 1941, the Russians had bottled up at Stalingrad on the Volga a German army rashly exposed by Hitler. The remnants surrendered in February in the most demoralizing defeat yet suffered by the Germans in Russia, and yet one which was only part of three splendid months of winter advance which marked the turning-point of the war on the eastern front.

The other great Allied victory has no specific date, but was as important as any of these. This was the Battle of the Atlantic. Its peak came in the early months of

7 December 1941: an American battleship stricken by the Japanese air attack on Pearl Harbour.

1942. In March nearly 850,000 tons of shipping were lost and six U-boats were sunk; six months later, the figures were 560,000 tons and eleven U-boats. The tide had turned, though there was still hard fighting ahead. At the end of the year nearly eight million tons of shipping had been lost for eighty-seven U-boats sunk. In 1943 the figures were three and a quarter million tons and 237 U-boats. This was the most crucial battle of all for the United Nations, for on it depended their ability to draw on American production.

Command of the sea also made possible re-entry to Europe. Roosevelt had agreed to give priority to the defeat of Germany, but the mounting of an invasion of France to take the strain off the Russian armies could not in the end be managed before 1944, and this angered Stalin. When it came, the Anglo-American invasion of northern France in June 1944 was the greatest seaborne expedition in history. Mussolini had by then been overthrown by Italians and Italy had already been invaded from the south; now German was fighting on three fronts. Soon after the landings in Normandy, the Russians entered Poland. Going faster than their allies, it still took them until next April to reach Berlin. In the west, Allied forces had broken out of Italy into central Europe and from the Low Countries into northern Germany. Almost incidentally, terrible destruction had been inflicted on German cities by a great air offensive which, until the last few months of the war, exercised no decisive strategic effect. When, on 30 April, the man who had ignited this conflagration killed himself in a bunker in the ruins of Berlin, historic Europe was literally as well as figuratively in ruins.

The war in the Far East took a little longer. At the beginning of August 1945 the Japanese government knew it must be defeated. Many of Japan's former conquests had been retrieved, her cities were devastated by American bombing and her sea-power, on which communications and safety from invasion rested, was in ruins. At this moment two nuclear weapons of a destructive power hitherto unapproached were dropped with terrible effect on two Japanese cities by the Americans. Between the explosions, the Russians declared war on Japan. The Japanese government abandoned its plan of a suicidal last-ditch stand and accepted terms of capitulation on 14 August. When a formal instrument of surrender was signed on 2 September, the Second World War had come to an end.

In its immediate aftermath it was difficult to measure the colossal extent of what happened. Only one clear and unambiguous good was at once visible, the overthrow of the Nazi regime. As the Allied armies advanced into Europe, the deepest evils of a system of terror and torture were revealed by the opening of the huge prison camps and the revelations of what went on in them. It was suddenly apparent that Churchill had spoken no more than the bare truth when he told his countrymen that 'if we fail, then the whole world, including the United States, including all that we have known and cared for, will sink into the abyss of a new Dark Age made more sinister, and perhaps more protracted, by the lights of perverted science'. The reality of this threat could be seen in Belsen and Buchenwald. Distinctions could hardly be meaningful between degrees of atrocity inflicted on political prisoners, slave labourers from other countries, or some prisoners of war. But the world's imagination was most struck by the systematic attempt which had been made to wipe out European Jewry, the so-called 'Final Solution' sought by Germans, an attempt which was carried far enough to change the demographic map: the Polish Jews were almost obliterated, and Dutch Jews, too, suffered terribly in proportion to their numbers. Overall, though complete figures may never be available, it is probable that between five and six million Jews were killed, whether in the gas-chambers and crematoria of the extermination camps, by shootings and extermination on the spot in east and south-east Europe, or by overwork and hunger.

Few people and no nations had engaged in the war because they saw it as a struggle against such wickedness, though no doubt many of them were heartened as it proceeded by the sense that the conflict had a moral dimension. Propaganda contributed to this. Even while England was the only nation in Europe still on her feet and fighting for her survival, a democratic society had sought to see in the struggle positive ends which went beyond survival and beyond the destruction of Nazism. Hopes of a new world of cooperation between great powers and social and economic reconstruction were embodied in the Atlantic Charter and United Nations. They were encouraged by sentimental goodwill towards allies and a tragic blurring of differences of interest and social ideals which were only too quickly to re-emerge. Much wartime rhetoric boomeranged badly with the coming of peace; disillusionment followed inspection of the world after the guns were silent. Yet for all this, the war of 1939–45 in Europe remains a moral struggle in a way, perhaps, in which no other has ever been. It is important to emphasize this. Too much has been heard of the regrettable consequences of Allied victory, and it is too easily forgotten that it crushed the worst challenge to liberal civilization which has ever arisen.

Some far-sighted men could see a deep irony in this. In many ways, Germany had been one of the most progressive countries in Europe; the embodiment of much that was best in its civilization. That Germany should fall prey to collective derangement on this scale suggested that something had been wrong at the root of that civilization itself. The crimes of Nazism had been carried out not in a fit of barbaric intoxication with conquest, but in a systematic, scientific, controlled, bureaucratic (though often inefficient) way about which there was little that was irrational except the appalling end which it sought. In this respect the Asian war was importantly different. Japanese imperialism replaced the old western imperialisms for a time, but often the subject peoples did not much regret the change. Propaganda during the war attempted to give currency to the notion of a 'fascist' Japan, but this was a distortion of so traditional a society's character. No such appalling consequences as faced European nations under German rule would have followed from a Japanese victory.

The second obvious result of the war was its unparalleled destructiveness. It was most visible in the devastated cities of Germany and Japan, where mass aerial bombing, one of the major innovations of the Second World War, proved much more costly to life and buildings than had been the bombing of Spanish cities in the Spanish civil war. Yet even those early essays had been enough to convince many observers that bombing alone could bring a country to its knees. In fact, although often invaluable in combination with other forms of fighting, the huge strategic bombing offensive against Germany built up by the British Royal Air Force from tiny beginnings in 1940 and steadily supplemented by the United States Air Force from 1942 onwards up to the point at which their combined forces could provide a target with continuous day and night bombing, achieved very little until the last few months of the war. Nor was the fiery destruction of the Japanese cities strategically so important as the elimination of her sea-power.

Not only cities had been shattered. The economic life and communications of central Europe had also been grievously stricken. In 1945, millions of refugees were wandering about in it, trying to get home. There was a grave danger of famine and epidemic because of the difficulty of supplying food. The tremendous problems of 1918 were upon Europe again, and this time confronted nations demoralized by defeat and occupation; only the neutrals and Great Britain had escaped those scourges. There were abundant arms in private hands, and a revolutionary threat could be feared. These conditions could also be found in Asia, but there the physical destruction was less severe and prospects of recovery better.

In Europe, too, the revolutionary political impact of the war was obvious. The power structure which had been a reality until 1914 and had an illusory prolongation of life between the two world wars, was doomed in 1941. Two great peripheral powers dominated Europe politically and were established militarily in its heart. This was evident at a meeting of the Allied leaders at Yalta in February 1945 which provided a basis for agreement, and the nearest thing to a formal peace settlement in Europe for decades. Its outcome was that old central Europe would disappear. Europe would be divided into eastern and western halves. Once again a Trieste-Baltic line became a reality, but now new differences were to be layered on top of old. At the end of 1945 there lay to the east a Europe of states which, with the exception of Greece, all had communist governments or governments in which communists shared power with others. The Russian army which had overrun them had proved itself a far better instrument for the extension of

Burma, 1944. Indian soldiers armed with American weapons fight other Asians on a battlefield dominated by the temples of Burmese Buddhism – which are themselves built from that most characteristic product of European industrialism, corrugated iron.

international communism than revolution had ever been. The pre-war Baltic republics did not emerge from the Soviet state, of course, and the Soviet Union now also absorbed parts of pre-war Poland and Romania.

Germany, the centre of the old European power structure, had effectively ceased to exist. A phase of European history which she had dominated was at an end, and Bismarck's creation was partitioned into zones occupied by the Russians, Americans, British and French. The other major political units of western Europe had reconstituted themselves after occupation and defeat, but were feeble; Italy, which had changed sides after Mussolini had been overthrown, had, like France, a much strengthened and enlarged communist party which, it could not be forgotten, was still committed to the revolutionary overthrow of capitalism. Only Great Britain retained her stature of 1939 in the world's eyes; it was even briefly enhanced by her stand in 1940 and 1941. She remained for a while the recognized equal of Russia and the United States. (Formally, this was true of France and China, too, but less attention was paid to them.) Yet Great Britain's moment

On 17 July 1945, at Potsdam, the heart of the Prussian military tradition, the Allied war leaders met for the first time after the defeat of Germany. President Truman had taken up his office automatically on the death of Roosevelt; Winston Churchill was soon afterwards to lose his after defeat in a general election. Only Stalin straddled war and peace.

was past. By a huge effort of mobilizing her resources and social life to a degree unequalled outside Stalin's Russia, she had been able to retain her standing. But she had been let out of a strategic impasse only by the German attack on Russia, and kept afloat only by American Lend-Lease. And this aid had not been without its costs: the Americans had insisted on the sale of British overseas assets to meet the bills before it was forthcoming. Moreover, the sterling area was dislocated. American capital was now to move into the old Dominions on a large scale. Those countries had learnt new lessons both from their new wartime strength and, paradoxically, from their weakness in so far as they had relied upon the mother country for their defence. From 1945, they more and more acted with full as well as formal independence.

It only took a few years for this huge change in the position of the greatest of the old imperial powers to become clear. Symbolically, when Great Britain made her last great military effort in Europe, in 1944, the expedition was commanded by an American general. Though British forces in Europe for a few months afterwards matched the Americans, they were by the end of the war outnumbered. In the Far East, too, though the British reconquered Burma, the defeat of Japan was the work of American naval and air power. For all Churchill's efforts, Roosevelt was by the end of the war negotiating over his head with Stalin, proposing *inter alia* the dismantling of the British empire. Great Britain, in spite of her victorious stand alone in 1940 and the moral prestige it gave her, had not escaped the shattering impact of the war on Europe's political structure. Indeed, she was in some ways the power which, with Germany, illustrated it best.

Thus was registered in Europe the passing of the European supremacy also evident at its periphery. British forces secured Dutch and French territories in Asia in time to hand them back to their former overlords and prevent the seizure of power by anti-colonial regimes. But fighting with rebels began almost immediately and it was clear that the imperial powers faced a difficult future. The war had brought revolution to the empires, too. Subtly and suddenly, the kaleidoscope of authority had shifted, and it was still shifting as the war came to an end. The year 1945 is not, therefore, a good place at which to pause; reality was then still masked somewhat by appearance and many Europeans still had to discover, painfully, that the European age of empire was over.

6
The Shaping of a New World

After the First World War, it had still been possible to embrace the illusion that an old order might be restored. In 1945, no one in authority could believe such a thing. This was one great and healthy contrast between the circumstances of the two great attempts of this century to re-order international life. The victory could not, of course, start with a clean sheet on which to plan. Events had closed off many roads, and even during the war crucial decisions had already been taken, some by agreement, some not, about what should follow victory. One of the most important had been that an international organization should be set up to maintain international peace. The fact that the great powers saw such an organization in different ways, the Americans as a beginning to the regulation of international life by law and the Russians as a means of maintaining the Grand Alliance, did not prevent them pressing forward. So the United Nations Organization (UNO) came into being at San Francisco in 1945.

Much thought, naturally, had been given to the League of Nations' failure to come up to expectations. One of its great defects was remedied in 1945: the United States and Russia belonged to the new organization from the start. Apart from this, the basic structure of the United Nations resembled that of the League in outline. Its two essential organs were a small Council and a large Assembly. Permanent representatives of all member states were to sit in the General Assembly. The Security Council had at first eleven members, of whom five were permanent; these were the representatives of the United States, Russia, Great Britain, France (at the insistence of Winston Churchill) and China. The Security Council was given greater power than the old League Council and this was largely the doing of the Russians. They saw that there was a strong likelihood that they would always be outvoted in the General Assembly – where, at first, fifty-one nations were represented – because the United States could rely not only on the votes of its allies, but also on those of its Latin American satellites. Naturally, not all the smaller powers liked this. They were uneasy about a body on which at any one moment any of them was likely not to sit, which would have the last word and in which the great powers would carry the main weight. Nevertheless, the structure the great powers wanted was adopted, as, indeed, it had to be if any organization was to work at all.

The other main issue which caused grave constitutional dispute was the veto power given to the permanent members of the Security Council. This was a necessary feature if the great powers were to accept the organization, though in the end the veto was restricted somewhat, in that a permanent member could not prevent investigation and discussion of matters which especially affected it unless they were likely to lead to action inimical to its interests.

In theory the Security Council possessed very great powers, but, of course, their operation was bound to reflect political reality. In its first decades, the importance of the United Nations proved to lie not in its power to act, but rather in the forum it provided for discussion. For the first time, a world public linked as never before by radio and film – and later, by television – would have to be presented with a case made at the General Assembly for what sovereign states did. This was something quite new. The United Nations at once gave a new dimension to international politics; it took much longer to provide effective new instrumentation for dealing with its problems. Sometimes, the new publicity of international argument led to feelings of sterility, as increasingly bitter and unyielding views were set out in debates which changed no one's mind. But an educational force was at work. It was important, too, that it was soon decided that the permanent seat of the General Assembly should be in New York; this drew American attention to it and helped to offset the historic pull of isolationism.

The United Nations General Assembly met for the first time, none the less, in London in 1946. Bitter debates began at once; complaints were made about the continued presence of Russian soldiers in Iranian Azerbaijan, occupied during the war, and the Russians promptly replied by attacking Great Britain for keeping her forces in Greece. Within a few days the first veto was cast, by the Soviet delegation. There were to be many more. The instrument which the Americans and British had regarded and continued to use as an extraordinary measure for the protection of special interests became a familiar piece of Soviet diplomatic technique. Already in 1946 the United Nations was an arena in which the USSR contended with a still inchoate western bloc which its policies were to do much to solidify.

The United Nations Building, New York.

Though the origins of conflict between the United States and Russia are often traced back a very long way, in the later years of the war the British government had tended to feel that the Americans made too many concessions and were over-friendly to the Soviet Union. Of course, there was always a fundamental ideological division; if the Russians had not always had a deep preconception about the roots of behaviour of capitalist societies, they would certainly have behaved differently after 1945 towards their wartime ally. It is also true that some Americans never ceased to distrust Russia and saw her as a revolutionary threat. But this did not mean that they had much impact on the making of American policy. In 1945, when the war ended, American distrust of Russian intentions was much less than it later became. Of the two states the more suspicious and wary was the Soviet Union.

At that moment, there were no other true great powers left. For all the legal fictions expressed in the composition of the Security Council, Great Britain was gravely overstrained, France barely risen from the living death of occupation and stricken by internal divisions (a large communist party threatened her stability), while Italy had discovered new quarrels to add to old ones. Germany was in ruins and under occupation. Japan was occupied and militarily powerless, while China had never yet been a great power in modern times. The Americans and Russians therefore enjoyed an immense superiority over all possible rivals. They were the only real victors. They alone had made positive gains from the war. All the other victorious states had, at the most, won survival or resurrection. To the United States and USSR, the war brought new empires.

Though that of the Soviet Union had been won at huge cost, she now had greater strength than she had ever known under the tsars. Her armies dominated a vast European glacis, much of which was sovereign Soviet territory; the rest was organized in states which were by 1948 in every sense satellites, and one of them was East Germany, a major industrial entity. Beyond the glacis lay Yugoslavia and Albania, the only communist states to emerge since the war without the help of Soviet occupation; in 1945 both seemed assured allies of Moscow. This advantageous position had been won by the fighting of the Red Army, but it also owed something to decisions taken by western governments and to their commander in Europe during the closing stages of the war, when General Eisenhower had resisted pressure to get to Prague and Berlin before the Russians. The resulting Soviet strategical preponderance in central Europe was all the more menacing because the old traditional barriers to Russian power had gone, in 1918 the Habsburg empire, and now a united Germany. An exhausted Great Britain and slowly-reviving France could not be expected to stand up to the Red Army, and no other conceivable counterweight on land existed if the Americans went home.

Russian armies also stood in 1945 on the borders of Turkey and Greece – where a communist rising was under way – and occupied northern Iran. In the Far East they had held much of Sinkiang, Mongolia, northern Korea and the naval base of Port Arthur as well as liberating the rest of Manchuria, though the only territory they actually took from Japan was the southern half of the island of Sakhalin and the Kuriles. The rest of these gains had been effectively at China's expense. Yet in China there was already visible at the end of the war the outline of a new communist state which could be expected to be friendly to Moscow. Stalin might have backed the wrong horse there in the past, but the Chinese communists could not hope for moral and material help from anyone else. So it seemed that in Asia, too, a new Soviet satellite might be in the making.

The new world power of the United States rested much less on occupation of territory than that of the USSR. She, too, had at the end of the war a garrison in the heart of Europe, but American electors in 1945 wanted it brought home as soon as possible. American naval and air bases round much of the Eurasian land mass were another matter. Though the USSR was a far greater Asian power than ever, the elimination of Japanese naval power, the acquisition of island airfields and technological changes which made huge fleet trains possible had together turned the Pacific Ocean into an American lake. Above all, Hiroshima and Nagasaki had demonstrated the power of the new weapon which the United States alone possessed, the atomic bomb. But the deepest roots of

American empire lay in her economic strength. Apart from the land-power of the Red Army, the overwhelming industrial power of the United States had been the decisive material factor behind the Allied victory. America had equipped not only her own huge forces but those of her allies. Moreover, by comparison with them, victory had cost her little. American casualties were fewer than theirs; even those of the United Kingdom were heavier and the Russian colossally so. The home base of the United States had been immune to enemy attack in any but a trivial sense and was undamaged; her fixed capital was intact, her resources greater than ever. Her citizens had seen their standard of living actually rise during the war; the armament programme ended a depression which had not been mastered by Roosevelt's New Deal. She was a great creditor country, with capital to invest abroad in a world where no one else could supply it. Finally, America's old commercial and political rivals were staggering under the troubles of recovery. Their economies drifted into the ambit of the American because of their own lack of resources. The result was a world-wide surge of indirect American power, its beginnings visible even before the war ended.

Something of the future implicit in the great power polarization could dimly be seen before the fighting stopped in Europe. It was made clear, for example, that the Russians would not be allowed to participate in the occupation of Italy or the dismantling of her colonial empire, and that the British and Americans could not hope for a Polish settlement other than the one wanted by Stalin. Yet (in spite of their record in their own hemisphere) the Americans were not happy about explicit spheres of influence; the Russians were readier to take them as a working basis. There is no need to read back into such divergences assumptions which became current a few years after the war, when conflict between the two powers was presumed to have been sought from the start by one or other of them. Appearances can be deceptive. For all the power of the United States in 1945, there was little political will to use it; the first concern of the American military after victory was to achieve as rapid a demobilization as possible. Lend-Lease arrangements with allies had already been cut off even before the Japanese surrender. This further reduced America's indirect international leverage; it simply weakened friends she would soon be needing who now faced grave recovery problems. They could not provide a new security system to replace American strength. Nor could the use of the atomic bomb be envisaged except in the last resort; it was too powerful.

It is much harder to be sure of what was going on in Stalin's Russia. Her peoples had clearly suffered appallingly from the war, more, possibly, than even the Germans. No one has been able to do more than provide estimates, but it seems likely that over twenty million Soviet citizens may have died. Stalin may well have been less aware of Soviet strength than of Soviet weakness when the war ended. True, his governmental methods relieved him of any need, such as faced western countries, to demobilize the huge land forces which gave him supremacy on the spot in Europe. But the USSR had no atomic bomb nor a significant strategic bomber force, and Stalin's decision to develop nuclear weapons put a further grave strain on the Soviet economy at a time when general economic reconstruction was desperately needed. The years immediately after the war were to prove as grim as had been those of the industrialization race of the 1930s. Yet in September 1949 an atomic explosion was achieved. In the following March it was officially announced that the USSR had an atomic weapon. By then much had changed.

Piecemeal, relations between the two major world powers had by then deteriorated very badly. This was largely the result of what happened in Europe, the area most in need of imaginative and coordinated reconstruction in 1945. The cost of the war's destruction there has never been accurately measured. Leaving out the Russians, about fourteen and a quarter million Europeans were dead. In the most-stricken countries those who survived lived amid ruins. One estimate is that about seven and a half million dwellings were destroyed in Germany and Russia. Factories and communications were shattered. There was nothing with which to pay for the imports Europe needed and currencies had collapsed; Allied occupation forces found that cigarettes and bully-beef were better than money. Civilized society had given way not only under the horrors of

Nazi warfare, but also because occupation had transformed lying, swindling, cheating and stealing into acts of virtue; they were not only necessary to survival, but they could be glorified as acts of 'resistance'. The struggles against German occupying forces had bred new divisions; as countries were liberated by the advancing Allied armies, the firing squads got to work in their wake and old scores were wiped out. It was said that in France more perished in the 'purification' of liberation than in the great Terror of 1793.

Above all, more finally than in 1918, the economic structure of Europe had disintegrated. The flywheel of much of European economic life had once been industrial Germany. But even if the communications and the productive capacity to restore the machine had been there, the Allies were at first bent on holding down German industrial production to prevent her recovery. Furthermore, Germany was divided. From the start the Russians had been carrying off capital equipment as 'reparations' to repair their own ravaged lands – as well they might; the Germans had destroyed 39,000 miles of railway track alone in their retreat in Russia. The Soviet Union may have lost a quarter of her gross capital equipment.

A political division between eastern and western Europe was coming to be evident before the end of the war. The British, in particular, had been alarmed by what happened to Poland. It seemed to show that Stalin would only tolerate governments in eastern Europe which were subservient. This was hardly what the Americans had envisaged as freedom for eastern Europeans to choose their own rulers, but until the war was over neither government nor public in the United States was much concerned or much doubted they could come to reasonable agreement with the Russians. Broadly speaking, Roosevelt had been sure that America could get on with the Soviet Union; they had common ground in resisting a revival of German power and supporting anti-colonialism. Neither he nor the American public showed any awareness of the historic tendencies of Russian policy. They disapproved strongly of British intervention in Greece against the communist guerrillas who sought to overthrow the monarchy after the German withdrawal. (Stalin did not object: he agreed with Churchill that Great Britain should have a virtually free hand in Greece in return for letting him have one in Romania.)

President Truman (who succeeded Roosevelt on his death in April 1945) and his advisers came to change American policies largely as a result of their experience in Germany. At first, the Russians had been punctilious in carrying out their agreement to admit British and American (and later French) armed forces to Berlin and share the administration of the city they had conquered. There is every indication that they wished Germany to be governed as a unit (as envisaged by the victors at Potsdam in July 1945), for this would give them a hand in controlling the Ruhr, potentially a treasure-house of reparations. Yet the German economy soon bred friction between West and East. Russian efforts to ensure security against German recovery led to the increasing practical separation of her zone of occupation from those of the three other occupying powers. Probably this was at first intended to provide a solid and reliable (that is, communist) core for a united Germany, but it led in the end to a solution by partition to the German problem which no one had envisaged. First, the western zones of occupation were for economic reasons integrated, without the eastern zone. Meanwhile Russian occupation policy aroused increasing distrust. The entrenchment of communism in the Soviet zone of Germany seemed to repeat a pattern seen elsewhere. In 1945 there had been communist majorities only in Bulgaria and Yugoslavia, and in other east European countries the communists only shared power in coalition governments. None the less, it increasingly looked as if those governments could, in fact, do little more than behave as Soviet puppets. Something like a bloc was already appearing in eastern Europe in 1946.

It is rash to dogmatize about Soviet motives. Stalin obviously feared any reunification of Germany except under a government he could control; Russia had too many memories of attacks from the west to trust a united Germany. It would always have a potential for aggression which a satellite could not have. This would have been true whatever the ideological character of the Russian regime; it only made it worse that a united Germany might be capitalist. Elsewhere, though, Soviet policy showed more flexibility. While anxiously organizing eastern Germany on the Russian side of a line slowly appearing across Europe, it was still in China officially supporting the KMT. In Iran, on the other hand, there was an obvious reluctance to withdraw Soviet forces as had been agreed. Even when they finally departed they left behind a satellite communist republic in Azerbaijan – to be later obliterated by the Iranians, to whom, by 1947, the Americans were giving military aid. In the Security Council the Soviet veto was more and more employed to frustrate her former allies and it was clear that the communist parties of western Europe were manipulated in Russian interests. Yet Stalin's calculations remain in doubt; perhaps he was waiting, expecting or even relying upon economic collapse in the capitalist world.

There had been and still was much goodwill for the USSR among her former allies. When Winston Churchill had drawn attention in 1946 to the increasing division of Europe by an 'Iron Curtain' he by no means spoke either for all his countrymen or for his American audience; some condemned him. Yet though the British Labour Government elected in 1945 was at first hopeful that 'Left could speak to Left', it quickly became more sceptical. British and American policy began to converge during 1946, as it became clear that the British intervention in Greece had in fact made possible free elections and as American officials had more experience of the tendency of Russian policy. Nor did President Truman have any prejudices in favour of Russia to shed. The British, moreover, were by now clearly embarked upon a deliberate policy of leaving India; that counted with American official opinion.

Truman took a momentous decision in February 1947. It was occasioned by a communication from the British government which, perhaps more than any other, conceded the long-resisted admission that Great Britain was no longer a world power. The British economy had been gravely damaged by the huge efforts made during the war; there was urgent need for investment at home. The first stages of decolonialization, too, were expensive. One outcome was that by 1947 the British balance of payments could only be maintained if forces were withdrawn from Greece. President Truman at once decided that the United States must fill the gap. Financial aid was given to Greece and Turkey, to enable them to survive the pressure they were under from Russia. He deliberately drew attention to the implication; much more than propping up two countries was involved. Although only Turkey and Greece were to receive aid, he deliberately offered the 'free peoples' of the world American leadership to resist, with American support, 'attempted subjugation by armed minorities or by outside pressures'. This was a reversal of the apparent return to isolation from Europe which the United States had seemed to pursue in 1945, and an enormous break with the historic traditions of American foreign policy. The decision to 'contain' Soviet power, as it was called, was possibly the most important in American diplomacy since the Louisiana Purchase. Ultimately, it was to lead to unrealistic assessments of the effective limits of American power, and, critics were to say, to a new American imperialism, as the policy was extended outside Europe, but this could not be seen at the time. It was provoked by Russian behaviour and the growing fears Stalin's policy had aroused over the previous eighteen months as well as by British weakness.

A few months later, the 'Truman Doctrine' was completed by another much more pondered step, an offer of American economic aid to European nations which would come together to plan jointly their economic recovery. This was the Marshall Plan, named after the American Secretary of State who announced it. Its aim was a non-military, unaggressive form of containment. It surprised everyone. The British Foreign Secretary, Ernest Bevin, was the first European statesman to grasp its implications. With the French, he pressed for the acceptance of the offer by western Europe. It was made, of course, to all Europe. But the Russians would not participate, nor did they allow their satellites to do so. Instead, they bitterly attacked the plan. When, with obvious regret, the Czechoslovakian coalition government also declined to adhere, that country, the only one in eastern Europe still without a fully communist government and not regarded as a Russian satellite, was visibly having to toe the Soviet line. Any residual belief in Czechoslovakia's independence was removed by a communist *coup* in February 1948. Another sign of Russian intransigence was an old pre-war propaganda device, the Comintern, revived as the Cominform in September 1947. It at once began the denunciation of what it termed a 'frankly predatory and expansionist course ... to establish the world supremacy of American imperialism'. Finally, when western Europe set up an Organization for European Economic Cooperation to handle the Marshall Plan, the Russians replied by organizing their own half of Europe in Comecon, a Council for Mutual Economic Assistance which was window-dressing for the Soviet integration of the command economies of the east.

Like fascism and Nazism before it in Italy and Germany, communism came to power in Czechoslovakia by management from above; the armed police and militia, the keys to power, march past the new government after the coup.

The Cold War (as it came to be called) had begun. The first phase of Europe's post-war history was over. The next, a phase in global history, too, was to continue well into the 1960s. In it, two groups of states, one led by the United States and one by Soviet Russia, strove throughout a succession of crises to achieve their own security by all means short of war between the principal contenders. Much of what was said was put into ideological terms. In some countries of what came to be a western bloc, the Cold War therefore also appeared as civil war or near-war, and as moral debate about values such as freedom, social justice and individualism. Some of it was fought in marginal theatres by propaganda and subversion or by guerrilla movements sponsored by the two great states. Fortunately, they always stopped short of the point at which they would have to fight with nuclear weapons whose increasing power made the notion of a successful outcome more and more unrealistic. The Cold War was also an economic competition by example and by offers of aid to satellites and uncommitted nations. Inevitably, in the process much opportunism got mixed up with doctrinaire rigidity. Probably it was unavoidable, but it was a blight which left little of the world untouched, and a seeping source of crime, corruption and suffering for more than thirty years.

In retrospect, for all the simple brutalities of the language it generated, the Cold War now looks somewhat like the complex struggles of religion in sixteenth- and seventeenth-century Europe, when ideology could provoke violence, passion, and even, at times, conviction, but could never wholly accommodate the complexities and cross-currents of the day. Above all, it could not contain those introduced by national interest. Like the religious struggles of the past, too, though, there was soon every sign that although specific quarrels might die down and disaster be avoided, its rhetoric and mythology could go rolling on long after they ceased to reflect reality.

The first important complication to cut across the Cold War was the emergence of a growing number of states which did not feel committed to one side or the other. Many new nations came into existence within a few years of 1945 as a result of decolonization, which caused just as great an upheaval in international relations as the Cold War. The United Nations General Assembly was much more important as a platform for anti-colonial than for Cold War propaganda (though they were sometimes confused).

There were many differences of circumstance and timing, but by and large the Asian nationalist movements had been guaranteed eventual success when the war flattened the card castle of European imperialism. This was first obvious in south-east Asia and Indonesia, but its repercussions were great even in areas (such as the Indian subcontinent) where the imperial power was not displaced by the Japanese. The surrender of sixty thousand British, Indian and Dominion troops at Singapore in 1942 had been the signal that European empire in Asia was doomed. No efforts could retrieve a disaster far worse than Yorktown. The loss of face had compromised the confidence and prestige of every European in the Far East. It did not matter that the Japanese often behaved badly to their new conquests. Parachuting arms to formerly subject peoples to resist the conquerors only made it likely that they would be used in due course against their former rulers in London, Paris or The Hague. Furthermore, though (by comparison with the upheavals caused by bombing, fighting, labour conscription, starvation and disease in Europe) life went on in most Asian villages undisturbed, there were notable side-effects. By 1945 there was a big potential for change in the East.

In the end, the former Asian empires were all but swept away within a few years. Yet territorial rule was not all that was at stake. Though Russian and American spheres of influence in Europe were (with the possible exception of Berlin) clearly enough demarcated in 1948 to remain unchanged for forty years, the settlement of great power relationships in the Far East was to be in doubt for much longer.

Some had always thought India might become a dominant Asian power once she achieved self-government. Even before 1939 it was plain that the survival of imperial rule in India was no longer in question. What was being discussed was the timetable and form of its replacement. Englishmen who favoured Indian independence hoped to keep it linked to the British Commonwealth of Nations, the name usually given to the empire since the Imperial Conference of 1926, which produced the first official definition of 'Dominion Status' as independent association to the Commonwealth in allegiance to the Crown, with complete control of internal and external affairs. This set a conceivable goal for India, though not one which British governments could concede as an immediate aim before 1940. Yet though unevenly, some progress was made before this, and this in part explains the absence in India of so complete a revulsion of anti-western feeling as had occurred in China.

Indian politicians had been deeply disappointed after the First World War. They had for the most part rallied loyally to the Crown; India had made big contributions of men and money to the imperial war effort, and Gandhi, later to be seen as the father of the Indian nation, had been one of those who had worked for it in the belief that this would bring a due reward. In 1917, the British government had announced that it favoured a policy of steady progress towards responsible government for India within the empire – Home Rule, as it were – though this was short of the Dominion status some

Indians were beginning to ask for. Reforms introduced in 1918 were none the less very disappointing, though they satisfied some moderates, and even such limited success as they had was soon dissipated. Economics came into play as international trading conditions worsened. In the 1920s the Indian government was already supporting Indian demands to put an end to commercial and financial arrangements favouring the United Kingdom, and soon insisted on the imperial government paying a proper share of India's contribution to imperial defence. Once into the world slump, it became clear that London could no longer be allowed to settle Indian tariff policy so as to suit British industry.

One influential factor hindering progress was the continuing isolation of the British community in India. Convinced that Indian nationalism was a matter of a few ambitious intellectuals, its members pressed merely for strong measures against conspiracy, a course not unattractive to administrators confronted with the consequences of the Bolshevik Revolution (though the Indian communist party was not founded until 1923). The result, against the wishes of all the Indian members of the legislative council, was the suspension of normal legal safeguards in order to deal with conspiracy. This provoked Gandhi's first campaign of strikes and pacifist civil disobedience. In spite of his efforts to avoid violence there were riots. At Amritsar in 1919, after some Englishmen had been killed and others attacked, a foolish general decided, as an example of his countrymen's determination, to disperse a crowd by force. When the firing stopped, nearly four hundred Indians had been killed and over a thousand wounded. An irreparable blow to British prestige was made worse when British residents in India and some members of parliament loudly applauded what had been done.

A period of boycott and civil disturbance followed, in which Gandhi's programme was adopted by Congress. Although Gandhi himself emphasized that his campaign was non-violent there was nevertheless much disorder and he was arrested and imprisoned for the first time in 1922 (and was soon released because of the danger that he might die in prison). This was the end of significant agitation in India for the next few years. In 1927 British policy began to move slowly forward again. A commission was sent to India to look into the working of the last series of constitutional changes (though this caused more trouble because no Indians had been included in it). Much of the enthusiasm which had sustained unity among the nationalists had by now evaporated and there was a danger of a rift bridged only by Gandhi's efforts and prestige between those who stuck to the demand for complete independence and those who wanted to work for Dominion status. Congress was, in any case, not so solid a structure as its rhetoric suggested. It was less a political party with deep roots in the masses than a coalition of local bigwigs and interests. Finally, a more grievous division still was deepening between Hindu and Moslem. The leaders of the two communities had watched the relations between their followers deteriorate rapidly in the 1920s into communal rioting and bloodshed. By 1930 the president of the Moslem political league was proposing that the future constitutional development of India should include the establishment of a separate Moslem state in the north-west.

That year was a violent one. The British viceroy had announced that a conference was to take place with the aim of achieving Dominion status, but this undertaking was made meaningless by opposition in Great Britain. Gandhi would not take part, therefore. Civil disobedience was resumed after a second conference foundered on the question of minorities' representation, and intensified as distress deepened with the world economic depression. The rural masses were now more ready for mobilization by nationalist appeals; although this alienated some elements in Congress, who saw their movement changing to take account of mass interests, it made Gandhi the first politician to be able to claim an India-wide following.

The wheels of the India Office were by now beginning to turn as they absorbed the lessons of the discussions and the 1927 commission. A real devolution of power and patronage came in 1935, when a Government of India Act was passed which took still further the establishment of representative and responsible government, leaving in the viceroy's sole control only such matters as defence and foreign affairs. Though the transfer of national power proposed in the Act was never wholly implemented, this was the culmination of legislation by the British. They had by now effectively provided a framework for a national politics. It was increasingly clear that at all levels the decisive struggles between Indians would be fought out within the Congress party, but it was already under grave strain. The 1935 Act once more affirmed the principle of separate communal representation and almost immediately its working provoked further hostility between Hindu and Moslem. Congress was by now to all intents and purposes a Hindu organization (though it refused to concede that the Moslem League should therefore be the sole representative of Moslems). But Congress had its internal problems, too, divided as it was between those who still wished to press forward to independence and those – some of them beginning to be alarmed by Japanese aggressiveness – who were willing to work the new institutions in cooperation with the imperial government. The evidence that the British were in fact devolving power was bound to be a divisive force. Different interests began to seek to insure themselves against an uncertain future.

Under the surface, the tide was running fast by 1941. Nearly two decades of representative institutions in local government and the progressive Indianization of the higher civil service had already produced a country which could not be governed except with the substantial consent of its élites. It was also one which had undergone a considerable preparatory education in self-government, if not democracy. Though the approach of war made the British increasingly aware of their need of the Indian army, they had already given up trying to make India pay for it and were by 1941 bearing the cost of its modernization. Then the Japanese attack forced the hand of the British government. It offered the nationalists autonomy after the war and a right of secession from the

The cremation of Gandhi's body after his murder by a Hindu fanatic.

Commonwealth, but this was too late; they now demanded immediate independence. Their leaders were arrested and the British Raj continued. A rebellion in 1942 was crushed much more rapidly than had been the Mutiny nearly a century earlier, but the sands were running out if the British wanted to go peacefully. One new factor was pressure from the

United States. President Roosevelt discussed confidentially with Stalin the need to prepare for Indian independence (as well as that of other parts of Asia, and the need for trusteeship for French Indo-China); the involvement of the United States implied revolutionary change in other people's wars just as it had done in 1917.

In 1945 the Labour Party, which had long had the independence of India and Burma as part of its programme, came to power at Westminster. On 14 March 1946, while India was torn with Hindu-Moslem rioting and its politicians were squabbling over the future, the British government offered full independence. Nearly a year later, it put a pistol to the head of the Indians by announcing that it would hand over power not later than June 1948. Thus the tangle of communal rivalries was cut, and the partition of the subcontinent followed. The only governmental unity it had ever enjoyed was ended and on 15 August 1947 two new Dominions appeared within it, Pakistan and India. The first was Moslem and was itself divided into two slabs of land at the extremities of northern India; the second was officially secular but was overwhelmingly Hindu in composition and inspiration.

Perhaps Partition was inevitable. India had never been ruled directly as one entity, even by the British, and Hindu and Moslem had been increasingly divided since the Mutiny. Nevertheless, its cost was enormous. The psychic wound to many nationalists was symbolized when Gandhi was murdered by a Hindu fanatic for his part in it. Huge massacres occurred in areas where there were minorities. Something like two million people fled to where their co-religionists were in control. Almost the only clear political gain on the morrow of independence was the solution, a bloody one, of the communal problem for the immediate future. Apart from this, the assets of the new states were the goodwill (arising from very mixed motives) shown to them by great powers, the inheritance of a civil service already largely native before independence, and an important infrastructure of institutions and services. These inheritances were not, however, equally shared, India tending to enjoy more of them than Pakistan.

Such advantages could not do much to deal with the grave problems of the subcontinent's economic and social backwardness. The worst problem was demographic. A steady rise in population had begun under British rule. Sometimes it was briefly mitigated by Malthusian disasters like the great influenza epidemic at the end of the First World War which struck down five million Indians or a famine in Bengal during the Second World War which carried off millions more. But in 1951 there was famine again in India, and in 1953 in Pakistan. The spectre of it lingered into the 1970s.

The subcontinent's industrialization, which had made important strides in the twentieth century (notably in the Second World War), did not offset this danger. It could not provide new jobs and earnings fast enough for a growing population. Though the new India had most of what industry there was, her problems were graver in this respect than those of Pakistan. Outside her huge cities, most Indians were landless peasants, living in villages where, for all the egalitarian aspirations of some of the leaders of the new republic, inequality remained as great as ever. The landlords who provided the funds for the ruling Congress party and dominated its councils stood in the way of any land reform which could have dealt with this. In many ways, the past lay heavy on a new state proclaiming the western ideals of democracy, nationalism, secularism and material progress, and it was to encumber the road of reform and development.

China had for a long time been engaged in fighting off a different sort of imperialism. Success against the Japanese and completion of her long revolution was made possible by the Second World War. The political phase of this transformation began in 1941, when the Sino-Japanese War merged in a world conflict. This gave China powerful allies and a new international standing. Significantly, the last vestiges of the 'unequal treaties' with Great Britain, France and the United States were then swept away. This was more important than the military help the Allies could give; for a long time they were too busy extricating themselves from the disasters of early 1942 to do much for China. A Chinese army, indeed, came instead to help to defend Burma from the Japanese. Still hemmed in to the west, though supported by American aircraft, the Chinese had for a long time to hold out as best they could, in touch with their allies only by air or the Burma Road. None the less a decisive change had begun.

China had at first responded to Japanese attacks with a sense of national unity long desired but never hitherto forthcoming except, perhaps, in the May 4th Movement. In spite of friction between the communists and the nationalists, sometimes breaking out into open conflict, this unity survived, broadly speaking, until 1941. Then, the new fact that the United States was now Japan's major enemy, and would eventually destroy her, subtly began to transform the attitude of the nationalist government. It came to feel that as ultimate victory was certain, there was no point in using up men and resources in fighting the Japanese when they might be husbanded for the struggle against the communists after the peace. Some of its members went further. Soon the KMT was fighting the communists again.

Two Chinas appeared. Nationalist China increasingly displayed the lethargy, self-seeking and corruption which had from the early 1930s tainted the KMT because of the nature of the support on which it drew. The regime was repressive and stifled criticism. It alienated the intellectuals. Its soldiers, sometimes badly officered and undisciplined, terrorized the peasant as much as did the Japanese. Communist China was different. In large areas controlled by the communists (often behind the Japanese lines) a deliberate attempt was being made to ensure the support of as wide a spectrum of interests as possible by moderate but unambiguous reform and disciplined behaviour. Outright attacks on landlords were usually avoided, but peasant goodwill was cultivated by enforcing lower rents and abolishing usury. Meanwhile, Mao published a series of theoretical writings designed to prepare the new communist cadres for the task that lay ahead. There was a need for political education as the party and the army grew steadily in numbers; when the Japanese collapsed in 1945 there were about a million Chinese communist soldiers.

The suddenness of victory was the second factor which shaped the last stage of the Chinese Revolution. Huge areas of China had suddenly to be reoccupied and reincorporated in the Chinese state. But many of them were already under communist control before 1945 and others could not possibly be reached by nationalist forces before the communists dug themselves in there. The Americans did what they could by sending soldiers to hold some of the ports until the nationalists could take them over. In some places the Japanese were told to hold on until the Chinese government could re-establish its authority. But when the final and military phase of the Revolution opened, the communists held more territory than they had ever done before and held it in the main with the support of the population who found that communist rule was by no means as bad as they had heard.

All unwittingly, the Japanese, by launching their attack on the KMT regime, had in the end brought about the very triumph of the Chinese Revolution they had long striven to avoid. It is at least possible that if the nationalists had been undistracted by foreign invasion and had not suffered the crippling damage it inflicted, they might have been able to master Chinese communism in the short run. In 1937 the KMT could still draw heavily on patriotic goodwill; many Chinese believed that it was the authentic carrier of the Revolution. The war destroyed the chance of exploiting this, if it were true, but also enabled China to resume for the last time her long march towards world power from which she had been deflected first by Europeans and then by fellow-Asians. The long frustration of Chinese nationalism was about to end, and the beneficiary would be communism.

The defeat of the KMT took three years. Although the Japanese usually sought to surrender to the KMT or Americans, the communists had acquired authority in new areas and often large stocks of arms when they gave in. The Russians, who had invaded Manchuria in the last days before the Japanese surrender, helped them by giving them access to the Japanese arms there. Mao made deliberately moderate policy pronouncements and continued to push forward with land reform. This conferred a further great advantage on the communists in the civil war which continued until 1949;

victory in that war was essentially a victory of the countryside over a city-based regime.

American policy was increasingly disillusioned by the revealed inadequacy and corruption of the Chiang K'ai-shek government. In 1947 American forces were withdrawn from China and the United States abandoned the efforts it had hitherto made to mediate between the two Chinas. In the following year, with most of the north in communist hands, the Americans began to cut down the amount of financial and military aid given to the KMT. From this time, the nationalist government ran militarily and politically downhill; it became obvious, and more and more employees of government and local authorities sought to make terms with the communists while they might still do so. The conviction spread that a new era was dawning. By the beginning of December, no important nationalist military force remained intact on the mainland and Chiang withdrew to Formosa (Taiwan). The Americans cut off their aid while this withdrawal was under way and publicly blamed the inadequacies of the nationalist regime for the débâcle. Meanwhile, on 1 October 1949, the People's Republic of China was officially inaugurated at Peking and the largest communist state in the world had come into existence. Once again, the Mandate of Heaven had passed.

In south-east Asia and Indonesia the Second World War was as decisive as elsewhere in ending colonial rule, though the pace was faster in Dutch and French colonies than British. The grant of representative institutions by the Dutch in Indonesia before 1939 had not checked the growth of a nationalist party, and a flourishing communist movement had appeared by then, too. Some nationalist leaders, among them one Achmed Sukarno, collaborated with the Japanese when they occupied the islands in 1942. They were in a favourable position to seize power when the Japanese surrendered, and did so by proclaiming an independent Indonesian republic before the Dutch could return. Fighting and negotiation followed for nearly two years until agreement was reached for an Indonesian republic still under the Dutch Crown; this did not work. Fighting went on again, the Dutch pressing forward vainly with their 'police operations' in one of the first campaigns by a former colonial power to attract the full blast of communist and anti-colonial stricture at the United Nations. Both India and Australia (which had concluded that she would be wise to conciliate the independent Indonesia which must eventually emerge) took the matter to the Security Council. Finally the Dutch gave in. The story begun by the East India Company of Amsterdam three and a half centuries before thus came to an end in 1949 with the creation of the United States of Indonesia, a mixture of more than a hundred million people scattered over hundreds of islands, of scores of races and religions. A vague union with the Netherlands under the Dutch Crown survived, but was dissolved five years later.

For a time the French in Indo-China seemed to be holding on better than the Dutch. That area's wartime history had been somewhat different from that of Malaysia or Indonesia because although the Japanese had exercised complete military control there since 1941 French sovereignty was not formally displaced until early 1945. The Japanese had amalgamated Annam, Cochin-China and Tongking to form a new state of Vietnam under the Emperor of Annam and as soon as the Japanese surrendered, the chief of the local communist party, the Viet Minh, installed himself in the government place at Hanoi and proclaimed the Vietnam republic. This was Ho Chi Minh, a man with long experience in the communist party and also in Europe. The revolutionary movement quickly spread. It was soon evident that if the French wished to re-establish themselves it would not be easy. A large expeditionary force was sent to Indo-China and a concession was made in that the French recognized the republic of Vietnam as an autonomous state within the French Union. But now there arose the question of giving Cochin-China separate status and on this all attempts to agree broke down. Meanwhile, French soldiers were sniped at and their convoys were attacked. At the end of 1946 there was an attack on residents in Hanoi and many deaths. Hanoi was relieved by French troops and Ho Chi Minh fled.

Thus began a war in which the communists were to struggle essentially for the nationalist aim of a united country, while the French tried to retain a diminished Vietnam which, with the other Indo-Chinese states, would remain inside the French Union.

The flag of the new state of Israel, raised at the port of Elath, on the gulf of Aqaba, during the 1948 war.

By 1949 they had come round to including Cochin-China in Vietnam and recognizing Cambodia and Laos as 'associate states'. But new outsiders were now becoming interested. The government of Ho Chi Minh was recognized in Moscow and Peking, that of the

Annamese emperor whom the French had set up by the British and Americans.

Thus in Asia decolonialization quickly burst out of the simple processes envisaged by Roosevelt. As the British began to liquidate their recovered heritage, this further complicated things. Burma and Ceylon had become independent in 1947. In the following year, communist-supported guerrilla war began in Malaya; though it was to be unsuccessful and not to impede steady progress towards independence in 1957, it was one of the first of several post-colonial problems which were to torment American policy. Growing antagonism with the communist world soon cut across the simplicities of anti-colonialism.

Only in the Middle East did things still seem clear-cut for the United States in 1948. In May that year, a new state, Israel, came into existence in Palestine. This marked the end of forty years during which only two great powers had needed to agree in order to manage the area. France and Great Britain had not found this too difficult. In 1939 the French still held mandates in Syria and the Lebanon (their original mandate had been divided into two), and the British retained theirs in Palestine. Elsewhere in the Arab lands the British exercised varying degrees of influence or power after negotiation with the new rulers of individual states. The most important were Iraq, where a small British force, mainly of air force units, was maintained, and Egypt, where a substantial garrison still protected the Suez Canal. The latter had become more and more important in the 1930s as Italy showed increasing hostility to Great Britain.

The war of 1939 was to release change in the Middle East as elsewhere, though this was not at first clear. After Italy's entry to the war, the Canal Zone became one of the most vital areas of British strategy and Egypt suddenly found herself with a battlefront for a western border. She remained neutral almost to the end, but was in effect a British base and little else. The war also made it essential to assure the supply of oil from the Gulf and especially from Iraq. This led to intervention when Iraq threatened to move in a pro-German direction after another nationalist *coup* in 1941. A British and Free French invasion of Syria to keep it, too, out of German hands led in 1941 to an independent Syria. Soon afterwards the Lebanon proclaimed its independence. The French tried to re-establish their authority at the end of the war, but unsuccessfully, and during 1946 these two countries saw the last foreign garrisons leave. The French also had difficulties further west, where fighting broke out in Algeria in 1945. Nationalists there were at that moment asking only for autonomy in federation with France and the French went some way in this direction in 1947, but this was far from the end of the story.

Where British influence was paramount, anti-British sentiment was still a good rallying-cry. In both Egypt and Iraq there was much hostility to British occupation forces in the post-war years. In 1946 the British announced that they were prepared to withdraw from Egypt, but negotiations on the basis of a new treaty broke down so badly that Egypt referred the matter (unsuccessfully) to the United Nations. By this time the whole question of the future of the Arab lands had been diverted by the Jewish decision to establish a national state in Palestine by force.

The Palestine question has been with us ever since. Its catalyst had been the Nazi revolution in Germany. At the time of the Balfour Declaration 600,000 Arabs had lived in Palestine beside 80,000 Jews – a number already felt by Arabs to be threateningly large. In some years after this, though, Jewish emigration actually exceeded immigration and there was ground for hope that the problem of reconciling the promise of a 'national home' for Jews with respect for 'the civil and religious rights of the existing non-Jewish communities in Palestine' (as the Balfour Declaration had put it) might be resolved. Hitler changed this.

From the beginning of the Nazi persecution the numbers of those who wished to come to Palestine rose. As the extermination policies began to unroll in the war years, they made nonsense of British attempts to restrict immigration which were the side of British policy unacceptable to the Jews; the other side – the partitioning of Palestine – was rejected by the Arabs. The issue was dramatized as soon as the war was over by a World Zionist Congress demand that a million Jews should be admitted to Palestine

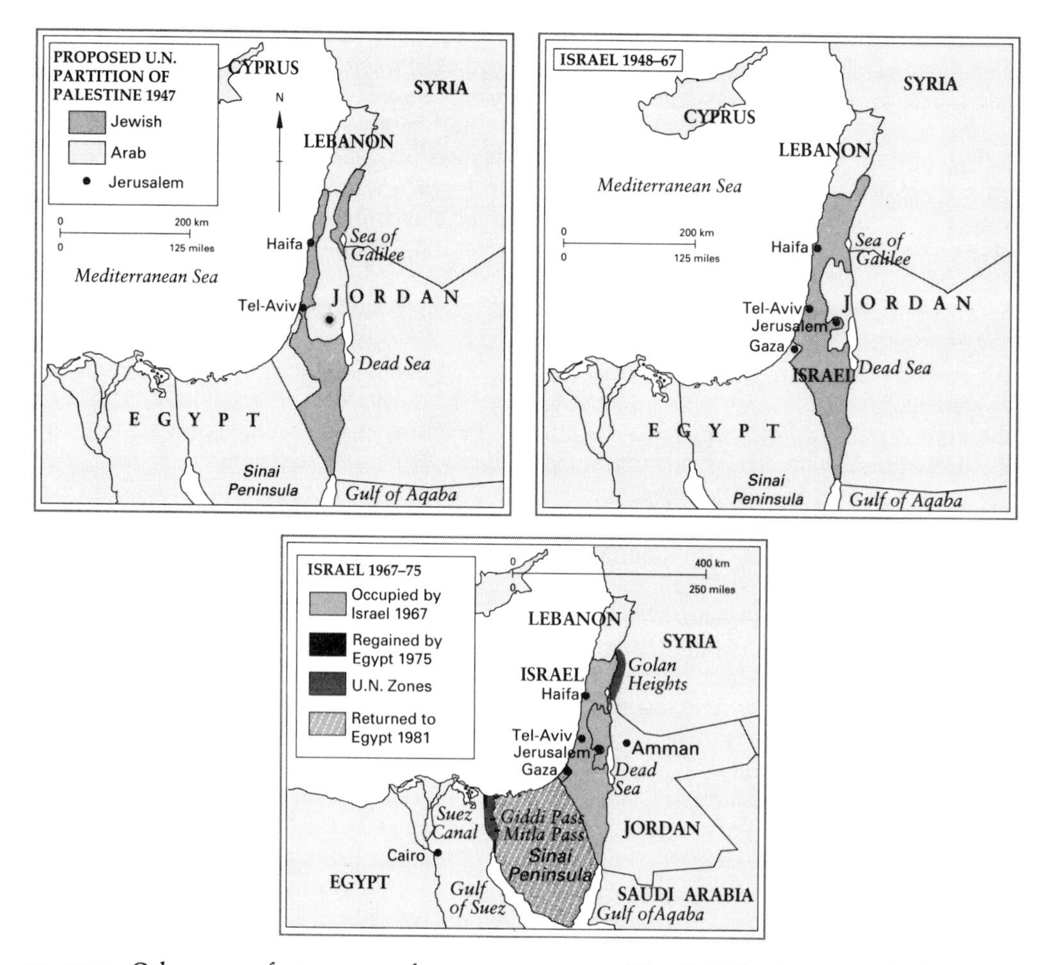

at once. Other new factors now began to operate. The British, in 1945, had looked benevolently on the formation of an 'Arab League' of Egypt, Syria, Lebanon, Iraq, Saudi Arabia, the Yemen and Transjordan. There had always been in British policy a strand of illusion – that pan-Arabism might prove the way in which the Middle East could be persuaded to settle down and that the coordination of the policies of Arab states would open the way to the solution of its problems. In fact the Arab League was soon preoccupied with Palestine to the virtual exclusion of anything else.

The other novelty was the Cold War. In the immediate post-war era, Stalin seems to have been impressed still with the old communist view that Great Britain was the main imperialist prop of the international capitalist system. Attacks on her position and influence therefore followed, and in the Middle East this, of course, coincided with traditional Russian interests, though the Soviet government had shown little interest in the area between 1919 and 1939. Pressure was brought to bear on Turkey at the Straits, and ostentatious Soviet support was given to Zionism, the most disruptive element in the situation. It did not need extraordinary political insight to recognize the implications of a resumption of Russian interest in the area of the Ottoman legacy. Yet at the same moment American policy turned anti-British, or, rather, pro-Zionist. This could hardly have been avoided. In 1946 mid-term congressional elections were held and Jewish votes were important. Since the Roosevelt revolution in domestic politics, a Democratic president could hardly envisage an anti-Zionist position.

Thus beset, the British sought to disentangle themselves from the Holy Land. From 1945 they faced both Jewish and Arab terrorism and guerrilla warfare in Palestine. Unhappy Arab, Jewish and British policemen struggled to hold the ring while the British government still strove to find a way acceptable to both sides of bringing the mandate to an end. American help was sought, but to no avail; Truman wanted a pro-Zionist solution. In the end the British took the matter to the United Nations. It recommended partition, but this was still a non-starter for the Arabs. Fighting between the two communities grew fiercer and the British decided to withdraw without more ado. On the day that they did so, 14 May 1948, the state of Israel was proclaimed. It was immediately recognized by the United States (sixteen minutes after the foundation act) and Russia, who were to agree about little else in the Middle East for the next quarter-century.

Israel was attacked almost at once by Egypt, whose armies invaded a part of Palestine which the United Nations proposal had awarded to Jews. Jordanian and Iraqi forces supported Palestinian Arabs in the territory proposed for them. But Israel fought off her enemies, and a truce, supervised by the United Nations, followed (during which a Zionist terrorist murdered the United Nations mediator). In 1949 the Israeli government moved to Jerusalem, a Jewish national capital again for the first time since the days of imperial Rome. Half of the city was still occupied by Jordanian forces, but this was almost the least of the problems left to the future. With American and Russian diplomatic support and American private money, Jewish energy and initiative had successfully established a new national state where no basis for one had existed twenty-five years before. Yet the cost was to prove enormous. The disappointment and humiliation of the Arab states assured their continuing hostility to it and therefore opportunities for great power intervention in the future. Moreover, the action of Zionist extremists and the far from conciliatory behaviour of Israeli forces in 1948–9 led to an exodus of Arab refugees. Soon there were 750,000 of them in camps in Egypt and Jordan, a social and economic problem, a burden on the world's conscience, and a potential military and diplomatic weapon for Arab nationalists. It would hardly be surprising were it true (as some students believe) that the first president of Israel quickly began to encourage his country's scientists to work on a nuclear energy programme.

Thus, many currents flowed together in a curious and ironical way to swirl in confusion in an area always a focus of world history. Victims for centuries, the Jews were in their turn now seen by Arabs as persecutors. The problems with which the peoples of the area had to grapple were poisoned by forces flowing from the dissolution of centuries of Ottoman power, from the rivalries of successor imperialisms (and in particular from the rise of two new world powers which dwarfed these in their turn), from the interplay of nineteenth-century European nationalism and ancient religion, and from the first effects of the new dependence of developed nations on oil. There are few moments in the twentieth century so soaked in history as the establishment of Israel. It is a good point at which to pause before turning to the story of the second half of the twentieth century.

Man's first steps on the moon, brought by television to approximately 500,000,000 people.

BOOK VIII

THE LATEST AGE

Whether we think 1917 a more important turning-point than 1919, or that the start of Japanese aggression in Manchuria in 1931 registers more of a new departure than the invasion of Poland in 1939, does not much matter; during the twentieth century history moved into a new phase. Politically, we could call it post-European, in that after 1945 no sensible person could deny that the days of political or military domination of the world by the great powers of Europe were over. But something else has also become clear: the world is one as never before. Events and trends interact more than ever; communications tie mankind together with a new closeness and rapidity. This partly explains and partly exemplifies a common human civilization, now shared more widely than any civilization hitherto, but still bearing the marks of its European origins.

This ought by now to be merely a commonplace, yet it can still surprise us. Motorists may have got used to sudden rises in the price of petrol if there is an emergency in the Middle East, but not long ago it would still have been incredible that – say – Indians should demonstrate about something going on in the South Pacific, or that a Chinese government should feel concerned by an Arab-Israeli war. Yet now almost anything that happens anywhere in the world may very rapidly produce effects elsewhere, and most of our leaders seem to recognize this, whether they do so because of ideology, calculation, or just simple fear.

Some of the world's new interconnectedness arises from greater economic and technological interdependence, and the faster exchange of information which goes with that. At a more fundamental level, there is a world-wide, still unrolling, change of attitudes. It marks a reversal of what went on for most of the thousands of years of prehistory and history already death with in this book. Once humanity was liberated in prehistoric times by its first primitive technologies from the simplest and most crushing restraints, it wandered for thousands of years along paths which diverged into increasingly different ways of life. The paths started slowly to converge again, a few centuries ago, when there began to spread from Europe a process which has various names and has been interpreted in many ways, but which is often called modernization. Since 1945, it looks as if it is the dominant theme of world history. The paths have come together again, and people are beginning to act accordingly.

Of course no absolute statements can be made in such general terms. We must not exaggerate the extent of change in many parts of the world. The richest and the poorest societies are now more obviously different than they have ever been. Great cultures and traditions still cling to their own values and ways of life, sensing threats in the individualism and hedonism which came from Europe even when often they were cloaked in otherwise desirable change. They seek, often successfully, to preserve social disciplines which, however harsh, they still find acceptable. Yet, when all is said and done, it remains true that because of the political supremacy Europeans established in the eighteenth and nineteenth centuries, and the technological advances, especially in communications, which followed, more common experiences and assumptions are now shared more widely than ever before.

Modernization has usually looked attractive because it promised wealth. Optimism about material progress and the possibility of improvement has now spread world-wide from its origins in eighteenth-century Europe, thanks to a golden age of wealth creation in the last couple of centuries. An immense increase in the consumption of resources provided unambiguous evidence of humanity's material betterment; there were more human beings, and most of them were living longer, than ever before. For all the notorious pockets of degradation and poverty which remained (or were created), the human race seemed to be able to make a greater success than ever before of the business of living on this planet. Only in the last couple of decades has anyone begun to worry much about the costs of this and what it might portend.

For most of the last half-century, issues other than those we now call 'environmental' have been far more worrying. One is the fragility of peace. Yet since 1945 the major powers avoided formally going to war with fighting likely though this sometimes seemed. On the other hand, for all the hopes and rhetoric swirling about the infant United Nations Organization, many of its members fought one another and sometimes did so more than once. There has been no return to what from this distance seems the Augustan calm of late nineteenth-century international relations. We have frequent enough reminders, moreover, of persistent divisions within mankind, and enough evidence of new ones appearing, to make the world still appear dangerously unstable. Nationality, ideology and economic interest still fragment it. Yet this, too, is a view from a particular perspective. There are others. The Second World War released many currents and some of them have yet to run their full course. Judgements in the last quarter-century have sometimes had to change almost day by day. The war in the Middle East in 1973 was from one point of view just another of many indecisive conflicts following the collapse of Ottoman empire; in another perspective, it was the first to reveal a reversal of the relations of some great industrial societies to the Islamic world because of dependence on its oil. After another decade, that reversal again seemed less permanent and important; the view had changed once more. Then came the Gulf crisis of 1990 and the most recent war of the Ottoman succession. 'Finally, three-quarters of a century of history seemed suddenly to vanish as the story opened in 1917 came to an end in the dissolution of the Soviet Union and the re-naming of Leningrad as St Petersburg.' All we can do in looking at near-contemporary history is to try to understand its changes contemporary history is to try to understand its changes historically. That may help to make the overwhelming rush of events just a little more comprehensible.

1
Perspectives

In 1974 the first world conference on population ever held met in Romania; uneasiness about its future had for the first time persuaded the human race to unite to consider the demographic outlook. The unwilled, seemingly uncontrollable and accelerating rise in world population which has gone on for the last couple of centuries is now seen to be a global problem, even if much about its exact nature remains uncertain.

POPULATION CHANGES

Accuracy in computing populations is still a highly relative business. We can only estimate to within one or two hundred millions how many people are now alive. None the less, the likely degree of error is not such that our estimates significantly distort what has happened. In round numbers, a world population of about seven hundred and twenty million in 1750 more than doubled by 1900, when it stood at about sixteen hundred million. Thus about eight hundred and fifty million had been added to it over a century and a half. It then took fifty years to add the next eight hundred and fifty million and somewhat more; by 1950 the population of the world stood at about two thousand five hundred million. Even more striking, the next eight hundred and fifty million were added by the middle of the 1960s, in less than twenty years. Now the total is over five thousand million. Though it had taken at least 50,000 years for *Homo sapiens* to number one thousand million (a figure reached in 1840 or thereabouts) the last one thousand million of his species has been added in only fifteen years or so to a total which has grown faster and faster. Though growth rates in some countries have fallen since the 1960s, up-to-date estimates still say the human race is growing by 1.63 per cent per year; that means it will probably number just under six thousand million before the end of this century.

Such figures are alarming. The spectre of Malthusian disaster has been revived. Fortunately, as Malthus himself observed, 'no estimate of future population or depopulation, formed from any existing rate of increase or decrease, can be depended upon'. We cannot be sure what might change or modify such an acceleration. Some societies have now accepted the possibility of conscious control of their shape and size, for instance. Strictly, this is not a new idea; in some places murder and abortion have long been customary ways of keeping down demands on scarce resources. Babies were exposed to die in medieval Japan; female infanticide was widespread in India a century and a half ago and has emerged again (or perhaps been acknowledged again) in recent years in China. What is new is that in some countries governments are now putting resources and authority behind more humane means of population control. Their aim is positive economic improvement instead of just the avoidance of family and personal disaster.

Not all governments make such efforts. This is a negative aspect of the complex truth that population growth, though world-wide, does not everywhere take the same form nor produce the same effects. Though many non-European countries have followed the pattern of nineteenth-century Europe (in first showing a fall in death rates without a corresponding fall in birth rates) it would be rash to prophesy that they will go on to repeat the next phase of the population history of developed countries. We cannot simply expect a pattern of declining natality shown in one place or one society to be repeated elsewhere. But we cannot be sure that they will not be, and there are even some signs that repetitions are beginning to emerge.

The dynamics of population growth or decline are exceedingly complex. They reflect limits set by ignorance, and by social and personal attitudes, and these are hard to measure, let alone manipulate.

At present it looks as if there is plenty of scope in many countries for things to go on for some time much as at present. At the very least, some poor countries cannot for a long time hope to achieve demographic equilibrium. Natality only began to drop in the last century when prosperity in a few countries made it attractive to large numbers of men and women to have smaller families; few of today's fast-growing countries are yet anywhere near that point. Further medical, nutritional and sanitary progress may make things much worse. The advances have been colossal since the nineteenth century, yet there are many places where they have yet to cut into mortality as dramatically as they did in Europe between 1800 and 1900. When and where they do so, humanity's numbers will probably rise faster still.

One rough indicator of potential for future growth is infant mortality. In the century before 1970 this fell dramatically from an average of about 225 per thousand live births to under 20 in developed countries; in 1988 the comparative figures for Bangladesh and Japan were 118 and 5. Such discrepancies continue to exist – and they are much greater than in the past – between poor and rich countries. There are comparable differences in life expectancy, too. In developed countries, life expectancy at birth rose from rather more than 40 in 1870 to slightly over 70 a hundred years later. It now shows a remarkable evenness. Life expectancy at birth in the United States, Great Britain and the USSR in 1987, for example, was 76, 75 and 70 years respectively; the differences were negligible by comparison with Ethiopia (41), or even India (58). Yet even the Indian baby faces prospects enormously better than those of Indian babies at the beginning of this century – let alone those of French babies in 1789. It is unlikely that the steam will go out of population growth while there are still such good prospects of improvement.

In the immediate future, this will present different problems in different places. For most of human history, all societies resembled pyramids, with very large numbers of young people and a few old. Now, developed societies are beginning to look like slowly tapering columns; the proportion of older people is bigger than in the past. The reverse is true in poorer countries. Over half Kenya's population is under 15, and two-thirds of China's is under 33. What this implies is too complicated for discussion here, but it shows that overall population growth is a somewhat obscuring notion. Although the world's population goes on growing mightily, it does so in ways which have different origins and produce very different historical effects.

Among them are big changes in the way population is shared. At the beginning of 1990, world population was distributed roughly as follows:

	Millions	*% of total*
Europe (excl. USSR	549	10.7
USSR	285	5.8
Asia (excl. USSR)	2943	57.7
Africa	610	11.9
South America and the Caribbean	430	8.4
North America	272	5.3
Australasia & Oceania	26	0.5

The fall from Europe's mid-nineteenth century quarter-share of world population is striking. Until the 1920s moreover, Europe was still exporting a lot of manpower overseas, notably to the Americas. This emigration was cut down by restrictions on entry to United States in that decade, dwindled even more during the world depression, and has never since recovered its former importance. On the other hand, immigration to the United States from the Caribbean, Central and South America, and Asia has surged upwards in the last two decades. Meanwhile, though some European countries still sent out many emigrants (in the early 1970s more Britons still left their country each year than were needed to balance an inflow from abroad), they also began from the 1950s onwards to attract North Africans, Turks, Asians and West Indians seeking work they could not find at home.

Present world patterns are not likely to be unchanged for long. Asia now contains over half mankind and China one-fifth of it, but huge as are the growth rates which have produced these populations, they are falling, though others remain threatening. Population growth in Brazil, for example, was running at more than twice the world rate in the early 1960s. In other Latin American countries, though, standards of living and life expectancy are still not very much better than the European levels early in this century, and growth has continued. Birth-rates are still very high. The long resistance of the Roman Catholic Church to contraception and legal abortion is not the whole explanation and is, in any case, increasingly ineffective. The attitudes of Latin American males and the social disciplines which impose large families on Latin American women – who, because of these same disciplines have long been unquestioningly complaisant – may have mattered more. The most threatening growth rates of all are to be found in the Islamic world; Jordan's annual rate is 3.9 per cent and present projections imply a doubling of population in sixteen years, while Iran grows at 3.5 and the much smaller Saudi population at 5.6 per cent.

The face of world poverty: Indian children, Delhi, January 1991.

There is nontheless now good evidence of major increases in contraceptive use in developing countries since 1970. Some of this can be attributed to conscious attempts by governments to stem population growth. For a long time, communist states did not warm to the ideas of population stabilization or reduction, but, though with very different historical backgrounds, both China and the Soviet Union began in the 1960s to try to control population growth by encouraging people to delay marriage and have smaller families. China has had some success, using legal regulation, tax incentives and social pressure, but found the unacceptable practice of female infanticide reappearing as a result; in India large sums have been spent on propaganda for contraception, but to uncertain effect, though it is clear that there was some increase in awareness of contraceptive methods. But neither revolutionized by industrialization (as was Japan) nor by a political attack on its traditional institutions (as was China), India is still a predominantly agrarian society. Both British and later Indian government has long been respectful of its social traditions. This has protected a conservatism in ideas and institutions which makes population control very difficult. One example is the survival, outside a tiny minority among India's élites, of a vast and traditional inequality in the status and employment prospects of men and women. Were attitudes towards women which are taken for granted in Europe or North America (and frequently denounced there as inadequate) even slightly more prevalent in India, they would be likely to raise dramatically the average age of women at marriage, and therefore to reduce the number of children in the average family. But such a change would presuppose fundamental and improbable revolution in Indian life, in the provision of new opportunities of employment and in the redistribution of authority inside the family and village, a much more radical break with India's past than independence in 1947. No country should be expected easily to shake off so much. No great tradition of civilization can be got rid of painlessly.

Perhaps, though, we need not be too gloomy. Knowledge of contraceptive possibilities and the use of them is spreading, and fertility has tended to fall in developing countries as economic well-being increases. If countries like India have not yet been able to generate much economic improvement in the lot of their peoples, elsewhere, as in parts of Latin America, there is evidence that economic growth has been followed by at least the beginning of a decline in natality. There is still a huge revolutionary power in the expanding, developing civilization of the European tradition; however it is packaged, it is the most powerful solvent of traditional structures history can show. A change in population structure in one way or another seems as unavoidable a concomitant of that civilization as the weakening of religious culture, the building of factories or the liberation of women – and the list can be hugely extended.

Meanwhile, differences of population and changes in those differences, affect the comparative strengths of nations. They are not, of course, simply translatable into differences of power. Resources and culture also come into the matter, and power for one purpose is not always power for another. None the less, population and power are related. At the end of 1988 the ten most populous states were

	million		million
China	1104	Brazil	144
India	796	Japan	123
USSR	283	Pakistan	105
USA	246	Nigeria	105
Indonesia	175	Bangladesh	104

West Germany, with 61 million, was then the most populous European country. On any reckoning, the list of ten contained the three most powerful countries in the world. China, now again part of a world which she long shunned, is bound to be a great power on grounds of population alone, for in a sense she is militarily unbeatable. Her social revolution has begun to increase her wealth, too, whereas the obstacle to some other highly-populated countries becoming very powerful is a poverty which looks unsurmountable, whether it is absolute, in the sense that natural resources are poor (Bangladesh), or relative, in that they are swallowed by population growth which is too fast (India and, until recently, Indonesia) and has overtaken the cashing of the cheque of aid from abroad and the reaping of the reward of improved technology and planned investment. Newly-generated wealth has simply been consumed. It is not easy, though, to generalize. In the early 1970s India was thought to be about to enter a period of self-sufficiency in food. Her agricultural output had doubled between 1948 and 1973. Yet this increase in wealth only just succeeded in holding the line for a population growing by a million a month.

Such population growth is itself revealing of another fact: greater world resources. The world's gross output of food has risen dramatically. Though many have starved, more have lived. Though millions have died in local famines, there has been (so far) no world-wide disaster. If the world had not been able to feed a growing population human numbers would be smaller. Whether this can continue is another matter. Yet experts have concluded that we shall be able for a considerable time to come to feed our growing numbers. Some hope that population policies could be effectively introduced to stabilize demand in relation to sustainable levels of supply while there is still time. In such matters, we enter the realms of speculation. Only the very existence of such hopes and aspirations should concern the historians, for they say something about what is – about the present and actual state of the world, where what is believed to be possible is a very important factor in determining what can happen. In considering how those hopes and aspirations have come to be we can direct our attention to another phenomenon of the last half-century: it was an age of unprecedented economic growth.

PLENTY

Many readers of this book will have seen, perhaps often, television pictures of harrowing details of famine and deprivation. Yet since 1945, at least in the developed world, economic growth has come to be taken for granted. It has there become the 'norm', in spite of hiccups and interruptions along the way, and the result is that even a slowing-down in its rate now occasions alarm. What is more, as population growth shows, there has been real economic growth in gross terms in much of the 'underdeveloped' world, too. Against the background of the 1930s, this can be considered a revolution, though the story does not in fact begin at the end of the Second World War. The decades since 1950 have been a golden age of growth, but the appropriate context in which to understand the unprecedented wealth production which has so far successfully carried the burden of world population growth is the whole twentieth century. Since 1900, there has been an increase in the world's wealth only briefly and locally interrupted by two world wars and, between them, a major world-wide setback, above all in the world economic depression of the 1930s. In 1945 wealth creation was resumed and has since barely ceased (though it greatly slowed in some countries after 1975), even under the impact of the most serious tests and even allowing for notable contrasts between different economies in the developed world. The overall rate of growth has now slowed from its peaks in the 1950s and 1960s, but the trend has continued upward.

In spite of huge disparities and setbacks affecting some countries more than others, this growth has been widely shared. Gross Domestic Product (GDP) has risen almost everywhere since 1960, and often it has risen *per capita*, too. Recent calculations provide the following examples of the change (in terms of 1988 dollars) over this century:

	1900	*1988*
Brazil	436	2451
Italy	1343	14432
Sweden	1482	21155
France	1600	17004
Japan	677	23325
UK	2798	14477
USA	2911	19815

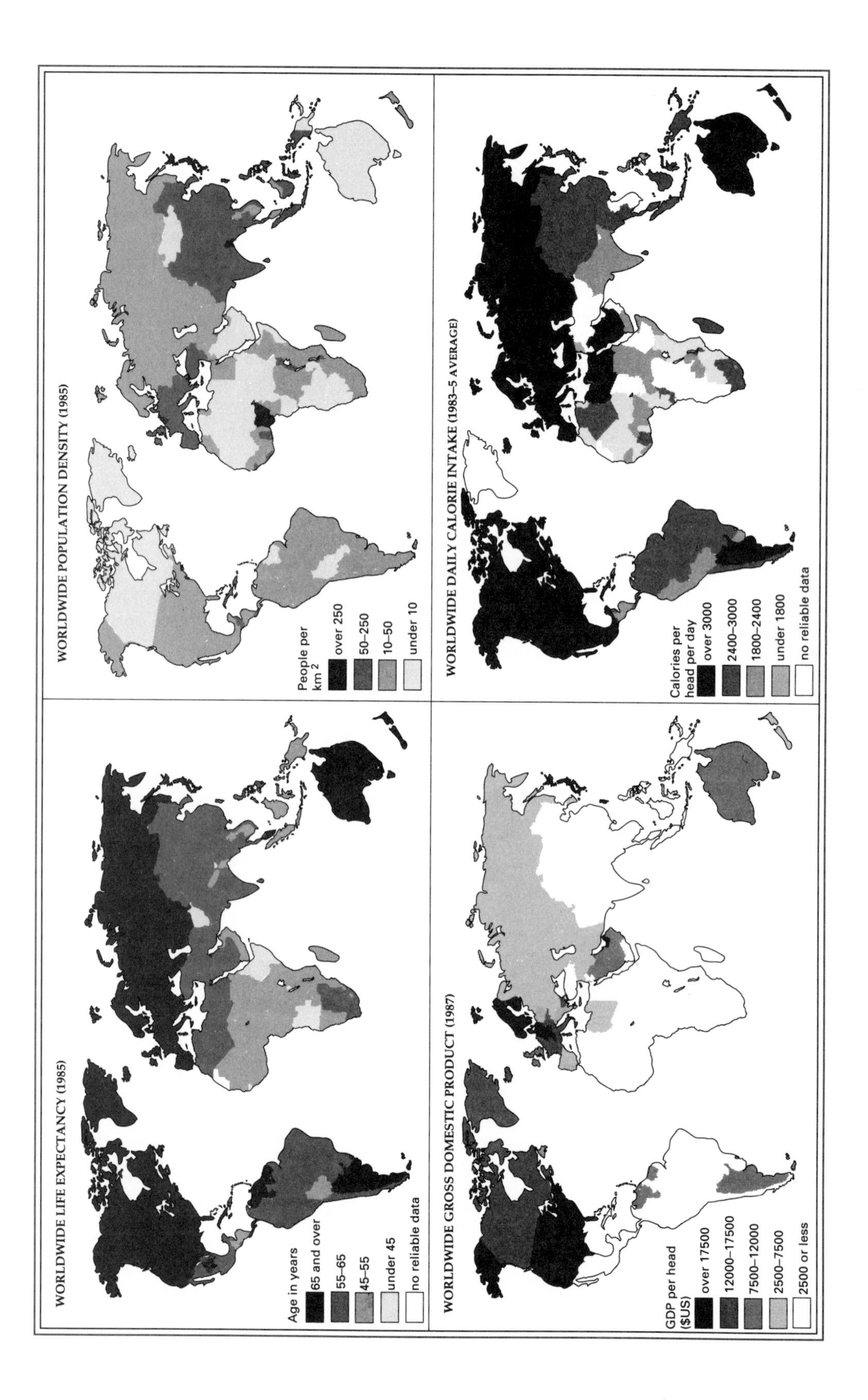
WORLDWIDE LIFE EXPECTANCY (1985)
Age in years
65 and over
55–65
45–55
under 45
no reliable data
WORLDWIDE POPULATION DENSITY (1985)
People per km 2
over 250
50–250
10–50
under 10
WORLDWIDE GROSS DOMESTIC PRODUCT (1987)
GDP per head ($US)
over 17500
12000–17500
7500–12000
2500–7500
2500 or less
WORLDWIDE DAILY CALORIE INTAKE (1983–5 AVERAGE)
Calories per head per day
over 3000
2400–3000
1800–2400
under 1800
no reliable data

Such selected figures require cautious interpretation, but they show how much richer the world has become, whatever the differential rates at which new wealth has been created and distributed. Some countries remain woefully poor. In 1988, Afghanistan, Madagascar, Laos, Tanzania, Ethiopia, Cambodia and Mozambique all had a GDP *per capita* of less than $150.

The major fact none the less remains that of wealth creation. The beginnings of a general explanation of a growing abundance must be the long peace between major industrial powers. The fact that the decades since 1945 have been studded by smaller-scale or incipient conflicts, that men and women have been killed in warlike operations on every day of them, that great powers have often had surrogates to do their fighting for them, and that there have been long periods of international tension, does not affect this. No such destruction of human and material capital as that of 1914–18 and 1939–45 has taken place. International rivalry, instead, has in fact sustained economic activity in many countries, provoked it in others. It has provided much technological spin-off, and has led to major capital investments and transfers for political motives, many of which have provided increased real wealth.

The first such transfers came about in the later 1940s, when American aid made possible the recovery of Europe as a major world centre of industrial production. For this to be successful, of course, the American dynamo had to be there to power recovery. The enormous wartime expansion of the American economy which had brought it out of the pre-war depression, together with the immunity of the American home base from physical damage by war, had ensured that it would be. They had rebuilt American economic strength. Explanation for its deployment as aid has to be sought in a complex of other factors. They include the international circumstances (of which the Cold War was an important part) which made it seem in America's interest to behave as she did; an imaginative grasp of opportunities by many of her statesmen and businessmen, the absence for a long time of any alternative source of capital on such a scale, and the wisdom of many men of different nations who, even before the end of the war, had tried to set in place institutions for cooperation in regulating the international economy. They were determined to avoid a return to the near-fatal economic anarchy of the 1930s. Their efforts produced the International Monetary Fund, the World Bank and the General Agreement on Tariffs and Trade (GATT). Much of the successful recovery of the non-communist world after 1945 (and therefore of the remarkable growth which followed) is attributable to these institutions. The economic stability which they provided underpinned two decades of growth in world trade at nearly 7 per cent per annum in real terms. Between 1945 and the 1980s the average level of tariffs on manufactured goods fell from 40 per cent to 5 per cent, and world trade multiplied more than fivefold. These institutions have substituted a degree of management and regulation for the reliance on 'natural' economic harmony which was the basis of the pre-1914 economic order, and for the virtual absence of order between 1914 and 1945. This does not mean they can solve all the world's economic problems, but that economic order can at least be improved.

Another decisive contribution of human agency to economic growth was less formal, often less visible, and has been made over a much longer term. It was provided by scientists and engineers. The continued application of scientific knowledge through technology, and the improvement and rationalization of processes and systems in the search for greater efficiency, were all visible before 1939. They came dramatically to the fore and began to exercise a quite new order of influence after 1945.

Agriculture, where improvement had begun long before industrialization itself was a recognizable phenomenon, may well be the best example. For thousands of years farmers edged their returns upwards almost entirely by clearing and breaking in new land. There is still a lot left which, with proper investment, can be made to raise crops (and much has been done in the last twenty-five years to use such land, even in a crowded country like India). Yet it is not by such means that world agriculture output has recently risen so dramatically. The root explanation is a continuation of the agricultural revolution which began in early modern Europe, and has been visible at least from the seventeenth century. Two hundred and fifty years later, it was speeding up at an accelerating rate.

Well before 1939, wheat was being successfully introduced to lands in which, for climatic reasons, it had not been grown hitherto. This was increasingly thanks to work by plant geneticists who evolved new strains of cereals, one of the first twentieth-century scientific contributions to agriculture comparable in effect to the trial-and-error 'improvement' of earlier times. Even greater contributions to world food supplies were to come from existing grain-growing areas, thanks to better chemical fertilizers. An unprecedented rate of replacement of nitrogen in the soil is the basic explanation of the huge yields which have now become commonplace in countries with advanced agriculture. They have their costs. A huge energy input is now needed to sustain the productivity of advanced Western agriculture, and some other ecological consequences are only just beginning to emerge.

Fertilizers are only one theme in the story of continuing agricultural innovation. Effective herbicides and insecticides began to be available commercially in the 1940s and 1950s. At the same time the use of machinery in agriculture grew enormously in developed countries. England had in 1939 the most mechanized farming in the world in terms of horsepower per acre cultivated. English farmers none the less then still did much of their work with horses, while combine harvesters (already common in the United States) were rare. Working horses are now preserved only as interesting curiosities in countries of advanced agriculture – though there has been some reintroduction of them for certain types of work where the rising cost of fuel does not justify the use of a tractor as a prime mover. But the fields are not the only part of the farm to be mechanized. The coming of electricity has brought automatic milking, grain-drying, threshing, the heating of animal sheds in winter. Now, the computer and automation have begun to reduce dependence on human labour even more; in both the United States and western Europe the agricultural workforce has continued to fall while production per acre has risen.

Yet, paradoxically, there are probably more subsistence farmers in the world today than in 1900. This is because there are more people. Their relative share of cultivated land and of the value of what is produced has fallen. In Europe the peasant – the cultivator of tiny farms – is fast disappearing, as he disappeared in Great Britain two hundred years ago. But this change is unevenly spread and easily disrupted. Russia was traditionally one of the great agricultural economies, but as recently as 1947 suffered famine so severe as once more to provoke outbreaks of cannibalism. Local disasters are not likely to cease in countries with large and rapidly-growing populations where subsistence agriculture is the norm and productivity remains low. Just before the First World War, the British yield of wheat per acre was already more than two and a half times that of India; by 1968 it was roughly five times. Over the same period the Americans raised their rice yield from 4.25 to nearly 12 tons an acre, while that of Burma, the 'rice bowl of Asia', rose only from 3.8 to 4.2. Another way of looking at these facts is to consider how many families can be provided for by one agricultural worker in different countries. In 1968 the answer for Egypt was slightly more than one, while for New Zealand the figure was over forty.

The most advanced agricultural practice is found in countries advanced in other ways. Unless they have a special mineral resource (like oil) or a particular agricultural speciality, countries where there is most need to improve agricultural productivity have often found it very difficult to produce crops more cheaply than can the leading industrial countries. Ironic paradoxes result; the Russians, Indians and Chinese, big grain and rice producers, have found themselves buying American and Canadian wheat. Disparities between developed and undeveloped countries have widened in the decades of plenty.

The most striking measure of this is relative consumption of the world's resources. Roughly half mankind consumes about six-sevenths of the world's production; the other half shares the rest. Moreover, even among the wealthy nations there are wide disparities. In 1970 there were about six Americans in every hundred human beings, but they used about forty of every hundred barrels of oil produced in the world. They each consumed roughly a quarter-ton of paper products a year; the corresponding figure for

China was about twenty pounds. The electrical energy used by China for all purposes in a year at that time would (it was said) just sustain the supply of power to the United States' air conditioners. Electricity production, indeed, is one of the best ways of making the point, since relatively little electrical power is traded internationally and most of it is consumed in the country where it is generated. At the end of the 1980s, the United States produced nearly 40 times as much electricity *per capita* as India, 23 times as much as China, but only 1.3 times as much as Switzerland.

In all parts of the world the disparity between rich and poor nations has grown more and more marked since 1945, not usually because the poor have grown poorer, but because the rich have grown much richer. Almost the only exception to this was to be found in the comparatively rich (by poor world standards) economies of the USSR and eastern Europe where mismanagement and the exigencies of a command economy held back growth and so narrowed the gap between those economies and poorer but developing countries. With these exceptions, even spectacular accelerations of production (some Asian countries, for example, pushed up their agricultural output between 1952 and 1970 proportionately more than Europe and much more than North America) have rarely succeeded in improving the position of poor countries in relation to that of the rich, because of the problems posed by rising populations – and rich countries, in any case, began at a higher level. As a result, though their ranking in relation to one another may have changed, those countries which enjoyed the highest standards of living in 1950 still, by and large, enjoy them today, though they have been joined by Japan. These are the major industrial countries. In the great acceleration of wealth production, manufacturing industry has played the major rôle. The manufacturing economies are today the richest *per capita*, and their example spurs poorer countries to seek their own salvation in industrialization. In 1970 three of the great industrial agglomerations of the world were still, as they had been in 1939, the United States, Europe and Russia; a fourth was Japan, in 1939 already the most important Asian industrial society. By 1990, though, the picture had changed somewhat. While the USSR was still one of the big four, it had fallen far behind the others and even behind one of the nations making up the west European industrial agglomeration, West Germany.

Comparisons have become more difficult, too. Major industrial economies today do not much resemble their nineteenth-century predecessors. The old heavy and manufacturing industries, which long provided the backbone of economic strength, are no longer a simple and satisfactory measure of it. Once-staple industries of leading countries have declined in importance. Of the three major steel-making countries of 1900, the first two (the USA and West Germany) were still among the first five world producers eighty years later, but in third and fifth places respectively; the United Kingdom (third in 1900) came tenth in the same world table – with Spain, Romania and Brazil close on her heels. In 1982, Poland made more steel than had the USA in 1900. What is more, newer industries often found a better environment for rapid growth in some developing countries than in the mature economies. It was in this way that the peoples of Taiwan and South Korea came by 1988 to enjoy *per capita* GDP of, in the first instance, nearly eighteen times that of India, and, in the second, more than fifteen times.

Modern industrial societies are much more than simple extrapolations in technology and structure from the past. Much of the economic growth of rich countries has been in industries – electronics and plastics are examples – which barely existed even in 1945. Coal, which replaced running water and wood in the nineteenth century as the major source of industrial energy, was long before 1939 joined by hydro-electricity, oil and natural gas; very recently, power generated by nuclear fission has been added to these.

Industrial growth has raised standards of living by providing cheaper power and materials for the production of goods. Improving transport has further lowered costs indirectly. An enormous growth in the production of commodities directly for the use and pleasure of the consumer was the result. Often the ramifications were enormous and one example must suffice. In 1885 there appeared the first vehicle propelled by an internal-combustion engine – one, that is to say, in which the energy produced by

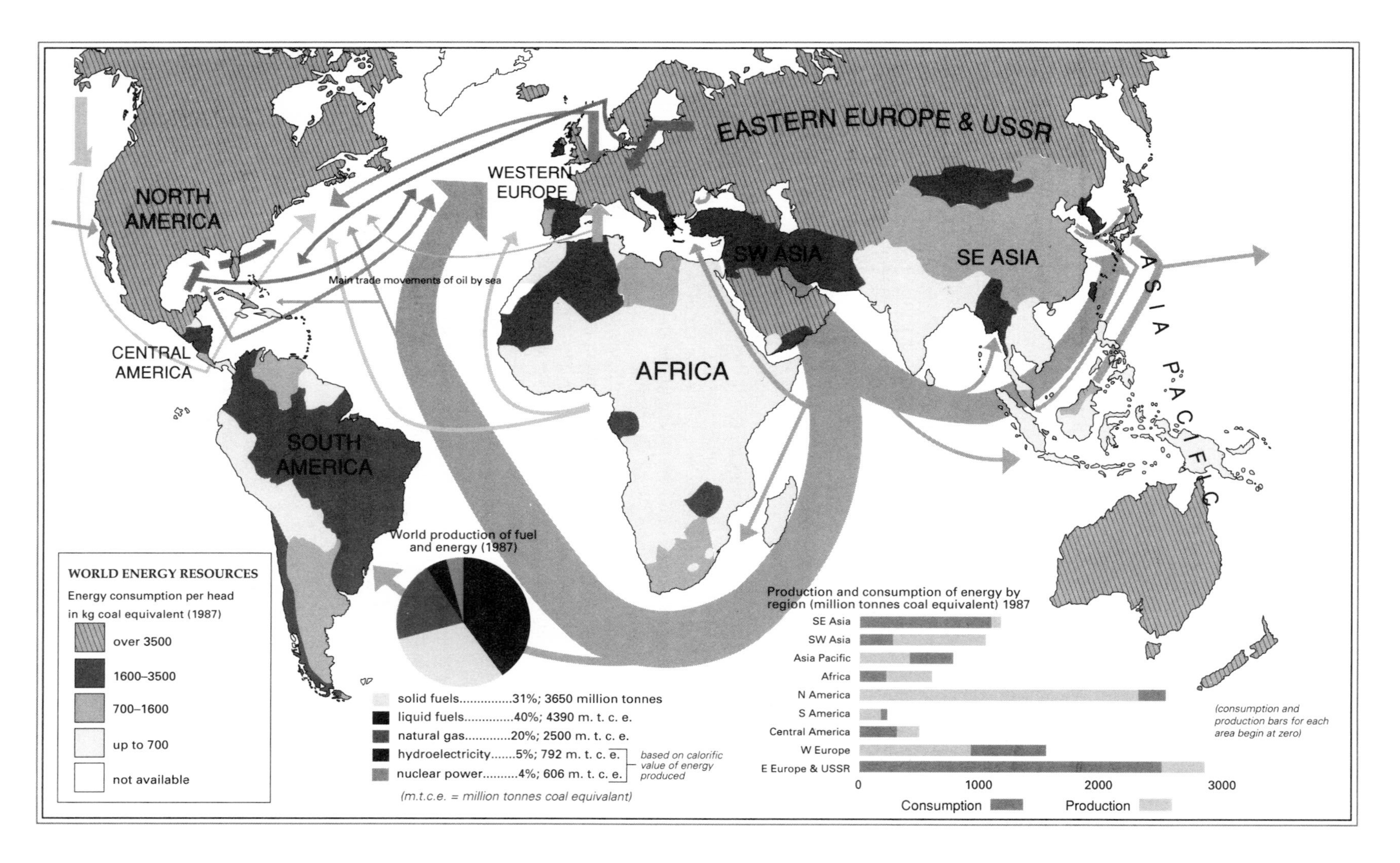
EASTERN EUROPE & USSR
NORTH AMERICA
WESTERN EUROPE
SW ASIA
SE ASIA
ASIA PACIFIC
Main trade movements of oil by sea
CENTRAL AMERICA
AFRICA
SOUTH AMERICA
World production of fuel and energy (1987)
WORLD ENERGY RESOURCES
Energy consumption per head in kg coal equivalent (1987)
over 3500
1600–3500
700–1600
up to 700
not available
solid fuels..............31%; 3650 million tonnes
liquid fuels.............40%; 4390 m. t. c. e.
natural gas............20%; 2500 m. t. c. e.
hydroelectricity.......5%; 792 m. t. c. e.
nuclear power..........4%; 606 m. t. c. e.
based on calorific value of energy produced
(m.t.c.e. = million tonnes coal equivalant)
Production and consumption of energy by region (million tonnes coal equivalent) 1987
SE Asia
SW Asia
Asia Pacific
Africa
N America
S America
Central America
W Europe
E Europe & USSR
0
1000
2000
3000
Consumption
Production
(consumption and production bars for each area begin at zero)

heat was used directly to drive a piston inside the cylinder of an engine instead of being transmitted via steam made with an external flame. Nine years later came a four-wheeled contraption which is a recognizable ancestor of the modern car. It was the French Panhard, and France and Germany kept a lead in producing cars in the next decade or so. When the first motor show was held in London in 1896 cars were still few, and rich men's toys. This was automobile prehistory. Automobile history began in the United States when Henry Ford in 1907 set up a production line for his Model T. He deliberately planned for a mass market. His car was to be sold at an unprecedently low price – $950. Demand rose so rapidly that by 1915 a million Ford cars were made each year. The price came down and by 1926 the Model T sold for less than $300. An enormous success was under way, and a great social change with it. Ford had provided for the masses something previously regarded as a luxury. There are now more cars than households in the United States. Ford changed the world, and perhaps as much as had the coming of the railways in the nineteenth century. His car gave people of modest incomes a mobility unavailable even to millionaires fifty years earlier. This huge increase in amenity was to spread round the globe.

The coming of the mass motor-car also changed history in other ways, and a long view is necessary to understand this. By the 1980s, a world-wide car manufacturing industry existed, in which international integration had gone very far and, in some countries, virtually dominated the manufacturing sector. Eight large producers now account for nearly three out of four of the world's cars. One country, Japan, could attribute a major part of its rise to economic ascendancy in the 1960s and 1970s to its car industry but was already, by 1990, consciously running it down at home in anticipation of new challengers abroad. Car manufacture stimulated major investment in other industries; today half the robots employed in the world's industry are welders in car factories (another quarter are painters in them). Meanwhile, over a much longer term, Ford's popularization of the car helped to stimulate a rising demand for oil, though this was already apparent before 1914, as more ships became oil-powered, and now transport is no longer oil's major consumer. Huge numbers of people came to be employed in the supply of fuel and services to car-owners. Investment in road-building – and later in arrangements for parking – became a major concern of government, both local and national, and profoundly affected the construction industry.

Finally, Ford can be credited (if that is the word) with a social and technical revolution by showing in his factories what could be done with mass-production. Like many great revolutionaries he brought other men's ideas to bear on his own. The result was the assembly line, the characteristic modern way of making consumer goods. On it, the article under manufacture is moved steadily from worker to worker, each one of them carrying out in the minimum necessary time the precisely delimited and, if possible, simple task in which he is skilled. Mass-production's psychological effect on the worker was soon deplored, but the technique was fundamental to a wider sharing of the wealth of the industrial economy. Ford saw that such work would be very boring and paid high wages to compensate for it (thereby incidentally contributing to another revolutionary economic change, the fuelling of economic prosperity by increasing purchasing power and, therefore, demand). Since then, good management practice has generated other solutions to this problem, and the diffusion of such practice in many countries through the automobile industries is, like the assembly line itself, a cultural force of incalculable significance in generalizing new attitudes, whether for good or ill.

Now, though, the assembly line may well be 'manned' by robots. The single greatest technological change since 1945 in the major industrial economies has come in Information Technology, the complex science of handling, managing, and devising electronically-powered machines to process information. Few innovatory waves have moved so fast. Applications of work, much of which was only done during the Second World War, were diffused in a couple of decades over a huge range of services and industrial processes. Rapid increases in the power and speed, reduction in the size, and improvement in the visual display capacity of computers only meant, in essence, that much more information could be ordered and processed than hitherto. But this was an example of a quantitative change bringing a qualitative transformation. Calculations which until recently would have required the lifetime of many mathematicians to complete can now be run through in a few minutes. Technical operations which would have had to wait decades for such calculations, or for the sorting and classifying of great masses of data, have now become feasible. Intellectual advance has never been so suddenly accelerated. At the same time as there was revolutionary growth in the capacity and power of computers, their technology made it easier and easier to pack their potential into smaller and smaller machines. Within thirty years, a 'microchip' the size of a credit card was doing the job which had first required a machine the size of an average British living-room. The transforming effects have been felt in every service – and, indeed, almost every human activity, from money-making to war-making.

Computers, soon able to transact business with one another, are the end of a long chain of development and innovation in communication. In its simplest form, this development was expressed in physical and mechanical movement of messages, goods and people whose major nineteenth-century achievements were most obviously manifest in such improvements as the application of first steam to land and sea transport, and then the production of the internal-combustion engine and electric tram. There had long been balloons, and the first 'dirigibles' existed before 1900 but the first flight by a man-carrying, powered, heavier-than-air machine was only made in 1903. Eighty years later, the value of goods imported and exported through Heathrow, London's biggest airport, was greater than that of any British seaport and aeroplanes are now the normal form of long-distance travel. Flight now offers a mobility to the individual which was hardly imaginable at the start of this century.

The communication of information had earlier been revolutionized even more by liberating it from physical interaction between the source and the recipient of a signal. By the middle decades of the nineteenth century, poles carrying lines for the electric telegraph were already a familiar sight, but by 1900, Marconi had exploited electro-magnetic theory to make possible the sending of the first 'wireless' messages. Transmitters and receivers no longer needed physical connexion. The first radio message to cross the Atlantic did so, appropriately, in 1901, the first year of a century to be revolutionized by his invention. By 1930 most people who owned 'wireless' receivers (and there were millions) had ceased to believe that windows had to be open to allow the broadcast 'waves' to reach them. There were by then large-scale radio broadcasting services in all major countries. A few years earlier still the first demonstration had been made of the devices on which television was based. In 1936 the BBC opened the first regular television broadcasting service; twenty years later the medium was commonplace in the leading industrial countries. Speedier information transmission had been going on over the whole century and the post-war foundation and explosive success of new industries devoted to the use of electronic methods of communication must be seen in that perspective.

THE MANAGEMENT OF NATURE

Technical progress was for a long time almost the only way in which science came home to most people, and perhaps it still is. In some industrial processes the science is very obvious: the use of nuclear fission to generate energy, or of computers to control machine-tools are examples. In others – and the production of plastic materials for almost every conceivable requirement is one of the most common – the basic science still lies concealed from the layman in obscure chemical processes. But in both it is easy to accept that the scientists' role has been paramount. In fact, by 1950 modern industry was already dependent on science, directly or indirectly, whether obviously so or not. Moreover, the transformation of fundamental science into an end-product is now often very rapid. The generalization of the use of the motor car after the principle of the internal combustion engine had been grasped took about half a century; the gap between the first use of penicillin and

Old and new: to the assembly line, long the typical form and symbol of industrial production, is now joined the robot. The end of the line, perhaps, if other car manufacturers' use of closely-involved and skilled labour proves to be the way ahead?

its large-scale manufacture was only about ten years.

One explanation is the spread of the idea of purposive research, and of willingness to spend money on it. In the nineteenth century most practical results of science were by-products of scientific curiosity. Sometimes they were accidental. By 1900 a change was under way. Some scientists had seen that consciously directed and focused research was sensible. Twenty years later, large industrial companies were beginning to see research as a proper call on their investment, albeit a small one. Some industrial research departments were in the end to grow into enormous establishments in their own right as petrochemicals, plastics, electronics and biochemical medicine made their appearance later in the century. Nowadays, the ordinary citizen of a developed country cannot lead a life which does not rely on applied science. This all-pervasiveness, coupled with its impressiveness in its most spectacular achievements, was one of the great reasons for the ever-growing recognition given to science. Money is one yardstick. The Cavendish Laboratory at Cambridge, for example, in which some of the fundamental experiments of nuclear physics were carried out before 1914, had then a grant from the university of about £300 a year – roughly $1500 at rates then current. When, during the war of 1939-45, the British and Americans decided that a major effort had to be mounted to produce nuclear weapons, the resulting 'Manhattan Project' (as it was called) is estimated to have cost as much as all the scientific research previously conducted by mankind from the beginnings of recorded time.

Such huge sums – and there were to be even larger bills to meet in the post-war world – mark another momentous change, a new importance of science to government. After being for centuries the object of only occasional patronage by the state, science is now a major political concern. Only governments can provide resources on the scale needed for some of the things done since 1945. One benefit they usually sought was better weapons, which explained much of the huge scientific investment of the United

States and the Soviet Union. The increasing interest and participation of governments has not, on the other hand, meant that science has grown more nationalistic; indeed, the reverse is true. The tradition of international communication among scientists is one of their most splendid inheritances from the first great age of science in the seventeenth century, but even without it, science would for theoretical and technical reasons have had to jump national frontiers.

Already before 1914 it was increasingly clear that boundaries between the individual sciences, some of them intelligible and usefully distinct fields of study since the seventeenth century, were tending first to blur and then to disappear. The full implications of this have only begun to appear very lately, though. In spite of the achievements of the great chemists and biologists of the eighteenth and nineteenth centuries, physics was the seed-bed of the major scientific achievements of the early twentieth century. Newtonian physics had provided a satisfying philosophical framework for a century and a half when James Clerk Maxwell, the first professor of experimental physics at Cambridge, published in the 1870s the work in electro-magnetism which broke effectively into fields and problems left untouched by seventeenth-century science. Maxwell's theoretical work and its experimental investigation profoundly affected the post-Newtonian view that the universe obeyed natural, regular and discoverable laws of a somewhat mechanical kind and that it essentially consisted of indestructible matter in various combinations and arrangements. Into this picture had now to be fitted a new component, the electro-magnetic fields whose technological possibilities quickly fascinated laymen and scientists alike.

The crucial work which followed and which founded modern physical theory was done between 1895 and 1914, by Rontgen who discovered X-rays, Becquerel who discovered radioactivity, Thomson who identified the electron, the Curies who isolated radium, and Rutherford who carried out the investigation of the atom's structure. They made it possible to see the physical world in a new way. Instead of lumps of matter, the universe began to look more like an aggregate of atoms which were tiny solar systems of particles in particular arrangements. These particles seemed to behave in a way which blurred the distinction between lumps of matter and electro-magnetic fields. Moreover, such arrangements of particles were not fixed, for in nature one arrangement might give way to another and thus elements could change into other elements. Rutherford's work, in particular, was decisive, when he established that atoms could be 'split' because of their structure as a system of particles. This meant that matter, even at this fundamental level, could be manipulated (though as late as 1935 he said that nuclear physics would have no practical implications – and no one rushed to contradict him). Two such particles were soon identified, the proton and the electron; others not until after 1932, when Chadwick discovered the neutron. But Rutherford, together with Bohr, provided the scientific world with an experimentally-validated picture of the atom's structure as a system of particles. Since then the discovery of new particles has continued as one of the main developmental lines along which physics has moved.

What this radically-important experimental work did not do was supply a new theoretical framework to replace the Newtonian. This was only achieved by a long revolution in theory, beginning in the last years of the nineteenth century and coming to its culmination in the 1920s. It was focused on two different sets of problems, which gave rise to the work designated by the terms relativity and quantum theory. The pioneers were Max Planck and Albert Einstein. By 1905 they had provided experimental and mathematical demonstration that the Newtonian laws of motion were an inadequate framework for explanation of a fact which could no longer be contested. This was that energy transactions in the material world took place not in an even flow but in discrete jumps – *quanta*, as they came to be termed. Planck showed that radiant heat (from, for example, the sun) was not, as Newtonian physics required, emitted continuously; he argued that this was true of all energy transactions. Einstein argued that light was propagated not continuously but in particles. Though much important work was to be done in the next twenty or so years, Planck's contribution had the most profound effect and it was again unsettling. Newton's views had been found wanting, but there was nothing to put in their place.

Meanwhile, after his work on quanta, Einstein had published in 1905 the work for which he was to be most famous, his statement of the theory of relativity. This was essentially a demonstration that the traditional distinctions of space and time, and mass and energy, could not be consistently maintained. Instead of Newton's three-dimensional physics, he directed men's attention to a 'space-time continuum' in which the interplay of space, time and motion could be understood. This was soon to be corroborated by astronomical observation of facts for which Newtonian cosmology could not properly account, but which could find a place in Einstein's theory. One strange and unanticipated consequence of the work on which relativity theory was based was his demonstration of the relations of mass and energy which he formulated as $E = mc^2$, where E is energy, m is mass, and c is the constant speed of light. The importance and accuracy of this theoretical formulation was not to be clear until much more nuclear physics had been done; it would then be apparent that the relationships observed when mass energy was converted into heat energy in the breaking up of nuclei also corresponded to this formula.

While these advances were absorbed, attempts continued to rewrite nuclear physics in the light of Planck's work. These did not get far until a major theoretical breakthrough in 1926 finally provided a mathematical framework for his observations and, indeed, for nuclear physics. So sweeping was the achievement of Schrödinger and Heisenberg, the two mathematicians mainly responsible, that it seemed for a time as if quantum mechanics might be of virtually limitless explanatory power in the sciences. The behaviour of particles in the atom observed by Rutherford and Bohr could now be accounted for. Further development of their work led to predictions of the existence of new nuclear particles, notably the positron, which was duly identified in the 1930s. Quantum mechanics seemed to have inaugurated a new age of physics.

By mid-century much more had disappeared in science than just a once-accepted set of general laws (and in any case it remained true that, for most everyday purposes, Newtonian physics was still all that was needed). In physics, from which it had spread to other sciences, the whole notion of a general law was being replaced by the concept of statistical probability as the best that could be hoped for. The idea, as well as the content, of science was changing. Furthermore, the boundaries between sciences collapsed under the onrush of new knowledge made accessible by new theories and instrumentation. Any one of the great traditional divisions of science was soon beyond the grasp of any single mind. The conflations involved in importing physical theory into neurology or mathematics into biology put further barriers in the way of attaining that synthesis of knowledge which had been the dream of the nineteenth century just as the rate of acquisition of new knowledge (some in such quantities that it could only be handled by the newly-available computers) became faster than ever.

Such considerations did nothing to diminish either the prestige of the scientists or the faith that they were mankind's best hope for the better management of its future. Doubts, when they came, arose from other sources than their inability to generate overarching theory intelligible to lay understanding as Newton's had been. Meanwhile, the flow of specific advances in the sciences continued. In some measure, the baton passed after 1945 from the physical to the biological sciences. Their current success and promise has, once again, deep roots in the origin of two paths of enquiry in nineteenth-century biology. The seventeenth-century invention of the microscope had first revealed the organization of tissue into discrete units called cells. In the nineteenth century investigators already understood that cells could divide and that they developed individually. Cell theory, widely accepted by 1900, suggested that individual cells, being alive themselves, provided a good approach to the study of life, and the application of chemistry to this became one of the main avenues of biological research. The other main line of advance in nineteenth-century biological science was provided by a new discipline, genetics, the study of the inheritance by offspring of characteristics from parents. Darwin

Albert Einstein
Old Grove Rd.
Nassau Point
Peconic, Long Island

August 2nd, 1939

F.D. Roosevelt,
President of the United States,
White House
Washington, D.C.

Sir:

Some recent work by E.Fermi and L. Szilard, which has been communicated to me in manuscript, leads me to expect that the element uranium may be turned into a new and important source of energy in the immediate future. Certain aspects of the situation which has arisen seem to call for watchfulness and, if necessary, quick action on the part of the Administration. I believe therefore that it is my duty to bring to your attention the following facts and recommendations:

In the course of the last four months it has been made probable - through the work of Joliot in France as well as Fermi and Szilard in America - that it may become possible to set up a nuclear chain reaction in a large mass of uranium,by which vast amounts of power and large quantities of new radium-like elements would be generated. Now it appears almost certain that this could be achieved in the immediate future.

This new phenomenon would also lead to the construction of bombs, and it is conceivable - though much less certain - that extremely powerful bombs of a new type may thus be constructed. A single bomb of this type, carried by boat and exploded in a port, might very well destroy the whole port together with some of the surrounding territory. However, such bombs might very well prove to be too heavy for transportation by air.

The post-war era has been dominated by the awareness that quite unprecedented dangers face mankind in the destructive powers of nuclear weapons. The technology of these was created with amazing swiftness. This letter, drafted by a Hungarian scientist, Leo Szilard, and signed by Einstein, was sent to President Roosevelt less than a month before war began in Europe. Six years and four days later, the first atomic bomb to be used was dropped at Hiroshima.

had invoked the principle of inheritance as the means of propagation of traits favoured by natural selection. The first steps towards understanding the mechanism which made this possible were those of an Austrian monk, Gregor Mendel. From a meticulous series of breeding experiments on pea plants Mendel concluded that there existed hereditary units controlling the expression of traits passed from parents to offspring. In 1909 a Dane gave them the name 'gene'.

Gradually the chemistry of cells became better understood and the physical reality of genes was accepted. In 1873 the presence in the cell nucleus of a substance which might embody the most fundamental determinant of all living matter was already established. Experiments then revealed a visible location for genes in chromosones, and in the 1940s it was shown that genes controlled the chemical structure of protein, the most important constituent of cells. In 1944 the first step was taken towards identifying the specific effective agent in bringing about changes in certain bacteria, and therefore in controlling protein structure. In the 1950s it was at last identified as 'DNA', whose physical structure (the double helix) was established in 1953. The crucial importance of this substance (its full name is deoxyribonucleic acid) is that it is the carrier of the genetic information which determines the synthesis of protein molecules at the basis of life. The chemical mechanisms underlying the diversity of biological phenomena were at last accessible. Physiologically, and perhaps psychologically, this implied a transformation of man's self-perception unprecedented since the general acceptance of Darwinian ideas in the last century.

The identification and analysis of the structure of DNA was the most conspicuous single step towards the manipulation of nature at the level of the shaping of life-forms. Once again, not only more scientific knowledge but new definitions of fields of study and new applications followed. 'Molecular biology', 'biotechnology' and 'genetic engineering' quickly became familiar terms. The genes of some organisms could, it was soon shown, be altered so as to give those organisms new and desirable characteristics. By manipulating their growth processes, yeast and other microorganisms could be made to produce novel substances, enzymes or other chemicals. This was one of the first extensions of the new science; the empirical technology which had been accumulated informally by millenia of experience in the making of bread, wine and cheese was at last to be overtaken. Genetic modification of bacteria can now be carried out so as to grow chemicals or hormones. By the end of the 1980s a world-wide collaborative investigation, the Human

Life Sciences: Milestones in Biotechnology and Genetic Engineering

1865	Mendel publishes first accounts of experiments with peas showing the presence of independent factors ('genes').
1909	First use of the term 'gene' in a book by Johannsen.
1928	Fleming discovers penicillin – publication of the first paper.
1941	First mass production of antibiotics (penicillin) in the USA (UK production insufficient for demand).
1947	First use of the word 'Biotechnology' (USA).
1953	Crick and Watson publish paper showing the structure of DNA.
1955	Structure of insulin published (first protein structure to be determined).
1973	Cohen and Boyer discover recombinant rDNA – the basis for genetic engineering.
1975	Milstein publishes the results of cell fusion experiments which lead to the development of monoclonal antibodies.
1982	Commercial production of first genetically engineered drug (insulin).
1983	First publication of gene mapping (Huntington's disease).
1986	Jefferys publishes method of 'DNA fingerprinting', first used forensically 1987.
1989	First isolation (structure determination) of a defective gene (cystic fibrosis).
1990	Goodfellow discovers structure of testis determining gene.
1991	First sex-reversed transgenic animal (mouse) produced.

Genome Project, whose almost unimaginable aim was the mapping of the human genetic apparatus so as to identify the position, structure and function of every human gene (there are from 50,000 to 100,000 in each cell, each of them having up to 30,000 pairs of the four basic chemical units that form the genetic code) was under way. Screening for the presence of certain defective genes, and even the replacement of some of them, is already achieved; the medical implications of this are enormous. At a more obvious level, that of day-to-day police work, what is called DNA 'fingerprinting' has been used to identify an individual from a blood or semen sample. Somewhat more eerily, a patent has now been acquired for a laboratory mouse genetically prone to cancer.

Progress in these matters has been startlingly rapid (and owes much to the availability of computers to handle very large quantities of information). It provides a remarkable instance of the accelerating tendency in scientific advance which in turn accelerates both new applications and the invasion of a world of settled assumptions and landmarks by new ideas. It is not easy to assess what this can mean. Once again, we face the old problem of establishing the level at which ideas have cultural, social, or political effect. For all the recent fundamental work in the 'life sciences', it is unlikely that even their approximate importance can be assessed by most of us.

For most human beings, the power of science is still most vividly demonstrated in its technological expression. For nearly twenty years that was above all visible in the spectacular achievements of the human exploration of space. The extension of our physical environment thus achieved may well turn out to have an ultimate significance dwarfing other historical processes which have been given more space in this book. It has been going on throughout most of the post-war era, the most exciting achievement of modern technology, suggesting that the capacity of our culture to meet unprecedented challenges is as great as ever, and perhaps the most obvious of all the manifestations of human domination of nature. Its huge psychological importance began to appear in October 1957. For most people, the space age began then, when an unmanned Soviet satellite called Sputnik I was launched by rocket and could soon be discerned in orbit around the earth emitting radio signals. The political impact was obvious: it shattered the belief that Russian technology lagged significantly behind American. The full significance of the event, though, was still obscured, because its entanglement with superpower rivalries swamped other considerations for most observers. In fact, it ended the era when the possibility of human travel in space could still be doubted, as events quickly showed. Almost incidentally and unnoticed, it marked a break in historical continuity as important as the European discovery of the Americas, or the Industrial Revolution. One of the futures dreamed of by writers of science fiction had begun.

Space exploration had deep roots. Visions of it could be found in the last years of the nineteenth century and the early years of the twentieth, when they were brought to the notice of the western public in fiction, notably, in the stories of Jules Verne and H.G. Wells. Its technology went back almost as far. A Russian scientist, K.E. Tsolikovsky, had designed multi-staged rockets and devised many of the basic principles of space travel (and he too had written fiction to popularize his obsession) well before 1914. The first Soviet liquid-fuelled rocket went up (three miles) in 1933, and a two-stage rocket six years later. The Second World War provoked a major German rocket programme, which the United States had drawn on to begin its own in 1955. It started with more modest hardware then the Russians (who already had a commanding lead) and the first American satellite weighed only three pounds (Sputnik I weighed 184 pounds). A much-publicized launch attempt was made at the end of December 1957, but the rocket caught fire instead of taking off and a sad little bleeping from the sands near the launching site was all that resulted. The Americans would soon do much better than this, but within a month of Sputnik I the Russians had already put up Sputnik II, an astonishingly successful machine, weighing half a ton and carrying the first passenger in space, a black-and-white mongrel called Laika. For nearly six months Sputnik II orbited the earth, visible to the whole inhabited world and an outrage to thousands of dog-lovers, for Laika was not to return.

The Russian and American space programmes had already by then somewhat diverged. The Russians, building on their pre-war experience, had put much emphasis on the power and size of their rockets, which could lift big loads, and here their strength was to continue to lie. This had military implications more obvious than those (equally profound but less spectacular) which flowed from American concentration on data-gathering and on instrumentation. A competition for prestige was soon under way, but although people soon spoke of a 'space race' the contestants were really running towards different goals. With one great exception (the wish to be first to put a man in space) their technical decisions were probably not much influenced by one another's performance. The contrast was clear enough when Vanguard, the American satellite which failed in December 1957, was successfully launched the following March. Tiny though it was, it went much deeper into space than any predecessor and has provided more valuable scientific information in proportion to its size than any other satellite. It is likely to be going around for another couple of centuries or so.

The Exploration and Use of Space: Major steps down to 1969

1903	Konstantin Tsolikovsky (1857–1935) publishes paper on rocket space travel using liquid propellants.
1933	1 May: Tsolikovsky predicts that many Soviet citizens will live to see the first space flights.
1944	German V.2 rockets used to bombard London and Antwerp.
1954	President Eisenhower announces a small scientific satellite, Vanguard, will be launched 1957–8.
1957	1 July: Launch of Sputnik 1 (USSR), weight 184 lbs. 3 November: Launch of Sputnik 2 (USSR), weight 1120 lbs., with the dog Laika as passenger.
1958	31 January: Launch of Explorer (USA) and discovery of Van Allen radiation belts. 17 March: Launch of Vanguard 1 (USA), weight 3.25 lbs. The first satellite with solar batteries.
1959	13 September: Luna 2 (USSR) crashes on Moon, the first man-made object to arrive there. 10 October: Luna 3 (USSR) photographs far side of Moon.
1960	11 August: Discoverer 13 (USA) recovered after first successful re-entry to atmosphere. 19 August: Sputnik 5 (USSR) orbits earth with two dogs which return unharmed.
1961	12 April: Major Yuri Gagarin (USSR) orbits the earth. 25 May: President Kennedy commits USA to landing man on moon by 1970. 6 August: Vostok 2 (USSR) makes seventeen orbits of earth.
1962	10 July: Launch of Telstar (USA) and first television pictures across the Atlantic. 20 February: First manned orbited space flight.
1965	18 March: On Voskhod 2 mission (USSR) Alexey Leonov makes ten-minute 'walk in space'. 2 May: Early Bird commercial communication satellite (USA) first used by television. 15 December: Launch of Gemini 6 (USA) which makes rendezvous with Gemini 7, the two craft coming within a foot of one another.
1966	July–November: Gemini mission 10, 11, 12 (USA) all achieve 'docking' with 'Agena' vehicle.
1967	27 January: first deaths in US programme.
1968	21–27 December: Apollo 8 (USA) makes first manned voyage round moon.
1969	14–17 January: Soyuz 4 and 5 (USSR) dock in space and exchange passengers. 16–24 July: Apollo 11 (USA) lands two men on the Moon.

New achievements then quickly followed; progress speeded up. At the end of 1958 the first satellite for communications purposes was successfully launched (it was American). In 1960 the Americans scored another 'first' – the recovery of a capsule after re-entry. The Russians followed this by orbiting and retrieving Sputnik V, a four-and-a-half-ton satellite, carrying two dogs, who thus became the first living creatures to have entered space and returned to earth safely. In the spring of following year, on 12 April, a Russian rocket took off carrying a man, Yuri Gagarin. He landed

108 minutes later after one orbit around the earth. The invasion of space by humanity had begun, four years after Sputnik I.

A twentieth-century Diaz or Columbus sets off: Yuri Gagarin preparing to leave on his globe-orbiting space mission, 12 April 1961. He reached a maximum altitude of 188 miles above the surface of the earth, which he circled in one hour and 48 minutes.

Possibly spurred by a wish to offset a recent publicity disaster in American relations with Cuba, President Kennedy proposed in May 1961 that the United States should try to land a man on the moon and return him safely to earth before the end of the decade. His reasons for recommending this were interestingly different from those which had led the rulers of fifteenth-century Portugal and Spain to back the Magellans and da Gamas. One was that such a project provided a good national goal, the next that it would be prestigious ('impressive to mankind' were the President's words); the third was that it was of great importance for the exploration of space; and the fourth was (somewhat oddly) that it was of unparalleled difficulty and expense. Kennedy said nothing of the advancement of science, of commercial or military advantage. Surprisingly, the project met virtually no opposition and the first money was soon allocated for the most expensive technological adventure in history.

During the early 1960s the Russians continued to make spectacular progress. The world was perhaps most struck when they sent a woman into space in 1963, but their technical competence continued to be best shown by the size of their vehicles – a three-man machine was launched in 1964 – and in the achievement the following year of the first 'space walk', when one of the crew emerged from his vehicle and moved about outside while in orbit (though reassuringly attached to it by a lifeline). The Russians were to go on to further important advances in achieving rendezvous for vehicles in space and in engineering their docking, but after 1967 (the year of the first death through space travel, when a Russian was killed during re-entry) the glamour passed to the Americans. In 1968, they achieved a sensational success by sending a three-man vehicle into orbit around the moon and transmitting television pictures of its surface. It was by now clear that 'Apollo', the moon-landing project, was going to succeed.

In May 1969 Apollo 10, a vehicle put into orbit with the tenth rocket of the project, approached to within six miles of the moon to assess the techniques of the final stage of landing. A few weeks later, on 16 July, a three-man crew was launched. Their lunar module landed on the moon's surface four days later. On the following morning, 21 July, the first human being to set foot on the moon was Neil Armstrong, the commander of the mission. President Kennedy's goal had been achieved with time in hand. Other landings were to follow. In a decade which had opened with humiliation for the United States in the Caribbean and was ending in the morass of an unsuccessful war in Asia, it was a triumphant reassertion of what America (and, by implication, capitalism) could do. It was also a sign of the latest and greatest extension by *Homo sapiens* of his environment, the beginning of a new phase of his history, that to be lived on other celestial bodies.

Even at the time, this wonderful achievement was decried. Its critics felt that the mobilization of resources the programme needed was unjustified, because irrelevant to the real problems of the earth. To some, the technology of space travel has seemed to be our civilization's version of the Pyramids, a huge investment in the wrong things in a world in crying out for money for education, nutrition, medical research – to name only a few of its pressing needs. It is difficult not to feel immediate sympathy for such a view, but it is more difficult to sustain it. In the first place, much of the scientific and economic effect of the space effort is hardly quantifiable (and has already been very far-reaching); the use of knowledge of miniaturization needed to make control systems, for example, rapidly spills over into application of obvious social and economic value. It cannot be said that this knowledge would necessarily have been available had not the investment in space come first. Nor, indeed, can we be confident that the resources lavished on space exploration would have been made available for any other great scientific or social goals, had they not been used in this way. Our social machinery does not operate like that.

The mythical importance of what has happened has also to be considered. However regrettable it may be, modern societies have shown few signs of being able to generate much interest and enthusiasm among their members for collective purposes except for brief periods or in war, whose 'moral equivalent' (as one American philosopher put it, well before 1914) is still to seek. The imagination of large numbers of people is not much spoken to by the prospect of adding marginally to the GDP or introducing one more refinement to a system of social services, however intrinsically desirable these things may be. Kennedy's identification of a national goal was shrewd; in the troubled 1960s, the Americans had much else to agitate and divide them, but they did not turn up to frustrate launchings of the space missions.

It is also important that space exploration became more of an international enterprise as it went on. Before the 1970s there was little cooperation between the two greatest nations concerned, the United States and Soviet Russia, and this undoubtedly led to wasteful duplication of effort and inefficiencies. Ten years before the Americans got there and planted the American flag on it, a Soviet mission had dropped a Lenin pennant on the moon. This seemed ominous; there was a basic national rivalry in the technological race itself and nationalism might provoke a Scramble for Space. But the dangers of competition were avoided; it was soon agreed that celestial objects were not subject to appropriation by any one state. In July 1975, some hundred and fifty miles above the earth, cooperation became a startling reality in a remarkable experiment in which Soviet and American machines connected themselves so that their crews could move from one to the other. In spite of doubts, exploration continued in a relatively benign international setting. The visual exploration of further space was carried beyond Jupiter by unmanned satellite, and 1976 brought the first landing of an unmanned exploration vehicle on the surface of the planet Mars. In 1977 the Space Shuttle, the first reusable space vehicle, made its maiden voyage. These achievements were tremendous, yet they are now hardly noticed, so jaded are our imaginations. So rapidly did there grow up a new familiarity with the idea of space travel that in the 1980s it seemed only mildly risible that people should have begun to make commercial bookings for it – and even for space burials (if that is the word), too. As the decade drew to a close what was the last big enterprise of the Soviet space effort took place in 1988, when a satellite was launched to prepare the way for a future manned voyage to Mars.

By then, space exploration had become somewhat overshadowed by uneasiness about mankind's interference with nature. Though successful adventure in space was a supreme example of technological manipulation to achieve human goals and satisfy human desires, within only a few years of Sputnik I misgivings were increasingly expressed about what lay at the very roots of so masterful a view of our relationship to the natural world. What is more, such uneasiness soon began to be expressed with a precision based on observed facts not hitherto considered in that light; science itself supplied the instrumentation and data which led to dismay about was going on. There began to be perceptible a conscious recognition of the damage interference with the environment might bring. It was, of course, the recognition that was new, not the phenomenon. Centuries before, migration southwards and the adoption of dryland crops from the Americas had devastated the great forests of south-west China, bringing soil erosion and the consequential silting of the Yangtze drainage system in its train, and so culminating in repeated flooding over wide areas. In the early Middle Ages, Islamic conquest had brought goat-herding and tree-felling to the North African littoral on a scale which destroyed a fertility once able to fill the granaries of Rome. But such great changes, though hardly unnoticed, were not then understood. The unprecedented rapidity of ecological interference initiated from the seventeenth century onwards by Europeans, however, was to bring things to a head. The unconsidered power of technology forced the dangers on the attention of mankind in the second half of this century. People began to reckon up damage as well as achievement, and by the middle of the 1970s it seemed to some of them that even if the story of growing human mastery of the environment was an epic, that epic might well turn out to be a tragic one.

Suspicion of science had never wholly disappeared in western societies. It had tended, nevertheless, to be confined to a few surviving primitive or reactionary enclaves as the majesty of what the scientific revolution of the seventeenth century implied gradually unrolled. History can provide much evidence of uneasiness about interference with Nature and attempts to control it, but until recently such uneasiness seemed to rest on non-rational grounds, such as the fear of provoking divine anger or nemesis. As time passed, it was steadily eroded by the palpable advantages and improvements which successful interference with Nature brought about, most obviously through the creation of new wealth expressed in all sorts of goods, from better medicine to better clothing and food. In the 1970s, though, it became clear that a new scepticism about science itself was abroad, even though only among a minority. Suspicion of what is one of the greatest tools of man-made change was marked, it is true, only in developed countries. There, a cynic might have said, the dividends on science had already been drawn. None the less, it manifested itself there first and in the 1980s there were signs that uneasiness was spreading as 'Green' political parties sought to promote policies protective of the 'environment'. They were not able to achieve much, but they proliferated; the established political parties and perceptive politicians therefore toyed with 'green' themes, too. Environmentalists, as the concerned came to be called, benefited from the modern communications which rapidly broadcast disturbing news even from the still uncommunicative USSR. In 1986,

A new peril. A Polish forest destroyed by 'acid rain' (originating in sulphur dioxide emissions by industrial plant).

an accident occurred at a Ukrainian nuclear power station. Suddenly and horribly, human interdependence was made visible. Grass eaten by lambs in Wales, milk drunk by Poles and Yugoslavs, and air breathed by Swedes, were all contaminated. An unknown number of Russians, it appeared, were going to die over the years from the slow effects of radiation. The alarming event was brought home to millions by television not long after other millions had watched on their screens an American rocket blow up with the loss of all on board. Chernobyl and *Challenger* showed to huge numbers of people for the first time both the limitations and the possible dangers of an advanced technological civilization.

These accidents reinforced and spread more widely the new concern with the environment. It is a complex matter and became tangled with other matters. Some of the doubts which have recently arisen accept that our civilization has been good at creating material wealth, but note that by itself that does not necessarily make men happy. This is hardly a new idea. Its application to society as a whole instead of to individuals is, none the less, a new emphasis. It has been pointed out that improvement of social conditions may not remove all human dissatisfactions and may, indeed, actually irritate some of them more acutely. Pollution, the oppressive anonymity of crowded cities and the nervous stress and strain of modern work conditions, easily erase satisfactions provided by material gain. Even in the 1960s, there was a recognition that things had come to a pretty pass in one of the most beautiful cities in the world when the noise level in the Place de l'Opéra in Paris was found to be greater than that at the Niagara Falls, and the Seine was carrying more sewer water than its natural flow. Scale had become a problem in its own right. Some difficulties posed by modern cities may even have grown to the point at which they are insoluble. Still greater misgivings began to be felt by more people about the threat of over-population, increasing pressure on diminishing resources, and the possibility of intensified competition for them in a world politically unstable.

It can at least be said that energy and material resources are now so wastefully and inequitably employed that a new version of the Malthusian peril is possible. We may well have by no means reached the end (or anything like it) of our capacity to produce food, and it is far more likely that other things than food will run out first. There would at once be an impossible situation if the whole world sought to consume other goods than foods at the level of developed countries today. There is a limit to what any one human being can eat, but virtually no limit to the goods he or she can consume in terms of a better environment, medicine, social services and the like. Yet the supply of such goods depends ultimately on energy and material resources which are finite. The social and political changes which must follow from this fact have hardly begun to be grasped. Mankind has to face the difficulty that to deal with such problems nothing like the knowledge or technique exists which is available, for example, to put someone on the moon.

In the 1970s, too, a new spectre began to be talked about, the possibility of man-made, irreversible climatic change. 1990 had hardly ended before it was pointed out that it had been the hottest year since climatic records began to be kept. Was this, some asked, a sign of 'global warming', the 'greenhouse effect' produced by the release into the atmosphere of the immense quantities of carbon dioxide produced by a huge population burning fossil fuels as never before? Nor was this the only contribution to the phenomenon of accumulating gases in the atmosphere whose presence prevents the planet from dissipating heat; methane, nitrous oxides and chlorofluorocarbons all make a contribution. And if global warming is not enough to worry about, then acid rain, ozone depletion leading to 'holes' in the ozone layer, and deforestation at unprecedented rates, all provided major grounds for new environmental concern.

It is easier to sense the scale of such problems, and so the danger, than to envisage feasible solutions, and while this remains so it is almost certain that uneasiness about interference with Nature will persist. To take only one problem: one estimate is that there is now some twenty-five per cent more carbon dioxide in the atmosphere than in pre- industrial times. It may be so (and as the world's output of the stuff is now said to be six thousand million tons a year, it is not for the layman to dispute the magnitudes).

The consequences, if the trend proved to be established and no effective countermeasures were forthcoming, could be enormous, expressing themselves in climatic change (average surface temperature on the earth might rise by between 1 and 4 degrees Celsius over the next century), agricultural transformation, rising sea-levels (6 centimetres a year has been suggested as possible and plausible), and major migrations. It was clear by 1990 that if the superpowers could eventually come round to cooperate rather than compete, there would be plenty of interests common to mankind to cooperate about-if they could agree on what had to be done. Beginnings have been made in controlling chlorofluorocarbons; and in February 1991, the officials from newly-created environment ministries met in Washington to begin to try to work out a treaty on global climatic change.

IDEAS, ATTITUDES, AUTHORITY

It should by now hardly need reiteration that it is hard for historians to say anything very confidently about what goes on in the minds of most people, and what effect it may have on the maelstrom of historical events. Yet there is evidence that changing or new ideas can have an impact on our collective life (for instance, in the recent political responses to environmental concern). Such impacts are expressed at a level implying a certain sophistication and refinement of mental processes; even now, only minorities worry about the ozone layer. But there are other impacts harder to assess. The 'cake of custom', to use a Victorian's phrase, formed by deep-seated, sometimes almost unrecognized assumptions and attitudes, is another matter altogether. To talk about changes in ideas which may affect it is much harder. But the effort has to be made. It is, for example, almost certainly true that, more than any other single fact, force, or idea, the abundance of commodities has shattered for millions a world of stable expectations. The effect of this is still spreading. Whatever has happened in developed countries, it is now most striking in poor countries, where cheap consumer goods – transistor radios, for example – can bring huge social changes in their train. Such goods confer status. They provide an incentive to work for higher wages to buy them (a spread of a cash economy is another aspect of the process). This often means leaving the villages in which so many still live and, in the process, the severing of ties with tradition and with ordered, stable communities.

The always-hastening onrush of a modernity rolling outwards from industrialized societies has other expressions and effects. Notably and paradoxically, the unprecedently dreadful tragedies and disasters of this century have left more people than ever before believing that human life and the condition of the world can be improved, perhaps indefinitely, and therefore that they should be. There is less willingness to recognize unavoidable, irredeemable tragedy as part of human destiny. The origins of such a new, optimistic, idea lie in Europe and until quite recently it was confined to cultures of European origins. For the majority of the human race, and in most parts of the globe, it is still unfamiliar, barely a half-century old. Few could yet formulate it clearly or consciously, even if asked. Yet it already affects behaviour everywhere.

It has almost certainly been spread less by conscious preaching (though there has been plenty of that) than by material changes which have since 1900 affected more and more people. The fact that they happened at all was the most important thing about many of them. Their psychological impact began to break up the cake of custom by setting out for the first time to many societies the implicit proposition that change is possible, that things need not always be what they have been. Of course the gulf which separates the modern European factory-worker from the Indian or Chinese peasant is vast. In terms of consumption it is certainly wider than that between their ancestors a thousand years ago. Yet to have implanted the notion that change is not only desirable but possible, perhaps even inevitable, is the most important of all the triumphs of the culture, European in origin, which we usually call 'western'.

Technical progress has almost always turned out to be a solvent of inherited ways. Sometimes this shows very directly and obviously; it is difficult, for instance, not to feel that radio and television imply a qualitative change in the history of culture greater even than the coming of print. Technical progress has also affected mentality less directly through the testimony it provides of the seemingly magical power of science to transform life by way of technology. In other ways, too, acquaintance with science and awareness of its importance is more widespread than ever before. There are more scientists about; more attention is given to science in education; scientific information is more widely diffused and more readily comprehensible through the media. Most of this has only come about since 1918; a significant break in scale becomes observable about then. Paradoxically, this has perhaps led to some diminution of the wonder felt in earlier times at what science could do. Success has provided diminishing returns in awe. When more and more things prove possible, there is less which is very surprising about the latest marvel. There is even (unjustifiable) disappointment and irritation when some problems prove recalcitrant. Yet the master idea of our age and perhaps the most widespread, the notion that purposive change can be imposed upon Nature if sufficient resources are made available, has grown stronger. It is a European idea, and the science now carried on around the globe is all based on the European experimental tradition. It continues to throw up ideas and implications disruptive of traditional, theocentric views of life and so we have reached the high phase of the long process of dethroning the idea of the supernatural.

Both the brute facts of a new economic order created by technology and a new appreciation of the power of science and the expectations it arouses have in the end a tendency to undermine traditional authority. Even if clear ideas about what science and technology might do have hardly filtered through to most laymen in developed societies, and even if much of humanity rests undisturbed in its traditional pieties, whether Christian, Hindu, Marxist or anything else, and whether diluted or not, well-being, science, technology and even mere improvement have made persistence in established ways harder. This has happened both at the levels of intellectual élites (who are, of course, disproportionately prominent in histories of thought and culture which stress documents they generate) and in shaping (even if sometimes in nonsensically-distorted ways) the climates of opinion in which most of mankind live. The second effect is perhaps more important in recent history than at earlier times because rising literacy and increasing rapidity of communication have pushed new ideas into mass culture more than ever before. But the impact of scientific ideas on élites is easier to trace. In the eighteenth century, Newtonian cosmology had been able to settle down into coexistence with Christian religion and other theocentric modes of thought without troubling the wide range of social and moral beliefs tied to them. As time passed, though, science has seemed harder and harder to reconcile with any fixed belief at all. It has appeared to stress relativism and the pressure of circumstance to the exclusion of any unchallengeable assumption or viewpoint.

A very obvious instance can be seen in some scientific attempts to investigate human behaviour, a topic particularly likely to raise questions about traditional attitudes. The most important of these were made by the practitioners of a new branch of study, psychology, which had evolved from a traditional interest in 'mental philosophy'. After 1900, psychology tended to take one of two paths. One was to be followed by, among others, Freud (the social impact of whose work has already been touched upon). Broadly speaking, it arose from clinical observation of mental disorder. It is, certainly, a clear example of something which a layman – though not all scientists – would call 'science', which has had an important and widespread effect. Comparatively rapidly, Freud's work became both famous and notorious, and influential. In the first place, it undermined confidence in many traditional assumptions. His ideas (or what were thought to be his ideas) and those of other teachers were diffused through advanced societies in derivative forms, such as in new attitudes to sexuality, or to responsibility and therefore to punishment. Secondly, Freudian ideas stimulated a body of work which, whatever its exact status and method, was based on the belief that therapy could be pursued and relevant material assembled by interrogating subjects about their wishes, feelings and thoughts, and interpreting this data in accordance with an appropriate body of theory.

This was the foundation of the activities which may conveniently be grouped under the name Freud applied to his own work, psychoanalysis, and was the inspiration of many artists, novelists, teachers and advertizing specialists.

The other notable early twentieth-century approach to psychology was more mechanistic and it had deeper roots in the past than psychoanalysis. Theoretically, it was quickly demolished by the philosophers. But this did not dispose of it; it went on to achieve much applause and apparent practical success. 'Behaviourism' (as it has been termed, though the word is often used imprecisely) appeared to generate a body of experimental data certainly as impressive (if not more so) as the successes claimed by psychoanalysis. The outstanding name associated with it is still that of the Russian, I.P. Pavlov. In a series of celebrated experiments in the first decade of this century he established generalizations which, with modification, seemed to be applicable to human beings. His most important single discovery was the 'conditioned reflex'. Essentially this was the manipulation of one of a pair of variables in an experiment in order to produce a predicted piece of behaviour by a 'conditioned stimulus'. The classical experiment provided for a bell to be sounded before food was given to dogs. After a time, the sounding of the bell produced in the dog the salivation which had earlier been produced by the appearance of food. Refinements and developments of such procedures followed. They provided new insights into human psychology which were exploited in many ways (one of the most depressing features of our age, perhaps, is the use made of them by torturers, though this did not produce successes which conclusively demonstrated superiority to earlier practitioners). Some beneficent results were forthcoming, notably in the treatment of mental illness and the improvement of teaching techniques. But if an attempt is to be made to grasp so shapeless a subject, the diffused effect of Behaviourism, though confined consciously to relatively few, seems curiously parallel to that of Freudianism, in that its bias lay towards the demolition of the sense of responsibility and individualism which was at the heart of European moral tradition. This, though, was a demolition also advanced by more empirical and experimental approaches to one aspect of the understanding of the mind, notably through the treatment of mental disorder by chemical, electrical and other physical interference.

It is difficult to believe that such studies, like those of other sciences, have not contributed importantly to a decline of religious belief simply by making more people conscious that something hitherto mysterious and inexplicable was now seen by respected men and women to be susceptible to scientific (as opposed to magical or religious) management, in however elementary and halting a way. But this decline, often alleged to be characteristic of the contemporary era, needs very careful qualification. When people talk about the waning power of religion, they often mean only the formal and organized authority of Christian churches. What people believe and do is quite a different matter. The last English monarch to take the precaution of consulting an astrologer about an auspicious day for her coronation was Elizabeth I, nearly four a half centuries ago. Yet in the 1980s the world was amused (and perhaps just a little alarmed) to hear that the wife of the President of the United States was in the habit of seeking astrological advice, while in 1947 the timing of the ceremony marking the achievement of Indian independence was only settled after appropriate consultation of astrologers. This is, perhaps, only superstition. But there are other important facts to be taken into account in considering the power of religion. India is now a republic which is non-confessional and secular in its constitution; to that extent it reflects the adoption of western ideas. More generally, confessional states and established religion are now unusual anywhere, outside Moslem countries. This does not mean though, that the real power of religions over their adherents has everywhere declined even if it indicates the movement of ideas among directing élites, whether or not they are able to impose them successfully. The founders of Pakistan were secular-minded, westernized men, but in a struggle with the conservative *ulema* after independence they lost. Pakistan became an orthodox Islamic state, and not a secular democracy on western lines which merely respected Islam as the religion of the majority of its people.

It may not be a paradox of our age that more people than ever before now give serious attention to what is said by religious authorities than have ever done so: there are more people alive, after all. Many people in England were startled in the 1980s when Iranian clergymen denounced an author as a traitor to Islam and pronounced a sentence of death upon him; it was a surprise to them to discover that, as it were, the Middle Ages were still in full swing in some parts of the world, without their having noticed it. They were even more startled when numbers of their Moslem fellow-citizens appeared to agree with the *fatwa*. Nevertheless, some believe that here as elsewhere western society has indicated a path which other societies will follow, and that conventional western liberalism will prevail. It may be so. Equally, it may not be. The interplay of religion and society is very complex and it is best to be cautious.

It is not obvious, for example, that even Islamic societies can avoid cultural corruption by the technology and materialism of the European tradition, though they appear to have been able to resist that tradition's ideological expression in atheistic communism. On the other hand, Islam is still an expanding and missionary faith and the notion of Islamic unity is not dead in Islamic lands. Recent events show only too well that it can still nerve men to political action, too. The example should suggest others elsewhere. United with strong social forces, religion produced terrifying massacres in the Indian subcontinent both during the months of partition in 1947 and in the struggles which led to the breaking away of East Bengal from Pakistan and the establishment of the new state of Bangladesh in 1971. In Ulster, a minority of Irishmen still mouth their hatreds and bitterly dispute the future of their country in the vocabulary of Europe's seventeenth-century religious wars. Though the hierarchies and leaders of different religions find it appropriate to exchange public courtesies and occasionally provide a gaping world with such bizarre spectacles as the hob-nobbing of Tibetan Lamas with Anglican bishops, it cannot be said that religion has ceased to be a divisive force because doctrine has become more amorphous. Whether, on the other hand, that means that the supernatural content of religion is losing its hold and that it is important today merely as a badge of group membership, is debatable. What is less debatable is that within the world whose origins are Christian, the decline of sectarian strife has gone along with a general decline of Christian belief and, often, of a loss of nerve. Ecumenism, the movement within Christianity whose most conspicuous expression was the setting up of a World Council of Churches (which Rome did not join) in 1948, owes much to Christians' growing sense in developed countries that they are living in hostile environments. It also owes something to greater uncertainty about what Christianity is, and what it ought to claim. The only unequivocally hopeful sign of unexhausted vigour in Christianity has been the growth (largely by natural increase) of the Roman Catholic Church. Most Roman Catholics are now non-Europeans, a change dramatized by the first papal visits to South America and Asia in the 1960s. By 1980 forty per cent of the world's Catholics lived in South America, and a majority of the College of Cardinals came from non-European countries.

As for the Papacy's historic position within the Roman Church, that appeared to be weakening in the 1960s, some symptoms being provided by an ecumenical council held at the Vatican in that decade. But 1978 (a year of three popes) brought to the throne of St Peter John Paul II, the first Polish pope, the first non-Italian for four and a half centuries, and the first whose investiture was attended by an Anglican archbishop of Canterbury. His pontificate soon showed a determination to exercise the historic possibilities of his office. It is hazardous to project trends in the history of an institution whose fortunes have so obviously fluctuated across the centuries (up with Hildebrandine reform, down with Schism and conciliarism; up with Trent, down with the Enlightenment; up with the first Vatican council ....) and safest simply to recognize that contraceptive practice – an issue sharpened by science and technology – may face it for the first time with a mortal threat to its unquestioned authority in the eyes of millions of Catholics.

Organized religion and the notion of fixed, unchanging moral law often linked to it, has always provided some of the most impressive underpinnings of mankind's

In 1964 Pope Paul VI became the most-travelled Pope in history. In December he was greeted by over a million people in Bombay where he had flown on the first papal visit ever to India to attend an international Eucharistic Congress.

achievements in social regulation. If they have both been much weakened in recent times, the state, the third great historic agent of social order, has kept its end up much better. It has probably never been so widely taken for granted as the normal form of political organization. There exist more states – that is, recognized and geographically-defined political units claiming legislative sovereignty and a monopoly of the use of force within their own borders – than ever before. More people than ever before look to government as their best chance of securing well-being rather than as an inevitable enemy. In many countries politics has apparently replaced religion (sometimes even appearing to eclipse market economics) as the focus of faith that can move mountains.

This success has been marked (and is partly caused) by governments' growing powers. For the greater part of a century the state has been given more and more power to do what was asked of it. Great wars and peacetime tensions helped. Both required huge mobilization of resources and this led to unprecedented extensions of governmental power. To such forces have also been added demands that governments indirectly promote the welfare of their subjects and undertake the provision of services either unknown hitherto, or left in the past to individuals or such 'natural' units as families and villages. There has also been the urge to modernize. Few countries outside Europe achieved this without direction from above and even in Europe some countries have owed much of their modernization to government. The twentieth century's outstanding examples, though, are Russia and China, two great agrarian societies which sought modernization by using the power of the state to impose it. Finally, there is technology. Better communications, more powerful weapons, more comprehensive information systems have, by and large, advantaged those who could spend most on them: governments.

At this point, though, a further explanatory factor, and one which is an important qualification, has to be introduced. The state has succeeded above all as the expression and instrument of nationalism. Where it has been in conflict with nationalism, the state has often been disadvantaged. This has been true even when, to all appearances, enormous power has been concentrated in the state apparatus. Shaped by the traditions of communist centralization though they have been, both the USSR and Yugoslavia have now disintegrated into national units. Quebec may separate from Canada. If speculation was the purpose of this book, it would be possible to cite other instances of disturbingly violent potential. In each instance, nationalism – whose bad name is tribalism – is the explanation. When it has found an expression in a state-form, on the other hand, nationalism has greatly reinforced the power of government and extended its real scope; so much so, that politicians in many countries are hard at work fostering new nationalisms where they do not exist in order to bolster shaky state structures which have emerged from decolonization.

The aggrandizement of the state – if we may so put it – long met with little effective resistance. Even in countries where governments have traditionally been distrusted and where institutions exist to check them, people tend to feel that they are much stronger and less resistible than even a few years ago. The strongest checks on the abuse of power remain those of habit and assumption; so long as electorates in liberal states can assume that governments will not quickly fall back on the use of force, they do not feel very alarmed. But the cause of liberal democracy world-wide does not always look very hopeful; there are now more dictators, and more authoritarian political regimes in the world than in 1939 (though, since changes in Greece, Portugal and Spain in the 1970s, and in eastern European countries in the 1980s, few in Europe). Like the undermining of other liberal assumptions, this is a measure of the narrowing base of what was once thought the cause of the future, but seemed to turn out to be only that of a few advanced societies of the nineteenth century. This does not mean that the forms of liberal politics have not survived. They have, indeed, in one sense prospered, for the rhetoric of democracy and constitutionalism has never been more widely employed, and nationalism is stronger than ever. Yet the substantial freedoms once associated with these ideas are often non-existent or conspicuously in danger and the lack of connexion between nationalism and liberalism is more obvious than ever.

One reason for this is that they have been exported to contexts inimical to them. It is unhistorical to deplore what has followed; as Burke pointed out long ago, political principles take their colour from circumstances. Often in the last half-century it has been shown that representative institutions and democratic forms cannot work properly in societies lacking solid foundations in habits coherent with them, or where there are powerful divisive influences. In such circumstances, the imposition of an authoritarian style of government has often been the best way of resisting social fragmentation once the discipline imposed by a colonial power is withdrawn. Only too obviously, this has not meant great freedom in most post-colonial countries. Whether greater happiness has followed may be more debatable, but is certainly not to be taken for granted.

The role played by the urge to modernize in strengthening the state – something prefigured long ago outside Europe in a Mehemet Ali or a Kemal – was an indication of new sources from which the state increasingly drew its moral authority. Instead of relying on personal loyalty to a dynasty or a supernatural sanction, it was to rest increasingly on the democratic argument that it was able to satisfy collective desires. Usually these were for material improvement, but sometimes not. If one value more than any other legitimizes state authority today it is that of nationalism, still the motive force of much of world politics. Its tenacity is remarked on page after page of this book and it has been successful in mobilizing allegiance as no other force has been able to do. Against it, ideologies have not proved effective; the forces working the other way, and helping to make the whole world one political system, have been economics, communications and technology, rather than comparably powerful moral ideas or mythologies.

Of course, the sheer technical prowess of the state has improved out of recognition since the days when even the great European monarchies could not carry out a census or demolish the obstacles to a unified internal market. Even the growth of international terrorism has left the state with a virtual monopoly of the main instruments of physical control. Already a hundred years ago, the police and armed forces of governments unshaken by war or uncorrupted by sedition gave them almost certain assurance of survival. Improved technology has only increased this near-certainty. New repressive techniques and weapons, moreover, are now only a small part of the story of state power. State intervention in the economy, through its power as consumer, investor or planner, and the improvement of mass communications in a form which leaves access to them highly centralized are also political facts of great importance. They did not appear suddenly in 1945. The welfare state was a reality in Germany and Great Britain before 1914; Hitler and Roosevelt made great use of radio (though for very different ends); and attempts to regulate economic life are as old as government itself. These were, however, all only slight pre-figurings of the Leviathan of our own age. When a British foreign secretary said in the late 1940s that he wished to see a state of affairs in which a man could set off for a foreign country with no need to worry about anything except buying a ticket, without passport, visa or other officially imposed documentation, he was virtually asking for the return of conditions taken for granted eighty years earlier. They still do not exist today – though now at least conceivable as practical goals within Europe.

Since 1945 four new factors have come into play which may indicate the start of a turn in the tide which began to run in favour of the state in sixteenth-century Europe, and has so run ever since. The first is more effective international organization. The United Nations is made up of sovereign states. It has, none the less, shown its capacity to organize collective action against its individual members as the League of Nations and earlier, even less substantial, associations and institutions never did. On a smaller, but still huge scale, the nations of Europe inch forward into an institutional framework which day by day encroaches upon their independent power to act as sovereign states. Nor are formal organizations, like the UN and the European Community, the whole story. Many non-economic problems today are tackled internationally and, given concerns over the environment, it is to be presumed that this tendency will continue.

The second force encroaching upon the state is the persistence of some supranational forces which from time to time eclipse by their magnitude the freedom of action of individual sovereign states. Islam is at times such a force; the modern state has never been so successful in the Middle East as elsewhere in assuring its authority. So, perhaps, is the racialist consciousness of pan-Africanism or what is called *négritude* an inhibition on some form of state action. A third curb (or determinant – words are not easy to choose at so general a level) for the state has been the re-integration of the world economy. Institutionalized by (sometimes) international agreement or by the simple organic growth of large companies, and driven by rising expectations, this has often dashed the hopes of politicians seeking to order the societies over which they are expected to preside. It is a factor which is closely linked, moreover, to the fourth emerging restraint on state power, the emergence of regional groupings of nations which require the observance of common disciplines. Some, like those of eastern Europe, have proved evanescent. Others, among which western Europe is the most obvious example, have slowly advanced their influence, even if the visions of some of their founders are as yet unrealized.

The effect of these forces will often be discernible in the background to the story told in the pages which follow. Perhaps they prefigure the end of the state's supremacy as the characteristic political institution of modern history. More probably, though, they may bring about some reduction in state power, while leaving forms largely intact (perhaps while power flowers and accumulates elsewhere instead). This is at least more probable than that radical forces will succeed in destroying state power. Such forces exist, and at times they draw strength and appear to prosper from new causes – ecology, feminism and a generalized anti-nuclear and 'peace' movement have all patronized them. But in forty years of activity such causes have only been successful when they have been able to influence and shape state policy, changing laws, and setting up new institutions. The idea that major amelioration can be achieved by by-passing so dominant an institution still seems as unrealistic as it was in the days of the anarchistic and utopian movements of the nineteenth century.

2
The Politics of the New World

The world is still unsettled by the global revolution in economic and political organization which began during the Second World War and has been going on ever since. Two and a half decades during which the central characteristics of the world political order seemed to be more and more frozen (though revolutionary developments were continuing elsewhere) were to be followed by a renewed quickening of the pace of change in the 1980s. By the end of that decade, landmarks taken for granted for thirty years and more had disappeared (sometimes almost overnight) and others, even more dominating, were already called in question. The whole process was so rapid and continuous as to make divisions in the story unusually artificial, but the historian has to try to analyze and uncover their structure. Some of the forces behind the turmoil were very profound and very mixed. Some, for instance, were rooted in the growing energy needs of industrialized societies already touched upon. Others can be traced ultimately much further back – to, for example, ideas first announced in the French Revolution. But though it is worth striving for as deep a historical perspective as possible, the long-term trends and forces cannot alone explain what happened. Like other great changes throughout history, many of those of the second half of the twentieth century arose almost from accident, circumstance, even personality. They are none the easier to explain for that. Often decisive evidence is unavailable, and almost always a proper perspective is difficult to achieve (as is usually the case when we stand close to events). Simply to trace the steps by which one sort of world changed into another is enough to aim for at this stage. The place to begin is with the alignments and structure which emerged in the 1950s and 1960s to provide the framework within power and policy were deployed.

COLD WAR BEGINNINGS

In those decades world history was dominated by a prolonged and bitter Soviet-American antagonism. The 'Cold War' made the rupture in the diplomatic history of the modern era which had first opened in 1917 the dominating fact of world affairs. The future great power which made its appearance through the Bolshevik coup was to approach international society in a new and uniquely troublesome way. Soviet Russia regarded diplomacy not just as a convenient way of doing business but as a weapon for the advance of ideology. However they qualified it in practice, Bolsheviks had said their aim was to overthrow the social institutions of non-communist societies, and they meant it, so far as the long run was concerned. After 1945 other communist states came into existence whose rulers agreed, at least in words. The result was a world seemingly divided into two camps, one led by the USSR, one by the United States. Each camp proclaimed itself as anti-imperialist. Each often, in fact, behaved (if effective domination is the test) in imperialist ways.

The original communist Europe of 1945 had been added to by other takeovers in 1947 as Hungary, Romania and Poland ceased to have any non-communists in their governments. Czechoslovakia followed in 1948. Then, the opening of the Marshall Aid programme was almost at once followed by what was to prove to be the first battle of the Cold War, over the fate of Berlin. It was decisive. It established the point at which, in Europe, the United States was prepared to fight. It does not seem that this outcome had been anticipated by the Russians, though they had provoked it by seeking to prevent the re-emergence of a reunited and economically powerful Germany which would not be under their control. The western powers had a different interest, to reanimate the German economy, at the very least in their own occupation zones. They sought to get on with this before Germany's future political shape was settled, sure that it was vital for the recovery of western Europe as a whole.

In 1948, without Russian agreement, the western powers introduced a currency reform in their own sectors. It had a galvanic effect, releasing the process of economic recovery in western Germany. Following on Marshall Aid, available (thanks to Soviet decisions) only to the western-occupied zones, this reform, more than any other step, cut Germany in two. It meant that the recovery of the eastern half would not be integrated with that of western Europe. A strong western Germany might now emerge by itself. That the western powers should get on with the business of putting German industry on its feet was undoubtedly economic sense, but eastern Germany was thenceforth decisively on the other side of the Iron Curtain. Currency reform divided Berlin, too, and thereby prejudiced communist chances of staging a popular putsch in the city.

The Soviet response was to disrupt communication between the western zones of Germany and Berlin. Whatever their original motives, the dispute escalated. Some western officials had already had it in mind that a severance of western Berlin from the three western occupation zones (Berlin was isolated inside the Soviet zone) might be attempted before this crisis; the word 'blockade' had been used and the Russians' acts were now interpreted in this sense. They had not interfered with the rights of the western allies to have access to their own forces in their own sectors of Berlin, but they were interfering with the traffic which ensured supply to the Berliners in those sectors. To supply West Berlin, the British and Americans organized an airlift to the city. The Russians wanted to demonstrate to the West Berliners that the western powers could not stay there if they did not want them to; they hoped thus to remove the obstacle which the presence of elected non-communist municipal authorities presented to Soviet control of Berlin. So, a trial of strength was under way. The western powers, in spite of the enormous cost of maintaining such a flow of food, fuel and medicine as would just keep West Berlin going, announced they were prepared to keep it up indefinitely. The implication was that they could be stopped only by force. For the first time since the war American strategic bombers moved back to bases in England. Neither side wanted to fight, but all hope of cooperation over Germany on the basis of wartime agreement was dead.

The blockade lasted over a year and defeating it was a remarkable technical achievement. Berlin's only airfield had to handle over a thousand aircraft a day for most of the time, with an average daily delivery of 5,000 tons of coal alone. Yet its real significance was political. Allied supply was not interrupted nor were the West Berliners intimidated. The Soviet authorities made the best of defeat by deliberately splitting the city and refusing the mayor access to his office. Meanwhile the western powers had signed a treaty setting up a new alliance, the North Atlantic Treaty Organization (NATO) in April 1949, a few weeks before the blockade was ended by agreement. NATO was the first Cold War creation to transcend Europe. The United States and Canada were members, as well as most western European states (only Sweden, Switzerland and Spain did not join). It was explicitly a defensive alliance, providing for the mutual defence of any member attacked and so yet another break with the now almost-vanished isolationist traditions of American foreign policy. In May a new German state, the Federal Republic, emerged from the three western zones of occupation and in the following October, a German Democratic Republic (the GDR) was set up in the east. Henceforth, there were to be two Germanies, it seemed, and the Cold War ran along an Iron Curtain dividing them, and not, as Churchill had suggested in 1946, further east, from Trieste to Stettin. But a particularly dangerous phase in Europe was over.

The foundation of NATO suggested perhaps that as well as two Europes, there might also be two worlds. This soon seemed more likely still, when the Cold War re-erupted in east Asia. In 1945 Korea had been divided along the 38th parallel, its industrial north being occupied by the Russians and the agricultural south by the Americans. The problem of reunification was eventually referred to the United Nations. After efforts to obtain elections for the whole country that organization recognized a government set up in the south as the only lawful government of the Republic of Korea. By then, the Soviet zone also had a government claiming sovereignty over the whole country. After Russian and American forces had both withdrawn, North Korean forces invaded the south in June 1950 (with, it now appears, Stalin's foreknowledge and approval). Within two days President Truman sent American forces to fight them, acting in the name of the United Nations. The Security Council had voted to resist aggression, and as the Russians were at that moment boycotting it, they could not veto United Nations action.

The Americans always provided the bulk of the UN forces in Korea, but other nations soon sent contingents. Within a few months the allied army was operating well north of the 38th parallel. It seemed likely that the North Koreans would be overthrown. When fighting drew near the Manchurian border, though, Chinese communist forces intervened and drove back the UN army. There was now a danger of a much bigger conflict. The question arose of direct action, possibly with nuclear weapons, by the United States against China. China was the second largest communist state in the world, and the largest in terms of population. Behind it stood the USSR; a man could walk from Erfurt to Shanghai without once leaving communist territory.

Prudently, Truman insisted that the United States must not become involved in a greater war on the Asian mainland. That much settled, further fighting showed that although the Chinese might be able to keep the North Koreans in the field, they could not overturn South Korea against American wishes. Armistice talks were started. The new American administration which came into office in 1953 was Republican and unequivocally anti-communist, but knew its predecessor had sufficiently demonstrated its will and capacity to uphold an independent South Korea and felt that the real centre of the Cold War was in Europe rather than in Asia. The armistice was signed in July 1953. Subsequent efforts to turn this into a formal peace have as yet failed; nearly forty years later, tension remains high between the two Koreas. In the Far East as well as in Europe the Americans had won the first battles of the Cold War, and in Korea they had been real battles; estimates suggest the war cost three million casualties in all, civilians included.

American bombers attacking a chemical works in North Korea. Until the coming of the intercontinental ballistic missile, American air-power was the first-line deterrent to what many people in the West regarded as an aggressive and expanding Communism; no comparable offsetting force existed in the East.

Shortly before the armistice, Stalin had died. It was very difficult to guess what this might mean. In due course, there seemed to have been something of a break in the continuity of Soviet policy, but that was not clear at the time. The American President Eisenhower remained distrustful of Russian intentions and in the middle of the 1950s, the Cold War was as intense as ever. Shortly after Stalin's death his successors revealed that they too had the improved nuclear weapon known as the hydrogen bomb. It was in a way

Stalin's final memorial. It guaranteed (if it had been in doubt) the USSR's status in the post-war world. Stalin had carried to their logical conclusion the repressive policies of Lenin, but he had done much more than his predecessor. He had rebuilt most of the tsarist empire and had given Russia the strength to survive (just, and with the help of powerful allies) her gravest hour of trial. What is not clear is that this could only have been achieved or was worth achieving at such cost, unless (as may well be thought) to have escaped German domination was justification enough. The Soviet Union was a great power, but, among the elements of that empire, Russia at least would doubtless have become one again without communism, and her peoples had been rewarded for their sufferings with precious little but an assurance of international strength. Domestic life after the war was harsher than ever; consumption was for years still held down and both the propaganda to which Soviet citizens were subjected and the brutalities of the police system seemed, if anything, to have been intensified after some relaxation during the war.

Another of Stalin's monuments was the division of Europe, clearer than ever at his death and confirmed in the next few years. The western half was by 1953 substantially rebuilt, thanks to American economic support. The Federal Republic and the GDR moved further and further apart. On two successive days in March 1954 the Russians announced that the eastern republic now possessed full sovereignty and the West German president signed the constitutional amendment permitting the rearmament of his country. In 1955 West Germany entered NATO; the Soviet riposte was the Warsaw Pact, an alliance of its satellites. Berlin's future was still in doubt, but it was clear that the NATO powers would fight to resist changes in its status except by agreement. In the east, the GDR agreed to settle with old enemies: the line of the Oder-Neisse was to be the frontier with Poland. Hitler's dream of realizing the greater Germany of the nineteenth-century nationalists had ended in the obliteration of Bismarckian Germany. Historic Prussia was now ruled by revolutionary communists, while the new West Germany was federal in structure, non-militarist in sentiment and dominated by Catholic and Social Democratic politicians whom Bismarck would have seen as 'enemies of the state'. So, without a peace treaty, the problem of containing the German power which had twice devastated Europe by war was settled for thirty-five years. Also in 1955 came the final definition of land frontiers between the European blocs, when Austria re-emerged as an independent state, the occupying allied forces being withdrawn, as were the last American and British troops from Trieste, with a settlement of the Italo-Yugoslav border dispute there.

We cannot, of course, attribute to Stalin or any one person another division which was already spread world-wide after the establishment of communism in China, that between what we may call capitalist and command (or would-be command) economies. Yet Stalin's policies helped to deepen it. Commercial relations between Soviet Russia and other countries had been encumbered by politics from the October Revolution onwards. There had followed the huge disruption of world trade after 1931 as the capitalist economies plunged into recession and sought salvation in protection (or, even, autarky). After 1945, though, all earlier divisions of the world market were transcended; two methods of organizing the distribution of scarce resources increasingly divided first the developed world and then a few other areas (of which the most important was constituted by China's stance in eastern Asia). The essential determinant of one system, the capitalist, was the market – though a market very different from that envisaged by the old liberal Free Trade ideology and in many ways a very imperfect one, tolerating a substantial degree of intervention through international agencies and agreement; in the communist-controlled group of nations (and some others) political authority was intended to be the decisive economic factor. Trade between the two systems continued, but on a severely restricted basis.

Neither system remained unchanged. Contacts between them multiplied as the years passed. None the less, they long offered alternative models for economic growth. Their competition was inflamed by the political struggles of the Cold War and actually helped to spread its antagonisms. This could not be a static situation. Before long the one system was much less completely dominated by the United States and the other somewhat less completely dominated by the Soviet Union than was the case in 1950. Both shared (though in far different degree) in the continuing economic growth of the 1950s and the 1960s, but were later to diverge as the market economies moved ahead more rapidly. The distinction between the two economic systems nevertheless remained a fundamental of world economic history from 1945 to the 1980s, shutting off some possibilities and suggesting others.

The entry of China to the world of what were called socialist economic systems was at first seen almost purely in Cold War terms, and as a shift in strategic balances. Yet by the time of Stalin's death there were many other signs that the prophecy made by the South African statesman Smuts more than a quarter-century before that 'the scene had shifted away from Europe to the Far East and the Pacific' had been realized. Although Germany continued to be the focus of Cold War strategy, Korea was the first dramatic evidence that the centre of gravity of world history was moving once again, this time from Europe to the Orient.

ASIAN REVOLUTION

The collapse of European power in Asia was bound to be followed by further changes as new Asian states came to be aware of their interests and power (or lack of it). Shape and unity given them by their former masters often did not long outlast the empires; the subcontinent of India lost its brief political unity at the very moment of decolonization; Malaysia and Indo-China were already before 1950 beginning to undergo important and not always welcome changes. Internal strains troubled some new nations; Indonesia's large Chinese communities had disproportionate weight and economic power and anything that happened in the new China might disturb them. Whatever their political circumstances, moreover, all these countries had fast-growing populations and were economically backward. For many of them, therefore, the end of European domination now seems less of a turning-point that once thought. The biggest changes came later.

Europe's control of their destinies had for the most part been fitful. Though Europeans had swayed the fate of millions of Asians, and had influenced their lives for centuries, their civilization had touched the hearts and minds of few but the ruling élites. In Asia that civilization had to contend with deeper-rooted and more powerful traditions than anywhere else in the world. Asian cultures had not been (because they could not be) swept aside like those of pre-Columban America. As in the Arab Islamic world, both the direct efforts of European and the indirect diffusion of European culture through self-imposed modernization faced formidable obstacles. The deepest layers of thought and behaviour often remained undisturbed even in some who believed themselves most emancipated from their past; nativities are still cast in educated Hindu families when children are born and marriages are contracted, and the Chinese Marxist draws on an unassailable sense of moral superiority grounded in age-old Chinese attitudes to the non-Chinese world.

For the purpose of understanding Asia's recent role in world history two zones of Asian civilization are as distinct and significant as they have been for centuries. A western Asian sphere is bounded by the mountain ranges of northern India, the Burmese and Siamese highlands and the huge archipelago of which Indonesia is the major component. Its centre is the Indian Ocean and in its history the major cultural influences have been three: Hindu civilization spreading from India to the south-east, Islam (which also spread eastward across it), and the European impact, felt for fairly long periods through commercial and religious activity, and then for a much shorter era of political domination. The other sphere is east Asian, and is dominated by China. In large measure this is a function of the simple geographical fact of that country's huge mass, but the numbers and, sometimes, the migration of her people, and, more indirectly and variably, China's cultural influence on the east Asian periphery – above all, Japan, Korea and Indo-China – all form part of the explanation. In this zone, direct European political domination had never compared with that further west in Asia, either in extent or duration.

It was easy to lose sight of such important differences, as of much else imposed by history, in the troubled years after 1945. In both zones there were countries which seemed to follow the same road. Both provided examples of angry rejections of the West, expressed in western language and appealing to world opinion on long-familiar lines. India, for example, absorbed within a few years both the princely states which had survived the Raj and the subcontinent's remaining French and Portuguese enclaves in the name of a truculent nationalism which owed little to Indian native tradition. Soon, the Indian security forces were energetically suppressing any threat of separatism or regional autonomy within the new republic. Perhaps this should not have been surprising. Indian independence was, on the Indian side, the work of a western-educated élite which had imported the ideas of nationalism – like those of equality and liberty – from the West, even if it had at first only sought equality and partnership with the Raj itself. A threat to that élite's position after 1947 could often be most easily (and sincerely) understood as a threat to an Indian nationhood which had in fact still to be created.

This was all the more true because the rulers of independent India had inherited, to a degree often overlooked, many of the aspirations and institutions of the British Raj. Ministerial structures, constitutional conventions, division of powers between central and provincial authorities, the apparatus of public order and security were all taken over, stamped with republican insignia, and continued to operate much as before 1947. The dominant and explicit ideology of government was a moderate and bureaucratic socialism in the current British mode, and this was not far removed in spirit from the public-works-and-enlightenment despotism-by-delegation of the Raj in its last years. The realities which faced India's rulers included a deep conservative reluctance among local notables who controlled votes to disturb traditional privilege at any level below that of the former princes. Yet awesome problems faced India – population growth, economic backwardness, poverty (the average annual *per capita* income of Indians in 1950 was $55), illiteracy, social, tribal, religious division, and great expectations of what independence ought to bring. It was clear that major change was needed.

The installation of the new constitution in 1950 did nothing to change these facts, some of which would not begin to exercise their full weight until at least the second decade of the new India's existence. Even at the end of the century, much of life of rural India still goes on virtually as it did in the past, when war, natural disaster, and the banditry of exploiting rulers allowed it to do so. This implies gross poverty for some. In 1960, over a third of the rural poor were living on less than a dollar a week (and, at the same time, half the urban population earned less than enough to maintain the official minimum daily calorie intake required for health). Economic progress had been swallowed by population growth. In the circumstances it is hardly surprising that the rulers of India should have incorporated in the constitution provisions for emergency powers as drastic as any ever enjoyed by a British viceroy, providing as they did for preventative detention and the suspension of individual rights, to say nothing of the suspension of state government and the submission of states to Union control under what was called 'President's Rule'.

The weaknesses and uneasiness of a 'new nation' made things worse when India quarrelled with her neighbour, Pakistan. It was first seriously evident over Kashmir, where a Hindu prince ruled over a majority of Moslem subjects. Fighting began there as early as 1947, when the Moslems tried to bring about union with Pakistan; the Maharajah asked for Indian help and joined the Indian republic. To complicate things further, the Moslem spokesmen of Kashmir were themselves divided. India refused to hold the plebiscite recommended by the United Nations Security Council; two-thirds of Kashmir then remained in Indian hands to ensure a running sore in Indo-Pakistan relations. Fighting stopped in 1949, only to break out again in 1965–6 and 1969–70. The issue had by then been further complicated by demarcation disputes and quarrels over the use of the Indus waters. In 1971 there was more fighting between the two states when East Pakistan, a Moslem, but Bengali-speaking, region broke away to form a new state, Bangladesh, under Indian patronage (thus showing that religious unity alone was not enough to constitute a viable nation). It soon faced economic problems even greater than those of India or Pakistan.

1965 brought fighting between India and Pakistan at many points on their frontiers from Tibet to the Arabian Sea, as well as in West Bengal and Assam. This Indian officer was captured in Kashmir.

In these troubled passages, India's leaders showed great ambitions (perhaps going at times so far as a wish to reunite the subcontinent) and sometimes blatant disregard of the interests of other peoples (such as the Nagas). The irritation aroused by Indian aspirations was moreover further complicated by the Cold War. India's leader, Nehru, had quickly insisted that India would not take sides. In the 1950s, this meant that India had warmer relations with Russia and communist China than with the United States; indeed, Nehru appeared to relish opportunities of criticizing American action, which helped to convince some sympathizers of India's credentials as a progressive, peaceful, 'non-aligned' democracy. It came as all the greater a shock, therefore, to them and to the Indian public to learn in 1959 that Nehru's government had been quarrelling with the Chinese about the northern borders for the previous three years without saying so. At the end of 1962, large-scale fighting began. Nehru took the improbable step of asking the Americans for military aid and, even more improbably, of receiving it at the same time as he also took it (in the form of aeroplane engines) from Russia.

Logically, the young Pakistan had not courted the same friends as India. She was in 1947 much weaker than her neighbour, with only a tiny trained civil service (Hindus had joined the old Indian Civil Service in much larger numbers than Moslems).

She was also divided geographically in two from the start, and almost at once had lost her ablest leader, Jinnah. Even under the Raj, Moslem leaders had always (and perhaps realistically) shown less confidence in democratic forms than Congress; usually, Pakistan has been ruled by authoritarian soldiers who have sought to ensure military survival against India, economic development, including land reform, and the safeguarding of Islamic ways.

Whereas India's friends in the West were sometimes disappointed by India's politics after independence, less had been expected of Pakistan. It always helped to distance her from India that she was formally Moslem while her neighbour was constitutionally secular and non-confessional (at first sight a seemingly 'western' stance, but one not hard to reconcile with India's predominantly Hindu and syncretic cultural tradition). This was to lead Pakistan towards increasing Islamic regulation of its internal affairs. Religious difference, though, affected Pakistan's foreign relations less than did the Cold War, which itself brought further confusion to Asian politics when a new association of professedly neutralist or 'non-aligned' nations emerged after a meeting of representatives of twenty-nine African and Asian states at Bandung in Indonesia in 1955. Most delegations other than China's were from lands which had been part of the colonial empires. From Europe, Yugoslavia was soon to join them. They were also poor and needy, more suspicious of the United States than of Russia, and more attracted to China than to either. These came to be called the 'Third World' nations, a term apparently coined by a French journalist in a conscious reminiscence of the legally-underprivileged French 'Third Estate' of 1789 which had provided much of the driving-force of the French Revolution. The implication was that such nations were disregarded by the great powers, and excluded from the economic privileges of the developed countries. Plausible though this might sound, the expression 'Third World' actually masked important differences between the members of that group and in their individual relations with the developed world. Not surprisingly, the coherence of Third World politics was not to prove very enduring. Since 1955 more people have been killed in wars and civil wars within that world than in conflicts external to it.

Nevertheless, ten years after the end of the Second World War Bandung forced the great powers to recognize that the weak had power if they could mobilize it. They bore this in mind as they looked for allies in the Cold War. By 1960 there were already clear signs that Russian and Chinese interests might diverge and each sought the leadership of the underdeveloped and uncommitted. At first this emerged obliquely in the disguise of differing attitudes to the Yugoslavs; it was in the end to be a world-wide contest. One early result was the paradox that as time passed Pakistan drew closer to China (in spite of a treaty with the United States) and Russia closer to India. When the United States declined to supply arms during her 1965 war with India, Pakistan asked for Chinese help. She got much less than she hoped, but this was early evidence of a new fluidity which began to mark international affairs in the 1960s.

No more than the USSR or China could the United States ignore this. Indeed, the Cold War was to produce an ironical change in the Americans' role in Asia; from being the patrons of the anti-colonialism which had done so much to dismantle their allies' empires, they began in some areas almost to appear as their successors, though in the east Asian rather than in the Indian Ocean sphere, where long and unrewarded efforts were made to placate an ungrateful India (before 1960 she received more economic aid from the United States than any other country).

One example of the new difficulties facing great powers was provided by Indonesia. Its vast sprawl encompasses many peoples, often with widely-diverging interests. The basic culture of the Sumatrans and Javanese is Hindu, but, at least formally, Indonesia also has the largest Moslem population under one government in the world. Arab traders had brought Islam to Indonesia's peoples from the thirteenth century onwards, and more than four-fifths of the Indonesian population is reckoned now to be Moslem, although traditional animism perhaps matters as much in determining their behaviour. Indonesia also has a well-entrenched Chinese community, enjoying in the colonial period a preponderant share of wealth and administrative jobs. The departure of the Dutch released communal tensions from the discipline an alien ruler had imposed just as the usual post-colonial problems – over-population, poverty, inflation – began to be felt.

In the 1950s the central government of the new republic was increasingly resented; by 1957 it had faced armed rebellion in Sumatra and unrest elsewhere. The time-honoured device of distracting opposition with nationalist excitement (directed against a continual Dutch presence in west New Guinea) did not work any more; popular support for Sukarno was not rebuilt. His government had already moved away from the liberal forms adopted at the birth of the new state and he leant more and more on Soviet support. In 1960 parliament was dismissed, and in 1963 Sukarno was named president for life. Yet the United States, fearing he might turn to China for help, long stood by him. This enabled him to swallow up (to the irritation of the Dutch) a would-be independent state which had emerged from west New Guinea. Sukarno then turned on the new federation of Malaysia, put together in 1957 from fragments of British empire in south-east Asia. With British help, Malaysia mastered Indonesian attacks on Borneo, Sarawak and the Malaysian mainland. Although he still enjoyed American patronage (at one moment, President Kennedy's brother appeared in London to support his cause), this setback seems to have been the turning-point for Sukarno. Exactly what happened is still obscure, but when food shortages and inflation went out of control, a coup was attempted (it failed) behind which, said the leaders of the army, were the communists. It is at least possible that Indonesia was intended by Mao to play a major part in the export of revolution; the communist party which Sukarno had tried to balance against other politicians was at one time alleged to be the third largest in the world. Whether or not a communist takeover was intended, though, the economic crisis was exploited by those who feared one. The popular and traditional Indonesian shadow theatres were for months seasoning the old Hindu epics which were their staple material with plentiful political allusions and overtones of coming change. When the storm broke, in 1965, the army stood back ostentatiously while popular massacre removed the communists to whom Sukarno might have turned. Estimates of the number killed vary between a quarter and a half a million. Sukarno himself was duly set aside the following year. A solidly anti-communist regime then took power which broke off diplomatic relations with China (they were not to be renewed until 1990). It has kept some of the losers of 1965 in jail until the present day, taking a few out from time to time to be hanged as evidence of resolute prosecution of the struggle against communism and, no doubt, *pour encourager les autres*.

Paradoxically (and almost incomprehensibly, given Indonesia's problems), American support for Sukarno had reflected the belief that strong, prosperous national states were the best bulwarks against communism. The history of the Far Eastern Asia in the last forty years can be read so as to offer support for that principle, but its successful expression in American policy was always far from unqualified. Difficulty lay in its practical application.

By 1960, the dominant strategical fact east of Singapore was the recreation of Chinese power. South Korea and Japan had successfully resisted communism but they benefited from the Chinese revolution; it gave them leverage against the West. Just as east Asians had always held off Europeans more successfully than the Indian Ocean countries, they have showed after 1947 an ability to buttress their independence in both communist and non-communist forms, and not to succumb to direct Chinese manipulation. It is difficult not to link this, ironically, to the deep and many-faceted conservatism of societies which had for centuries drawn on Chinese example. In their discipline, capacity for constructive social effort, disregard for the individual, respect for authority and hierarchy, and deep self-awareness as members of civilizations proudly distinct from the West, the east Asians drew on something much deeper than the Chinese Revolution; indeed, that revolution is only comprehensible against the background dominated by that something.

We may nevertheless begin with the Chinese Revolution's victory and installation in power in 1949. Peking was once more the capital of China. Some thought

In China, 1949, the year of the establishment of the Communist People's Republic, a great French photographer caught the atmosphere as the long-drawn-out struggle of traditional and revolutionary China came to a climax. The result was this picture of an old gentleman and a column of new recruits to the army.

this showed that China's leaders might again be more aware of pressure from her land frontiers in the north than of the threat from the sea which had faced her for more than a century. However this may be, the Soviet Union was the first state to recognize the new China, closely followed by the United Kingdom, India and Burma. Given Cold War preoccupations elsewhere and the circumstances of the nationalist collapse, the new China in fact faced no real threat from the outside. Her rulers could concentrate on the long overdue and immensely difficult task of modernization; the nationalists, cooped up in Taiwan, could be disregarded, though for the moment irremovable. When a major threat appeared, as the United Nations forces approached the Yalu river frontier of Manchuria in 1950, the Chinese reaction was strong and immediate: they sent a large army to Korea. But the main preoccupation of China's new rulers was the internal state of the country. Poverty was universal. Disease and malnutrition were widespread. Material and physical construction and reconstruction were overdue, population pressure on land was as serious as ever, and the moral and ideological void presented by the collapse of the *ancien régime* over the preceding century had to be filled.

The peasants were the starting-point. Here 1949 is not a very significant date. Since the 1920s land reform had been carried out largely by the peasants themselves in areas the Communists dominated. By 1956 China's farms were collectivized in a social transformation of the villages which was intended to give control of the new units to their inhabitants. The essential change was the overthrow of local village leaders and landlords; it was often violent and such persons must have made up a large number of the 800,000 Chinese later reported by Mao to have been 'liquidated' in the first five years of the People's Republic. Meanwhile industrialization was also pressed forward, with Soviet help, the only source from which China could draw. The model chosen, too, was the Soviet one: a Five Year Plan was announced and launched in 1953. It was a remarkable success. By 1956 it had produced an increase in the Chinese net domestic product proportionately greater than the increase in the production of food. A contrast with India was emerging which was to grow much more striking.

By then China was once more a major influence abroad, though her independence was long masked by the superficial unity of the communist bloc and her continued exclusion from UNO at the insistence of the United States. A Sino-Soviet treaty in 1950 was interpreted – especially in the United States – as evidence that China was entering the Cold War. Certainly, the regime was communist and talked revolution and anti-colonialism, and its choices were bound to be confined by the parameters of the Cold War. Yet in a longer perspective more traditional concerns now seem evident in Chinese communist policy from the start. At a very early point, there was visible a concern to re-establish Chinese power within the area it had always tended to fill up.

The security of Manchuria is by itself enough to explain Chinese military intervention in Korea, but the peninsula had also long been an area of dispute between imperial China and Japan. A Chinese occupation of Tibet in 1951 was another incursion into an area which had for centuries been under Chinese suzerainty. But from the start the most vociferous demand made for regaining control of the Chinese periphery was for the eviction of the KMT government from Taiwan, seized in 1895 by the Japanese and only briefly restored in 1945 to control by the mainland. By 1955, a United States government was so deeply committed to the support of the KMT regime there that the president announced that the United States would protect not merely the island itself but the smaller islands near the Chinese coast which were thought essential to its defence. About this issue and against a psychological background provided by a sense of inexplicable rebuff from a China long patronized by American philanthropy and missionary effort, the interest of Americans in Chinese affairs tended to crystallize for over a decade. So obsessively did it do so, that the KMT tail seemed at times to wag the American dog. Conversely, during the 1950s, both India and Russia supported Peking over Taiwan, insisting that the matter was one of Chinese internal affairs; it cost them nothing to do so. Sensation was therefore all the greater when the next decade brought fighting between China and these two countries.

The quarrel with India grew out of the Chinese occupation of Tibet. When the Chinese further tightened their grasp on that country in 1959, Indian policy still seemed basically sympathetic to China. An attempt by Tibetan exiles to set up a government on Indian soil was stifled. But by then territorial disputes had begun, and had already led to clashes. The Chinese announced that they did not recognize a border with India along lines drawn by a British-Tibetan negotiation in 1914 and never formally accepted by any Chinese government. Forty-odd years' usage was hardly significant against China's millenial historical memory. As a result, there was much heavier fighting in the autumn of 1962, when Nehru demanded a Chinese withdrawal from the disputed zone. The Indians did badly, though fighting ceased, at the end of the year, on the initiative of the Chinese.

It was at this moment that, early in 1963, a startled world suddenly heard the Soviet Union bitterly denounced by the Chinese communists. On the one hand, said the Chinese, the Russians had helped India, and on the other, they had, in a hostile gesture, cut off economic and military aid to China three years earlier. The second charge showed there were complex origins to this quarrel, but by no means went to the root of the matter. Some Chinese communists (Mao among them) could, after all, remember what had happened when Chinese interests had been subordinated to the international interest of communism, as interpreted by Moscow, in the 1920s. Since that time there had always been a tension in the leadership of the Chinese party between Soviet and native forces. Mao himself represented the latter. Unfortunately, such subtleties were difficult to disentangle because Chinese resentment of Soviet policy had to be presented to the rest of the world in Marxist jargon. Since the new leadership in Russia was engaged at the time in the dismantling of the Stalin myth, this almost accidentally led the Chinese to sound more Stalinist than Stalin in their public pronouncements even when they were pursuing non-Stalinist policies.

In 1963, it would also have been of utility to non-Chinese observers to recall an even more remote past to Sino-Soviet relations. Long before the foundation of the CCP, the Chinese Revolution had been a movement of national regeneration. One of its primary aims had been the recovery from the foreigners of China's control over her own destiny. Among these foreigners, the Russians were pre-eminent. Their record of encroachment upon the Chinese sphere went back to Peter the Great. It had continued all through the tsarist to the Soviet era. A protectorate over Tannu Tuva had been established in 1914 by the tsars, but the area was annexed by the Soviet Union in 1944. In 1945 Russian armies entered Manchuria and north China and thus reconstituted the tsarist Far East of 1900; they remained in Sinkiang until 1949, in Port Arthur until 1955. In Mongolia they left behind a satellite Mongolian People's Republic they had set up in the 1920s. With something like 4,500 miles of shared frontier (if Mongolia is included), the potential for dispute along its huge length was immense.

Mao Tse-tung's personal experience must also have counted for much. Although his later intellectual formation had been Marxist and although he found its categories helpful in explaining his country's predicament, he appears always to have diluted them with a certain pragmatism. He escaped the Marxist dogma of the Bolshevik period because of a firm belief in the lessons of experience, and advocated a sinicized Marxism which envisaged a society unlike that of Soviet Russia as well as non-capitalist. His attitude to knowledge and ideas was predominantly utilitarian and moralistic, and thus very much in the Chinese tradition. The basis of his world view, rather than the bloodless categories of the dialectic, appears to have been a classification of phenomena into contending forces on which human will-power could at any moment play to bring about morally valuable and creative change.

Mao's relationship with his party had not always been smooth. His first opportunity had come when his policy towards the peasantry provided a way ahead after disaster had overtaken urban communism. After a temporary setback in the early 1930s he was from about 1935 virtually supreme within the party. Rural influences were on top. A way was also opened for Mao to future enormous international influence; the notion of a protracted revolutionary war, waged from the countryside and carried into the towns, looked promising in other parts of the world where the orthodox Marxist belief that industrial development was needed to create a revolutionary proletariat was not persuasive.

The Sino-Soviet conflict announced in 1963 in the end entangled the whole communist world. It was inflamed by Russian tactlessness. The Soviet leaders seem to have been as careless as any western imperialists of the emotions of Asiatic allies: Krushchev once revealingly remarked that when touring in China, he and other Russians 'used to laugh at their primitive forms of organization'. The withdrawal of Soviet economic and technical help in 1960 had been a grave affront just when natural disasters – floods are said by official Chinese sources to have drowned a hundred and fifty million acres of agricultural land – had followed the collapse of an economic offensive launched in 1958 by Mao, the 'Great Leap Forward'. The object of the 'leap' appears to have been to decentralize the economy into 'communes', of which there were some twenty-five thousand, thus repudiating centralized planning on the Russian model with its bureaucratic dangers, as well as calling directly upon the participation of local forces from which the regime had previously benefited. It failed badly. Mao's standing suffered.

His rivals came together to put the economy back on the road to modernization again. One striking symbol was the explosion of a Chinese nuclear weapon in 1964, an expensive admission card to a very exclusive club. It was probably more important, at least immediately, that the regime managed to avoid crippling famine and kept the loyalty of the people. Though more slowly than in the past, the Chinese population had continued to rise to even more colossal totals. Five hundred and ninety million has been thought a reasonable estimate for 1950; twenty-five years later, it was 835 million. Though China's share of world population may have sometimes been higher in the past – perhaps she contained nearly 40 per cent of mankind on the eve of the Taiping rebellion – she was in the 1960s stronger than ever before. Her leaders even talked as if they were unmoved by the possibility of nuclear war; Chinese would survive in greater numbers than the peoples of other countries. There were signs that the USSR was alarmed by the presence of such a demographic mass on the border of her most thinly-populated regions. China, after all, is not only one of the most highly, but also one of the most densely populated countries in the world.

Meanwhile, the revolution seemed not to have lost the dynamism which the party strove to keep alive. Much turned on the fear of what had happened in the USSR, where substantial change and relaxation seemed to follow in the decade after Stalin's death and, as was later to become common knowledge, deep corruption and conservatism gradually took hold of the party and the bureaucrats. The fear that something similar might happen in China lay behind the 'Cultural Revolution' of 1966–9. This huge upheaval, sweeping through party and administration, was an attempt to offset the danger that a new ruling class would emerge, and a consequence also of the re-establishment of Mao's prestige. The cult of Maoism was to be revitalized for a new generation. One way in which Cultural Revolution expressed itself was, oddly, by closing universities. Physical labour was demanded of all citizens in order to change traditional attitudes towards intellectuals. The new emphasis was upon self-sacrifice and subordination to the thought of Mao Tse-tung. By 1968 the country had been turned upside-down by young 'Red Guards' fighting entrenched officialdom and there were signs that Mao himself believed things had gone too far. After the army re-established order and instituted new cadres a party congress reconfirmed his leadership, but he had again failed.

The moral preoccupation of this mysterious episode is very striking. It was, in a way, an attempt (Mao's last, as it turned out) to give his ideas a spiritual meaning for a new generation. Clearly Mao personally felt a danger that the Revolution might congeal and lose the moral élan which had carried it so far. In seeking to protect it, old ideas had to go. This should not seem surprising. Of the great revolutions of world history, the Chinese has been unquestionably one of the most far-reaching. It had to be. Society, government and economy were enmeshed and integrated with one another into a whole system in China as nowhere else. The traditional prestige of intellectuals and scholars still embodied the old order, just as had the examination system whose abolition more than fifty years earlier had been one of the announcements that a real revolution and not just a change of regime was under way in the most unchanging society on earth. The 'demotion' of intellectuals was urged as a necessary consequence of making a new China. It is in this perspective that Chinese communism's achievement and direction stand out from the mystifying welter of events which bemuse the foreign observer. Deliberate attacks on family authority, for example, were not merely attempts by a suspicious regime to encourage informers and delation, but attacks on the most conservative of all Chinese institutions. Similarly, the advancement of women and propaganda to discourage early marriage had dimensions going beyond 'progressive' feminist ideas or population control; they were an assault on the past such as no other revolution had ever made, for in China the past meant a rôle for women far inferior to anything to be found in pre-revolutionary America, France or Russia. The attacks on party leaders which accused them of flirtation with Confucian ideas were much more than the jibes which comparable attacks would have been in western countries; indeed, they could not have occurred in the West, where no vision of a past to be rejected which was so total could exist after centuries of cultural pluralism.

Yet rejection is only half the story. More than two thousand years of a continuity stretching back to the Ch'in and perhaps further also lives behind the Chines Revolution. One clue is the role of authority in it. For all its cost and cruelty, it was a heroic endeavour; in scale it is matched only by such gigantic social efforts as the spread of Islam, or Europe's assault on the world in early modern times. Yet it was different from such movements because it was at least in intention centrally controlled and directed. It is a paradox of the Chinese Revolution that it rests on popular fervour but is unimaginable without conscious direction from a state inheriting all the mysterious prestige of the traditional bearers of the Mandate of Heaven. Chinese tradition respects authority and

'Bombard the Headquarters!' A poster from 1966, launching the Great Proletarian Cultural Revolution.

gives it a moral endorsement which has long been hard to find in the West. No more than any other great state could China shake off its history, and as a result the communist regime sometimes has a paradoxically conservative appearance. No large society had for so long driven home to its peoples the lessons that the individual mattered less than the collective whole, that authority could rightfully command the services of millions at any

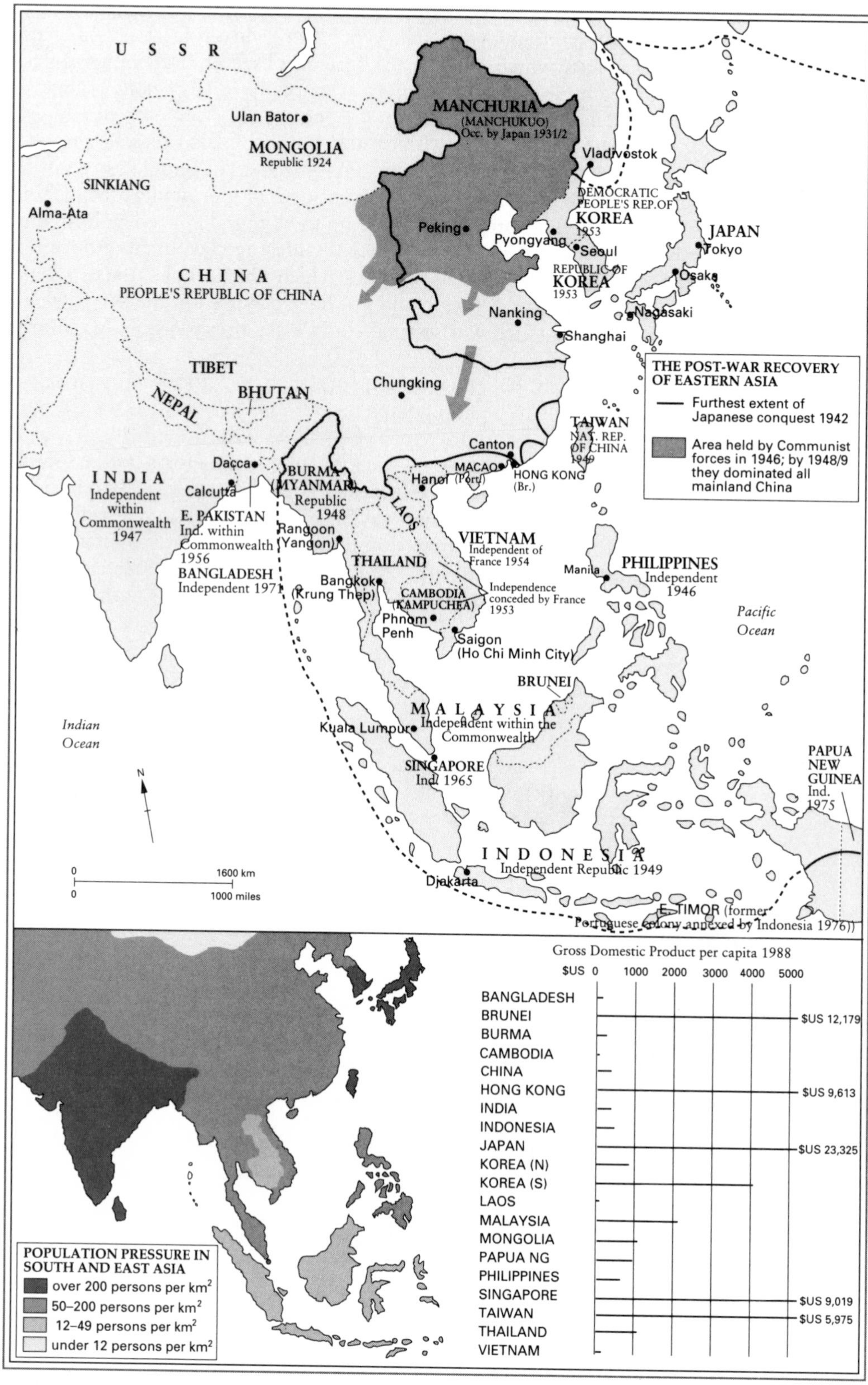

THE POST-WAR RECOVERY OF EASTERN ASIA

Furthest extent of Japanese conquest 1942

Area held by Communist forces in 1946; by 1948/9 they dominated all mainland China

POPULATION PRESSURE IN SOUTH AND EAST ASIA

over 200 persons per km²

50–200 persons per km²

12–49 persons per km²

under 12 persons per km²

Gross Domestic Product per capita 1988

$US 0 1000 2000 3000 4000 5000

BANGLADESH
BRUNEI — $US 12,179
BURMA
CAMBODIA
CHINA
HONG KONG — $US 9,613
INDIA
INDONESIA
JAPAN — $US 23,325
KOREA (N)
KOREA (S)
LAOS
MALAYSIA
MONGOLIA
PAPUA NG
PHILIPPINES
SINGAPORE — $US 9,019
TAIWAN — $US 5,975
THAILAND
VIETNAM

cost to themselves in order to carry out great works for the good of the state, that authority is unquestionable so long as it is exercised for the common good. The notion of opposition is distasteful to the Chinese because it suggests social disruption; that implies the rejection of the kind of revolution involved in the adoption of western individualism, though not of collective radicalism.

Mao fitted Chinese popular tradition. The regime over which he presided benefited from the Chinese past as well as destroying it, because Mao was easily comprehensible within its idea of authority. He was presented as a ruler-sage, as much a teacher as a politician; Western commentators were amused by the status given to his thoughts by the omnipresence of the Little Red Book (in the West after the Protestant Reformation similar extravagant adulation was sometimes given to the Bible), but the utterances of great teachers have always commanded respect in China. Mao was the spokesman of a moral doctrine which was presented as the core of society, just as Confucianism had been. There was also something traditional in Mao's artistic interests; he was admired by the people as a poet and his poems won the respect of qualified judges. In China power has always been sanctioned by the notion that the ruler did good things for his people and sustained accepted values. Mao's actions could be read in such a way.

The weight of the past was evident in Chinese foreign policies, too. Though it came to patronize revolution all over the world, China's main concern was with the Far East and, in particular, with Indo-China, a traditionally tributary country. There, Russian and Chinese policy again diverged. After the Korean War the Chinese began to supply arms to the communist guerrilla forces in Vietnam for what was less a struggle against colonialism – that was decided already – than about what should follow it. In 1953 the French had given up both Cambodia and Laos. In 1954 they lost at a base called Dien Bien Phu a battle decisive both for French prestige and for the French electorate's will to fight. After this, it was impossible for the French to maintain themselves in the Red River delta. A conference at Geneva agreed to partition Vietnam between a South Vietnamese government and the communists who had come to dominate the north, pending elections which might reunite the country. The elections never took place. Instead, there soon opened in Indo-China what was to become the fiercest phase since 1945 of the Asian war against the West begun in 1941.

The western contenders were no longer the former colonial powers (the French had gone home and the British had problems enough elsewhere), but the Americans; on the other side was a mixture of Indo-Chinese communists, nationalists and reformers supported by the Chinese and Russians. American anti-colonialism and the belief that the United States should support indigenous governments led it to back the South Vietnamese as it backed South Koreans and Filipinos. Unfortunately neither in Laos nor South Vietnam, nor, in the end, in Cambodia, did there emerge regimes of unquestioned legitimacy in the eyes of those they ruled. American patronage merely identified them with the western enemy so disliked in east Asia. American support also tended to remove the incentive to carry out reforms which would have united people behind these regimes, above all in Vietnam, where de facto partition did not produce good or stable government in the south. While Buddhists and Roman Catholics quarrelled bitterly and the peasants were more and more alienated from the regime by the failure of land reform, an apparently corrupt ruling class seemed able to survive government after government. This benefited the communists. They sought reunification on their own terms and maintained from the north support for the communist underground movement in the south, the Vietcong.

By 1960 the Vietcong had won control of much of the south. This was the background to a momentous decision taken by the American president, John Kennedy, in 1962, to send not only financial and material help but also 4,000 American 'advisers' to help the South Vietnam government put its military house in order. It was the first step towards what Truman had been determined to avoid, the involvement of the United States in a major war on the mainland of Asia (and, in the end, the loss of more than 50,000 American lives).

Another of Washington's responses to Cold War in Asia had been to safeguard as long as possible the special position arising from the American occupation of Japan. This was virtually a monopoly, with only token participation by British Commonwealth forces. It had been possible because of the Soviet delay in declaring war on Japan and the speed of Japan's surrender. Stalin was taken by surprise. The Americans firmly rejected later Soviet requests for a share in an occupation Soviet power had done nothing to bring about, and the results were to be startling. The United States provided the last great example of western paternalism in Asia and the Japanese once more demonstrated their astonishing gift for learning from others only what they wished to learn while safeguarding their own society against unsettling change.

1945 forced Japan spiritually into a twentieth century she had already entered technologically. Defeat confronted her people with deep and troubling problems of national identity and purpose. The westernization of the Meiji era had led to the dream of 'Asia for the Asians'. In its Japanese version, this idea was now blown away by defeat. It had really been a kind of Japanese Monroe doctrine, underpinned by the anti-western sentiment so widespread in the Far East and cloaking imperialism. The slogan was given meaning in a different sense after 1945 by the rolling back of colonialism. That left Japan with no obvious and creditable Asian role. True, at that moment she seemed unlikely for a long time to have the power for one. Moreover, the war's demonstration of Japan's vulnerability had been a great shock; like the United Kingdom she had depended on control of the surface of the sea, and the loss of it had doomed her. Then there were the other results of defeat; the loss of territory to Russia on Sakhalin and the Kurile islands and the occupation by the Americans. Finally, there was vast material and human destruction to repair.

On the asset side, the Japanese in 1945 still had the unshaken central institution of the monarchy, whose prestige was undimmed and, indeed, had made the surrender possible. Many Japanese saw in the emperor Hirohito the man who had saved them from annihilation. The American commander in the Pacific, General MacArthur, wanted to maintain the monarchy as an instrument of a peaceful occupation. He took care to have a new Japanese constitution adopted before republican enthusiasts in the United States could interfere; he found it effective to argue that Japan should be helped economically in order to get it more quickly off the back of the American taxpayer. Japanese social cohesiveness and discipline was a great help, though for a time it seemed that the Americans might undermine this by the determination with which they pressed democratic institutions upon the country. Some problems must have been eased by a major land reform in which about a third of the cultivated area of Japan passed from landowners' to cultivators' ownership. By 1951 that democratic education and careful demilitarization was deemed adequate for a peace treaty between Japan and most of her former opponents except the Russians and nationalist Chinese (terms with them followed within a few years). Japan regained her full sovereignty, including a right to arms for defensive purposes, but gave up virtually all her former overseas possessions. Thus the Japanese emerged from the post-war era to resume control of their own affairs. An agreement with the United States provided for the maintenance of American forces on her soil. Confined to her own islands, and facing a China stronger and much better consolidated than for a century, Japan's position was not necessarily a disadvantageous one. In less than twenty years this much reduced status was, as it turned out, to be transformed again.

The Cold War had changed the implications of the American occupation even before 1951. Japan was separated from Russians and Chinese by, respectively, 10 and 500 miles of water. Korea, the old area of imperial rivalry, was only 150 miles away. The spread of the Cold War to Asia guaranteed Japan even better treatment from the Americans, now anxious to see her working convincingly as an example of democracy and capitalism, and also gave her the protection of the United States nuclear 'umbrella.' The Korean War made her important as a base, and galvanized the Japanese economy. The index of industrial production quickly went back up to the level of the 1930s. The United States promoted Japanese interests abroad through diplomacy. Finally, Japan long had no defence costs, since she was until 1951 forbidden to have any armed forces.

Japan's close connexion with the United States, proximity to the communist world, and her advanced and stable economy and society, all made it natural that she should eventually take her place in the security system built up by the United States in Asia and the Pacific. Its foundations were treaties with Australia, New Zealand and the Philippines (which had become independent in 1946). Others followed with Pakistan and Thailand; these were the Americans' only Asian allies other than Taiwan. Indonesia and (much more important) India remained aloof. These alliances reflected, in part, the new conditions of Pacific and Asian international relations after the British withdrawal from India. For a little longer there would still be British forces east of Suez, but Australia and New Zealand had discovered during the Second World War that the United Kingdom could not defend them and that the United States could. The fall of Singapore in 1942 had been decisive. Though British forces still sustained the Malaysians against the Indonesians in the 1950s and 1960s, their important colony at Hong Kong survived, they knew, only because it suited the Chinese that it should. On the other hand, there was no question of sorting out the complexities of the new Pacific by simply lining up states in the teams of the Cold War. The peace treaty with Japan itself caused great difficulty because United States policy saw Japan as a potential anti-communist force while others – notably Australia and New Zealand – remembered 1941 and feared a revival of Japanese power.

Thus American policy was not made only by ideology. None the less, it was long misled by what was believed to be the disaster of the communist success in China and by Chinese patronage of revolutionaries as far away as Africa and South America. There had certainly been a transformation in China's international position and it would go further. Yet the crucial fact was China's re-emergence as a power in her own right and in the end this did not reinforce the dualist, Cold War system but made nonsense of it. Though at first only within the former Chinese sphere, it was bound to bring about a big change in relative power relationships; the first sign of this was seen in Korea, where the United Nations armies were stopped and it became necessary to consider bombing China. But the rise of China was also of the most acute importance to the Soviet Union which from one pole of a bipolarized system became the corner of a triangle, as well as losing its unchallenged pre-eminence in the world revolutionary movement. And it was in relation to the Soviet Union, perhaps, that the wider significance of the Chinese Revolution most readily appeared. Overwhelmingly the most important though it might be, the Chinese Revolution was only the outstanding instance of a rejection of western domination which was Asia-wide. Paradoxically, of course, that rejection in all Asian countries was most obviously expressed in forms, language and mythology borrowed from the West itself, whether they were those of industrial capitalism, nationalism or Marxism.

INHERITORS OF THE EMPIRE: THE MIDDLE EAST AND AFRICA

The survival of Israel, the coming of the Cold War and a huge rise in the demand for oil revolutionized the politics of the Middle East after 1948. Israel focused Arab feeling more sharply than Great Britain had ever done. It enabled pan-Arabism to look plausible. On the injustice of the seizure of what were regarded as Arab lands, the plight of the Palestine refugees and the obligations of the great powers and the United Nations to act on their behalf, the Arab masses could brood bitterly and Arab rulers were able to agree as on nothing else.

None the less, after the defeat of 1948-9, the Arab states were not for some time disposed again to commit their own forces openly. A formal state of war persisted but a series of armistices established for Israel de facto borders with Jordan, Syria and Egypt which lasted until 1967. There were continuing border incidents in the early 1950s, and raids were carried out upon Israel from Egyptian and Syrian territory by bands of young guerrilla soldiers recruited from the refugee camps, but immigration, hard work and money from the United States steadily consolidated the new Israel. A siege psychology helped to stabilize Israel's politics; the prestige of the party which had brought about the very existence of the new state was scarcely troubled while the Jews transformed their new country. Within a few years they could show massive progress in bringing barren land under cultivation and establishing industries. The gap between Israel's *per capita* income and that of the more populous among the Arab states steadily widened.

Here was another irritant for the Arabs. Foreign aid to their countries produced nothing like such dramatic change. Egypt, the most populous of them, faced particularly grave problems posed by high rates of population growth. While the oil-producing states were to benefit from growing revenue and a higher GDP, this often led to further strains and divisions. Contrasts between different Arab states, and between classes within them both deepened. Most of the oil-producing countries were

In 1954 an attempt was made on the life of Gamal Nasser, already the hero of Egyptian nationalists and a rising star in the political firmament of the whole Arab world. The photograph was taken as he was on his way back to Cairo after the incident.

ruled by small, wealthy, sometimes traditional and conservative, sometimes nationalist and westernized, élites, usually uninterested in the poverty-stricken peasants and slum-dwellers of more populous neighbours. The contrast was exploited by a new Arab political movement, founded during the war, the Ba'ath party. It attempted to synthesize Marxism and pan-Arabism, but the Syrian and Iraqi wings of the movement (it was always strongest in those two countries) had fallen out with one another almost from the start.

Pan-Arabism had much to overcome, for all the impulse to united action stemming from anti-Israeli and anti-western feeling. The Hashemite kingdoms, the Arabian sheikdoms, and the Europeanized and urban states of North Africa and the Levant all had widely divergent interests and very different historical traditions. Some of them, like Iraq or Jordan, were artificial creations whose shape was dictated by the needs and wishes of European powers after 1918; some were social and political fossils. Even Arabic was in many places a common language only within the mosque (and not all Arabic-speakers were Moslems). Though Islam was a tie between many Arabs, it for a long time seemed little more; in 1950 few Moslems talked of it as a militant, aggressive faith. It was only the Israeli issue which provided a common enemy. Hopes were awoken among Arabs in many countries, nevertheless, by a revolution in Egypt from which there eventually emerged a young soldier, Gamal Abdel Nasser. For a time he seemed likely both to unite the Arab world against Israel and to open the way to social change. In 1954 he became the leader of the military junta which had overthrown the Egyptian monarchy two years previously. Egyptian nationalist feeling had for decades found its main focus and scapegoat in the British, still garrisoning the country, and now blamed for their part in allowing the establishment of Israel. The British government, for its part, did its best to cooperate with Arab rulers because of its fears of Russian influence in an area still thought crucial to British communications and oil supplies. The Middle East, ironically (given the motives which had taken the British there in the first place), did not seem to matter less after withdrawal from India.

It was a time of strong anti-western currents elsewhere in the Arab world. In 1951 the King of Jordan had been assassinated; in order to survive, his successor had to make it clear that he had severed the old special tie with Great Britain. Further west, the French, who had been forced to recognize the complete independence of Morocco and Tunisia soon after the war, faced in 1954 the start of an Algerian national rebellion. It was to become a full-scale war; no French government could easily abandon a country where there were over a million settlers of European stock. Moreover oil had just been discovered in the Sahara. In such a context Nasser's rhetoric of social reform and nationalism had wide appeal. His anti-Israeli feelings were not in doubt and he quickly had to his credit the success of an agreement with Great Britain for the evacuation of the Suez base. The Americans, too, increasingly aware of Russian menace in the Middle East, looked on him for a while with favour as a spotless anti-colonialist and potential client.

He soon came to appeal to them less. The guerrilla raids on Israel from Egyptian territory (the 'Gaza Strip', where many Palestinian refugee camps lay) provoked irritation in Washington. In 1950, the British, French and Americans had already said they would provide only limited supplies of arms to the Middle East states and only on such conditions as would keep a balance between Israel and the Arabs. When Nasser made an arms deal with Czechoslovakia and Egypt recognized communist China, second thoughts about him hardened. By way of showing displeasure, an American and British offer to finance a cherished development project, a high dam on the Nile, was withdrawn. As a riposte, Nasser seized the assets of the private company which owned and ran the Suez Canal, saying its profits should finance the dam; this touched an old nerve of British imperial sensibility. Instincts only half-disciplined by imperial withdrawal seemed for once to be coherent both with anti-communism and with friendship towards more traditional Arab states whose rulers were beginning to look askance at Nasser as a revolutionary radical. The British prime minister, too, was obsessed with a false analogy which led him to see Nasser as a new Hitler, to be checked before he embarked upon a career of successful aggression. As for the French, they were aggrieved by Nasser's support for the Algerian insurrection. Both

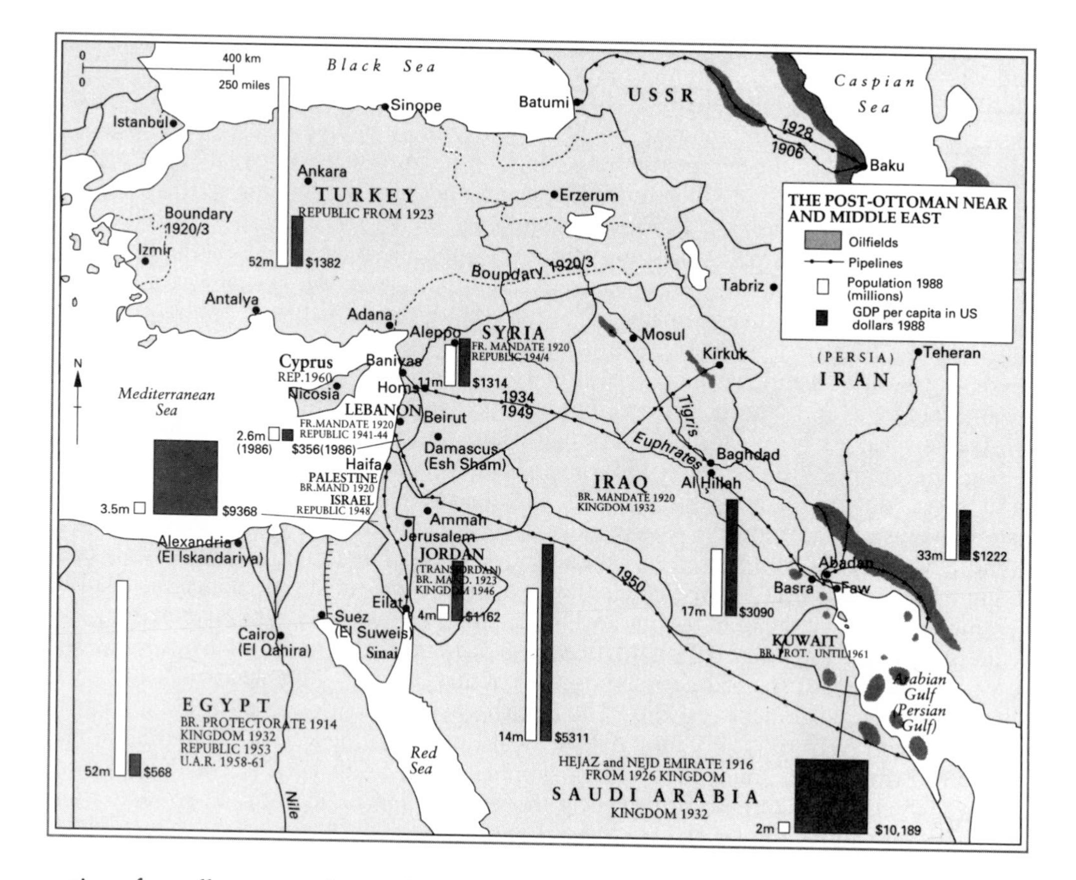

nations formally protested over the Canal's seizure and, in collusion with Israel, began to plan Nasser's overthrow.

In October 1956, the Israelis suddenly invaded Egypt to destroy, they announced, bases from which guerrillas harassed their settlements. The British and French governments at once said freedom of movement through the Canal was in danger. They called for a cease-fire; when Nasser rejected this they launched (on Guy Fawkes' Day) first an air attack and then a seaborne assault on Egypt. Collusion with Israel was denied but the denial was preposterous. It was a lie, and, worse still, from the first incredible. Soon, the Americans were thoroughly alarmed; they feared advantage for the USSR in this renewal of imperialism. They used financial pressure to force a British acceptance of a cease-fire negotiated by the United Nations. The Anglo-French adventure ended in humiliation.

The Suez affair looked (and was) a western disaster, but in the long run its main importance was psychological. The British suffered most; it cost them much goodwill, particularly within the Commonwealth, and squandered confidence in the sincerity of their retreat from empire. It confirmed the Arabs' hatred of Israel; the suspicion that she was indissolubly linked to the West made them yet more receptive to Soviet blandishment. Nasser's prestige soared still higher. Some thought, too, that Suez had badly distracted the West at a crucial moment from eastern Europe (where a revolution in Hungary against its Soviet satellite government had been crushed by the Russian army while the western powers quarrelled with one another). Nevertheless, the essentials of the region's affairs were left by the crisis much as before, animated by a new wave of pan-Arab enthusiasm though they might be. Suez did not change the balance of the Cold War, or of the Middle East.

In 1958 an attempt was made by Ba'ath sympathizers to unite Syria and Egypt in a United Arab Republic. It was briefly to bear fruit in 1961. The pro-western government of the Lebanon was overthrown that year, and the monarchy of Iraq was swept aside by revolution. But differences between Arab countries soon reasserted themselves. The world watched curiously when American forces were summoned to the Lebanon and British to Jordan to help maintain their governments against pro-Nasser forces. Meanwhile, fighting went on sporadically on the Syrian-Israeli border, though the guerrillas were for a time held in check. From Suez to 1967 the most important development in the Arab world was none of these, but the Algerian revolution. The intransigence of the French settlers and the bitterness of many soldiers who felt they were asked to do an impossible job there nearly brought about a coup détat in France itself. The government of General de Gaulle nevertheless opened secret negotiations with the Algerian rebels and in July 1962, after a referendum, France formally granted independence to a new Algeria. As Libya had emerged from United Nations trusteeship to independence in 1951, the entire North African coast outside the tiny Spanish enclaves was now clear of European supremacy. Yet external influences still bedeviled the history of the Arab lands as they had done ever since the Ottoman conquests centuries before, but now indirectly, through aid and diplomacy, as the United States and Russia sought to buy friends.

The United States laboured under a disadvantage: no American president or Congress could abandon Israel. The importance of Jews among American voters was too great, though President Eisenhower had been brave enough to face them down over Suez, even in an election year. In spite of America's clean hands therefore, Egyptian and Syrian policy continued to sound anti-American and prove irritating. The USSR, on the other hand, had dropped its early support of Israel as soon as it ceased to be a useful weapon with which to embarrass the British. Soviet policy now took a steady pro-Arab

The Six-Day war recovered for the Israeli those parts of the city of Jerusalaem previously held by the Jordanians. Once more Jews were able to pray at the Wailing Wall, the last substantial portion of the ancient Temple to survive.

line and assiduously fanned Arab resentment over survivals of British imperialism in the Arab world. Marginally, too, the Russians earned a cheap bonus of Arab approval in the later 1960s by harassing their own Jews.

Meanwhile the terms of the Middle Eastern problem were slowly changing because of oil. In the 1950s there were two important developments. One was a much greater rate of oil discovery than hitherto, particular on the southern shores of the Persian Gulf, in the small sheikdoms then still under British influence, and in Saudi Arabia. The second was a huge acceleration of energy consumption in western countries, especially in the United States. The prime beneficiaries of the oil boom were Saudi Arabia, Libya, Kuwait and, some way behind, Iran and Iraq, the established major producers. This had two important consequences. Countries dependent upon Middle Eastern oil – the United States, Great Britain, Germany and, soon, Japan – had to give greater weight to Arab views in their diplomacy. It also meant big changes in the relative wealth and standing of Arab states. None of the three leading oil producers was either heavily populated or traditionally very weighty in international affairs.

The importance of the oil factor was still not very evident in the last Middle East crisis of the 1960s, which began when a much more extreme government took power in Syria with Russian support. The King of Jordan was threatened if he did not act in support of the Palestinian guerrillas (organized since 1964 as the Palestine Liberation Organization, or PLO). Jordanian forces therefore began to prepare to join in an attack on Israel with Egypt and Syria. But in 1967, provoked by an attempt to blockade their Red Sea port, the Israelis struck first. In a brilliant campaign they destroyed the Egyptian air force and army in Sinai and hurled back the Jordanians, winning in six days' fighting new borders on the Suez Canal, the Golan Heights, and the Jordan. For defence, these were far superior to their former boundaries and the Israelis announced that they would keep them. This was not all. Defeat had ensured the eclipse of the glamorous Nasser, the first plausible leader of pan-Arabism. He was left visibly dependent on Russian power (a Soviet naval squadron arrived at Alexandria as the Israeli advance-guards reached the Suez Canal), and on subsidies from the oil states. Both demanded more prudence from him, and that meant difficulties with the radical leaders of the Arab masses.

Yet the Six Day War of 1967 had solved nothing. There were new waves of Palestinian refugees; by 1973 about 1,400,000 Palestinians were said to be dispersed in Arab countries, while a similar number remained in Israel and Israeli-occupied territory. When the Israelis began to establish settlements in their newly-won conquests, Arab resentment grew even stronger. Even if time, oil, and birth rates seemed to be on the Arab side, not much else was clear. In the United Nations, a 'Group of 77' supposedly non-aligned countries achieved the suspension of Israel (like South Africa) from certain international organizations and, perhaps more important, a unanimous resolution condemning the Israeli annexation of Jerusalem. Another called for Israel's withdrawal from Arab lands in exchange for recognition by her neighbours. Meanwhile, the PLO turned to terrorism outside the disputed lands to promote their cause. Like the Zionists of the 1890s, they had decided that the western myth of nationality was the answer to their plight: a new state should be the expression of their nationhood, and like Jewish militants in the 1940s, they determined that assassination and indiscriminate murder would be their weapons. It was clear that in time there would be another war, and therefore a danger that, because of the identification of American and Russian interests with opposing sides, a world war might suddenly blow up out of a local conflict, as in 1914.

The danger became imminent when Egypt and Syria attacked Israel on the Jewish holy day of Yom Kippur, in October 1973. The Israelis for the first time faced the prospect of military defeat by the greatly improved and Soviet-armed forces of their opponents. Yet once again they won, though only after the Russians were reported to have sent nuclear weapons to Egypt and the Americans had put their forces on the alert round the world. This grim background, like the possibility that the Israelis themselves might have nuclear weapons they would be prepared to use in extremity, was not fully discernible to the public at the time. More immediately obvious was the impact of the acts of other Arab states, led by Saudi Arabia. By announcing restrictions on oil supply to Europe, Japan and the United States they raised the possibility that the diplomatic support on which Israel had always been able to rely might not be available to her for ever: Israel might not always be able to count on guilt about the Holocaust, on sympathy and admiration for a civilized and liberal state in a backward region, and on the weight of Jewish votes in the United States. Though for the moment a UN force was put into Sinai to separate the Israelis and Egyptians, none of the region's fundamental problems had been solved.

Nor was this all. The impact of the new 'oil diplomacy' had been immediate: oil prices shot up. A major change in world relationships had taken place. In the 1950s and most of the 1960s the United States and United Kingdom-had been able to assure stable and cheap oil supplies through their informal influence on Iraq (down to 1963, when a Ba'ath regime seized power), the Gulf States and Saudi Arabia. In the 1970s this broke down under the strain, primarily, of the Israeli question. Overnight, economic problems which had gone grumbling along but were tolerable in the 1960s became acute. Dependence on oil imports played havoc with balance-of-payment problems. Even the United States was badly shaken, while Japan and Europe were soon showing signs of economic recession. There was even talk of a new world depression like that of the 1930s. The golden age of economic growth which had begun with post-war recovery came to an end, and poorer countries which depended on imported oil were the worst hit of all. Many of them were soon facing renewed price inflation and some a virtual obliteration of the surplus of earnings they needed in order to pay interest on their large debts to foreign creditors.

This was so in many African countries. In the 1950s and 1960s Africa underwent a dramatic decolonization which left some fragile new states behind. This was especially true south of the Sahara. In spite of the long European influence in the former Ottoman lands, direct European rule had always been less radically innovative in the Islamic lands of the Atlantic and Mediterranean coasts than in Black Africa, where the map and the legal status of most of the area was now transformed in two decades. France and Great Britain were the major colonial powers concerned in what was for a long time a surprisingly peaceful process (Italy had lost her last African territories in 1943). Only in Algeria did the making of a decolonialized Africa cost much bloodshed (though there was plenty to come in the post-colonial era, when African set about African). Though Portugal only gave in after a domestic revolution in 1974, colonialism elsewhere in Africa was replaced fairly peacefully well before that date. Both French and British politicians were anxious to retain, if they could, some sort of influence by ostentatiously benevolent interest in their former subjects; settlers, rather than imperial rulers, were the usual brake on withdrawal.

As a result, black Africa owes its present form in the main to decisions of nineteenth-century Europeans (just as much of the Middle East owes its framework to their successors in this century). New African 'nations' were usually defined by the boundaries of former colonies. These boundaries often enclosed peoples of many languages, stocks and customs, over whom colonial administrations had provided little more than a formal unity. As Africa lacked the unifying influence of great indigenous civilizations such as those of Asia to offset the colonial fragmentation of the continent, imperial withdrawal from the continent was followed by its Balkanization. The doctrine of nationalism which appealed to the westernized African élites (Senegal, a Moslem country, had a president who was a writer of poetry in French and an expert on Goethe) confirmed a continent's fragmentation, often ignoring important realities which colonialism had contained or manipulated. The sometimes strident nationalism of new rulers was often a response to the dangers of centrifugal forces. West Africans combed the historical record – such as it was – of ancient Mali and Ghana, and east Africans brooded over the past which might be hidden in relics such as the ruins of Zimbabwe in order to forge national mythologies like those of earlier nation-makers in Europe as they strove to find unifying and rallying influences.

In 1966 the British colony Beuchuanaland became independent. The Union Jack was hauled down in a ceremony at which British administrators were by then already practiced hands. Its successor in this case was the flag of Botswana, Africa's fortieth independent state at that time.

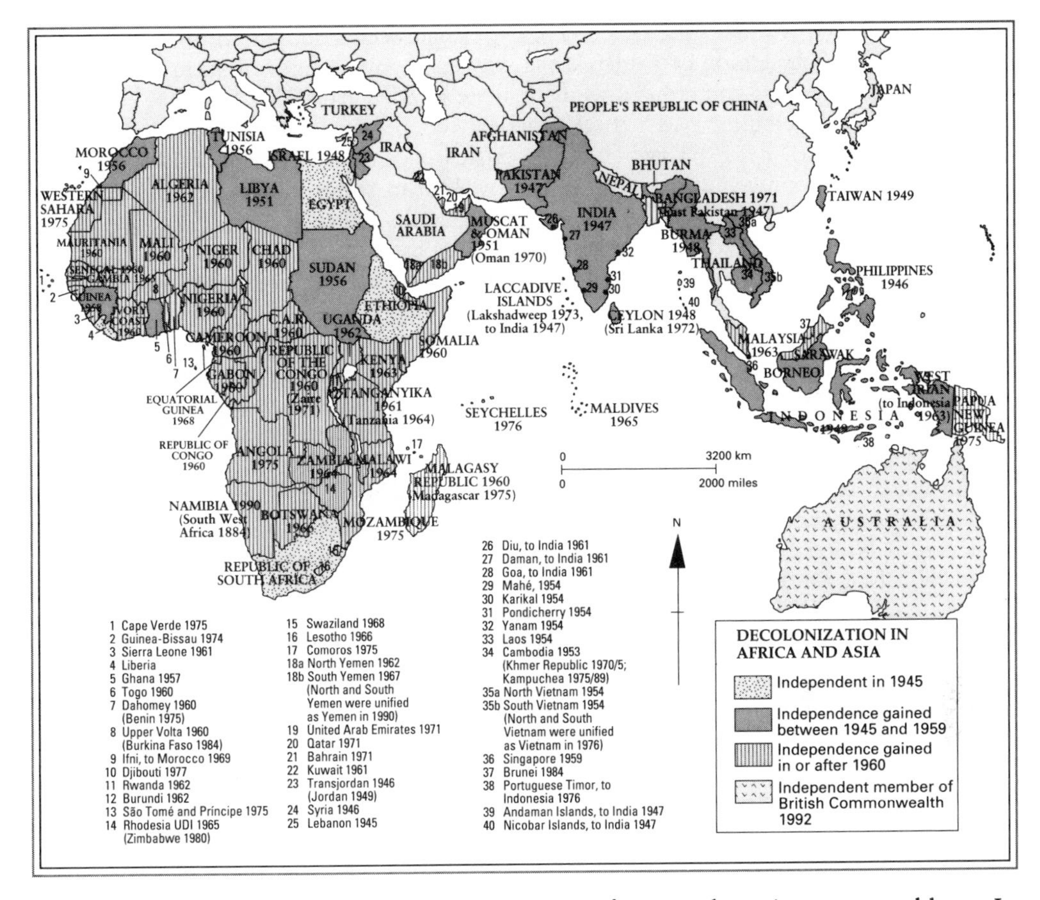

Its new internal divisions were not Africa's only or its worst problems. In spite of the continent's great economic potential, the economic and social foundations for a prosperous future were shaky. Once again, the imperial legacy was significant. Colonial regimes in Africa left behind feebler cultural and economic infrastructures than in Asia. Rates of literacy were low and trained cadres of administrators and technical experts were small. Africa's important economic resources (especially in minerals) required for their exploitation skills, capital and marketing facilities which could only come in the near future from the world outside (and white South Africa long counted as 'outside' to many black politicians). What was more, some African economies had only recently undergone disruption and diversion because of European needs and in European interests. During the war of 1939-45, agriculture in some of the British colonies had shifted towards the growing of cash crops on a large scale for export. Whether this was or was not in the long-term interests of peasants who had previously raised crops and livestock only for their own consumption is debatable, but what is certain is that the consequences were rapid and profound. One was an inflow of cash in payment for produce the British and Americans needed. Some of this came through in higher wages, but the spread of a cash economy often had disturbing local effects. Unanticipated urban growth and regional development took place. Many African countries were thus tied to a particular pattern of development which was soon to show its vulnerabilities and limitations in the post-war world. Even the benevolent intentions of a programme like the British Colonial Development and Welfare Fund, or many international aid programmes, unintentionally helped to shackle African producers to a world market. In such circumstances, as populations rose more and more rapidly after 1960 and as disappointment with the reality of 'freedom' from the colonial powers set in, discontent was inevitable.

None the less, in spite of the difficulties experienced along the way, the process of decolonization in black Africa was hardly interrupted. In 1945 the only wholly independent countries in Africa other than Egypt had been Ethiopia and Liberia, though in reality and law the Union of South Africa was a self-governing Dominion of the British Commonwealth and is therefore only formally excluded from this category (a slightly vaguer status also cloaked the virtual practical independence of the British colony of Southern Rhodesia). By 1961 (when South Africa became a fully independent republic and left the Commonwealth), twenty-four new African states had come into existence. There are now over fifty. Ten years later, Portugal was the only former colonial power still hanging on to black African possessions and by the end of 1975 they had gone, too. The Iberians who had led the European adventure of overseas dominion were almost the last to abandon it.

As Africans shook off colonialism, dangers quickly came to the surface. Ghana, in 1957, was the first ex-colonial new nation to emerge in sub-Saharan Africa. In the next 27 years twelve wars were to be fought in Africa and thirteen heads of state would be assassinated. There were two especially bad outbreaks of strife. In the former Belgian Congo an attempt by the mineral-rich region of Katanga to break away provoked a civil war in which rival Soviet and American influences quickly became entangled, while the United Nations strove to restore peace. Then, at the end of the 1960s, came a distressing civil war in Nigeria, hitherto one of the most stable and promising of the new African states, which again drew non-Africans to dabble in the blood-bath (one reason was that Nigeria had joined the ranks of the oil producers). In other countries, there were less bloody but still fierce struggles between factions, regions and tribes which distracted the small westernized élites of politicians and quickly led them to abandon democratic and liberal principles much talked of in the heady days when a colonial system was in retreat.

In many of the new nations, the need, real or imaginary, to prevent disintegration, suppress open dissent and strengthen central authority, had led by the 1970s to one-party, authoritarian government or to the exercise of political authority by soldiers (it was not unlike the history of South America after the Wars of Liberation). Nor did the surviving regimes of an older independent Africa escape. Impatience with an ancien régime seemingly incapable of providing peaceful political and social change led in 1974 to revolution in Ethiopia. The setting aside of the 'Lion of Judah' was almost incidentally the end of the oldest Christian monarchy in the world (and of a line of kings supposed in one version of the story to run back to the son of Solomon and the Queen of Sheba). A year later, the soldiers who had taken power seemed just as discredited as their predecessors. From similar changes elsewhere in Africa there sometimes emerged tyrant-like political leaders who reminded Europeans of earlier dictators, but this comparison may be misleading. Africanists have gently suggested that many of the 'strong men' of the new nations can be seen as the inheritors of the mantle of pre-colonial African kingship, rather than in western terms. Some were simply bandits, though.

Their own troubles did not diminish the frequent irritation with which many Africans reacted to the outside world. Some of the roots of this may lie very deep, in (for example) the mythological drama built on the old European slave trade, which Africans were encouraged to see as a supreme example of racial exploitation, or in the sense of political inferiority which lay near the surface in a continent of relatively powerless states (some with populations of less than a million). In political and military terms, a disunited Africa could not expect to have much weight in international affairs. Attempts were therefore made to overcome the weakness which arose from division. An abortive attempt in 1958 to found a United States of Africa opened an era of alliances, partial unions, and essays in federation which had culminated in the emergence in 1963 of the Organization for African Unity (the OAU), largely thanks to the Ethiopian emperor, Haile Selassie.

Politically, the OAU had little success, though it concluded in 1975 a beneficial trade negotiation with Europe in defence of African producers. The very disappointment of much of the early political history of independent Africa directed thoughtful politicians towards cooperation in economic development, above all in relation to Europe, whose former colonial powers remained Africa's most important source of capital, skill and counsel. Unfortunately, the economic record of black Africa has been dreadful. Unsuitable schemes have led to agricultural decline. Political concern with urban voters, corruption and prestigious investment have played havoc with commercial and industrial policy. Meanwhile, population inexorably rises, and famine has inexorably recurred. The onset of world economic recession after the 1973 oil revolution had a shattering effect on Africa, one worsened within a few years by the impact of repeated drought. In Black Africa annual *per capita* growth in GDP has turned downward since 1960, and fell by 1.7 per cent from 1980 to 1985.

Against this background, political cynicism flourished and the leaders of the independence era often seemed to lose their way. Yet they showed an almost total lack of self-criticism and often expressed their frustration in the encouragement of new resentments (sometimes exacerbated by external attempts to entangle Africans in the Cold War). These could be disappointing, too. Marxist revolution had little success. Paradoxically, it was only in Ethiopia, most feudally backwards of independent African states, and the former Portuguese colonies, the least-developed former colonial territories, that formally Marxist regimes took root. Former French and British colonies were hardly affected.

Scapegoats, inevitably, were sought. Increasingly, but perhaps explicably, given the completeness and rapidity of decolonization in Africa and the geographical remoteness of much of it, they tended to be found at hand; resentments came to focus more on the racial division of black and white in Africa itself. This was flagrant in the most powerful of African states, the Union of South Africa.

The Afrikaans-speaking Boers who by 1945 dominated that country cherished against the British grievances which went back to the Great Trek, had been intensified by defeat in the Boer War and had then led to the progressive destruction of ties with the British Commonwealth after the First World War. This had been made easier by the concentration of voters of Anglo-Saxon origin in the provinces of Cape Town and Natal; the Boers were entrenched in the Transvaal and the major industrial areas as well as the rural hinterland. South Africa, it is true, entered the war in 1939 on the British side and supplied important forces to fight in it, but even then intransigent 'Afrikaners', as they increasingly called themselves, supported a movement favouring cooperation with the Nazis. Its leader became prime minister in 1948, after defeating South Africa's senior statesman, Jan Smuts, in a general election. As the Afrikaners had steadily engrossed power inside the Union, and had built up their economic position in the industrial and financial sectors, the prospect of imposing a policy towards the black African which diverged from their deep prejudices was soon inconceivable. The eventual result was the construction of a system of separation of the races: *apartheid*. It systematically embodied and reinforced the reduction of the black African to the inferior status he occupied in Boer ideology. Its aim was to guarantee the position of the whites in a land where industrialism and market economies had done much to break down the regulation and distribution of the growing black population by the old tribal divisions.

Apartheid had some appeal – on even less excusable grounds than the primitive superstitions or supposed economic necessities of the Afrikaners – to white people elsewhere in Africa. The only country where a similar balance of black and white population to that of South Africa and a similar concentration of wealth existed was Southern Rhodesia which, to the great embarrassment of the British government, seceded from the Commonwealth in 1965. The aim of the secessionists, it was feared, was to move towards a society more and more like South Africa's. The British government dithered and missed its chance. There was nothing that the black African states could immediately do about Rhodesia, and not much that the United Nations could do either, though 'sanctions' were invoked in the form of an embargo on trade with the former colony; many black African states ignored them and the British government winked at the steps taken by major oil companies to ensure their product reached the rebels. In one of the most shameful episodes in the history of a feeble government Great Britain's stock sank in the eyes of Africans who, understandably, did not see why a British government could not intervene militarily to suppress a colonial rebellion as flagrant as that of 1776. Many British reflected that precisely that precedent made the outlook for intervention by a remote and militarily weak imperial sovereign look discouraging.

Though South Africa (the richest and strongest state in Africa, and growing richer and stronger all the time) seemed secure, she was, together with Rhodesia and the Portuguese colonies, the object of mounting black African anger as the 1970s began. The drawing of the racial battlelines was hardly offset by minor concessions to South Africa's blacks and her growing economic ties with some black states. There was a danger, too, that outside powers might soon be involved. In 1975, after the Portuguese withdrawal from Angola, a Marxist regime took power there. When civil war followed, foreign communist soldiers arrived to support the government, while South African support was soon given to rebels against it.

The South African government soon showed it was taking thought. It sought to detach itself from the embarrassment of association with an unyielding independent Rhodesia (whose prospects had sharply worsened when Portuguese rule also came to an end in Mozambique and a guerrilla campaign was launched from it against her). The American government contemplated the outcome if Rhodesia collapsed at the hands of black nationalists depending on communist support. It applied pressure to the South Africans, who, in turn, applied it to the Rhodesians. In September 1976 the Rhodesian prime minister sadly told his countrymen that they had to accept the principle of black majority rule. The last attempt to found an African country dominated by whites had failed. It was another landmark in the recession of European power. Yet the guerrilla war continued, worsening as black nationalists sought to achieve unconditional surrender. At last, in 1980 Rhodesia briefly returned to British rule before re-emerging into independence, this time as the new nation of Zimbabwe, with a black prime minister.

This left South Africa alone as the sole white-dominated state in the continent, the richest of its economies, and the focus of black (which, in this context, meant non-white) resentment the world around. Although the OAU had been split by civil war in Angola, African leaders could usually find common ground against South Africa. In 1974 the General Assembly of the United Nations forbade South Africa to attend its sessions because of *apartheid*, and in 1977 the UN Commission of Human Rights deftly side-stepped demands for the investigation of the horrors perpetrated by black against blacks in Uganda, while castigating South Africa (along with Israel and Chile) for its misdeeds. From Pretoria, the view northwards looked more and more menacing. The arrival of Cuban troops in Angola showed a new power of strategic action against South Africa by the USSR. Both that former Portuguese colony and Mozambique also provided bases for South African dissidents who fanned unrest in the black townships and sustained urban terrorism in the 1980s.

These were no doubt among the factors which brought about changes in the position of the South African government. By the middle of that decade, the issue seemed to be no longer whether the more obnoxious features of *apartheid* should be dismantled, but whether black majority rule could be conceded by South African whites, and whether it could happen without armed conflict. A change was apparent when a new prime minister took office in 1978. To the dismay of many Afrikaners, Mr P.W. Botha began slowly to unroll a policy of concession. Before long, though, his initiative slowed; continuing signs of hostility to South Africa in the United Nations, urban terrorism at home, an increasingly-dangerous and militarily-demanding situation on the northern frontiers in Namibia (allocated to South Africa years before as a UN trusteeship territory), and increased distrust of Botha among his Afrikaner supporters (shown in elections), all led him back towards repression. His last gesture to relaxation was a new constitution in 1983 which provided representation for non-white South Africans in a way which outraged black political leaders by its inadequacy, and disgusted white conservatives by conceding the principle of non-white representation at all.

Meanwhile, the pressure of what were called 'sanctions' against South Africa by other countries was growing. In 1985 even the United States imposed them to a limited extent; by then, international confidence in the South African economy was falling, and the effects were showing at home. Straws before the wind of change in domestic opinion could be discerned in the decision of the Dutch Reformed Church that *apartheid* was at least a 'mistake' and could not (as had been claimed) be justified by Scripture, and in growing division among Afrikaner politicians. It probably helped, too, that in spite of its deepening isolation, South African military action successfully mastered the border threats, though it was incapable of defeating the Angola government so long as Cuban forces remained there. In 1988 Namibia came to independence on terms South Africa found satisfactory and peace was made with Angola.

This was the background against which Mr Botha (President of the republic since 1984) reluctantly and grumpily stepped down in 1989 and Mr F.W.de Klerk succeeded him. He soon made it clear that the movement towards liberalization was to continue and would go much further than many thought possible, even if this did not mean the end of *apartheid* in all respects. Political protest and opposition were allowed much more freedom. Meetings and marches were permitted, imprisoned black Nationalist leaders were released. Meanwhile, an important change in the relations between the superpowers had produced agreement between the United States and the Soviet Union over ending the struggles in Angola and Mozambique and giving freedom to Namibia.

Suddenly, the way ahead opened up dramatically. In February 1990 Mr de Klerk announced a 'new South Africa'. Nine days later the symbolic figure of Mr Nelson Mandela, leader of the African National Congress, emerged at last from jail. He was before long engaged in discussion with the government about what might come next. For all the intransigence of his language, there were hopeful signs of a new realism that the task of reassuring the white minority about a future under a black majority must be attempted. Just such signs, of course, prompted other black politicians to greater impatience.

1990: a scene startling to many South Africans, black and white alike – Mr. Mandela meets the South African Prime Minister, Mr. F. W. de Klerk, and the Foreign Minster, Mr. R. ('Pik') Botha.

By the end of 1990 Mr de Klerk had gone a long way. He had taken his followers further than Mr Mandela had taken his. He had even said he would rescind the land legislation which was the keystone of *apartheid*. In 1991, at last, the other *apartheid* laws were repealed. It was an interesting indicator of the pace with which events had moved in South Africa that the interest of the world was focused by then less on the sincerity or insincerity of white South African leaders, than on the realism (or lack of it) of their black counterparts and their ability (or inability) to control their followers. The hopes surrounding Mr Mandela at the time of his release had soon given way to misgivings and there were plentiful signs of division between black South Africans. It was only clear that a stony path still lay ahead, even if once unthinkable steps towards a democratic South Africa had already been taken.

LATIN AMERICA

By 1900, some Latin American countries were beginning to settle down not only to stability but to prosperity. To the original colonial implantation of culture had been added the influence of nineteenth-century Europe, especially of France, to which Latin American élites had been drawn in the post-colonial period. Their upper classes were highly Europeanized and the modernity of many of the continent's great cities reflected this, as their populations reflected recent European immigration which was swallowing the old colonial élites. As for the descendants of the aboriginal Americans, they were hardly to be taken into account. Their suppression had been so complete in one or two countries as to approach even their extinction.

Almost all Latin American states were primary producers, whether of agricultural or mineral resources. Although some were relatively highly urbanized, their manufacturing sectors were inconsiderable, and for a long time did not seem to be troubled by the social and political problems of nineteenth-century Europe. Capital had flowed into the continent, only briefly and occasionally checked by periodic financial disasters and disillusionments. The only social revolution in a Latin American state before 1914 (as opposed to countless changes in governmental personnel) overthrew the Mexican dictator Porfirio Diaz in 1911. It opened the way to nearly ten years' fighting and a million deaths, but the primary role in it was played by a middle class which felt excluded from the benefits of the regime, not by an industrial proletariat. Latin American countries did not display the class conflict of industrialized Europe in spite of large-scale nineteenth-century immigration, though in rural areas there was class conflict aplenty.

These promising-looking societies survived the First World War prosperously. It brought important changes in their relations with Europe and North America. Before 1914, though she was the predominant political influence in the Caribbean, the United States did not yet exercise much economic weight in South America's affairs. In 1914 she supplied only seventeen per cent of all foreign investment south of the Rio Grande; Great Britain was far ahead of her. The liquidation of British holdings in the Great War changed this; by 1929 the United States was the largest source of investment in South America, providing about forty per cent of the continent's foreign capital. Then came the world economic crisis; 1929 was the doorway to a new and unpleasant era for the Latin American states, the true beginning of their twentieth century. Many defaulted on their payments to foreign investors. It became almost impossible to borrow further capital abroad. The collapse of prosperity led to growing nationalist assertiveness, sometimes against other Latin American states, sometimes against the North Americans and Europeans; foreign oil companies were expropriated in Mexico and Bolivia. The traditional Europeanized oligarchies were compromised by their failure to meet the problems posed by falling national incomes. From 1930 onwards there were military *coups* in every country except Mexico.

1939 again brought prosperity as commodity prices rose because of wartime demand (in 1950 the Korean War prolonged this trend). In spite of the notorious admiration of Argentina's rulers for Nazi Germany, most of the republics were sympathetic to the Allies, who courted them; most joined the United Nations' side before the war ended, and one, Brazil, sent a small expeditionary force to Europe, a striking gesture. The most important effects of the war on Latin America, though, were economic. One, of great significance, was that the old dependence on the United States and Europe for manufactured goods now became apparent in shortages. An intensive drive to industrialize gathered speed in several countries. On the urban work-forces which industrialization built up was founded a new form of political power which entered the lists as a competitor with the military and the traditional élites in the post-war era. Authoritarian, semi-fascist, but popular mass movements brought to power a new kind of strong man. Peron in Argentina was the most famous, but Colombia in 1953 and Venezuela in 1954 threw up similar rulers. Communism had no such conspicuous success among the masses.

A significant change also came about (though not as a result of war) in the way the United States used its preponderant power in the Caribbean. Twenty times in the first twenty years of the century American armed forces had intervened directly in neighbouring republics, twice going so far as to establish protectorates. Between 1920 and 1939 there were only two such interventions, in Honduras in 1924 and Nicaragua two years later. Indirect pressure also declined. In large measure this was a sensible recognition of changed circumstances. There was nothing to be got by direct intervention in the 1930s and President Roosevelt made a virtue of this by proclaiming a 'Good Neighbour' policy which stressed non-intervention by all American states in one another's affairs. Yet after 1950 there was another change. While American policy was dominated by European concerns in the early phase of the Cold War, after Korea it began slowly to look southwards again. Washington had not been unduly alarmed by manifestations of Latin American nationalism which tended to find a scapegoat in American policy, but became increasingly concerned lest the hemisphere provide a lodgement for Russian influence. The Cold War had come to it, and there followed greater selectivity in giving support to Latin American governments and, at times, to covert operations: for example, to the overthrow in 1954 of a government in Guatemala which had communist support.

At the same time United States policy-makers were anxious that the footholds provided for communism by poverty and discontent should be removed. They provided more economic aid (Latin America had only a tiny fraction of what went to Europe and Asia in the 1950s but much more in the next decade) and applauded governments which said they sought social reform. Unfortunately, whenever the programmes of such governments moved towards the eradication of American control of capital by nationalization, American policy tended to veer away again, demanding compensation on such a scale as to make reform very difficult. On the whole, therefore, while it might deplore the excesses of an individual authoritarian regime, such as that of Cuba before 1958, the American government tended to find itself, as in Asia, supporting conservative interests in Latin America. This was not invariably so; some governments acted effectively, notably Bolivia, which carried out land reform in 1952. But it remained true that, as for most of the previous century, the worst-off Latin Americans had virtually no hearing from either populist or conservative rulers, in that both listened only to the towns – the worst-off, of course, were the peasants, for the most part American Indians by origin.

Yet for all the Americans' nervousness the revolutionary change in Latin America was small. This was one of the lessons of the only victorious revolution, that in Cuba, of which so much was hoped and feared at the time. That country was in a number of respects very exceptional. Its island position in the Caribbean, within a relatively short distance of the United States, gave it special significance. The approaches to the Canal Zone had often been shown to have even more importance in American strategical thinking than Suez in the British. Secondly, Cuba had been especially badly hit in the depression; it was virtually dependent on one crop, sugar, and that crop had only one outlet, the United States. This economic tie, moreover, was only one of several which gave Cuba a closer and more irksome 'special relationship' with the United States than had any other Latin American state. There were historic connexions which went back before 1898 and the winning of independence from Spain. Until 1934 the Cuban constitution had included special provisions restricting Cuba's diplomatic freedom. The Americans kept a naval base on the island (as they still do). There was heavy American investment in urban property and utilities, and Cuba's poverty and low prices made it an attractive holiday resort for Americans. All in all, it should not have been surprising that Cuba produced, as it did, a strongly anti-American nationalist movement with much popular support.

The United States was seen as the real power behind the conservative post-war Cuban regime. Under the dictator Batista who came to power in 1952 this in fact ceased to be so; the State Department disapproved of him and cut off help to him in 1957. By the time this happened a young nationalist doctor, Fidel Castro, had already begun a guerrilla campaign against his government. In two years he was successful. In 1959, as

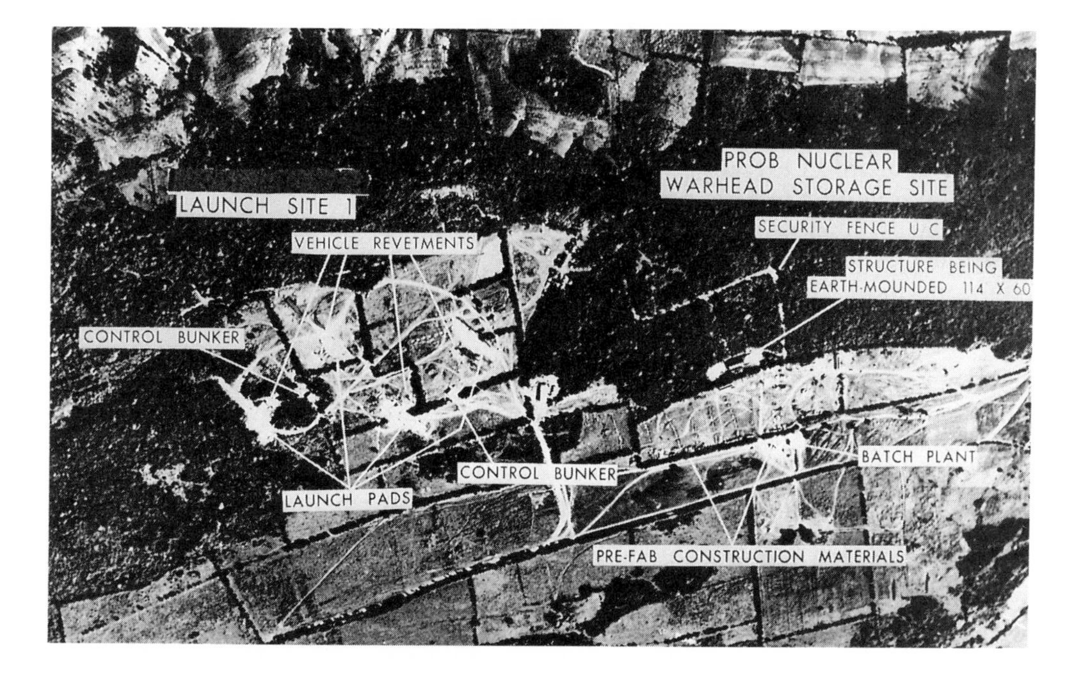

The importance to modern military intelligence of photographic and television observation from aircraft and satellites was dramatically illustrated when the United States government produced its photographs (of which this is one) of missile sites under construction in Cuba in 1962.

prime minister of a new, revolutionary, Cuba, he described his regime as 'humanistic' and, specifically, not communist.

Castro's original aims are still not known. Perhaps he was himself not clear what he thought. From the start he worked with a wide spectrum of people who wanted to overthrow Batista, from liberals to Marxists. This helped to reassure the United States, which patronized him as a Caribbean Sukarno; American public opinion idolized him as a romantic figure and beards became fashionable among American radicals. The relationship quickly soured once Castro turned to interference with American business interests, starting with agrarian reform and the nationalization of sugar concerns. He also denounced publicly those Americanized elements in Cuban society which had supported the old regime. Anti-Americanism was a logical means – perhaps the only one – open to Castro for uniting Cubans behind the revolution. Soon the United States broke off diplomatic relations with Cuba and began to bring to bear other kinds of pressure as well. The American government became convinced that the island was likely to fall into the hands of the communists upon whom Castro increasingly relied. It did not help when the Soviet leader Khrushchev warned the United States of the danger of retaliation from Soviet rockets if it acted militarily against Cuba and declared the Monroe doctrine dead; the State Department quickly announced that reports of its demise were greatly exaggerated. Finally the American government decided to promote his overthrow by force.

It was agreed that this should be done by Cuban exiles. When the presidency changed hands in 1961 Kennedy inherited this policy. Exiles were already training with American support in Guatemala, and diplomatic relations with Cuba had been broken off. He had not initiated it, but he was neither cautious nor thoughtful enough to impede it. This was the more regrettable because there was much else which promised well in the new president's attitude to Latin America, where it had been obvious for some time that the United States needed to cultivate goodwill. As it was, the possibilities of a more positive approach were almost at once blown to pieces by the fiasco known as the 'Bay of Pigs' operation, when an expedition of Cuban exiles supported by American money and arms came to a miserable end in April 1961. Castro now turned in earnest towards Russia, and at the end of the year declared himself a Marxist-Leninist.

A new and much more explicit phase of the Cold War thus began in the western hemisphere, and began badly for the United States. The American initiative incurred disapproval everywhere because it was an attack on a popular, solidly-based regime. Henceforth, Cuba was a revolutionary magnet in Latin America. Castro's torturers replaced Batista's and his government pressed forward with policies which, together with American pressure, badly damaged the economy, but embodied egalitarianism and social reform (in the 1970s, Cuba claimed to have the lowest child mortality rates in Latin America).

Almost incidentally and as a by-product of the Cuban revolution there soon took place the most serious confrontation of the whole Cold War and perhaps its turning-point. It is not yet known why or how the Soviet government decided to install in Cuba missiles capable of reaching anywhere in the United States and thus roughly to double the number of American bases or cities which were potential targets. Nor is it known whether the initiative came from Havana or Moscow. Though Castro asked the USSR for arms, it seems likeliest that it was the second. But whatever the circumstances, American photographic reconnaissance confirmed in October 1962 that the Russians were building missile sites in Cuba. President Kennedy waited until this could be shown to be incontrovertible and then announced that the United States Navy would stop any ship delivering further missiles to Cuba and that those already in Cuba would have to be withdrawn. One Lebanese ship was boarded and searched in the days that followed; Soviet ships were only observed. The American nuclear striking force was prepared for war. After a few days and some exchanges of personal letters between Kennedy and Khrushchev, the latter agreed that the missiles should be removed.

This crisis by far transcended the history of the hemisphere, and its repercussions for the rest of the world are best discussed elsewhere. So far as Latin American history is concerned, though the United States promised not to invade Cuba, it went on trying to isolate it as much as possible from its neighbours. Unsurprisingly, the appeal of Cuba's revolution nevertheless seemed for a while to gain ground among the young of other Latin American countries. This did not make their governments more sympathetic towards Castro, especially when he began to present Cuba as a revolutionary centre for the rest of the continent. In the event, as an unsuccessful attempt in Bolivia showed, revolution was not likely to prove easy. Cuban circumstances had been very atypical. The hopes entertained of mounting a peasant rebellion elsewhere proved illusory. Local communists, indeed, appear to have deplored Castro's efforts. While there were in some places plenty of materials about for revolution they turned out to be urban rather than rural and it was in the major cities that guerrilla movements were within a few years taking the headlines. Though spectacular and dangerous, it is not clear that these movements enjoyed wide popular support, though the brutalities practised in dealing with them alienated middle-class support from authoritarian governments in some countries. Anti-Americanism continued to run high. Kennedy's hopes for a new American initiative, based on social reform, an 'Alliance for Progress' as he termed it, made no headway against the animosity aroused by American treatment of Cuba. His successor as president, Johnson, did no better, perhaps because he was less interested in Latin America than in domestic reform and tended to leave hemisphere policy to fellow-Texans with business interests there. The initiative was never recaptured after the initial flagging of the Alliance. Worse still, it was overtaken in 1965 by fresh evidence of the old Adam of intervention, this time in the Dominican Republic, where, four years before, American help had assisted the overthrow and assassination of a corrupt and tyrannical dictator and his replacement by a reforming democratic government. When this was pushed aside by soldiers acting in defence of the privileged who felt threatened by reform, the Americans cut off aid; it looked as if, after all, the Alliance for Progress might be used discriminately. But aid was restored – as it was to other right-wing regimes – when Johnson came to power. A rebellion against the soldiers in 1965 resulted in the arrival of 20,000 American troops to put it down.

By the end of the decade the Alliance had virtually been forgotten, in part because of the persistent fears of communism which led American policy to put its weight behind conservatives, in part because the United States had plenty of more pressing problems elsewhere. One result was a new wave of attacks on American property by governments which did not have to fear the loss of American support while the communist threat seemed to endure. Chile nationalized the largest American copper company, the Bolivians took over oil concerns and the Peruvians American-owned plantations. In 1969 there was a historic meeting of Latin American governments at which no United States representative was present and *Yanqui* behaviour was explicitly and implicitly condemned. A tour undertaken by a representative of the president of the United States that year led to protest, riots, the blowing up of American property and requests to stay away from some countries. It was rather like the end of the previous decade, when a 'goodwill' tour by Eisenhower's vice-president ended in him being mobbed and spat upon. All in all, it looked by 1970 as if Latin American nationalism was entering a new and vigorous period. If Cuba-inspired guerrillas had ever presented a danger, they appeared to do so no longer. Once the spur of an internal fear was gone there was little reason for governments not to try to capitalize on anti-American feeling.

Yet it was clear that the real problems of Latin America were not being met. The 1970s, and still more, the 1980s revealed chronic economic troubles and, by 1985 it was reasonable to speak of an apparently insoluble crisis. There were several sources for this. For all its rapid industrialization, the continent was threatened by appalling population growth. The hundred or so million Latin Americans and Caribbean islanders of 1950 will probably have become five hundred millions by the year 2000. This huge rate of growth began to be obvious just as the difficulties of the Latin American economies were again beginning to show their intractability. The aid programme of the Alliance for Progress patently failed to cope with them, and failure spawned quarrels over the use of American funds. Mismanagement produced huge foreign debts which crippled attempts to sustain investment and achieve better trade balances. Social divisions remained menacing. Even the most advanced Latin American countries displayed vast discrepancies of wealth and education. Constitutional and democratic process, where they existed, seemed increasingly impotent to confront such problems. In the 1960s and 1970s, Peru, Bolivia, Brazil, Argentina and Paraguay all underwent prolonged authoritarian rule by soldiers and some of the upholders of those regimes undoubtedly believed that only authoritarianism could bring about changes of which civilian government had proved incapable.

Latin Americans' problems were vividly brought to the notice of the world by reports of torture and violent repression from countries like Argentina, Brazil and Uruguay, all once regarded as civilized and constitutional states. Then Chile, a country with a more continuous history of constitutional democracy than other Latin American states, underwent a military coup which overthrew in 1973 a government many Chileans believed to be under the control of communists. The counter-revolutionary movement had approval and probably support from the United States, but many Chileans acquiesced because they had been frightened by the revolutionary tendencies of the displaced elected regime. There followed the installation of a very authoritarian government. It long seemed unable to extract the country from economic disaster, though in the end it rebuilt the economy and even, in the late 1980s, began to look as if it might be able to liberalize itself.

It was on a troubled and distracted continent that there fell, to cap its troubles, the oil crisis of the 1970s which finally sent the foreign debt problems of its oil-importing countries out of control. By 1990, most orthodox economic remedies had been tried, in one country or another, but had proved unworkable or unenforceable in dealing with runaway inflation, interest charges on external debt, the distortion in resource allocation arising from bad government in past, and simple administrative and cultural inadequacy for the support of good fiscal policies. It remains impossible to guess how the complex and consequent economic crisis can be surmounted. While it is not, Latin

1973 in Chile. An early episode in the coup overthrowing the left-wing government of President Allende (second from right).

America remains an explosive, disturbed continent of nations less and less like one another, for all their shared roots, except in their distress. Most Latin Americans are now poorer, if *per capita* income is the measure, than ten years ago. Even culture is still a divisive force. To the layers laid down by Indian, slave, colonial and post-colonial experiences, all of which were reflected strongly in differences of economic level, have now been added the differences brought by the arrival in the 1950s and 1960s of the assumptions of developed, high-technology societies, whose benefits are available to the better-off but not to the poor who observe them. Just as in Asia, though the world's imagination has hardly begun to grasp it, the strains of the impact of modern civilization on a historically deep-rooted society are now more obvious than ever before. Latin America has been undergoing some of them since the sixteenth century, but they are now expressed additionally through the terrorism displayed by radical revolutionaries and reactionary soldiers alike, and they continue to undermine civilized and constitutional standards already achieved.

BELL SCOTT *William Bell Scott, a not very successful Scottish painter and poet, was for twenty years from 1843 a master in the new government schools of design at Newcastle-on-Tyne, where he played a major part in the organizing of art schools in the region. While he did so he found time to paint a series of pictures celebrating themes in the history of Northumberland and the borders, among them this canvas of* 'The Industry of the Tyne', *an idealization of one of the core areas of the second phase of British industrialization, the age of coal and iron. The nineteenth century brought the burgeoning of the new staple industries which were the base of 'the workshop of the world'. By the 1870s it was a joke in Newcastle that in a local shipyard iron colliers were built by the mile and then chopped into lengths as required.*

SAMUEL SCOTT *The Old Custom House Quay of the East India Wharf, London, in the 1750s as painted by 'the English Canaletto', Samuel Scott. The huge expansion of overseas trade which was the major feature of the British economy in the eighteenth century was based above all on textile exports and a switch from traditional (and only slowly-growing) European markets colonial customs. Between the 1720s and 1780s, exports to British North America quadrupled; those to the 'East Indies' (which included India) rose eightfold in volume. It was the main engine of new industrial growth and the source of prosperity of the provincial ports which were to challenge London's old supremacy. It laid the foundation of the world-wide British commercial leadership of the nineteenth century.*

3
Crumbling Certainties

In the middle of the 1970s, the two giants still dominated the world as they had done since 1945, and they still often talked as if they divided it, too, into adherents or enemies. But there had been changes in they way they were regarded. The United States was widely believed to have lost its once overwhelming military preponderance over the Soviet Union and those easily frightened by signs of instability wondered what would happen if there was to be a new Cuban crisis. Others found reassurance in such a shift: perhaps a more even balance would make such a crisis unlikely. The superpowers had to live in a changing environment, moreover. The two once more-or-less-disciplined blocs surrounded by small fry trying to escape from being swallowed by them were showing signs of strain. New quarrels were beginning to cut across old ideological divisions. More interesting still, there were signs that new aspirants to the role of superpower might be emerging. Some people had even begun to talk about an era of *détente*.

SUPERPOWER DIFFICULTIES

Once again, the roots of change go back some way; there are no sharp dividing lines between phases even if things looked more frozen then ever after the Cuba crisis. The death of Stalin, for instance, could hardly have been without effect, though it brought no obvious immediate change in Russian policy, and even more difficulty in interpreting it. The subsequent changes of personnel, whose outcome after nearly two years was the emergence of Nikita Khrushchev in a directing rôle, and the retirement in 1956 of Molotov, Stalin's old henchman and veteran of Cold War diplomacy, from his post as foreign minister were followed, too, by a sensational speech made by Khrushchev at a secret session of the twentieth congress of the Soviet communist party. In it he denounced the misdeeds of the Stalin era and declared 'coexistence' the goal of Russian foreign policy. Given announcements of Soviet reductions in armaments it might be said that 1956 had already seemed a promising year for international change for the better, until the atmosphere was fouled by the Suez invasion and a contemporaneous revolution in Hungary. The first led to Soviet threats to Great Britain and France; the Russians were not going to risk Arab goodwill by failing to support Egypt openly. The second event operated against a deeper background, for ever since 1948 Soviet policy had been almost morbidly sensitive to signs of deviation or dissatisfaction among its satellites. In that year, Soviet advisers had been recalled from Yugoslavia, which was expelled from the Cominform. Her treaties with Russia and other communist states were denounced by them and there followed five years of vitriolic attacks on 'Titoism' (though, in the end, the Russians were to climb down and ask to reopen diplomatic relations).

Yugoslavia's damaging and embarrassing survival as a socialist state outside the Warsaw Pact left Russia sensitive to any tremor in the eastern camp. Anti-Soviet riots in East Berlin in 1953 and in Poland in the summer of 1956 had seemed to show that nationalism was still stronger than communism. It was against this background that disturbances in Budapest in October 1956 grew into a nation-wide movement which led to the withdrawal of Russian forces from the city, a new Hungarian government and a promise of free elections and the end of one-party rule. Unfortunately, the new

The briefly successful Hungarian revolution of 1956 released the frustration and bitterness of years, often with brutal results for members of the regime's former security police, many of whom were lynched.

regime soon went further, withdrawing from the Warsaw Pact, declaring Hungary's neutrality, and asking the United Nations to take up the Hungarian question. At this, the Russian army returned. The Hungarian revolution was crushed. The UN General Assembly twice condemned the intervention, and the episode hardened attitudes on both sides. The Russians were once more made aware of how little they were liked by the peoples of eastern Europe, and therefore became even more distrustful of western talk of 'liberating' them. Western Europe was once again reminded of the real face of Soviet power, and sought to consolidate itself further.

Seeing danger in a rearmed West Germany, the Russians were anxious to strengthen their satellite, the German Democratic Republic. The continued existence of West Berlin inside its territory was a grave weakness. The city's frontiers were open and easily crossed. Its prosperity and freedom drew more and more East Germans – especially skilled workers – to the West. In 1958, this led the Russians to denounce the arrangements under which Berlin had been run since 1948; they said they would hand over their sector to the GDR if better ones could not be found. In August 1961, after some two years of drawn-out diplomacy, the East Germans suddenly erected a wall to cut off the Soviet sector of Berlin from the western. They felt driven to do this by a huge increase in the outflow of refugees as the atmosphere of crisis over Berlin had deepened. 140,000 crossed in 1959, 200,000 in 1960 and more than 100,000 in the first six months of 1961. The new wall raised tension in the short run, but probably lowered it in the long. The GDR had succeeded in stopping emigration and Khrushchev quietly dropped his more extreme demands when it was clear that the United States was not prepared to give way over the legal status of Berlin even at the risk of war.

Substantially, this rhythm was repeated the following year over Cuba, although the risk was then far greater. The allies of the United States were not so directly interested as they had been over a possible change in the German settlement, nor did the Russians seem to pay much attention to Cuba's interests. In a virtually 'pure' confrontation of the superpowers the Soviet Union appeared to have been forced to give way. While avoiding action or language which might have been dangerously provocative, and while leaving a simple route of retreat open to his opponent by confining his demands to essentials, President Kennedy none the less made no apparent concessions, though, quietly, the withdrawal of American missiles from Turkey followed after a little while. Immediately, Khrushchev had to be satisfied with an undertaking that the United States would not invade Cuba.

It is difficult to believe that this was not a major turning-point. The prospect of nuclear war as the ultimate price of geographical extension of the Cold War had been faced and found unacceptable. The subsequent setting-up of direct telephone communication between the heads of the two states – the 'hot line' – recognized that the danger of conflict through misunderstanding made necessary some more intimate connexion than the ordinary channels of diplomacy. It was also clear that in spite of Soviet boasting to the contrary, American preponderance in weapons was as great as ever. The new weapon which mattered for purposes of direct conflict between the two superpowers was the inter-continental rocket missile; at the end of 1962 the Americans had a superiority in this weapon of more than six to one over the Russians, who set to work to reduce this disparity. The choice was made of rockets before butter and once again the Soviet consumer was to bear the burden. Meanwhile, the Cuban confrontation had probably helped to achieve the first agreement between Great Britain, the United States and the Soviet Union on the restriction of testing nuclear weapons in space, the atmosphere or underwater. Disarmament would still be pursued without success for many years, but this was the first positive outcome of any negotiations about nuclear weapons.

In 1964 Khrushchev was removed from office. As head of both government and party since 1958 it seems likely that his personal contribution to Soviet development had been to provide a great shaking-up. This brought qualified 'de-Stalinization', a huge failure over agriculture, and a change in the emphasis of the armed services (towards the strategic rocket services which became their élite arm). Khrushchev had himself undertaken initiatives in foreign policy besides the disastrous Cuban adventure, and they may have been the fundamental cause of the decision to remove him. Yet though he was set aside with the connivance of the army by colleagues whom he had offended and alarmed, he was not killed, sent to prison or even to run a power-station in Mongolia. Evidently the Soviet Union was improving its techniques of political change. The contrast with earlier Soviet politics was striking and the nature of the regime such that people accounted this progress. Soviet society had relaxed a little after Stalin's death. The speech at the Twentieth Congress could not be unsaid, even if much of it was aimed at diverting criticism from those who (like Khrushchev himself) had been participants in the crimes of which Stalin was accused. (Symbolically, Stalin's body had been removed from Lenin's tomb, the national shrine.) In the next few years there was what some called a 'thaw'. Marginally-greater freedom of expression was allowed to writers and artists, while the regime appeared briefly to be a little more concerned about its appearance in the eyes of the world over such matters as its treatment of Jews. But this was personal and sporadic: liberalization depended on who had Khrushchev's ear. It seems clear only that after Stalin's death, particularly during the era of Khrushchev's ascendancy, the party had re-emerged as a much more independent factor in Russian life. The authoritarian nature of the Russian government, though, seemed unchanged – which is what might have been expected.

This makes it now seem odd that some came to think that the United States and Soviet Russia were growing more and more alike, and that this would help to make Russian policy less menacing. This theory of 'convergence' gave a distorted emphasis to an undoubted truth: the Soviet Union was a developed economy. In the 1960s some still thought socialism a plausible road to modernization because of that. It was often overlooked that the Soviet economy was also inefficient and distorted. Soviet industrial growth, though faster than that of the United States at least in the 1950s, had long been

The last years of the USSR: military might on parade, May Day 1982.

overwhelmingly a matter of heavy manufacture. The private consumer in the Soviet Union remained poor by comparison with the American, and would have been even more visibly so but for a costly system of subsidies. Russian agriculture, which had once fed the cities of Central Europe and paid for the industrialization of the tsarist era, had been a continuing failure; paradoxically, the Soviet Union often had to buy American grain. The official Soviet communist party programme of 1961 proposed that by 1970 the USSR should have outstripped the United States in industrial output, but the proposal was not made reality (unlike President Kennedy's of the same year to put a man on the moon). Yet the USSR, in comparison with many undeveloped countries, was undoubtedly rich. In spite of the obvious disparity between them as consumer societies, to the poor the USA and USSR sometimes looked alike. Many Soviet citizens, too, were more aware of the contrast between their stricken and impoverished country in the 1940s with one much less so in the 1960s, than of comparison with the United States. Moreover, the contest of the two systems was not always one-sided. Soviet investment in education, for example, was thought to have achieved literacy rates as good as, and even slightly higher at times, than the American. Such comparisons, which fall easily over the line from quantitative to qualitative judgement, nevertheless do not alter the basic fact that the *per capita* GDP of the Soviet Union still in the 1970s lagged far behind that of the United States. If its citizens had at last been given old age pensions in 1956 (nearly half a century after the British people), they also had to put up with health services which fell further and further behind those available in the West. There had been a long legacy of backwardness and disruption to eliminate; only in 1952 had real wages in Russia even got back to their 1928 level. The theory of 'convergence' was always too optimistic.

None the less, by 1970 a scientific and industrial base existed in the Soviet Union which in scale and at its best rivalled that of the United States. Its most obvious expression, and a great source of patriotic pride to the Soviet citizen, was the exploration of space. By 1980 there was so much ironmongery in orbit that it was difficult to recapture the startling impression made in 1957 by the first Soviet satellites. Although American successes had speedily followed, Soviet space achievements remained of the first rank. There seemed to be something in space exploration which fed the patriotic imagination and rewarded patience with other aspects of daily life in the USSR. It is not too much to say that for some Soviet citizens their space technology justified the Revolution; Russia was shown by it to be able to do almost anything another nation could do, and much that only one other could do, and perhaps one or two things which, for a while, no other could do. She was modernized at last.

Whether this meant that she was in some sense becoming a satisfied nation, with leaders more confident and less suspicious of the outside world and less prone to disturb the international scene, is a different matter. Soviet responses to Chinese resurgence were not encouraging; there was talk of a pre-emptive nuclear attack on the Chinese border. Soviet society was beginning to show new signs of strain, too, by 1970. Dissent and criticism, particularly of restraints upon intellectual freedom, had become obvious for the first time in the 1960s. So did symptoms of anti-social behaviour, such as hooliganism, corruption and alcoholism. But such defects probably held both as much and as little potential for significant change as in other large countries. Less spectacular facts may turn out to be more important in the long run; later perspectives may reveal that one important watershed was passed in the 1970s, when native Russian speakers for the first time became a minority in the Soviet Union. Meanwhile, the Soviet Union remained a police state where the limits of freedom and the basic privileges of the individual were defined in practice by an apparatus backed up by administrative decisions and political prisons. The real difference between the Soviet Union and the United States (or any west European nation) was still best shown by such yardsticks as her enormous expenditure on jamming foreign broadcasting.

For obvious reasons, changes in the United States were more easily observed than those in the Soviet Union, but this did not always make it easier to discern fundamentals. Of the sheer growth of American power there can be no doubt, nor of its importance to the world. In the middle of the 1950s the United States contained about 6 per cent of the world's population but produced more than half the world's manufactured goods. In 1968 the American population passed the two hundred million mark (in 1900 it had been 76 million), only one in twenty of whom were not native-born (though within ten years there would be worries about a huge Spanish-speaking immigration from Mexico and the Caribbean). Numbers of births went up while the birth rate dropped after 1960; the United States was unique among major developed countries in this respect. Nearly a quarter of all births in 1987 were to unmarried women. More Americans than ever lived in cities or their suburbs, and the likelihood that they would die of some form of malignancy had trebled since 1900; this, paradoxically, was a sure sign of improvement in public health, because it showed a growing mastery of other diseases.

The immensely-successful American industrial structure was dominated in 1970 by very large corporations, some of them commanding resources and wealth greater than those of some nations. Concern was often expressed for the interests of the public and the consumer, given the weight in the economy of these giants. But no doubts existed about the economy's ability to create wealth and power. Though it was to be shown that it could not do everything that might be asked of it, American industrial strength was the great constant of the post-war world. It sustained the huge military potential upon which the conduct of American foreign policy inevitably rested.

American economic success was a good generator of political mythologies. President Truman's second administration and those of President Eisenhower were marked by noisy debate and shadow-boxing about the danger of governmental interference with free enterprise. This was beside the point. Ever since 1945 the federal government has held and indeed increased its importance as the first customer of the American economy. Government spending has been the primary economic stimulant and the goal of hundreds of interest groups, and hopes of balanced budgets and cheap, business-like administration always ran aground upon this fact. What was more, the United States was a democracy; whatever the doctrinaire objections to it, and however much rhetoric might be devoted to attacking it, the Welfare State slowly advanced because voters wanted it that way. These facts made the old ideal of free enterprise unchecked and

Storage of new cars from General Motors in the 1950s, a symbol of huge productive power and problems of traffic management and pollution to come.

uninvaded by the influence of government more unrealistic than ever in the 1950s and 1960s. They also helped to prolong the Democratic coalition. The Republican presidents who were elected in 1952 and 1968 on each occasion benefited from war-weariness and neither was able to persuade Americans that they should also elect Republican congresses. On the other hand, signs of strain were to be seen in the Democratic bloc before 1960 – Eisenhower appealed to many southern voters – and by 1970 something a little more like a national conservative party had appeared under the Republican banner because some Southerners had been offended by Democratic legislation on behalf of the blacks. Twenty years later, though, the Democratic-voting 'Solid South' created by the Civil War had disappeared as a political constant.

Presidents could sometimes shift emphasis. The Eisenhower years leave an impression that little happened in the domestic history of the United States during them; it was not part of that president's vision of his office that he should provide a strong policy lead at home. Partly because of this, Kennedy's election by a narrow margin of the popular vote in 1960 – the arrival of a new man (and a young one, too) – produced a sense of striking change. Too much was made at the time of the more superficial aspects of this, but in retrospect it can be agreed that both in foreign and domestic affairs, the eight years of renewed Democratic rule from 1961 brought great change to the United States, though not in the way in which Kennedy or his vice-president, Lyndon Johnson, hoped when they took office.

One issue already there in 1960 was what could still then be called the Negro question. A century after emancipation, the black American was likely to be poorer, more often on relief, more often unemployed, less well-housed and less healthy than the white American. Thirty years later, this was still, and, sadly, even more the case. In the 1950s, though, there was widespread optimism about changing things. The position of blacks in American society suddenly began to appear intolerable and became a great political question because of three facts. One was migration. This turned a local Southern question into a national problem. Between 1940 and 1960 the black population of Northern states almost trebled. New York became the state with the biggest black population of the Union. This brought blacks into view not only in new places, but in new ways. It revealed that the problem facing them was not only one of legal rights, but more complex; it involved economic and cultural deprivation, too. The second fact pushing the question forward on to the national stage lay outside the United States. Many of the new nations which were becoming a majority at the UN were nations of coloured peoples. It was an embarrassment – of which communist propaganda always made good use – for the United States to display at home so flagrant a contravention of the ideals she espoused abroad as was provided by the plight of many of her blacks. Finally, the action of blacks themselves under their own leaders, some inspired by Gandhian principles of passive resistance to oppression, won over many whites. In the end the legal and political position of black Americans was radically altered for the better as a result. Bitterness and resentment were not eliminated in the process, though, but in some places actually increased and more blacks than before remained evidently poor and actually deprived.

The first and most successful phase of the campaign for equal status of the black was a struggle for 'Civil Rights', of which the most important were the unhindered exercise of the franchise (always formally available, but actually not in some Southern states) and access to equality of treatment in other ways. The success stemmed from decisions of the Supreme Court in 1954 and 1955. The process thus began not with legislation but with judicial interpretation. These important first decisions provided that the segregation of different races within the public school system was unconstitutional and that where it existed it should be brought to an end within a reasonable time. This challenged the social system in many Southern states, but by 1963 some black and white children were attending public schools together in every state of the Union, even if some all-black and all-white schools survived.

Legislation was not really important until after 1961. After the inauguration of a successful campaign of 'sit-ins' by black leaders (which itself achieved many important local victories) Kennedy initiated a programme going beyond the securing of voting rights to attack segregation and inequality of many kinds. It was to be continued by his successor. Poverty, poor housing, bad schools in run-down urban areas were symptoms of deep dislocations inside American society, and were inequalities made more irksome by the increasing affluence in which they were set. The Kennedy administration appealed to Americans to see their removal as one of the challenges of a 'New Frontier'.

Even greater emphasis was given to legislation to remove them by Johnson, who succeeded to the presidency when Kennedy was murdered in November 1963. Unhappily, laws did not help; the deepest roots of the American black problem appeared to lie beyond their reach in what came to be called the 'ghetto' areas of great American cities. The perspective from 1990 is again helpful. In 1965 (a hundred years after emancipation became law throughout the whole United States) a ferocious outbreak of rioting in a black district of Los Angeles was estimated to have involved at its height as many as 75,000 people. Other troubles followed in other cities, but not on the same scale. Twenty-five years later, all that had happened in Watts (where the Los Angeles outbreak took place) was that conditions had further deteriorated. By 1990 Los Angeles police were regarded there as members of an occupying army: they were in that year authorized to add dum-dum bullets to their already formidable weaponry. Over the United States as a whole a young black male American of the 1990s was seven times more likely to be murdered than his white contemporary (and it would probably be by a fellow-black). He was more likely to have a term in prison than to go to a university. By then, too, two-thirds of black babies were born to unmarried mothers. The problem was (it was usually agreed) one of economic opportunity, but none the easier to solve for that. It not only remained unsolved but appeared to be running away from solution. The poisons it secreted burst out in crime, a major collapse in health standards in some black communities, and in ungovernable and virtually unpoliceable inner-city areas. In the culture and politics of white America they seemed at times to have produced a near-neurotic obsession with colour and racial issues.

A demonstration by black campaigners for civil rights. Washington, 1963. The memorial to Lincoln – the president who issued the Emancipation Proclamation a hundred years earlier – was a frequent focus of such rallies.

His own poor Southern background had made Johnson a convinced and convincing exponent of the 'Great Society' in which he discerned America's future and perhaps this might have held promise for the handling of the black economic problem had he survived. Potentially one of America's great reforming presidents, Johnson nevertheless experienced tragic failure, for all his aspirations, experience and skill. His constructive and reforming work was soon forgotten (and, it must be said, set aside) when his presidency came to be overshadowed by an Asian war disastrous enough before it ended to be called by some America's Sicilian Expedition.

American policy in south-east Asia under Eisenhower had come to rest on the dogma that a non-communist South Vietnam was essential to security, and that it had to be kept in the western camp if others in the area – perhaps as far away as India and Australia – were not to be subverted. So, the United States had become the backer of its conservative government. President Kennedy did not question this view. He began to back up American military aid with 'advisers'. When he was murdered there were 23,000 of them in South Vietnam, and, in fact, many of them were in action in the field. President Johnson followed the course already set, believing that pledges to other countries had to be shown to be sound currency. But government after government in Saigon turned out to be broken reeds. At the beginning of 1965 Johnson was advised that South Vietnam might collapse; he had the authority to act (given him by Congress after North Vietnamese attacks on American ships the previous year) and air attacks were launched against targets in North Vietnam. Soon after, the first official American combat units were sent to the South. American participation quickly soared out of control. In 1968 there were over

500,000 American servicemen in Vietnam; by Christmas that year a heavier tonnage of bombs had been dropped on North Vietnam than fell on Germany and Japan together in the entire Second World War.

The outcome was politically disastrous. It was almost the least of Johnson's worries that the American balance of payments was wrecked by the war's huge cost, which also took money from badly-needed reform projects at home. Worse was the bitter domestic outcry which arose as casualties mounted and attempts to negotiate seemed to get nowhere. Rancour grew, and with it the alarm of moderate America. It was small consolation that Russia's costs in supplying arms to North Vietnam were heavy, too.

More was involved in domestic uproar over Vietnam than the agitation of young people rioting in protest and distrust of their government or the anger of conservatives who found their ideals outraged by ritual desecrations of the symbols of patriotism and refusals to carry out military service. Vietnam was bringing about a transformation in the way many Americans looked at the outside world. It was in south-east Asia that it was at last borne in on the thoughtful that even the United States could not obtain every result she wanted, far less obtain it at any reasonable cost. The late 1960s were a sunset era, though the sunset was not of American power but rather of the illusion that American power was limitless and irresistible. Americans had approached the post-war world with this illusion intact. Their country's strength had, after all, decided two world wars. Beyond them there stretched back a century and a half of virtually unchecked and unhindered continental expansion, of immunity from European intervention, of the growth of an impressive hegemony in the American hemisphere. There was nothing in American history which was wholly disastrous or irredeemable, hardly anything in which there was, ultimately, failure, and nothing over which most Americans felt any guilt. It had been easy and natural for that background to breed a careless assumption of limitless possibility and for that to be carried over from domestic to foreign concerns. It was understandable that Americans should forget the special conditions of continental isolation and the supremacy of British sea-power on which their success story had long been built.

The reckoning had begun to be drawn up in the 1950s, when many Americans had to be content with a lesser victory in Korea than they had hoped for. There had then opened twenty years of frustrating dealings with nations often enjoying not a tenth of the power of the United States but apparently able to thwart her. At last, in the Vietnam disaster, both the limits of power and its full costs were revealed. In March 1968 the strength of domestic opposition to the war was shown clearly in the primary elections. Johnson had already drawn the conclusion that the United States could not win, had restricted bombing and asked the North to open negotiations again. Dramatically, he also announced that he would not stand for re-election in 1968. Just as the casualties of the Korean War won Eisenhower election in 1952, so the casualties of Vietnam, on the battlefield and at home, helped to elect another Republican president in 1968 (only four years after a huge Democratic majority for Johnson) and to re-elect him in 1972. Vietnam was not the only factor, but it was one of the most important in dislocating the old Democratic coalition.

The new president, Mr Richard Nixon, began to withdraw American ground forces from Vietnam soon after his inauguration. Peacemaking took three years. In 1970 secret negotiations began between North Vietnam and the United States, accompanied by further withdrawals but also by renewed and intensified bombing of the North by the Americans. The diplomacy was tortuous and difficult. The United States could not admit it was abandoning its ally, though in fact it had to do so, nor would the North Vietnamese accept terms which did not leave them with power to harass the Southern regime through their sympathizers in the south. Amid considerable public outcry in the United States, bombing was briefly resumed at the end of 1972, but for the last time. Soon afterwards, on 27 January 1973, a cease-fire was signed in Paris.

The war had cost the United States vast sums of money and 57,000 dead. It had gravely damaged American prestige, eroded American diplomatic influence, had

War in Vietnam: a Vietnamese woman pleads with an American soldier as her home blazes behind her.

ravaged domestic politics and had frustrated reform. What had been achieved was a brief preservation of a shaky South Vietnam saddled with internal problems which made its survival uncertain, while terrible destruction had been inflicted on the peoples of Indo-China. Perhaps the abandonment of the illusion of American omnipotence offsets these costs.

It was at least a domestic success to disentangle the United States from the morass and President Nixon reaped the benefit. The liquidation of the venture had followed other signs of his recognition of how much the world had already changed since the Cuban crisis. The most striking was a new policy of normal and direct relations with China. It came to a climax with a visit of the president to China in February 1972. It was not only an attempt to start to bridge what he described as '16,000 miles and twenty-two years of hostility' (he might have added 'and 2500 years of history'), but the first visit by an American president to mainland Asia. Diplomatic relations with the United States at the highest level were paralleled by another gain for China, her entry to the United Nations Organization and a seat on the Security Council.

When Mr Nixon followed his Chinese trip by becoming also the first American president to visit Moscow (in May 1972) and this was followed by an interim agreement on arms limitation, the first of its kind, it seemed that another important change had come about. The stark polarized simplicities of the Cold War were clearly gone, however doubtful might be what came next. The Vietnam settlement followed and can hardly have been unrelated to it; Moscow and Peking both had to be squared if there was to be a cease-fire. China's attitude to the Vietnamese struggle was, we may guess, by no means simple; it was complicated by the potential danger from Soviet policy, by the United States' use of its power elsewhere in Asia, notably Taiwan and Japan, and by Vietnamese nationalism. For all China's help, its Indo-Chinese communist satellite could not be trusted. The Vietnamese had a history of struggle against Chinese as well as French imperialism. In the immediate aftermath of the American withdrawal, too, the nature of the struggle going on in Vietnam (and now spread to Cambodia) was more and more clearly revealed as a civil war.

The North Vietnamese did not wait long before resuming operations. For a time the United States government had to pretend not to see this; there was too much relief at home over the liquidation of the Asian commitment for scruples to be expressed over the actual observation of the peace terms which had made withdrawal possible. When a political scandal forced Nixon's resignation, his successor faced a Congress suspicious of what it saw as dangerous foreign adventures and determined to thwart them. This meant that there would be no attempt to uphold the peace terms of 1972 in so far as they guaranteed the South Vietnamese regime against overthrow.

Early in 1975 American aid to Saigon came to an end. A government which had lost virtually all its other territory was reduced to a backs-to-the-wall attempt to hold the capital city and lower Mekong with a demoralized and defeated army. At the same time, communist forces in Cambodia were destroying another regime earlier supported by the United States. Congress prevented the sending of further military and financial help. The pattern of China in 1947 was being repeated; the United States was eventually to cut her losses at the expense of those who had relied on her (though 117,000 Vietnamese left with the Americans).

Such an outcome was doubly ironical. In the first place it seemed to show that the hardliners on Asian policy had been right all along – that only the knowledge that the United States was in the last resort prepared to fight for them could guarantee the post-colonial regimes' resistance to communism. Secondly, the swing back to isolationism in the United States was accentuated, not muffled, by defeat and disaster; those who reflected on the American dead and missing and the huge cost now saw the whole Indo-China episode as a pointless and unjustifiable waste on behalf of peoples who would not fight to defend themselves. Yet an alternative reading of the American position in east Asia was possible. It was arguable that better relations with China mattered much more than the loss of Vietnam.

The opening of a new chapter of history, President and Mrs Nixon visit the Great Wall near Peking in 1972.

By the decade's end, America and her allies were confused and worried. The situation was not easy to read. Objectively, though, there were good grounds for reassurance. The American democratic system showed no sign of breaking down, or of not meeting many of the country's needs, even if it could not find answers to all its problems. The economy had, astonishingly, been able to continue for years to pay for a hugely expensive war, a space exploration programme that put men on the moon, and for the maintenance of garrisons around the world on an unprecedented scale. True, the black American's plight continued to worsen, and some of the country's greatest cities seemed stricken by urban decay, while a deep psychological wound had been inflicted by the Vietnam war on those who had hitherto believed unquestioningly in the traditions of American patriotism. Fewer Americans, though, seemed to find such facts so worrying as their country's supposed military inferiority (essentially, in missiles) to the Soviet Union (it was to be an issue in the presidential election of 1980). Concern was exacerbated by the undermining in the early 1970s of confidence in the traditional leadership of the president in foreign affairs by domestic scandal and a new distrust of the executive power. President Ford (who had taken office in 1974 on the resignation of his predecessor) had already had to face a Congress unwilling to countenance further aid to its allies in Indo-China. When Cambodia collapsed, and South Vietnam quickly followed, questions began to be asked at home and abroad about how far what looked like a world-wide retreat of American power might go. If the United States would no longer fight over Indo-China, would she, then, do so over Thailand? More alarmingly still, would she fight over Israel – or even Berlin? There were good reasons to think the Americans' mood of resignation and dismay would not last for ever, but while it lasted, their allies looked about them and felt uneasy.

TWO EUROPES

Europe was the birthplace of Cold War and for a long time its main theatre. Yet well before 1970 there had been signs that the terrible simplifications institutionalized in NATO and (more rigidly still) in the Warsaw Pact might not be all that was shaping history there. Even in eastern Europe, seemingly long insulated by Soviet power from external stimuli to change and by its command economies, there were signs of division. The violence with which Albania, the tiniest of Europe's communist countries, condemned the Soviet Union and applauded China when the two fell out in the 1960s had to be endured by the Russians; Albania had no frontier with other Warsaw Pact countries and so was not likely to have to take account of the Red Army. It was more striking when Romania, with Chinese support, successfully contested the direction of her economy by Comecon, asserting a national right to develop it in her own interest. She even took up a vaguely neutralist position on questions of foreign policy – though remaining inside the Warsaw Pact – and did so, oddly enough, under a ruler who imposed on his countrymen one of the most rigidly dictatorial regimes in eastern Europe. But Romania (unlike Albania) had no land frontier with a NATO country, and one eight hundred kilometres long with Russia; her skittishness could be tolerated, therefore. That there were clear limits to the dislocation of the old monolithic unity of communism was none the less shown in 1968. When a communist government in Czechoslovakia set about liberalizing its internal structure and developing trade relations with West Germany, tolerance was not forthcoming. After a series of attempts to bring her to heel Czechoslovakia was invaded in August 1968 by Warsaw Pact forces. To avoid a repetition of what had happened in Hungary in 1956, the Czech government did not resist and a brief attempt to provide an example of 'socialism with a human face', as a Czech politician had put it, was obliterated.

None the less, Sino-Soviet tension combined with tremblings within the eastern bloc and the uneasiness of the United States over relations with Latin American countries to lead to suggestions that the world as a whole was abandoning bipolarity for 'polycentrism' as an Italian communist put it. The loosening of Cold War simplicities had indeed been surprising. Another complicating development had meanwhile emerged in western Europe, the region where the evolution of a unified world history had actually begun. In 1980 it was clear that one of the historic rôles of the peoples of that part of the continent was over. Western Europeans by then ruled no more of the world's surface than their ancestors had done five hundred years earlier. Huge transformations had taken place, and irreversible things had been done in those five centuries. But Europe's imperial past was over. Yet the discovery of a new rôle for her was already well under way. Thirty years earlier, in the 1950s, western Europe had begun to show some of the earliest signs, feeble though they were, that nationalism's grip on the human potential for large-scale organization might be loosening in the very place where nationalism had been born.

The legacies of common experience which have shaped western Europe have been, not unreasonably, traced by enthusiasts back to the Carolingians, but 1945 is far enough to go back at this point. From that date, the decisive agents for more than forty years were the outcome of the war and Soviet policy. They had ended by partition the German problem. This quietened the old fears of France. The likelihood of another great civil war in the West over the German question had receded. Soviet policy had also given the western countries many new reasons to cooperate more closely; what happened in eastern Europe in the late 1940s struck them as a warning of what might happen if the Americans ever went home and they remained divided. The Marshall Plan and NATO therefore turned out to be only the first two of many important steps towards an integration whose heart was to be found in a new Europe.

That integration had more than one source. The initiation of the Marshall Plan was followed by the setting-up of an Organization (at first of sixteen countries, but later expanded) of European Economic Cooperation in 1948, but the following year, a month after the signing of the treaty setting up NATO, the first political bodies representing ten different European states were also set up under a new Council of Europe. The economic forces making for integration were developing more rapidly,

A dramatic confrontation in Prague during the Russian invasion in 1968.

though. Customs Unions had already been created in 1948 between the 'Benelux' countries (Belgium, the Netherlands, and Luxembourg), and (in a different form) between France and Italy. The most important of the early steps towards economic integration, though, emerged from a French proposal for a Coal and Steel Community. This came into existence formally in 1951 and embraced France, Italy, the Benelux countries and, most significantly, West Germany. It made possible the rejuvenation of the industrial heartland of western Europe and was the major step towards the integration of western Germany into a new international structure. Through economic arrangement, there came into existence the means of containing while reviving West Germany, whose strength, it was becoming clear, was needed in a western Europe menaced by Soviet land-power. Under the influence of events in Korea, American official opinion (to the consternation of some Europeans) was in the early 1950s rapidly coming round to the view that Germany had to be rearmed.

Other facts, too, helped to ease the way to supra-national organization in Europe. The political weakness symptomized by their domestic communist parties subsided in both France and Italy, mainly thanks to economic recovery. Communists had ceased to take part in their governments as early as 1947 and the danger that French and Italian democracy might suffer a fate like Czechoslovakia's had disappeared by 1950. Anti-communist opinion tended to coalesce about parties whose integrating forces were either Roman Catholic politicians or social democrats well aware of the fate of their comrades in eastern Europe. Broadly speaking, these changes meant that western Europe governments of a moderate right-wing complexion pursued similar aims of economic recovery, welfare service provision, and western European integration in practical matters during the 1950s.

Further institutions emerged. In 1952 a European Defence Community formalized West Germany's military position. This was to be replaced by German membership of NATO, but a major thrust towards greater unity, as before, was economic. The crucial step came in 1957: the European Economic Community (EEC or 'Common Market') then came into being when France, Germany, Belgium, the Netherlands, Luxembourg and Italy joined in signing the Treaty of Rome. Some enthusiasts spoke of the reconstitution of Charlemagne's heritage. It spurred countries which had not joined the EEC to set up their own, looser and more limited, European Free Trade Association (EFTA) two-and-a-half years later. By 1986, the six countries of the original EEC (by then it was usually called the EC – the word 'Economic', significantly, being dropped) were twelve; while EFTA had lost all but four of its members to it. Five years later still, and what was left of EFTA was envisaging merging with the EC.

Western Europe's slow but accelerating movement towards political unity, demonstrated the confidence of those who made the arrangements that armed conflict could never again be an acceptable alternative to cooperation and negotiation between their countries. The era of western European international war, rooted in the beginnings of the national state system, was over. Tragically, though recognizing that fact, Great Britain's rulers did not seize at the outset the chance to join in giving it institutional expression; their successors were twice to be refused admission to the EEC before finally joining it. Meanwhile, the Community's interests were steadily cemented together by a Common Agricultural Policy which was, to all intents and purposes, a huge bribe to the farmers and peasants who were so important a part of the German and French electorates, and, later, to those of poorer countries as they became members.

For a long time determined opposition to further integration, at the political as opposed to the economic level, came from France. It was expressed strongly by General de Gaulle, who returned to politics in 1958 to become president when the French republic was threatened with civil war over Algeria. His first task was to negotiate these rapids by carrying through important constitutional reforms. His next service to France was as great as any in his wartime career, the liquidation of her Algerian commitment in 1961. The legions came home, some disgruntled. The act freed both him and his country for a more vigorous international role, though a very negative one. De Gaulle's view of European consolidation was limited to cooperation between independent nation-states; he saw the EEC as above all a way of protecting French economic interests. He was quite prepared to strain the new organization badly to get his way. Further, he in effect twice vetoed British application to join it. Wartime experience had left de Gaulle with a deep distrust of the 'Anglo-Saxons' and a belief, by no means ill-founded, that the British still hankered after integration with an Atlantic community embracing the United States, rather than with continental Europe. In 1964 he annoyed the Americans by exchanging diplomatic representatives with communist China. He insisted that France go ahead with her own nuclear weapons programme, declining to be dependent on American patronage. Finally, after causing it much trouble, he withdrew from NATO. This could be seen as the coming of 'polycentrism' to the western bloc. When de Gaulle resigned after an unfavourable referendum on further constitutional change in 1969, a major political force making for uncertainty and disarray in western Europe disappeared. Those who led France for the next couple of decades, while trying to sound like him, proved less intransigent, although French policy still opposed the evolution towards a more complete cohesion both of a true Common Market, free from internal restraints on trade, and of a politically united Europe.

Great Britain finally joined the EEC in 1973, a recognition, at last, of the facts of twentieth-century history by the most conservative of the historic nation-states. The decision complemented the withdrawal from empire and acknowledged that the British strategic frontier lay no longer on the Rhine, but on the Elbe. It was the climax, though not the end, of an era of uncertainty which had lasted almost since 1945. For a quarter-century all British governments had tried and failed to combine economic growth, increased social service provision and a high level of employment. The second depended ultimately on the first, but when difficulty arose, the first had always been sacrificed to the other two. The United Kingdom was, after all, a democracy whose voters, greedy and gullible, had to be placated. The vulnerability of the traditional British economy's commitment to international trade was a handicap, too. Others were its old staple industries, starved of investment, and the deeply conservative attitudes of the people. Though the United Kingdom grew richer (in 1970 virtually no British manual worker had four weeks' paid holiday a year and ten years later a third of them did), it fell behind more and more other developed countries both in its wealth and its rate of creating it. If the British had managed a decline in international power and the achievement of a rapid decolonization without the violence and domestic bitterness visible elsewhere, it remained unclear whether they could shake off the past in other ways and ensure themselves even a modest prosperity as a second-rank nation.

One obvious and symptomatic threat to order and civilization was posed in Northern Ireland. Protestant and Catholic hooligans alike seemed bent on destroying their homeland rather than cooperate with their rivals. Lunatic nationalism was as alive in Ulster as ever and cost the lives of thousands of British citizens – soldiers, policemen and civilians, Protestant and Catholic, Irish, Scotch, and English alike – in the 1970s and 1980s. Fortunately it did not disrupt British party politics as Irish questions had done in the past. The electorate remained preoccupied, rather, by material concerns. Inflation ran at unprecedented levels (the annualized rate 1970–80 was over 13 per cent) and gave new fierceness to industrial troubles in the 1970s, especially in the wake of the oil crisis. There was speculation about whether the country was 'ungovernable' as a miners' strike brought down one government, while many leaders and interpreters of opinion seemed obsessed with the themes of social division. Even the question whether the United Kingdom should remain in the EEC, which was submitted to the revolutionary device of a referendum in June 1975, was often put in these terms. It was therefore all the more surprising to many politicians when the outcome was unambiguously favourable to continued membership. It was the first sign for a decade or so that the views of the country at large were not necessarily represented by those who were considered their spokesmen; it was also possible that it was a decision which would in the end prove a turning-point in the history of a once-great nation, by closing off one sequence of choices which would have flowed ineluctably from a reassertion of insularity.

None the less, more bad times (economically speaking) lay immediately ahead; inflation (in 1975 running at 26.9 per cent in the wake of the oil crisis) was at last identified by government as the major threat. Wage demands by trades unions

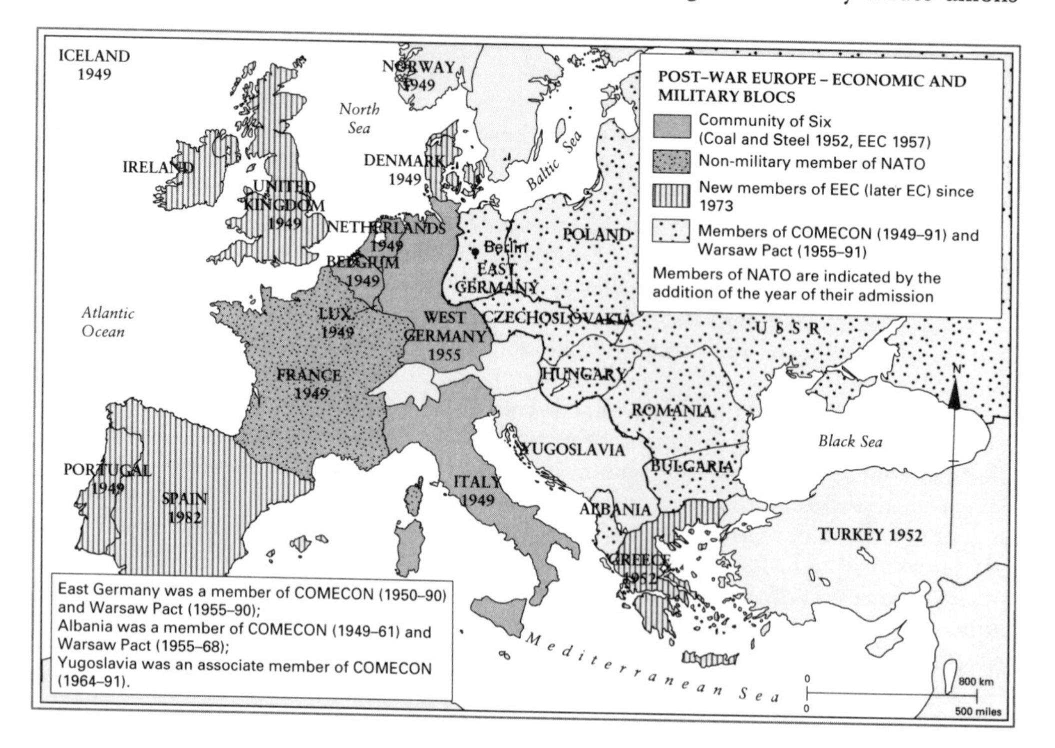

were anticipating inflation still to come and it began to dawn on some that the era of unquestioned growth in consumption was over. There was a gleam of light; a few years earlier vast oil fields had been discovered under the sea-bed off the coasts of northern Europe. In 1976 the United Kingdom became an oil-exporting nation. That did not at once help much; in the same year, a loan from the International Monetary Fund was required. When Mrs Thatcher, the country's (and Europe's) first woman prime minister and the first woman leader of a major political party (the Conservative), took office in 1979 she had, in a sense, little to lose; her opponents were discredited. So, many felt, were ideas which had been long accepted uncritically as the determinants of British policy. A radical new departure for once really did seem to be a possibility. To the surprise of many and the amazement of some among both her supporters and her opponents, that is exactly what Mrs Thatcher was to provide in what was to prove the longest tenure of power of any British prime minister in this century.

Not far into her premiership, Mrs Thatcher found herself in 1982 presiding unexpectedly over what may well prove to have been Great Britain's last colonial war. The reconquest of the Falkland Islands after their brief occupation by Argentinian forces was a great feat of arms as well as a major psychological and diplomatic success. The prime minister's instincts to fight for the principles of international law and territorial sovereignty, and for the islanders' right to say by whom they should be governed, were well-attuned to the popular mood (whose more vociferous and vulgar manifestations displeased the more fastidious among her opponents). She also correctly judged the international possibilities. After an uncertain start (unsurprising, given its traditional sensitivity over Latin America) the United States provided important practical and clandestine help. Chile, by no means easy with her restive neighbour, was not disposed to object to British covert operations on the mainland of South America. More important, most of the EC countries supported the isolation of Argentina in the UN, and resolutions which condemned the Argentinian action. It was especially notable that the British had from the start the support (not traditionally offered so readily) of the French government, which knew a threat to international security and vested rights under international law when it saw one.

It now seems clear that Argentinian action had been encouraged by the misleading impressions of likely British reactions gained from British diplomacy in previous years (for this reason, the British foreign secretary resigned at the outset of the crisis). Happily, one political consequence was the fatal wounding in its prestige and cohesiveness of the military regime which had ruled Argentina since 1976. It was replaced at the end of 1983 by a constitutional and elected government. In the United Kingdom, Mrs Thatcher's prestige rose with national morale; abroad, too, her standing was enhanced, and this was important. For the rest of the decade it provided the country with an influence with other heads of state (notably the American president) which the raw facts of British strength could scarcely have sustained by themselves. Not everyone agreed that this influence was always advantageously deployed; in this sphere (and as perhaps in some others) a comparison with General de Gaulle is to the point. Like his, Mrs Thatcher's personal convictions, preconceptions and prejudices were always very visible and she, like him, was no European, if that meant allowing emotional or even practical commitment to Europe to blunt their respective visions of national interest.

In domestic affairs, the effects of Mrs Thatcher's policies will require a long perspective for their full evaluation. What can be discerned even now is that she transformed the terms of British politics, and perhaps of cultural and social debate, dissolving a long *bien-pensant* consensus about national goals. This, together with the undoubted radicalism of many of her specific policies, awoke both enthusiasm and an unusual animosity. For all the rhetoric, though, she failed to achieve two of her most important aims, the reduction of public spending, and the withdrawal of central government from as much as possible of the national life. Though former pretensions to direct the economy were at last set aside, ten years after she took up office government was playing a greater, not a smaller, part in many areas of society, and the public money spent on health and social security had gone up in real terms by a third since 1979 (unfortunately for her, without satisfying greatly increased demands).

In 1990, many of her political colleagues had come to believe that although she had led the Conservative Party to three general election victories in a row (a unique achievement in British politics), she would be a vote-loser in the next contest, which could not be far away. Faced with the erosion of loyalty and support, she resigned. Her successor, Mr Major, was something of an unknown quantity, having not had a long exposure in the front ranks of politics. It seemed likely, though, that British policy might now become less obstructive in its approach to the Community and its affairs (and would certainly be less rhetorical in style).

1984: the British Prime Minister, Mrs. Thatcher, enjoying a conference of the European Economic Community. Her deep distrust of its possible implications both brought her popularity and led to the alienation of many Conservative politicians by 1989.

The 1970s had been difficult years for all the members of the EEC. Growth fell away and their individual economies reeled under the impact of the oil crisis. This contributed to institutional bickering and squabbling (particularly on economic and financial matters) which had reminded Europeans of the limits to any transcendence of bipolarity so far achieved. It continued in the 1980s and, coupled with uneasiness about the success of the Far Eastern economic sphere, dominated by Japan, and a growing realization that other nations would wish to join the ten, led to further crystallization of ideas about the Community's future. Many Europeans saw more clearly that greater unity, a habit of cooperation and increasing prosperity were prerequisites of Europe's political independence, but there was also an emerging sense that such independence would always remain hollow unless Europe, too, could turn herself into a superpower.

Comfort could be drawn from further progress in integration. In 1979, a month after Mrs Thatcher formed her government, the first direct elections to the European parliament were already being held. Greece in 1981, Spain and Portugal in

1986, were soon to join the Community. In 1987 the foundations of a common European currency and monetary system were laid (though the United Kingdom did not agree) and it was settled that 1992 should be the year which would see the inauguration of a genuine single market, across whose national borders goods, people, capital and services were to move freely. Members even endorsed in principle the idea of European political union, though the British and French had notable misgivings. This by no means made at once for greater psychological cohesion and comfort as the implications emerged, but it was an indisputable sign of progress of some sort.

In December 1991 at Maastricht another step forward was taken when the EC members agreed on measures to integrate the Community further, though, again with reservations and special arrangements for the cautious British. By then a common monetary system restricting the independence of members in managing the devaluation or revaluation of their own currencies was already in place. This was a significant stride towards a Common European currency. As at every point in the Community's history, satisfaction about what has been achieved was almost lost to sight amid quarrels, misgivings, misinterpretations, ambiguities about what should or might lie ahead. Yet many other European nations were now knocking at or approaching the door of entry, a good testimony to what they saw as the advantages of membership. Since 1957, western Europe had come a very long way, further, perhaps, than was always grasped by men and women born and grown to maturity since the Treaty of Rome had been signed. Underlying the institutional changes, too, were growing similarities – in politics, social structure, consumption habits and beliefs about values and goals. The old disparities of economic structure had greatly diminished, as the decline in numbers and increase in prosperity of French and German farmers showed. On the other hand, new problems had presented themselves as poorer and perhaps politically less stable countries had joined the EC. That there had been huge convergences could not, in 1991, be contested. What was still unclear was what this might imply for the future.

NEW CHALLENGES TO THE COLD WAR WORLD ORDER

In December 1975 Mr Ford became the second American president to visit China. The adjustment of his country's deep-seated attitudes towards the People's Republic had begun with the slow recognition of the lessons of the Vietnam disaster. On the Chinese side, changes were to be understood also in a deeper perspective. They were a part of China's resumption of the international and regional rôle appropriate to her historical stature and potential. This can be said to have been going on since 1911, but could only come to fruition after 1949. In the 1970s it was completed. An approach could then be made towards establishing normal relations with the United States. It helped that the Chinese were concerned over Soviet policy, which was seen as expansionist and threatening. Formal recognition of what had been achieved came in 1978. In a Sino-American agreement the United States made the crucial concession that its forces should be withdrawn from Taiwan and that official diplomatic relations with the island's KMT government should be ended.

Mao had died in 1976. The threat of the ascendancy of a 'gang of four' of his coadjutors (one was his widow) who had promoted the policies of the cultural revolution was quickly averted by their arrest (and, eventually, trial and punishment in 1981). Under new leadership dominated by party veterans, it soon became clear that the excesses of the cultural revolution were to be corrected. In 1977 there rejoined the government as a vice-premier the twice-previously disgraced Deng Xiaoping, firmly associated with the contrary trend. Scope was now to be given to individual enterprise and the profit motive, and economic connexions with non-communist countries would be encouraged. The aim was to resume the process of technological and industrial modernization. The major definition of the new course was undertaken in 1981 at the plenary session of the central committee of the party which met that year. It undertook, too, the delicate task of distinguishing the positive achievements of Mao, a 'great proletarian revolutionary', from what it called his 'gross mistakes' and his responsibility for the setbacks of the Great Leap Forward and, more importantly, the cultural revolution.

For all the comings-and goings in CCP leadership, and the mysterious debates and sloganizing which continued to obscure political realities, and though Deng Xiaoping and his associates had to work through a collective leadership which included conservatives, the 1980s were dominated by the new current. They settled the question which for thirty years had been at the heart of the party's history, and therefore at the heart of China's, too. Modernization had at last been given precedence over Marxist socialism, even if that could hardly be said aloud (the secretary-general of the party pronounced in 1986 the amazing judgement that 'Marx and Lenin cannot solve our problems', but he was dismissed soon afterwards), and even if much Marxist language still pervaded the rhetoric of government. Some said China was resuming the 'capitalist road'. This, too, was obscuring, though natural. There persisted in the party and government a clear grasp of the need for positive planning of the economy; what was new was a recognition of its practical limits and a willingness to try to discriminate more carefully between what was and was not within the scope of effective regulation in the pursuit of the major goals of economic and national strength, the improvement of living standards, and a broad egalitarianism.

One remarkable change was, to all intents and purposes, agriculture was privatized in the next few years. New slogans – 'to get rich is glorious' – were coined to encourage the development of village industrial and commercial enterprise, and a pragmatic road to development was signposted with 'four modernizations'. Special economic areas, enclaves for free trade with the capitalist world, were set up; the first was at Canton, the historic centre of Chinese trade with the West. It was not a policy without costs – grain production fell at first, inflation began to show itself in the early 1980s and foreign debt rose. Some blamed the growing visibility of crime and corruption on the new line.

Yet of its economic success there can be no doubt. Mainland China began in the 1980s to show that perhaps an economic 'miracle' like that of Taiwan was within her grasp. By 1986 she was the second largest producer of coal in the world, and the fourth largest of steel. GDP rose at more than ten per cent a year between 1978 and 1986, while industrial output had doubled in value in that time. *Per capita* peasant income nearly tripled, and by 1988 the average peasant family was estimated to have about six months' income in the savings bank. Taking a longer perspective, the contrasts are even more striking, for all the damage done by the Great Leap Forward and the cultural revolution. The value of foreign trade multiplied roughly twenty-five times in *per capita* terms between 1950 and the middle of the 1980s. The social benefits which have accompanied such changes are also clear: increased food consumption and life expectancy, a virtual end to many of the great killing and crippling diseases of the old regime, and a huge inroad into mass illiteracy. China's continuing population growth was alarming and prompted stern measures of intervention, but it had not, as had India's, devoured the fruits of economic development.

The new line specifically linked modernization to strength. Thus it reflected the aspirations of China's reformers ever since the May 4th Movement, and of some even earlier. But China's international weight had already been apparent in the 1950s; what now happened was that it began to show itself in different ways. One important sign was agreement with the British in 1984 over terms for the reincorporation of Hong Kong on the expiration of the lease covering some of its territories in 1997. A later agreement with the Portuguese provided for the resumption of Macao, too. It was a blemish on the general recognition of China's due standing that Vietnam (with which China's relations at one time degenerated into open warfare, when the two countries were rivals for the control of Cambodia, traditionally part of the old imperial zone of Chinese hegemony) remained hostile to her among neighbouring countries; but the Taiwanese were somewhat reassured by Chinese promises that the reincorporation of the island in the territory of the republic in due course would not endanger its economic system. Similar assurances were given over Hong Kong. Like the establishment of special trading enclaves on the mainland where external commerce could flourish, such statements underlined the importance China's new rulers attached to commerce as a channel of modernization. China's sheer size gave such a policy direction importance over a wide area. By 1985 the whole of east and south-east Asia constituted a single trading zone of unprecedented potential.

Within it, new centres of industrial and commercial activity were developing so fast in the 1980s as to justify by themselves the view that the old global balance of economic power had disappeared. South Korea, Taiwan, Hong Kong and Singapore had all shed any aura of undeveloped economies; Malaysia, Thailand and Indonesia were, by 1990, clearly moving up rapidly towards joining them. Their success was part of that of east Asia as a whole, and Japan had been indispensable to this outcome. The rapidity with which she, like China, recovered her former status as a power (and surpassed it) had obvious implications for her place both in the Asian and the world balance. By 1970 the Japanese had the second-highest GDP in the non-communist world. They had renewed their industrial base and had moved with great success into new areas of manufacture. Only in 1951 did a Japanese yard launch the country's first ship built for export; twenty years later, Japan had the largest shipbuilding industry in the world. At the same time she took a commanding position in consumer industries such as electronics and motorcars, of which Japan made more than any country except the United States. This caused resentment among American manufacturers, the supreme compliment. In 1979 it was agreed that Japanese cars should be made in England, the beginning of penetration of the EEC market. The debit side of this account was provided by a fast-growing population and by the ample evidence of the cost of economic growth in the destruction of the Japanese environment and the wear and tear of urban life.

Japan was long favoured by circumstances. Vietnam, like Korea, was a stroke of luck. The American enforcement of a bias towards investment rather than consumption during the occupation years also helped. Yet human beings must act to take advantage of favourable circumstances, and Japanese attitudes were crucial. Post-war Japan could deploy intense pride and an unrivalled willingness for collective effort among her people; both sprang from the deep cohesiveness and capacity for subordinating the individual to collective purposes which had always marked Japanese society. Strangely, such attitudes seemed to survive the coming of democracy. It may be too early to judge how deeply democratic institutions are rooted in Japanese society; after 1951 there soon appeared something like a consensus for one-party rule (though irritation with this quickly expressed itself in the emergence of more extreme groupings, some anti-liberal). Mounting uneasiness was shown, too, over what was happening to traditional values and institutions. The costs of economic growth loomed up not only in huge conurbations and pollution, but in social problems which strained even Japanese custom. Great firms still operated with success on the basis of group loyalties buttressed by traditional attitudes and institutions. None the less, at a different level, even the Japanese family seemed to be under strain.

Economic progress also helped to change the context of foreign policy, which moved away in the 1960s from the simplicities of the preceding decade. Economic strength made the yen internationally important and drew Japan into the monetary diplomacy of Europe. Prosperity involved her in the affairs of many parts of the world. In the Pacific basin, she was a major consumer of other countries' primary produce; in the Middle East she became a large buyer of oil. In Europe, Japan's investment was thought alarming by some (even though her aggregate share was not large), while imports of her manufactured goods threatened European producers. Even food supply raised international questions; in the 1960s 90 per cent of Japan's requirements for protein came from fishing and this led to alarm that the Japanese might be over-fishing important grounds.

As these and other matters changed the atmosphere and content of foreign relations, so did the behaviour of other powers, especially in the Pacific area. Japan increasingly assumed in the 1960s a position in relation to other Pacific countries not unlike that of Germany towards central and eastern Europe before 1914. She became the world's largest importer of resources, too. New Zealand and Australia found their economies increasingly and profitably tied in to Japanese consumption rather than to the old British market. Both of them supplied meat, and Australia minerals, notably coal and iron ore. On the Asian mainland the Russians and the South Koreans complained about the Japanese fishing. This added a new complication to an old story of economic involvement there. Korea was Japan's second biggest market (the United States was the biggest) and the Japanese started to invest there again after 1951. This at first revived a traditional distrust; it was ominous to find that South Korean nationalism had so anti-Japanese a tone that in 1959 the president of South Korea could urge his countrymen to unite 'as one man' against not their northern neighbour, but Japan. Within twenty years, too, Japanese car manufacturers were looking askance at the vigorous rival they had helped create. As in Taiwan, so in Korea industrial growth had been built on technology diffused by Japan. Furthermore, Japan's dependence on imported energy had meant that she underwent a nasty economic shock when oil prices shot up in the 1970s. Yet these causes for concern did not seem to affect her progress. Japanese exports to the United States in 1971 were worth $6,000,000,000; by 1984, that total had grown tenfold. By the end of the 1980s Japan was the world's second largest economic power in terms of GDP. As her industrialists turned to advanced information technology and biotechnology, and talked of running down car manufacturing, there was no sign that she had lost her power of disciplined self-adaptation.

Greater strength had already meant greater responsibilities. The withdrawal of American direction was logically rounded off twenty years later when Okinawa (one of the first of her overseas possessions to be re-acquired) was returned to Japan. There remained the question of the Kuriles, still in Russian hands, and of Taiwan, in the possession of the Chinese nationalists and claimed by the Chinese communists, but Japanese attitudes on all these matters remained – no doubt prudently – reserved. There was also the possibility that the question of Sakhalin might be reopened. All such issues began to look much more susceptible to revisions or at least reconsideration in the wake of the great changes brought to the Asian scene by Chinese and Japanese revival. The Sino-Soviet quarrel gave Japan much greater freedom for manoeuvre, both towards the United States, her erstwhile patron, and towards China and Russia. The embarrassment which too close a tie with the Americans might bring became clearer as the Vietnam War unrolled and political opposition to it grew in Japan. Her freedom was limited, in the sense that all three of the other great powers of the area were by 1970 equipped with nuclear weapons (and she, of all nations, had most reason to know their effect), but there was little doubt that Japan could produce them within a relatively brief time if she had to. Altogether, the Japanese stance had the potential to develop in various directions; in 1978 the Chinese vice-president visited Tokyo. In that year trade between China and Japan was worth as much as China's trade with the United States and West Germany combined. Indisputably, Japan was once more a world power.

If the test of that status is the habitual exercise of decisive influence, whether economic, military or political, outside a country's own geographical area, then by the 1980s India was clearly not one. This is perhaps one of the surprises of the second half of the century. India moved into independence with many advantages enjoyed neither by other former European dependencies, nor by Japan in the aftermath of defeat. She had in 1947 an effective administration, well-trained and dependable armed forces, a well-educated élite, thriving universities (some seventy of them), international benevolence and goodwill to draw upon and, soon, the advantages of Cold War polarization to exploit. She had also then had poverty, malnutrition and major public health problems to confront, but so did China, and the contrast between the two countries by the 1980s was very great and even visible; the streets of Chinese cities were by 1970 filled by serviceably (though drably) dressed and well-nourished people, while those of India still displayed horrifying examples of poverty and disease.

In considering India's poor development performance it is easy to be pessimistically selective. There were sectors where growth was substantial and impressive.

Separatism is seen as a major threat by Indian governments. Here, soldiers in Amritsar, where the Golden Temple is the chief shrine of the Sikh religion, prepare for a march by Sikh sextremists.

But such achievements are overshadowed by the fact that economic growth was followed closely by population increase; most Indians remained as poor, or only a little better-off, as those who welcomed independence in 1947.

It has been argued that to have kept India together at all was a great achievement, given the country's fissiparous nature and potential divisions. But in the 1980s there were signs that even this success might prove not much longer durable. Sikh particularism brought itself vividly to the world's notice by the assassination of a prime minister in 1984, after the Indian army had carried out an attack on the foremost shrine of Sikh faith at Amritsar. In the next seven years, more than ten thousand Sikh militants, innocent bystanders, and members of the security forces were to be killed. Fighting with Pakistan over Kashmir, too, broke out again in the later part of the decade. In 1990 it was officially admitted that 890 people had died that year in Hindu-Moslem riots, the worst since 1947. It was an ominous symptom of reaction to the dangers of division that an orthodox and deeply conservative Hindu party made its appearance in Indian politics as the first plausible threat to the hegemony of Congress.

That hegemony persisted, nonetheless. Congress forty years after independence was more visibly than ever not so much a political party in the European sense as an India-wide coalition of interest groups, notables and controllers of patronage, and this gave it, even under the leadership of Nehru, for all his socialist aspirations and rhetoric, an intrinsically conservative character. It was never the function of Congress, once the British were removed, to bring about change, but, rather, to accommodate it. This was in a manner symbolized by the dynastic nature of Indian government. Nehru had been succeeded as prime minister by his daughter, Mrs Gandhi, and she, after her murder, by her son, Rajiv Gandhi. When he, in turn, was blown up by an assassin (he was not in office at the time), Congress leaders almost at once showed an almost-automatic reflex in seeking to persuade his widow to take up the leadership of the party.

Democracy nevertheless succeeded briefly in giving India governments other than those of Congress. Mrs Gandhi had been defeated in 1977 by the Janata party, which then provided India with its first non-Congress government since independence. Her return to power in 1980 gave government back to Congress. Her son won an overwhelming electoral victory in 1985 and Congress continued in power until 1989 when elections produced a minority coalition (it, too, contained two members of the Nehru family) dependent on Janata and left-wing parties for its votes. On the other side of the coin, India's democratic record looked less firm after the authoritarianism showed by Mrs Gandhi in the 1970s (though it must be remembered that this led to her exclusion from office, even if not for long), or the recurrent use of president's powers to suspend normal constitutional government in specific areas, or the frequent brutalities of the police and security forces towards minorities. Under the Raj, Congress had sought for westernized Indians the constitutional rights of Englishmen: it was not very successful in guaranteeing them after independence, far less in extending them to the non-westernized.

Once again, it is difficult not to return to the banal reflexion that the weight of the past was very heavy in India, and that no dynamic force emerged to throw it off. As memories of pre-independence India faded, the reassertion of India's tradition was always likely. Symbolically, when the moment for independence had come in 1947 it had been at midnight, because the British had not consulted the astrologers to provide an auspicious day and a moment between two days had therefore to be chosen for the birth of a new nation: it was an assertion of the power of Indian ways which were to lose little of their force in the next forty years. Partition had then redefined the community to be governed in much more dominantly Hindu terms. By 1980 the last Indian civil service officer recruited under the British had retired. India has still not reached the point at which it can feel assured of its modernization. It lives still with a conscious disparity between its ingrafted western political system and the traditional society on which that has been imposed. For all the great achievements of many of its leaders, devoted men and women, the entrenched past, with all that means in terms of privilege, injustice and inequity, still stands in India's way. Perhaps those who believed in her future in 1947 were simply failing to recognize how difficult and painful fundamental change must be – and it is not for those of us who have found it so hard to accomplish much less fundamental change in our own society to be supercilious about that.

India's neighbour Pakistan turned more consciously to some aspects of her own tradition – the Islamic – and in so doing participated in a movement of renewal which was visible across most of the Moslem world. Not for the first time, western politicians have again had occasion in recent decades to recall that Islam is the creed of lands which stretch from Morocco in the west to China in the east. Indonesia, the largest south-east Asian country, Pakistan, Malaysia and Bangladesh between them contain nearly half the world's Moslems. Beyond those countries and the Arabic lands, both the Soviet Union and Nigeria, the most populous African country, also had large numbers of Moslem subjects (as long ago as 1906, the tsarist government of Russia had been alarmed by revolution in Iran because of its possibly disturbing effect on the Tsar's Moslem peoples). But new perceptions of the Islamic world took time to appear, and well into the 1970s the rest of the world thought mainly of the Arab countries of the Middle East, and especially of the oil-rich states among them, when it thought of Islam much at all.

So far as it went – and sensitivity to the area's problems was much heightened after the oil crisis of 1974 – this limited perception was also for a long time obscured and confused by the perspectives of the Cold War. The shape of that conflict sometimes blurred into older frameworks, too; to some observers a traditional Russian desire for influence in the area seemed to be a strand in Soviet policy now nearer satisfaction than at any time in the past. The Soviet Union had by 1970 a world-wide naval presence rivalling that of the United States and established even in the Indian Ocean, the only great maritime region in which Cold War confrontations had yet to occur. Following British withdrawal from Aden in 1967, that base had been used by the Russians with the concurrence of the South Yemen government. All this was taking place at a time when further south, too, there had been strategic setbacks for the Americans. The coming of the Cold War to the Horn of Africa and the former Portuguese colonies had added significance to events taking place further north.

Yet Soviet policy, in a longer perspective, does not appear to have benefited much within the Moslem world from the disarray of the Americans, marked though that was in the mid-1970s in the Middle East. Egypt had by then fallen out with Syria. She had turned to the United States in the hope of making a face-saving peace with Israel. When in 1975 the General Assembly of the United Nations denounced Zionism as a form of racism and granted the PLO 'observer' status in the Assembly it was a consequence that Egypt was more and more isolated from other Arab states. By this time, the PLO's activity across the northern border was not only harassing Israel, but was steadily driving Lebanon, once a bastion of western values, into ruin and disintegration. In 1978 Israel invaded southern Lebanon in the hope of ending the PLO raids. Though the non-Islamic world applauded when the Israeli and Egyptian prime ministers the following year met in Washington to agree a peace providing for Israel's withdrawal from Sinai, the Egyptian paid the price of assassination three years later by those who felt he had betrayed the Palestinian cause.

The limited settlement between Israel and Egypt owed much to President Carter, the Democratic candidate who had won the American presidential election of 1976. American morale was by then suffering from setbacks other than those in the Middle East. The Vietnam War had destroyed one president and his successor's presidency had been built on the management of American withdrawal and the 1973 settlement (though it was soon clear how little that settlement was worth). There was in the background, too, the fear many Americans shared of the rising strength of the USSR in ballistic missiles. All this affected American reactions to an almost wholly unforeseen event in 1979, which not only dealt a damaging blow to the United States, but revealed a potentially huge new dimension to the troubles of the Middle East. This was the overthrow of the Shah of Iran.

Long the recipient of American favour as a reliable ally, in January 1979 the Shah was driven from his throne and country by a coalition of outraged liberals and Islamic conservatives. An attempt to secure constitutional government soon collapsed as popular support rallied to the Islamic faction. Iran's traditional ways and social structure had been dislocated by a policy of modernization in which the Shah had followed – with less caution – his father Reza Khan. Almost at once, there emerged a Shi'ite Islamic Republic, led by an elderly and fanatical cleric. The United States quickly recognized the new regime, but unavailingly. It was tarred with guilt by association, as the patron of the former Shah and the outstanding embodiment of capitalism and western materialism. It was small consolation that the Soviet Union was soon undergoing similar vilification by the Iranian religious leaders, as the second 'Satan' threatening the purity of Islam. Some Americans were encouraged, though, when the particularly ferocious Ba'ath regime in Iraq, already viewed with favour for its ruthless execution and pursuit of Iraqi communists, fell out with the new Iran in a conflict inflamed (in spite of Ba'athist secularism) by the traditional animosity of Mesopotamian Suni and Persian Shi'ite Moslems. It was in July 1979 that Saddam Hussein took over as president in Baghdad, and that looked encouraging to the State Department: he was likely to offset the Iranian danger in the Gulf, it seemed.

The Iranian revolution nevertheless implied more than just the American loss of a client state. Though a coalition of grievances had made possible the overthrow of the Shah, a speedy reversion to archaic tradition (strikingly, in the treatment of women) showed that more than a ruler had been repudiated. The Iranian Islamic republic was also an expression of the rage shared by many Moslems world-wide (especially in Arab lands) at the onset of secular westernization and the failure of the promise of modernization. In the Middle East, as nowhere else, nationalism, socialism and capitalism had failed to solve the region's problems – or at least to satisfy passions and appetites they had aroused. Moslem fundamentalists thought that Ataturk, Reza Khan and Nasser had all led their people down the wrong road. Islamic societies had successfully resisted the contagion of atheistic communism but to many Moslems the contagion of the West

Rioting in Teheran and many other places in Iran in 1978 led to the overthrow of a government, the imposition of martial law in all major cities, and promises to end corruption – all to no avail. The religious leaders of the mobs remained determined to overthrow the Shah.

now seemed even more threatening. Paradoxically, the western revolutionary notion of capitalist exploitation helped to feed this revulsion of feeling.

The roots of Islamic fundamentalism (to use a crude blanket term) were varied and very deep. They could tap centuries of struggle against Christianity. They were refreshed from the 1960s onwards by the obviously growing difficulties of western powers (including the USSR) in imposing their will on the Middle East and Persian Gulf, given their Cold War divisions. There was the mounting evidence for many Arabs that the western principle of nationality advocated since the 1880s as an organizational remedy for the instability which followed Turkish decline had not worked (only too evidently in the 1980s, it would be clear that the wars of the Ottoman Succession were not over). A favourable conjunction of embarrassments was made more promising still by the recent revelation of the potency of the oil factor. But then there was also, since 1945, the growing awareness of pious Moslems that western commerce, communications and the simple temptations offered to those rich with oil, were more dangerous to Islam than any earlier (let alone purely military) threat had been. When the descendants of Mehemet Ali (as it were) sent their sons to Harvard and Oxford, they did not, it was feared, acquire only academic instruction there but bad habits as well. This made for strain within Islam.

This was not all that divided Moslems. Suni and Shi'ite hostility went back centuries. In the post-1945 period, the Ba'ath socialist movement which inspired many Moslems and which was nominally entrenched in Iraq, had become anathema to the Moslem Brotherhood, which deplored the 'godlessness' of both sides even in the Palestinian quarrel. Popular sovereignty was a goal fundamentalists rejected; they sought Islamic control of society in all its aspects, so that, before long, the world began to be used to hearing that Pakistan forbade mixed hockey, that Saudi Arabia punished crime by stoning to death and amputation of limbs, that Oman was building a university in which men and women students were to be segregated during lectures – and much, much more. By 1980, the fundamentalists were powerful enough to secure their goals in some Islamic countries. Even students in a comparatively 'westernized' Egypt had already by 1978 been voting for the fundamentalists in their own elections, while some of the girls among them were refusing in medical school to dissect male corpses and demanded a segregated, dual system of instruction.

To put such attitudes in perspective (and at first sight it is not obvious to western eyes why student 'radicals' should happily espouse such obviously reactionary causes), they have to be understood in the context of a long absence within Islam of any state or institutional theory such as that of the West. Even in orthodox hands, and even if it delivered some desirable goods, the state as such is not self-evidently a legitimate authority in Islamic thought – and, on top of that, the very introduction of state structures in Arab lands since the nineteenth century had been in imitation, conscious or unconscious, of the West. Youthful radicalism which had tried and found wanting the politics of left-wing socialism (or what was thought to be that, and was in any case another western import) felt that no intrinsic value resided in states or nations; it looked elsewhere, and that, in part, explains the efforts shown first in Libya, and then in Iran, to arrive at new ways of legitimating authority. Whether the age-old Islamic bias against public institutions and towards tribalism and the brotherhood of Islam can be sustained remains to be seen.

The violence of politics in many Arab states frequently exhibits a simple polarization between repressive authoritarianism on the one hand and the fundamentalist wave on the other. In the 1980s both Morocco and Algeria were to find their domestic order thus troubled. The situation was made the more dangerous and explosive by the demography of the Arab lands. The average age of most Islamic societies is said to be between 15 and 18, and they are growing at very fast rates. There is just too much youthful energy and frustration about for the outlook to be promising for peace.

Soon after the Iran revolution, students in Teheran worked off some of their exasperation by storming the American embassy and seizing diplomats and others as hostages. A startled world suddenly found the Iranian government supporting the students, taking custody of the hostages and endorsing the students' demands for the return of the Shah. President Carter could hardly have faced a more awkward situation at a moment when American policy in the Islamic world was above all preoccupied with Soviet intervention in Afghanistan. A severance of diplomatic relations with Iran and the imposition of economic sanctions were the first responses. Then came an attempted rescue operation, which failed dismally. The unhappy hostages were in the end to be recovered by negotiation (and, in effect, a ransom: the return of Iranian assets in the United States which had been frozen at the time of the revolution), but the humiliation of the Americans was by no means the sole or even the major importance of the episode. Besides its wide policy repercussions, the retention of the hostages was a symbolic moment. It was a shock (registered in a unanimous vote of condemnation at the UN) to the convention evolved first in Europe and then developed over more than three centuries throughout the civilized world that diplomatic envoys should be immune from interference. The Iranian government's action announced that it was not playing by the accepted rules. That was a blatant rejection of western assumptions which made some in the West wonder what else Islamic revolution might imply.

4
The End of an Era

The historian has to distil his account from events which flow uninterruptedly. Now, as never before, those events quickly affect other events the world round and that, too, makes organizing the story of their flow harder. What is more, though they are likely to show greater and faster interaction than in past ages, the current does not run evenly. In one place it hesitates, or is blocked, while elsewhere it rushes on. The phasing of one part of the story does not always match that of other parts. Nevertheless, there are sometimes sweeping waves of change; almost kaleidoscopic rearrangements take place in long-accepted facts and assumptions about very important matters and over wide areas. Suddenly, new mental maps seem to be needed and we talk of turning-points and new eras. The weight of the past seems for a moment to be overcome, or something within it to be breaking out in a new way. This is the sort of thing which justifies historians in attaching special importance to such markers as the christianizing of the Roman empire, the destruction of the unity of Christendom in the sixteenth or seventeenth centuries, or the launching of new political ideas on the world by the French Revolution. And now it begins to look as if we may have been experiencing recently the start of just such a big change. Its meaning and even its outlines, at least so far as the world order to which we have grown used is concerned, are still far from clear. But it is a global change at least in political arrangements and assumptions, and (though we must not be too rigid in distinguishing sequences) it has become apparent only since 1985, though more and more dramatically as the end of the decade approached.

The story can be taken up ten years earlier, in a region where great change then seemed imminent but was, in fact, not to follow. This was the Middle East. Though tension there was high in 1980, great disappointment lay ahead for those who hoped for emergence from the Arab-Israeli impasse. For a time, the Iranian revolution looked as if it might transform the rules of the game played hitherto. Ten years later, it was still very difficult to assess in a balanced way what had actually changed, or what was the true significance of Islamic fundamentalism. What had looked for a time like a unified Islamic resurgence can now be seen, at least in part, as one of the recurrent waves of puritanism which have across the centuries helped from time to time to regenerate the Faithful. Clearly, too, it owed much to circumstance; Israel's occupation of the third of Islam's Holy Places in Jerusalem had suddenly enhanced the sense of Islamic solidarity. Yet, turning to the Islamic past brought its own difficulties. An attack by Iraq on Iran in 1980 led to a bloody war lasting eight years and costing a million lives. Once more, Islamic peoples were divided along ancient lines: Iraq was Sunnite, Iran Shi'ite.

It soon appeared, too, that although she could irritate and alarm the superpowers (the USSR especially, because of its millions of Moslem subjects), Iran could not thwart them. At the end of 1979, her rulers had to watch helplessly while a Russian army went to Afghanistan to support a puppet communist regime there against Moslem rebels. This was one reason why they backed terrorists and kidnappers; it was the best (or worst) they could do. Nor, in spite of their success over the American hostages, could the Iranians get back the former Shah to face Islamic justice.

By successfully tweaking the eagle's tail-feathers in the hostage affair, Iran, it is true, humiliated the United States, but in perspective this seems much less significant than it did at the time. In retrospect, a declaration by President Carter in 1980 that the

THE SOVIET UNION AND ITS SUCCESSORS
Territory acquired 1939–45
Border disputes with China and Mongolia after 1945
Independent republics emerging 1990–91
N
0 1000 km
0 600 miles
Arctic Ocean
Bering Sea
Sea of Okhotsk
Sea of Japan
Khabarovsk
Khabarovsk
RUSSIAN FEDERATION
MONGOLIA
CHINA
KAZAKHSTAN
UZBEKISTAN
KYRGYZSTAN
TAJIKISTAN
TURKMENISTAN
Alma-Ata
Bishkek
Dushanbe
Tashkent
Ashkhabad
Aral Sea
Caspian Sea
Baku
AZERBAIJAN
Yerevan
ARMENIA
Tbilisi
GEORGIA
Black Sea
Kishinev
MOLDOVA
UKRAINE
Kiev
Minsk
BELARUS
Moscow
ESTONIA
LATVIA
Tallinn
Riga
Vilnius
LITHUANIA
PART OF RUSSIAN FEDERATION
Baltic Sea

United States regarded the Persian Gulf as an area of vital interest looks more important. It was an early sign of the ending of a dangerously exaggerated mood of American uncertainty and defeatism. A central reality of international politics was about to reassert itself. For all the dramatic changes since the Cuban crisis, the American republic was still in 1980 one of the only two states whose might gave them unquestioned status as (to use an official Soviet definition) 'the greatest world powers, without whose participation not a single international problem can be solved'. This participation in some instances would be implicit rather than explicit, but it was a fundamental *datum* of the way the world worked. Even a spectacular economic challenge in the Far East and oil blackmail could not affect it, let alone international terrorism.

History, moreover, has no favourites for long. Though some Americans had been alarmed by the growth in Soviet strength after the Cuban missile crisis, there were plentiful signs by the early 1970s that the Soviet rulers were in difficulties. They had to face a truism that Marxism itself proclaimed, that consciousness evolves with material conditions. Among other results of the real but limited rewards Soviet society had given its citizens were two: an evident dissidence, trivial in scale but suggesting a growing demand for greater spiritual freedom, and a less explicit, but real, groundswell of opinion that further material gains should be forthcoming. The Soviet Union nevertheless continued to spend colossal sums on armaments (of the order of a quarter of its GDP in the 1980s). Yet these could hardly suffice, it appeared. To carry even this burden, western technology, management techniques and, possibly, capital, would be needed. What change might follow on that was debatable, but that there would be change was certain. If nothing else, it seemed unlikely that the huge Soviet military-industrial complex would passively accept a real diminution of its role in the USSR.

Reassuringly, there had grown even stronger by 1980 a compelling tie between the two superpowers. For all the huge effort to give the Soviet Union superior nuclear striking power over the United States, superiority in these matters at such a level is a rather notional affair. The Americans, with their gift for the arresting slogan, concisely summed up the situation as MAD; that is to say, both countries had the capacity to produce 'Mutually Assured Destruction', or, more precisely, a situation in which each of two potential combatants had enough striking power to ensure that even if a surprise attack deprived it of the cream of its weapons, what remained would be sufficient to ensure a reply so appalling as to leave an opponent's cities smoking wildernesses and its armed forces capable of little but attempting to control the terrorized survivors.

This bizarre possibility was a great conservative force. Even though madmen (to put the matter simply) are occasionally to be found in seats of power, Dr Johnson's observation that the knowledge that you are to be hanged wonderfully concentrates the mind is applicable to collectivities threatened with disaster on this scale: the knowledge that a blunder may be followed by extinction is a great stimulus to prudence. Here may well lie the most fundamental explanation of a new degree of cooperation which began to be shown in the 1970s by the United States and the Soviet Union in spite of specific quarrels. A 1972 treaty on missile limitation was one of its first fruits; it owed something, too, to a new awareness on both sides that science could now monitor infringements of such agreements (not all military research made for an increase of tension). In the following year talks began on further arms limitations while another set of discussions began to explore the possibility of a comprehensive security arrangement in Europe.

In return for the implicit recognition of Europe's post-war frontiers (above all, that between the two Germanies), the Soviet negotiators finally agreed in 1975 at Helsinki to increase economic intercourse between eastern and western Europe and to sign a guarantee of human rights and political freedom. The last was, of course, unenforceable. Yet it may well have much outweighed the symbolic gains of frontier recognition to which the Soviet negotiations had attached such importance. Western success over human rights was not only to prove a great encouragement to dissidents in communist Europe and Russia, but silently swept aside old restraints on what had been deemed interference in the internal affairs of communist states. Though very slowly, there began to flow public criticisms which were in the end to help to force change on communist governments in eastern Europe. Meanwhile, the flow of trade and investment between the two Europes began almost at once to rise, though slowly. It was the nearest approach so far to a peace treaty ending the Second World War, and it gave the Soviet Union what its leaders most desired, assurance of the territorial security which was one of the major spoils of victory in 1945.

For all that, Americans were very worried about world affairs as 1980, the year of a presidential election, approached. Eighteen years before, the Cuban crisis had shown the world that the United States was top dog. She had then enjoyed superior military strength, the (usually dependable) support of allies, clients and satellites the world round, and the public will to sustain a world diplomatic and military effort while grappling with huge domestic problems. By 1980, many of her citizens felt the world had changed and were unhappy about it. When the new Republican president, Ronald Reagan, took office in 1981, many of his supporters looked back on a decade of what seemed increasing American powerlessness. He inherited an enormous budgetary deficit, disappointment over what looked like recent successful initiatives by the Russians in Africa and Afghanistan, and dismay over what was believed to be the disappearance of the superiority in nuclear weapons the United States had enjoyed in the 1960s.

In the next five years Mr Reagan was to restore the morale of his countrymen by remarkable (even if often cosmetic) feats of leadership. Symbolically, on the day of his inauguration, the Iranians released their American hostages, the close of a humiliating and frustrating episode (many Americans believed the timing of the release to have been stage-managed by the new administration's supporters). This was by no means the end of the problems American policy faced in the Middle East and the Gulf. Two fundamental difficulties did not go away – the threat posed to international order in that area while Cold War attitudes endured, and the question of Israel. The war between Iran and Iraq was evidence of the danger, many people thought. Soon, the dangerous instability of some Arab countries became more obvious. Ordered government virtually disappeared in the Lebanon, which collapsed into an anarchy disputed by bands of gunmen patronized by the Syrians and Iranians. This gave the revolutionary wing of the PLO a much more promising base for operation against Israel than in the past. Israel, therefore, took to increasingly violent and expensive military operations on and beyond her northern borders. There followed in the 1980s a heightening of tension and ever more vicious Jewish-Palestinian conflict.

The United States was not the only great power troubled by these enduring ills. When the Soviet Union sent its soldiers to Afghanistan in 1979 (where they were to stay bogged down for most of the next decade), Iranian and Moslem anger elsewhere was bound to affect Moslems inside the Soviet Union. Some thought this a hopeful sign, believing the growing confusion of the Islamic world might induce caution on the part of the two superpowers, and perhaps lead to less unconditional support for their satellites and allies in the region. This mattered most, of course, to Israel. Meanwhile, the more alarming manifestations and rhetoric of the Iranian revolution made it look for a moment as if a true conflict of civilizations was beginning. Its aggressive puritanism, though, also caused shivers among conservative Arabs and in the oil-rich kingdoms of the Gulf – above all, Saudi Arabia.

There were indeed numerous signs of what looked like spreading sympathy for the radical conservativism of the Iranian revolution in other Islamic countries. Fundamentalists murdered the president of Egypt in 1981. The government of Pakistan continued to proclaim (and impose) its Islamic orthodoxy, and winked at assistance to the anti-communist Islamic rebels in Afghanistan. (Yet by the end of the decade it had accepted a woman as prime minister, uniquely among Islamic countries, and even, in 1989, rejoined the British Commonwealth.) North Africa presented more alarming evidence of radical Islamic feeling as the decade drew on. This was less a matter of the bizarre sallies and pronouncements of the excited dictator of Libya (he called upon other oil-producing states to stop supplying the United States while one third of Libyan oil continued to find a market there, and in 1980 briefly 'united' his country to Ba'athist Syria) than of political developments further west. Algeria had made a promising start after winning its independence, but by 1980 its economy was flagging, the consensus which had sustained the independence, movement was crumbling, and emigration to look for work in Europe seemed the only outlet available for the energies of many of her young men. In the 1990 elections in Algeria, for the first time in any Arab country, an Islamic fundamentalist party won a majority of votes. In the previous year a military *coup* had brought a military and militant Islamic fundamentalist regime to power in the Sudan which at once suppressed the few remaining civic freedoms of the people of that unhappy land.

Nonetheless, for all the attractions of Islamic radicalization, there were plentiful signs by 1990 that more moderate and conservative Arab politicians had been antagonized, and that indigenous opposition to the fundamentalists was significant and sometimes effective. It remains hard to believe that sufficient leverage is available to the would-be revolutionaries, even after setting aside both such political realities and deeper questions about the feasibility of successful Islamic revolution, when so many of its would-be supporters sought, unknowingly, to realize goals of power and modernization systematically incompatible with Islamic teaching and custom. Libya could destabilize other African countries and arm Irish terrorists, but achieved little else. Because of preoccupations of changing circumstances elsewhere, the old Soviet-American rivalry was decreasingly available for exploitation. All that was left for the fundamentalists to look to were two potentially rich countries, Iraq and Iran, and for most of the 1980s they were fatally entangled in a costly struggle with one another.

The Intifada revolt of 1989 began with this demonstration to celebrate the twenty-fith anniversary of the founding of the PLO.

There was also growing evidence that the ruler of Iraq, patronized by the Americans and the major trouble-maker of the Middle East, was only tactically and pragmatically a supporter of Islam. Saddam Hussein was a Moslem by upbringing, but led a formally secular Ba'athist regime actually based on patronage, family and the self-interest of soldiers. He sought power, and technological modernization as a way to it, and there is no evidence that the welfare of the Iraqi people ever concerned him. When he launched his war on Iran, the prolongation of the struggle and evidence of its costs were greeted with relief by other Arab states – notably the other oil-producers of the Gulf – because it appeared at the same time to pin down both a dangerous bandit and the Iranian revolutionaries whom they feared. It was, however, less pleasing to them that the war distracted attention from the cause of the Palestinians question and unquestionably made it easier for Israel to deal with the PLO.

During nearly a decade of alarums and excursions in the Gulf, some of which raised the spectre of further interference with Western oil supplies, incidents seemed at times to threaten a widening of armed conflict, notably between Iran and the United States. Meanwhile, events in the Levant embittered the stalemate there. Israel's annexation of the Golan Heights, her vigorous operations in Lebanon against Palestinian guerilla bands and their patrons, and her government's encouragement of further Jewish immigration (notably from the USSR) all helped to buttress her against the day when she might once again face united Arab armies. At the end of 1987, though, there came the first outbreaks of violence among Palestinians in the Israeli-occupied territories. They persisted and grew into an enduring insurrection, the *intifada*. The PLO, though it won further international sympathy by officially recognizing Israel's own right to exist was none the less in a disadvantaged position in 1989, when the Iraq-Iran war finally ended. In the following year the Ayatollah Khomeni died and there were signs that his successor might be less adventurous in support of both the Palestinian and the fundamentalist Islamic cause.

During the Iraq-Iran war, the United States had favoured Iraq, seeing Iran as a danger. This was in part because of American over-estimation of the fundamentalist threat. When, nevertheless, the Americans found themselves at last face-to-face at war in the Gulf with a declared enemy, it was with the Iraqis, not the Iranians. In 1990, after making a generous peace with Iran, Hussein took up an old border dispute with the sheikdom of Kuwait. He had also quarrelled with its ruler over oil quotas and prices. It is not easy to believe in the reality of these grievances; whatever they may have meant symbolically to Hussein himself, what seems to have moved him most was a simple determination to seize the immense oil wealth of Kuwait. During the summer of 1990, his threats increased. Then, on 2 August, the armies of Iraq invaded Kuwait, and in a few hours subdued it.

There followed a remarkable mobilization of world opinion against Iraq in the UN. Hussein sought to play both the Islamic and the Arab cards by confusing the pursuit of his own predatory ambitions with Arab hatred for Israel. Except in the streets of Middle Eastern cities, where there were many demonstrations in his favour, they proved to be cards of very low value. Only the PLO and Jordan spoke up for him. No doubt to his shocked surprise, Saudi Arabia, Syria and Egypt, on the other hand, actually became improbable partners in the alliance which rapidly formed against him. Almost equally surprising to him must have been the acquiescence of the USSR in what followed. Finally, and most startlingly of all, the United Nations Security Council produced (with overwhelming majorities) a series of resolutions condemning Iraq's actions and, finally, authorizing the use of force against her to ensure the liberation of Kuwait. Once again, a much-feared second oil crisis did not follow.

Huge forces were assembled in Saudi Arabia under American command. On 16 January 1991 they went into action. Within a month Iraq gave in and withdrew, after suffering considerable loss (allied losses were by comparison insignificant). It is too soon to weigh all the consequences of that humiliation, though it did not obviously threaten Hussein's survival. It was not the turning-point in the Middle East which so many had longed for. Yet, in spite of Hussein's attempts to inspire an anti-Israel Islamic crusade, he had not found takers; it was no turning-point for Arab revolutionaries, either. The greater losers were the PLO. Israel was the greatest gainer. Arab military success at her expense was inconceivable for the near future. Yet at the end of one more war of the Ottoman succession, the Israel-Arab problem was still there. Syria and Iran had already before the Kuwait crisis begun to show signs that, for their own reasons, they intended

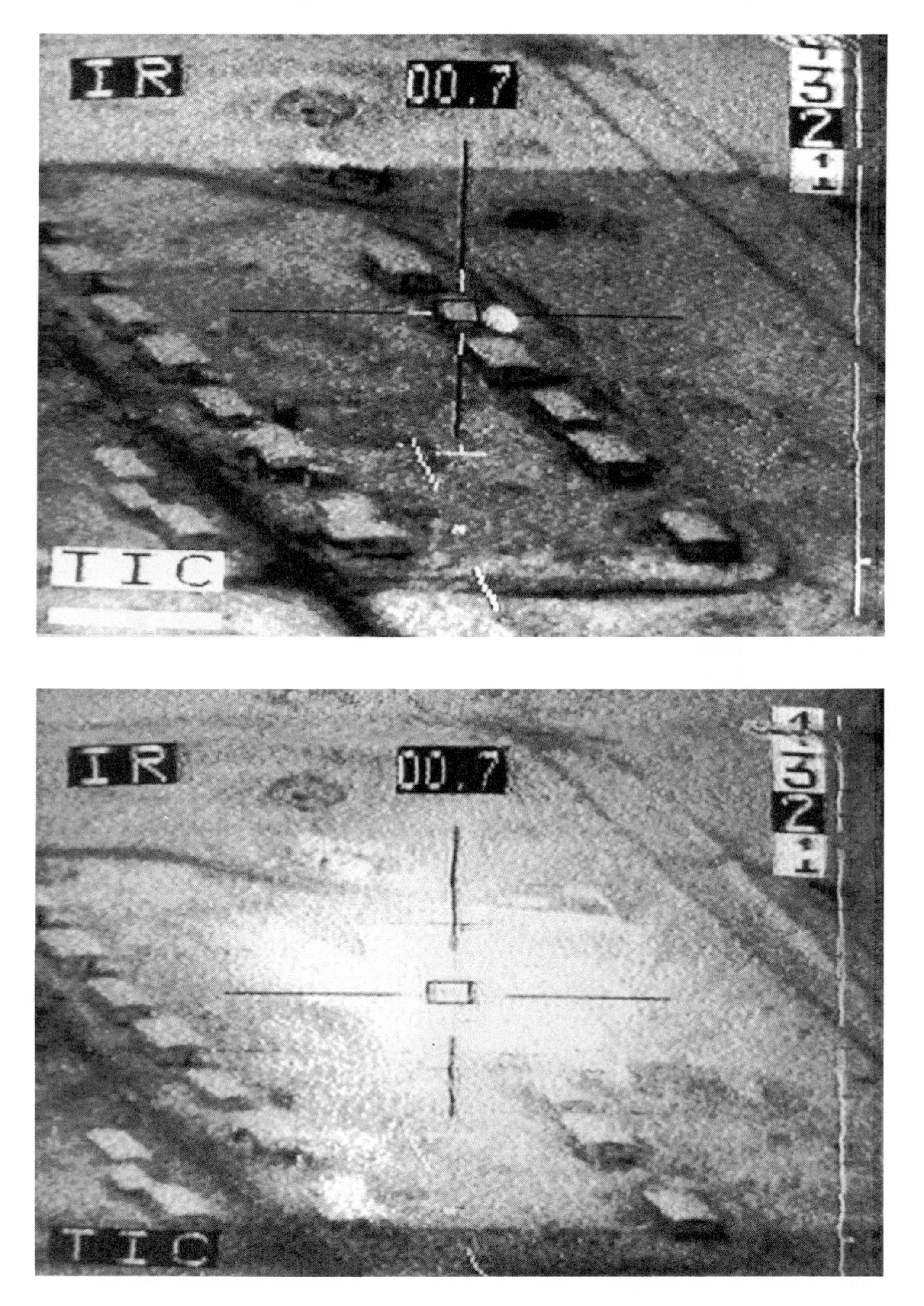

A new kind of war. Photographs of a laser-guided missile fired at a munitions dump in Kuwait, January 1991.

to make attempts to get a negotiated settlement, but whether one would emerge was another matter. For the United States, it was clearly more of a priority than ever to get one and that encouraged hopes that Israel might, at last, show less intransigence. Meanwhile, the alarming spectre of a radical and fundamentalist pan-Islamic movement had been dissipated. For all the unrest and discontent with which Islamic countries faced the West, there was virtually no sign that their resentments could yet be coordinated in an effective response, and less than ever that they could do without the subtly corrosive means of modernization which the West offered. Almost incidentally, too, crisis in the Gulf appeared to reveal that the oil weapon had lost much of its power to damage the developed world. It was against this background that American diplomacy was at last successful in 1991 in persuading Arabs and Jews to take part in a conference on the Middle East.

The great transformations of the 1980s had been elsewhere. They bore upon events in the Middle East only because they shaped what the USA and USSR did there. In 1979–80, the American presidential election campaign had been deliberately fought so as to play on the public's fears of the Soviet Union. Unsurprisingly, this re-awoke animosity at the official level; the conservative leaders of the Soviet Union showed renewed suspicion of the trend of United States policy. It seemed likely that promising steps towards disarmament might be swept aside – or even worse. In the event, the American administration showed a remarkable pragmatism in foreign affairs, while, on the Soviet side, internal change was to open the way to greater flexibility.

One crucial event was the death, in November 1982 of Leonid Brezhnev, Khrushchev's successor and for eighteen years general secretary of the Party. His immediate replacement (the head of the KGB) soon died and a septuagenarian whose own death followed even more quickly succeeded him before there came to the office of general secretary in 1985 the youngest member of the Politburo, Mr Mikhail Gorbachev. He was fifty-four. Virtually the whole of his political experience had been of the post-Stalin era. His impact upon his country's, and the world's, history cannot yet be properly assessed. His personal motivation and the conjunction of forces which propelled him to the succession remains unclear. The KGB, presumably, did not oppose his promotion, and his first acts and speeches were orthodox (although he had already, in the previous year, made an impression on the British prime minister as 'someone with whom business could be done'). He soon articulated a new political tone. The word 'communism' was heard less in his speeches and 'socialism' was re-interpreted to exclude egalitarianism (though from time to time he reminded his colleagues that he *was* a communist). For want of a better term, his aim was seen as liberalization, which was an inadequate western attempt to sum up two Russian words he used a great deal: *glasnost* (openness) and *perestroika* (restructuring). The implications of the new course were to be profound and dramatic, and for the remainder of the decade Mr Gorbachev was to grapple with them. In the end there became clear his recognition that the Soviet economy could no longer provide its former military might, sustain its commitments to allies abroad, improve (however slowly) living standards at home, and assure self-generated technological innovation without radical modernization. The implications of that were vast.

Mr Reagan drew the first dividends on Mr Gorbachev's arrival in power. In the sphere of foreign affairs, the Soviet leader's new course soon became clear in their meetings. Discussion of arms reduction was renewed. Agreements were reached on other issues, and this was made easier by the decision of the Soviet leadership in 1989 to withdraw their forces from Afghanistan. In America's domestic politics, a huge and still growing budgetary deficit, and a flagging economy which would under most presidents have produced political uproar were for years virtually lost to sight in the euphoria produced by a seeming transformation of the international scene. The alarm and fear with which the 'evil empire' (as Mr Reagan had termed it) of the Soviet Union was regarded by many Americans began to evaporate. Optimism and confidence grew as the USSR showed signs of growing division and difficulty in reforming its affairs, while Americans were promised wonders by their government in the shape of new defensive measures in outer space. Though thousands of scientists said the project was unrealistic, the Soviet government could not face the costs of competing with that. Americans were heartened, too, in 1986 when American bombers were launched from England on a punitive mission against Libya, whose unbalanced ruler had been supporting anti-American terrorists (significantly, the Soviet Union expressed less concern about this than did many west Europeans). The President was less successful, though, in convincing many of his countrymen that more enthusiastic assertions of American's interests in Central America were truly in their interests. But he survived as a remarkably popular figure; only after he had left office did it begin to dawn that for most Americans the decade had been one in which they had got poorer. Only rich Americans got richer during the Reagan presidency.

In 1987, the fruits of negotiation on arms control were gathered in an agreement over intermediate range nuclear missiles. In spite of so many shocks and its erosion by the emergence of new foci of power, the nuclear balance had held long enough for the first stand-downs by the superpowers. They had shown they could still manage their conflicts and the world's crises without all-out war, after all. They at least, if not other countries seeking to acquire nuclear weapons, appeared to have recognized that nuclear war, if it came, held out the prospect of virtual extinction for mankind, and were slowly beginning to do something about it. In 1991 there were further dramatic developments as the USA and USSR agreed to major reductions in existing weapons stocks.

In 1986 President Reagan and Chairman Mikhail Gorbachev held a 'summit' meeting in Iceland. There were no formal results but the meeting was important in establishing personal relations between the two men.

Such a huge change in the international scene cannot be disentangled from its many consequences for others than the superpowers. They have to be artificially separated to be expounded, but one could not have occurred without the other. At the end of 1980 there had been little reason to believe that the peoples of eastern Europe and the Soviet Union were about to enter a new phase of their destiny, a decade bringing changes unmatched since the 1940s. All that then could be seen was the growing difficulty of European communist countries in maintaining even the modest growth rates they attained. Comparison with the market economies of the non-communist world had become more and more unfavourable to them. Yet this hardly suggested any challenge to the verdicts of 1952, 1956 and 1968: in 1980 Soviet power seemed to hold eastern Europe as firmly as ever in its grip.

What had been, in a measure, lost to sight within the carapace provided by the Warsaw Pact was social and political change which had been going on for thirty years (and more, if one counts the great unwilled changes of the Second World War and its aftermath). At first sight, the outcome of a long experiment with a particular model of development had been a remarkable uniformity. In each communist-ruled country, the party was supreme; careerists built their lives round it as, in earlier centuries, men on the make clustered about courts and patrons. In each, (and above all in the USSR itself) there was also an unspeakable and unexaminable past which could not be mourned or deplored, whose weight overhung and corrupted intellectual life and political discussion, so far as there was any. As for the east European economies, investment in heavy industrial and capital goods had produced a surge of early growth (in some of them more vigorous than in others) and then an international system of trading arrangements with other communist countries, dominated by the USSR and rigidified by aspirations to central planning. Increasingly and obviously, a growing thirst for consumer goods could not be met; commodities taken for granted in western Europe remained luxuries in the east European countries, cut off as they were from the advantages of international economic specialization. On the land, private ownership had been much reduced by the middle of the 1950s, usually to be replaced by a mixture of co-operatives and state farms, though, within this broadly uniform picture, different patterns had later emerged. In Poland, for instance, peasants were already moving back into smallholdings by 1960; eventually, something like four-fifths of Polish farmland was to return to private ownership even under communist government. Output remained low; in most east European countries agricultural yields were only from half to three-quarters those of the European Community. By the 1980s all of them, in varying degree, were in a state of economic crisis with the possible exception only of the GDR. Even there, *per capita* GDP stood at only $9,300 a year in 1988, against $19,500 in the Federal Republic.

What had come to be called the 'Brezhnev doctrine' (after a speech that functionary made in Warsaw in 1968) said that developments within eastern bloc countries might require – as in Czechoslovakia that year – direct intervention to safeguard the interests of the USSR and its allies against any attempts to turn socialist economies back towards capitalism. Yet Brezhnev had also been interested in pursuing *détente* and it was not unreasonable to interpret his doctrine as a recognition of the possible dangers presented to international stability by breakaway developments in communist Europe, and a way of limiting them by drawing clearer lines. Since then, internal change in western countries, steadily growing more prosperous, and with memories of the late 1940s and the seeming possibility of subversion far behind them, had not increased east-west tension. By 1980, after revolutionary changes in Spain and Portugal, not a dictatorship survived west of the Trieste-Stettin line and democracy was everywhere triumphant. For thirty years, the only risings by industrial workers against their political masters had been in East Germany, Hungary, Poland and Czechoslovakia – all communist countries (conspicuously, when Paris was in uproar in 1968, and student riots destroyed the prestige of de Gaulle's government, the Parisian working-class had done nothing).

After 1970, and even more after the Helsinki agreement of 1975, as awareness of contrasts with western Europe grew in the eastern *bloc*, dissident groups emerged, survived and even strengthened their positions, in spite of severe repression. Gradually, too, a few officials or economic specialists, and even some party members, began to show signs of scepticism about the effects of detailed centralized planning and there was increasing discussion of the advantages of utilizing market mechanisms. The key to stability in the east, nevertheless, remained the Soviet Army. There was no reason to believe that fundamental change was possible in any of the Warsaw Pact countries if the Brezhnev doctrine held, and continued to provide support to governments subservient to the USSR.

1980, Gdansk. Lech Walesa addresses his supporters in the Lenin shipyard.

The first clear sign that this might not always be the case came in the early 1980s, in Poland. The Polish nation had retained, to a remarkable degree (but not for the first time in its history), its collective integrity by following its priests and not its rulers. The Roman Catholic church had an enduring hold on the affections and minds of most Poles as the embodiment of the nation, and was often to speak for them – all the more convincingly when a Polish pope had been enthroned. It did so on behalf of workers who protested in the 1970s against economic policy, and condemned their ill-treatment. This, together with the worsening of economic conditions, was the background to 1980, a year of crisis for Poland. A series of strikes then came to a head in an epic struggle in the Gdansk shipyard. From them emerged a new and spontaneously organized federation of trades unions, 'Solidarity'. It added political demands to the economic goals of the strikers, among them one for free and independent trades unions. Solidarity's leader was a remarkable, often-imprisoned, electrical union leader, Lech Walesa, a devout Catholic, closely in touch with the Polish hierarchy. The shipyard gates were decorated with a picture of the Pope and open-air masses were held by the strikers.

The world was surprised to see, soon, a shaken Polish government, troubled as strikes spread, making historic concessions. The crucial step was that Solidarity was recognized as an independent, self-governing trade union. Symbolically, regular broadcasting of the Catholic Mass on Sundays was also conceded. But disorder did not cease, and with the winter, the atmosphere of crisis deepened. Threats were heard from Poland's neighbours of possible intervention; forty Soviet divisions were said to be ready in the GDR and on the Russian frontier. But the dog did not bark in the night; the Soviet army did not move and was not ordered by Brezhnev to do so, nor by his successors in the turbulent years which followed. It was the first sign of changes in Moscow which were the necessary premise of what was to follow in eastern Europe in the next ten years.

In 1981, tension continued to rise, the economic situation worsened, but Walesa strove to avert provocation. On five occasions the Russian commander of the Warsaw Pact forces came to Warsaw. On the last occasion, the Solidarity radicals broke away from Walesa's control and called for a general strike if emergency powers were taken by the government. On 13 December, martial law was declared. There followed fierce repression (opposition may have cost hundreds of lives). But the Polish military's action may also have helped to make Russian invasion unnecessary. Solidarity went underground, to begin seven years of struggle, during which it became more and more evident that the military government could neither prevent further economic deterioration, nor enlist the support of the 'real' Poland, the society alienated from communism, for the regime. A moral revolution was taking place. As one Western observer put it, Poles began to behave '*as if* they lived in a free country'; clandestine organization and publication, strikes and demonstrations, and continuing ecclesiastical condemnation of the regime sustained what was at times an atmosphere of civil war.

Although after a few months the government cautiously abandoned formal martial law, it still continued to deploy a varied repertoire of overt and undercover repression. Meanwhile, the economy declined further, western countries offered no help and little sympathy, and, after 1985, the change in Moscow began to produce its effects. Yet the climax came only in 1989, for Poland her greatest year since 1945, as it was for other countries, too, thanks to her example. It opened with the regime's acceptance that other political parties and organizations, including Solidarity, had to share in the political process. As a first step to true political pluralism, elections were held in June in which some seats were freely contested. Solidarity swept the board in them. Soon the new parliament denounced the German-Soviet agreement of August 1939, condemned the 1968 invasion of Czechoslovakia, and set up investigations into political murders committed since 1981.

In August Walesa announced that Solidarity would support a coalition government; the communist diehards were told by Mr Gorbachev that this would be justifiable (and some Soviet military units had already left the country). In September a coalition dominated by Solidarity and led by the first non-communist Prime Minister since 1945 took office as the government of Poland. Western economic aid was soon promised. By Christmas 1989 the Polish People's Republic had passed from history and, once again, the historic Republic of Poland had risen from its grave.

Poland led eastern Europe to freedom. The importance of events there had quickly been perceived in other communist countries, whose leaders were much alarmed. In varying degree, too, all eastern Europe was exposed to the new factor of a steadily increasing flow of information about non-communist countries, above all, through television (which was especially marked in the GDR). More freedom of movement, more access to foreign books and newspapers had imperceptibly advanced the process of criticism there as in Poland. In spite of some ludicrous attempts to go on controlling information (Romania still required that typewriters be registered with the state authorities), a change in consciousness was under way.

Mr Gorbachev had come to power during the early stages of these developments. Five years later, it was clear that his assumption of office had released revolutionary institutional change in the Soviet Union too, first, as power was taken from the party, and then as the opportunities so provided were seized by newly emerging opposition forces, above all in republics of the Union which began to claim greater or lesser degrees of autonomy. Before long, it began to look as if he might be undermining his own authority. Paradoxically, too, and alarmingly, the economic picture looked worse and worse. It became clear that a transition to a market economy, whether slow or rapid, was likely to impose far greater hardship on many – perhaps most – Soviet citizens than had been envisaged. By 1989 it was clear that the Soviet economy was out of control and running down. As always before in Russian history, modernization had been launched from the centre and flowed out to the periphery through authoritarian structures. But that was precisely what could not now be relied upon to happen, first because of the resistance of the *nomenklatura* and the administration of the command economy, and then, at the end of the decade, because of the visibly and rapidly crumbling power of the centre.

By 1990 much more information was available to the rest of the world about the true state of the Soviet Union and its people's attitudes than ever before. Not only were there many overt signs of popular feeling, but *glasnost* had brought to the Soviet Union its first surveys of public opinion through polls. Some rough-and-ready judgements could be made: the discrediting of the party and *nomenklatura* was profound, even if it had not by 1990 gone so far as in some other Warsaw Pact countries (more surprisingly, the long unprotesting Orthodox Church appeared to have retained more respect and authority than other institutions of the Marxist-Leninist *ancien régime*).

But it was clear that economic failure everywhere hung like a cloud over any liberalizing of political processes. Soviet citizens as well as foreign observers began to talk by 1989 of the possibility of civil war. The thawing of the iron grip of the past had revealed the power of nationalist and regional sentiment when excited by economic collapse and opportunity. After seventy years of efforts to make Soviet citizens, the USSR was revealed to be a collection of peoples as distinct as ever, organized in fifteen republics, some of which (above all the three Baltic republics of Latvia, Estonia and Lithuania) were quick to show dissatisfaction with their lot and, in the end, were to lead the way to political change. Azerbaijan and Soviet Armenia posed problems which were complicated by the shadow of Islamic unrest which hung over the whole Union. To make matters worse, some believed there was a danger of a military coup; commanders who were as discontented by the Soviet failure in Afghanistan as some American soldiers had been by withdrawal from Vietnam were talked about as potential Bonapartes – a danger long flourished as a bogey of Bolshevik mythology.

The signs of disintegration multiplied, although Mr Gorbachev succeeded in clinging to office and, indeed, in obtaining formal enhancements of his nominal powers. (But this had the disadvantage of focusing responsibility for failure, too). One dramatic moment came in March 1990, when a declaration of the Lithuanian parliament declared the annexation of 1939 invalid and reasserted Lithuania's independence, though this was followed by complicated negotiation to avoid provoking the armed suppression of the revived republic by Soviet forces. Latvia and Estonia also claimed their independence, though in slightly different terms. The upshot was that Mr Gorbachev did not seek to revoke the fact of secession, but in return won agreements that the Baltic republics should guarantee the continued existence of certain practical services to the USSR. Yet this proved to be the beginning of the end for Mr Gorbachev. A period of increasingly rapid manoeuvring between reforming and conservative groups, allying himself first to one and then, to redress the balance, to the other, led by the end of 1990 to the compromise of the previous summer already looking out of date and unworkable. Connivance at repressive action by the soldiers and KGB in Vilnius and Riga early in the new year did not stem the tide. Parliaments in nine of the Soviet republics had already by then either declared they were sovereign or had asserted a substantial degree of independence from the Union government. Some of them had made local languages official; and some had transferred Soviet ministries and economic agencies to local control. The Russian republic – the most important – set out to run its own economy separately from that of the Union. The Ukrainian republic proposed to set up its own army. In March, elections led Mr Gorbachev once more back to the path of reform and a search for a new Union treaty which could preserve some central rôle for the State. The world looked on, bemused.

The Polish example had growing weight in other countries as they realized that an increasingly divided and paralyzed USSR would not (perhaps could not) intervene to uphold its creatures in the communist party bureaucracies of the other Warsaw Pact countries. This shaped what happened in them after 1986. The Hungarians had moved almost as rapidly in economic liberalization as the Poles, even before overt political change, but their most important contribution to the dissolution of Communist Europe came in August 1989. Germans from the GDR were then allowed to enter Hungary freely as tourists, though their purpose was known to be to present themselves to the embassy and consulates of the Federal Republic for asylum. A complete opening of Hungary's frontiers came in September (when Czechoslovakia followed suit) and a flow became a flood. In three days 12,000 East Germans crossed from these countries to the west. The Soviet authorities remarked that this was 'unusual'. For the GDR it was the beginning of the end. On the eve of the carefully-planned and much-vaunted celebration of forty years' 'success' as a socialist country, and during a visit by Mr Gorbachev (who, to the dismay of the German communists, appeared to urge the east Germans to seize their chance), riot police had to battle with anti-government demonstrators on the streets of east Berlin. The government and party threw out their leader, but this was not enough. November opened with huge demonstrations in many cities against a regime whose corruption was becoming evident; on 9 November came the greatest symbolic act of all, the breaching of the Berlin Wall. The East German Politburo caved in and the demolition of the rest of the Wall followed.

More than anywhere else, events in the GDR showed that even in the most advanced communist countries there had been a massive alienation of popular feeling from the regime. 1989 had brought it to a head. All over eastern Europe,

After the Berlin Wall was spontaneously broken through by individuals, within a few days it was undergoing official demolition.

it was suddenly clear that communist governments had no legitimacy in the eyes of their subjects, who either rose against them or turned their backs and let them fall down. The institutional expression of this alienation was everywhere a demand for free elections, with opposition parties freely campaigning. The Poles had followed their own partially-free elections in which some seats were still reserved to supporters of the existing regime, with the preparation of a new constitution; in 1990, Lech Walesa became President. A few months earlier, Hungary had elected a parliament from which emerged a non-communist government. Soviet soldiers began to withdraw from the country. In June 1990, Czechoslovakian elections produced a free government and it was soon agreed that the country was to be evacuated of Soviet forces by May 1991. In none of these countries did the former communist politicians get more than 16 per cent of the vote. Free election in Bulgaria was less decisive: there, the contest was won by communist party members turned reformers and calling themselves socialists.

In two countries, events turned out differently. Romania underwent a violent revolution (ending in the killing of its former communist dictator) after a rising in December 1989 which revealed uncertainties about the way ahead and internal divisions ominously foreshadowing further strife. By June 1990 a government some believed still to be heavily influenced by former communists had turned on some of its former supporters, now critics, and crushed student protest with the aid of vigilante squads of miners at some cost in lives and in disapproval abroad. The GDR was the other country where events took a special turn. It was bound to be a special case, because the question of political change was inescapably bound up with the question of German re-unification. The breaching of the Wall revealed that not only was there no political will to support communism, there was no will to support the GDR either. A general election there in March 1990 gave a majority of seats (and a 48 per cent vote) to a coalition dominated by the Christian Democrat party – the ruling party of the western German Federal Republic. Unity was no longer in doubt; only the procedure and timetable remained to be settled.

In July the two Germanies joined in a monetary, economic and social union. In October they united, the former territories of the GDR becoming provinces of the Federal Republic. The change was momentous, but no serious alarm was openly expressed, even in Moscow, and Mr Gorbachev's acquiescence was his second great service to the German nation. Yet alarm there must have been in the USSR. The new Germany would be the greatest European power to the west. Russian power was now in eclipse as it had not been since 1918. The reward for Mr Gorbachev was a treaty with the new Germany promising economic help with Soviet modernization. It might also be said, by way of reassurance to those who remembered 1941–5, that the new German state was not just an older Germany revived. Germany was now shorn of the old east German lands (had, indeed, formally renounced them) and was not dominated by Prussia as both Bismarck's *Reich* and the Weimar republic had been. More reassuring still (and of importance to west Europeans who felt misgivings), the Federal Republic was a federal and constitutional state seemingly assured of economic success, with nearly forty years' experience of democratic politics to build on, and embedded in the structures of the EC and NATO. She was given the benefit of the doubt by west Europeans with long memories, at least for the time being.

At the end of 1990, the condition of what had once seemed the almost monolithic east European bloc already defied generalization or brief description. As former communist countries (Czechoslovakia, Poland, Hungary) applied to join the EC, or got ready to do so (Bulgaria), some observers speculated about a potentially wider degree of European unity than ever before. More cautious judgements were made by those who noted the virulent emergence of new – or re-emergence of old – national and communal division to plague the new East. Above all, over the whole area there gathered the storm-clouds of economic failure and the turbulence they might bring. Liberation might have come, but it had come to peoples and societies of very different levels of sophistication and development, and with very different historical origins. Prediction was clearly unwise.

How unwise became clear in 1991. In that year, a jolt was given to optimism over the prospects of peaceful charge when two of the constituent republics of Yugoslavia announced their decision to separate from the federal state. Their action was influenced by deep-rooted national animosities which had for years been increasingly evident in the Serb-dominated republic. In August, sporadic fighting by both air and ground forces began between Serbs and Croats. Precedents for intervention by outsiders did not ever seem promising – though different views were held by different EC countries – and such a prospect became even less attractive when the USSR in July uttered a warning about the dangers of spreading local conflict to the international level. By the end of the year Macedonia, Bosnia-Herzegovina and Slovenia had all declared themselves independent of the Yugoslav federal republic, and had joined Croatia in doing so.

The Soviet warning was the last diplomatic *démarche* of the regime. It was soon eclipsed by a much more momentous event. On 19 August a still mysterious attempt was made by conservatives to set aside Mr Gorbachev by *coup d'état*. It failed, and three days later he was again in occupation of the presidency. Nonetheless, his position was not the same; continual changes of side in a search for compromise had ruined his political credibility. He had clung too long to the Party and the Union; Soviet politics had taken a further lurch forward. To many it seemed as if it was toward disintegration. The circumstances of the *coup* had given an opportunity which he seized to Mr Boris Yeltsin, the leader of the Russian republic, the largest in the Union. The army, the only conceivable threat to his supporters, did not move against him. He now appeared both as the strong man of the Soviet scene without whose concurrence nothing could be done, and as a possible standard-bearer for a Russian chauvinism which might threaten other republics. While foreign observers waited to understand, the purging of those who had supported or acquiesced in the *coup* was developed into a determined replacement of Union officialdom at all levels, the redefinition of roles for the KGB and a redistribution of control over it between the Union and the republics. The most striking change of all was the demolition of the Communist Party of the Soviet Union, which began almost at once. Almost bloodlessly, at least to begin with, it appeared, the huge creation which had grown out of the Bolshevik *coup* of 1917 was coming to an end. There seemed at first good grounds for rejoicing over that, though it was not clear that nothing but good would follow.

Nor was that easier to see as the year came to an end. With the decision to abandon price controls in the Russian republic in the near future, it seemed likely that not only inflation unparalleled since the earliest days of the Soviet system, but perhaps starvation, too, would soon face millions of Russians. In another republic, Georgia, fighting had already broken out between the supporters of the president elected after the first free elections there and the discontented opposition. Dwarfing all such facts, though, was the end of the great Soviet Union itself. The giant superpower which had emerged from the bloody experiments of the Bolshevik revolution to be, almost to the end, for nearly seventy years the hope of revolutionaries around the world, and the generator of military strength that had fought and won the greatest land campaigns in history, dissolved suddenly and helplessly into a set of successor states. The last of the great European multi-national empires had gone. Russian, Ukrainian and Belorussian leaders met at Minsk on 8 December and announced the end of the Soviet Union and the establishment of a new 'Commonwealth' of Independent States. On 21 December 1991, a gathering of representatives from eleven of the former republics met briefly at Alma-Ata to confirm this. They agreed that the formal end of the Union would come on the last day of the year. Almost immediately, Mr Gorbachev resigned.

It was the climax of one of the most startling and important changes of modern history. Of what lay ahead, no one could be sure – except that it would be a period of danger, difficulty and, for many former Soviet citizens, misery. In other countries, politicians were rarely tempted to express more than caution over the turn events had taken. There was too much uncertainty ahead. As for the USSR's former friends, they were silent. A few of them had deplored the turn of events earlier in the year so much that they had expressed approval or encouragement for the failed *coup* of August. Understandably, Libya and the PLO did so, since any return to anything like Cold War groupings was bound to arouse their hopes of renewed possibilities of manoeuvre in an international arena newly constricted first by *détente* between the USA and USSR and then by the growing powerlessness of the latter.

Events must have been followed with special interest in China. Her rulers had their own reasons for uneasiness about the direction in which events across their longest land frontier appeared to be going after the collapse of communism in eastern Europe. At the end of 1991, they were the rulers of the only multinational empire still intact. This was so although China had been engaged for most of the 1980s in a continuing process of controlled modernization, based on substantial liberalization (at least in the eyes of old-fashioned communists) of the economy. This did not imply any weakening of the will to power of the regime. China's rulers remained firmly in control and intended to do so. They were helped by the persistence of the old Chinese social disciplines, by the relief felt by millions that the cultural revolution had been left behind, and by the policy (contrary to that of Marxism as still expounded in Moscow until 1980) that economic rewards should flow through the system to the peasant. This built up rural purchasing power, and made for contentment in the countryside. There was a major swing of power away from the rural communes, which in many places practically ceased to be relevant, and by 1985 the family farm was back as the dominant form of rural production over much of China. At the same time, it was apparent to many Chinese that China enjoyed respect and status abroad; one striking, if paradoxical, sign was an official state visit by Queen Elizabeth II in 1985.

A few years later, nevertheless, China was clearly experiencing major difficulties. Foreign debt had shot up. Inflation was running at a annual rate of about 30 per cent by the end of the decade. There was anger over evidence of corruption, and divisions in the leadership (some following upon the deaths and illness among the gerontocrats who dominated the Party) were widely known to exist. Those believing that a reassertion of political control was needed began to gain ground, and there were signs that they were manoeuvring to win over Deng Xiaoping. Yet western observers and perhaps some Chinese had been led by the policy of economic liberalization to take unrealistic and over-optimistic views about the possibility of political relaxation. The exciting changes in eastern Europe stimulated further hopes of this. But the illusions suddenly crumbled.

In the early months of 1989, China's city-dwellers were feeling the pressures both of the acute inflation and of an austerity programme which had been imposed to deal with it. This was the background to a new wave of student demands. Encouraged by the presence of sympathizers with liberalization in the governing oligarchy, they demanded that the Party and government should open a dialogue with a newly formed and unofficial Student Union about corruption and reform. Posters and rallies began to champion calls for greater 'democracy'. The regime's leadership was alarmed, refusing to recognize the Union which, it was feared, might be the harbinger of a new Red Guards movement. There were demonstrations in many cities and as the seventieth anniversary of the May 4th Movement approached the student activists invoked its memory so as to give a broad patriotic colour to their campaign. They were not able to arouse much support in the countryside, or in the southern cities, but, encouraged by the obviously sympathetic attitude of the general secretary of the CCP, Zhao Ziyang, began a mass hunger strike. It won widespread popular sympathy and support in Peking. It started only shortly before Mr Gorbachev arrived in the capital; his state visit, instead of providing further reassuring evidence of China's international standing, only served to remind people of what was going on in the USSR as a result of policies of liberalization. This cut both ways, encouraging the would-be reformers and frightening the conservatives. By this time the most senior members of the government, including Deng Xiaoping, seem to have been thoroughly alarmed. Widespread disorder might be in the offing; they believed China faced a major crisis. Some feared a new cultural revolution if things got out of control (and Deng Xiaoping's own son, they could have remarked, was still a cripple as a result of the injuries inflicted on him by Red Guards). On 20 May martial law was declared.

There were signs for a moment that the government might not be able to impose its will, but the army's reliability was soon assured. The repression which followed was ruthless. The student leaders had moved the focus of their efforts to an encampment in Peking in Tiananmen Square, where, thirty years before, Mao had proclaimed the foundation of the People's Republic. From one of the gates of the old Forbidden City a huge portrait of him looked down on the symbol of the protesters: a plaster figure of a 'Goddess of Democracy', deliberately evocative of New York's Statue of Liberty. On 2 June the first military units entered the suburbs of Peking on their way to the square. There was resistance with extemporized weapons and barricades. They forced their way through. On 4 June the students and a few sympathizers were overcome by rifle-fire, teargas, and a brutal crushing of the encampment under the treads of tanks which swept into the square. Killing went on for some days, mass arrests followed (perhaps as many as ten thousand in all). Much of what happened took place before the eyes of the world, thanks to the presence of film-crews in Peking which had for days familiarized television audiences with the demonstrators' encampment. Foreign disapproval was almost universal.

As so often in China, it is hard to know what had really happened. Obviously her rulers felt they faced a grave threat. It is probable, too, that they acted

Liberty, a western icon, rises in Tienanmen Square, Peking, in May 1989.

in a way deplored and opposed by many of their fellow-Chinese. Yet the rural masses did not sympathize with the protesters; rather, they were against them. Changes in the ruling hierarchy and vigorous attempts to impose political orthodoxy followed. Economic liberalization was reined in. Neo-Marxist slogans were heard again. China, it was clear, was not going to go the way of eastern Europe or the USSR. But where was she going? Perhaps the safest conclusion to be drawn at this stage is that she was once again moving to her own rhythms and stimulants, not at once to be interpreted in categories drawn from the western world, for all the rhetoric of regime and protesters alike. The students in Tiananmen Square met the tanks not only rallied around the statue which was their icon of Liberty, but with another gesture that showed what they owed to another non-Chinese and western inspiration: they sang the *Internationale*. That may suggest both the complexity (and even the incoherence) of the opposition movement and its alienation from much that was still influential in China. As recently as 1987, a poll had reported that even among urban Chinese, the moral defect which was most strongly deplored was that of 'filial disobedience'. Transformed though so much of the world already was, China after Tiananmen Square still baffled observers and futurologists by her seemingly massive immunity to currents outside her borders. One of the traditional roles of her governments has always been to act as the guardian of Chinese values. If, anywhere, modernization might turn out in the end not to mean 'westernization', it could be in China.

Epilogue: In the Light of History

History cannot have an end unless we extinguish the human race; if we do not, someone will always be there to think about its past, and about his or her forerunners. History is unrolling (at a remarkably brisk pace, too) as this is being written and will go on doing so, so far as we can guess. A history of the world cannot pull up at a clean chronological boundary, nor be pinned down at a single date as a neat pattern of 'significant' topics and events. The flow of actuality will in due course be plotted, studied and analyzed by historians, and they will make choices and draw distinctions which make sense of it. The historically significant is always what one age thinks worth noting about another. The events of our age will acquire new meanings when people in the future start to wonder about what has made their world.

What is more, today's judgements about what is important may look eccentric even in a few months if changes come as quickly as they have been coming in the last few years. More than any of his colleagues, the historian of the recent past faces the old problem of perspective – a good image, which we too easily take for granted, forgetting its pictorial and visual origins. It means getting things in the right relationships to one another.

In the aftermath of the huge political transformations of the last decade we all know only too well that events can alter perspective with startling rapidity. Other changes, less immediately observable, are going on, too, and may be even more fundamental. If, as some say, the startling figure of a total world population of five thousand million was probably passed on the first Monday in July 1986, and that of six thousand million is likely to be reached even before the year 2000, what are the implications of that? What can be said about the degree of damage already done to the ozone layer by the enormous increase in the burning of fossil fuels since 1945? Faced with such changes and the difficulty of assessing them, it will be best not to try to do too much, and to be content with having a last word at all. Soon, no doubt, the world will have changed still more and new surprises will make even the last few years look very different.

At the very least, prophecy must be avoided; it is not the historian's business, even if disguised as extrapolation. It may, admittedly, serve as a pedagogic or rhetorical device. It can sharpen awareness and clarify possibilities with hypothetical projections. It can sometimes demonstrate trends which cannot continue (if the world's scientists, for example, had gone on increasing in number at the rate of the two centuries down to the 1960s, the population of the world would before long consist of nothing but scientists). Guesses, too, are legitimate if they throw light on the scale of present facts. Perhaps fossil fuels will go the way that the larger prehistoric mammals went at the hands of human hunters – or perhaps they will not. Thinking about the question, though, throws some light on the way the oil crisis of the 1970s, working through the blind forces of the market expressed in costs, awoke more prudent attitudes among some of the biggest consumers. But it remains vital always to recall that the historian's subject-matter is the past. It is all he has to talk about. When it is the recent past, what he can try to do is to see consistency or inconsistency, continuity or discontinuity with what has gone before. Here we are back with the difficulties posed by the mass of facts which crowd in on us. The very confusion they present suggests a much more revolutionary period than any earlier one

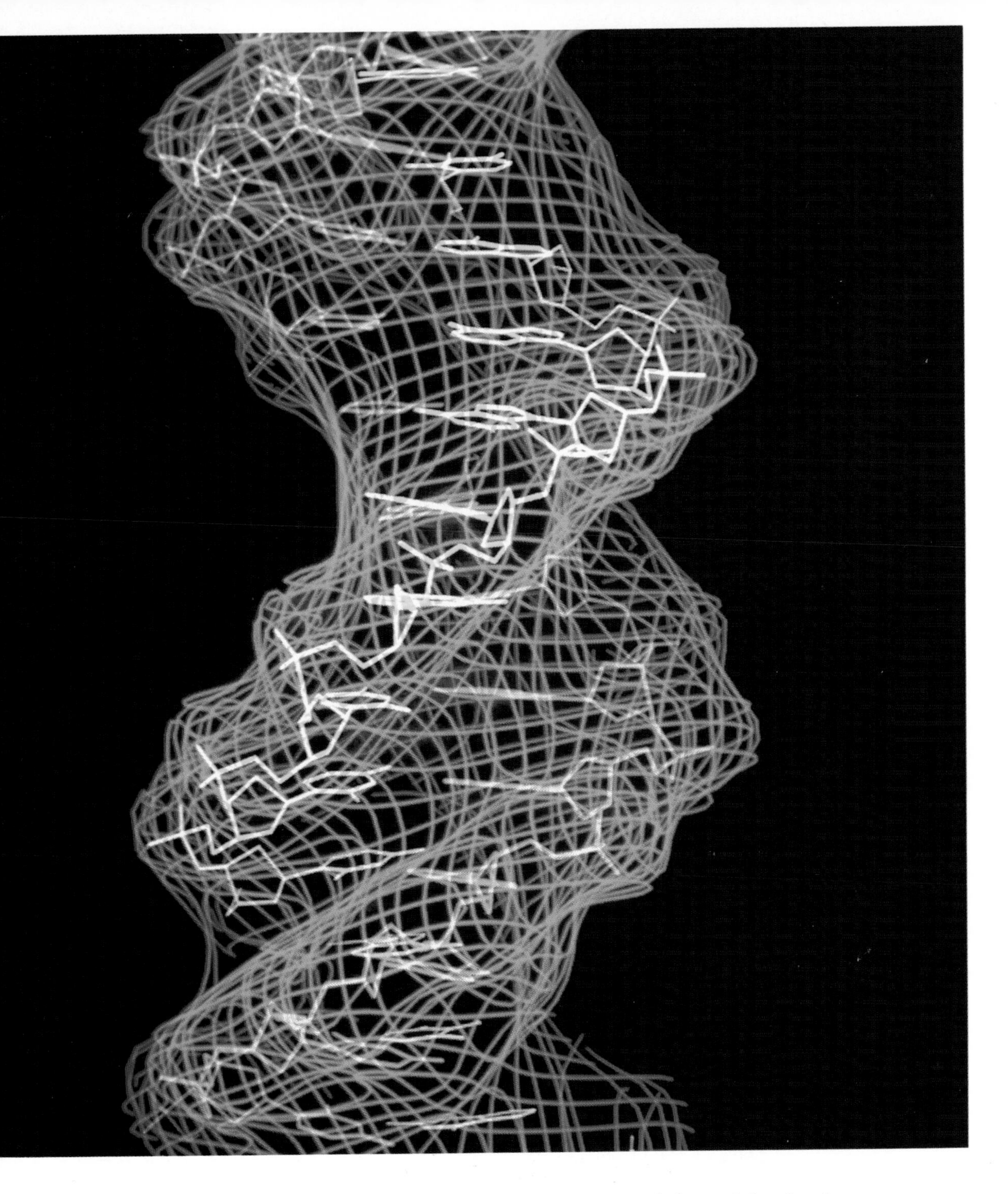

DNA – THE SHAPING OF LIFE *Computer graphics have now provided a tool for showing structures previously only visible by ultra-microscopical means. This is a representation of a short section of DNA (deoxyribonucleic acid) which forms the basis of the genetic material of most living organisms. The DNA molecule contains all the information from which a particular living organism is built and maintained. It consists of a double helix of strands of sugar phosphate residues linked by pairs of organic, purine and pyrimidine, bases arranged as the rungs of a spiral ladder. The precise sequence of the bases determines genetic inheritance by forming genes which may be transcribed to produce RNA (ribonucleic acid) which is translated via specific amino acids into different protein chains. These ultimately produce the molecules which constitute and maintain living organisms. The blue lines in the model show the surface of the structure – the three-dimensional form of the double helix.*

MODERN JAPAN *Cultural borrowings are sometimes mystifying at first sight. It is not immediately easy to see why Japanese girls should be paid to don and pose in the traditional ballet-skirt of Europe. It becomes slightly more thought-provoking to learn that they do so as part of an advertisement. Western commercial and industrial technique has had a huge impact on Japanese life, especially since 1960, and through the big domestic market must have influenced widely-held attitudes and views. At first sight messages used outside Japan may seem to promote the same results and fuel the same responses. Yet scholars still argue fiercely about the degree and depth of 'westernization' in one of the most resilient, traditional and insular cultures in the world nearly a hundred and fifty years after the start of the changes released by Japan's opening to the West in the nineteenth century.*

and all that has been said so far about the continuing acceleration of change confirms this. This does not, on the other hand, imply that these more violent and sweeping changes are not consistent with what has gone before, and so do not emerge from the past in a way which is explicable and for the most part understandable.

I have argued in these pages that if there is any general trend at all in history, it is twofold, towards a growing unity of human experience and a growing human capacity to control the environment. Let us start there. It can be argued, for instance, that the expression 'one world' is now little more than a cant term. In so far as the people who first used it hoped to ease the way to common political action, too, it seems exploded. There is just too much conflict and quarrelling about. Since 1931 there has hardly ever been more than a few weeks at a time when human beings were not fighting one another somewhere in the world. Moreover, political divisions, even when they do not break out in overt violence, can none the less be expensive and dangerous, as the Cold War showed only too well. Although there emerged from victory in 1945 an international institution called the United Nations and although it has shown some potential for collective action, it is based, ironically, on the theory that the whole surface of the globe is divided into territories belonging to sovereign states; there are now nearly two hundred of them and a strong likelihood that their number will increase. A civil war had been going on in Yugoslavia for some months before this book went to press with precisely that as the aim of some of the contenders in it. None of this can be denied and much, much more could be said along the same lines.

Yet it does not quite meet the thesis already set out. The point could be made by thinking about a more remote period. The world of Islam during the European Middle Ages was the scene of constant strife between different Islamic rulers. Yet Islam had, none the less, a cultural unity, though not one which was complete, let alone homogeneous. All over the Islamic world, institutions and behaviour expressed this in uniformities and similarities which were not observable between other civilizations or cultures, and the societies over which they exercised their sway. That is still to a considerable extent true. One argument of this book, though, is that, for all the conflicts of the modern world, they often more and more resemble – though still far from completely – civil wars between contestants sharing a common background. A creeping unity has seized mankind. Conflicts of civilization and clashes of culture are rarer than in the past. At this level, though, the argument can hardly be put without it being overstated. More humdrum matters may make the point more exactly.

Curiously enough, it is most obvious at the level of personal experience, though this is usually the level at which men feel most acutely the distinctions between them. In the days when men of neighbouring villages all over the world spoke significantly different dialects even if they shared a common language, when in the whole of their lives most of them would only exceptionally travel ten miles from their homes, when even their clothes and tools might provide in their shape and workmanship evidence of big differences of technology, style and custom, then human experience was in important ways much more differentiated than it is now. The great physical, racial and linguistic divisions of the past were in their day much harder to overcome than are their equivalents today, thanks to improved communication, mass education, mass production of commonly-required artifacts, and so on. The results are obvious to a traveller in any part of the modern world. Though we can still see exotic or unfamiliar clothes in some countries, more people over most of the globe now dress alike than ever before. Only rural districts or consciously nationalist regimes now cling to traditional local costume instead of the near universal shirt-and-trousers of males, for example. Kilts, kaftans, kimonos are becoming tourist souvenirs, or the carefully-preserved relics of a sentimentalized past; less picturesque traditional clothing, meanwhile, is more and more the sign of poverty and backwardness. The efforts of a few self-consciously conservative and nationalist regimes only bear this out. When Rousseau encouraged the eighteenth-century Poles to national regeneration, he urged them to guard zealously their national dress, pastimes and amusements and to reject foreign fashion. There are still regimes which are his disciples today. Uganda legislated against the mini-skirt and the Iranian revolutionaries put women back into the chador, because the common experience pouring in from the world outside was felt to be corrosive of tradition. With exactly opposite motives, though, Ataturk forbade Turks to wear the fez and Peter the Great dressed his courtiers in Western European clothes, in order to destroy tradition. More was involved than fashion. What was at stake for them was orientation towards a new range of experience, not simply taste or fashion. They had grasped the importance of symbolic immersion in what they saw as a common, advancing, progressive culture.

The wider basis of shared experience now available to many human beings is rarely a matter of such conscious commitment to the world culture of our day. If they live in modernizing societies, they are increasingly liberated from differences of climate by electricity, air-conditioning and medicine. All over the world millions of them live in cities with similar street lighting and traffic signals, policemen on point duty, banks and bus stops. In the shops there is a growing likelihood that the same goods will be available as in other countries; toothpaste (though only invented a short time ago) can be bought world-wide. Men who do not understand one another's languages can service the same machines and derive the same or very similar advantages and disadvantages from their use. Everywhere in the world the motor car has imposed in greater or lesser degree the same demands on urban living and threatens it with similarly intolerable strains and stresses. Country districts still escape many of these blessings, but the shared experience is clearly to be seen in the cities – and in some of the world's greatest cities it is for millions an experience of uniform squalor, economic precariousness and comparative deprivation. Cairo, Calcutta and Rio can offer similar spectacles of misery, for all the important differences between Moslem, Hindu and Christian origins.

The point can be made by an effort of historical imagination. A traveller who went from imperial Rome to the Han capital of Loyang would have wondered at everything he saw; not only would clothes have been cut differently and made of different materials, but food would have been different, the animals in the streets would have been of different breeds, the weapons of the soldiers and their armour would have been shaped differently. Even wheelbarrows would have had a different design. A modern American or European finds much less to surprise him. China is perhaps still one of the cultures most resistant to external trends, but even if Chinese cuisine retains its distinctiveness, a Chinese airliner now looks like any other. Yet not so long ago, when junks were the only ocean-going Chinese shipping, they did not at all look like contemporary European cogs or caravels.

It can be said that such shared experience is peripheral to the lives of the masses of mankind who still live in villages and struggle to get a living from the soil, often with traditional tools and ideas. It is also true that the all-too-visible difference between life in rich and poor countries far transcends any difference which existed in the past. A thousand years ago all societies were by modern standards poor and consequently closer to one another in their experience than they are today. The difficulty of winning one's daily bread and the fragility of man's life before the mysterious, implacable forces which cut him down like grass were things all men had in common, whatever language they spoke or creed they followed. Now, a large minority of mankind lives in countries with an average *per capita* annual income of over $3000, and millions of others in counties where the corresponding figure is less than one-tenth of this sum. There are even colossal distinctions among the poor; people have begun to talk of a 'Fourth World' as countries of the Third World which enjoy natural resources in great demand have begun to show a potential for self-sustained development virtually inconceivable to the truly poor.

This is all part of the complex reality of our world, and is not to be overlooked, yet the importance of such contrasts can be exaggerated. They are, in the first place, the product of a relatively recent and brief historical era; we should no more assume they will endure for centuries than we may assume they will easily or swiftly disappear. Perhaps they will dwindle in a world which becomes still more homogenized. The leading classes and elites even in the poorest countries almost always look to some version of modernization as a way out of their troubles. They look, that is to say, to the West. Their societies, wherever they display vitality and an urge to change, thus provide new confirmation of the pervasive influence of the civilization which has proved so triumphant elsewhere.

It is not a counter-argument to say that modernization is only a matter of technology and that more fundamental matters of belief, institutions and attitudes are the real determinants of social behaviour. Not only does that side-step questions about the way material experience shapes culture; it is easy also to point to the evidence that ideas and institutions, too, as well as material artifacts and techniques, have become more generally spread among mankind. Our world has been slow to give much practical respect to such documents as the United Nations Declaration of Human Rights, but the interest shown in drawing them up and signing them has symptomatically been intense. Although many of the signatories have little intention of respecting them, a 'decent regard for the opinion of mankind' – to borrow a historic phrase – compels them to pay lip-service to certain principles. Such principles usually turn out to be derived from the European tradition of civilization (as, indeed, the widespread acceptance of the notion of a 'Declaration of Rights' suggests) and this is only a recent example of the tendency of the last three centuries during which that civilization has extended its influence around the whole world. We sometimes now call that influence 'Western', rather than European, but its ultimate origin and the heartland of the civilization which created it is still Europe, even if North America now so importantly shapes it. The great age of western political domination has now passed, but the reasonable grounds for talking of the first world civilization have already been set out. It is not culturally arrogant to remark that Aztec and Inca civilizations could not stand up to the Spanish, and that Hindu and Chinese civilizations were somewhat more successful against later 'Franks'. Such things are true or untrue: they are neither admirable nor repugnant. Whether we regard the European tradition as greedy, oppressive, brutal and exploitative, or as objectively improving, beneficent and humane, is neither here nor there. It was either the master-source of the modern world or it was not and that is all our concern here.

As this book has argued at what some may regard as perhaps excessive length, European ideas and institutions have by no means everywhere displaced native tradition. That is not the point. Our world is, indeed, still shaped by many deeply different traditions. Women are not treated in the same way – whether for good or ill is irrelevant – in Islamic and Christian societies. Indians still take into account astrology in fixing the day of a wedding, while English people may find train timetables or imperfect weather information which they believe to be 'scientific' more relevant. Though the philosophy (or what is taken to be the philosophy) of ancient Asia may have a cult attractiveness for a minority of modern Americans, the roots of American behaviour are still to be found, if in any ideological source, in the confidence of the Enlightenment and the conviction felt by many early puritan settlers that they were a people set apart, freemen of a city builded on a hill, or that of later emigrants that they were truly entering a New World. One could go on and on with such contrasts. Differing traditions make even the use of shared technology and ideas different. Japanese capitalism does not work in the same way as British, and any explanation must lie deep in the different histories of two peoples otherwise similar in a few respects (as islanders for example). What remains true in spite of this is that no other tradition has shown the same vigour and attractiveness in alien settings as the European: it has no competitors as a world shaper.

Even its grossest manifestations, its material greed and rapacity, show this. Societies once rooted in changeless acceptance of things as they are have taken up the belief that limitless improvement in material well-being is a proper goal for them. They have thus taken aboard much of the mental heritage of an expanding Europe. The very idea that willed change is possible is itself deeply subversive. Many other such European ideas have imposed a layer of assumptions and myths drawn from the experience of European liberalism on top of social institutions of great antiquity and toughness in many countries. There are republics the world round nowadays, and everyone speaks the language of democracy and the rights of man. There is an effort, too, to bring to bear the rationalizing and utilitarian approach to government and administration and to replicate elsewhere models of institutions which have been found successful in countries in the European tradition. One reason why many black men clamour vociferously against the white-dominated societies they live in is that they in fact wish to realize the ideals of human rights and dignity evolved by European civilization. Meanwhile, their black cousins in new nations wish to participate in the benefits of the rising wealth made possible by industrialization and hope to realize the values it generates even when preaching the merits of 'negritude' or the timeless truths of Islam.

Very few cultures, if any, have been able altogether to resist this vigorous European tradition: even China has kow-towed to Marx and science. Some have resisted more successfully than other, but almost everywhere the individuality of other great civilizations has been in some measure sapped. When modernizers have sought to take some things but not others from the dominant mode, they have not found it easy to do so. It is possible, at a certain cost, to get a selective modernity, but it usually comes in a package, some of whose other contents may be unwelcome, as *sotto voce* protests about pressure to give more attention to human rights showed at the 1991 conference of British Commonwealth prime ministers.

The emancipation of women makes the point: its impact on family structure and authority, on economic life and culture, and through them upon the emotions and attitudes of individuals, would be bound to prove revolutionary in many parts of the world, even were it carried only so far as to produce the state of affairs often decried by feminists in advanced Western countries as oppressive. Few non-European countries and cultures have made any but partial concessions to the principle; in some Arabic-speaking countries girls now go to universities but they are segregated in the lecture rooms, from which they will emerge to take their places in an Islamic world where the relations of the sexes are ruled by tradition. As for uneducated women – above all the millions of peasant women of Africa and Asia – they often remain wholly untouched even by such changes. Yet they, too, may benefit directly and practically from the difference made to their lot by the arrival of modernity in the shape of pumped water or electricity in their villages, and such things have ultimately a power to transform. Not all such disparities are the result of conscious selection from the European tradition. Organic change always comes about only slowly and sometimes indirectly.

Female liberation, indeed, has taken a long time to come as far as it has done even in western countries. Christianity had from the start a fundamental (even if at first sight barely visible) bias towards the improvement of the lot of women, because it took for granted that they, like men, had souls of infinite value in the eyes of God. On this was to be built the modern freedom of women in societies in the Christian tradition. However theologians might chip away at the idea in the interests of male prejudice they could not in the end gainsay the principle. Yet it still took some seventeen centuries to produce the first advocates of feminism and another two to arrive at the substantial legal equality of men and women in the West today. It needed, too, the reinforcement provided by industrialization, and its economically liberating effects upon women, before this could be achieved. The difference made to western women's lives by technical change in a huge diversity of forms, from the coming of running hot water, to the perfection of detergents, synthetic fibres, prepared foods (to name only a few), has been as great or greater than that brought by the franchise.

Conscious rejection of some elements in the package, the sheer inertia of well-established cultural patterns and the relative accessibility of the material as opposed to the moral products of the European tradition, sufficiently explain why modernizing change still sometimes seems superficial. It is a little like Hellenization in the Near East in the centuries after Alexander. Borrowed institutions, ideas and styles often show something of the same lack of real spirit and inspiration as the borrowings of Greek ideas in the East. There is the same spread of the signs and symbols of material achievement – nowadays dams, universities, steelworks, airlines, instead of baths, theatres, temples –

the same standard forms of government and administration, the same literary and artistic fashion and even the same aesthetic. English has become the *lingua franca* of business and intellectual life as Greek once was. Yet the results often fail to ring true; successfully transplanted to other traditions though such things seem to have been, they are somehow not the same.

What this means is not that the experience of civilization now so widely diffused is unreal or a sham, but that it is incomplete and selective, though hardly ever in the way would-be selectors would like. This does not invalidate the claim that for the first time in human history the variety of human experience is converging instead of diverging. The most obvious sign of this is what all parts of the world have found most seductive and compelling about European culture, its promise of material success. The real triumph of European civilization came when it began to convince people in other cultures that through it lay the road to success for themselves. Europeanization came to be synonymous with modernization. What were then seized upon were not only techniques and institutions, not merely such superficialities as clothes and hairstyles, but goals and patterns of behaviour, and these sometimes produced incongruous and unanticipated results. The ideas of progress as limitless material improvement, of the right of the individual to assert himself, of nationalism as the proper basis for political organization, have all produced consequences going far beyond what was expected by those who so confidently passed on to others the recipes which they believed to underlie their own success. Meanwhile, the introduction of new machines, the building of railways and mines, the coming of banks and newspapers, and much, much else, transformed social life in ways no one had willed or envisaged. The process, once begun, was irreversible. Once European methods and goals were accepted (as they have been in greater or less degree, consciously or unconsciously, by élites in almost all countries), then an uncontrollable evolution had begun. Even in the most tightly controlled essays in modernization, new and unexpected needs and demands would erupt from time to time. There now looms up a new spectre – that modernization's successes may have communicated to mankind goals which are materially and psychologically unobtainable, limitlessly expanding and unsatisfiable in principle.

The revolution in the minds of men which has been going on with increasing vigour for a long time – the acceptance of the idea that continuous material improvement is possible – was the climax of centuries of growing success in the manipulation of the environment. Ironically, the idea took root over all the world almost at the moment when the first misgivings were beginning to be felt about it in its birthplace. They have prompted some to pessimism. To say whether pessimism or optimism about the future is justified, though, is not (it must be repeated) the historian's job. What he can do is say whether history has or has not been taken into account in making a judgement about the present or future. Substantially, it is now often urged that humanity's undoubted success in manipulating its world is threatened by danger of disappearing because of two interconnected and observable tendencies. One is that success creates new problems, and does so perhaps too rapidly for answers to be found before irreversible damage is done. Thus we face the problems of depleted natural resources, disturbance of existing ecologies, the creation of new stresses and dangers such as those involved in gathering people into great cities. The second tendency is that the achievements of science outrun our capacity to manage the power they give us. Examples are the dangers inherent in already irreversible damage to the environment, in the proliferation of nuclear weapons or the reduction in death-rate through medical advance which leads to rapid population growth, in interference with the genetic process through *in vitro* fertilization (the first 'test-tube baby' was born in 1978, in Lancashire), or genetic engineering, or in the provision of mass communication without adequate cultural preparation. If such misgivings prove justified, then the present age will indeed seem to our successors to mark a break in historical continuity. It might imply that the progressive power to control environment which marked the whole of earlier history and all known prehistory was at an end.

This would certainly be a prospect not to be lightly regarded, but once again we must recall how little we can say about the future. There are still no reasons to believe that the ways of discovering techniques to meet problems in the past cannot again be brought to bear successfully. We have no grounds either logical or empirical for thinking that the steady accretion of control over nature which has marked all history until now will not continue. All that is different in that change is quite simply, more sweeping and faster than ever before. But this applies to the search for solutions as well as to the emergence of problems. We know of nothing in the nature of the problems now facing the human race which in principle renders them incapable of solution. They may be more urgent and potentially more damaging, but this is only to say that their solution may require more urgent and radical methods, more drastic political and social change, not that they are insoluble. We may have to decide to live in a different way, but we need not assume mankind will be extinguished. There is no reason to conclude that this series of tests must prove fatal to mankind when earlier ones (the onset of the Ice Ages, for example) did not, though they had to be faced with far poorer resources, both mental and technological. We have plenty of evidence of human adaptability in the past. The only clear warning which does stand out is that, whatever we do, we are likely to be gravely misled about the future if we simply extrapolate present trends. We must prepare for discontinuity as well as continuity.

The greatest discontinuity of recent times was the successful penetration of space and the landing of men on the moon over twenty years ago. But besides marking a break, that achievement embodied great continuities, too. Landing on the moon was the most complete and dazzling affirmation to that date of the belief that Man lives in a universe he can manage. The instruments for doing so were once magic and prayer; they are now science and technology. But there is a continuity in the growing confidence of Man through history that he can manipulate the natural world. It cannot be said that landing on the moon is more or less of a landmark in that continuity than, say, the mastery of fire, the invention of agriculture or the discovery of nuclear power. But it is emphatically an event of that order.

It can properly be compared also to the great age of terrestrial discovery. The timescales are interestingly different. Something like about eighty years of exploration were required to take the Portuguese round Africa and India; there were eight between the launching of the first man into space and the arrival of men on the moon. The target set in 1961 was achieved with about eighteen months to spare. Exploration in space is safer, too. It long had no fatalities; in spite of a few spectacular accidents, in terms of deaths per passenger-mile travelled it is still the safest form of transport known to man, while fifteenth-century seafaring was a perilous business at best; if you did not die of shipwreck there was a good chance that tropical disease, scurvy or irritated natives might take you off. Actuarially, the risk of travelling in the *Santa Maria* – or even the *Mayflower* – must have been much greater than that faced by the crew of Apollo 11. The comparison suggests another instance of continuity. The age of oceanic discovery was for a long time mainly dominated by one people, the Portuguese. They built on a slow accumulation of knowledge. Cumulatively, the base of exploration widened as data was added, piece by piece, to what was known. Five hundred years later, Apollo was launched from a far broader base, nothing less than the whole scientific knowledge of mankind. The distance to the moon was already known; so were the conditions which would greet men arriving there, most of the hazards they might encounter, the quantities of power, supplies and the nature of the other support systems they would need to return, the stresses their bodies would undergo. In part this diminished the tremendous impact of the event. Though things might have gone wrong, there was a widespread feeling they would not. In its predictable, as in its cumulative, quality, space exploration epitomizes our science-based civilization.

The increasing mastery of the nature of which space exploration is the latest step mainly the achievement of only the last seven or eight millennia. Behind them lay the hundreds of thousands of years during which prehistoric technology inched forwards from the discovery that a cutting edge could be put on a stone chopper and that fire could be mastered. The weight of genetic programming and environmental pressure then still loomed much larger then than did conscious control. The dawning of consciousness that more than this was possible was the major step in man's evolution after his physical structure had settled into more or less what it is today. The control and use of experience became possible with it, and then experiment and analysis. There is no need to conclude that they will not still provide mankind with a tool-kit to survive in a world which it has made so different from that of even our recent ancestors. But neither, of course, must we conclude that such a tool-kit will be found, or if found, properly and usefully employed. Pessimism is understandable – and probably temperamentally ineradicable.

One danger which does not seem quite so threatening as it did even a few years ago, certainly, is political. It no longer is solely a matter of the superpowers themselves and their ability to manage their relationships, but of the context in which the international balance must be preserved. The responsible leaders of the United States and the Soviet Union appeared even a couple of decades ago to have accepted that they could not be sure of imposing their will on the other by threat of war. They were also aware that victory in any real sense would be impossible in a full-scale nuclear war. Both began to seek agreements to limit or reduce their nuclear arsenals and both eventually, in 1991, and thanks to major earlier political changes, announced huge reductions. These developments may well be thought reassuring. But even superpowers have to live in a world which is changing and threatened still by destabilization. Indeed, that may have become even more of a threat now that the giants no longer confront one another on an otherwise almost empty stage. It is filling up rapidly, and the bit-part players have their own lines they long to deliver. Some of them have been waiting a long time to do so and are impatient; others are frightened, and that may be more dangerous. Several countries feel they must seek to acquire nuclear weapons of their own; others, emerging from a crumbling Soviet Union, find themselves *de facto* in possession of them. The realization that a nuclear war might mean destruction for mankind is still not shared by rulers in all countries. As long ago as 1968, Great Britain, the Soviet Union and the United States signed a Nuclear Non-Proliferation Treaty. More than a hundred and forty other states subsequently adhered to it, but France and China (already then possessing nuclear weapons) have not done so, nor have Israel, South Africa, India (the first 'non-aligned' country to detonate a nuclear device successfully, in 1974), Pakistan, Argentina and Brazil (all of whom either have effective devices, rudimentary though some of them probably are, or will have them in the near future). Grave uncertainty clocks the nuclear future of the successor states of the USSR. In addition, Iraq and North Korea, though signatories to the 1968 treaty, are known to be trying to acquire a nuclear weapons production capacity.

This must provoke sobering reflexions in a world still horribly violent, as countless deaths of men, women and children from political or quasi-political conflict have shown since 1945. And there is plenty of rubbish for new bonfires lying about. The economic ills which feed unrest in the underdeveloped world and keep it in ferment show few signs of disappearing. Covert means, subversion and terrorism are still used by governments to achieve inadmissible ends, and nationalism has lost none of its power to exploit human differences. Many historical ghosts some thought long laid to rest have been released in eastern and central Europe in the last two or three years to harrass and distract us once again.

And yet much has changed for the better. In the longest term and the perspectives of environmental damage, there is now at least recognition of problems where, even recently, none existed; there is the outline of some agreed agenda in sight. At the end of 1990 a UN conference met in Geneva with representatives of 137 nations present and concluded that global warming was a real threat to humanity. It also set to work to have a convention on man-made change in the climate ready for signature in 1992. Something, at least, could be done. Developed nations have already begun to show their capacity to restrict consumption of fossil fuels. In face of a different danger, the United Nations has now at least begun to try to police international society as was hoped by some of its founders and as the League never did. In the Far East and Europe astonishing political changes have taken place which must have reduced the likelihood of war from *some* long-established causes.

Again, the historian must not prophesy (happily for him). If it seems disappointing to some that history can offer no grounds for unequivocal assertions one way or the other about what may be the outcome of our uncertainties, that is because they expect too much. At the end of the day, the only advantage a historian has in considering such problems is that he may be a little less surprised by the outcome, whatever it is, than those who have not reflected on the history behind it. Only two general truths emerge from the study of history. One is that things tend to change much more, and more quickly, than one might think. The other is that they tend to change much less, and much more slowly, than one might think. The past hangs around longer and is more difficult to keep peacefully buried, even by strenuous efforts, than we believe. That ought to be obvious enough today, after the many reminders given us in the Middle East that the wars of the Ottoman succession which began long ago in the eighteenth century in south-east Europe are still unfinished. Innovation and inertia tend to be exemplified in any specific historical situation. It was the conclusion of the first version of this book that because of this, for good or ill, we shall always find what happens somewhat surprising. Nearly twenty years later, there seems to be no reason to say otherwise.

Illustration Sources

Book I

CHAPTER 1

2 *Homo-erectus pekinensis* skull found near Peking. © The Natural History Museum, London

8 Scrapers and hand-axes. © Cambridge University, Museum of Archaeology and Anthropology, Cambridge

14 The Laussel Venus, Dordogne. Museé d'Aquitaine Bordeaux

CHAPTER 2

19 Wall painting from a cave in Lascaux, Dordogne. © Caisse Nationale des Monuments Historiques et des Sites, Paris

22 Cave at Lascaux, Dordogne. © Caisse Nationale des Monuments Historiques et des Sites, Paris

Book II

CHAPTER 1

30 Statue of Cretan woman. © Ronald Sheridan, (Harrow), London

33 Limestone tablet found in Mesopotamia, dated 3500 BC. © Ashmolean Museum, Oxford

35 Scene at Sumerian dairy. By permission of the trustees of The British Museum, London

CHAPTER 2

39 Mesopotamian statues found at Tell Asmar, Iraq. Courtesy of The Oriental Institute, University of Chicago, Chicago

43 Mesopotamian cylinder seal, c 2300 BC. © Michael Holford, London

46 Statue of Ibihil, a steward, found at Mari, Syria. © Louvre, Paris

50 The great ziggurat at Ur, Iraq. © Roger-Viollet, Paris

CHAPTER 3

55 Tomb painting from Thebes, Egypt. By permission of the trustees of The British Museum, London

57 The pyramid at Saqqara, Egypt. © Barnaby's Picture Libary, London. Photograph Gerald Clyde

59 Painting on the tomb of Pharaoh Rekhmaza, Eighteenth Dynasty. Photograph Andre Held, Paris

63 Portion of the court of Ramses II at Luxor, Egypt. © Barnaby's Picture Library, London. Photograph Marie Mattson

66 Akhnaton or Amenhotep IV, New Kingdom. © Ronald Sheridan, (Harrow), London

67 Musicians and dancers, Egypt. By permission of the trustees of The British Museum, London

69 The colossi of Memnon at Thebes, Egypt. © Ronald Sheridan, (Harrow), London

CHAPTER 4

73 Mortuary temple of Ramses III, Exterior detail of the sea battle with the 'sea-peoples', New Kingdom, Dynasty XX. Courtesy of The Oriental Institute, University of Chicago, Chicago

76 Reconstructed fresco at Knossos showing the Minoan distinctive 'bull-leaping'. Photograph C M Dixon Photo Resources, Kent

79 Theseus and the Minotaur, fourth century BC, Greece. © Ronald Sheridan, (Harrow), London

81 Painted vase from Gournia, Minoan neo-palatial period, 1700–1450 BC. © Ronald Sheridan, (Harrow), London

83 Silver bowl from Cyprus. By permission of the trustees of The British Museum, London

87 Egyptian wall painting, Thebes, Dynasty XVIII, Tomb of Rekhirma. All rights reserved, The Metropolitan Museum of Art, New York

91 Jewish king bowing to Assyrian ruler. By permission of the trustees of The British Museum, London

CHAPTER 5

96 Soapstone figure from Mohenjo-Daro, Pakistan. The Government of Pakistan, Department of Archaeology

101 Ruins of Mohenjo-Daro, 3000 BC, Pakistan. Photograph J Allan Cash Ltd, London

106 The Great Wall of China. © Barnaby's Picture Library, London. Photograph Hubertus Kanus

110 Lid of clay pot from pre-Shang Honan, China. © Museum of Far Eastern Antiquities, Stockholm

115 Oracle on tortoise shell, China. Institute of History and Philology, Academia Sinica, Taiwan

CHAPTER 6

120 Iron-age bronze cauldron. Naturhistorisches Museum, Vienna

122 Gold plaque from Peru. The Dumbarton Oaks Research Library and Collections, Washington D.C.

123 Pre-historic tomb, Brittany. Zodiaque, France

124 Torc from Norfolk, England. By permission of the trustees of The British Museum, London

125 Crested helmet and body armour, Villanovan culture, Italy, seventh century BC. The University Museum, University of Pennsylvania, Philadelphia

Book III

CHAPTER 1

132 Statue from the Acropolis. © V and N Tombazi. Courtesy the National Tourist Organization of Greece, Athens

135 Pot by Exekias of Athens. Photograph Staatliche Antikensammlungen und Glypothek, Munich

CHAPTER 2

141 Hector and Achilles, fifth century BC vase, Greece. By permission of the trustees of The British Museum, London

144 Spartan pot, Greece. Bibliotheque Nationale, Paris

CHAPTER 3

150 The Parthenon, Athens. Photograph Werner Forman Archive, London

155 Bust of Thales. The Mansell Collection, London

158 Dionysus on a pot of about 500 BC. Staatliche Antikensammlungen und Glypothek, Munich. Photograph Christa Koppermann

161 The Erechtheum, Athens. The Mansell Collection, London

165 Statue of youth, found in Milos, about 550 BC. Courtesy the National Tourist Organization of Greece, Athens

CHAPTER 4

169 Part of a mosaic from Pompeii depicting Alexander the Great in the National Museum of Naples. Photograph Andre Held, Paris
172 The Venus de Milo, Louvre. Photograph Roger-Viollet, Paris
176 Statue of Alexander the Great in Egyptian style, Luxor. © Ronald Sheridan, (Harrow), London

CHAPTER 5

183 Etruscan couple on sarcophagus, about 500 BC. The Mansell Collection, London
185 Bronze vessel used for cremated remains, about 500 BC. Photograph Museo Civico Archeoogico Bologna
188 Julius Caesar, Fotomas Index, Kent. Propaganda coin. By permission of the trustees of The British Museum, London

CHAPTER 6

194 Onyx gem, Italy. Kunsthistorisches Museums, Vienna
196 Roman aqueduct in Segovia. Photograph J Allan Cash Ltd, London
202 Bronze head, Gallic chief. Bernisches Historisches Museum, Bern

CHAPTER 7

209 Carving from a tomb of the Menorah, Rome. The Mansell Collection, London
210 The Pontius Pilate Stone, found at Caesarea. © Ronald Sheridan, (Harrow), London
211 Masada, Herod's The Great's fortress above the Dead Sea. Photograph Weidenfeld & Nicolson Ltd, London
212 Dead Sea Scrolls, Israel. © Ronald Sheridan, (Harrow), London
215 The Arch of Titus in the Forum, Rome. The Mansell Collection Ltd, London
217 Coptic fresco, wall painting, Karanis. © The University of Michigan, Kelsey Museum Archives, Ann Arbor 1

CHAPTER 8

220 Stilicho, the son of a Vandal and last general of the Western Roman Empire. Photograph Hirmer Fotoarchiv, Munich
227 The head of the Emperor Constantine, part of a huge seated statue. Photograph Hirmer Fotoarchiv, Munich

CHAPTER 9

238 The Skelling, Kerry. Bord Fáilte (Irish Tourist Board), Dublin
241 Saxon barbarian warrior, 700 AD. Archiv für Kunst und Geschichte, Berlin
243 Mosaics from the Church of San Vitale, Ravenna. Courtesy the Italian State Tourist Office, London
246 Leo the Great meets Attila the Hun in a tapestry. © Ronald Sheridan, (Harrow), London
248 Ivory of St Gregory, tenth century. Kunsthistorisches Museums, Vienna

Book IV

CHAPTER 1

250 Hagia Sophia, looking towards the east end from the West. © A F Kersting, London
256 Detail of Scythian gold spherical cup, Kul Oba, 300–500 AD. Hermitage Museum, St Petersburg. © Werner Forman Archive, London

CHAPTER 2

266 Leaf from manuscript of *Automata* by Abu 'l'Izz Isma'il al Jazari, colours and gilt on paper, probably Syrian, 1315, Mamluk School. The Metropolitan Museum of Art, Bequest of Cora Timken Burnett, 1957, New York
269 Illumination from a book on chess, Spain. MAS, Barcelona
271 The Mosque of Cordoba, Spain. Instituto Amatller de Arte Hispanico. MAS, Barcelona

CHAPTER 3

276 Pope Honorius III. © Ronald Sheridan, (Harrow), London
282 Psalter for Basil II, eleventh century. Hirmer Fotoarchiv, Munich
284 Barkeinni psalter. Weidenfeld & Nicholson Ltd, London

CHAPTER 4

296 Mongol warrior, Persian painting. Topkapi Palace Museum, Istanbul
302 Suleiman the Magnificent, Turkish painting. Topkapi Palace Museum, Istanbul
305 Timur Lang's tomb, Samarkand, Uzbekistan. Robert Harding Picture Library, London
307 Mehmet II, by Gentile Bellini, 1480. National Gallery, London

CHAPTER 5

314 Codex Vigilanus, El Escorial, Madrid. MAS, Barcelona
318 Psalter, ninth century, made in France, Wurttemberg Landesbibliotek. Bildarchiv Foto Marburg
320 Psalter, Emperor Otto III, Staatsbibliothek, Munich. Hirmer Verlag, Munich
323 Viking ship, Norway. Fotograph Mittet Foto A/S, Oslo
325 The Witanagemot and king, Anglo-Saxon, England. By permission of The British Library, London
327 Mozarabic art, eleventh century illuminated manuscript. By permission of The British Library, London
332 Miniature of King Edgar, tenth century, England. By permission of The British Library, London

CHAPTER 6

342 Taj Mahal, Agra, India. © Barnaby's Picture Library, London. Photograph Hubertus Kanus
345 Painted cotton hanging of Indians' vision of Europeans, Shahibag, Ahmedabad, India. Calico Museum of Textiles, Ahmedabad, India

CHAPTER 7

355 Bronze figure of horse, Eastern Han Dynasty, second century, China. Robert Harding Picture Library, London
358 Painting of an examination, eighteenth century, China. Bibliotheque Nationale, Paris
362 Temple of Heaven, Peking, China. Sally and Richard Greenhill, London
366 Scholars on a scroll, China. Collection of the National Palace Museum, Republic of China

CHAPTER 8

370 Deity from Matsumo shrine, Japan. Sakamoto Photo Research Laboratory, Tokyo
372 Minamoto Yoritomo, Japan. Sakamoto Photo Research Laboratory, Tokyo
375 The Lady Murasaki, Japan. By permission of the trustees of The British Museum, London
377 Japanese swordmanship, Japan. © The Board of the Trustees of the Victoria & Albert Museum, London

CHAPTER 9

380 Inscribed votive tablet of red slate from temple at Meroe, second century BC , Africa. © Werner Forman Archive, London
383 Macchu Pichu, Peru. © Barnaby's Picture Library, London. Photograph Peter Larsen
386 Mayan relief. By permission of the trustees of The British Museum, London
387 Mixtec manuscript. By permission of the trustees of The British Museum, London

CHAPTER 10

392 St Etienne, Nevers, France. Photo Jean Roubier, Neuilly
394 Ivory of the martyrdom of St Thomas a Becket, English, circa 1400. The Metropolitan Museum of Art, The Cloisters Collection, Purchase, 1970, New York
396 Heretics being burned, France, fifteenth century. Bibliotheque Nationale, Paris
400 Stone carving of Eve, Autun Cathedral, France. Photo Jean Roubier, Neuilly
403 Manuscript of warfare, fifteenth century, English. By permission of The British Library, London

410 December in the fifteenth century, English. The British Libary, London

CHAPTER 11
418 The Reconquest, fifteenth-century altar-piece of James I of Aragon defeating the Moors, Spain. © The Board of the Trustees of the Victoria & Albert Museum, London
419 The Krak des Chevaliers, from the southwest, Syria. A F Kersting, London
424 Map of Columbus' discoveries. New York Public Library, New York
426 Magellan's discovery of the straits. Fotomas Index, Kent
429 Stone carved relief on tomb, Bologna, Italy. Museo Civico Medievale di Bologna

Book V
CHAPTER 1
434 Vasco da Gama. By permission of The British Library, London
440 Title-page of French business manual, seventeenth century. The Mansell Collection, London
442 Drawing of the hall of the Hanse merchants, sixteenth century, Antwerp. The Mansell Collection, London
447 A view of the House of Commons, London, eighteenth century. Christine Vincent, Hertfordshire
450 An engraving of the Frankfurt ghetto, Germany, seventeenth century. The Mansell Collection, London

CHAPTER 2
453 Titian painting of Charles V, El Prado, Madrid. The Mansell Collection, London
457 Luther by Cranach, circa 1525. The Mansell Collection, London
461 Elizabeth I of England by M Gheeraerts. The National Portrait Gallery, London
463 Rubens painting of Ignatius Loyola, founder of the Jesuits. The Mansell Collection, London
465 Adam Schaliger, a German Jesuit missionary in China, seventeenth century. The Mansell Collection, London
468 Medallion of Oliver Cromwell by Thomas Simon, 1650. National Portrait Gallery, London
471 Title-page of *Leviathan* by Thomas Hobbes, 1651. The Mansell Collection, London

CHAPTER 3
476 Painting by Gentile Bellini of the reception of the Venetian ambassador at Cairo, sixteenth century,in the Louvre, Paris. The Mansell Collection, London
479 Charles V and Pope Clement VII, Rome. The Mansell Collection, London
481 Medal struck to celebrate the defeat of the Armada, 1588. The Mansell Collection, London
491 Monument of Peter the Great at St Petersburg,Russia. © The Hulton Picture Companuy, London
495 Catherine the Great, painting at the Winter Palace, St Petersburg, Russia. The Mansell Collection, London
498 State of Poland, 1773, allegorical print showing Catherine the Great, Joseph II and Frederick II by J E Nilson. The Mansell Collection, London

CHAPTER 4
502 Salt cellar in Ivory from Benin, Africa. Werner Forman Archive, London
510 Drake's route to the Caribbean. Fotomas Index, Kent
514 John Smith captures the King of Pamaunkee. The Mansell Collection, London

CHAPTER 5
526 Samuel Pepys. The Mansell Collection, London
529 Gravestone of black servant, England. Christine Vincent, Hertfordshire
531 Painting of William Penn in North America by Benjamin West. The Mansell Collection, London
532 Captain Cook's mean shooting sea horses. Christine Vincent, Hertfordshire

CHAPTER 6
536 Engraving of The Royal Observatory, Greenwich, London in 1675. The Science Museum, London
540 Title-page of Vesalius' work on human anatomy. The Mansell Collection, London
542 Illustration from Harvey's treatise on the circulation of blood. The Mansell Collection, London
545 English Freemasons, eighteenth century. The Mansell Collection, London
547 Joseph Wright's painting, 'The Air-Pump'. National Gallery, London
550 Voltaire by Hubert Carnavalet. The Mansell Collection, London
551 Portrait of Jean-Jacques Rousseau by Quentin la Tour. The Mansell Collection, London

Book VI
CHAPTER 1
554 Statue of Queen Victoria, Calcutta, India. © Popperfoto, Northampton
557 George Cruikshank drawing, 'Taking the Census', 1851. Mary Evans Picture Library, London
559 Print of a soup-kitchen, England, nineteenth century. Christine Vincent, Hertfordshire
561 Early Sydney, Australia. © The Hulton Picture Company, London
563 The mechanical reaper, North America, nineteenth century. Science Museum, London
565 Coalbrookdale, Shropshire, 1758. © The Hulton Picture Company, London
566 Mill in 1840, Silesia, Germany. Mary Evans Picture Library, London
568 Isambard Kingdom Brunel viewing his last ship, *The Great Eastern*. Photograph Institute of Agricultural History and Museum of English Rural Life, University of Reading
570 Russian workers in St Petersburg, Russia, 1890s. © Novosti Press Agency, London
572 The 'Free Trade' hat, England. The Mansell Collection, London

CHAPTER 2
575 Engraving of the Boston Massacre, Paul Revere, 1770, Boston, USA. The Metropolitan Museum of Art, Gift of Mrs Russell Sage, 1909, New York
578 The Declaration of Independence, USA. The Mansell Collection, London
582 The Estates General at Versailles, France, 5 May 1789. The Mansell Collection, London
586 Delacroix's painting of Liberty, Paris, France. Louvre, Paris
589 David's painting of Napoleon, France. The Mansell Collection, London

CHAPTER 3
596 Giuseppe Mazzini. The Mansell Collection, London
598 *Punch* cartoon showing the frightened rulers in 1848, London. The Mansell Collection, London
605 Mr and Mrs Karl Marx, London. The Mansell Collection, London
609 War in the Crimea, attack on Malakoff and Little Redan, Russia. John Hillelson Agency, London

CHAPTER 4
614 *Punch* cartoons. Punch Publications Ltd, London
621 Cotton pickers, USA, nineteenth century. Culver Pictures Inc, New York
624 Evicted British labourers, 1870s. Photograph Institute of Agricultural History and Museum of English Rural Life, University of Reading
626 Great Exhibition, Crystal Palace, London, England, 1851. Mary Evans Picture Library, London
628 Engraving of well-fed British lion of free trade, London. Fotomas Index, Kent

CHAPTER 5
632 Christian missionaries in Africa. Bodleian Library, Oxford
634 The Maxim gun, England. National Army Museum, London
636 Belgian magnate with his chauffeur, Katanga, Zaire. Werner Forman Archive, London
641 Europe's cultural superiority

as depicted in an 1890s advertisement. Christine Vincent, Hertfordshire
646 Boer fighters, 1899, South Africa. © The Hulton Picture Company, London
649 Battery in Gibraltar, Highland Regiment, circa 1868. John Hillelson, London
659 Destruction of the US Battleship *Maine* in Havana Harbor, 1898. Chicago Historical Society, Chicago
660 The *Missouri* is the first battleship to pass through the Panama Canal. © The Hulton Picture Company, London

CHAPTER 7
663 English cartoon of the reception in at the court in China of Lord Macartney's deputation, 1792. Fotomas Index, Kent
666 General Gordon's Bodyguard trained on half of the Chinese government to resist the Taipings, China, 1860. John Hillelson Agency, London. Photograph Felice Beato
668 Pu-Yi, the last emperor of China, 1911. The Mansell Collection, London
670 *Punch* cartoon on an anti-Boxer rebellion theme, 1900. The Mansell Collection, London
671 Yuan Skih-k'ai at his inauguration, 1912, China. © The Hulton Picture Company, London
673 The Japanese capture of Tientsin in China, 1894, from a contemporary print. © The Board of the Trustees of the Victoria & Albert Museum, London
674 The first railway station in Japan, wood block print, 1872. Christine Vincent, Hertfordshire
677 Japanese parliament house, 1891. © The Hulton Picture Company, London
681 A young Indian prince and his advisers. Oriental and India Office Collections, The British Library, London
682 King George V and Queen Mary show themselves to the people, Delhi, Durbar, India, 1911. © Popperfoto, Northampton
686 Operating theatre in Bombay, India. Oriental and India Office Collections,The British Library, London

Book VII
CHAPTER 1
688 The Red Banner of victory over the Reichstag,1945. © Novosti Press Agency, London
692 Ballooning at Ranelagh, England, 1907 and distressing scenes in the East End of London, 1912. © The Hulton Picture Company, London
696 Woman chaining herself to the railings of Buckingham Palace, 1914. © Popperfoto, Northampton
699 German postage stamp of the coronation of Kaiser William II. Edimedia, Paris
702 Charles Darwin, a photograph by Captain Darwin. The Mansell Collection, London

CHAPTER 2
707 Reactionaries marching in Odessa, Ukraine, 1905. © The Hulton Picture Company, London
710 A group of De Havilland long-distance bombing machines, Serny Aerodrome, France, 1918. Photograph the Imperial War Museum, London
713 Middlesex battalion, 1916, France. © Popperfoto, Northampton
714 Indian soldiers on the British Western Front. © Popperfoto, Northampton
716 French tank, First World War. The Imperial War Museum, London
718 Lenin, Krupskaya, Zagorsky, Litvinov, Ludvinskaya in Red Square, May Day, 1919. © Novosti Press Agency, London
721 Stanley Baldwin and Ramsey MacDonald during the Coalition government, 1931. John Hillelson, London
724 Famine in Russia, post-Revolution. © Roger-Viollet, Paris

CHAPTER 3
732 Nanking arsenal, 1867, China. © The Hulton Picture Company, London
738 Chian K'ai-shek on Black Dragon, 1929. © Barnaby's Picture Library, London
740 Japanese marines behind sandbags in China, 1932. © Popperfoto, Northampton
741 Low cartoon from the *Evening Standard*, 1933. © John Appleton. Photograph Centre for the Study of Cartoons and Caricature, University of Kent at Canterbury, Kent
743 Japanese soldiers in Shanghai, 1937. Foto Ullstein Bilderdienst, Berlin

CHAPTER 4
746 Opening of the Suez Canal, 1869, a French vessel passes through. © The Hulton Picture Company, London
748 RAF armoured cars in Palestine. © Popperfoto, Northampton
750 Day of destiny in Iraq: an uncontrollable fountain of oil gushes 140 ft above the derrick at Kirkuk. © The British Petroleum Company plc, London
754 Kemal Ataturk. Photograph the Turkish Embassy, London
758 'Die Traumdentung' title page. Mary Evans Picture Library, London

CHAPTER 5
760 Franklin Delano Roosevelt, Warm Springs, Georgia, USA, 1939. © Associated Press, London
764 Hitler and the Party—massing of the standards at Nuremberg, 1934. © Barnaby's Picture Library, London
769 Nazi atrocities in Russia, 1942. © Novosti Press Agency, London
770 South Lancashire battalion lands in Normandy, 1944. Photograph The Imperial War Museum, London
771 The attack on Pearl Harbour, 1941. © Popperfoto, Northampton
774 The Battle for Fort Dufferin, Burma, Second World War. © Popperfoto, Northampton
775 Churchill, Stalin and Truman at the Potsdam Conference in 1945. © Popperfoto, Northampton

CHAPTER 6
778 The UN building in 1985 on the occasion of its fortieth anniversary. © UN Photo 166456/Lois Conner, London
783 Aftermath of Czech Crisis—Police and militia parade in Prague, 1948. © Hulton Picture Company, London
786 Gandhi's funeral, 1948, India. © Magnum Photos, London. Photograph Henri Cartier-Bresson
790 Raising the flat in 1948 in Israel. © Israel Defense Department, Tel Aviv

Book VIII
CHAPTER 1
794 Man on the Moon, 1969. Photograph NASA, Houston
799 Street children in Delhi, India after a snowfall in January, 1991. © Popperfoto, Northampton
809 The assembly line of a Nissan car factory showing robots at work. © Picturepoint, London
812 Letter to President Franklin Roosevelt from Albert Einstein, 1939. Photograph courtesy of Franklin D Roosevelt Library, New York
816 'Goodbye comrades'—Yuri Gagarin waves goodbye before entering space capsule which took him around the world, 15 April 1961. © Topham Picture Source, Kent
819 Polish Greens leader, Andrzej Pietrowski, shows a lone pine in thousands of hectares of forest destroyed by sulphur dioxide emissions from power plants in nearby Czechoslovakia, German and Poland itself, 1991. © Popperfoto, Northampton
825 Pope Paul VI in Bombay in 1964. © Popperfoto, London

CHAPTER 2
832 USAF bombers over Korea. Robert Hunt Library, London
836 Indian army officer seen blindfolded after being taken prisoner by Pakistan troops during

fighting in Kashmir, September, 1965. Topham Picture Source, Kent
839 Old man amongst Nationalist recruits, Beijing, 1944. Magnum Photos, London. Photograph Henri Cartier-Bresson
843 Poster which introduced the Cultural Revolution in 1966, it shows Mao inspiring the people. © Camera Press, London
848 Nasser greets the masses, Cairo, 1954. © Popperfoto, Northampton
851 Prayers at the Wailing Wall, Jerusalem, Israel. © Camera Press, London. Photographer J N Reicher
854 The Union Jack makes way for the new country of Botswana's flag, 1966. © Camera Press, London
859 F W de Klerk and Nelson Mandela confer in Davos, Switzerland, 1992. © Popperfoto, Northampton
863 Missile site in Cuba, 1962. © Popperfoto, London
866 Salvador Allende, 1973 in Chile. © Topham Picture Source, Kent

CHAPTER 3

868 Street scene in Budapest, Hungary in November 1956. © The Hulton Picture Company, London
870 Brezhnev's last November parade 1982, Moscow. © Camera Press, London
872 Hundreds of new cars await delivery at General Motor's storage area in New Jersey, USA. © Camera Press, London
874 Black demonstrators in Washington D.C. 1963. © Topham Picture Source, Kent
876 US soldier in Vietnam with woman and child. © Camera Press, London. Photograph James Pickerell
878 Mrs and Mrs Richard Nixon in China, 1972. © Topham Picture Source, Kent
880 Prague, Czechoslovakia, 1968. © Magnum Photos, London
884 Mrs Thatcher at European Economic Community, 1984. © Topham Picture Source, Kent
889 Armed soliders guard downtown Amritsar in preparation for marches on the Golden Temple by disgruntled Sikhs, India. © Popperfoto, Northampton
892 An effigy of the Shah is burned in Tehran, 1978. © Camera Press, London. Photographer Michel Giannoulatos

CHAPTER 4

899 An Israeli soldier fires a rubber-coated steel ball towards Palestinian demonstrators, September 1989. © Popperfoto, Northampton
901 French Jaguar fighter during bombing of a munitions dump in Kuwait, January, 1991. © Popperfoto, Northampton
903 Gorbachev and Reagan share a laugh at the Geneva Summit concluding ceremony, November 1985. © Popperfoto, Northampton
905 Lech Walesa addressing the workers in Gdansk shipyard, Poland, 1980. © Popperfoto/UPI, Northampton
908 The Berlin Wall comes down near the Brandenberg Gate in East Berlin, Germany, February, 1990. © Camera Press, London
912 A statue of Liberty is raised in Tienamen Square, Peking, China, May, 1989. © Popperfoto, Northampton

Colour Plates

Maiden Castle (Robert Estall, Suffolk); Sphinx (Robert Estall/Alistair Scott, Suffolk); Tholos, East India Company official, Benin head (Michael Holford, London); Stela, Easter Island, Lion-strangling, Japanese Scroll, Mozarabic Art (Werner Forman Archive, London); Roman Urbanism (John Hillelson/Brian Brake, London); The Medieval World Order (SCALA, Florence); Reliquary (Hirmer Verlag, Munich); Samuel Scott, Bell Scott (ET Archive, London,); DNA (Science Photo Library, London); Modern Japan (Hutchison, London)

Index

Aachen:
  Charlemagne's court at, 317–18
  Ottonian use of, 321
Abbasid caliphate, 261
  seizure of power, 264
  nature of rule, 267–8
  achievements of civilization, 268–72
  Charlemagne and, 272, 316
  end of rule, 272, 295, 300
Abbas 'the Great' (*c.* 1557–1628/9), shah of Persia (1587–1628):
  built Isfahan, 310
  religious intolerance, 310
  English employees, 311
Abdul Hamid II (1842–1918), sultan of Turkey (1876–1909), 747
Abgar VIII or IX, Christian king of Osrhoene (179–214), 206
aborigines, Australian, 529, 647
Abraham, Hebrew patriarch, 87–8, 270
Abu-al-Abbas, first Abbasid caliph (749–754), 267
Abu-Bakr (573–634), first Arab caliph, 261
Abu Simbel, 53, 128
Abyssinia *see* Ethiopia
Academy of Athens, first university, 163, 175
  abolished, 243
Achaean peoples, 80
Achaemenid dynasty of Persia, 94, 130–1, 147
  evoked by Sassanids, 222–3, 252
  Scythians and, 255
Achilles, Greek hero, 140, 171
Acts of Parliament (British):
  Enclosure, 439, 561
  Navigation (1651), 525
  Corn Laws, 572–3, 625
  Quebec (1774), 576
  Stamp (1765), 576
  Reform (1832), 624
  Factory (1801), 627
  Union (with Ireland, 1801), 629
  Government of India (1935), 785–6
Aden:
  Turks in, 309
  British naval base, 649
  British withdrawal, 890
Aegean:
  islands, 68, 76, 77, 137–8, 287, 304
  Dark Ages of, 82–4, 85
Aeschylus (525–456 BC), Greek tragedian, 164
Aesop (mid-6th cent. BC), Greek slave, 140
Afghanistan:
  Alexander and, 171
  revolt from Persia, 311
  USSR and, 893–4, 895, 898, 902, 907
Africa:
  prehistory, 7–9, 12, 15–16, 19, 118–21
  Roman province, 187, 222, 229
  Christianity in, 226, 228, 234–5 (*see also* missionary activity)
  Vandal invasion, 232, 242
  Islam and, 272, 273
  early native cultures, 379–85
  European trade bases, 503
  slave trade in, 530–1, 856
  imperialism in, 656–8
  World War I and, 715
  World War II and, 771, 855
  négritude, 827, 918
  economic deterioration, 853–6
  Cold War and, 857
  See also individual areas
Agni, Aryan god, 99
agriculture:
  in America, prehistoric, 24–5, 28, 121
  invention, 25
  social effects, 26
  in Middle East, 45, 272
  in Egypt, 52, 56–7
  Mediterranean, 77
  in India, 97, 801
  in China, 106–7, 840
  in Africa, 119–20, 379–80, 855
  in western Europe, 124, 881
  Greece, 138, 152–3
  basis of Roman economy, 183, 205, 221
  of early medieval economy, 333–4
  Indian, 383, 387–8
  productivity increases, 409–11, 412, 438–41, 559–62
  setbacks, 412
  in market economy, 441, 560, 564
  in Russia, 493–4, 562–3, 727, 870
  Caribbean, 512–13
  crop and animal transplants, 532–3
  worldwide productivity increases, 564, 803–5
  large-scale northern, 623
  in Japan, 675
  eastern European, 904
  See also animals, crops, plough
Ahura Mazda, Persian god, 253
air travel and warfare:
  aeroplanes, 704, 808
  airships, 716, 808
  in World War II, 767–8, 770, 773
Akbar (1542–1605), Moghul emperor of India (1556–1605), 347–50, 351
Akhnaton (Amenhotep IV), king of Egypt (*c.* 1375–1358 BC), 67, 88
Akkad, Akkadians, 45–6, 51
Alaska, sold to US, 648
Albania:
  Ottoman conquest, 308
  state, 697
  Chinese affiliations, 879
Albert of Saxe-Coburg-Gotha (1819–61), prince consort of UK, 629
Albertus Magnus (1193 or 1206–80), scholastic philosopher, 428
Albigensian heretics, persecuted, 397
alchemy, 538
alcohol:
  etymology, 40
  in Mesopotamia, 45
  in Egypt, 62
  early medieval beer, 333
  Indian abstention, 339
  trading commodity, 411, 532
Alcuin (735–804), English scholar, 317
Alembert, Jean le Rond d' (1717–83), French encyclopaedist, 548
Alexander 'the Great' (356–323 BC), king of Macedon (336–323 BC) and conqueror in East, 170–3, 337
Alexander I (1777–1825), tsar of Russia (1801–25), 607
Alexander II (1818–81), tsar of Russia (1855–81), 611, 706
Alexandria:
  Egypt, founded, 171
  culture at, 174–5, 204
  Jews in, 208
  falls to Arabs, 281
  French community, 745
Alfred 'the Great' (849–99), king of Wessex (871–99) and England (886–899), 326
Algeria:
  French imperialism, 653, 656
  nationalism in, 791, 849
  independence of, 851, 881
  Islamic fundamentalism and, 893, 899
Al-Khwarizmi (d. 835), Islamic astronomer, 269
Al-Kindi (d. *c.* 870), Islamic philosopher, 270
Alliance for Progress, in S. America, 864
Al-Mansur (939–1002), Arab conqueror of Spain, 328
Alva, Ferdinand, duke of (1508–82), Spanish general, 466
Amarna, 67
Amboyna, spice trade base at, 504

Ambrose, St (*c*. 340–97), bishop of Milan, 234, 235
Amenemhet I, king of Egypt (*c*. 2000–1970 BC, 65
Amenhotep III, king of of Egypt (*c*. 1411–1375 BC), 67
Amenhotep IV *see* Akhnaton
America *see* Central America, North America, South America, West Indies, individual states
American Revolution, 574–6
issues, 576–81
war, 576–7
constitution, 577–81
Declaration of Independence, 577
domestic divisions, 577–80
France and, 577, 581–2
Amon-Re, Egyptian god, 60, 65, 67
Amorites, 48, 50–1, 90
Amos (8th cent, BC), Hebrew prophet, 91–2
Amsterdam, 438, 441, 447, 527
anarchists, 693
Anatolia, 25, 38, 42, 73, 76
Hittites in, 37, 48, 51, 86
Huns in, 232
Arabs and, 281
Seljuks in, 296
in Ottoman empire, 308
See also Ottoman empire, Turkey
Anaximander (610–547 BC), Greek philosopher, 160
Anglican Church:
development, 459–60
dissenters, 467
Anglo-Persian Oil Company, 755
Angola, 858
animals:
domesticated, earliest, 26
in India, 97, 101, 102, 103, 339
in Europe, early medieval, 333–4
in S. America, 388
species spread by Europeans, 533
selective breeding, 561
in World War I, 715
See also horses, rabbits
Anna Ivanovna (1693–1740), tsarina of Russia (1730–40), 494
Antigonid dynasty of Macedon, 173
Antiochus IV, king of Syria (175–164 BC), 207
anti-Semitism:
medieval, 404, 413
early modern, 441
Nazi, 763–5, 791
Antoninus Pius (86–161), Roman emperor (138–61), 197
Antony, St (*c*. 250–350), Christian hermit, 245
Antwerp, 441, 445, 464
Anu, Sumerian god, 42, 51
apartheid, 857–9
Aphrodite, female deity, 79, 142
Apollo:
Greek god, 140
See also Delphi
Apollonius of Tyana (*fl*. AD 50), Greek philosopher, 204
Aquinas, St Thomas (*c*. 1226–74), Christian philosopher, 416, 428, 430, 433
Arab League, formation, 792
Arabs:
conquests under Islam, 258, 261–2
nomadic pastoralists, 258
empires, 264–74
Islamic civilization, 268–72
in Europe, 272, 321, 323–4
in India, 346
sources for early Africa, 381–2
European trade rivalry, 503, 504
World War I and, 715
national states emerge, 748–55
Israel and, 791–3, 847–50, 852–3, 895, 900
oil and, 848–9, 887
Pan-arabism, 849
national states and nationalism in, 892–3
See also Islam, Umayyads, wars (Arab-Israeli), individual areas
Aramaeans, 92
Archimedes (*c*. 287–212 BC), Greek mathematician, 175, 189, 543
architecture:
Mesopotamian models, 40, 49
Egyptian organization, 57–8
in ancient China, 117
megaliths, 126–7
Greeks set standards, 166
Roman technical skill, 199
Islamic, 270
Taj Mahal, 350
ecclesiastical, 395–6
See also building
Ardashir I (Artaxerxes), king of Persia (224–40), 222, 252, 341
Argentine:
expansion of food production, 563, 640, 715–16
Indians in, 640
dictatorship in, 865
Falklands War, 883
nuclear weapons, 921
Argos, 82
Aristarchus of Samos (*c*. 310–264 BC), Greek astronomer, 175
Aristophanes (*c*. 448–380 BC), Greek comedian, 166
Aristotle (384–322 BC), Greek philosopher, 331, 416
collected data, 159, 163
on Pythagoras, 160
influence, 161, 163, 175, 268
on city-state, 163, 175
founder of deductive logic, 163
tutor of Alexander, 171
medieval prestige, 416, 428, 538
Arius (*c*. 256–336), Arianism, Christian heresy of, 228, 234, 238, 239, 247
Armenia:
Alexander and, 173
Romans and, 197–8
focus of conflict, 222
Persian conquest, 252
Arabs in, 281
and Mongols, 300, 304
Russian gains in, 648
1923 settlement, 753
in USSR, 907
Arminius (17 BC–AD 21), Ger, chieftain, 197
arms control, 903
Armstrong, Neil (1930–), on moon, 817
Arsacid dynasty of Parthia, 252
art:
Palaeolithic, 18–21, 28–9
Sumerian, 44, 47, 48
Egyptian, 53, 61–2
Chinese, 117, 354, 361–2, 368
Greek, 141–2, 166–7
Islamic, 270
court arts in Japan, 373–4
in black Africa, 382
Renaissance Europe, 429
changing 20th-cent. patterns, 756–7
See also painting, pottery, sculpture
Artaxerxes *see* Ardashir I
Arthur (6th cent.), legendary Romano-British king, 237, 403, 414
Aryan peoples:
in India, 99–103
in Persia, 129
Ashurbanipal, king of Assyria (668–626 BC), 85, 93, 129
Asia Minor:
Mycenaean colonies, 82
Achaeans in, 83, 86
Hittites in, 86
Greek cities, 147
Alexandrine empire, 168–9, 170, 173
Romans in, 185–7, 229
Byzantine empire and, 254, 281
See also individual areas Asoka, Maurya emperor of India (264–228 BC), 339–40
Assassins, Moslem sect, 300
Assur, Assyrian god, 93
Assyria:
Assyrians, 38, 48–9, 51
conquers Israel, 92
conquests in Mesopotamia, 92–3
defeated, 93
in Egypt, 129
Astarte, Semitic goddess, 79
astrology, 177, 823, 834, 890, 917
astronomy:
Babylonian, 49
Egyptian, 58
Chinese, 112
and megaliths, 127
erroneous Greek theorizing, 162
Arabic, 269
Ptolemaic primacy, 428–9, 543
early modern scientific views, 543
observatories, 703
Einstein and, 811
See also space
Athanasius (*c* 298–373), bishop of Alexandria, against Arianism, 229
Athena, Greek goddess, 139
Athens:
settled by Achaeans, 80, 82
economy, 142, 144–5, 152, 154–6
government, 144, 154–7
against Persia, 147–8
exploited Delian League, 149
Peloponnesian War and, 149–51
social institutions, 152–3, 167
cultural and political dominance, 154, 157, 175
atheism as impiety, 159
defeat by Macedon, 168–9, 173
Atlantic Charter (1941), 768, 773
atomic theory, 810
Greek, 160, 538
Aton, Egyptian god, 61, 67
Attalid dynasty of Pergamon, 173, 189
Attica, 80, 82, 143, 145, 148, 152
See also Athens
Attila (*c*. 406–53), Hun chieftain, 232, 247
Augsburg, Peace of (1530), 458, 480, 481
Augustine:
St (354–430), bishop of Hippo, Christian philosopher, 234–6, 253, 317, 430–1
St (d. 604), archbishop of Canterbury, 247
Augustus (63 BC–AD 14), first Roman emperor (27 BC–AD 14):
maintained republican forms, 194–5, 219
rise to power, 195
successors, 195

military aims, 196–8
use of religion, 203
Aurelian (c. 214–75), Roman emperor (270–75), 224
Aurungzebe (1618–1707), Moghul emperor of India (1658–1707), 350, 351
Australia:
European settlement, 500, 521, 533, 623
aborigines, 529, 647
food production in, 563, 645
'White Australia' policy, 638
federal structure, 643
and New Guinea, 661
in World War II, 770
and Indonesia, 789
foreign policy, 847
economy, 887
Australopithecus, 7–9, 12
Austria:
Dual Monarchy *see* Habsburgs
French revolutionary wars, 585, 588
Venice acquired, 592
Vienna settlement and, 595
1848 revolts and, 599–600
Slav nationalism in, 697, 719
Anschluss, 765
post 1945, 833
See also Habsburgs
avant-garde, idea of, 757
Avars, 280
in Balkans, 254
Persian allies, 254
use of stirrup, 257
and Byzantium, 288
Charlemagne and, 316
Averroes (1126–98), Islamic philosopher, 270
Avicenna *see* Ibn-Sina
Avignon, papal residence at, 399
Ayyubid dynasty in Egypt and Levant, 298
Azanian culture in Africa, 382
Aztecs:
tributary empire of, 385–6
Spanish conquest, 509, 647

Babur (1483–1530), founder of Moghul empire (1526–30), 347
Babylon:
first empire, 48–51
Kassites in, 71, 72
Hittite raid, 86
Assyrians and, 92
conquers Jerusalem, 92, 206–7
last empire, 94
Bacon, Francis (1561–1626), English philosopher, 539–41, 548
Bactria:
Bactrians, Greek kingdom of, 173
in India, 341
Baghdad:
Abbasid capital, 267, 268
Varangians in, 290
sacked by Mongols, 300
taken by Turks, 310
balance of power:
established as aim, 484
modern, 897
Balboa, Vasco Nuñez de (c. 1475–1517), Spanish explorer, 509
Balfour Declaration (1917) on Jewish homeland, 751, 791
Balkans:
Slavs and Avars in, 254
Ottoman empire and, 308, 423
independent states emerge, 697
See also individual areas
Baluchistan, 96, 97
Bandung Conference (1955), 837
Bangladesh, 824, 835–6
Bantu peoples, 381
Basil II (c. 958–1025), Byzantine emperor (976–1025), 280, 281
Basques, 470
Bath, thermal springs at, 237
Batista, Fulgencio (1901–73) Cuban dictator, 862
battles:
Marathon (490 BC), 148
Mycale (480 BC), 148
Plataea (480 BC), 148
Salamis (480 BC), 148
Thermopylae (480 BC), 148
Leuctra (371 BC), 151
Cannae (216 BC), 185
Trasimene (217 BC), 185
Zama (202 BC), 186
Actium (31 BC), 195
Adrianople (378), 231
Manzikert (1071), 286, 297
Mohács (1526), 308
Tours (732), 321,315
Stamford Bridge (1066), 326
Hastings (1066), 327
Agincourt (1415), 405
Crécy (1346), 405
Tannenberg (1410), 422
Lepanto (1571), 480
Plassey (1757), 507
Saratoga (1777), 577
Yorktown (1781), 577, 582
Somme (1916), 585, 715, 767
Aboukir (1798), 590
Trafalgar (1805), 590
Waterloo (1815), 593
Omdurman (1898), 633
Verdun (1916), 715, 767
Passchendaele (1917), 717
Atlantic, the (1939–45), 767, 771–2
Britain (1940), 767–8
El Alamein (1942), 771
Stalingrad (1942), 771
Dien Bien Phu (1954), 845
Bayle, Pierre (1647–1706), French philosopher, 545
Beccaria, Cesare (1738–94), Italian penologist, 549
Becket, St Thomas (1118–1170), English bishop, politician, martyr, 395
Becquerel, Antoine Henri (1852–1908) French scientist, 810
Bede (c. 673–735), English scholar monk, 239
Beethoven, Ludwig van (1770–1827), German composer, 757
Belgium, Belgians:
Franks in, 237–8
area adumbrated, 466
independence, 598
in Congo, 656, 658, 856
World War II and, 767, 772
See also Flanders, Netherlands
Belisarius (c. 505–65), Roman general, 242
Belshazzar (6th cent. BC), Babylonian prince, 94
Benedictine monasteries, 246, 331
Benedict, St (c. 480–c. 544):
monasticist, 245–6
Rule of, 317, 318, 331
Benelux *see* Belgium, Netherlands
Bengal:
British entry, 507
partition, 681
famine in, 787
secession of East, 824, 835–6
Benin, 382
Bentham, Jeremy (1748–1832), British philosopher, 549, 705
Berber peoples, 118, 234, 262
Berlin:
growth, 568
Congress of (1878), 710
division and blockade of, 829–30
Wall of, 868, 908
Bernard of Clairvaux, St (1090–1153), Cistercian reformer, 391, 398, 417
Bevin, Ernest (1881–1951), British statesman, 782
Bhagavad Gita, Hindu text, 340
Bible:
diffusion, 87, 538
Carolingian copies, 317
printed, 430, 431
literal truth questioned, 701
and Red Book, 845
See also New Testament, Old Testament
biology, biological sciences, 811–14
Bismarck, Otto von (1815–98), Prussian statesman, 833, 909
German unification and, 602, 604, 774
imperialism and, 697
foreign policy of, 706, 711
Boadicea (d. 62), British queen, 204
Boeotia, 152
Boers:
on Great Trek, 644, 645
settlement, 645
wars with British, 645
See also South Africa
Boethius, Anicius Manlius Severinus (c. 480–524), Roman philosopher, 239, 317, 326, 331
Bogomil heresy, 287
Bohr, Niels (1885–1962), Danish physicist, 810, 811
Boleslav I, king of Poland (992–1025), 294
Bolivar, Simon (1783–1830), S. American statesman, 640
Bolivia, 864, 865
Bologna, university of, 427
Bolsheviks, 694
October Revolution and, 717
former allies and, 719, 751
Stalin and, 727
as Chinese inspiration, 734
in Middle East, 753, 755
conservative fears of, 762
aims, 829
bomb, atomic, hydrogen *see* nuclear, weapons
Bonaparte, Napoleon *see* Napoleon I
Boniface, St (680–754), evangelist of Germany, 315, 331
Boniface VIII, pope (1294–1303), 398–9
Bosnia:
in Ottoman empire, 308
annexed by Austria-Hungary, 711, 747
declares independence, 910
Boston:
'Massacre' (1770), 576
'Tea Party' (1773), 576
Botha, P.W. (1916–), S. African prime minister, 858
Bougainville, Louis de (1729–1811), French explorer, 533
Bourbon dynasty of France, 460–1, 477
against Habsburgs, 481
in exile, 589
restored, 590, 593, 597
Boxer movement in China (1900):

xenophobic, 669, 678, 731
Germans and, 699
indemnities for, 734, 738
Brahe, Tycho (1546–1601), Danish astronomer, 543
brahmanas, 99, 100, 103, 104, 204
Hindu thought and, 103
in Maurya empire, 339, 340–1
evolved into Hinduism, 343–4
Brahmans, Indian texts, 101–2
brain, growth of, 9, 10
Brandenburg:
electorate of, 482, 496
See also Prussia
Brazil:
Portuguese in, 425, 511
slaves emancipated, 631, 640
independence of, 640
society, 640
republican, 641
resources of, 641
military rule, 865
nuclear weapons, 921
Brezhnev, Leonid (1906–1982), Russian statesman, 902, 906
Britain:
Roman province of, 197, 204, 237
See also England, United Kingdom
British Broadcasting Corporation (BBC), 808
British Commonwealth of Nations:
support UK in World War II, 767, 775
Imperial Conference (1926), 784
Indian constitutional reforms, 785–6
Indian partition, 787
in southern Africa, 855
British Empire and Commonwealth:
seeds, 507, 513–19, 524
American Revolution and, 574–6
19th-cent. attitudes to, 643
Boers and, 644–5, 857
'Dominion status' created, 644
native populations and, 646–7, 838, 850
in India, 649, 678–84, 687, 781, 784–7
naval bases, 649
in black Africa, 656–8
in Egypt, 656, 712
in S.E. Asia, 659, 685, 686, 687
1914 extent, 661
World War I and, 715
World War II and, 767, 775
internal divisions, 850
effect of decolonization on UK, 882
bronze:
working of, 27
in Sumer, 45
in Egypt, 72
in India, 102
in China, 107, 117, 354
in Japan, 369
Bronze Age:
term, 27, 40
fades, 72, 80, 85
stable pattern of, 128
Bruno, Giordano (*c.* 1548–1690), Italian philosopher, 544
Buddha (Siddhartha Gautama) (*c.* 563–*c.* 483 BC), 104
deified, 341, 344
sculpture of, 341, 374
Buddhism:
in India, 102–5, 339–40, 344
in China, 116, 358–9, 364–5
among Mongols, 300
Mahayana type, 344
in Japan, 369, 675
See also Zen
building:
in brick, 32, 97
in Sumer, 45
in Egypt, 57–8 (*see also* pyramids)
of neolithic Crete, 75–6
by Solomon, 84, 90–1
Harappan, 97
in China, 109
early African, 382
Mayan, 384–5
See also architecture, cities, technology
Bulgaria:
Bulgars, movements, 242, 281
Byzantine conquest, 281–2, 286–7
first Slav state, 288
Russia and, 291, 292
Ottoman conquest, 308
independent, 697, 747
World War I and, 714, 721
communist state, 781
Burgundy, duchy acquired by Habsburgs, 407
Burke, Edmund (1729–97), British political philosopher, 552, 588, 826
Burma:
Chinese conquest (1294), 364
Portuguese trade with, 504
British in, 658, 667, 685, 687
World War II and, 770, 776, 787
independence, 787, 791
Buwayhid dynasty in Persia, 272, 295, 297
Byblos, 47, 84–5
Byzantium:
Byzantine, Constantinople founded, 229
centre of later Roman empire, 241, 242, 243, 245
socio-economic strains, 242, 244
Persia and, 253
divergence from West, 277, 285
religion, 277–8, 283
character of, 279, 285
struggle for survival, 280
Arabs and, 281
recovery, 281
decline and fall, 285–8, 306
Bulgars and, 288
threatening Russia converted, 291
clerical intransigence in, 306
Constantinople taken by Turks, 308, 422
See also Constantinople, Rome

Cadiz, 85
Caesar, Gaius Julius (*c.* 102–44 BC):
in Gaul, 192, 197
Roman general and dictator, 192–3
adopted Octavian, 195
deified, 195
Calais, English capture of, 405
Calcutta:
British trade post, 506
'Black Hole' of, 507
capital city, 680
calendar:
Egyptian, 58
in China, 112, 365
Julian, 192
Moslem, 260, 754
Mayan, 384
in Japan, 674
Calvin, John (1509–64), Protestant theologian, 458–9, 462
Cambodia:
French imperialism in, 685–6
'associate state', 790
communist victory, 877
Cambridge, university of, 427, 536, 680, 809, 810
Canaan:
Philistines in, 86
Hebrews in, 88, 89–90, 91
Canada:
French settlement, 515, 516–17
British conquest, 517, 623
expansion of food production, 563, 645
Quebec Act (1774), 576
Dominion created, 643
member of NATO, 831
Canute (*c.* 995–1035), king of England (1016–35), Denmark and Norway, 326
Capetian dynasty of France, 320, 404–5, 452
capitalism:
origins, 412, 442
applied to industrialization, 567
Marx's view, 606, 693
associated with US democracy, 621
unemployment and, 693
women and, 695
Chinese hostility, 733, 737
target of Hitler, 763
world economic division and, 763
See also Economic organization, financial services Cappadocia, 74, 173, 226
Caracalla (186–217), Roman emperor (211–17), 219
Caribbean *see* West Indies
Carolingian empire:
territorial acquisitions, 316
intellectual life, 317
subordination of Church, 317, 330
partition, 319, 320–1
See also Charlemagne
Carter, James Earl (1924–), US president, 891, 892, 895–6
Carthage:
foundation, 85–6, 147
western Greeks and, 147, 173, 179
Athens and, 151
Rome and, 185–7
Germanic capture, 232
child sacrifice in, 234
Arab capture, 262
cartography, 134, 424, 426
Casas, Bartolomé de las (1474–1566), Spanish defender of Indians, 512
caste-system in India, 95, 100, 339, 340–1, 343
irrelevant to Buddha, 104
Castro, Fidel, (1927–), Cuban dictator, 862–4
Çatal Hüyük, Turkey, 27, 32
Catherine II (1729–96), tsarina of Russia (1762–96), 494–6, 497–9, 549
Cato, Marcus (234–149 BC), Roman statesman, 186
Cavendish Laboratory, Cambridge, 809
Cavour, Camillo Benso di (1810–61), Italian statesman, 602, 604
Celestine V (d. 1296), pope (1294), 398–9
Celtic peoples, 124–5, 173, 189
on British fringes, 237, 312
Central America:
human immigration, 21
agriculture in, 121, 384, 387–8
early cultures of, 121–2, 384–5, 388
European discovery of, 425
territorial occupation, 509–13

See also Aztecs, Mayan civilization, individual areas
Central Asia:
nomads from, 93, 255–6, 341
topography of, 255
See also Huns, Mongols
cereal crops, 26
in Mesopotamia, 38, 45
in Crete, 77
in peasant diet, 409–10
trade in, 439, 443, 493
Russian production, 493–4, 564
in New World, 563
tariffs on, 572–3
new strains of, 804
yields, 804
(i) barley, 25, 45, 56, 134
(ii) emmer, 25–6, 56
(iii) maize, 121, 351, 389
to Africa, 532
(iv) millet, 25, 45
in China, 107
in Africa, 120
(v) rice, 25
in India, 98, 101, 339
in Africa, 120
in China, 360, 363
in Italy, 439
in Caribbean, 513
in Japan, 675
(vi) sesame, 45
(vii) wheat, 45, 134
international trade in, 532
Cetewayo (d. 1884), Zulu chieftain, 647
Ceylon:
British in, 508
Dutch in, 508
independent, 791
Chadwick, Sir James (1891–1974), British physicist, 810
Chalcolithic, 24
Chaldees:
Chaldaeans, 49, 92, 93
See also Ur
Champlain, Samuel de (1567–1635), French explorer, 516
Chandra Gupta, first Gupta emperor of India (4th cent.), 342
Chandragupta, Maurya ruler of India (*c.* 321–296 BC), 337–9
Ch'ang-an, T'ang capital, 361–2
chariot:
in Egypt, 65
development of, 72
used by Achaeans, 80–1
value to Hittites, 86
used by Aryans, 99
used by Chinese, 108
Charlemagne (742–814), king of Franks and emperor (800–814), 272
against Vikings, 316, 325
Christian empire-builder, 316
cultural interests, 317–18
dominated Church, 317, 330
personality of, 317–18
Charles Martel (*c.* 689–741), Christian warrior, 315, 328
Charles the Bald (823–77), king of W. Franks (843–77), 319, 326
Charles III (832–88), king of W. Franks (885–7), 319
Charles I and V (1500–1558), king of Spain (1516–56), Holy Roman emperor 1519–56), 464, 466, 503
and Luther, 458
dynastic aims of, 478–9
Charles I (1600–1649), king of England and Scotland (1625–49), 467, 469
Charles II (1630–85), king of England and Scotland (1660–85), 467, 470, 505, 523
Charles X (1757–1836), king of France (1824–30), 597
Chartism, British radical movement, 627, 628
Chaucer, Geoffrey (*c.* 1340–1400): English poet, 404, 429
Canterbury Tales, 431
Chavin culture in Peru, 121
chemistry:
early progress in, 542
chemical industry, 690
industrial applications, 704
agriculture and, 804
Cheng Ho, (*c.*1371–1435), Chinese admiral, 365
Ch'en Tu-hsiu (1879–1942), Chinese revolutionary, 733, 735
Cheops, pyramid of, 58
Chernobyl, disaster at, 820
Chiang K'ai-shek (1887–1975), Chinese statesman:
in Moscow, 737
KMT leader, 739, 742, 789
USA and, 742
children:
allowances for, 559
labour exploited, 571
marriage of, 679
Chile, 858, 883
Europeans in, 509
natural resources of, 642
military coup in, 865
and Falklands War, 883
China:
earliest civilization, 33, 95
rice in, 98
continuity of civilization, 105
(i) ancient, 105–17
topography of, 105–7
origins of civilization, 106–7
growth under Shang and Chou, 107–14
period of the Warring States, 109–10, 113, 114, 352–3
ancestor-worship, 111
social distinctions in, 111, 113–14
writing, 112
growth of cities, 113
thought, 114–17
art, 117
Huns and, 255–6
Mongol period, 299–300, 301, 303
continuity of tradition, 352
(ii) classical, 352–68
Ch'in unification, 353
Han unity, 353, 354
state and kingship under T'ang, 354–5
centralized bureaucracy based on examinations, 356–7, 364
Confucian values dominant, 356–8, 359
pressure on peasants, 359
classical arts, 361–2, 368
fails to exploit technical ingenuity, 363–4, 368
European presence, 364, 503–4
Mongols reluctantly assimilated, 364
(iii) Manchu empire, 365–8
Japanese cultural dependence, 369, 373
cultural impact on West, 527, 548
in world trade, 635, 653
attitudes to foreigners, 662–3, 664–5, 667, 669
traditional framework threatened, 662, 731–4
early revolutionary movements, 664–70
Revolution, 670–1, 732–9, 787–9
Japan and, 676–7, 739–41, 742
(iv) Republic:
European imperialism and, 685, 731
nationalism, 732–3, 740–1
Marxist communism, 735–9, 742, 788–9
UNO and, 777
Russia and, 779, 782
KMT collapse, 788
(v) People's Republic, 789
population policy, 797, 800, 842
Korean war and, 831, 832, 840
balance of power and, 833
foreign policies, 836, 845, 847, 885
relations with India, 836, 837
domination of E. Asia, 838
economy, 840, 842, 886–7, 911
relations with US, 840, 877
relations with USSR, 840, 885
Maoism, 842
nuclear weapons, 842, 921
government after Mao, 885–7, 911–13
relations with Japan, 888
Tiananmen Square, 910
social attitudes, 913
See also Mao
Ch'in dynasty in China, 353, 355–6, 359
Ch'ing dynasty in China *see* Manchu
Chinghis Khan (1162–1227), Mongol conqueror and emperor (1206–27), 299–300, 301, 360
Chosroes I, Sassanid king of Persia (531–79), 254, 316
Chosroes II, Sassanid king of Persia (590–628), 254
Chou culture in China, 107–14, 356
Christ *see* Jesus
Christianity:
Hebrew influence on, 91, 208, 212
Zoroastrian influence on, 131, 204
in Mediterranean, 134
soul-body schism, 162
impact on Europe, 206
origin and spread, 209–15
Roman response, 214–17
maintenance of doctrine, 216–18
takes root in Europe despite persecution, 225–6
officially established by Constantine, 226–9
success against Julian, 233
shaped by Augustine, 234–6
temper, 234
doctrinal divergences, 244
assimilated pagan practices, 249
in Persia, 253
Islam and, 260
in Japan, 375, 376
in Africa, 380–1
European mainspring, 390–1, 417
Protestant Reformation, 455–60
Counter-Reformation, 462–4
and spread of European values, 633
declining political force, 700
ecumenism, 824
women and, 918
See also Jesus, missionary activity, Orthodox Church, Roman Catholic Church
Christian kingdoms, 326–7

Ch'u culture China, 110
Churchill, Winston Spencer (1874–1965), British statesman:
  leader in war, 767, 768, 772, 776
  UNO and, 777
  Greece and, 781
  'Iron Curtain' speech, 782, 831
Church and intellectual activity:
  educational dominance, 331–2, 429
  belief in Creator, 542
  Church conservatism, 543–5
  'enlightened' revolt, 546–7, 550–2
Church and state:
  distinction foreshadowed, 229, 234
  in Byzantium, 277–8
  under Charlemagne, 316–17, 330
  under Otto, 321
  medieval tensions, 391–5, 397–8
  lay authority in England, 459–60
  unity in Spain, 462–4
  in Habsburg lands, 474
  in Russia, 489–90
  state support for Church, 536
  Church conservatism, 549–50
  in France, 587, 589–90
  in S. America, 641
Chu Yüan-chang (1328–98), first Ming emperor of China (1368–98), 365
Cicero, M. Tullius (106–43 BC), Roman orator, 203, 235
Cilicia, 93, 189, 226
cinema, for propaganda, 716
Cistercian monastic order, 391
cities:
  development of, 34, 40
  foster civilization, 75
  growth in China, 112–13
  transmit Hellenistic culture, 173
  outside feudal structure, 415–16
  industrial centres, 567–9
  social effects of, 569, 690
  bombing of, 772, 773
  20th century growth, 820
  'ghettos' in, 873, 916
  See also urbanization
city-state:
  Greek *polis*, 143, 149, 170
  Aristotle on, 163, 175
  in early Africa, 381
  Italian, 407
civilization:
  origins of, 28, 32–3, 35–6
  definitions, 32, 34, 535
  early, 32–7
  variety of, 34–6
  diffusion of, 85–6, 118
  integration in Near East, 128–9, 131
  Indian world-view, 343, 344, 345–6
  medieval Church domination, 390–1, 430–1
  post-medieval world-view, 430–1, 432
  Europeanizing of world, 500, 630–3, 635
  revolutionized by industrialization, 564
  modern standardization, 915–19
civil rights campaign, US, 387
class structure and consciousness:
  in Japan, 374–6, 675
  in medieval Europe, 414–15 (*see also* feudalism)
  in early modern Europe, 446–9
  in Russia, 490, 496
  revolutionary element in France, 582–3
  socialist egalitarianism, 604–6
  Marx's view, 606–7
  British stratification, 623, 627–8
  Comintern doctrine on, 762
Claudius (10 BC–AD 54), Roman emperor (AD 41–54), 197
Clemenceau, Georges (1841–1929), French statesman, 722–3
Clement of Alexandria, St (*fl. c.* AD 96), Christian Platonist, 218, 225
Cleopatra (*c.* 68–30 BC), queen of Egypt (51–30 BC), 173, 192, 195
clergy:
  early medieval, 329–31
  medieval, 391–3
  legal immunity of, 395
  training revivified, 427, 431
  anti-clericalism, 456–7, 512, 536, 587, 641, 701
climate:
  effect on evolution, 5–6, 18, 21, 36
  Egypt, 52 (*see also* Nile)
  India, 95–6
  China, 105
  Africa, 118–19
  fears over, global warming, 820, 921
  See also Ice Ages
Clive, Robert (1725–74), British imperialist, 507
clocks, 365, 368, 541, 662, 672
Clovis (465–511), king of the Franks (481–511), 238–9, 249
Cluny, Cluniac monasteries, 331, 391
coal:
  in China, 363
  for smelting iron, 565
  Coal and Steel Community, Europe, 880
Cobden, Richard (1804–65), British statesman, 572–3
Cold War, 911
  ideological content, 783–4, 829
  Berlin blockade, 831
  Korean War, 831–2
  split in world economy, 833–4
  Asia and, 837
  See also Cuba
Colombia, 861
  US intervention, 642
  and Panama, 660
colonization:
  Phoenician, 84–5, 147
  Greek, 138, 142, 147
  Norse, 324–6
  demography and, 441, 513, 637–8
  early modern European, 500–19
  in Africa and Asia, (Portuguese), 501, 503–4; (Dutch), 504–5; (British), 505–9
  in Americas, (Spanish) 509–13, 519; (French), 513, 515, 516, 517; (British) 513–19; (Dutch), 516
  economic effects of, 519
  political effects of, 519
  early economic theory and, 521–2
  cultural effects of, 630–2, 687
  of Australasia, 643, 647
  of S. Africa, 644, 647
  See also British Empire, imperialism
colour problems *see* racial issues
Columbus, Christopher (1451–1506), Italian explorer, 423, 425, 509
Comecon, 783, 879
Cominform, 783, 867
Comintern (Third International):
  communist spearhead, 724, 726, 783
  in China, 735, 737, 739, 742
  preached class struggle, 762
commodities *see* Economic organization, trade, individual commodities
Commodus (161–92), Roman emperor (180–92), 197, 219
communications:
  telegraph, 567, 604, 636, 679
  postal services, 625
  diminish isolation, 636, 915
  population movements and, 637
  telephone, 704
  satellite, 817
  television, 906
  See also railways, shipping, transport
communism:
  fears of, 724–6
  economic policy split, 726–7
  Chinese Revolution and, 737–9, 742, 788
  in Arab states, 753
  in W. Europe, 762, 773–4
  in E. Europe, 773–4, 781, 782–3, 902–10 (*see also* Russia)
  in S.E. Asia, 789–91
  See also Bolsheviks, Comintern, Marx, socialism
Comneni dynasty in Byzantium, 287
compass, magnetic, 363, 368, 423–4
Comte, Auguste (1797–1857), French sociologist, 705
concentration camps at Belsen and Buchenwald, 772
Confucius (*c.* 550–478 BC), shaper of Chinese thought:
  Confucianism, 114–17
  as conservative force, 116, 732, 742
  values dominated Chinese state, 356–8, 360
  in Japan, 369–70, 675
  tablets attacked by Taiping, 665
Congo (Zaire), 856
  Belgians in, 656, 658
conquistadores, 509, 528
Conrad I, king of E. Franks (911–18), 320
Constantine I 'the Great' (*c.* 274–337), Roman emperor (306–37):
  established Christianity, 226–9, 279
  reunited and reformed empire, 226–7, 275
Constantine V (718–75), Byzantine emperor (740–75), 283
Constantine XI (1449–53), Byzantine emperor (1449–53), 308
Constantinople:
  Christian capital, 229, 241, 280
  fall to Turks, 275, 308
  Arab sieges, 280, 281
  fall to Crusaders, 287, 418
Constantius (*c.* 225–300), Roman emperor in W. (305–6), 226
contraception:
  in Egypt, 62
  early modern, 436
  modern, 558, 695, 799, 800, 824
Cook, James (1728–79), British explorer, 533
Copernicus, Nicolaus (1473–1543), Polish astronomer, 543
copper, 24, 27, 45, 53, 65
  in Yugoslavia, 72
  in Near East, 90–1
  in India, 97, 102
  in Europe, 124
Coptic Church, 244
Corinth, 147, 150
Corn Law repeal, 572–3, 625, 697
Cortés, Herman (1485–1547), Spanish soldier, 509
Cossacks, semi-autonomous, 487, 488
cotton:
  in ancient India, 97, 222
  trade in, 532
  slavery and, 563, 616–17

factory processing of, 567
Japanese spinning, 675
councils:
Chalcedon (451), 244
Basle (1431–49), 400
Church, 824
first ecumenical (1123), 395
Nicaea (325), 228, 279
Florence (1439), 306, 422
Constance (1414–18), 400
Pisa (1409), 400
Siena, 400
Trent (1545–63), 462, 464, 531
Counter-Reformation, 462–4
and politics, 466, 474, 486–7
and missionary enterprise, 531
creoles, 511, 638
Crete:
ancient, 75, 76–9, 86 (*see also* Minoan civilization)
taken by Turks, 308, 485
autonomous, 697
united with Greece, 747
World War II and, 768
Croatia, 910
Croesus (d. *c.* 546 BC), king of Lydia, 142, 147
Cromwell, Oliver (1599–1658), English soldier-statesman, 467, 522–3
crops *see* cereal, plantation, olive, vegetables, vine
Crusades, 273, 287, 417–19
fourth, 287, 418–19
first, 298, 395, 417, 419
response to Seljuk power, 298
third, 298, 418
second, 417–18
evoked by French imperialists, 751
Cuba:
Europeans in, 509
independent, 641, 642, 660
US interest, 659, 862–4
supports communist regimes in Africa, 858
Castro's revolution, 862–3
missile bases in, 864, 868–9
Curie, Marie (1867–1934) and Pierre (1860–1906), French scientists, 810
currency *see* money
Cyclades, 82, 287
Cynic school of philosophy, 177
Cyprus, 93, 190
raided by Arabs, 262
taken by Turks, 308, 485
Cyril, St (9th cent.), apostle of Slavs, 289–90, 294
Cyrus (d. 529 BC), king of Persia, 94, 129–31
Czechoslovakia:
Palaeolithic sites in, 21
state established, 721, 722
democracy in, 729
German seizure, 765
Russian influence and, 782–3
liberalizing movement crushed, 879
end of communist government, 909

Dada movement in art, 757
Damascus:
Amorites in, 48
Umayyad capital, 264
mosque at, 270, 304
Mongol capture (1257), 300–1
British capture (1917), 751
Dampier, William (1652–1715), English explorer, 533
Danegeld, English tribute to Danes, 326–7
Dante Alighieri (1265–1321), Italian poet, 270, 404, 430
Dardanelles, 1915
campaign, 715, 751
Darius I (*c.* 558–486 BC), king of Persia (522–486 BC), 130–1, 147
Darius III, king of Persia (335–330 BC), 171, 173
Darwin, Charles (1809–82), British scientist, 701–2, 705, 757, 813
David (*c.* 1060–970 BC), king of Israel, 90
Dead Sea Scrolls (Qumran writings), 209
Debs, Eugene (1855–1926), US socialist, 759
Decembrist movement in Russia, 608
Decius (201–51), Roman emperor (249–51), 226
Delhi, 346–7, 684
Delian League, 149, 154
Delphi, 140, 147, 166, 177
Demosthenes (*c.* 384–322 BC), Athenian orator, 169, 170
Deng Xiaoping (1904–), Chinese communist politician, 885, 911
Denmark:
raiding *see* Norsemen
Danes, amber source, 85
megaliths, in, 127
Charlemagne and, 316
settlement in England, 326
World War II and, 767
Descartes, René (1596–1650), French philosopher, 472, 544–5, 546
Devi, Hindu goddess, 344
Dewey, John (1859–1952), US philosopher, 733
Dhamma, Asoka's social philosophy, 339–40
diadochi, Alexander's successors, 173
diamonds in S. Africa, 644
diaspora, dispersion of Jews, 216, 218
Diaz, Porfirio (1830–1915), Mexican dictator, 860
Diderot, Denis (1713–84), French encyclopaedist, 548
Diocletian (245–313), Roman emperor (284–305), 224, 226
diplomacy:
professionalized, 477, 711
Cold War, 829
term détente used, 867
'hot line', 869
disease:
Black Death (1348–50), 306, 413
epidemics, 438
spread by imperialism, 509, 512, 528, 530
reduction in incidence, 557–8
influence, 787
See also medicine, mortality rates
DNA, 813–14
Domesday Book, Norman economic survey, 404, 416, 433
Dominican order of friars, 396–7
Dominican Republic, 641, 642, 864
'Donation of Constantine', 330
Donatists:
Christian sect, 228, 234
attacked by Augustine, 235, 236, 381
Dorians, 82
drama:
Greek innovations, 164–6
Indian, 343
Japanese, 374, 376, 378
Dravidian rule in Deccan, 341
drugs:
import to Europe, 527
Chinese opium trade, 665
aspirin, 703
chemical research and, 704
See also plantation crops (tobacco)
Dupleix, Joseph (1697–1763), French governor in India, 507
Dutch *see* Netherlands

East Africa:
Swahili-speakers settle, 381–2
illiterate artistry, 382
technology, 382
European trade, 503
See also individual areas
Eastern Question:
Europe against Islam, 485–6
See also Ottoman Empire East India Company (British):
in Persia, 311
founded, 349, 505
political interventions in India, 507
growth, 527
to China, 533
'dual control', 650
East India Company (Dutch), 504
East India Company (French), 351, 508
Economic organization:
(i) in ancient world:
obscurity, 76
in India, 97–8
in China, 112–13
in Mediterranean lands, 134
in Greece, 152
in Hellenistic areas, 171, 173–5
in Roman republic, 183–4
in Roman empire, 204–5, 221–2, 224, 226–7, 229, 230–1, 232
in Byzantine empire, 280
(ii) medieval:
early medieval regression, 331–2
money economy in Japan, 377–8
European, 409–12
(iii) modern:
increased emphasis on commerce, 441–3
early industrialization, 444–5
early modern Russia, 494
plantation economies, 512–13, 517
effects of imperialism, 526, 528–9, 532
effects of population growth, 556
in industrial societies, 566–7, 570–1
state intervention debated, 571–3, 606, 694, 729
French Revolution and, 581–2
US expansion, 616, 622, 708–9
UK 19th-cent. prosperity, 628
trans-oceanic, 633–7
S. American resources, 641–2
Japanese expansion, 671–2, 675, 731–2
20th-cent. Russian, 706, 724, 726–7
World War I and, 715, 720, 723
slump, 729–30, 739–40, 756, 759
state intervention accepted in capitalist areas, 756, 760–1
US aid to Europe, 768, 775, 782
US postwar strength, 779–80, 804
European postwar weakness, 780–1, 803–4
in modern India, 787
international disparities, 801–3
two world models, 833–4
in modern S. America, 865–6
European organizations, 879–80
in modern China, 883, 885–6
South-East Asian, 887

See also financial services, GATT, IMF, inflation, money, trade, World Bank
Economic theory:
trade limited, 510–11
on colonies, 521–2
Adam Smith, 546, 549, 635
laissez-faire, 571–2, 756
state intervention debated, 571–3, 606, 694, 729, 730
socialism, 604–6, 726
See also Free Trade
Edo (historic name of Tokyo), 376, 377, 378, 671, 672, 673
Education:
organized in Sumer, 43–4
of Egyptian scribes, 55–6
clerical domination, 332, 427
early modern Russia, 492
increase in literacy, 537–9
Rousseau's influence, 552
in British India, 651–2, 679–80, 682
in Japan, 674–5
scientific, 703
Turkish reforms, 754
See also literacy, universities
Edward III (1312–77), king of England (1327–77), 405
Egypt:
(i) ancient civilization, 32, 52–70
use of chariot, 65
Hebrews in, 68, 87, 89
foreign relations, 78, 82, 119, 120
decline of, 168
Ptolemaic state, 173, 175, 177, 192, 195
Fatimid caliphate, 272, 295, 298
(ii) modern:
Napoleon in, 653
British-French intervention, 654–5, 656, 712, 746, 752–3
modernization and nationalism, 745–6, 752–3
economy, 746
World War II and, 770, 791
in Arab League, 792
Israel and, 792–3, 891
Suez crisis, 849–50
and Syria, 891
and Gulf War, 900
Einhard (*c.* 770–840), Carolingian intellectual, 317
Einstein, Albert (1879–1955), German-Swiss physicist, 810–11
Eisenhower, Dwight David (1890–1969), US president, 872
Russia and, 779, 832
election, 832
S.E. Asia and, 832
and Jews, 851
economy and, 871
Elam, Elamites, 40, 47, 48, 51, 93, 94, 129
electricity:
use developed, 690, 704
household use, 695–6
Faraday and, 703
hydro-electric schemes, 761
production and consumption, 805
Elijah (*c.* mid-9th cent. BC), Hebrew prophet, 91
Elizabeth I (1533–1603), queen of England (1558–1603):
India and, 348, 349–50
Church and, 460
national idol, 466–7
against Spain, 480–1, 522
Russia and, 488, 492
navy and, 524
consults astrologer, 823
Elizabeth (1709–62), tsarina of Russia (1741–62), 494
Elizabeth II (1926–), queen of UK, 911
enclosure, agricultural, 439, 445, 561
energy:
human, 74, 412
steam, used by Hero, 175
animal, 412
water, 412, 690
wind, 412, 690
in agriculture, 562
for railways and shipping, 564, 565, 655
hydro-electric schemes, 761
ecological problem, 804
wasteful use, 820
acceleration of consumption, 852
See also electricity, oil
Engels, Frederick (1820–95), German socialist, 569
England, English:
(i) Anglo-Saxon:
regeneration under Alfred, 326–7
Norman administration, 404
(iii) early modern:
trade with Persia, 311
first trade posts in India, 349, 350–1, 505
Papacy and, 395, 398–9, 400
wars with France, 405
royal power, 408
peasant rebellions, 413
trade growth, 441–2, 443, 445, 492
class structure, 446–8
social changes, 446–7
Tudor administrative techniques, 454, 460, 477–8
Church in, 460
Crown and parliament, 467–9, 470–2, 490
wars with Spain, 480–1
Dutch and, 482, 484, 522–3, 525
colonization in N. America, 601–7
See also Britain, United Kingdom
Enki, Sumerian god, 42
enlightened despotism, 473–5, 552
Enlightenment:
definition, 545
ideas and attitudes, 545–53, 701
Enlil, Sumerian god, 42, 51
Environmental concern and change, 818–20, 921
Epicurus (341–270 BC), Epicurean school of philosophy, 177, 199
Erasmus, Desiderius (*c.* 1466–1536), Christian humanist scholar, 430, 433, 456, 457
Eratosthenes (276–194 BC), Greek scientist, 175, 178
Eridu, Sumerian settlement, 40
Erigena, Johannes Scotus (*c.* 815–77), Irish philosopher, 428
Eskimoes, 28, 324, 384, 646
Estonia, 907
Ethiopia:
hominid sites in, 7
Moslems in, 260
early kingdom, 379, 382
Portuguese and, 503
retained independence, 633, 655, 656, 855
Italian invasion, 761
World War II and, 770
revolution, 856
Etruscans:
Villanovan culture, 127, 147, 180
culture, 180–1
origins, 180–1
Euclid (323–283 BC), Greek geometrician, 175, 268, 331, 428
Euripides, (*c.* 484–407 BC), Greek tragedian, 165–6
Europa, legendary Greek heroine, 79
European Economic Community (Common Market), 881, 882, 884
European Economic Co-operation, Organization for, 783
Europe, Eastern:
German expansion, 421
socio-economic contrasts with W, 423, 447, 449–50
Turkish confrontation, 485–7
nationalism in, 697
revolutionary movements in, 717–19, 725–6
Versailles settlement and, 722
authoritarian regimes, 729
polycentrism, 879
post-1945 economy, 904–5
See also Byzantium, Slavs, individual countries Europe, Western:
(i) prehistory:
Homo sapiens prototypes, 15–16
Palaeolithic cultures, 17
Palaeolithic art, 18–21
first agriculture, 26
prehistoric cultures, 122–3
racial groups, 122–7
prehistoric insignificance, 124–6, 127
(ii) *See* Rome
(iii) early medieval:
Arab incursions, 262
post-Roman Germanic cultures, 237–40, 247–9
medieval civilization shaped, 312–36
(iv) the Middle Ages:
interpretations, 390–1
role of Church, 391–402
political pattern, 402–8
social and economic pattern, 408–16
aggressive Christianity, 417–21
navigational enterprise, 423–7
mental enlargement, 427–33
(v) early modern:
demography, 436–9, 637–8, 690–3
economy, 438–46
class structure, 446–9
women, 450–1
growth of state authority, 452–75
dynastic political geography, 477–84
Ottoman threats, 485–7
imperial expansion, 500–19
(vi) modern:
world-wide spread of values, 500–3, 630–3, 635
19th-cent. revolutionary politics, 549–607, 693–4
demography, 556–9, 798–9
agricultural revolution, 559–63
industrialization, 564–7, 690–2
urbanization and labour, 567–71, 692–3
economic ideologies, 572–3, 693–4
political geography after Napoleon, 592
centre of world trade, 633–6
female emancipation, 695–6
mental attitudes, 700–5
(vii) 20th cent.:
tensions, 706–7

build-up to World War I, 713–17, 719
peace settlement, 721–3
political regimes, 725–6
economic slump, 729–30
changing patterns of thought, 757
World War II, 765–76
(vii) supranational organizations, 879–80
Ezekiel (7th cent. BC), Hebrew prophet, 206
Ezra (mid-5th cent. BC), Hebrew prophet, 207

factories:
19th-cent. growth, 566–7
new social patterns, 571
legislation on, 625, 627
assembly-line technique, 807
Falkland Islands, 883
famine:
in Roman empire, 243
medieval, 413
early modern outbreaks, 438, 439
in Ireland, 559, 599, 629
in Russia, 559, 727
in India, 679, 787
World War I and, 717
World War II and, 773
in Africa, 856
Faraday, Michael (1791–1867), British scientist, 703
fascism:
authoritarianism, 724, 728
in Italy, 727–8
intellectual impetus, 757
Fatimid caliphate in Egypt, 272, 295, 298
Ferdinand V (1452–1516), king of Castile (1474–1516) and Aragon (1479–1516), 405
Fertile Crescent, 25–6
cultural melting-pot, 36, 37, 38, 47–8, 71
Hittite hegemony, 86
nomad intrusions, 87, 93–4
See also individual areas
feudalism:
chain of obligation, 334–6
in Japan, 374, 671, 673
increasingly unreal structure, 413–16, 446
Tudors and, 454
abolished in France, 562, 584, 587
survival of values, 562, 587
ended in Central Europe, 600
financial services:
banking, 411, 412, 441
joint-stock companies, 441–2, 635
Stock Exchanges, 441–2, 635
insurance schemes, 442–3
expansion of, 527, 635
facilitate industrial growth, 567
focused on London, 634–5
damaged by World War I, 720, 729
Finland, 695, 768
fire, earliest use of, 11–12, 107
fish, fishing, trade in, 443, 887–8
Flanders:
textile centres, 411
metallurgy, 412
drainage in, 439
independent tendencies, 464
Florence, 411, 412, 479
Council of, 306, 422
food supply and diet:
earliest storing, 8–9
meat-eating, 12
extension of sources, 18
Neolithic revolution, 24
in Mesopotamia, 38
Egyptian diet, 56
in Africa, 119–20
limited Dark Age diet, 333–4
medieval European, 409–11
demographic dependence on, 441, 556
increased variety of, 527
transplantation of commodities, 532–3
revolutionary increases in supply, 560–1
processing and preserving, 696
India, 801
world, 801, 820, 888
S.E. Asia, 838, 840
Ford, Gerald (1913–), US president, 878, 885
Ford, Henry (1863–1947), US automobile manufacturer, 805–7
France:
Neanderthal population, 21
Roman province (Gaul), 187, 189, 191, 192, 197
Germanic invasions, 231–2, 237–40
(i) Frankish settlement, 239
repel Arabs, 262, 315, 328
support Papacy, 313
Charlemagne's empire, 316–19
divided, 319
territorial units, 320, 322
(ii) medieval political pattern, 404–5
peasant rebellion, 413
(iii) early modern:
class structure, 446
wars of religion, 460–1
Frondes, 469
Louis XIV's absolute monarchy, 472, 474–5, 490
against Habsburgs, 477, 479–80, 481
Louis XIV's foreign policy, 482
in S.E. Asia, 505
in India, 507, 508
in Caribbean, 513
in Canada, 515, 516–17
in black Africa, 530
philosophes, 548, 549–53
'feudal system' abolished, 562, 583, 587
industrialization, 567
American Revolution and, 576, 581–2
(iv) Revolution (1789) *see* French Revolution
constitutional monarchy, 597–8
Second Republic, 599, 602
Parisian revolutions, 600, 604, 607
(v) Third Republic:
in N. Africa, 653, 656, 657
in Indo-China, 658, 684–5, 687, 740, 742, 745
income tax, 694
nationalism, 698
foreign policies, 708, 712, 759, 761, 765
World War I and, 714, 717
Versailles settlement, 721, 722–3
in Syria, 751, 753
World War II and, 766–7, 772, 781
(vi) Fourth Republic, 774, 779
UNO and, 777
in Indo-China, 789, 845
in Near and Middle East, 791
in N. Africa, 851
Gaullist policies, 881
nuclear weapons, 921
Francis of Assisi, St (1182–1226), founder of Franciscan order, 396–7, 399
Francis Xavier, St (1506–52), Jesuit missionary, 375
Franciscan order of friars, 396–7, 399
Franco, General Francisco (1892–1975), Spanish dictator, 765
Franks *see* Germanic peoples
Frederick II (1194–1250), Holy Roman emperor (1212–50), 427, 428
Frederick II 'the Great' (1712–86), king of Prussia (1740–86), 474, 497, 549, 602–3
Frederick William I (1688–1740), king of Prussia (1713–40), 496–7
freemasons, 537
Free Trade:
19th-cent. defences, 573, 625, 635
questioned, 708
abandoned, 756
French Revolution:
stimuli, 581–2, 583–4
National Assembly formed, 584
political divisions, 585
Republican Convention, 585–6
Terror, 585
ideological debates, 587–8
institutions rebuilt, 587
consolidated by Napoleon, 588–90, 592–3
influence, 595, 640
Freud, Sigmund, originator of psychoanalysis, 757, 822–3
friars *see* Dominicans, Franciscans
Fujiwara clan in Japan, 369, 371
arts under, 374

Gagarin, Yuri (1934–68), Russian cosmonaut, 815–16
Galen (*c.* 130–*c.* 200), Greek physician, 268, 416, 429
Galerius, Roman emperor (305–11), 226
Galilei, Galileo (1564–1642), Italian astronomer, 541, 543
Gama, Vasco da (*c.* 1460–1524), Portuguese navigator, 425, 500
gambling, 75
Gandhi, Indira (1917–84), Indian stateswoman, 889, 890
Gandhi, Mohandas (Mahatma) (1869–1948), Indian social and political leader:
in S. Africa, 645
nationalism, 784–5
death, 787
Gandhi, Rajiv (1944–1991), Indian statesman, 888, 890
Ganges, 95, 99, 101, 346
Garibaldi, Giuseppe (1807–84), Italian nationalist, 681–2
GATT, 803
Gaul *see* France
Gaulle, Charles de (1890–1970), French statesman, 767, 851, 881
Gauls, sacked Rome, 184
General Agreement on Tariff and Trade, 803
Geneva, theocratic state under Calvin, 458
Genghis Khan *see* Chinghis Khan
Genoa:
commercial city-state, 306, 408
end of republic, 595
geology, study of, 701
George, St, patron of England, 403
George III (1738–1820), king of UK (1760–1820), 404, 576, 662

George IV (1762–1830), king of UK (1820–30), 452
George V (1865–1936), king of UK and emperor of India (1910–36), 684
George, Henry (1839–97), American economist, 669
German Democratic Republic, 830, 833, 867, 868
collapse, 908
Germanic peoples:
pressure on Rome, 223–4
folk-movements, 231–2
Dark Age culture, 249
(i) in East:
Goths, 223, 231, 232, 242, 249
Alamanni, 223, 231, 238
Franks, 223, 231, 237–8, 249, 313
Ostrogoths, 231, 239
Visigoths, 231–2, 237, 239
Vandals, 232
(ii) in North:
Angles and Jutes, 237
Saxons, 237, 315, 320–1
See also Carolingians, Charlemagne, Pepin
Germany:
resistance to Rome, 197
Frankish settlement, 237–8, 311, 319–20
(i) medieval:
political fragmentation, 320
Ottonian empire, 321–2
individual states, 406–7
expansion E, 421
(ii) modern:
Protestant Reformation in, 456–8
Habsburg interest in, 478–9, 481, 497
rise of Prussia, 497, 499
Enlightenment in, 549
industrialization, 567
Napoleonic reorganization, 591–2, 593
1848
revolts, 599, 600
(iii) Second Reich:
unification under Prussia, 602–3
African imperialism, 656
rise of SDP, 693, 762
nationalism, 698–9
foreign policies, 706, 709–14, 716–17, 719
economy, 708, 729
World War I, 713–17, 719–21
Versailles treaty, 721, 726, 759
(iv) Weimar Republic:
liberal constitution, 723, 759
fascism and Nazi origins, 728, 761–3
(v) Nazi rule:
autocratic terrorism, 763, 772, 791–2
expansion E, 765
World War II strategies, 765–72
shattered, 774
(vi) post-war:
division, 781–2, 830, 866
Federal Republic of, 830, 833, 881
(vi) post-war, 881
re-unification, 909
See also Berlin, German Democratic Republic (GDR)
Ghana, 381, 853, 856
Ghazan (1271–1304), Mongol Il-khan, (1295–1304), 303
Ghaznavid dynasty in Turkey, 297, 346
Gibbon, Edward (1737–94), English historian, 262, 400, 549
Gibraltar, 262
Gilgamesh, Epic of, 41–2, 49, 93
Giza, pyramids at, 58
glass, 45
Gnostic Christian heresy, 218, 234
gods *see* religion, individual deities
Gogol, Nicolai Vasilyevich (1809–52), Russian writer, 165
gold:
from Africa, 381, 382, 424
financial effects of, 446, 604
as imperialist motive, 509, 644, 646
as exchange standard, 635, 729, 730
Golden Bull on jurisdiction of German princes, 406
Gorbachev, Mikhail Sergevich (1931–), Soviet statesman, 902, 906–8, 911
Goths *see* Germanic peoples
government and administration:
in Minoan Crete, 79–80
in Mycenae, 80–1
in Near East, 92–3
in archaic Greece, 143–7 (*see also* city–state)
in Hellenistic world, 173–4
of Roman republic, 181–4, 187, 195
Roman imperial bureaucracy, 198, 219–20, 224
of Arab empire, 265
of Seljuk Turks, 297
Mongol, 301–2
'feudalism', 334–6
in Moghul empire, 348–9
Chinese bureaucracy, 356–7, 364
Japanese shogunate, 370–3, 376
early African kingdoms, 382
early modern Europe, 454–5
strengthening of state power, 454–5, 460–7, 595, 625–6
in N. American colonies, 517
colonial governments, 533–4
US constitutionalism, 577–80
French revolutionary, 584, 587
under Napoleon, 589–90, 592
19th-cent. French, 597, 612
Tsarist Russia, 608–10, 612
19th-cent. UK, 625
S. Africa, 645
British India, 651, 678–9, 680–1, 683, 785–7
French Indo-China, 685
modern state authority, 825–8
See also feudalism, political institutions, individual countries
Gracchus, Tiberius, (163–133 BC) and Gaius (153–121 BC), Roman agrarian reformers, 190
Great Britain *see* England, United Kingdom
Great Wall of China, 105, 256, 359, 741
Greece, 80
Mycenaean culture, 80–2
Greeks in Egypt, 129, 138
contribution to civilization, 136
(i) civilization, 137–48
conscious Hellenic identity, 137–9
topography, 137–8
Greeks overseas, 138, 142, 147, 150–1, 255
aristocratic society, 140
economic life, 140–2
political development, 143–7
struggle against Persia, 147–8
expansion, 148
(ii) classical civilization:
political history, 149–52
nature of culture, 151–2, 157, 166–7
limited economic patterns, 152
social structure, 152–4
dominance of Athens, 154–7
growth of rational thought, 158–63
literary achievements, 163–6
supreme artistic quality, 166–7
(iii) Hellenistic age, 168–78
influence on Macedon, 168
Philip II's expansion, 168–9
Alexander spreads culture, 170–3
changed spirit, 173–4
(iv) and Rome, 186, 187, 189
(v) in Ottoman empire, 309
revolt, 598–9
independence, 697
World War I and, 714
Turkish concessions, 753
World War II and, 768
Russians on borders, 779
British intervention, 781, 782
US aid, 782
joins EEC, 884
Greenland, 324
Gregory, St (538–94), bishop of Tours, 239
Gregory I (*c.* 540–604), pope (590–604), 247, 249
Gregory VII (Hildebrand) (*c.* 1023–85), pope (1073–85), 393, 399
Guatemala, 726
Gupta empire in India, 342–4
Gutenberg, Johannes (*c.* 1400–1468), German printer, 431
Gutian peoples, 47, 48

Habsburg dynasty in Europe:
Holy Roman emperors, 407
Church and state, 474
wars against France, 477, 479–80, 481
territories, 478
against Prussia, 497, 602
and Napoleon, 591–2
security and danger, 595, 600
Dual Monarchy, 602, 697, 706, 719, 722
foreign policies, 709, 710, 711, 712, 713
Ottoman parallels, 749–51
Hadrian (76–138), Roman emperor (117–38), 197, 204, 216
Hadrian's Wall, 197
Haile Selassie (1891–1976), Ethiopian emperor, 856
Haiti, 641
Hamdanid dynasty, 295
Hammurabi, king of Babylon (*c.* 1792–1750 BC), 48–9, 50
code of, 48–9
Han dynasty in China, 353–6, 359
Later Han, 353–6
Hannibal (247–182 BC), Carthaginian general, 186
Hanoi, 789
Hanse, German trade league, 292, 411, 415, 422
Harappan culture in India, 97–9, 102–3
Haroun-al-Raschid (*c.* 764–809), caliph of Baghdad, 268, 272, 316, 360
Harvey, William (1578–1657), English physician, 541
Hatshepsut (*c.* 1490–1470 BC), queen of Egypt, 65
Hawaii, 659
See also Pearl Harbor
Hebrews:
Egypt and, 68, 87, 89
origins, 87
monotheism, 88–9, 91
settle in Palestine, 90

prophets, 91–2
See also Jews, Judaism
Heisenberg, Werner (1901–76), German physicist, 811
Hellenistic civilization:
Alexander's empire, 171–3
government by cities, 173–5
intellectual activity, 174–8
influence on Byzantium, 279
Persian damage, 281
Turkish annihilation, 308
Helsinki agreement (1975), 897, 904
Henry 'the Fowler' (c. 876–936), king of E. Franks (912–36), 321
Henry II (973–1024), Ottonian emperor (1002–24), 321–2
Henry IV (1050–1106), Holy Roman emperor (1056–1106), 393–4
Henry VII (1457–1509), king of England (1485–1509), 408, 477
Henry VIII (1491–1547), king of England (1509–47), 402, 460, 524, 561
Henry IV (1553–1610), king of France (1589–1610), 462, 477, 535
Henry 'the Navigator' (1394–1460), prince of Portugal, 424–5
Heracles, Greek hero, 82, 168
Heraclius (c. 575–641), Roman emperor in E. (610–41), 254, 257, 279, 281, 288
Hero of Alexandria (2nd cent. AD), Greek engineer, 175
Herod Agrippa I (d. AD 44), king of Palestine, 216
Herod the Great (74–4 BC), king of Judaea (40–4 BC), 208–9
Herodotus (c. 484–c. 424 BC):
Greek historian, 58–9, 94, 130
on Africa, 119
'father of history', 163
on Scythians, 255
Herrero tribe in S.W. Africa, 656
Hesiod (c. 700 BC), Greek poet, 164
Hideyoshi, (1536–98), Japanese general and dictator, 375, 376
Hildebrand see Gregory VII
Hinduism:
in ancient India, 102–3, 104
development of, 340, 344–5, 349
social practices, 343–4, 679, 682, 787
Hindu-Moslem rivalry in India, 507, 680–3, 785–7, 824, 834, 889
essential forms revived, 680
Hippocrates of Cos (c. 460–c. 357 BC), Greek physician, 162, 428, 429
Hirohito, emperor of Japan (1901–1989), 846
historiography:
evidence, methods and problems, 436–7
Greek, 163–4
Gibbon, 549
Hitler, Adolf (1889–1945), German dictator:
racial views, 762–3, 768, 791
rise to power, 762–3
allied with Mussolini, 765
foreign policies, 765–6
war strategies, 766–7, 768
death, 772
use of radio, 827
Hittites, 48, 51
influence on Egypt, 67, 68
and Mycenae, 82
in Asia Minor, 86
collapse, 86, 92
Hobbes, Thomas (1588–1679), English political philosopher, 470
Ho Chi Minh (1892–1969), N. Vietnamese statesman, 789–90
Hohenzollern dynasty in Germany, 474, 496–7, 549, 602, 719
Holy Roman Empire:
foundations in Ottonian empire, 322
Habsburg domination, 407, 474, 478–9
power limited, 452, 478
Napoleonic revision, 591–2
Homer, author of *Iliad* and *Odyssey*, 75, 83, 137, 164
evidence of, 139–40, 144, 157
importance of, 139
hominids, 7–9
fossil sites, 16
Homo erectus, 9–14
Homo sapiens:
prototypes, 9–10, 13, 15–17
racial groups, 21, 36–7
physical development, 25
adaptable, 27, 29
Homo sapiens sapiens, 9, 16–17, 25
Honduras, 862
Hong Kong, British base, 635, 737, 886
Honorius I, pope (625–38), 279
horses:
exploited for war, 65, 72, 80, 86, 99, 129, 258
in Africa, 379
in Americas, 384, 533
used for agriculture, 412, 804
declining use, 704
Horus, Egyptian god, 55, 60
Hottentots, 528
Hsiung-Nu, 255–7
expelled from China, 354
See also Huns
Hudson, Henry (d. 1611), English explorer, 516
Hudson's Bay Company, 523
Hugh Capet (c. 938–96), king of France (987–96), 320
Hugo, Victor (1802–85), French writer, 571
Huguenots, Protestants in France, 469, 470, 472, 521
Hulugu, Mongol khan, 303
humanism, embodied by Erasmus, 456
Hume, David (1711–76), Scottish philosopher, 549
Hunas in India, 346
Hungary:
Dual Monarchy see Habsburgs
defeat by Mongols, 300
defeat by Turks, 308, 485
acquired by Habsburgs, 407
freed from Turks, 486
revolt, 599, 601
Slav nationalism in, 697, 719
Magyar nationalism in, 706
anti-Soviet revolution, 867–8
end of communist regime, 908
Hung Hsiu-ch'uan (1812–64), Taiping leader, 666, 667
Huns:
Asian nomads, assault Roman empire, 231, 232, 247
Mongolian power, 255–6
as types of brutality, 699
hunting, 12, 18, 21
origin of aristocracy, 27
Hus, John (c. 1373–1415), Church reformer, 401
Hussein (c. 1854–1931), sherif of Mecca (1908–24), 749, 751, 752
Huxley, Thomas Henry (1825–95), British thinker, 733
Hyksos, Semitic peoples, 65, 72, 88

Ibn-Sina (979–1037), Arab philosopher, 270
Ibsen, Henrik Johan (1828–1906), Norwegian dramatist, 165, 695, 733
Ice Ages, 5, 15, 17, 18, 20, 24
Iceland, 324
Iconoclasm movement in Orthodox Church, 279, 283–5, 315, 329
iconography:
Orthodox Christianity and, 283–5
of the Buddha, 374
IMF, see International Monetary Fund
Imhotep (*fl. c.* 2800 BC), Egyptian architect, 58
Imperialism:
popular in Athens, 156
by suction, 256
crusades as examples of, 419
impetus for European, 500–1, 521–2, 526–7, 647, 655
commercial empires of Portuguese and Dutch, 503–5, 640
British in India, 505–9, 650–3, 678–84
Spanish in America, 509–13, 638–40
N. American settlement, 513–20
economic effects of, 526–7, 529–31, 633–6
intellectual impact of, 528
social effects of, 528–31, 533, 630–2, 651–2
Russian expansion, 648
British naval bases, 649
Africa, 654–8
S.E. Asia, 658, 685–7, 740
US expansion, 658–60
Mandates developed, 722, 752–3, 791
in Ottoman territories, 745–54
Japanese, 846
British and Suez, 850
British, Arab resentment, 852
legacy for Africa, 855
See also colonization
Inca society, 388–9
Pizarro and, 509
India:
earliest civilization, 33, 71
regions, 95
(i) ancient, 95–105
origins of civilization, 95–7
Harappan culture, 97–9
Aryan culture, 99–103
leadership of north, 101
religions, 102–4
Alexander halted, 171, 173, 337
(ii) medieval:
attacked by Timur Lang, 304
Mauryan empire, 337–41
western accounts, 337
political disunity, 340–1
Gupta civilization, 342–4
religious cross-currents, 344, 346
Islamic settlement, 346–7
Moghul empire, 346–51, 506–7
European commercial presence, 349, 350–1, 505
(iii) British rule:
British political intervention, 507–9
Hindu-Moslem clash, 507, 681–4, 785–7
British control extended, 650–2, 678–9
social attitudes, 651–2, 679–80, 787
Mutiny, 652–3, 678, 679
economy, 653, 787
nationalism, 680–1, 785–7

religion, 687
civil disobedience, 785
(iii) self-government:
British withdrawal, 782, 784–7
constitution, 785–6, 835, 890
partition, 787, 890
food supplies, 805
religion in, 823
foreign policy, 835, 889
politics, 888–9
Sikh terrorism, 889
nuclear weapons, 921
Indians, N. American:
early colonial relations, 517–18, 532–3, 574
extermination and eviction, 529, 533, 658
revolts, 574
depressed status, 623
in Canada, 646
Indo-China:
China and, 354, 845
French in, 658, 667, 685, 687, 845
World War II and, 770, 789
communists in, 789–91, 845, 877–8
See also individual areas
Indo-European peoples, 37, 48, 80, 82, 86, 99, 128, 178–9, 180
Indonesia:
Dutch imperialism, 504–5
nationalism in, 686–7, 789
World War II and, 770, 784
internal tensions, 837–8, 890
economy, 887
Indra, Aryan god, 99
industrialization, 598, 606, 805
medieval power sources, 412
slow growth, 445
early modern Russian, 493–4, 611, 634
'Industrial Revolution', use of mineral power, 564–6
technology, 564–6
factories, 566–7
transport, 566
topography, 567
social effects, 568–71
socialism and, 604–5
in India, 679, 787
social impact of widespread ind., 690–2
political tensions and, 706
stimulated by war, 715
Stalin and, 726–7
China, 840
See also technology
inflation:
in Roman empire, 221
early modern European, 445–6
19th-cent. Chinese, 664
post-World War I, 729
International Monetary Fund (IMF), 803
in UK, 882
information technology, 808
Innocent III (1161–1216), pope (1198–1216), 395, 397, 418
Inquisition:
Spanish, 462–4, 512, 536
intellectual censorship, 538, 544
intellectual activity:
growth of rational thought in Greece, 157–66
Hellenistic culture, 174–8
Roman, 199
Islamic, 268–70
Carolingian, 317
English under Alfred, 326
Gupta advances in India, 343
Islamic influence on Europe, 428–30
medieval scholasticism, 428
Renaissance, 430–1, 546
broadens via imperial growth, 528
promoted by literacy, 537–9
attitudes to science, 538–45
Enlightenment ideas, 545–53
19th-cent. movements, 701–5
changing 20th-cent. patterns, 757
acceleration, 808
élites, 822
China, 842–3
Soviet Union, 871, 904
See also literature, philosophy, science
International, Second Socialist, 693
International Monetary Fund (IMF), 803
Investiture contest, 392–3
Ionia:
Ionians, 82, 131, 147–8
philosophers, 159–60
islands, Ottoman conquest, 308, 485
British acquisition, 648
Iran:
ancient *see* Persia
modern state:
oil and, 852
and Russia, 648, 755, 778, 782
Reza Khan's coup, 755
World War II and, 770
revolution, and relations with USA, 891, 893, 898
war with Iraq, 895, 900
Iraq:
in Arab Empire, 261, 262, 265, 267
nationalism in, 749
British mandate, 752
oil source, 752–3, 791, 852
World War II and, 770, 791
in Arab League, 792
revolution, 850–1
war with Iran, 895, 900
See also Baghdad, Mesopotamia
Ireland:
Norse trade with, 324
raid, 325
famines in, 559, 599, 629
Act of Union with UK, 629
demand for Home Rule, 629, 697–8, 722
Irish Free State, 723
civil war in Ulster, 824, 882
religion in politics, 824
Irenaeus (*c.* 130–202), bishop of Lyons, 218
Irene (*c.* 752–803), Byzantine empress (769–802), 283, 285
iron, 28, 72, 73
in China, 34, 112–13, 363
military advantages, 72
spread of, 86
value to Hittites, 86
in India, 99, 100
in Africa, 120, 379
in Persia, 129
in Italy, 180
in Japan, 369
Russian exports, 494
smelted, 565
in Germany, 690
'Iron Curtain' division in Europe, 782, 831
Isaac, Hebrew patriarch, 87, 88
Isabella (1451–1504), queen of Castile (1474–1504), 405
Isaiah (8th cent. BC), Hebrew prophet, 91–2, 94
Isaurian dynasty in Byzantine empire, 281
Isfahan, 310, 311
Isis, Egyptian goddess, 60, 217
Islam:
Hebrew influence on, 91
origins, 257–60
conquests, 261–2
spread of, 261–2, 273–4
lack of records, 264
schism, 264, 267
non-Arab converts, 265
nature of civilization, 268–70, 421
Orthodox view of, 277
cultural stability of, 295
political role in Middle East, 297–8
Mongol attacks, 300–1
in India, 345–7, 682–3, 785–7
in Africa, 379, 381, 382
intellectual contribution to Europe, 428–30
pan-Islamic movement, 747, 749
modern force of, 824, 849, 890
fundamentalism in, 891–2, 895
See also Crusades, Shi'ites, Suni
Ismail (1486–1524), Safavid ruler of Persia (1500–1524), 310, 311
Ismail (1830–95), first Egyptian khedive (1866–79), 746
Israel:
(i) religious vision of, 88
tribes in, 89–90, 91, 92
defeated by Assyrians, 92
(ii) Jewish national state, 789–93
Arab hostility to, 791–3, 847, 852, 898
Six Day War (1967), 852
Yom Kippur War (1973), 852
and Egypt, 891
and Gulf War, 900
nuclear weapons, 921
See also Jews, Palestine, Zionism
Italy:
Ostrogothic kingdom, 239, 242
linked with Germany under Ottonians, 321
disunity, 329–30, 479–80, 484
city-states emerge, 407
revival of trade, 411, 412
vanguard of literacy, 537
Enlightenment in, 549
Napoleonic division, 592, 593
1848 revolts, 599, 600
unification under Sardinia, 602
(i) 1900–1945:
civil unrest, 693
foreign policy, 706, 711, 714, 747, 753, 761, 765
fascism in, 728
World War II and, 767, 771, 791
(ii) post-1945:
internal divisions, 779
African territories, 853
communism in, 880
See also Etruscans, Rome
Ivan III 'the Great' (1440–1505), tsar of Russia (1462–1505), 422–3, 487, 492
Ivan IV 'the Terrible' (1530–84), tsar of Russia (1544–84), 487–8, 727
Ivan VI, Antonovich (1740–64), tsar of Russia (1740–41), 494

Jackson, Andrew (1767–1845), US president, 618
Jacob, Hebrew patriarch, 87, 88
Jacobin club, French revolutionary association, 597
jade, 107, 117

Jagiellon dynasty in E. Europe, 485, 486
Jahangir (1569–1627), Moghul emperor of India (1605–27), 350
Jainism, 104
James, St (d. AD 44), apostle of Jesus, 214, 328
James I and VI (1566–1625), king of England (1603–25) and Scotland (1567–1625), 350, 467, 481
James II and VII (1633–1701), king of England and Scotland (1685–8), 467, 470, 523
James Stuart 'the Pretender' (1688–1766), 484
Januarius, St (d. AD 305), 535
Japan:
  (i) earliest pottery, 28
  (ii) medieval, 369–78
    resists Kubilai, 364
    clan power and shoguns, 369–78
    cultural dependence on China, 369, 373
    topography, 369
    court arts, 373–4
    European presence in, 374–5, 377
    military ideal, 374, 376
    economy, 377–8
  (iii) modern to 1945:
    western contacts, 504, 632, 672–5
    industrialization, 634
    westernization, 637, 674
    retained independence, 655
    China and, 667, 669, 676–7
    economy, 671–2, 675, 731, 734–5, 739–40
    feudal disintegration, 671, 673
    technical receptivity, 672
    aggressive nationalism, 676, 708
    World War I and, 713–14, 715
    Asian domination, 731, 733–7, 741–4
    War II and, 769–70, 772, 773, 786, 787–8
    occupied, 779, 846
  (iii) modern to 1945:
    World War II and, 784, 788
    foreign policy, 846, 887–8
    postwar reconstruction, 846
    economy, 887–8
    society, 887
Jaroslav 'the Wise', prince of Rus (1015–54), 292
Java, 364, 504, 532
Jefferson, Thomas (1743–1826), US president, 619
Jeremiah (*fl.* 626–586 BC), Hebrew prophet, 91–2
Jericho, prehistoric settlements at, 32
Jerome, St (*c.* 340–420), Christian churchman, 236
Jerusalem, 90, 92
  Temple of, 84, 91, 129
  destroyed by Babylonians, 94, 206–7
  Christians at, 214
  sacked by Romans, 215
  sacked by Persians, 254
  fall to Arabs, 262
  crusader kingdom, 298, 417
  British capture (1917), 751
  Israeli capital, 793, 895
Jesus (*c.* 4 BC–AD 29/30):
  Christ, life, 209–12
  teaching, 210–13
  Jewish reactions to, 213
Jesus, Society of (Jesuits):
  in India, 348
  in China, 365, 367, 368, 527, 633, 662
  in Japan, 375
  founded, 462
  dissolved, 473, 536, 552
  as missionaries, 531
Jews:
  origins, 86–7, 91–2
  literary influence, 136
  spiritual nationalists under Rome, 198, 204, 208–9, 213, 215–16
  theocratic view of history, 206
  transformed by Exile, 207
  spread outside Judaea, 208, 209, 215–16, 218, 260
  alienated from Christians, 233, 243–4, 254
  immigration to Palestine, 749, 751, 752, 791–2
  Nazi persecution of, 752, 763, 772, 791
  in Russia, 852
  See also anti-Semitism, Hebrews, Israel, Judaism, Zionism
Jinnah, 837
Joan of Arc, St (1412–31), French heroine, 404
John (1st cent. BC/AD), pre-Christian prophet, 212
John XXIII, pope (1410–15), deposed, 400
John VIII, Byzantine emperor (1425–8), 306
John (1167–1216), king of England (1199–1216), 399
John Paul II pope (1920–), 905
Johnson, Lyndon Baines (1908–73), US president:
  S. American and, 864
  domestic policies, 872, 873
  Vietnam and, 874
Johnson Dr Samuel, (1709–84), English man of letters, 3, 897
Jordan:
  prehistory, 25
  Transjordan, 752
  Israel and, 852
  See also Arabs, Ottoman empire, Palestine
Joseph, Hebrew patriarch, 88
Joseph II (1741–90), archduke of Austria and Holy Roman emperor (1765–90), 474, 497, 550
Josephus, Flavius (37–*c.* 100), Jewish historian, 215
journalism *see* press
Juan-Juan *see* Avars
Judaea, Roman province of, 207, 208–9, 215
Judah, Hebrew kingdom of, 92, 93, 94, 207
Judaism:
  Zoroastrian influence, 131
  source of Christianity, 206
  shaped by Exile, 207
  attractions of, 208
  Pharisaic interpretation, 208
  Khazar conversion, 280–1
  See also Hebrews, Old Testament, prophets, Yahweh
Julian 'the Apostate' (*c.* 331–63), Roman emperor (361–3), 233
Justinian I (483–562), Roman emperor in E. (527–62), 241–3
  administration, 242
  attempts to restore empire, 242–5, 280
  failed to reunite Church, 244, 247
Justin Martyr (*c.* 100–*c.* 163), Christian churchman, 225
Kalidasa (*fl. c.* 375), Indian poet, 343
Kamakura period in Japan, 371–3, 374
Kama Sutra, Indian text, 343
K'ang-hsi (1654–1722), Manchu emperor of China (1662–1722), 366–8, 573, 662
Kanishka (*fl. c.* 200), Kushan ruler of India, 344
Kant, Immanuel (1724–1804), German philosopher, 545, 549
Kashmir, 102, 650, 835
Kassites:
  rule in Babylon, 71–2, 92
  use of horse, 72
Katanga, 856
Kemal Ataturk (1881–1938), first president of Turkish republic, 754, 916
Kennedy, John Fitzgerald (1917–63), US president:
  space exploration and, 816
  and S.E. Asia, 845
  Cuba and, 863, 864, 869
  domestic policies, 873
Kenya, British settlement of, 658
Kepler, Johannes (1571–1630), German astronomer, 541, 543
Keynes, John Maynard (1883–1946), British economist, 720
Khazars, central Asian tribe, 257, 262, 280, 288, 290
Khomeni, Ayatollah Ruholla Ruholla (1902–), Iranian statesman, 900
Khorsabad, Assyrian palace at, 93
Kiev:
  Rus centre, 290–4, 328
  Mongol capture of, 294, 300, 421–2
Kipling, J. Rudyard (1865–1936), British writer, 655
Klerk, F.W. de (1936–), S.African statesman, 859
knighthood, orders of:
  Teutonic Knights, 300, 421, 422, 496
  St John of Jerusalem (Templars), 420
Knossos:
  palace of, 76–7, 78
  collapse, 80
Kochba, Simon bar (2nd cent. AD), Jewish messianic claimant, 216
Koran, 259–60, 264, 428
Korea:
  Chinese in, 354, 365, 667, 677
  Japanese concern, 369, 373
  division and war, 831
Kremer, Gerhard *see* Mercator
Krishna, Hindu god, 340
Kruger, Paul (1825–1904), president of Transvaal Republic, 645
Krushchev, Nikita Sergeyevich (1894–1971), Soviet statesman, 867, 868
  in China, 842
  and Cuban missile bases, 863, 868
  and Berlin Wall, 868
  fall, 869
Kubilai Khan (1216–94), Mongol emperor of China (1259–94), 301, 303, 364
Kuomintang (KMT):
  Chinese National People's Party, 732–3
  Twenty-One Demands and, 734
  CCP and, 737–9, 741–2, 788–9
  in Taiwan, 840
Kurile islands, 888
Kush, African kingdom of, 65, 120–1, 128–9, 379–80
Kushan period in India, 341

Kuwait, 749, 852, 900

labour:
  Byzantine, 281
  agricultural wage-labour, 409, 439, 561–2
  demographic influences, 413
  European craftsmen, 444–5
  children exploited, 571
  unemployment, 692–3, 730
  women and, 695, 715
  assembly-line use, 807
  robots, 808
  See also serfdom, slavery, trades unionism
Laika, 814
Langland, William (*c.* 1332–*c.* 1440), English writer, 404
language:
  development of, 13
  Doric dialect, 82
  Sanskrit, 103, 343, 680
  Aramaic, 130
  Greek, 136, 137, 166, 173, 199, 278
  Latin, 136, 199, 240, 278, 428, 433
  Arabic, 264, 268, 270, 849
  Norse influences, 324
  Romance, 324
  Urdu, 349
  French as international lang, 537–8, 549
  English in India, 651, 680
  Hindi, 680
  impact of Freud, 757
  western in Asia, 835
  in Soviet Union, 907
Laos:
  French imperialism, 685
  'associate state', 790
  internal divisions, 845
Lao-Tse (*c.* 604–531 BC), Chinese thinker, 116
Lateran treaties, 728
Later Han dynasty in China, 353–6
Latin America *see* Central America, South America, individual states
Latvia, 907
law and legal institutions:
  Sumerian, 44
  code of Hammurabi, 48–9
  idea of law, 75
  Chinese theory, 114
  in Greek *polis*, 143, 155–6
  Roman jurisprudence, 199, 242
  Germanic codifications, 240
  idea of fundamental laws, 453
  areas of legal immunity, 454
  privileges abolished, 583, 584, 587
  Napoleonic code, 592, 754
League of Nations:
  created, 722
  insecurity of, 723
  Asian policies, 734, 741, 742
  Arab mandates granted, 752
  mediation of, 755, 761
  and UNO, 777
Lebanon:
  Ottoman rule *see* Ottoman empire
  Sumerians and, 47
  Phoenicians in, 84
  French interest, 749, 791
  independent, 791
  in Arab League, 792
  internal tensions, 850, 851, 898
  invasion by Israel, 889
  and PLO, 891
  See also Levant
Lend-Lease system of US aid, 768, 775, 780
Lenin, Vladimir Ilyich (V. I. Ulyanov) (1870–1924), Russian revolutionary statesman, 833
  Bolshevik revolution and, 717, 719, 726
  death, 726
Leo I 'the Great', pope (440–61), 247
Leo III, pope (795–816), 316, 330
Leo IX, pope (1049–54), 391, 393
Leo III, Byzantine emperor (717–41), 281, 283
Leo IV, Byzantine emperor (775–80), 283
Leonidas (d. 480 BC), king of Sparta, 148
Levant:
  man in, 16
  early agriculture, 25–6
  focus of dispute, 65, 71
  trading posts, 82, 138
  Christian, 245
  Islamic rule, 295, 298
  French presence in, 654–5, 687
  Ottoman rule, 749
  and Israel, 900
  See also individual states
liberalism:
  contribution of Enlightenment, 548–52
  and industrialism, 572–3
  associated with republicanism, 580
  creed of market society, 594–5, 730
  associated with nationalism, 595
  in Russia, 607
  social weaknesses, 694
  threatened, 694
Liberia, 656, 855
Libya:
  Egypt and, 86
  independent, 851, 893, 898
  oil and, 852
  anti-Americanism, 899, 903
Lincoln, Abraham (1809–65), US president, 620–1, 622
Lindisfarne, English monastery raided by Vikings, 325
Lister, Joseph (1827–1912), British surgeon, 703, 704
Li Ta-chao, Chinese Marxist, 735
literacy:
  in Sumer, 40–1
  spreads in Near East, 75
  in Minoan Crete, 80
  in India, 97–8
  in China, 109
  in Mediterranean, 136, 181
  rare in Dark Age Europe, 240
  Islamic, 268
  early medieval clerical monopoly, 332
  absent in E. Africa, 382
  effects of European mastery, 537–9, 569, 697, 716
  spread in Japan, 672
  and declining religious faith, 701
  low in black Africa, 855
literature:
  fostered by cities, 35
  in Mesopotamia, 41–2, 49, 93
  in India, 99–100, 101–2, 343
  flowers in Mediterranean, 136
  range and primacy of Greek, 164–6
  civic content of Roman, 199
  Arabic, 268
  Russian, 292–3
  Carolingian Renaissance, 317
  Norse Sagas, 324, 328
  See also drama, historiography, poetry, individual writers
Lithuania:
  medieval duchy, 422
  union with Poland, 486
  break with USSR, 907
Livy, Titus Livius (59 BC–AD 17), Roman historian, 199, 203
Lloyd George, David (1863–1945), British statesman, 722
Locke, John (1632–1704), English philosopher, 527, 546, 549, 581
London:
  growth, 411, 438, 568
  port, 441–2, 527
  Great Exhibition, 567, 625
  squalor, 569
  19th-cent. focus of world economy, 634–5
Long March (1934), 742
Lothair I (795–855), Holy Roman emperor (817–55), 319
Louis I 'the Pious' (778–840), Holy Roman emperor (814–40), 319, 325
Louis II 'the German' (805–76), king of E. Francia (817–76), 319
Louis XIV (1638–1715), king of France (1643–1715), 366, 549
  absolute monarchy of, 472, 490
  foreign policy, 484
  navy, 525
  compared with Napoleon, 590–1
Louis XVI (1754–93), king of France (1774–92), 581, 584–5, 597
Louis XVII (1785–95), nominal king of France, 597
Louis XVIII (1755–1824), king of France (1814–24), 597
Louis Philippe (1773–1850), king of France (1830–48), 597, 599
Louisiana Purchase, 615, 618, 782
Loyola, St Ignatius (1491–1556), founder of Jesuits, 462
Luther, Martin (1483–1546), Protestant reformer, 456–8, 460
Lycurgus (? 9th cent. BC), Spartan legislator, 145
Lydia, ancient kingdom of, 142, 147
Lyell, Charles (1797–1875), Scottish geologist, 701
Lyons, persecuted Christians at, 217, 218

Macao, 856, 886
MacArthur, Douglas (1880–1964), US general, 846
Macartney, George, Lord (1737–1806), British ambassador to China, 662, 665
Macedon:
  Spartan ally, 149
  expansion by Philip II, 168–70
  empire created by Alexander, 170–3
  Roman conquest of, 186–7
Macedonia (part of Yugoslavia), 910
Magadha, Indian kingdom of, 102, 337
Magellan, Ferdinand (*c.* 1480–1521), Portuguese navigator, 425
magic:
  in Egypt, 60, 62
  and art, 75
Magna Carta, 455
Magyars:
  nationalism of, 291, 320–1, 711, 722
  See also Hungary
Mahabharata, Indian text, 101, 340
Mahdi, the (Mohammed Ahmed) (1848–85), Sudanese leader, 633
Mahmud, Afghan ruler of Persia, 311
Mahrattas, Hindu opposition to Moghuls, 350–1, 507
Maitland, Frederick William

(1850–1906), English historian, 581
Malaysia:
  Dutch power in, 505
  British interest, 650, 658, 685
  World War II and, 770
  post-war and communists, 791
  and Indonesia, 838
  economy, 887
Mali, African kingdom of, 381, 382, 853
Malta:
  temples in, 126, 127
  British acquisition, 648
Malthus, Thomas R. (1766–1834), British economist, 556, 559, 690, 701–2, 797, 820
Mamelukes, Turkish mercenaries, 296, 298–9, 301, 303, 304
Manchu dynasty in China (1644–1912), 360
  enter China, 365
  empire of K'ang-hsi, 367–8
  decline, 662–70
  popular resentment of, 665, 669
  against Taiping, 666–7
  fall, 670
Manchukuo, Japanese puppet state, 741
Manchuria:
  home of Manchu, 365
  Japanese invasion, 678, 740–1
  western interests, 734
  Russian invasion, 779, 788
  Chinese policy and, 840
  Russia and, 841
Mandela, Nelson (1918–), S. African politician, 859, 860
Manhattan Project on nuclear weapons, 809
Manichaeism, Christian heresy, 234, 235, 253
Mani (Manichaeus) (*c.* 216–*c.* 276), founder of Manichaeism, 253
Maori peoples of New Zealand, 643, 647
Mao Tse-tung (1893–1976), Chinese revolutionary statesman:
  and formation of CCP, 735, 737
  Marxist studies of, 735
  peasant revolution and, 739, 742, 788, 841–2
  writings, 788
  domination, 841
  'Cultural revolution' and, 842, 886
  views, 842–5
  death, 885
Marathas *see* Mahrattas
Marconi, Guglielmo (1874–1937), Italian scientist, 808
Marco Polo (*c.* 1254–*c.* 1324), Venetian commercial traveller, 303, 363, 364, 429
Marcus Aurelius Antoninus (121–80), Roman emperor (161–80), 197
Marduk, Babylonian god, 49–50, 51, 88, 94, 129
Mari:
  Sumerian religious centre, 44
  Babylonian centre, 48, 49
Maria Theresia (1717–80), archduchess of Austria and empress (1740–80), 474, 497, 549
Marius, Gaius (155–86 BC), Roman consul, 190
Mark Antony (83–30 BC), Roman general, 195, 197
Marlborough, John Churchill, duke of (1650–1722), English soldier, 712
marriage and family, 798, 918
  in Sumer, 44
  Pharaonic incest, 64
  in China, 111, 799–800, 843
  in Greece, 145–6, 152–4
  Romans and barbarians, 240
  Hindu practices, 344, 679
  clan power, 354–5, 735, 737
  medieval death-rate and, 437
  effects of female emancipation, 695
  polygamy forbidden in Turkey, 754
  Latin-America, 799
  in Soviet Union, 799–800
  India, 800, 834
  Japan, 887
  See also women
Marshall Plan of US aid to Europe, 782, 783, 829
Martin V, pope (1417–31), 400
Marx, Karl Heinrich (1818–83), German socialist philosopher:
  Communist Manifesto, 606
  worldwide veneration, 637
  and social science, 705
Marxism, Marxists:
  Chinese, 107, 637, 735–6, 738, 739
  quasi-religious doctrine, 607, 694
  and Bolsheviks, 694, 719
  and Stalin, 727
  intellectual impetus, 757
  target of Hitler, 763
  in Africa, 857
Mary Tudor (1516–58), queen of England (1553–8), 460, 464
Masada, 215
mathematics:
  Babylonian, 44
  Sumerian, 44
  Egyptian, 58
  mystical approach of Pythagoras, 160–1
  Euclid, 175
  Arabic, 268, 270, 428–9
  decimal system in India, 343, 429
  techniques develop, 541, 543–4
  quantum theory, 810–11
Maurice (582–602), Byzantine emperor, 254
Maurya empire in India, 337–41
Maximian (d. 310), Roman emperor (286–305), 224
Maximilian I (1459–1519), Holy Roman emperor (1493–1519), 405
Maximinus (d. 388), Roman emperor in W, 221
Maxwell, James Clerk (1831–79), British physicist, 810
May 4th Movement in China, 734–5, 788
Mayan civilization in Meso-america, 28, 384–5
Mazzini, Giuseppe (1805–72), Italian nationalist, 595, 604
  influence on India, 680, 681–2
Mecca, holy city of Islam, 258, 260, 295, 310, 351
Medes, 93, 129
Medici family in Florence, 479
medicine:
  in Egypt, 62–3
  Greek empiricism, 162
  Arabic textbooks, 270
  medieval European, 416, 429
  progress against disease, 557–8, 655
  women practitioners, 601
  antiseptics, 703, 704
  anaesthetics, 704
  and genetics, 814
  See also disease
Medina, City of the Prophet, 260, 295, 310, 751
Mediterranean:
  early civilization, 134–6
  topography, 134
  Ottoman empire, 309
  commercial activity in, 333, 408, 636
  See also individual areas
Megasthenes (*fl. c.* 300 BC), Greek writer on India, 337–9
Mehemet Ali (1769–1849), pasha of Egypt (1805–49), 654, 745, 754
Mehmet II (1429–81), sultan of Turkey (1451–81), 308, 309–10
Meiji Restoration in Japan, 673
  authoritarian, 675–6
  westernization, 675
  foreign policy, 676–7
Memel, 765
Memphis, Egypt, 54, 65
Mencius (*c.* 372–289 BC), Chinese thinker, 117
Mendel, Gregor (1822–84), Austrian biologist, 812–13
Menes (35th cent. BC), king of Egypt, 53, 54
Mercator, Gerhardus (1512–94), Flemish cartographer, 427
Meroe, Sudanese kingdom, 121, 379
Merovingian dynasty of Franks, 238, 240, 249, 313–15
Merton, Walter de (d. 1277), bishop of Rochester and Lord Chancellor of England, 427
Meso-american cultures, 384–5
  See also Aztecs, Mayan civilization, Olmecs
Mesolithic era, 17, 24
Mesopotamia:
  first civilization, 32, 34, 38–51
  contact with Egypt, 53
  literacy in, 75
  confusion in, 85, 86, 88
  trade with Harappans, 97
  Parthians in, 178–9
  Arab conquest of, 257, 281
  Turks in, 298, 749
  Mongols in, 304
  See also Iraq
metallurgy:
  Neolithic origins, 25
  social and economic effects, 27–8
  Mesopotamian imports, 45
  European skills, 72–3, 124
  in ancient China, 112–13, 117
  of Etruscans, 180
  European expansion, 412, 445
  See also bronze, copper, industrialization, iron, technology, tin
Methodius, St (9th cent.), apostle of Slavs, 289–90, 294
Metternich-Winneburg, Prince Clemens von (1773–1859), Austrian statesman, 599
Mexico:
  maize in, 121
  Aztec culture, 385–6
  urban growth, 438
  Spanish in, 509–10, 511, 512, 528, 640
  territorial losses to US, 616, 618
  oil, 641
  social divisions, 642
  US intervention, 642
  social divisions and revolution, 860
Michael (1596–1645), tsar of Russia (1613–45), 489
Michael II, Byzantine emperor (820–29), 283

Mieszko I, king of Poland (960–92), 293–4
Miletus:
wool a speciality, 152
early thinkers in, 159–60
militarism:
samurai ideals, 374, 376, 671
restrictions on barbarity, 699–700
military techniques:
cavalry, 72
Indo-European advances, 72
of Assyrians, 93
in China, 113
Greek hoplites, 142
Macedonian phalanx, 168
Parthian skills, 178, 197
Roman army, 190, 205, 224, 231–2
Scythian skills, 255
Mongol cavalry, 301
knightly class, 335–6
British army in India, 679
modern technology and, 699–700
in World War I, 714–15
World War II bombing, 773
See also weapons
Mill, John Stuart (1806–73), British philosopher, 669, 680
Minamoto clan in Japan, 373
Ming dynasty in China (1368–1644), 360, 365
Minoan civilization in Crete:
earliest, 32, 76
character, 76–7
destruction, 77
naval power, 77–8
art, 78
religion, 78–9
technology, 78
trade, 78
government, 79–80
achievements, 85
See also Knossos
Minos, legendary king of Crete, 77, 79
Minotaur, legendary monster, 77, 79
missionary activity (Christian):
early medieval, 290, 294, 315, 324, 331
in Japan, 376–7
impetus to imperialism, 501, 687
in Latin America, 511, 512
effect of Counter-Reformation, 531–2
in European cities, 569
and Europeanization, 633, 636
in Australasia, 643
in India, 651, 687
in China, 665, 666–7
in Asia, 687
See also Jesus, Society of
Missouri Compromise, 619
Mitanni peoples, 92
Mithras, Zoroastrian deity, 204, 217
Mithridates I (171–138 BC), king of Parthia, 178–9
Mithridates II (*c.* 124–87 BC), king of Parthia, 179
Mithridates VI (132–63 BC), king of Pontus (120–63 BC), 189
Mitsui, Japanese traders, 671–2
Moghul empire in India:
established by Babur, 347
stabilized by Akbar, 347–50
religious eclecticism, 348
European trading posts, 349, 350–1
arts in, 350
popular revolts, 350–1
collapse, 507
formally ended, 653
Mohammed Reza Shah Pahlavi (1919–80), shah of Iran, 891
Mohenjo-Daro, 102, 105
Harappan site, 97, 99
Molotov, Vyacheslav Mikhailovich (1890–1986), Soviet politician, 867
monasticism:
in India, 103
Christian, 245–6
in India, 138–9, Christian, 278
in England, 326, 331, 460
early medieval reform, 331
money:
invented, 74
in China, 109, 113, 356, 361
encouraged Greek trade, 142, 145
terminology, 183
inflation in Roman empire, 221
Chinese use of paper spreads, 301
Ottoman, 308
little Dark Age coinage, 333
medieval growth of money economy, 409–11
decimal systems, 587
See also gold, inflation
Mongolia, 299, 737, 741, 841
Mongols, 294
(i) nomad warriors, 298–9
inspired aggression of Chinghis Khan, 299–300
assault on Islam, 300–1
khanates, 301
reasons for success, 301
Il-khans, 303–4
religion, 303
Timur Lang, 304, 346
(ii) ascendancy in China (1264–1368), 360, 364–5
repelled by Japanese, 373
Monophysites, Christian heretics, 244, 279
Monroe doctrine:
US hegemony in N. America, 616, 640
intervention in S. America, 642, 660
deterred European imperialism, 649–50
stimulated US expansion, 658
alive and well, 863
Montenegro, 697
Montesquieu, Charles Louis de Secondat, baron (1689–1755), French philosopher, 451, 548
Morocco:
French interest, 654, 656, 712, 751
independent, 849
Islamic fundamentalism in, 893
mortality rates, 416, 437
(i) fall in Europe, 437, 557–8, 559, 690
of slaveships, 444
in World War I, 715, 716, 717, 719–20
in World War II, 780
(ii) infant:
in Egypt, 62–3
in medieval Europe, 416
fall in Europe, 558
in war, 716
world disparities, 798–9
in Cuba, 864
(iii) life expectancies, 798
Moscow:
medieval, 293, 422, 423
early modern, 488, 491, 494
Moses (*c.* 14th cent. BC), Hebrew patriarch, 68, 89, 207, 544
Moslems:
Moslem-Hindu clash in India, 507, 681–4, 785–6, 824, 834, 889
Moslem brotherhood, 893
See also Islam
motor cars:
spread of, 636, 704, 807–8
early use, 690
cheap American, 703, 730, 805, 807
Japanese production, 887
Mo-Tzu (5th cent. BC), Chinese thinker, 116
Mozambique, 858, 859
Mu-Awiyah, first Umayyad caliph (661–80), 264
Muhammad (570–632), 258–60
background, 258
opposed by Quraysh, 260, 264
teaching, 260
Munich agreements (1938), 765
Murasaki, Lady (978–*c.* 1015), Japanese novelist, 374
Muscovy, 294, 328, 608, 422, 449, 487, 490, 492
music:
Arabic, 270–2
20th-cent. movements, 757
Mussolini, Benito (1883–1945), Italian dictator:
rise to power, 728
aggressive foreign policy, 761, 765
overthrow, 772
Mycenae, Mycenaeans, 80–5

Nadir Kali (1688–1747), shah of Persia, 311
Namibia, 858, 859
Napier, John (1550–1617), Scot, inventor of logarithms, 544
Naples, 306, 427, 479
Napoleon I (Bonaparte) (1769–1821), emperor of France (1804–15):
consolidated French Revolution, 588–90
rise, 588–9
redrew European map, 591
awakened nationalism, 593
influence on Russians, 607–8
Egyptian expendition, 653
Napoleon III (1808–73), emperor of France (1852–71), 602, 604
Nasser, Gamal Abdel (1918–70), Egyptian statesman, 849, 852
nationalism:
growth in Europe, 402–4, 466, 484
stimulated by Napoleon, 593
19th-cent. growth, 594–5, 698–9
associated with liberalism, 595
1848
revolts and, 599
in Eastern Question, 599, 601, 611, 697, 745–55
Slav, 599, 697, 711
in Italy and Germany, 602
Chinese, 669, 671
in Japan, 676, 740
in India, 680–4, 785–7
in S.E. Asia, 687, 789
in Ireland, 697
in World War I, 733–7
in World War II, 766–7, 784
space exploration and, 814–17
modern political force, 826
Arab, 849
naval power and tactics:
in Crete, 77–8
of Solomon, 90
of Athens, 150–1, 154
Byzantine, 242

English, 481, 522, 524–8, 617, 635, 649, 698, 712
Spanish, 481, 522–4
Turkish, 485
warships, 485, 501, 525, 770
Dutch, 522–3
French, 525
German, 698, 712
Japanese, 734
See also shipping
navigation:
medieval advances in Europe, 423–4, 426–7
astrolabe, 428–9
Chinese and Arab, 500
See also shipping
Nazism *see* Germany, Hitler
Neanderthal man:
sites, 15–16, 107
characteristics, 16–17
art, 18
Near East:
Homo sapiens sapiens found, 16
crucial zone of early development, 25–6, 28, 32, 36–7, 76
first civilizations, 34, 38–51, 85–94
racial interplay, 36–7, 85–7
papyrus in, 62
literacy in, 75
iron in, 86
religious forces, 88
Assyrian influences, 92–3
integration of civilizations, 128–9
See also Fertile Crescent, Islam, Ottoman empire, individual areas
Nebuchadnezzar II, king of Babylon (604–562 BC), 94, 129
Negroid peoples *see* racial issues (ethnic groups) Negro in US *see* racial issues, United States of America Nehru, Pandit Jawaharlal (1889–1964), Indian statesman, 836, 889
Neolithic era, 17
matrix of civilization, 24–5
pottery, 32
Mesopotamian settlements, 38–40
Egyptian settlements, 53
in China, 107
neo-Platonism, 199
Nero (37–68), Roman emperor (54–68), 214, 216
Nestorians:
Christian heretics, 244, 254
among Mongols, 300, 303, 304
in China, 364
Netherlands:
shipping, 441, 525, 527
commercial interests, 443, 447, 672
Dutch revolt from Spain, 464, 480
commercial imperialism, 482, 504–5, 507, 513, 522, 686–7, 789
in N. America, 516
Anglo-Dutch wars, 522, 524
slave trading, 530
World War II and, 767, 772
Nevsky, Alexander (1200–63), grand duke of Novgorod, 421
'New Deal' in US, 760–1, 780
New Testament:
Gospel record of Jesus, 209–12
Acts of the Apostles, 214
Erasmus and, 456
Newton, Sir Isaac (1642–1727), British natural philosopher, 543–4, 548–9, 552, 757, 810
New York, 516
New Youth, Chinese revolutionary journal, 733, 735
New Zealand:
discovered by Europeans, 533
expansion of food production, 563, 645
Maoris, 643, 647
self-government, 643–4
foreign policy, 847
economy, 887
Nicaea, Council of (325), 228, 279
Nicaragua, 642, 862
Nicholas I, pope (858–67), 330
Nicholas I (1796–1855), tsar of Russia (1825–55), 608–10, 611
Nigeria:
early kingship in, 382
civil war in, 856
Nightingale, Florence (1820–1910), British hospital reformer, 601
Nile River:
basis of Egyptian civilization, 52–3, 54–5
cultural influence, 58, 60, 62, 70
marshes reclaimed, 70
Nimrud, Babylonian city, 48
Nineveh, Babylonian city, 48, 92, 93
Nixon, Richard M. (1913–), US president, 875–7
Normans:
Normandy, in Mediterranean, 287, 417
duchy granted, 291, 319–20
loss of Scandinavian character, 328
conquest of England, 404–5
duchy acquired by Fr, 405
Norsemen:
Slav tribes and, 290–1
colonizing, 324–6, 383
shipping, 324, 423
raids on Christendom, 325–6
absorbed into native cultures, 326
See also Varangians
North America:
human immigration, 18, 21, 383
primitive agriculture, 26, 28, 383
Norse settlement, 324, 383
European settlement, 513–19, 524, 611
War of American independence, 576–81
See also American Revolution, Canada, United States of America
North Atlantic Treaty Organization (NATO):
established, 831
Germany and, 833
strains in, 881
Norway:
early colonizing, 324
women's vote, 695
World War II and, 767
See also Norsemen
Novgorod, 421, 422
Nubia, 65, 86, 120
See also Kush, Sudan
nuclear fission, 808

oar, 74
OAU *see* Organization for African Unity
Odoacer (d. 493), German chieftain, 232, 240
Oghuz Turks, migrate to Iran, 296, 297
oil (mineral):
in S. America, 642
energy source, 690
political factor, 751, 752–3, 755, 853, 884
demand for, 807
in Middle East, 850
'energy crisis' and, 853, 884, 914
Okinawa, 888
Old Testament:
(i) evidence of, 87, 88, 89–90, 91, 92
translated into Greek, 208
Newton's endorsement, 544
(ii) books:
Genesis, 89
Joshua, 89
Daniel, 94
Isaiah, 94, 208
Pentateuch, 207
Old Testament, 75, 87, 207
Olduvai Gorge:
earliest home base, 8–9
fossil relics found, 8–9, 12, 16
Oleg (10th cent.), prince of Russia, 290
Olga (d. *c.* 969), Christian ruler of Kiev, Russia (945–62), 291
olive, 77, 134, 152
Athenian oil exported, 145
Olmec culture in Meso-america, 121–2, 384
Olympic Games, 137
Oman, 893
'Open Door' policy in China, 731, 734
opium *see* drugs, wars oracles:
in China, 112
in Greece, 140, 147, 177
at Siwah, 171
Organization for African Unity (OAU), 856, 858
Origen (*c.* 185–*c.* 254), Christian churchman, 218, 225
origins of Man, 19th-cent. theories, 701–2
origins of Man, 4–14
primates, 6–7
prosimians, 6–7
Australopithecus, 7–9
first hominids, 7
Homo erectus, 9–14
physiological development, 9–10
slowness of development, 24
See also Homo sapiens
Orkhan, first Ottoman sultan (1326–59), 307–8
Orthodox Church:
characteristics, 277
divergence from R.C. Church, 278–9, 285, 419
dispute over icons, 283–5
conversion of Slavs, 290, 292, 293–4
centred in Russia, 422
Osiris, Egyptian god, 55, 60
Osman I (1259–1326), Ottoman sultan (1299–1326), 306–7
Osmanlis *see* Ottomans
Osrhoene, Syrian kingdom, 206
Otto I (912–73), Holy Roman emperor (936–73):
German territories organized, 321
Italian crown assumed, 321
Otto II (955–83), Holy Roman emperor (973–83), 321
Otto III (980–1002), Holy Roman emperor (983–1002), and 'renewal of Roman Empire', 321
Ottoman empire (of Osmanli Turks), 273, 304–9
identity, 306
military organization, 307–8
capture of Constantinople, 308
further conquests, 308–9
multi-racial policy, 309
perpetuated E-W distinction, 309–10
rivalry with Safavid Persians, 310
European campaigns, 480, 485–6
Russian threat to, 488, 496, 497–9, 600–1, 611

subjects' nationalism, 598–9, 654–5, 697, 745, 749–52
Indian sympathy, 684
World War I and, 711, 713–14, 715
disintegration, 745–51, 753–4
Ottonian empire:
territorial acquisitions, 321–2
Papacy and, 323, 330, 331
Ovid, Publius Ovidius Naso (43 BC–AD 17), Roman poet, 195
Oxford, university of, 427, 436, 536, 680

Pacific basin and islands:
Tahiti, 533
US interests, 659
See also Hawaii, Philippines
painting:
Palaeolithic, 18–21, 29, 62
Indian, 347, 350
Chinese, 354, 361–2
Japanese, 374
perspective, 429, 539
abstract, 757
Pakistan:
Moslem state created, 787, 835, 890, 898
East Bengal secession, 835–6
India and, 835–7
foreign policy, 836–7
government, 836–7
nuclear weapons, 921
Palaeolithic era:
Lower, 17
Upper, 17–23, 29
art, 18–21, 29, 61
hunting, 18
physiological development of man, 18–21, 24–5
tools, 18
cultural distinctions arise, 27
excavations of, 701
Palestine, 65
focus of dispute, 71, 90
Philistines in, 86
crusaders in, 417–21
Jewish immigration, 749, 752, 791
Ottoman rule, 749
British policy, 751, 752, 791
Arab nationalists, 753, 852, 900
refugee problem, 852
See also Israel, Palestine Liberation Organization
Palestine Liberation Organization (PLO), 852, 891, 900, 911
Panama:
Congress (1826), 640
canal, 642, 660
Pan-American congresses (1889), 642
Papacy:
rise of, 246–7
Frankish involvement, 315–16, 321–2, 323, 330
religious and political focus, 319
diplomatic contacts with China, 354
defence against lay power, 391–5, 397–8
support friars, 397
Avignon exile, 399
Protestant critics, 455–60
Counter-Reformation, 462–4, 473
Enlightened critics, 552
loss of Temporal Power, 700
Lateran treaties, 728
modern trends, 824–5
paper:
Chinese invention, 362
used in W, 402, 412, 429, 431
papyrus, 53, 62, 75, 175
use declined, 333
Paracelsus (1493–1541), Swiss physician, 539
Paraguay, 865
parchment, 175
enabled minuscule writing, 333
gave way to paper, 402
Paris:
capital of Clovis, 239
core of France, 320, 405
growth, 411, 438, 568
university, 427
revolutions, 597, 600, 604
Commune, 604, 607
Parliament:
English, early growth, 408
composition, 446, 623
use by Tudors, 460
quarrel with Crown (1640–60), 467–9
sovereignty questioned in colonies, 576, 577
flexibility of, 627
buildings, 628
House of Lords limited, 694
parliaments *see* political institutions
Parthenon, Athenian temple, 154, 166
Parthia:
Parthians, threaten Hellenistic world, 179
resist Romans, 197–8, 222
overrun by Scythians, 255
in India, 341
Pasiphaë, legendary wife of Minos, 77, 79
Pasteur, Louis (1822–95), French chemist, 704
Patna, Maurya palace at, 337
Paul, St (*c.* 3–*c.* 64 or 68), Christian apostle, 214, 215–16, 225
Paul III, pope (1534–49), missionary activity of, 531
Pavlov, Ivan Petrovich (1849–1936), Russian biologist, 823
Pearl Harbor, Hawaii, US base at, 659, 770
Peasants' Revolt, England, 413
Pechenegs, S. Russian nomads, 287, 291, 293
Peel, Sir Robert (1788–1850), British statesman, 573, 625
Peking:
Chinese capital, 364, 365, 367
diplomats, 667
in Chinese Revolution, 732
activist students, 733, 734, 735
Tiananmen incident, 912
Peking man, 107
Pelagianism, Christian heresy, 234, 235
Peloponnese, 80, 170, 308
See also Mycenae, Sparta
Peloponnesian League, in Spartan interests, 148
penology:
transportation, 533
early modern savagery, 535
reformers, 549
guillotine, 585
Pepin 'the Short' (714–68), king of Franks (751–68), 315, 329–30
Pergamon, 173, 186, 189
library at, 174–5
Pericles (*c.* 490–429 BC), Athenian statesman, 150,156, 157
Péron, Juan Domingo (1895–1973), Argentine statesman, 861
Perry, Matthew (1794–1858), US naval officer, 672, 677
Persepolis, 131, 171
Persia, Persians:
(i) ancient:
Indo-European origins, 72
invade Babylon, 94
under Cyrus, 129–31, 207
under Darius, 130–1
wars with Greece, 147–8
Spartan ally, 151
decline, 168
hostility of Macedon, 170–1
rule of Alexander, 171–3
(ii) Sassanid rule, 222, 252–4, 257, 261–2
wars with Justinian, 240–1, 242
(iii) Safavid dynasty, 310–11
cultural flowering, 310
English trade, 311
(iv) modern *see* Iran
Peru:
Chavin culture, 121
Inca society, 388–9
Spanish conquest, 509
military rule, 865
Peter, St (d 64), Christian apostle, 214, 216
papal authority attributed to, 246, 315, 330
Peter I 'the Great' (1672–1725), tsar of Russia (1682–1725), 473–4, 916
modernizing autocrat, 490–4, 499, 727
territorial expansion, 490–2, 497
Pharaohs:
role of, 54–5
marriages of, 64
Phidias (*c.* 500–*c.* 431 BC), Greek sculptor, 157, 167
Philip II (382–336 BC), king of Macedon (359–336 BC), 168–70
Philip II (1527–98), king of Spain (1556–98), 464, 480
Philippines:
US interests, 659–60, 687
World War II and, 770
independent, 847
philosophy:
Confucianism, 114–17
Taoism, 116–17
Greeks achieve rational thought, 157–60
and Plato, 160, 162–3
Hellenistic quietism, 177–8
Indian thought, 343, 344, 345–6
medieval scholasticism, 428
Enlightenment, 545–53, 701
Romanticism, 552–3
impact of science, 705
Phoenicia:
Phoenician people, alphabet, 75, 84, 130, 137
art, 84
location, 84–5
trade, 84–5, 127, 189
under Hebrews, 90
Phrygian people, 86, 93
physics:
study of, early modern advances, 542, 543–4
modern discoveries, 810–13
relativity, 810–11
Pilate, Pontius, Roman governor of Judaea (26–36), 209, 215
Pilgrim Fathers, The, 517
piracy:
Norse, 324–6
Caribbean, 480, 513, 522, 525
privateers, 525
Pizarro, Francisco (*c.* 1475–1541), Spanish conqueror of Peru, 509

Planck, Max (1858–1947), German physicist, 810–11
plantation crops:
(i) coffee, 527
trade in, 505
in Caribbean economy, 512, 532
in Java, 532
in Brazil, 642
(iii) sugar, 527, 532
(ii) rubber, 532
in Caribbean economy, 512, 513
object of war, 524
duties on, 576
(v) tobacco, 351, 527
trade in, 505, 527
(v) tobacco:
in Caribbean economy, 512, 513
in N. America, 516, 517
in Ceylon, 532
See also cotton
Plato (*c.* 428–*c.* 348 BC), Greek philosopher, 156
on Socrates, 156, 159
insistence on rationalism, 158, 159
reality immaterial, 160, 162, 163
first textbook writer, 162
transmitted to Islam, 175, 268
influence on Christinanity, 218, 225, 546
Pleistocene period, 5
Early, 7–9
Late, 9–10, 15–17
Middle, 9–14
See also Neolithic era, Palaeolithic era
Pliny the Elder, (23–79 AD), Roman naturalist and writer, 201
Pliny the Younger (62–113 AD), Roman administrator, 331, 342
plough, 28
in Europe, 334, 412
in Africa, 379
in Americas, 384, 387
in Russia, 493
poetry:
Greek achievement in drama, 164
poet as teacher in Greece, 164
Japanese, 374
Renaissance forms, 431
See also individual writers
Poland:
Slav origins, 293–4
conversion to Catholicism, 294
defeat by Mongols, 300
government, 473, 485, 487
Russia and, 486–7, 496, 781, 878
first partition, 499, 611
revolts, 598, 599, 603, 611
Versailles settlement and, 722, 759
authoritarian regime, 729
fourth partition, 765–6
post-1945, 829, 867, 904–5
Solidarity, 905, 906
collapse of communist regime, 906
police forces, 604
communist use of, 726
agent of government, 827
political ideas:
legislative sovereignty, 453, 454, 472, 475
enlightened despotism, 473–5
dynasticism, 475, 481, 484–5
popular sovereignty in US, 581
in France, 584, 587, 597
'Right' and 'Left', 587
attacked, 599–600
See also communism, fascism, government administration, liberalism, Marxism, socialism
political institutions:
(i) monarchy:
Sumerian, 43, 45–7
Pharaonic, 55
Greek, 84, 145
divine kingship in Rome, 203, 224, 225
Byzantine autocracy, 275
Mongol autocracy, 303
royal power in Europe, 336, 402–3, 454–5, 469, 470, 472–5
Tsarist autocracy, 490
Napoleonic autocracy, 590–1
(ii) aristocracy:
early Greek, 140
legal privileges of, 446–8, 525–7
(*see also* class structure)
(iii) oligarchy:
in Greece, 145, 156, 170
of Roman republic, 182–3
of Italian city-states, 407
(iv) parliaments and representative bodies:
Athenian democracy, 144–5, 155–6
medieval European, 408
pre-revolutionary France, 469, 582–3, 584
French National Assembly, 584, 587, 597
Indian National Congress, 681, 684, 785–6
demand for, 595
US democratization, 618, 620, 622
British democratization, 623–4 (*see also* Parliament, English)
use by socialists, 693–4
franchise extensions, 695, 731
hostility of Bolsheviks, 718–19
hostility of Nazis, 763
republicanism:
in US constitution, 580–1, 618
in French constitutions, 585, 587
(vi) federalism, 580–1, 602, 643
See also government and administration, individual countries
political party systems:
strengthened in US, 622
organized in U K, 624–5
SDP in Germany, 693
Russian socialists, 694
spread of, 697
KMT and CCP in China, 732–3
See also Bolsheviks, communism
pollution *see* environmental concern and change
Polybius (*c.* 201–120 BC), Roman historian, 189
Pompadour, Jeanne de (1721–64), mistress of Louis XIV, 451
Pompey (106–46 BC), Roman general, 192, 208
Pontiac's rebellion, 574
Pope, Alexander (1688–1744), English poet, 544
population and demographic patterns:
(i) prehistoric, 21–3, 26–7, 36, 83
(ii) Europe:
Dark Ages, 334
medieval, 409–11, 413
early modern trends, 436–9, 441, 443–4, 493
modern, 556–9, 569, 583, 690
(iii) India, 102, 378, 663, 697, 787, 360, 363, 368, 369, 377
and food supply, 441, 556
Americas, 513, 516, 616, 622–3, 641, 657
(iv) Malthusian theory, 556, 797, 820
emigration and, 557, 637–8
Africa, 657
(v) worldwide growth and structure, 797–801
See also contraception, mortality rates
Portugal:
Portuguese, trading ships, 309
interest in Persia, 311
trade with India, 349, 351
in China, 365, 501, 667
in Japan, 374–5, 376
maritime expeditions, 424–5
Spain and, 425, 470, 484, 504
commercial imperialism, 501, 503–5, 511, 533
slave trading, 530
government in exile, 640
World War I and, 714
authoritarian regime, 728
World War II and, 770
and decolonization, 835
joins EEC, 884–5
Poseidon, Greek god, 139
Postumus (d. 267), Roman imperial claimant, 221
Potocki family in Poland, 487
Potsdam Conference (1945), 781
Pottery:
Neolithic, 28, 32
Sumerian, 40, 45
Indian, 97
Chinese, 107, 117, 354, 361, 362, 368
Athenian, 152
Arabic, 268, 270
Praxiteles (4th cent. BC), Greek sculptor, 167
press:
censorship, 472, 590
rise of periodicals, 538
imperialism and, 655
in Japan, 674
growth of cheap press, 697
Chinese revolutionary, 733, 735
Arabic, 749
Prester John (*c.* 12th–13th cent.), legendary Christian king, 298, 424
Primates, development of, 6–7
printing:
spreads Bible, 87, 432
movable type in China, 362–3
stimulus to papermaking, 412
impact on Europe, 431–2, 528
diffusion of ideas, 537, 541, 701
progress:
idea of, developed in Europe, 432–3, 503, 535
justifies European imperialism, 647, 689
accepted worldwide, 919
prophets, Hebrew, 91–2, 207
prosimians, 6
Protestant Reformation:
origins and spread, 456–60
undermined authority, 469, 547, 553
force for literacy, 538
Protestants:
Lutheranism, 456–60
Anabaptists, 458
Calvinism, 458–9, 460
in New England, 516
See also Anglican Church
Proust, Marcel, (1851–1922), French writer, 442
Prussia:
enlightened despotism in, 474
Hohenzollern state, 496–7, 499

French revolutionary wars and, 585, 590, 593
German unification and, 599, 602–3
psychology:
science of, psychoanalysis, 757, 822–3
behaviourism, 823
Ptolemy of Alexandria (2nd cent. AD), Egyptian astronomer, 161–2, 425, 426, 428
Ptolemy I 'Soter' (*c.* 367–283 BC), ruler of Egypt (323–285 BC), 173, 175
Puerto Rico, US acquisition of, 660
Pugachev, Russian rebel (1773), 496
Punch, British journal, 643
puritanism:
soul-body schism and, 162
influenced by Augustine, 235
17th-cent. force, 469, 516
Pylos, 80
pyramids, 57–8, 60
Pyramid of Cheops, 58
Step Pyramid, 58
Mayan, 384
Pyrrhus (*c.* 318–272 BC), king of Epirus (295–272 BC), 179, 184–5
Pythagoras (*c.* 570–500 BC), Greek mystical mathematician, 160–1
Quebec, 516, 826
Quetzalcoatl, Aztec god, 388
Quraysh, Bedouin tribe:
Muhammad a member, 258
caliphs, 261, 264

rabbits, 621
racial issues:
(i) ethnic groups, 36–7, 119, 124–6
(ii) imperialism and, 529, 532–3
multi-racial Latin America, 511, 640–1
in southern Africa, 645
natives and colonists, 646–7
(iii) in USA, results of slave emancipation, 622
(iv) Nazi theories, 763
(v) *négritude*, 827
See also anti-Semitism, Germanic peoples, Indo-Europeans
radio:
early use, 703, 808
Nazi monopoly, 763
political use, 827
Soviet jamming of Western, 871
Radischev, Alexander Nikolaievich (1749–1802), Russian political exile, 496
Radziwill family in Poland, 487
railways:
facilitate trade and communications, 563–4, 566, 604, 636, 637, 690
British, 567
in war, 601
and nationalism, 643
Indian, 651, 679
Japanese, 674
Rama, Hindu god, 340
Ramayana, Indian epic, 101, 340
Ramses III, king of Egypt (*c.* 1200–1168 BC), 68
Ramses XI, king of Egypt (12th cent. BC), 68
Ravenna, 232, 247, 315, 317, 329
Reagan, Ronald (1911–), US president, 898, 902
Red Guards in China, 842, 911, 912
Reformation *see* Protestant Reformation
religion:
(i) origins, 20
divinity of kings, 54–5, 175–7, 203, 224, 225
diversity, 75
and morality, 91–2
mystery cults, 204, 217
Byzantine theological disputes, 278–9, 283–5
legal support of, 535–6
awe mitigated by Enlightenment, 546, 701
social cement dissolves, 569, 756
declining political force, 700
impact of science on, 822
as modern social force, 822, 823–4
(ii) Sumerian, 40, 41, 42
Egyptian, 58–61
Hebrew vision, 89
Babylonian, 94
Indian, 99, 102–4
Chinese ancestor-worship, 111
Greece, common factor in, 139, 177
and Roman civic life, 203
Mongol, 303
Aztec, 385, 387
See also individual religions and deities
Renaissance:
classical inspiration and arts, 430–1, 546
technology, 444–5
descriptive science, 539–41
Renan, Ernest (1823–92), French scholar, 701
revolution, idea of:
accepted into political thought, 574, 595
'Right' and 'Left' debate, 587
in Marxist thought, 606–7
British reluctance, 627
Chinese peasant emphasis, 739
Reza Khan (1878–1944), shah of Persia (1925–41), 755
Rhodesia:
Zimbabwe buildings, 382
British settlement, 656, 658
independence, 856
See also Zimbabwe
Richelieu, Cardinal Armand de (1585–1642), French statesman, 469, 472, 481
Rig-Veda, Indian text, 99–100, 103
Risorgimento, 714, 727
See also Italy
Rollo (d. 921), first duke of Normandy, 291, 320
Roman Catholic Church:
(i) growth:
early temper, 234
political importance in Dark Ages, 238, 245
rise of Papacy, 246–7
estranged from E, 279, 285, 419
early medieval insecurity, 328–30
cultural monopoly, 331–2
(ii) medieval role:
basis of European civilization, 391–2
reform movements in, 391–3, 398–401
role of Papacy, 391–5, 397–400
friars, 397
intolerance, 397–8
social role, 401–2, 449
legal immunity, 455
(iii) modern:
Protestant Reformation and, 456–60, 469, 538, 547, 553
Counter-Reformation, 460–4, 466, 474, 486, 531
in Latin America, 512
in Canada, 576
in France, 584, 590
authority questioned, 701
anti-Marxism of, 728
fascist concessions, 728
world-wide strength, 824
See also Church and state, clergy, Papacy
Romania:
established, 601, 697
World War I and, 714
Versailles settlement and, 722, 759
Russia and, 781, 879
collapse of communist regime, 909
Romanov dynasty in Russia, 489, 492
See also individual rulers
Romanticism, 553
idealization of artist, 757
Rome, Romans, 179
(i) Republic, 180–93
Etruscan inheritance, 180–1
political structure, 182–3, 187–9
expansion, 184–7
wars with Carthage, 184–7
army enters politics, 190
civil wars, 191–2, 195
(ii) Empire:
241–54, Augustan settlement, 195
Augustan successors, 195
territorial expansion, 197–8
administrative and cultural cosmopolitanism, 198
technical skills, 199
brutality, 201
social divisions, 201
civic and mystery religions, 203–4
economic and military basis, 204–5
(iii) decline and fall, 218–32
weakness of government, 219
economic weakness, 221
border troubles in E, 222
barbarian invasions in W, 223, 224
Diocletian's reorganization, 224
Christianity established, 226–9
Constantine's administrative reforms, 226–7
persecution fails to eradicate Christians, 226
empire collapses in W, 229–31
(iv) Empire in E, 229–31
Justinian's efforts, 296–9
See Byzantine empire
Romulus and Remus, legendary founders of Rome, 180
Röntgen, Wilhelm (1845–1923), German physicist, 811
Roosevelt, Franklin Delano (1882–1945), US president:
'New Deal', 759–61
World War II and, 768–9, 776
Russians and, 781
Asian policies, 787
use of radio, 827
Roosevelt, Theodore (1858–1919), US president, 660
Rosetta stone, 61
Rousseau, Jean Jacques (1712–78), Swiss philosopher, 552–3, 915
Roxana (d. 311 BC), wife of Alexander the Great, 171, 172
Rum, sultanate of, 286, 297, 300
Rurik (d. 879), prince of Russia, 291
Russia, Russians:
origins, 290
(i) Kiev Russia:

rise, 290–1
conversion to Orthodoxy, 292
links with W, 292–3
Kiev supremacy wanes, 293, 294
Mongol attacks, 300, 421
dismemberment of Persia, 311
rivalry with China, 367, 368, 488
(ii) emergence of state, 422–3, 487–93
modernization of state, 474, 491, 492–6, 611
territorial gains, 491–2
abolition of serfdom, 563, 611
grain trade, 564
and Eastern Question, 598–9, 601
19th-cent. revolutionary tradition, 607–8
authoritarian bureaucracy, 608–10, 706
and Japan, 677–8
1905 revolutions, 693
economy, 706–7, 726–7, 729
foreign policy, 709, 710, 711–12, 714
(iii) Bolshevik revolution (October), USSR, 717–27, 735
former allies and, 722
civil wars, 726
Asian policies, 731, 737, 739, 740
Turkey and, 749–51, 753
Persia and, 755
foreign policies, 759, 765–6
internal crisis, 762
World War II and, 768, 769, 771, 773–4, 776, 788–9
(iv) post-1945:
UNO and, 777–9
strength of, 779, 780
domestic policies, 780, 867–70
E. European policies, 781, 782–3, 830–1, 867–8, 902, 904–5, 906–7
in Middle East, 781–2, 850, 851–2, 891
Cold War and, 783–4
space exploration, 814–17
China and, 841, 842, 879
economy, 869, 907
in balance of power, 871, 895–6
African policies, 890
in Afghanistan, 895, 902
Asian policies, 895
and arms control, 897, 900, 906
end of USSR, 907–9, 910–11
Russian Republic, 907
nuclear weapons, 921
Rutherford, Ernest, baron (1871–1937), New Zealand scientist, 810, 811

Saddam Hussein (1937–), Iraqi politican, 891, 899–900
Safavid dynasty in Persia, 310–11
Sahara:
animal husbandry, 118
racial groups, 118–19
French and, 656
St Peter, Rome, church of, 456
St Petersburg, tsarist capital, 492, 494, 568
Saint-Simon, Claude de (1760–1825), French social scientist, 606
St Sophia, Constantinople, church of, 242, 275, 285, 287, 290, 308
Sakhalin, 846
Saladin, Yusuf ibn-Ayyub (1138–93), ruler of Egypt and Syria (1174–93), 270, 287
chivalrous repute, 298
triumph over crusaders, 298, 418
salt, from Africa, 381
Salvation Army, 569
Samanid dynasty, 295, 296–7
Samaria, 92
Samarkand, 16, 255, 300, 347
Samoa, naval base at, 659
Samuel, Hebrew prophet, 90, 92
samurai, 374, 376–8, 671, 672, 675, 676
Saqqara, 58
Sarajevo, assassination at, 713
Sardinia:
growth, 484, 592, 595
leader of Italian unity, 602
Sargon I (*c.* 2400–*c.* 2350 BC), king of Akkad and Sumer, 73
45–7, 97
Sargon II, king of Assyria (721–705 BC), 93
Sassanid dynasty in Persia:
pressure on Rome, 222
basis of power, 252–3
aggression under Chosroes II, 254
fall, 257, 261–2
Satsuma rebellion, 676
Saudi Arabia:
kingdom established, 752
oil and, 852
Gulf War and, 900
Saul (d. *c.* 1010 BC), king of Israel, 90
Savonarola, Girolamo (1452–98), Italian reformer, 401
Savoy:
dynasty, 484
duchy lost, 602
Saxons *see* England, Germanic peoples
Scandinavia *see* Norsemen, individual areas
Schrödinger, Erwin (1887–1961), Austrian physicist, 811
science:
Greek, 159
Hellenistic, 175
Arabic, 268–70
transmitted to W, 429
introduced to Russia, 492
medieval weaknesses, 539
modern impact, 701–5, 803, 809–15, 822
distrust of, 818–21
See also chemistry, physics, technology
Scotland:
Norsemen and, 324
backward, 449
wars with England, 467, 482
union with England (1707), 484
intellectual flowering, 549
scripts:
cuneiform types, 40–1, 43, 49, 51, 75
Egyptian hieroglyphic, 61–2, 75
Phoenician alphabet, 75, 84, 130, 137
Minoan, 80
Hebrew, 90
Cyrillic, 290, 292
Carolingian minuscule, 317
Turkish latinization, 754
See also literacy, writing
sculpture:
Egyptian, 61, 65
Persians, 131
Greek supermacy in, 166–7
Hellenistic, 177–8
Indian classical, 343
Scythian peoples:
advance to Iran, 93, 130
hold off Persians, 147
identity and culture, 255
in Punjab, 341
Sebastopol, 601
Seleucus (*c.* 325–281 BC):
Seleucid dynasty, 173
united Hellenism and orient, 173–4, 175
Hellenizing of Jews, 207
Seljuk Turks, 295–6, 297–9
second Turkish empire, 297
response to crusades, 298
Serbia:
Ottoman conquest of, 308
state established, 745
World War I and, 711, 715, 722
serfdom:
in later Roman empire, 221
in feudal Europe, 335
recedes in medieval Europe, 409, 413, 439
in Russia, 492–3, 496
abolished in France, 562
abolished, 563, 611
Severus, L. Septimus (146–211), Roman emperor (193–211), 204, 219, 222
Sforza family, 479
Shah Jahan (1592–1666), Moghul emperor of India (1627–58), 350–1
Shakespeare, William (1564–1616), English dramatic poet, 477, 539
Shang China, 107–9, 111–12
Shi'ites:
Islamic sect, 264, 267, 272
Buwayhid members, 297
in Persia, 301
Safavid members, 310
Iran/Iraq war, 891, 895
Shinto, Japanese religion, 371, 675
shipping:
sail used, 74, 412
in Mediterranean, 134
Scandinavian, 324
medieval advances, 423–4
Dutch, 441, 443, 525, 527
early modern developments, 501–2, 524–5
increased by colonies, 519
steam-driven, 563–4, 565–6, 637, 690
in war, 601
submarine warfare, 717, 772
Soviet, 890
See also naval power and tactics, navigation
Shiva, Hindu god, 102, 344
Shriners, religious sect, 59
Siam (Thailand):
early European presence, 504, 505
retained independence, 655, 658, 685
economic development, 887
Siberia, Russian colonization of, 488, 648
Sicily:
disastrous Athenian expedition to, 150–1, 179
Roman conquest of, 185, 186
Normans in, 287, 417
Arab conquest, 323
Sidon, Phoenician site, 84, 85–7
Sierra Leone, Slave refuge, 530
Sikhs:
empire in N.W. India, 505
in Indian politics, 889
silk:
from China, 135, 178, 222, 303, 352, 354
manufacture, 429
silver:
for currency, 74, 142, 333
from Spain, 85
mines, 152
from S. America, 509, 510

Sinai:
copper in, 65
focus of dispute, 71
Hebrews in, 89
to Egypt, 749
Singapore:
Japanese capture of, 784
economic development, 887
slavery:
Babylonian, 49
Egyptian, 64
Greek, 140–1
helots, 146
critics, 177, 512
omnipresent in Rome, 201, 204
under Islam, 272
in early Russia, 293
use by Turks, 297
in India, 339
black African, 443, 530–1
in W. Indies, 512, 513
and abolition in USA, 563, 612, 616–22
worldwide abolition, 631–2, 640, 644
See also serfdom
slave trade:
by Arabs, 333, 381
by Italians, 411
commercial value, 443
to W. Indies, 513, 524, 527
growth and effects on Africa, 530–1
worldwide abolition, 631–2
and African consciousness, 856
Slav peoples, 278–91
in central Europe and Balkans, 254, 294
Greek influence, 278
topography and culture, 288–90
in Russia, 290, 291
in central Poland, 293–4
conflict with Teutons, 294, 421, 768
and Arab slavers, 333
nationalism, 599, 697
Hitler and, 768
Slovenia, 910
Smith, Adam (1723–90), Scottish economist, 546, 549, 556, 571–2
and Free Trade, 635
Smuts, Jan Christian (1870–1950), S. African statesman, 857
Smyrna, 82, 217
Sobieski, John (1624–96), king of Poland (1674–96), 486, 487
socialism:
economic theory, 604–6
Marx and, 606–7
English paradox, 627
Second International, 693
revisionism, 694
See also communism, Marxism
social services and welfare:
child allowances, 559
British Poor Laws, 625
taxation and, 694
World War I and, 715
British 20th cent., 881, 884
societies and clubs:
(i) intellectual:
Académie des Sciences, 536
Royal Society, 536, 541
popularity of, 536–7, 541
(ii) political:
des Jacobins, 599
working-class, 627
(iii) secret:
Chinese, 359, 365, 663, 665, 732
freemasons, 537
Turkish, 747
Arab, 749
society:
social institutions, variety, 34–6, 75
(i) ancient:
stimulated by coordinated efforts, 38–40
ordered by Hammurabi's code, 48–9
in Egypt, 53, 63–4, 70
in India, 99–102
in China, 107–9, 110–11
in Greece, 140, 152–4
Roman, 182, 201, 243
of Arab empire, 264–5
(ii) modern:
feudal Europe, 334–6
Hindu practices, 344, 679, 682
church in medieval Europe, 401–2
early modern European structure, 446–51
decline of medieval institutions, 454, 561
westernization in Russia, 492
racially mixed in S. America, 511, 533
effects of missionary activity, 531–2
effects of agricultural transplants, 532
industrial rhythms, 570
19th-cent. British patterns, 627
Anglo-Indian attitudes, 651–2, 678–9
Indian, 800
USA, 807, 872
British 20th cent., 881, 884
Iran, 891
Socrates (*c.* 469–399 BC), Athenian philosopher, 156, 159, 162
Solomon (*c.* 974–*c.*. 937 BC): king of Israel, 84
Ethiopian descent from, 380
Solon (*c.* 638–558 BC), Athenian legislator, 145
Song of Roland, 404, 419
Sophocles (495–406 BC), Greek tragedian, 164, 166, 167
South Africa:
Hottentot fate, 529
European settlement, 613
Boer wars, 644–5
black peoples, 647
republic of, 855, 857–60
nuclear weapons, 921
South America:
human immigration, 18, 21
early cultures, 121–2, 388–9
European discovery, 425
territorial acquisitions, 509–13, 638–40
growth of independent states, 638–40
economy, 860, 865–6
USA and, 860–2, 864–5
Cold War and, 862–3
20th-cent. nationalism, 865
authoritarian governments, 865
population, 865
See also individual states
South-east Asia:
earliest agriculture, 25–6
Dutch in, 504–5, 686–7, 838
Portuguese trade with, 504
French in, 505, 658, 685
British in, 650, 658, 685, 687
modernization and nationalism, 685–7, 784, 789, 837–8
US involvement, 837, 844
See also individual states
South Korea, 832, 888
'South Sea Bubble', 442
South Yemen, 890
Soviet Union (USSR) *see* Russia
space, exploration of, 815–18
Spain:
Spaniards, site of minerals, 85, 126
Roman province, 185, 187
Vandal state, 232
Visigothic state, 232, 239, 242
Umayyad control, 272, 316
Viking raids, 325
in Americas, 385, 425, 509–13, 514–15, 638–40
Arabs expelled from, 404, 413, 417
Catholic monarchy, 407, 462–6
revolts, 471
war with England, 480
Bourbon monarchy, 484
intellectual lag, 549
US war, 656, 659–60
disorders, 693
authoritarian regime, 728
Civil War, 765, 773
World War II, 770
joins EEC, 884–5
Sparta, 82
conservative structure, 142, 145–6
against Persia, 148
rivalry with Athens, 149–51
Spectator, British journal, 451, 538
spices:
from India, 135, 222
from China, 303
Dark Age rarity, 333
from Africa, 425
European trade in, 503, 504, 505
sport:
boxing, 75–6
bull-leaping, 76, 79
brutality of Roman, 201
early modern brutality, 535
Sputnik(s), 814–16
Stalin (Joseph Vissarionovich Djugashvili) (1879–1953), Russian statesman:
economic policy, 727
Asian policy, 739, 787, 831
European policy, 765–6, 781, 782
World War II strategies, 768, 773–4, 776
nuclear weapons developed by, 780
Middle Eastern policy, 792
death and legacy, 832–3
de-Stalinization, 869
statistics:
censuses, 436–7
collection of, 436, 437
steel, 807
Steinheim man, 15
Stephen (d. *c.* AD 30), first Christian martyr, 214, 216
Stephen II, pope (752–7), 315, 329–30
stirrup, 72, 257, 336
Stoics:
Stoic school of philosophy, 177
in Rome, 199, 204
stone, for tools, 17, 18
Stone Age *see* Neolithic era, Palaeolithic era
Stonehenge, megalithic monument, 127
Stuart dynasty in England and Scotland, 467, 469, 470–2, 484
and navy, 523, 524–5
Sudan, 86, 120, 376, 381, 899
Mahdi's war, 633
Suez Canal, 636, 745–6, 849–50
W. trade with Asia and, 653
in World War I, 715
British protection of, 751, 752–3, 791
Egyptian nationalization of, 849

'suffragette' movement, 697
 See also women
sugar, 862, 863
Sukarno, Achmed (1901–70), Indonesian statesman, 789, 838
Sulla, L. Cornelius (138–78 BC), 191, 192
Sumer:
 earliest civilization in, 40
 religion in, 40, 41, 42–4
 legacy of, 42, 47–9, 51
 art in, 44, 47, 48
 technology in, 44–5
 women and law, 44, 48
 history of, 45–8
 influence on Egypt, 52, 53
Sung dynasty in China (960–1126), 360, 364
 Mongol threat to, 360
 art under, 362
 technology under, 363–4
Suni, Islamic sect, 264, 267, 893, 895
 Seljuk members, 297
 Ottoman members, 310
 and Iran/Iraq war, 895
Sun Yat-sen (1866–1925), Chinese revolutionary, 669, 670, 685
 KMT and, 732–3, 737–8
 Japanese and, 734
 death, 737, 739
superstition:
 Greek, 158, 159, 160, 177
 in Roman empire, 203, 216
 early modern, 535
Surrealism, 757
Sviatoslav, prince of Russia (962–72), 291
Swanscombe man, 15
Sweden:
 Swedes, in Poland, 486
 against Russia, 491, 496
 against Prussia, 496–7
 power declines, 496
 colonization, 515
 World War II and, 770
 NATO and, 830
 See also Norsemen, Varangians
Switzerland, 770
Symmachus, Q. Aurelius (*c.* 340–*c.* 416), Roman aristocrat, 234
Syracuse, 150–1, 179, 186
Syria:
 Minoan connexions with, 78
 defeat by Assyria, 92
 in Alexandrine empire, 170–1, 173
 invaded by Persia, 222
 Christian, 226, 229, 244, 245
 Arab conquest, 261
 Umayyad seat, 264, 270
 Ottoman rule, 367, 749
 crusades in, 419
 French interest in, 654, 751, 752, 753
 World War II and, 770
 independent, 791
 Arab League and, 792
 in United Arab Republic, 850
 Egypt and, 891
 Gulf War and, 900

Taiping rebellion in China (1850–64), 665–7
Taiwan (Formosa), 789
Taj Mahal, 350
Tamberlane *see* Timur Lang
T'ang dynasty in China:
 collapse of, 296
 family and state under, 354–5
 Confucian civil service, 356–8
 population pressure under, 360
 classical age of art, 361–2
Tantrism, semi-magical cult, 346
Taoism, 116–17, 358–9, 364–5
Tariq (d. *c.* 720), Berber commander, 262
Tatars, Mongolian tribe, 299, 422
technology:
 earliest tools, 7, 9, 13
 advanced by fire, 11–12
 Palaeolithic artifacts and materials, 17, 18
 Neolithic, 24, 25
 in Sumer, 44–5
 in Egypt, 53, 62
 Minoan, 78
 in India, 97
 in China, 107, 112–13, 362–4
 Roman technical skill, 199
 early medieval European, 334
 early E. African, 382
 agricultural advances, 412, 439, 561–2
 industrial advances, 444–5, 563–4
 domestic machines, 696
 social impact of, 702–3
 space explorations and, 814, 817
 See also industrialization, metallurgy, plough, science, telescope, tools
telescope, 541, 543
Tengri, Mongol god, 303
Tennessee Valley scheme, 761
Tenochtitlan, Aztec city, 385
Teotihuacán, early American city, 384
Tertullian (*c.* 152–222), Christian churchman, 225, 234, 331, 430
textiles:
 Asian, 445
 E-W trade in, 505
 factory industry, 567
 See also cotton, silk, wool
Thailand *see* Siam
Thales (*c.* 640–546 BC), Greek philosopher, 160
Thatcher, Margaret (1925–), British prime minister, 883–4
Thebes:
 Egypt, 54, 67
 Greece, 151, 168
Themistocles (*c.* 514–449 BC), Athenian statesman, 156
Theodoric (455–526), Ostrogothic king of Italy (474–526), 239, 240
Theodosius I (*c.* 346–95), Roman emperor in E (379–95), 233
Theseus, legendary Athenian hero, 77
Thespis (6th cent. BC), Greek actor, 164
'Third World', political grouping of, 837
Thomas à Kempis (*c.* 1380–1471), Christian mystic, 397
Thomson, Sir Joseph John (1856–1940), British physicist, 810
Thor, Norse god, 290
Thotmes III, king of Egypt (*c.* 1500–*c.* 1446 BC), 65
Thrace, 86, 138, 169, 231
Thucydides (*c.* 460–400 BC), Greek historian, 151, 156, 163–4
Tiberius, Claudius Nero (42 BC–AD 37), Roman emperor (AD 14–37), 195–6
Tibet:
 Chinese invasions of, 367, 368
 British invasion of, 633, 658
 Chinese and, 840
Timur Lang (*c.* 1336–1405), Mongol emperor:
 aspired to rival Chinghis Khan, 304, 308
 sacked Delhi, 346
tin, 85, 126, 135
Tito (Josip Broz) (1892–1980), Yugoslav statesman, 867
Tokugawa shogunate, 375–6, 378
 problems and responses, 671–3
Tokyo (formerly Edo), 673
Toledo, Spain, 428
Toltec peoples, 385
tools *see* technology, stone, metallurgy
trade:
 in Mesopotamian economy, 45
 in Egypt, 65
 increasing complexity of, 73, 86
 money and barter, 74
 in India, 97–9
 'degrading', 140, 154, 415
 simple Greek structure, 152
 expands in Hellenistic cities, 174
 Mongol exploitation of, 301–3, 364
 Dark Age decline, 333–4
 early western links with India, 343, 349, 351
 stimulated urbanization, 411
 world growth, 441–3, 503–5, 512, 519, 527, 633–6
 Russia and W, 492, 494
 motive for imperialism, 503–5, 510–11, 512
 British imperial concern, 648–9, 653
 Europe and China, 662–3, 665
 Japanese, 671–2, 731–2, 739
 post-1919 tariff walls, 729
 and ideology, 833
 See also Economic organization, individual commodities
trade routes:
 Silk Road, 178, 270, 279, 361
 E-W contacts via, 222, 256–7, 301–3, 309
 encourage development in Africa, 381
 stimulate towns, 411
 See also Suez Canal
trades unionism:
 in France, 587
 inspired by Marxism, 607
 in UK, 627, 694
 and fascism, 728
Trajan, M. Ulpius (*c.* 52–117), Roman emperor (98–117), 197, 198, 204
transport and travel:
 prehistoric patterns, 73–4
 Roman, 237
 Mongol stimulus, 301–3
 Chinese canals, 356, 359
 medieval European, 412
 international, 443, 563–4, 565–6
 effects on industrial growth, 565–6, 690
 in British India, 651
 safety, 920
 See also air travel, motor car, railways, shipping, Suez Canal
Transvaal Republic, 645
treaties:
 Verdun (843), 319
 Tordesillas (1494), 425
 Westphalia (1648), 481
 Kutchuk Kainardji (1774), 497
 Breda (1667), 523
 Paris (1783), 577
 Vienna (1815), 595, 598
 Japanese 'unequal', 672–3, 677, 740, 787
 'Reinsurance' (1887), 711
 Brest-Litovsk (1918), 719
 Versailles (1919), 721, 723, 726, 729, 765

Locarno (1925), 759
Maastricht (1991), 885
Non-proliferation (1968), 921
Trent, Council of (1545–63), 462, 464, 531
Trieste, 833
Tripoli, Italian attack on, 656, 661, 747
Trotsky, Leon Davidovich (1879–1940), Russian revolutionary, 727
Troy, 86
siege of, 82, 83, 275
Truman, Harry S. (1884–1972), US president:
and Europe, 781, 782
and Middle East, 793
and Far East, 831
Truman Doctrine, 782
Tsolikovsky, Konstantin (1857–1935), Russian scientist, 814
Tudor dynasty in England, 477
administrative techniques, 454, 460, 477–8
rejection of Papacy, 460
rebellions against, 470
See also individual rulers
Tunisia, 849
Turkey:
Ottoman power *see* Ottoman empire
Turks, first empire, 257, 295–7
pressure in Asia Minor, 286
tribal movements, 295
contacts with East, 296–7
Seljuk power, 297–9
Versailles settlement and, 721
nationalist rebellion, 747–9
Kemalist modernization, 754
Russians on borders, 779
US aid to, 782
US missiles withdrawn, 869
Tutankhamon, king of Egypt (1360–1350 BC), 68
Tyre:
Phoenician site, 84, 85, 90
destroyed by Alexander, 171

Uganda, 656
Ulster *see* Ireland
Umayyad caliphate, 260–2, 264–7
art and architecture under, 270, 304
in Spain, 272, 316
United Arab Republic, formed, 850
United Kingdom:
(i) 1707–1900:
formed (1707), 484
in India, 505–9
commercial orientation of foreign policy, 524
seapower of, 524–5
slave trading, 524, 530
Enlightenment thinkers, 548–9
America and, 556–9, 613–15
agricultural revolution, 561–2
industrial growth, 566–9
wars with France, 590, 593
Eastern Question and, 601
political democratization, 623–5
social legislation, 625
non-revolutionary politics, 627, 697, 708
Irish problem, 629, 697–8, 722
19th-cent. centre of world economy, 634–5
imperial policies, 643–4, 648–9, 678–84
Boers and, 644–7
Japan and, 677, 712
(ii) 1900–1945:
European policies, 708, 712, 761, 765
World War I and, 714–17, 719
Versailles settlement and, 722
Asian policies, 731, 734, 738, 742
India, 740
Middle East and, 747, 749–53
domestic reforms, 762
World War II and, 765–9, 770–6
(iii) post-1945:
UNO and, 777
European policies, 781
India and, 782, 784–7
in Near and Middle East, 782, 791–3, 849
USA and, 782
S.E. Asia and, 791
Rhodesia and, 855
EEC and, 881, 882
economy, 882–3
Irish problem, 882
Falklands War, 883
nuclear weapons, 921
United Nations Organization:
importance, 777–8, 784
(UNO), constitution, 777
use of veto, 777, 782
anti-colonialism in, 789
mediation of, 793
Israel and, 852, 891
PLO and, 852
S. Africa and, 852, 858
Hungary and, 868
and Gulf War, 900
Declaration of Human Rights, 917
United Provinces, 466, 484
United States of America:
(i) expansion of food production, 563–4, 623
slavery and abolition in, 563, 612, 616–22
industrialization, 567
independence sought and gained, 574–7, 585
constitution, 577–80, 616, 618
compared with UK, 613–15
foreign and imperialist policy, 613–17, 642, 658–60, 672, 687
territorial expansion, 615–16, 622–3, 658–60
economic growth, 616, 622, 634, 635, 658
population growth, 616, 622
democratization of politics, 618, 621
North against South (Civil War), 618–22
party growth, 622
immigration, 637, 708
(ii) 1900–45:
economy, 708–9, 715, 729–30, 756
Europe and, 709, 715, 730, 756
investment in Europe, 730
Asian policies, 734, 737, 742, 744
Middle Eastern policies, 752
'New Deal', 760–1
World War II and, 768–72, 774, 775, 776
(iii) post-1945:
in UNO, 777–8
sources of strength, 779–80
aid to Europe, 782, 803
in Middle East, 782, 791
Cold War and, 784, 864
Asian policies, 787, 789, 837, 845, 846–7, 877
economy, 803, 871–2
space exploration and, 815–16
and Israel, 851, 898, 902
S. American policies, 860–2
in balance of power, 867, 871, 894, 897, 898, 902
negro question, 872–3
domestic effects of Vietnam war, 874–6
and Iran, 891, 893
political patterns, 902
nuclear weapons, 921
See also American Revolution
universities:
Plato's Academy, 163, 175, 243
medieval foundations, 397, 427–8
clerical emphasis in, 536
Indian, 651
in Vietnam, 686
science departments, 703
political activism at, 733, 734
in China, 842
women in, 893, 918
See also Cambridge, Oxford
Upanishads, Indian texts, 103
Ur, 42, 44–5, 47, 48, 49
home of Abraham, 88
Urban II, pope (1088–99), 395
urbanization:
origins, 34, 40
new scale of Roman, 199, 201
decline in times of economic weakness, 222, 231, 333–4
European towns grow, 411, 438
industrial concomitant, 567–9
social effects of, 567–9, 693
See also cities
Uruguay, expansion of food production in, 563
Uruk, 40
USSR (Union of Socialist Soviet Republics) *see* Russia

Valens (*c.* 328–78), Roman emperor (364–78), 231
Valerian, Roman emperor (253–60), 222, 226
Valois dynasty in France, 477–8, 480
Varangians, Norsemen in East, 290–1, 324
Varuna, Aryan god, 99
Vedanta, 344
vegetables, 56
potatoes, 121, 351, 389
mustard, 237
pulses in Europe, 409
in Europe, 439, 527
Venezuela, 642
blockade of, 642
dictatorship in, 861
Venice:
Venetians, commercial empire of, 287, 306, 408
banking in, 411
small scale of, 412
capture Constantinople, 418
diplomacy of, 476, 522
Turks and, 480, 485
end of republic, 592, 595
Versailles:
palace of, 473, 482, 603
See also treaties
Vesalius, Andreas (1514–64), Belgian anatomist, 539, 543
Vespasian, T. Flavius (9–79), Roman emperor (69–79), 197
Vespucci, Amerigo (1454–1512), Italian navigator, 425
Victoria (1819–1901), queen of UK (1837–1901), empress of India (1876–1901), 625, 629
on Boer War, 645

and medicine, 704
Vienna:
Turkish sieges, 308, 485, 486
growth of, 568
revolution in, 599
See also treaties
Viet Cong, 845
Vietnam:
Chinese invasions, 364
French in, 685–6, 740
republic formed, 789
partition and war, 845
US policy and, 845, 874–7
and China, 877, 886
Vikings *see* Norsemen
Villanovan culture *see* Etruscans vine, 77, 134, 147, 152
Virgil, P. Virgilius, Maro (70–19 BC), Roman poet, 199
and Dante, 430
Vishnu, Hindu god, 103, 340, 344
Vladimir (*c.* 956–1015), Christian prince of Russia (980–1015), 292
Vladivostok, 611, 648, 667, 677
trans-Siberian railway to, 741
Voltaire (François Marie Arouet) (1694–1778), French writer, 341, 546, 549, 552

Walesa, Lech (1943–), Polish politician, 905, 906, 909
warlords, Chinese, 733, 735, 737, 738, 741
Warring States period in China, 109–10, 113, 114
regional cultures evolve, 352
wars:
Persian (490 BC, 480 BC), 147–8
Peloponnesian (431–404 BC), 149–51, 156
Punic (3rd cent. BC), 185–6
Roman Civil (49–31 BC), 191–2, 195
crusades (1096–1291) *see* separate entry
Hundred Years' (1337–1453), 405–6
of the Roses (1455–85), 408, 477
Dutch Revolt (1554–1648), 464, 480
English Civil (1642–51), 467, 468, 469, 470
Thirty Years' (1614–48), 469, 481, 486, 521
Spanish Succession (1701–14), 482–4, 523
Austrian Succession (1740–48), 497, 524
Seven Years' (1756–63), 507, 524
Anglo-Dutch (1652–4, 1665–7), 522, 524–5
American Independence (1773–83), 574–7, 585
French revolutionary (1792–1802), 585
Napoleonic (1804–15), 590–3, 653
Crimean (1854–6), 601
US Civil (1861–5), 618–22, 709–10
Opium (1840, 1841), 635, 664–5
S. American Independence (1810–22), 638–40
Boer (1899–1902), 644–5
Sino-Japanese (1894–1902), 667
Java (1825), 686–7
of 1812, 709–10
First World, 'the Great' (1914–18), 709, 714–17, 719–20
Second World (1939–45), 765–76
Spanish Civil (1936–9), 765, 773
Korean (1950–3), 831–2
Vietnam (1964–73), 845, 875–6
Arab-Israeli (1967, 1973), 852
Falkland Islands (1982), 883
Gulf (1990), 898
See also Alexander, Arabs, Chinghis Khan, Cold War, Ottoman empire
Washington, George (1732–99), US president, 577, 579
foreign policy, 615
weapons:
Palaeolithic, 17, 18
Chinese, 113, 353, 363
'Greek fire', 280, 281, 291
guns and gunpowder, 308, 363, 368, 376
warships, 485, 524–5
19th-cent. European superiority, 633
and modern technology, 699–700
in World War I, 715, 716
nuclear, 772, 780, 810, 832, 897
bombs in World War II, 773
restrictions on tests and numbers, 869, 903
in balance of power, 897
See also chariot, military techniques
weights and measures:
standardized in India, 97
in China, 356
metric systems, 587, 592
Weimar Republic *see* Germany
Wessex, Anglo-Saxon kingdom, 326
West Indies:
West Indians, discovery by Europeans, 425, 509
colonial economy, 510, 512–14
English acquisitions in, 523, 524, 639
USA, 862, 871
population growth, 865
wheel, 34, 45
unknown to Africa, 380
to America, 385
Wilhelm II (1854–1941), emperor of Germany (1888–1918), 698–9
William I (1027–87), duke of Normandy and king of England (1066–87), 404, 416
William III (1650–1702), king of England (1689–1702), 472, 484, 523
William of Roebuck (13th cent.), Franciscan friar, 303
William 'the Silent' (1533–84), prince of Orange (1544–84), 466, 480
Willibrord, St (658–739), Anglo-Saxon missionary, 331
Wilson, T. Woodrow (1856–1924), US president:
and Versailles settlement, 721, 723
and League of Nations, 722, 759
and Ottoman empire, 751
wine *see* alcohol
Woden, Norse god, 290
woman:
role of, human sexual specialization, 9, 10, 13, 21
Sumerian, 44
Egyptian, 64
subordinated in Greece, 152–4
under Islam, 273
Hindu subordination, 344
Chinese disregard for, 361
medieval European, 415–16
early modern European, 450–1
labour of, 571, 695, 715
in professions, 601
equality, 666
emancipation by technology, 695–6
franchise and, 695, 697
Turkish emancipation, 755
Indian attitudes, 800
See also contraception
wool:
in Crete, 77
speciality of Miletus, 152
English, 445
from Australia, 645
World Bank, 804
World Economic Conference (1933), 756
writing:
earliest forms, 40
in Sumer, 40–1, 43
in Babylon, 49
in Egypt, 53, 61–2
spread of, 75, 142
Harappan, 97
in ancient China, 112
in Americas, 121
in Persia, 130
in Greece, 137
Turkish, 296, 301
See also literacy, scripts
Wyclif, John (*c.* 1320–84), English Church reformer, 401

Xenophon (*c.* 430–*c.* 354 BC), Greek soldier and historian, 168
Ximenes, Francisco (1437–1517), Spanish cardinal, 457
X-rays, 704, 810

Yahweh:
Temple of, 84, 91
Hebrew god, 88–9
Yeltzin, Boris (1931–), Russian politician, 910
yoga, 104
Young Turk movement, 747–8, 749
Yuan Shih-kai (1859–1916), Chinese dictator, 671, 732, 734
Yueh-Chih people, 256
Yugoslavia:
emergence of state, 711, 721, 722
World War II and, 768
communist state, 779, 781
Russian hostility, 867
dissolution, 910

Zen Buddhism in Japan, 374
Zeno (*c.* 340–265 BC), Greek founder of Stoicism, 177
Zeus, Greek god, 79
Zhao-Ziyang (1919–), Chinese politician, 911
Ziggurats, 42, 45, 47
Zimbabwe, 382
state of, 858
Zionism:
Congress, 749
Balfour Declaration, 751
World Zionist Congress, 791–2
Russian attitude, 792
US attitudes, 792
See also Israel, Jews, Palestine
Zoroastrianism, 131, 222, 252, 253, 257
Zulu peoples, 644, 647, 656
Zwingli, Ulrich (1484–1531), Swiss Protestant theologian, 458